Lecture Notes in Computer Science   7031

Commenced Publication in 1973
Founding and Former Series Editors:
Gerhard Goos, Juris Hartmanis, and Jan van Leeuwen

Lora Aroyo   Chris Welty
Harith Alani   Jamie Taylor
Abraham Bernstein   Lalana Kagal
Natasha Noy   Eva Blomqvist (Eds.)

The Semantic Web – ISWC 2011

10th International Semantic Web Conference
Bonn, Germany, October 23-27, 2011
Proceedings, Part I

 Springer

Volume Editors

Lora Aroyo
VU University Amsterdam, The Netherlands; l.m.aroyo@vu.nl

Chris Welty
IBM Research, Yorktown Heights, NY, USA; cawelty@gmail.com

Harith Alani
The Open University, Milton Keynes, UK; h.alani@open.ac.uk

Jamie Taylor
Google, Mountain View, CA, USA; jamietaylor@google.com

Abraham Bernstein
University of Zurich, Switzerland; Bernstein@ifi.uzh.ch

Lalana Kagal
Massachusetts Institute of Technology, Cambridge, MA, USA; lkagal@csail.mit.edu

Natasha Noy
Stanford University, CA, USA; noy@stanford.edu

Eva Blomqvist
Linköping University, Sweden; eva.blomqvist@liu.se

ISSN 0302-9743   e-ISSN 1611-3349
ISBN 978-3-642-25072-9   e-ISBN 978-3-642-25073-6
DOI 10.1007/978-3-642-25073-6
Springer Heidelberg Dordrecht London New York

Library of Congress Control Number: 2011939851

CR Subject Classification (1998): C.2, H.4, H.3, H.5, J.1, K.4

LNCS Sublibrary: SL 3 – Information Systems and Application, incl. Internet/Web and HCI

Typesetting: Camera-ready by author, data conversion by Scientific Publishing Services, Chennai, India

Printed on acid-free paper

Springer is part of Springer Science+Business Media (www.springer.com)

Preface

Ten years ago, several researchers decided to organize a workshop to bring together an emerging community of scientists who were working on adding machine-readable semantics to the Web, the Semantic Web. The organizers were originally planning for a few dozen researchers to show up. When 200 of them came to Stanford in August 2001, the Semantic Web Workshop became the Semantic Web Working Symposium, and the International Semantic Web Conference (ISWC) was born. Much has changed in the ten years since that meeting. The Semantic Web has become a well-recognized research field in its own right, and ISWC is a premier international research conference today. It brings together researchers, practitioners, and users in artificial intelligence, databases, social networks, distributed computing, Web engineering, information systems, human–computer interaction, natural-language processing, and others. Companies from Facebook to Google to the *New York Times* rely on Semantic Web technologies to link and organize their data; governments in the United States, United Kingdom, and other countries open up their data by making it accessible to Semantic Web tools; scientists in many domains, from biology, to medicine, to oceanography and environmental sciences, view machine-processable semantics as key to sharing their knowledge in today's data-intensive scientific enterprise; semantic technology trade shows attract more than a thousand attendees. The focus of Semantic Web research has moved from issues of representing data on the Web and the growing pains of figuring out a common format to share it, to such challenges as handling billions of statements in a scalable way to making all this data accessible and usable to regular citizens.

This volume contains the main proceedings of the 10th International Semantic Web Conference (ISWC 2011), which was held in Bonn, Germany, in October 2011. We received tremendous response to our calls for papers from a truly international community of researchers and practitioners. Indeed, every track of the conference received a record number of submissions this year. The careful nature of the review process, and the breadth and scope of the papers finally selected for inclusion in this volume, speak to the quality of the conference and to the contributions made by researchers whose work is presented in these proceedings.

The Research Track of the conference attracted 264 submissions. Each paper received at least three, and sometimes as many as five, reviews from members of the Program Committee. After the first round of reviews, authors had the opportunity to submit a rebuttal, leading to further discussions among the reviewers, a meta-review and a recommendation from a member of the Senior Program Committee. Every paper that had at least one recommendation for acceptance was discussed in a virtual meeting of the Senior Program Committee.

As the Semantic Web develops, we find a changing variety of subjects that emerge. This year the keywords of accepted papers were distributed as follows (frequency in parentheses): ontologies and semantics (15), database, IR, and AI technologies for the Semantic Web (14), management of Semantic Web data (11), reasoning over Semantic Web data (11), search, query, integration, and analysis on the Semantic Web (10), robust and scalable knowledge management and reasoning on the Web (10), interacting with Semantic Web data (9), ontology modularity, mapping, merging, and alignment (8), languages, tools, and methodologies for representing and managing Semantic Web data (8), ontology methodology, evaluation, reuse, extraction, and evolution (7), evaluation of Semantic Web technologies or data (7), specific ontologies and ontology patterns for the Semantic Web (6), new formalisms for the Semantic Web (4), user interfaces to the Semantic Web (3), cleaning, assurance, and provenance of Semantic Web data, services, and processes (3), social Semantic Web (3), evaluation of Semantic Web technology (3), Semantic Web population from the human Web (3).

Overall, the ISWC Program Committee members adopted strict standards for what constitutes high-quality Semantic Web research and what papers must deliver in terms of theory, practice, and evaluation in order to be accepted to the Research Track. Correspondingly, the Program Committee accepted only 50 papers, 19% of the submissions.

The Semantic Web In-Use Track received 75 submissions. At least three members of the In-Use Program Committee provided reviews for each paper. Seventeen papers were accepted – a 23% acceptance rate. The large number of submissions this year demonstrated the increasingly diverse breadth of applications of Semantic Web technologies in practice. Papers demonstrated how semantic technologies could be used to drive a variety of simulation and test systems, manage distributed content and operate within embedded devices. Several papers tapped the growing amount of semantically enriched environmental data available on the Web allowing communities to visualize, organize, and monitor collections for specific purposes.

The Doctoral Consortium has become a key event at the conference over the years. PhD students get an opportunity to present their thesis proposals and to get detailed feedback on their research topics and plans from the leading academic and industrial scientists in the field. Out of 31 submissions to the Doctoral Consortium, 6 were accepted as long papers for presentation at the conference, and 9 were accepted for presentation at the special Consortium-only poster session. Each student was assigned a mentor who led the discussion following the presentation of their proposal, and provided extensive feedback and comments.

A unique aspect of the ISWC conference is the Semantic Web Challenge. In this competition, the ninth to be held at the conference, practitioners and scientists showcase useful and leading-edge applications of Semantic Web technology. Diana Maynard and Chris Bizer organized the Semantic Web Challenge this year.

The keynote talks given by leading scientists in the field further enriched the ISWC program. Alex (Sandy) Pentland, the director of the Human Dynamics Laboratory and the Media Lab Entrepreneurship Program at the Massachusetts

Institute of Technology, discussed the New Deal on Data—a new data ecosystem that can allow personal data to become an accessible asset for the new generation of systems in health, finance, logistics, and transportation. Gerhard Weikum, a Research Director at the Max Planck Institute for Informatics, discussed the issues and approaches to extending and enriching linked data, in order to improve its scope, quality, interoperability, cross-linking, and usefulness. Frank van Harmelen, a professor at the VU University Amsterdam, and a participant and leader in Semantic Web research, provided his analysis of the past ten years, discussing whether any universal patterns have emerged in the way we built the Semantic Web. Nigel Shadbolt, Deputy Head of the School of Electronics and Computer Science at the University of Southampton, gave a lively dinner talk.

As in previous ISWC editions, the conference included an extensive Tutorial and Workshop program. Tania Tudorache and Heiner Stuckenschmidt, the Chairs of this track, created a stellar and diverse collection of 7 tutorials and 16 workshops, where the only problem that the participants faced was which of the many exciting workshops to attend.

We would like to thank Marta Sabou and Guilin Qi for organizing a lively Poster and Demo Session. This year, the Posters and Demos were introduced in a Minute Madness Session, where every presenter got 60 seconds to provide a teaser for their poster or demo. Marco Neumann coordinated an exciting Industry Track with presentations both from younger companies focusing on semantic technologies and software giants, such as Yahoo! and Microsoft.

As we look forward to the next ten years of Semantic Web research, we organized an Outrageous Ideas Session, with a special award sponsored by the Computing Community Consortium. At this track, we invited scientists to submit short papers describing unconventional and innovative ideas that identify new research opportunities in this field. A Program Committee of established Semantic Web researchers judged the submissions on the extent to which they expand the possibilities and horizons of the field. After presentation of shortlisted papers at the conference both the PC members and the audience voted for the prize winners.

We are indebted to Eva Blomqvist, our Proceedings Chair, who provided invaluable support in compiling the volume that you now hold in your hands (or see on your screen) and exhibited super-human patience in allowing the other Chairs to stretch deadlines to the absolute limits. Many thanks to Jen Golbeck, the Fellowship Chair, for securing and managing the distribution of student travel grants and thus helping students who might not have otherwise attended the conference to come to Bonn. Mark Greaves and Elena Simperl were tireless in their work as Sponsorship Chairs, knocking on every conceivable virtual 'door' and ensuring an unprecedented level of sponsorship this year. We are especially grateful to all the sponsors for their generosity.

As has been the case in the past, ISWC 2011 also contributed to the linked data cloud by providing semantically annotated data about many aspects of the conference. This contribution would not have been possible without the efforts of Lin Clark, our Metadata Chair.

Juan Sequeda, our Publicity Chair, was tirelessly twittering, facebooking, and sending old-fashioned announcements on the mailing lists, creating far more lively 'buzz' than ISWC ever had.

Our very special thanks go to the Local Organization Team, led by Steffen Staab and York Sure-Vetter. They did a fantastic job of handling local arrangements, thinking of every potential complication way before it arose, often doing things when members of the Organizing Committee were only beginning to think about asking for them. Special thanks go to Ruth Ehrenstein for her enormous resourcefulness, foresight, and anticipation of the conference needs and requirements. We extend our gratitude to Silke Werger, Holger Heuser, and Silvia Kerner.

Finally, we would like to thank all members of the ISWC Organizing Committee not only for handling their tracks superbly, but also for their wider contribution to the collaborative decision-making process in organizing the conference.

October 2011

Lora Aroyo
Chris Welty
Program Committee Co-chairs
Research Track

Harith Alani
Jamie Taylor
Program Committee Co-chairs
Semantic Web In-Use Track

Abraham Bernstein
Lalana Kagal
Doctoral Consortium Chairs

Natasha Noy
Conference Chair

Conference Organization

Organizing Committee

Conference Chair

Natasha Noy   Stanford University, USA

Program Chairs–Research Track

Lora Aroyo   VU University Amsterdam, The Netherlands
Chris Welty   IBM Watson Research Center, USA

Semantic Web In-Use Chairs

Harith Alani   KMI, Open University, UK
Jamie Taylor   Google, USA

Doctoral Consortium Chairs

Abraham Bernstein   University of Zurich, Switzerland
Lalana Kagal   Massachusetts Institute of Technology, USA

Industry Track Chair

Marco Neumann   KONA, USA

Posters and Demos Chairs

Guilin Qi   Southeast University, China
Marta Sabou   MODUL University, Austria

Workshops and Tutorials Chairs

Heiner Stuckenschmidt   University of Mannheim, Germany
Tania Tudorache   Stanford University, USA

Semantic Web Challenge Chairs

Christian Bizer   Free University Berlin, Germany
Diana Maynard   University of Sheffield, UK

Metadata Chair

Lin Clark   DERI Galway, Ireland

Local Organization Chairs

Steffen Staab	University of Koblenz-Landau, Germany
York Sure-Vetter	GESIS and University of Koblenz-Landau, Germany

Local Organization

Ruth Ehrenstein	University of Koblenz-Landau, Germany

Sponsorship Chairs

Mark Greaves	Vulcan, USA
Elena Simperl	Karlsruhe Institute of Technology, Germany

Publicity Chair

Juan Sequeda	University of Texas at Austin, USA

Fellowship Chair

Jen Golbeck	University of Maryland, USA

Proceedings Chair

Eva Blomqvist	Linköping University and ISTC-CNR, Sweden/Italy

Webmaster

Holger Heuser	GESIS, Germany

Senior Program Committee – Research

Mathieu d'Aquin	Open University, UK
Philippe Cudré-Mauroux	University of Fribourg, Switzerland
Jérôme Euzenat	INRIA and LIG, France
Aldo Gangemi	STLab, ISTC-CNR, Italy
Jeff Heflin	Lehigh University, USA
Ian Horrocks	University of Oxford, UK
Geert-Jan Houben	Delft University of Technology, The Netherlands
Aditya Kalyanpur	IBM Research, USA
David Karger	MIT, USA
Manolis Koubarakis	National and Kapodistrian University of Athens, Greece
Diana Maynard	University of Sheffield, UK
Peter Mika	Yahoo! Research, Spain
Peter F. Patel-Schneider	Bell Labs, USA
Axel Polleres	Siemens AG/DERI Galway, Austria/Ireland

Jie Tang	Tsinghua University, China
Paolo Traverso	FBK, Italy
Lei Zhang	IBM China Research, China

Program Committee – Research

Fabian Abel	Xiaoyong Du
Sudhir Agarwal	Dieter Fensel
Faisal Alkhateeb	Achille Fokoue
Yuan An	Fabien Gandon
Melliyal Annamalai	Zhiqiang Gao
Grigoris Antoniou	Fausto Giunchiglia
Kemafor Anyanwu	Birte Glimm
Knarig Arabshian	Bernardo Grau
Lora Aroyo	Alasdair Gray
Manuel Atencia	Paul Groth
Sören Auer	Tudor Groza
Christopher Baker	Michael Grüninger
Jie Bao	Gerd Gröner
Michael Benedikt	Nicola Guarino
Sonia Bergamaschi	Volker Haarslev
Eva Blomqvist	Peter Haase
Piero Bonatti	Willem van Hage
Kalina Bontcheva	Harry Halpin
Aidan Boran	Andreas Harth
John Breslin	Michael Hausenblas
Paul Buitelaar	Sandro Hawke
Diego Calvanese	Tom Heath
Mari Carmen	Nathalie Hernandez
Enhong Chen	Stijn Heymans
Key-Sun Choi	Michiel Hildebrand
Benoit Christophe	Kaoru Hiramatsu
Lin Clark	Pascal Hitzler
Oscar Corcho	Aidan Hogan
Isabel Cruz	Laura Hollink
Richard Cyganiak	Matthew Horridge
Claudia D'Amato	Wei Hu
Theodore Dalamagas	Jane Hunter
Mike Dean	David Huynh
Stefan Decker	Eero Hyvönen
Renaud Delbru	Giovambattista Ianni
Ian Dickinson	Lalana Kagal
Stefan Dietze	Yevgeny Kazakov
Li Ding	Anastasios Kementsietsidis
John Domingue	Teresa Kim

Yasuhiko Kitamura
Vladimir Kolovski
Kouji Kozaki
Thomas Krennwallner
Markus Krötzsch
Oliver Kutz
Ora Lassila
Georg Lausen
Juanzi Li
Shengping Liu
Carsten Lutz
Christopher Matheus
Jing Mei
Christian Meilicke
Alessandra Mileo
Knud Moeller
Boris Motik
Yuan Ni
Daniel Oberle
Jacco van Ossenbruggen
Sascha Ossowski
Jeff Pan
Bijan Parsia
Alexandre Passant
Chintan Patel
Terry Payne
Jorge Perez
Giuseppe Pirrò
Valentina Presutti
Guilin Qi
Yuzhong Qu
Anand Ranganathan
Riccardo Rosati
Marie-Christine Rousset
Sebastian Rudolph
Tuukka Ruotsalo
Alan Ruttenberg
Marta Sabou
Uli Sattler
Bernhard Schandl
François Scharffe
Ansgar Scherp
Daniel Schwabe
Luciano Serafini

Yidong Shen
Amit Sheth
Pavel Shvaiko
Michael Sintek
Sergej Sizov
Spiros Skiadopoulos
Kavitha Srinivas
Giorgos Stamou
Johann Stan
George Stoilos
Umberto Straccia
Markus Strohmaier
Heiner Stuckenschmidt
Rudi Studer
Xingzhi Sun
Valentina Tamma
Sergio Tessaris
Philippe Thiran
Christopher Thomas
Cassia Trojahn dos Santos
Raphaël Troncy
Christos Tryfonopoulos
Tania Tudorache
Anni-Yasmin Turhan
Octavian Udrea
Victoria Uren
Stavros Vassos
Maria Vidal
Ubbo Visser
Denny Vrandecic
Holger Wache
Haofen Wang
Fang Wei
Chris Welty
Max L. Wilson
Katy Wolstencroft
Zhe Wu
Guotong Xie
Bin Xu
Yong Yu
Ondrej Zamazal
Ming Zhang
Antoine Zimmermann

Additional Reviewers – Research

Saminda Abeyruwan
Pramod Ananthram
Marcelo Arenas
Ken Barker
Thomas Bauereiß
Domenico Beneventano
Meghyn Bienvenu
Nikos Bikakis
Holger Billhardt
Victor de Boer
Georgeta Bordea
Stefano Borgo
Loris Bozzato
Volha Bryl
Carlos Buil-Aranda
Federico Caimi
Delroy Cameron
Davide Ceolin
Melisachew Wudage Chekol
Yueguo Chen
Gong Cheng
Alexandros Chortaras
Mihai Codescu
Anna Corazza
Minh Dao-Tran
Jérôme David
Brian Davis
Jianfeng Du
Liang Du
Songyun Duan
Michel Dumontier
Jinan El-Hachem
Cristina Feier
Anna Fensel
Alberto Fernandez Gil
Roberta Ferrario
Pablo Fillottrani
Valeria Fionda
Haizhou Fu
Irini Fundulaki
Sidan Gao
Giorgos Giannopoulos
Kalpa Gunaratna

Claudio Gutierrez
Ollivier Haemmerlé
Karl Hammar
Norman Heino
Cory Henson
Ramon Hermoso
Daniel Herzig
Matthew Hindle
Joana Hois
Julia Hoxha
Matteo Interlandi
Prateek Jain
Ernesto Jiménez-Ruiz
Mei Jing
Fabrice Jouanot
Martin Junghans
Ali Khalili
Sheila Kinsella
Szymon Klarman
Matthias Knorr
Srdjan Komazec
Jacek Kopecky
Adila Alfa Krisnadhi
Sarasi Lalithsena
Christoph Lange
Danh Le Phuoc
Ning Li
John Liagouris
Dong Liu
Nuno Lopes
Uta Lösch
Theofilos Mailis
Alessandra Martello
Michael Martin
Andrew Mccallum
William J. Murdock
Vijayaraghava Mutharaju
Nadeschda Nikitina
Andriy Nikolov
Philipp Obermeier
Marius Octavian Olaru
Matteo Palmonari
Catia Pesquita

Rafael Peñaloza
Marco Pistore
Laura Po
Martin Przyjaciel-Zablocki
Behrang Qasemizadeh
Paraskevi Raftopoulou
Padmashree Ravindra
Yuan Ren
Alexandre Riazanov
Francesco Ricca
Karl Rieb
Mariano Rodriguez-Muro
Marco Rospocher
Silvia Rota
Luigi Sauro
Rafael Schirru
Florian Schmedding
Michael Schmidt
Patrik Schneider
Thomas Schneider
Alexander Schätzle
Murat Sensoy
Arash Shaban-Nejad
Elena Simperl
Giorgos Siolas
Evren Sirin
Sebastian Speiser
Campinas Stephane
Ljiljana Stojanovic

Cosmin Stroe
Vojtěch Svátek
Stuart Taylor
Manolis Terrovitis
Andreas Thalhammer
Ioan Toma
Thanh Tran
Despoina Trivela
Giovanni Tummarello
Geert Vanderhulst
Tassos Venetis
Luis Vilches-Blázquez
Andreas Wagner
Cong Wang
Wenbo Wang
Xiaoyuan Wang
Zhe Wang
Zhichun Wang
Jens Wissmann
Fei Wu
Kejia Wu
Tobias Wunner
Guohui Xiao
Hong Qing Yu
Valentin Zacharias
Massimo Zancanaro
Dmitriy Zheleznyakov
Ming Zuo

Program Committee – Semantic Web In-Use

Sofia Angeletou
Anupriya Ankolekar
Sören Auer
Christian Bizer
Ivan Cantador
Oscar Corcho
Gianluca Correndo
Mike Dean
Salman Elahi
Lee Feigenbaum
Nicholas Gibbins
John Goodwin
Mark Greaves

Tudor Groza
Siegfried Handschuh
Michael Hausenblas
Manfred Hauswirth
Tom Heath
Ivan Herman
Pascal Hitzler
Bo Hu
David Huynh
Krzysztof Janowicz
Matthias Klusch
Jens Lehmann
Libby Miller

Knud Möller	Manuel Salvadores
Lyndon Nixon	Milan Stankovic
Daniel Olmedilla	Nenad Stojanovic
Jeff Z. Pan	Martin Szomszor
Massimo Paolucci	Edward Thomas
Alexandre Passant	Tania Tudorache
Carlos Pedrinaci	Mischa Tuffield
H. Sofia Pinto	Giovanni Tummarello
Axel Polleres	Denny Vrandecic
Yves Raimond	Holger Wache
Matthew Rowe	Shenghui Wang
Marta Sabou	Fouad Zablith

Additional Reviewers – Semantic Web In-Use

Charlie Abela	Meenakshi Nagarajan
Lorenz Bühmann	Vít Nováček
Toni Cebrián	Kunal Sengupta
Juri Luca De Coi	William Smith
Gianluca Demartini	Claus Stadler
Laura Dragan	Vinhtuan Thai
Philipp Frischmuth	Christian Von Der Weth
Benjamin Grosof	Josiane Xavier Parreira
Nicholas Humfrey	Zi Yang
Nophadol Jekjantuk	Fouad Zablith
Amit Krishna Joshi	Yuting Zhao

Program Committee – Doctoral Consortium

Abraham Bernstein	David Karger
Philip Cimiano	Diana Maynard
Fabio Ciravegna	Enrico Motta
Philippe Cudré-Mauroux	Natasha Noy
Jen Golbeck	Guus Schreiber
Tom Heath	Evren Sirin
Jeff Heflin	Tania Tudorache
Lalana Kagal	

Additional Reviewers – Doctoral Consortium

Andriy Nikolov

Sponsors

Platinum Sponsors

AI Journal
Elsevier
fluid Operations
OASIS
THESEUS

Gold Sponsors

Microsoft Research
ontoprise
Ontotext
PoolParty
SoftPlant
Yahoo! Research

Silver Sponsors

Computing
  Community
  Consortium
IBM Research
IEEE Intelligent
  Systems
IGI Global
IOS Press
LATC
Monnet
Planet Data
RENDER
Springer

Table of Contents – Part I

Research Track

Table of Contents – Part II

Semantic Web In-Use Track

Doctoral Consortium

Invited Talks—Abstracts

Leveraging the Semantics of Tweets for Adaptive Faceted Search on Twitter

Fabian Abel[1], Ilknur Celik[1], Geert-Jan Houben[1], and Patrick Siehndel[2]

[1] Web Information Systems, Delft University of Technology
{f.abel,i.celik,g.j.p.m.houben}@tudelft.nl
[2] L3S Research Center, Leibniz University Hannover
siehndel@l3s.de

Abstract. In the last few years, Twitter has become a powerful tool for publishing and discussing information. Yet, content exploration in Twitter requires substantial effort. Users often have to scan information streams by hand. In this paper, we approach this problem by means of faceted search. We propose strategies for inferring facets and facet values on Twitter by enriching the semantics of individual Twitter messages (tweets) and present different methods, including personalized and context-adaptive methods, for making faceted search on Twitter more effective. We conduct a large-scale evaluation of faceted search strategies, show significant improvements over keyword search and reveal significant benefits of those strategies that (i) further enrich the semantics of tweets by exploiting links posted in tweets, and that (ii) support users in selecting facet value pairs by adapting the faceted search interface to the specific needs and preferences of a user.

Keywords: faceted search, twitter, semantic enrichment, adaptation.

1 Introduction

The broad adoption and ever increasing popularity of Social Web have been re-shaping the world we live in. Millions of people from all over the world use social media for sharing masses of (user-generated) content. This data, for example from social tagging or (micro-)blogging, is often unstructured and is not in compliance with the Semantic Web standards. Research efforts aiming at transforming social data into RDF data such as DBpedia [3], and services like revyu.com do exist along with other pioneer exceptions like Semantic Media Wiki[1], semantic tagging [1], and semantic (micro-)blogging (SMOB [2]). However, the big players often do not adhere to Semantic Web principles. For instance, on Twitter, the most popular microblogging service on the Web, the content of Twitter messages (tweets) is not semantically described, which has a negative impact on search. Even though Twitter does allow for metadata[2], this metadata is for describing the *context* of a tweeting activity; e.g. location of the user, Twitter client from

---

[1] http://semantic-mediawiki.org/wiki/Semantic_MediaWiki
[2] http://dev.twitter.com/pages/annotations_overview

L. Aroyo et al. (Eds.): ISWC 2011, Part I, LNCS 7031, pp. 1–17, 2011.
© Springer-Verlag Berlin Heidelberg 2011

which the user tweeted, date and time of the tweet and so on. Yet, there is still a lack of tools to automatically enrich the semantics of tweets and fill those meta-data fields with semantics in order to describe the *content* of a Twitter message. The lack of semantics and structure makes searching and browsing on Social Web applications like Twitter a really challenging task.

Although considerable amount of research has been directed towards Twitter recently, search on Twitter has not been studied extensively yet which motivates, for example, the TREC 2011 track on Microblogs that defines the first search tasks on Twitter[3]. In line with the TREC research objectives, we investigate ways to enhance search and content exploration in the microblogosphere by means of faceted search. In an open and enormous network like Twitter, users may get lost, become de-motivated and frustrated easily with the information overload. Hence, there is a need for an effective personalized searching option from the users' point of view that would assist them in following the optimal path through a series of facets to find the information they are looking for, while providing a structured environment for relevant content exploring. In this paper we propose and evaluate an adaptive faceted search framework for Twitter. We investigate how to extract facets from tweets, how to design appropriate faceted search strategies on Twitter and analyze the impact of the faceted search strategy building blocks on the search performance by means of an automated evaluation framework for faceted search. Our main contributions can be summarized as follows[4].

Semantic Enrichment. To allow for faceted search on Twitter, we present methods for enriching the semantics of tweets by extracting facets from tweets and related external Web resources that describe the content of tweets.

Adaptive Faceted Search Framework. We introduce different building blocks that allow for various faceted search strategies for content exploration on Twitter and propose methods that adapt to the interests/context of a user.

Evaluation Framework. We present an evaluation environment based on an established model for simulating users' click behavior to evaluate different strategies of our adaptive faceted search engine for Twitter. Given this, we prove the effectiveness of our methods on a large Twitter dataset of more than 30 million tweets. We reveal the benefits of faceted search over keyword search and investigate the impact of the different building blocks of our adaptive faceted search framework on the search performance.

2   Background

Twitter is the second most popular social media application which has experienced exponential growth over the last few years in terms of number of users and

---

[3] http://sites.google.com/site/trecmicroblogtrack/

[4] Our adaptive faceted search framework, the code of our evaluation framework and the dataset are available via: http://wis.ewi.tudelft.nl/iswc2011/

tweets published. A recent report shows that one billion tweets are published in a week which corresponds to an average of 140 million tweets per day[5]. This astounding growth and popularity of the microblogging service have naturally been attracting significant amount of research from various perspectives and fields lately. In this section, we present the background and related work that help to understand the usage dynamics and semantics of Twitter messages and motivations for faceted search on Twitter.

2.1   Understanding Twitter Semantics

Tweets are distinctively short text messages of maximum 140 characters that do not explicitly feature facets, in addition to being too short to extract meaningful semantics from. Furthermore, the language and syntax of tweets are significantly different than other Web documents, since Twitter users tend to use abbreviations and short-form for words to save space, as well as colloquial expressions, which make it even harder to infer semantics from tweets.

Mining the semantics of tweets could lead to interesting applications. For instance, Twitris 2.0, a Semantic Web application, facilitates understanding perceptions from social media by capturing semantics with spatial, temporal, thematic dimensions, user intentions and sentiments, and networking behavior from Twitter [8]. Following a top-down approach, Stankovic et al. mapped tweets to conference talks and exploited metadata of the corresponding research papers to enrich the semantics of tweets in order to better understand the semantics of the tweets published in conferences [9]. We follow a similar approach to this where we try to leverage the semantics of tweets for enhancing search on Twitter. Therefore, instead of a restricted domain like scientific conferences, we try to enrich the tweets in general.

Studies on the social network of Twitter and information diffusion dynamics show that tweets are often news related. For instance, Kwak et al. showed that the majority of the trending topics and 85% of all the posted tweets in Twitter are related to news [4]. Sankaranarayanan et al. investigated the use of Twitter to build a news processing system from tweets, called TwitterStand, by capturing tweets that correspond to late breaking news [10]. Some researchers differentiated between news and casual information by investigating the credibility of news propagated through Twitter, while others studied the information propagation via re-tweeting during emergency events [11]. Building on such studies which revealed that Twitter is used more as a news media than a social network [4], and identified "information seekers" as a primary category of Twitter users [7], we try to map tweets to news articles on the Web over the same time period in order to enrich them and to allow for extracting more entities to generate richer facets for search.

Another distinct characteristic of the Twitter syntax is the use of hashtags. Hashtags are meant to be identifiers for related messages of the same topic. By including a hashtag in a message, users indicate to which conversations their

---

[5] http://blog.twitter.com/2011/03/numbers.html

message is related to. Due to the unorganized and fragmented streams of information in Twitter, the use of hashtags has become the means of creating threads of conversations and gathering those serving for a particular interests. When used appropriately, searching on hashtags would return messages that belong to the same conversation. Huang et al. studied the use of hashtags and tagging behavior in Twitter in comparison to Delicious, where they found that hashtags are often just meaningful for a short period of time, and described tagging in Twitter as "conversational" [14].

2.2   Search on Twitter

Since Twitter has become an important source of information for late-breaking news, Twitter posts are already being exploited by major search engines such as Google and Bing. The simplicity of Twitter is one of its powers that has played an important role in its success. However this simplicity brings about negative effect when it comes to searching, browsing or mining the Twitterverse for various uses. Aggregating functions are limited to filtering tweets by users or hashtags, or restricting by keywords, organized by time and not by relevance [12]. Our work is motivated by the inaccuracies of the current keyword search option and the lack of semantics in tweets that hinders a better browsing experience.

Searching and browsing are indeed limited in Twitter. For example, one can search for tweets by a keyword or by a user in a timeline that would return the most recent posts. So, if a user wants to see the different tweets about a field of sports, and were to search for "sports" in Twitter, only the recent tweets that contain the word "sports" would be listed to the user. Many tweets that do not contain the search keyword, but are about different sport events, sport games and sport news in general, would be filtered out. This keyword search is not only imprecise, but is also missing out on a number of messages that do not contain the particular keyword. As tweets are unconventionally short and do not contain explicit meanings, searching microblogging platforms and making sense of the streams of messages passing through the system become even more challenging.

On the other hand, semantic search augments and improves traditional search results by using data from the Semantic Web. Guha et al. described two semantic search systems and outlined how the semantics of search terms can be used for improving search results [15]. We follow a similar approach to adding explicit semantics in order to improve search by extracting entities from tweets and linking external Web sources to tweets in order to enhance their semantics.

A systematic overview of search behavior on Twitter and what differentiates it from Web search was presented by [16]. Researchers investigated why people search Twitter and found out that people mainly search socially generated content to find temporally relevant information (e.g. breaking news, traffic jams etc.) and social information (e.g. opinion, general sentiment, information related to people of interest), as well as to "monitor" the associated results. It was also noted that Twitter search queries are shorter, more popular and less likely to evolve as part of a session, whereas Web queries change and develop during a session to "learn" more about a topic. We take the search behavior of the users into account when we develop strategies for our faceted search framework, such as time sensitive or personalized rankings of facet values.

2.3 Faceted Search

Faceted search is becoming a popular method to allow users to interactively search and navigate complex information spaces. Faceted search systems help people find what they are looking for by allowing them to specify not just keywords related to their information needs, but also metadata which is used for query refinement. Hearst defined facets as "a set of meaningful labels organized in such a way as to reflect the concepts relevant to a domain" [18]. Koren et al. defined three common characteristics for faceted search interfaces; (i) facets and facet-values, (ii) previous search results, and (iii) the current query [17]. By choosing from the suggested facet-values, a user can interactively refine the query. Traditional faceted search interfaces allow users to search for items by specifying queries regarding different dimensions and properties of the items (facets) [19]. For example, online stores such as eBay[6] or Amazon[7] enable narrowing down their users' search for products by specifying constraints regarding facets such as the price, the category or the producer of a product. In contrast, information on Twitter is rather unstructured and short, which does not explicitly feature facets. This puts constraints on the size and the number of keywords, as well as facets that can be used as search parameters without risking to filter out many relevant results.

As a solution, we enrich the semantics of tweets by extracting facets and assigning semantics to them, which allows for a rather semantic faceted search than a keyword search. For instance, given a tweet like *"Off to BNP Paribas at Indian Wells"*, entities such as *"BNP Paribas"* and *"Indian Wells"* are extracted and assigned to facet types such as "SportsEvents" and "Locations" respectively, which allows for searching in different dimensions (multiple facets) even though the words like "sport", "event" or "location" are not included in the tweet (see Figure 1(a)).

2.4 Problem Formalization

On Twitter, facets describe the properties of a Twitter message. For example, persons who are mentioned in a tweet or events a tweet refers to. Oren et al. [19] formulate the problem of faceted search in RDF terminology. Given an RDF statement $(subject, predicate, object)$, the faceted search engine interprets (i) the subject as the actual resource that should be returned by the engine, (ii) the predicate as the facet type and (iii) the object as the facet-value (restriction value). We follow this problem formulation proposed by Oren et al. [19] and interpret tweets as the actual resources (subjects) which the faceted search engine should return, entities that are mentioned in a tweet as facet value and the type of an entity as facet type.

---

[6] http://ebay.com/
[7] http://amazon.com/

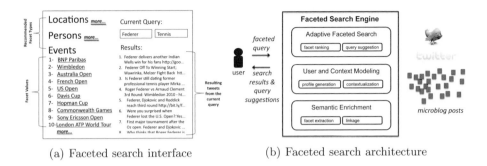

(a) Faceted search interface   (b) Faceted search architecture

Fig. 1. Adaptive faceted search on Twitter: (a) example interface and (b) architecture of the faceted search engine

Figure 1(a) illustrates how we envision the corresponding faceted search interface that allows users to formulate faceted queries. Given a list of facet values which are grouped around facet types such as locations, persons and events, users can select facet-value pairs such as $(URI_{event}, URI_{wimbledon})$ to refine their current query $((URI_{person}, URI_{federer}), (URI_{sportsgame}, URI_{tennis}))$. A faceted query thus may consist of several facet-value pairs. Only those tweets that match all facet-value constraints will be returned to the user. The ranking of the tweets that match a faceted query is a research problem of its own (cf. [16]). In this paper, we rank matching tweets according to their creation time, i.e. the older a tweet the lower its ranking. The core challenge of the faceted search interface is to support the facet-value selection as good as possible. Hence, the facet-value pairs that are presented in the faceted search interface (see left in Figure 1(a)) have to be ranked so that users can quickly narrow down the search result lists until they find the tweets they are interested in. Therefore, the *facet ranking problem* can be defined as follows.

Definition 1 (Facet Ranking Problem). *Given the current query F_{query}, which is a set of facet-value pairs $(predicate, object) \in F_{query}$, the hit list H of resources that match F_{query}, a set of candidate facet-value pairs $(predicate, object) \in F$ and a user u, who is searching for a resource r at time t via the faceted search interface, the core challenge of the faceted search engine is to rank the facet-value pairs F. Those pairs should appear at the top of the ranking that restrict the hit list H so that u can retrieve t with the least possible effort.*

3   Framework for Adaptive Faceted Search

The architecture of the engine that we propose for faceted search on Twitter is depicted in Figure 1(b) and features three main components. The *semantic enrichment* layer aims to extract facets from tweets and generates RDF statements that describe the semantic meaning of a Twitter message. In order to adapt the

Table 1. Building blocks of faceted search framework: strategies for extracting facet-value pairs (FVPs) from tweets, inferring user interest in FVPs and ranking FVPs

Building Block	Description
Semantic Enrichment	Enriching tweets to extract FVPs for representing tweets: (1) tweet-based enrichment (2) tweet-based and link-based enrichment
User/Context Modeling	Strategies for generating profiles that represent (current) user demands in FVPs: (1) user modeling based on tweets published by a user [6] (2) context modeling based on user context when issuing a query (here: query time)
Adaptive Faceted Search	Strategies for adapting the faceted search interface to the user and context and for ranking FVPs in particular: (1) Occurrence Frequency: ranking based on frequency of a FVP in the tweets (2) Personalization: adapting the FVP ranking to a given user profile (3) Time Sensitivity: adapting the FVP ranking to temporal context (4) Diversification: strategy to increase variety among the top-ranked FVPs

faceted search engine to the people who are using it, we propose *user modeling and context modeling* strategies that infer interests of the users in facets. Based on the semantically enriched tweets and the user profiles inferred by the user modeling layer, the *adaptive faceted search* layer solves the actual facet ranking problem. It provides methods that adapt the facet-value pair ranking to the given context and user. Table 1 lists the components and different strategies of the three main building blocks. Below, we explain these building blocks in detail.

3.1 Semantic Enrichment

Twitter messages are short text messages that do not feature facets describing the content of the message. Twitter messages such as "Federer is great http://bit.ly/2fRds1t" can be represented in RDF using, for example, SIOC vocabulary[8], the semantic meaning of such messages is however not explicitly defined:

```
<http://twitter.com/bob/statuses/48748435752333312>
    a <sioc:Post> ;
    dcterms:created "2011-07-08T15:52:51+00:00" ;
    sioc:content "Federer is great http://bit.ly/2fRds1t";
    sioc:has_creator <http://twitter.com/bob> ;
    sioc:links_to <http://bit.ly/2fRds1t> ;
```

While the above RDF representation specifies the tweet's metadata such as the creator of the tweet or the creation time, it requires further enrichment so that

---

[8] http://rdfs.org/sioc/spec/

the content of a tweet is semantically described as well. Representing the semantics of Twitter messages will allow for semantic search strategies such as faceted search (for casual users) or SPARQL queries (for advanced users and application developers). Our faceted search framework features two core strategies for extracting the semantics from tweets: (i) tweet-based enrichment where named entities are extracted from Twitter messages and (ii) tweet-based and link-based enrichment where tweets are further enriched with entities that are extracted from external Web resources that are referenced from the tweets. Therefore, our framework connects to three named entity recognition services: OpenCalais[9], DBpedia spotlight[10] and Alchemy[11]. Using our semantic enrichment infrastructure, we can represent the semantics of the above Twitter message:

```
<http://twitter.com/bob/statuses/48748435752333312>
    a <sioc:Post> ;
    ...
    sioc:has_topic <http://dbpedia.org/resource/Roger_Federer> ;
    sioc:has_topic <http://dbpedia.org/resource/Tennis> ;
    sioc:has_topic <http://dbpedia.org/resource/2009_Wimbledon_Championships> .
```

While the relation to *dbpedia:Roger_Federer* can be inferred by merely analyzing the tweet (tweet-based enrichment), inferring that the tweet refers to Federer's achievements at the Wimbledon tournament 2009 is possible when following the link that is posted in the tweet (link-based enrichment). Our engine uses the identified entities as facet values and exploits the type of the entities to group facet values into facet types. In our evaluation, we process tweets by means of the OpenCalais API which allows us to infer 39 different facet types.

3.2   User and Context Modeling

The goal of the user modeling module is to create a user profile that represents the current demands of the user so that the faceted search interface can be adapted to the inferred profile. Therefore, we define a user profile as a list of weighted facet values (entities):

Definition 2 (User Profile). *The profile of a user $u \in U$ is a set of weighted entities where with respect to the given user u for an entity $e \in E$ its weight $w(u, e)$ is computed by a certain function w.*

$$P(u) = \{(e, w(u, e)) | e \in E, u \in U\}$$

Here, E and U denote the set of entities and users respectively.

In this paper, we apply a lightweight user modeling strategy that weights the entities according to their occurrence frequency in the complete history of tweets which have been published by the user u before she is performing the search

---

[9] http://opencalais.com/

[10] http://dbpedia.org/spotlight

[11] http://alchemyapi.com/

activity. The time of a search activity is considered as context and in the subsequent section we introduce a strategy that exploits this feature to adapt the faceted search engine to the temporal context. For more detailed information on user and context modeling strategies that are part of our search framework, we refer the reader to [6].

3.3   Adaptive Faceted Search

Given the strategies for enriching the semantic descriptions of tweets as well as user and context modeling strategies, the module for adaptive faceted search can operate on semantically rich Twitter items and profiles to solve the ranking task specified above (see Definition 1). Below, we present four ranking strategies that order the facet-value pairs to adapt the faceted search interface to the current context and user.

Occurrence Frequency. A lightweight approach is to rank the facet-value pairs $(p, e) \in F$ based on their occurrence frequency in the current hit list H, the set of tweets that match the current query (cf. Definition 1):

$$rank_{frequency}((p, e), H) = |H_{(p,e)}| \qquad (1)$$

$|H_{(p,e)}|$ is the number of (remaining) tweets that contain the facet-value pair (p, e) which can be applied to further filter the given hit list H. By ranking those facet values high that appear in most of the tweets, $rank_{frequency}$ minimizes the risk of ranking relevant facet values low. However, this might increase the effort a user has to invest to narrow down search results: by selecting facet values which occur in most of the remaining tweets the size of the hit list is reduced slowly.

Personalization. The personalized facet ranking strategy adapts the facet ranking to a given user profile that is generated by the user modeling layer depicted in Figure 1(b). Given the set of facet-value pairs $(p, e) \in F$ (cf. Definition 1), the personalized facet ranking strategy utilizes the weight $w(u, e)$ in $P(u)$ (cf. Definition 2) to rank the facet-value pairs:

$$rank_{personalized}((p, e), P(u)) = \begin{cases} w(u, e) & \text{if } w(u, e) \in P(u) \\ 0 & \text{otherwise} \end{cases} \qquad (2)$$

Diversification. The main idea of the diversification strategy is to produce facet rankings for which the highly ranked facet-values lead to diverse subsets of the current hit list H. For example, if a user is searching for news on "Egypt", based on the frequency, the highly ranked facet-values would be entities such as "Cairo" or "Middle East", because they appear in most of the resources in the hit list. However, these facet-values may refer to very similar items, i.e. issuing the query "Cairo" on top of "Egypt" will not filter out many more items. Hence, to drill down to a small result set as quickly as possible, it might be more appropriate to display facet value pairs which (i) are more selective and (ii) are diverse from the other facet-value pairs so that users with diverse information needs can be satisfied.

The diversification algorithm that we propose uses occurrence frequency as basis ranking strategy and then reorders the FVPs according to the number of items in the current hit list that (1) match the given FVP and (2) do not match the higher ranked FVPs (see Equation 3 and Equation 4).

$$rank_{diversify}((p, e), H) = rank_{frequency}((p, e), H) + d \cdot diversify((p, e), H) \quad (3)$$

$$diversify((p, e), H) = |H_{(p,e)} \setminus \cup_{i=1}^{N} H_{(p_i, e_i)}| \quad (4)$$

Here, $d \in \mathbb{R}$ allows for adjusting the influence of the diversification – in this paper we set $d = 1$. N is the number of higher ranked facet value pairs. All items in the hit list which contain higher ranked FVPs are not taken into account for the scoring of the remaining facet value pairs.

Time Sensitivity. The time sensitive ranking strategy takes into account the current temporal context (query time) and the publishing time of the tweets that match a facet-value pair. The core idea is to rank those FVPs that recently occurred in tweets (*trending FVPs*) higher than FVPs that constantly are mentioned in tweets. To achieve this, we take the creation times of all tweets that match a facet-value pair (p, e) and calculate the average age of these tweets, i.e. the average distance to the actual query time. For each FVP, we therefore obtain a score that describes how recently the tweets are that match the FVP. The smaller the score – i.e. the younger the matching tweets – the higher the rank. In practice, we combine the time sensitive ranking score with one of the above ranking strategies such as occurrence frequency:

$$rank_{time}((p, e), H) = \frac{d}{\overline{avg}_{age}(H_{p,e})} \cdot rank_{frequency}((p, e), H) \quad (5)$$

Here, $\overline{avg}_{age}(H_{p,e})$ is the average, normalized age of the tweets in H that match FVP (p, e). Normalization is done by dividing $avg_{age}(H_{p,e})$ by the maximum average age associated with a FVP in H. The dampen factor $d \in \mathbb{R}$ allows to adjust the influence of the time sensitive score with respect to the ranking score $rank_{frequency}((p, e), H)$. In our experiments, we set $d = 1$ and test the time sensitive scoring method also with other ranking strategies such as $rank_{personalized}((p, e), H)$.

4   Analysis of Facet Extraction

In this section, we analyze the characteristics of a large Twitter corpus of more than 30 million Twitter messages, and investigate how the semantic enrichment of tweets impacts the facet-value pair extraction so that tweets are discoverable by means of faceted search. We collected those tweets by monitoring the Twitter activities of more than 20,000 Twitter users over a period of more than four months starting on November 15, 2010. We started the crawling process by monitoring popular Twitter accounts in the news domain such as the New York Times (*nytimes*) and CNN Breaking News (*cnnbrk*) and then extended the set of accounts in a snowball manner with users who replied or re-tweeted messages

Characteristics	Tweet-based enrichment	Tweet & Link-based enrichment
avg. num. of facet values per tweet	1.85	5.72
avg. num. of discoverable tweets	61161.23	75782.76
avg. num. of FVP-selects to filter results	1.95	2.25
avg. size of filtered result set	1685.320	189.48

(a) Impact of link-based enrichment

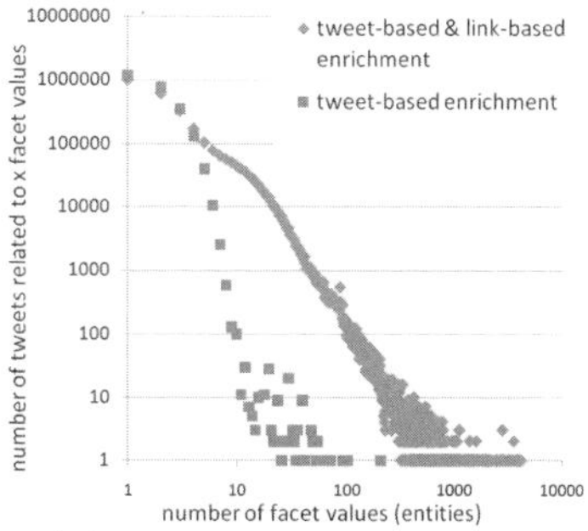

(b) Facet values per tweet

Fig. 2. Impact of link-based enrichment on (a) the characteristics of tweets and the faceted search settings and (b) the number of facet values per tweet

of Twitter users whom we followed already. Figure 2(b) shows the distribution of identified entities per tweet. The distribution shows that for most tweets only a very small number of related entities are identified. Nearly 50% of tweets contain only one entity. Around 92% of the tweets contain 3 or less entities. Moreover, Figure 2(b) shows the impact of the semantic enrichment for the tweets. While the majority of the tweets still contain one, two or three different entities, a significant increase is observed for the number of tweets containing five or more different facet values with the aid of enrichment. In this scenario the number of tweets which are related to more than four facet-value pairs is around 14 times larger when using the semantic enrichment based on link exploitation.

Figure 2(a) overviews some of the characteristics of tweets and the faceted search scenarios for both tweet-based and tweet- and link-based enrichment. It reveals that the number of facet value pairs related to each tweet increases on average when using the semantic enrichment functionality which exploits links to external Web resources. While the tweets contain on average 1.85 facet values, the link-based enrichment strategy features 5.72 facet values per tweet, thus allowing the end-user to find a tweet via many more alternative search paths and faster by drilling down to a smaller set of resulting tweets. The numbers showing discoverable tweets, FVP-selects and the size of the result set are related to the simulated search scenario. Figure 2(a) shows that the link-based-enrichment increases the number of discoverable tweets significantly. This suggests more FVP-selects to drill down the result list. However our evaluations show that the slight increase of 15% more click actions result in a much smaller result set. When using the link exploration strategy, the size of the result set is 9 times smaller, this helps the user to find the tweet(s) of interest faster.

5   Evaluation of Faceted Search Strategies

Having analyzed the characteristics of the facet extraction, we now evaluate the performance of the faceted search strategies proposed in Section 3 and answer the following research questions.

1. How well does faceted search that is supported by the semantic enrichment perform in comparison to keyword search?
2. What strategy performs best in ranking facet-value pairs that allow users to find relevant tweets on Twitter?
3. How do the different building blocks of our faceted search framework (see Table 1) impact the performance?

5.1   Evaluation Methodology

Our evaluation methodology extends an approach introduced by Koren et al. [17] that simulates the clicking behavior of users in the context of faceted search interfaces.

The core evaluation setup consists of parameters describing the user interface itself and algorithms characterizing the simulated user behavior. In general, faceted search user interfaces share some common characteristics and feature at least two parts: an area displaying the facets and a part showing the search results (see Figure 1(a)). Based on such an interface, a user can perform different actions, where the goal is to find a relevant tweet. We consider a tweet as relevant for a user if it was re-tweeted by the user. In a faceted search interface, a user can perform different actions and we focus on the following types of actions: (1) selection of a facet-value pair to refine the query and drill down the search result list, (2) if no appropriate facet-value pair is shown to the user then she can ask for more facet-value pairs and (3) if the user cannot select further facet-value pairs then she has to scan the result set until she finds the relevant tweet. In our simulation, we assume that the user knows the tweet she is looking for and only selects facet-value pairs that match the target tweet.

We model the user's facet-value pair selection behavior by means of a *first-match user* that selects the first matching facet-value pair. To evaluate the performance, we generated search settings by randomly selecting 1000 tweets that have been re-tweeted. Each search setting consists of (i) a target tweet (= the tweet that was re-tweeted), (ii) a user that is searching for the tweet (= the user who re-tweeted the tweet) and (iii) the timestamp of the search activity (= the time when the user re-tweeted the message). The set of candidate items was given by all those tweets which have been published within the last 24 hours before the search activity. On average, the number of candidate items is 61161.23 for the tweet-based enrichment strategy and 75782.76 for the tweet-based and link-based enrichment strategy (see Figure 2(a)) while there is only one single relevant tweet (target tweet) per search session.

For measuring the performance of our facet ranking strategies, we utilize *Success@k* which is the probability that a relevant facet-value pair, the user selects to narrow down the search result list, appears within the top k of the facet-value pair ranking. This metric is a direct indicator for the effort a user needs to spends using the search interface: the higher Success@k, the faster the user will find a relevant facet-value pair when scanning the facet-value pair ranking.

For evaluating the performance of faceted search in comparison with keyword search, we use hashtags as keyword queries and measure the performance by

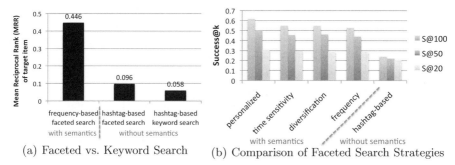

(a) Faceted vs. Keyword Search   (b) Comparison of Faceted Search Strategies

Fig. 3. Overview on results: (a) mean reciprocal rank (MRR) of target item in the search result ranking for faceted search and keyword search and (b) performance of the faceted search strategies for ranking FVPs

means of the mean reciprocal rank (MRR) which indicates at which position we find the target item in the search result ranking.

5.2 Results

Using the evaluation method presented above, we analyze the quality of the search strategies. Figure 3 overviews the results that allow us to answer the research questions raised at the beginning of this section.

Faceted Search vs. Keyword Search. Figure 3(a) shows that our approach to faceted search clearly outperforms faceted search based on hashtags as well as keyword search. Using tweet-based semantic enrichment for extracting FVPs and occurrence frequency as weighting scheme for ranking FVPs (*frequency-based faceted search*), we achieve an improvement regarding MRR of more than 360% (from 0.096 to 0.446) over *hashtag-based faceted search* where hashtags mentioned in the tweets are exploited as facets. Comparing the semantic faceted search strategy to *hashtag-based keyword search*, where a user issues a single hashtag as a query, shows actually an improvement regarding MRR of more than 660% (from 0.058 to 0.446). Furthermore, it is important to state that the results shown in Figure 3(a) are based on those 28% of the search settings for which the target tweet contains at least one hashtag[12]. For the remaining search settings, hashtag-based strategies fail which further proves that the semantic enrichment of our faceted search framework is highly beneficial and important for search on Twitter.

Comparison of Strategies for Ranking FVPs. Figure 3(b) gives an overview of the performance of the different facet ranking strategies measured by Success@100, Success@50 and Success@20. Again, we observe that the hashtag-based faceted search strategy, which exploits hashtags as FVPs and applies occurrence frequency as weighting scheme, is clearly outperformed by the semantic faceted search strategies provided by our framework. For example, the

---

[12] Given the more than 30 million tweets of our dataset, we actually observe that just 19.82% of the tweets mention a hashtag.

tweet-based semantic enrichment in combination with occurrence frequency as weighting scheme (*frequency*) improves over the *hashtag-based* baseline by 121.5, 99.5% and 42.4% regarding S@100, S@50 and S@20 respectively. For the faceted search strategies that make use of semantic enrichment, we observe that the *personalized* ranking strategy outperforms the other strategies for all metrics. When looking at the Success@100, the personalized strategy performs approximately 12% better than the other three strategies. Knowing the preferences of a user for certain topics, which are modeled via the FVPs, thus brings advantages for adapting the faceted search interface to the user who is searching for a tweet. Furthermore, knowing the user's temporal context also improves the search performance slightly (see *time sensitivity* in Figure 3(b)). However, the differences between the ranking strategies based on time, diversification and frequency are very small which might be caused by the fact that both the time sensitive strategy and the diversification use occurrence frequency as basic weighting function. Hence, reducing the influence of the frequency-based scoring on these strategies could possibly lead to further improvements.

Impact of the different Building Blocks on Faceted Search. Figure 4 illustrates the impact of some of the building blocks of our framework on the faceted search performance. In Figure 4(a), we compare the performance of the frequency-based strategy and personalized strategy when doing (i) semantic enrichment solely on tweets (*tweet-based enrichment*) or (ii) semantic enrichment by analyzing both the content of the tweets and Web resources that are linked from the tweets (*tweet-based & link-based enrichment*). It shows that the link-based enrichment significantly improves the Success@100 for both strategies. Hence, while semantic enrichment by means of named entity recognition in tweets improves already the faceted search performance over faceted search based on hashtags, we achieve further improvements if we follow the links posted in Twitter messages to further describe the semantic meaning of a Twitter message. Furthermore, the improvement gained by personalization are consistent through the different enrichment strategies. For example, the personalized strategy improves over the frequency-based strategy by 8.3% when link-based enrichment is conducted.

Figure 4(b) shows how the Success@100 rates for the frequency and the diversification based rankings increase when the temporal context of the search activity is taken into account to adjust the ranking of the FVPs. One can see improvements for both strategies with 3.8% for the frequency-based strategy and 5.3% for the diversification strategy.

Synopsis. Given these observations, we now revisit our research questions raised at the beginning of this section. We showed that faceted search clearly outperforms hashtag-based keyword search. Using the our faceted search framework, we achieve a more than eight times higher MRR than keyword search. In response to which strategy performs the best for ranking facet-value pairs, we revealed that the personalization strategy – i.e. adapting the facet-value pair ranking to the interest profile of the user who is searching for a tweet – performs best for faceted search on Twitter. Furthermore, we showed that the different building

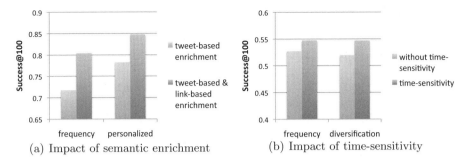

Fig. 4. Impact of (a) semantic enrichment strategy on the frequency-based and the personalized strategies for search settings where the target tweet contains a link and (b) impact of time-sensitive re-ordering on the performance of the frequency-based and the diversification-based strategies

blocks of our faceted search framework all have positive impact on the facet-value pair ranking in order to answer our second question. Semantic enrichment by means of exploiting both tweets and Web resources that are referenced from the tweets increases the number of tweets that are discoverable via faceted search. It also increases the number of facet-value pairs per tweet so that users have more alternatives in narrowing down the search result list. Moreover, we showed that time-sensitivity – i.e. adapting the facet-value pair ranking to the temporal context – improves the performance of the facet-value pair ranking so that users can find their intended tweets faster and with less effort.

6   Conclusions

In this paper, we tackled the problem of searching for relevant messages on Twitter. We introduced an adaptive faceted search framework that features semantic enrichment of tweets as a solution to this problem. Our framework allows adding semantics to tweets by extracting entities and enriching them with external resources in order to create facets (e.g. persons, locations, organizations etc.) and facet-values that describe the content of tweets. To support users in selecting facet-value pairs during their faceted search activities, we studied different strategies that adapt the ranking of facet-value pairs to the user and context (e.g. temporal context).

We presented an evaluation framework that allows for simulating users' search behavior and applied this simulator on a large Twitter dataset of more than 30 million tweets. Our evaluation proves the effectiveness of our strategies and reveals that our faceted search framework achieves tremendous improvements in comparison with hashtag-based keyword search. Moreover, we see that personalization and context-adaptation gain the best performance among the faceted search strategies. Our analysis of semantic enrichment strategies also showed that the exploitation of links that are posted in Twitter messages is beneficial for describing the semantic meaning of tweets and therefore improves the search performance as well.

Acknowledgements. The research leading to these results has received funding from the European Union Seventh Framework Programme (FP7/2007-2013) in context of the ImREAL project[13] and the SYNC3 project[14].

References

1. Passant, A., Laublet, P.: Meaning Of A Tag: A collaborative approach to bridge the gap between tagging and Linked Data. In: Workshop on Linked Data on the Web, Beijing, China (2008)
2. Passant, A., Hastrup, T., Bojars, U., Breslin, J.: Microblogging: A Semantic Web and Distributed Approach. In: Workshop on Scripting For the Semantic Web, Tenerife, Spain, vol. 368. CEUR-WS.org (2008)
3. Auer, S., Bizer, C., Kobilarov, G., Lehmann, J., Cyganiak, R., Ives, Z.G.: DB-pedia: A Nucleus for a Web of Open Data. In: Aberer, K., Choi, K.-S., Noy, N., Allemang, D., Lee, K.-I., Nixon, L.J.B., Golbeck, J., Mika, P., Maynard, D., Mizoguchi, R., Schreiber, G., Cudré-Mauroux, P. (eds.) ASWC 2007 and ISWC 2007. LNCS, vol. 4825, pp. 722–735. Springer, Heidelberg (2007)
4. Kwak, H., Lee, C., Park, H., Moon, S.: What is twitter, a social network or a news media? In: WWW, pp. 591–600. ACM (2010)
5. Bernstein, M., Kairam, S., Suh, B., Hong, L., Chi, E.H.: A torrent of tweets: managing information overload in online social streams. In: CHI Workshop on Microblogging: What and How Can We Learn From It? (2010)
6. Abel, F., Gao, Q., Houben, G.-J., Tao, K.: Analyzing user modeling on twitter for personalized news recommendations. In: Konstan, J.A., Conejo, R., Marzo, J.L., Oliver, N. (eds.) UMAP 2011. LNCS, vol. 6787, pp. 1–12. Springer, Heidelberg (2011)
7. Java, A., Song, X., Finin, T., Tseng, B.: Why we twitter: understanding microblogging usage and communities. In: Workshop on Web Mining and Social Network Analysis, pp. 56–65. ACM (2007)
8. Jadhav, A., Purohit, H., Kapanipathi, P., Ananthram, P., Ranabahu, A., Nguyen, V., Mendes, P.N., Smith, A.G., Cooney, M., Sheth, A.: Twitris 2.0: Semantically empowered system for understanding perceptions from social data. In: Semantic Web Challenge (2010)
9. Stankovic, M., Rowe, M., Laublet, P.: Mapping Tweets to Conference Talks: A Goldmine for Semantics. In: Workshop on Social Data on the Web, Shanghai, China, vol. 664. CEUR-WS.org (2010)
10. Sankaranarayanan, J., Samet, H., Teitler, B.E., Lieberman, M.D., Sperling, J.: Twitterstand: news in tweets. In: SIGSPATIAL, pp. 42–51. ACM (2009)
11. Sakaki, T., Okazaki, M., Matsuo, Y.: Earthquake shakes Twitter users: real-time event detection by social sensors. In: WWW, pp. 851–860. ACM (2010)

---

[13] http://imreal-project.eu
[14] http://sync3.eu

12. Laniado, D., Mika, P.: Making Sense of Twitter. In: Patel-Schneider, P.F., Pan, Y., Hitzler, P., Mika, P., Zhang, L., Pan, J.Z., Horrocks, I., Glimm, B. (eds.) ISWC 2010, Part I. LNCS, vol. 6496, pp. 470–485. Springer, Heidelberg (2010)
13. Letierce, J., Passant, A., Breslin, J., Decker, S.: Understanding how Twitter is used to widely spread scientific messages. In: Web Science Conference (2010)
14. Huang, J., Thornton, K.M., Efthimiadis, E.N.: Conversational Tagging in Twitter. In: Hypertext, pp. 173–178. ACM (2010)
15. Guha, R., Mccool, R., Miller, E.: Semantic search. In: WWW, pp. 700–709. ACM (2003)
16. Teevan, J., Ramage, D., Morris, M.R.: #twittersearch: a comparison of microblog search and web search. In: WSDM, pp. 35–44. ACM (2011)
17. Koren, J., Zhang, Y., Liu, X.: Personalized interactive faceted search. In: WWW, pp. 477–486. ACM (2008)
18. Hearst, M.A.: Design recommendations for hierarchical faceted search interfaces. In: Workshop on Faceted Search Co-located with SIGIR, pp. 26–30 (2006)
19. Oren, E., Delbru, R., Decker, S.: Extending Faceted Navigation for RDF Data. In: Cruz, I., Decker, S., Allemang, D., Preist, C., Schwabe, D., Mika, P., Uschold, M., Aroyo, L.M. (eds.) ISWC 2006. LNCS, vol. 4273, pp. 559–572. Springer, Heidelberg (2006)

ANAPSID: An Adaptive Query Processing Engine for SPARQL Endpoints

Maribel Acosta[1], Maria-Esther Vidal[1], Tomas Lampo[2], Julio Castillo[1],
and Edna Ruckhaus[1]

[1] Universidad Simón Bolívar, Caracas, Venezuela
{macosta,mvidal,ruckhaus}@ldc.usb.ve, julio@gia.usb.ve
[2] University of Maryland, College Park, USA
tlampo@cs.umd.edu

Abstract. Following the design rules of Linked Data, the number of available SPARQL endpoints that support remote query processing is quickly growing; however, because of the lack of adaptivity, query executions may frequently be unsuccessful. First, fixed plans identified following the traditional optimize-then-execute paradigm, may timeout as a consequence of endpoint availability. Second, because blocking operators are usually implemented, endpoint query engines are not able to incrementally produce results, and may become blocked if data sources stop sending data. We present ANAPSID, an adaptive query engine for SPARQL endpoints that adapts query execution schedulers to data availability and run-time conditions. ANAPSID provides physical SPARQL operators that detect when a source becomes blocked or data traffic is bursty, and opportunistically, the operators produce results as quickly as data arrives from the sources. Additionally, ANAPSID operators implement main memory replacement policies to move previously computed matches to secondary memory avoiding duplicates. We compared ANAPSID performance with respect to RDF stores and endpoints, and observed that ANAPSID speeds up execution time, in some cases, in more than one order of magnitude.

1 Introduction

The Linked Data publication guideline establishes the principles to link data on the Cloud, and make Linked Data accessible to others[1]. Based on these rules, a great number of available SPARQL endpoints that support remote query processing to Linked Data have become available, and this number keeps growing. Additionally, the W3C SPARQL working group is defining a new SPARQL 1.1 query language to respect the SPARQL protocol and specify queries against federations of endpoints [19]. However, access to the Cloud of Linked datasets is still limited because many of these endpoints are developed for very lightweight use. For example, if a query posed against a linkedCT endpoint[2] requires more than 3 minutes to be executed, the endpoint will timeout without producing any answer. Thus, to successfully execute real-world queries, it

---

[1] http://www.w3.org/DesignIssues/LinkedData.html
[2] Clinical Trials data produced by the ClinicalTrials.gov site available at
http://linkedCT.org. and http://hcls.deri.org/sparql

L. Aroyo et al. (Eds.): ISWC 2011, Part I, LNCS 7031, pp. 18–34, 2011.
© Springer-Verlag Berlin Heidelberg 2011

may be necessary to decompose them into simple sub-queries, so that the endpoints will then be capable of executing these sub-queries in a reasonable time. Additionally, since endpoints may unpredictably become blocked, execution engines should modify plans on-the-fly to contact first the available endpoints, and produce results as quickly as data arrives.

Several query engines have been developed to locally access RDF data [1,10,12,17,24]. The majority have implemented optimization techniques and efficient physical operators to speed up execution time [12,17,24]; others have defined structures to efficiently store and access RDF data [17,25], or have developed strategies to reuse data previously stored in cache [1,10,17]. However, none of these engines are able to gather Linked Data accessible through SPARQL endpoints, or hide delays from users.

Recently several approaches have addressed the problem of query processing on Linked Data [2,7,9,13,14,15,16,21]; some have implemented source selection techniques to identify the most relevant sources for evaluating a query [7,14,21], while others have developed frameworks to retrieve and manage Linked Data [2,8,9,13,15,16], and to adapt query processing to source availability [9]. Additionally, Buil-Aranda et al. [4] have proposed optimization techniques to rewrite federated queries specified in SPARQL 1.1, and reduce the query complexity by generating well-formed patterns. Finally, some RDF engines[18,20] have been extended to query federations of SPARQL endpoints. Although all these approaches are able to access Linked Data, none of them can simultaneously provide an adaptive solution to access SPARQL endpoints.

In this paper we present ANAPSID, an engine for SPARQL endpoints that extends the adaptive query processing features presented in [22], to deal with RDF Linked Data accessible through SPARQL endpoints. ANAPSID stores information about the available endpoints and the ontologies used to describe the data, to decompose queries into sub-queries that can be executed by the selected endpoints. Also, adaptive physical operators are executed to produce answers as soon as responses from available remote sources are received. We empirically analyze the performance of the proposed techniques, and show that these techniques are competitive with state-of-the-art RDF engines which access data either locally or remotely.

The paper is comprised of six additional sections. We start with a motivating example in the following section. Then, we present the ANAPSID architecture in section 3 and describe the query engine in section 4. Experimental results are reported in section 5, and section 6 summarizes the related work. Finally, we conclude in section 7 with an outlook to future work.

2   Motivating Example

LinkedSensorData[3] is a dataset that makes available sensor weather data of approximately 20,000 stations around the United States. Each station provides information about weather observations; the ontology O&M-OWL[4] is used to describe these

---

[3] http://wiki.knoesis.org/index.php/LinkedSensorData

[4] http://knoesis.wright.edu/ssw/ont/sensor-observation.owl

observations; a Virtuoso endpoint is provided to remotely access the data. Further, each station is linked to its corresponding location in *Geonames*[5].

Consider the acyclic query: *Retrieve all sensors that detected freezing temperatures on April 1st, 2003, between 1:00am and 3:00am*[6]. The answer comprises 1,600 sensors.

```
prefix om-owl:<http://knoesis.wright.edu/ssw/ont/sensor-observation.owl#>
prefix rdf:<http://www.w3.org/1999/02/22-rdf-syntax-ns#>
prefix weather:<http://knoesis.wright.edu/ssw/ont/weather.owl#>
prefix sens-obs:<http://knoesis.wright.edu/ssw/>
prefix xsd:<http://www.w3.org/2001/XMLSchema#>
prefix owl-time:<http://www.w3.org/2006/time#>
prefix gn:<http://www.geonames.org/ontology#>
SELECT DISTINCT ?sensor
WHERE {
?sensor om-owl:generatedObservation ?observation .
?observation rdf:type weather:TemperatureObservation .
?observation om-owl:samplingTime ?time .?time owl-time:inXSDDateTime ?xsdtime .
?observation om-owl:result ?result .?result om-owl:floatValue ?value .
?result om-owl:uom weather:fahrenheit .FILTER(?value <= "32.0"^^xsd:float).
FILTER(?xsdtime >= "2003-04-01T01:00:00-07:00"^^http://www.w3.org/2001/XMLSchema#dateTime")
FILTER(?xsdtime <= "2003-04-01T03:00:00-07:00"^^http://www.w3.org/2001/XMLSchema#dateTime").
?sensor om-owl:hasLocatedNearRel ?location .?location om-owl:hasLocation ?ga. ?ga gn:name ?name}
```

Using the *LinkedSensorData* endpoint[7], we executed several versions of the former query with different date ranges. Table 1 reports on the observed execution time values. Different instantiations of the SPARQL endpoint parameter SPARQL SPONGE[8] were set up to indicate the type of dereferences to be executed during query processing.

Table 1. Execution Time (secs.) of Queries Against the *LinkedSensorData* SPARQL Endpoint; SPONGE parameter: Local, Grab All, Grab All *seeAlso*, Grab Everything

	Local	Grab All	Grab All (*seeAlso*)	Grab Everything
Average	0.35	100.78	Timeout	Timeout
Standard Deviation	0.04	38.32	Timeout	Timeout
Minimum	0.30	58.95	Timeout	Timeout
Maximum	0.45	155.59	Timeout	Timeout

We can observe that if the query is run on data locally stored in the endpoint, i.e., SPONGE is equal to Local and only one endpoint is contacted, the queries can be executed in less than one second. However, if IRI's are dereferenced by using the *Grab All* option, the execution time increases in average two orders of magnitude. Moreover, if the *seeAlso* references are considered and the corresponding endpoints are contacted (Grab All *seeAlso*), the execution reaches a timeout of 86,400 secs. (one day). Similarly, if all the referred resources are downloaded (Grab Everything), the endpoint reaches the timeout without producing any answer. These results suggest that when the *LinkedSensorData* endpoint requires to download data from remote endpoints, it may become blocked waiting for answers; this may be caused by a blocking query processing model executed by existing endpoints.

---

[5] http://www.geonames.org/ontology

[6] Time is specified in Mountain Time; temperature in Fahrenheit scale.

[7] http://sonicbanana.cs.wright.edu:8890/sparql

[8] http://docs.openlinksw.com/virtuoso/virtuososponger.html

Traditionally, query processing engines are built on top of a blocking iterator model that fires a query execution process from the root of the execution plan to the leaves, and does not incrementally produce any result until its corresponding children have completely produced the answer. Thus, if any of the intermediate nodes becomes blocked while producing answers, the root of the plan will also be blocked. We consider plans whose leaves are endpoints; however, similar problems may occur, if leaves corresponds to URIs that need to be dereferenced.

To overcome limitations of existing execution models when Linked Data is dereferenced, some state-of-the-art approaches have proposed adaptive query engines that are able to produce answers as data becomes available [9,14]. For example, Hartig et al. [9] extend the traditional iterator model and provide an adaptive query engine that hides delays that occur when any linked dataset becomes blocked. This adaptive iterator detects when a URI look-up stops responding, and resumes the query execution process executing other iterators; results can be incrementally produced, and delays in retrieving data during URI look-ups are hidden from the users. Further, Ladwig et al. [14] use a non-blocking operator to opportunistically produce answers as soon as dereferenced data is available. However, none of these approaches support the access to a federation of SPARQL endpoints. Finally, some RDF engines have been extended to deal with SPARQL queries against federations of endpoints[4,18,20], but no adaptive query techniques have been implemented, and queries are frequently unsuccessful when endpoints become blocked. In this paper we present an adaptive engine that makes use of information about endpoints, to decompose the query into simple sub-plans that can be executed by the remote endpoints. Also, we propose a set of physical operators that gather data generated by the endpoints, and quickly produce responses.

3   The ANAPSID Architecture

ANAPSID is based on the architecture of wrappers and mediators [26] to query federations of SPARQL endpoints (Figure 1).

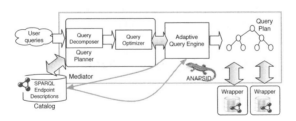

Fig. 1. The ANAPSID Architecture

Lightweight *wrappers* translate SPARQL sub-queries into calls to endpoints as well as convert endpoint answers into ANAPSID internal structures. *Mediators* maintain information about endpoint capabilities, statistics that describe their content and performance, and the ontology used to describe the data accessible through the endpoint.

Following the approach developed in previous work [11], the Local As View (LAV) approach is used to describe endpoints in terms of the ontology used in the endpoint dataset. Further, mediators implement query rewriting techniques, decompose queries into sub-queries against the endpoints, and gather data retrieved from the contacted endpoints. Currently, only SPARQL queries comprised of joins are considered; however, the rewriting techniques have been extended to consider all SPARQL operators, but this piece of work is out of the scope of this paper. Finally, mediators hide delays, and produce answers as quickly as data arrives; they are composed of the following components:

- *Catalog*: maintains a list of the available endpoints, their ontology concepts and capabilities. Contents are described as views with bodies comprised of predicates that correspond to ontology concepts; execution timeouts indicate endpoint capabilities. Statistics are updated on-the-fly by the adaptive query engine.
- *Query Decomposer*: decomposes user queries into multiple simple sub-queries, and selects the endpoints that are capable of executing each sub-query. Simple sub-queries are comprised of a list of triple patterns that can be evaluated against an endpoint, and whose estimated execution time is less than the endpoint timeout. Vidal et al. [24] suggest that the cardinality of the answer of sub-queries comprised of triple patterns that share exactly one variable, may be small-sized; so the query decomposer will try to identify low cost sub-queries that meet this property.
- *Query Optimizer*: identifies execution plans that combine sub-queries and benefits the generation of bushy plans composed of small-sized sub-queries. Statistics about the distribution of values in the different datasets are used to identify the best combination of sub-queries. These statistics and capabilities of the endpoints are collected by following an Adaptive Sampling Technique [3,24], or on-the-fly during query execution.
- *Adaptive Query Engine*: implements different physical operators to gather tuples from the endpoints. These physical operators are able to detect when endpoints become blocked, and incrementally produce results as the data arrives. Additionally, the query engine can modify an execution plan on-the-fly to execute first the requests against the endpoints that are available; information gathered during runtime is used to update catalog statistics, and to re-optimize delayed queries.

4   The ANAPSID Query Processing Engine

The ANAPSID query engine provides a set of operators able to gather data from different endpoints. Opportunistically, these operators produce results by joining tuples previously received even when endpoints become blocked. Additionally, the physical operators implement main memory replacement policies to move previously computed matches to secondary memory, ensuring no duplicate generation. Each join operator maintains a data structure called Resource Join Tuple (RJT), that records for each instantiation of the join variable(s), the tuples that have already matched. Suppose that for the instantiation of the variable $?X$ with the resource r, the tuples $\{T_1, ..., T_n\}$ have matched, then the RJT will be the pair $(r, \{T_1, ..., T_n\})$, where the first argument, *head* of the RJT, corresponds to the resource and the second, *tail* of the RJT, is the list of tuples.

4.1 The Adaptive Group Join (agjoin)

The **agjoin** operator is based on the Symmetric Hash Join [5] and XJoin [22] operators, defined to quickly produce answers from streamed data accessible through a wide-area network. Basically, the Symmetric Hash Join and XJoin are non-blocking operators that maintain a hash table for the data retrieved from sources A and B. Execution requests against A and B are submitted in parallel, and when a tuple is generated from source A (resp. B), it is inserted in the hash table of A (resp. hash table of B) and probed against the hash table of B (resp. hash table of A). An output is produced each time a match is found. Further, the XJoin implements a main memory replacement policy that flushes portions of the hash tables to secondary memory when they become full, and ensures that no duplicates are generated. Even though these operators produce results incrementally, results are produced one-by-one because tuples are first inserted in the corresponding hash table and then probed against the other hash table to find one match at a time. To speed up query answering, we propose the **agjoin** operator. The **agjoin** maintains for source A (resp. B) a list L_A (resp. L_B) of RJTs, which represents for each instantiation, $\mu(?X)$, of the tuples already received from source A, the tuples received from B that match $\mu(?X)$. L_A (resp. L_B) is indexed by the values of $\mu(?X)$ that correspond to the heads of the RJTs in L_A (resp. L_B); thus, **agjoin** provides a direct access to the RJTs. When a new tuple t with instantiation $\mu(?X)$ arrives from source A, **agjoin** probes against L_A to find an RJT whose head corresponds to $\mu(?X)$; if there is a match, the **agjoin** quickly produces the answer as the result of combining t with all the tuples in the tail of RJT of $\mu(?X)$; if not, nothing is added to L_A. Independently of the success of the probing process, t is inserted in its corresponding RJT in L_B. Figure 2 illustrates main memory contents during the execution of **agjoin** between sources A and B.

Fig. 2. agjoin between sources A and B: (a) L_A and L_B current state; (b) effects of arriving a tuple A5 from source A, three tuples are immediately produced, and RJT (r2,{A5}) is inserted in L_B

L_A and L_B in Figure 2 (a) indicate that tuples (B1,A1), (B1,A2), (B2,A1), (B2,A2), (B3,A1), (B3,A2), (B7,A3), (B7,A4), (B8,A3), (B8,A4) have been already produced; also, at this time, no tuples with $\mu(?X) = r_2$ have been received from A, while three of these tuples have arrived from source B. Figure 2 (b) shows the current state of L_A and L_B after a tuple A5 with $\mu(?X) = r_2$ arrives from source A, i.e., shows the effects in L_B of arriving a new tuple A5 with $\mu(?X) = r_2$ from source A. In this case, A5 is probed against L_A and three outputs are produced immediately; concurrently, the insert process is fired, and RJT (r2,{A5}) is inserted in L_B.

Property 1. *Consider the current state of lists L_A and L_B in an instant t, the number of answers produced until t, NAP_t, is given by the following formula:*

$$NAP_t = \sum_{RJT_a \in L_A \,\wedge\, RJT_b \in L_B \,\wedge\, head(RJT_a)=head(RJT_b)} (|tail(RJT_a)| \times |tail(RJT_b)|).$$

A three-stage policy is implemented to flush RJTs; completeness and no duplicates are ensured. A first stage is performed while at least one source sends data; a second stage is fired when both sources are blocked, and the third is only executed when all data have completely arrived from both sources. Note that the same operator can execute first or second stages at different times and depending on the availability of the sources, it can move from one stage to the other; however, the third stage is executed only once.

In a first stage, when a tuple t arrives from source A, it is inserted in an RJT in L_B; the probe time of t in L_A and the insert time of t in L_B are stored with t. Further, if a portion of the main memory assigned to A becomes full, an RJT victim is chosen based on the time of the last probe; thus, the least recently probed RJT is selected, flushed to secondary memory, and annotated with the flush time. In case RJTs with the same head are chosen as victims at different times, only one RJT will be stored to secondary memory; the tail will be comprised of the tails of the different victim RJTs; these tails will be annotated with the respective flush time. Figure 3 illustrates the process performed when a main memory failure occurs and the timestamps of the stored tuples.

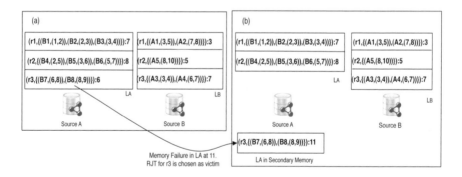

Fig. 3. Timestamp annotations and Main Memory Failures: (a) L_A and L_B timestamps; (b) effects of a main memory failure in L_A; RJT for r3 is flushed

Figure 3 (a) illustrates the RJTs in Figure 2, annotated with the probe and insert times of the tuples, and the RJTs probe times[9]. Thus, we can say that B1 was probed at time 1 and inserted in L_A at 2; also, timestamp 7 associated with the RJT of r1 in L_A, indicates that the last probe of a tuple from source B was performed against this RJT at time 7. Further, suppose that a failure of memory occurs at time 11 in the portion of main memory assigned to source A, then the RJT with head r3, is flushed to secondary memory and its flush time is annotated with 11. Figure 3 (b) illustrates the final state of

---

[9] An RJT probe time corresponds to the most recent probe time of the tuples in the RJT.

L_A (main and secondary memory) and L_B after flushing the RJT to secondary memory. Definition 1 states the conditions to meet when tuples are joined during a first stage.

Definition 1. *Let RJT_i and RJT_j be Resource Join Tuples in L_A and L_B, respectively, such that, head(RJT_i)=head(RJT_j). Suppose RJT_j has been flushed to secondary memory. Then, a tuple $B_j \in$ tail(RJT_i) was matched to tuples of tail(RJT_j) during a first stage of the* **agjoin***, i.e., before RJT_j was flushed, if and only if:*

$$probeTime(B_j) < flushTime(RJT_j).$$

A second stage is fired when both sources become blocked; Definition 2 establishes the conditions to be satisfied by tuples that are matched in a second stage.

Definition 2. *Let RJT_i and RJT_j be Resource Join Tuples in L_A and L_B, respectively, such that, head(RJT_i)=head(RJT_j). Suppose RJT_j has been flushed to secondary memory. Then, a tuple $B_j \in$ tail(RJT_i) was matched to tuples of tail(RJT_j) during a second stage of the* **agjoin***, i.e., before RJT_i was flushed to secondary memory[10], if and only if, there is a second state ss:*

$$flushTime(RJT_j) < insertTime(B_j) < TimeSecondStage(ss) < flushTime(RJT_i).$$

To produce new answers during a second stage, the **agjoin** selects the largest RJTs in secondary memory, and probes them against their corresponding RJTs in main memory. To avoid duplicates, conditions in Definitions 1 and 2 are checked. The execution of a second stage is finished, when one source becomes unblocked, and all the RJTs in secondary memory are checked to find new matches. A global variable named *Time-LastSecondStage*, is maintained and updated when a second stage finishes; also, for each second stage, we maintain the time it was performed.

Suppose tuple t from RJT_i matches tuples in RJT_j in the second stage at time st, then the probe time of t and the probe time of its RJT in main memory are updated to st. To illustrate this process, consider the current state of L_A and L_B reported in Figure 3 (b); also suppose that the last second stage was performed at time 14. Following the policy to select RJTs in secondary memory, (r3,{(B7,(6,8)),(B8,(8,9))}) in the secondary memory version of L_A, is chosen and probed against (r3,{(A3,(3,4)),(A4,(6,7))}) in L_B; the RJT in secondary memory was chosen because it has the longest tail. Since conditions in Definition 1 hold for tuples B7 and B8, no new answers are produced and their timestamps are not changed. Finally, one of the sources becomes available at time 15, then the second stage finalizes, and *TimeLastSecondStage* is updated to 15.

The third stage is fired when data has been completely received. Tuples that do not satisfy conditions in Definitions 1 and 2 are considered to produce the rest of the answers. First, RJTs in main memory are probed with RJTs in secondary memory. Then, RJTs in secondary memory are probed to produce new results. Figure 4 illustrates states of L_A and L_B right after all the tuples have been received at time 100 and the third stage is fired; the last second stage was performed at time 60. First, **agjoin** tries to combine RJT of r3 in secondary memory of source A with RJT of r3 in main memory of source B. Because A21 was inserted in the RJT at time 63, i.e., after the last second stage was

---

[10] If an RJT is in main memory, then its flush time is ∞.

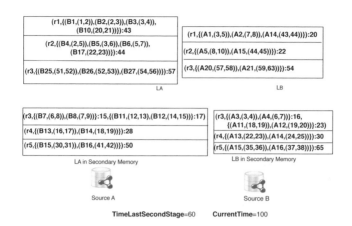

Fig. 4. The **agjoin** third stage at time 100, after having the last second stage at time 60

performed, the combination of A21 with all the tuples of RJT of r3 in secondary memory of source A, must be output. The rest of the combinations between tuples in these RJTs were already produced. Then, RJT of r3 in secondary memory of B and RJT of r3 in main memory of source A are considered, and no answers are produced because all the tuples satisfy conditions in Definition 2. Next, RJTs in secondary memory are combined, but no answers are produced: (a) tuples of RJTs of r3 in secondary memory were matched in a first stage, (b) tuples of RJTs of r4, and tuples of RJTs of r5, were matched in a first stage; at this point **agjoin** finalizes.

Property 2. *Let A and B be sources joined with the* **agjoin** *operator, no duplicates are generated. Additionally, if A and B send all the tuples, the output is complete.*

4.2   The Adaptive Dependent Join (adjoin)

The **adjoin** extends the Dependent join operator [6] with the capability to hide delays to the user. The Dependent join is a non-commutative operator, that is required when instantiations of input attributes need to be bound to produce the output. Similarly, the **adjoin** is executed when a certain binding is required to execute part of a SPARQL query. For example, suppose triple pattern $t_1 = \{s\ p_1\ ?X\}$ is part of an outer sub-query, triple pattern $t_2 = \{?X\ p_2\ o\}$ is part of the inner sub-query, and the predicate p_1 is foaf:page, rdfs:seeAlso, or owl:sameAs. For each instantiation μ of variable $?X$, dereferences of μ must be performed before executing the inner sub-query, i.e., the **adjoin** is used when instantiations from the outer sub-query need to be dereferenced to execute the inner sub-query. Also, the clause BINDINGS in SPARQL 1.1 represents this type of dependencies. We implemented the **adjoin** as an extension of the **agjoin** operator, but instead of asynchronously accessing sources A and B, accesses to source B are only fired when tuples from source A are inserted in L_B. The rest of the operator remains the same.

5   Experimental Study

We empirically analyze the performance of the proposed query processing techniques, and report on the execution time of plans comprised of ANAPSID operators versus queries posed against SPARQL endpoints, and state-of-the-art RDF engines.

Dataset	Number of triples
LinkedSensorData-blizzards	56,689,107
linkedCT	9,809,330
DBPedia	287,524,719

(a) Dataset Cardinality

Benchmark	#patterns	answer size
1	24-30	1,298-9,008
2	13-17	1-99
3	16-20	0-7

(b) Query Benchmarks

Fig. 5. Experiment Configuration Set-Up

Datasets and Query Benchmarks[11]**:** LinkedSensorData-blizzards[12], linkedCT[13], and DBPedia (english articles)[14] were used; datasets are described in Table of Figure 5(a). Sensor data[15] was accessed through a Virtuoso SPARQL endpoint; the timeout was set to 86,400 secs. We could not execute our benchmark queries against existing endpoints for clinical trials because of timeout configuration, so we implemented our own Virtuoso endpoint with timeout equal to 86,400 secs.[16] Three sets of queries were considered (Table of Figure 5(b)); each sub-query was executed as a query against its corresponding endpoint. Benchmark 1 is a set of 10 queries against LinkedSensorData-blizzards; each query can be grouped into 4 or 5 sub-queries. Benchmark 2 is a set of 10 queries over linkedCT with 3 or 4 sub-queries. Benchmark 3 is a set of 10 queries with 4 or 5 sub-queries executed against linkedCT and DBPedia endpoints.

Evaluation Metrics: We report on runtime performance, which corresponds to the *user time* produced by the `time` command of the Unix operation system. Experiments were executed on a Linux CentOS machine with an Intel Pentium Core2 Duo 3.0 GHz and 8GB RAM. Experiments in RDF-3X were run in both cold and warm caches; to run cold cache, we cleared the cache before running each query by performing the command sh -c "sync ; echo 3 > /proc/sys/vm/drop_caches"; to run on warm cache, we executed the same query five times by dropping the cache just before running the first iteration of the query. Each query executed by ANAPSID and SPARQL endpoints was run ten times, and we report on the average time.

Implementations: ANAPSID was implemented in Python 2.6.5.; the SPARQL Endpoint interface to Python (1.4.1)[17] was used to contact endpoints. To be able to configure delays and availability, we implemented an endpoint simulator in Python 2.6.5. This simulator is comprised of servers and proxies. Seven instances of this script were run and listened on different ports, simulating seven endpoints. Servers

[12] http://wiki.knoesis.org/index.php/LinkedSensorData
[13] http://linkedCT.org
[14] http://wiki.dbpedia.org/Datasets
[15] http://sonicbanana.cs.wright.edu:8890/sparql
[16] http://virtuoso.bd.cesma.usb.ve/sparql
[17] http://sparql-wrapper.sourceforge.net/

materialize intermediate results of queries in Benchmark 2, and were implemented using the Twisted Network framework 11.0.0[18]. Proxies send data between servers and RDF engines, following a particular transfer delay and respecting a given size of messages; they were implemented using the Python low level networking socket interface.

5.1 Performance of the ANAPSID Query Engine

We compare ANAPSID performance with respect to Virtuoso SPARQL endpoints, ARQ 2.8.8. BSD-style[19], and RDF-3X 0.3.4.[20]. RDF-3X is the only engine that accessed data stored locally, so we ran queries in both cold and warm caches. Execution times in warm caches indicate a lower bound on the execution time, and correspond to a best scenario when all the datasets are locally stored and physical structures are created to efficiently access the data. Datasets linkedCT and DBPedia were merged; RDF-3X ran queries in Benchmark 3 against this dataset. Queries ran in ANAPSID were comprised of sub-queries combined using the **agjoin** and **adjoin** operators. To facilitate the execution of queries against the Virtuoso endpoints, the SPONGE parameter was set to *Local*, i.e., the endpoint only considered data locally stored in its database; the rest of the configurations of SPONGE failed, reporting the errors: `server stopped responding` and `proxy error 502`. Table 2 reports on execution times and geometric means for Benchmarks 1, 2 and 3.

We can observe that RDF-3X is able to improve cold cache execution time by a factor of 1.37 in the geometric mean when the Benchmark 1 queries were run in warm cache, by a factor of 1.8 for Benchmark 2, and by a factor of 2.85 for Benchmark 3. This is because RDF-3X exploits compressed index structures and caching techniques to efficiently execute queries in warm cache. ANAPSID accesses remote data and does not implement any caching technique or compressed index structures; however, it is able to reduce the execution time geometric means of the other RDF engines. For queries in Benchmark 1, Virtuoso SPARQL endpoint execution time is reduced by a factor of 19.31, and RDF-3X warm cache execution time is improved by a factor of 3.62; ARQ failed evaluating these queries.

Further, queries in Benchmark 2 timed out in all linkedCT SPARQL endpoints. Similarly, queries q4 to q9 timed out after 12 hours in ARQ. However, ANAPSID was able to run all the Benchmark 2 queries, and overcome RDF-3X in warm cache and ARQ by a factor of 1.1 and 4,160.56, respectively. Finally, for queries in Benchmark 3, which combine data from linkedCT and DBPedia, we observed that RDF-3X did not exhibit a good performance, while the SPARQL endpoints as well as ARQ, failed executing all the queries. Bad performance of RDF-3X may be because the dataset result of mixing linkedCT and DBPedia has around 18GB, and this size impacts on the aggregated index structures needed to be accessed during both optimization and query execution. Furthermore, the endpoints were not able to execute these queries, because they could not dereference the URIs in the queries before meeting the timeout. Finally, ARQ executed all the joins as *Nested Loop joins*, and invoked many times the different endpoints,

---

[18] http://twistedmatrix.com
[19] http://sourceforge.net/projects/jena/
[20] http://www.mpi-inf.mpg.de/~neumann/rdf3x/

Table 2. Execution Time (secs) Different RDF Engines; Virtuoso Endpoint Sponge Local

	q1	q2	q3	q4	q5	q6	q7	q8	q9	q10	Geom. Mean
Benchmark 1											
					Cold Caches						
RDF-3X	7.83	7.12	8.47	7.45	6.36	523.89	551.20	462.77	472.42	473.20	60.60
					Warm Caches						
	4.40	4.14	4.09	4.18	4.05	466.79	465.26	464.65	475.95	463.96	44.10
SPARQL Endpoint	380.71	147.03	129.40	141.06	93.86	374.56	464.02	330.16	466.62	198.86	234.86
ANAPSID	16.60	9.22	9.54	6.80	9.59	21.48	14.34	13.48	11.08	16.19	12.16
Benchmark 2											
	q1	q2	q3	q4	q5	q6	q7	q8	q9	q10	Geom. Mean
					Cold Caches						
RDF-3X	6.35	3.55	4.13	1,543.82	3.71	4.36	1,381.9	2.75	3.83	0.51	10.62
					Warm Caches						
	2.44	2.28	2.41	1,385.09	2.71	1.75	1,321.05	1.74	1.73	0.14	5.87
SPARQL Endpoint	Timeout	Timeout	Timeout	Timeout	Timeout	Timeout	Timeout	Timeout	Timeout	Timeout	Timeout
ANAPSID	6.21	6.11	6.67	7.27	6.94	6.24	6.89	6.76	4.28	1.10	5.30
ARQ	21,043.34	17,686.52	18,936.85	43,200+	43,200+	43,200+	43,200+	43,200+	43,200+	593.36	22,051.01+
Benchmark 3											
	q1	q2	q3	q4	q5	q6	q7	q8	q9	q10	Geom. Mean
					Cold Caches						
RDF-3X	6.84	4.15	4.12	34,037.8	2,954.76	2,447.02	35,497.11	2,403.11	2,402.71	0.33	268.49
					Warm Caches						
	0.88	0.92	0.90	27,779.41	2,468.83	2,416.54	26,420.77	2,374.60	2,374.51	0.003	94.01
SPARQL Endpoint	Timeout	Timeout	Timeout	Timeout	Timeout	Timeout	Timeout	Timeout	Timeout	Timeout	Timeout
ANAPSID	12.54	11.66	12.97	18.17	10.41	9.79	12.60	12.87	6.68	7.03	11.03

which failed executing the queries because the maximum number of allowed requests was exceeded. However, ANAPSID showed a stable behavior along all the queries, overcoming RDF-3X in warm caches by a factor of 8.52. ANAPSID performance relies on the operators and the shape of plans; they are composed of small-sized sub-queries that can be executed very fast by the endpoints. These results indicate that even in the best scenarios where data is locally stored and state-of-the-art RDF engines are used to execute the queries, ANAPSID is able to remotely access data and reduce the execution time.

5.2 Adaptivity of ANAPSID Physical Operators

We also conducted an empirical study to analyze adaptivity features of ANAPSID operators in presence of unpredictable data transfers or data availability. We implemented an endpoint simulator, and ran different types of physical join operators to analyze the impact on the query execution time, of different data transfer distributions. We considered three join implementations: (a) Blocking corresponds to a traditional Hash join which produces all the answers at the end of the execution, (b) SHJ implements a Symmetric Hash Join, and (c) the ANAPSID **agjoin** operator. All the operators were implemented in Python 2.6.5. We measured the time to produce the first tuple, and time to completely produce the query answer. To run the simulations, queries of Benchmark 2 were executed and all intermediate results were stored in files, which were accessed by the endpoint simulator server during query execution simulations; five different simulated

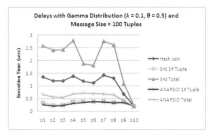

(a) Gamma(*k*=0.1;*θ*=0.5), 100 tuples.

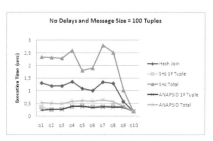

(b) No Delays, 100 tuples.

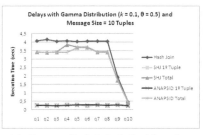

(c) Gamma(*k*=0.1,*θ*=0.5), 10 tuples.

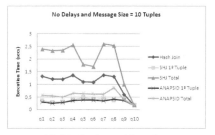

(d) No Delays, 10 tuples.

Fig. 6. Execution time (secs.) of Hash Join, Symmetric Hash Join (SHJ), and ANAPSID operators

endpoints were executed. Data transfer rates were configured to respect a Gamma distribution with $k = 0.1$ and $\theta = 0.5$; message sizes were set to 100 and 10 tuples. Finally, the performance of all the operators in an ideal environment with no delays, was also studied.

Figure 6 reports on the performance of the proposed operators. We can observe that the usage of RJTs in ANAPSID, benefits a faster generation of the first tuple as well as the output of the complete answer, even considering the cost of managing asynchronous processes in the non-blocking operators. In case that the tuple transfer delays are high (Figure 6 (c)), SHJ and ANAPSID operators exhibit a similar behavior; this is because the savings produced by using the RJTs are insignificant with respect to the time spent in receiving the data. Based on these results, we can conclude that ANAPSID operators overcome blocking operators, and that their performance may be affected by the distribution data transfer rate.

Finally, we ran ARQ, Hash join, SHJ, and ANAPSID against the endpoint simulator, and evaluated their performance in the following SPARQL 1.1. query:

```
SELECT DISTINCT ?fn3 ?fn5 ?C WHERE
{     {SERVICE <http://127.0.0.1:9000> {
        ?A4 <http://data.linkedct.org/resource/linkedct/intervention_name> "Coenzyme Q10" .
        ?A3 <http://data.linkedct.org/resource/linkedct/intervention> ?A4 .
        ?A3 <http://data.linkedct.org/resource/linkedct/condition> ?C .
        ?A3 <http://xmlns.com/foaf/0.1/page> ?fn3 .}} .
      {SERVICE <http://127.0.0.1:9001> {
        ?A6 <http://data.linkedct.org/resource/linkedct/intervention_name> "Niacin" .
        ?A5 <http://data.linkedct.org/resource/linkedct/intervention> ?A6 .
        ?A5 <http://data.linkedct.org/resource/linkedct/condition> ?C .
        ?A5 <http://xmlns.com/foaf/0.1/page> ?fn5 .}} .}
```

Intermediate results to answer the query were loaded in 15 files which were accessed through two simulated endpoints. We considered three types of delay distributions as well as no delays; Figure 7 reports on execution time (secs. log-scale).

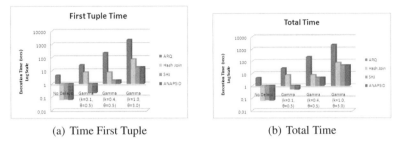

(a) Time First Tuple   (b) Total Time

Fig. 7. ANAPSID Physical Operators versus state-of-the-art Join Operators. Execution time in (secs. log-scale).

We observe that SHJ and ANAPSID operators are able to produce the first tuple faster than ARQ or Hash join, even in an ideal scenario with no delays; further, ARQ performance is clearly affected by data transfer distribution and its execution time can be almost two orders of magnitude greater than the time of SHJ or ANAPSID. We notice that SHJ and ANAPSID are competitive, this is because the number of intermediate results is very small, and the the benefits of the RJTs cannot be exploited. This suggests that the performance of ANAPSID operators depends on the selectivity of the join operator and the data transfer delays.

6   Related Work

Query optimization has emphasized on searching strategies to select the best sources to answer a query. Harth et al. [7] present a Qtree-based index structure which stores data source statistics that have been collected in a pre-processing stage. A Qtree is a combination of histograms and an R-tree multidimensional structure; histograms are used for source ranking, while regions determine the best sources to answer a join query. Li and Heflin [16] build a tree structure which supports the integration of data from multiple heterogeneous sources. The tree is built in a bottom-up fashion; each triple pattern is rewritten according to the annotations on its corresponding datasets. Kaoudi et al. [13] propose a technique that runs on Atlas, a P2P system for processing RDF distributed data that are stored in hash tables. The purpose of this technique is to minimize the query execution time and the bandwidth consumed; this is done by reducing the cardinality of intermediate results. A dynamic programming algorithm was implemented that relies on message exchange among sources. None of these approaches use information about the processing capacity of the selected sources; in consequence, they may select endpoints that will time out because the submitted query is too complex.

The XJoin [22] is a non-blocking operator based on the Symmetric Hash Join, and it follows two principles: incremental production of answers as sources become available, and continuous execution including the case when data sources present delays; access to the sources is not done through SPARQL endpoints, and the XJoin operator can only be applied when its arguments are evaluated independently. The Tukwila integration system [5] executes queries through several autonomous and heterogeneous sources. Tukwila decomposes original queries into a number of sub-queries on each source, and uses adaptive techniques to hide delays. We consider dependency between arguments and define operators able to respect binding pattern restrictions while delays are hidden.

Urhan et al. [23] present the algorithm of scrambling query plans that aims to hide delays; in case a source becomes blocked and all the previously gathered data have already been considered, the execution plan is reordered to produce at least partial results. Hartig et al. [9] rely on an adaptive iterator model that is able to detect when a dereferenced dataset stops responding, and submits other query requests to alive datasets; also, heuristics-based techniques are proposed to minimize query intermediate results [8]. Ludwig and Tran [14] propose a mixed query engine; sources are selected using aggregated indexes that keep information about triple patterns and join cardinalities for available sources; these statistics are updated on-the-fly. Execution ends when all relevant sources have been processed or a stop condition given by the user is hold; additionally, the Symmetric Hash Join is implemented to incrementally produce answers; recently, this approach was extended to also process Linked Data locally stored [15]. Avalanche [2] produces the first k results, and sources are interrogated to obtain statistics which are used to decompose queries into sub-queries that are executed based on their selectivity; sub-queries results are sent to the next most selective source until all sub-queries are executed; execution ends when a certain stop condition is reached. Finally, some RDF engines are able to process federated SPARQL queries[4,18,20]. Although these approaches are able to access Linked data, none of them provide an adaptive solution to query SPARQL endpoints.

7 Conclusions and Future Work

We have defined ANAPSID, an adaptive query processing engine for RDF Linked Data accessible through SPARQL endpoints. ANAPSID provides a set of physical operators and an execution engine able to adapt the query execution to the availability of the endpoints and to hide delays from users. Reported experimental results suggest that our proposed techniques reduce execution times and are able to produce answers when other engines fail. Also, depending on the selectivity of the join operator and the data transfer delays, ANAPSID operators may overcome state-of-the-art Symmetric Hash Join operators. In the future we plan to extend ANAPSID with more powerful and lightweight operators like *Eddy* and *MJoin* [5], which are able to route received responses through different operators, and adapt the execution to unpredictable delays by changing the order in which each data item is routed.

References

1. Atre, M., Chaoji, V., Zaki, M.J., Hendler, J.A.: Matrix "Bit" loaded: a scalable lightweight join query processor for RDF data. In: Proceedings of the WWW, pp. 41–50 (2010)
2. Basca, C., Bernstein, A.: Avalanche: Putting the Spirit of the Web back into Semantic Web Querying. In: The 6th International Workshop on SSWS at ISWC (2010)

3. Blanco, E., Cardinale, Y., Vidal, M.-E.: A sampling-based approach to identify qos for web service orchestrations. In: iiWAS, pp. 25–32 (2010)
4. Buil-Aranda, C., Arenas, M., Corcho, O.: Semantics and Optimization of the SPARQL 1.1 Federation Extension. In: Antoniou, G., Grobelnik, M., Simperl, E., Parsia, B., Plexousakis, D., De Leenheer, P., Pan, J. (eds.) ESWC 201. LNCS, vol. 6644, pp. 1–15. Springer, Heidelberg (2011)
5. Deshpande, A., Ives, Z.G., Raman, V.: Adaptive query processing. Foundations and Trends in Databases 1(1), 1–140 (2007)
6. Florescu, D., Levy, A.Y., Manolescu, I., Suciu, D.: Query optimization in the presence of limited access patterns. In: SIGMOD Conference, pp. 311–322 (1999)
7. Harth, A., Hose, K., Karnstedt, M., Polleres, A., Sattler, K.-U., Umbrich, J.: Data summaries for on-demand queries over linked data. In: WWW, pp. 411–420 (2010)
8. Hartig, O.: Zero-Knowledge Query Planning for an Iterator Implementation of Link Traversal Based Query Execution. In: Antoniou, G., Grobelnik, M., Simperl, E., Parsia, B., Plexousakis, D., De Leenheer, P., Pan, J. (eds.) ESWC 2011, Part I. LNCS, vol. 6643, pp. 154–169. Springer, Heidelberg (2011)
9. Hartig, O., Bizer, C., Freytag, J.C.: Executing SPARQL Queries Over the Web of Linked Data. In: Bernstein, A., Karger, D.R., Heath, T., Feigenbaum, L., Maynard, D., Motta, E., Thirunarayan, K. (eds.) ISWC 2009. LNCS, vol. 5823, pp. 293–309. Springer, Heidelberg (2009)
10. Idreos, S., Kersten, M.L., Manegold, S.: Self-organizing tuple reconstruction in column-stores. In: SIGMOD Conference, pp. 297–308 (2009)
11. Izquierdo, D., Vidal, M.-E., Bonet, B.: An Expressive and Efficient Solution to the Service Selection Problem. In: Patel-Schneider, P.F., Pan, Y., Hitzler, P., Mika, P., Zhang, L., Pan, J.Z., Horrocks, I., Glimm, B. (eds.) ISWC 2010, Part I. LNCS, vol. 6496, pp. 386–401. Springer, Heidelberg (2010)
12. Jena TDB (2009), http://jena.hpl.hp.com/wiki/TDB
13. Kaoudi, Z., Kyzirakos, K., Koubarakis, M.: SPARQL Query Optimization on Top of DHTs. In: Patel-Schneider, P.F., Pan, Y., Hitzler, P., Mika, P., Zhang, L., Pan, J.Z., Horrocks, I., Glimm, B. (eds.) ISWC 2010, Part I. LNCS, vol. 6496, pp. 418–435. Springer, Heidelberg (2010)
14. Ladwig, G., Tran, T.: Linked Data Query Processing Strategies. In: Patel-Schneider, P.F., Pan, Y., Hitzler, P., Mika, P., Zhang, L., Pan, J.Z., Horrocks, I., Glimm, B. (eds.) ISWC 2010, Part I. LNCS, vol. 6496, pp. 453–469. Springer, Heidelberg (2010)
15. Ladwig, G., Tran, T.: SIHJoin: Querying Remote and Local Linked Data. In: Antoniou, G., Grobelnik, M., Simperl, E., Parsia, B., Plexousakis, D., De Leenheer, P., Pan, J. (eds.) ESWC 2011, Part I. LNCS, vol. 6643, pp. 139–153. Springer, Heidelberg (2011)
16. Li, Y., Heflin, J.: Using Reformulation Trees to Optimize Queries Over Distributed Heterogeneous Sources. In: Patel-Schneider, P.F., Pan, Y., Hitzler, P., Mika, P., Zhang, L., Pan, J.Z., Horrocks, I., Glimm, B. (eds.) ISWC 2010, Part I. LNCS, vol. 6496, pp. 502–517. Springer, Heidelberg (2010)
17. Neumann, T., Weikum, G.: Scalable join processing on very large rdf graphs. In: SIGMOD International Conference on Management of Data, pp. 627–640 (2009)
18. Quilitz, B., Leser, U.: Querying Distributed RDF Data Sources with SPARQL. In: Bechhofer, S., Hauswirth, M., Hoffmann, J., Koubarakis, M. (eds.) ESWC 2008. LNCS, vol. 5021, pp. 524–538. Springer, Heidelberg (2008)
19. Harris, S., Andy Seaborne, E.P.: SPARQL 1.1 Query Language (June 2010)
20. Stoker, M., Seaborne, A., Bernstein, A., Keifer, C., Reynolds, D.: SPARQL Basic Graph Pattern Optimizatin Using Selectivity Estimation. In: WWW (2008)

21. Tran, T., Zhang, L., Studer, R.: Summary Models for Routing Keywords to Linked Data Sources. In: Patel-Schneider, P.F., Pan, Y., Hitzler, P., Mika, P., Zhang, L., Pan, J.Z., Horrocks, I., Glimm, B. (eds.) ISWC 2010, Part I. LNCS, vol. 6496, pp. 781–797. Springer, Heidelberg (2010)
22. Urhan, T., Franklin, M.J.: Xjoin: A reactively-scheduled pipelined join operator. IEEE Data Eng. Bull. 23(2), 27–33 (2000)
23. Urhan, T., Franklin, M.J., Amsaleg, L.: Cost based query scrambling for initial delays. In: SIGMOD Conference, pp. 130–141 (1998)
24. Vidal, M.-E., Ruckhaus, E., Lampo, T., Martínez, A., Sierra, J., Polleres, A.: Efficiently Joining Group Patterns in SPARQL Queries. In: Aroyo, L., Antoniou, G., Hyvönen, E., ten Teije, A., Stuckenschmidt, H., Cabral, L., Tudorache, T. (eds.) ESWC 2010. LNCS, vol. 6088, pp. 228–242. Springer, Heidelberg (2010)
25. Weiss, C., Karras, P., Bernstein, A.: Hexastore: sextuple indexing for semantic web data management. PVLDB 1(1), 1008–1019 (2008)
26. Wiederhold, G.: Mediators in the architecture of future information systems. IEEE Computer 25(3), 38–49 (1992)

Modelling and Analysis of User Behaviour in Online Communities

Sofia Angeletou, Matthew Rowe, and Harith Alani

Knowledge Media institute, Open University, UK
{s.angeletou,m.c.rowe,h.alani}@open.ac.uk

Abstract. Understanding and forecasting the health of an online community is of great value to its owners and managers who have vested interests in its longevity and success. Nevertheless, the association between community evolution and the behavioural patterns and trends of its members is not clearly understood, which hinders our ability of making accurate predictions of whether a community is flourishing or diminishing. In this paper we use statistical analysis, combined with a semantic model and rules for representing and computing behaviour in online communities. We apply this model on a number of forum communities from Boards.ie to categorise behaviour of community members over time, and report on how different behaviour compositions correlate with positive and negative community growth in these forums.

1 Introduction

Online communities form a fundamental part of the web today where a large portion of the Internet's traffic is driven by and through them [16]. These communities are where the majority of web users share content, seek support, and socialise. On the one hand, for companies and businesses, such online communities tend to yield much value in terms of idea generation, customer support, problem solving, etc. [15]. On the other hand, managing and hosting these communities can be very costly and time consuming, and hence their owners and managers have a great vested interest in ensuring that these communities continue to flourish, and that their members remain active and productive.

One of the main metrics often used by community managers to measure community health is the number of members and posts. These numbers give a good indication of community popularity. However, for deeper assessment and forecasting, other more complex qualitative and behavioural parameters need to be considered [6, 13]. For example, behavioural analytics complement other community assessment tools and increase the value of the data [8].

Health of online communities is a relatively new and complex concept that is codependent on the emergence and evolution of user behaviour in those communities. Domination of any type of behaviour, whether positive or negative, could encourage others to change their behaviour or even abandon the community [10]. Therefore, monitoring and analysis of behaviour and its evolution over time, in addition to straightforward metrics such as post and user counts, can provide valuable information on how healthy an online community currently is or will

L. Aroyo et al. (Eds.): ISWC 2011, Part I, LNCS 7031, pp. 35–50, 2011.

be in the near future. Behaviour in online communities is usually associated with various social and technical parameters which influence the roles users hold in different settings [11]. Associating users with behavioural categories involves identifying and applying constraints, expectations and frameworks to categorise and follow user behaviour in the community [4].

To support community owners and managers in observing and maintaining the good health of their communities, we first need to (a) model, capture and monitor the activities of community members, (b) analyse emergent behaviours and their change over time, (c) understand the correlation of certain types of behaviour with community evolution, and (d) learn how and when to intervene to influence the interactions and behaviour of community members. In this work we focus on the first three tasks; the first task is concerned with producing a semantic model for representing user activities in online communities and the attention they generate in those communities. The second task focuses on comparing the emergence and change in patterns of behaviour with the evolution of those communities. And the third task explores how community composition, i.e. a macro-analysis of the community, is correlated with the health of the community. By monitoring activities in online communities we will be able to better understand and predict their evolution directions; i.e. whether they are flourishing (positive evolution) or diminishing (negative evolution). The main contributions in this paper are as follows:

1. *Method to infer user roles in online communities:* We employ semantic rules to label community users with their role and utilise dynamic feature binning to account for the dynamic nature of communities and their propensity to evolve.
2. *Ontology to model behavioural features and support community role inference:* Allowing user features to be captured in a common machine-readable format across communities and platforms.
3. *Analysis of community health through role composition:* We demonstrate the utility of our approach by analysing three communities over a 3 year period, showing the effects of behaviour composition changes on community health and compositions that are key signifiers of healthy or unhealthy communities.

In the following section we report on various related works in the area of behaviour and community analysis. In section 3 we present our methodology for user behaviour analysis and how we utilise Semantic Web technologies to infer the role that a user has within an online community. Section 4 describes our analysis of community health in the three sample forums, followed by discussions and future plans in section 5, and finishes with conclusions in section 6.

2   Related Work

In this section we report on existing works on analyses of behaviour patterns and roles in online communities. The identification of behaviour is often based on features which reflect the intensity, persistence, focus, reciprocity and polarity of user activities.

For instance, users who contribute with high intensity, reciprocity and persistence, positive polarity and are focused on supporting and contributing to the community are characterised as *moderators, mediators* [10], *captains* and *pillars* [14]. When such users are able to set the standard for community interactions, they get labelled as *celebrities* [4]. *Popular initiator, popular participant* and *joining conversationalist* [3] are three roles very similar to the celebrity type since their intensity, persistence and reciprocity are also quite high. Another type of prolific, but not as widely popular, user is the *elitist*, who demonstrates high values for the above dimensions but communicates with a smaller group of users. On the lower end of the activity scale the *lurker* is the most frequently observed role and is defined as a participant who consumes but does not contribute and usually has a strong personal focus [4, 10, 14]. Similarly described roles are those of *content consumers* [9], *grunts* and *taciturns* [3] who do contribute but with low intensity. The polarity of the user contribution has also been used to distinguish the negative roles of *troll* and *flamer* who exhibit disruptive behaviour similar to the *ranter*. Like celebrities, ranters also demonstrate high intensity and persistence yet their primary goal is to raise discussions on the topic of their interest for some personal goals, same as *over-riders* and *generators* [14].

Although there is no commonly agreed set of behaviour patterns and labels, the social and technical features considered by the above works when categorising behaviour do share some characteristics, albeit sometimes tailored to suit the online communities under investigation. The approaches followed in the above references are normally based on correlating a set of features taken from a specified snapshot of a community, then labelling users with behaviour roles that fits the results from that snapshot.

Our analysis extends these approaches by introducing a framework for representing, computing, and monitoring users' behaviour over time. We extend the state of the art by demonstrating how various features can be modelled and articulated into semantic rules to automate the detection and categorisation of users with specific types of behaviour.

Furthermore, is it often the case that fixed ranges of feature-values are calculated when associating them with behaviour types, so if the feature value for a given user falls within that range, then that user will be labelled with the corresponding behaviour. However, it is often the case that such value ranges seize to apply if the time window or community changes. Here we present a framework that enables a more dynamic association of features to roles (Section 3.2) and allows for on-the-fly value threshold assignment that takes context into account.

Community health indicators are normally dependent on the goals and characteristics of the community [11]. Straightforward measures such as number and frequency of posting are often used as an index of community health. For example, it has been shown that the activity of a group can be maintained in high levels by long term members, who help keep the group together [1, 10]. On the other hand, it has been found that having *lurkers* in a community does not necessarily have a negative influence [12]. Our framework allows us to investigate the influence that various user behaviours and interactions have on the overall health of communities over time. This helps in understanding what the optimum compositions of behaviour should be for a given community, and in forecasting community evolution. We analyse the influence and predictability of a wide set of behaviours on community health.

3   Methodology for Behaviour Analysis

In this section we describe our approach for labelling users in online communities with the roles they hold in the context of these communities and in a specified timeframe. To perform such labelling we first need to capture the activity of users in online communities, define what sort of behaviour is associated with particular roles and compare it to a user's activity. For capturing users' activities we define an *ontology* (Section 3.1) that represents all involved entities and their interactions. We also define *"rule skeletons"* (Section 3.2) which provides high-level descriptions of how certain *features* are associated with various behaviour roles. Our community analysis then *fleshes out* these rules with *dynamic and automatically computed value-ranges* that will eventually determine which users will be categorised with which behaviours. Finally we apply these rules on all community members to infer their behaviour types (Section 3.3).

3.1   Ontology

Capturing a user's activity in online communities is a primary step to analysing his behaviour. Fig. 1 presents a portion of our Behaviour Ontology[1] which represents online community users and their interactions. The ontology extends SIOC [2] to refine the representation of low level user activities and interactions. It also extends the Social Reality [5] ontology which provides an abstract representation of social roles and their contexts. The main concepts and properties of the ontology are:

- **sioc:UserAccount** is a SIOC class to represent online community users.
- **oubo:Post** represents users' main activities; writing and replying to posts.
- **oubo:PostImpact** summarises a post's replies, comments, forwards, etc.
- **oubo:UserImpact** encodes the user impact (behaviour)
- **oubo:TimeFrame** is the temporal context during which the analysis is carried out and the association of a user to a role holds.
- **social-reality:C** represents context, such as time period (oubo:TimeFrame) and a forum (sioc:Forum).
- **oubo:Role** represents the roles we derive for users based on their activities in the community.
- **oubo:belongsToContext** links context-related concepts, such as oubo: TimeFrame, sioc:Forum, and social-reality:C.
- **social-reality:counts_as** associates a user with a oubo:Role.
- **social-reality:context** associates a user role with its context.

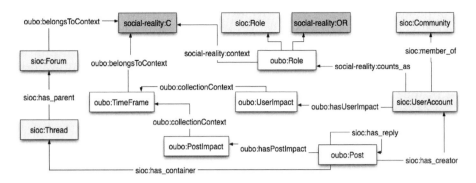

Fig. 1. Behaviour Ontology

The ontology also defines the rules for the population of features and classification of the users in different roles using dynamically populated feature weights (Fig. 3). This is discussed in Section 3.3.

3.2   Behaviour Roles

To derive the behaviour roles of community users, their activity patterns need to be compared against the behavioural characteristics of each role type. In the literature, the features (e.g. number of posts and replies, in/out degrees) associated with behaviour are often given static value ranges (i.e. min and max values for corresponding behaviours) which are calculated based on the community snapshot under analysis (e.g. [3, 9, 17]). However, a common characteristic of online communities is their propensity to evolve and develop as new users participate and the dynamic of the community changes. An effect of such dynamism is that should we learn a static value for the maximum and minimum values for each role's features then applying such values at a later point in time will lead to users being omitted from the labelling process.

To counteract such effects we use the notion of *skeleton rules*, where each rule contains a mapping between a given feature and the *level* that the value of the feature should take to indicate a certain type of behaviour: *low*, *medium* or *high*. In using this method we can shift the bounds that constitute such levels as the dynamics of the community changes, thereby allowing more users to be labelled with behaviour roles.

Many different behaviours and associated features have been proposed in the literature (section 2). Our framework for modelling and computing behaviour is not tied to any specific community or behaviour types or feature compositions. To demonstrate our framework, we selected the behaviour roles defined in [3], which covers a range of common activity and participation roles. In [3], Chan and Hayes performed clustering over users within Boards.ie community forums and then carried out manual analysis to derive the behaviour labels for each cluster. They clustered the users using a list of key features that covered (a)

---

[1] http://purl.org/net/oubo/0.3

the structural network properties of a user within the community, (b) the user's popularity amongst the community, (c) their propensity to initialise discussions, and (d) their persistence in discussions. These features are:

- **In-degree Ratio:** The proportion of users U that reply to user v_i, thus indicating the concentration of users that reply to v_i.
- **Posts Replied Ratio:** Proportion of posts by user v_i that yield a reply, used to gauge the popularity of the user's content based on replies.
- **Thread Initiation Ratio:** Proportion of threads that have been started by v_i. This feature captures the propensity of a user to instigate discussions and generate fresh content for the community.
- **Bi-directional Threads Ratio:** Proportion of threads where user v_i replies to a user and receives a reply, thus forming a *reciprocal* communication.
- **Bi-directional Neighbours Ratio:** The proportion of neighbours where a *reciprocal* interaction has taken place - e.g. v_i replied to v_j and v_j replied to v_i. This can be thought of as the intersection between the set of *repliers* and *recipients*. This measure allows the *reciprocal* characteristics of the user to be captured and their participation with users in the community, where higher values demonstrate a tendency to interact.
- **Average Posts per Thread:** The average number of posts made in every thread that user v_i has participated in. Allows the level of discussion that the user participates in to be gauged.
- **Standard Deviation of Posts per Thread:** The standard deviation of the number of posts in every thread that user v_i has participated in. This gauges the distribution of the discussion lengths, for example, one would expect that a user who often discusses at length with other users would have a high *Average Posts per Thread* and a low *Standard Deviation of Posts per Thread*, while someone who varies their participation will have a higher *Standard Deviation of Posts per Thread*.

Based on the feature-behaviour compositions in [3] and in other literature, we deduce a mapping of these common feature to value ranges for each behaviour role (Table 1).

3.3   Constructing and Applying Behaviour Rules

Our approach for constructing and applying rules is shown in Fig. 2 and is composed of four stages that function in a cyclical manner: *First*, we construct features for all users who participated in the given community at a specific point in time. *Second*, we derive bins for features in the community, thus providing the bounds for the *low*, *medium* or *high* levels of each feature. *Third*, the rule base is constructed using the skeleton rule base and the levels from the binning. *Fourth* we apply the rules to each member of the community and derive a role label, this provides the role composition of the community at a given time snapshot. As Fig. 2 shows, the process is repeatable over time, thereby allowing the composition of a given community to be monitored by inferring the role of each community user at a given point in time. We now explain the four steps in greater detail.

Table 1. Roles and the feature-to-level mappings

Role	Feature	Level
Elitist	In-Degree Ratio	low
	Bi-directional Threads Ratio	high
	Bi-directional Neighbours Ratio	low
Grunt	Bi-directional Threads Ratio	med
	Bi-directional Neighbours Ratio	med
	Average Posts per Thread	low
	STD of Posts per Thread	low
Joining Conversationalist	Thread Initiation Ratio	low
	Average Posts per Thread	high
	STD of Posts per Thread	high
Popular Initiator	In-Degree Ratio	high
	Thread Initiation Ratio	high
Popular Participants	In-Degree Ratio	high
	Thread Initiation Ratio	low
	Average Posts per Thread	med
	STD of Posts per Thread	med
Supporter	In-Degree Ratio	med
	Bi-directional Threads Ratio	med
	Bi-directional Neighbours Ratio	med
Taciturn	Bi-directional Threads Ratio	low
	Bi-directional Neighbours Ratio	low
	Average Posts per Thread	low
	STD of Posts per Thread	low
Ignored	Posts Replied Ratio	low

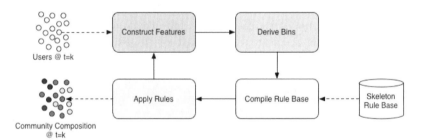

Fig. 2. Overview of the approach to analyse user behaviour, label users with behaviour roles and derive the community composition

Step 1: Constructing Features. The previously defined statistical features are constructed for each user at a given point in time. For our experiments, as we describe in the following section, we use a window function to extract all posts made within a given community during that time period. Using the reply-to structure of the posts, we then compile the above features and create an instance of oubo:UserImpact that contains the features for the given user measured at a given point in time.

Step 2: Deriving Bins. Our skeleton rule base contains mappings of features-to-levels, allowing the bounds of the levels to be altered depending on the dynamics of the community. We set the bounds through *binning*, a process that *discretizes* continuous feature values into three bins (*low, medium* or *high*) using Weka's[2] discretization filter together with equal frequency binning. Using a naive binning approach - e.g. splitting a feature range into thirds - can result in a large frequency skew within a single bin, equal frequency binning avoids this and provides a distribution-dependent notion of levels. The process of deriving the bins is performed when we analyse the community at a different point in time, in doing so our intuition is that we will reduce the number of unclassified users and account for changes in the community's dynamics - we show this empirically in the following section.

Step 3: Compiling the Rule Base. The association of users to roles is inferred by analysing the captured data for each user against features that each role embodies. To perform such inferences our approach employs the SPARQL Inference Notation (SPIN)[3] framework, allowing the encoding of rules as SPARQL queries. The benefits of such an approach is that the rules are embedded in an ontological model and can, therefore, be shared and executed across platforms that support SPARQL Extensions and Jena.

To compile the rule base we create a rule for each behaviour role within the community. For each role a new instance of the oubo:RoleClassifier Class is created and associated with a set of features as shown in Fig. 3. Each feature has a minimum and maximum value which specify the range of feature values a user should have for this feature in order to be assigned to this role. We use the skeleton rule of the role to provide the rule's syntax and then replace the levels with the necessary bounds produced by our binning procedure. In the majority of cases a combination of features is required for the association of a user to a role. In these cases, all the feature values of the user should belong to the ranges specified by each feature belonging to the relative RoleClassifier instance.

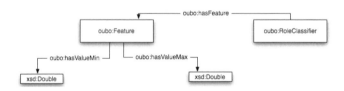

Fig. 3. Association of Roles with Features

Step 4: Applying Rules. The produced rules are SPARQL Construct queries that exploit the power of SPIN Functions by testing each instance of sioc:UserAccount against each instance of oubo:RoleClassifier, and then assigning the user to the role whose associated classifier is matched. Fig. 4 presents an example of one such query that is encoded as a spin:rule for the class sioc:UserAccount. For each instance of sioc:UserAccount (represented by the variable ?this) a set of triples are inferred representing the association of the user with a new instance of a specific role ?role. The ?role is an instance

---

[2] http://www.cs.waikato.ac.nz/ml/weka/
[3] http://spinrdf.org

of a subclass of oubo:Role, `?t` depending on the classification of the user as described in Fig. 4. The spin function oubo:fn_getRoleType executes the process described above and returns the appropriate classifier. Then using the smf:buildURI SPARQL Motion function `?t` is built so that it is associated to the correct subclass of oubo:Role. Finally, the `?context` is created and connected to the particular `?role` via the relation `social-reality:hasContext`, and is also associated to the temporal and forum relevant contexts in which it makes sense that the user holds this particular role - i.e. a given user can have multiple roles within different communities and time periods.

```
CONSTRUCT {
    ?role a ?t .
    ?this social-reality:count_as ?role .
    ?context a social-reality:C .
    ?role social-reality:content ?context .
    ?temp a oubo:TemporalContext .
    ?forum a sioc:Forum .
    ?forum oubo:belongsToContext ?context .
    ?temp oubo:belongsToContent ?context
} WHERE {
    BIND (oubo:fn_getRoleType(?this) AS ?type) .
    BIND(smf:buildURI("oubo:Role{?type}") AS ?t) .
    .....
}
```

Fig. 4. SPARQL CONSTRUCT encoded as a spin:rule in the class sioc:UserAccount

4   Analysis of Community Health

Healthy communities provide users with the resources from which information can be sought, interactions made and discussions participated in. In this section we explore the relation between the composition of a community, i.e. the various roles that users have within a community and the proportion to which such roles make up the community (e.g. 20% elitists, 10% taciturns, etc.), and the activity in the community. Through experiments and analysis of the subsequent results, we seek to identify key community compositions that are associated with both an increase and decrease in community activity. In doing so, we are provided with an understanding of how certain behaviour types are correlated with community evolution and what compositions are signifiers of healthy and unhealthy communities. We here consider the level of activity as a proxy of community health, but other parameters (e.g. reciprocity, sentiment) could also be considered.

4.1   Experimental Setup

For our experiments we used a dataset collected from the Irish community discussion forums, Boards.ie. We extracted all posts from the beginning of 2004 through to the end of 2006 for our analysis - thereby capturing a 3 year period over which we could perform our analyses. Rather than analysing the entire site, we selected 3 forums that showed a variance in activity throughout the analysed period - the plots of post activity are shown in Fig. 6.

– *Forum 246 (Commuting and Transport):* Demonstrates a clear increase in activity over time.
– *Forum 388 (Rugby):* Exhibits periodic increase and decrease in activity and hence it provides good examples of healthy/unhealthy evolutions.
– *Forum 411 (Mobile Phones and PDAs):* Increase in activity over time with some fluctuation - i.e. reduction and increase over various time windows.

In order to compile a dataset for each forum we used the following process: beginning on 1st January 2004 we used a window from 13 weeks prior to this date up until the date as our feature window. Within this window we extracted all the posts made within the forum, and used the posts to compile the statistics for each unique user who had made a post within that window. Once we had finished building the statistics for each user at that collection date, we then rolled the date forward 84 days, leaving a 12 week gap between our last collection date. The window was compiled once again: going 13 weeks back, returning all posts within the window, and then building the user features for each unique user within the window. We repeated this process until the end of 2006. To provide a coarse measurement of the community's *health* we also counted the number of posts made in the forum during that window - allowing the activity at one point in time to be contrasted against earlier activity.

Following the compilation of our user statistics at the incremental time steps (13 time steps in total) and the instantiation of oubo:UserImpact, we categorised each user using our previously described rules. In doing so we were able to measure the composition of the community over time as differing percentages of users that have taken on such roles within the community. We then correlated this composition with the health of the community at that point in time, seeking patterns that describe a healthy and unhealthy community in terms of either an increase or decrease in activity, e.g. having many users of a certain role type reduces community activity. We also report on how our approach greatly reduces the percentage of users that could not be classified by the current behaviour rules.

4.2   Results

Fig. 5 shows the correlograms from the individual forums. The upper panel shows the extent to which a correlation exists between two features and the polarity of the correlation (i.e. positive or negative). The greater the portion of the circle that is filled then the greater the correlation. The colour indicates the polarity: blue indicates a positive correlation and red indicates a negative correlation.

For forum 246 (Commuting and Transport), shown in Fig 5(a), a positive correlation exists between the post count and both the number of *elitists* and *popular participants* (*'partic'* in the chart), indicating that as more users assume such roles within the community then activity increases. This is due to the *popular participants* driving discussions and joining in with the community to make it more vibrant. Meanwhile, *elitists* communicate a lot with their own group and thus drive its activity. In forum 246 we also observe a negative correlation between the post count and the proportion of *taciturns* within the community, indicating that users who communicate very little with others can reduce the overall interactions and dynamic of the community. Fig 5(a) also shows that an increased number of *ignored* users has a negative effect on community activity, which follows intuition. As forum 246 concerns transport discussion, many users post questions regarding travel situations and modes of transport, and hence if a large portion of those users are ignored then activity in this community diminishes as questions remain unanswered.

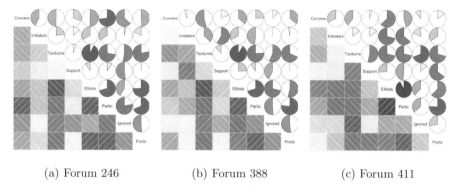

(a) Forum 246   (b) Forum 388   (c) Forum 411

Fig. 5. Correlation between the various features within each forum

Similarly to forum 246, in forum 388 (Rugby) we also find a positive correlation between post counts and number of *elitists* and *popular participants* (Fig. 5(b)). There is also a slight correlation between post count and the proportion of *conversationalists* within the community, which demonstrates the value of conversations and debates in driving this community. There is also a negative correlation between the post count and the number of *taciturns* and *ignored* within the community.

For forum 411, (Mobile Phones and PDAs), the correlogram in Fig. 5(c) demonstrates similar patterns to the previous two forums in terms of positive correlations. Once again, we find that the post count has a positive correlation with the proportion of *elitists* and *popular participants*. We also find that the post count has a negative correlation with the proportion of *taciturns* and, in this individual forum, with the proportion of *supporters*. Supporters have a mid-level range of 'Bi-directional Threads Ratio', indicating that conversation is one of the drivers behind this role. However in forum 411, debate between users appears to be limited, as users require information regarding support and are less inclined to chat with other users repeatedly. This is also supported by the lack of positive correlation between the post count and the proportion of *conversationalists* in this forum.

Within Fig. 6 we show the composition changes in each of the analysed forums over time, plotted together with the post count within each forum. For all forums we find that activity increases where the proportion of *ignored* users decreases. For forum 246 we see that as the proportion of *joining conversationalists* and *popular initiators* increases so does activity. The same applies to forums 388 and

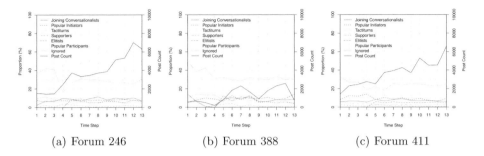

(a) Forum 246   (b) Forum 388   (c) Forum 411

Fig. 6. Changes in composition over time plotted with forum post counts

411, although for the latter the number of *popular initiators* was more important, confirming our earlier correlation analysis for that forum where conversations are not driving activity in this forum.

An interesting factor in each forum is the effect of composition stability on community activity. In the case of forum 246 and forum 411 we find that the composition converges on relatively stable proportions (i.e. with no extreme fluctuations in role types) and leads to a large rise in activity over time. Conversely for forum 388, Fig. 6(b) shows that the lack of stability in the community's composition leads to fluctuations in activity. This suggests that although limiting the number of ignored users and taciturns in the community would be beneficial, a stable mix of user types actually improves community health.

Unclassified Users. Our rule-based approach for inferring the role of a given user at a given point in time utilises dynamic binning to update the low, mid and high bounds for each rule's features. In utilising dynamic binning our intuition was that our approach could adapt to the changing dynamics of the community as it evolves over time, i.e. the *low* bound for a feature during one year will differ from a year later. Therefore to demonstrate the utility of our approach we measured the proportion of users that are unclassified in each forum. We contrast this against the proportion of users who are unclassified when the feature bounds are not updated at each time step (40.05% unclassified users for forum 246, 37.92% for forum 388 and 39.84% for forum 411). Results show that our method of dynamically updating the bins for feature bounds enables a greater coverage of the users (29.06% unclassified users for 246, 28.06% for 388 and 28.68% for 411), and therefore enables, on average, a greater proportion of users to be labelled with a given role. Additional behaviour rules can be added to increase the percentage of classified users even more.

Predicting Community Health. Thus far we have concentrated on identifying correlations between the post count within single forums and the proportion of roles within such communities. An important aspect of undertaking such analysis is the ability to forecast community health should the composition of the community change. To demonstrate the utility of such an approach we performed a binary classification task to identify, based on the composition of the community, whether the activity had either *increased* or *decreased* since the previous time window. We built a dataset for each forum and constructed an instance for each of the 13 time windows. Each instance contained the features describing the 7 behaviour roles in the community together with the percentage of users allocated with such roles and a class label denoting the activity in the forum as having either increased (*pos*) or decreased (*neg*) since the previous time window. For our classification task we used the J48 decision tree classifier in a 10-fold cross validation setting (due to the limited size of the datasets) by: *first*, identifying increases and decreases in each of the forums, and *secondly*, identifying activity changes across communities, by combining forum datasets together into a single dataset. To report on the performance of our approach we used precision, recall, f-measure (setting $\beta = 1$) and the area under the Receiver Operator Characteristic Curve (ROC).

Table 2. Results from detecting changes in activity using community composition

Forum	P	R	F_1	ROC
246	0.799	0.769	0.780	0.800
388	0.603	0.615	0.605	0.775
411	0.765	0.692	0.714	0.617
All	0.583	0.667	0.607	0.466

Table 2 presents the results from our classification experiments. For forum 246 we achieve the highest F_1 value due to the activity in the forum steadily increasing over time and the precision value indicating that in this forum the composition patterns account for fluctuations in activity. For forum 388 we return the lowest F_1 value, indicating that the variance in activity renders the prediction of activity increase difficult within this forum, this could possibly be due to the seasonal fluctuations in interest surrounding the rugby season. For forum 411 we achieve high precision, indicating that activity can be precisely detected based on the composition in this forum. When performing cross-community health predictions we achieve lower F_1 values than those for forums 246 and 411 and the lowest ROC value. This indicates that cross-community patterns are not as reliable as individual community analysis, where patterns in compositions for single forums account for the idiosyncratic behaviour.

For our next task we induced Linear Regression models by regressing the post count on the community composition, using each of the role proportions as our predictor variables, seeking a relationship between the change in the overall composition of the community and the health of the forum. We now report on the model learnt for forum 388 (Commuting and Transport), given that this model achieved the highest coefficient of determination while forums 246 and 411 achieved R^2 values of 0.649 and 0.793 respectively.

Table 3 shows the results from the induced model.[4] The model indicates that should a community increase in its proportion of *popular initiators* and *popular*

---

[4] We found no multicolinearity between variables in the model when testing using the Variance Inflation Factor, suggesting that the roles are distinct in this forum and there are no clear dependencies between them.

participants while decreasing in the proportion of *supporters* and *ignored* then the community's activity will increase. However, an increase in *ignored* and *supporters* will yield a reduction in the post count and therefore a reduction in the "health" of the community. Such patterns can be used to alert a community manager of the current state of their community and its projected evolution. Managers could then use this information to decide what action to take to influence the evolution of their community in a positive way.

Our analysis of community forums has explored the correlation between community composition and health, and how predictions can be performed. Through this analysis we have identified four key *take-home* messages:

1. Healthy communities contain more elitists and popular participants.
2. Unhealthy communities contain many taciturns and ignored users.
3. Communities exhibit idiosyncratic compositions, thus reflecting the differing dynamics that are required/exhibited by individual communities.
4. A stable composition, with a mix of roles, increases community health.

Table 3. Linear regression model induced from the forum composition of f388

Role	Est' Coefficient	Standard Error	t-Value	P($x >$t)
Joining Conversationalist	69.20	43.82	1.579	0.1751
Popular Initiators	173.41	54.72	3.169	0.0248 **
Taciturns	-135.97	101.91	-1.334	0.2397
Supporters	-266.53	109.60	-2.432	0.0592 *
Elitists	-105.19	55.88	-1.882	0.1185
Popular Participants	372.44	103.24	3.608	0.0154 **
Ignored	-75.69	33.39	-2.267	0.0727 *

Summary: Res. St Err: 311.5, Adj R^2: 0.8514, $F_{7,5}$: 10.82, p-value: 0.0092

Signif. codes: p-value < 0.001 *** 0.01 ** 0.05 * 0.1 . 1

5 Discussion and Future Work

The communities we chose to analyse in this paper were forums from Boards.ie. It is possible of course that different behavioural patterns could emerge when analysing different communities. However, there is no reason to assume that our current behaviour types would not apply, since the basic statistics that underpin them are not specific to any community. As for the features we chose to measure users' value, we have already started comparing them with results from Twitter and highlighting variations in their influence from Boards.ie.

Churn - i.e. the loss of community members - is a risk posed to online communities and one that community operators wish to avoid. Churn is normally affected by various community features [7]. By analysing the community composition that is correlated with a healthy community that evolves into an unhealthy community, we will be able to learn patterns that could then be used to pre-empt such changes, and thus warn community managers of the possibility of such decline. Our future work will also seek to identify key users within online communities and monitor their behaviour to predict their churn which would have a detrimental effect on the community. The combination of such *micro* and *macro*-level analysis would enable community managers to identify which users to pay more attention to in order to maintain a healthy community.

The emergence and evolution of certain types of behaviour could be dependent on the rise and fall of other behavioural types. Such possible correlations need further investigation and can support prediction of community evolution. On the other hand, negative user behaviour could badly influence health of the community, and an unhealthy community could foster negative behaviour. Many studies have shown that certain features (e.g. sentiment, time, user popularity) influence the spread of some types of behaviour, or increase response to posts. Some of these findings differ from one online community to another. The ideal mix and spread of behavioural types that boosts health in communities is still unknown, and it is likely to be dependent on the characteristics and goals of the communities in question.

Our approach for inferring user roles accounts for the dynamic nature of communities by utilising the repeated binning of feature values and using skeleton rules that map features to value levels. In doing so we have shown the ability of this approach to reduce the proportion of unclassified users when compared with an approach that does not utilise such updating. However, on average our approach still misses ~29% of users and is unable to associate those users with behaviour types. Our future work will explore methods to reduce this percentage by exploring the use of clustering and outlier detection techniques to account for new emerging roles within the community.

6   Conclusions

In this paper we have presented an approach to label the users of online communities with their role based on the behaviour they exhibit. We presented an ontology to capture the behavioural characteristics of users as numeric attributes and explained how semantic rules can be employed to infer the role that a given user has. There is currently no standard or agreed list of behaviour types for describing activities of users in online communities. Behaviour categories suggested in the literature are sometimes based on different observations and conceptions. In this paper our aim was not to identify the ultimate list of behavioural types, but rather to demonstrate a semantic model for representing and inferring behaviour of online community members.

A key contribution of this paper is the analysis of community composition over time and the correlation of such compositions with the health of communities, characterised by the number of posts made within a given community. Our empirical analysis of such correlations identified patterns in community composition that lead to both healthy and unhealthy communities, where a greater proportion of *elitists* and *popular participants* lead to an increase in activity, while a greater proportion of *taciturns* and *ignored* users lead to a decrease in activity. We also found that a stable community composition of role proportions lead to an increase in activity within the community, suggesting that wide fluctuations in role types could reduce community health.

Acknowledgment. This work was supported by the EU-FP7 projects WeGov (grant 248512) and Robust (grant 257859). Also many thanks to Boards.ie for providing data.

References

1. Backstrom, L., Kumar, R., Marlow, C., Novak, J., Tomkins, A.: Preferential behavior in online groups. In: Proc. Int. Conf. on Web Search and Web Data Mining (WSDM), New York, NY, USA (2008)
2. Breslin, J.G., Harth, A., Bojars, U., Decker, S.: Towards semantically-interlinked online communities. In: Gómez-Pérez, A., Euzenat, J. (eds.) ESWC 2005. LNCS, vol. 3532, pp. 500–514. Springer, Heidelberg (2005)
3. Chan, J., Hayes, C., Daly, E.: Decomposing discussion forums using common user roles. In: Proc. Web Science Conf (WebSci 2010), Raleigh, NC, US (2010)
4. Golder, S.A., Donath, J.: Social roles in electronic communities. In: Association of Internet Researchers (AoIR) 5.0 (2004)
5. Hoekstra, R.: Representing social reality in OWL 2. In: Proc. of OWLED (2010)
6. Lithium Technologies Inc. Community health index for online communities (2009), http://pages.lithium.com/community-health-index.html
7. Karnstedt, M., Rowe, M., Chan, J., Alani, H., Hayes, C.: The effect of user features on churn in social networks. In: Proc. ACM Web Science Conf. (WebSci 2011), Koblenz, Germany (2011)
8. LeClaire, J., Rushin, J.: Behavioral Analytics for Dummies. Wiley (2010)
9. Maia, M., Almeida, J., Almeida, V.: Identifying user behavior in online social networks. In: Proceedings of the 1st Workshop on Social Network Systems, SocialNets 2008, pp. 1–6. ACM, New York (2008)
10. Preece, J.: Online Communities - Designing Usability, Supporting Sociability. John Wiley & Sons, Ltd. (2000)
11. Preece, J.: Sociability and usability in online communities: Determining and measuring success. Behavior and Information Technology Journal 20(5), 347–356 (2001)
12. Soroka, V.: Invisible participants: how cultural capital relates to lurking behavior. In: Proceedings of the 15th International Conference on World Wide Web, pp. 163–172 (2006)
13. Sterne, J.: Social Media Metrics: How to Measure and Optimize Your Marketing Investment. John Wiley & Sons (2010)
14. Strijbos, J.-W., De Laat, M.F.: Developing the role concept for computer-supported collaborative learning: An explorative synthesis. Computers in Human Behavior 26(4), 495–505 (2010); Emerging and Scripted Roles in Computer-supported Collaborative Learning
15. Tapscott, D., Williams, A.: Wikinomics. Atlantic Books (2007)
16. Wire, N.: Led by facebook, twitter, global time spent on social media sites up 82% year over year (2010), http://blog.nielsen.com/nielsenwire/global/led-by-facebook-twitter -global-time-spent-on-social-media-sites-up-82-year-over-year/
17. Zhu, T., Wang, B., Wu, B., Zhu, C.: Role defining using behavior-based clustering in telecommunication network. Expert Syst. Appl. 38(4), 3902–3908 (2011)

Alignment-Based Trust for Resource Finding in Semantic P2P Networks

Manuel Atencia[1,2], Jérôme Euzenat[1],
Giuseppe Pirrò[3], and Marie-Christine Rousset[2]

[1] INRIA, Grenoble, France
{Manuel.Atencia,Jerome.Euzenat}@inrialpes.fr
[2] University of Grenoble, Grenoble, France
Marie-Christine.Rousset@imag.fr
[3] Free University of Bolzano-Bozen, Bolzano, Italy
giuseppe.pirro@unibz.it

Abstract. In a semantic P2P network, peers use separate ontologies and rely on alignments between their ontologies for translating queries. Nonetheless, alignments may be limited —unsound or incomplete— and generate flawed translations, leading to unsatisfactory answers. In this paper we present a trust mechanism that can assist peers to select those in the network that are better suited to answer their queries. The trust that a peer has towards another peer depends on a specific query and represents the probability that the latter peer will provide a satisfactory answer. In order to compute trust, we exploit both alignments and peers' direct experience, and perform Bayesian inference. We have implemented our technique and conducted an evaluation. Experimental results showed that trust values converge as more queries are sent and answers received. Furthermore, the use of trust improves both precision and recall.

1 Introduction

Peer-to-peer (P2P) systems have received considerable attention because their underlying infrastructure is very appropriate for scalable and flexible distributed applications over Internet. In P2P systems, there is no centralised control or hierarchical organisation: each peer is equivalent in functionality and cooperates with other peers in order to solve a collective task. P2P systems have evolved from simple keyword-based file sharing systems such as Napster and Gnutella to semantic data management systems such as EDUTELLA [14], PIAZZA [8] or SOMEWHERE [1].

In this paper, by a *semantic P2P network* we refer to a fully decentralised overlay network of people or machines (peers) sharing and searching for resources (documents, videos, photos, data, services) based on their semantic annotations using ontologies. In semantic P2P systems, every peer is free to organise her local resources as instances of classes of her own ontology serving as query interface for other peers. Alignments between ontologies make possible to reformulate queries from one local peer vocabulary to another. The result of a query is a set

L. Aroyo et al. (Eds.): ISWC 2011, Part I, LNCS 7031, pp. 51–66, 2011.

of resources (*e.g.*, documents) which are instances of some classes corresponding, possibly via subsumption or equality, to the initial query posed to a specific peer.

Trust is widely acknowledged as a central factor when considering networks of autonomous interacting entities and notably in the context of the Semantic Web. When referring to the notion of trust, T. Berners-Lee advocates for a user to be able to search for reasons why he or she should be confident of a returned answer [3]. Trust is helpful to select, from a given set of peers, those that are expected to answer with most satisfactory instances. Peers may use this information for broadcasting their queries to a reduced set of peers and to have an approximation of the reliability of provided answers. Furthermore, peers may preventively send selected queries in order to improve the trust they have towards another peer. Finally, by identifying "weak correspondences", peers may signal faulty alignments and trigger new matching of the ontologies.

Several proposals have been made that do not share the same meaning for trust [15,2]. Many are user/agent/peer centred and rely on the assumption that all peers share similar implicit goals. Trust is then closely related to the notion of reputation in a community.

In contrast, in the context of semantic P2P systems, each peer may have her own view on how categorising the resources that are exchanged between peers. For this reason, we rather promote the computation of subjective trust values based on direct experiences between peers. We also argue for a finer grained approach to trust in order to take into account the fact that, for answers provided by the same peer, the trust into these answers may vary according to which class they are instance of within the peer ontology.

An Illustrative Scenario

Consider a semantic P2P system for exchanging bookmarks, in which a peer *Alice* organises her bookmarks according to two main categories: *FavouriteMusic* and *GoodRestaurants*. These in turn are divided into subcategories: *Jazz*, *PopRock* and *Folk* for *FavouriteMusic*, and *Italian* and *Chinese* for *GoodRestaurants*. Within the Semantic Web, this can be implemented as a lightweight ontology that can be expressed in RDFS, in which categories and subcategories correspond to classes and subclasses, and the URLs identifying bookmarks correspond to URIs declared as instances of some classes.

Suppose that *Alice* is acquainted with *Bob* and *Chris* with whom she shares some interests in music and restaurants. This is captured by *correspondences* between her ontology and *Bob*'s and *Chris*'s ontologies. If *Bob* organises his best-of songs according to his favourite singers (*e.g.*, the classes *MichaelJackson* and *LouisArmstrong* are declared as subclasses of *BestSongs* in his ontology), the following correspondence expresses that any URL bookmarked by *Bob* as an instance of his class *MichaelJackson* can be bookmarked by *Alice* as an instance of her own class *PopRock*:[1]

$$Bob : MichaelJackson \leqslant Alice : PopRock$$

---

[1] We make use of the notation $P : A$ for identifying a class A of peer P's ontology.

An alignment between two peer ontologies is a set of correspondences between some classes used by these peers. Figure 1 shows the ontologies and alignments between *Alice*'s, *Bob*'s and *Chris*'s ontologies. It must be seen as a (small) part of a semantic P2P system that can be queried for resource finding.

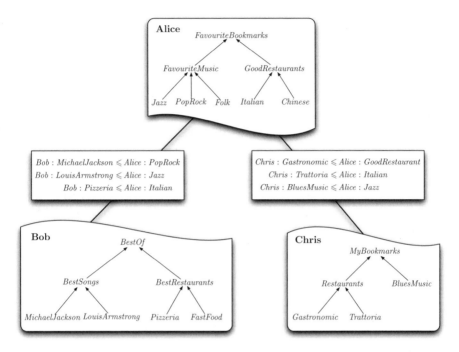

Fig. 1. Three semantic peers in a P2P semantic network

Suppose *Alice* wants to get bookmarks from her acquaintances in the network to enrich her bookmarks about *Italian* restaurants. The alignments between her ontology and *Bob*'s and *Chris*'s ontologies allow to reformulate her initial query about *Italian* restaurants into the query *Pizzeria* asked to *Bob*, and *Trattoria* asked to *Chris*. As his answer set, *Bob* will return to her the set of instances in the extension of his class *Pizzeria*, and *Chris* the set of instances in the extension of his class *Trattoria*.

Alice notices that *Chris* has some bookmarks in common with her, and thus tends to trust *Chris* for providing her with instances that fits well with her taste in terms of *Italian* restaurants. Subsequently, she may be inclined to add new bookmarks in her class *Italian* when they come from *Chris*.

However, although she trusts *Chris* for restaurants, she may not trust him for his musical tastes. For instance, for getting new bookmarks about *Jazz*, she can discover by choosing a sample of the set of URLs returned by *Chris* as the extension of his class *BluesMusic* that very few corresponding music files fit well with her taste in terms of Jazz music. For music, she will later tend not to trust *Chris* and will prefer to query *Bob* on this topic.

Contributions

In this paper we propose a probabilistic model to handle trust in a semantic P2P setting. We define the trust of a peer P towards another peer P' regarding a class C (belonging to P's ontology) as the probability that an instance returned by P' as an answer to the query asking for instances of C is satisfactory for P. In order to compute trust, we exploit the information provided by peers' ontologies and alignments, along with the information that comes from direct experience. Trust values are refined over time as more queries are sent and answers received.

We have designed an experimental protocol to study the convergence of trust, and to measure the gain of using trust for resource finding in practice.

Finally, a by-product of our trust model is a probabilistic setting for resource finding, in which the instances returned as answer for a given query are associated with a probability. This is in line with the recent trends towards probabilistic databases [4].

The paper is organised as follows. The background of our work is presented firstly. Then we introduce the notion of probabilistic populated ontology and the definition of trust. Later we explain the computation of trust and update of probabilistic populated ontologies. We discuss experimental results, and finally give some concluding remarks.

2   Preliminaries

In this section the components of a semantic peer-to-peer network are presented: populated ontologies, alignments and acquaintance graphs. The kind of queries that we take into account is also described.

2.1   Ontologies and Populated Ontologies

We draw a distinction between the ontological structure and the instances used to populate it. We deal with lightweight ontologies: classes linked by means of a less-general-than relation and a disjointness relation.

Definition 1. *An* ontology *is a tuple $O = \langle C, \leqslant, \perp \rangle$ where C is a non-empty finite set of* class symbols*; $\leqslant$ is a partial order on C; $\perp$ is an irreflexive and symmetric relation on C; and for all $c, c', d, d' \in C$,*

$$\text{if } c \perp d, \ c' \leqslant c \text{ and } d' \leqslant d \text{ then } c' \perp d'$$

A populated ontology is the result of adding instances to an ontology in accordance to the intended meaning of the two ontological relations.

Definition 2. *A* populated ontology *$\mathcal{O}$ is a tuple $\langle O, I, ext \rangle$, where O is an ontology, I is a set of* instances*, and ext is a function that maps each class c of O with a subset $ext(c)$ of I called the* extension *of c, in such a way that the family of class extensions covers I, and for all classes c, d the following hold:*

1. *if $c \leqslant d$ then $ext(c) \subseteq ext(d)$, and*
2. *if $c \perp d$ then $ext(c) \cap ext(d) = \emptyset$.*

2.2   Alignments

In an open and dynamic environment as a P2P network, the assumption of peers sharing the same ontology is not realistic. But if peers fall back on different ontologies, there must be a way to connect ontologies and translate queries so that their addressees are able to process them. Typically this is done by means of alignments —sets of correspondences between semantically related ontological entities— and finding alignments is what ontology matching is aimed at (see [7]).

A correspondence between two classes c and c' of two ontologies O and O', respectively, is usually defined as a tuple $\langle c, c', r \rangle$ with $r \in \{\leqslant, =, \geqslant, \perp\}$, where $c \leqslant c'$ (or $\langle c, c', \leqslant \rangle$) is read "$c$ is less general than c'", $c = c'$ is read "c is equal to c'", $c \geqslant c'$ is read "c is more general than c'", and $c \perp c'$ is read "c is disjoint from c'". Here, however, we deal with a more general notion of a correspondence inspired from [6].

Definition 3. *Let O and O' be two ontologies, and let c and c' be two classes of O and O', respectively. A* correspondence *between c and c' is a tuple $\langle c, c', R \rangle$ with $R \in 2^\Gamma$ where Γ is the set $\{=, >, <, \emptyset, \perp\}$. An* alignment *$\mathcal{A}$ between O and O' is a set of correspondences between classes of O and O'.*

In such correspondences, a class is connected to another through a set of base relations to be thought of as an exclusive disjunction. For instance, $c\{>, <\}c'$ (*i.e.*, $\langle c, c', \{>, <\} \rangle$) is read "either c is more general than c' or less general than c'". In this way, we can express uncertainty with regard to the alignment relation. Note that the relations '$\geqslant$' and '$\leqslant$' can be seen as abbreviations for $\{=, >\}$ and $\{=, <\}$, respectively. Secondly, a nonstandard symbol '$\emptyset$' is introduced. It reflects the idea of overlapping: classes the extensions of which share some instances but no one is equal to or contained into the other. Finally, $c\,\Gamma\,c'$ states total uncertainty about the relation between c and c'.

According to Definition 3, an alignment may include correspondences that link the same two classes through different relations, or no one connecting two particular classes. However, one would like alignments to relate any pair of classes and to do it in one way. If there exists no correspondence between c and c' in an alignment $\mathcal{A}$, we can simply add $\langle c, c', \Gamma \rangle$. If $\langle c, c', R \rangle, \langle c, c', S \rangle \in \mathcal{A}$ with $R \neq S$, we can replace both with $\langle c, c', R \cap S \rangle$. This follows the interpretation of alignments as a set of correspondences which all hold. The resulting alignment is said to be normalised.

Definition 4. *Let $\mathcal{A}$ be an alignment between two ontologies O and O'. The* normalisation *of $\mathcal{A}$ is the alignment $\overline{\mathcal{A}}$ made up of all correspondences $\langle c, c', R \rangle$ with $c \in C$, $c' \in C'$ and $R = \bigcap \mathcal{R}_\mathcal{A}(c, c')$ where $\mathcal{R}_\mathcal{A}(c, c') = \{S : \langle c, c', S \rangle \in \mathcal{A}\}$. The alignment $\mathcal{A}$ is said to be* normalised *providing $\mathcal{A} = \overline{\mathcal{A}}$.*

Remark 1. Recall that if $\mathcal{R}_\mathcal{A}(c, c') = \emptyset$ then $\bigcap \mathcal{R}_\mathcal{A}(c, c') = \Gamma$.

All alignments considered in this work are assumed to be normalised.

2.3 Peers and Acquaintance Graphs

We consider a finite set $\mathcal{P} = \{P_i\}_{i=1}^n$ of peers. In this work, P_i will be identified by i. We assume that each peer P_i is associated with one populated ontology $\mathcal{O}_i = \langle O_i, I_i, ext_i \rangle$ (where $1 \leq i \leq n$).

An acquaintance graph stands for peers' acquaintances (or neighbours) in the network. As usual, a link between two peers reflects the fact that they know the existence of each other. In addition, we assume that there exists one alignment between their respective ontologies.

Definition 5. *An* acquaintance graph *is a labelled directed graph* $\langle \mathcal{P}, \mathrm{ACQ} \rangle$ *where* $\mathcal{P} = \{P_i\}_{i=1}^n$ *is the set of vertices and any edge in* ACQ *is of the form* $\langle i, j \rangle$ *with* $i \neq j$, *and it is labelled with an alignment* $\mathcal{A}_{ij}$ *between ontologies* O_i *and* O_j. *Moreover, if* $\langle i, j \rangle \in \mathrm{ACQ}$ *then* $\langle j, i \rangle \in \mathrm{ACQ}$ *and* $\mathcal{A}_{ji}$ *is the inverse of* $\mathcal{A}_{ij}$.[2]

Peer P_j *is said to be an* acquaintance *of peer* P_i *if* $\langle i, j \rangle \in \mathrm{ACQ}$. *The set of* acquaintances *of* P_i *is denoted by* $\mathrm{ACQ}(P_i)$.

Remark 2. Note that, given two ontologies O and O', we can always consider the trivial alignment, that is, the one that is made up of all correspondences $\langle c, c', \Gamma \rangle$ with $c \in C$ and $c' \in C'$.

2.4 Queries and Query Translations

Peers pose queries to obtain information concerning other peers' populated ontologies. We deal with a simple query language, as peers can only request class instances: if peer P_j is an acquaintance of peer P_i, she may be asked

$$\mathcal{Q} = c(X)? \tag{1}$$

by P_i with $c \in O_i$. Now, since we do not assume that all peers share the same ontology, queries may require to be translated for their recipients to be able to process them.

Query translations are determined by correspondences of the alignments of the network. Specifically, if peer P_i wants to send $\mathcal{Q}$ to P_j, she will first choose one correspondence $\langle c, d, R \rangle \in \mathcal{A}_{ij}$ (typically R is equal to '$=$' or '$>$') and then send P_j the translation

$$\mathcal{Q}' = d(X)? \tag{2}$$

The answer to (1) through its translation (2) is the set of instances of class d in P_j's populated ontology. Unlike queries, we assume that no translation of instances is ever required. Since alignments may be unsound and incomplete, this answer may contain *unsatisfactory* instances, *i.e.*, instances which are not considered instances of c by P_i.

A peer cannot foresee whether the answer that another peer provides to one of her queries contains satisfactory instances or not, but this uncertainty can be estimated with the help of a trust mechanism.

---

[2] Given an alignment $\mathcal{A}$ between O and O', the inverse of $\mathcal{A}$ is the alignment $\mathcal{A}^{-1} = \{\langle c', c, R^{-1} \rangle : \langle c, c', R \rangle \in \mathcal{A}\}$ between O' and O, where $R^{-1} = \{r^{-1} : r \in R\}$ and r^{-1} is '$>$' and '$<$' if r is '$<$' and '$>$', respectively, and $r^{-1} = r$ otherwise.

3   The Trust Mechanism

As mentioned above we look at trust as a way to estimate the proportion of satisfactory instances in a peer answer. The notion of satisfactory instance can be faithfully captured by an ideal populated ontology $\mathcal{O}_i^*$ that corresponds to a hypothetical situation in which peer P_i classified all instances of the network according to her ontology O_i. In this way we can express the fact that P_i considers an arbitrary instance a as an instance of $c \in O_i$ by $a \in ext_i^*(c)$. It is assumed that $O_i = O_i^*$ and $ext_i(c) \subseteq ext_i^*(c)$ for every class $c \in C_i$. This populated ontology is referred to as the *reference populated ontology* of peer P_i.

If peer P_i receives a set B as an answer to the query (2), the proportion of satisfactory instances is given by the conditional probability $p(ext_i^*(c)|B)$. The probability space under consideration here is the triple $(\Omega, \mathfrak{A}, p(\cdot))$ where Ω is the set of instances of the network (a finite set), the σ-algebra $\mathfrak{A}$ is the power set of Ω, and $p(\cdot)$ is Laplace's definition of probability. Our approach for trust aims at finding approximations to these conditional probabilities. Before the definition of trust we introduce the notion of a probabilistic populated ontology.

3.1   Probabilistic Populated Ontologies

Once an answer is received, it can be (partly) added or not to the extension of the queried class. In order to capture the evolution of class extensions in the network, we consider a time variable $t \in \mathbb{N}$, and we will write $\mathcal{O}_i^t$ to denote peer P_i's populated ontology at instant t (beginning with $\mathcal{O}_i$):

$$\mathcal{O}_i = \mathcal{O}_i^0, \mathcal{O}_i^1, \ldots, \mathcal{O}_i^t, \ldots \tag{3}$$

We assume that the underlying ontology never changes, *i.e.*, $O_i = O_i^t$ $(t \in \mathbb{N})$, and that the sequence of class extensions $\{ext_i^t(c)\}_{t \in \mathbb{N}}$ is monotonically increasing for all $c \in C_i$.

Nonetheless, since we deal with probabilities new instances may not be 100% satisfactory. For this reason, at $t \in \mathbb{N}$, peer P_i is associated with a *probabilistic populated ontology*.

Definition 6. *Peer P_i's probabilistic populated ontology at time t is a triple*

$$\widetilde{\mathcal{O}}_i^t = \langle O_i, I_i^t, \widetilde{ext}_i^t \rangle$$

where I_i^t is a set of instances and $\widetilde{ext}_i^t$ is a function that maps each class c of O_i with its probabilistic extension

$$\widetilde{ext}_i^t(c) = \langle A^*, \mathcal{F} \rangle \text{ with } \mathcal{F} = \{\langle A^k, [p^k, q^k] \rangle\}_{k \in K} \text{ where}$$

- A^* *is a (possibly empty) subset of $ext_i^*(c)$, that is, a set of instances which are certain to be instances of the class c, and all*
- A^k *are pairwise disjoint subsets of I_i^t which are also disjoint from A^*, and all $[p^k, q^k]$ are distinct subintervals of the unit interval $[0,1]$, where $k \in K$ and K is a (possibly empty) index set of integers.*

Furthermore, the tuple $\mathcal{O}_i^t = \langle O_i, I_i^t, ext_i^t \rangle$ with $ext_i^t(c) = A^* \uplus \biguplus_{k \in K} A^k$ must be a populated ontology (so that the axioms that relate classes with their extensions are fulfilled).

A probabilistic extension $\widetilde{ext}_i^t(c)$ can be seen as a classical extension $ext_i^t(c)$ partitioned into a number of subsets $A^*, A^1, \ldots, A^n$. All instances of A^* are sure to be instances of the class c and then $p(ext_i^*(c)|A^*) = 1$. However, the set A^k ($1 \leq k \leq n$) may contain instances that are actually not instances of c. The idea behind the interval $[p^k, q^k] \subseteq [0,1]$ is that there exists some statistical evidence for $p^k \leq p(ext_i^*(c)|A^k) \leq q^k$.[3] The way probabilistic extensions are built is explained in Section 3.5.

Remark 3. Every populated ontology $\mathcal{O}_i$ can be seen as a probabilistic populated ontology $\widetilde{\mathcal{O}}_i = \langle O_i, I_i, \widetilde{ext}_i \rangle$ where $\widetilde{ext}_i(c) = \{\langle ext_i(c), \emptyset \rangle\}$ for all $c \in C_i$.

Peers build probabilistic populated ontologies as more queries are sent and answered (starting with the "probabilistic version" of $\mathcal{O}_i$):

$$\widetilde{\mathcal{O}}_i = \widetilde{\mathcal{O}}_i^0, \widetilde{\mathcal{O}}_i^1, \ldots, \widetilde{\mathcal{O}}_i^t, \ldots \tag{4}$$

And what was said about (3) at the beginning of this section holds for the underlying populated ontologies of (4).

3.2   Definition of Trust

With the new terminology, P_j's answer to query (1) via its translation (2) at time t is the extension $ext_j^t(d)$, and an arbitrary instance $a \in ext_j^t(d)$ is qualified as satisfactory provided that $a \in ext_i^*(c)$. The proportion of satisfactory instances in $ext_j^t(d)$ is given by the conditional probability $p(ext_i^*(c)|ext_j^t(d))$. Our proposal is that the higher this value is, the more P_i trusts P_j.

Definition 7. *Let us consider two peers P_i and P_j ($i \neq j$) and two classes c and d of O_i and O_j, respectively. The trust that P_i has towards P_j with respect to the translation $\langle c, d \rangle$ at time t is the conditional probability $p(ext_i^*(c)|ext_j^t(d))$ and it is denoted by $trust^t(P_i, P_j, \langle c, d \rangle)$.*

This idea is slightly different from most of the existing approaches for trust. In our setting cheating is not directly addressed: unsatisfactory answers are seen as the result of peers' incapacity to understand each other. In addition, trust is dependent on translations: peers may be very trustworthy in regard with some translations but not with others. In the following section, we explain our approach for computing trust. It exploits the information provided by alignments and revises it with direct experience.

---

[3] The use of intervals follows Lukasiewicz's notation for conditional constraints in probabilistic knowledge bases [11].

3.3 Computation of Trust

Our approach for trust aims at approximating $trust^t(P_i, P_j, \langle c, d \rangle)$ by Bayesian inference. A probability distribution $T^t(P_i, P_j, \langle c, d \rangle)$ represents P_i's belief about $\theta = p(ext_i^*(c)|ext_j^t(d))$. If there is no direct experience, alignments are taken to construct prior beliefs. Answers are later used to revise these beliefs. As they can be of a size that cannot be processed manually, we propose to perform sampling with replacement in order to estimate the number of satisfactory instances. We work with beta distributions as they are typically used to describe the parameter of a binomial distribution.

No direct experience: alignment-based trust. If $T^t(P_i, P_j, \langle c, d \rangle)$ is not defined (this is the case when, for instance, $t = 0$), we fall back on alignments. Peers P_i and P_j's ontologies are linked through $\mathcal{A}_{ij}$. Since this alignment is normalised then there exists a unique $R \subseteq \Gamma$ such that $\langle c, d, R \rangle \in \mathcal{A}_{ij}$. The intending meaning of correspondences is

$$
\begin{aligned}
R = \{=\} \quad &\text{iff} \quad ext_i^*(c) = ext_j^*(d) \\
R = \{>\} \quad &\text{iff} \quad ext_i^*(c) \supset ext_j^*(d) \\
R = \{<\} \quad &\text{iff} \quad ext_i^*(c) \subset ext_j^*(d) \\
R = \{\perp\} \quad &\text{iff} \quad ext_i^*(c) \cap ext_j^*(d) = \emptyset \\
R = \{\lozenge\} \quad &\text{iff} \quad \text{none of the above holds}
\end{aligned}
$$

Hence, provided that $ext_j^t(d) \subseteq ext_j^*(d)$,

$$
\begin{aligned}
\text{if } R \text{ is `=' or `>'} \quad &\text{then} \quad p(ext_i^*(c)|ext_j^t(d)) = 1 \\
\text{if } R \text{ is `}\perp\text{'} \quad &\text{then} \quad p(ext_i^*(c)|ext_j^t(d)) = 0 \\
\text{if } R \text{ is `<' or `}\lozenge\text{'} \quad &\text{then} \quad p(ext_i^*(c)|ext_j^t(d)) \in [0, 1]
\end{aligned}
$$

In the general case, R is a set $\{r_1, \ldots, r_n\} \subseteq \Gamma$ (with $n \le 5$). If we assume that all relations in R are equiprobable, by the law of total probability, we have

$$
trust^t(P_i, P_j, \langle c, d \rangle) \in [u, v] \quad \text{with} \quad u = \frac{1}{n}\sum_{k=1}^{n} a_k \quad v = \frac{1}{n}\sum_{k=1}^{n} b_k
$$

where $[a_k, b_k] = [1, 1]$ if r_k is `=' or `>', $[a_k, b_k] = [0, 0]$ if r_k is `$\perp$', and finally $[a_k, b_k] = [0, 1]$ if $r_k$ is `<' or `$\lozenge$' $(k = 1, \ldots, n)$.

The information above can be used to construct P_i's prior belief about the parameter $\theta = p(ext_i^*(c)|ext_j^t(d))$ by means of a beta distribution $\text{Beta}(\alpha, \beta)$. If $[u, v] = [0, 1]$, we take the uniform distribution $U[0, 1] = \text{Beta}(1, 1)$. If not, and $u < v$, we equal $[u, v]$ with $[\mu - 2\sigma, \mu + 2\sigma]$ and then find $\text{Beta}(\alpha, \beta)$ whose mean and deviation are μ and σ. This is the standard way to find a confidence interval based on the normal distribution. If $u = v$, we proceed with $\sigma = 0.005$ and $\mu = u$ unless $u = 1$ and $u = 0$, in which cases $\mu = 0.99$ and $\mu = 0.01$, respectively.[4] In this way, we define the trust distribution $T^t(P_i, P_j, \langle c, d \rangle) = \text{Beta}(\alpha, \beta)$.

---

[4] The values for α and β can be found by solving $\mu = \frac{\alpha}{\alpha+\beta}$ and $\sigma = \sqrt{\frac{\mu(\mu-1)}{\alpha+\beta+1}}$.

Example 1. If $R = \{<, =\}$ then $[u, v] = [.5, 1]$. If we make $[.5, 1] = [\mu - 2\sigma, \mu + 2\sigma]$ then $\mu = .75$ and $\sigma = .125$. This leads to Beta$(8.25, 2.75)$ whose shape is depicted in Figure 2(a). Figure 2 is completed with the shapes of beta distributions for the relations $R = \{=\}$ and $R = \Gamma$. The latter corresponds to Beta$(0.4, 0.8)$.[5]

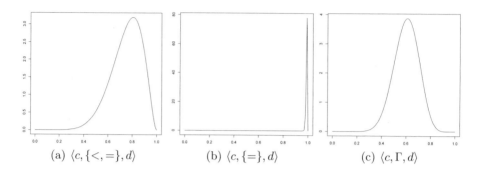

(a) $\langle c, \{<, =\}, d\rangle$ (b) $\langle c, \{=\}, d\rangle$ (c) $\langle c, \Gamma, d\rangle$

Fig. 2. Beta distributions for different correspondences

Direct experience. Trust at time t is used to choose a peer to which to send a query, as well as a class through which to translate it. This is explained in detail in Section 3.4. Let us imagine that P_i receives $B = ext_j^t(d)$. A sampling with replacement is performed over B in order to estimate the number of satisfactory instances. Let $S \subseteq B$ be a sample (strictly speaking, S is a multiset). We assume that every peer can call an *oracle* (typically the user) to find out whether an instance is satisfactory or not. More specifically, given $a \in S$, P_i's oracle provides a yes/no response to the question: "$a \in ext_i^*(c)$?". Even this, nonetheless, may be a high burden for P_i's oracle. We can benefit from peers' populated ontologies to process some instances automatically without the need to call oracles. Recall that P_i is associated with a probabilistic populated ontology $\tilde{\mathcal{O}}_i^t$, and that the probabilistic extension of class c includes a set A^* of instances which are certain to be instances of $ext_i^*(c)$. So if $a \in B \cap A^*$, $a \in ext_i^*(c)$. We can also identify unsatisfactory instances automatically: if $a \in S$ is such that there exists c' in O_i with $c \perp c'$ and $a \in A'^*$ then $a \notin ext_i^*(c)$. The remaining instances are processed by peer P_i's oracle.

Assume that $T^t(P_i, P_j, \langle c, d\rangle) = \text{Beta}(\alpha, \beta)$. If s is the sample size, s^+ is the number of successes (satisfactory instances), and $s^- = s - s^+$ is the number of failures, peer P_i's posterior belief about $\theta = p(ext_i^*(c)|ext_j^t(d))$ is summarised in Beta$(\alpha + s^+, \beta + s^-)$. Thus we define

$$T^{t+1}(P_i, P_j, \langle c, d\rangle) = \text{Beta}(\alpha + s^+, \beta + s^-) \tag{5}$$

---

[5] Although $c\Gamma d$ stands for total uncertainty about the relation between c and d, the mean of its associated beta distribution, Beta$(0.4, 0.8)$, is not 0.5 but 0.6. However, our aim is not to find out the correct relation between c and d, but to estimate the probability $p(ext_i^*(c)|ext_j^t(d))$. In this sense, total uncertainty arises with $c < d$, $c \lozenge d$ or $c\{<, \lozenge\}d$, which are all modelled with a uniform distribution (whose mean is 0.5).

3.4   Use of Trust

Imagine that peer P_i wants to query $c(X)$? $(c \in C_i)$ at time $t \in \mathbb{N}$. Then P_i chooses an element from the set

$$\mathcal{P}_0 = \{\langle P_j, d_j \rangle : P_j \in \text{ACQ}(P_i) \text{ and } d_j \in O_j\}$$

so that, if $\langle P_{j_0}, d_{j_0} \rangle$ is the preferred tuple, P_i will send $d_{j_0}(X)$? to P_{j_0}. This choice depends on trust: P_i opts for $\langle P_{j_0}, d_{j_0} \rangle$ iff

$$\text{E}(T^t(P_i, P_{j_0}, \langle c, d_{j_0} \rangle)) = \max_{\langle P_j, d_j \rangle \in \mathcal{P}_0} \{\text{E}(T^t(P_i, P_j, \langle c, d_j \rangle))\}$$

where $\text{E}(\cdot)$ denotes the expected value of a distribution.

3.5   Updating Probabilistic Populated Ontologies

In the end, trust is used for class extensions to be increased with new satisfactory instances. If peer P_i receives $B = ext_j^t(d)$ as an answer to "$c(X)$?" then B will be (partly) added to $ext_i^t(c)$. In line with the computation of trust based on direct experience (see Section 3.3), the set B is partitioned into three subsets:

$$B = B_{aut}^+ \uplus B_{aut}^- \uplus B_{\overline{aut}}$$

- $B_{aut}^+ = \{a \in B : a \in ext_i^t(c)\} = B \cap ext_i^t(c)$
- $B_{aut}^- = \{a \in B : \text{there exists } c' \in O_i \text{ with } a \in ext_i^t(c') \text{ and } c \perp c'\}$
- $B_{\overline{aut}} = \{a \in B : a \notin B_{aut}^+ \text{ and } a \notin B_{aut}^-\} = B \setminus (B_{aut}^+ \uplus B_{aut}^-)$

The set B_{aut}^+ contains the instances in B that already belong to $ext_i^t(c)$, and B_{aut}^- comprises those instances that, if added to $ext_i^t(c)$, would yield to a logical inconsistency. The set $B_{\overline{aut}}$ embodies the new information that can be included in $ext_i^t(c)$.[6] Since the answer B was received as the result of a comparison of trust values, it seems reasonable to add all instances of $B_{\overline{aut}}$ to $ext_i^t(c)$. The fact that these instances may not be 100% satisfactory, though, should be reflected in P_i's populated ontology. As described in Section 3.1, probabilistic populated ontologies are designed for this purpose.

The set $B_{\overline{aut}}$ will be included in $ext_i^t(c)$ along with an interval $[p, q]$ such that $p \leq p(ext_i^*(c)|B_{\overline{aut}}) \leq q$ on the basis of statistical evidence. Again, we propose to perform Bayesian inference, but, instead of weighing more on P_i's oracle, we lean on the previous sampling and make use of the formula

$$p(ext_i^*(c), B_{\overline{aut}}|ext_j^t(d)) = p(ext_i^*(c)|B_{\overline{aut}}) \cdot p(B_{\overline{aut}}|ext_j^t(d)) \tag{6}$$

Let us explain this in detail. The probability $p(B_{\overline{aut}}|ext_j^t(d))$ represents the proportion of instances of $B_{\overline{aut}}$ in $ext_j^t(d)$ and its computation is straightforward. By monotonicity, we have

$$0 \leq p(ext_i^*(c), B_{\overline{aut}}|ext_j^t(d)) \leq p(B_{\overline{aut}}|ext_j^t(d))$$

---

[6] The subscript "aut" stands for "automatic", as both instances from B_{aut}^+ and B_{aut}^- can be automatically processed, whereas this is not the case for $B_{\overline{aut}}$.

In order to compute a prior about $\vartheta = p(ext_i^*(c), B_{\overline{aut}}|ext_j^t(d))$, we proceed as the computation of alignment-based trust (Section 3.3): we equate the interval $[u, v]$, where $u = 0$ and $v = p(B_{\overline{aut}}|ext_j^t(d))$, with $[\mu - 2\sigma, \mu + 2\sigma]$, and then find $Beta(\alpha, \beta)$ whose mean and deviation are μ and σ, respectively. A posterior is computed with the same sampling S used for Equation 5, but this time we count as a success any satisfactory instance that also belongs to $B_{\overline{aut}}$.

Let $Beta(\alpha', \beta')$ be the resulting posterior, and let μ' and σ' be its mean and deviation. The set $B_{\overline{aut}}$ is included in $ext_i^*(c)$ along with the interval $[p, q]$ where

$$p = \frac{1}{p(B_{\overline{aut}}|ext_j^t(d))}(\mu' - 2\sigma') \quad q = \frac{1}{p(B_{\overline{aut}}|ext_j^t(d))}(\mu' + 2\sigma')$$

Hence, $p \leq p(ext_i^*(c)|B_{\overline{aut}}) \leq q$ with 95% probability, which is based on the normal approximation to the posterior density for ϑ and Equation 6. Actually, if S^+ denotes the set of satisfactory instances in the sample S, $B_{\overline{aut}}$ is partitioned into $B_{\overline{aut}} \cap S^+$ and $B_{\overline{aut}} \setminus S^+$, which are added to $ext_i^t(c)$ separately. Thus $[p, q]$ must be resized accordingly, and then replaced by another interval $[p', q']$. Below we explain explicitly how probabilistic populated ontologies are built.

As remarked in Section 3.1, $\widetilde{\mathcal{O}}_i^0$ is defined as the probabilistic version of P_i's initial populated ontology $\mathcal{O}_i$, that is,

- $\widetilde{\mathcal{O}}_i^0 = \widetilde{\mathcal{O}}_i$, and
- at time $t \in \mathbb{N}$, if $\widetilde{ext}_i^t(c) = \langle A^*, \mathcal{F} \rangle$ then we define

$$\widetilde{ext}_i^{t+1}(c) = \langle A^* \uplus (B_{\overline{aut}} \cap S^+), \mathcal{F} \uplus \langle B_{\overline{aut}} \setminus S^+, [p', q'] \rangle \rangle$$

In order for $\widetilde{\mathcal{O}}_i^{t+1}$ to be a probabilistic populated ontology, though, $B_{\overline{aut}}$ must be included in the extension of any superclass c' of c. For the sake of space, we give a brief explanation of how this is done. Notice first that no instance in $B_{\overline{aut}}$ belongs to the extension of a class disjoint from c' as $B_{aut}^- \cap B_{\overline{aut}} = \emptyset$ and c is a subclass of c'. All instances in $B_{\overline{aut}} \cap S^+$ are certainly instances of c' since $ext_i^*(c) \subseteq ext_i^*(c')$. Instead of $B_{\overline{aut}} \setminus S^+$ we include $B_{\overline{aut}} \setminus (S^+ \cup ext_i^t(c'))$ as some instances of $B_{\overline{aut}} \setminus S^+$ may already belong to $ext_i^t(c')$. In order to find an interval with which to estimate $p(ext_i^*(c')|B_{\overline{aut}} \setminus (S^+ \cup ext_i^t(c')))$, we proceed as before to approximate $p(ext_i^*(c)|B_{\overline{aut}} \setminus (S^+ \cup ext_i^t(c')))$ and then apply the monotonicity of probability. In this way, the upper bound that we obtain is equal to 1.

By construction, $\widetilde{\mathcal{O}}_i^{t+1}$ is a probabilistic populated ontology.

4   Experimental Analysis

This section reports on a preliminary experimental campaign that has been conducted to test the viability of the trust mechanism described in this paper.

We set out to answer two research questions:

1. Do trust values converge as more queries are sent and answers received?
2. Is there any gain in query-answering performance —measured in precision and recall— by using the trust technique?

In what follows we first describe the experimental setting and then explain the execution and evaluation.

4.1   Experimental Setting

The trust mechanism presented in this work has been implemented in a simulator written in Java. The simulator also deals with aspects indirectly related to trust, such as generation of P2P networks, populated ontologies and alignments. In the remainder of the section we elaborate more on these aspects.

P2P network topology. Social networks are well-known to exhibit small-world characteristics [5]. For this reason, a small-world topology was used for the entire evaluation. To generate this topology, we ran Kleinberg's algorithm included in the JUNG Java library.[7] A node in the network represents a peer associated with a populated ontology. The total number of peers in our evaluation was 20.

Populated ontologies. All populated ontologies in the evaluation had the same underlying ontology $O_i = O$. More specifically, we chose the ontological scheme described in [10] (with 64 classes). The semantic heterogeneity was reproduced by the way classes were populated with instances. The simulator implements an ontology population module which was utilised for both reference populated ontologies $\mathcal{O}_i^*$ and initial populated ontologies $\mathcal{O}_i = \mathcal{O}_i^0$. First, a set S of abstract instances is generated. In our evaluation, the size of S was 6000. Second, for each peer P_i, a sample S_i is taken from S. Furthermore, this sampling is performed in a way that S_i and S_j overlap for each pair i, j. The size of each S_i is determined with a Zipfian distribution, which is often used to approximate data in physical and social sciences [12]. The skewing factor considered was 0.5. Third, the top class of $\mathcal{O}_i^*$ is populated with S_i and a top-down population process is carried out by removing instances randomly for the remainder of classes. During this process, we check that all ontological axioms —subclass and disjoint relations— are fulfilled. Initial populated ontologies are generated in a similar way, starting this time with a sample of S_i instead of S to populate the top class in $\mathcal{O}_i$.

Alignment generation. A connection between peers P_i and P_j in the network (edge between nodes) is labelled with an alignment $\mathcal{A}_{ij}$ between their respective ontologies. This is seen as a declined version of a reference alignment $\mathcal{A}_{ij}^*$ which is never available to the peers. Thus we can capture the real practice of ontology matching. Reference alignments are built by comparing class extensions in the reference populated ontologies (for instance, $c < d$ is included in $\mathcal{A}_{ij}^*$ iff $ext_i^*(c) \subset ext_j^*(d)$). To build initial alignments, correspondences in reference alignments are discarded or replaced randomly in accord with global values for precision and recall. In our evaluation, we chose 0.6 for both measures.

4.2   Execution and Evaluation

From all peers and classes in the network we chose a subset $\mathcal{P}_0 \subseteq \mathcal{P}$ of 15 peers and a subset $C_0 \subseteq C$ of 25 classes randomly and ran 100 simulations. At each round $n \leq 100$ of the execution, a peer $P_i \in \mathcal{P}_0$ and a class $c \in C_0$ are randomly chosen. Then an acquaintance P_j of P_i and a class $d \in C_j$ are selected by using

---

[7] http://jung.sourceforge.net

the trust mechanism (Section 3.4). Notice that $C_i = C_j = C$ as we chose a single ontological scheme O. To process answers, the maximum number of oracle calls allowed was 40. The subset $B_{\overline{aut}} \subseteq ext_j^t(d)$ is included in peer P_i's probabilistic populated ontology if the expected value $\mathrm{E}(T^n(P_i, P_j, \langle c, d \rangle))$ is greater than a given threshold. In our evaluation, this threshold was 0.6.

In order to test the convergence of trust, we analysed the difference

$$\Delta^n = |\mathrm{E}(T^n(P_i, P_j, \langle c, d \rangle)) - p(ext_i^*(c)|ext_j^*(d))|$$

over the 10 most occurred queries. Figure 3 shows the experimentation results. After a number of rounds, Δ^n approached 0. Actually, in most of the cases, no more than 5 rounds were needed for Δ^n to be close to 0.1.

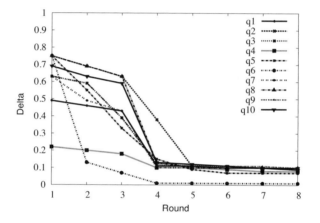

Fig. 3. Test of convergence of trust

In order to test the gain in query-answering performance, we compared the use of the trust mechanism with a naive strategy. In the latter, peers randomly choose acquaintances and always accept their answers. For the evaluation to be fair, the same set of queries was used in both strategies. This time we analysed precision and recall measured by

$$Precision(n) = \frac{|ext_i^*(c) \cap ext_i^n(c)|}{|ext_i^n(c)|} \qquad Recall(n) = \frac{|ext_i^*(c) \cap ext_i^n(c)|}{|ext_i^*(c)|}$$

Figure 4 depicts the average precision and recall over the 100 rounds for the 20 most occurred queries. As expected, the naive strategy produced lower values for both measures. Furthermore, the use of the trust mechanism ensured high precision. However, this was not the case for recall. The reason is that peers only ask their neighbours, and these ones never change. As instances are spread all over the network, many instances may be unaccessible to peers. It is expected that if instances were more accessible, recall would be higher, but this remains to be experimented. The theoretical model presented in this paper is general enough to cover the case where peers receive answers from non-neighbour peers.

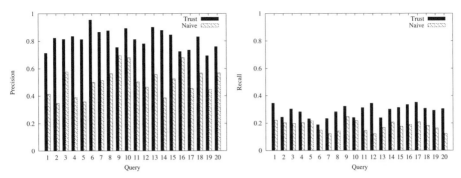

Fig. 4. Comparison between the use of trust and the naive strategy

5 Concluding Remarks

We have proposed a trust mechanism in semantic P2P systems. The trust that a peer has towards another peer depends on a specific query and represents the probability that the latter will provide a satisfactory answer. In order to compute trust, we exploit alignments and peers' direct experience, and perform Bayesian inference. Preliminary experimental results show that trust values converge as more queries are sent and answers received, and that there is a gain in query-answering precision and recall when peers make use of the trust mechanism.

The notion of probabilistic populated ontology has been introduced. This is a by-product of trust computation that allows to store and process the instances obtained from query answers in the same way as it is done in probabilistic databases [4]. More precisely, a probabilistic populated ontology can be seen as a probabilistic database in which each fact $C(i)$ is associated with a (lower bound of) probability. As a result, query answers can be ranked, and only top-k answers can be returned to interested users. In addition, since trust evolves over time as more queries are spread over the network and their answers are processed and stored with their probabilities, the resulting probabilistic populated ontologies somehow *capture* and *compile* the results of a trust propagation.

Many different probabilistic approaches to trust can be found in the literature [16,13]. Some also perform Bayesian inference over feedback on past interactions. However, to the best of our knowledge, our model is the only one which explicitly benefits from ontological content and alignments.

EigenTrust [9] is a peer-to-peer algorithm which, like ours, has a direct trust computation. Direct trust is then propagated among peers and aggregated to calculate global trust which can be very costly. As remarked above, we avoid this computation by exploiting the information on global trust stored and compiled in the probabilistic populated ontologies of acquaintance peers.

As future work, we plan to extend our trust model in order to deal with more expressive ontology and query languages. Although witness peers are not considered in this paper, the use of witness information is another future research line. Witness peers can help to find new trustworthy acquaintances. In this way, recall values can increase. Furthermore, the impact of malicious peers that hide or bias information, or lie, will be studied too.

Regarding the experimentation, we aim to perform a thorough experimental analysis concerning different network configurations in terms of number of peers, instances and oracle calls. Moreover, we want to investigate the relation between the quality of alignments and the speed of convergence of trust values.

Acknowledgements. This work is supported under the Dataring project, which is sponsored by the Agence Nationale de Recherche (ANR-08-VERS-007), and the Webdam project, sponsored by the European Research Council (FP7-226513).

References

1. Adjiman, P., Chatalic, P., Goasdoué, F., Rousset, M.C., Simon, L.: Distributed reasoning in a peer-to-peer setting: application to the Semantic Web. Journal of Artificial Intelligence Research 25, 269–314 (2006)
2. Artz, D., Gil, Y.: A survey of trust in computer science and the Semantic Web. Journal of Web Semantics 5(2), 58–71 (2007)
3. Berners-Lee, T.: Cleaning up the user interface (1997), http://www.w3.org/DesignIssues/UI.html
4. Dalvi, N., Re, C., Suciu, D.: Query evaluation on probabilistic databases. IEEE Data Engineering Bulletin 29(1), 25–31 (2006)
5. Watts, D.J.: Networks, dynamics, and the small-world phenomenon. American Journal of Sociology 105(2), 493–527 (1999)
6. Euzenat, J.: Algebras of ontology alignment relations. In: Sheth, A.P., Staab, S., Dean, M., Paolucci, M., Maynard, D., Finin, T., Thirunarayan, K. (eds.) ISWC 2008. LNCS, vol. 5318, pp. 387–402. Springer, Heidelberg (2008)
7. Euzenat, J., Shvaiko, P.: Ontology matching. Springer, Heidelberg (2007)
8. Halevy, A., Ives, Z., Tatarinov, I., Mork, P.: Piazza: data management infrastructure for Semantic Web applications. In: Proceedings of the 12th International Conference on World Wide Web, WWW 2003, pp. 556–567 (2003)
9. Kamvar, S.D., Schlosser, M.T., Garcia-Molina, H.: The EigenTrust algorithm for reputation management in P2P networks. In: Proceedings of the 12th International Conference on World Wide Web, WWW 2003, pp. 640–651 (2003)
10. Lorenz, B.: Ontology of transportation networks. REWERSE project, IST-2004-506779, EU FP6 Network of Excellence (NoE). Deliverable (2005)
11. Lukasiewicz, T., Straccia, U.: Managing uncertainty and vagueness in description logics for the Semantic Web. Journal of Web Semantics 6(4), 291–308 (2008)
12. Manning, C.D., Schütze, H.: Foundations of Statistical Natural Language Processing. The MIT Press (1999)
13. Mui, L., Mohtashemi, M., Ang, C., Szolovits, P., Halberstadt, A.: Ratings in distributed systems: a bayesian approach. In: Proccedings of the 11th Workshop on Information Technologies and Systems, WITS 2001 (2001)
14. Nejdl, W., Wolf, B., Qu, C., Decker, S., Sintek, M., Naeve, A., Nilsson, M., Palmér, M., Risch, T.: EDUTELLA: a P2P networking infrastructure based on RDF. In: Proceedings of the 11th International Conference on the World Wide Web, WWW 2002, pp. 604–615 (2002)
15. Sabater, J., Sierra, C.: Review on computational trust and reputation models. AI Review 24(1), 33–60 (2005)
16. Schillo, M., Funk, P., Rovatsos, M.: Using trust for detecting deceitful agents in artificial societies. Applied Artificial Intelligence 14(8), 825–848 (2000)

The Justificatory Structure of
the NCBO BioPortal Ontologies

Samantha Bail, Matthew Horridge, Bijan Parsia, and Ulrike Sattler

The University of Manchester
Oxford Road, Manchester, M13 9PL
{bails,bparsia,sattler@cs.man.ac.uk}

Abstract. Current ontology development tools offer debugging support
by presenting justifications for entailments of OWL ontologies. While
these minimal subsets have been shown to support debugging and understanding tasks, the occurrence of *multiple* justifications presents a significant cognitive challenge to users. In many cases even a single entailment
may have many distinct justifications, and justifications for distinct entailments may be critically related. However, it is currently unknown how
prevalent significant numbers of multiple justifications per entailment are
in the field. To address this lack, we examine the justifications from an
independently motivated corpus of actively used biomedical ontologies
from the NCBO BioPortal. We find that the majority of ontologies contain multiple justifications, while also exhibiting structural features (such
as patterns) which can be exploited in order to reduce user effort in the
ontology engineering process.

1 Introduction

Debugging and repair of an OWL ontology is a crucial step in the ontology
development process in order to ensure the correctness and quality of the ontology. Finding the source of an error and modifying it to remove the fault can
be a tedious and error-prone task in large and often complex OWL ontologies.
Adequate *explanation* support for arbitrary entailments is therefore an essential
component of OWL ontology editors.

Justifications, minimal subsets of an ontology that are sufficient for an entailment to hold, are currently the prevalent form of explanation in OWL ontology
development tools such as Protégé 4. Previous research has mainly dealt with
improving the comprehensibility of single justifications for an individual entailment [16,8,13], as well as optimising the performance of computing justifications
[14,3,24]. We are now attempting to tackle the issue of coping with *multiple
justifications*.

Multiple justifications for a single entailment occur in a large number of OWL
ontologies, regardless of the size or description logic expressivity of the ontology,
often reaching up to several hundred justifications per entailment [4]. However,
even small numbers of multiple justifications can cause a cognitive overload
for the ontology engineer. Choosing a *minimal repair*, i.e. a smallest possible
modification to remove the entailment without affecting the remainder of the

L. Aroyo et al. (Eds.): ISWC 2011, Part I, LNCS 7031, pp. 67–82, 2011.

ontology, requires significant cognitive effort from the user when faced with not just one, but multiple justifications.

Further, repairing entailments in isolation might also cause non-minimal repairs, as demonstrated in [15]. Considering multiple entailments to repair simultaneously almost certainly[1] requires the user to deal with multiple justifications. As in the case of multiple justifications for a single entailment, the ontology engineer must recognize *relationships* between the justifications in order to find a suitable repair. While the root and derived justifications described in [15] point out one type of relations (namely subset relations), there exist many other structural aspects of justifications which have not been explored yet.

While there is a clear use case for improved coping mechanisms, we may also consider gathering additional knowledge about an ontology from the justifications occurring in it. A user may want to learn about the modelling of an ontology by considering not only its explicitly asserted structure and metrics, but also its implicit structure, which is described by the relations between entailments and their entailing axiom sets.

To date, there has been no systematic investigation into the problem of multiple justifications in an independently motivated corpus of OWL ontologies. In this paper, we analyse the relationships between justifications in a set of ontologies from the biomedical domain which were extracted from the NCBO BioPortal.[2] This analytical work constitutes the first step on the road to developing an explanation tool with improved coping strategies for multiple justifications with the aim of supporting ontology engineers in the debugging process.

To facilitate the description of justificatory structure, we introduce a graph-based framework for capturing and analysing relationships between justifications in OWL ontologies. Using these JGraphs, we outline different aspects of the justificatory structure and perform an analysis of an ontology corpus. The contributions of this paper are: 1) A framework to compute and describe the justificatory structure as the foundation for improved explanation support for multiple justifications. 2) Metrics for OWL ontologies that describe implicit features of the ontology. 3) A survey of a representative set of bio-ontologies that demonstrates the concept of justificatory structure and provides insight into structural aspects of the corpus.

2   Preliminaries

In the following section we provide a brief overview of the Web Ontology Language OWL and discuss the notion of entailment sets. We then introduce justifications as a form of explanation for entailments.

2.1   OWL

The Web Ontology Language OWL 2,[3] may be regarded as a syntactic variant of the expressive description logic (DL) [2] $\mathcal{SROIQ}$, with an OWL 2 ontology

---

[1] A set of axioms can be a minimal entailing set for multiple entailments.
[2] http://bioportal.bioontology.org
[3] http://www.w3.org/TR/2009/REC-owl2-overview-20091027

corresponding to a set of $\mathcal{SROIQ}$ [11] axioms. These axioms, i.e. statements about the entities in the ontology, can take the form of subsumptions (denoted by the symbol $\sqsubseteq$ in DL and SubClassOf in OWL Manchester Syntax[4]) and equivalent classes (denoted by $\equiv$ in DL and EquivalentClasses in Manchester Syntax). They may involve complex class expressions to describe the relationships between the classes in the ontology, which are based on a wide range of available constructors in OWL 2. For example, the equivalent class axiom

$$\mathsf{DNA} \equiv \mathsf{NucleicAcid} \sqcap \exists \mathsf{hasPart.Deoxynucleotide}$$

defines the class DNA as a NucleicAcid that has some (at least one) part which is a Deoxynucleotide. In addition to subsumptions and equivalences between classes, we can also make statements about the *individuals* and the *roles* in the ontology. In the remainder of this paper we will use the term OWL interchangeably with OWL 2 when referring to OWL 2 ontologies.

2.2   Entailments

Any statement which holds in all models of an ontology $\mathcal{O}$ is considered an *entailment* of the ontology. For example, an ontology $\mathcal{O}$ containing the above axiom entails that DNA is a subclass of NucleicAcid, which is expressed as $\mathcal{O} \models$ DNA $\sqsubseteq$ NucleicAcid.

   We consider the entailment set of an ontology to be a set of entailments of interest, given by a function $\varepsilon(\mathcal{O})$:

Definition 1 (Entailment set). *Let $\mathcal{O}$ be an ontology and $\varepsilon(\mathcal{O})$ a function that returns a finite set of axioms $\{\eta_1 \ldots \eta_n\}$ such that $\mathcal{O} \models \eta_i$; this set is an entailment set of $\mathcal{O}$.*

While the entailment relation $\mathcal{O} \models \eta$ is well defined through the semantics of description logics, the term is often used in an ambiguous way [6]. In order to specify a particular *finite* subset of the set of all entailments of an ontology $\mathcal{O}$, multiple variables need to be fixed: For the purpose of analysing the justificatory structure of ontologies, we focus on entailments that are *direct subsumptions* between atomic (named) classes, including $\top$ and $\bot$. This set includes asserted entailments, as there may be additional (other than the axiom itself) reasons for the entailment to hold, which may be missed when excluding asserted entailments from the analysis. Tautologies such as $A \sqsubseteq A$, $A \sqsubseteq \top$ and $\bot \sqsubseteq A$ for a named class A are omitted, as they hold no information value.

2.3   Justifications

Justifications [23,20] are a form of explanation of entailments of OWL ontologies, which is used in OWL ontology editors such as Protégé 4 to provide explanation support to the user. A justification is a minimal subset of an ontology $\mathcal{O}$ that causes an entailment η to hold.

---

[4] http://www.w3.org/TR/owl2-manchester-syntax

Definition 2 (Justification). $\mathcal{J}$ *is a justification for* $\mathcal{O} \models \eta$ *if* $\mathcal{J} \subseteq \mathcal{O}, \mathcal{J} \models \eta$ *and, for all* $\mathcal{J}' \subset \mathcal{J}$, *it holds that* $\mathcal{J}' \not\models \eta$.

For every axiom which is asserted in the ontology, the axiom itself naturally is a justification. We are, however, only interested in *non-trivial entailments*, i.e. justifications which have some (at least one) justification which is not the axiom itself; in this case, the justification is also called a *non-trivial justification*.

A justification is defined with respect to a single entailment η and an ontology $\mathcal{O}$; in order to describe the set of all justifications for all entailments in an entailment set $\varepsilon(\mathcal{O})$, we introduce the notion of *justification sets*:

Definition 3 (Justification set). *Given an ontology* $\mathcal{O}$ *and a function* $\varepsilon(\mathcal{O})$, *the justification set* $Justs(\mathcal{O}, \varepsilon)$ *is the set of all justifications* $\{\mathcal{J}_1 \ldots \mathcal{J}_m\}$, $\mathcal{J}_i \subseteq \mathcal{O}$, *for the axioms in the entailment set* $\varepsilon(\mathcal{O})$.

Further, we define the set of all axioms occurring in all justifications for a particular entailment set:

Definition 4 (Justification axioms)

$$JustAx(\mathcal{O}, \varepsilon) = \{\alpha \mid \text{there is a } \mathcal{J} \in Justs(\mathcal{O}, \varepsilon) \text{ s.t. } \alpha \in \mathcal{J}\}$$

With respect to debugging unwanted entailments (i.e. $\varepsilon(\mathcal{O})$ is the set of all unwanted entailments, e.g. unsatisfiable classes) of an ontology, a *repair* is a subset of $JustAx(\mathcal{O}, \varepsilon)$ which, if removed from the ontology, would *break* all these unwanted entailments. We are particularly interested in finding a *minimal repair*, which corresponds to a *minimal hitting set* [21] for the justifications in $Justs(\mathcal{O}, \varepsilon)$.

3   JGraphs

In this section we provide the necessary definitions for justification graphs which capture the relations between axioms in the ontology, justifications (sets of axioms) and entailments of interest. Based on the above definitions of entailments sets and justification sets, we can now define the justification graph of an ontology $\mathcal{O}$. A justification graph (JGraph) $G_{\mathcal{J}}$ is a directed graph whose set of vertices is the union of the set of axioms $\varepsilon(\mathcal{O})$ which are entailed by $\mathcal{O}$ and the set $JustAx(\mathcal{O}, \varepsilon)$ of axioms that participate in justifications for these entailments, together with the set of all justifications $Justs(\mathcal{O}, \varepsilon)$. The edges indicate whether an axiom is an element of a justification, and whether a justification is a justification for a particular entailed axiom; hence, the graph is bipartite.

Definition 5 (Justification graph)

$$G_{\mathcal{J}} = (\varepsilon(\mathcal{O}) \cup JustAx(\mathcal{O}, \varepsilon) \cup Justs(\mathcal{O}, \varepsilon), E_1 \cup E_2) \ where$$
$$E_1 = \{(u, v) \in \varepsilon(\mathcal{O}) \cup JustAx(\mathcal{O}, \varepsilon) \times Justs(\mathcal{O}, \varepsilon) \mid u \in v\},$$
$$E_2 = \{(v, w) \in Justs(\mathcal{O}, \varepsilon) \times \varepsilon(\mathcal{O}) \mid v \in Justs(\mathcal{O}, w)\}.$$

Side remarks: 1) The set of justifications for a given set of entailments is unique, so is the set of axioms in the justifications, and the edges follow from these unambiguous relations; therefore the JGraph is unique. 2) $\varepsilon(\mathcal{O})$ and $JustAx(\mathcal{O}, \varepsilon)$ are not disjoint; i.e. an axiom in $\mathcal{O}$ may have a non-trivial justification, while also being an element of a justification for some other entailment. 3) Any axiom vertex in the graph with an in-degree of at least one is an entailment in $\varepsilon(\mathcal{O})$. (4) Similarly, any axiom vertex in the graph with an out-degree of at least one is in $JustAx(\mathcal{O}, \varepsilon)$.

The principle of JGraphs is demonstrated by the following example ontology:

$$\mathcal{O} = \{A \sqsubseteq \exists R.B, \qquad\qquad (a1)$$
$$\exists R.B \sqsubseteq C \sqcap D, \qquad (a2)$$
$$A \sqsubseteq D, \qquad\qquad (a3)$$
$$F \sqsubseteq G\} \qquad\qquad (a4)$$

The entailment set $\varepsilon(\mathcal{O})$ comprising the direct and indirect atomic subsumptions that are entailed by $\mathcal{O}$ contains the following axioms:

$$\varepsilon(\mathcal{O}) = \{A \sqsubseteq C, \qquad\qquad (a5)$$
$$A \sqsubseteq D, \qquad\qquad (a3)$$
$$F \sqsubseteq G\} \qquad\qquad (a4)$$

Only the first two entailments $a5$ and $a3$ in $\varepsilon(\mathcal{O})$ have a non-trivial justification $j1 = \{a1, a2\}$, and the entailed axiom $a3$ has an additional trivial justification, which is the axiom itself: $j2 = \{a3\}$. The set of vertices in JGraph $G_{\mathcal{J}}$ therefore is: $\{a1, a2, a3, a5, j1, j2\}$. The respective sets of edges in the graph are: $E_1 = \{(a1, j1), (a2, j1), (a3, j2)\}$ and $E_2 = \{(j1, a5), (j1, a3), (j2, a3)\}$.

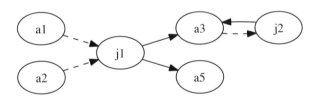

Fig. 1. Example of a JGraph

4   Justificatory Structure

Using the JGraphs defined above, we outline different aspects of the justificatory structure, which allows us to examine and describe implicit structural properties of an OWL ontology.

4.1   Number and Size of Justifications

Justificatory Redundancy. The number of justifications per entailment is an indicator of *justificatory redundancy* in the ontology; it demonstrates "how often

the same thing is expressed in different ways". The in-degree of entailments in
the JGraph corresponds to this metric. We must point out that, while the term
redundancy has mostly negative connotations, this measurement can also be a
criterion for the *richness* and *inferential power* of an ontology. It is also not
to be confused with logical redundancy, as this would imply that it is possible
to remove a set of axioms from the ontology without breaking any entailments.
Furthermore, we cannot make any claims about the purpose of the axioms in
the justifications; they might have been added without the intention of causing
the entailment.

Activity. Finally, with respect to the user effort required when dealing with an
ontology, we define the *activity* of an OWL ontology. This is the total number
of axioms that occur in justifications for non-trivial entailments, that is, the
size of the subset of the ontology which actively participates in inference. An
ontology in which the set of asserted axioms is the same as the set of inferred
ones (with respect to some definition of $\varepsilon(\mathcal{O})$) has an activity value of 0. This
measurement does not take into account equivalent sets of axioms where the
axioms have different sizes; for example, a subsumption of the form $A \sqsubseteq B \sqcap D$,
$A \sqsubseteq C \sqcap D$ could be re-written into a single axiom $A \sqsubseteq B \sqcap C \sqcap D$, which is
logically equivalent, but has a lower activity value.

4.2   Self-Justifications

Any justification which is simply the entailed axiom itself is classified as a self-
justification. In the JGraph this is expressed as a cycle between an axiom node
and a justification node, where the in-degree of the justification node is one. The
justification $j2$ in Figure 1 illustrates a self-justification for the axiom $a3$. There
are different reasons for the existence of self-justifications: (1) A conscious design
decision to improve reasoner performance or a tailoring towards a particular
ontology browser interface which does not support reasoning, i.e. the inferred
subsumptions were added back into the ontology. (2) The absence of a reasoner
during the ontology engineering process and the modeller not being aware that
the subsumption is already entailed. (3) The entailed subsumption could simply
be a side-effect of axioms that were added to the ontology without the aim of
causing the entailment.

4.3   Justification Overlap

Arbitrary Overlap. Justifications for both single and multiple entailments
may share some axioms, i.e. the justifications *overlap* to a certain extent. With
respect to coping with multiple justifications, this overlap is an indicator for a
common *lemma*, i.e. an intermediate entailment caused by a subset of a justification [9]. A suitable lemma may support understanding structural similarities
between multiple justifications, which in turn reduces the task of understanding
multiple seemingly distinct justifications to understanding a smaller number of
lemmas.

Root and Derived Justifications. A special case of justification overlap are *root* and *derived* unsatisfiable classes [15], which describes the containment of one justification in another. An unsatisfiable class C is *derived* if one of its justifications is a superset of the justification of another unsatisfiable class D; all other unsatisfiable classes are *roots*. Meyer et al. [17] propose a repair strategy that extends root and derived unsatisfiable classes to arbitrary sets of entailments. These relationships are captured by the edge set E_3 in the JGraph:

Definition 6 (Root and derived justifications). *Given a JGraph $G_{\mathcal{J}} = (\varepsilon(\mathcal{O}) \cup JustAx(\mathcal{O}, \varepsilon) \cup Justs(\mathcal{O}, \varepsilon), E_1 \cup E_2)$ and a vertex $v \in Justs(\mathcal{O}, \varepsilon)$; v is a derived justification if there is a justification v' such that $\{w \mid (w, v) \in E_1\} \supset \{w \mid (w, v') \in E_1\}$. Else, v is a root justification.*

This definition aligns with the root and derived unsatisfiable classes defined in [15]; in this case, the entailment set $\varepsilon(\mathcal{O})$ is comprised of all unsatisfiable classes in an ontology $\mathcal{O}$.

Equality. Equality is another special case of justification overlap, where the justifications for different entailments contain the same axioms. In terms of the JGraph, these justifications are represented by a single vertex which has an out-degree greater than one. The equality of justification is illustrated by the following two axioms from the above example ontology:

Example 1 (Multiple Entailments)

$$A \sqsubseteq \exists R.B$$

$$\exists R.B \sqsubseteq C \sqcap D$$

This minimal set of axioms entails the two atomic subsumptions $A \sqsubseteq C$ and $A \sqsubseteq D$ and therefore represents a justification for both entailments. In the graph shown in Figure 1, this justification is represented by the vertex labelled $j1$, which entails the two axioms $a3$ ($A \sqsubseteq D$) and $a5$ ($A \sqsubseteq C$).

We may consider the number of axioms that a justification entails as the *inferential power* of the justification, answering the question "how much can be expressed with how little?". In the context of repairing unwanted entailments, users can benefit from examining multiple unwanted entailments at the same time if they share some justifications rather than looking at each justification in isolation, as this reduces the total number of justifications to repair.

Axiom Power. The *power* of an axiom, also denoted as *arity* [23], is the out-degree of any axiom in $JustAx(\mathcal{O}, \varepsilon)$. The respective axiom occurs in multiple justification, which corresponds to a justification overlap of size one. *Key axioms* are those axioms with the maximal out-degree. These provide informations about the ontology in two ways: With respect to repairing an unwanted entailment, the user may focus on removing or weakening the key axioms first, as they have the highest repair powers. Secondly, key axioms are those statements that contribute to a large number of entailments of the ontology, and are therefore structurally (not necessarily in the context of domain knowledge) relevant to understanding the ontology.

A set of axioms can be a justification for more than one entailment. Therefore, we need to consider how many entailments would be removed from the ontology through the removal or weakening an axiom in the repair process. The *impact* of an axiom is the number of entailments that would break in addition to the entailment we focus on, which corresponds to the out-degree of all justifications that the axiom has edges to.

4.4   Patterns

Two types of patterns can be identified in the context of the justificatory structure: (1) Graph surface patterns, and (2) isomorphism between justifications. A surface pattern is a structural similarity in the JGraph, such as matching subgraphs. Surface patterns in the JGraph reveal modelling similarities in the ontology, regardless of whether the justifications and axioms in the pattern also interact in a similar way. Highlighting a pattern of this type may support user understanding of the modelling in the ontology, while it may also be an indicator for isomorphic justifications.

Two justifications $\mathcal{J}_1$ and $\mathcal{J}_2$ are isomorphic [7] if there is an injective renaming from $\mathcal{J}_1$ to $\mathcal{J}_2$, i.e. the axioms in the justifications have the same structure while using different class and property names. It can be claimed that if a person is able to understand $\mathcal{J}_1$ (from a structural point of view, not considering domain knowledge), they can also understand $\mathcal{J}_2$. Making this sameness explicit may reduce the user effort required when faced with a large number of justifications in the debugging process, as it reduces the number of justifications that need to be examined.

4.5   Components

The number of components of the graph provides a measure for the disjointness of justifications in the whole ontology. The disjointness of justifications strongly affects the justification computation process, which makes use of Reiter's Hitting Set Tree (HST) algorithm for diagnosis [21,22]. In the HST, the vertices are labelled with justifications and the paths constitute hitting sets, i.e. minimal repairs for the justifications. Optimisations for the HST algorithm are mainly based on closing a branch in the tree if the path to it is labelled with a superset of an existing path, which is not possible if the justifications are disjoint. This leads to a rapid growth of the HST and has significant negative effects on the performance of computing all justifications for an entailment.

5   BioPortal Ontology Survey

In this section we apply the JGraph metrics on a corpus of ontologies used in bio-health applications. By analysing selected aspects of justificatory structure, we draw conclusions regarding the occurrence and the nature of multiple justifications in OWL ontologies. In a nutshell, the ontologies surveyed cover a wide range of justificatory structure from very lean to very rich, with size or DL flavour of the ontology being no indicator of its implicit structure.

5.1   Test Corpus: The NCBO BioPortal

The purpose of this study was to analyse the justificatory structure of a realistic and representative set of OWL ontologies. Thus, the test corpus was selected based on the following criteria: 1) The ontologies had to be publicly accessible. 2) In order to avoid hand-picking "suitable" ontologies that may not be representative of naturally occurring ontologies, the choices were between a random sample of web ontologies, or an existing set from a web-based repository. 3) For the same reason, we focused on realistic ontologies that were actively used, i.e. ontologies that were built simply for training purposes (such as the *Koala* or *Pizza* ontologies) would be excluded. The NCBO BioPortal [18] repository meets all the above criteria while containing a large number of OWL and OWL-compatible ontologies; it was therefore selected as the test corpus for our survey.

The BioPortal provides ontologies from various groups from the biomedical domain, including the full set of daily updated OBO Foundry[5] ontologies, which are built based on common design principles. OBO ontologies use a flat-file format, which can be translated into OWL 2 and were therefore included in the test corpus.

Dataset. At the time of downloading (12 March 2011), the BioPortal repository listed 226 latest versions of ontologies in OWL and OBO format, out of which 218 could be downloaded and parsed with the OWL API[6] parsers [10]. 8 ontologies could not be processed due to the file being not available under the given URL, or parsing errors. For each ontology the imports closure was downloaded and merged with the root ontology, while missing imports were ignored. We then extracted all entailed atomic subsumptions from the parseable OWL files and excluded those ontologies that did not contain any non-trivial entailments, which left us with 72 ontologies. At this stage, 5 ontologies were removed from the set as they could not processed by the justification generation and the JGraph framework due to their large size and number of entailments.

The structural analysis of the remaining 67 ontologies was further restricted to only include ontologies with coherent TBoxes, i.e. ontologies that contained only statements about the concept hierarchy and no unsatisfiable classes. Both justifications for unsatisfiable classes and ABox entailments need to be treated differently from justifications for subsumptions between named classes; a separate investigation of the justificatory structure of both incoherent ontologies and ontologies with ABoxes are omitted due to space limitations, but are part of future work. This filtered out 20 ontologies with an ABox, and 5 ontologies with

---

[5] http://obofoundry.org
[6] http://owlapi.sourceforge.net

unsatisfiable classes, leaving us with 42 ontologies that contained non-trivial entailments for which the justifications could be generated and processed in the JGraph generator.

Imported Entailments. In the next stage of data pruning, the import structure of the 42 ontologies was examined in order to determine to which extent entailments were imported from external ontologies. An entailment whose justifications contain only axioms from an imported ontology is classified as an "imported entailment", whereas an entailment whose justifications contain only axioms from the importing ontology is called a "native entailment" [6]. 28 ontologies did not have any imported entailments at all, which could be either due to them having no imports, the imported ontology having no entailments that matched our criteria, or missing imports, which were ignored in the pre-processing stage.

We found that 7 ontologies in BioPortal import the Basic Formal ontology (BFO), an "upper" ontology for biological data which itself is contained in the BioPortal corpus. 3 of these ontologies had 70 imported entailments each, which all stemmed exclusively from BFO, and no native entailments. The remaining 4 ontologies had the 70 imported entailments from BFO, plus additional entailments which were either native or imported from ontologies other than BFO. A further 7 ontologies had imported entailments from other ontologies, which, in most cases, could be attributed to the ontology intentionally being split up over several files. For example, the Chemical Information ontology (76 entailments) had 1 native entailment, 72 imported entailments from an ontology titled "cheminf-external", and 3 entailments from "cheminf-core".

In order to prevent skewed results due to the dominance of BFO, the ontologies which imported BFO were also removed from the set (while BFO itself remained in the corpus). This left us with 35 ontologies which had only native (or mixed) non-trivial entailments from a coherent TBox, which could be processed by the JGraph framework.[7]

5.2   Results and Analysis

Entailments. From the 35 ontologies, 12,010 non-trivial entailments were extracted, with a total of 7,176 distinct non-trivial justifications. In addition, 2,340 self-justifications were found, i.e. 2,340 entailments were asserted in the ontologies while also having additional justifications. The average number of non-trivial entailments across all ontologies is 343, which is mainly affected by the large number of entailments in the NCI Thesaurus (7,862), and the Experimental Factor ontology (1,787), which deviates by an order of magnitude from the other ontologies.

---

[7] Due to space limitations and the wealth of data obtained in the experiments, we have to omit tables and graphs displaying the results. The raw data from the graph analysis as well as detailed overview tables are available from
http://owl.cs.manchester.ac.uk/research/publications/
supporting-material/iswc2011-juststruct

The majority of ontologies (74.3%) that produce non-trivial entailments has 100 or less entailments. Surprisingly, small numbers of non-trivial entailments occurred in some of the largest ontologies with high DL expressivity, such as the MaHCO ontology ($\mathcal{ALCHIQ(D)}$, 13,844 axioms, 26 non-trivial entailments) and Cardiac Electrophysiology ($\mathcal{SHF(D)}$, 176,113 axioms, 19 non-trivial entailments).

As we can see from these examples, the size of an ontology does not necessarily affect the number of non-trivial entailments. This is confirmed by the Spearman's rank coefficient[8] of $\rho = 0.18$, which indicates that there is no correlation between the two values.

Multiple Justifications. The average in-degree for the JGraph vertices representing entailments is 1.3 (standard deviation $\sigma = 4.4$), not counting the self-justifications (i.e. the justification which is the asserted axiom itself). In particular, only 10 out of the 35 ontologies (28.6%) have exactly one justification for each of its entailments;[9] 27 ontologies (57.1%) have between one and 2 justifications per entailment; and 5 (14.3%) ontologies have an average of more than 2 justifications per entailment.

The largest number of justifications for a single entailment (236) can be found in the Gene Ontology Extension, an ontology that has several entailments with more than 100 justifications each, followed by the Experimental Factor ontology, which has 20 entailments with 16 or more justifications each. There exists no correlation between the size of an ontology and the number of justifications per entailments ($\rho = -0.21$), so neither size nor expressivity of an ontology are an indicator for multiple justifications.

The occurrence of multiple justifications in 71.4% of the surveyed ontologies, as well as the large maximal number of justifications per entailment in some ontologies shows that both the computational requirements as well as the cognitive complexity of multiple justifications can pose a significant challenge when dealing with OWL ontologies found in practice.

Self-Justifications. In the surveyed corpus, 2,340 of the 12,010 entailments have self-justifications, which means that the entailments are asserted as well as inferred through additional reasons in the ontology. While the occurrence of self-justifications is common (e.g. due to the entailment simply being a side-effect of some axioms in the ontology), it is surprising that 8 ontologies do not contain any self-justifications, despite large numbers of both entailments and justifications. On the other extreme, one ontology, the Software Ontology, has a self-justification for each of its 332 entailments; a detailed discussion of this ontology follows in Section 5.2.

---

[8] Values for Spearman's rank coefficient range from -1, which indicates a perfect negative correlation between two variables, to +1 (perfect positive correlation), with 0 indicating no correlation.

[9] Only 11 out of the 800 justifications in the NCI Thesaurus have multiple justifications (2 and 3 respectively), which yields an average of 1 justification per entailment after rounding.

Size of Justifications. The average size of justifications in the ontology is 5.9 axioms ($\sigma = 4.0$), with a maximum size of 20 axioms for a justification in the NCI Thesaurus. We have to emphasize that the justifications were only computed for direct subsumptions; this means that large justifications cannot stem from long subsumption chains. The largest justifications were found in the Gene Ontology Extension (average size: 9.5 axioms) and the NCI Thesaurus (9.6 axioms), with the latter having over 300 entailments that have justifications of size 10 or larger.

All 35 ontologies have an average justification size of less than 10 axioms, which gives us a useful indicator of the potential cognitive complexity of understanding justifications: While there are some extreme cases, such as the NCI Thesaurus, most ontologies produce justifications with a size that can be deemed "manageable" by human users. As with the number of justifications per entailment, we cannot detect any correlation between the size of an ontology and the average size of its justifications ($\rho = 0.11$).

Activity. Regarding the number of axioms of the ontology that participate in justifications, the largest total numbers can be found in the NCI Thesaurus (6,479 axioms) and the Experimental Factor ontology (3,813 axioms). Interestingly, when taking into account the total size of the ontology, the NCI Thesaurus only uses 4.4% of its axioms in justifications, whereas the axioms occurring in justifications in Experimental Factor make up 53.7% of the whole ontology, with both ontologies having relatively large numbers of entailments (7,862 and 1,787 entailments respectively). This state is reflected by the average number of entailments per justification, i.e. the inferential power of the justifications in the ontology: While the NCI Thesaurus has an average of 9.8 entailments per justification, the justifications in Experimental Factor only have 1.2 entailments each.

The majority of ontologies in the corpus (21 out of 35) have comparatively small numbers (less than 100) of axioms that participate in inference, regardless of the size of the ontology. Not surprisingly, there is a strong correlation between the total number of non-trivial entailments of an ontology and the number of axioms that participate in the justifications for these entailments ($\rho = 0.9$).

Axiom Power. The average axiom power across all ontologies in the corpus is 3.0, which means that, on average, an axiom occurs in 3 justifications. While only 4 ontologies have an average axiom power of exactly 1.0, all remaining ontologies have surprisingly high-power axioms. More than half of the ontologies (54.3%) contain at least one axiom which occurs in 9 or more justifications, and 31.4% have axioms with a power of 20 or higher, peaking at one axiom that occurs in 510 justifications in the Experimental Factor ontology.

Again, this shows that considering justifications in isolation when debugging an ontology can lead to non-minimal repairs, e.g. by removing a different axiom from each justification rather than removing a shared axiom, which also causes an unnecessary overhead in terms of effort required.

Since the analysis of impact depends on a particular justification/entailment pair in the context of ontology repair, a general discussion on this metric is omitted.

Patterns. One interesting case of patterns in the form of isomorphic justifications is the Software ontology (2,080 axioms, DL expressivity $\mathcal{ALCHIN}$). Almost all of the 332 entailments have exactly one self-justification and one additional non-trivial justification which contains 2 axioms; in fact, only one of the justifications contains 3 axioms. 318 of these entailments are of the form $X \sqsubseteq R\ Software$ for X being some class name in the ontology, with all of the justifications of size two being structurally isomorphic and the equivalence axiom occurring in all 318 justifications: $\mathcal{J}_i = \{R\ Software \equiv \exists is\_encoded\_in.R\ Language$, $X \sqsubseteq \exists is\_encoded\_in.R\ Language\}$.[10]

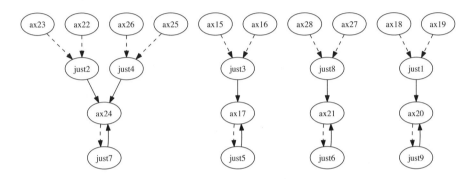

Fig. 2. JGraph of the Biopax ontology, with 3 isomorph subgraphs

A similar pattern can be found in the Biopax ontology (Figure 2). The 3 isomorphic subgraphs are justifications for atomic subsumptions of the type $A \sqsubseteq B$ which themselves are asserted in the ontology (i.e. they have self-justifications, $just5$, $just6$ and $just9$). 2 of the 3 justifications ($just3$ and $just8$) contain an axiom of the type $A \sqsubseteq \exists d.x$, where d is a data property and x a data property value (in this case a string), together with a data property domain axiom $Domain(d, B)$. Justification $just1$ contains axioms of the same structure with an object property and object property domain axiom instead. These 3 justifications have exactly the same surface structure in the graph, while not being strictly isomorphic.

This phenomenon highlights the importance of structural analysis of multiple justifications, as we can easily understand the nature of several hundred justifications (as in the case of the Software ontology) by examining only one justification and understanding the isomorphic structure, rather than dealing with every single justification independently.

6   Related Work

Structural analysis of (OWL) ontologies is an actively researched topic in the areas of ontology quality measurement, ontology integration, and ontology

---

[10] The ontology uses IDs for the class names, which we here display with their human-readable labels instead.

matching. Most of the frameworks developed for these purposes focus on analysing the asserted class hierarchy using basic metrics, such as the numbers of root classes, numbers of leaf classes and depth of the inheritance tree, to determine structural metrics of the ontology [25,27,1,19]. A more in-depth approach to analysing the asserted class graph is proposed in [12], which applies methods from Social Network Analysis for the purpose of comparing ontologies.

The research area of exploring structural relationships between the *reasons* for entailments of OWL ontologies is in its early stages. In one of the first approaches to analysing this *implicit* structure of ontologies, Vrandecic et al. [26] introduce ontology metrics that also consider the semantics of the ontology language, mainly focusing on the entailed statements of an ontology.

Root and *derived* justifications [15] provide a way of exploiting structural relations between justifications for multiple entailments. Focusing on root justifications can drastically reduce the number of justifications a user has to examine when attempting to repair unsatisfiable classes. In [17], the authors describe an extension for root and derived justifications to cover *arbitrary* sets of entailments beyond unsatisfiable classes. The focus of the work of root and derived justifications is to provide improved ontology debugging and repair support; there is no indication of inferring information about the ontology from its root and derived structure.

In [5], we introduce the notion of justificatory structure and conduct a preliminary survey on a set of ontologies. We analyse a small number of properties, such as the number of justifications per entailment, with findings indicating that the problem of multiple justifications for entailments is common in real-life OWL ontologies.

The characteristics of justifications in bio-ontologies in particular are the focus of a survey of the NCBO BioPortal [10]. In this study, the number of non-trivial entailments and justifications per entailment are analysed against the background of the ontology size and expressivity. It is found that a large number of ontologies contains non-trivial entailments, which indicates the use of inference in the ontology engineering process.

7   Conclusion and Future Work

We have presented a graph-based framework that captures the relations between axioms, entailments and justifications of an OWL ontology. Analysing this *justificatory structure* of OWL and OWL-compatible ontologies from the NCBO BioPortal has clearly shown that multiple justifications do occur in a large proportion (71.4%) of the surveyed ontologies.

Furthermore, some ontologies have very large numbers of multiple justifications per entailment (up to several hundred), which poses a significant computational and cognitive challenge against the background of explanation support for ontology debugging tasks. We have found no correlations between the DL expressivity and size of an ontology and the complexity of its justificatory structure. With improved ontology development tool support and modelling patterns, we expect OWL ontologies to become more complex in the future, which may lead to a more complex justificatory structure. We can therefore conclude that it is necessary to focus attention on developing improved coping mechanisms for multiple justifications in order to reduce user effort and limit the computational load.

In our presentation of justificatory structure, we have also discussed and demonstrated structural relations between the justifications in an ontology, such as axiom power, patterns and justification overlap, which may be exploited for these coping mechanisms. Based on the axiom power analysis and the patterns found in the surveyed ontologies (e.g. the Software ontology), we have shown how structural analysis can drastically reduce the number of seemingly distinct justifications that have to be examined when attempting to understand the justifications in an ontology.

For future work, we plan to further explore justification overlap and lemmas, using approaches from formal concept analysis. This will provide us with the necessary structural information which can then be used to suggest to the user a suitable repair strategy when confronted with multiple justifications. Finally, as the current basic visualization of JGraphs is clearly limited in terms of scalability and the representation of most structural aspects, we aim to investigate approaches to developing a user oriented interaction framework.

References

1. Alani, H., Brewster, C.: Metrics for ranking ontologies. In: Proc. of EON-2006 (2006)
2. Baader, F., Calvanese, D., McGuinness, D.L., Patel-Schneider, P., Nardi, D.: The description logic handbook: theory, implementation, and applications. Cambridge University Press (2003)
3. Baader, F., Peñaloza, R., Suntisrivaraporn, B.: Pinpointing in the Description Logic EL+. In: Hertzberg, J., Beetz, M., Englert, R. (eds.) KI 2007. LNCS (LNAI), vol. 4667, pp. 52–67. Springer, Heidelberg (2007)
4. Bail, S., Parsia, B., Sattler, U.: JustBench: A framework for OWL Benchmarking. In: Patel-Schneider, P.F., Pan, Y., Hitzler, P., Mika, P., Zhang, L., Pan, J.Z., Horrocks, I., Glimm, B. (eds.) ISWC 2010, Part I. LNCS, vol. 6496, pp. 32–47. Springer, Heidelberg (2010)
5. Bail, S., Parsia, B., Sattler, U.: The justificatory structure of OWL ontologies. In: Proc. of OWLED-2010 (2010)
6. Bail, S., Parsia, B., Sattler, U.: Extracting finite sets of entailments from OWL ontologies. In: Proc. of DL 2011 (2011)
7. Horridge, M., Bail, S., Parsia, B., Sattler, U.: The cognitive complexity of OWL justifications. In: Proc. of DL 2011 (2011)
8. Horridge, M., Parsia, B., Sattler, U.: Laconic and Precise Justifications in OWL. In: Sheth, A.P., Staab, S., Dean, M., Paolucci, M., Maynard, D., Finin, T., Thirunarayan, K. (eds.) ISWC 2008. LNCS, vol. 5318, pp. 323–338. Springer, Heidelberg (2008)
9. Horridge, M., Parsia, B., Sattler, U.: Lemmas for justifications in OWL. In: Proc. of DL 2009 (2009)

10. Horridge, M., Parsia, B., Sattler, U.: The state of bio-ontologies. In: Proc. of ISMB 2011 (2011)
11. Horrocks, I., Kutz, O., Sattler, U.: The even more irresistible SROIQ. In: Proc. of KR 2006, pp. 57–67 (2006)
12. Hoser, B., Hotho, A., Jäschke, R., Schmitz, C., Stumme, G.: Semantic Network Analysis of Ontologies. In: Sure, Y., Domingue, J. (eds.) ESWC 2006. LNCS, vol. 4011, pp. 514–529. Springer, Heidelberg (2006)
13. Ji, Q., Qi, G., Haase, P.: A relevance-based algorithm for finding justifications of DL entailments. Technical report, University of Karlsruhe (2008)
14. Kalyanpur, A., Parsia, B., Horridge, M., Sirin, E.: Finding All Justifications of OWL DL Entailments. In: Aberer, K., Choi, K.-S., Noy, N., Allemang, D., Lee, K.-I., Nixon, L.J.B., Golbeck, J., Mika, P., Maynard, D., Mizoguchi, R., Schreiber, G., Cudré-Mauroux, P. (eds.) ASWC 2007 and ISWC 2007. LNCS, vol. 4825, pp. 267–280. Springer, Heidelberg (2007)
15. Kalyanpur, A., Parsia, B., Sirin, E., Hendler, J.: Debugging unsatisfiable classes in OWL ontologies. J. of Web Semantics 3(4), 268–293 (2005)
16. Lam, J.S.C., Sleeman, D., Pan, J.Z., Vasconcelos, W.W.: A Fine-Grained Approach to Resolving Unsatisfiable Ontologies. In: Spaccapietra, S. (ed.) Journal on Data Semantics X. LNCS, vol. 4900, pp. 62–95. Springer, Heidelberg (2008)
17. Meyer, T., Moodley, K., Varzinczak, I.: First steps in the computation of root justifications. In: Proc. of ARCOE 2010 (2010)
18. Noy, N.F., Shah, N.H., Whetzel, P.L., Dai, B., Dorf, M., Griffith, N., Jonquet, C., Rubin, D.L., Storey, M.-A., Chute, C.G., Musen, M.A.: Bioportal: ontologies and integrated data resources at the click of a mouse. Nucleic Acids Research 37, W170–W173 (2009)
19. Orme, A.M., Yao, H., Etzkorn, L.H.: Complexity metrics for ontology based information. Int. J. of Technology Management 47(1/2/3), 161–173 (2009)
20. Parsia, B., Sirin, E., Kalyanpur, A.: Debugging OWL ontologies. In: Proc. of WWW-2005, pp. 633–640 (2005)
21. Reiter, R.: A theory of diagnosis from first principles. Artificial Intelligence 32(1), 57–95 (1987)
22. Schlobach, S.: Diagnosing terminologies. In: Proc. of AAAI-2005 (2005)
23. Schlobach, S., Cornet, R.: Non-standard reasoning services for the debugging of description logic terminologies. In: Proc. of IJCAI-2003, pp. 355–362 (2003)
24. Suntisrivaraporn, B., Qi, G., Ji, Q., Haase, P.: A Modularization-Based Approach to Finding All Justifications for OWL DL Entailments. In: Domingue, J., Anutariya, C. (eds.) ASWC 2008. LNCS, vol. 5367, pp. 1–15. Springer, Heidelberg (2008)
25. Tartir, S., Arpinar, I., Moore, M., Sheth, A., Aleman-Meza, B.: OntoQA: Metric-based ontology quality analysis. In: Proc. of KADASH 2005 (2005)
26. Vrandečić, D., Sure, Y.: How to Design Better Ontology Metrics. In: Franconi, E., Kifer, M., May, W. (eds.) ESWC 2007. LNCS, vol. 4519, pp. 311–325. Springer, Heidelberg (2007)
27. Yao, H., Orme, A., Etzkorn, L.: Cohesion metrics for ontology design and application. J. of Computer Science 1(1), 107–113 (2005)

Effective and Efficient Entity Search in RDF Data

Roi Blanco[1], Peter Mika[1], and Sebastiano Vigna[2]

[1] Yahoo! Research
Diagonal 177, 08018 Barcelona, Spain
{roi,pmika}@yahoo-inc.com
[2] Università degli Studi di Milano
via Comelico 39/41, I-20135 Milano, Italy
vigna@acm.org

Abstract. Triple stores have long provided RDF storage as well as data access using expressive, formal query languages such as SPARQL. The new end users of the Semantic Web, however, are mostly unaware of SPARQL and overwhelmingly prefer imprecise, informal keyword queries for searching over data. At the same time, the amount of data on the Semantic Web is approaching the limits of the architectures that provide support for the full expressivity of SPARQL. These factors combined have led to an increased interest in semantic search, i.e. access to RDF data using Information Retrieval methods. In this work, we propose a method for effective and efficient entity search over RDF data. We describe an adaptation of the BM25F ranking function for RDF data, and demonstrate that it outperforms other state-of-the-art methods in ranking RDF resources. We also propose a set of new index structures for efficient retrieval and ranking of results. We implement these results using the open-source MG4J framework.

1 Introduction

The amount of data published on the Semantic Web has grown at increasing rates in the past years due to the activities of the Linked Data community and the adoption of RDFa by major web publishers. The amount of data to be managed is stretching the scalability limitations of triple stores that are conventionally used to manage Semantic Web data. At the same time, the Semantic Web is increasingly reaching end users who need efficient and effective access to large subsets of this data. Such end users prefer simple, but ambiguous natural language queries over highly selective, formal graph queries in SPARQL, the query language of triple stores. In a web search scenario, formulating SPARQL queries may not be feasible altogether due to the heterogeneity of data.

These requirements are spurring interest in the field of Semantic Search, in particular the adaptation of Information Retrieval methods to data access. IR-style indexing is efficient in that it scales well with respect to the size of text collections in both index construction and retrieval. The field has also developed a number of methods for effective ranking of documents that match user

L. Aroyo et al. (Eds.): ISWC 2011, Part I, LNCS 7031, pp. 83–97, 2011.
© Springer-Verlag Berlin Heidelberg 2011

queries. The challenge in Semantic Search is adapting these results in indexing and ranking to exploit the inherent structure and semantics of RDF data, and expanding them to support the user tasks common in RDF retrieval. The most basics of these tasks is entity-search or Ad-hoc Object Retrieval (AOR) as described by [15], i.e. the retrieval of RDF resources that are representation of an entity described in a keyword query.

This problem has direct relevance to the operation of Web search engines, which increasingly incorporate structured data in their search results pages. Figure 1 shows a search result page from Yahoo! Search for the query *vienna, austria*. Besides the ten blue links representing document results, we can see on the left-bar suggestions for points of interest in Vienna. This requires an understanding that this query represents the city of Vienna, and a ranking over the points of interest. Similarly, an information box above the result pages shows relevant travel information such as the current weather and the geographic location of the city. This again requires a decision that travel information might be relevant to this query, and to execute a top-1 query for the most relevant city in the travel database, and to retrieve the location of the city and the current weather.

In this paper, we describe our system for entity search that adapts a state-of-the-art IR ranking model by taking into consideration the structure and semantics of RDF data. We show that this ranking model outperforms in effectiveness all 14 submissions that have been evaluated on the task of entity-search at the Semantic Search workshop in 2010. We also discuss the combination of index structures that allow this system to be efficient even on large and heterogenous datasets collected from the Web.

2   Related Work

Besides the core problem of document retrieval, ranking models from Information Retrieval have been applied in the past to the problem of retrieval over XML [9] and the relational data model [1,8,10]. However, adaptations to the RDF model are relatively new.

The typical way of providing online access to RDF collections is by using triple stores (or quad stores) that implement database-style indexing of the structure of RDF graphs. Triple stores (such as OWLIM[1] and 4store[2]) allow the option to index the text values of literals in an inverted index on the side (e.g. using Lucene), or rely on text-indexing of the underlying DBMSs (such as Oracle[3] and Virtuoso[4]), but these indices are only used for matching (filtering candidate solutions). As SPARQL does not have a built in query language for full text search in literals, this functionality is typically exposed using 'magic predicates'

---

[1] http://www.ontotext.com/owlim

[2] http://4store.org/

[3] http://www.oracle.com/technetwork/database/options/semantic-tech/index.html

[4] http://virtuoso.openlinksw.com/dataspace/dav/wiki/Main/VOSIntro

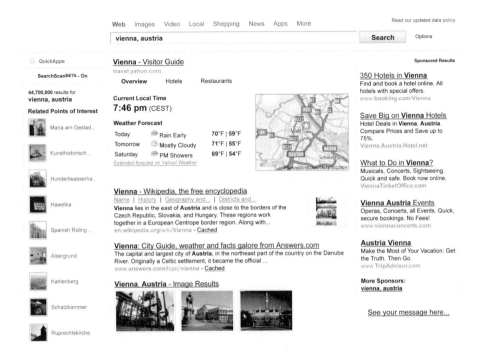

Fig. 1. Examples of structured data in the search result page

that are specific to the triple store. SPARQL 1.0 allows matching using regular expressions, but this is typically not supported by IR engines. Triple stores in general do not perform ranking.

The work of Wang et al. [18] on the Semplore system considers a deeper integration of DB and IR technology, where a set of inverted indices are used for matching limited forms of conjunctive queries, in particular tree-shaped queries with a single target variable at the root of the tree. Such queries may include "keyword concepts", i.e. the set of resources that have a predicate-value that contains a given set of keywords. They show that resolving such queries can be more efficient than the combination of a triple-store with a full-text index on the side. They also propose a simple propagation algorithm to transfer the scores from the keyword matching along the relations back to the root node of the query tree, thereby obtaining a ranking over the results. The relations themselves do not change the scoring.

All of the above systems consider an expert user who is familiar with the structure of the data and is able to denote his information need using a structured query, i.e. providing graph patterns for matching. The scenario we consider in our work is one of ad-hoc retrieval, i.e. retrieval by users who are not assumed to have prior knowledge of the system, including the representation of data. This scenario is typical for open search systems with inexperienced users who are not aware of the schema of the data, but also for systems that contain heterogeneous collections of data that don't conform to any single schema. As an example, the web dataset considered in this paper consists of over 30,000 unique RDF properties. In such cases, it is impossible to translate the user's information need into a single correct structured representation of the query as suggested by [17].

More specifically, we consider the task of Ad-hoc Object Retrieval (AOR) defined by [15]. Pound et al. point out that over 40% of searches in a typical web search usage are looking for a single object or entity. The task –also commonly called entity search– is thus to provide a ranking over RDF resources in terms of their relevance to an entity that is explicitly named in the query (though the query may contain more information than just the name of the entity). Though this task is basic, it requires solving basic problems in ad-hoc retrieval, in particular, dealing with multiple potential interpretations, and ranking partially matching resources based on their degree of relevance.

This task has been evaluated in a campaign run at the SemSearch 2010 workshop, where six participants entered 14 submissions [7]. The submissions represent a wide range of retrieval approaches, including the ones used in existing Semantic Web search engines such as Sindice [13]. We show that our method outperforms these systems in retrieval effectiveness by as much as 40%.

The method we use is closest in nature to the one proposed by Pérez-Agüera [14]. We also adapt BM25F, a scoring function that is considered state-of-the-art in text retrieval. The index structure used in their system is similar to our horizontal index, but it considers five fields: text (all text from property values), title (words from the URI), object (tokens from the URIs of objects), inlinks (tokens from predicates of 'incoming' triples). The main difference to our work is that BM25F scoring is applied to this five-field structure, while in our work BM25F is used on the vertical index where there is one field per predicate in the data. In other words, while Pérez-Agüera et al. consider all predicates with equal weight, we design a system where it is possible to assign different weights to different predicates. We will show that such weight assignment can improve retrieval performance.

Alternative evaluations exist for other important tasks in Semantic Search. The TREC Entity Track is focusing on entity search over text or hybrid collections (text with metadata).[5] The 1st Workshop on Question Answering over Linked Data (QALD)[6] focused on natural language question-answering over selected RDF datasets, where ranking is not required. The evaluation campaign organized by the European SEALS project focuses on user experience and employs user studies in addition to automated testing [19]. We do not expect that our system could be applied directly in all these scenarios, but some of the techniques described may be useful in designing solutions for them.

---

[5] http://krisztianbalog.com/files/trec2010-entity-overview.pdf
[6] http://www.sc.cit-ec.uni-bielefeld.de/qald-1

3   Ranking Model

The basis for our ranking is the ranking function BM25F [16] that has been originally developed for text retrieval. It is an extension of the BM25 probabilistic model that weights query terms differently depending on which document *fields* they appear in. Originally, BM25F was employed to weigh occurrences of terms in the *title, body,* or *anchor text* of Web pages, whereas we will break down the description of an RDF resource by the property, and consider as values the literals that appear for each unique datatype-property (see Section 4).

The features that BM25F uses are the field term frequency tf_{si} (number of times term i appears in field s), the field length l_s (number of tokens in the field s) and the field weights v_s. The ranking function does not exploit proximity information or term dependencies.

Using BM25F, a document D is scored against a query Q using a summation over individual scores of query terms $q \in Q$:

$$score^{BM25F}(Q,D) = \sum_{q \in Q} w_i^{BM25F} \tag{1}$$

First, BM25F computes a document length normalization factor as

$$B_s = \left((1 - b_s) + b_s \cdot \frac{l_s}{avl_s} \right) , \tag{2}$$

where avl_l is the average length of field l and b_s is a tunable parameter ($0 \leq b_s \leq 1$) that controls the amount of normalization. Next, BM25F aggregates the weighted term frequencies over all the fields S, normalizing them using B_s as

$$\tilde{tf}_i = \sum_{s=1}^{S} v_s \frac{tf_{si}}{B_s} , \tag{3}$$

and finally these frequencies are normalized using a sigmoid function as

$$w_i^{BM25F} = \frac{\tilde{tf}}{k_1 + \tilde{tf}_i} \cdot w_i^{IDF} , \tag{4}$$

where k_1 is a parameter and w_i^{IDF} is the *inverse document frequency* of term i, calculated as $\log \left(\frac{D - n_i + 0.5}{n_i + 0.5} \right)$ (n_i is the number of documents i occurs in).

The ranking function as described in its most general form requires information of all field lengths (l_s), which is infeasible to index for very large collections. Instead, we use a simplified version of the ranking function where the size of the document D is used as the length of all fields ($l_s = l$). An additional problem of RDF collections is that many objects are very short and are promoted by the normalization component. In order to mitigate this problem, we select a threshold l_{max} so that if $l > l_{max} \rightarrow l = l_{max}$, and set $l_{max} = 10$ for all the experiments.

Standard document retrieval models also allow for incorporating document *query-independent* features, which might come from different sources such as the Web graph. Two examples are the document PageRank values or the number of inlinks that point to a particular Web page. In our case, we classify documents based on their domain into three classes, just like the field weights. We add this document-weights w_D to compute a final retrieval score as [2]:

$$score(Q, D) = w_D \cdot score^{BM25F}(Q, D) \tag{5}$$

4   Indexing

Information Retrieval engines rely on indices for efficient access to the information required for computing scores at query time [11]. Indexing in IR is a basic process of *inversion* (hence the name *inverted index*) in which a document is made accessible by the *term(s)* appearing in the content rather than by some identifier of the document. In more detail, an inverted index provides, for each term that appears in a collection of documents, a *posting list*, that is a list of numbers identifying the documents in which the term appears. The posting lists can be richer, providing, for instance, also the number of occurrences and possibly the exact positions (always expressed as offsets from the start of the document).

Current off-the-shelf retrieval packages allow references to multiple indices or fields within a single query. In addition, state-of-the-art packages provide support for an *alignment* operator. Alignment of queries is useful for *parallel texts*. The need for handling parallel texts comes originally from the area of natural language indexing, e.g. storing part-of-speech information. For example, a text parallel to "Washington won several battles" could be "PERSON VERB - NOUN". Once parallel texts have been indexed, an alignment operator between terms of two different indices returns just the documents in which two terms appear *in the same positions*. For instance, an alignment between "Washington" and "PERSON" would return the document associated to the parallel texts above, but an alignment between "Washington" and "PLACE" would not (even if "Washington" does appear in the document).

This technique is implemented in MG4J [5], an open-source engine for text indexing. MG4J provides, for each query, a *minimal-interval semantics*—a set of regions of text satisfying the query which are incomparable by containment (i.e., no region is contained inside another region). The resulting semantics are an extension of the Clarke–Cormack–Burkowski lattice [6] that handles multiple indices (e.g., title and main text) particularly suited to parallel texts. Indeed, the alignment operator can align any set of regions, and since the set of regions associated to a term is exactly given by the positions in which the term appears, we obtain the alignment of parallel texts we described. Other possibilities are also available, such as operators that are weaker than exact alignment.

These functionality allow two main alternatives to implement structured retrieval.

The first option is illustrated in Table 2 using the sample data shown in Table 1. For simplicity, we will call this a *horizontal index* on the basis that RDF resources are represented using only three fields, one field for the tokens from values, one for the properties and one for the tokens from the subject URI. The token and property indices are aligned in that there is a correspondence between the positions in the token and property fields, i.e. the value in the token field at a given position is (part of) the value for the property written to the same position in the property index. (Note that we write the complete predicate in each position of the property field.) The alignment operator is used to align the matches in the token and property fields where the query specifies a token to match in a particular field.

The second option, which we will call a *vertical index* is shown in Table 3 using the same data. Here we create a field for each property occurring in the data. In this case performing matching on particular properties only requires the ability to restrict matches by field. Positions can be still useful, e.g. to make sure the first and last name are matched as consecutive words. Note that structured retrieval can also be implemented using a single field, e.g. by encoding fields as a post-fix of tokens or storing field information as payload. These alternatives are much less appealing. Post-fixing, for example storing terms like *peter_foaf:name*, leads to an explosion in dictionary size, especially when using a large number of fields. On the other hand, encoding fields as payload makes it inefficient to restrict searches to particular fields.

Table 1. Sample RDF data in Turtle format

```
@prefix foo: <http://example.org/ns#> .
@prefix foaf: <http://xmlns.com/foaf/0.1/> .
@prefix vcard: <http://www.w3.org/2006/vcard/ns#> .
foo:peter foaf:name "peter mika" .
foo:peter foaf:age "32" .
foo:peter vcard:location "barcelona" .
```

Table 2. Horizontal index of the data in Table 1

Field	pos1	pos2	pos3	pos4	pos5
token	peter	mika	32	barcelona	
property	foaf:name	foaf:name	foaf:age	vcard:location	
subject	http	example	org	ns	peter

In previous work [12], we have shown that both of these index structures can be efficiently built in a distributed fashion using a single MapReduce job. Since indexing can be efficiently parallelized, the index building time is linear in the

Table 3. Vertical index of the data in Table 1

Field	pos1	pos2	pos3	pos4	pos5
foaf:name	peter	mika			
foaf:age	32				
vcard:location	barcelona				

Table 4. R-vertical index of the data in Table 1

Field	pos1	pos2	pos3	pos4	pos5
w_{imp}	peter	mika	barcelona		
w_{neut}					
w_{uni}	32				

size of the input given the same number of machines in the cluster, and also linear in the number of machines given the same input (up to the natural limit where the cost of distribution outweighs the cost of indexing). The resulting indices are similar in size for the horizontal and vertical case and a small fraction of the size of the input data. Note that the vertical index alone does not contain all the information we need for ranking, in particular only the horizontal index provides direct access to term frequencies and the document sizes that is used in our ranking. Thus in practice we can either use the horizontal index on its own or use a combination of the vertical and the horizontal index, where the vertical index is used for faster matching, but the horizontal index is also accessed when computing resource scores.

In our current work, we propose a third additional index structure for improved performance. For purposes of ranking, we only need to distinguish fields that have different weights assigned. In our ranking function, we will use three different weight levels for important, neutral, and unimportant properties so that we can index all properties with the same weight using only three fields, instead of the much larger number of fields we build for the regular vertical index. We call this reduced version of the vertical index the *r-vertical* index. Table 4 shows how we would index our sample data using this index structure, assuming that we classify *foaf:name* and *vcard:location* as important, and *foaf:age* as unimportant. The disadvantage of the r-vertical index is the loss in functionality: using this index it is not possible any more to issue queries that explicitly restrict matches to particular properties, e.g. to retrieve resources where the word *peter* matches in *foaf:name* and not in other fields. Note that using the r-vertical index instead of the vertical index does not change the way ranking is performed, it merely provides faster access and therefore speeds up the ranking process. We investigate this next. We refer the reader to [12] for more discussion on how we build these indices, the time spent and the distributed methods used to scale up indexing.

5 Evaluation

5.1 Evaluation of Efficiency

To measure the efficiency of these structures, we index the Billion Triples Challenge 2009 dataset.[7] It contains RDF data collected by various Semantic Web crawlers, and as such the data is highly heterogeneous. It contains 2,680,081 classes and 33,164 properties, and therefore it is unlikely that any user could be aware of the complete structure of the data and compose formal queries to match this structure. The collection contains 1.14 billion quads, which is 249GB of data in uncompressed N-Quads format. The usage of predicates is highly skewed, and fits an exponentially decaying distribution (refer to the webpage for other statistics). Note that the scale of the data justifies the use of distributed indexing, i.e. a single-machine setup would have been much slower in indexing this amount of data.

For indexing, we grouped the quads by subject URI, and considered as virtual documents the quads with the same subject. We subdivided each document into fields by considering each unique predicate as a separate field. We only indexed datatype-properties, i.e. quads with literals in the object position. In case of multiple values for the same subject and predicate, we simply considered the concatenation of values. We performed a minimal processing of values at indexing time, namely we removed stop-words using a list of 389 common English terms and lower-cased terms. We also indexed the subject URIs by replacing delimiters with blank spaces and applying the same processing to the resulting string. The version of the BTC 2009 dataset used in the evaluation does not include blank nodes, i.e. all blank node identifiers have been replaced by URIs. We index this data using all three index structures. For the vertical index, we select the top three-hundred most common datatype-properties for indexing.

In the experiments, we measure the efficiency of retrieval, i.e. the time it takes to process queries including matching and ranking, but not result rendering. We consider two execution modes: *AND* where we require all keywords to be present in a document to be scored and *OR* where only a single term is needed. The former execution mode resembles Web search engines whereas the latter is the mode by default in our ranking model. To show the additional cost of structured retrieval, we also include a plain BM25 run using the *token* index.

We sample 150K queries taken from Yahoo!'s query logs with the restriction that they lead to a click in Wikipedia, in order to ensure there is an entity focus in the user intent. Among those queries 68% are unique and the average query length is 2.2 terms. Table 5 presents the average running times, which converge after a couple of thousand queries are being executed. The Table shows results for the baseline BM25 retrieval, all three index configurations (horizontal, vertical and reduced-vertical) and the two query execution modes.

Our tests show that the vertical approach is about eight time faster than the horizontal approach when queries are executed in AND mode, while it is only slightly faster in OR mode.

---

[7] http://vmlion25.deri.ie/

Table 5. Retrieval efficiency using different index structures and execution modes

	AND mode	OR mode
BM25	46 ms	80 ms
Horizontal	819 ms	847 ms
Vertical	97 ms	780 ms
R-Vertical	48 ms	152 ms

In general, AND queries execute faster than OR queries. In AND mode, it is necessary to compute the intersection of two (or more) posting lists; for OR queries, it is necessary to compute the union. Clearly, in the second case we always need to read the full posting list of each term involved. In the AND case, instead, it is often possible to skip over documents that are not necessary using *skip-pointers* [11]. For instance, when computing the AND of a very common and a very rare term, most of the postings of the very common term are not needed to compute the result, as the rare term doesn't appear there.

The difference between AND or OR execution modes is small in the horizontal case, because the alignment operator dominates execution times. Further, we can see that the r-vertical index is almost as fast as the vertical index. This proves that it is possible to trade-off query expressivity for faster execution times and apply our scoring at execution times comparable to the current state-of-the-art in Web search engines.

5.2 Evaluation of Effectiveness

We evaluate the effectiveness of our ranking using the data set, the queries and the relevance assessments that have been made available as part of the Semantic Search Challenge of 2010 [7]. All of the data has been made publicly available for research use.[8]

The collection used in this evaluation is the Billion Triples Challenge 2009 data set that we have described in Section 4. The query set consists of 92 queries with an entity focus selected from the query logs of Microsoft Live Search and Yahoo! Search (see [7] for details.) We use a proprietary, state-of-the-art spell corrector to fix a small number of user mistakes in the queries and apply the same term-processing as on the collection.

For ranking, we use the ranking function described in Section 3. We classify manually the properties into three classes (important, unimportant and neutral) and assign the same v_s for each class. In principle, we could learn or select a different v_s for each field, but in practice this would lead to an excessive number of parameters. Table 6 shows the list of important and unimportant properties.

Similarly, we do not assign a weight w_D individually to each document, but manually classify a small number of domains into the three classes. Table 7 shows the list of important and unimportant domains, while all other domains are considered neutral. We then set w_D to w_D^i, w_D^u, w_D^n for documents coming

---
[8] http://km.aifb.kit.edu/ws/semsearch10/

from domains classified as important, unimportant and neutral respectively. It is future work to look at how we could automatically learn these lists, i.e. based on the likelihood of the fields matching in relevant documents or domains vs. the likelihood of matching in irrelevant documents or domains. Similarly, we use a single b parameter for all b_s. We choose a separate weight for the *subject* field, which plays a special role as the identifier of the resource. We score the documents after matching in OR execution mode.

Table 6. Manually selected list of important and unimportant properties. URIs are abbreviated using known prefixes provided by the prefix.cc web service

important	dbp:abstract, rdfs:label, rdfs:comment, rss:description, rss:title, skos:prefLabel, akt:family-name, wn:lexicalForm, nie:title
unimportant	dc:date, dc:identifier, dc:language, dc:issued, dc:type, dc:rights, rss:pubDate, dbp:imagesize, dbp:coorDmsProperty, dbo:birthdate, foaf:dateOfBirth,foaf:nick, foaf:aimChatID, foaf:openid, foaf:yahooChatID, georss:point, wgs84:lat, wgs84:long

We use the official relevance assessments for evaluation, which were gathered using Amazon Mechanical Turk and used a three-scale grading for excellent results, fair results and irrelevant results [3]. We report the retrieval performance using Mean Average Precision (MAP) [11] which is more robust to noise perturbations than the P@10 measure [4] and check for statistical significant differences against the baseline using Wilcoxon's signed rank test (significance level set to 0.01).

Table 7. Manually selected list of important and unimportant domains

important	dbpedia.org, netflix.com
unimportant	www.flickr.com, www.vox.com, ex.plode.us

Our results are shown in Table 8. We perform two rounds of parameter tuning, in each round using a linear search over the individual parameter spaces. First, we select a default configuration for the parameters and tune the performance of each one of the features individually and report on their individual contribution to the increase in performance. Next, given the parameter list ordered as displayed in the table, we report the performance increase when adding a new parameter to the model, one at a time. This allows us to determine what is the benefit of adding each parameter over the best configuration found for the model so far.

We report the contribution of each of the features described Section 3. We start with the plain BM25 function with no structure ($v_s = 1$). We then investigate the effect of tuning BM25's b parameter. We then look at the result of assigning field weights other the default $v_s = 1$, in particular the effect of finding an optimal weight for the subject field (v_{sjc}), and for important and unimportant fields according to Table 6. Last, we look at changing document weights to other than the default $w_D^n = 1$. In particular, we assign a higher weight to documents that are from important domains as given by Table 7, and then decrease the weights of documents from unimportant domains. We omit the results for the k_1 parameter as it has little effect in retrieval performance.

Table 8. Feature importance measured with MAP. Improvements are statistically significant against plain BM25 using Wilcoxon's pairwise sign rank test (p-value < 0.01). The *Individual Features* column computes the improvement of each feature independently, on top of the untuned baseline, whereas the *Combination* column shows cumulative gain as we add features in the listed order, one at a time.

Feature	Description	Individual Features	Combination
BM25	BM25	0.1805	0.1805
b	BM25's b parameter	0.2450 (+35.7%)	0.2450 (+35.7%)
v_{sjc}	weighting for the subject field	0.2279 (+26.26%)	0.2512 (+2.5%)
v_{imp}	weighting for *important* properties	0.2261 (+25.25%)	0.2565(+2.1%)
v_{uni}	weighting for *unimportant* properties	0.2160 (+19.72%)	0.2590 (+1%)
w_D^i	weighting for *important* domains	0.2229 (+23.49%)	0.2730 (+5.4%)
w_D^u	weighting for *unimportant* domains	0.2319 (+28.47%)	0.2754 (+1%)

The first column of results shows that all features are able to improve significantly the baseline, even adding them individually. It is interesting to note that property field weighting (v_{sjc}, v_{imp}, v_{uni}) is able to improve the MAP score by more than 20%. This is a promising result given that we only took a few properties into account, and potentially adding more parameters to the ranking function could boost the performance by a larger margin. Adding query-independent domain-based weights (w_D^i, w_D^u) is also beneficial, despite the fact that we only included a limited number of site domains. This indicates that there is still room for improvement, given enough training data available and further analysis of which fields and properties should be weighted differently.

The second column of the table shows the accumulated improvement when we introduce one parameter at a time in the model. The total improvement using this one-step linear tuning of features is around 53% over the untuned baseline. 35% of the improvement is due to the b parameter, and on top of that, the field and site features are able to boost the performance another 18%, which is an encouraging result. The fact that the document normalization component plays an important role in the performance (controlled by b) goes accordingly to results in document retrieval. This indicates that the model is able to incorporate many different signals and boost up the performance significantly by combining them in a suitable way.

Next, we perform a 2-fold cross validation splitting the query set in two halves in order to determine the performance of the combination of features and what would be the effectiveness of the system in a real search environment, with

Table 9. Cross-validated results comparing our ranking function against the BM25 baseline and the best performing submission at SemSearch 2010 (percentage improvements are relative to SemSearch 2010)

Method	MAP	NDCG
SemSearch'10	0.1909	0.3134
BM25	0.1805 (-6%)	0.3869 (+23%)
BM25F	0.2705 (+42 %)	0.4800 (+52%)

limited training data available. We tune the parameters performance with a linear search and the promising directions algorithm [16] on each one of the halves separately. The algorithm starts with an initial set of parameter values, and performs one independent linear search over each parameter. Then, it selects the vector going from the initial set of parameter values and the best found values, which defines a *promising direction* in the parameter space. The algorithm explores the parameter space over this vector and repeats the whole process until convergence to a local minimum or when a maximum number of iterations is reached. We report the results averaged over the two halves in Table 9 using both the MAP and the NDCG metric, where the latter exploits graded relevance judgments. Our method improves 50% over the BM25 baseline, and 42% over the best run submitted to SemSearch 2010 using MAP [7]. These results are extremely significant and would necessarily translate to a qualitative jump in user experience.

Looking at the results in more detail, we could conclude that we did poorly on long queries such as *the morning call lehigh valley pa*. We also did poorly on queries with only one relevant result that we didn't find such as *kaz vaporizer*, in this case because the single result came from ex.plode.us domain which we marked as unimportant due to poor quality data (a flat list of tags).[9] We also performed low on the query *hospice of cincinnati*, which is a long-term care provider in Cincinnati that has no directly relevant resource in the BTC dataset. In this case, our system favored blog posts from RSS feeds that mentioned all three words and in general talked about hospice care in Cincinnati. However, the assessors marked as fair results other institutes in Cincinnati, such as the University of Cincinnati, the Hyde Park in Cincinnati and the Cincinnati Police Department. Conversely, we did well on queries that were short but highly selective such as *mst3000*, which stands for Mystery Science Theater 3000, an American cult television comedy. We also did well on queries where there was only one relevant result that we did manage to find, e.g. *fitzgerald auto mall chambersburg pa*. This auto mall has no relevant information in the BTC dataset, but the City of Chambersburg, Pennsylvania was accepted as a fair result by the assessors. All other queries fell in between these extremes, and typically had more than one relevant result.

---

[9] http://ex.plode.us is a social aggregator that is not in service any more.

6 Conclusions

Ad-hoc object retrieval is one of the most basic tasks in semantic search and it has direct applications in search engines that incorporate structured data in their result pages. In this paper, we have proposed an adaptation of the BM25F ranking function to the RDF data model that incorporates both field weights, document priors and a separate field for the subject URIs. We have shown that each of these features contributes to effectiveness on its own and in combination with other features. In cross-validation, the combination of these features outperforms in effectiveness the baseline BM25 method that ignores RDF structure and semantics by 50% in MAP score. It also improves on other state-of-the-art methods on the ad-hoc object retrieval task by 42% in MAP and 52% in NDCG scores.

We have shown two basic index structures, which we called the horizontal and vertical indices, for efficient retrieval of the information required for scoring. Both provide the same query expressivity, but represent different trade-offs in effectiveness. The vertical index becomes ineffective as the number of properties grow, while the horizontal index is able to capture all our data, but requires the slower alignment operator to resolve queries. We also proposed a modified version of the vertical index, which groups properties with the same weight, and thereby trades off query expressivity for a performance that is comparable to retrieval over text. In previous work, we have shown that both basic structures can be efficiently built using MapReduce.

In future work, we plan to explore the combination of retrieval with data integration to reduce the redundancy in current object search results. For this, we need to find co-referent objects in search results and integrate the information that different sources provide. A second problem we would like to address is the ranking of information that is provided about each object. As some objects may have several hundreds of triples associated with them, it is necessary to select only those triples for display that are most descriptive of the object and at the same time pertinent to the user query.

References

1. Bhalotia, G., Hulgeri, A., Nakhe, C., Chakrabarti, S., Sudarshan, S.: Keyword Searching and Browsing in Databases using BANKS. In: ICDE, pp. 431–440 (2002)
2. Blanco, R., Barreiro, Á.: Probabilistic Document Length Priors for Language Models. In: Macdonald, C., Ounis, I., Plachouras, V., Ruthven, I., White, R.W. (eds.) ECIR 2008. LNCS, vol. 4956, pp. 394–405. Springer, Heidelberg (2008), http://portal.acm.org/citation.cfm?id=1793274.1793322
3. Blanco, R., Halpin, H., Herzig, D.M., Mika, P., Pound, J., Thompson, H.S., Tran, D.T.: Repeatable and reliable search system evaluation using crowdsourcing. In: Proceeding of the 34th International ACM SIGIR Conference on Research and Development in Information Retrieval. SIGIR, ACM (2011)
4. Blanco, R., Zaragoza, H.: Beware of relatively large but meaningless improvements. Yahoo! Research Technical Report (2011)

5. Boldi, P., Vigna, S.: MG4J at TREC 2005. In: Voorhees, E.M., Buckland, L.P. (eds.) The Fourteenth Text REtrieval Conference (TREC 2005) Proceedings. No. SP 500-266 in Special Publications, NIST (2005), http://mg4j.dsi.unimi.it/
6. Clarke, C.L.A., Cormack, G.V., Burkowski, F.J.: An algebra for structured text search and a framework for its implementation. The Computer Journal 38(1), 43–56 (1995), http://comjnl.oxfordjournals.org/content/38/1/43.abstract
7. Halpin, H., Herzig, D., Mika, P., Blanco, R., Pound, J., Thompon, H., Duc, T.T.: Evaluating ad-hoc object retrieval. In: Proceedings of IWEST (2010)
8. Hristidis, V., Papakonstantinou, Y.: DISCOVER: Keyword Search in Relational Databases. In: VLDB, pp. 670–681 (2002)
9. Kamps, J., Geva, S., Trotman, A., Woodley, A., Koolen, M.: Overview of the Inex 2008 Ad Hoc Track. In: Geva, S., Kamps, J., Trotman, A. (eds.) INEX 2008. LNCS, vol. 5631, pp. 1–28. Springer, Heidelberg (2009)
10. Luo, Y., Wang, W., Lin, X.: SPARK: A Keyword Search Engine on Relational Databases. In: ICDE, pp. 1552–1555 (2008)
11. Manning, C.D., Raghavan, P., Schütze, H.: Introduction to Information Retrieval. Cambridge University Press, Cambridge (2008)
12. Mika, P.: Distributed indexing for semantic search. In: SEMSEARCH 2010 Proceedings of the 3rd International Semantic Search Workshop, pp. 1–4. ACM (2010), http://portal.acm.org/citation.cfm?id=1863879.1863882
13. Oren, E., Delbru, R., Catasta, M., Cyganiak, R., Stenzhorn, H., Tummarello, G.: Sindice.com: {A} Document-oriented Lookup Index for Open Linked Data. International Journal of Metadata, Semantics and Ontologies 3(1) (2008), http://www.sindice.com/pdf/sindice-ijmso2008.pdf
14. Pérez-Agüera, J.R., Arroyo, J., Greenberg, J., Iglesias, J.P., Fresno, V.: Using BM25F for semantic search. In: Proceedings of the 3rd International Semantic Search Workshop on - SEMSEARCH 2010, pp. 1–8. ACM Press, New York (2010), http://portal.acm.org/citation.cfm?doid=1863879.1863881, http://km.aifb.kit.edu/ws/semsearch10/Files/bm25f.pdf
15. Pound, J., Mika, P., Zaragoza, H.: Ad-hoc Object Ranking in the Web of Data. In: Proceedings of the WWW, pp. 771–780. Raleigh, USA (2010)
16. Robertson, S., Zaragoza, H.: The probabilistic relevance framework: BM25 and beyond, foundations and trends in information retrieval. Foundations and Trends in Information Retrieval 3(4), 333–389 (2009), http://dx.doi.org/10.1561/1500000019
17. Tran, T., Wang, H., Haase, P.: Hermes: Data Web search on a pay-as-you-go integration infrastructure. Web Semantics: Science, Services and Agents on the World Wide Web 7(3), 189–203 (2009), http://linkinghub.elsevier.com/retrieve/pii/S1570826809000213
18. Wang, H., Liu, Q., Penin, T., Fu, L., Zhang, L., Tran, T., Yu, Y., Pan, Y.: Semplore: A scalable IR approach to search the Web of Data. Web Semantics: Science, Services and Agents on the World Wide Web 7(3), 177–188 (2009), http://www.sciencedirect.com/science/article/B758F-X1SBDK-1/2/8efe2a494e75791c8b333a1abdfc4188
19. Wrigley, S.N., Reinhard, D., Elbedweihy, K., Bernstein, A., Ciravegna, F.: Methodology and campaign design for the evaluation of semantic search tools. In: Proceedings of the 3rd International Semantic Search Workshop on - SEMSEARCH 2010, pp. 1–10. ACM Press, New York (2010), http://portal.acm.org/citation.cfm?doid=1863879.1863889

An Empirical Study of Vocabulary Relatedness and Its Application to Recommender Systems

Gong Cheng, Saisai Gong, and Yuzhong Qu

State Key Laboratory for Novel Software Technology, Nanjing University,
Nanjing 210093, China
{gcheng,yzqu}@nju.edu.cn, saisaigong@gmail.com

Abstract. When thousands of vocabularies having been published on the Semantic Web by various authorities, a question arises as to how they are related to each other. Existing work has mainly analyzed their similarity. In this paper, we inspect the more general notion of relatedness, and characterize it from four angles: well-defined semantic relatedness, lexical similarity in contents, closeness in expressivity and distributional relatedness. We present an empirical study of these measures on a large, real data set containing 2,996 vocabularies, and 15 million RDF documents that use them. Then, we propose to apply vocabulary relatedness to the problem of post-selection vocabulary recommendation. We implement such a recommender service as part of a vocabulary search engine, and test its effectiveness against a handcrafted gold standard.

Keywords: Ontology, recommendation, relatedness, vocabulary.

1 Introduction

The Semantic Web enriches data with machine-readable, unambiguous meaning by advising different applications to use common vocabularies (a.k.a. ontologies), and to adhere strictly to the term descriptions provided. It would enable an even wider range of applications that operate on integrated data when vocabularies from different communities are interconnected, e.g. aligned. A large body of work has been devoted to this problem of matching [9], which aims at finding terms (i.e. classes or properties) in different vocabularies that have the same intensional meaning. Accordingly, approaches thus far mainly follow a paradigm that measures the *similarity* between terms [9] or between vocabularies [17,5]. In fact, similarity is just a specific kind of *relatedness*. As other forms of relatedness, one vocabulary may extend another by defining more specific subclasses, and two vocabularies may describe closely related domains so that they are often used together, etc. However, this more general notion of relatedness has been addressed by only few work [19,23,11], and none of these approaches has been evaluated on a representative sample of real-world vocabularies. In this regard, whereas our previous work [4] has analyzed only explicit relations between terms, in this paper, we will characterize several different aspects of relatedness between vocabularies via an empirical study of many real-world, diverse vocabularies.

L. Aroyo et al. (Eds.): ISWC 2011, Part I, LNCS 7031, pp. 98–113, 2011.

Vocabulary relatedness can find many applications. For example, it could be employed to rank and find central vocabularies [7]. Here we conceive another application called *post-selection vocabulary recommendation*. Assume that a user has shown an interest in a vocabulary, or in other words, she has *selected* a vocabulary. Such selection widely exists in many scenarios, e.g. having selected a vocabulary for further exploration when interacting with a vocabulary search engine, or having selected a vocabulary for use when developing an application. Then, a recommender system will automatically suggest several other vocabularies that the user might also be interested in, e.g. one as an alternative or complementary to the selected one for a particular use. Naturally, such recommendation mainly relies on the features of the selected vocabulary, and thus we call it post-selection recommendation. We will discuss how this specific task can be supported by the study of vocabulary relatedness.

To summarize, the contribution of this paper is threefold:

- Rather than similarity, we study the more general notion of relatedness between vocabularies on the Semantic Web. We discuss four kinds of relatedness: (a) semantic relatedness defined by vocabulary (meta-)descriptions, (b) content similarity which exploits lexical features, (c) expressivity closeness according to the language constructs adopted, and (d) distributional relatedness derived from vocabulary usage.
- We apply six proposed relatedness measures to a real-world data set crawled by a Semantic Web search engine, which contains 2,996 vocabularies instantiated by other 15 million RDF documents (collectively containing 4 billion RDF triples). We analyze and compare the effects of our measures, and report many statistical findings that help characterize real-world vocabularies.
- We consider the problem of post-selection vocabulary recommendation, and propose to solve it by using relatedness measures. We also examine the popularity of vocabularies for recommendation. We evaluate our approach based on a handcrafted gold standard, and also develop such a recommender system and incorporate it into a vocabulary search engine.

In the remainder of this paper, Sect. 2 characterizes our data set, in particular the vocabularies identified from it. Section 3 describes and compares several relatedness measures. Section 4 introduces and evaluates a solution to the problem of post-selection vocabulary recommendation. Finally, Sect. 5 compares related work, and Sect. 6 concludes the paper.

2   Vocabularies in the Real World

2.1   Data Set

The data set investigated in this work is the one — at the time of writing — used by the Falcons search engine.[1] As summarized in Table 1, it comprises

---

[1] http://ws.nju.edu.cn/falcons/

15 million RDF (including RDF/XML and RDFa) documents, which collectively contain 4 billion RDF triples, crawled from 5 thousand pay-level domains[2] between February 2010 and May 2011.

Table 1. Data set statistics

Number of RDF documents	15,947,721
Number of pay-level domains hosting RDF documents	5,805
Aggregate number of RDF triples	4,099,414,887
Number of vocabularies	2,996
Number of pay-level domains hosting vocabularies	261
Aggregate number of classes	396,023
Aggregate number of properties	59,868

To characterize the data set, Figure 1 presents the distribution of the number of pay-level domains over the number of RDF documents hosted on a log-log scale. The distribution approximates a power law, but having a long tail to the right which corresponds to several large data sources including hi5.com, 13s.de, geonames.org, dbpedia.org, etc. This power law phenomenon has also been observed on other data sets such as the one crawled by Swoogle [8].

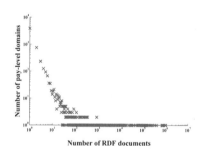

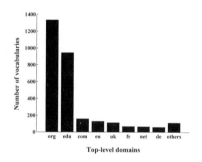

Fig. 1. Distribution of the number of pay-level domains over the number of RDF documents hosted

Fig. 2. Distribution of the number of vocabularies hosted over top-level domains, where "others" represents an aggregate of all the ones not presented

2.2 Identifying Vocabularies

We study only the vocabularies that are published by applying best practice.[3] Accordingly, since a vocabulary description may be distributed among multiple

---

[2] A *pay-level domain* is a domain that requires payment at a (country-code) top-level domain [14]. For instance, the URI http://ws.nju.edu.cn/falcons/ belongs to the pay-level domain nju.edu.cn. We use the Apache Nutch package (nutch.apache.org) to identify the pay-level domain of a URI.

[3] http://www.w3.org/TR/swbp-vocab-pub/

documents, we employ a bottom-up strategy to identify vocabularies from the data set. That is, firstly we identify a *term* as a dereferenceable URI that refers to a class or a property in the RDF document retrieved via dereferencing the URI. Then, terms sharing a common namespace URI are grouped into a *vocabulary*, using this namespace URI as its identification. In this way, we may miss some old-fashioned vocabularies that are not dereferenceable, and may also fail to find all the terms for some vocabulary, but we believe that the results obtained would accurately reflect real-world conditions at our best.

As summarized in Table 1, we have identified 396,023 classes and 59,868 properties, which are grouped into 2,996 vocabularies. They come from 261 pay-level domains or 33 top-level domains. That is, among 5,805 pay-level domains in our data set that serve RDF documents, only a small portion (4.50%) publish their own vocabularies. Figure 2 depicts the distribution of the number of vocabularies hosted over top-level domains, in which `org` and `edu` dominate with 44.53% and 31.58%, respectively, followed by `com` and several country-code ones. This distribution is also close to the one for Swoogle [8].

These vocabularies vary considerably in size and composition. The largest ones, in terms of the number of terms, are some versions of YAGO and Cyc which comprise tens of thousands of terms, whereas most of the others (72.30%) contain not more than 25. Even among large vocabularies, some (e.g. YAGO) mainly provide classes when some others (e.g. SUMO) are rich in both classes and properties.

3   Characterizing Relatedness between Vocabularies

In this section, we discuss, from different points of view, four kinds of relatedness between vocabularies, and formalize them as numerical measures. In particular, we assume that relatedness measures are symmetric. We perform an empirical analysis of these measures, and finally make a comparison.

3.1   Semantic Relatedness

Vocabularies on the Semantic Web are described in a structured way. When one vocabulary is connected to another via a typed link, it naturally indicates certain kind of relatedness having well-defined semantics, and this leads to our first kind of relatedness measure.

Explicit Relation. Major vocabulary languages such as OWL provide mechanisms for describing information about a vocabulary itself. For instance, `owl:imports` references another vocabulary whose meaning will be included in the present one. Since such relation between vocabularies is directly given in the meta-description of a vocabulary, we call it *explicit relation*. Further, when there are explicit relations between vocabulary v_1 and v_2, and between v_2 and v_3, we observe some kind of relation between v_1 and v_3, which looks "longer" and thus is probably weaker than the two original relations.

These observations could be represented as an edge-weighted graph G_E, where nodes correspond to vocabularies, and every pair of explicitly related vocabularies v_i and v_j are connected by an undirected edge, associated with a weight w indicating how weak the relation is:

$$w(v_i, v_j) = \begin{cases} 2 & \text{if } v_i \text{ references } v_j \text{ or } v_j \text{ references } v_i, \\ 1 & \text{if } v_i \text{ references } v_j \text{ and } v_j \text{ references } v_i. \end{cases} \quad (1)$$

Then, the relatedness (denoted by R_S^E) between two vocabularies is defined as the multiplicative inverse of the weight of a shortest path between their corresponding nodes in G_E, which is thus inside (0,1], or 0 when unreachable. Note that we actually ignore the specific types of relations, as we will see later that most relations observed in practice are quite homogeneous.

Implicit Relation. In a vocabulary, the description of a term may refer to terms in other vocabularies, e.g. via rdfs:subClassOf or complex OWL constructs, which suggests a kind of *implicit relation* between vocabularies, in the sense that they are revealed by term-level descriptions but might not be mentioned in the meta-description of vocabulary. Analogous to G_E, here we devise another edge-weighted graph G_I to convey such relations, which differs from G_E in only one respect that: implicit but not explicit relation is considered. Then, a relatedness measure, denoted by R_S^I, is defined based on G_I analogously.

Hybrid Relation. When we take both explicit and implicit relations into consideration, we obtain a kind of *hybrid relation* between vocabularies. Analogously, it could be characterized as an edge-weighted graph G_{E+I}, based on which a relatedness measure, denoted by R_S^{E+I}, is defined.

Empirical Analysis. Among 2,996 vocabularies in the data set, explicit, implicit and hybrid relations are observed between 2,968, 2,845 and 4,691 pairs of vocabularies, respectively. According to Table 2 which summarizes several statistical properties of G_E, G_I and G_{E+I}, whereas G_E and G_I are similar in terms of the number of edges, G_E seems more fragmented, suggested by the percentages of isolated nodes and the metrics below for characterizing reachability. On the other hand, there are far more edges in G_{E+I} than in G_E, indicating that many implicit relations between vocabularies are not captured by the meta-descriptions thereof.

In particular, only 17 types of explicit relations are observed in our data set, and only 6 occur in the meta-descriptions of more than one vocabulary. As shown in Table 3, when owl:imports dominates largely, most others are negligible.

3.2   Content Similarity

In a vocabulary description, terms are not only interconnected but also usually associated with human-readable contents, e.g. labels. Given two vocabularies

Table 2. Statistical properties of G_E, G_I and G_{E+I}

	G_E	G_I	G_{E+I}
Number of nodes	2,996	2,996	2,996
Number of edges	2,968	2,845	4,691
Average degree	1.98	1.90	3.13
Maximum degree	786	684	848
Percentage of isolated nodes	56.88%	36.72%	32.31%
Number of connected components	1,763	1,143	1,007
Percentage of nodes in the largest connected component	32.78%	57.44%	62.18%
Percentage of pairs of connected nodes	5.40%	16.50%	19.33%

Table 3. Relations used in the highest percentages of vocabulary meta-descriptions

`http://www.w3.org/2002/07/owl#imports`	36.58%
`http://www.daml.org/2001/03/daml+oil#imports`	1.60%
`http://www.w3.org/2000/01/rdf-schema#seeAlso`	0.30%
`http://www.w3.org/2002/07/owl#priorVersion`	0.10%
`http://purl.org/dc/terms/requires`	0.07%
`http://www.openlinksw.com/schema/attribution#isDescribedUsing`	0.07%

modeling the same or related domains, their textual descriptions often overlap. By detecting this aspect, we present our second kind of relatedness measure.

Specifically, the relatedness (denoted by R_C) between two vocabularies v_i and v_j combines the *content similarity* between their classes (denoted by C_i and C_j) and the one between their properties (denoted by P_i and P_j):

$$
R_C(v_i, v_j) = \begin{cases} \frac{\text{SetSim}(C_i,C_j)+\text{SetSim}(P_i,P_j)}{2} & \text{if } C_i \times C_j \neq \emptyset \text{ and } P_i \times P_j \neq \emptyset, \\ \text{SetSim}(C_i, C_j) & \text{if } C_i \times C_j \neq \emptyset \text{ and } P_i \times P_j = \emptyset, \\ \text{SetSim}(P_i, P_j) & \text{if } C_i \times C_j = \emptyset \text{ and } P_i \times P_j \neq \emptyset, \\ 0 & \text{if } C_i \times C_j = \emptyset \text{ and } P_i \times P_j = \emptyset, \end{cases}
$$
(2)

where SetSim is a similarity measure for term sets that determines the extent to which the lexical features of both sets are covered by each other:

$$
\text{SetSim}(T_i, T_j) = \text{HMean}(\frac{1}{|T_i|} \sum_{t_i \in T_i} \max_{t_j \in T_j} \text{LS}(t_i, t_j), \frac{1}{|T_j|} \sum_{t_j \in T_j} \max_{t_i \in T_i} \text{LS}(t_i, t_j)),
$$
(3)

where HMean returns the harmonic mean of the two parameters, and $\text{LS}(t_i, t_j)$ gives the lexical similarity between terms. As one implementation of LS, we apply a string metric [24] to all pairs of the respective labels of the two terms, normalize each result to be inside the interval [0,1], and finally take the maximum.

Empirical Analysis. In our data set, to exploit term descriptions for labels, we retrieve property values from `rdfs:label`, `dc:title` and their subproperties (e.g. `skos:prefLabel`) that are defined via or can be inferred from the `rdfs:subPropertyOf` relation, which collectively amount to 86 types of properties. In this way, at least one label can be found for 63.67% of all the terms, which are distributed among 36.21% of all the vocabularies. Since the absence of label is still commonly observed, the local name of each term URI is also employed.

Another thing we would like to point out is: computing content similarity is the most expensive task performed in our experiments, which costs a multithreading program running on a multi-core server several weeks. This is not surprising because all pairs of 2,996 vocabularies are compared, and for each pair, every class (resp. property) in one vocabulary is compared with every class (resp. property) in another, which is again time-consuming in particular for large vocabularies, as illustrated in Sect. 2.2.

3.3   Expressivity Closeness

Vocabularies vary from lightweight taxonomies to heavyweight ones with complex constraints. In this regard, two vocabularies are close when they are similar in expressivity. Accordingly, we develop our third kind of relatedness between vocabularies based on their *expressivity closeness.*

The expressivity of a vocabulary is mainly (though not fully) captured by the language constructs (e.g. `rdfs:subClassOf` vs. `owl:complementOf`) adopted for describing terms. Besides, other meta-level terms may also be employed for description, e.g. Dublin Core metadata terms and those for meta-modeling. Therefore, we propose to characterize the expressivity of a vocabulary v by MetaTerms(v) — the set of all meta-level terms that are instantiated in v's description. Then, given two vocabularies v_i and v_j, we define their relatedness (denoted by R_E) as follows:

$$R_E(v_i, v_j) = \text{J}(\text{MetaTerms}(v_i), \text{MetaTerms}(v_j)), \qquad (4)$$

where J returns the Jaccard similarity coefficient of the two sets.

Empirical Analysis. We observe 4,978 meta-level terms that are instantiated in at least one vocabulary's description, 469 (9.42%) of which are used in at least two, showing a wide variety. In particular, the meta-level terms instantiated in the highest percentages of vocabulary descriptions are all language constructs, led by `rdf:type`, `rdfs:domain` and `rdfs:range`. Excluding these, Table 4 presents the top-ranked ones remaining, which are all not widely used.

On the other hand, describing a vocabulary needs to instantiate an average of 10.13 types of meta-level terms. In fact, 92.96% of all the vocabularies in our data set use not more than 20 types. However, we still recognize hundreds of types of meta-level terms in some complex vocabularies such as Cyc.

Table 4. Meta-level terms (excluding those in RDF, RDFS, OWL or DAML) instantiated in the highest percentages of vocabulary descriptions

http://purl.org/dc/elements/1.1/description	1.50%
http://purl.uniprot.org/core/encodedIn	0.90%
http://www.w3.org/2004/02/skos/core#definition	0.73%
http://purl.org/dc/terms/modified	0.67%
http://www.swop-project.eu/ontologies/pmo/product.owl#unit	0.67%
http://purl.org/dc/terms/issued	0.63%
http://www.w3.org/2003/06/sw-vocab-status/ns#term_status	0.63%

3.4 Distributional Relatedness

Whereas all the previous notions of relatedness rely on the *intensional* descriptions of vocabularies, our fourth kind of measure looks at the *extensional* side, i.e. to investigate vocabulary usage in practice.

Recall that on the fruitful topic of relatedness in the field of computational linguistics, among others, *distributional relatedness* [20] defines close words as those that are used in similar contexts, e.g. having many co-occurring words in common. Accordingly, a "distributional profile" is created for each word, which characterizes the strength of association between the word and every other word that co-occurs with it, commonly by using conditional probability. Then, the similarity (e.g. cosine) between distributional profiles is calculated, as a proxy for relatedness between words.

Inspired by this line of research, we study vocabulary co-occurrence in use, which in the context of the Semantic Web amounts to vocabulary co-instantiation. We conceive an RDF document as the context from which co-instantiation is observed, and let $IV(d)$ be the set of all vocabularies instantiated in RDF document d. Then, given the set of all vocabularies V and $v \in V$, the distributional profile of v is represented by a $|V|$-dimensional vector, denoted by $DP(v)$, where:

$$DP_i(v) = \frac{|\{d \in D|\, v, v_i \in IV(d)\}|}{|\{d \in D|\, v \in IV(d)\}|}, \qquad (5)$$

where D is the set of all RDF documents under investigation. In particular, $DP(v)$ is defined as $\mathbf{0}$ when no instantiation of v can be observed in any $d \in D$. Finally, the relatedness between vocabulary v_i and v_j, denoted by $R_D(v_i, v_j)$, is given by the cosine similarity between $DP(v_i)$ and $DP(v_j)$.

This straightforward implementation is improved in two ways. Firstly, language-level vocabularies (e.g. RDF) are trivially and widely instantiated, which function as stop words in computational linguistics. Hence they are filtered out prior to processing. Otherwise, they may undesirably, even largely, increase the relatedness between many pairs of vocabularies. Secondly, as discussed in Sect. 2.1, considering the distribution of the number of pay-level domains over the number of RDF documents hosted, a large data source in the long tail of the distribution may unfairly affect the computation of relatedness. To avoid this, we limit the effects that could be caused by a single pay-level domain. Specifically, we define $\mathrm{PLD}(D)$ as a partition of D such that each element of $\mathrm{PLD}(D)$ corresponds to all the RDF documents in D that are hosted by one particular pay-level domain. Then, we rewrite (5) as follows:

$$\mathrm{DP}_i(v) = \frac{|\{S \in \mathrm{PLD}(D)| \exists d \in S, v, v_i \in \mathrm{IV}(d)\}|}{|\{S \in \mathrm{PLD}(D)| \exists d \in S, v \in \mathrm{IV}(d)\}|}. \tag{6}$$

Empirical Analysis. In our data set, instantiation is observed for 1,874 (62.55%) vocabularies. Table 5 shows the most widely instantiated ones, led by Dublin Core metadata vocabularies and FOAF. Further, among 9,763 pairs of vocabularies that have co-instantiation, Table 6 presents the most frequent ones.

Table 5. Vocabularies (excluding RDF, RDFS, OWL and DAML) instantiated in RDF documents hosted by the highest percentages of pay-level domains

http://purl.org/dc/elements/1.1/	37.45%
http://xmlns.com/foaf/0.1/	22.79%
http://purl.org/dc/terms/	15.90%
http://www.icra.org/rdfs/vocabularyv03#	10.65%
http://www.w3.org/2003/01/geo/wgs84_pos#	5.22%
http://purl.org/vocab/bio/0.1/	2.76%
http://www.w3.org/2000/10/swap/pim/contact#	2.76%
http://rdfs.org/sioc/ns#	2.20%
http://usefulinc.com/ns/doap#	1.67%
http://purl.org/vocab/relationship/	1.38%

Table 6. Pairs of vocabularies (excluding those involving RDF, RDFS, OWL or DAML) co-instantiated in RDF documents hosted by the highest percentages of pay-level domains

http://purl.org/dc/elements/1.1/ http://purl.org/dc/terms/	14.42%
http://purl.org/dc/elements/1.1/ http://www.icra.org/rdfs/vocabularyv03#	10.65%
http://purl.org/dc/terms/ http://www.icra.org/rdfs/vocabularyv03#	10.61%
http://xmlns.com/foaf/0.1/ http://purl.org/dc/elements/1.1/	9.42%
http://www.w3.org/2003/01/geo/wgs84_pos# http://xmlns.com/foaf/0.1/	5.05%

3.5   Comparison

Now we study the levels of *agreement* between different relatedness measures. We apply, to all pairs of 2,996 vocabularies in our data set, each of our six relatedness measures, namely R_S^E, R_S^I, R_S^{E+I}, R_C, R_E and R_D. Each measure will induce a ranking of these pairs, and we leverage the Spearman's rank correlation coefficient, denoted by ρ, to measure the correspondence between these rankings and assess its significance. ρ is inside the interval [-1,1], and an increasing value implies increasing agreement.

The results are summarized in Fig. 3. All the values are positive, i.e., all these measures are positively correlated. Larger values are found between R_S^I and R_S^{E+I} (0.88), and between R_S^E and R_S^{E+I} (0.53), which are not surprising since R_S^{E+I} comprises R_S^E and R_S^I. In particular, the second largest value (0.66) is observed between R_S^E and R_D, indicating that explicitly related vocabularies are also most likely to be instantiated together, and vice versa.

	R_S^I	R_S^{E+I}	R_C	R_E	R_D
R_S^E	0.39	0.53	0.21	0.19	0.66
R_S^I	-	0.88	0.26	0.38	0.35
R_S^{E+I}	-	-	0.30	0.26	0.43
R_C	-	-	-	0.32	0.23
R_E	-	-	-	-	0.24

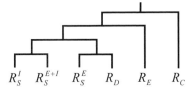

Fig. 3. Levels of agreement between individual relatedness measures

Fig. 4. Dendrogram showing the single-link hierarchical clustering of individual relatedness measures based on their levels of agreement

Further, based on ρ values, we employ the single-link hierarchical clustering technique to depict the relationships between measures. As shown in Fig. 4, R_C is relatively far from the other measures. One reason might be that for describing the same domain, different authorities may publish their own vocabularies, which vary considerably in expressivity and are rarely connected to each other.

4   Post-selection Vocabulary Recommendation

In this section, we describe an application that can be enriched by the study of vocabulary relatedness. Recall that when we browse online book stores or movie databases, some of these applications will provide recommendations to avoid overloading users with information. For instance, when we look at the introduction of a book, several "related items" are also presented, e.g. books written by the same author. Analogously, when interacting with a vocabulary repository, e.g. a vocabulary search engine, after a vocabulary has been selected for examining details, the system is expected to recommend several related vocabularies. In the next, we address this problem of *post-selection vocabulary recommendation*. We describe an approach as well as an extension, and present evaluation results.

4.1   Relatedness-Based Ranking

In Sect. 3, we have introduced six measures of relatedness between vocabularies, namely $\mathfrak{R} = \{R_S^E, R_S^I, R_S^{E+I}, R_C, R_E, R_D\}$, all returning values inside the interval [0,1]. For a selected vocabulary v_0, we argue that a vocabulary v_i is more likely to be recommended if it is more *related* to v_0, in terms of some $R_j \in \mathfrak{R}$. That is, we rank recommendation candidates by $R_j(v_i, v_0)$. Here, which R_j to use is specified by users according to their specific needs. $\mathfrak{R}$ can also be extended to include other metrics developed in the future.

When users intend to receive recommendations featuring several different characteristics, it requires employing multiple measures. Further, users may attach different degrees of importance to different measures. To this end, we allow ranking recommendation candidates by a *linear combination* of all the measures in $\mathfrak{R}$, i.e. $\sum_{R_j \in \mathfrak{R}} \alpha_j R_j(v_i, v_0)$, where $\alpha_j \in [0, 1]$ is a group of weightings.

We implement such a *recommender service* in Falcons Ontology Search.[4] When exploring a retrieved vocabulary, users could enquire about related ones after specifying a weighting for each relatedness measure.

4.2   Popularity-Based Re-ranking

Besides relatedness, another factor we would like to consider in vocabulary recommendation is *popularity*. Recall that the Semantic Web could facilitate data integration on the semantic level exactly because different Semantic Web applications produce and consume data adhering to common vocabularies. Hence, we argue that a recommender service should return more popular vocabularies, i.e. those having been used by more applications. To incorporate popularity into the criteria for ranking, given $\text{Pop}(v)$ — the number of pay-level domains hosting RDF documents that instantiate v, we extend our approach to rank recommendation candidates by the following metric:

$$\sum_{R_j \in \mathfrak{R}} \alpha_j R_j(v_i, v_0) \cdot (1 + \log_b (1 + \text{Pop}(v_i))), \qquad (7)$$

where b is a parameter that tunes the degree of influence of popularity on recommendation. When decreasing b from $+\infty$ to a small value (e.g. 2), the degree of influence increases. But apparently, popularity is achieved at the relative cost of relatedness. A trade-off needs to be studied for specific applications.

4.3   Evaluation

Firstly, without considering popularity, we examine which $R_j \in \mathfrak{R}$ is more useful for recommendation. To achieve this, we compare generated rankings thereof to the gold standard given by human experts. We identify 1,302 vocabularies from our data set for this experiment, each containing 5–25 terms, being neither

---

[4] http://ws.nju.edu.cn/falcons/ontologysearch/

too small to be significant nor too large for manual investigation. We choose 20 from them at random as "selections" for testing post-selection recommendation. For each selection, we can hardly ask experts to give a ranking of all the other 1,301 vocabularies, but rather, we apply the depth-10 pooling technique, which is widely adopted for evaluating information retrieval (IR) systems. To be specific, we apply each $R_j \in \mathfrak{R}$ to score all the other 1,301 vocabularies, retain only those having positive relatedness values, and collect the top-10 ones. For all $R_j \in \mathfrak{R}$, these top-ranked vocabularies collectively form a pool to be used in the experiment. The pool is randomly divided up and assigned to two experts. For each assignment, the expert is asked to assess the relatedness between the assigned vocabulary and the selection, and report (a) "closely related", (b) "somewhat related", or (c) "unrelated", corresponding to ratings 2, 1 and 0, respectively. In particular, 5 vocabularies in each pool are assigned to both experts.

We receive 739 assessments in total, of which 81.60% are unrelated, 10.55% somewhat related and 7.85% closely related. Unrelated vocabularies take the largest proportion, which in fact is quite common under pooling methods. Besides, among 100 (20 × 5) vocabularies assessed by both experts, agreement is reached on 80%. If we consider only binary ratings by taking closely and somewhat related as "related", agreement is reached on 91%, suggesting a high quality of the assessments. Finally, to form one single gold standard, when two experts give different assessments on a vocabulary, we take the higher rating.

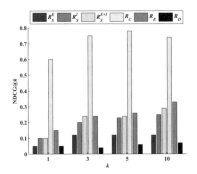

Fig. 5. NDCG of individual measures

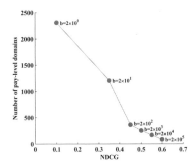

Fig. 6. Relationship between relatedness and popularity under different b values

For each selection, we evaluate each $R_j \in \mathfrak{R}$ by calculating its normalized discounted cumulative gain (NDCG) — a widely used metric for IR evaluation. NDCG@k, inside the interval [0,1], measures the quality of the k top-ranked vocabularies against their gold-standard ratings. Figure 5 summarizes the results averaged over all the 20 selections, under different settings of k. R_C noticeably outperforms the others, showing that our experts assess relatedness between vocabularies mainly based on the overlap between their contents. On the other hand, we attribute the bad performances of R_S^E and R_D to the fact that, as presented in Sect. 3, 56.88% of vocabularies in our data set are not explicitly related to any other ones, and that 37.45% have no instantiation. Thereby, R_S^E and R_D fail to find any related vocabularies for 13 and 11 selections, respectively. In these cases, NDCG is defined as 0, which largely hurt their overall performances.

Secondly, we look at combinations of measures. Since R_C performs the best in the first experiment, we combine it with every other measure in $\mathfrak{R}$ to see whether better results can be achieved. Figure 7 illustrates the evaluation results of several combinations. Actually, for each kind of combination, we show only one group of weightings with which the best result is obtained. We find that under different settings of k, better or equal results are consistently observed when R_C is combined with R_S^E, R_E or R_D, whereas R_S^I and R_S^{E+I} seem only helpful when $k = 1$, i.e. in generating the top-ranked vocabulary. However, the reader is reminded that these results only reflect the bias of our experts, whereas our flexible approach indeed allows task-oriented combination.

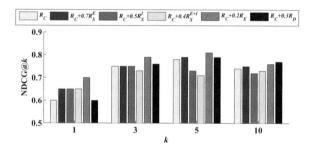

Fig. 7. NDCG of several combinations of measures

Finally, we illustrate, with R_C, the relationship between relatedness and popularity. Under different b values in (7), we evaluate relatedness by NDCG@1 and evaluate popularity by the number of pay-level domains hosting RDF documents that instantiate the top-ranked vocabulary. As shown in Fig. 6, when NDCG decreases from 0.60 to 0.45 (averaged over all the 20 selections), the number of pay-level domains increases linearly. A much higher popularity can also be achieved, which however loses relatedness considerably. It reveals that looking for a good trade-off is not an easy task, but has to rely on specific applications.

5   Related Work

5.1   Relatedness

In computational linguistics, a substantial amount of research has been conducted on the *measurement of relatedness* between words (or senses) [3]. Most existing methods exploit semantic networks such as WordNet, and operate on shortest paths or information theory. These ideas have also been transplanted to the Semantic Web for measuring relatedness between terms in a vocabulary [19,23,11], by treating a vocabulary description as a semantic network.

Complementary to this, another line of research studies the co-occurrence of words to measure their distributional relatedness [20].

Differently, we look at *relatedness on the vocabulary but not the term level*. In an earlier work [4], we derive a vocabulary dependence graph from the relations between terms, and perform a complex network analysis, which reveals its scale-free nature. In [7], several types of relations between terms and between vocabularies are identified, to characterize a random surfer's behavior for ranking. In [25], vocabularies are clustered based on their use of language constructs. Whereas each of these studies investigates very few kinds of relatedness, the work in this paper characterizes it in four aspects and compares six measures.

As a special kind of relatedness, *similarity* between terms [9] and between vocabularies [17,5] have attracted extensive research. Further, the similarities among a collection of vocabularies can be represented as a graph, on which statistical analysis [10,21] and complex network analysis [12] have been carried out. Besides, more sophisticated measures of similarity have been established based on such graph [6]. In our work, we also implement content-based similarity as one aspect, when we deal with the more general notion of relatedness.

5.2 Recommendation

Recommender systems have become an important research area [1]. In particular, collaborative approaches have been applied to vocabulary recommendation [22,15], which are grounded on *user-generated ratings*. A closely related problem is vocabulary search, which usually takes a keyword query as input and in fact performs *query-biased recommendation* [2,13,18]; these approaches mainly investigate how well a vocabulary is relevant to a keyword query. Inspired by [16], the problem of post-selection recommendation addressed in our work is in a different context that takes a selected vocabulary as input and demands *selection-biased recommendations*, to which relatedness measurement is the natural solution.

6   Conclusions and Future Work

In this paper, we have discussed vocabulary-level relatedness from four aspects. Our empirical analysis on a large, real data set compares six developed measures, and also, reports many statistical findings, which help characterize vocabularies on the real Semantic Web. In particular, we observe that many cross-vocabulary relations between terms are not embodied in their vocabulary meta-descriptions, and vocabularies having explicit relations tend to be instantiated together. After that, we have proposed to apply relatedness measures to the problem of post-selection vocabulary recommendation. The evaluation results demonstrate the effectiveness of our measures in recommendation, particularly when they are combined appropriately. We have enriched our Falcons Ontology Search system with such a flexible recommender service.

In fact, our relatedness measures have not fully exploited vocabularies. As future work, textual descriptions and provenance information in vocabulary meta-description still need investigation. About vocabulary recommendation, it would be interesting to combine our relatedness measures with collaborative methods and ontology evaluation techniques.

Acknowledgments. This work was supported in part by the NSFC under Grant 60973024 and 61021062, and in part by ZTE Corp. (R&Dcon1105160003). We thank Min Liu for his time and effort in the experiments.

References

1. Adomavicius, G., Tuzhilin, A.: Toward the Next Generation of Recommender Systems: A Survey of the State-of-the-Art and Possible Extensions. IEEE Trans. Knowl. Data Eng. 17(6), 734–749 (2005)
2. Alani, H., Brewster, C.: Ontology Ranking Based on the Analysis of Concept Structures. In: 3rd International Conference on Knowledge Capture, pp. 51–58. ACM, New York (2005)
3. Budanitsky, A., Hirst, G.: Evaluating WordNet-based Measures of Lexical Semantic Relatedness. Comput. Linguist. 32(1), 13–47 (2006)
4. Cheng, G., Qu, Y.: Term Dependence on the Semantic Web. In: Sheth, A., Staab, S., Dean, M., Paolucci, M., Maynard, D., Finin, T., Thirunarayan, K. (eds.) ISWC 2008. LNCS, vol. 5318, pp. 665–680. Springer, Heidelberg (2008)
5. David, J., Euzenat, J.: Comparison Between Ontology Distances (Preliminary Results). In: Sheth, A., Staab, S., Dean, M., Paolucci, M., Maynard, D., Finin, T., Thirunarayan, K. (eds.) ISWC 2008. LNCS, vol. 5318, pp. 245–260. Springer, Heidelberg (2008)
6. David, J., Euzenat, J., Šváb-Zamazal, O.: Ontology Similarity in the Alignment Space. In: Patel-Schneider, P.F., Pan, Y., Hitzler, P., Mika, P., Zhang, L., Pan, J.Z., Horrocks, I., Glimm, B. (eds.) ISWC 2010, Part I. LNCS, vol. 6496, pp. 129–144. Springer, Heidelberg (2010)
7. Ding, L., Pan, R., Finin, T., Joshi, A., Peng, Y., Kolari, P.: Finding and Ranking Knowledge on the Semantic Web. In: Gil, Y., Motta, E., Benjamins, V.R., Musen, M.A. (eds.) ISWC 2005. LNCS, vol. 3729, pp. 156–170. Springer, Heidelberg (2005)
8. Ding, L., Finin, T.: Characterizing the Semantic Web on the Web. In: Cruz, I., Decker, S., Allemang, D., Preist, C., Schwabe, D., Mika, P., Uschold, M., Aroyo, L.M. (eds.) ISWC 2006. LNCS, vol. 4273, pp. 242–257. Springer, Heidelberg (2006)
9. Euzenat, J., Shvaiko, P.: Ontology Matching. Springer, Heidelberg (2007)
10. Ghazvinian, A., Noy, N.F., Jonquet, C., Shah, N., Musen, M.A.: What Four Million Mappings Can Tell You about Two Hundred Ontologies. In: Bernstein, A., Karger, D.R., Heath, T., Feigenbaum, L., Maynard, D., Motta, E., Thirunarayan, K. (eds.) ISWC 2009. LNCS, vol. 5823, pp. 229–242. Springer, Heidelberg (2009)
11. Hawalah, A., Fasli, M.: A Graph-based Approach to Measuring Semantic Relatedness in Ontologies. In: International Conference on Web Intelligence, Mining and Semantics, pp. 29:1–29:12. ACM, New York (2011)
12. Hu, W., Chen, J., Zhang, H., Qu, Y.: How Matchable Are Four Thousand Ontologies on the Semantic Web. In: Antoniou, G., Grobelnik, M., Simperl, E., Parsia, B., Plexousakis, D., De Leenheer, P., Pan, J. (eds.) ESWC 2011, Part I. LNCS, vol. 6643, pp. 290–304. Springer, Heidelberg (2011)
13. Jonquet, C., Musen, M.A., Shah, N.H.: Building a Biomedical Ontology Recommender Web Service. J. Biomed. Semant. 1(suppl.1), S1 (2010)

14. Lee, H.-T., Leonard, D., Wang, X., Loguinov, D.: IRLbot: Scaling to 6 Billion Pages and Beyond. In: 17th International Conference on World Wide Web, pp. 427–436. ACM, New York (2008)

15. Lewen, H., d'Aquin, M.: Extending Open Rating Systems for Ontology Ranking and Reuse. In: Cimiano, P., Pinto, H.S. (eds.) EKAW 2010. LNCS(LNAI), vol. 6317, pp. 441–450. Springer, Heidelberg (2010)

16. Lv, Y., Moon, T., Kolari, P., Zheng, Z., Wang, X., Chang, Y.: Learning to Model Relatedness for News Recommendation. In: 20th International Conference on World Wide Web, pp. 57–66. ACM, New York (2011)

17. Maedche, A., Staab, S.: Measuring Similarity between Ontologies. In: Gómez-Pérez, A., Benjamins, V.R. (eds.) EKAW 2002. LNCS (LNAI), vol. 2473, pp. 251–263. Springer, Heidelberg (2002)

18. Martínez Romero, M., Vázquez -Naya, J.M., Munteanu, C.R., Pereira, J., Pazos, A.: An Approach for the Automatic Recommendation of Ontologies Using Collaborative Knowledge. In: Setchi, R., Jordanov, I., Howlett, R.J., Jain, L.C. (eds.) KES 2010. LNCS, vol. 6277, pp. 74–81. Springer, Heidelberg (2010)

19. Mazuel, L., Sabouret, N.: Semantic Relatedness Measure Using Object Properties in an Ontology. In: Sheth, A., Staab, S., Dean, M., Paolucci, M., Maynard, D., Finin, T., Thirunarayan, K. (eds.) ISWC 2008. LNCS, vol. 5318, pp. 681–694. Springer, Heidelberg (2008)

20. Mohammad, S., Hirst, G.: Distributional Measures of Concept-distance: A Task-oriented Evaluation. In: 2006 Conference on Empirical Methods in Natural Language Processing, pp. 35–43. ACL, Sydney (2006)

21. Nikolov, A., Motta, E.: Capturing Emerging Relations between Schema Ontologies on the Web of Data. In: 1st International Workshop on Consuming Linked Data. CEUR Workshop Proceedings (2010)

22. Noy, N.F., Guha, R., Musen, M.A.: User Ratings of Ontologies: Who Will Rate the Raters? In: 2005 AAAI Spring Symposium, pp. 56–63. The AAAI Press, Menlo Park (2005)

23. Pirró, G., Euzenat, J.: A Feature and Information Theoretic Framework for Semantic Similarity and Relatedness. In: Patel-Schneider, P.F., Pan, Y., Hitzler, P., Mika, P., Zhang, L., Pan, J.Z., Horrocks, I., Glimm, B. (eds.) ISWC 2010, Part I. LNCS, vol. 6496, pp. 615–630. Springer, Heidelberg (2010)

24. Stoilos, G., Stamou, G., Kollias, S.: A String Metric for Ontology Alignment. In: Gil, Y., Motta, E., Benjamins, V.R., Musen, M.A. (eds.) ISWC 2005. LNCS, vol. 3729, pp. 624–637. Springer, Heidelberg (2005)

25. Tempich, C., Volz, R.: Towards a Benchmark for Semantic Web Reasoners-An Analysis of the DAML Ontology Library. In: 2nd International Workshop on Evaluation of Ontology-based Tools. CEUR Workshop Proceedings (2003)

RELIN: Relatedness and Informativeness-Based Centrality for Entity Summarization

Gong Cheng[1], Thanh Tran[2], and Yuzhong Qu[1]

[1] State Key Laboratory for Novel Software Technology, Nanjing University,
Nanjing 210093, China
[2] Institute AIFB, Karlsruhe Institute of Technology, D-76131 Karlsruhe, Germany
{gcheng,yzqu}@nju.edu.cn, ducthanh.tran@kit.edu

Abstract. Linked Data is developing towards a large, global repository for structured, interlinked descriptions of real-world entities. An emerging problem in many Web applications making use of data like Linked Data is how a lengthy description can be tailored to the task of quickly identifying the underlying entity. As a solution to this novel problem of entity summarization, we propose RELIN, a variant of the random surfer model that leverages the relatedness and informativeness of description elements for ranking. We present an implementation of this conceptual model, which captures the semantics of description elements based on linguistic and information theory concepts. In experiments involving real-world data sets and users, our approach outperforms the baselines, producing summaries that better match handcrafted ones and further, shown to be useful in a concrete task.

Keywords: Distributional relatedness, entity summarization, informativeness, PageRank, random surfer model.

1 Introduction

Linked Data can be conceived as a large collection of entity descriptions. As descriptions evolve on the Linked Data Web, they are linked to others. The result is that descriptions become increasingly lengthy. Already today, lengthy descriptions can be found in many existing data sets. For instance, the latest version of the well-known DBpedia data set[1] describes 3.5 million entities with 672 million facts (i.e. RDF triples). This means each entity description is associated with an average of 192 RDF triples. Lengthy descriptions take long time for human users to read, which is unacceptable in tasks that require quick identification of the underlying entities. For example during entity search [5,14], users want to quickly browse through search results to identify the ones that match a given information need. Another task is pay-as-you-go data integration [11,19], where users evaluate entity mappings computed by the matching system by identifying the referred entities and judging whether they denote the same thing. To improve the efficiency of these tasks, we aim at solving this novel problem that we

---

[1] http://dbpedia.org/

L. Aroyo et al. (Eds.): ISWC 2011, Part I, LNCS 7031, pp. 114–129, 2011.

call *entity summarization* to produce a version of the original description that is more concise, yet containing sufficient information for users to quickly identify the underlying entity.

The more general problem of *data summarization* has been studied by different communities. For example, *database* [2] and *graph summarization* [13] compute compact representations of data that generalize the original data elements (e.g. cells in a data cube, or a graph) to a more coarse-grained level (e.g. dimension-based regions, or an aggregated graph). That is, data elements are categorized, and then are compactly represented using the resulting categories. However, this is proposed for lossless or lossy (but with bounded errors) data representation, which is distinct from the summary pursued in our problem of entity summarization that is for facilitating quick identification of the underlying entity, or in other words, for helping to efficiently distinguish one entity from others. Thereby, rather than categorization, a solution needed here could be a way of selecting a few central data elements that are most useful in characterizing an entity. This is more similar to *extractive text* [8] and *ontology summarization* [20], the goal of which is to find the central topics of the given data (e.g. a document or an ontology). Unlike categories in database and graph summaries, a topic here is an element extracted from the original data, e.g. a text sentence or an ontology element. To find central elements, the notion of centrality is often employed. Existing approaches [8,20] mainly simulate a random surfer's behavior (as in PageRank [15]), and incorporate data elements that are most likely to be visited by the surfer into the summary. We follow this line of research in our work.

To summarize, we propose to look at this novel (1) *problem of entity summarization*. In this first (to the best of our knowledge) solution to the problem, we elaborate on (2) *a variant of the random surfer model*. This well-known model is used as the basis to support the idea of incorporating central elements into the summary. However, it is revised by a more specific notion of centrality, called RELIN, where the computation of central elements involves relatedness (or similarity) between elements as well as their informativeness, i.e. the amount of information carried that helps to identify the entity. It extends the previous idea of capturing the main themes [8,20] that describe the data, to find more specific central elements that identify the data. To this end, instead of a traditional random surfer, we simulate a rather goal-directed surfer that explores an entity description with the aim of identifying the underlying entity. We model two kinds of action, namely relational move and informational jump, that follow non-uniform probability distributions. The surfer, to achieve her goal, prefers related elements when she moves, and prefers informative elements when she jumps. We propose a simple but effective (3) *implementation of these notions of relatedness and informativeness* that exploits the semantic information captured by the graph structure of the data (as in [20]) as well as the labels of nodes and edges. For the latter, we apply well-known linguistic and information theory concepts. We carried out an extensive (4) *empirical study* of the proposed approach. The results show that it significantly outperformed the baseline approaches, both in an intrinsic evaluation based on a comparison with handcrafted summaries, and in an extrinsic evaluation where the computed summaries are used for the task of confirming entity mappings.

The remainder of this paper is organized as follows. The problem is defined in Sect. 2. The approach is detailed in Sect. 3, and an implementation is given in Sect. 4. Related work is discussed in Sect. 5. Experimental results are presented in Sect. 6 before we conclude in Sect. 7.

2   Problem Statement

For the investigated problem, we employ a graph-structured data model corresponding to RDF, which describes entities in the form of attribute values and relations to other entities (collectively called property values). Let E be the set of all *entities*, L the set of all *literals*, and P the set of all *properties*.

Definition 1 (Data Graph). *A* data graph *is a digraph* $G = \langle V, A, \mathrm{Lbl}_V, \mathrm{Lbl}_A \rangle$, *where* V *is a finite set of nodes,* A *is a finite set of directed edges where each* $a \in A$ *has a source node* $\mathrm{Src}(a) \in V$ *and a target node* $\mathrm{Tgt}(a) \in V$, *and* $\mathrm{Lbl}_V : V \mapsto E \cup L$ *and* $\mathrm{Lbl}_A : A \mapsto P$ *are labeling functions that map nodes and edges to entities or literals, and properties, respectively.*

Definition 2 (Feature). *A* feature f *is a property-value pair where* $\mathrm{Prop}(f) \in P$ *and* $\mathrm{Val}(f) \in E \cup L$ *denote the property and the value, respectively. An entity* e *has a feature* f *in a data graph* $G = \langle V, A, \mathrm{Lbl}_V, \mathrm{Lbl}_A \rangle$ *if there exists* $a \in A$ *such that* $\mathrm{Lbl}_A(a) = \mathrm{Prop}(f)$, $\mathrm{Lbl}_V(\mathrm{Src}(a)) = e$ *and* $\mathrm{Lbl}_V(\mathrm{Tgt}(a)) = \mathrm{Val}(f)$.

That is, a feature of an entity corresponds to one of its associated edges in the data graph. We actually consider both incoming and outgoing edges (i.e. where e appears as target and source node, respectively). Without loss of generality, we focus on outgoing edges for the sake of clear presentation.

A feature is regarded as the smallest meaningful description element for an entity, based on which we characterize an entity description as a set of features:

Definition 3 (Feature Set). *Given a data graph* G, *the* feature set *of an entity* e, *denoted by* $\mathrm{FS}(e)$, *is the set of all features of* e *that can be found in* G.

The left part of Fig. 1 depicts the data graph for our running example, which describes a person and one of his publications. Given this data graph, the feature set of the entity ex:Rudi_Studer is shown in the right part of Fig. 1.

Finally, the problem of entity summarization is defined as extracting a subset from a lengthy feature set, subject to a cardinality constraint.

Definition 4 (Entity Summarization). *Given* $\mathrm{FS}(e)$ *and a positive integer* $k < |\mathrm{FS}(e)|$, *the problem of* entity summarization *is to select* $\mathrm{Summ}(e) \subset \mathrm{FS}(e)$ *such that* $|\mathrm{Summ}(e)| = k$. $\mathrm{Summ}(e)$ *is called a* summary *of* e.

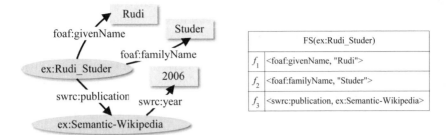

Fig. 1. The feature set of the entity `ex:Rudi_Studer` (on the right), given the data graph (on the left) containing two entities (ellipses) and three literals (rectangles)

In the running example, valid summaries of `ex:Rudi_Studer` include $\{f_1, f_2\}$, $\{f_1, f_3\}$ and $\{f_2, f_3\}$ given $k = 2$. In the next sections, we will introduce an approach to finding a summary from $\binom{|\,\mathrm{FS}(e)|}{k}$ candidates that best characterizes e for quick identification.

It is worth noting that we impose a length constraint on the summary based on the number of features. In fact, the content of individual features may also be a factor that deserves consideration because, for instance, features may contain literals that significantly vary in length. However, this will not be investigated in our work. Besides, we actually concentrate on *what* information (i.e. which features) should be presented, but will not address *how* this information should be presented (e.g. by using visualization or natural language generation methods), although the latter is also an important part of the summarization task.

3   Entity Summarization

We conceive the problem of entity summarization as the one of ranking, i.e. selecting the k top-ranked features from the feature set for a summary. In this sense, entity summarization and feature ranking refer to the same task.

3.1   Centrality-Based Ranking

Centrality-based ranking has been successfully applied to text [8] and ontology summarization [20], and we follow this direction to solve entity summarization. This paradigm requires constructing a graph where nodes correspond to the data elements to be ranked, i.e. sentences in text summarization [8], RDF sentences in ontology summarization [20], and features in entity summarization. Every pair of related nodes are connected by undirected [8] or directed edges [20], and such pairs could be defined based on some numerical relatedness measures with a predefined threshold [8] or problem-specific heuristics [20]. Finally, nodes are ranked according to their centralities in the graph, often computed by using PageRank [15]. Basically, PageRank simulates a surfer, who navigates from node to node, choosing with a uniform probability which edge to follow at each step, and with a small probability, occasionally jumps to a random node; the ranking of nodes is obtained by considering the stationary distribution of such a Markov chain, and a node with a higher probability of being reached by the surfer is ranked higher. In this way, top-ranked nodes (i.e. data elements) are believed to capture the main themes of the original data, since they are central to the original data with regard to the relatedness among data elements.

However, applying the random surfer model like this to our scenario yields two problems. (1) The supported notion of centrality may be too general. Capturing the main themes of the original entity description is not the only goal pursued here. Recall that the summary we are looking for is the one that can best characterize the underlying entity and help to distinguish the entity from others. That is, the measurement of centrality should also give consideration to *how much information a feature carries that can contribute to the identification of the entity.* (2) To apply this random surfer model, edges are added between "significantly related" nodes, where relatedness is actually defined as a boolean-valued function: nodes are either related (and thus connected by an edge) or not (and thus not adjacent). Then, all the adjacent nodes of a node are treated as being equally related to it, since the surfer chooses from them with a uniform probability which one to visit. In other words, the model does not *represent the degree of relatedness on a more fine-grained level.* This imprecision may lead to suboptimal results, particularly when such a boolean-valued function is derived from a relatedness threshold, as it is often the case.

3.2   RELIN: Relatedness and Informativeness-Based Centrality

To remedy the flaws pointed out above, we extend the standard random surfer model as follows. For the first issue, inspired by [10], we propose to embed the measurement of informativeness in the random surfer model. Recall that in the standard model, the surfer jumps to a random node with a given probability. We replace this uniform probability distribution with a non-uniform one that is *dependent on the amount of information carried by each target node that helps to identify the entity.* As a result, a feature that is informative in terms of distinguishing the underlying entity from others will more likely to be reached by the surfer, and thus will be ranked higher. For the second issue, we propose to construct an edge-labeled complete graph, as illustrated in Fig. 2 (solid lines). Then the surfer at a node chooses which edge to follow not with a uniform probability but with a probability (derived from the label of the edge) *proportional to the relatedness between the two associated nodes* (i.e. the current node and the target). In this way, we avoid the problem of finding the most appropriate threshold (which is shown to be difficult [8]) and can also fully exploit the computed numerical relatedness values.

To be specific, we propose RELIN, a variant of the random surfer model that measures RELatedness and INformativeness-based graph centrality for entity summarization. Similar to the standard model in PageRank, we simulate a random surfer's behavior using two kinds of action, one called *relational move* and

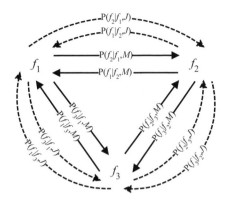

Fig. 2. A graph under the RELIN random surfer model, where nodes represent features, solid lines represent relational moves (i.e. edges) between features, and dashed curves represent informational jumps between features. Each action is associated with a non-uniform probability.

the other *informational jump*. This hypothetical surfer is a goal-directed one that navigates through a feature set in order to identify the underlying entity. To this end, the surfer either performs a relational move — more likely to a feature that carries related information about the theme currently under investigation, or performs an informational jump — more likely to a feature that provides a large amount of new information for clarifying the identity of the underlying entity. These choices are represented by two non-uniform probability distributions, one given by the relatedness between features and the other by the informativeness of features. For the running example, Fig. 2 illustrates the graph under this new random surfer model.

Now we formalize our solution using a general probabilistic framework [7]. The surfer's behavior in RELIN, namely relational move (M) and informational jump (J), is defined with respect to the current feature f_q:

- $P(M|f_q)$: the probability of performing a relational move from f_q, and
- $P(J|f_q)$: the probability of performing an informational jump from f_q.

There exist only two kinds of action, and thus they satisfy $P(M|f_q)+P(J|f_q) = 1$. Then both actions are defined with targets:

- $P(f_p|f_q, M)$: the probability of performing a relational move from feature f_q to feature f_p, and
- $P(f_p|f_q, J)$: the probability of performing an informational jump from feature f_q to feature f_p.

These sets of probabilities must satisfy the following normalization constraints for each $f_q \in$ FS, where FS is the feature set under consideration:

- $\sum_{f_p \in \text{FS}} P(f_p|f_q, M) = 1$, and

$$- \sum_{f_p \in FS} P(f_p | f_q, J) = 1.$$

Let $\mathbf{x}(t)$ be a $|FS|$-dimensional vector where $\mathbf{x}_p(t)$ is the probability that the surfer visits feature f_p at step t. By taking all the possibilities of the surfer's behavior into account, the probability $\mathbf{x}_p(t+1)$ is updated as follows:

$$\mathbf{x}_p(t+1) = \sum_{f_q \in FS} \mathbf{x}_q(t) \cdot \left(P(M | f_q) \cdot P(f_p | f_q, M) \right.$$
$$\left. + P(J | f_q) \cdot P(f_p | f_q, J) \right). \tag{1}$$

All the above probabilities defining the RELIN random surfer model can be organized into the following $|FS| \times |FS|$ matrices:

- $\mathbf{M}$, where $\mathbf{M}_{p,q} = P(f_p | f_q, M)$,
- $\mathbf{J}$, where $\mathbf{J}_{p,q} = P(f_p | f_q, J)$,
- $\mathbf{\Delta}$, a diagonal matrix where $\mathbf{\Delta}_{q,q} = P(M | f_q)$, and
- $\mathbf{\Lambda}$, a diagonal matrix where $\mathbf{\Lambda}_{q,q} = P(J | f_q)$.

Then (1) can be rewritten as:

$$\mathbf{x}(t+1) = (\mathbf{M} \cdot \mathbf{\Delta} + \mathbf{J} \cdot \mathbf{\Lambda}) \cdot \mathbf{x}(t). \tag{2}$$

It has been proved [7] that:

$$\lim_{t \to \infty} \mathbf{x}(t) = \mathbf{x}^*, \tag{3}$$

where $\mathbf{x}^*$ is a constant vector that does not depend on the initial distribution $\mathbf{x}(0)$, if $P(J | f_q) \neq 0$ and $P(f_p | f_q, J) \neq 0$ for every $f_p, f_q \in FS$. In practice, the iterative computation of (2) is usually configured to stop after a certain number of iterations.

Finally, features in FS are ranked by $\mathbf{x}^*$. That is, feature f_p will be ranked higher than feature f_q if $\mathbf{x}_p^* > \mathbf{x}_q^*$.

To implement this model, we need to give $\mathbf{M}$, $\mathbf{J}$, $\mathbf{\Delta}$ and $\mathbf{\Lambda}$, i.e., to define $P(f_p | f_q, M)$, $P(f_p | f_q, J)$, $P(M | f_q)$ and $P(J | f_q)$ for every $f_p, f_q \in FS$. This will be discussed in the next section.

4   Implementation

Firstly, we define $\mathbf{\Delta}$ and $\mathbf{\Lambda}$ as follows:

$$\mathbf{\Delta}_{q,q} = 1 - \lambda, \quad q = 1, \ldots, |FS|,$$
$$\mathbf{\Lambda}_{q,q} = \lambda, \quad q = 1, \ldots, |FS|, \tag{4}$$

where $\lambda \in [0, 1]$ is regarded as a parameter to be tuned and tested in experiments. Here we actually follow PageRank to assume that the surfer has a consistent probability of choosing between move and jump.

Next, as a general strategy for computing $\mathbf{M}$, the relatedness between features, and $\mathbf{J}$, the informativeness of features, we propose to exploit the information captured by the labels of nodes and edges in the original data graph. We now discuss one possible implementation.

4.1 Relatedness

Two features, i.e. property-value pairs, are related when they have related properties, e.g. "has paper" and "has research interest" which are both about academic information, or they have related values, e.g. a paper titled "Semantic Wikipedia" and a research interest "Semantic Web". Thus, we define the relatedness between two features as a combination of the relatedness (denoted by Rel) between their properties and the relatedness between their values (subjected to the mentioned probability normalization):

$$\mathbf{M}_{p,q} = \sqrt{\mathrm{Rel}(\mathrm{Prop}(f_p), \mathrm{Prop}(f_q)) \cdot \mathrm{Rel}(\mathrm{Val}(f_p), \mathrm{Val}(f_q))}. \qquad (5)$$

As potential implementations of Rel, various notions of relatedness have been proposed in the literature. Among others, a well-known line of research [3] employs semantic network such as WordNet[2] to measure the relatedness between text phrases, usually based on the length of the shortest path between their corresponding nodes in the network. However, it is difficult to find such a source of background knowledge that has a good coverage of all the properties and values in our problem that might be encountered in practice. Therefore, we employ another notion called *distributional relatedness* [12], i.e., two phrases are more related if they more often co-occur in certain contexts (e.g. documents). We use an implementation called *Pointwise Mutual Information* (PMI). Let $\mathrm{P}(s_i)$ be the probability that phrase s_i occurs in a document, which could be estimated by counting throughout a corpus. Analogously, let $\mathrm{P}(s_i, s_j)$ be the joint probability of phrase s_i and phrase s_j. Their PMI is defined as follows:

$$\mathrm{PMI}(s_i, s_j) = \log \frac{\mathrm{P}(s_i, s_j)}{\mathrm{P}(s_i) \cdot \mathrm{P}(s_j)}. \qquad (6)$$

Note that to estimate this for every edge in a graph under the RELIN random surfer model, we need to obtain probabilities for every entity, literal and property that might be mentioned in the nodes (i.e. features). Achieving this coverage requires a large and diverse corpus. To this end, we leverage the Google search engine to obtain the contexts in which phrases may co-occur. Let $\mathrm{Hits}(s_i)$ be the number of documents returned by the search engine that match phrase s_i (which could be the name of a property or an entity, or the lexical form of a literal), and N a predefined normalizing constant. Then, we estimate $\mathrm{P}(s_i, s_j)$, $\mathrm{P}(s_i)$ and $\mathrm{P}(s_j)$ by computing $\frac{\mathrm{Hits}(s_i, s_j)}{N}$, $\frac{\mathrm{Hits}(s_i)}{N}$ and $\frac{\mathrm{Hits}(s_j)}{N}$, respectively.

For instance, in the running example, f_1 is more related to f_2 than to f_3, mainly because their property names "given name" and "family name" have a higher PMI than "given name" and "publication" have.

It is worth noting that the use of a Web search engine could become a performance bottleneck that limits a practical summarization system. Solutions include completing all (or most) potential queries prior to placing the system in service, or using a local corpus instead.

---

[2] http://wordnet.princeton.edu/

4.2   Informativeness

We use a well-known information theory concept to measure the informativeness of features. Given o an outcome of a random variable with probability $P(o)$, its *self-information* is defined as:

$$\text{SelfInfo}(o) = -\log(P(o)).\tag{7}$$

That is, the smaller probability an outcome has, the more information its occurrence provides. In this sense, according to the RELIN random surfer model, we should look at $P(f_p|f_q)$ — the probability that feature f_p belongs to a feature set, given f_q belongs to the same one. This can be estimated via a statistical analysis of the original data graph:

$$P(f_p|f_q) = \frac{|\{e \in E \mid f_p, f_q \in \text{FS}(e)\}|}{|\{e \in E \mid f_q \in \text{FS}(e)\}|}.\tag{8}$$

Then, the amount of new information that the surfer obtains by performing an informational jump from f_q to f_p is measured by (subjected to the mentioned probability normalization):

$$\mathbf{J}_{p,q} = \text{SelfInfo}(f_p|f_q) = -\log(P(f_p|f_q)).\tag{9}$$

For instance, given f_1 in the running example, in terms of distinguishing the underlying entity from others, f_3 is more informative than f_2 because there is only one Rudi that is an author of the publication "Semantic Wikipedia", but there are probably more than one Rudi whose family name is Studer.

It is worth noting that, when computing $\text{SelfInfo}(f_p|f_q)$ between all pairs of features is too costly in practice, $\text{SelfInfo}(f_p)$ can be used as an approximation. This would assume that the informativeness of one feature is independent from the information provided by other features such that for informational jump, only the information of the targets plays a role.

5   Related Work

Summarization methods can be classified into *extractive* and *non-extractive* ones. Most text [18] and ontology summarization [16,20] work employs the more popular extractive strategies, which produce a summary by choosing a subset from the original data elements. We follow it by conceiving an entity description to be summarized as a feature set. On the other hand, database [2] and graph summarization [13] usually adopt the non-extractive paradigm, and define the notion of summary on a level that is more coarse-grained than the original data.

The adoption of a PageRank-like [15] graph centrality measure for entity summarization here is motivated by research in related fields. Firstly, for the closely related problem of text summarization, *centrality*-based methods (e.g. [8]) have proven to be superior to those simple *centroid*-based statistical methods (e.g. [17]), which basically rank data elements according to their relatedness to

the "centroid" of the entire data. A detailed theoretical and empirical comparison of these two styles is given in [8]. Secondly, among various notions of graph
centrality, we prefer a *principled PageRank-like measure* mainly because PageR-
ank has shown a better performance than its competitors (e.g. degree-based
measures) in the experiments on previous summarization tasks [8,20].

Although the proposed RELIN approach also builds upon the random surfer
model [15] as in previous methods [8,16,20], it is about computing central elements (i.e. features) that not just represent the main themes of the original
data, but rather, can best *identify the underlying entity*. Thus, instead of a traditional random surfer, we simulate a goal-directed one that has a *preference for
related and informative features*. Further, different from the standard model, the
surfer's behavior in RELIN is characterized by *non-uniform probability distributions*. This idea of assigning different weights to different actions in the random
surfer model has been investigated, among others, for dealing with the problem of Web search [10]. Besides dealing with a different problem, the proposed
approach also uses information that is completely different from the notions of
relatedness and informativeness implemented in our work.

Complementary to centrality, *diversity* [4] is another popular metric for evaluating a summary, by measuring its coverage of themes in the original data. In
fact, this diversity aspect is partially supported by our implementation, since
according to Sect. 4.2, informational jumps are dependent on the amount of
"new" information. However, this matter is not elaborated in the paper because
it is orthogonal to our concern in the sense that it can be easily incorporated
into our approach as a re-ranking step, as proposed before [4,20].

From another point of view, our work is also related to the topic of *ranking in RDF graphs*, for which different ideas have been studied. For instance,
[6] performs a hierarchical link analysis for ranking entities; [9] applies tensor
decomposition to find latent aspects of the data and generate aspect-specific
rankings; [1] ranks associations (i.e. paths) between entities by means of a wide
range of customizable metrics. However, none of these approaches is well fit for
the problem of entity summarization here, which requires ranking data elements
according to *how much they help identify the underlying entity*. In this respect,
[16,20] are the most related work: basically, an RDF graph is decomposed into
a set of subgraphs called "RDF sentences"; based on the common nodes they
share, links are defined between them, from which a new graph with nodes
representing RDF sentences is derived; then, various graph centrality measures
(e.g. PageRank) are applied to this new graph for ranking. In comparison, our
work leverages the information contained in the *labels of nodes and edges in the
original data graph*. This goes beyond [16,20], which mainly employ the *graph
structure* for ranking.

6   Experiments

In the experiments, two real-world data graphs were used: (1) the English version
of the DBpedia 3.4 core data sets, which collectively contain 124,404,962 RDF

triples,[3] and (2) the December 2009 Link Export of the Freebase data set,[4] which contains 325,158,504 RDF triples.[5] Both are domain-independent encyclopedic data sets, which belong to the largest that have been made publicly available on the Web as part of the Linked Data initiative. They cover a broad range of descriptions of entities such as people, cities and music albums.

We implemented RELIN as described previously in the paper. For the parameter λ in (4), which assigns the importance of informational jump relative to relational move, we tested 5 different values, namely 0.00, 0.15, 0.50, 0.85 and 1.00. Particularly, with $\lambda = 0.00$, our approach relies only on relatedness between features, which can then be regarded as an application of traditional text summarization methods (e.g. [8]) to entity summarization. On the contrary, with $\lambda = 1.00$, it amounts to one that employs informativeness of features only. Besides, the iterative computation in RELIN was set to stop after 10 iterations.

We intend to compare our approach with other work on ranking RDF data. However, as discussed in Sect. 5, no existing method is well-suited to our problem. Thus, to establish baselines, on the one hand, we implemented *OntoSum*, which is an adaptation of the most related approach given in [20] to our problem of ranking features.[6] To be specific, given the data graph comprising all the features of the entity under consideration, the notion of RDF sentence proposed in [20] amounts to one single RDF triple, which further corresponds to a feature of the entity. Thereby, a ranking of RDF sentences produced by [20] would naturally induce a ranking of features. On the other hand, we also implemented *RandomRank* that always produces a random ranking of features.

We ran two independent evaluations. In an *intrinsic* one, automatically computed summaries were compared with ideal ones. The other *extrinsic* evaluation aimed at investigating the usefulness of the summaries in a practical task.

6.1   Intrinsic Evaluation

In the intrinsic evaluation, 24 participants (comprising graduate and undergraduate students majoring in computer science) were invited to manually construct ideal entity summaries as the gold standard. A sample of 149 entities were selected at random from DBpedia under the constraint that the cardinality of each one's feature set is inside the interval [20,40], such that it is neither too small to be significant for a summarization task nor too lengthy for manual investigation. Then, each entity was randomly assigned to an average of 4.43 participants; given an entity description presented as a list of features sorted in random order, a participant was asked to return two ideal summaries — one containing 5 features and the other containing 10 — that could best clarify the identity

---

[3] The data sets *Links to Wikipedia Article* and *External Links* were not imported since they are less relevant to the summarization task.
[4] http://www.freebase.com/
[5] The RDF triples that involve non-English literals were removed since our participants cannot read.
[6] Among several centrality measures compared in [20], we chose the best-preforming one, namely PageRank.

of the underlying entity. That is, given $k \in \{5, 10\}$ and an entity e, we would receive, from n different participants, n ideal summaries, denoted by $\mathrm{Summ}_i^I(e)$ for $i = 1, \ldots, n$.

Firstly, we report the level of agreement between ideal summaries. Given $k \in \{5, 10\}$, an entity e and n ideal summaries received, their *agreement* is defined by their average overlap:

$$\frac{2}{n(n-1)} \sum_{i=1}^{n} \sum_{j=i+1}^{n} |\mathrm{Summ}_i^I(e) \cap \mathrm{Summ}_j^I(e)| . \tag{10}$$

In our experiments, when $k = 5$, the agreement averaged over all the entities is 2.91 features, and when $k = 10$, the overall agreement is 7.86. These indicate a *significant level of agreement between participants about ideal summaries.*

Secondly, based on ideal summaries, for a summary automatically computed, denoted by $\mathrm{Summ}(e)$, its *quality* is evaluated by looking at the average overlap between $\mathrm{Summ}(e)$ and each $\mathrm{Summ}_i^I(e)$:

$$\mathrm{Quality}(\mathrm{Summ}(e)) = \frac{1}{n} \sum_{i=1}^{n} |\mathrm{Summ}(e) \cap \mathrm{Summ}_i^I(e)| . \tag{11}$$

Table 1 presents the quality of summaries computed under each approach setting, averaged over all the entities. The best quality values are highlighted.

Table 1. Quality of summaries computed under each approach setting

	$k = 10$	$k = 5$
OntoSum	3.69	1.01
RandomRank	3.36	0.76
RELIN, with $\lambda = 0.00$	3.58	1.61
RELIN, with $\lambda = 0.15$	3.84	1.73
RELIN, with $\lambda = 0.50$	4.40	1.99
RELIN, with $\lambda = 0.85$	**4.88**	2.29
RELIN, with $\lambda = 1.00$	4.86	**2.40**

Our approach, under almost all the tested values of λ, outperformed the two baselines. When $k = 10$, compared with *OntoSum* and *RandomRank*, the quality achieved by RELIN increases by up to 32.2% and 45.2%, respectively. When $k = 5$, the highest increases are 137.6% and 215.8%, respectively. These results suggest that w.r.t. entity summarization, *our approach is clearly superior to the most related competitor in the literature, and both are better than a random selection.*

By testing different values of λ, we found that informativeness (i.e. when $\lambda = 1.00$) is more effective than relatedness (i.e. when $\lambda = 0.00$), particularly in generating extremely short summaries ($k = 5$), where it achieved the best result. That means, it seems the participants preferred only to jump from one informative feature to one another. However, it changed when the summaries became longer. For instance, when $k = 10$, the best result was achieved when $\lambda = 0.85$, i.e. the use of informativeness in combination with relatedness. Thus, these results suggest that *whereas both actions are useful, the choice of λ should be tested and tuned in experiments, and the defined length of summary is one important factor that determines this trade-off.*

6.2   Extrinsic Evaluation

In the extrinsic evaluation, 19 participants (comprising graduate and under-graduate students majoring in computer science) were invited to confirm entity mappings. That is, given two (summaries of) descriptions, a participant was asked to judge whether they refer to the same underlying entity. The accuracy and efficiency of these judgments would reflect the usefulness of automatically computed summaries when being applied to assist users in this particular task. A sample of 47 pairs of entities were used in the experiments. Each pair, consisting of one entity from DBpedia and the other from Freebase, is either correct (i.e. referring to the same real-world entity) or incorrect (i.e. referring to different real-world entities). These pairs were constructed as follows. Firstly, a sample of 47 entities were selected at random from DBpedia under the constraint that when submitting each one's name as a keyword query against the Freebase search engine, at least two entities could be retrieved. Then, the DBpedia entity and one Freebase entity randomly selected from the top-2 search results formed an entity mapping. If such a mapping could be found in the DBpedia extended data set *Links to Freebase*, which explicitly defines entity mappings (in the form of `owl:sameAs` relation) between DBpedia and Freebase, it was deemed correct, or otherwise incorrect. In this way, we obtained 24 correct mappings and 23 incorrect ones. These judgments were used as gold-standard answers.

For each mapping, under each of the five approach settings as shown in Table 2, the two entity descriptions were summarized and then were randomly and blindly assigned to an average of 3.62 participants to judge. Each summary was presented as a list of features sorted by their ranking values. In particular, under the *ReturnsAll* setting, all the features in a description would be presented in random order without summarization. To compare different approach settings, we examined the (1) *accuracy* of the judgments made by the participants, and the (2) *time* they spent. The accuracy of a judgment is 1.0 if it coincides with the gold standard, or otherwise 0.0. To eliminate the difference in participants' intrinsic efficiency,[7] before aggregation, every time value spent by a participant was normalized by the average time per judgment spent by this participant. In this sense, 1.0 would mean medium efficiency, when smaller values indicate higher efficiency. Table 2 summarizes the experimental results averaged over all the mappings and participants, where better results are highlighted.

By looking at the three settings under the same $k = 5$, we found *our approach achieved the highest accuracy*, whereas *OntoSum* performed even worse

---

[7] For example, an inefficient participant would unfairly increase the overall time spent under those approach settings that she was involved in.

Table 2. Accuracy and time for judgments using summaries computed under each approach setting

	k	Accuracy	Time
OntoSum	5	0.56	**0.84**
RandomRank	5	0.60	**0.87**
RELIN, with $\lambda = 0.85$	5	**0.70**	**0.92**
RELIN, with $\lambda = 0.85$	10	**0.68**	1.12
ReturnsAll	n/a	0.60	1.41

than *RandomRank*. Considering the other two settings as well, we can see the effect of summary length on the time spent: with summarization enabled (i.e. other than *ReturnsAll*), the time is significantly shorter. This corresponds to our expectation that *participants' efficiency in carrying out tasks can be improved when using concise entity descriptions.* By comparing the results of RELIN under $k = 5$ and $k = 10$, we further found that even when being generated by the same approach, *longer summaries required noticeably more time.*

An interesting finding reflected by the last three rows of Table 2 is that, with longer summary length, although the time increases as expected, the accuracy actually decreases. That is, the *accuracy of judgments does not positively correlate with the amount of presented data.* Many participants reported in post-experiment interviews that it was because they got rather lost when facing a large amount of (often low-quality and confusing) information. This indicates that *providing a concise entity description could also improve the user experience and effectiveness (e.g. accuracy here).*

6.3   Discussion

Although our approach performed better than the baselines, the results are still far from perfect. For example, in the intrinsic evaluation, the overall levels of agreement between a computed summary and an ideal summary (i.e. quality) are 4.88 and 2.40 at best when $k = 10$ and $k = 5$, respectively, which are much lower than the ones between ideal summaries (7.86 and 2.91, respectively). That means, *automatically computed summaries still cannot replace handcrafted ones.*

The experimental results also revealed some factors that deserve consideration when further improving our approach. Firstly, although some features were ranked high because of their high informativeness and notable relatedness, e.g. features that stand for the longitude and latitude of a city, they were not preferred by most participants because the information they carry were deemed too "domain-specific" to be exploited. That is, these features are highly informative for domain experts that can deal with this particular kind of knowledge, but are not as valuable when presented to average users. This suggests a *user-specific notion of informativeness*, which could be implemented by leveraging user profiles or feedback. Secondly, *information redundancy* was observed in entity descriptions, which should be reduced during summarization. For example, the location of a city in DBpedia is usually not only described by the properties "longitude" and "latitude" but also redundantly described by an additional "point" property. Although our implementation has partially addressed the issue of diversity (as discussed in Sect. 5), other strategies are still needed to cope with more general cases. Thirdly, as described in Sect. 2, we focus on ranking and selecting features, rather than presenting. However, several participants reported that they could hardly understand some features, whereas some others suggested that they would prefer to see summaries presented using a richer widget, as opposed to simply a list of features as we did in the experiments. Thus, we can conclude that besides selecting the best features, *methods used for presenting entity summaries also have an impact on the user-perceived quality.*

7   Conclusions and Future Work

We have studied the problem of entity summarization, which is related to but different from the problems of extractive text and ontology summarization, since it is more about identifying the entity that underlies a lengthy description. To this novel problem, we have proposed a solution called RELIN. As a variant of the random surfer model, it is based on non-uniform probability distributions, and embeds informativeness into the traditional relatedness-based centrality measure. We have presented an implementation that rests on the information captured by the labels of nodes and edges in the data graph. It goes beyond related methods for ontology summarization which mainly build upon the graph structure. The experimental results of applying our approach to entity summarization are quite promising. It performs better than the baselines in terms of producing summaries that are closer to handcrafted ideal summaries, and that assist users in confirming entity mappings more accurately.

The experimental results and feedback obtained from the participants have indicated directions for future research. We will study "human factors" in the context of entity summarization. For instance, we will look at user feedback. We are also interested in application-specific entity summaries, such as query-biased summaries for entity search.

Acknowledgments. This work was supported in part by the NSFC under grant 61003018 and 61021062, and from the AIFB part, by the German Federal Ministry of Education and Research (BMBF) under the CollabCloud project grant (01IS0937A-E). We would like to thank Dr. Xiang Zhang, Saisai Gong, and all the participants in the experiments.

References

1. Aleman-Meza, B., Halaschek-Wiener, C., Budak Arpinar, I., Ramakrishnan, C., Sheth, A.P.: Ranking Complex Relationships on the Semantic Web. IEEE Internet Comput. 9(3), 37–44 (2005)

2. Bu, S., Lakshmanan, L.V.S., Ng, R.T.: MDL Summarization with Holes. In: 31st International Conference on Very Large Data Bases, pp. 433–444. ACM, New York (2005)
3. Budanitsky, A., Hirst, G.: Evaluating WordNet-based Measures of Lexical Semantic Relatedness. Comput. Linguist. 32(1), 13–47 (2006)
4. Carbonell, J., Goldstein, J.: The Use of MMR, Diversity-Based Reranking for Reordering Documents and Producing Summaries. In: 21st Annual International ACM SIGIR Conference on Research and Development in Information Retrieval, pp. 335–336. ACM, New York (1998)
5. Cheng, G., Qu, Y.: Searching Linked Objects with Falcons: Approach, Implementation and Evaluation. Int. J. Semant. Web Inf. Syst. 5(3), 49–70 (2009)
6. Delbru, R., Toupikov, N., Catasta, M., Tummarello, G., Decker, S.: Hierarchical Link Analysis for Ranking Web Data. In: Aroyo, L., Antoniou, G., Hyvönen, E., ten Teije, A., Stuckenschmidt, H., Cabral, L., Tudorache, T. (eds.) ESWC 2010, Part II. LNCS, vol. 6089, pp. 225–239. Springer, Heidelberg (2010)
7. Diligenti, M., Gori, M., Maggini, M.: A Unified Probabilistic Framework for Web Page Scoring Systems. IEEE Trans. Knowl. Data Eng. 16(1), 4–16 (2004)
8. Erkan, G., Radev, D.R.: LexRank: Graph-based Centrality as Salience in Text Summarization. J. Artif. Intell. Res. 22, 457–479 (2004)
9. Franz, T., Schultz, A., Sizov, S., Staab, S.: TripleRank: Ranking Semantic Web Data by Tensor Decomposition. In: Bernstein, A., Karger, D.R., Heath, T., Feigenbaum, L., Maynard, D., Motta, E., Thirunarayan, K. (eds.) ISWC 2009. LNCS, vol. 5823, pp. 213–228. Springer, Heidelberg (2009)
10. Haveliwala, T.H.: Topic-Sensitive PageRank: A Context-Sensitive Ranking Algorithm for Web Search. IEEE Trans. Knowl. Data Eng. 15(4), 784–796 (2003)
11. Jeffery, S.R., Franklin, M.J., Halevy, A.Y.: Pay-as-you-go User Feedback for Dataspace Systems. In: 2008 ACM SIGMOD International Conference on Management of Data, pp. 847–860. ACM, New York (2008)
12. Mohammad, S., Hirst, G.: Distributional Measures of Concept-distance: A Task-oriented Evaluation. In: 2006 Conference on Empirical Methods in Natural Language Processing, pp. 35–43. ACL, Sydney (2006)
13. Navlakha, S., Rastogi, R., Shrivastava, N.: Graph Summarization with Bounded Error. In: 2008 ACM SIGMOD International Conference on Management of Data, pp. 419–432. ACM, New York (2008)
14. Nie, Z., Ma, Y., Shi, S., Wen, J.-R., Ma, W.-Y.: Web Object Retrieval. In: 16th International World Wide Web Conference, pp. 81–90. ACM, New York (2007)
15. Page, L., Brin, S., Motwani, R., Winograd, T.: The PageRank Citation Ranking: Bringing Order to the Web. Technical report, Stanford InfoLab (1999)
16. Penin, T., Wang, H., Tran, T., Yu, Y.: Snippet Generation for Semantic Web Search Engines. In: Gómez-Pérez, A., Yu, Y., Ding, Y. (eds.) ASWC 2009. LNCS, vol. 5926, pp. 493–507. Springer, Heidelberg (2009)
17. Radev, D.R., Jing, H., Styś, M., Tam, D.: Centroid-based Summarization of Multiple Documents. Inf. Process. Manag. 40(6), 919–938 (2004)
18. Spärck Jones, K.: Automatic Summarising: The State of the Art. Inf. Process. Manag. 43(6), 1449–1481 (2007)
19. Tran, T., Wang, H., Haase, P.: Hermes: Data Web Search on a Pay-as-you-go Integration Infrastructure. J. Web Semant. 7(3), 189–203 (2009)
20. Zhang, X., Cheng, G., Qu, Y.: Ontology Summarization Based on RDF Sentence Graph. In: 16th International World Wide Web Conference, pp. 707–716. ACM, New York (2007)

Decomposition and Modular Structure of BioPortal Ontologies

Chiara Del Vescovo[1], Damian D.G. Gessler[2], Pavel Klinov[2], Bijan Parsia[1],
Ulrike Sattler[1], Thomas Schneider[3], and Andrew Winget[4]

[1] University of Manchester, UK
{delvescc,bparsia,sattler}@cs.man.ac.uk
[2] University of Arizona, AZ, USA
dgessler@iplantcollaborative.org, pklinov@email.arizona.edu
[3] Universität Bremen, Germany
tschneider@informatik.uni-bremen.de
[4] St. John's College, NM, USA
andrewwinget@gmail.com

Abstract We present the first large scale investigation into the modular
structure of a substantial collection of state-of-the-art biomedical ontologies, namely those maintained in the NCBO BioPortal repository.[1] Using
the notion of Atomic Decomposition, we partition BioPortal ontologies
into logically coherent subsets (atoms), which are related to each other
by a notion of dependency. We analyze various aspects of the resulting
structures, and discuss their implications on applications of ontologies.
In particular, we describe and investigate the usage of these ontology decompositions to extract modules, for instance, to facilitate matchmaking
of semantic Web services in SSWAP (Simple Semantic Web Architecture
and Protocol). Descriptions of those services use terms from BioPortal so
service discovery requires reasoning with respect to relevant fragments
of ontologies (i.e., modules). We present a novel algorithm for extracting
modules from decomposed BioPortal ontologies which is able to quickly
identify atoms that need to be included in a module to ensure logically
complete reasoning. Compared to existing module extraction algorithms,
it has a number of benefits, including improved performance and the possibility to avoid loading the entire ontology into memory. The algorithm
is also evaluated on BioPortal ontologies and the results are presented
and discussed.

Keywords: OWL, modularity, atomic decomposition, semantic Web
services, SSWAP.

1 Introduction

State-of-the art biomedical ontologies, e.g., those provided by the NCBO Bio-
Portal, are often maintained as monolithic collections of axioms in single files or

---

[1] http://bioportal.bioontology.org/

L. Aroyo et al. (Eds.): ISWC 2011, Part I, LNCS 7031, pp. 130–145, 2011.
© Springer-Verlag Berlin Heidelberg 2011

in a few files. This is not ideal for applications which require access to individual fragments of ontologies, for example, axioms relevant for a particular term. One example is use of ontology terms in descriptions of Semantic Web services or requests for their discovery. In such cases it is undesirable to load the entire ontology into memory (or transfer it over the network) in order to reason about a limited signature.

Semantic Web Services, such as SSWAP[2] (Simple Semantic Web Architecture and Protocol [8]) or SADI (Semantic Automated Discovery and Integration [14]), offer particular challenges for monolithic ontologies. In this application, semantic Web services reference (and dereference) ontological terms at transaction time–often requiring only a few terms from numerous ontologies in order to complete a transaction between two agents. This creates two challenges specific to ontology decomposition and modularity: 1) Semantic Web services operate under both AAA (Anyone can say Anything about Anything[3]) and the OWA (Open World Assumption). Thus even if service providers had complete knowledge of all BioPortal ontologies before transaction, this could become incomplete at transaction time because service providers could be presented with new terms from new ontologies where said terms could imply arbitrarily complex relations with cached ontologies (e.g., class subsumption or equivalence). This implies new, on-demand reasoning, which places a premium on minimizing the size and complexity of relevant ontologies to those components necessary and sufficient for the transaction at hand; 2) memory and hard disk resources are not limiting for virtually all biological ontologies. But network bandwidth and latency is limiting: large, monolithic ontologies can exceed 10 Mbytes when serialized as RDF/XML. Therefore it is important to investigate the possibility of maintaining ontologies in a more flexible form which supports reasoning over small (from the network's or the reasoner's viewpoint) fragments.

This paper presents the first, to our knowledge, large-scale investigation into decomposability and modular aspects of the NCBO BioPortal ontologies and demonstrates that most of them can be split into *small* logically coherent parts (atoms), from which modules can be efficiently assembled before reasoning. We discuss such good (on average) decomposability of BioPortal ontologies, its implications for applications, and also comment on occasional poor decomposability (Section 3). Finally, we describe a novel algorithm for decomposition-based module extraction (and the auxiliary algorithm for computing minimal seed signatures) and present evaluation results in Section 4.

2   Modularity and Atomic Decomposition

We assume the reader to be familiar with OWL and the underlying Description Logics [1], and sketch here some of the central notions around locality-based modularity [2] and Atomic Decomposition [6]. We use $\mathcal{L}$ for a Description Logic,

---

[2] http://sswap.info
[3] For details see paragraph 2.2.6 of the "RDF: Concepts and Abstract Syntax' document at http://www.w3.org/TR/rdf-concepts#section-anyone

e.g., $\mathcal{SHIQ}$, and $\mathcal{O}, \mathcal{M}$, etc., for an ontology, i.e., a finite set of axioms. Moreover, we respectively use $\widetilde{\alpha}$ or $\widetilde{\mathcal{O}}$ for the signature of an axiom α or of an ontology $\mathcal{O}$, i.e., the set of class, property, and individual names used in α or in $\mathcal{O}$.

Given a set of terms, or *seed signature*, Σ, a Σ-module $\mathcal{M}$ based on deductive-Conservative Extensions [9] is a minimal subset of an ontology $\mathcal{O}$ such that, for all axioms α with terms only from Σ, we have that $\mathcal{M} \models \alpha$ iff $\mathcal{O} \models \alpha$, i.e. $\mathcal{O}$ and $\mathcal{M}$ have the same entailments over Σ. Deciding if a set of axioms is a module in this sense is hard or even impossible for expressive DLs [12], but if we drop the minimality requirement we can define "good sized" approximations, as in the case of *syntactic locality*, or *locality* for short, which can be efficiently extracted. Such modules provide strong logical guarantees by capturing *all* the relevant entailments about Σ, despite not necessarily being minimal subsets of $\mathcal{O}$ with this property [11]. A module extractor is implemented in the OWL API.[4]

Given an ontology $\mathcal{O}$ and a seed signature Σ, we say that an axiom $\alpha \in \mathcal{O}$ is $\perp$-*local* w.r.t. Σ if we can "clearly identify" the result of replacing all terms in α not in Σ with $\perp$ as a tautology; see [2] for a formal definition. Then, a $\perp$-module for Σ contains all axioms that are non-$\perp$-local w.r.t. Σ, plus all those needed to preserve the meaning of terms occurring in these axioms. Similarly we can define $\top$-modules. Additionally, by nesting these two notions until a fixpoint is reached we obtain $\top\perp^*$-modules. Hence, locality-based modules come in 3 flavours, namely $\top, \perp$, and $\top\perp^*$: roughly speaking, a $\top$-module for Σ gives a view "from above" because it contains all subclasses of class names in Σ; a $\perp$-module for Σ gives a view "from below" since it contains all superclasses of class names in Σ; a $\top\perp^*$-module is a subset of both the corresponding $\top$- and $\perp$-modules, containing all entailments to imply that two classes in Σ are in the subclass relation, but not necessarily all their sub- or super-classes. Given a module notion $x \in \{\top, \perp, \top\perp^*\}$, we denote by x-mod$(\Sigma, \mathcal{O})$ the x-module of $\mathcal{O}$ w.r.t. Σ.

In [6] we have introduced a new approach to represent the whole family $\mathfrak{F}_{\mathcal{O}}^x$ of locality-based x-modules of an ontology $\mathcal{O}$. The key point is observing that some axioms appear in a module only if other axioms do. In this spirit, we have defined a notion of "logical dependence" as follows: an axiom α depends on another axiom β if, whenever α occurs in a module $\mathcal{M}$, then β belongs to $\mathcal{M}$, too. Next, we observe that, for each axiom α, the x-module for the signature $\widetilde{\alpha}$ is the smallest x-module containing α; we call α-*module* a module x-mod$(\widetilde{\alpha}, \mathcal{O})$ and denote it by $\mathcal{M}_{\alpha}^x$.

The dependence between axioms allows us to identify clumps of highly inter-related axioms that are never split across two or more modules [6]; these clumps are called *atoms*. More precisely, for $x \in \{\top, \perp, \top\perp^*\}$ an x-*atom* of an ontology $\mathcal{O}$ is a maximal subset of $\mathcal{O}$ which is either contained in, or disjoint with, any x-module of $\mathcal{O}$. The family of x-atoms of $\mathcal{O}$ is denoted by $\mathcal{A}(\mathfrak{F}_{\mathcal{O}}^x)$ and is called x-*Atomic Decomposition (x-AD)*. If x is clear from the context, we drop it.

Since every atom is a set of axioms, and atoms are pairwise disjoint, the AD is a partition of the ontology $\mathcal{O}$. Hence, the number of atoms is at most linear

---

[4] http://owlapi.sourceforge.net

w.r.t. the size of $\mathcal{O}$. Moreover, atoms are the building blocks of all modules [7]. For an atom $\mathfrak{a} \in \mathcal{A}(\mathfrak{F}_{\mathcal{O}}^x)$, the module $\mathcal{M}_{\mathfrak{a}}^x = x\text{-mod}(\tilde{\mathfrak{a}}, \mathcal{O})$ is called *compact*.

Proposition 1. *Let $\mathfrak{a}$ be an atom in the AD $\mathcal{A}(\mathfrak{F}_{\mathcal{O}}^x)$ of an ontology $\mathcal{O}$ and $\alpha \in \mathfrak{a}$; then, for any selection of axioms $\mathcal{S} = \{\alpha_1, \ldots, \alpha_\kappa\} \subseteq \mathfrak{a}$ we have that $x\text{-mod}(\tilde{\mathcal{S}}, \mathcal{O}) = \mathcal{M}_{\alpha}^x$. In particular, for each $\alpha_i \in \mathfrak{a}$, $\mathcal{M}_{\alpha_i}^x = \mathcal{M}_{\alpha}^x$. Vice versa, if $\mathcal{M}_{\alpha}^x = \mathcal{M}_{\beta}^x$, then there exists some $\mathfrak{a}$ such that $\alpha, \beta \in \mathfrak{a}$.*

As a consequence of Prop. 1, the set of compact modules coincides with the set of α-modules, and we denote by $\mathcal{M}_{\mathfrak{a}}$ the module $\mathcal{M}_\alpha$ for each $\alpha \in \mathfrak{a}$. Now, we are ready to extend the definition of logical dependence to atoms. Let $\mathfrak{a}$ and $\mathfrak{b}$ be two distinct atoms of an ontology $\mathcal{O}$. Then, $\mathfrak{a}$ is *dependent* on $\mathfrak{b}$ (written $\mathfrak{a} \succeq \mathfrak{b}$) if $\mathcal{M}_{\mathfrak{b}} \subseteq \mathcal{M}_{\mathfrak{a}}$. The dependence relation $\succeq$ on AD is a partial order (i.e., dependence is transitive, reflexive, and antisymmetric) and thus can be represented by means of a Hasse diagram, i.e. a graph showing the dependencies between its nodes. Moreover, $\succeq$ provides the basis for a polynomial-time algorithm for computing the AD, since they allow us to construct $\mathcal{A}(\mathfrak{F}_{\mathcal{O}}^x)$ via α-modules only [6].

Given the Hasse diagram of an AD, it is easy to get all compact modules of an ontology by considering the *principal ideal* of an atom $\mathfrak{a}$, i.e. the set $(\mathfrak{a}] = \{\alpha \in \mathfrak{b} \mid \mathfrak{a} \succeq \mathfrak{b}\} \subseteq \mathcal{O}$.

Example 2. Consider the ontology $\{\alpha_1, \ldots, \alpha_7\}$ and its $\bot$-AD:

$\alpha_1 = $ 'Animal $\sqsubseteq (= 1\text{hasGender}.\top)$',
$\alpha_2 = $ 'Animal $\sqsubseteq (\geq 1\text{hasHabitat}.\top)$',
$\alpha_3 = $ 'Person $\sqsubseteq$ Animal',
$\alpha_4 = $ 'Vegan $\equiv$ Person $\sqcap \forall\text{eats}.(\text{Vegetable} \sqcup \text{Mushroom})$',
$\alpha_5 = $ 'TeeTotaller $\equiv$ Person $\sqcap \forall\text{drinks}.\text{NonAlcoholicThing}$',
$\alpha_6 = $ 'Student $\sqsubseteq$ Person $\sqcap \exists\text{hasHabitat}.\text{University}$',
$\alpha_7 = $ 'GraduateStudent $\equiv$ Student $\sqcap \exists\text{hasDegree}.\{\text{BA}, \text{BS}\}$'

Here the $\bot$-atoms in the AD contain the following axioms respectively: $\mathfrak{a}_1 = \{\alpha_1, \alpha_2\}$, $\mathfrak{a}_2 = \{\alpha_3\}$, $\mathfrak{a}_3 = \{\alpha_4\}$, $\mathfrak{a}_4 = \{\alpha_5\}$, $\mathfrak{a}_5 = \{\alpha_6\}$, $\mathfrak{a}_6 = \{\alpha_7\}$. The compact module for the atom $\mathfrak{a}_6$ is $\mathcal{M}_{\mathfrak{a}_6} = \mathfrak{a}_1 \cup \mathfrak{a}_2 \cup \mathfrak{a}_5 \cup \mathfrak{a}_6$.

Next, we are interested in modules that do not "fall apart", and thus can be said to have an internal logical coherence. A module is called *fake* if there exist two $\succeq$-uncomparable modules $\mathcal{M}_1, \mathcal{M}_2$ with $\mathcal{M}_1 \cup \mathcal{M}_2 = \mathcal{M}$; a module is called *genuine* if it is not fake. Interestingly, the notions of α-modules, principal ideals of atoms, and genuine modules coincide [6], so from now on we refer to them simply as Genuine Modules (GMs). Note that fake modules are represented in the Hasse diagram of an AD as union of principal ideals of atoms; the converse does not hold: not all combinations of principal ideals of atoms are fake modules.

Whilst getting GMs is an easy task to perform via ADs, extracting a module for a general signature is more complicated. This happens because axioms can pull in a module terms that are not "strictly necessary" for them to be non-local. For example, only axiom α_4 in Ex. 2 is non-$\bot$-local w.r.t. $\Sigma = \{\text{Vegan}\}$. However, each module containing α_4 contains also α_1, α_2, and α_3, because in order to preserve the meaning of Vegan we need first to preserve the meaning of the other terms occurring in this axioms. To guarantee this condition, we need to enlarge Σ with the terms pulled in by relevancy, and then re-check the axioms against relevancy w.r.t. the new signature.

We formalize this idea as follows. We define a *minimal seed signature* for a module $\mathcal{M} = x\text{-mod}(\Sigma, \mathcal{O})$ to be a $\subseteq$-minimal signature Σ' such that $\mathcal{M} = x\text{-mod}(\Sigma', \mathcal{O})$. We denote the set of all minimal seed signatures of a module by $x\text{-mssig}(\mathcal{M}, \mathcal{O})$. We call an atom $\mathfrak{a}$ *relevant for a signature* Σ if there exists $\Sigma' \in x\text{-mssig}(\mathcal{M}_\mathfrak{a}, \mathcal{O})$ such that $\Sigma' \subseteq \Sigma$.

Proposition 3. *Let* $x \in \{\bot, \top\}$ *and* Σ_0 *the input signature. Let us consider* $\mathcal{M}_0^x = \{\alpha \in (\mathfrak{a}] \mid \mathfrak{a}$ *is relevant for* $\Sigma_0\}$ *and, for* $i \geq 1$, $\mathcal{M}_i^x = \{\alpha \in (\mathfrak{a}] \mid \mathfrak{a}$ *is relevant for* $\widetilde{\mathcal{M}_{i-1}^x} \cup \Sigma_0\}$. *Then, the chain of inclusions* $\mathcal{M}_0^x \subsetneq \mathcal{M}_1^x \subsetneq \ldots$ *eventually stops, and denoted by* $\mathcal{M}_*^x$ *the fixpoint, we have that* $\mathcal{M}_*^x = x\text{-mod}(\Sigma_0, \mathcal{O})$.

The procedure described in Prop. 3 is equivalent to the standard extraction of a module only for the two notions $\top$ and $\bot$, because the $\top\bot^*$-AD only partially reflecting dependencies between atoms; see [3] for an example.

In summary, x-atoms and related genuine modules form a basis for all x-locality-based modules. Next, we analyse ADs of existing ontologies and discuss their decomposability.

3   Decomposability of BioPortal Ontologies

Decomposing ontologies into suitable parts is clearly beneficial when it comes to processing, editing, and analyzing them, or to reusing their parts. When ontologies are decomposed automatically, e.g., by computing an AD, it is interesting to discuss and evaluate the suitability of such decomposition for different scenarios, and whether all or which ontologies decompose "well", what it means to decompose well, and which properties of an ontology lead to "good" decomposability.

In this paper we discuss and evaluate the performance of ADs w.r.t. a specific task, i.e. Fast Module Extraction. Suitable application and maintenance scenarios for this task are, as stated before, semantic Web services, or SADI services. We prove that in these cases the AD is generally a good decomposition. On the other hand, "good decomposability" may have a different meaning in other scenarios.

A first such scenario, called Collaborative Ontology Development and Reuse, involves different ontology engineers working on different modules of an ontology. The aim is to minimize the risk of conflicts which could result from two or more ontology engineers making changes to logically related parts of the ontology (i.e., one engineer could be changing the semantics of terms used by another). Modularity provides the notion of "safety" which defines conditions under which there is no such risk [2]. We assume that each engineer works within *their* module and uses other terms in a safe way, and that modules different engineers work on do not overlap. Here, a fine-grained decomposition is desirable.

Another scenario, called Topicality for Ontology Comprehension, is based on the assumption that, in order to enable the understanding of what the ontology deals with, we can search for its "topics" and their interrelations [4]. In this case, a good decomposition should provide a "bird's-eye" view of the topical structure of an ontology. This means that a very fine-grained decomposition is undesirable because it does little to help understanding. On the other hand, large clumps of axioms could aggregate, hence hide, specific topical relations. In this scenario, a good decomposition should be only modestly fine-grained.

We now present the results of decomposing BioPortal ontologies w.r.t. our notions of locality. Due to space restrictions we present only summaries of this results, but full decompositions, spreadsheets with metrics and other data is available online at http://tinyurl.com/modbioportal.

The 3 notions of ADs we use are strongly related since $\top\!\perp^*$-AD is a refinement w.r.t. set inclusion of both $\perp$- and $\top$-AD, see [3]. As a consequence, we expect ontologies to have more, smaller $\top\!\perp^*$-atoms than $\perp$- or $\top$-atoms.

Proposition 4. *The $\top\!\perp^*$-AD is finer than both the $\perp$-AD and $\top$-AD, i.e., for any $\top\!\perp^*$-atom $\mathfrak{a}$, there exists a $\perp$-atom $\mathfrak{b}$ and a $\top$-atom $\mathfrak{c}$ with $\mathfrak{a} \subseteq \mathfrak{b}$ and $\mathfrak{a} \subseteq \mathfrak{c}$.*

The NCBO BioPortal ontology repository contains over 250 bio-medical ontologies, of which 218 are OWL or OBO ontologies. Among these, we filtered out those whose file was corrupted, those that do not contain any logical axioms, and some very large ontologies.[5] The result is a corpus of 181 ontologies, designed and built by domain experts, that vary greatly in size and expressivity [10].

We have decomposed these 181 BioPortal ontologies according to all three notions of syntactic locality: $\perp$, $\top$, and $\top\!\perp^*$. For each decomposition, we compute a basic set of metrics: for each ontology, we compute the average and maximal size of atoms and Genuine Modules (GM) measured in numbers of axioms (axs. in the table), and then we take the average of the resulting numbers over all 181 ontologies. The results are presented in the following table.

Notion of locality	Average average axs./atom	Average maximum axs./atom	Average average axs./GM	Average maximum axs./GM	Average nr. of conn. components
$\top\!\perp^*$	1.73	86	66	143	826
$\perp$	2.19	93	73	156	45
$\top$	330.45	1,417	1,166	2,093	1.64

It can be seen that the $\top\!\perp^*$-AD is generally quite fine-grained: the average size of an atom is less than 2 axioms; indeed, only 54 ontologies out of 181 have at least one atom greater than 10 axioms. Next, $\perp$-AD is fairly, even suprisingly close in granularity to $\top\!\perp^*$-AD as the average atom is only slightly larger than 2 axioms, and all other metrics are surprisingly close. This remark is supported by the Spearman's coefficient [13] comparing the number of atoms per ontology in the $\perp$-AD with the one in the $\top\!\perp^*$-AD. It has a value of $\rho \cong 0.9946$, showing a strong, monotonic correlation between the two measures. Moreover, closer inspection reveals that these two ADs even coincide in 34/181 ontologies. This is interesting for FME, as we will see later.

---

[5] See the technical report [3] for statistics for ontologies with over 20K axioms.

In contrast, $\top$-AD is substantially coarser than both $\top\!\perp^*$ and $\perp$-ADs as the average atom is two orders of magnitude larger, and all other metrics are much larger as well. Given the nature of $\top$-locality [2], this is not surprising, and it supports our general understanding that $\top$-ADs are not a good choice when small size of atoms and modules are relevant. Also, observe that the connectivity of $\top\!\perp^*$-AD is much looser than that of the other two ADs: this reflects the fact that the dependency relation, for $\top\!\perp^*$-AD, only reflects one kind of dependency, which is the reason why a $\top\!\perp^*$ version of Prop. 3 does not hold.

In the majority of the ontologies investigated, we observe rather good decomposability in terms of atom size. There are, however, ontologies that contain abnormally huge atoms even for $\top\!\perp^*$-AD, e.g., over 6K axioms. This is of concern since a module of these ontologies is likely to be of at least that size. For example, in the context of Web services, an attempt to discover a service whose description uses terms from such an atom may require transmitting and reasoning with thousands of axioms, which is undesirable. We observe these huge atoms both in absolute terms, i.e., with more than 200 axioms, and in relative terms, i.e., with more than 50% of axioms of the ontology. In the following table, we list ontologies whose $\top\!\perp^*$-ADs have a huge atom, absolute, relative, or both. We report their size, the size of the maximal atoms, plus some other data that is explained in what follows.

Ontology $\mathcal{O}$ (ID in BioPortal)	#$\mathcal{O}$	#max Atom	#Eq. axs.	#Disj. axs.
Nanoparticle Ontology (1083)	16, 267	6, 425	42	6, 106
Breast Tissue Cell Lines Ontology (1438)	2, 734	2, 201	0	7
IMGT Ontology (1491)	1, 112	729	38	594
SNP Ontology (1058)	3, 481	598	30	210
Amino Acid Ontology (1054)	477	445	8	190
Comparative Data Analysis (1128)	804	434	8	190
Family Health History (1126)	1, 091	378	0	1
Neural Electromagnetic Ontologies (1321)	2, 286	259	21	0
Computer-based Patient Record Ontology (1059)	1, 454	238	18	20
Basic Formal Ontology (1332)	95	89	13	41
Ontology of Medically-related Social Entities (1565)	138	100	17	41
Ontology for General Medical Science (1414)	194	102	17	41
Cancer Research and Mgmt Acgt Master (1130)	5, 435	3, 796	16	42

We carried out a preliminary investigation of ontologies with huge atoms, trying to understand the reasons for the existence of huge atoms. It turns out that some huge atoms are due to the abundance of Disjoint Covering Axioms (DCAs) and we assume that their abundance is due to a specific usage pattern of ontology editors. More precisely, one version of DCAs is a pair of axioms of the form $\{A \equiv (B_0 \sqcup \ldots \sqcup B_n), \text{PairwiseDisjoint}(B_0, \ldots, B_n)\}$. Since our notion of modularity is based on axioms and subsets of an ontology and is self-contained, any module that mentions B_i contains both axioms, and thus pulls in all axioms about B_j as well. When DCAs occur on many classes on all levels in the class hierarchy of an ontology, then this results, unsurprisingly, in a huge atom. Moreover, note that not only disjointness causes axioms to tie together, as the explicit covering axiom shows the same behaviour. For disjointness, however, this "pulling-in" effect does not occur if we rewrite the n-ary disjointness axiom into equivalent pairwise disjointness axioms or even make the disjointness implicit, as in the following example: $\{B_0 \sqsubseteq A \sqcap (= 0\,R.\top), \ldots, B_{n-1} \sqsubseteq A \sqcap (= n-1\,R.\top), B_n \sqsubseteq A \sqcap (\geq n\,R.\top)\}$.

In the previous table, we see that ontologies with huge atoms often have a large number of DCAs in these atoms, as indicated by the number of equivalence class and disjointness axioms in the last two columns: e.g., in the first ontology, which also has the largest atom, almost all axioms in this atom are disjointness axioms; additionally, upon inspection, it turns out that some of the equivalence axioms in this atom are covering axioms involving 10 or more classes. Also, in the second ontology, even though the largest atom only contains 7 disjointness axioms, it turns out that one disjointness axiom contains 52 terms.

The numbers for Comparative Data Analysis and Amino Acid ontologies look very similar because the first ontology imports the second. Trivially, large atoms persist also in the imports closure of an ontology: they can only grow. This is particularly relevant for ontologies that are used as base for others. In our corpus, we indeed find such a basis, which causes other ontologies to decompose badly in the sense described above: the Basic Formal Ontology consists of 95 axioms, 89 of which form an atom, which is due to the abundant usage of DCAs. Among the "relative huge atoms" ontologies, two import the Basic Formal Ontology, and their decomposability is affected.

Other patterns also lead to huge atoms, and an investigation of possible patterns is part of future work.

The last remark about this data concerns its analysis under the viewpoint of scenarios different from semantic Web services. For Collaborative Ontology Development and Reuse, these results are promising since they show a seemingly good decomposability of ontologies for $\top\bot^*$-AD and $\bot$-AD, i.e., the existence of small, disjoint sets of axioms that can be safely updated in parallel. In contrast, in the Topicality for Ontology Comprehension scenario we observe that, when the number of atoms is comparable with the number of axioms, then atoms do not provide any summarization over axioms and we cannot hope that considering atoms can provide any summarization benefit. In this case, the atoms reflect only very fine-grained topics of an ontology [4]. However, the dependency structure reflects the logical dependency between atoms, and thus can be used to consider, e.g., dependent components which, in turn, may better reflect the topics of an ontology. Of course, to really support ontology comprehension, we might have to consider "most relevant" atoms of an ontology [5] and, definitely, suitable labeling of modules. Both directions are part of future work.

4   Labeled Atomic Decomposition and Decomposition-Based Module Extraction

One particular application of atomic decomposition explored in this paper is module extraction. In this section we describe a module extraction algorithm, called FME for "Fast Module Extraction", which is (a) usually faster than the standard ME algorithm and (b) does not require loading the entire ontology into memory.

As explained in Section 2, every module is a union of atoms, however, not every union of atoms is a module. In general, it is non-trivial to determine which atoms the module for a given seed signature Σ consists of. In particular, a seemingly irrelevant atom, whose signature is disjoint with Σ, may turn out to be a part of the module. One way to help determining relevant atoms is to *label* them, i.e., associate them with extra information regarding seed signatures. In this paper we consider a particular kind of labels which, for each atom $\mathfrak{a}$, contains the set of the *Minimal Seed Signatures* $\mathsf{MSS}((\mathfrak{a}])$ (recall that each $(\mathfrak{a}]$ is a module).

Labelling each atom $\mathfrak{a}$ with the minimal seed signatures of its module $\mathsf{MSS}((\mathfrak{a}])$ can have several uses. First, every $\Sigma \in \mathsf{MSS}((\mathfrak{a}])$ can be regarded as a (minimal) topic that determines $(\mathfrak{a}]$ and $\mathfrak{a}$. In this sense, all MSSs of all atoms constitute all relevant minimal topics about which the ontology speaks. This can be exploited for comprehension. The case where atoms have too many MSSs—$(\mathfrak{a}]$ could have up to $2^{\#(\mathfrak{a}]}$ many—is the subject of a representation method that allows the adjustment of granularity and is deferred to future work. Second, the collection of all MSSs guides the extraction of a single module by suggesting possible topics (MSSs as inputs of the extraction algorithm). Again, the number of topics needs to be controlled by adjusting the granularity of the presentation.

4.1   Labeling Algorithm and Evaluation

First, we present an AD-driven algorithm for computing, for each atom $\mathfrak{a}$ in the decomposition, the set of its minimal seed signatures $\mathsf{MSS}((\mathfrak{a}])$. Currently, the algorithm is limited to $\top$ or $\bot$-locality. We plan to extend it to $\top\bot^*$-locality in the future.

Note: in Algorithm 1 the symbol $\cup^*$ means "union and minimization w.r.t. set inclusion". This operator guarantees that every set S of seed signatures does not contain Σ' if $\Sigma \subseteq \Sigma'$ for some $\Sigma \in S$. For example, $\{\Sigma_1, \Sigma_2\} \cup^* \{\Sigma_3, \Sigma_4\}$, where $\Sigma_2 \subset \Sigma_3$, is equal to $\{\Sigma_1, \Sigma_2, \Sigma_4\}$.

Algorithm 1 first computes the set $\mathsf{MGS}(\mathfrak{a})$ (minimal globalizing signatures) for all axioms in $\mathfrak{a}$ (Line 4). For an axiom α and a given notion of locality x, $\mathsf{MGS}(\alpha)$ is the set of all $\Sigma \subseteq \widetilde{\alpha}$ such that α is x-non-local w.r.t. Σ and α is x-local w.r.t. all proper subsets of Σ. For bottom atoms $\mathfrak{a}$ (i.e., atoms which do not depend on other atoms) the sets $\mathsf{MSS}((\mathfrak{a}])$ and $\mathsf{MGS}((\mathfrak{a}])$ coincide.

Now, every signature $\Sigma \in \mathsf{MGS}(\mathfrak{a})$ is necessarily a seed signature for $(\mathfrak{a}]$ but, unless $\mathfrak{a}$ is a bottom atom, is not necessarily minimal. The reason is that $\Sigma' \subset \Sigma$ could be a seed signature for a module $(\mathfrak{b}]$, for some atom $\mathfrak{b} \preceq \mathfrak{a}$ if $\Sigma \subseteq \Sigma' \cup \widetilde{(\mathfrak{b}]}$. In that case, informally, Σ' first "pulls" $(\mathfrak{b}]$ into the module (Σ' being a seed signature for $(\mathfrak{b}]$) and then the extended seed signature $\Sigma' \cup \widetilde{(\mathfrak{b}]}$ "pulls" the axioms of $\mathfrak{a}$ and the rest of $(\mathfrak{a}]$. With "extended seed signature", we mean the seed signature against which locality is checked at some iteration of the standard ME algorithm. Even worse, there could be MSSs for $(\mathfrak{a}]$ which are not subsets of any signature in $\mathsf{MGS}(\mathfrak{a})$ – or not even subsets of $\widetilde{\mathfrak{a}}$, as illustrated in Example 5.

Algorithm 1. Computing MSSs for a principal ideal

1: **Input: Ontology** $\mathcal{O}$**; its AD** x**-mod-AD,** $x \in \{\top, \bot\}$**; atom** $\mathfrak{a}$
2: **Output: MSS($\mathfrak{a}$), the set of all MSSs for** $(\mathfrak{a}]$
3: MSS($\mathfrak{a}$), PreMSS($\mathfrak{a}$) $\leftarrow \emptyset$
4: MGS($\mathfrak{a}$) $\leftarrow \bigcup_{\alpha \in \mathfrak{a}}^* \text{MGS}(\alpha)$
5: DD($\mathfrak{a}$) $\leftarrow$ the set of atoms that $\mathfrak{a}$ non-transitively depends on
6: **if** DD($\mathfrak{a}$) $= \emptyset$ **then**
7:    **return** MGS($\mathfrak{a}$)
8: **end if**
9: **for** each $\mathfrak{b} \in \text{DD}(\mathfrak{a})$ **do**
10:    MSS($\mathfrak{b}$) $\leftarrow$ recursively compute MSSs for $(\mathfrak{b}]$
11: **end for**
12: **for** each $\Sigma \in \text{MGS}(\mathfrak{a})$ **do**
13:    $\text{RC}_\Sigma(\mathfrak{a}) \leftarrow \{\mathfrak{b} \in \text{DD}(\mathfrak{a}) \mid \Sigma \cap \widetilde{(\mathfrak{b}]} \neq \emptyset\}$
14:    **for** each $\{\mathfrak{b}_1, \ldots, \mathfrak{b}_n\} \in \wp(\text{RC}_\Sigma(\mathfrak{a}))$ **do**
15:       $\Sigma_\mathfrak{a} \leftarrow \Sigma \setminus \bigcup_{i=1,\ldots,n} \widetilde{(\mathfrak{b}_i]}$
16:       **for** each $X \in \text{MSS}(\mathfrak{b}_1) \times \cdots \times \text{MSS}(\mathfrak{b}_n)$ **do**
17:          PreMSS($\mathfrak{a}$) $\leftarrow$ PreMSS($\mathfrak{a}$) $\cup^* \{\Sigma_\mathfrak{a} \cup X\}$
18:       **end for**
19:    **end for**
20: **end for**
21: **for** each $\Sigma \in \text{PreMSS}(\mathfrak{a})$ **do**
22:    MSS($\mathfrak{a}$) $\leftarrow$ MSS($\mathfrak{a}$) $\cup^* \{\{\Sigma'\} \mid \Sigma' \subseteq \Sigma$ and $x\text{-mod}(\Sigma', \mathcal{O}) = (\mathfrak{a}]\}$
23: **end for**
24: **return** MSS($\mathfrak{a}$)

Example 5. Let $\mathcal{O} = \{\alpha, \beta, \gamma\}$ with $\alpha = $ 'A $\equiv$ B $\sqcap$ C', $\beta = $ 'B $\equiv$ D $\sqcup$ E', and $\gamma = $ 'C $\equiv$ F $\sqcup$ G)'. Then the following hold:

$$\bot\text{-mod}(\{\mathtt{A}\}, \mathcal{O}) = \bot\text{-mod}(\{\mathtt{B}, \mathtt{C}\}, \mathcal{O}) \qquad\qquad = \{\alpha, \beta, \gamma\}$$
$$\bot\text{-mod}(\{\mathtt{B}\}, \mathcal{O}) = \bot\text{-mod}(\{\mathtt{D}\}, \mathcal{O}) = \bot\text{-mod}(\{\mathtt{E}\}, \mathcal{O}) = \{\beta\}$$
$$\bot\text{-mod}(\{\mathtt{C}\}, \mathcal{O}) = \bot\text{-mod}(\{\mathtt{F}\}, \mathcal{O}) = \bot\text{-mod}(\{\mathtt{G}\}, \mathcal{O}) = \{\gamma\}$$

Therefore, there are three atoms $\mathfrak{a} = \{\alpha\}$, $\mathfrak{b} = \{\beta\}$, $\mathfrak{c} = \{\gamma\}$ with the dependencies $\mathfrak{b} \preceq \mathfrak{a}$ and $\mathfrak{c} \preceq \mathfrak{a}$. Now take the MSS $\{\mathtt{B}, \mathtt{C}\}$ for $\mathfrak{a}$ and replace B and C, which occur in $\mathfrak{b}$ and $\mathfrak{c}$, with the MSSs $\{\mathtt{D}\}$ and $\{\mathtt{F}\}$ for $\mathfrak{b}$ and $\mathfrak{c}$. Then $\{\mathtt{D}, \mathtt{F}\}$ is an MSS for $(\mathfrak{a}] = \mathcal{O}$ although obviously $\{\mathtt{D}, \mathtt{F}\}$ is disjoint with $\widetilde{\mathfrak{a}}$ and with any member of MGS($\mathfrak{a}$).

Despite these complications, axioms of $\mathfrak{a}$ can only be pulled into the module once the extended seed signature includes at least one of the members of MGS($\mathfrak{a}$). The algorithm next recursively computes MSS for all direct children of $\mathfrak{a}$ (Line 10) and then proceeds to discover other MSSs of $(\mathfrak{a}]$ by combining the sets MSS for direct children of $\mathfrak{a}$ with the set MGS($\mathfrak{a}$) (Lines 12–20).

   It does so by "elaborating" each $\Sigma \in \text{MGS}(\mathfrak{a})$. It selects those atoms $\mathfrak{b} \preceq \mathfrak{a}$ which behave as described above, i.e., $\widetilde{(\mathfrak{b}]}$ overlaps with Σ. The set of all such direct children of $\mathfrak{a}$ w.r.t. Σ is stored as $\mathsf{RC}_\Sigma(\mathfrak{a})$ (Line 13). Then the algorithm removes from Σ (the signature being "elaborated") the terms in the "lower" atoms ($\bigcup_{i=1,\dots,n} \widetilde{(\mathfrak{b}_i]}$) and stores the result in $\Sigma_\mathfrak{a}$ (Line 15). Lines 16–18 go through all seed signatures X which are guaranteed to pull every atom in $\mathsf{RC}_\Sigma(\mathfrak{a})$. Then, $X \cup \Sigma_\mathfrak{a}$ is a seed signature (not necessarily minimal) for $(\mathfrak{a}]$, as explained above. All such $X \cup \Sigma_\mathfrak{a}$ are collected in $\mathsf{PreMSS}(\mathfrak{a})$.

The members $\Sigma \in \mathsf{PreMSS}(\mathfrak{a})$ are not guaranteed to be a *minimal* seed signature for $(\mathfrak{a}]$ because of possible weak dependencies between direct children of $\mathfrak{a}$. Informally, there could be a subset of Σ which first pulls some $\mathfrak{b}_i$, then some child of $\mathfrak{b}_i$ and only then $\mathfrak{b}_j$. Therefore, the algorithm has to "minimize" every $\Sigma \in \mathsf{PreMSS}(\mathfrak{a})$ by checking whether any of its subsets are, by themselves, already seed signatures of $(\mathfrak{a}]$ (Lines 21–23). However, entries of $\mathsf{PreMSS}(\mathfrak{a})$ are usually good approximations of truly minimal seed signatures; in particular, they are much better approximations than just the signature of $(\mathfrak{a}]$.

4.2   Properties of the Labeling Algorithm

The correctness of Algorithm 1 is established in [3]. It requires time exponential in the size of the ontology in the worst case, see the discussion in [3]. Despite the worst-case intractability the algorithm has the *anytime* property: the loops for elaborating (Lines 14–21) and minimizing (Line 21–23) a seed signature could be interrupted upon time-out, which will result in computing some subset of the MSS set for an atom.[6] This allows for practical approximations in the case when computing all MSS takes too long. We call atoms whose labels do not contain all MSS *dirty* (other atoms are called *clean*).

Dirty atoms require special handling during module extraction because their relevance may not be determinable due to missing of some MSS. In other words, if a dirty atom $\mathfrak{a}$ is not relevant to a signature, it could mean two things: first, the atom is not a part of the module or, second, the atom *is* part of the module but a seed signature, which would indicate the relevance of $\mathfrak{a}$, has not been computed due to the time-out. Therefore, in order for the FME algorithm to remain correct it is forced to include dirty atoms into the module even though they may be irrelevant. This means, in particular, that performance of the FME algorithm directly depends on whether the MSS algorithm has been able to compute all MSS for every atom. This is subject of the evaluation which we discuss next. The open-source Java implementation used for our experiments is available at `http://tinyurl.com/bioportalFME`.

4.3   Evaluation of the Labeling Algorithm

We evaluated the labeling algorithm on the same BioPortal ontologies as used in Sect. 3. The main goal of the evaluation is to assess the practical feasibility of computing all MSS for atoms in the BioPortal ontologies. We set the time-out for

---

[6] Minimization has to be interrupted carefully to make sure that all produced signatures are minimal w.r.t. inclusion even though some signatures could be missing.

computing labels for every atom to be 5 seconds, so the algorithm is guaranteed to finish in 5 times the number of atoms in seconds. The results are presented in the following table.

Total no. of ont.s	Avg. size of MSS($\mathfrak{a}$)	Avg. number of terms in all MSS($\mathfrak{a}$)	Max. size of MSS($\mathfrak{a}$)	Number of ont. with dirty atoms	Max. number of dirty atoms
181	1.4	2.1	4, 252	5	554

For the vast majority of ontologies (176 out of 181) the algorithm was able to compute all MSS for all atoms. Also, the average label size (that is, the number of MSSes per atom) and the average number of terms in all MSSes per atom are small: 1.4 and 2.1, respectively (when averaged first within an ontology then over all ontologies). This is yet another consequence of the simplicity of the BioPortal ontologies: their atoms are relevant to only a small number of terms which implies a small average number of atoms (and consequently, axioms) per module, see the next subsection. This observation might suggest that the BioPortal ontologies, in contrast to those examined by Del Vescovo et al. in [5], do not have exponential numbers of modules, but it is no firm evidence because it does not tell us about the asymptotic growth of their module numbers relative to their sizes.

Regarding the few ontologies with dirty atoms, they either do not decompose well or have an interesting property of the AD graph: certain atoms *non-transitively* depend on a high number of other atoms. Both reasons are true, e.g., for the Nanoparticle ontology, for which the MSS algorithm left 554 atoms dirty and managed to compute 1, 019 MSS sets for one atom, and the International Classification for Nursing Practice ontology (72 dirty atoms and 4, 252 MSS sets, respectively). We leave it for future research to investigate such cases, where a subset of an ontology turns out to be relevant for such a high number of distinct, but overlapping, seed signatures.

4.4   Fast Module Extraction Algorithm and Evaluation

Finally, we present a LAD-based FME algorithm, which extracts modules based on Prop. 3 (i.e., by examining MSS sets in labels), and its evaluation. Similarly to the labeling algorithm, the current version of the FME algorithm is restricted to $\top$- or $\bot$-locality.

The relevance check at Line 6 takes into account the possible dirtiness of an atom. More formally, the atom is *possibly relevant* to Σ if it is clean and there exists $\Sigma' \in \mathsf{MSS}(\mathfrak{a})$ such that $\Sigma' \subseteq \Sigma$ or it is dirty and $\widetilde{(\mathfrak{a}]} \cap \Sigma \neq \emptyset$ and there is no $\Sigma' \in \mathsf{MSS}(\mathfrak{a})$ such that $\Sigma \subset \Sigma'$.[7]

The FME algorithm has two important advantages over the standard ME algorithm. First, it should be faster for most of ontologies because it benefits from the labeled AD in two ways: i) it exploits labels to quickly detect relevant atoms,

---

[7] Observe that if a subset of $\mathsf{MSS}(\mathfrak{a})$ contains a proper superset of Σ, then, since all seed signatures are minimal, the full set $\mathsf{MSS}(\mathfrak{a})$ cannot contain a subset of Σ.

Algorithm 2. Atomic decomposition-based module extraction algorithm (FME)

1: **Input:** LAD for FME of an ontology $\mathcal{O}$, a seed signature Σ
2: **Output:** The module $x\text{-mod}(\Sigma, \mathcal{O})$, where $x \in \{\top, \bot\}$
3: $\mathcal{M} \leftarrow \emptyset$
4: **repeat**
5:   enlarged $\leftarrow$ false
6:   $\mathcal{M} \leftarrow \mathcal{M} \cup$ "all atoms that are *possibly* relevant to Σ"
7:   **if** $\widetilde{\mathcal{M}} \setminus \Sigma \neq \emptyset$ **then**
8:     enlarged $\leftarrow$ true
9:   **end if**
10: $\Sigma \leftarrow \Sigma \cup \widetilde{\mathcal{M}}$
11: **until** enlarged = false
12: **return** $\mathcal{M}$

---

ii) once an atom $\mathfrak{a}$ is established to be relevant the corresponding module $(\mathfrak{a}]$ is added to the module without further checks. Second, it consumes substantially less memory since only relevant atoms (and their principal ideals) need to be loaded. The second advantage is especially important when modules are small comparing to the size of the ontology. This is the case with most of the BioPortal ontologies where the median module's size for small seed signatures is under 1%, as illustrated by the FME evaluation results, which we show next.

We ran the FME algorithm on the same set of BioPortal ontologies, which were used for decomposition and the labeling evaluation. Seed signatures are generated by a random selection of class names. For each size both FME and ME algorithms were run 100 times on different seed signatures and the results are averaged over all runs. The results are averaged over all 181 ontologies and presented in the following table. Correctness of the FME algorithm was also verified empirically by checking that the resulting modules contain all axioms extracted by the standard ME algorithm.[8]

Size of seed sig.	Avg. (median) rel. module size (%)	Number of positive cases	Avg. ME runtime (ms)	Avg. FME speed-up	Max. FME speed-up
2	0.77 (0.04)	173	1.09	7.33	37.28
5	0.91 (0.08)	169	1.15	3.86	27.12
10	0.99 (0.13)	150	1.18	2.48	8.34

"Relative module size" = size of the module divided by the size of the ontology
"Positive cases" = ontologies for which FME is faster than ME
"Avg. (max.) speed-up" = average (max.) value of ME time divided by FME time

Several conclusions can be drawn from the results. First, good decomposability of BioPortal ontologies indeed implies small modules on average (column 2).

---

[8] The converse is only guaranteed to be true when there is no dirty atoms. Otherwise an FME module could be a superset (i.e., an approximation) of the ME module for the same seed signature. Of course, the irrelevant atoms can easily be removed by running the ME algorithm on the FME module, i.e., by refining the approximation.

Second, even the standard ME is very fast (around 1 millisecond). Third, the FME algorithm is typically faster than the standard ME algorithm, however, this depends on several factors: i) decomposability of the ontology, ii) average number of atoms' labels, and iii) size of the seed signature. The first factor is important for both FME and ME algorithms as it effects the size of the module. The second factor determines how quickly the FME algorithm can perform the relevance check on an atom. In the worst case, the algorithm has to examine each MSS for an atom to decide if it is relevant.[9] The seed signature's size determines the number of relevant atoms. When the seed signature gets larger, the algorithm has to examine more atoms for relevancy. Finally, note that the results include ontologies with dirty atoms on which the FME algorithm could be up to 5 times slower than the ME algorithm because of considering possibly irrelevant atoms (this illuminates the importance of efficient labeling).

We also investigated the cases in which the FME algorithm runs an order of magnitude faster than the standard ME algorithm. This seems to be the case with ontologies which decompose into small atoms with a low number of MSS set, and small seed signatures. This is fairly typical for BioPortal ontologies, including some well-known ones. We illustrate this by comparing the running time of FME and ME on randomly generated samples of size between 10K and 60K axioms of GO (the Gene Ontology) and ChEBI (Chemical Entities of Biological Interest Ontology).[10] Seed signatures of size 2, 5 and 10 are generated as in the previous experiment. The results are shown in the two figures below (GO on the left, ChEBI on the right).

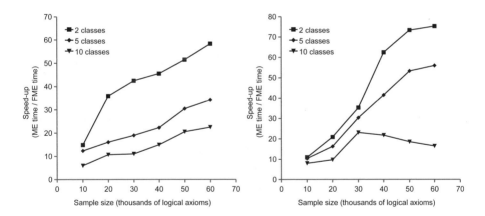

The graphs show that FME time tends to grow more slowly with the size of the ontology than ME time. This is unsurprising because the ratio of the module's size to the ontology size is decreasing (provided the seed signature's size remains constant) and the FME is usually able to quickly locate relevant

---

[9] In fact, this depends on the data structure used to store sets of MSS. We use simple hash sets, so the check takes $O(|\mathsf{MSS}(\mathfrak{a})| \times |\Sigma|)$, where Σ is the seed signature.

[10] Both ontologies are slightly over 60K logical axioms.

atoms while the ME algorithm has to examine each axiom.[11] As in the previous experiments, the speed-up is greater for smaller seed signatures. Note, however, that for seed signatures of 10 terms the relative speed-up of FME decreases after $30K$ axioms for ChEBI. Although it is still an order of magnitude, the behavior suggests that additional optimizations might be necessary for FME. For example, the relevance check could be made much quicker if MSS sets are stored in a data structure tuned for testing set inclusion.

In addition, the FME algorithm can work when only labels and the graph structure of the AD (but not axioms) are loaded into memory. This could be important for maintaining large ontologies, or even large collections of large ontologies, such as ontology repositories. In that case, contrary to the standard ME, the FME algorithm could still extract modules by loading axioms of only relevant atoms (plus possibly some dirty atoms for which irrelevance cannot be proved). For example, if BioPortal ontologies were maintained in the decomposed form, it would be possible to provide clients, such as SSWAP, with modules for a required seed signature in a scalable (from the memory perspective) way.

5   Summary and Future Directions

In this paper we have presented results of decomposing and extracting modules from most of BioPortal ontologies. We showed that the majority of ontologies decompose well, discussed possible reasons for poor decomposability, and implications of decomposability for possible use cases, in particular, semantic Web service annotation and discovery. In addition, we presented novel AD-based algorithms for computing minimal seed signatures for compact modules and module extraction.

Overall, the reported results show the utility of ontology modularity and decomposition for such tasks as semantic Web service matchmaking. In particular, it is likely that only small portion of a biomedical ontology is relevant for terms used in a Web service description, e.g., on SSWAP or SADI (see the average module size in the table on Page 142). Therefore, reasoning required to discover the service could be (efficiently) performed on a small set of OWL axioms. Furthermore, decomposition helps to get that set (module) faster than the standard module extraction and without the necessity to keep the ontology in memory.

We intend to continue our work on decomposition in several directions. First, we will investigate the possibility of maintaining ontologies in a decomposed form. This is more scalable from the memory perspective, enables faster ME, and is also potentially useful for comprehension and collaborative development of the ontology. However, it will require the possibility of incremental updates to the AD since its computation can be time consuming. Second, we will extend our algorithms to $\top\bot^*$-modules and, possibly, to semantic locality. Third, we

---

[11] For space reasons our description of the FME algorithm does not show how labels serve as *indexes* by enabling us to perform the relevance test only on atoms whose MSS sets overlap with the seed signature. We must mention that a syntactic indexing (but coarser and less efficient) could be used for the standard ME as well.

will keep on investigating modeling guidelines for developing well decomposable ontologies and will seek to improve understanding of poor decomposability.

Acknowledgements. We thank the anonymous reviewers for their comments and Evan Lane for discussions with D.G. and A.W. This material is based upon work supported by the National Science Foundation (NSF) under grant #0943879 and the NSF Plant Cyberinfrastructure Program (#EF-0735191).

References

1. Baader, F., Calvanese, D., McGuinness, D., Nardi, D., Patel-Schneider, P.F.: The Description Logic Handbook: Theory, Implementation, and Applications. Cambridge University Press (2003)
2. Cuenca Grau, B., Horrocks, I., Kazakov, Y., Sattler, U.: Modular reuse of ontologies: Theory and practice. J. of Artif. Intell. Research 31, 273–318 (2008)
3. Del Vescovo, C., Gessler, D., Klinov, P., Parsia, B., Sattler, U., Schneider, T., Winget, A.: Decomposition and modular structure of bioportal ontologies. Tech. rep. (2011), http://tinyurl.com/modbioportal
4. Del Vescovo, C., Parsia, B., Sattler, U.: Topicality in logic-based ontologies. In: Proc. of ICCS-2011, pp. 187–200 (2011)
5. Del Vescovo, C., Parsia, B., Sattler, U., Schneider, T.: The modular structure of an ontology: an empirical study. In: Proc. of DL 2010 (2010), http://ceur-ws.org
6. Del Vescovo, C., Parsia, B., Sattler, U., Schneider, T.: The modular structure of an ontology: Atomic decomposition. In: Proc. of IJCAI-2011, pp. 2232–2237 (2011)
7. Del Vescovo, C., Parsia, B., Sattler, U., Schneider, T.: The modular structure of an ontology: atomic decomposition. Tech. rep., University of Manchester (2011), http://bit.ly/i4olYO
8. Gessler, D., Schiltz, G.S., May, G.D., Avraham, S., Town, C.D., Grant, D.M., Nelson, R.T.: SSWAP: A simple semantic web architecture and protocol for semantic web services. BMC Bioinformatics 10, 309 (2009)
9. Ghilardi, S., Lutz, C., Wolter, F.: Did I damage my ontology? A case for conservative extensions in description logics. In: KR-2006, pp. 187–197 (2006)
10. Horridge, M., Parsia, B., Sattler, U.: The state of bio-medical ontologies. In: Proc. of 2011 ISMB Bio-Ontologies SIG (2011)
11. Jiménez-Ruiz, E., Cuenca Grau, B., Sattler, U., Schneider, T., Berlanga Llavori, R.: Safe and Economic Re-Use of Ontologies: A Logic-Based Methodology and Tool Support. In: Bechhofer, S., Hauswirth, M., Hoffmann, J., Koubarakis, M. (eds.) ESWC 2008. LNCS, vol. 5021, pp. 185–199. Springer, Heidelberg (2008)
12. Konev, B., Lutz, C., Walther, D., Wolter, F.: Formal Properties of Modularisation. In: Stuckenschmidt, H., Parent, C., Spaccapietra, S. (eds.) Modular Ontologies. LNCS, vol. 5445, pp. 25–66. Springer, Heidelberg (2009)
13. Spearman, C.: The proof and measurement of association between two things. Amer. J. Psychol. 15, 72–101 (1904)
14. Wilkinson, M.D., Vandervalk, B., McCarthy, L.: SADI Semantic Web services – cause you can't always GET what you want! In: Proc. of APSCC, pp. 13–18 (2009)

A Clustering-Based Approach to Ontology Alignment

Songyun Duan[1], Achille Fokoue[1], Kavitha Srinivas[1], and Brian Byrne[2]

[1] IBM T.J. Watson Research Center, Hawthorne, New York, USA
{sduan,achille,ksrinivs}@us.ibm.com
[2] IBM Software Group, Information Management, Austin, Texas, USA
byrneb@us.ibm.com

Abstract. Ontology alignment is an important problem for the linked data web, as more and more ontologies and ontology instances get published for specific domains such as government and healthcare. A number of (semi-)automated alignment systems have been proposed in recent years. Most combine a set of similarity functions on lexical, semantic and structural features to align ontologies. Although these functions work well in many cases of ontology alignments, they fail to capture alignments when terms or structure varies vastly across ontologies. In this case, one is forced to rely on manual alignment. In this paper, we study whether it is feasible to re-use such expert provided ontology alignments for new alignment tasks. We focus in particular on many-to-one alignments, where the opportunity for re-use is feasible if alignments are *stable*. Specifically, we define the notion of a cluster as being made of multiple entities in the source ontology S that are mapped to the same entity in the target ontology T. We test the *stability hypothesis* that the formed clusters of source ontology are stable across alignments to different target ontologies. If this hypothesis is valid, the clusters of an ontology S, built from an existing alignment with an ontology T, can be effectively exploited to align S with a new ontology T'. Evaluation on both manual and automated high-quality alignments show remarkable stability of clusters across ontology alignments in the financial domain and the healthcare and life sciences domain. Experimental evaluation also demonstrates the effectiveness of utilizing the stability of clusters in improving the alignment process in terms of precision and recall.

1 Introduction

Ontology alignment is an important problem for the linked data web, as more and more ontologies get published for specific domains such as government and healthcare. A number of (semi-)automated alignment systems have been developed in recent years (*e.g.,* Lily [16], ASMOV [8], Anchor-Flood [11], RiMOM [13]). Most systems combine a large set of similarity functions on lexical, semantic and structural features to align ontologies (for surveys, see [2], [14]). While these similarity functions are important and effective for many cases of ontology alignments, there are also cases where none of the similarity functions

L. Aroyo et al. (Eds.): ISWC 2011, Part I, LNCS 7031, pp. 146–161, 2011.

adequately capture the nature of the alignment; this is particularly true when the two ontologies of extremely different modeling granularities are involved. For instance, one alignment exercise frequently conducted by IBM consultants in the field is to align models that describe assets at an IT level (*e.g.*, the IBM Information FrameWork model used to describe IT assets in the banking industry) to models that describe the same assets at a business level (*e.g.*, the IBM Component Business Model for banking). Because the two models describe the same assets in different terms and different structures, traditional approaches to automated ontology alignment fail abysmally (the mapping precision we have measured can be as low as 1% in these cases). In fact, the only alternative in such cases is to rely on a domain expert who can provide the alignment between these types of models. However, if the expert has actually done the hard work of mapping the models once, is it feasible to re-use these high-quality manual ontology alignments, to improve the alignment process for new alignments when the two models evolve, or when the same model needs to be aligned to new models? This is the research focus of our paper.

For the purpose of investigating the re-use of manual mappings, we direct our attention in this paper to many-to-one (or conversely, one-to-many) mappings, because this is where mapping re-use can be readily applied while similarity functions fail to produce valuable information for alignments. In many-to-one mapping scenarios, multiple entities in one ontology S get mapped to a single entity in a target ontology T_1. The grouping of multiple entities in S can be viewed as user-specified clustering of source entities. In principle, there is a chance that prior ontology alignments can provide some guidance for the current alignment task in hand, if there is some *stability* in mapping certain entities in one ontology to the same entity in target ontologies. Put it another way, the question is whether the user-specified clusters based on the alignment of S to T_1 tend to appear when S is aligned with ontology T_2 different from T_1. If the user-specified clustering in S is in fact stable, then the clustering information can be exploited when alignment needs to be performed from S to T_2. Specifically, a mapping provided by an expert on one of the entities in a cluster of S can be automatically generalized to map other members of this cluster.

To evaluate the *stability hypothesis*, we define two novel metrics to measure the similarity of clusters constructed for an ontology S based on its alignment results to different ontologies. These metrics are conceptually similar to Levenshtein and Jaccard measures of string similarity. We also design a mapping strategy that utilizes the clustering information for new alignments and study the effectiveness of this strategy in terms of the classical mapping quality metrics such as precision and recall. Furthermore, we characterize mapping *efficiency* in terms of the amount of saving in human effort required in the alignment task with and without the clustering information. We apply these metrics to compare two independent alignments that were performed by IBM consultants in the field. The first alignment involved the mapping of the IBM Component Business Model (CBM), a flat model of business functions expressed in business terms to Information FrameWork (IFW), a structured and detailed model of enterprise processes described from an IT perspective. The second alignment involved a very different version of the CBM model which was aligned to a mostly unchanged IFW model. The alignment process was conducted about a year apart, by different consultants. As mentioned earlier, applying any of the standard similarity functions to either model pair fails to detect any meaningful alignments. Manual mappings produced by IBM consultants had most CBM entities mapped to multiple IFW entities. Using these expert created mappings as reference, we tested whether the user defined clusters of IFW entities stayed stable when it was mapped to a very different version of CBM. Our evaluation of the previously defined metrics showed remarkable stability of clustering of IFW entities (the average similarity of clusters is 0.89, within the range of $[0, 1]$). For the same dataset, the improvement in mapping precision is 0.4, and the efficiency is 0.95 within a scale of 0 to 1; the higher the better. For repeatability purposes, we evaluated these same metrics for 2312 ontology comparisons publicly available on the BioPortal web site[1] with again remarkable stability of clustering of source ontology entities (the average similarity of clusters is 0.84). From these positive stability results, the opportunity for re-use is quite clear: clustering information generated from existing alignments is very helpful for new alignment tasks. For instance, if entities a, and b in ontology $\mathcal{S}$ are mapped to entity c in ontology $\mathcal{T}_1$, and a is mapped to entity d in another ontology $\mathcal{T}_2$, we know b should be mapped to d in $\mathcal{T}_2$ as well.

Our main contributions in this paper are as follows:

- We present a novel technique to uncover, from existing many-to-one (or conversely, one-to-many) alignments, internal structures of related entities (*i.e.*, clusters of entities) in ontologies.
- We show the stability of those clusters across alignments in two different domains (finance and healthcare & life sciences) and on both manually created mappings and automatically generated high-quality mappings.
- We describe how clusters discovered in existing many-to-one and one-to-many alignments can be exploited for performing new alignments, and evaluate the impact on both mapping quality (precision/recall) and mapping efficiency (saving in human effort).

The remainder of the paper is organized as follows. In the next section, we present an overview of our clustering-based ontology alignment approach and the fundamental stability hypothesis it relies on. In Section 3, we describe cluster similarity measures needed to validate the stability hypothesis. The evaluation results on many-to-one alignments are presented in Section 4. Finally, after discussing related work in Section 5, we conclude in Section 6.

2   Overview of Clustering-Based Ontology Alignment

In many-to-one alignment scenarios, multiple entities in the source ontology $\mathcal{S}$ get matched to the same entity in the target ontology $\mathcal{T}$. One way to interpret

---

[1] http://bioportal.bioontology.org

the alignment result of $\mathcal{S} \rightarrow \mathcal{T}$ is that the entities in $\mathcal{S}$ are partitioned into clusters (*i.e.*, groups) such that each cluster of entities are matched to the same entity in $\mathcal{T}$. Consider a simple example.

Source ontology $\mathcal{S} = \{a, b, c, d\}$, target ontology $\mathcal{T} = \{e, f\}$, and their alignment result: $\mathcal{S} \rightarrow \mathcal{T} = \{a \rightarrow e, b \rightarrow e, c \rightarrow f, d \rightarrow f\}$. In this case, ontology $\mathcal{S}$ is partitioned into 2 clusters: $\mathcal{P}_s = \{\{a, b\}, \{c, d\}\}$.

It naturally follows that a source ontology could be partitioned in different ways based on its alignment results with different target ontologies. Our clustering-based ontology alignment approach relies on the following fundamental hypothesis:

Hypothesis (H): *The partitions of a source ontology (based on alignment results with different target ontologies) are stable across ontology alignments.*

If this hypothesis is valid, it is feasible to leverage the alignment result of ontology $\mathcal{S}$ to ontology $\mathcal{T}_1$ to help a new alignment of $\mathcal{S}$ to ontology $\mathcal{T}_2$ as follows:

- Generate a partition (*i.e.*, a set of clusters) of $\mathcal{S}$, denoted as $\mathcal{P}_s$, from the alignment result of $\mathcal{S} \rightarrow \mathcal{T}_1$;
- To perform the alignment task of $\mathcal{S} \rightarrow \mathcal{T}_2$, instead of matching individual entities in $\mathcal{S}$ independently with the entities in $\mathcal{T}_2$, it may be more efficient and more accurate to match a cluster of entities in $\mathcal{P}_s$ to the entities in $\mathcal{T}_2$. The intuition is that the entities in one cluster are expected to match to the same entity in $\mathcal{T}_2$.

This approach would be particularly valuable to maintain alignments as ontologies evolve. For example, if a high-quality alignment from ontology $\mathcal{S}$ to ontology $\mathcal{T}_1$ has been produced through a manual or semi-automated process and ontology $\mathcal{T}_1$ then evolves to a new version $\mathcal{T}_1$', this approach would significantly reduce the amount of pairwise mappings to consider in order to build an alignment from $\mathcal{S}$ to $\mathcal{T}_1$'.

Table 1. Example of an IFW cluster based on manual alignment to CBM

IFW	CBM
Provide FMO Transaction Reconciliation	Account Reconciliation
Request Amended Counterparty Confirmation	Account Reconciliation
Accumulate Futures Transaction Values	Account Reconciliation
Analyze FMO Transaction Details	Account Reconciliation
Compare FMO Transaction Details	Account Reconciliation
Verify FMO Transaction Details	Account Reconciliation

Tables 1 and 2 show two examples of clusters obtained respectively through manual alignment and through automated alignment.

In Table 1, most entities in the IFW cluster (*i.e.*, 'Provide FMO Transaction Reconciliation', 'Request Amended Counterparty Confirmation', 'Accumulate Futures Transaction Values', and 'Analyze FMO Transaction Details') show little to no lexical or structural similarity between themselves or with the target

Table 2. Example of a Mouse Anatomy cluster based on lexical alignment to Brenda Tissue Ontology

Mouse Anatomy	Brenda Tissue
intestine (no synonym)	intestine (synonyms: bowel, gut)
bowel (no synonym)	intestine (synonyms: bowel, gut)
gut (no synonym)	intestine (synonyms: bowel, gut)

CBM entity, 'Account Reconciliation'. In fact, applying standard similarity functions to directly map IFW to CBM produce extremely poor results because, as mentioned in Section 1, the two models are very different from almost all perspectives: different vocabularies (IT vocabulary for IFW vs. business vocabulary for CBM), very different structures (deep nested structure for IFW vs. flat structure for CBM), modeling at different levels of abstraction (modeling at the IT process level for IFW vs. modeling at the business functions level for CBM). The semantic similarity between IFW entities in the cluster, which could not be computed from information present in both models, was indirectly identified by the domain experts (IBM consultants) when they map these IFW entities to the same CBM entity.

Table 2 shows partial results of aligning the adult Mouse Anatomy Ontology (MA) and Brenda Tissue Ontology (BTO) using the automated process[2] described in [9]. Like the IFW-CBM case, entities in the cluster of MA ontology do not exhibit any meaningful similarity that could be computed based only on information in MA ontology. However, as opposed to the previous IFW case, entities in MA are lexically similar to the mapped entity (*i.e.*, *intestine* which has as explicit synonyms *bowel* and *gut*) in the target ontology. In this case, the alignment to BTO serves as a dictionary look up that uncovers the semantic similarity between *intestine*, *bowel*, and *gut*. This uncovered semantic similarity could then be used in the next alignment involving MA ontology.

3   Measures of Cluster Similarity

To test our stability hypothesis (H), we need to evaluate the similarity between the partitions of the same ontology, which requires a similarity measure on a pair of partitions (*i.e.*, sets of clusters). To ease presentation, consider two alignments, $\mathcal{S}$ to $\mathcal{T}_1$ and $\mathcal{S}$ to $\mathcal{T}_2$. Based on their alignment results, we can generate two partitions of $\mathcal{S}$: $\mathcal{P}_{s,1} = \{C_1, C_2, \ldots, C_m\}$ and $\mathcal{P}_{s,2} = \{C'_1, C'_2, \ldots, C'_n\}$, where each cluster C_i or C'_j is a collection of entities in the source ontology $\mathcal{S}$. So the real challenge is to define an appropriate measure to evaluate the similarity of $\mathcal{P}_{s,1}$ and $\mathcal{P}_{s,2}$. A good measure needs to be symmetric and have a fixed range of values, preferably $[0, 1]$, such that similarity values computed for different pairs of partitions are comparable. Here we consider two similarity measures which are conceptually similar to similarity metrics for strings.

---

[2] The ontologies and the alignments are available at
  http://bioportal.bioontology.org/

3.1 Measure I: Jaccard Similarity on Entity Pairs

For each cluster C in the partition $\mathcal{P}_s$ of ontology $\mathcal{S}$, we can generate all pairs of entities in cluster C. Thus, the partition $\mathcal{P}_s$ can be represented as the union of all the sets of entity pairs (one set per cluster in $\mathcal{P}_s$). The generated set of entity pairs is *equivalent* to the original partition in the sense that we can re-generate the partition from the set of entity pairs. For instance, consider a partition $\mathcal{P}_1 = \{\{a, b\}, \{c, d, e\}\}$. The corresponding set of entity pairs is $\mathcal{P}_1' = \{\{a, b\}, \{c, d\}, \{c, e\}, \{d, e\}\}$. Note that given $\mathcal{P}_1'$, we can re-generate the original partition $\mathcal{P}_1$. For another partition $\mathcal{P}_2$ (say, $\mathcal{P}_2 = \{\{a, b, c\}, \{d, e\}\}$), we can also generate a set of entity pairs as $\mathcal{P}_2' = \{\{a, b\}, \{a, c\}, \{b, c\}, \{d, e\}\}$. The similarity of the two sets P_1' and P_2' can then be computed with the standard Jaccard similarity [1] by treating each entity pair (without considering the sequence of entities) as the basic element of a set. Therefore, the similarity of $\mathcal{P}_1$ and $\mathcal{P}_2$ can be computed as follows:

$$PSim_1(\mathcal{P}_1, \mathcal{P}_2) = \frac{|\mathcal{P}_1' \cap \mathcal{P}_2'|}{|\mathcal{P}_1' \cup \mathcal{P}_2'|} \tag{1}$$

where the numerator is the size of set intersection, and the denominator is the size of set union, with each entity pair as a basic unit in the set. The similarity measure $PSim_1$ has the desired property that it is symmetric (*i.e.,* $Sim_1(P_1, P_2) = Sim_2(P_2, P_1)$) and the range of the similarity value is $[0, 1]$. Furthermore, $PSim_1$ captures the effect of big clusters in a partition, because big clusters will generate entity pairs that are exponential in size to cluster size; thus reflecting the natural preference for big clusters.

3.2 Measure II: Partition Edit Distance

One measure that is closely related to similarity is distance. The distance between two partitions can be intuitively characterized by the minimum amount of work to transform one partition into the other, which is conceptually similar to edit distance (*i.e.,* the minimum number of edits, including insertion, deletion, and substitution) between two strings. The basic operations for partitions we consider include *Split* and *Merge*. A Split operation on a cluster C_1 creates two non-overlapping clusters C_2 and C_3, with the union of C_2 and C_3 including all the elements in C_1. Merge is an inverse operation of Split. To continue with the previous example, to transform partition $\mathcal{P}_1$ into partition $\mathcal{P}_2$, we need 2 operations: a Split operation on the cluster $\{c, d, e\}$ generates two clusters $\{c\}$ and $\{d, e\}$; and a Merge operation of the two clusters $\{a, b\}$ and $\{c\}$ creates a new cluster $\{a, b, c\}$, thus resulting in partition $\mathcal{P}_2$. So the edit distance between $\mathcal{P}_1$ and $\mathcal{P}_2$ is 2.

Definition: The edit distance between two partitions $\mathcal{P}_1$ and $\mathcal{P}_2$, denoted as $ED(\mathcal{P}_1, \mathcal{P}_2)$, is the length of the shortest edit path composed of Splits and Merges from $\mathcal{P}_1$ to $\mathcal{P}_2$. A nice property of the partition edit distance is that it is symmetric, *i.e.,* $ED(\mathcal{P}_1, \mathcal{P}_2) = ED(\mathcal{P}_2, \mathcal{P}_1)$. Although the edit path from $\mathcal{P}_1$ to $\mathcal{P}_2$ is different from the path of transforming $\mathcal{P}_2$ to $\mathcal{P}_1$, these two paths have the same length, since the two basic operations of Merge and Split are inverse of each other.

Because the edit distance between two partitions of the same ontology is dependent on ontology size, we need a normalization factor to transform edit distance into a similarity measure. The normalization factor we consider here is ontology size, *i.e.*, the number of entities in a source ontology. Thus, the similarity measure derived from edit distance of two partitions $\mathcal{P}_1$ and $\mathcal{P}_2$ is:

$$PSim_2(\mathcal{P}_1, \mathcal{P}_2) = 1 - \frac{1}{|\mathcal{S}|}ED(\mathcal{P}_1, \mathcal{P}_2) \qquad (2)$$

where $|\mathcal{S}|$ is the size of the source ontology. The similarity measure $PSim_2$ is also symmetric.

3.3   Measure III: Mapping Quality

The above two measures reflects the stability of partitions from the similarity perspective. We also propose another measure to evaluate the actual quality of mappings which are generated based on the clustering information. To this end, we simulate the procedure of generating partitions of a source ontology and applying the clustering information for a new alignment that involves the same source ontology:

- Generate a partition $\mathcal{P}_1$ of the source ontology $\mathcal{S}$ based on the mapping result from $\mathcal{S}$ to a target ontology $\mathcal{T}_1$;
- For a new alignment task from $\mathcal{S}$ to another target ontology $\mathcal{T}_2$, generate the mappings as follows:
  - For each cluster C in the partition $\mathcal{P}_1$, randomly pick one entity s from C and find the mapped entity t in $\mathcal{T}_2$;
  - Generalize the mapping to other entities in the same cluster, with the mappings being $\{\langle s', t \rangle | s' \in C\}$.

Since in this paper we focus on many-to-one mappings, we exclude the one-to-one mappings from the estimation of precision and recall, the two classical metrics for measuring mapping quality.

$$precision = \frac{|M \cap M_{GS}|}{|M|}, recall = \frac{|M \cap M_{GS}|}{|M_{GS}|}$$

where M is the mappings generated using the strategy described above, and M_{GS} is the gold-standard (*i.e.*, reference) mappings. Note that $|M|$ and $|M_{GS}|$ are equal in this scenario, so precision and recall are equal, and we will only report results in precision in the experiment section. In addition to the mapping quality, we also measure the amount of saving of human effort to generate the mappings, compared to the baseline approach of independently generating mappings for each entity in the source ontology from scratch. The human effort is estimated as the number of mappings that require human input. We thus define mapping efficiency with utilization of clustering information as:

$$\texttt{efficiency} = 1 - |\mathcal{P}_{m2o}|/|\mathcal{S}_{m2o}|$$

where $|\mathcal{P}_{m2o}|$ is the number of *non-singleton* clusters (*i.e.*, clusters with more than one entity) in the partition $\mathcal{P}$ of the source ontology $\mathcal{S}$, and $|\mathcal{S}_{m2o}|$ is the total number of entities in the non-singleton clusters. Intuitively, the bigger the clusters, the more efficient the approach based on clustering. At the same time, however, bigger clusters tend to be more *impure* (*i.e.*, meaning entities in the same cluster are mapped to different entities in the target ontology). Therefore, clusters of size exceeding the optimal value will adversely affect mapping quality.

3.4   Discussion

The three measures described above reflect different information aspects for the hypothesis testing about partition stability. The Jaccard similarity indicates whether the partitions generated based on mappings to different target ontologies are at the same granularity. For example, if the target ontology $\mathcal{T}_1$ is more fine-grained than another target ontology $\mathcal{T}_2$, we expect that the Jaccard similarity of the two partitions of the source ontology is relatively low. A simple example will illustrate this fact. Suppose we have one partition containing just one cluster $\{a, b, c, d\}$, and the other partition is $\{\{a, b\}, \{c, d\}\}$. It is easy to see that the target ontologies in the two alignments are at different granularity. The Jaccard similarity of the two partitions is $1/3$, which is relatively low. The partition edit distance, on the other hand, is insensitive to such partition granularity. Continue that simple example. We can see that the edit distance between the two partitions is 1, so the normalized similarity based on the edit distance is $1 - 1/4 = 0.75$. The advantage of edit distance is that it can capture both the cases where entities mapped to the same entity are mapped to different entities in another target ontology, and the cases where entities mapped to different entities in one target ontology are mapped to the same entities in another ontology. The third measure, mapping quality, provides information about whether the partition information is reliable for end use. That is, how much the users can trust the partition information provided by one alignment task, when they perform a related alignment task in the same domain with the same source ontology. In some sense, mapping quality is a hybrid measure of Jaccard similarity and partition edit distance, and can provide an estimate of usefulness of the clustering information for end users.

4   Evaluating Partitioning Stability

In this section, we evaluate the stability of partitioning, using the three measures defined in Section 2, on one dataset from the financial domain and one from the life sciences domain that is publicly available on the BioPortal website.

4.1 IFW - CBM: Ontology Evolution Scenario

As discussed earlier in Section 1, we first studied the case where we had two high-quality manual alignments: IFW-CBM$_1$ and IFW-CBM$_2$, where CBM$_2$ reflects an evolution of CBM$_1$. CBM$_1$ has 65 entities, and CBM$_2$ has 120 entities; they overlap in 37 entities. There are 2165 entities in IFW that are involved in many-to-one mappings. The partition of IFW based on the mappings from IFW to CBM$_1$ consists of 62 clusters, and the partition based on the mappings from IFW to CBM$_2$ consists of 111 clusters. The average cluster size in both partitions is 25. Recall that the average cluster size determines the mapping efficiency, *i.e.*, the amount of human effort that can be saved by leveraging the clustering information. Therefore, the mapping efficiency in this case is expected to be high; the actual efficiency value is 0.95. We also calculated the similarity of the two partitions: (1) The similarity based on partition edit distance is 0.89; and (2) the Jaccard similarity is 0.53. The low Jaccard similarity is likely due to the fact that the number of clusters in two partitions is quite different (62 vs. 111), as is the cluster size. As a consequence, the number of entity pairs generated from the clusters of IFW entities changes significantly. Because Jaccard similarity is quite sensitive to the size of the sets of entity pairs, the two partitions have a low Jaccard similarity. Jaccard similarity clearly reflects the actual change in granularity of the two versions of CBM.

The mapping precision metric is not symmetric, which means using the clustering information based on the mappings from IFW to CBM$_1$ to generate mappings for IFW to CBM$_2$ may have a precision quite different from that in the other direction. Therefore, we estimated mapping precision in both directions, and the average precision is 0.78. To determine the improvement in precision due to the use of clustering information, we measured the overlap between the two alignment results (*i.e.*, IFW-CBM$_1$ and IFW-CBM$_2$) as the baseline. The intuition is that, if we directly use one alignment result to generate mappings for the other alignment, only the overlap of the two alignments can generate correct mappings; the precision for this approach is 0.38. So through the utilization of clustering information from one alignment for the other alignment, we improve the mapping precision by 0.4; which is statistically significant. As mentioned in Section 1, the lexical and structural similarity between IFW and CBM is extremely low; we actually ran our alignment algorithm [5] for the two alignments IFW-CBM$_1$ and IFW-CBM$_2$ and got a precision around 0.01. In this scenario, manual mapping is therefore a must, and improving precision by 0.4 by alignment re-use is a significant saving.

4.2 Large Scale Evaluation on BioPortal Ontologies

The BioPortal website contains 149 ontologies, 9.3K ontology comparisons, and 1.75 million matchings of elements in various ontologies that were largely lexically generated.

Recall that we can create one partition of the source ontology from one ontology alignment result. For a given source ontology $\mathcal{S}$, there could be multiple

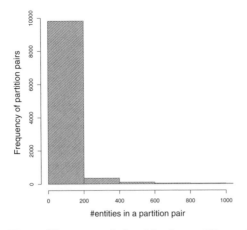

Fig. 1. Histogram of #entities in partition pairs

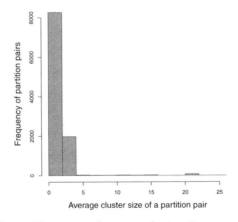

Fig. 2. Histogram of average cluster size per partition

partitions of $\mathcal{S}$, based on the alignment results with respect to different target ontologies. We used the two similarity measures (described in Section 3) to estimate pairwise similarity of the partitions on the same source ontology. If an ontology $\mathcal{S}$ is aligned with k ontologies, we will generate k partitions of $\mathcal{S}$, and there will be $\binom{k}{2}$ similarity computations of the pairs of partitions. Therefore, the total number of pairwise comparison of partitions is $\sum_{i=1}^{K} \binom{k_i}{2}$, where K is the number of ontologies, and k_i is the number of times an ontology $\mathcal{S}_i$ is aligned with other ontologies. In this setting, we have altogether 24K similarity computations between generated partitions.

Since we were focused on many-to-one matching scenarios, we needed to preprocess the expected matchings from the BioPortal website before analyzing the similarity of partitions on the same source ontology, using the following steps:

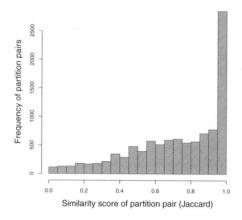

Fig. 3. Histogram of Jaccard similarity values

1) For a pair of partitions on the same source ontology, we identified the entities that appeared in both partitions. We removed from further analysis those entities that only appeared in one of the partitions. The rationale for this pruning was that these entities were really analogous to missing observations. That is, if an entity is missing from the partition it could be either due to incomplete alignment by domain experts, or because it is a singleton in this alignment, or because it should have been mapped to a different cluster. Since we had no way of knowing which of the three cases these entities fell into, we basically eliminated the entities from the analysis.

2) For any entity that is a singleton cluster in both partitions, we also removed them from further analysis; although the singleton clusters common in two partitions do not actually affect the similarity values, due to the robustness of our similarity measures.

3) To make the analysis meaningful, we also removed ontology comparisons that contained less than 10 entities involved in many-to-one matchings.

After preprocessing the expected matchings, we had 10.4K pairs of partitions for the similarity comparison. Figure 1 shows the distribution of the number of entities in each pair of partitions. The average number of entities involved in many-to-one matching scenarios is 64, which ensures that our analysis of partitioning stability is based on a reasonable number of data points and is reliable. Figure 2 shows the distribution of average cluster size per partition. It is clear that a majority of the partitions have small clusters, with a size of 2 or 3; note that we have excluded singleton clusters generated from one-to-one mappings. Since the mapping result is incomplete and often only covers a small part of the ontology, we made only considered the entities mentioned in both matchings, which partially explains small clusters.

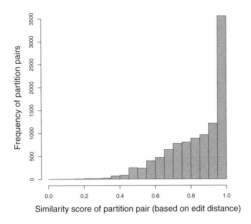

Fig. 4. Histogram of similarity values based on partition edit distance

Figure 3 shows the distribution of the similarity values of all pairs of partitions using Jaccard similarity on entity pairs (see Section 3.1). The mean of the similarity values is 0.72, and the standard deviation is 0.26. Figure 4 shows the distribution of the similarity values of all pairs of partitions based on partition edit distance (see Section 3.2). The mean of the similarity values is 0.84, and the standard deviation is 0.16.

Both Figure 3 and Figure 4 show that the partition of an ontology S is reasonably stable based on the results of aligning S with different ontologies. This observation indicates that we can leverage the partition of ontology S constructed from an existing alignment result to help new ontology alignments, which can be done in the following way: (1) Given the result of aligning S to T_1, we can generate a partition (*i.e.*, clusters) of S, denoted as P_s; (2) For a new alignment from S to T_2, we match each cluster of entities in P_s to the same entity in T_2. This alignment strategy has two benefits: (i) it improves alignment quality, since the alignment tool can aggregate the information from all entities in a cluster to make alignment decisions rather than make decisions based on individual entities independently; and (ii) it improves alignment efficiency, because the alignment of one entity in a cluster can be easily generalized to the other entities in the same cluster. Figure 5 shows the distribution of precision when we apply the mapping strategy to the 10.4K ontology pairs. The average precision is 0.92, with a standard deviation of 0.11. This result verifies that it is viable to utilize the clustering information from one ontology pair for the alignment of another pair in the same domain, certainly with the same source model. Figure 6 shows the distribution of mapping efficiency in terms of the percentage of mappings that can be automatically generated by leveraging the partition information. The average efficiency is 0.37, with a standard deviation of 0.19. As explained in the previous section, the efficiency is highly dependent on the average cluster

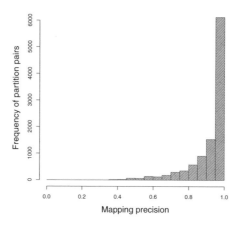

Fig. 5. Histogram of mapping precision values

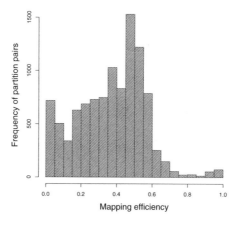

Fig. 6. Histogram of mapping efficiency values

size; the bigger the average cluster size, the higher the efficiency. Since the average cluster size of the partitions is small (see Figure 2), the efficiency is thus modest.

We also note that the observed stability of clusters for the BioPortal ontologies is not simply an artifact of the fact that the mappings were computed using lexical matching. For instance, the concepts *COO:F0005386 hyaluronidase activity*, *CCO:F0004395 hyaluronate lyase activity*, and *CCO:F0000824 hyalurononglucosaminidase activity* are all mapped to *PHI:0000199 hyaluronidase activity* based on their broad synonyms. Yet, each of the 3 concepts is mapped to different concepts in the Gene Ontology (GO). It is clear that stability is independent of whether or not lexical similarity drives the alignment process, which was also shown earlier with the IFW-CBM alignments.

Can the use of clustering information improve the alignment for BioPortal ontologies? Unfortunately, we do not have the luxury of having overlaps between two versions of the same model, like the IFW-CBM case, that can be used as a baseline. We do note, however, that there were a substantial number (48,261) of mappings, generated by our clustering-based approach, that are missing from the BioPortal website. Since the mappings provided by BioPortal are incomplete [9], it is unclear whether some entities in part of a cluster should not be mapped to any entity in the target ontology or the extra mappings we found are valid. Although we were unable to verify the validity of all the mappings due to lack of expertise, a number of them seemed correct based on their synonyms (see Table 3 for a few examples below). In the table, CLL is missed because it is an acronym for chronic lymphocytic leukemia, lung neoplasms is missed because it is a synonym of lung cancer, and similarly, RB1 is missed because it is an acronym for retinoblastoma. This observation indicates that our clustering-based alignment approach can improve the recall for the alignments of BioPortal ontologies; note that the average precision estimated with the existing mappings is 0.92.

Table 3. Examples of missed matches as defined by clustering

Concept 1	Concept 2
estrogen receptor alpha (CDR0000322904)	estrogen receptor (PRO_000007204)
retinoblastoma (MPATH:378)	RB1 (CDR0000043571)
non-small cell lung cancer (CDR0000040862)	Lung neoplasms (D008175)
renal cell carcinoma (CDR0000038140)	carcinoma, renal cell (C1534)
B-cell chronic lymphocytic leukemia (CDR0000039824)	CLL (LP34550-1)

5   Related Work

The alignment technique we proposed in this paper, which exploits internal structures of ontologies discovered through existing high-quality alignments, can be contrasted with previous work in terms of its singular focus on many-to-one and one-to-many alignments and in terms of the novelty of its approach to learning from existing alignments.

Although many approaches have been proposed to perform ontology alignment in the literature, there have been, to the best of our knowledge, no significant efforts to tailor the alignment process for alignments with cardinality different from one-to-one. After computing an aggregate similarity score for each candidate matching, most state-of-the art systems (*e.g.,* AgreementMaker [3] and BLOOMS [15]) simply return the matchings above a given threshold under a given alignment cardinality constraint (*e.g.,* one-to-one, one-to-many, many-to-one) without any consideration for the internal structures implied by one-to-many or many-to-one alignments. Other systems (*e.g.,* ASMOV [12]) have been optimized for one-to-one alignments to the point of considering multiple entity correspondences, where the same entity in one ontology is matched with multiple entities in the other ontology, as an inconsistency check in the final semantic verification step. This bias for one-to-one alignments also transpires from the relatively large collection of mostly one-to-one ontology alignments used to evaluate and systematically characterize the performance of state-of-the-art ontology alignment systems at the annual Ontology Alignment Evaluation Initiative[3] event.

Prior work on learning from existing high-quality alignments (*e.g.*, [5], [4], [7] and [6]) has typically taken a machine learning approach to customize the alignment process either for a given pair of ontologies, for which a partial reference alignment is available, or for a domain where multiple reference alignments are available. The outcome of this traditional learning approach is the specification of the optimal value for each parameter of the alignment process for a particular alignment or for alignments in a given domain. However, the learning approach does not work well when there is little or no lexical/structural similarity between the ontologies to align; in which cases the similarity functions can provide little signal for the learning process. Furthermore, no information is learned about the intrinsic structure of ontologies and then used to help new alignment. In contrast, in this paper, we describe how existing many-to-one (or one-to-many) alignments can be used to discover internal structures (*i.e.*, grouping entities within an ontology); such structures can then be leveraged in new ontology alignment as discussed in Section 2.

To the best of our knowledge, [9] and [10] are the only related work which attempts to learn structural characteristics of ontologies from matchings. However, our work is different from [9] in terms of its goals. The main goal of [9] is to uncover the network structure of the set of ontologies, and learn from their links (*i.e.*, entity matchings) the interesting properties of the ontologies in the particular domain; for example, which ones are the most relevant and most appropriate to serve as background knowledge for domain-specific tools. Our goal is to uncover internal ontological structures to enhance future alignments. Reference [10] proposed an alignment technique to generate mappings between source ontology and target ontology by composing previously determined mappings that involve intermediate ontologies. Our work differs from [10] in that we evaluated the soundness of the hypothesis that the partition (*i.e.*, clustering of entities) of the source ontology is stable across ontology alignments, which validates the underlying assumption made by [10]; so our work is complementary to [10].

6   Conclusions

In this paper we proposed the hypothesis that the internal structure of an ontology, *i.e.*, clusters of its entities discovered from the many-to-one alignment scenario, is stable across ontology alignments in the same domain. To evaluate this hypothesis, we defined two novel metrics to measure the similarity of clusters generated for one ontology based on its alignments with different target ontologies. Experimental evaluation with datasets from the financial domain and

---

[3] http://oaei.ontologymatching.org/2010/

the healthcare and life sciences domain demonstrated that the stability hypothesis is valid. In addition, we designed a mapping strategy that can leverage the clustering information for new alignment tasks, and characterized the effectiveness of this mapping strategy in terms of the impact on mapping quality and mapping efficiency. Experimental evaluation showed that clustering information discovered from one alignment can help improve, with a statistical significance, the mapping quality and mapping efficiency of a new alignment.

References

1. Bishop, C.M.: Pattern Recognition and Machine Learning. Springer, Heidelberg (2007)
2. Choi, N., Song, I.-Y., Han, H.: A survey on ontology mapping. SIGMOD Rec. (2006)
3. Cruz, I.F., Antonelli, F.P., Stroe, C.: Agreementmaker: efficient matching for large real-world schemas and ontologies. In: Proc. VLDB Endow., vol. 2, pp. 1586–1589 (August 2009)
4. Doan, A., Madhavan, J., Domingos, P., Halevy, A.: Ontology matching: A machine learning approach. In: Handbook on Ontologies in Information Systems. Springer, Heidelberg (2003)
5. Duan, S., Fokoue, A., Srinivas, K.: One size does not fit all: Customizing ontology alignment using user feedback. In: Patel-Schneider, P.F., Pan, Y., Hitzler, P., Mika, P., Zhang, L., Pan, J.Z., Horrocks, I., Glimm, B. (eds.) ISWC 2010, Part I. LNCS, vol. 6496, pp. 177–192. Springer, Heidelberg (2010)
6. Eckert, K., Meilicke, C., Stuckenschmidt, H.: Improving Ontology Matching Using Meta-Level Learning. In: Aroyo, L., Traverso, P., Ciravegna, F., Cimiano, P., Heath, T., Hyvönen, E., Mizoguchi, R., Oren, E., Sabou, M., Simperl, E. (eds.) ESWC 2009. LNCS, vol. 5554, pp. 158–172. Springer, Heidelberg (2009)
7. Ehrig, M., Staab, S., Sure, Y.: Bootstrapping Ontology Alignment Methods with APFEL. In: Gil, Y., Motta, E., Benjamins, V.R., Musen, M.A. (eds.) ISWC 2005. LNCS, vol. 3729, pp. 186–200. Springer, Heidelberg (2005)
8. Jean-Mary, Y.R., et al. : ASMOV: Results for OAEI 2009. In: OM (2009)
9. Ghazvinian, A., Noy, N.F., Jonquet, C., Shah, N., Musen, M.A.: What Four Million Mappings Can Tell You About Two Hundred Ontologies. In: Bernstein, A., Karger, D.R., Heath, T., Feigenbaum, L., Maynard, D., Motta, E., Thirunarayan, K. (eds.) ISWC 2009. LNCS, vol. 5823, pp. 229–242. Springer, Heidelberg (2009)
10. Gross, A., Hartung, M., Kirsten, T., Rahm, E.: Mapping composition for matching large life science ontologies. In: International Conference on Biomedical Ontology (2011)
11. Hanif, M.S., Aono, M.: Anchor-Flood: Results for OAEI 2009. In: OM (2009)
12. Jean-Mary, Y.R., Patrick Shironoshita, E., Kabuka, M.R.: Ontology matching with semantic verification. Web Semantics: Science, Services and Agents on the World Wide Web 7(3), 235–251 (2009)
13. Li, J., Tang, J., Li, Y., Luo, Q.: RiMOM: A dynamic multistrategy ontology alignment framework. IEEE Trans. Knowl. Data Eng. (2009)
14. Noy, N.F.: Semantic integration: a survey of ontology-based approaches. SIGMOD Rec. (2004)
15. Pesquita, C., Stroe, C., Cruz, I., Couto, F.M.: Blooms on agreementmaker: Results for oaei 2010. In: OM (2010)
16. Wang, P., Xu, B.: Lily: Ontology alignment results for OAEI 2009. In: OM (2009)

Labels in the Web of Data

Basil Ell, Denny Vrandečić, and Elena Simperl

KIT
Karlsruhe, Germany
{basil.ell,denny.vrandecic,elena.simperl}@kit.edu

Abstract. Entities on the Web of Data need to have labels in order to be exposable to humans in a meaningful way. These labels can then be used for exploring the data, i.e., for displaying the entities in a linked data browser or other front-end applications, but also to support keyword-based or natural-language based search over the Web of Data. Far too many applications fall back to exposing the URIs of the entities to the user in the absence of more easily understandable representations such as human-readable labels. In this work we introduce a number of label-related metrics: completeness of the labeling, the efficient accessibility of the labels, unambiguity of labeling, and the multilinguality of the labeling. We report our findings from measuring the Web of Data using these metrics. We also investigate which properties are used for labeling purposes, since many vocabularies define further labeling properties beyond the standard property from RDFS.

Keywords: Web of Data, labels, human interfaces, data quality.

1 Introduction

A growing number of applications is expected to use the Web of Data. They will discover descriptions of interesting entities on the Web, load these descriptions, and improve the user experience by being smarter, or enable completely new scenarios, by building on the knowledge found in the Semantic Web [8]. These applications often need to expose the entities and the data they have gathered about these entities from the Web to the end user. In order to do so, labels are often used as human-readable names for the entities. Labels can be utilized for a number of different purposes:

– displaying the data to end-users, instead of displaying the URIs,
– for searches over the Web of Data, be they keyword-based or question-based,
– for indexing purposes, or
– for training and using annotation tools with a given knowledge base, etc.

In order to be able to utilize labels, they need to be made accessible to the application. In the general case it is assumed that labels will be made available, among other information, by dereferencing the URI of an entity using the HTTP protocol, following Linked Open Data principles [21].

L. Aroyo et al. (Eds.): ISWC 2011, Part I, LNCS 7031, pp. 162–176, 2011.
© Springer-Verlag Berlin Heidelberg 2011

In reality, the situation is slightly more complicated. Issues such as internationalization, multiple labels for an entity, the computational costs associated with dereferencing, or the use of alternative labeling properties make the task of finding a label for a given entity much harder than expected. In this paper we investigate how labeling on the Web of Data is actually used. The findings of our analysis allow us to derive a number of recommendations for data publishers. We define a number of metrics that provide a baseline for a quantitative analysis of the state of labeling on the Web. We finally come up with some suggestions on how to improve the current situation. The suggestions are aimed at simplifying the usage of data from the Semantic Web in any application.

The paper is structured as follows. Section 2 describes related work, especially how current applications (mostly browsers for linked data) deal with the issue of labeling. Section 3 draws the distinction between information resources and non-information resources, and how they are currently dealt with by data publishers with regards to labels. In Section 4 we investigate which properties are actually used to provide labels. Even though there is a property defined in the RDFS standard, a number of vocabularies define alternative properties to provide labels. Based on those properties, we define metrics in Section 5 in order to assess the current state of labeling in the Web of Data, followed by the results of applying those metrics on a sample of the Web in Section 6. We close with a number of recommendations and conclusions in Section 7.

2   Related Work

Applications enabling human users to exploit the Web of Data can be classified into three categories: Linked Data browsers, Linked Data search engines, and domain specific Linked Data applications [20].

Linked data browsers, such as Disco [9], Tabulator [5] or Marbles [4] to name just a few, enable human users the exploration of linked data similar to how HTML browsers enable exploration of the traditional Web of documents. Instead of navigating between HTML pages, they allow navigation between RDF documents following links in the data by following RDF links. Since applications consuming linked data such as linked data browsers are intended to be used by a broad audience if the Web of Data becomes widely used, hiding technical details such as URIs when displaying facts to human users becomes crucial. For annotating entities with human-readable descriptions, the property `rdfs:label` from the RDF vocabulary is commonly used to provide a human-readable version of a resource's name besides its URI [10].

For example when displaying data available in the linked data cloud for the artist Sidney Bechet using the linked data browser *Sig.ma*, the list of information items for his affiliation contains, amongst other items, the following three items:

- `http://rdf.freebase.com/ns/m.049jnng`
- `http://rdf.freebase.com/ns/m.043j22x`
- Sidney Bechet and His Orchestra

For the first two items no human-readable labels are available to Sig.ma, therefore the URI is displayed which does not represent anything meaningful to the user besides the information that Freebase contains information about Sidney Bechet.

If for a resource no label is known or an unexpected property is used for labeling or the label is not retrieved by resolving the URI, developers of linked data browser came up with a set of options when dealing with the problem of missing human-readable labels:

1. The URI itself is displayed to the user. The URI can be meaningful for some users that do not regard it as noise and that are capable of deriving the meaning from some readable strings in the URI. However, this requires URIs that have been created by following a convention to use meaningful names for URIs.[1] Displaying the URI also often leads to an overly technical feel of the interface.
2. The last part of the URI is used, i.e. the local name or the fragment identifier. For example for the URI
   `http://www.example.com/about#bob` the fragment identifier `bob` is used, and for the URI
   `http://www.example.com/people/alice` the last part of the path is used, i.e. `alice`.
3. A more complex mechanism, as e.g. used in Protégé [18] which allows the user to specify which property values to display.

Human-oriented search engines such as Falcons, Sindice, MicroSearch, Watson, SWSE, and Swoogle provide keyword-based search services. Keyword search on graphs relies on the existence of nodes that are labeled thus allowing to match keywords to nodes via their labels [19,30], or on meaningful URIs .

While measurements of the Web of Data have been performed before [14,32,13], an analysis of labels in the Web of Data has not been performed. However, Azlinayati et al. [24] analyzed identifiers and labels in 219 OWL ontologies. Given that the Web of Data mainly consists of instance data, their analysis regarding schema data can be seen as complementing our approach which analyses instance data.

3   Information Resources and Non-information Resources

URIs are used to identify resources, where a resource might be anything from a person over an abstract idea to a simple document on the Web [23]. *Information resources* (IR) are resources that consist of information and therefore all of their essential characteristics can be conveyed in a message and be transported over

---

[1] However, `http://www.w3.org/Provider/Style/URI` recommends not to use topic names in a URI since thereby an URI's creator binds herself to some classification that can be subject to change, and would therefore require a renaming of the URI, which is considered undesired.

protocols such as HTTP. IRs can be copied from and downloaded via the Internet given their URLs. Disjoint from this set of resources is the set of *non-information resources* (NIR) – resources that cannot be accessed and downloaded via the Internet – such as a person or a country. Nevertheless, a non-information resource can be identified with a URI. Resolving the URI should result in metadata that describes the non-information resource. This idea is part of the Linked Open Data principles [21].

The distinction between information and non-information resources is relevant for the further investigation of labeling behaviour on the Web of Data: whereas NIRs are not directly accessible to the machine (i.e. the machine can talk *about* a resource, but not access or transform it), IRs can be downloaded, displayed, and further processed. IRs do not necessarily require labels in order to be useful to the end-user, whereas for NIRs there is not much else that can be used to represent them in the user interface. IRs can be represented by themselves (in case of a picture), or by a hyperlink to the document, or by the document title (in case of an HTML page or Office document). Applications such as Linked Data browsers should thus be aware of the difference, and treat NIRs and IRs differently. Indeed, some browsers do so. Tabulator [6], Explorator [2], and Graphite[2] display, for instance, images inline with the other data in the browser.

Whether a URI refers to an information resource or to a non-information resource should be determined as follows: Non-information resources should have a hash URI or, if they have a slash URI, resolving the URI should lead to an HTTP 303 See also response. Hash URIs include a fragment, with a special part that is separated from the rest of the URI by a hash symbol # [27].[3] URIs of information resources on the other hand should ultimately resolve with the given information resource, which means with an HTTP response code 200 OK (after following redirects). When we receive an error when resolving a URI (i.e. a response in the 4xx or 5xx range), we cannot infer whether this URI refers or has referred to an information resource or a non-information resource.

Even though URIs are supposed to be opaque [7], an analysis performed on URIs with extensions from the BTC 2010 corpus revealed that URIs with file name extensions such as .html or .jpg often refer to information resources. In order to test this hypothesis, we collected all URIs ending in extensions from the BTC 2010 corpus. The Billion Triple Challenge (BTC) 2010 corpus[4] is a dataset consisting of linked data crawled from the web which is stored as *ntriples*. Here, each of the 3,167,799,445 ntriples is a quad constituted by a subject, a predicate, an object, and a context, where the context is the URI of the resource the triple has been crawled from. When ignoring the context, thus reducing the quads to triples, the dataset contains 1,441,499,718[5] distinct triples. Looking through the corpus, we found 75,6 Million distinct URIs that were either in

---

[2] http://graphite.ecs.soton.ac.uk/
[3] e.g. http://www.example.com/about#alice
[4] Available at http://km.aifb.kit.edu/projects/btc-2010/, (accessed May 2011)
[5] http://gromgull.net/blog/2010/09/redundancy-in-the-btc2010-data-its-only-1-1b-triples/ (accessed 2011-06-29)

the subject or the object position.[6] Of these, 10,3 Million URIs ended in an extension (13,6%). For each extension, we selected a random sample of 50 URIs, and issued HTTP HEAD requests. The aim of the request was not to retrieve the whole resource, but only the HTTP header information. If the response to the request was a 303 See other, the URI would have been a non-information resource even though the URI ended in a file extension. Extensions that appear more than 100,000 times in the BTC 2010 corpus are .jpg, .html, .rdf, .bml, .do, .json, .ttl, .jsp, .xml, .php, .htm, .png, and .gif. The percentage of NIRs among those resources is 0% – indeed not a single URI returned a 303 among these extensions. A complete list of all extensions and results can be found online[7]. The results show that almost all URIs ending with an extension are indeed information resources, as expected. The only surprising number we encountered was among .svg files, which were encountered 3,287 times. Of these SVG URIs, 31% gave a 303 See other response. We further investigated the matter, and found that all those URIs came from DBpedia [3], and can be traced back to DBpedia's extraction mechanism, which transforms infobox links to local SVG files on Wikipedia articles as properties of a given entity.

The BTC 2010 corpus also provides a file that contains all URIs that had a 303 See other response when they have been resolved, and the URIs they have been redirected to.[8] This list contains about 6 Million URIs. Some of them point to HTML documents, and not only to RDF files, but in general we assume that this list contains a subset of the NIRs that are within the BTC 2010 corpus.

4   Labeling Properties

The RDFS standard defines the property label, which can be used to connect an entity to a name aimed at human consumption [11]. But rdfs:label is only one of the many means that are actually used on the Web to assign a human-readable name to an entity. There are several different reasons for using alternative labeling properties. Some vocabularies prefer to use more specific properties to assign names. For example, the FOAF vocabulary [12] defines foaf:name to assign a name to a person, as it sounds much more acceptable to give a person a name than a label. The SWRC ontology [29] provides swrc:name as well. SKOS even provides a set of properties for preferred and alternative labels [26], as the simple label property from RDFS is not sufficient for the needs of SKOS. Other vocabularies might provide an alternative labeling property due to legacy reasons. FOAF introduces a foaf:LabelProperty class for labeling such properties, but this is not used even within FOAF itself.

In order to find out which properties are used for labeling, we examined the BTC 2010 corpus. From the corpus we extracted the property from all quads with

---

[6] We also looked at the URIs in the property positions, but within a sample of ca. 40 Million triples we only found a single URI with an extension, and subsequently ignored this case.

[7] http://km.aifb.kit.edu/sites/label/btc/

[8] The file redirects.nx in the BTC 2010 corpus.

Table 1. Most often used properties for labeling purposes

Number of quads	Property URI
184,848,373	http://www.w3.org/2000/01/rdf-schema#label
71,742,600	http://xmlns.com/foaf/0.1/nick
17,005,858	http://purl.org/dc/elements/1.1/title
7,107,149	http://purl.org/rss/1.0/title
6,083,581	http://xmlns.com/foaf/0.1/name
2,914,013	http://purl.org/dc/terms/title
2,808,455	http://www.geonames.org/ontology#name
2,413,957	http://xmlns.com/foaf/0.1/nickname
1,649,940	http://swrc.ontoware.org/ontology#name
1,506,497	http://sw.cyc.com/CycAnnotations_v1#label
1,133,192	http://rdf.opiumfield.com/lastfm/spec#title
1,021,985	http://www.proteinontology.info/po.owl#ResidueName
713,219	http://www.proteinontology.info/po.owl#Atom
713,219	http://www.proteinontology.info/po.owl#Element
713,219	http://www.proteinontology.info/po.owl#AtomName
663,485	http://www.proteinontology.info/po.owl#ChainName
541,038	http://purl.uniprot.org/core/fullName
488,528	http://purl.uniprot.org/core/title
452,537	http://www.aktors.org/ontology/portal#has-title
434,237	http://www.w3.org/2004/02/skos/core#prefLabel
404,950	http://www.aktors.org/ontology/portal#name
391,730	http://xmlns.com/foaf/0.1/givenName
358,077	http://www.w3.org/2000/10/swap/pim/contact#fullName
337,650	http://xmlns.com/foaf/0.1/surName
336,063	http://swrc.ontoware.org/ontology#title
317,076	http://swrc.ontoware.org/ontology#booktitle
290,178	http://www.aktors.org/ontology/portal#has-pretty-name
283,754	http://purl.uniprot.org/core/orfName
253,034	http://purl.uniprot.org/core/name
211,193	http://www.daml.org/2003/02/fips55/fips-55-ont#name
186,984	http://www.geonames.org/ontology#alternateName
157,019	http://purl.uniprot.org/core/locusName
132,317	http://www.w3.org/2004/02/skos/core#altLabel
126,250	http://creativecommons.org/ns#attributionName
126,126	http://www.aktors.org/ontology/portal#family-name
126,086	http://www.aktors.org/ontology/portal#full-name

a literal with the datatype xsd:string or without a given datatype. We counted the number of occurrences for each such property. From the set of 178 properties that occurred at least 100,000 times[9] we manually assessed whether the property is used for the purpose of labeling. To do so we performed a URI lookup on the property itself, checking the label and the description of the property, and then looked at instance data. This resulted in a list of 36 properties shown in Table 1 that are used for the purpose of labeling. Note that the numbers in Table 1 should not be read as the number of labeled entities, since an entity can have multiple labels or an entity can be labeled several times in multiple contexts.

Most of these properties are not connected to rdfs:label in a way that would allow for machines to automatically discover the alternative labeling property. From the given list, only FOAF [12], SKOS [26], and Geonames[10] explicitly connect their labeling properties to rdfs:label via the rdfs:subPropertyOf property. Under both RDFS [11] and OWL 2 semantics [17], this would allow to automatically infer that any label connected with the alternative labeling property is also a valid value for rdfs:label.[11] Also, the pattern occurs so frequently that it might be worthwhile to hard-code it into an application, to avoid the overhead implied by the usage of a reasoner. Note that proteinontology contains multiple properties used for labeling. This is due to the fact, that these properties are annotated as functional properties with a given domain. For example the domain of the property po:Atom is the class po:Atoms. That means that when using such a property, besides labeling an entity this, this entity can be uniquely referred to via that label and it can be inferred that this entity belongs to class po:Atoms.

5   Metrics

In this section we define a number of metrics that help study the properties of labeling within a dataset. In the following section we will discuss the results of measuring the Web of Data along these metrics.

5.1   Completeness

All non-information resources should have labels. The labeling completeness metric LC tells us if this is indeed the case. It is the ratio of all URIs with at least one value for a labeling property to all URIs in a given knowledge base. The metric is extended with three parameters: the actual properties used to assign the label, the entities to be regarded by the metric, and the dataset.

Labeling properties are indicated by the subscript of the metric. They may be defined strictly as only rdfs:label (LC_{rdfs}), or including any formally defined

---

[9] The whole set is available at http://km.aifb.kit.edu/sites/label/btc/

[10] http://www.geonames.org

[11] Note that this was not true for the OWL 1 Lite and DL semantics since rdfs:label is an owl:AnnotationProperty [28], but OWL 2 was extended to enable this pattern.

subproperty of rdfs:label (LC_{rdfs+}), or as any other set of labeling properties lp (LC_{lp}) (such as the set presented in Section 4, which we call BTC).

The regarded entities are defined by the superscript. Most often, we will only want to consider the non-information resources (LC^{NIR}). For an automatic assessment of this metric we also must devise a method to decide whether a given URI is an information resource, or a non-information resource, as discussed in Section 3. One might also argue that some non-information resources actually do not require labels, as some resources are basically artifacts of the knowledge representation (LC^-). In RDFS and OWL this would most prominently include nodes that model n-ary relations [25].

The third parameter is given as the argument of the metric. Thus $LC(D)$ is the labeling completeness of the dataset D. We expect $LC(D)$ to always be 1 for a good knowledge base D.

Note that a dataset may include data from several RDF files, and indeed most of the time LC is defined over the merged data from a whole site. In this paper, for example, we regard the BTC as a whole, the merged data from several million look-ups.

5.2 Efficient Accessibility

A wide-spread method to work with data from the Semantic Web is called *follow your nose*, and it works due to the Linked Open Data principles [21]: whenever an application encounters an unknown URI, it can simply dereference the URI in order to access information about the entity identified by the URI. This will usually include a label for the entity of interest, and also links to other entities to which the given entity is connected, so that the application can further dereference these as well.

Assume that for the URI ex:Berlin the result of this exercise looks as follows:

```
ex:Berlin ex:location ex:Germany .
ex:Berlin rdfs:label "Berlin" .
```

A linked data browser can display the string *Berlin* to represent the resource of interest, but it has to look up both ex:location and ex:Germany before it can represent the single fact that is included in the response. If an RDF graph contains 50 triples, with about 60-80 different URIs, the application actually needs to make several dozens of HTTP requests in order to display the facts within that single resource. This turns out to be the main reason for the slow performance of linked data browsers [31]: a single browsing step can fire dozens, if not hundreds, of requests.

Imagine that the response would instead be:

```
ex:Berlin ex:location ex:Germany .
ex:Berlin rdfs:label "Berlin" .
ex:location rdfs:label "Location" .
ex:Germany rdfs:label "Germany" .
```

Now the application can display the fact without any additional lookup. This approach has nevertheless several disadvantages: it implies redundancy, and leads to larger data files. In general it is expected to nevertheless *reduce* the load and bandwidth of serving linked open data as the amount of requests would be significantly reduced.

We define the metric LE as the ratio of all mentioned URIs with at least one value for a labeling property to all mentioned URIs in a given RDF graph. The subscript and superscript are defined as for LC, the superscript can further define a background set of known labels (e.g. for a widely deployed vocabulary like FOAF or GoodRelations [22]). For example, the following graph would have a LE_{rdfs}^{foaf-} of 1, but a LE_{rdfs}^{-} of 0.5 (since the foaf:img property has no label). Note that for brevity RDF and RDFS are always assumed to be known.

```
ex:Basil foaf:img ex:basil.jpg .
ex:Basil rdfs:label "Basil" .
```

Whereas for the LC metric we can always look up a given URI, this is not allowed for LE. Nevertheless, LE with sensible parameters should always be 1 in order to increase the utility of any given response for inquiring applications.

5.3   Unambiguity

Each entity can have a whole set of different labels attached to it. This will likely yield meaningful results if the application can distinguish between these labels: SKOS includes different properties for denominating preferred and alternative labels [26], and given a multi-lingual knowledge base we expect to have several labels for a given entity, one in each language (see the following section).

But an entity can also have several labels that are not at all differentiated. In this case an application has to select one of the labels. And unless it does not have a deterministic selection procedure, the application might end up being inconsistent, displaying different labels every time the entity is displayed – which might easily lead to confusion for the user of the application. Even if the application provides a deterministic selection procedure, as long as this procedure is not common among all applications the user uses to interact with a given knowledge base, the user will be exposed to confusing inconsistencies in the interface.

We introduce the metric LU_f which is the ratio of all URIs that have exactly one preferred label according to a selection procedure f to all URIs with any label in a given knowledge base. The superscript is the same as for LC, but the subscript is replaced by the selection procedure f, which might be, in the simplest case, just selecting any value of rdfs:label (LC_{rdfs}), but could also include a more sophisticated preference function (e.g. if there is a skos:prefLabel take that, otherwise any rdfs:label).

As with all the other metrics in this paper, a good knowledge base should have a LU of 1.

5.4   Multilinguality

Language tags can be used on plain literals to state the natural language used by the literal. This enables applications to select the most appropriate literals based on their user's language preferences. An example for a literal with a language tag is `"university"@en` or `"Universität"@de`.

In order to measure multilinguality we define two metrics: LLN, the number of languages used with a labeling property, and LLC^{lang}, the completeness for a given language, i.e. the ratio of URIs with at least one label tagged with the given language or a less specific one to all URIs in a given knowledge base. The same sub- and superscripts apply as for LC (note that there are two different superscripts). If no superscript defines the language, then the average over all used languages is supposed.

6   Results

We used the metrics defined in the previous section on the BTC 2010 corpus. For measuring, we did not consider entailments as defined by the formal semantics of RDFS, OWL, or RIF. In particular we did not mush entities together through `owl:sameAs` statements or inverse functional properties, but regarded them URI by URI.

The BTC2011 corpus consists of 219 chunks. From each chunk we extracted the URIs from the first 100 nquads which resulted in 7195 URIs. For each URI we performed a lookup and identified 1376 NIRs by 303 `See other` redirect. By following the redirect and analyzing the RDF data we found that for 526 NIRs at least one label exists given the properties in Table 1. This means that only 38.2% of the analyzed NIRs have a label. Table 2 shows which properties are used to assign labels.

Table 2. Completeness of NIR labels

Number of NIRs	Labeling property
451	http://www.w3.org/2000/01/rdf-schema#label
73	http://xmlns.com/foaf/0.1/name
53	http://purl.org/dc/elements/1.1/title
20	http://xmlns.com/foaf/0.1/givenName
13	http://purl.org/dc/terms/title
5	http://xmlns.com/foaf/0.1/nick
4	http://www.w3.org/2004/02/skos/core#prefLabel

In order to measure the efficient accessibility, we looked through a sample of five random graphs from each second level domain in the BTC 2010 corpus. The results are given in Figure 1. In order to define a set of known vocabularies, we took the ten most widely used vocabularies in the BTC 2010 corpus (see Table 3).

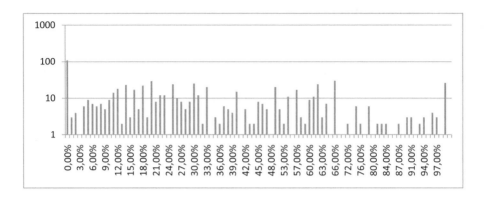

Fig. 1. Histogramm of the LE_{BTC}^{top} of up to five random graphs from each of the domains in the BTC 2010 corpus, for a total of 741 graphs

Table 3. Top ten most occurring vocabulary namespaces in the BTC 2010 corpus (according to http://gromgull.net/2010/10/btc/explore.html)

Vocabulary namespace	Number of occurences
http://www.w3.org/2000/01/rdf-schema#	845,952,387
http://data-gov.tw.rpi.edu/vocab/p/90/	651,432,324
http://www.w3.org/1999/02/22-rdf-syntax-ns#	567,247,265
http://purl.org/goodrelations/v1#	527,323,224
http://xmlns.com/foaf/0.1/	209,249,423
http://purl.uniprot.org/core/	41,961,030
http://purl.org/dc/elements/1.1/	29,596,285
http://www.proteinontology.info/po.owl#	13,661,605
http://purl.org/dc/terms/	12,579,646
http://www.w3.org/2002/07/owl#	12,362,503

We measured the unambiguity of the corpus. From the set of 57,532 NIRs that have at least one label in the corpus, 903 NIRs have multiple labels – either multiple labels for at least one of the labeling properties shown in Table 1, or multiple labels for at least one property and language. This results in an unambiguity ratio of 0.98.

Finally, we measured the multilinguality of the Web of Data. In general, most data sources contained at most one language (2.2%), if any was specified. A merry few (0.7%) contained several language tags, but even they did not have a high completeness. The most commonly used language tags are en (44.72%), de (5.22%), and fr (5.11%).

Labels are used in order to provide a human-readable names for entities. Every entity should have labels in all relevant languages. Almost none of the datasets on the Web have a full set of labels in more than one language, i.e. most ontologies are not multi-lingual. Thus they miss a potential benefit of the Semantic Web, i.e. the language-independence of the Web of Data.

7   Conclusion

Our work has investigated the current state of labeling the Web of Data, and some problems that need to be considered in the future in order to optimize the ways application developers and potential end-users interact with the data. We have defined metrics to assess the completeness, efficient accessibility, unambiguity, and multilinguality. These metrics address issues that were problematic during the development of applications. The list is not complete, but sound given that they are all based in previous experience. While defining the metrics, we noticed that we had to include a number of parameters that depend on the application that will use the knowledge. This is not surprising: data on the Web of Data is hardly ever evaluable by itself – it greatly benefits from knowing the context of an application that will use the data. The parameters in the evaluation metrics allow to customize the metrics based on the given application, on the labeling properties the application understands, and on the set of entities that are expected to play a role when using the application.

Based on our findings and the argumentation leading to the definition our metrics, we can nevertheless make a number of suggestions on how to improve the quality and usefulness of labels in the data:

- Provide labels for all URIs *mentioned* in a given RDF graph, not only for the *main entities*, as this will considerably speed displaying the data with human-readable names and reduce the number of requests significantly.
- Provide a complete set of labels in all supported languages. One of the biggest advantages of the Web of Data is its inherent multilinguality, but currently this is a tremendously underused feature of the architecture.
- If you are using a labeling property of your own, connect your labeling property to `rdfs:label` explicitly with the `rdfs:subPropertyOf` property. Use `rdfs:label` redundantly as well, since many tools will not provide the inferencing needed to understand your labeling property. If possible, simply avoid using your own labeling property.
- Do not provide more than one obvious preferred label for each entity, in order to decrease the possible confusion for the end-user when using an application over your data.

The suggestions given above lead to an obvious problem: even a moderately small RDF graph with about 100 triples will include about 150 entities. Labeling all these entities in, e.g. ten languages will lead to an extra 1500 triples – a huge overhead (and not even considering the costs creating those labels, a task that would highly benefit from automation). While one could devise new protocols to deal with these problems, there is also an under-utilized existing solution: HTTP allows to set the `Accept-Language` header, that defines a set of natural languages the response should cover [16]. By using the HTTP headers a data provider could both provide all labels necessary for an efficient exposure of the data, as well as not unnecessarily inflate the size of the response by only providing the requested languages.

Labels should follow a style guide and be used consistently. A style guide should define if classes are labeled with plural or singular noun, if properties are labeled with nouns or verbs, etc. Labels should never use camel case or similar escape mechanisms for multi word terms, but instead simply use space characters (or whatever is most suitable for the given language). I.e. an URI `http://example.org/LargeCity` should have a label `"large city"@en`. External dictionaries such as WordNet [15] can be used to check consistency with regards to a style guide.

In an environment where datasets are assembled on the fly from multiple datasets [1], the assembled parts may follow different style guides. The assembled dataset will then not adhere to a single style guide and thus offer an inconsistent user interface. It is not expected that a single style guide will become ubiquitous on the whole Web. Instead, a dataset may specify explicitly what style guide it follows, and even provide labels following different style guides. This would allow to introduce a subproperty of label that is style guide specific, which would in return allow for the consistent display of assembled datasets.

Even when subproperties of `rdfs:label` are defined, there should always be one label (per supported language) given explicitly by using `rdfs:label` itself. Even though this is semantically redundant, many tools (especially visualization tools) do not apply reasoning for fetching the labels of an entity but simply look for the explicit triple stating the entity's label.

Many of the problems described in this paper are a consequence of publishing data using the Linked Open Data principles. It is not clear if this is indeed the best way to publish data on the Web of Data. Serving data through a SPARQL endpoint provides a viable alternative, with the big advantage that the application can, in a very fine-grained way, describe exactly what kind of information, labels, and language it needs. The SPARQL endpoint can then try to understand the query and do its best effort to provide a viable response.

The Linked Open Data principles have spread widely due to their obvious advantages derived from being part of the Web architecture. But the principles are meeting their limitations, as this investigation on labels shows. The Semantic Web has long struggled with the chicken and egg problem of data vs. applications. Now that the data is there, we see that applications don't yet follow with the same force that the datasets had. One of the reasons is the lack of quality in some of the published datasets.

Labeling may be just a small, but at the same time it is an absolutely essential piece of the puzzle that is needed for the Web of Data to finally become widely used.

Acknowledgements. We thank Andreas Harth for his support on how to crawl the data. The work presented in this paper is supported by the European Union's 7th Framework Programme (FP7/2007-2013) under Grant Agreement 257790.

References

1. Alani, H.: Position paper: ontology construction from online ontologies. In: Carr, L., Roure, D.D., Iyengar, A., Goble, C.A., Dahlin, M. (eds.) Proceedings of the 15th International Conference on World Wide Web (WWW 2006), Edinburgh, Scotland, pp. 491–495. ACM (May 2006)
2. Araújo, S.F.C., Schwabe, D., Barbosa, S.D.J.: Experimenting with explorator: a direct manipulation generic rdf browser and querying tool
3. Auer, S., Lehmann, J.: What Have Innsbruck and Leipzig in Common? Extracting Semantics from Wiki Content. In: Franconi, E., Kifer, M., May, W. (eds.) ESWC 2007. LNCS, vol. 4519, pp. 503–517. Springer, Heidelberg (2007)
4. Becker, C., Bizer, C.: Marbles (2009)
5. Berners-Lee, T., Chen, Y., Chilton, L., Connolly, D., Dhanaraj, R., Hollenbach, J., Lerer, A., Sheets, D.: Tabulator: Exploring and Analyzing linked data on the Semantic Web. In: Proceedings of the 3rd International Semantic Web User Interaction Workshop, vol. 2006 (2006)
6. Berners-Lee, T., Chen, Y., Chilton, L., Connolly, D., Dhanaraj, R., Hollenbach, J., Lerer, A., Sheets, D.: Tabulator: Exploring and analyzing linked data on the Semantic Web. In: Rutledge, L., Schraefel, M.C., Bernstein, A., Degler, D. (eds.) Proceedings of the Third International Semantic Web User Interaction Workshop SWUI 2006 at the International Semantic Web Conference ISWC (2006)
7. Berners-Lee, T., Fielding, R., Masinter, L.: Uniform Resource Identifier (URI): Generic Syntax. Technical Report 3986, Internet Engineering Task Force, RFC 3986 (June 2005), http://www.ietf.org/rfc/rfc3986.txt
8. Berners-Lee, T., Hendler, J., Lassila, O.: The semantic web. Scientific American 2001(5) (2001), http://www.sciam.com/2001/0501issue/0501berners-lee.html
9. Bizer, C., Gau, T.: Disco - hyperdata browser (January 2007)
10. Brickley, D., Guha, R.: RDF Vocabulary Description Language 1.0: RDF Schema. W3C Recommendation (2004)
11. Brickley, D., Guha, R.V.: RDF Vocabulary Description Language 1.0: RDF Schema. W3C Recommendation (February 2004)
12. Brickley, D., Miller, L.: The Friend Of A Friend (FOAF) vocabulary specification (July 2005)
13. d'Aquin, M., Baldassarre, C., Gridinoc, L., Angeletou, S., Sabou, M., Motta, E.: Characterizing knowledge on the semantic web with watson. In: Garcia-Castro, R., Vrandecic, D., Gómez-Pérez, A., Sure, Y., Huang, Z. (eds.) EON. CEUR Workshop Proceedings, vol. 329, pp. 1–10. CEUR-WS.org (2007)
14. Ding, L., Finin, T.: Characterizing the Semantic Web on the Web. In: Cruz, I., Decker, S., Allemang, D., Preist, C., Schwabe, D., Mika, P., Uschold, M., Aroyo, L.M. (eds.) ISWC 2006. LNCS, vol. 4273, pp. 242–257. Springer, Heidelberg (2006)
15. Fellbaum, C.: WordNet: An Electronic Lexical Database (Language, Speech, and Communication). MIT Press (May 1998)
16. Fielding, R., Gettys, J., Mogul, J., Frystyk, H., Masinter, L., Leach, P., Berners-Lee, T.: Hypertext Transfer Protocol – HTTP/1.1. RFC 2616 (June 1999)
17. Grau, B.C., Horrocks, I., Motik, B., Parsia, B., Patel-Schneider, P., Sattler, U.: OWL 2: The next step for OWL. Web Semantics: Science, Services and Agents on the World Wide Web 6(4), 309–322 (2008)

18. Grosso, W.E., Eriksson, H., Fergerson, R.W., Tu, O.S.W., Musen, M.A.: Knowledge modeling at the millennium: the design and evolution of PROTEGE-2000. In: Proceedings of the 12th International Workshop on Knowledge Acquisition, Modeling and Management (KAW-1999), Banff, Canada (October 1999)
19. He, H., Wang, H., Yang, J., Yu, P.S.: Blinks: ranked keyword searches on graphs. In: Chan, C.Y., Ooi, B.C., Zhou, A. (eds.) SIGMOD Conference, pp. 305–316. ACM (2007)
20. Heath, T.: How Will We Interact with the Web of Data? IEEE Internet Computing 12, 88–91 (2008)
21. Heath, T., Bizer, C.: Linked Data: Evolving the Web into a Global Data Space. In: Synthesis Lectures on the Semantic Web: Theory and Technology. Morgan & Claypool (2011)
22. Hepp, M.: GoodRelations: An Ontology for Describing Products and Services Offers on the Web. In: Gangemi, A., Euzenat, J. (eds.) EKAW 2008. LNCS (LNAI), vol. 5268, pp. 329–346. Springer, Heidelberg (2008)
23. Jacobs, I., Walsh, N.: Architecture of the World Wide Web 2004. W3C Recommendation 1 (December 15, 2004), http://www.w3.org/TR/webarch/
24. Manaf, N.A.A., Bechhofer, S., Stevens, R.: A Survey of Identifiers and Labels in OWL Ontologies. In: Proceedings of the 6th International Workshop on OWL Experiences and Directions, OWLED 2010 (2010)
25. Noy, N., Rector, A.: Defining n-ary relations on the semantic web. W3C Working Group Note (April 2006), http://www.w3.org/TR/swbp-n-aryRelations/
26. Miles, A., Bechhofer, S.: SKOS Simple Knowledge Organization System Reference. W3C Recommendation (August 18, 2009), http://www.w3.org/TR/skos-reference/ (2009)
27. Sauermann, L., Cyganiak, R.: Cool URIs for the Semantic Web. W3C Interest Group Note (December 2008)
28. Smith, M.K., Welty, C., McGuinness, D.: OWL Web Ontology Language Guide. W3C Recommendation (February 10, 2004), http://www.w3.org/TR/owl-guide/
29. Sure, Y., Bloehdorn, S., Haase, P., Hartmann, J., Oberle, D.: The SWRC Ontology - Semantic Web for Research Communities. In: Bento, C., Cardoso, A., Dias, G. (eds.) EPIA 2005. LNCS (LNAI), vol. 3808, pp. 218–231. Springer, Heidelberg (2005)
30. Tran, D.T., Wang, H., Rudolph, S., Cimiano, P.: Top-k exploration of query candidates for efficient keyword search on graph-shaped (rdf) data. In: Proceedings of the 25th International Conference on Data Engineering (ICDE 2009), Shanghai, China (März 2009)
31. Vrandečić, D., Ratnakar, V., Krötzsch, M., Gil, Y.: Shortipedia: Aggregating and curating semantic web data. In: Proceedings of the ISWC 2010, Shanghai, China (November 2010)
32. Wang, T.D., Parsia, B., Hendler, J.: A Survey of the Web Ontology Landscape. In: Cruz, I., Decker, S., Allemang, D., Preist, C., Schwabe, D., Mika, P., Uschold, M., Aroyo, L.M. (eds.) ISWC 2006. LNCS, vol. 4273, pp. 682–694. Springer, Heidelberg (2006)

Semantic Search: Reconciling Expressive Querying and Exploratory Search

Sébastien Ferré[1] and Alice Hermann[2]

[1] IRISA/Université de Rennes 1, Campus de Beaulieu, 35042 Rennes cedex, France
ferre@irisa.fr
[2] IRISA/INSA de Rennes, Campus de Beaulieu, 35708 Rennes cedex 7, France
alice.hermann@irisa.fr

Abstract. Faceted search and querying are two well-known paradigms to search the Semantic Web. Querying languages, such as SPARQL, offer expressive means for searching RDF datasets, but they are difficult to use. Query assistants help users to write well-formed queries, but they do not prevent empty results. Faceted search supports exploratory search, i.e., guided navigation that returns rich feedbacks to users, and prevents them to fall in dead-ends (empty results). However, faceted search systems do not offer the same expressiveness as query languages. We introduce *Query-based Faceted Search* (QFS), the combination of an expressive query language and faceted search, to reconcile the two paradigms. In this paper, the LISQL query language generalizes existing semantic faceted search systems, and covers most features of SPARQL. A prototype, Sewelis (aka. Camelis 2), has been implemented, and a usability evaluation demonstrated that QFS retains the ease-of-use of faceted search, and enables users to build complex queries with little training.

1 Introduction

With the growing amount of available resources in the Semantic Web (SW), it is a key issue to provide an easy and effective access to them, not only to specialists, but also to casual users. The challenge is not only to allow users to retrieve particular resources (e.g., flights), but to support them in the exploration of a knowledge base (e.g., which are the destinations? Which are the most frequent flights? With which companies and at which price?). We call the first mode *retrieval search*, and, following Marchionini [10], the second mode *exploratory search*. Exploratory search is often associated to *faceted search* [5,13], but it is also at the core of Logical Information Systems (LIS) [4,2], and Dynamic Taxonomies [12]. Exploratory search allows users to find information without *a priori* knowledge about either the data or its schema. Faceted search works by suggesting restrictions, i.e., selectors for subsets of the current selection of items. Restrictions are organized into facets, and only those that share items with the current selection are suggested. This has the advantage to provide guided navigation, and to prevent dead-ends, i.e., empty selections. Therefore, faceted search is *easy-to-use* and *safe*: *easy-to-use* because users only have to

L. Aroyo et al. (Eds.): ISWC 2011, Part I, LNCS 7031, pp. 177–192, 2011.
© Springer-Verlag Berlin Heidelberg 2011

choose among the suggested restrictions, and *safe* because, whatever the choice made by users, the resulting selection is not empty. The selections that can be reached by navigation correspond to queries that are generally limited to conjunctions of restrictions, possibly with negation and disjunction on values. This is far from the expressiveness of query languages for the semantic web, such as SPARQL[1]. There are *semantic faceted search* that extend the expressiveness of reachable queries, but still to a small fragment of SPARQL (e.g., SlashFacet [7], BrowseRDF [11], SOR [9], gFacet [6]). For instance, none of them allow for cycles in graph patterns, unions of complex graph patterns, or negations of complex graph patterns.

Querying languages for the semantic web are quite expressive but are difficult to use, even for specialists. Users are asked to fill an empty field (problem of the writer's block), and nothing prevents them to write a query that has no answer (dead-end). Even if users have a perfect knowledge of the syntax and semantics of the query language, they may be ignorant about the data schema, i.e., the *ontology*. If they also master the ontology or if they use a graphical query editor (e.g., SemanticCrystal [8], SCRIBO Graphical Editor[2]) or an auto-completion system (e.g., Ginseng [8]) or keyword query translation (e.g., Hermes [14]), the query will be syntactically correct and semantically consistent w.r.t. the ontology but it can still produce no answer.

The contribution of this paper, *Query-based Faceted Search* (QFS), is to define a semantic search that is (1) easy to use, (2) safe, and (3) expressive. Ease-of-use and safeness are retained from existing faceted search systems by keeping their general principles, as well as the visual aspect of their interface. Expressiveness is obtained by representing the current selection by a *query* rather than by a set of items, and by representing navigation links by *query transformations* rather than by set operations (e.g., intersection, crossing). In this way, the expressiveness of faceted search is determined by the expressiveness of the query language, rather than by the combinatorics of user interface controls. In this paper, the query language, named LISQL, generalizes existing semantic faceted search systems, and covers most features of SPARQL. The use of queries for representing selections in faceted search has other benefits than navigation expressiveness. The current query is an intensional description of the current selection that complements its extensional description (list of items). It informs users in a precise and concise way about their exact position in the navigation space. It can easily be copied and pasted, stored and retrieved later. Finally, it allows expert users to modify the query by hand at any stage of the navigation process, without loosing the ability to proceed by navigation.

The paper is organized as follows. Section 2 discusses the limits of set-based faceted search by formalizing the navigation from selection to selection. Section 3 introduces LISQL queries and their transformations. In Section 4, navigation with QFS is formalized and proved to be *safe* and *complete* w.r.t. LISQL, and *efficient*. Section 5 reports about a usability evaluation, and Section 6 concludes.

---

[1] see http://www.w3.org/TR/rdf-sparql-query/

[2] http://www.scribo.ws/xwiki/bin/view/Blog/SparqlGraphicalEditor

2   Limits of Set-Based Faceted Search

The principle of faceted search [13] is to guide users from *selection* of items to selection of items. At each navigation step, a new selection is derived by applying a set operation between the current selection S and a *restriction* R. A restriction is a *feature* that applies to at least one item of the current selection, i.e., $S \cap R \neq \emptyset$. Typically, a feature is a pair facet-value, and the set operation is intersection: $S := S \cap R$. The new selection is the set of items that belong to the current selection, and that belong to the restriction. Extensions of faceted search may allow for the exclusion of a restriction ($S := S \setminus R$), or the union with a restriction ($S := S \cup R$). Restrictions can also be tags or item names.

In the context of the Semantic Web, items and values are resources, facets are properties, and tags are classes. Because of the relational nature of semantic data, new kinds of restrictions and set operations have been introduced in semantic faceted search (e.g., /facet [7], BrowseRDF [11], SOR [9], gFacet [6]). A restriction can be the set of items that are subject of some property (the domain of the property), or that are object of some property (the range of the property) (e.g., BrowseRDF). A facet can be defined as a path of properties. Finally, a property p can be crossed forwards ($S := p(S,.)$) or backwards ($S := p(.,S)$) (e.g., /facet, SOR, gFacet).

Both in theory and in practice, it is useful to distinguish between syntax and semantics. For example, we should distinguish between a pair facet-value (syntax), and the set of items it matches (semantics). In the following table, we define the syntax and semantics of the various kinds of restrictions: r denotes any RDF resource (URI, literal), c denotes a RDFS class, p denotes a RDF property, and S_0 denotes the set of all items (possibly all resources of a RDF dataset).

restriction	syntax	semantics	examples
name	r	$\{r\}$	`<JohnSmith>`, `"John"`, `2011`
tag	`a` c	$\mathrm{rdf{:}type}(.,\{c\})$	`a person`
(facet, value)	p `:` r	$p(.,\{r\})$	`year : 2011`
(facet, value)	p `of` r	$p(\{r\},.)$	`mother of <JohnSmith>`
domain	p `: ?`	$p(.,S_0)$	`year : ?`
range	p `of ?`	$p(S_0,.)$	`mother of ?`

The same distinction can be made for complex selections, and we introduce in the following table a syntax for the various set operations that can be applied between selections and restrictions: S denotes a selection, and R denotes a restriction that is relevant to S: i.e., $S \cap R \neq \emptyset$.

selection	syntax	semantics
initial	`?`	S_0
intersection	S `and` R	$S \cap R$
exclusion	S `and not` R	$S \setminus R$
union	S `or` R	$S \cup R$
crossing backwards	p `:` S	$p(.,S)$
crossing forwards	p `of` S	$p(S,.)$

The syntactic form of restrictions are *features*. The syntactic form of selections are *queries* whose answers are sets of items, i.e., subsets of S_0. The above tables implicitly define a grammar for features and queries:

$$S \to ? \mid S \text{ and } R \mid S \text{ and not } R \mid S \text{ or } R \mid p \; : \; S \mid p \text{ of } S$$
$$R \to r \mid a \; c \mid p \; : \; r \mid p \text{ of } r \mid p \; : \; ? \mid p \text{ of } ?.$$

This grammar already defines a rich language of accessible queries, but it has strong limits in terms of flexibility and expressivity, as we discuss now. To reach some selections requires a precise ordering in navigation steps, which hinders the flexibility of the search, and assumes that the user has a clear idea of his query in advance. For example, to reach the query `father of (mother of (name : "John") and name : "Jane")`, the user has first to select `name : "John"` (*people named John*), then to cross forward `mother` (*their mothers*), then to intersect with `name : "Jane"` (*...whose name is Jane*), and finally to cross forward `father` (*their fathers*). Any other ordering will fail; starting from the expected result (grand-fathers) will lead to the set of grand-children instead.

Some useful selections that can be defined in terms of set operations are not reachable by set-based faceted search. For example, the following kinds of selections are not reachable: unions of complex selections. e.g., $(R_1 \cap R_2) \cup (R_3 \cap R_4)$; or intersection of crossings from complex selections, e.g., $p_1(., R_1 \cap R_2) \cap p_2(., R_3 \cap R_4)$. Note that a selection $S_1 \cap p(., S_2)$ cannot in general be obtained by first navigating to S_1, then crossing forwards p, navigating to S_2, and finally crossing backwards p, because it is not equivalent to $p(., p(S_1, .) \cap S_2)$ unless p is inverse functional. Therefore, not all combinations of intersection, union, and crossing are reachable, which is counter-intuitive and limiting for end users.

Existing approaches to semantic faceted search often have additional limitations, which are sometimes hidden behind a lack of formalization. A same facet (a property path) cannot be used several times, which is fine for functional properties but not for relations such as "child": $p(., f_1 \cap f_2)$ is reachable but not $p(., f_1) \cap p(., f_2)$ (e.g., BrowseRDF, gFacet). A property whose domain and range are the same cannot be used as a facet (e.g., /facet), which includes all family and friend relationships for instance.

3   Expressive Queries and Their Transformations

The contribution of our approach, *Query-based Faceted Search* (QFS), is to significantly improve the expressivity of faceted search, while retaining its properties of safeness (no dead-end), and ease-of-use. The key idea is to define navigation steps at the syntactic level as query transformations, rather than at the semantic level as set operations. The navigation from selection to selection, as well as the computation of restrictions related to the current selection, are retained by defining the semantics of features and queries, i.e., the mapping from a feature f or a query q to a set of items: $R = items(f)$ and $S = items(q)$. Transformations at the syntactic level are necessary because there exist useful navigation steps that cannot be obtained by applying set operations on the current selection. For example, given $S = R_1 \cap R_2$, the set of items $S' = R_1 \cap (R_2 \cup R_3)$ cannot be derived from S and R_3. On the contrary, the query f_1 and (f_2 or f_3) can be derived from the query f_1 and f_2 and the feature f_3 because enough information is retained at the syntactic level.

In this section, we generalize in a natural way the set of queries compared to Section 2. This defines a query language, which we call LISQL (LIS Query Language). We then define a set of query transformations so that every LISQL query can be reached in a finite sequence of such transformations. This is in contrast with previous contributions in faceted search that introduce new selection transformations, and leave the query language implicit. We think that making the language of reachable queries explicit is important for reasoning on and comparing different faceted search systems. In Section 3.3, we give a translation from LISQL to SPARQL, the reference query language of the Semantic Web. This provides both a way to compute the answers of queries with existing tools, and a way to evaluate the level of expressivity achieved by LISQL.

3.1   The LIS Query Language (LISQL)

A more general query language, LISQL, can be obtained simply by merging the syntactic categories of features and queries in the grammar of Section 2, so that every query can be used in place of a feature.

Definition 1 (LISQL queries). *The syntax and semantics of the LISQL constructs is defined in the following table, where r is a resource, c is a class, p is a property, S_0 is the set of all items, and q_1, q_2 are LISQL queries s.t. $S_1 = items(q_1)$ and $S_2 = items(q_2)$.*

query	syntax (q)	semantics $(items(q))$
resource	r	$\{r\}$
class	a c	$rdf{:}type(., \{c\})$
all	?	S_0
crossing backwards	p : q_1	$p(., S_1)$
crossing forwards	p of q_1	$p(S_1, .)$
complement	not q_1	$S_0 \setminus S_1$
intersection	q_1 and q_2	$S_1 \cap S_2$
union	q_1 or q_2	$S_1 \cup S_2$

The definition of LISQL allows for the arbitrary combination of intersection, union, complement, and crossings. In order to further improve the expressiveness of LISQL from tree patterns to graph patterns, we add variables (e.g., ?X) as an additional construct. Variables serve as co-references between distant parts of the query, and allows for the expression of cycles. For example, the query that selects people who are an employee of their own father can be expressed as a person and father : ?X and employee of ?X, or alternately as a person and ?X and employee of father of ?X. The semantics of queries with variables is given with the translation to SPARQL in Section 3.3, because it cannot be defined like in the table of Definition 1.

Syntactic constructs are given in increasing priority order, and brackets are used in concrete syntax for disambiguation. The most general query ? is a neutral element for intersection, and an absorbing element for union. In the following, we use the example query q_{ex} = a person and birth : (year : (1601 or 1649) and place : (?X and part of England)) and father : birth : place : not ?X, which uses all constructs of LISQL, and selects the set of "persons born in 1601 or 1649 at some place in England, and whose father is born at another place".

3.2 Query Transformations

We have generalized the query language by allowing complex selections in place of restrictions: e.g., $S_1 \cap S_2$ instead of $S \cap R$. However, because the number of suggested restrictions in faceted search must be finite, it is not possible to suggest arbitrarily complex restrictions. More precisely, the *vocabulary* of features must be finite. In QFS, we retain the same set of features as in Section 2, which is a finite subset of LISQL for any given dataset.

The key notion we introduce to reconcile this finite vocabulary, and the reachability of arbitrary LISQL queries is the notion of *focus* in a query.

Definition 2 (focus). *A focus of a LISQL query q is a node of the syntax tree of q, or equivalently, a subquery of q. The set of foci of q is noted $\Phi(q)$; the root focus corresponds to the root of the syntax tree, and represents the whole query. The subquery at focus $\phi \in \Phi(q)$ is noted $q[\phi]$; and $q[\phi := q_1]$ denotes the modified query q, where the subquery at focus ϕ has been replaced by q_1.*

In the following, when it is necessary to refer to a focus in a query, the corresponding subquery is underlined with the focus name as a subscript, like in mother of $\underline{?}_\phi$. Foci are used in QFS to specify on which subquery a query transformation should be applied. For example, the query (f_1 and f_2) or (f_3 and f_4) can be reached from the query (f_1 and f_2) or f_3 by applying the intersection with restriction f_4 to the subquery f_3, instead of to the whole query. Similarly, the query p_1 : (f_1 and f_2) and p_2 : (f_3 and f_4) can be reached by applying the intersection with restriction f_4 to the subquery f_3. This removes the problem of unreachable selections in set-based faceted search presented in Section 2. Moreover, this removes the need for a strategy in the ordering of navigation steps. For example, the query a woman and mother of name : "John" can be reached by first selecting a woman, then by selecting mother of $\underline{?}_\phi$, then by inserting name : "John" at the focus ϕ.

Definition 3 (query transformation). *The different kinds of LISQL query transformations are listed in the following table, where each transformation is paramaterized by a focus ϕ and a query q_1. The expression $q[t]$ is the query that results from the application of transformation t to query q.*

transformation	notation (t)	result query ($q[t]$)
intersection	ϕ and q_1	$q[\phi := q[\phi]$ and $q_1]$
exclusion	ϕ and not q_1	$q[\phi := q[\phi]$ and not $q_1]$
union	ϕ or q_1	$q[\phi := q[\phi]$ or $q_1]$

We show in the following equations how the intersection with an arbitrary LISQL query can be recursively decomposed into a finite sequence of intersections with features, and exclusions and unions with the most general query ?.

$$
\begin{aligned}
q[\phi \text{ and } (?)] &= q \\
q[\phi \text{ and } (p : q_1)] &= q[\phi \text{ and } p : \underline{?}_{\phi_1}][\phi_1 \text{ and } q_1] \\
q[\phi \text{ and } (p \text{ of } q_1)] &= q[\phi \text{ and } p \text{ of } \underline{?}_{\phi_1}][\phi_1 \text{ and } q_1] \\
q[\phi \text{ and } (\text{not } q_1)] &= q[\phi \text{ and not } \underline{?}_{\phi_1}][\phi_1 \text{ and } q_1] \\
q[\phi \text{ and } (q_1 \text{ and } q_2)] &= q[\phi \text{ and } \underline{q_1}_{\phi_1}][\phi_1 \text{ and } q_2] \\
q[\phi \text{ and } (q_1 \text{ or } q_2)] &= q[\phi \text{ and } \underline{q_1}_{\phi_1}][\phi_1 \text{ or } \underline{?}_{\phi_2}][\phi_2 \text{ and } q_2]
\end{aligned}
$$

For example, the complex query q_{ex} = a person and birth : (year : (1601 or 1649) and place : (?X and part of England)) and father : birth : place : not ?X can be reached through the navigation path: $\underline{?}_{\phi_0}[\phi_0$ and a person] $[\phi_0$ and birth : $\underline{?}_{\phi_1}]$ $[\phi_1$ and year : $\underline{1601}_{\phi_2}]$ $[\phi_2$ or $\underline{?}_{\phi_3}]$ $[\phi_3$ and 1649] $[\phi_1$ and place : $\underline{?}_{\phi_4}]$ $[\phi_4$ and ?X] $[\phi_4$ and part of England] $[\phi_0$ and father : $\underline{?}_{\phi_5}]$ $[\phi_5$ and birth : $\underline{?}_{\phi_6}]$ $[\phi_6$ and place : $\underline{?}_{\phi_7}]$ $[\phi_7$ and not $\underline{?}_{\phi_8}]$ $[\phi_8$ and ?X]. The classical facet-value features appear to be redundant for navigation as their intersection can be decomposed, but they are still useful for visualization in a faceted search interface.

Sequences of query transformations are analogous to the use of graphical query editors, but the key difference is that answers and restrictions are returned at each step, providing feedback, understanding-at-a-glance, no dead-end, and all benefits of exploratory search. Despite the syntax-based definition of navigation steps, those have a clear semantic counterpart. Intersection is the same as in standard faceted search, only making it available on the different entities involved in the current query. In the above example, intersection is alternately applied to the person, his birth, his birth's place, his father, etc. The set of relevant restrictions is obviously different at different foci. The union transformation introduces an alternative to some subquery (e.g., an alternative birth's year). The exclusion transformation introduces a set of exceptions to the subquery (e.g., excluding some father's birth's place). In Section 4, we precisely define which query transformations are suggested at each navigation step, and we prove that the resulting navigation graph is safe (no dead-end), and complete (every "safe" query is reachable).

3.3 Translation to and Comparison with SPARQL

We here propose a (naive) translation of LISQL queries to SPARQL queries. It involves the introduction of variables that are implicit in LISQL queries. As this translation applies to LISQL queries with co-reference variables, it becomes possible to compute their set of items.

Definition 4 (SPARQL translation). *The* SPARQL translation *of a LISQL query q is* $sparql(q) = $ SELECT DISTINCT $?x$ WHERE $\{ S_0(x)\ GP(x, q) \}$, *where* the graph pattern $S_0(x)$ binds x to any element of the set of all items S_0, and the function GP inductively defines the graph pattern of q with variable x representing the root focus.

$$
\begin{aligned}
GP(x, ?v) &= S_0(v) \text{ FILTER } (?x = ?v) \\
GP(x, r) &= \text{FILTER } (?x = r) \\
GP(x, \texttt{a } c) &= ?x \text{ rdf:type } c \\
GP(x, p : q_1) &= ?x \; p \; ?y. \; GP(y, q_1) \quad \text{where } y \text{ is a fresh variable} \\
GP(x, p \text{ of } q_1) &= ?y \; p \; ?x. \; GP(y, q_1) \quad \text{where } y \text{ is a fresh variable} \\
GP(x, ?) &= \{ \; \} \\
GP(x, \texttt{not } q_1) &= \text{NOT EXISTS } \{ \; GP(x, q_1) \; \} \\
GP(x, q_1 \text{ and } q_2) &= GP(x, q_1) \; GP(x, q_2) \\
GP(x, q_1 \text{ or } q_2) &= \{ \; GP(x, q_1) \; \} \text{ UNION } \{ \; GP(x, q_2) \; \}
\end{aligned}
$$

We now discuss the translations of LISQL queries compared to SPARQL in general. They have only one variable in the SELECT clause because of the nature of faceted search, i.e., navigation from set to set. From SPARQL 1.0, LISQL misses the optional graph pattern, and the named graph pattern. Optional graph patterns are mostly useful when there are several variables in the SELECT clause. LISQL has the NOT EXISTS construct of SPARQL 1.1. If we look at the graph patterns generated for intersection and union, the two subpatterns necessarily share at least one variable, x. This is a restriction compared to SPARQL, but one that makes little difference in practice as disconnected graph patterns are hardly useful in practice.

4   A Safe and Complete Navigation Graph

In this section, we formally define the navigation space over a RDF dataset as a graph, where vertices are navigation places, and edges are navigation links. A navigation place is made of a query q and a focus ϕ of this query. The focus determines the selection of items to be displayed, and the corresponding restrictions at this focus. A navigation link is defined by a query transformation and, possibly, a focus move. Before defining the navigation graph itself, we first define the set of items and the set of restrictions for some query q and some focus $\phi \in \Phi(q)$. The set of items is defined as the set of items of the query $flip(q, \phi)$ that is the reformulation of q from the point of view of the focus ϕ. For example, the reformulation, called the *flip*, of the query a woman and mother of name : "John"$_\phi$ is the query name : "John" and mother : a woman.

Definition 5 (flip at focus). *The* flip *of a query q at a focus $\phi \in \Phi(q)$ is defined as $flip(q, \phi) = flip'(?, q, \phi)$, where the function $flip'(k, q', \phi)$ is inductively defined, with k representing the context of q' in q, by (only main cases are given):*

$$
\begin{aligned}
flip(k, p : q_1, \phi) &= flip(p \text{ of } k, q_1, \phi) & \text{if } \phi \in q_1 \\
flip(k, q_1 \text{ and } q_2, \phi) &= flip(k \text{ and } q_2, q_1, \phi) & \text{if } \phi \in q_1 \\
flip(k, q_1 \text{ or } q_2, \phi) &= flip(k, q_1, \phi) & \text{if } \phi \in q_1 \\
flip(k, \texttt{not } q_1, \phi) &= flip(k, q_1, \phi) & \text{if } \phi \in q_1 \\
flip(k, q', \phi) &= q' \text{ and } k & \text{otherwise}
\end{aligned}
$$

When the focus is in the scope of an union, only the alternative that contains the focus is used in the flipped query. This is necessary to have the correct set of restrictions at that focus, and this is also useful to access the different subselections that compose an union. For example, in the query a man and (firstname : "John"$_\phi$ or lastname : "John"), the focus ϕ allows to know the set of men whose firstname is John without forgetting the second alternative. When the focus is in the scope of a complement, this complement is ignored in the flipped query. This is useful to access the subselection to be excluded. For example, in the query a man and not father : ?$_\phi$, the focus ϕ allows to know the set of men who have a father, i.e., those who are to be excluded from the selection of men.

Definition 6 (items at focus). *The* items *of a query q at focus ϕ is defined as the items of the flip of q at focus ϕ, i.e., $items(q, \phi) = items(flip(q, \phi))$.*

This enables the definition of the set of restrictions at each focus in the normal way. The navigation graph can then be formally defined.

Definition 7 (restrictions at focus). *The* restrictions *of a query q at focus ϕ is defined as the features that share items with the query q at focus ϕ:*

$$restr(q, \phi) = \{f \mid items(q, \phi) \cap items(f) \neq \emptyset\}.$$

Definition 8 (navigation graph). *Let D be a RDF dataset. The* navigation graph $G_D = (V, E)$ *of D has its set of vertices defined by*

$$V = \{(q, \phi) \mid q \in LISQL, \phi \in \Phi(q)\},$$

and its set of edges defined by the following table for every vertice (q, ϕ). The notation $(q', \phi') = (q, \phi)[l]$ denotes the navigation place obtained by traversing the navigation link l from the navigation place (q, ϕ).

navigation link	notation (l)	target ((q', φ'))	conditions
focus change	focus φ'	(q, ϕ')	for every focus $\phi' \in \Phi(q)$
intersection	and f	$(q[\phi$ and $f_{\phi'}], \phi')$	for every $f \in restr(q, \phi)$
exclusion	and not ?	$(q[\phi$ and not $?_{\phi'}], \phi')$	
union	or ?	$(q[\phi$ or $?_{\phi'}], \phi')$	
name	name ?v	$(q[\phi$ and $?v_{\phi'}], \phi')$	for some fresh variable v
reference	ref ?v	$(q[\phi$ and $?v_{\phi'}], \phi')$	for every $v \in vars(q)$ s.t. $items(q', \phi') \neq \emptyset$
delete	delete	$(q[\phi := ?], \phi)$	

The number of navigation places (vertices) is infinite because there are infinitely many LISQL queries, but the number of outgoing navigation links (edges) is finite at each navigation place because the vocabulary of features is finite, and the number of foci and variables in a query is finite. By default, the initial navigation place is $v_0 = (?_\phi, \phi)$. The following lemma shows that intersection navigation links behave as in standard faceted search.

Lemma 1. *For every query q, focus $\phi \in \Phi(q)$, and feature f, the following equality holds: $items((q, \phi)[$and $f]) = items(q, \phi) \cap items(f)$.*

4.1 Safeness and Completeness

From the formal definition of navigation graphs, we can now formally state safeness and completeness theorems. Those theorems have subtle conditions w.r.t. focus change, and the main purpose of this section is to discuss them. For reasons of space, lemmas and proofs have been removed, but they are fully available in a research report [3] (the presentation is slightly different but equivalent).

Theorem 1 (safeness). *Let D be a RDF dataset. The navigation graph G_D is safe except for some focus changes, i.e., for every path of navigation links without focus change from (q, ϕ) to (q', ϕ'), $items(q, \phi) \neq \emptyset$ implies $items(q', \phi') \neq \emptyset$.*

We justify to allow for unsafe focus changes by considering the following navigation scenario. The current query has the form $q = \underline{f_1 \text{ or } f_2}_\phi$, i.e., the union of two restrictions. The feature f_3 is a restriction of q such that $items(f_2) \cap items(f_3) = \emptyset$, i.e., only items of f_1 match f_3. The intersection with f_3 leads to the query $q' = (f_1 \text{ or } f_2) \text{ and } \underline{f_3}_{\phi'}$, and a focus change on f_2 leads to an empty selection. We could prevent intersection with f_3 but this would be counter-intuitive because it is a valid restriction for (q, ϕ). We could simplify the query q' by removing the second alternative f_2 ($q' = f_1 \text{ and } f_3$), or forbid the focus change, but we think users should have full control on the query they have built. Finally, allowing for the unsafe focus change is a simple way to inform users that no item of f_2 matches the new restriction feature f_3.

Theorem 2 (completeness). *Let D be a RDF dataset. The navigation graph G_D is complete except for some queries having an unsafe focus change, i.e., for every query q s.t. for every $\phi \in Phi(q)$, $items(q, \phi) \neq \emptyset$, there is a navigation path from v_0 to the navigation place $(\underline{q}_\phi, \phi)$.*

In the above scenario, it was possible to navigate to $(f_1 \text{ or } f_2) \text{ and } f_3$ that has an unsafe focus change on f_2, but it is not possible to navigate to the equivalent $f_3 \text{ and } (f_1 \text{ or } f_2)$ because $f_2 \notin restr(f_3 \text{ and } (f_1 \text{ or } \underline{?}_{\phi_2}), \phi_2)$. Fortunately, a query that is not a dead-end but has unsafe focus changes can be simplified into an equivalent query (same set of items) without unsafe focus changes. It suffices to delete from the query empty alternatives ($S \cup \emptyset = S$), and empty exclusions ($S \setminus \emptyset = S$).

4.2 Efficiency

Each navigation step from a navigation place (q, ϕ) requires the computation of the set of items $items(q, \phi)$, the set of restrictions $restr(q, \phi)$, and the set of navigation links as specified in Definition 8. In many cases, the set of items can be obtained efficiently from the previous set of items, and the last navigation link. If the last navigation link was an intersection, Lemma 1 shows that the set of items is the result of the intersection that is performed during the computation of restrictions, like in standard faceted search. For an exclusion or a naming, the set of items is unchanged. For a reference, the set of items was already computed at the previous step. Otherwise, for an union or a focus change, the set of items is computed with a LISQL query engines, possibly reusing existing query engines for the Semantic Web (see Section 3.3).

Computing the set of restrictions is equivalent to set-based faceted search, i.e., amounts to compute set intersections between the set of items and the precomputed set of items of features. The same datastructures and algorithms can therefore be used. As features are LISQL queries, their set of items can be computed like for queries, possibly with optimizations given features are simple queries. Finally, determining the set of navigation links requires little additional computation. A navigation link is available for each focus of the query (focus change), and each restriction (intersection). Three navigation links for exclusion, union, and naming are always available. Only for reference navigation links it is necessary, for each variable in the query, to compute the set of items of the target navigation place, in order to check it is not empty. This additional cost is limited as the number of variables in a LISQL query is very small in practice, and is bounded by the number of foci of the query.

5   Usability Evaluation

This section reports on the evaluation of QFS in terms of usability[3]. We have measured the ability of users to answer questions of various complexities, as well as their response times. Results are strongly positive and demonstrate that QFS offers expressiveness and ease-of-use at the same time.

Prototype. QFS has been implemented as a prototype, Sewelis[4] (aka. Camelis 2). Figure 1 shows a screenshot of Sewelis. From top to bottom, and from left to right, it is composed of a menu bar (M), a toolbar (T), a query box (Q), query controls (QC), feature controls (FC), an answer list or extension box (E), a facet hierarchy (F), and a set of value boxes (V). A query engine can be derived from Sewelis by retaining only the components Q and E. A standard faceted search system can be derived by retaining only the components E, F, and V. Navigation links, i.e., suggested query transformations, are available on all components. Whenever a navigation control is triggered, the corresponding navigation link is applied, and components (Q,E,F,V) are refreshed accordingly. The query box (Q) is clickable for setting the focus on any subquery. Query controls (QC) provide buttons for *naming, union, exclusion* (and a few others). Every element of components (E,F,V) can be used as an argument for *intersection*, with the guarantee that the resulting query does have answers. Restriction are dispatched between components (E,F,V) according to their types. The facet hierarchy (F) contains variables of the current query (e.g., ?X), classes (e.g., a person), and property paths (e.g., father of ?, birth : year : ?). Each value box (V) contains a list or hierarchy of relevant values for some property path facet (e.g.,

---

[3] Details can be found on http://www.irisa.fr/LIS/alice.hermann/camelis2.html
[4] See http://www.irisa.fr/LIS/softwares/sewelis/ for a presentation, screencasts, a Linux executable, and sample data.

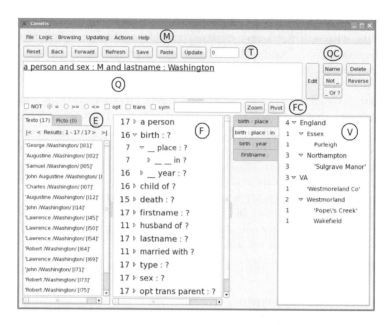

Fig. 1. A screenshot of the user interface of Sewelis. It shows the selection of male persons whose lastname is Washington.

father of 'George Washington', birth : year : 1601). The extension box (E) contains resources (e.g., England). The hierarchical organization of facets in (F) is based on RDFS class and property hierarchies. A value box (V) is hierarchically organized according to the last property of its property path, if that property is transitive (here, in = part of).

Dataset. The datasets were chosen so that subjects had some familiarity with the concepts, but not with the individuals. We found genealogical datasets about former US presidents, and converted them from GED to RDF. We used the genealogy of Benjamin Franklin for the training, and the genealogy of George Washington for the test. The latter describes 79 persons by their birth and/or death events, which are themselves described by their year and place, by their firstname, lastname, and sex, and by their relationships (father, mother, child, spouse) to other persons. Places are linked by a transitive *part-of* relationship, allowing for the display of place hierarchies in Sewelis.

Methodology. The subjects consisted of 20 graduate students in computer science. They had prior knowledge of relational databases but neither of Sewelis, nor of faceted search, nor of Semantic Web. None was familiar with the dataset used in the evaluation. The evaluation was conducted in three phases. First, the subjects learned how to use Sewelis through a 20min tutorial, and had 10 more minutes for free use and questions. Second, subjects were asked to answer a set of questions, using Sewelis. We recorded their answers, the queries they built,

Table 1. Questions of the test, by category, and the minimum number of navigation links to answer them

Category	Question (# navig. links)
Visualization	1 How many persons are there? (0) 2 How many men are there? (0) 3 How many persons have a birth's place in the base? (0)
Selection	4 How many women are named Mary? (4) 5 Who was born at Stone Edge? (4) 6 Which man was born in 1659? (5) 7 Who is married with Edward Dymoke? (3)
Path	9 Which man has his father married with Alice Cooke? (5) 11 Which man is married with a woman born in 1708? (7)
Disjunction	8 Which women have for mother Jane Butler or Mary Ball? (6) 12 Which men are married with a woman whose birth's place is Cuck-fields or Stone Edge? (9)
Negation	10 How many men were born in the 1600 or 1700 years, and not in Norfolk? (12) 13 How many women have a mother whose death's place is not Warner Hall? (7)
Inverse	14 Who was born in the same place as Robert Washington? (6) 15 Who died during the year when Augustine Warner was born? (6)
Cycle	16 Which persons died in the same area where they were born? (9) 17 How many persons have the same firstname as one of their parent? (8) 18 Which persons were born the same year as their spouse? (10)

and the time they spent on each question. Finally, we got feedback from subjects through a SUS questionnaire and open questions [1]. The test was composed of 18 questions, with smoothly increasing difficulty. Table 1 groups the questions in 7 categories: the first 2 categories are covered by standard faceted search, while the 5 other categories are not in general. For category *Visualization*, the exploration of the facet hierarchy was sufficient. In category *Selection*, we asked to count or list items that have a particular feature. In category *Path*, subjects had to follow a path of properties. Category *Disjunction* required the use of unions. Category *Negation* required the use of exclusions. Category *Inverse* required the crossing of the inverse of properties. Category *Cycle* required the use of co-reference variables (naming and reference navigation links).

Results. Figure 2 shows the number of correct queries and answers, the average time spent on each question and the number of participants who had a correct query for at least one question of each category. For example, in category "Visualization", the first two questions had 20 correct answers and queries; the third question had 10 correct answers and 13 correct queries; all the 20 participants had a correct query for at least one question of the category; the average response times were respectively 43, 21, and 55 seconds. The difference between the number of correct queries and correct answers is explained by the fact that some subjects forgot to set the focus on the whole query after building the query.

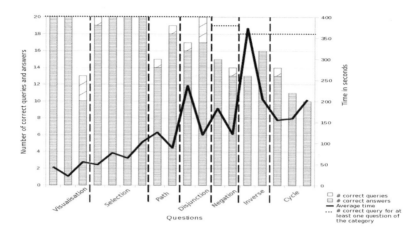

Fig. 2. Average time and number of correct queries and answers for each question

All subjects but one had correct answers to more than half of the questions. Half of the subjects had the correct answers to at least 15 questions out of 18. Two subjects answered correctly to 17 questions, their unique error was on a disjunction question for one and on a negation question for the other. All subjects had the correct query for at least 11 questions. For each question, there is at least 50 percent of success. The subjects spent an average time of 40 minutes on the test, the quickest one spent 21 minutes and the slowest one 58 minutes.

The first 2 categories corresponding to standard faceted search, visualization and selection, had a high success rate (between 94 and 100) except for the third question. The most likely explanation for the latter is that the previous question was so simple (a man) that subjects forgot to reset the query between the questions 2 and 3. All questions of the first two categories were answered in less than 1 minute and 43 seconds on average. Those results indicate that the more complex user interface of QFS does not entail a loss of usability compared to standard faceted search for the same tasks.

For other categories, all subjects but two managed to answer correctly at least one question of each category. Within each category, we observed that response times decreased, except for the *Cycle* category. At the same time, for *Path*, *Disjunction* and *Inverse*, the number of correct answers and queries increased. Those results suggest a quick learning process of the subjects. The decrease in category *Negation* is explained by a design flaw in the interface. For category *Cycle*, we conjecture some lassitude at the end of the test. Nevertheless, all but two subjects answered correctly to at least one of *Cycle* questions. The peak of response time in category *Inverse* is explained by the lack of inverse property examples in the tutorial. It is noticeable that subjects, nevertheless, managed to solve the *Inverse* questions with a reasonable success rate, and a decreasing response time.

Table 2. Results of SUS questions

SUS Question	Score (on a 0-4 scale)	
I think that I would like to use this system frequently	2.8	Agree
I found the system unnecessarily complex	0.8	Strongly disagree
I thought the system was easy to use	2.6	Agree
I think that I would need the support of a technical person to be able to use this system	1.5	Disagree
I found the various functions in this system were well integrated	2.9	Agree
I thought there was too much inconsistency in this system	0.6	Strongly disagree
I would imagine that most people would learn to use this system very quickly	2.5	Agree
I found the system very cumbersome to use	1.0	Disagree
I felt very confident using the system	2.8	Agree
I needed to learn a lot of things before I could get going with this system	1.7	Neutral

SUS Questionnaire. Table 2 shows the answers to the SUS questions, which are quite positive. The first noticeable thing is that, despite the relative complexity of the user interface, subjects do not find the system *unnecessarily complex* nor *cumbersome to use.* We think this is because the principles of QFS are very regular, i.e., they follow few rules with no exception. The second noticeable thing, which may be a consequence of the first, is that subjects *felt confident using the system* and found no *inconsistency.* Finally, even if it is necessary for subjects to learn how to use the system, they *thought that the system was easy to use,* and that they *would learn to use it very quickly.* The results of the test demonstrate that they are right, even for features that were not presented in the tutorial (the Inverse category).

6   Conclusion

We have introduced *Query-based Faceted Search* (QFS) as a search paradigm for Semantic Web knowledge bases, in particular RDF datasets. It combines most of the expressiveness of the SPARQL query language, and the benefits of exploratory search and faceted search. The user interface of QFS includes the user interface of other faceted search systems, and can be used as such. It adds a query box to tell users where they are in their search, and to allow them to change the focus. It also adds a few controls for applying some query transformations such as the insertion of disjunction, negation, and variables.

QFS has been implemented as a prototype, Sewelis. Its usability has been demonstrated through a user study, where, after a short training, all subjects were able to answer simple questions, and most of them were able to answer complex questions involving disjunction, negation, or co-references. This means QFS retains the ease-of-use of other faceted search systems, and gets close to the expressiveness of query languages such as SPARQL.

Acknowledgments. We would like to thank the 20 students, from the University of Rennes 1 and the INSA engineering school, for their volunteer participation to the usability evaluation.

References

1. Brooke, J.: SUS: A quick and dirty usability scale. In: Jordan, P., Thomas, B., Weerdmeester, B., McClelland, A. (eds.) Usability Evaluation in Industry, pp. 189–194. Taylor and Francis, London (1996)
2. Ferré, S.: Camelis: a logical information system to organize and browse a collection of documents. Int. J. General Systems 38(4) (2009)
3. Ferré, S., Hermann, A., Ducassé, M.: Semantic faceted search: Safe and expressive navigation in RDF graphs. Research report, IRISA (2011), http://hal.inria.fr/inria-00410959/PDF/PI-1964.pdf
4. Ferré, S., Ridoux, O.: A File System Based on Concept Analysis. In: Palamidessi, C., Moniz Pereira, L., Lloyd, J.W., Dahl, V., Furbach, U., Kerber, M., Lau, K.-K., Sagiv, Y., Stuckey, P.J. (eds.) CL 2000. LNCS (LNAI), vol. 1861, pp. 1033–1047. Springer, Heidelberg (2000)
5. Hearst, M., Elliott, A., English, J., Sinha, R., Swearingen, K., Yee, K.P.: Finding the flow in web site search. Communications of the ACM 45(9), 42–49 (2002)
6. Heim, P., Ertl, T., Ziegler, J.: Facet Graphs: Complex Semantic Querying Made Easy. In: Aroyo, L., Antoniou, G., Hyvönen, E., ten Teije, A., Stuckenschmidt, H., Cabral, L., Tudorache, T. (eds.) ESWC 2010. LNCS, vol. 6088, pp. 288–302. Springer, Heidelberg (2010)
7. Hildebrand, M., van Ossenbruggen, J., Hardman, L.: /facet: A Browser for Heterogeneous Semantic Web Repositories. In: Cruz, I., Decker, S., Allemang, D., Preist, C., Schwabe, D., Mika, P., Uschold, M., Aroyo, L.M. (eds.) ISWC 2006. LNCS, vol. 4273, pp. 272–285. Springer, Heidelberg (2006)
8. Kaufmann, E., Bernstein, A.: Evaluating the usability of natural language query languages and interfaces to semantic web knowledge bases. J. Web Semantics 8(4), 377–393 (2010)
9. Lu, J., Ma, L., Zhang, L., Brunner, J., Wang, C., Pan, Y., Yu, Y.: SOR: A practical system for ontology storage, reasoning and search (demo). In: Int. Conf. Very Large Databases (VLDB), pp. 1402–1405. VLDB Endowment, ACM (2007)
10. Marchionini, G.: Exploratory search: from finding to understanding. Communications of the ACM 49(4), 41–46 (2006)
11. Oren, E., Delbru, R., Decker, S.: Extending Faceted Navigation for RDF Data. In: Cruz, I., Decker, S., Allemang, D., Preist, C., Schwabe, D., Mika, P., Uschold, M., Aroyo, L.M. (eds.) ISWC 2006. LNCS, vol. 4273, pp. 559–572. Springer, Heidelberg (2006)
12. Sacco, G.M.: Dynamic taxonomies: A model for large information bases. IEEE Transactions Knowledge and Data Engineering 12(3), 468–479 (2000)
13. Sacco, G.M., Tzitzikas, Y. (eds.): Dynamic taxonomies and faceted search. The information retrieval series. Springer, Heidelberg (2009)
14. Tran, T., Wang, H., Haase, P.: Hermes: Data web search on a pay-as-you-go integration infrastructure. Web Semantics: Science, Services and Agents on the World Wide Web 7, 189–203 (2009)

Effectively Interpreting Keyword Queries on RDF Databases with a Rear View

Haizhou Fu and Kemafor Anyanwu

Semantic Computing Research Lab, Department of Computer Science,
North Carolina State University, Raleigh NC 27606, USA
{hfu,kogan}@ncsu.edu

Abstract. Effective techniques for keyword search over RDF databases incorporate an explicit interpretation phase that maps keywords in a keyword query to structured query constructs. Because of the ambiguity of keyword queries, it is often not possible to generate a unique interpretation for a keyword query. Consequently, heuristics geared toward generating the top-K likeliest user-intended interpretations have been proposed. However, heuristics currently proposed fail to capture any user-dependent characteristics, but rather depend on database-dependent properties such as occurrence frequency of subgraph pattern connecting keywords. This leads to the problem of generating top-K interpretations that are not aligned with user intentions. In this paper, we propose a context-aware approach for keyword query interpretation that personalizes the interpretation process based on a user's query context. Our approach addresses the novel problem of using a sequence of structured queries corresponding to interpretations of keyword queries in the query history as contextual information for biasing the interpretation of a new query. Experimental results presented over DBPedia dataset show that our approach outperforms the state-of-the-art technique on both efficiency and effectiveness, particularly for ambiguous queries.

Keywords: Query Interpretation, Keyword Search, Query Context, RDF Databases.

1 Introduction

Keyword search offers the advantage of ease-of-use but presents challenges due to their often terse and ambiguous nature. Traditional approaches [1][2][8] for answering keyword queries on (semi)structured databases have been based on an assumption that queries are explicit descriptions of semantics. These approaches focus on merely matching the keywords to database elements and returning some summary of results i.e., IR-style approaches. However, in a number of scenarios, such approaches will produce unsatisfactory results. For example, a query like *"Semantic Web Researchers"* needs to be *interested* as a list of people, many of which will not have all keywords in their labels and so will be missed by IR-style approaches. For such queries, each keyword needs to be interpreted and the entire query needs to be mapped to a set of conditional expressions,

L. Aroyo et al. (Eds.): ISWC 2011, Part I, LNCS 7031, pp. 193–208, 2011.

i.e., *WHERE* clause and return clause. It is not often easy to find a unique mapping, therefore this problem is typically done as a top-K problem with the goal of identifying the *K* likeliest user intended interpretations.

Existing top-K query interpretation approaches [11] for RDF databases employ a cost-based graph exploration algorithm for exploring schema and data to find connections between keyword occurrences and essentially fill in the gaps in a keyword query. However, these techniques have the limitation of using a *"one-size-fits-all"* approach that is not *user-dependent* but rather more *database-dependent*. The heuristics used are based on the presumption that the likeliest intended interpretation is the interpretation that has the most frequent support in the database, i.e., the interpretation is related to classes of high-cardinality. Unfortunately, since such metrics are not user-dependent, the results generated do not always reflect the user intent.

In this paper, we address the problem of generating *context-aware query interpretations* for keyword queries on RDF databases by using information from a user's query history. The rationale for this is that users often pose a series of related queries, particularly in exploratory scenarios. In these scenarios, information about previous queries can be used to influence the interpretation of a newer query. For example, given a keyword query *"Mississippi River"*, if a user had previously queried about *"Mortgage Rates"*,then it is more reasonable to select the interpretation of the current query as being that of a *financial institution "Mississippi River Bank"*. On the other hand, if a user's previous query was *"Fishing Techniques"*, it may make more sense to interpret the current query as referring to a *large body of water*: the *"Mississippi River"*. Two main challenges that arise here include (i) effectively capturing and efficiently representing query history and (ii) effectively and efficiently exploiting query history during query interpretation. Towards addressing these challenges we make the following **contributions**:

i) Introduce and formalize the problem of *Context-Aware keyword query interpretation* on RDF databases.

ii) Propose and implement a *dynamic weighted summary graph model* that is used to concisely capture essential characteristics of a user's query history.

iii) Design and implement an efficient and effective top-K *Context-Aware graph exploration algorithm* that extends existing cost-balanced graph exploration algorithms, with support for biasing the exploration process based on context as well as with early termination conditions based on a notion of dominance.

iv) Present a comprehensive evaluation of our approach using a subset of the DBPedia dataset [3], and demonstrate that the proposed approach outperforms the state-of-the-art technique on both efficiency and effectiveness.

2 Foundations and Problem Definition

Let *W* be an alphabet of database tokens. An RDF database is a collection of subject-property-object triples linking RDF resources. These triples can be represented as a graph $G_D = V_D, E_D, \lambda_D, \varphi_D$, where subject or object is represented as node in V_D while property is represented by edge in E_D. An object node can either represent another entity (RDF resource) or literal value. λ_D is a labeling function $\lambda_D : (V_D \cup E_D) \to 2^W$ that captures the *rdfs:label* declarations and returns a set of all distinct tokens in the label of any resource or property in the data graph. In addition, for any literal node $v_l \in V_D$, $\lambda_D(v_l)$ returns all distinct tokens in the literal value represented by v_l. φ_D is the incidence function: $\varphi_D : V_D \times V_D \to E_D$.

An RDF schema is also a collection of subject-property-object triples, which can also be represented as a graph: $G_S = (V_S, E_S, \lambda_S, \varphi_S, \pi)$, where the nodes in V_S represent classes and edges in E_S represent properties. λ_S is a labeling function $\lambda_S : V_S \cup E_S \to 2^W$ that captures the *rdfs:label* declarations and returns a set of all distinct tokens in the label of any class or property in the schema graph. φ_S is an incidence function: $\varphi_S : V_S \times V_S \to E_S$. $\pi : V_S \to 2^{V_D}$ is a mapping function that captures the predefined property *rdf:type* maping a schema node representing a class C to a set of data graph nodes representing instances of C. Nodes/edges in a schema can be organized in a subsumption hierarchy using predefined properties *rdfs:subclass* and *rdfs:subproperty*.

We define some special nodes and edges in the schema graph that are necessary for some of the following definitions:

- let $V_{LITERAL} \subset V_S$ be a set containing all literal type nodes (i.e., a set of literal nodes representing literal types such as "XSD:string");
- let $V_{LEAF\_CLASS} \subseteq (V_S - V_{LITERAL})$ be a set containing all leaf nodes (i.e. those nodes representing classes who do not have sub-classes) which are not literal type nodes;
- let $E_{LEAF\_PROPERTY} \subseteq E_S$ be a set containing all leaf edges (i.e. those edges representing properties which do not have sub-properties);
- let $V_{LEAF\_LITERAL} \subseteq V_{LITERAL}$ be a set containing all literal type nodes who are joined with leaf edges, for example, in Fig 1, literal type node $v_{string1}$ is in $V_{LEAF\_LITERAL}$ but $v_{string2}$ is not because the edge e_{name} connecting v_{Place} and $v_{string2}$ is not a leaf edge.

We define a keyword query $Q = \{w_1, w_2, \ldots, w_n | w_i \in W\}$ as a sequence of keywords, each of which is selected from the alphabet W. Given a keyword query Q, an RDF schema and data graphs, the traditional problem that is addressed in relation to keyword queries on RDF databases is how to translate an keyword (unstructured) query Q into a set of conjunctive triple patterns (structured query) that represents the intended meaning of Q. We call this process as **keyword query "structurization"/interpretation**. To ensure that the structured query has a defined semantics for the target database, the translation process is done on the basis of information from the data and schema graphs. For example, given a keyword query "*Mississippi River Bank*", the schema graph and the data graph shown in Fig 1, we can find a structured query with conjunctive triple patterns listed at the top of Fig 1. Because of the class hierarchy defined in the schema, there could be many equivalent triple patterns for a given keyword query. For instance, in the schema graph of Fig 1, "*Organization*" is the

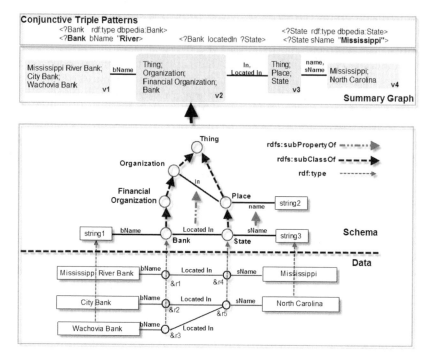

Fig. 1. Graph Summarization

super class of *"Bank"*. Assuming that only *"Bank"* has the property *"bName"*, thus, the two pattern queries:

⟨ *?x bName "river"* ⟩, ⟨ *?x rdf:type Organization* ⟩, and
⟨ *?x bName "river"* ⟩, ⟨ *?x rdf:type Bank* ⟩

are equivalent because the domain of the property *"bName"* requires that the matches of ?x can only be the instances of *"Bank"*. To avoid redundancy and improve the performance, usually a summary graph structure is adopted that concisely summarizes the relationships encoded in the subsumption hierarchies and the relationships between tokens and the schema elements they are linked to.

Recall that our goal is to enable context-awareness for keyword query interpretation, we would also like this summary graph structure to encode information about a user's query history such as which classes have been associated with recent queries. This leads to a notion of a *context-aware summary graph* which is defined in terms of the concept of *"Upward Closure"*:

DEFINITION 1 (*Upward closure*): Let v_C be a node in a schema graph G_S that represents a class C. The *upward closure* of v_C is $v_C^\wedge$, which is a set containing v_C and all the nodes representing super classes of C. For example, in Fig 1, the upward closure of the node v_{State} is: $v_{State}^\wedge = \{v_{Thing}, v_{Place}, v_{State}\}$. The *upward closure* of an edge $e_P \in E_S$ denoted by $e_P^\wedge$ is similarly defined.

DEFINITION 2 (*Context-aware Summary Graph*) : Given an RDF schema graph G_S, a data graph G_D and a query history QH: $QH = \{Q_1, \ldots, Q_T\}$, where Q_T is the most recent query, a *context-aware summary graph* can be defined as $SG = (V_{SG}, E_{SG}, \theta, \lambda_{SG}, \Psi_{SG}, \omega)$, where

- $\theta : V_{SG} \cup E_{SG} \rightarrow 2^{(V_S \cup E_S)}$ is an injective mapping function that maps any node or edge in SG to a set of nodes or edges in G_S.
- $V_{SG} = \{v_i | \exists u \in V_{LEAF\_CLASS} \cup V_{LEAF\_LITERAL}$ such that $\theta(v_i) = u^{\wedge}\}$.
  For example, the context-aware summary graph in Fig 1 contains $\{v1, v2, v3, v4\}$ four nodes, each of which can be mapped to the upward closure of one of the leaf nodes in $\{v_{string1}, v_{Bank}, v_{State}, v_{string3}\}$ in the schema graph respectively.
- $E_{SG} = \{e_i | \exists u \in E_{LEAF\_PROPERTY}$ such that $\theta(e_i) = u^{\wedge}\}$.
  For example, the summary graph in Fig 1 contains $\{e1, e2, e3\}$ three edges, each of which can be mapped to the upward closure of one of the leaf edges in $\{e_{bName}, e_{locatedIn}, e_{sName}\}$ respectively.
- λ_{SG} is a labeling function: $\lambda_{SG} : (V_{SG} \cup E_{SG}) \rightarrow 2^W$.
  - $\forall v \in V_{SG}$ where $\theta(v) = v_C^{\wedge}$ and $v_C \in V_S$ representing class C,
    $$\lambda_{SG}(v) = \{\bigcup_{v_i \in v_C^{\wedge}} \lambda_S(v_i)\} \cup \{\bigcup_{r_j \in \pi(v_C)} \lambda_D(r_j)\},$$
    i.e., union of all distinct tokens in the labels of the super classes of C and distinct tokens in labels of all instances of C. For example, in Fig 1,
    $\lambda_{SG}(v4) = \{$ *"Mississippi"*, *"North"*, *"Carolina"*$\}$;
    $\lambda_{SG}(v3) = \{$ *"Thing"*, *"Place"*, *"State"*$\}$.
  - $\forall e \in E_{SG}$ where $\theta(e) = e_P^{\wedge}$ and $e_P \in E_S$ representing property P,
    $$\lambda_{SG}(e) = \bigcup_{e_i \in e_P^{\wedge}} \lambda_S(e_i),$$
    which is a union of all distinct tokens in the labels of all super classes of P. For example,
    $\lambda_{SG}(e2) = \{$ *"LocatedIn"*, *" In"*$\}$.
- Ψ_{SG} is the incidence function: $\Psi_{SG} : V_{SG} \times V_{SG} \rightarrow E_{SG}$, such that if $\theta(v_1) = v_{C1}^{\wedge}, \theta(v_2) = v_{C2}^{\wedge}, \theta(e) = e_P^{\wedge}$, then $\Psi_{SG}(v_1, v_2) = e$ implies $\varphi_S(v_{C1}, v_{C2}) = e_P$.
- $\omega : (QH, V_{SG} \cup E_{SG}) \rightarrow \mathbf{R}$ is a query history dependent weighting function that assigns weights to nodes and edges of SG. For a query history QH_{T-1} and $QH_T = QH_{T-1} + Q_T$, and $m \in SG$, $\omega(QH_{T-1}, m) \geq \omega(QH_T, m)$ if $m \in Q_T$.

Note that, we only consider user-defined properties for summary graph while excluding pre-defined properties. Further, we refer to any node or edge in a context-aware summary graph as a *summary graph element*.

DEFINITION 3 (*Hit*): Given a context-aware summary graph SG and a keyword query Q, a *hit* of a keyword $w_i \in Q$ is a summary graph element $m \in SG$ such that $w_i \in \lambda(m)$ i.e., w_i appears in the label of m. Because there could be multiple hits for a single keyword w, we denote the set of all hits of w as $HIT(w)$. For example, in Fig 1, $HIT(\text{"bank"}) = \{v1, v2\}$.

DEFINITION 4 (*Keyword Query Interpretation*): Given a keyword query Q and a context-aware summary graph SG, a *keyword query interpretation* QI

is a connected sub-graph of SG that connects at least one hit of each keyword in Q.

For example, the summary graph shown in Fig 1 represents the interpretation of the keyword query "*Mississippi, River, Bank*" which means "*Returning those banks in the Mississippi State whose name contains the keyword 'River'* ". The equivalent conjunctive triple patterns are also shown at the top of Fig 1. Note that for a given keyword query Q, there could be many query interpretations due to all possible combinations of hits of all keywords. Therefore, it is necessary to find a way to rank these different interpretations based on a cost function that optimizes some criteria which captures relevance. We use a fairly intuitive cost function in the following way: $cost(QI) = \sum_{m_i \in QI} \omega(m_i)$,

which defines the cost of an interpretation as a combination function of the weights of the elements that constitute the interpretation. We can formalize the *context-aware top-k keyword query interpretation problem* as follows:

DEFINITION 5 (*Context-aware Top-k Keyword Query Interpretation Problem*): Given a keyword query Q, and a context-aware summary graph SG, let $[[Q]] = \{QI_1, \ldots, QI_n\}$ be a set of all possible keyword query interpretations of Q, the *context-aware top-K keyword query interpretation problem* is to find the top K keyword query interpretations in IS: $TOPK = \{QI_1, \ldots, QI_K\} \subseteq [[Q]]$ such that

(i.) $QI_i \in TOPK$ and $QI_j \in ([[Q]] - TOPK)$, $cost(QI_i) \leq cost(QI_j)$.
(ii.) If $1 \leq p < q \leq k$, $cost(QI_p) \leq cost(QI_q)$, where $QI_p, QI_q \in TOPK$.

This problem is different from the traditional top-k keyword query interpretation problem in that the weights are dynamic and are subject to the evolving context of query history. Because some queries are more ambiguous than others, keyword query interpretation problem requires effective techniques to deal with large interpretation space. We propose a concept called ***Degree of Ambiguity (DoA)*** for characterizing the ambiguity of queries: The DoA of the keyword query Q is defined as $DoA(Q) = \prod_{w_i \in Q} |HIT(w_i)|$, which is the number of all combinations of keyword matches. It will be used as a performance metric in our evaluation.

Overview of our approach. Having defined the problem, we start with an overview of our approach. It consists of the following key steps as shown in Fig 2:

- Find keyword hits using an inverted index for a given keyword query Q. (Step (1)–(3)).
- The query interpreter takes the hits and utilizes a graph exploration algorithm to generate a set of top-K interpretations of Q. (Step (4)–(5))
- The top-1 interpretation of the top-K interpretation is passed to a cost model to update the weights of the context-aware summary graph. (Step (6)–(7))
- Steps involved in Fig 2 only capture one of the iteration cycles of the interactions between user and our interpretation system. The new weights of context-aware summary graph will be used to bias the graph exploration in the next iteration when user issues a new query

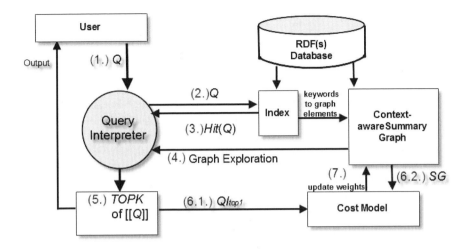

Fig. 2. Architecture and workflow

The cost model will be discussed in the next section and the graph exploration algorithm will be discussed in section 4.

3   Representing Query History Using a Dynamic Cost Model

The implementation of the dynamic cost model for representing query history consists of two main components: i) data structures for implementing a labeled dynamic weighted graph i.e., the context-aware summary graph; ii) a dynamic weighting function that assigns weights to summary graph elements in a way that captures their relevance to the current querying context. To understand what the weighting function has to achieve, consider the following scenario. Assuming that we have the following sequence of queries $Q1 =$ "*Ferrari, price*" and $Q2 =$ "*F1, calendar*", with $Q2$ as the most recent query. If $Q1$ is interpreted as "*Car*", the relevance score of the concept "*Car*" as well as related concepts (concepts in their immediate neighborhood such as "*Auto Engine*") should be increased. When $Q2$ arrives and is interpreted as "*Competition*", the relevance score for it should be increased. Meanwhile, since "*Auto Engine*" and all the other concepts that are not directly related to $Q2$, their relevance scores should be decreased. Then ultimately, for a new query $Q3 =$ "*Jaguar, speed*", we will prefer the concept with higher relevance score as its interpretation, for example, we prefer "*Car*" than "*Mammal*".

To achieve this effect, we designed the dynamic weighting function to be based on a *relevance function* in terms of two factors: *historical impact factor* (hif) and *region factor* (rf).

Let T indicate the historical index of the most recent query Q_T, t be the historical index of an older keyword query Q_t, i.e., $t \leq T$, and m denote a summary graph element. Assume that the top-1 interpretation for Q_t has already been generated : QI_t. *Region factor* is defined as a monotonically decreasing function of the graph distance $d(m, QI_t)$ between m and QI_t:

$rf(d(m, QI_t)) = \frac{1}{\alpha^{d(m,QI_t)}}$

$(rf(d(m, QI_t)) = 0 \ if \ d(m, QI_t) \geq \tau)$ τ is a constant value, and $d(m, QI_t)$ is the shortest distance between m and QI_t, i.e., among all the paths from m to *any* graph element in the sub-graph QI_t, $d(m, QI_t)$ is the length of the shortest path. Here, $\alpha > 1$ is a positive integer constant. The region factor represents the relevance of m to QI_t . *Historical impact factor* captures the property that the relevance between a query and a graph element will decrease when that query ages out of the query history. hif is a monotonically decreasing function: $hif(t) = 1/\beta^{T-t}$, where $\beta > 1$ is also a positive integer constant. We combine the two factors to define the relevance of m to query interpretation QI_t as $hif(t) * rf(d(m, QI_t))$. To capture the aggregate historical and region impacts of all queries in a user's query history, we use the combination function as the relevance function γ :

$$\gamma(m, T) = \sum_0^T hif(t) * rf(d(m, QI_t)) = \sum_0^T (1/\beta^{T-t})(1/\alpha^{d(m,QI_t)}) \quad (1)$$

To produce a representation of (1) for a more efficient implementation, we rewrite a function as recursive:

$$\gamma(m, T) = \gamma(m, T-1)/\beta + 1/\alpha^{d(m,QI_T)} \quad (2)$$

The consequence of this is that, given the relevance score of m at time $T - 1$, we can calculate $\gamma(m, T)$ simply by dividing $\gamma(m, T-1)$ by β then adding $1/\alpha^{d(m,QI_T)}$. In practice, we use $d(m, QI_T) < \tau = 2$, so that, only m and the neighboring nodes and edges of m will be have their scores updated.

Boostrapping. At the initial stage, there are no queries in the query history, so the relevance score of the summary graph elements can be assigned based on the $TF - IDF$ score, where each set of labels of a summary graph element m i.e., $\lambda_{SG}(m)$ is considered as a document. User-feedback is allowed at every stage to select the correct interpretation if the top-1 query interpretation generated is not the desired one.

Since top-K querying generation is based on finding the smallest cost connected subgraphs of the summary, the definition of our weighting function for the dynamic weighted graph model is defined as the following function of the relevance function γ.

$$\omega(m, t) = 1 + 1/\gamma(m, t) \quad (3)$$

This implies that a summary graph element with a higher relevance value will be assigned a lower weight. In the next section, we will discuss how to find interpretations with top-k minimal costs.

4 Top-K Context-Aware Query Interpretation

The state of the art technique for query interpretation uses cost-balanced graph exploration algorithms [11]. Our approach extends such an algorithm [11] with a novel context-aware heuristic for biasing graph exploration. In addition, our approach improves the performance of the existing algorithm by introducing an early termination strategy and early duplicate detection technique to eliminate the need for duplicate detection as a postprocessing step. *Context Aware Graph Exploration (CoaGe)* algorithm shown in Fig 3.

CoaGe (*Q, SG, K*)		TopKCombination (*TOPK, CL,K*)	
1	Initialize priority queues *TOPK, CQ*;	1	Initialize the combination enumerator *Enum=CL.CEnum*
2	Create cursor for each hit of each keyword;	2	Initialize the threshold list *TL*
3	Insert each cursor to *CQ*;	3	**while** (*cur_comb = Enum.current()* AND cur_comb != NULL)
4	**while** *CQ* is not empty	4	**if** exist *h* in *TL*, **Dominate**(*cur_comb,h*) == TRUE
5	*c* = *CQ*.ExtractMin(); //get cheapest cursor	5	**if** (*Enum.**DirectNext**(h)* == NULL) break;
6	*v* = *cursor.path*[0]; //the visiting node	6	**else**
7	**if** (*v* is a root)	7	**if** (**DuplicateDetection**(*cur_comb, TOPK*) == TRUE)
8	**TopKCombination** (*TOPK, v.CL*)	8	**if** (*Enum.**Next**()* == NULL) break;
9	**if** (*TOPK.count* >= *K* AND	9	**else** continue;
	TOPK.Max() < *CQ*.Min())	10	**else** //not duplicate combination
10	TERMINATE;	11	**if** (*TOPK.count* >= *K* AND
11	**if** (*c.depth* is less than the threshold)		*TOPK*.Max() < *cur_comb.cost*)
12	**foreach** neighbor *n* of *v*	12	*TL*.Add(*cur_comb*);
13	if *n* is not visited by *c*	13	**if** (*Enum.**DirectNext**(v)* == NULL) break;
14	create new cursor *new_cur* for *n*;	14	**else**
15	if *c.topN*== FALSE	15	**if** (*TOPK.count* >= *K*) *TOPK*.ExtractMin();
16	*c.cost* *= **penalty_factor**;	16	*TOPK*.Insert(*cur_comb*);
17	*n.CL*[*c.keyword*].Add(*new_cur*);	17	**if** (*Enum.**Next**()* == NULL) break;

Fig. 3. Pseudocodes for CoaGe and TopCombination

4.1 CoaGe

CoaGe takes as input a keyword query *Q*, a context-aware summary graph *SG* and an integer value *K* indicating the number of candidate interpretations that should be generated. In *CoaGe*, a max binomial heap *TOPK* is used to maintain top-K interpretations and a min binomial heap *CQ* is used to maintain cursors created during the graph exploration phase (line 1). At the initialization stage, for each hit of each keyword, *CoaGe* generates a cursor for it. A cursor originates from a hit m_w of a keyword w is represented as *c*(*keyword, path, cost, topN*), where *c.keyword* = *w*; *c.path* contains a sequence of summary graph elements in the path from m_w to the node that *c* just visited; *c.cost* is the cost of the path, which is the sum of the weights of all summary graph elements in *c.path*;

$c.topN$ is a boolean value that indicates whether m_w is among the *top-N hits* of $HIT(w)$. The Top-N hit list contains the N minimum weighted hits of all hits in $HIT(w)$.

Each node v in the context-aware summary graph has a cursor manager CL that contains a set of lists. Each list in CL is a sorted list that contains a sequence of cursors for keyword w that have visited v, we use $CL[w]$ to identify the list of cursors for keyword w. The order of the elements in each list is dependent on the costs of cursors in that list. The number of lists in CL is equal to the number of keywords: $|CL| = |Q|$. During the graph exploration, the cursor with minimal cost is extracted from CQ (line 5). Let v be the node just visited by this "*cheapest*" cursor (line 6). $CoaGe$ first determines whether v is a root (line 7). This is achieved by examining if all lists in $v.CL$ is not empty, in other words, at least one cursor for every keyword has visited v. If v is a root, then, there are $\prod_{w_i \in Q} |v.CL[w_i]|$ combinations of cursors. Each combination of cursors can be used to generate a sub-graph QI. However, computing all combinations of cursors as done in the existing approach [11] does is very expensive. To avoid this, we developed an algorithm $TopCombination$ to enable early termination during the process of enumerating all combinations. $TopCombination$ algorithm (line 8) will be elaborated in the next subsection. A second termination condition for the $CoaGe$ algorithm is if the smallest cost of CQ is larger than the largest cost of the top-K interpretations (line 9). After the algorithm checks if v is a root or not, the current cursor c explores the neighbors of v if the length of $c.path$ is less than a threshold (line 11). New cursors are generated (line 14) for unvisited neighbors of c (not in $c.path$, line 13). New cursors will be added to the cursor manager CL of v (line 17). The cost of new cursors are computed based on the cost of the path and if c is originated from a top-N hits.

Unlike the traditional graph exploration algorithms that proceed based on static costs, we introduce a novel concept of '*velocity*' for cursor expansion. Intuitively, we prefer an interpretation that connects keyword hits that are more relevant to the query history, i.e., lower weights. Therefore, while considering a cursor for expansion, it penalizes and therefore "*slows down*" the velocity of cursors for graph elements that are *not* present in the top-N hits (line 16). By so doing, if two cursors have the same cost or even cursor c_A has less cost than cursor c_B, but c_B originates from a top-N hit, c_B may be expanded first because the cost of c_A is penalized and $c_A.cost * penalty\_factor > c_B.cost$. The space complexity is bounded by $O(n \cdot d^D)$, where $n = \sum_{w_i \in Q} |HIT(w_i)|$ is the total number of keyword hits, $d = \Delta(SG)$ is the maximum degree of the graph and D is the maximum depth a cursor can explore.

4.2 Efficient Selection of Computing Top-k Combinations of Cursors

The $TopCombination$ algorithm is used to compute the top-K combinations of cursors in the cursor manager CL of a node v when v is a root. This algorithm avoids the enumeration of all combinations of cursors by utilizing a notion of *dominance* between the elements of CL. The **dominance relation-**

ship between two combinations of cursors $Com_p = (CL[w_1][p_1], ...CL[w_L][p_L])$ and $Com_q = (CL[w_1][q_1], ...CL[w_L][q_L])$ is defined as follows: Com_p *dominates* Com_q, denoted by $Com^p \succ Com^q$ if for all $1 \le i \le L = |Q|$, $p_i \ge q_i$ and exists $1 \le j \le L$, $p_j > q_j$. Because every list $CL[w_i] \in CL$ is sorted in a non decreasing order, i.e., for all $1 \le s \le L$, $i \ge j$ implies that $CL[w_s][i].cost \ge CL[w_s][j].cost$. Moreover, because the scoring function for calculating the cost of a combination Com is a monotonic function: $cost(Com) = \sum_{c_i \in Com} c_i.cost$, which equals to the sum of the costs of all cursors in a combination, then we have:

$Com_p = (CL[w_1][p_1], CL[w_2][p_2], ..., CL[w_L][p_L])$

$\succ$

$(Com_q = CL[w_1][q_1], CL[w_2][q_2], ..., CL[w_L][q_L])$

implies that for all $1 \le i \le L$,

$CL[w_i][p_i].cost \ge CL[w_i][q_i].cost$ and $cost(Com_p) \ge cost(Com_q)$.

In order to compute top-k minimal combinations, given the combination Com_{max} with the max cost in the top-k combinations, we can ignore all the other combinations that dominate Com_{max}. Note that, instead of identifying all non-dominated combinations as in line with the traditional formulation, our goal is to find top-K minimum combinations that require dominated combinations to be exploited.

The pseudocodes of the algorithm $TopKCombination$ is shown in Fig 3. $TopKCombination$ takes as input a max binomial heap $TOPK$, a cursor manager CL and an integer value K indicating the number of candidate interpretations that should be generated. The algorithm has a combination enumerator $Enum$ that is able to enumerate possible combinations of cursors in CL (line 1). TL is initialized to contain a list of combinations as thresholds (line 2). The enumerator starts from the combination

$Com_0 = (CL[w_1][0], CL[w_2][0], ..., CL[w_L][0])$,

which is the *"cheapest"* combination in CL. Let

$Com_{last} = (CL[w_1][l_1], CL[w_2][l_2], ..., CL[w_L][l_L])$, be the last combination, which is the most *"expensive"* combination and $l_i = CL[w_i].length - 1$, which is the last index of the list $CL[w_i]$.

The enumerator outputs the next combination in the following way: if the current combination is

$Com_{current} = (CL[w_1][s_1], CL[w_2][s_2], ..., CL[w_L][s_L])$,

from 1 to L, $Enum.Next()$ locates the first index i, where $1 \le i \le L$ such that $s_i \le l_i$, and returns the next combination as $Com_{next} =$

$(CL[w_1][0], ..., CL[w_{i-1}][0], CL[w_i][s_i + 1], ..., CL[w_L][s_L])$,

where, for all $1 \le j < i$, s_j is changed from $l_j - 1$ to 0, and $s_j = s_j + 1$. For example, for $(CL[w_1][9], CL[w_2][5])$, if $CL[w_1].length$ equals to 10 and $CL[w_2].length > 5$, then, the next combination is $(CL[w_1][0], CL[w_2][6])$. The enumerator will terminate when $Com_{current} == Com_{last}$.

Each time $Enum$ move to a new combination cur_comb, it is compared with every combination in TL to check if there exists a threshold combination $h \in TL$ such that $cur\_comb \succ h$ (line 4). If so, instead of moving to the next combination using $Next()$, $Enum.DirectNext()$ is executed (line 5) to directly return the next combination that does not dominate h and has not been not enumerated before. This is achieved by the following steps: if the threshold combination is

$Com_{threshold} = (CL[w_1][s_1], CL[w_2][s_2], ..., CL[w_L][s_L])$,

from 1 to L, $Enum.DirectNext()$ locates the first index i, where $1 \leq i \leq L$ such that $s_i \neq 0$, and from $i + 1$ to L, j is the first index such that $s_j \neq l_j - 1$, then the next generated combination is $Com_{direct\_next} =$

$(CL[w_1][0], ..., CL[w_i][0], ..., CL[w_{j-1}][0], CL[w_j][s_j + 1], ..., CL[w_L][s_L])$

where for all $i \leq r < j$, s_r is changed to 0, and $s_j = s_j + 1$. For example, for

$com_{threshold} = (CL[w_1][0], CL[w_2][6], CL[w_3][9], CL[w_4][2])$,

assume that the length of each list in CL is 10, then its next combination that does not dominate it is

$com_{direct\_next} = (CL[w_1][0], CL[w_2][0], CL[w_3][0], CL[w_5][3])$.

In this way, some combinations that could be enumerated by "$Next()$" function and will dominate $com_{threshold}$ will be ignored. For instance, $com_{next} = Next(com_{threshold}) =$

$(CL[w_1][1], CL[w_2][6], CL[w_3][9], CL[w_4][2])$,

and the next combination after this one: $Next(com_{next}) =$

$(CL[w_1][2], CL[w_2][6], CL[w_3][9], CL[w_4][2])$

will all be ignored because they dominate $com_{current}$.

If a new combination is "$cheaper$" than the max combination in $TOPK$, it will be inserted to it (line 16), otherwise, this new combination will be considered a new threshold combination, and inserted to TL (line 12) such that all the other combinations that dominate this threshold combination will not be enumerated. The time complexity of $TopKCombination$ is $O(K^k)$, where $K = |TOPK|$ is the size of $TOPK$, $k = |Q|$ is the number keywords. Because, for any combination

$com = (CL[w_1][s_1], ..., CL[w_L][s_L])$, where for all s_i, $1 \leq i \leq L$, $s_i \leq K$

$com_K = (CL[w_1][K + 1], ..., CL[w_L][K + 1]) \succ com$

In the worst case, any combinations that dominates com_K will be ignored and K^k combinations are enumerated. Consequently, the time complexity of $CoaGe$ is $O(n \cdot d^D \cdot K^k)$, where n is the total number of keyword hits, $d = \Delta(SG)$ is the maximum degree of the graph, D is the maximum depth. The time complexity of the approach in [11] (we call this approach $TKQ2S$) is $O(n \cdot d^D \cdot S^{D-1})$, where $S = |SG|$ is the number of nodes in the graph.

5   Evaluation

In this section, we discuss the experiments including efficiency and effectiveness of our approach. The experiments were conducted on a machine with Intel duel core 1.86GHz and 3GB memory running on Windows 7 Professional. Our test bed includes a real life dataset DBPedia, which includes 259 classes and over 1200 properties. We will compare the efficiency and effectiveness with $TKQ2S$.

5.1   Effectiveness Evaluation

Setup. 48 randomly selected college students were given questionnaires to complete. The questionnaire contains 10 groups of keyword queries (To minimize the cognitive burden on our evaluators we did not use more than 10 groups in this questionnaire). Each group contains a short query log consisting of a sequence of up to 5 keyword queries from the oldest one to the newest one. For each group, the questionnaire provides English interpretations for each of the older queries. For the newest query, a list of candidate English interpretation for it is given, each interpretation is the English interpretation representing a structured query generated by either $TKQ2S$ or $CoaGe$. Therefore, this candidate interpretation list provided to user is a union of the results returned by the two algorithm. Then users are required to pick up to 2 interpretations that they think are the most intended meaning of the newest keyword query in the context of the provided query history. A consensus interpretation (the one that most people pick) was chosen as the desired interpretation for each keyword query.

Metrics. Our choice of a metric of evaluating the query interpretations is to evaluate how relevant the top-K interpretations generated by an approach is to the desired interpretation. Further, it evaluates the quality of the ranking of the interpretations with respect to their relative relevance to the desired interpretation. Specifically, we adopt a standard evaluation metric in IR called "Discounted cumulative gain (DCG)" with a refined relevance function: $DCG_K = \sum_{i=1}^{K} \frac{2^{rel_i}-1}{\log_2(1+i)}$, where K is the number of top-K interpretations generated, rel_i is the graded relevance of the resultant interpretation ranked at position i. In IR, the relevance between a keyword and a document is indicated as either a match or not, rel_i is either zero or one. In this paper, the relevance between a resultant interpretation QI and a desired interpretation QI_D for a given keyword query Q cannot be simply characterized as either a match or not. QI and QI_D are both sub-graphs of the summary graph and could have some degree of overlapping, which means $rel_i \in [0,1]$. Of course, if $QI == QI_D$, QI should be a perfect match. In this experiment, we define the relevance between a candidate interpretation QI and the desired interpretation QI_D as: $rel_i = \frac{|QI_i \cup QI_D|-|QI_i \cap QI_D|}{|QI_i \cup QI_D|}$, where QI_i is the interpretation ranked at position i. rel_i returns the fraction of those overlapping summary graph elements in the union of the the two sub-graphs. Large overlapping implies high similarity between QI_i and QI_D, and therefore, high relevance score. For example, if QI_i has 3 graph elements representing a class *"Person"*, a property *"given name"* and a class *"XSD:string"*. The desired interpretation QI_D also has 3 graph elements, and it represents class *"Person"*, a property *"age"* and a class *"XSD:int"*. Therefore, the relevance between QI_i and QI_D is equal to $1/5 = 0.2$ because the union of them contains 5 graph elements and they have 1 common node.

On the other hand, we use another metric precision to evaluate the results. The precision of a list of top-K candidate interpretation is:

$P@K = |$ relevant interpretations in $TOPK|/|TOPK|$,

which is the proportion of the relevant interpretations that are generated in $TOPK$. Because sometimes, when the user votes are evenly distributed, the consensus interpretation cannot represent the most user intended answer, our evaluation based on DCG may not be convincing. The precision metric can

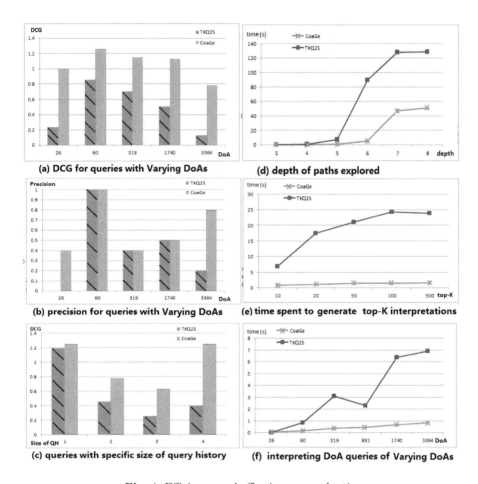

Fig. 4. Efficiency and effectiveness evaluation

overcome this limitation by consider the candidate interpretations that over 10% people have selected as desired interpretations.

Discussion. We compared the DCG of the results returned by our approach and $TKQ2S$. Top-8 queries are generated for each algorithm for each keyword query. The result shown in Fig 4 (a) illustrates the quality of interpreting queries with varying DoA values. From (a), we can observe that $TKQ2S$ does not generate good quality interpretations for queries with high DoA. The reason is that they prefer the popular concepts and rank the desired interpretation which is not popular but is higher relevant to the query history at a low position. It also does not rank higher relevant interpretations higher. Fig 4 (b) illustrates the precision of the top-5 queries generated by each algorithm. In most of cases, $TKQ2S$ generates same number of relevant queries as $CoaBe$, but it fails to generate enough relevant interpretations for the last query with $DoA = 3364$. For the first query in (b), $TKQ2S$ does not output any relevant interpretations, therefore, the precision is 0. Fig 4 (c) illustrates how different lengths of query history will affect results. In this experiment, 4 groups of queries are given, the ith group contains i queries in the query history. Further, the ith group contains all the queries in the $(i-1)$th group plus a new query. Given the 4 different query histories, the two algorithms are to interpret another query Q. (c) illustrates the quality of interpreting Q given different query histories. We can observe that our approach will do better with long query history. But for the first group, both algorithms generate a set of interpretations that are very similar to each other. Both the DCG values are high because user have to select from the candidate list as the desired interpretation, even though they may think none of them is desired. For the third group, that difference in performance is due to a transition in context in the query history. Here the context of query changed in the 2nd or 3rd query. This resulted in a lower DCG value which started to increase again as more queries about new context were added.

5.2   Efficiency Evaluation

From the result of the efficiency evaluation in Fig 4 (d)–(f), we can see that, our algorithm outperforms $TKQ2S$ especially when the depth (maximum length of path a cursor will explore) and the number of top-K interpretations and the degree of ambiguity DoA are high. The performance gain is due to the reduced search space enabled by early termination strategy using the $TopKCombination$ algorithm.

6   Related Work

There is a large body of work supporting keyword search for relational databases [1][2][7][8] based on the interpretation as "match" paradigm. Some recent efforts such as Keymantic[5] QUICK[13], SUITS[6], Q2Semantics[12] and [11] have focused on introducing an explicit keyword interpretation phase prior to answering the query. The general approach used is to find the "*best*" sub-graphs (of the schema plus a data graph) connecting the given keywords and represent the intended query meaning. However, these techniques are based on fixed data-driven heuristics and do not adapt to varying user needs. Alternative approaches [6][13] use techniques that incorporate user input to incrementally construct queries by providing them with query templates. The limitation of these techniques is the extra burden they place on users.

Query history has been exploited in IR [4][9][10]. These problems have different challenges from the ones addressed in this work. The similarity measures are based on mining frequent query patterns. However, we need to exploit the query history to identify similar search intent, which may not necessarily be the most frequent query patterns. Our problem requires unstructured queries, their intended interpretations (structured queries) and the ontology to be managed.

7 Conclusion and Future Work

This paper presents a novel and effective approach for interpreting keyword queries on RDF databases by integrating the querying context. In addition to the techniques proposed in this paper, we plan to explore the idea of exploiting the answers to a priori queries for query interpretation.

Acknowledgement. The work presented in this paper is partially funded by NSF grant IIS-0915865.

References

1. Aditya, B., Bhalotia, G., Chakrabarti, S., Hulgeri, A., Nakhe, C., Parag, Sudarshan, S.: Banks: Browsing and keyword searching in relational databases. In: VLDB, pp. 1083–1086 (2002)
2. Agrawal, S., Chaudhuri, S., Das, G.: Dbxplorer: enabling keyword search over relational databases. In: SIGMOD Conference, p. 627 (2002)
3. Auer, S., Bizer, C., Kobilarov, G., Lehmann, J., Cyganiak, R., Ives, Z.G.: DBpedia: A Nucleus for a Web of Open Data. In: Aberer, K., Choi, K.-S., Noy, N., Allemang, D., Lee, K.-I., Nixon, L.J.B., Golbeck, J., Mika, P., Maynard, D., Mizoguchi, R., Schreiber, G., Cudré-Mauroux, P. (eds.) ASWC 2007 and ISWC 2007. LNCS, vol. 4825, pp. 722–735. Springer, Heidelberg (2007)
4. Bar-Yossef, Z., Kraus, N.: Context-sensitive query auto-completion. In: WWW, pp. 107–116 (2011)
5. Bergamaschi, S., Domnori, E., Guerra, F., Lado, R.T., Velegrakis, Y.: Keyword search over relational databases: a metadata approach. In: SIGMOD Conference, pp. 565–576 (2011)
6. Demidova, E., Zhou, X., Zenz, G., Nejdl, W.: SUITS: Faceted User interface for Constructing Structured Queries from Keywords. In: Zhou, X., Yokota, H., Deng, K., Liu, Q. (eds.) DASFAA 2009. LNCS, vol. 5463, pp. 772–775. Springer, Heidelberg (2009)
7. He, H., Wang, H., Yang, J., Yu, P.S.: Blinks: ranked keyword searches on graphs. In: SIGMOD Conference, pp. 305–316 (2007)
8. Hristidis, V., Papakonstantinou, Y.: Discover: Keyword search in relational databases. In: VLDB, pp. 670–681 (2002)
9. Kelly, D., Gyllstrom, K., Bailey, E.W.: A comparison of query and term suggestion features for interactive searching. In: SIGIR, pp. 371–378 (2009)
10. Pound, J., Paparizos, S., Tsaparas, P.: Facet discovery for structured web search: a query-log mining approach. In: SIGMOD Conference, pp. 169–180 (2011)
11. Tran, T., Wang, H., Rudolph, S., Cimiano, P.: Top-k exploration of query candidates for efficient keyword search on graph-shaped (rdf) data. In: ICDE, pp. 405–416 (2009)
12. Wang, H., Zhang, K., Liu, Q., Tran, T., Yu, Y.: Q2Semantic: A Lightweight Keyword Interface to Semantic Search. In: Bechhofer, S., Hauswirth, M., Hoffmann, J., Koubarakis, M. (eds.) ESWC 2008. LNCS, vol. 5021, pp. 584–598. Springer, Heidelberg (2008)
13. Zenz, G., Zhou, X., Minack, E., Siberski, W., Nejdl, W.: From keywords to semantic queries - incremental query construction on the semantic web. J. Web Sem. 7(3), 166–176 (2009)

Extracting Semantic User Networks from Informal Communication Exchanges

Anna Lisa Gentile, Vitaveska Lanfranchi, Suvodeep Mazumdar,
and Fabio Ciravegna

Department of Computer Science
University of Sheffield
Sheffield, United Kingdom
{a.l.gentile,v.lanfranchi,s.mazumdar,f.ciravegna}@dcs.shef.ac.uk

Abstract. Nowadays communication exchanges are an integral and time consuming part of people's job, especially for the so called *knowledge workers*. Contents discussed during meetings, instant messaging exchanges, email exchanges therefore constitute a potential source of knowledge within an organisation, which is only shared with those immediately involved in the particular communication act. This poses a knowledge management issue, as this kind of contents become "buried knowledge". This work uses semantic technologies to extract buried knowledge, enabling expertise finding and topic trends spotting. Specifically we claim it is possible to automatically model people's expertise by monitoring informal communication exchanges (email) and semantically annotating their content to derive dynamic user profiles. Profiles are then used to calculate similarity between people and plot *semantic knowledge-based networks*. The major contribution and novelty of this work is the exploitation of semantic concepts captured from informal content to build a semantic network which reflects people expertise rather than capturing social interactions. We validate the approach using contents from a research group internal mailing list, using email exchanges within the group collected over a ten months period.

1   Introduction

Email is a common tool for quick exchange of information between individuals and within groups, especially in formal organisations [10], via the use of official or unofficial mailing lists. Mailing lists exchanges are often used to reach a wider audience that includes both the initiator's personal networks and other individuals with shared interests [31] and who may be potential sources of expertise. This poses a knowledge management issue, as the emails' knowledge content is not shared with the whole organisation but only with those included in the recipient list, thus implicitly creating and/or reinforcing the existence of dynamic communities inside organisations. Whilst this is positive as it increases flexibility and innovation, the drawback is that knowledge remains implicit or not shared with the rest of the organisation, becoming "buried knowledge" [32]. Moreover as recognised by [25] the lines between inter-communication on professional and

L. Aroyo et al. (Eds.): ISWC 2011, Part I, LNCS 7031, pp. 209–224, 2011.

social levels are increasingly blurred: people tend to share more aspects of their social life with co-workers and these exchanges often lead to establishing professional coooperations or sharing topics of interests. Extracting information from emails could prove useful in a knowledge management perspective, as it would provide means to build social networks, determine experts and communities of practice, taking into account not only professional content but also social and emerging topics, that may highlight emerging cooperations, interests etc. As proved by [14] email traffic can be analysed and used to identify COINS (Collaborative Innovation Networks), groups of individuals inside organisations that are self-motivated and work together on a new idea. In this paper we propose an approach to automatically model people's expertise and dynamic communities interests by monitoring informal communication exchanges. The content of communication exchanges is semantically annotated and used to derive user profiles. Profiles are then used to calculate similarity between people and plot *semantic knowledge-based networks*, highlighting groups of users with shared knowledge or with complementary knowledge. The main novelty concerns the profile generation; with respect to the state of the art in the Social Network Analysis, where profiles are mostly based (i) on information declared by users in their static profiles, (ii) on rates of communication exchanges between users and (iii) on the morphology of the social graph, our work proposes a profile generation which is based on semantic concepts extracted from user generated content; these profiles have the advantage of being both dynamic, as they are created from user generated content and semantic, in the sense that unique and meaningful concepts are extracted. To confirm its quality and validity we experimented the proposed approach on a dataset consisting of a research group internal mailing list exchanges (as in [13]), extracting profiles with different degrees of semantics. The similarity between users has then been calculated for every type of profile and compared against users subjective similarity judgements, to understand if increasing the level of semantics in the user profiles increases the accuracy of similarity, therefore increasing the potential for exploiting the technique for different tasks, such as expert finding, knowledge sharing or replicated expertise detection within an organisation. To demonstrate the usefulness of the approach, this has been integrated into a knowledge management system aimed at capturing and sharing knowledge exchanged during informal communication exchanges (meetings, chats, etc.): this has provided clear scenarios of usage that will be evaluated in the coming months to understand whether semantic dynamic user profiling is beneficial to increase efficiency and efficacy in knowledge management.

The paper is structured as follows. Section 2 presents a review of the state of the art in knowledge capture from informal communication exchanges, on semantic user profiling and on measures for user similarity. Section 3 describes the proposed approach to automatically model people's expertise profiles and calculate similarity between them. Section 4 presents the experiments used to extract and evaluate user profiles and similarities, whilst section 5 introduces the applications of the approach and possible scenarios of use. Section 6 discusses the results and the next stages of our research.

2   State of the Art

2.1   Knowledge Capture from Informal Communication Exchanges

Capturing, representing and sharing knowledge from informal communication exchanges has been a topic of research in the knowledge management and in the information seeking behaviour communities for many years, as this type of knowledge is tacit, often very specilized and precise and is not shared with anyone else than the immediate recipient of the informal communication exchange. Previous researches on knowledge workers information seeking behaviours proved how engineers spend 40-66% of their time sharing information [18,30]. Different types of communication exchange can be recognised at different levels of formality, with structured planning meetings or corporate mailing lists at one side of the spectrum and informal chats at the coffee machine or chats over internet messaging programs at the other. Independently from the degree of formality all these communication exchanges contain invaluable knowledge that is often buried [32]. Different techniques have been explored in previous works to extract, represent and share this buried knowledge, often focusing on one specific type of communication exchange, such as emails, meeting recordings etc. In this work we focus on email exchanges. Email content, and addressee and recipient, often provide clues about the interests and expertise of participants [5] and they are used as a source for automatic expertise identification and knowledge elicitation [7,31]. The main techniques adopted to extract information and build social networks from emails are usually quantitative data analysis, such as frequency of exchanges between individuals, and data mining over the email content.

Exchange Frequency. In the panorama of work on extracting social networks from email, the frequency of email exchange has been widely used as the main indicator of relevance of a connection. In some cases the effort is on determining frequency thresholds [33,11,3,9], while in others time-dependent threshold conditions are defined to detect dynamic networks [6,19]. Diesner et al. [10] construct a social network via weighted edges over a classical dataset, the *Enron* corpus[1], a large set of email messages made public during the legal investigation of the Enron corporation. They reported the emergence of communication sub-groups with unusually high email exchange in the period prior to the company becoming insolvent in 2001, when email was a key tool for obtaining information especially across formal, inter-organisational boundaries. Diesner et al. [10] also observed that variations in patterns of email usage were influenced by knowledge about and reputation of, in addition to, formal roles within the organisation. Our approach differs from the above cited works as we choose to not take into account the frequency of individual communication exchanges but to perform content-based analysis of mailing list archives, where the emphasis is not on individual recipients of an email but on reaching a wider base of colleagues to find experts that could answer very specific questions.

---

[1] http://www.cs.cmu.edu/~enron

Content-Based Analysis. Email content analysis has been used for different purposes: determining expertise [31], analysing the relations between content and people involved in email exchanges [5,17,25,39], or simply extracting useful information about names, addresses, phone numbers [21]. Our approach takes inspiration from Schwartz et al. [31] in trying to derive expertise and common interests within communities from email exchange but moves on from the keyword-based extraction approach, to consider McCallum et al. [25] contribution in applying Machine Learning (ML) and Natural Language Processing (NLP) to retrieve the rich knowledge content of the information exchanged in automatically gathered social networks, and better interpret the attributes of nodes and the types of relationships between them.

Laclavík et al. [21] observe that enterprise users largely exploit emails to communicate, collaborate and carry out business tasks. They also adopt a cont-based approach and exploit pattern-based Information Extraction techniques to analyse enterprise email communications, and exploit the data obtained to create social networks. The test sets (one in English containing 28 emails, and a second in Spanish with 50 emails) consist of mainly formal emails exchanged between different enterprises. The results obtained indicate that emails are a valid means for obtaining information across formal, inter-organisational boundaries. Lin et al. [24,23]propose a framework for social-oriented Knowledge Management within an organization. They exploit the content of emails for corporate search and for providing expert finding and social network facilities. The focus of these work is rather on the framework description in terms of provided functionalities than on the description of how the content is processed and how the user profiles are generated. The work we present in this paper, on the other hand, makes use of a test set containing informal email exchanges from an internal mailing list for an academic research group, for a pilot, exploratory set of experiments. We adopt multiple approaches for extracting information at different levels of semantic granularity to aid the understanding of the content of the conversations carried out via email, and depict the variety of topics discussed using this communication medium and ultimately derive effective user profiles.

2.2 Semantic User Profiles

Using semantic technologies to derive user profiles has been a topic of research in recent years within different communities, especially in the field of Information Retrieval, with the objective of providing customized search results and in the Recommender Systems community, with the aim of generating effective customized suggestions to the users. In the IR community the focus is on building a user profile which reflects the user interests more than the user expertise, to customize search results. Daoud et al. [8] represent semantic user profiles as graphs of concepts derived from a pre-defined ontology and represent documents as vector of concepts of the same ontology. Then profiles are exploited to determine the personalized result ranking, using a graph-based document ranking model. In the field of Recommender Systems the goal is improving the accuracy recommendations. Abel et al. [1] build semantic user profiles for news recommendations: they enrich users' Twitter activities with semantics extracted from news articles and then use these profiles to suggest users articles to read.

The input for the semantic user profile generation can be gathered in different ways: Kramar [20] e.g. proposes to monitor user activities on the web end extracting metadata (tags, keywords, named entities) from the visited documents. Another direction is the usage of user generated content to extract interests and knowledge levels of users. The most similar work to ours in this direction is the one proposed by Abel et al. [2] who use Twitter posts to generate User Models. Semantic user profiles are extracted from the messages people post on Twitter; in particular they generate three types of profiles: hashtag-based, topic-based or entity-based profiles. The main focus of their work is how to extract semantic from short text like tweets rather than extensively comparing the different types profiles. Our work also propose to generate three types of profiles, with increasing level of semantics (keyword, named entities and concepts), but we also investigate how increasing the levels of semantic in the user profile improves the quality of the profile for a certain task; specifically we show how a more semantic profile better reflects the user perceived similarity with other users in the community.

2.3   Measures for User Similarity

Calculating similarity between user is a key research question in many fields. In Social Networks the similarity between two users is usually a binary function, indicating the "friendship" of two users who are either connected or not. Non-binary similarities have also been proposed [35]. The relationship between two users can be explicit, as stated by the two users, or predicted by means of automatic techniques. Typical features which are used to infer the similarity are attributes from the user's profile like geographic location, age, interests [27]. Social connections already present in the graph are also used as features to predict new possible connections. Other commonly used features (for example in Facebook) are interaction-counters top-friend, and picture graphs. Also spatio-temporal data pertaining to an individuals trajectories has been exploited to geographically mine the similarity between users based on their location histories [22,37].

In our study we take inspiration from the way content-based recommender systems create user profiles as vectors of defining keywords [4,28]; we build user profiles by exploiting the content of emails belonging to each user and we represent them as vectors. The novelty of our technique is using semantic concepts as feature representation rather than keywords. With respect to existing techniques for user similarities in Social Networks, the novelty of our technique is exploiting user generated content to build the feature space, rather than exploiting static features from user profiles. We calculate similarity between users within a network but we use dynamic and semantic features rather that static user-defined ones. The main advantage is that we capture the dynamicity and evolutionary nature of the user interaction.

3 Semantic Network Extraction from Informal Mail Exchanges

The proposed approach models people's expertise and dynamic communities interests by analysing informal communication exchanges. The approach performs the generation of content-based user profiles by analysing user generated contents. User profiles are then used to derive similarity between users. The value of similarity can then be exploited to plot semantic network between users, which will reflect the similarity on the basis of shared knowledge. The following sections will discuss the profile generation and similarity derivation in details.

3.1 Building User Profiles

User profiles are built by extracting information from email content using three techniques, with varying degrees of semantics. This is done to ascertain the quality of user profiles and their suitability to model people's interests and expertise.

Keyword-based profile. Each email e_i in the collection E is reduced to a Bag of Keywords representation, such as $e_i = \{k_1, \ldots, k_n\}$. Each user keyword-based profile consists of Bag of Keywords, extracted from their sent emails.

Entity-based profile. Each email e_i in the collection E is reduced to a Bag of Entities representation, such as $e_i = \{ne_1, \ldots, ne_k\}$. Entities are elements in text which are recognized as belonging to a set of predefined categories (classical categories are persons, locations, organizations, but more fine grained classification is typically used). Each user entity-based profile consists of Bag of Entities, extracted from their sent emails.

Concept-based profile. Each email e_i in the collection E is reduced to a Bag of Concepts representation, such as $e_i = \{c_1, \ldots, c_n\}$. Concepts are elements in text which are identified as unique objects and linked to an entry in a reference Knowledge Base (Wikipedia in this case). Each user concept-based profile consists of Bag of Concepts, extracted from their sent emails.

Implementation details. The keyword extraction process has been performed using Java Automatic Term Recognition Toolkit (JATR v1.0[2]). JATR implements a voting mechanism to combine the results from different methods for terminology recognition (dealing with single- and multi-word term recognition) into an integrated output, improving results of integrated methods taken separately [38].

The Named Entity extraction has been performed using the Open Calais web service[3]. Open Calais is an ontology-based service which returns extraction results in RDF. Together with named entity recognition, the service performs instance recognition (concept identification) and facts extraction. For the purposes of this work we only exploited the Named Entities returned by OpenCalais.

---

[2] http://staffwww.dcs.shef.ac.uk/people/Z.Zhang/resources/
    tools/jatr_v1.0.zip
[3] http://www.opencalais.com/

The Concept extraction process has been performed using the Wikify web service [26]. Wikify uses a machine-learning approach to annotate Wikipedia concepts within unstructured text. The disambiguation procedure uses three features: a priori probability of each Wikipedia concept, weighted context terms in the target text and a measure of the goodness of the context. The context terms weights are calculated by using the average semantic relatedness with all other context terms. The measure of goodness of the context reflects its homogeneity and it is used to dynamically balance a priori probability and context term weights, instead of using a fixed heuristic. The candidate terms are generated by gathering all n-grams in the text and discarding those below a low threshold (to discard non-sense phrases and stopwords). Experiments show that the method perform as well on non-Wikipedia texts as on Wikipedia ones, with an F-measure of 75% on non-Wikipedia texts.

3.2 Deriving People Similarity

The obtained user profiles are then used to calculate the similarity strength between users, measured on a [0,1] range. Similarity values reflect the amount of knowledge shared between two users. When used to plot a network of users similarity values can be useful to identify central users, small communities with shared knowledge (users with higher values of similarity among each other).

Following [15] similarity score is calculated using Jaccards index. The Jaccard similarity coefficient measures similarity between sample sets, and is defined as the size of the intersection divided by the size of the union of the sample sets. Sample sets for our user similarity are concepts (or keywords or Named Entities respectively) in each user profile. Moreover the similarity calculated over semantic user profiles is compared with the same measure calculated over the keyword based profiles and the named entity based profiles, to prove that increasing the semantic value of a profile increases its quality and suitability for modelling people's expertise. The semantic profiles amongst the others, better mimic the users' perceived similarities between each other. Results for similarities calculated over the three different types of profiles are shown in section 4.

4 Experiments

The aim of the experiments was to validate the hypothesis that increasing the level of semantic in the user profile generation improves the quality of profiles. For such purpose we used the task of inferring similarities between users and assessed the correlation with human judgment. The corpus used for analysis and knowledge content extraction is an internal mailing list of the OAK Group in the Computer Science Department of the University of Sheffield[4]. The mailing list is used for quick exchange of information within the group on both professional and social topics. We use the same corpus as [13], but with a broader period

---

[4] http://oak.dcs.shef.ac.uk

coverage: we selected all emails sent to the mailing list in the ten month period from July 2010 to May 2011, totalling 1001 emails. We will refer to this corpus as *mlDataset*. For each email we extracted the subject and the email body. The number of users in the mailing list is 40; 25 of them are active users (users sending email to mailing list). The average message length is 420 characters. The average message length per user is shown in table 1, together with number of concepts in each user profile. Table 1 reports statistic about all 25 active users, even if only 15 of those 25 participated to the evaluation exercise.

Table 1. Corpus statistics for *mlDataset*. Column ID contains an anonymous identifier for the users. Column AvgMsg contains the average message length for that user, expressed in number of characters. Column Conc contains the number of concepts in the user profile.

ID	AvgMsg	n. sent email	Conc	ID	AvgMsg	n. sent email	Conc
1	445	62	42	16	612	121	71
3	453	209	90	18	170	27	35
5	1192	9	10	21	290	24	25
6	543	5	5	22	155	27	14
7	489	13	11	23	345	53	39
8	330	9	14	24	271	10	14
9	462	74	67	25	399	65	79
10	237	37	23	27	236	36	33
11	282	90	80	28	523	20	23
12	841	23	40	33	102	8	5
13	766	12	12	36	227	30	15
14	338	17	18	40	224	3	3
15	516	17	22				

We compared user similarity obtained with the three different profile types (section 3) against the users perceived similarity. Pearson correlation has been calculated for user judgments compared against the automatic generated similarities using Keyword based profiles, Named Entity based profiles and Concepts based profiles.

The evaluation was conducted as a paper-based exercise in which the participants were asked to fill in a table containing a list of people within the working group and rate their perceived similarity on a scale from 1 to 10, with *1 = not similar at all* and *10 = absolutely similar*. If the user was not known they were instructed to leave the rating blank. The participants were asked to score the similarity in terms of topics or interest shared with the other person, both from a professional and social point of view (e.g. hobbies or other things which emerge within the working time) to concentrate the user thoughts towards "general" similarity without thinking about what specific type of similarity they share. The exercise was repeated twice for a small subset of users, with the second one seven days after the first one. The inter-annotator agreement was calculated by comparing for each user his/her perceived similarities about other participants and similarities perceived by all the rest of participants for that particular user. Inter-annotator agreement is shown in the last column of table 2.

A total of 15 users took part in the evaluation exercise, all providing valid questionnaires.

Table 2. Correlation (C) of similarity with user judgment at (S) significance level, obtained using Keyword based profiles (K), Named Entity based profile (NEs), Concept based profiles (Conc). Column (Agr) reports Inter-annotator agreement for each user at significance < 0.001.

ID	K		NEs		Conc		Agr
	C	S	C	S	C	S	C
14	0.55	0	0.41	0.04	0.68	0	0.91
7	0.48	0.02	0.39	0.06	0.58	0	0.87
28	0.5	0.01	0.41	0.04	0.57	0	0.89
10	0.47	0.02	0.39	0.05	0.57	0	0.94
27	0.32	0.11	0.29	0.16	0.48	0.02	0.92
21	0.34	0.11	0.42	0.04	0.42	0.04	0.91
1	0.35	0.02	0.32	0.11	0.42	0.04	0.94
3	0.3	0.14	0.31	0.14	0.38	0.06	0.86
9	0.28	0.18	0.36	0.07	0.38	0.06	0.9
18	0.5	0.01	0.5	0.01	0.36	0.07	0.87
8	0.17	0.53	0.19	0.48	0.35	0.18	0.82
11	0.59	0	0.42	0.04	0.34	0.1	0.83
25	0.25	0.22	0.33	0.11	0.3	0.14	0.73
23	0.21	0.32	0.33	0.1	0.19	0.36	0.86

Table 2 shows the Pearson correlation between automatically generated similarity with user judgment. For the three types of user profiles, Keyword based profiles (K), Named Entity based profile (NEs) and Concept based profiles (Conc), the table shows the correlation value (C) and the respective significance level (S). The inter-annotator agreement for each user reported in column Agr, has been calculated with Pearson correlation at significance < 0.001. Results are presented in descending order of correlation for similarities over concept based profiles. Figures show that the correlation almost always improves by the usage of Concept based profiles over Keyword based profiles, except for three users (18, 11, 23). Moreover the significance level for the correlation on concept based similarity is lower than the one on keyword based similarity (except for user 23). For the three users not confirming the trend, the inter-annotator agreement average (0.83) is lower than the general average (0.88).

5   Applications and Scenarios of Usage

The approach presented in this paper allows to automatically extract content and user similarity from an email corpus to build networks which reflects people expertise rather than capturing users' social interactions. This approach could prove particularly useful for knowledge management, in particular for tasks such as expertise finding, trend spotting and identification of dynamic communities inside an organisation. In order to prove its usefulness for knowledge management the approach has been integrated into a knowledge management system aimed at capturing and sharing knowledge exchanged during informal communication exchanges (meetings, chats, etc.): this has provided clear scenarios of usage, which will be helpful to understand whether semantic dynamic user profiling increases efficiency and efficacy in expert finding and trend spotting tasks. The knowledge management framework adopts semantic user profiling to capture content from numerous informal communication exchanges such as emails, meeting recordings and minutes etc.; these are then visualised using SimNET (Similarity and Network Exploration Tool), a tool for exploring and searching over socio-centric content, that interactively displays content and users networks as part of the searching and browsing capabilities provided by the knowledge management system. Section 5.1 introduces SimNET, whilst section 5.2 introduces two scenarios of usage of the knowledge management system within a large organisation. The scenarios are presented to highlight the capabilities and usefulness of the approach; an evaluation of the expert finding and trend analysis tasks is scheduled for the coming months.

5.1   Semantic Network Visualisation

SimNET is a dynamic and real-time filtering interface that offers multiple visualisations for knowledge extracted from user generated content, using classical techniques such as node-Link diagrams [12,29], force-directed layouts [16],[34], radial layouts [36] (as shown in Figure 1) and tag clouds (detail of a tag cloud in Figure 3). SimNET has been built as a flexible visualisation tool to explore socio-centric content using the most suitable visualisation for the undertaken task. Radial visualization is provided to focus on the interactions between users, while force-directed layout is provided to support by-topic knowledge exploration and trends observation. SimNET has two main interaction paradigms - email visualisation and similarity visualisation. The users are initially presented with an interface that visualises email interactions and a tag cloud describing all the concepts. The users can then choose between a radial or a force-directed layout according to the task and can use filtering and searching widgets to focus the exploration of the dataset. For example, when clicking on concepts in the tag cloud relevant emails are highlighted in the radial layout and vice versa. Users can also select to visualise similarities by clicking on 'Plot Similarity'. The radial graph is updated to show the similarities as edges between nodes. These edges are colour coded to signify the similarity among the connecting nodes. The interface provides the users with a *similarity slider*, which can be dragged to set a similarity threshold for showing edges and a *date range selection bar*, to explore the network evolution over time.

Providing visualisation capabilities is critical for making sense of user profiles, topics and similarities as it allows users to access and manipulate knowledge.

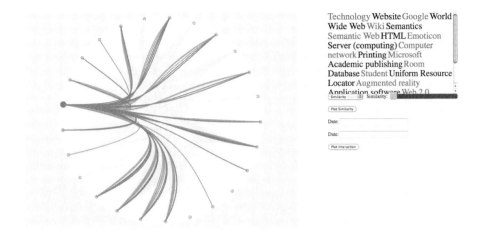

Fig. 1. SimNET, Similarity and Network Exploration Tool

5.2   Scenarios of Use

Expertise Finding. Our system supports the task of finding an expert for a given topic in multiple ways. For example a user may browse the tag cloud or perform a specific query. The system will then plot the results using a force-directed or a radial layout and the user will be able to interact with the visualisation to explore it and highlight users that are experts on that specific topic. This is very important as it addresses the long tail of information, allowing to discover expertise for topics that are not well known or for which not many people are knowledgeable. It is even more important when applied to large and dynamic organisation where the number of users is very high and it is very likely that they do not know each other and they are not aware of who are the current experts on certain topics, who are similar users to involve in a certain activity. Having a system that highlights the long-tail of information allows sharing knowledge in a more efficient manner; for example if a user working in a big organization is looking for information of a common and well-known topic within her/his department almost everyone in her/his group will be able to answer her request, whilst if looking for more obscure information that is interdisciplinary, people in the same department may not know the answer. In such a case a system that allows to discover expertise tailoring the short and long tail of information is invaluable as it quickly highlights people in the organisations for quick help.

Expertise Trend Analysis. When wanting to understand the expertise of a group of people or emerging topic trends inside a dynamic organisation it could be helpful to plot all the topics accordingly to their relevance. Figure 2 refers to the 25 users of the experiment shown in section 4. It shows a number of topics discussed within the research group (as extracted from *mlDataset*). The concepts closer to the central node *Oak* are the ones shared by the majority of users, while nodes on the outer circle (randomly picked) are concepts shared by

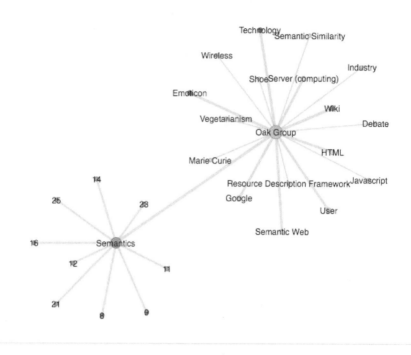

Fig. 2. A force-directed layout to highlight the expertise of the group

July 2010

Technology **World Wide Web**
Website **Wiki** Printing **Server**
(computing) Emoticon **Google** Uniform
Resource Locator **Microsoft** Computer network
Room Wireless **Application software** Augmented reality
Privacy Time Web 2.0 Semantic Web **Semantics** Laboratory **Facebook**
Database **Week** Student HTML Hewlett-Packard Institute of Electrical and Electronics Engineers Internet Innovation

March 2011

Technology **Website** Semantics **HTML**
Semantic Web **Database** Uniform Resource
Locator **Academia** Internet Server (computing) World Wide Web
Microsoft Innovation Printing Google Wiki Internet Innovation Hyperlink **Application**
software Portable Document Format **Facebook** Thought **Privacy**
Computer network **Information technology** Web page **Wireless**

Fig. 3. Tag clouds generated over two different period of time in the *mlDataset*

a small number of people (2 or 3). For example, Figure 2 clearly highlights the
emerging topics in the group as {*Semantics, Wiki, Emoticon, Technology, User
(computing), Week, Semantic Web, Server (computing), HTML, Google*}, but
also allows to identify topics that are emerging or have less wide-spread and more
specific expertise such as {*Industry, Semantics, Semantic similarity, Javascript,
Debate, Vegetarianism, Marie Curie, resource Description Framework, Wireless,
Shoe, Chocolate*}.

By using temporal filtering on the data it is also possible to study the trends
evolution over time and the hot topics during a certain period. Figure 3 shows
the topics discussed in July 2010 and in March 2011 (from *mlDataset*); the data

tag cloud visualisation helps discovering that in one period there were discussions e.g. about *Augmented Reality* and about *Facebook* on the other.

6 Conclusions and Future Work

This paper proposed an approach to automatically and dynamically model user expertise from informal communication exchanges. The main novelty of our approach consists of generating semantic user profiles from emails (and more generally from any textual user generated content) guaranteeing flexibility, dynamicity and providing ways to connect these data with Linked Open Data (LOD). Whilst linked data have not been exploited in the current work, future work will consider semantic concept expansions, enriching user profiles by exploring the LOD cloud starting from concepts within the profiles. Indeed, the actual concepts in each profile are dbpedia[5] objects, therefore already highly connected to the LOD cloud[6].

Extracting information from informal communication exchanges could be hugely beneficial for knowledge management inside an organisation, as it offers means to recover buried knowledge without any additional effort from the individuals and respecting their natural communication patterns. In order to prove and evaluate the possible benefits for knowledge management the approach has been integrated in a knowledge management system aimed at capturing and sharing knowledge gathered from informal communication exchanges (meetings, chats, etc.) that dynamically builds user networks, determines experts and communities of practice and identifies emerging topics, mirroring the natural evolution of the organisational communities and displays it in an interactive interface that provides means to access and manipulate knowledge. Further evaluations of the approach will be conducted shortly with users to test whether the approach provides advantages over standard knowledge management practices. During this evaluation, attention will be given as whether the approach increases efficiency and efficacy for a given task by presenting in a visual way trends and topics of expertise and providing means to search for experts inside the organisation.

Acknowledgments. A.L Gentile and V. Lanfranchi are funded by SILOET (Strategic Investment in Low Carbon Engine Technology), a TSB-funded project. S. Mazumdar is funded by Samulet (Strategic Affordable Manufacturing in the UK through Leading Environmental Technologies), a project partially supported by TSB and from the Engineering and Physical Sciences Research Council.

References

1. Abel, F., Gao, Q., Houben, G.J., Tao, K.: Analyzing User Modeling on Twitter for Personalized News Recommendations. In: Konstan, J., Conejo, R., Marzo, J., Oliver, N. (eds.) UMAP 2011. LNCS, vol. 6787, pp. 1–12. Springer, Heidelberg (2011)

---

[5] http://dbpedia.org/
[6] http://linkeddata.org/

2. Abel, F., Gao, Q., Houben, G.J., Tao, K.: Semantic Enrichment of Twitter Posts for User Profile Construction on the Social Web. In: Antoniou, G., Grobelnik, M., Simperl, E.P.B., Parsia, B., Plexousakis, D., Leenheer, P.D., Pan, J.Z. (eds.) ESWC 2011, Part II. LNCS, vol. 6644, pp. 375–389. Springer, Heidelberg (2011)
3. Adamic, L., Adar, E.: How to search a social network. Social Networks 27(3), 187–203 (2005)
4. Balabanović, M., Shoham, Y.: Fab: content-based, collaborative recommendation. Commun. ACM 40, 66–72 (1997)
5. Campbell, C.S., Maglio, P.P., Cozzi, A., Dom, B.: Expertise identification using email communications. In: Proceedings of the Twelfth International Conference on Information and Knowledge Management, CIKM 2003, pp. 528–531. ACM, New York (2003)
6. Cortes, C., Pregibon, D., Volinsky, C.: Computational methods for dynamic graphs. Journal Of Computational And Graphical Statistics 12, 950–970 (2003)
7. Culotta, A., Bekkerman, R., McCallum, A.: Extracting social networks and contact information from email and the web. In: CEAS 2004: Proc. 1st Conference on Email and Anti-Spam (2004)
8. Daoud, M., Tamine, L., Boughanem, M.: A Personalized Graph-Based Document Ranking Model Using a Semantic User Profile. In: De Bra, P., Kobsa, A., Chin, D. (eds.) UMAP 2010. LNCS, vol. 6075, pp. 171–182. Springer, Heidelberg (2010)
9. De Choudhury, M., Mason, W.A., Hofman, J.M., Watts, D.J.: Inferring relevant social networks from interpersonal communication. In: Proceedings of the 19th International Conference on World Wide Web, WWW 2010, pp. 301–310. ACM, New York (2010)
10. Diesner, J., Frantz, T.L., Carley, K.M.: Communication networks from the enron email corpus "it's always about the people. enron is no different". Comput. Math. Organ. Theory 11, 201–228 (2005)
11. Eckmann, J., Moses, E., Sergi, D.: Entropy of dialogues creates coherent structures in e-mail traffic. Proceedings of the National Academy of Sciences of the United States of America 101(40), 14333–14337 (2004)
12. Freeman, L.C.: Visualizing Social Networks. JoSS: Journal of Social Structure 1(1) (2000)
13. Gentile, A.L., Basave, A.E.C., Dadzie, A.S., Lanfranchi, V., Ireson, N.: Does Size Matter? When Small is Good Enough. In: Rowe, M., Stankovic, M., Dadzie, A.-S., Hardey, M. (eds.) Proceedings, 1st Workshop on Making Sense of Microposts (#MSM 2011), pp. 45–56 (May 2011)
14. Gloor, P.A., Laubacher, R., Dynes, S.B.C., Zhao, Y.: Visualization of communication patterns in collaborative innovation networks - analysis of some w3c working groups. In: Proceedings of the Twelfth International Conference on Information and Knowledge Management, CIKM 2003, pp. 56–60. ACM, New York (2003)
15. Guy, I., Jacovi, M., Perer, A., Ronen, I., Uziel, E.: Same places, same things, same people?: mining user similarity on social media. In: Proceedings of the 2010 ACM Conference on Computer Supported Cooperative Work, CSCW 2010, pp. 41–50. ACM, New York (2010)
16. Heer, J., Boyd, D.: Vizster: Visualizing online social networks. In: Proceedings of the 2005 IEEE Symposium on Information Visualization, p. 5. IEEE Computer Society, Washington, DC, USA (2005)
17. Keila, P.S., Skillicorn, D.B.: Structure in the Enron email dataset. Computational & Mathematical Organization Theory 11, 183–199 (2005)

18. King, D.W., Casto, J., Jones, H.: Communication by Engineers: A Literature Review of Engineers' Information Needs, Seeking Processes, and Use. Council on Library Resources, Washington (1994)
19. Kossinets, G., Watts, D.J.: Empirical analysis of an evolving social network. Science 311(5757), 88–90 (2006)
20. Kramár, T.: Towards Contextual Search: Social Networks, Short Contexts and Multiple Personas. In: Konstan, J., Conejo, R., Marzo, J., Oliver, N. (eds.) UMAP 2011. LNCS, vol. 6787, pp. 434–437. Springer, Heidelberg (2011)
21. Laclavik, M., Dlugolinsky, S., Seleng, M., Kvassay, M., Gatial, E., Balogh, Z., Hluchy, L.: Email analysis and information extraction for enterprise benefit. Computing and Informatics, Special Issue on Business Collaboration Support for Micro, Small, and Medium-Sized Enterprises 30(1), 57–87 (2011)
22. Li, Q., Zheng, Y., Xie, X., Chen, Y., Liu, W., Ma, W.y.: Mining User Similarity Based on Location History. Architecture (c) (2008)
23. Lin, C.Y., Cao, N., Liu, S.X., Papadimitriou, S., Sun, J., Yan, X.: SmallBlue: Social Network Analysis for Expertise Search and Collective Intelligence. In: IEEE 25th International Conference on Data Engineering, ICDE 2009, pp. 1483–1486 (2009)
24. Lin, C.Y., Ehrlich, K., Griffiths-Fisher, V., Desforges, C.: Smallblue: People mining for expertise search. IEEE Multimedia 15, 78–84 (2008)
25. McCallum, A., Wang, X., Corrada-Emmanuel, A.: Topic and role discovery in social networks with experiments on Enron and academic email. Journal of Artificial Intelligence Research 30, 249–272 (2007)
26. Milne, D., Witten, I.H.: Learning to link with wikipedia. In: Proceeding of the 17th ACM Conference on Information and Knowledge Management, CIKM 2008, pp. 509–518. ACM, New York (2008)
27. Mislove, A., Viswanath, B., Gummadi, K.P., Druschel, P.: You are who you know: inferring user profiles in online social networks. In: Proceedings of the Third ACM International Conference on Web Search and Data Mining, WSDM 2010, pp. 251–260. ACM, New York (2010)
28. Pazzani, M., Billsus, D.: Learning and revising user profiles: The identification of interesting web sites. Machine Learning 27, 313–331 (1997)
29. Reingold, E.M., Tilford, J.S.: Tidier drawings of trees. IEEE Transactions on Software Engineering 7, 223–228 (1981)
30. Robinson, M.A.: Erratum: Correction to robinson, m.a, an empirical analysis of engineers' information behaviors. Journal of the American Society for Information Science and Technology 61(4), 640–658 (2010); J. Am. Soc. Inf. Sci. Technol. 61, 1947–1947 (September 2010)
31. Schwartz, M.F., Wood, D.C.M.: Discovering shared interests using graph analysis. Communications of the ACM 36(8), 78–89 (1993)
32. Tuulos, V.H., Perkiö, J., Tirri, H.: Multi-faceted information retrieval system for large scale email archives. In: Proceedings of the 28th Annual International ACM SIGIR Conference on Research and Development in Information Retrieval, SIGIR 2005, pp. 683–683. ACM, New York (2005)
33. Tyler, J., Wilkinson, D., Huberman, B.: E-Mail as spectroscopy: Automated discovery of community structure within organizations. The Information Society 21(2), 143–153 (2005)
34. Viégas, F.B., Donath, J.: Social network visualization: can we go beyond the graph. In: Workshop on Social Networks for Design and Analysis: Using Network Information in CSCW 2004, pp. 6–10 (2004)
35. Xiang, R., Lafayette, W., Lafayette, W.: Modeling Relationship Strength in Online Social Networks. North, 981–990 (2010)

36. Yee, K.P., Fisher, D., Dhamija, R., Hearst, M.: Animated exploration of dynamic graphs with radial layout. In: Proceedings of the IEEE Symposium on Information Visualization 2001 (INFOVIS 2001), p. 43. IEEE Computer Society Press, Washington, DC, USA (2001)
37. Ying, J.J.C., Lu, E.H.C., Lee, W.C., Tseng, V.S.: Mining User Similarity from Semantic Trajectories. Cell, 19–26 (2010)
38. Zhang, Z., Iria, J., Brewster, C., Ciravegna, F.: A comparative evaluation of term recognition algorithms. In: Calzolari, N., Choukri, K., Maegaard, B., Mariani, J., Odjik, J., Piperidis, S., Tapias, D. (eds.) Proceedings of the Sixth International Language Resources and Evaluation (LREC 2008). European Language Resources Association (ELRA), Marrakech (2008)
39. Zhou, Y., Fleischmann, K.R., Wallace, W.A.: Automatic text analysis of values in the Enron email dataset: Clustering a social network using the value patterns of actors. In: HICSS 2010: Proc., 43rd Annual Hawaii International Conference on System Sciences, pp. 1–10 (2010)

Verification of the OWL-Time Ontology

Michael Grüninger

Department of Mechanical and Industrial Engineering,
University of Toronto, Toronto, Ontario, Canada M5S 3G8

Abstract. Ontology verification is concerned with the relationship between the intended structures for an ontology and the models of the axiomatization of the ontology. The verification of a particular ontology requires characterization of the models of the ontology up to isomorphism and a proof that these models are equivalent to the intended structures for the ontology. In this paper we provide the verification of the ontology of time introduced by Hobbs and Pan, which is a first-order axiomatization of OWL-Time. We identify five modules within this ontology and present a complete account of the metatheoretic relationships among the modules and between other time ontologies for points and intervals.

1 Introduction

Over the years, a number of first-order ontologies for time have been proposed. In addition to ontologies for timepoints and ontologies for time intervals ([2]), there are also ontologies that axiomatize both timepoints and time intervals together with the relationships between them. More recently, Hobbs and Pan ([10], [12]) have proposed a first-order axiomatization $T_{owltime}$ of OWL-Time[1] as an ontology of time for the Semantic Web that also includes both timepoints (referred to as instants) and intervals.

   The primary objective of this paper is to provide a characterization of the models of $T_{owltime}$ up to isomorphism using the notion of theory reducibility from [6]. This will lead to a modularization of $T_{owltime}$ and allow us to identify incorrect or missing axioms in the current axiomatization of $T_{owltime}$. Finally, we will also use this reduction to compare $T_{owltime}$ to other time ontologies for points and intervals, and address the question as to whether $T_{owltime}$ forms an adequate core theory for time ontologies, or whether it is too weak or too strong to play such a role.

2 Ontology Verification

Our methodology revolves around the application of model-theoretic notions to the design and analysis of ontologies. The semantics of the ontology's terminology can be characterized by a set of structures, which we refer to as the set of intended structures for the ontology. Intended structures are specified with respect to the models of well-understood mathematical theories (such as partial orderings, lattices, incidence structures, geometries, and algebra). The extensions of the relations in an intended structure are then specified with respect to properties of these models.

---

[1] http://www.w3.org/TR/owl-time/

L. Aroyo et al. (Eds.): ISWC 2011, Part I, LNCS 7031, pp. 225–240, 2011.
© Springer-Verlag Berlin Heidelberg 2011

Why do we care about ontology verification? The relationship between the intended models and the models of the axiomatization plays a key role in the application of ontologies in areas such as semantic integration and decision support. Software systems are semantically integrated if their sets of intended models are equivalent. In the area of decision support, the verification of an ontology allows us to make the claim that any inferences drawn by a reasoning engine using the ontology are actually entailed by the ontology's intended models. If an ontology's axiomatization has unintended models, then it is possible to find sentences that are entailed by the intended models, but which are not provable from the axioms of the ontology. The existence of unintended models also prevents the entailment of sentences or a possible barriers to interoperability.

With ontology verification, we want to characterize the models of an ontology up to isomorphism and determine whether or not these models are elementarily equivalent to the intended structures of the ontology. From a mathematical perspective this is formalized by the notion of representation theorems. The primary challenge for someone attempting to prove representation theorems is to characterize the models of an ontology up to isomorphism. For this we use the following notion from [6]:

Definition 1. *A class of structures $\mathfrak{M}$ can be represented by a class of structures $\mathfrak{N}$ iff there is a bijection $\varphi : \mathfrak{M} \rightarrow \mathfrak{N}$ such that for any $\mathcal{M} \in \mathfrak{M}$, $\mathcal{M}$ is definable in $\varphi(\mathcal{M})$ and $\varphi(\mathcal{M})$ is definable in $\mathcal{M}$.*

The key to ontology verification is that a theorem about the relationship between the class of the ontology's models and the class of intended structures can be replaced by a theorem about the relationship between the ontology (a theory) and the theory axiomatizing the intended structures (assuming that such axiomatization is known). We can use automated reasoners to prove this relationship and thus verify an ontology in a (semi-)automated way.

The relationship between theories T_A and T_B is the notion of interpretation ([5]), which is a mapping from the language of T_A to the language of T_B that preserves the theorems of T_A. If there is an interpretation of T_A in T_B, then there exists a set of sentences (referred to as translation definitions) in the language $L_A \cup L_B$ of the form

$$(\forall \overline{x}) \, p_i(\overline{x}) \equiv \varphi(\overline{x})$$

where $p_i(\overline{x})$ is a relation symbol in L_A and $\varphi(\overline{x})$ is a formula in L_B. Translation definitions will be used extensively in the proofs of theorems later in the paper.

We will say that two theories T_A and T_B are definably equivalent iff they are mutually interpretable, i.e. T_A is interpretable in T_B and T_B is interpretable in T_A.

The key to using theorem proving and model finding to support ontology verification is the following theorem ([6]):

Theorem 1. *A theory T_1 is definably equivalent with a theory T_2 iff the class of models $Mod(T_1)$ can be represented by $Mod(T_2)$.*

Let $\mathfrak{M}^{intended}$ be the class of intended structures for the ontology, and let T_{onto} be the axiomatization of the ontology. The necessary direction of a representation theorem (i.e. if a structure is intended, then it is a model of the ontology's axiomatization) can be stated as

$$\mathcal{M} \in \mathfrak{M}^{intended} \Rightarrow \mathcal{M} \in Mod(T_{onto})$$

If we suppose that the theory that axiomatizes $\mathfrak{M}^{intended}$ is the union of some previously known theories $T_1, ..., T_n$, then by Theorem 1 we need to show that T_{onto} interprets $T_1 \cup ... \cup T_n$. If Δ is the set of translation definitions for this interpretation, then the necessary direction of the representation theorem is equivalent to the following reasoning task:

$$T_{onto} \cup \Delta \models T_1 \cup ... \cup T_n$$

The sufficient direction of a representation theorem (any model of the ontology's axiomatization is also an intended structure) can be stated as

$$\mathcal{M} \in Mod(T_{onto}) \Rightarrow \mathcal{M} \in \mathfrak{M}^{intended}$$

In this case, we need to show that $T_1 \cup ... \cup T_n$ interprets T_{onto}. If Π is the set of translation definitions for this interpretation, the sufficient direction of the representation theorem is equivalent to the following reasoning task:

$$T_1 \cup ... \cup T_n \cup \Pi \models T_{onto}$$

Proving these two entailment problems constitutes the reducibility theorem for T_{onto}; the set of theories $T_1, ..., T_n$ form the reduction of T_{onto}.

All of the theories introduced in this paper are being investigated in the context of the COLORE (Common Logic Ontology Repository) project, which is building an open repository of first-order ontologies that serve as a testbed for ontology evaluation and integration techniques, and that can support the design, evaluation, and application of ontologies in first-order logic. We identify the theories for the reduction by searching through the COLORE ontology repository for ontologies that are definably equivalent to the ontology that we are verifying, and then prove that the mappings between the ontologies are correct.

3   Modularization of OWL-Time

The verification of the OWL-Time Ontology $T_{owltime}$ will also provide a decomposition of the ontology into a set of subtheories which are related by conservative extension. We will not present the verification of the entire OWL-Time Ontology in this paper; rather we will focus on the subtheory $T_{timespan}$ of $T_{owltime}$ that omits the axioms for durations and dates. Furthermore, throughout the paper we will identify additional axioms that are required to prove the representation theorems, leading to the specification of a theory $T_{timespan}^*$ as an extension of $T_{timespan}$. By the end of the paper, we will show that $T_{timespan}^*$ is definably equivalent to the following theories (which will be introduced and discussed in detail in the following sections) from COLORE:

$$T_{linear\_order} \cup T_{pseudo\_complete} \cup T_{log} \cup T_{weak\_planar} \cup T_{diamond}$$

In Figure 1, we can see which subsets of these theories interpret subtheories of $T_{owltime}$, and this set of subtheories [2] constitutes the modules within $T_{owltime}$. In this way, modules within OWL-Time are obtained by identifying ontologies within the COLORE ontology repository that are definably equivalent to subtheories of OWL-Time. In the following sections, we will specify the reduction theorems for each of these subtheories, culminating in the representation theorem for $T_{owltime}^*$.

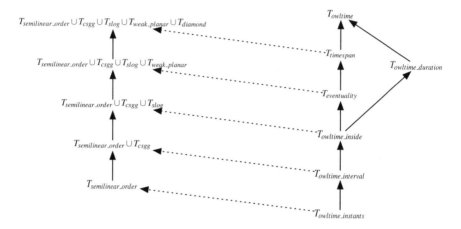

Fig. 1. Relationships between the subtheories of $T_{owltime}$ and the theories from COLORE which are used to prove the representation theorem for $T_{owltime}$. Solid lines denote conservative extension and dotted lines denote interpretability between theories.

4   Instants and Intervals

The first two modules within OWL-Time that we will consider are $T_{owltime\_instant}$ and $T_{owltime\_interval}$, which axiomatize fundamental intuitions about timepoints (Instants) and time intervals (Intervals). Besides the objective of ontology verification, we will also be interested in using the reduction of $T_{owltime\_interval}$ to determine the relationship to other ontologies for timepoints and time intervals. Given an existing ontology T of timepoints and intervals, we will identify the theory that is a common extension of both $T_{owltime\_interval}$ and T, and also identify the common subtheory of $T_{owltime\_interval}$ and T.

---

[2] The original first-order axiomatization of $T_{owltime}$ can be found at
http://www.cs.rochester.edu/~ferguson/daml/
daml-time-nov2002.txt
The CLIF (Common Logic Interchange Format) axiomatization of the subtheories of $T_{owltime}$ discussed in this paper can be found at
http://stl.mie.utoronto.ca/colore/time/owltime_instants.clif
http://stl.mie.utoronto.ca/colore/time/owltime_interval.clif
http://stl.mie.utoronto.ca/colore/time/owltime_inside.clif
http://stl.mie.utoronto.ca/colore/time/eventuality.clif
http://stl.mie.utoronto.ca/colore/time/timespan.clif

4.1 Time Ontologies for Points

Ontologies for time points have been studied in [2] and [9], and all axiomatize an ordering relation over the points. Although Hobbs and Pan rightly argue that a restriction to linear orders is too strong, allowing arbitrary partial orderings is arguably too weak. We consider the following extension:

Definition 2. $T_{owltime\_instant\_s}$ *is the extension of* $T_{owltime\_instant}$ *with the axiom*

$$(\forall x, y)\, Instant(x) \wedge Instant(y) \supset (\exists z)(before(z,x) \vee (z = x)) \wedge (before(z,y) \vee (z = y)))$$

It is straightforward to see that the resulting ordering on the set of instants is a semilinear ordering [4] as axiomatized by the theory $T_{semilinear\_ordering}$[3].

4.2 Time Ontologies for Points and Intervals

In his Catalog of Temporal Theories [9], Hayes introduced three time ontologies that axiomatize both timepoints and time intervals. The *endpoints* theory combines the language of intervals and points by defining the functions *beginof*, *endof*, and *between* to relate intervals to points and vice-versa. This theory imports the axioms of *linear_point* that define the binary *before* relation between timepoints as transitive and irreflexive, and impose the condition that all timepoints are linearly ordered and infinite in both directions. The *vector_continuum* theory is a theory of timepoints and intervals that introduces the notion of orientation of intervals. It also imports theory *linear_point*. In this theory single-point intervals, known as *moments*, are defined as intervals whose *beginof* and *endof* points are the same. The *point_continuum* theory combines intervals and points by defining the relation *in* that relates a point to the interval it is contained in. The verification of these three time ontologies is given in [7].

   The theory $\mathcal{IQ}$ in [13] (which we will refer to as T_{vila} in this paper) is a common subtheory for all of the ontologies of time points and intervals. Later in the paper, we will characterize the relationship between T_{vila} and $T_{owltime\_interval}$. To lay the groundwork for this characterization, we review the classes of structures that will be used for the reducibility theorem for these theories.

Graphical Incidence Structures. The basic building blocks for the models presented in this paper are based on the notion of incidence structures ([3]).

Definition 3. *A k-partite incidence structure is a tuple* $\mathbb{I} = (\Omega_1, ..., \Omega_k, \mathbf{in})$, *where* $\Omega_1, ..., \Omega_k$ *are sets with*
$$\Omega_i \cap \Omega_j = \emptyset, i \neq j$$
and
$$\mathbf{in} \subseteq (\bigcup_{i \neq j} \Omega_i \times \Omega_j)$$
Two elements of $\mathbb{I}$ *that are related by* **in** *are called incident.*

---
[3] http://stl.mie.utoronto.ca/colore/ordering/
semilinear_ordering.clif

The models of the time ontologies in this paper will be constructed using special classes of incidence structures.

Definition 4. *An strong graphical incidence structure is a bipartite incidence structure*

$$\mathbb{S} = \langle X, Y, \mathbf{in}^{\mathbf{S}} \rangle$$

such that all elements of Y are incident with either one or two elements of X, and for each pair of points $\mathbf{p}, \mathbf{q} \in X$ there exists a unique element in Y that is incident with both $\mathbf{p}$ and $\mathbf{q}$.

The class of strong graphical incidence structures is axiomatized by $T_{strong\_graphical}$[4]

Definition 5. *A loop graphical incidence structure is a bipartite incidence structure*

$$\mathbb{S} = \langle X, Y, \mathbf{in}^{\mathbf{S}} \rangle$$

such that all elements of Y are incident with either one or two elements of X, and for each pair of points $\mathbf{p}, \mathbf{q} \in X$ there exists a unique element in Y that is incident with both $\mathbf{p}$ and $\mathbf{q}$, and for each point $\mathbf{r} \in X$ there exists a unique element in Y that incident only with $\mathbf{r}$.

The class of loop graphical incidence structures is axiomatized by $T_{loop\_graphical}$[5].

These classes of incidence structures get their names from graph-theoretic representation theorems of their own.

Definition 6. *A graph $G = (V, E)$ consists of a nonempty set V of vertices and a set E of ordered pairs of vertices called edges.*

An edge whose vertices coincide is called a loop. A graph with no loops or multiple edges is a simple graph.

A complete graph is a graph in which each pair of vertices is adjacent.

Theorem 2. *Let $G = (V, E)$ be a complete graph.*

A bipartite incidence structure is a strong graphical incidence structure iff it is isomorphic to $\mathbb{I} = (V, E, \in)$.

Theorem 3. *Let $G = (V, E)$ be a complete graph with loops.*

A bipartite incidence structure is a loop graphical incidence structure iff it is isomorphic to $\mathbb{I} = (V, E, \in)$.

These representation theorems show that there is a one-to-one correspondence between the particular class of incidence structures and the given class of graphs; in so doing, we have a characterization of the incidence structures up to isomorphism.

---

[4] http://stl.mie.utoronto.ca/colore/incidence/
strong-graphical.clif

[5] http://stl.mie.utoronto.ca/colore/incidence/
loop-graphical.clif

Reducibility Theorems for Points and Intervals. We will ultimately be interested in determining the relationship between the theory T_{vila} and $T_{owltime\_interval}$. On the one hand, one of the design objectives for OWL-Time was to be a relatively weak theory that could be consistently extended to other time ontologies. On the other hand, T_{vila} is the common subtheory of existing time ontologies for points and intervals.

The approach taken in this paper is to compare different time ontologies by comparing the theories in their reductions. Using the axiomatizations of these classes of incidence structures, we can prove the following reducibility theorem for T_{vila}.

Theorem 4. T_{vila} is definably equivalent to

$$T_{linear\_order} \cup T_{strong\_graphical}$$

Proof.

Using the same set of translation definitions, we can also prove the following:

Theorem 5. Let T_{moment} be the extension of T_{vila} with the axiom

$$(\forall t)\ timepoint(t) \supset (\exists i)\ timeinterval(i) \wedge (beginof(i) = t) \wedge (endof(i) = t)$$

T_{moment} is definably equivalent to

$$T_{linear\_order} \cup T_{loop\_graphical}$$

In the next section we will show how this theory T_{moment} is definably equivalent to an extension of $T_{owltime\_interval}$.

4.3   Reducibility Theorems for Extensions of $T_{owltime\_interval}$

Given that the time ontologies in [9] and [13] use linear orderings over timepoints, We introduce the following extensions and subtheories of $T_{owltime\_interval}$ that impose a linear ordering on instants.

$T_{owltime\_linear}$ is the extension of $T_{owltime\_interval}$ with the axiom

$$(\forall t_1, t_2)\ Instant(t_1) \wedge Instant(t_2) \supset (before(t_1, t_2) \vee before(t_2, t_1) \vee (t_1 = t_2))$$

$T_{owltime\_e}$ is the extension of $T_{owltime\_interval}$ with the axiom

$$(\forall i)\ Interval(i) \supset (\exists t_1, t_2)\ begins(t_1, i) \wedge ends(t_2, i)$$

$T_{owltime\_le}$ is the extension of $T_{owltime\_linear}$ with $T_{owltime\_e}$.
$T_{owltime\_leu}$ is the extension of $T_{owltime\_le}$ with the axiom

$$(\forall t_1, t_2, i_1, i_2)\ begins(t_1, i_1) \wedge ends(t_2, i_1) \wedge begins(t_1, i_2) \wedge ends(t_2, i_2) \supset (i_1 = i_2)$$

$T_{owltime\_m}$ is the subtheory of $T_{owltime}$ without the axioms

$$(\forall t)\ (Instant(t) \equiv begins(t, t))$$

$$(\forall t)\ (Instant(t) \equiv ends(t, t))$$

The relationships between these theories are shown in Figure 2.

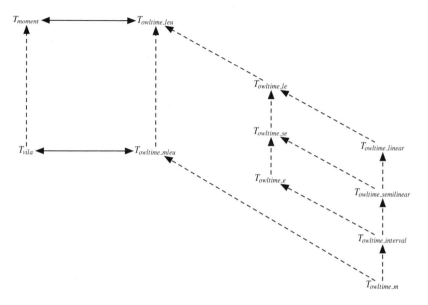

Fig. 2. Relationships between extensions of $T_{owltime\_interval}$ and other time ontologies for points and intervals. Dotted lines denote nonconservative extension and solid lines denote definable equivalence.

Theorem 6. $T_{owltime\_leu}$ *is definably equivalent to*

$$T_{linear\_order} \cup T_{loop\_graphical}$$

Proof. Let Δ_1 be the following set of translation definitions:

$$(\forall x) \, point(x) \equiv Instant(x)$$

$$(\forall x) \, line(x) \equiv Interval(x)$$

$$(\forall x, y) \, in^G(x, y) \equiv (begins(x, y) \lor ends(x, y))$$

$$(\forall x, y) \, before(x, y) \equiv lt(x, y)$$

Using Prover9 [11], we have shown that[6]

$$T_{owltime\_le} \cup \Delta_1 \models T_{linear\_order} \cup T_{loop\_graphical}$$

Let Π_1 be the following set of translation definitions:

$$(\forall x) \, Instant(x) \equiv point(x)$$

---

[6] All proofs in this paper that have been generated by Prover9 can be found at
http://stl.mie.utoronto.ca/colore/time/mappings/proofs/

$$(\forall x)\ Interval(x) \equiv line(x)$$

$$(\forall x, y)\ begins(x, y) \equiv ((in^G(x, y) \wedge ((\forall z)\ in^G(z, y) \supset lt(x, z)))$$

$$(\forall x, y)\ ends(x, y) \equiv ((in^G(x, y) \wedge ((\forall z)\ in^G(z, y) \supset lt(z, x)))$$

$$(\forall x, y)\ before(x, y) \equiv lt(x, y)$$

Using Prover9, we have shown that

$$T_{linear\_order} \cup T_{loop\_graphical} \cup \Pi_1 \models T_{owltime\_le}$$

Combining Theorem 6 and Theorem 4 gives us

Corollary 1. $T_{owltime\_leu}$ *is definably equivalent to* T_{moment}.

Since all subtheories and extensions of $T_{owltime\_interval}$ have the same nonlogical lexicon, we can use the same translation definitions as in the proof of Theorem 6 to prove the following theorem:

Theorem 7. $T_{owltime\_mleu}$ *is definably equivalent to*

$$T_{linear\_order} \cup T_{strong\_graphical}$$

Combining Theorem 7 and Theorem 5 gives us

Corollary 2. $T_{owltime\_mleu}$ *is definably equivalent to* T_{vila}.

4.4   Semilinear Ordering on Instants

If we now generalize these results to semilinear orderings on instants, then we need to consider a different class of structures for the reducibility theorem.

Definition 7. *A semilinear betweenness relation* **B** *is a ternary relation that is definable in a semilinear ordering* $\langle X, < \rangle$ *by the formula*

$$(\forall x, y, z)\ B(x, y, z) \equiv (((x < y) \wedge (y < z)) \vee ((z < y) \wedge (y < x)))$$

The semilinear betweenness relation captures the notion of comparability among instants within the semilinear ordering. The key axiom of $T_{owltime\_interval}$ is the one which states that for every two comparable points, there exists an interval of which they are the endpoints. This relationship between intervals and the ordering over instants is then captured by the following class of structures:

Definition 8. *A closed semilinear graphical geometry is a structure* $\mathbb{S} = \langle X, Y, \mathbf{B}, in^G \rangle$ *such that*

1. $\mathbb{B} = \langle X, \mathbf{B} \rangle$ *is a semilinear betweenness relation;*
2. $\mathbb{I} = \langle X, Y, in^G \rangle$ *is a graphical incidence structure;*

3. *triples of points in the extension of the betweenness relation* **B** *are incident with the same line in Y ;*
4. *any triple of points that are incident with the same line in Y are ordered by the betweenness relation* **B**.

The class of closed semilinear graphical geometries is axiomatized by T_{csgg}[7].

We did not need this class for extensions of $T_{owltime\_interval}$ with linear orderings because all instants in a linear ordering are comparable. This is not the case with semilinear orderings – elements on different branches are not comparable.

We can use translation definitions similar to those from the proof of Theorem 6 to prove the reducibility theorem for $T_{owltime\_se}$:

Theorem 8. *The theory* $T_{owltime\_se} = T_{owltime\_interval\_e} \cup T_{owltime\_instant\_s}$ *is definably equivalent to*

$$T_{semilinear\_order} \cup T_{csgg}$$

4.5   Intervals without Begin and Ends

The preceding sections have shown the reducibility of theories that are extensions of $T_{owltime\_e}$ (which requires that all intervals have endpoints) rather than $T_{owltime\_interval}$ (which allows intervals that do not have beginning or end instants). Nevertheless, we can use these results to provide a representation theorem for $T_{owltime\_interval}$.

Definition 9. *Let* $\mathbb{I} = \langle P, I, \mathbf{in} \rangle$ *be an incidence structure.*
   A line $l \in I$ *is solitary iff it is incident with a unique point.*
   A line $l \in I$ *is isolated iff it is not incident with any point.*

Theorem 9. *Any model* $\mathcal{M}$ *of* $T_{owltime\_interval}$ *contains as substructures a unique model* $\mathcal{N}$ *of* $T_{owltime\_e}$ *and a bipartite incidence structure* $\mathbb{I} = \langle P, L \cup R \cup C, \mathbf{in} \rangle$ *such that*

$$\mathcal{M} \cong \mathcal{N} \cup \mathbb{I}$$

where L and R are disjoint sets of solitary lines and C is a set of isolated lines.

4.6   Discussion

If we consider the summary of results in Figure 2, we can see that $T_{owltime\_leu}$ is the theory which is a common extension of both $T_{owltime\_interval}$ and T_{vila}. In addition, $T_{owltime\_mleu}$ is the common subtheory of $T_{owltime\_interval}$ and T_{vila}; as such, it is intepretable by all existing time ontologies for points and intervals.

As a consequence of these results, we can see that $T_{owltime\_interval}$ is not interpretable by all of the existing ontologies for timepoints and intervals; in other words, there are ontologies that are not definably equivalent to any consistent extension of $T_{owltime\_interval}$. In particular, it is not interpretable by any time ontology that prevents the existence of moments [1], such as $T_{endpoints}$ from [9].

---

[7] http://stl.mie.utoronto.ca/colore/geometry/csgg.clif

5   Inside Intervals

Moving on to the next module $T_{owltime\_inside}$ of $T_{owltime}$, we encounter a new relation, $inside$, between instants and intervals. Before we introduce the classes of structures required to characterize the models of this module, we briefly discuss the problem of unintended models of $T_{owltime\_inside}$.

5.1   A Critique of the Axioms

Ordering on Instants Inside an Interval. Mace can be used to construct a model of $T_{owltime\_interval}$ that satisfies the sentence

$$(\exists t_1, t_2, i) inside(t_1, i) \wedge inside(t_2, i) \wedge \neg before(t_1, t_2) \wedge \neg before(t_2, t_1) \vee (t_1 \neq t_2))$$

that is, a model in which the Instants in an Interval are not linearly ordered, even though the axioms do entail the condition that the beginning and end instants themselves are linearly ordered.

We add the following axiom to $T_{owltime}$ to guarantee that all Instants in an Interval are linearly ordered:

$$(\forall t_1, t_2, i) \; inside(t_1, i) \wedge inside(t_2, i)$$

$$\supset (before(t_1, t_2) \vee before(t_2, t_1) \vee (t_1 = t_2)) \tag{1}$$

Which Instants are Inside? Although Hobbs and Pan assert:

> The concept of inside is not intended to include the beginnings and ends of intervals.

the axiomatization in $T_{owltime\_inside}$ does not quite capture these intended models. While we can use Prover9 to show that the following sentence is entailed by $T_{owltime}$

$$(\forall i, t_1, t_2) \; ProperInterval(i) \wedge begins(t_1, i) \wedge ends(t_2, i) \supset \neg inside(t_1, i) \wedge \neg inside(t_2, i)$$

Mace can be used to construct models of $T_{owltime\_inside}$ that falsify each of the following sentences:

$$(\forall i, t_1) \; ProperInterval(i) \wedge begins(t_1, i) \supset \neg inside(t_1, i)$$
$$(\forall i, t_1) \; ProperInterval(i) \wedge ends(t_1, i) \supset \neg inside(t_1, i)$$
$$(\forall i, t_1) \; Interval(i) \wedge begins(t_1, i) \supset \neg inside(t_1, i)$$
$$(\forall i, t_1) \; Interval(i) \wedge ends(t_1, i) \supset \neg inside(t_1, i)$$

In other words, $T_{owltime\_inside}$ is not strong enough to eliminate models in which only the beginnings or ends of intervals are included as instants inside the interval.

If we are to entail these sentences (which should follow from the original intuition), we need to extend $T_{owltime}$ with the following two sentences:

$$(\forall i, t_1, t_2) \; inside(t_1, i) \wedge begins(t_2, i) \supset before(t_2, t_1) \tag{2}$$

$$(\forall i, t_1, t_2) \; inside(t_1, i) \wedge ends(t_2, i) \supset before(t_1, t_2) \tag{3}$$

236   M. Grüninger

5.2   Semilinear Ordered Geometries

When we used incidence structures to represent the models of $T_{owltime\_interval}$, we only needed to worry about two instants that are incident with an interval, namely, the beginning and the end. For models of $T_{owltime\_inside}$, intervals may be incident with a larger set of instants that are linearly ordered. We therefore need to introduce a new class of structures.

Definition 10. *A semilinear ordered geometry is a structure* $\mathbb{L} = \langle X, Y, \mathbf{B}, \mathbf{in^L} \rangle$ *such that*

1. $\mathbb{B} = \langle X, \mathbf{B} \rangle$ *is a semilinear betweenness relation;*
2. $\mathbb{I} = \langle X, Y, \mathbf{in^L} \rangle$ *is a weak bipartite incidence structure;*
3. *any triple of points that are incident with the same line in* Y *are ordered by the betweenness relation* $\mathbf{B}$.

The class of semilinear ordered geometries is axiomatized by T_{slog}[8].

5.3   Reducibility Theorem for $T_{owltime\_inside}$

Let $T^*_{owltime\_inside}$ be the axioms in $T_{owltime\_inside} \cup T_{owltime\_se}$ together with Axioms 1, 2, and 3.

Theorem 10. $T^*_{owltime\_inside}$ *is definably equivalent to*

$$T_{semilinear\_order} \cup T_{csgg} \cup T_{slog}$$

Proof. Let Δ_2 be the set of translation definitions in Δ_1 together with:

$$(\forall x, y) \, in^L(x, y) \equiv inside(x, y)$$

$$(\forall x, y, z) B(x, y, z) \equiv ((before(x, y) \wedge before(y, z)) \vee (before(z, y) \wedge before(y, z)))$$

Using Prover9, we have shown that

$$T^*_{owltime\_inside} \cup \Delta_2 \models T_{semilinear\_order} \cup T_{csgg} \cup T_{slog}$$

Let Π_2 be the set of translation definitions in Π_1 together with:

$$(\forall x, y) \, inside(x, y) \equiv in^L(x, y)$$

Using Prover9, we have shown that

$$T_{semilinear\_order} \cup T_{csgg} \cup T_{slog} \cup \Pi_2 \models T^*_{owltime\_inside}$$

---

[8] http://stl.mie.utoronto.ca/colore/geometry/slog.clif

6   Eventualities

Hobbs and Pan introduce the class of eventualities to *"cover events, states, processes, propositions, states of affairs, and anything else that can be located with respect to time."* In this section, we characterize the models of the two subtheories, $T_{eventuality}$ and $T_{timespan}$, of $T_{owltime}$ which axiomatize the intuitions for eventualities and their relationships to instants and intervals.

6.1   Weak Planar Geometries

Since $T_{eventuality}$ extends $T_{owltime\_interval}$, a natural approach is to extend the geometries that are the underlying structures for the intended models.

Definition 11. *A weak planar geometry is a tripartite incidence structure*

$$\mathbb{E} = \langle X, Y, Z, \mathbf{in^E} \rangle$$

in which $N(\mathbf{q})$ is a linear ordered geometry for each $\mathbf{q} \in Z$.

Examples of weak planar geometries can be found in Figure 3(a) and 3(b). In each example, the incidence relation between the eventuality e_1 and the intervals i_j corresponds to the **during** relation; the incidence relation between e_1 and the instants t_k corresponds to the **atTime** relation.

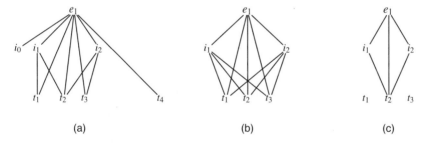

(a)   (b)   (c)

Fig. 3. Examples of weak planar geometries and diamond geometries

The class of weak planar geometries is axiomatized by T_{weak_planar}[9].

6.2   Reducibility Theorem for $T_{eventuality}$

Let $T^*_{eventuality}$ be the axioms in $T_{eventuality} \cup T^*_{inside}$ together with

$$(\forall e)\ Eventuality(e) \supset \neg TemporalEntity(e) \tag{4}$$

$$(\forall e)\ atTime(e) \supset Eventuality(e) \tag{5}$$

$$(\forall e)\ during(e) \supset Eventuality(e) \tag{6}$$

$$(\forall t, e)\ timeSpan(t, e) \supset Eventuality(e) \wedge TemporalEntity(e) \tag{7}$$

---

[9] http://stl.mie.utoronto.ca/colore/geometry/weak_planar.clif

Theorem 11. $T_{eventuality}$ *is definably equivalent to*

$$T_{semilinear\_order} \cup T_{csgg} \cup T_{slog} \cup T_{weak\_planar}$$

Proof. Let Δ_3 be the set of translation definitions in Δ_2 together with:

$$(\forall x, y)\ in^E(x, y) \equiv (inside(x, y) \vee during(x, y) \vee atTime(x, y))$$

Using Prover9, we have shown that

$$T_{eventuality} \cup \Delta_3 \models T_{semilinear\_order} \cup T_{csgg} \cup T_{slog} \cup T_{weak\_planar}$$

Let Π_3 be the set of translation definitions in Π_2 together with:

$$(\forall x)\ eventuality(x) \equiv plane(x)$$

$$(\forall x, y)\ during(x, y) \equiv in^E(x, y) \wedge plane(x) \wedge line(y)$$

$$(\forall x, y)\ atTime(x, y) \equiv in^E(x, y) \wedge plane(x) \wedge point(y)$$

Using Prover9, we have shown that

$$T_{semilinear\_order} \cup T_{csgg} \cup T_{slog} \cup T_{weak\_planar} \cup \Pi_3 \models T_{eventuality}$$

6.3 Diamond Semilattices

We can now proceed to the final class of structures that we will need to characterize models of $T_{eventuality}$ as part of the full representation theorem.

Definition 12. *A diamond semilattice is a bounded semilattice in which all nonextremal elements are incomparable.*

Definition 13. *A diamond geometry is a tripartite incidence structure*

$$\mathbb{D} = \langle X, Y, Z, \mathbf{in^D} \rangle$$

in which $N(\mathbf{q})$ is isomorphic to a diamond semilattice for each $\mathbf{q} \in Z$.

The class of diamond geometries is axiomatized by $T_{diamond}$[10].

Figure 3(c) is an example of a diamond geometry, which is isomorphic to the extension of the **timespan** relation; note that it is a substructure of the weak planar geometry in Figure 3(b). Figure 3(a) is an example of a weak planar geometry that does not contain a diamond geometry as a substructure; this corresponds to a model of $T_{timespan}$ in which the extension of the **timespan** relation is empty.

---

[10] http://stl.mie.utoronto.ca/colore/geometry/diamond.clif

6.4 Reducibility Theorem for $T_{timespan}$

Theorem 12. $T^*_{timespan} = T_{timespan} \cup T^*_{eventuality}$ *is definably equivalent to*

$$T_{semilinear\_order} \cup T_{csgg} \cup T_{slog} \cup T_{weak\_planar} \cup T_{diamond}$$

Proof. Let Δ_4 be the set of translation definitions in Δ_3 together with:

$$(\forall x, y) \, in^D(x, y) \equiv timeSpan(x, y)$$

Using Prover9, we have shown that

$$T^*_{timespan} \cup \Delta_4 \models T_{semilinear\_order} \cup T_{csgg} \cup T_{slog} \cup T_{weak\_planar} \cup T_{diamond}$$

Let Π_4 be the set of translation definitions in Π_3 together with:

$$(\forall x, y) \, timeSpan(x, y) \equiv (in^D(x, y) \wedge plane(x) \wedge ((\exists z) \, line(z) \wedge in^E(z, x)$$

$$\wedge((\forall w) \, point(w) \supset (in^E(w, x) \equiv in^E(w, z)))))$$

Using Prover9, we have shown that

$$T_{semilinear\_order} \cup T_{csgg} \cup T_{slog} \cup T_{weak\_planar} \cup T_{diamond} \cup \Pi_4 \models T^*_{timespan}$$

7 Representation Theorem for $T_{owltime}$

The reducibility of $T^*_{timespan}$ in Theorem 12 is the first step in the verification of the ontology The second step is to define the class of intended models:

Definition 14. $\mathfrak{M}^{owltime}$ *is the following class of structures:* $\mathcal{M} \in \mathfrak{M}^{owltime}$ *iff*

1. $\mathcal{M} \cong \mathcal{P} \cup \mathbb{G} \cup \mathbb{L} \cup \mathbb{E} \cup \mathbb{D}$, *where*
   (a) $\mathcal{P} = \langle P, < \rangle$ *is a linear ordering;*
   (b) $\mathbb{G} = \langle P, I, in^G \rangle$ *is a closed semilinear graphical geometry;*
   (c) $\mathbb{L} = \langle P, I, \mathbf{B}, in^L \rangle$ *is a semilinear ordered geometry;*
   (d) $\mathbb{E} = \langle P, I, E, in^E \rangle$ *is a weak planar geometry;*
   (e) $\mathbb{D} = \langle P, I, E, in^D \rangle$ *is a diamond geometry.*
2. $\langle \mathbf{p_1}, \mathbf{p_2} \rangle \in \mathbf{before}$ *iff* $\mathbf{p_1} < \mathbf{p_2}$;
3. $\langle \mathbf{p}, \mathbf{i} \rangle \in \mathbf{begins}$ *iff* $\langle \mathbf{p}, \mathbf{i} \rangle \in in^G$
   and for any $\mathbf{p}' \in P$ *such that* $\langle \mathbf{p}', \mathbf{i} \rangle \in in^G$, *we have* $\mathbf{p} < \mathbf{p}'$;
4. $\langle \mathbf{p}, \mathbf{i} \rangle \in \mathbf{ends}$ *iff* $\langle \mathbf{p}, \mathbf{i} \rangle \in in^G$
   and for any $\mathbf{p}' \in P$ *such that* $\langle \mathbf{p}', \mathbf{i} \rangle \in in^G$, *we have* $\mathbf{p}' < \mathbf{p}$;
5. $\langle \mathbf{p}, \mathbf{i} \rangle \in \mathbf{inside}$ *iff* $\langle \mathbf{p}, \mathbf{i} \rangle \in in^L$;
6. $\langle \mathbf{e}, \mathbf{p} \rangle \in \mathbf{atTime}$ *iff* $e \in E$, $\mathbf{p} \in P$ *and* $\langle \mathbf{e}, \mathbf{p} \rangle \in in^E$;
7. $\langle \mathbf{e}, \mathbf{i} \rangle \in \mathbf{during}$ *iff* $e \in E$, $\mathbf{i} \in I$ *and* $\langle \mathbf{e}, \mathbf{i} \rangle \in in^E$;
8. $\langle \mathbf{t}, \mathbf{e} \rangle \in \mathbf{timeSpan}$ *iff* $\langle \mathbf{t}, \mathbf{e} \rangle \in in^D$ *and* $N(\mathbf{e}) \cong K_{1,m,n}$.

We can now state the Representation Theorem for $T^*_{timespan}$:

Theorem 13. $\mathcal{M} \in \mathfrak{M}^{owltime}$ iff $\mathcal{M} \in Mod(T^*_{timespan})$.

Proof. The theorem follows from Theorem 12 and Theorem 1, together with the fact that each of the substructures of $\mathcal{M} \in \mathfrak{M}^{owltime}$ corresponds to a theory in the reduction – $T_{semilinear\_order}$ axiomatizes the class of semilinear orderings, T_{csgg} axiomatizes the class of closed semilinear graphical geometries, T_{slog} axiomatizes the class of semilinear ordered geometries, T_{weak_planar} axiomatizes the class of weak planar geometries, and $T_{diamond}$ axiomatizes the class of diamond geometries.

8 Summary

The first-order time ontology for the Semantic Web proposed by Hobbs and Pan aims to be a core ontology for specifying temporal concepts for a wide variety of web applications and services. If it is to play this role effectively, we need a characterization of the models of the ontology and a guarantee that these models are equivalent to the intended models of the ontology's concepts. In this paper, we have provided a characterization of the models of $T_{owltime}$ up to isomorphism. This verification of $T_{owltime}$ has also led to modularization of the ontology and the identification of additional axioms to capture intuitions for intended models. We have also shown that two axioms of $T_{owltime}$ make it inconsistent with some existing time ontologies for points and intervals.

The next step is the proof of representation theorems for all of $T_{owltine}$, including the axioms for duration and dates, based on the duration and dates ontology in [8].

References

1. Allen, J., Hayes, P.: Moments and points in an interval-based temporal logic. Computational Intelligence 5, 225–238 (1989)
2. van Benthem, J.F.A.K.: The Logic of Time: A Model-Theoretic Investigation into the Varieties of Temporal Ontology and Temporal Discourse, 2nd edn. Springer, Heidelberg (1991)
3. Buekenhout, F.: Handbook of Incidence Geometry. Elsevier (1995)
4. Droste, M., Holland, W., MacPherson, H.: Automorphism groups of infinite semilinear orders. Proceedings of the London Mathematical Society 58, 454–478 (1989)
5. Enderton, H.: Mathematical Introduction to Logic. Academic Press (1972)
6. Gruninger, M., Hashemi, A., Ong, D.: Ontology Verification with Repositories. In: Formal Ontologies and Information Systems 2010, Toronto, Canada (2010)
7. Gruninger, M., Ong, D.: Verification of Time Ontologies with Points and Intervals. In: 18th International Symposium on Temporal Representation and Reasoning (2011)
8. Gruninger, M.: Ontologies for Dates and Duration. In: Twelfth International Conference on Principles of Knowledge Representation and Reasoning (2010)
9. Hayes, P.: A Catalog of Temporal Theories, Tech. Report UIUC-BI-AI-96-01, University of Illinois (1996)
10. Hobbs, J., Pan, F.: An Ontology of Time for the Semantic Web. ACM Transactions on Asian Language Information Processing 3, 66–85 (2004)
11. McCune, W.: Prover9 and Mace4 (2005-2010),
    http://www.cs.unm.edu/~mccune/Prover9
12. Pan, F., Hobbs, J.: Time in OWL-S. In: Proceedings of AAAI Spring Symposium on Semantic Web Services. Stanford University (2004)
13. Vila, L.: Formal Theories of Time and Temporal Incidence. In: Fisher, Gabbay, Vila, (eds.) Handbook of Temporal Reasoning in Artificial Intelligence. Elsevier (2005)

The Cognitive Complexity of OWL Justifications

Matthew Horridge, Samantha Bail, Bijan Parsia,
and Ulrike Sattler

The University of Manchester
Oxford Road, Manchester, M13 9PL
{matthew.horridge,bails,
bparsia,sattler}@cs.man.ac.uk

Abstract. In this paper, we present an approach to determining the cognitive complexity of justifications for entailments of OWL ontologies. We introduce a simple cognitive complexity model and present the results of validating that model via experiments involving OWL users. The validation is based on test data derived from a large and diverse corpus of naturally occurring justifications. Our contributions include validation for the cognitive complexity model, new insights into justification complexity, a significant corpus with novel analyses of justifications suitable for experimentation, and an experimental protocol suitable for model validation and refinement.

1 Introduction

A justification is a minimal subset of an ontology that is sufficient for an entailment to hold. More precisely, given $\mathcal{O} \models \eta$, $\mathcal{J}$ is a justification for η in $\mathcal{O}$ if $\mathcal{J} \subseteq \mathcal{O}$, $\mathcal{J} \models \eta$ and, for all $\mathcal{J}' \subsetneq \mathcal{J}$, $\mathcal{J}' \not\models \eta$. Justifications are the dominant form of explanation in OWL,[1] and justification based explanation is deployed in popular OWL editors. The primary focus of research in this area has been on explanation for the sake of debugging problematic entailments [8], whether standard "buggy" entailments, such as class unsatisfiability or ontology inconsistency, or user selected entailments such as arbitrary subsumptions and class assertions. The debugging task is naturally directed toward "repairing" the ontology and the use of "standard errors" further biases users toward looking for problems in the logic of a justification.

As a form of explanation, justifications are a bit atypical historically. While they present the ultimate, ontology specific reasons that a given entailment holds, they, unlike proofs, do not *articulate* how those reasons support the entailment, at least, in any detail. That is, they correspond to the *premises* of a proof, but do not invoke any specific proof calculus. Clearly, this brings advantages, as justifications are calculus independent, require nothing more than knowledge of OWL, and do not involve a host of knotty, unresolved issues of long standing (such as what to do about "obvious" steps [2]). Furthermore, justifications are highly manipulable: Deleting an axiom breaks the entailment, which allows for a very active, ontology related form of experimentation

---

[1] Throughout this paper, "OWL" refers to the W3C's Web Ontology Language 2 (OWL 2).

L. Aroyo et al. (Eds.): ISWC 2011, Part I, LNCS 7031, pp. 241–256, 2011.

by users. However, in spite of their field success, justifications are held to be lacking *because* they don't articulate the connection and thus are too hard to understand.[2]

The Description Logic that underpins OWL, $\mathcal{SROIQ}$, is N2ExpTime-complete [10], which suggests that even fairly small justifications could be quite challenging to reason with. However, justifications are highly successful in the field, thus the computational complexity argument is not dispositive. We do observe often that certain justifications are difficult and frustrating to understand for ontology developers. In some cases, the difficulty is obvious: a large justification with over 70 axioms is going to be at best cumbersome however simple its logical structure. However, for many reasonably sized difficult justifications (e.g. of size 10 or fewer axioms) the source of cognitive complexity is not clearly known.

If most naturally occurring justifications are easy "enough" to understand, then the need for auxilliary explanation faculties (and the concomitant burden on the user to master them and the tool developer to provide them) is reduced. In prior work [5,3,6], we proposed a predictive complexity model based on an exploratory study plus our own experiences and intuitions. However, in order to deploy this metric reliably, whether to assess the state of difficulty of justifications or to deploy an end-user tool using it, the model needed validation.

In this paper, we present the results of several experiments into the cognitive complexity of OWL justifications. Starting from our cognitive complexity model, we test how well the model predicts error proportions for an entailment assessment task. We find that the model does fairly well with some notable exceptions. A follow-up study with an eye tracker and think aloud protocol supports our explanations for the anomalous behaviour and suggests both a refinement to the model and a limitation of our experimental protocol.

Our results validate the use of justifications as the primary explanation mechanism for OWL entailments as well as raising the bar for alternative mechanisms (such as proofs). Furthermore, our metric can be used to help users determine when they need to seek expert help or simply to organise their investigation of an entailment.

2   Cognitive Complexity and Justifications

While there have been several user studies in the area of debugging [11,9], ontology engineering anti-patterns [16], and our exploratory investigation into features that make justifications difficult to understand [5], to the best of our knowledge there have not been any formal user studies that investigate the cognitive complexity of justifications.

Of course, if we had a robust theory of how people reason, one aspect of that robustness would be an explanation of justification difficulty. However, even the basic mechanism of human deduction is not well understood. In psychology, there is a long standing rivalry between two accounts of human deductive processes: (1) that people

---

[2] See, for example, a related discussion in the OWL Working Group
http://www.w3.org/2007/OWL/tracker/issues/52. Also, in [1], the authors rule out of court justifications as a form of explanation: "It is widely accepted that an explanation corresponds to a formal proof. A formal proof is constructed from premises using rules of inference".

apply inferential rules [15], and (2) that people construct mental models [7].[3] In spite of a voluminous literature (including functional MRI studies, e.g., [14]), to date there is no scientific consensus [13], even for propositional reasoning.

Even if this debate were settled, it would not be clear how to apply it to ontology engineering. The reasoning problems that are considered in the literature are quite different from understanding how an entailment follows from a justification in a (fairly expressive) fragment of first order logic. For example, our reasoning problems are in a regimented, formalised language for which reasoning problems are far more constrained than deduction "in the wild." Thus, the artificiality of our problems may engage different mechanisms than more "natural" reasoning problems: e.g. even if mental models theory were correct, people *can* produce natural deduction proofs and might find that doing so allows them to outperform "reasoning natively". Similarly, if a tool gives me a justification, I can use my knowledge of justifications to help guide me, e.g., that justifications are minimal means that I must look at all the axioms presented and I do not have to rule any out as irrelevant. As we will see below, such meta-justificatory reasoning is quite helpful.

However, for ontology engineering, we do not need a *true account of human deduction*, but just need a way to determine how *usable* justifications are for our tasks. In other words, what is required is a theory of the *weak cognitive complexity* of justifications, not one of *strong cognitive complexity* [17].

A similar practical task is generating sufficiently difficult so-called "Analytical Reasoning Questions" (ARQs) problems in Graduate Record Examination (GRE) tests. ARQs typically take the form of a "logic puzzle" wherein an initial setup is presented, along with some constraints, then the examinee must determine possible solutions. Often, these problems involve positioning entities in a constrained field (e.g., companies on floors in a building, or people seated next to each other at dinner). In [13], the investigators constructed and validated a model for the complexity of answering ARQs via experiments with students. Analogously, we aim to validate a model for the complexity of "understanding" justifications via experiments on modellers.

In [13], Newstead et al first build a preliminary complexity model, as we did, based on a small but intense pilot study using think aloud plus some initial ideas about the possible sources of complexity. Then they validated their model in a series of large scale controlled experiments wherein a set of students were given sets of questions which varied systematically in complexity (according to their model) and in particular features used. One strong advantage Newstead el al have is that the problems they considered are very constrained and comparatively easy to analyse. For example, the form of ARQ question they consider have finite, indeed, easily enumerable, sets of models. Thus, they can easily determine how many possible situations are ruled out by a given constraint which is a fairly direct measure of the base line complexity of the problem. Similarly, they need merely to *construct* problems of the requisite difficulty, whereas we need to *recognise* the difficulty of arbitrary inputs. Finally, their measure of difficulty is exactly what proportion of a given cohort get the questions right, whereas we are dealing with a more nebulous notion of understanding.

---

[3] (1) can be crudely characterised as people use a natural deduction proof system and (2) as people use a semantic tableau.

Of course, the biggest advantage is that their problems are expressed in natural language and reasonably familiar to millions of potential participants, whereas our investigations necessarily require a fair degree of familiarity with OWL — far more than can be given in a study-associated training session. Nevertheless, the basic approach seems quite sound and we follow it in this paper.

3 A Complexity Model

We have developed a cognitive complexity model for justification understanding. This model was derived partly from observations made during an exploratory study (see [5,3,6] for more details) in which people attempted to understand justifications from naturally occurring ontologies, and partly from intuitions on what makes justifications difficult to understand.

Please note that reasonable people may (and do!) disagree with nigh every aspect of this model (the weights are particularly suspect). For each factor, we have witnessed the psychological reality of their causing a reasonably sophisticated user difficulty in our exploratory study. But, for example, we cannot warrant their orthogonality, nor can we show that some combinations of factors is easier than the sum of the weights would indicate. This should not be too surprising especially if one considers the current understanding of what makes even propositional formulae difficult for automated reasoners. While for extremely constrained problems (such as propositional kCNF), we have long had good predictive models for reasoning difficulty for key proving techniques, more unconstrained formulae have not been successfully analysed. Given that the complexity of a given algorithm is intrinsically more analysable than human psychology (consider simply the greater ease of controlled experiments), the fact that we do not have good predictive models for automated reasoners should be a warning for theorists of cognitive complexity. However, while daunting, these facts do not mean we should give up, as even a fairly crude model can be useful, as we have found. Furthermore, we can hope to improve the predictive validity of this model, even without determining the structure of the phenomena.

Table 1 describes the model, wherein $\mathcal{J}$ is the justification in question, η is the focal entailment, and each value is multiplied by its weight and then summed with the rest. The final value is a complexity score for the justification. Broadly speaking, there are two types of components: (1) structural components, such as **C1**, which require a syntactic analysis of a justification, and (2) semantic components, such as **C4**, which require entailment checking to reveal non-obvious phenomena.

Components **C1** and **C2** count the number of different kinds of axiom types and class expression types as defined in the OWL 2 Structural Specification.[4] The more diverse the basic logical vocabulary is, the less likely that simple pattern matching will work and the more "sorts of things" the user must track.

Component **C3** detects the presence of universal restrictions where *trivial satisfaction* can be used to infer subsumption. Generally, people are often surprised to learn that if $\langle x,y\rangle \notin R^{\mathcal{I}}$ for all $y \in \Delta^{\mathcal{I}}$, then $x \in (\forall R.C)^{\mathcal{I}}$. This was observed repeatedly in the exploratory study.

---

[4] http://www.w3.org/TR/owl2-syntax/

Table 1. A Simple Complexity Model

Name	Base value	Weight
C1 AxiomTypes	Number of axiom types in $\mathcal{J}$ & η.	100
C2 ClassConstructors	Number of constructors in $\mathcal{J}$ & η.	10
C3 UniversalImplication	If an $\alpha \in \mathcal{J}$ is of the form $\forall R.C \sqsubseteq D$ or $D \equiv \forall R.C$ then 50 else 0.	1
C4 SynonymOfThing	If $\mathcal{J} \models \top \sqsubseteq A$ for some $A \in \text{Signature}(\mathcal{J})$ and $\top \sqsubseteq A \notin \mathcal{J}$ and $\top \sqsubseteq A \neq \eta$ then 50 else 0.	1
C5 SynonymOfNothing	If $\mathcal{J} \models A \sqsubseteq \bot$ for some $A \in \text{Signature}(\mathcal{J})$ and $A \sqsubseteq \bot \notin \mathcal{J}$ and $A \sqsubseteq \bot \neq \eta$ then 50 else 0.	1
C6 Domain&NoExistential	If $\text{Domain}(R, C) \in \mathcal{J}$ and $\mathcal{J} \not\models E \sqsubseteq \exists R.\top$ for some class expressions E then 50 else 0.	1
C7 ModalDepth	The maximum modal depth of all class expressions in $\mathcal{J}$.	50
C8 SignatureDifference	The number of distinct terms in $\text{Signature}(\eta)$ not in $\text{Signature}(\mathcal{J})$.	50
C9 AxiomTypeDiff	If the axiom type of η is not the set of axiom types of $\mathcal{J}$ then 50 else 0	1
C10 ClassConstructorDiff	The number of class constructors in η not in the set of constructors of $\mathcal{J}$.	1
C11 LaconicGCICount	The number of General Concept Inclusion axioms in a laconic version of $\mathcal{J}$	100
C12 AxiomPathLength	The number of maximal length expression paths in $\mathcal{J}$ plus the number of axioms in $\mathcal{J}$ which are not in some maximal length path of $\mathcal{J}$, where a *class (property) expression subsumption path* is a list of axioms of length n where for any $1 \leq i < n$, the axiom at position i is $C_i \sqsubseteq C_{i+1}$.	10

Components **C4** and **C5** detect the presence of synonyms of $\top$ and $\bot$ in the signature of a justification where these synonyms are *not explicitly* introduced via subsumption or equivalence axioms. In the exploratory study, participants failed to spot synonyms of $\top$ in particular.

Component **C6** detects the presence of a domain axiom that is not paired with an (entailed) existential restriction along the property whose domain is restricted. This typically goes against peoples' expectations of how domain axioms work, and usually indicates some kind of non-obvious reasoning by cases. For example, given the two axioms $\exists R.\top \sqsubseteq C$ and $\forall R.D \sqsubseteq C$, the domain axiom is used to make a statement about objects that have R successors, while the second axiom makes a statement about those objects that do not have any R successors to imply that C is equivalent to $\top$. This is different from the typical pattern of usage, for example where $A \sqsubseteq \exists R.C$ and $\exists R.\top \sqsubseteq B$ entails $A \sqsubseteq B$.

Component **C7** measures maximum modal depth of sub-concepts in $\mathcal{J}$, which tend to generate multiple distinct but interacting propositional contexts.

Component **C8** examines the signature difference from entailment to justification. This can indicate confusing redundancy in the entailment, or synonyms of $\top$, that may not be obvious, in the justification. Both cases are surprising to people looking at such justifications.

Components **C9** and **C10** determine if there is a difference between the type of, and types of class expressions in, the axiom representing the entailment of interest and the types of axioms and class expressions that appear in the justification. Any difference can indicate an extra reasoning step to be performed by a person looking at the justification.

Component **C11** examines the number of subclass axioms that have a complex left hand side in a *laconic*[5] version of the justification. Complex class expressions on the left hand side of subclass axioms in a laconic justification indicate that the conclusions of several intermediate reasoning steps may interact.

Component **C12** examines the number of obvious syntactic subsumption paths through a justification. In the exploratory study, participants found it very easy to quickly read chains of subsumption axioms, for example, $\{A \sqsubseteq B, B \sqsubseteq C, C \sqsubseteq D, D \sqsubseteq E\}$ to entail $A \sqsubseteq E$. This complexity component essentially increases the complexity when these kinds of paths are lacking.

The weights were determined by rough and ready empirical twiddling, without a strong theoretical or specific experimental backing. They correspond to our sense, esp. from the exploratory study, of sufficient reasons for difficulty.

4   Experiments

While the model is plausible and has behaved reasonably well in applications, its validation is a challenging problem. In principle, the model is reasonable if it successfully predicts the difficulty an arbitrary OWL modeller has with an arbitrary justification sufficiently often. Unfortunately, the space of ontology developers and of OWL justifications (even of existing, naturally occurring ones) is large and heterogeneous enough to be difficult to randomly sample.

4.1   Design Challenges

To cope with the heterogeneity of users, any experimental protocol should require minimal experimental interaction, i.e. it should be executable over the internet from subjects' own machines with simple installation. Such a protocol trades access to subjects, over time, for the richness of data gathered. To this end, we adapted one of the experimental protocols described in [13] and tested it on a more homogeneous set of participants—a group of MSc students who had completed a lecture course on OWL. These students had each had an 8 hour lecture session, once a week, for five weeks on OWL and ontology engineering, and had completed 4 weeks of course work including having constructed several ontologies. The curriculum did not include any discussion of justifications or explanation per se, though entailment and reasoning problems had been covered.[6] Obviously, this group is not particularly representative of all OWL ontologists: They are young, relatively inexperienced, and are trained in computer science. However, given their inexperience, especially with justifications, things they find easy should be reliably easy for most trained users.

While the general experimental protocol in [13] seems reasonable, there are some issues in adapting it to our case. In particular, in ARQs there is a restricted space of

---

[5] Laconic justifications [4] are justifications whose axioms do not contain any superfluous parts.

[6] See http://www.cs.manchester.ac.uk/pgt/COMP60421/ for course materials.

possible (non-)entailments suitable for multiple choice questions. That is, the wrong answers can straightforwardly be made plausible enough to avoid guessing. A justification inherently has one statement for which it is a justification (even though it will be a minimal entailing subset for others). Thus, there isn't a standard "multiple set" of probable answers to draw on. In the exam case, the primary task is successfully answering the question and the relation between that success and predictions about the test taker are outside the remit of the experiment (but there is an established account, both theoretically and empirically). In the justification case the standard primary task is "understanding" the relationship between the justification and the entailment. Without observation, it is impossible to distinguish between a participant who really "gets" it and one who merely acquiesces. In the exploratory study we performed to help develop the model, we had the participant rank the difficulty of the justification, but also used think aloud and follow-up questioning to verify the success in understanding by the participant. This is obviously not a minimal intervention, and requires a large amount of time and resources on the part of the investigators.

To counter this, the task was shifted from a justification understanding task to something more measurable and similar to the question answering task in [13]. In particular, instead of presenting the justification/entailment pair *as* a justification/entailment pair and asking the participant to try to "understand" it, we present the justification/entailment pair as a set-of-axioms/candidate-entailment pair and ask the participant to *determine* whether the candidate is, in fact, entailed. This diverges from the standard justification situation wherein the modeller knows that the axioms entail the candidate (and form a justification), but provides a metric that can be correlated with cognitive complexity: *error proportions*.

4.2 Justification Corpus

To cope with the heterogeneity of justifications, we derived a large sample of justifications from ontologies from several well known ontology repositories: The Stanford BioPortal repository[7] (30 ontologies plus imports closure), the Dumontier Lab ontology collection[8] (15 ontologies plus imports closure), the OBO XP collection[9] (17 ontologies plus imports closure) and the TONES repository[10] (36 ontologies plus imports closure). To be selected, an ontology had to (1) entail one subsumption between class names with at least one justification that (a) was not the entailment itself, and (b) contains axioms in that ontology (as opposed to the imports closure of the ontology), (2) be downloadable and loadable by the OWL API (3) processable by FaCT++.

While the selected ontologies cannot be said to generate a *truly representative* sample of justifications from the full space of possible justifications (even of those on the Web), they are diverse enough to put stress on many parts of the model. Moreover, most of these ontologies are actively developed and used and hence provide justifications that a significant class of users encounter.

---

[7] http://bioportal.bioontology.org
[8] http://dumontierlab.com/?page=ontologies
[9] http://www.berkeleybop.org/ontologies/
[10] http://owl.cs.manchester.ac.uk/repository/

For each ontology, the class hierarchy was computed, from which direct subsumptions between class names were extracted. For each direct subsumption, as many justifications as possible in the space of 10 minutes were computed (typically all justifications; time-outs were rare). This resulted in a pool of over 64,800 justifications.

While large, the actual logical diversity of this pool is considerably smaller. This is because many justifications, for different entailments, were of exactly the same "shape". For example, consider $\mathcal{J}_1 = \{A \sqsubseteq B, B \sqsubseteq C\} \models A \sqsubseteq C$ and $\mathcal{J}_2 = \{F \sqsubseteq E, E \sqsubseteq G\} \models F \sqsubseteq G$. As can be seen, there is an injective renaming from $\mathcal{J}_1$ to $\mathcal{J}_2$, and $\mathcal{J}_1$ is therefore *isomorphic* with $\mathcal{J}_2$. If a person can understand $\mathcal{J}_1$ then, with allowances for variations in name length, they should be able to understand $\mathcal{J}_2$. The initial large pool was therefore reduced to a smaller pool of 11,600 *non-isomorphic justifications*.

4.3   Items and Item Selection

Each experiment consists of a series of test items (questions from a participant point of view). A test *item* consists of a *set of axioms*, one *following axiom*, and a *question*, "Do these axioms entail the following axiom?". A participant *response* is one of five possible answers: "Yes" (it is entailed), "Yes, but not sure", "Not Sure", "No, but not sure", "No" (it is not entailed). From a participant point of view, any item may or may not contain a justification. However, in our experiments, every item was, in fact, a justification.

It is obviously possible to have non-justification entailing sets or non-entailing sets of axioms in an item. We chose against such items since (1) we wanted to maximize the number of actual justifications examined (2) justification understanding is the actual task at hand, and (3) it is unclear how to interpret error rates for non-entailments in light of the model. For some subjects, esp. those with little or no prior exposure to justifications, it was unclear whether they understood the difference between the set merely being entailing, and it being minimal and entailing. We did observe one person who made use of this metalogical reasoning in the follow-up study.

Item Construction: For each experiment detailed below, test items were constructed from the pool of 11,600 non-isomorphic justifications. First, in order to reduce variance due primarily to size, justifications whose size was less than 4 axioms and greater than 10 axioms were discarded. This left 3199 (28%) justifications in the pool. In particular, this excluded large justifications that might require a lot of reading time, cause fatigue problems, or intimidate, and excluded very small justifications that tended to be trivial.[11]

For each justification in the pool of the remaining 3199 non-isomorphic justifications, the complexity of the justification was computed according to the model presented in Table 1, and then the justification was assigned to a complexity bin. A total of 11 bins were constructed over the range of complexity (from 0 to 2200), each with a complexity interval of 200. We discarded all bins which had 0 non-isomorphic justifications of size 4-10. This left 8 bins partitioning a complexity range of 200-1800.

---

[11] Note that, as a result, nearly 40% of all justifications have no representative in the pruned set (see Figure 3). Inspection revealed that most of these were trivial single axiom justifications (e.g. of the form $\{A \equiv B\} \models A \sqsubseteq B$ or $\{A \equiv (B \sqcap C)\} \models A \sqsubseteq B$, etc.

Figure 1 illustrates a key issue. The bulk of the justifications (esp. without the trivial), both with and without isomorphic reduction, are in the middle complexity range. However, the model is not sophisticated enough that small differences (e.g. below a difference of 400-600) are plausibly meaningful. It is unclear whether the noise from variance in participant abilities would wash out the noise from the complexity model. In other words, just from reflection on the model, justifications whose complexity difference is 400 or less do not seem reliably distinguishable by error rates. Furthermore, non-isomorphism does not eliminate all non-significant logical variance. Consider a chain of two atomic subsumptions vs. a chain of three. They have the same basic logical structure, but are not isomorphic. Thus, we cannot yet say whether this apparent concentration is meaningful.

Since we did not expect to be able to present more than 6 items and keep to our time limits, we chose to focus on a "easy/hard" divide of the lowest three non-empty bins (200-800) and the highest three non-empty bins (1200-1800). While this limits the claims we can make about model performance over the entire corpus, it, at least, strengthens negative results. If error rates overall do not distinguish the two poles (where we expect the largest effect) then either the model fails or error rates are not a reliable marker. Additionally, since if there is an effect, we expect it to be largest in this scenario thus making it easier to achieve adequate statistical power.

Each experiment involved a fixed set of test items, which were selected by randomly drawing items from preselected spread of bins, as described below. Please note that the selection procedure changed in the light of the pilot study, but only to make the selection more challenging for the model.[12]

The final stage of item construction was justification obfuscation. All non-logical terms were replaced with generated symbols. Thus, there was no possibility of using domain knowledge to understand these justifications. The names were all uniform, syntactically distinguishable (e.g. class names from property names) and quite short. The entailment was the same for all items, i.e. $c_1 \sqsubseteq c_2$. It is possible that dealing with these purely symbolic justifications distorted participant response from response in the field, even beyond blocking domain knowledge. For example, they could be alienating and thus increase error rates or they could engage less error prone pattern recognition.

5   Results

The test items that were selected by the above sampling methodology are shown below. Every set of axioms is a justification for $c_1 \sqsubseteq c_2$. There was no overlap in participants across the studies. For the main study, none of the authors were involved in facilitating the study, though Bail and Horridge participated in recruitment.

5.1   Pilot study

Participants: Seven members of a Computer Science (CS) Academic or Research Staff, or PhD Program, with over 2 years of experience with ontologies and justifications.

---

[12] The selections are available from http://owl.cs.manchester.ac.uk/research/publications/supporting-material/iswc2011-cog-comp

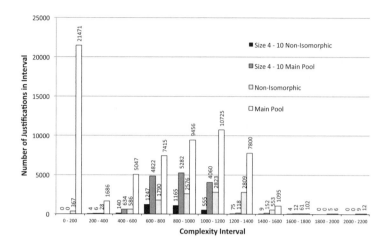

Fig. 1. Justification Corpus Complexity Distribution

Materials and procedures: The study was performed using an in-house web based survey tool, which tracks times between all clicks on the page and thus records the time to make each decision.

The participants were given a series of test items consisting of 3 practice items, followed by 1 common easy item (**E1** of complexity 300) and four additional items, 2 ranked easy (**E2** and **E3** of complexities 544 and 690, resp.) and 2 ranked hard (**H1** and **H2** of complexities 1220 and 1406), which were randomly (but distinctly) ordered for each participant. The easy items were drawn from bins 200-800, and the hard items from bins 1200-1800. The expected time to complete the study was a maximum of 30 minutes, including the orientation, practice items, and brief demographic questionnaire (taken after all items were completed).

Results: Errors and times are given in Table 2. Since all of the items were in fact justifications, participant responses were recoded to success or failure as follows: Success = ("Yes" | "Yes, but not sure") and Failure = ("Not sure" | "No, Not sure" | "No"). Error proportions were analysed using Cochran's Q Test, which takes into consideration the pairing of successes and failures for a given participant. Times were analysed using two tailed paired sample t-tests.

Table 2. Pilot Study Failures and Response Times

Item	Failures	Mean Time (ms)	Time StdDev. (ms)
E1	0	65,839	39,370
E2	1	120,926	65,950
E3	2	142,126	61,771
H1	6	204,257	54,796
H2	6	102,774	88,728

An initial Cochran Q Test across all items revealed a strong significant difference in error proportions between the items $[Q(4) = 16.00, p = 0.003]$. Further analysis using Cochran's Q Test on pairs of items revealed strong statistically significant differences in error proportion between: **E1/H1** $[Q(1) = 6.00, p = 0.014]$, **E1/H2** $[Q(1) = 6.00, p = 0.014]$ **E2/H2** $[Q(1) = 5.00, p = 0.025]$ and **E3/H2** $[Q(1) = 5.00, p = 0.025]$. The differences in the remaining pairs, while not exhibiting differences above $p = 0.05$, were quite close to significance, i.e. **E2/H1** $[Q(1) = 3.57, p = 0.059]$ and **E3/H1** $[Q(1) = 5.00, p = 0.10]$. In summary, these error rate results were encouraging.

An analysis of times using paired sample t-tests revealed that time spent understanding a particular item is not a good predictor of complexity. While there were significant differences in the times for **E1/H1** $[p = 0.00016]$, **E2/H1** $[p = 0.025]$, and **E3/H1** $[p = 0.023]$, there were no significant differences in the times for **E1/H2** $[p = 0.15]$, **E2/H2** $[p = 0.34]$ and **E3/H2** $[p = 0.11]$. This result was anticipated, as in the exploratory study people gave up very quickly for justifications that they felt they could not understand.

5.2   Experiment 1

Participants: 14 volunteers from a CS MSc class on OWL ontology modelling, who were given chocolate for their participation.[13] Each participant had minimal exposure to OWL (or logic) before the class, but had, in the course of the prior 5 weeks, constructed or manipulated several ontologies, and received an overview of the basics of OWL 2, reasoning, etc. They did not receive any specific training on justifications.

Materials and procedures: The study was performed according to the protocol used in the pilot study. A new set of items were used. Since the mean time taken by pilot study participants to complete the survey was 13.65 minutes, with a standard deviation of 4.87 minutes, an additional hard justification was added to the test items. Furthermore, all of the items with easy justifications ranked easy were drawn from the highest easy complexity bin (bin 600-800). In the pilot study, we observed that the lower ranking easy items were found to be quite easy and, by inspection of their bins, we found that it was quite likely to draw similar justifications. The third bin (600-800) is much larger and logically diverse, thus is more challenging for the model.

The series consisted of 3 practice items followed by 6 additional items, 3 easy items(**EM1**, **EM2** and **EM3** of complexities: 654, 703, and 675), and 3 hard items (**HM1**, **HM2** and **HM3** of complexities: 1380, 1395, and 1406). The items were randomly ordered for each participant. Again, the expectation of the time to complete the study was a maximum of 30 minutes, including orientation, practice items and brief demographic questionnaire.

Results. Errors and times are presented in Table 3. The coding to error is the same as in the pilot. An analysis with Cochran's Q Test across all items reveals a significant difference in error proportion $[Q(5) = 15.095, p = 0.0045]$.

A pairwise analysis between easy and hard items reveals that there are significant and, highly significant, differences in errors between **EM1/HM1** $[Q(1) = 4.50, p =$

---

[13] It was made clear to the students that their (non)participation did not affect their grade and no person with grading authority was involved in the recruitment or facilitation of the experiment.

Table 3. Experiment 1 Failures and Response Times

Item	Failures	Mean Time (ms)	Time StdDev. (ms)
EM1	6	103,454	68,247
EM2	6	162,928	87,696
EM3	10	133,665	77,652
HM1	12	246,835	220,921
HM2	13	100,357	46,897
HM3	6	157,208	61,437

0.034], **EM1/HM2** [$Q(1) = 7.00$, $p = 0.008$], **EM2/HM1** [$Q(1) = 4.50$, $p = 0.034$], **EM2/HM2** [$Q(1) = 5.44$, $p = 0.02$], and **EM3/HM2** [$Q(1) = 5.44$, $p = 0.02$].

However, there were no significant differences between **EM1/HM3** [$Q(1) = 0.00$, $p = 1.00$], **EM2/HM3** [$Q(1) = 0.00$, $p = 1.00$], **EM3/HM3** [$Q(1) = 2.00$, $p = 0.16$] and **EM3/HM1** [$Q(1) = 0.67$, $p = 0.41$].

With regards to the nonsignificant differences between certain easy and hard items, there are two items which stand out: An easy item **EM3** and a hard item **HM3**, which are shown as the last pair of justifications in Figure 2.

In line with the results from the pilot study, an analysis of times using a paired samples t-test revealed significant differences between some easy and hard items, with those easy times being significantly less than the hard times **EM1/HM1** [$p = 0.023$], **EM2/HM2** [$p = 0.016$] and **EM3/HM1** [$p = 0.025$]. However, for other pairs of easy and hard items, times were not significantly different: EM1/HM1 [$p = 0.43$], **EM2/HM1** [$p = 0.11$] and **EM3/HM2** [$p = 0.10$]. Again, time is not a reliable predictor of model complexity.

Anomalies in Experiment 1: Two items (**EM3** and **HM3**) did not exhibit their predicted error rate relations. For item **EM3**, we conjectured that a certain pattern of superfluous axiom parts in the item (not recognisable by the model) made it harder than the model predicted. That is, that the *model* was wrong.

For item **HM3** we conjectured that the model correctly identifies this item as hard,[14] but that the MSc students answered "Yes" because of misleading pattern of axioms at the start and end of item **HM3**. The high "success" rate was due to an error in reasoning, that is, a *failure* in understanding.

In order to determine whether our conjectures were possible and reasonable, we conducted a followupup study with the goal of observing the conjectured behaviours in situ. Note that this study does *not* explain what happened in Experiment 1.

5.3   Experiment 2

Participants: Two CS Research Associates and one CS PhD student, none of whom had taken part in the pilot study. All participants were very experienced with OWL.

---

[14] It had been observed to stymie experienced modellers in the field. Furthermore, it involves deriving a synonym for $\top$, which was not a move this cohort had experience with.

Materials and procedures: Items and protocol were exactly the same as Experiment 1, with the addition of the think aloud protocol [12]. Furthermore, the screen, participant vocalisation, and eye tracking were recorded.

Results: With regard to **EM3**, think aloud revealed that all participants were distracted by the superfluous axiom parts in item **EM3**. Figure 3 shows an eye tracker heat map for the most extreme case of distraction in item **EM3**. As can be seen, hot spots lie over the superfluous parts of axioms. Think aloud revealed that all participants initially tried to see how the $\exists$ prop1.C6 conjunct in the third axiom contributed to the entailment and struggled when they realised that this was not the case.

EM1

$C1 \sqsubseteq \exists\,prop1.C3$

$prop1 \sqsubseteq prop2$

$prop2 \sqsubseteq prop3$

$C3 \sqsubseteq C4$

$C4 \sqsubseteq C5$

$C5 \sqsubseteq C6$

$C6 \sqsubseteq C7$

$C7 \sqsubseteq C8$

$C2 \equiv \exists\,prop3.C8$

HM1

$C1 \equiv \exists\,prop1.C3$

$prop1 \equiv prop2^-$

$prop2 \sqsubseteq prop3$

$prop3 \equiv prop4^-$

$C3 \equiv (\exists\,prop5.C4) \sqcap (\exists\,prop2.C1)$
$\sqcap\;(\forall prop5.C4) \sqcap (\forall\,prop2.C1)$

$prop6 \equiv prop5^-$

$\exists\,prop6.\top \sqsubseteq C5$

$C6 \sqsubseteq C7$

$C6 \equiv (\exists\,prop5.C5) \sqcap (\forall prop5.C5)$

$C2 \equiv \exists\,prop4.C7$

EM2

$C1 \equiv C3 \sqcap (\exists\,prop1.C4) \sqcap (\exists\,prop2.C5)$

$C1 \sqsubseteq C6$

$C6 \sqsubseteq C7$

$C7 \sqsubseteq C8$

$C8 \equiv C9 \sqcap (\exists\,prop1.C10)$

$C2 \equiv C9 \sqcap (\exists\,prop1.C4) \sqcap (\exists\,prop2.C5)$

HM2

$C3 \equiv (\exists\,prop1.C5) \sqcup (\forall\,prop1.C5)$

$C3 \sqsubseteq C4$

$\exists\,prop1.\top \sqsubseteq C4$

$C4 \sqsubseteq C2$

EM3

$C1 \sqsubseteq C3$

$C3 \sqsubseteq C4$

$C4 \equiv C5 \sqcap (\exists\,prop1.C6)$

$C5 \equiv C7 \sqcap (\exists\,prop2.C8)$

$C1 \sqsubseteq \exists\,prop1.C9$

$C9 \sqsubseteq C10$

$C2 \equiv C7 \sqcap (\exists\,prop1.C10)$

HM3

$C1 \sqsubseteq \forall\,prop1.C3$

$C6 \equiv \forall\,prop2.C7$

$C6 \sqsubseteq C8$

$C8 \sqsubseteq C4$

$C4 \sqsubseteq \exists\,prop1.C5$

$\exists\,prop2.\top \sqsubseteq C4$

$C2 \equiv (\exists\,prop1.C3) \sqcup (\forall\,prop3.C9)$

Fig. 2. Justifications Used in Experiment 1. All justifications explain the entailment $C1 \sqsubseteq C2$.

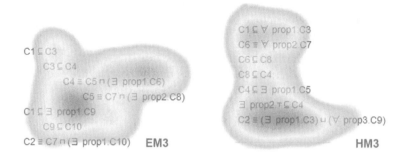

Fig. 3. Eye Tracker Heat Maps for **EM3** & **HM3**

In the case of **HM3**, think aloud revealed that none of the participants understood how the entailment followed from the set of axioms. However, two of them responded correctly and stated that the entailment did hold. As conjectured, the patterns formed by the start and end axioms in the item set seemed to mislead them. In particular, when disregarding quantifiers, the start axiom C1 ⊑ ∀prop1.C3 and the end axiom C2 ⊑ ∃prop1.C3 ⊔ . . . look very similar. One participant spotted this similarity and claimed that the entailment held as a result. Hot spots occur over the final axiom and the first axiom in the eye tracker heat map (Figure 3), with relatively little activity in the axioms in the middle of the justification.

6   Dealing with Justification Superfluity

Perhaps the biggest issue with the current model is that it does not deal at all with superfluity in axioms in justifications. That is, it does not penalise a justification for having axioms that contain, potentially distracting, superfluous parts—parts that do not matter as far as the entailment is concerned. Unfortunately, without a deeper investigation, it is unclear how to rectify this in the model. Although it is possible to identify the superfluous parts of axioms using laconic and precise justifications [4], throwing a naive superfluity component into the model would quite easily destroy it. This is because there can be justifications with plenty of superfluous parts that are trivial to understand. For example consider $\mathcal{J} = \{A \sqsubseteq B \sqcap C\} \models A \sqsubseteq B$, where C is along and complex class expression, and yet there can be justifications with seemingly little superfluity (as in the case of **EM3**) which causes complete distraction when trying to understand an entailment. Ultimately, what seems to be important is the location and shape of superfluity, but deciding upon what "shapes" of superfluity count as non-trivial needs to be investigated as part of future work.

One important point to consider, is that it might be possible to deal with the problems associated with superfluity by presentation techniques alone. It should be clear that the model does not pay any attention to how justifications are presented. For example, it is obvious that the ordering (and possibly the indentation) of axioms is important. It can make a big difference to the readability of justifications and how easy or difficult they are to understand, yet the model does not take into consideration how axioms will be ordered when a justification is presented to users. In the case of superfluity, it is conceivable that *strikeout* could be used to cross out the superfluous parts of axioms and this would dispel any problems associated with distracting superfluity. Figure 4 shows the helpful effect of strikeout on **EM3**. As can be seen, it immediately indicates that the problematic conjunct, $\exists$ prop1.C6, in the third axiom should be ignored. Some small scale experiments, carried out as part of future work, could confirm this.

EM3

C1 $\sqsubseteq$ C3

C3 $\sqsubseteq$ C4

C4 $\equiv$ C5 $\sqcap$ ($\exists$ prop1.C6)

C5 $\equiv$ C7 $\sqcap$ ($\exists$ prop2.C8)

C1 $\sqsubseteq$ $\exists$ prop1.C9

C9 $\sqsubseteq$ C10

C2 $\equiv$ C7 $\sqcap$ ($\exists$ prop1.C10)

EM3

C1 $\sqsubseteq$ C3

C3 $\sqsubseteq$ C4

C4 $\equiv$ C5 ~~$\sqcap$ ($\exists$ prop1.C6)~~

C5 $\equiv$ C7 ~~$\sqcap$ ($\exists$ prop2.C8)~~

C1 $\sqsubseteq$ $\exists$ prop1.C9

C9 $\sqsubseteq$ C10

C2 $\equiv$ C7 $\sqcap$ ($\exists$ prop1.C10)

Fig. 4. EM3 with and without strikeout

7 Discussion and Future Work

In this paper we presented a methodology for validating the predicted complexity of justifications. The main advantages of the experimental protocol used in the methodology is that minimal study facilitator intervention is required. This means that, over time, it should be possible to collect rich and varied data fairly cheaply and from geographically distributed participants. In addition to this, given a justification corpus and population of interest, the main experiment is easily repeatable with minimal resources and setup. Care must be taken in interpreting results and, in particular, the protocol is weak on "too hard" justifications as it cannot distinguish a model mislabeling from people failing for the wrong reason.

The cognitive complexity model that was presented in this paper fared reasonably well. In most cases, there was a significant difference in error proportion between model ranked easy and hard justifications. In the cases where error proportions revealed no difference better than chance, further small scale follow-up studies in the form of a more expensive talk-aloud study was used to gain an insight into the problems. These inspections highlighted an area for model improvement, namely in the area of superfluity. It is unclear how to rectify this in the model, as there could be justifications with superfluous parts that are trivial to understand, but the location and shape of superfluity seem an important factor.

It should be noted that the goal of the experiments was to use error proportion to determine whether two justifications come from different populations—one from the set of easy justifications and one from the set of hard justifications. This is rather different than being able to say, with some level of statistical confidence, that the model generalises to the whole population of easy or hard justifications. For the former the statistical toolbox that is used is workable with very small sample sizes. Ultimately the sample size depends on the variance of the sample, but sample sizes of less than 10 can work, where sample size is the number of outcomes (successes or failures) per justification. For the latter, sample sizes must be much larger. For example, by rule of thumb, around 400 justifications would be needed from the hard category to be able say with 95% confidence that all of hard justifications are actually hard justifications. While being able to generalise to the whole population would be the best outcome, the fact that participants would have to answer 400 items means that this is not achievable, and so the focus is on using error proportion to determine the actually hardness of a justification.

The refinement and validation of our model is an ongoing task and will require considerably more experimental cycles. We plan to conduct a series of experiments with different cohorts as well as with an expanded corpus. We also plan to continue the analysis of our corpus with an eye to performing experiments to validate the model over the whole (for some given population).

References

1. Borgida, A., Calvanese, D., Rodriguez-Muro, M.: Explanation in the DL-lite family of description logics. In: Chung, S. (ed.) OTM 2008, Part II. LNCS, vol. 5332, pp. 1440–1457. Springer, Heidelberg (2008)
2. Davis, M.: Obvious logical inferences. In: IJCAI-1981 (1981)
3. Horridge, M., Parsia, B.: From justifications towards proofs for ontology engineering. In: KR-2010 (2010)
4. Horridge, M., Parsia, B., Sattler, U.: Laconic and Precise Justifications in OWL. In: Sheth, A.P., Staab, S., Dean, M., Paolucci, M., Maynard, D., Finin, T., Thirunarayan, K. (eds.) ISWC 2008. LNCS, vol. 5318, pp. 323–338. Springer, Heidelberg (2008)
5. Horridge, M., Parsia, B., Sattler, U.: Lemmas for justifications in OWL. In: DL 2009 (2009)
6. Horridge, M., Parsia, B., Sattler, U.: Justification Oriented Proofs in OWL. In: Patel-Schneider, P.F., Pan, Y., Hitzler, P., Mika, P., Zhang, L., Pan, J.Z., Horrocks, I., Glimm, B. (eds.) ISWC 2010, Part I. LNCS, vol. 6496, pp. 354–369. Springer, Heidelberg (2010)
7. Johnson-Laird, P.N., Byrne, R.M.J.: Deduction. Psychology Press (1991)
8. Kalyanpur, A., Parsia, B., Sirin, E., Cuenca-Grau, B.: Repairing Unsatisfiable Concepts in OWL Ontologies. In: Sure, Y., Domingue, J. (eds.) ESWC 2006. LNCS, vol. 4011, pp. 170–184. Springer, Heidelberg (2006)
9. Kalyanpur, A., Parsia, B., Sirin, E., Hendler, J.: Debugging unsatisfiable classes in OWL ontologies. Journal of Web Semantics 3(4) (2005)
10. Kazakov, Y.: $\mathcal{RIQ}$ and $\mathcal{SROIQ}$ are harder than $\mathcal{SHOIQ}$. In: KR 2008. AAAI Press (2008)
11. Lam, S.C.J.: Methods for Resolving Inconsistencie In Ontologies. PhD thesis, Department of Computer Science, Aberdeen (2007)
12. Lewis, C.H.: Using the thinking-aloud method in cognitive interface design. Research report RC-9265, IBM (1982)
13. Newstead, S., Brandon, P., Handley, S., Dennis, I., Evans, J.S.B.: Predicting the difficulty of complex logical reasoning problems, vol. 12. Psychology Press (2006)
14. Parsons, L.M., Osherson, D.: New evidence for distinct right and left brain systems for deductive versus probabilistic reasoning. Cerebral Cortex 11(10), 954–965 (2001)
15. Rips, L.J.: The Psychology of Proof. MIT Press, Cambridge (1994)
16. Roussey, C., Corcho, O., Vilches-Blázquez, L.: A catalogue of OWL ontology antipatterns. In: Proc. of K-CAP-2009, pp. 205–206 (2009)
17. Strube, G.: The role of cognitive science in knowledge engineering. In: Contemporary Knowledge Engineering and Cognition (1992)

Visualizing Ontologies: A Case Study

John Howse[1], Gem Stapleton[1], Kerry Taylor[2], and Peter Chapman[1]

[1] Visual Modelling Group, University of Brighton, UK
{John.Howse,g.e.stapleton,p.b.chapman}@brighton.ac.uk
[2] Australian National University and CSIRO, Australia
Kerry.Taylor@csiro.au

Abstract. Concept diagrams were introduced for precisely specifying ontologies in a manner more readily accessible to developers and other stakeholders than symbolic notations. In this paper, we present a case study on the use of concept diagrams in visually specifying the Semantic Sensor Networks (SSN) ontology. The SSN ontology was originally developed by an Incubator Group of the W3C. In the ontology, a sensor is a physical object that implements sensing and an observation is observed by a single sensor. These, and other, roles and concepts are captured visually, but precisely, by concept diagrams. We consider the lessons learnt from developing this visual model and show how to convert description logic axioms into concept diagrams. We also demonstrate how to merge simple concept diagram axioms into more complex axioms, whilst ensuring that diagrams remain relatively uncluttered.

1 Introduction

There is significant interest in developing ontologies in a wide range of areas, in part because of the benefits brought about by being able to reason about the ontology. In domains where a precise (formal) specification of an ontology is important, it is paramount that those involved in developing the ontology fully understand the syntax in which the ontology is defined. For instance, one formal notation is description logic [3], for which much is known about the complexity of reasoning over its fragments [4].

Notations such as description logics require some level of mathematical training to be provided for the practioners using them and they are not necessarily readily accessible to all stakeholders. There have been a number of efforts towards providing visualizations of ontologies, that allow their developers and users access to some information about the ontology. For example, in Protégé, the OWLViz plugin [10] shows the concept (or class) hierarchy using a directed graph. Other visualization efforts provide instance level information over populated ontologies [11]. To the best of our knowledge, the only visualization that was developed as a direct graphical representation of description logics is a variation on existential graphs, shown to be equivalent to $\mathcal{ACL}$ by Dau and Eklund [7]. However, existential graphs, in our opinion, are not readily usable since their syntax is somewhat restrictive: they have the flavour of a minimal first-order logic

L. Aroyo et al. (Eds.): ISWC 2011, Part I, LNCS 7031, pp. 257–272, 2011.

with only the $\exists$ quantifier, negation and conjunction; the variation developed as an equivalent to $\mathcal{ACL}$ uses ? to act as a free variable.

In previous work, Oliver et al. developed concept diagrams (previously called ontology diagrams) as a formal logic for visualizing ontology specifications [12,13], further explored in Chapman et al. [5]. Whilst further work is needed to establish fragments for which efficient reasoning procedures can be devised, concept diagrams are capable of modelling relatively complex ontologies since they are a second-order logic.

The contribution of this paper is to demonstrate how concept diagrams can be used to model (part of) the Semantic Sensor Networks (SSN) ontology, in its current version, which was developed over the period February 2009 to September 2010 by an Incubator Group of the W3C, called the SSN-XG [6]. We motivate the need for accessible communication of ontology specifications in section 2, ending with a discussion around why visualization can be an effective approach. Section 3 presents a formalization of some of the SSN ontology's axioms using concept diagrams and using description logic, contrasting the two approaches. Section 4 demonstrates how to translate description logic axioms to concept diagrams and some inference rules. Section 5 concludes the paper.

2 Motivation

Complex ontologies are often developed by groups of people working together, consistent with their most important application: to support the sharing of knowledge and data. The most common definition of ontology refers to "an explicit representation of a shared conceptualisation" [9]. A *shared* conceptualisation is usually needed for the purposes for which ontologies are most used: for representation of data to be shared amongst individuals and organisations in a community. The "sharing" is necessary when domain-knowledge capture through an ontology requires modelling of either commonly-held domain knowledge or the common element of domain knowledge across multiple domains. This needs to take account of instances that are asserted to exist in the ontology, and also instances that *might* exist, or come into existence when the ontology is applied to describe some data.

The W3C's Web Ontology Language (OWL 2.0) is a very expressive but decidable description logic: a fragment of first order predicate calculus. A brief and incomplete introduction is given here: the reader is referred to [1] for a complete treatment. In common with all ontology languages, a hierarchical taxonomy of concepts (called *classes* in OWL) is the primary modelling notion. *Individuals* can be members (or *instances*) of concepts and all individuals are members of the predefined concept Thing, no individuals are members of the predefined Nothing. Binary relations over concepts, called *roles*, are used to relate individuals to others, and concept constructors comprising complex logical expressions over concepts, roles and individuals are used to relate all these things together. Most important here are role restrictions: expressions that construct a concept by referring to relations to other concepts. There are also a range of *role characteristics* that can constrain the relations wherever they occur: such as domain, range, transitive, subproperty and inverse. Two key features of OWL designed for the semantic web applications is that all entities: classes (concepts), properties (roles) and individuals are identified by URI (a Web identifier that can be a URL), and that it has an RDF/XML serialization (commonly considered unreadable).

In the experience of these authors, when people meet to develop conceptual structures, including models of knowledge intended to become an OWL ontology, they very quickly move to sketching 2D images to communicate their thoughts. At the beginning, these may be simple graph structures of labelled nodes connected by labelled or unlabelled arcs. For example, figure 1 shows the whiteboard used at the first Face-to-face meeting of the W3C Semantic Sensor Networks Incubator Group, in Washington DC, USA, November 2009. Unlike some modelling languages, OWL does not have a heritage in visual representations, and modellers struggle with different interpretations of the visualizations used in the group. For example, in OWL, it is very important to know whether a node represents an individual or a class. In a more advanced example, modellers need to know whether a specified subsumption relationship between concepts is also an equivalence relationship. As we shall see, concept diagrams are capable of visualizing exactly these kinds of definitions.

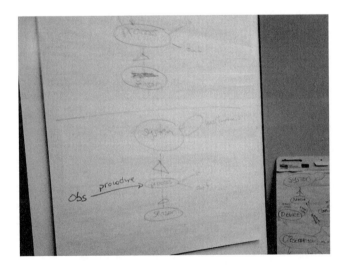

Fig. 1. Whiteboard used at face-to-face meeting; Photo: M. Hauswirth

The major feature of OWL as a modelling language is also its greatest hindrance for shared development: the formal semantics and capability for *reasoning*. The examples we give later, in our case study, demonstrate that information that would sometimes need to be inferred is actually explicitly visible on concept

diagrams (so-called *free-rides* which we explain later). It is commonplace for
ontology developers, as domain experts, to be unaware of the formal semantics underlying OWL, and if they are aware it remains very difficult to apply the knowledge of the semantics in practice while developing in a team environment. For
example, even the simple difference between universal and existential role restrictions are difficult to represent and to assimilate in diagrammatic form. As
another example, the semantic difference between rdfs:domain and rdfs:range constraints on properties, as opposed to local restrictions on those properties in the
context of class definitions, is difficult to represent diagrammatically and very
hard to take into account when studying spatially-disconnected but semantically-
connected parts of an ontology sketch. There is a need for semantically-informed
sketching tools that help ontology developers to better understand their modelling
in real time.

3   Visualizing the SSN Ontology

In this section we walk through parts of the SSN ontology, showing how to
express it in our concept diagrams. At the end of section 3.2 we will give
a summary of the concept diagram syntax. The SSN ontology is available at
purl.oclc.org/NET/ssnx/ssn and extensive documentation and examples of its
use are available in the final report of the SSN-XG [2]. An alternative visualization of the SSN ontology was created using CMAP from IHMC (see
www.ihmc.us/groups/coe/) and may be compared with the visualisation presented here. The ontology imports, and is aligned with, the Dolce Ultra-Lite
upper ontology [14] from which it inherits upper concepts including Event, Object, Abstract, Quality, PhysicalObject, SocialObject, InformationObject, Situation,
Description, Method, and Quality.

3.1   Concept Hierarchy Axioms

To represent the concept hierarchy, concept diagrams use Euler diagrams [8],
which effectively convey subsumption and disjointness relationships. In particular, Euler diagrams comprise closed curves (often drawn as circles or ellipses) to
represent sets (in our case, concepts). Two curves that have no common points
inside them assert that the represented sets are disjoint whereas one curve drawn
completely inside another asserts a subsumption relationship. In addition, Euler
diagrams use shading to assert emptiness of a set; in general, concept diagrams
use shading to place upper bounds on set cardinality as we will see later.

In the SSN ontology, descriptions of the concepts are given as comments in
the ssn.owl file [2], which we summarize here. The SSN ontology is defined over
a large vocabulary of which we present the subset required for our case study.
At the top level of the SSN hierarchy are four concepts, namely Entity, Feature-
OfInterest, Input, and Output. The concept Entity is for anything real, possible
or imaginary that the modeller wishes to talk about. Entity subsumes five other
concepts which, in turn, may subsume further concepts. The five concepts are:

1. **Abstract** These are entities that cannot be located in space and time, such as mathematical concepts.
2. **Event** These are physical, social, or mental processes, events, or states. Event is, therefore, disjoint from Abstract.
3. **Object** These are physical, social or mental objects or substances. Therefore, Object is disjoint from Abstract and Event.
4. **FeatureOfInterest** A feature of interest is an abstraction of real world phenomena and is subsumed by the union of Event and Object.
5. **Quality** This is any aspect of an entity that cannot exist without that entity, such as a surface of a solid object. Quality is also disjoint from Abstract, Event, and Object.

An Euler diagram asserting these subsumption and disjointness properties as a single axiom, alongside the axioms expressed using description logic, can be seen here:

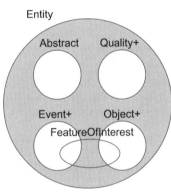

1. Abstract $\sqsubseteq$ Entity
2. Quality $\sqsubseteq$ Entity
3. Event $\sqsubseteq$ Entity
4. Object $\sqsubseteq$ Entity
5. Abstract $\sqcap$ Quality $\sqsubseteq \perp$
6. Abstract $\sqcap$ Event $\sqsubseteq \perp$
7. Abstract $\sqcap$ Object $\sqsubseteq \perp$
8. Quality $\sqcap$ Event $\sqsubseteq \perp$
9. Quality $\sqcap$ Object $\sqsubseteq \perp$
10. Event $\sqcap$ Object $\sqsubseteq \perp$
11. FeatureOfInterest $\sqsubseteq$ Event $\sqcup$ Object
12. Entity $\equiv$ Abstract $\sqcup$ Object $\sqcup$ Event $\sqcup$ Quality

The Euler diagram has a certain succinctness over the description logic in that there are 6 DL axioms asserting disjointness properties, for example.

Notice, in the figure above, the concept Object is annotated with a plus symbol, as are Event and Quality. Whilst not part of the formal syntax, this plus symbol is used to indicate that there are concepts subsumed by each of these concepts that are not displayed in this diagram; with tool support, one could imagine clicking on this plus to 'expand' the diagram, to show the subsumed concepts. In the SSN ontology, Object is the union of two disjoint concepts, PhysicalObject and SocialObject:

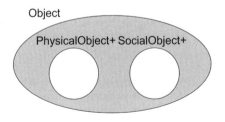

1. Object $\equiv$ PhysicalObject $\sqcup$ SocialObject
2. PhysicalObject $\sqcap$ SocialObject $\sqsubseteq \perp$

A PhysicalObject is an object that has a proper space region whereas a SocialObject exists only within some communication Event, in which at least one PhysicalObject participates. Again, as indicated by the plus sign, PhysicalObject subsumes various other concepts: Sensor, System, Device, and SensingDevice. A Sensor can do sensing: that is, a Sensor is any entity that can follow a sensing method and thus observe some Property of a FeatureOfInterest. A System is a unit of abstraction for pieces of infrastructure for sensing, namely Device and SensingDevice. A Device is a physical piece of technology, of which SensingDevice is an example. Additionally, SensingDevice is an example of Sensor. This information about the SSN ontology is axiomatized by the single Euler diagram below, and equivalently by the adjacent description logic axioms:

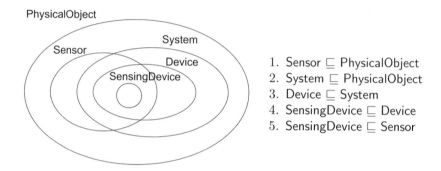

1. Sensor ⊑ PhysicalObject
2. System ⊑ PhysicalObject
3. Device ⊑ System
4. SensingDevice ⊑ Device
5. SensingDevice ⊑ Sensor

It should be clear that the diagram just given makes some informational content explicit, whereas it needs to be derived from the description logic axioms. For instance, one can easily read off, from the diagram, that SensingDevice is subsumed by PhysicalObject, since the closed curve representing the former is contained by the closed curve representing the latter. From the description logic axioms, one must use the transitive property of ⊑ to extract this information: SensingDevice ⊑ Sensor ⊑ PhysicalObject. To make this deduction, one has to identify appropriate description logic axioms from the list given, which requires a little more effort than reading the diagram. This example, illustrating the inferential properties of the diagram, is a typical example of a *free-ride* (sometimes called a *cheap ride*), the theory of which was developed by Shimojima [16], later explored by Shimojima and Katagiri [17]. In general, a free-ride is a piece of information that can be readily 'seen' in a diagram that would typically need to be inferred from a symbolic representation.

SocialObject also subsumes various other concepts, which we do not describe in full here. Three will be of use to us later: a Situation is a view on a set of entities; an Observation is a Situation in which a SensingMethod has been used to estimate or calculate a value of a Property of a FeatureOfInterest; and Sensing is a process that results in the estimation, or calculation, of the value of a phenomenon. The following Euler diagram defines an axiom from the SSN ontology, and the adjacent description logic statements capture the same information:

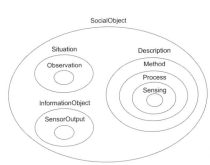

1. Situation ⊑ SocialObject
2. Observation ⊑ Situation
3. InformationObject ⊑ SocialObject
4. SensorOutput ⊑ InformationObject
5. Description ⊑ SocialObject
6. Method ⊑ Description
7. Process ⊑ Method
8. Sensing ⊑ Process
9. Situation ⊓ Description ⊑ ⊥
10. Situation ⊓ SocialObject ⊑ ⊥
11. InformationObject ⊓ Description ⊑ ⊥

This diagram, presenting information about the concepts subsumed by SocialObject, also has many free-rides, such as Sensing is subsumed by Description, and that Process is disjoint from Observation since the two curves do not overlap. For the latter, to deduce this from the given description logic axioms, one would need to use axiom numbers 2, 6, 7, and 9.

We saw earlier that Quality was subsumed by one of the top-level concepts, Entity. In turn, Quality subsumes Property, which is an observable quality of an event or object. Property subsumes many concepts, but we only make use of one of them later: MeasurementCapability. This concept collects together measurement properties (accuracy, range, precision, etc) as well as the environmental conditions in which those properties hold, representing a specification of a sensor's capability in those conditions:

1. Property ⊑ Quality
2. MeasurementCapability ⊑ Property

The last part of the concept hierarchy that we demonstrate concerns Event, which was subsumed by the top-level concept Entity. Two of the concepts subsumed by Entity are Stimulus and SensorInput. A sensor input is an event that triggers the sensor and the concept SensorInput is equivalent to Stimulus:

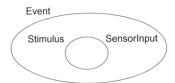

1. Stimulus ⊑ Event
2. SensorInput ⊑ Event
3. SensorInput ≡ Stimulus

Notice here that, in the Euler diagram, we have asserted equivalence between concepts by drawing two circles on top of one another.

We have represented 24 of the SSN concepts using Euler diagrams to assert subsumption and disjointness relationships. The discussions around free-rides indicate that Euler diagrams (the basis of concept diagrams) can be an effective method of axiomatizing concept hierarchies. We refer the reader to [2] for further information on the hierarchy.

3.2   Role Rescrictions

Moving on to role restrictions, concept diagrams extend Euler diagrams by incorporating more syntax to increase their expressiveness. In particular, arrows are used to represent role restrictions. The source of the arrow is taken to restrict the domain of the role, and the target provides some information about the image of the role under the domain restriction. The nature of the information given is determined by the arrow's type: arrows can be dashed or solid. Given a solid arrow, a, sourced on C and targeting D, representing the role R, a asserts that the image of R when it's domain is restricted to C is *equal* to D, that is:

$$image(\mathsf{R}|_\mathsf{C}) = \mathsf{D} \qquad \text{where } image(\mathsf{R}|_\mathsf{C}) = \{y : \exists x \in \mathsf{C} \ (x,y) \in \mathsf{R}\}.$$

By contrast, if a were instead dashed then it would assert that the image of R when its domain is restricted to C includes at least the elements in D, that is:

$$image(\mathsf{R}|_\mathsf{C}) \supseteq \mathsf{D}.$$

As we shall see in our examples, the syntax that can be used as sources and targets of arrows, including closed curves (both labelled, as in Euler diagrams, or unlabelled), or dots. Unlabelled closed curves represent anonymous concepts and dots represent individuals. As with closed curves, dots can be labelled to represent specific individuals, or unlabelled to represent anonymous individuals. The syntax and semantics will be more fully explained as we work through our examples.

Our first example of some role restrictions concerns the concept Sensor, since this is at the heart of the SSN ontology. There are various restrictions placed on the roles detects, observes, hasMeasurementCapability and implements. The first of these, detects, is between Sensor and Stimulus: sensors detect only stimuli. Next, there is a role observes between Sensor and Property: sensors observe only properties. Thirdly, every sensor hasMeasurementCapability, the set of which is subsumed by MeasurementCapability. Finally, every sensor implements some sensing. That is, sensors have to perform some sensing. The concept diagram below captures these role restrictions:

Here, we are quantifying over the concept Sensor, since we have written 'For all Sensor s' above the bounding box of the diagram (in the formal abstract syntax of concept diagrams, this would be represented slightly differently, the details of which are not important here). The dot labelled s in the diagram is then the source of four arrows, relating to the role restrictions just informally described. The solid arrow labelled detects is used to place the following restriction on the detects role:

$$image(\mathsf{detects}|_{\{s\}}) \subseteq \mathsf{Stimulus},$$

For all Sensor *s*

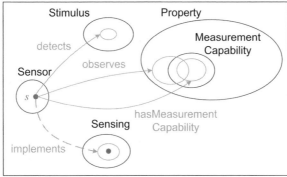

treating the individual *s* as a singleton set. The unlabelled curve is acting as an existentially quantified anonymous set so, strictly, the arrow and the unlabelled curve assert:

$$\exists X \, image(\mathsf{detects}|_{\{s\}}) = X \wedge X \subseteq \mathsf{Stimulus}.$$

Earlier, we defined an axiom that asserted MeasurementCapability is subsumed by Property, along with axioms that give the disjointness information conveyed in the diagram above. We have made use of that information in the diagram above, by drawing the curves with appropriate containment and disjointness properties. A further point of note is that, in this diagram, we have not asserted anything about whether $image(\mathsf{observes}|_{\{s\}})$ and $image(\mathsf{hasMeasurementCapability}|_{\{s\}})$ are disjoint, or whether one subsumes the other. All we know is that the former is subsumed by Property and the latter is subsumed by MeasurementCapability. Finally, the dashed arrow provides partial information:

$$\exists X \exists y \, image(\mathsf{implements}|_{\{s\}}) \supseteq X \wedge X \subseteq \mathsf{Sensing} \wedge y \in X$$

where X arises from the unlabelled curve targeted by the arrow and y arises from the unlabelled dot placed inside this curve; we are using unlabelled dots to assert the existence of individuals.

The role restrictions just given, together with the disjointness information, can also be expressed using the following description logic axioms:

1. Sensor $\sqsubseteq \forall$ detects.Stimulus
2. Sensor $\sqsubseteq \forall$ observes.Property
3. Sensor $\sqsubseteq \exists$ implements.Sensing
4. Sensor $\sqsubseteq \forall$ hasMeasurementCapability.MeasurementCapability
5. Sensor $\sqcap$ Stimulus $\sqsubseteq \perp$
6. Sensor $\sqcap$ Property $\sqsubseteq \perp$
7. Sensor $\sqcap$ Sensing $\sqsubseteq \perp$
8. Stimulus $\sqcap$ Property $\sqsubseteq \perp$
9. Stimulus $\sqcap$ Sensing $\sqsubseteq \perp$
10. Property $\sqcap$ Sensing $\sqsubseteq \perp$

We can see that the concept diagram has free-rides arising from the use of the unlabelled curves. For example, it is easy to see that $image(\text{detects}|_{\{s\}})$ is disjoint from Property, but this information is not immediately obvious from the description logic axioms: one must make this deduction from axioms 1 and 8.

Our second collection of role restrictions concerns the concept Observation. Here, an observation includes an event, captured by the role includesEvent, which is a Stimulus, illustrated by the dashed arrow in the diagram immediately below. In addition, observations are observedBy exactly one (unnamed) individual, which is a Sensor. Similarly, observations have exactly one observedProperty and this is a Property, exactly one sensingMethodUsed and this is a Sensing object, and a set of observationResults all of which are SensorOutputs. Finally, observations have exactly one featureOfInterest (role), which is a FeatureOfInterest (concept). All of these role restrictions are captured in the diagram below, where again we have used previous information about disjointness to present a less cluttered diagram:

For all Observation *o*

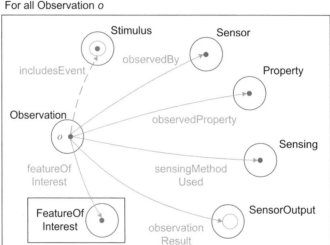

Here, of note is the use of a rectangle around the closed curve labelled Feature-OfInterest. The rectangle is used to assert that we are not making any claim about the disjointness of FeatureOfInterest with respect to the other concepts appearing in the diagram.

To allow the reader to draw contrast with description logic, the role restrictions just given are expressed by 21 description logic axioms, comprising 15 disjointness axioms and the following 6 axioms that correspond to the information provided by the six arrows:

1. Observation $\sqsubseteq \exists$ includesEvent.Stimulus
2. Observation $\sqsubseteq (= 1$ observedBy) $\sqcap$ ($\forall$ observedBy.Sensor)
3. Observation $\sqsubseteq (= 1$ observedProperty) $\sqcap$ ($\forall$ observedProperty.Property)
4. Observation $\sqsubseteq (= 1$ sensingMethodUsed) $\sqcap$ ($\forall$ sensingMethodUsed.Sensing)
5. Observation $\sqsubseteq \forall$ ObservationResult.SensorOutput
6. Observation $\sqsubseteq (= 1$ FeatureOfInterest) $\sqcap$ ($\forall$ FeatureOfInterest.FeatureOfInterest)

Consider axiom 2, which corresponds to the arrow labelled observedBy. From the description logic axiom, a little reasoning is required to see that every observation is related to exactly one individual, which must be a sensor: one must deduce this from the information that Observation is subsumed by the set of individuals that are related to exactly one thing under observedBy intersected with the set of individuals that are related to only properties under observedBy. In our opinion, the diagram more readily conveys the informational content of the axioms than the description logic syntax and in a more succinct way (although this, of course, could be debated).

To conclude this section, we summarize main syntax of concept diagrams:

1. **Rectangles.** These are used to represent the concept Thing.
2. **Closed Curves.** These are used to represent concepts. If the curve does not have a label then the concept is anonymous. The spatial relationships between the curves gives information about subsumption and disjointness relationships.
3. **Dots.** These are used to represent individuals. As with closed curves, unlabelled dots represent anonymous individuals. The location of the dot gives information about the type of the individual. Distinct dots represent distinct individuals. When many dots are present in a region, we may use $\leq$, $=$, and $\geq$ as shorthand annotations (this will be demonstrated later).
4. **Shading.** Shading in a region asserts that the concept represented by the region contains only individuals represented by dots.
5. **Solid Arrows.** These are used to represent role restrictions. In particular, the image of the role whose label appears on the arrow has an image, when the domain is restricted to (the concept or individual represented by) the source, is *equal* to the target.
6. **Dashed Arrows.** These are used to represent role restrictions. In particular, the image of the role whose label appears on the arrow has an image, when the domain is restricted to the source, which is a *superset* of the target.

In addition, quantifiers and connectives can be used in the standard way.

4   Discussion

We will now extract, from the case study that we have presented, some general constructions of diagrams, from description logic axioms. Moreover, we will show how to take these simple axioms and merge them into more complex axioms, by providing inference rules. These inference rules are inspired by the manner in which we produced our visualization of the SSN ontology, aiming for diagrams with minimal clutter, without compromising their informational content.

With regard to subclass and disjointness information, where C and D are concepts, we have the following translations:

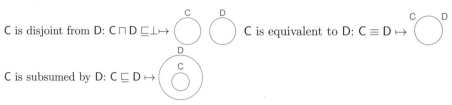

C is disjoint from D: $C \sqcap D \sqsubseteq \bot \mapsto$

C is equivalent to D: $C \equiv D \mapsto$

C is subsumed by D: $C \sqsubseteq D \mapsto$

However, using these translations would give one diagram for every disjoint-ness, subsumption and equivalence axiom in the ontology. As we have seen, it is possible to produce readable diagrams that correspond to many axioms of the kind just given (all of our diagrams that conveyed concept hierarchy information expressed more than one description logic axiom). There is clearly a requirement on the ontology deverloper to determine a balance between the number of ax-ioms like these conveyed in a single diagram and the clutter in a diagram. Our diagrams were drawn in a manner that concepts were only in the same diagram if we wanted to assert something about their relationship. Later, we will give some general rules for merging these simple diagrams into larger diagrams.

As we saw earlier, we can readily translate information about 'only' or 'some' role restrictions into diagrammatic form. For example, in the Sensor concept, we have Sensor detects only Stimulus and Sensor implements some Sensing. Abstract-ing from this, and including more general constraints, we have role restrictions of these forms, where C and D are concepts and R is a role:

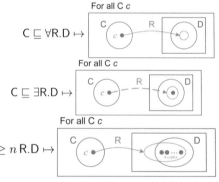

C is subsumed by the anonymous concept containing individuals related, under R, to only individuals in D: $C \sqsubseteq \forall R.D \mapsto$

C is subsumed by the anonymous concept containing individuals related, under R, to an individual in D: $C \sqsubseteq \exists R.D \mapsto$

C is subsumed by the anonymous concept containing individuals related, under R, to at least n individuals in D: $C \sqsubseteq \geq n R.D \mapsto$

In the above, instead of drawing n dots, we could use one dot annotated with $\geq n$ as shorthand which is sensible if n gets beyond, say, 4. We can also adopt this shorthand for $\leq n$; we recall that shading is used to place upper bounds on set cardinality, generalizing the use of shading in Euler diagrams, in a shaded region all elements must be represented by dots. We now give two further translations:

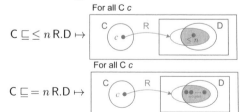

C is subsumed by the anonymous concept containing individuals related, under R, to at most n individuals in D: $C \sqsubseteq \leq n R.D \mapsto$

C is subsumed by the anonymous concept containing individuals related, under R, to exactly n individuals in D: $C \sqsubseteq = n R.D \mapsto$

Again, in the diagram just given, we could have used the shorthand $= n$.

The above translations demonstrate how to produce concept diagrams from role restrictions defined using description logic. These translations are sound and, in fact, information preserving. As with the hierarchy information, there are often more succinct, elegant diagrams that can be created by representing many of these axioms in a single diagram. We will call the diagrams obtained by applying the transformations just given *atomic axioms*. We now demonstrate how to produce non-atomic axioms (like the diagrams given in the SSN ontology) from atomic axioms.

We begin by giving some inference rules that allow us to replace some axioms with others; in many cases there are obvious generalizations of the inference rules. We adopt a traditional presentation style, where axioms written above a line can be used to infer those written below the line. Each rule has a name, displayed in shorthand: Dis for Disjunction, Sub for Subsumption, and Mer for Merge. All of these rules are equivalences (no informational content is lost) and can be formalized and proved sound, but here we just present them informally using illustrative diagrams. First we have, concerning hierarchy information:

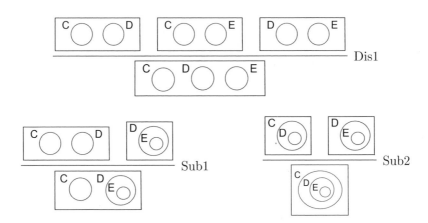

For instance, Dis1 says three axioms that tell us three concepts are pairwise disjoint is equivalent to a single diagram telling us that the concepts are pairwise disjoint. Sub1, tells us, roughly speaking, that if D appears in one axiom, a, and we know that E is subsumed by D then we can copy E into a, placing it inside D. Sub2 is another instance of this kind of inference.

Regarding role restrictions, we have seen that it is possible to use information about disjointness when creating these kinds of axioms. For instance, if we know that C and D are disjoint then we can simplify an axiom that tells us C is subsumed by $\forall R.D$: we do not have to place D in a separate box. This intuition is captured by our first role restriction rule, Dis2. Our second role restriction rule, Mer1, takes atomic axioms arising from $C \sqsubseteq \forall R.D$ and $C \sqsubseteq \forall S.E$ and merges them into a single diagram. Rule Mer2 is similar.

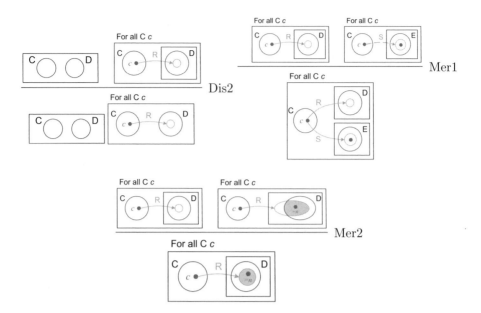

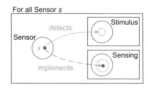

To demonstrate the use of these inference rules, we consider the example from the SSN network concerning Sensor on page 265. Translating the associated description logic axioms numbered 1, 3, 5, 7, and 9, using the techniques of the previous subsection, we get the following five diagrams:

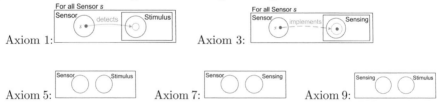

Using the rule Mer2, from axioms 1 and 3 we obtain:

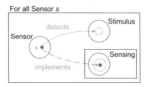

Using axiom 5, and a generalization of Dis2, we can delete the rectangle around Stimulus (since Stimulus and Sensor are disjoint):

Using axioms 7 and 9, we further deduce:

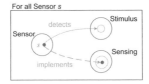

We leave it to the reader to use these kinds of manipulations to obtain the single diagram given for the role restrictions imposed over the Sensor concept.

5   Conclusion

In this paper we have discussed the need for sophisticated ontology visualization techniques that will allow disparate groups of ontology developers and users to communicate effectively. Concept diagrams are a visual notation that were developed with this need in mind. We have used concept diagrams to produce a visualization of (part of) the Semantic Sensor Network ontology, including information about the concept hierarchy and role restrictions. Thus, this paper demonstrates that concept diagrams can be applied to modelling real-world ontologies. Concept diagrams may undergo further refinement as more case studies are developed and as they are applied in other domains.

An important future development is the implementation of tool support. We envisage developing tools which allow the automatic conversion of symbolically specified ontologies to concept diagrams. This will involve solving challenging problems, such as identifying what constitutes an effective diagram (as shown in this paper, there are different diagrams that convey the same information) and how to automatically draw chosen diagrams from abstract descriptions of them. This functionality could build on recent advances in automated Euler diagram drawing [15,18,19], although the layout problem for concept diagrams is more challenging. In addition, we want to allow ontology creators to be able to specify the axioms directly with concept diagrams, which may require a sketch recognition engine to be devised; this could also build on recent work that recognizes sketches of Euler diagrams [20]. These automatically drawn sketches can be translated into symbolic form, so that we can make use of sophisticated tool support that already exists for ontology development.

References

1. W3C   OWL   Working   Group,   OWL   2   Web   Ontology   Language Document   Overview,   W3C   Recommendation   (October   27,   2009), http://www.w3.org/TR/2009/REC-owl2-overview-20091027/ (accessed June 2011)
2. Lefort, et al.: The W3C Semantic Sensor Network Incubator Group Final Report, http://www.w3.org/2005/Incubator/ssn/XGR-ssn-20110628/ (accessed June 2011)

3. Baader, F., Calvanese, D., McGuinness, D., Nadi, D., Patel-Schneider, P. (eds.): The Description Logic Handbook. CUP (2003)
4. Baader, F., Calvanese, D., McGuinness, D., Nadi, D., Patel-Schneider, P. (eds.): The Description Logic Handbook, ch. 3. CUP (2003)
5. Chapman, P., Stapleton, G., Howse, J., Oliver, I.: Deriving sound inference rules for concept diagrams. In: IEEE Symposium on Visual Languages and Human-Centric Computing. IEEE (2011)
6. Compton, et al.: The SSN Ontology of the Semantic Sensor Network Incubator Group. Submitted to The Journal of Web Semantics (July 2011)
7. Dau, F., Eklund, P.: A diagrammatic reasoning system for the description logic $\mathcal{ACL}$. Journal of Visual Languages and Computing 19(5), 539–573 (2008)
8. Euler, L.: Lettres a une princesse d'allemagne sur divers sujets de physique et de philosophie. Letters 2, 102–108 (1775); Berne, Socit Typographique
9. Gruber, T.: A translation approach to portable ontology specifications. Knowledge Acquisition 5(2) (1993)
10. Horridge, M.: OWLViz, http://www.co-ode.org/downloads/owlviz/ (accessed June 2009)
11. Jambalaya: http://www.thechiselgroup.org/jambalaya
12. Oliver, I., Howse, J., Stapleton, G., Nuutila, E., Torma, S.: A proposed diagrammatic logic for ontology specification and visualization. In: International Semantic Web Conference (Posters and Demos) (2009)
13. Oliver, I., Howse, J., Stapleton, G., Nuutila, E., Torma, S.: Visualising and specifying ontologies using diagrammatic logics. In: 5th Australasian Ontologies Workshop, vol. 112, pp. 87–104. CRPIT (2009)
14. Presutti, V., Gangemi, A.: Content Ontology Design Patterns as Practical Building Blocks for Web Ontologies. In: Li, Q., Spaccapietra, S., Yu, E., Olivé, A. (eds.) ER 2008. LNCS, vol. 5231, pp. 128–141. Springer, Heidelberg (2008)
15. Rodgers, P., Zhang, L., Fish, A.: General Euler Diagram Generation. In: Stapleton, G., Howse, J., Lee, J. (eds.) Diagrams 2008. LNCS (LNAI), vol. 5223, pp. 13–27. Springer, Heidelberg (2008)
16. Shimojima, A.: Inferential and expressive capacities of graphical representations: Survey and some generalizations. In: Blackwell, A.F., Marriott, K., Shimojima, A. (eds.) Diagrams 2004. LNCS (LNAI), vol. 2980, pp. 18–21. Springer, Heidelberg (2004)
17. Shimojima, A., Katagiri, Y.: An Eye-Tracking Study of Exploitations of Spatial Constraints in Diagrammatic Reasoning. In: Stapleton, G., Howse, J., Lee, J. (eds.) Diagrams 2008. LNCS (LNAI), vol. 5223, pp. 74–88. Springer, Heidelberg (2008)
18. Simonetto, P., Auber, D.: Visualise undrawable Euler diagrams. In: 12th International Conference on Information Visualization, pp. 594–599. IEEE (2008)
19. Stapleton, G., Zhang, L., Howse, J., Rodgers, P.: Drawing Euler diagrams with circles: The theory of piercings. IEEE Transactions on Visualisation and Computer Graphics 17, 1020–1032 (2011)
20. Wang, M., Plimmer, B., Schmieder, P., Stapleton, G., Rodgers, P., Delaney, A.: SketchSet: Creating Euler diagrams using pen or mouse. In: IEEE Symposium on Visual Languages and Human-Centric Computing 2011. IEEE (2011)

LogMap:
Logic-Based and Scalable Ontology Matching

Ernesto Jiménez-Ruiz and Bernardo Cuenca Grau

Department of Computer Science, University of Oxford
{ernesto,berg}@cs.ox.ac.uk

Abstract. In this paper, we present LogMap—a highly scalable ontology matching system with 'built-in' reasoning and diagnosis capabilities. To the best of our knowledge, LogMap is the only matching system that can deal with semantically rich ontologies containing tens (and even hundreds) of thousands of classes. In contrast to most existing tools, LogMap also implements algorithms for 'on the fly' unsatisfiability detection and repair. Our experiments with the ontologies NCI, FMA and SNOMED CT confirm that our system can efficiently match even the largest existing bio-medical ontologies. Furthermore, LogMap is able to produce a 'clean' set of output mappings in many cases, in the sense that the ontology obtained by integrating LogMap's output mappings with the input ontologies is consistent and does not contain unsatisfiable classes.

1 Introduction

OWL ontologies are extensively used in biology and medicine. Ontologies such as SNOMED CT, the National Cancer Institute Thesaurus (NCI), and the Foundational Model of Anatomy (FMA) are gradually superseding existing medical classifications and are becoming core platforms for accessing, gathering and sharing bio-medical knowledge and data.

These reference bio-medical ontologies, however, are being developed independently by different groups of experts and, as a result, they use different entity naming schemes in their vocabularies. As a consequence, to integrate and migrate data among applications, it is crucial to first establish correspondences (or *mappings*) between the vocabularies of their respective ontologies.

In the last ten years, the Semantic Web and bio-informatics research communities have extensively investigated the problem of automatically computing mappings between independently developed ontologies, usually referred to as the *ontology matching problem* (see [8] for a comprehensive and up-to-date survey).

The growing number of available techniques and increasingly mature tools, together with substantial human curation effort and complex auditing protocols, has made the generation of mappings between real-world ontologies possible. For example, one of the most comprehensive efforts for integrating bio-medical ontologies through mappings is the UMLS Metathesaurus (UMLS) [2], which integrates more than 100 thesauri and ontologies.

L. Aroyo et al. (Eds.): ISWC 2011, Part I, LNCS 7031, pp. 273–288, 2011.

However, despite the impressive state of the art, modern bio-medical ontologies still pose serious challenges to existing ontology matching tools.

Insufficient scalability. Although existing matching tools can efficiently deal with moderately sized ontologies, large-scale bio-medical ontologies such as NCI, FMA or SNOMED CT are still beyond their reach. The largest test ontologies in existing benchmarks (e.g., those in the OAEI initiative) contain around 2000-3000 classes (i.e., with several million possible mappings); however, to the best of our knowledge, no tool has been able to process ontologies with tens or hundreds of thousands of classes (i.e., with several billion possible mappings).

Logical inconsistencies. OWL ontologies have well-defined semantics based on first-order logic, and mappings are commonly represented as OWL class axioms. Hence, the ontology $\mathcal{O}_1 \cup \mathcal{O}_2 \cup \mathcal{M}$ resulting from the integration of $\mathcal{O}_1$ and $\mathcal{O}_2$ via mappings $\mathcal{M}$ may entail axioms that don't follow from $\mathcal{O}_1$, $\mathcal{O}_2$, or $\mathcal{M}$ alone. Many such entailments correspond to logical inconsistencies due to erroneous mappings in $\mathcal{M}$, or to inherent disagreements between $\mathcal{O}_1$ and $\mathcal{O}_2$. Recent work has shown that even the integration of ontologies via carefully-curated mappings can lead to thousands such inconsistencies [9,5,16,13] (e.g., the integration of FMA-SNOMED via UMLS yields over 6,000 unsatisfiable classes). Most existing tools are based on lexical matching algorithms, and may also exploit the structure of the ontologies or access external sources such as WordNet; however, these tools disregard the semantics of the input ontologies and are thus unable to detect and repair inconsistencies. Although the first reasoning-based techniques for ontology matching were proposed relatively early on (e.g., S-Match [10]), in practice reasoning is known to aggravate the scalability problem (e.g., no reasoner known to us can classify the integration NCI-SNOMED via UMLS). Despite the technical challenges, there is a growing interest in reasoning techniques for ontology matching. In particular, there has been recent work on 'a-posteriori' mapping debugging [12,13,14,15], and a few matching tools (e.g., ASMOV [11], KOSIMap [21], CODI [19,20]) incorporate techniques for 'on the fly' semantic verification.

In this paper, we present LogMap—a novel ontology matching tool that addresses both of these challenges. LogMap implements highly optimised data structures for lexically and structurally indexing the input ontologies. These structures are used to compute an initial set of *anchor mappings* (i.e., 'almost exact' lexical correspondences) and to assign a confidence value to each of them. The core of LogMap is an iterative process that, starting from the initial anchors, alternates *mapping repair* and *mapping discovery* steps. In order to detect and repair unsatisfiable classes 'on the fly' during the matching process, LogMap implements a sound and highly scalable (but possibly incomplete) ontology reasoner as well as a 'greedy' diagnosis algorithm. New mappings are discovered by iteratively 'exploring' the input ontologies starting from the initial anchor mappings and using the ontologies' *extended class hierarchy*.

To the best of our knowledge, LogMap is the only matching tool that has shown to scale for rich ontologies with tens (even hundreds) of thousands of classes. Furthermore, LogMap is able to produce an 'almost clean' set of output

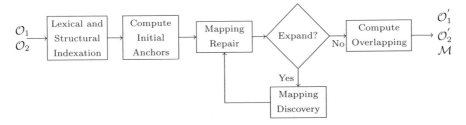

Fig. 1. LogMap in a nutshell

mappings between FMA, SNOMED and NCI; as shown 'a posteriori' using a fully-fledged DL reasoner, LogMap only failed to detect one unsatisfiable class (out of a total of several thousands) when integrating these large-scale ontologies.

2   The Anatomy of LogMap

We next provide an overview of the main steps performed by LogMap, which are schematically represented in Figure 1.

1. **Lexical indexation.** The first step after parsing the input ontologies is their lexical indexation. LogMap indexes the labels of the classes in each ontology as well as their lexical variations, and allows for the possibility of enriching the indexes by using an external lexicon (e.g., WordNet or UMLS-lexicon).
2. **Structural indexation.** LogMap uses an *interval labelling schema* [1,4,18] to represent the extended class hierarchy of each input ontology. Each extended hierarchy can be computed using either simple structural heuristics, or an off-the-shelf DL reasoner.
3. **Computation of initial 'anchor mappings'.** LogMap computes an initial set of equivalence *anchor mappings* by intersecting the lexical indexes of each input ontology. These mappings can be considered 'exact' and will later serve as starting point for the further discovery of additional mappings.
4. **Mapping repair and discovery.** The core of LogMap is an iterative process that alternates *repair* and *discovery* steps.
   - In the *repair step*, LogMap uses a sound and highly scalable (but possibly incomplete) reasoning algorithm to detect classes that are unsatisfiable w.r.t. (the merge of) both input ontologies and the mappings computed thus far. Then, each of these undesirable logical consequences is automatically repaired using a 'greedy' diagnosis algorithm.
   - To *discover new mappings*, LogMap maintains two *contexts* (sets of 'semantically related' classes) for each anchor. Contexts for the same anchor are expanded in parallel using the class hierarchies of the input ontologies. New mappings are then computed by matching the classes in the relevant contexts using ISUB [23]—a flexible tool that computes a similarity score for any pair of input strings. This mapping discovery strategy is based on a *principle of locality*: if classes C_1 and C_2 are

Table 1. Fragment of the lexical indexes for NCI and FMA ontologies

Inverted index for NCI labels		Index for NCI class URIs	
Entry	*Cls ids*	*Cls id*	*URI*
secretion	49901	49901	NCI:CellularSecretion
cellular,secretion	49901	37975	NCI:Trapezoid
cellular,secrete	49901	62999	NCI:TrapezoidBone
trapezoid	37975,62999	60791	NCI:Smegma
trapezoid,bone	62999		
smegma	60791		
Inverted index for FMA labels		**Index for FMA class URIs**	
Entry	*Cls ids*	*Cls id*	*URI*
secretion	36792	36792	FMA:Secretion
bone,trapezoid	20948,47996	47996	FMA:Bone_of_Trapezoid
trapezoid	20948	20948	FMA:Trapezoid
smegma	60947	60947	FMA:Smegma

correctly mapped, then the classes semantically related to C_1 in $\mathcal{O}_1$ are likely to be mapped to those semantically related to C_2 in $\mathcal{O}_2$.

LogMap continues the iteration of repair and discovery steps until no context is expanded in the discovery step. The output of this process is a set of mappings that are likely to be 'clean'—that is, it will not lead to logical errors when merged with the input ontologies (c.f., evaluation section).

5. **Ontology overlapping estimation.** In addition to the final set of mappings, LogMap computes a fragment of each input ontology, which intuitively represent the 'overlapping' between both ontologies. When manually looking for additional mappings that LogMap might have missed, curators can restrict themselves to these fragments since 'correct' mappings between classes not mentioned in these fragments are likely to be rare.

2.1 Lexical Indexation

LogMap constructs an 'inverted' lexical index (see Table 1) for each input ontology. This type of index, which is commonly used in information retrieval applications, will be exploited by LogMap to efficiently compute an initial set of anchor mappings.

The English name of ontology classes as well as their alternative names (e.g., synonyms) are usually stored in OWL in label annotations. LogMap splits each label of each class in the input ontologies into components; for example, the NCI class 'cellular_secretion' is broken into its component English words 'cellular' and 'secretion'. LogMap allows for the use of an external lexicon (e.g., UMLS lexicon[1] or WordNet) to find both their synonyms and lexical variations; for example, UMLS lexicon indicates that 'secrete' is a lexical variation of 'secretion'.

---

[1] UMLS Lexicon, unlike WordNet, provides only normalisations and spelling variants.

LogMap groups the component words of each class label and their variations into sets, which will then constitute the key of an inverted index. For example, the inverted index for NCI contains entries for the sets 'cellular, secretion' and 'cellular, secrete'. The range of the index is a numerical ID that LogMap associates to each corresponding class (see Table 1). Thus, in general, an entry in the index can be mapped to several classes (e.g., see 'trapezoid' in Table 1) .

The use of external lexicons to produce a richer index is optional and LogMap allows users to select among well-known lexicons depending on the application.

These indexes can be efficiently computed and bear a low memory overhead. Furthermore, they only need to be computed once for each input ontology.

2.2 Structural Indexation

LogMap exploits the information in the (extended) class hierarchy of the input ontologies in different steps of the matching process. Thus, efficient access to the information in the hierarchies is critical for LogMap's scalability.

The basic hierarchies can be computed by either using structural heuristics, or an off-the-shelf DL reasoner. LogMap bundles HermiT [17] and Condor [22], which are highly optimised for classification. Although DL classification might be computationally expensive, it is performed only once for each ontology.

The class hierarchies computed by LogMap are *extended*—that is, they contain more information than the typical classification output of DL reasoners. In particular, LogMap exploits information about explicit *disjoint classes*, as well as the information in certain complex class axioms (e.g., those stating subsumption between an intersection of named classes and a named class).

These extended hierarchies are indexed using an interval labelling schema—an optimised data structure for storing DAGs and trees [1]. The use of an interval labelling schema has been shown to significantly reduce the cost of computing typical queries over large class hierarchies [4,18].

In this context, the ontology hierarchy is treated as two DAGs: the *descendants* DAG representing the descendants relationship, and the *ancestors* DAG, which represents the ancestor relationship. Each named class C in the ontology is represented as a node in each of these DAGs, and is associated with the following information (as in [18]).

- **Descendants preorder number:** predesc(C) is the order in which C is visited using depth-first traversal of the descendants DAG.
- **Ancestors preorder number:** preanc(C) is the preorder number of C in the ancestors DAG.
- **Topological order:** deepest associated level within the descendants DAG.
- **Descendants interval:** the information about descendants of C is encoded using the interval [predesc(C), maxpredesc(C)], where maxpredesc(C) is the highest preorder number of the children of C in the descendants DAG.
- **Ancestors interval:** the information about ancestors of C is encoded using the interval [preanc(C),maxpreanc(C)] where maxpreanc(C) is the highest (ancestor) preorder number of the parents of C in the ancestors DAG.

Anatomy ⊑ ¬BiologicalProcess

TransmembraneTransport ⊑ ∃BP_hasLocation.CellularMembrane

∃BP_hasLocation.⊤ ⊑ BiologicalProcess

⊤ ⊑ ∀BP_hasLocation.Anatomy

CellularSecretion ⊑ TransmembraneTransport

ExocrineGlandFluid ⊑ ∃AS_hasLocation.ExocrineSystem

⊤ ⊑ ∀AS_hasLocation.Anatomy

∃AS_hasLocation.⊤ ⊑ Anatomy

Smegma ⊑ ExocrineGlandFluid

ExocrineGlandFluid ⊓ ExfoliatedCells ⊑ Smegma

(a) NCI ontology fragment

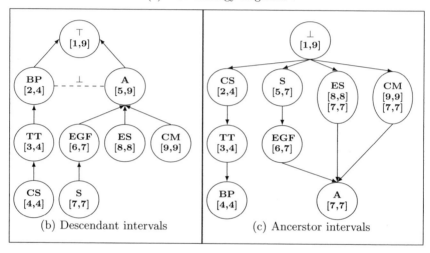

(b) Descendant intervals   (c) Ancerstor intervals

Fig. 2. NCI extended hierarchies. Abbreviations: *BP*=BiologicalProcess, *A*=Anatomy, *TT*=TransmembraneTransport, *CM*=CellularMembrane, *EGF*=ExocrineGlandFluid, *CS*=CellularSecretion, *ES*=ExocrineSystem, *S*=Smegma

Figure 2 shows a fragment of NCI and its labelled (entailed) hierarchy. Disjointness and complex class axioms are represented in a separate structure that also uses integer intervals.

The interval labelling schemas provides LogMap with an interface to efficiently answer queries about taxonomic relationships. For example, the following typical queries over ontology hierarchies only require simple integer operations (please, refer to Figure 2 for the class label abbreviations):

- 'Is Smegma *a subclass of* Anatomy?': check if predesc(S)=7 is contained in descendants interval [predesc(A), maxpredesc(A)]=[5,9];
- 'Do Smegma *and* CellularSecretion *have ancestors in common?*': check if the intersection of ancestors intervals [preanc(S), maxpreanc(S)]=[5,7] and [preanc(CS), maxpreanc(CS)]=[2,4] is non-empty.

Table 2. Fragment of the intersection between the inverted indexes for FMA and NCI

Entry	FMA ids	NCI ids	Mappings
secretion	36792	49901	FMA:Secretion $\equiv$ NCI:CellularSecretion
smegma	60947	60791	FMA:Smegma $\equiv$ NCI:Smegma
trapezoid	20948	37975, 62999	FMA:Trapezoid $\equiv$ NCI:Trapezoid FMA:Trapezoid $\equiv$ NCI:TrapezoidBone
trapezoid,bone	20948, 47996	62999	FMA:Trapezoid $\equiv$ NCI:TrapezoidBone FMA:Bone_of_Trapezoid $\equiv$ NCI:TrapezoidBone

2.3 Computing Anchor Mappings

LogMap computes an initial set of anchor mappings by simply intersecting the inverted indexes of the input ontologies (i.e., by checking whether two lexical entries in the indexes of the input ontologies contain exactly the same strings). Anchor computation is hence extremely efficient. Table 2 shows the result of intersecting the inverted indexes of Table 1, which yields five anchor mappings.

Given an anchor $m = (C_1 \equiv C_2)$, LogMap uses the string matching tool ISUB to match the neighbours of C_1 in the ontology hierarchy of $\mathcal{O}_1$ to the neighbours of C_2 in the hierarchy of $\mathcal{O}_2$. LogMap then assigns a confidence value to m by computing the proportion of matching neighbours weighted by the ISUB similarity values. This technique is based on a principle of locality: if the hierarchy neighbours of the classes in an anchor match with low confidence, then the anchor may be incorrect. For example, LogMap matches classes FMA:Trapezoid and NCI:Trapezoid (see Table 2). However, NCI:Trapezoid is classified as a polygon whereas FMA:Trapezoid is classified as a bone. LogMap assigns a low confidence to such mappings and hence they will be susceptible to be removed during repair.

2.4 Mapping Repair and Discovery

The core of LogMap is an iterative process that alternates *mapping repair* and *mapping discovery* steps. In each iteration, LogMap maintains two structures.

- A working set of *active mappings*, which are mappings that were discovered in the immediately preceding iteration. Mappings found in earlier iterations are *established*, and cannot be eliminated in the repair step. In the first iteration, the active mappings coincide with the set of anchors.
- For each anchor, LogMap maintains two *contexts* (one per input ontology), which can be expanded in different iterations. Each context consists of a set of classes and has a distinguished subset of *active classes*, which is specific to the current iteration. In the first iteration, the contexts for an anchor $C_1 \equiv C_2$ are $\{C_1\}$ and $\{C_2\}$ respectively, which are also the active classes.

Thus, active mappings are the only possible elements of a repair plan, whereas contexts constitute the basis for mapping discovery.

Table 3. Propositional representations of FMA, NCI, and the computed mappings

Propositional FMA ($\mathcal{P}_1$)		Propositional NCI ($\mathcal{P}_2$)	
(1)	Smegma → Secretion	(8)	Smegma → ExocrineGlandFluid
(2)	Secretion → PortionBodySusbstance	(9)	ExocrineGlandFluid → Anatomy
(3)	PortionBodySusbstance → AnatomicalEntity	(10)	CellularSecretion → TransmembraneTransport
Computed mappings ($\mathcal{P}_M$)		(11)	TransmembraneTransport → TransportProcess
(m_4)	FMA:Secretion → NCI:CellularSecretion	(12)	TransportProcess → BiologicalProcess
(m_5)	NCI:CellularSecretion → FMA:Secretion	(13)	Anatomy ∧ BiologicalProcess → false
(m_6)	FMA:Smegma → NCI:Smegma	(14)	ExocrineGlandFluid ∧ ExfolCells → Smegma
(m_7)	NCI:Smegma → FMA:Smegma		

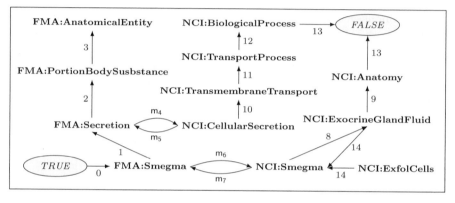

Fig. 3. Graph representation of Horn-clauses in Table 3

Mapping Repair. LogMap uses a Horn propositional logic representation of the extended hierarchy of each ontology together with all existing mappings (both active and established). As an example, Table 3 shows Horn clauses obtained from the extended hierarchies of FMA and NCI (which have been computed using a DL reasoner), and the anchor mappings computed by LogMap. As shown in the table, LogMap splits each equivalence mapping into two Horn clauses.

The use of a propositional Horn representation for unsatisfiability detection and repair is key to LogMap's scalability since DL reasoners do not scale well with the integration of large ontologies via mappings. The scalability problem is exacerbated by the number of unsatisfiable classes (more than 10,000 found by LogMap when integrating SNOMED and NCI using only anchors) and the large number of additional reasoner calls required for repairing each unsatisfiability.

Unsatisfiability checking. LogMap implements the well-known Dowling-Gallier algorithm [7] for propositional Horn satisfiability, and calls the Dowling-Gallier module once (in each repair step) for each class. Our implementation takes as input a class C (represented as a propositional variable) and determines the satisfiability of the propositional theory $\mathcal{P}_C$ consisting of

- the rule (true $\rightarrow C$);
- the propositional representations $\mathcal{P}_1$ and $\mathcal{P}_2$ (as in Table 3) of the extended hierarchies of the input ontologies $\mathcal{O}_1$ and $\mathcal{O}_2$; and
- the propositional representation $\mathcal{P}_M$ of the mappings computed thus far.

We make the following important observations concerning our encoding of the class satisfiability problem into propositional logic.

- Our encoding is *sound*. If the propositional theory $\mathcal{P}_C$ is unsatisfiable, then the class C is indeed unsatisfiable w.r.t. the DL ontology $\mathcal{O}_1 \cup \mathcal{O}_2 \cup \mathcal{M}$, where $\mathcal{O}_1$ and $\mathcal{O}_2$ are the input ontologies and $\mathcal{M}$ is the set of mappings computed so far by LogMap (represented as DL concept inclusions)
- Due to the properties of the Dowling-Gallier algorithm, our encoding is worst-case linear in the size of $\mathcal{P}_C$. Furthermore, the total number of calls to the Dowling-Gallier module is also linear in the number of classes of $\mathcal{O}_1$ and $\mathcal{O}_2$. As shown in the evaluation section, these favourable computational properties are key to the scalability of LogMap.
- Our encoding is *incomplete*, and hence we might be reporting unsatisfiable classes as satisfiable. Incompleteness is, however, mitigated by the following facts. First, the extended hierarchies of $\mathcal{O}_1$ and $\mathcal{O}_2$ have been computed using a complete reasoner and many consequences that depend on non-propositional reasoning have already been pre-computed. Second, mappings computed by LogMap (and by most ontology matching tools) correspond to Horn rules. For example, as shown in our experiments, LogMap only failed to report one unsatisfiable class for FMA-NCI (from more than 600).

A complete description of the Dowling and Gallier algorithm can be found in [7]. As an example, consider Figure 3, which shows the graph representation of all propositional clauses that are involved in the unsatisfiability of the class Smegma in FMA. Each node represents a propositional variable in Table 3; furthermore, the graph contains a directed edge labelled with a propositional rule r from variable C to variable D if the head of r is D and C occurs in the body of r. Note that there is a path from true to NCI:BiologicalProcess and a path from true to NCI:Anatomy which involve only rules with a single variable in the antecedent; furthermore, the variables NCI:BiologicalProcess and NCI:Anatomy constitute the body of rule (13), whose head is precisely false.

Computing repair plans. LogMap computes a *repair* for each unsatisfiable class identified in the input ontologies. Given an unsatisfiable class C and the propositional theory $\mathcal{P}_C$, a *repair* $\mathcal{R}$ of $\mathcal{P}_C$ is a minimal subset of the active mappings in $\mathcal{P}_M$ such that $\mathcal{P}_C \setminus \mathcal{R}$ is satisfiable.

To facilitate computation of repairs, LogMap extends Dowling-Gallier's algorithm to record all *active mappings* ($\mathcal{P}_{act}$) that may be involved in each unsatisfiability. For our example in Figure 3, LogMap records the active mappings $\mathcal{P}_{act} = \{m_4, m_5, m_6, m_7\}$, which may be relevant to the unsatisfiability of FMA:Smegma. This information is used in the subsequent repair process.

To improve scalability, repair computation is based on the 'greedy' algorithm in Table 4. Unsatisfiable classes in each ontology are ordered by their topological

Table 4. Repair in LogMap. A call to DowlingGallier returns a satisfiability value *sat* and, if *sat* = false, it optionally returns the relevant *active mappings* ($\mathcal{P}_{act}$).

Procedure Repair
Input: *List*: Ordered classes; $\mathcal{P}_1$, $\mathcal{P}_2$ and $\mathcal{P}_M$ Horn-propositional theories.
Output: $\mathcal{P}_M$: set of repaired mappings
1: **for each** $C \in List$ **do**
2:       $\mathcal{P}_C := \mathcal{P}_1 \cup \mathcal{P}_2 \cup \mathcal{P}_M \cup \{\text{true} \rightarrow C\}$
3:       $\langle sat, \mathcal{P}_{act} \rangle := \text{DowlingGallier}(\mathcal{P}_C)$
4:       **if** sat = false **then**
5:             $Repairs := \emptyset$
6:             $repair\_size := 1$
7:             **repeat**
8:                   **for each** subset $\mathcal{R}$ of $\mathcal{P}_{act}$ of size $repair\_size$ **do**
9:                         $sat := \text{DowlingGallier}(\mathcal{P}_C \setminus \mathcal{R})$
10:                         **if** sat = true **then** $Repairs := Repairs \cup \{\mathcal{R}\}$
11:                   **end for**
12:                   $repair\_size := repair\_size + 1$
13:             **until** $|Repairs| > 0$
14:             $\mathcal{R} :=$ element of $Repairs$ with minimum confidence.
15:             $\mathcal{P}_M := \mathcal{P}_M \setminus \mathcal{R}$
16:       **end if**
17: **end for**
18: **return** $\mathcal{P}_M$

level in the hierarchy. Since subclasses of an unsatisfiable class are unsatisfiable, repairing first classes high-up in the hierarchy is a well-known repair strategy.

Given each unsatisfiable class C and the relevant active mappings $\mathcal{P}_{act}$ computed using Dowling-Gallier, the algorithm identifies subsets of $\mathcal{P}_{act}$ of increasing size until a repair is found. Thus, our algorithm is guaranteed to compute all repairs of smallest size. In our example, our algorithm computes repairs $\mathcal{R}_1 = \{m_4\}$ and $\mathcal{R}_2 = \{m_6\}$ consisting of only one mapping. If more than one repair is found, LogMap selects the one with the minimum confidence value.

Finally, each equivalence mapping is split into two propositional rules, which are treated independently for repair purposes. Hence, a repair may include only one such rule, thus 'weakening' the mapping, as in the case of $\mathcal{R}_1$ and $\mathcal{R}_2$.

Mapping Discovery. LogMap computes new mappings by first expanding the contexts $\mathbf{C}_1^m$ and $\mathbf{C}_2^m$ for each anchor m, and then (incrementally) matching the classes in $\mathbf{C}_1^m$ to those in $\mathbf{C}_2^m$ using ISUB, as described next.

Context expansion. LogMap only expands contexts that are *open* (i.e., with at least one active class). The expansion of an open context is performed by adding each neighbour (in the corresponding class hierarchy) of an active class in the context. The set of active classes in each context is then reset to the empty set.

Context matching using ISUB. LogMap makes a call to ISUB for each pair of classes $C \in \mathbf{C}_1^m$ and $D \in \mathbf{C}_2^m$, but only if the same call has not been performed in previous discovery steps (for these or other contexts). Thus, LogMap never calls ISUB twice for the same input classes. We call *relevant* those 'new' lexical correspondences found by ISUB (in the current iteration) with a similarity value exceeding a given *expansion* threshold.

LogMap uses these relevant correspondences to determine the set of active classes of $\mathbf{C}_1^m$ and $\mathbf{C}_2^m$ for the next iteration as well as the set of new mappings.

- The new active classes of $\mathbf{C}_1^m$ and $\mathbf{C}_2^m$ are those that participate in some relevant correspondence.
- The current set of mappings is expanded with those relevant correspondences with similarity value exceeding a *mapping threshold* (which is higher than the expansion threshold). These new mappings will constitute the set of *active mappings* for the next repair step.

The use of ISUB allows LogMap to discover new mappings that, unlike anchors, are not lexically 'exact' (but with similarity higher than the mapping threshold). The number of ISUB tests performed is relatively small: only contexts for the same anchor are matched using ISUB, the same ISUB call is never performed twice, and context growth is limited by the expansion threshold.

2.5 Overlapping Estimation

In addition to the mappings, LogMap also returns two (hopefully small) fragments $\mathcal{O}_1'$ and $\mathcal{O}_2'$ of $\mathcal{O}_1$ and $\mathcal{O}_2$, respectively. Intuitively, $\mathcal{O}_1'$ and $\mathcal{O}_2'$ represent the 'overlapping' between $\mathcal{O}_1$ and $\mathcal{O}_2$, in the sense that each 'correct' mapping not found by LogMap is likely to involve only classes in these fragments. Thus, domain experts can focus only on $\mathcal{O}_1'$ and $\mathcal{O}_2'$ when looking for missing mappings between $\mathcal{O}_1$ and $\mathcal{O}_2$. The computation of $\mathcal{O}_1'$ and $\mathcal{O}_2'$ is performed in two steps.

1. *Computation of 'weak' anchors.* Recall that LogMap computed the initial anchors by checking whether two entries in the inverted index of $\mathcal{O}_1$ and $\mathcal{O}_2$ contained *exactly the same* set of strings (c.f., Section 2.3). For the purpose of overlapping estimation (only), LogMap also computes new anchor mappings that are 'weak' in the sense that the relevant entries in the inverted index are only required to contain *some* common string. Thus, weak anchors represent correspondences between classes that have a common lexical component.
2. *Module extraction.* The sets S_i of classes in $\mathcal{O}_i$ involved in either a weak anchor or a mapping computed by LogMap are then used as 'seed' signatures for module extraction. In particular, $\mathcal{O}_1'$ (resp. $\mathcal{O}_2'$) are computed by extracting a locality-based module [6] for S_1 in $\mathcal{O}_1$ (resp. for S_2 in $\mathcal{O}_2$).

Note that, unlike anchors, 'weak anchors' are not well-suited for mapping computation since they rarely correspond to real mappings, and hence they introduce unmanageable levels of 'noise'. For example, the discovered correspondence NCI:CommonCarotidArteryBranch ~ FMA:BranchOfCommonCochlearArtery is a weak anchor between NCI and FMA because both classes share the terms 'branch', 'common' and 'artery'; however, such correspondence is clearly not a standard mapping since none of the involved classes is subsumed by the other.

Table 5. Repairing *Gold Standards*. The ⊑ column indicates subsumption mappings. The % of total mappings includes those 'weakened' from equivalence to subsumption.

Ontologies	GS Mappings		Repaired Mappings		
	Total	Unsat.	Total	⊑	Time (s)
FMA-NCI	3,024	655	(96%) 2,898	78	10.6
FMA-SNOMED	9,072	6,179	(89%) 8,111	1,619	81.4
SNOMED-NCI	19,622	20,944	(93%) 18,322	837	812.4
Mouse-NCI$_{Anat.}$	1,520	0	1,520	-	-

3 Evaluation

We have implemented LogMap in Java and evaluated it using a standard laptop computer with 4 Gb of RAM.

We have used the following ontologies in our experiments: SNOMED CT Jan. 2009 version ($306, 591$ classes); NCI version 08.05d ($66, 724$ classes); FMA version 2.0 ($78, 989$ classes); and NCI Anatomy ($3, 304$ classes) and Mouse Anatomy ($2, 744$ classes), both from the OAEI 2010 benchmark [8]. Classification times for these ontologies were the following: 89s for SNOMED, 575s for NCI, 28s for FMA, 1s for Mouse Anatomy, and 3s for NCI Anatomy.[2] We have performed the following experiments,[3] which we describe in detail in the following sections.

1. *Repair of gold standards.* We have used LogMap's mapping repair module (c.f. Section 2.4) to automatically repair the mappings in two gold standards:
   - The mappings FMA-NCI, FMA-SNOMED and SNOMED-NCI included in UMLS Metathesaurus [2] version 2009AA;[4] and
   - the OAEI 2010 anatomy track gold standard [3].
2. *Matching large ontologies.* We have used LogMap to match the following pairs of ontologies: FMA-NCI, FMA-SNOMED, SNOMED-NCI, and Mouse Anatomy-NCI Anatomy. To the best of our knowledge, no tool has so far matched FMA, NCI and SNOMED; hence, we only compare our results with other tools for the case of Mouse Anatomy-NCI Anatomy.
3. *Overlapping estimation.* We have used LogMap to estimate the overlapping between our test ontologies as described in Section 2.5.

3.1 Repairing Gold Standards

Table 5 summarises our results. We can observe the large number of UMLS mappings between these ontologies (e.g., almost $20, 000$ for SNOMED-NCI). Using LogMap we could also detect a large number of unsatisfiable classes (ranging from 655 for FMA-NCI to $20, 944$ for SNOMED-NCI), which could be repaired efficiently (times range from 10.6s for FMA-NCI to 812.4s for SNOMED-NCI).

---

[2] We used ConDOR [22] to classify SNOMED, and HermiT [17] for the others.

[3] Output resources available in: http://www.cs.ox.ac.uk/isg/projects/LogMap/

[4] The mappings are extracted from the UMLS distribution files (see [13] for details).

Table 6. Mappings computed by LogMap

| Ontologies | Found Mapp. | | Output Mapp. | | Time (s) | |
	Total	Unsat.	Total	⊑	Anchors	Total
FMA-NCI	3,185	597	(94%) 3,000	43	28.3	69.8
FMA-SNOMED	2,068	570	(99%) 2,059	32	35.6	92.2
SNOMED-NCI	14,250	10,452	(95%) 13,562	1,540	528.6	1370.0
Mouse-NCI$_{Anat}$	1,369	32	(99%) 1,367	3	1.8	15.7

Finally, the repair process was not aggressive, as it resulted in the deletion of a small number of mappings;[5] for example, in the case of NCI and FMA LogMap preserved 96% of the original mappings, and also managed to 'weaken' 78 equivalence mappings into subsumption mappings (instead of deleting them).

We have used the reasoners HermiT and ConDOR to classify the merge of the ontologies and the repaired mappings, thus verifying the results of the repair. For FMA-NCI, we found one unsatisfiable class that was not detected by LogMap's (incomplete) reasoning algorithm. Unsatisfiability was due to a complex interaction of three 'exact' lexical mappings with axioms in NCI and FMA involving existential and universal restrictions. For FMA-SNOMED and SNOMED-NCI we could not classify the merged ontologies, so we extracted a module [6] of the mapped classes in each ontology. For FMA-SNOMED we could classify the merge of the corresponding modules and found no unsatisfiable classes. For SNOMED-NCI no reasoner could classify the merge of the modules.

In the case of the Mouse Anatomy and NCI Anatomy ontologies from OEAI, we found no unsatisfiable class using both LogMap and a DL reasoner.

3.2 Matching Large Ontologies

Table 6 summarises the results obtained when matching our test ontologies using LogMap for a default expansion threshold of 0.70 and mapping threshold of 0.95.

The second and third columns in Table 6 indicate the total number of mappings found by LogMap (in all repair-discovery iterations), and the total number of detected unsatisfiable classes, respectively. The fourth and fifth columns provide the total number of output mappings (excluding those discarded during repair) and shows how many of those mappings were 'weakened' from equivalence to simple subsumption during the repair process. We can observe that, despite the large number of unsatisfiable classes, the repair process was not aggressive and more than 94% (in the worst case) of all discovered mappings were returned as output. Finally, the last two columns show the times for anchor computation and repair, and the total matching time.[6]

Total matching time (including anchor computation and repair-discovery iterations) was less than two minutes for FMA-NCI and FMA-SNOMED. The

---

[5] The repair process in our prior work was much more aggressive [13]; for example, 63% of UMLS for SNOMED-NCI were deleted.

[6] Excluding only indexation time, which is negligible.

Table 7. Precision and recall w.r.t. Gold Standard

Ontologies	Found Mappings			Output Mappings		
	Precision	Recall	F-score	Precision	Recall	F-score
FMA-NCI	0.767	0.843	0.803	0.811	0.840	0.825
FMA-SNOMED	0.767	0.195	0.312	0.771	0.195	0.312
SNOMED-NCI	0.753	0.585	0.659	0.786	0.582	0,668
Mouse-NCI$_{Anat}$	0.917	0.826	0.870	0.918	0.826	0.870

Table 8. Missed mappings by LogMap with respect to repaired gold standard

Ontologies	GS ISUB ≥ 0.95		GS ISUB ≥ 0.80		GS ISUB ≥ 0.50	
	% Mapp.	Recall	% Mapp.	Recall	% Mapp.	Recall
FMA-NCI	88%	0.96	93%	0.90	97%	0.87
FMA-SNOMED	21%	0.95	64%	0.30	92%	0.21
SNOMED-NCI	62%	0.94	75%	0.77	89%	0.65
Mouse-NCI$_{Anat}$	75%	0.99	87%	0.95	95%	0.88

slowest result was obtained for SNOMED-NCI (20 minutes) since repair was costly due to the huge number of unsatisfiable classes. We could only compare performance with other tools for Mouse-NCI$_{Anat}$ (the largest ontology benchmark in the OAEI). LogMap matched these ontologies in 15.7 seconds, whereas the top three tools in the 2009 campaign (no official times in 2010) required 19, 23 and 10 minutes, respectively; furthermore, the CODI tool, which uses sophisticated logic-based techniques to reduce unsatisfiability, reported times between 60 to 157 minutes in the 2010 OAEI [20].

Table 7 shows precision and recall values w.r.t. our Gold Standards (the 'clean' UMLS-Mappings from our previous experiment and the mappings in the anatomy track of the OAEI 2010 benchmark). The left-hand-side of the table shows precision/recall values for the set of all mappings found by LogMap (by disabling the repair module), whereas the right-hand-side shows precision/recall for the actual set of output mappings. Our results can be summarised as follows:

- Although the main benefit of repair is to prevent logical errors, the table shows that repair also increases precision without harming recall.
- In the case of Mouse-NCI$_{Anat}$ we obtained an F-score in line with the best systems in the 2010 OAEI competition [8].
- Results for FMA-NCI were very positive, with both precision and recall exceeding 0.8. Although precision was also high for SNOMED-NCI and FMA-SNOMED, recall values were much lower, especially for FMA-SNOMED.

We have analysed the reason for the low recall values for FMA-SNOMED and SNOMED-NCI. Our hypothesis was that SNOMED is 'lexically incompatible' with FMA and NCI since it uses very different naming conventions. Results in Table 8 support this hypothesis. Table 8 shows, on the one hand, the percentage of gold standard mappings with an ISUB similarity exceeding a given threshold and, on the other hand, the recall values for LogMap w.r.t. such mappings only.

Table 9. Overlapping computed by LogMap

Ontologies	Overlapping for $\mathcal{O}_1$			Overlapping for $\mathcal{O}_2$		
$\mathcal{O}_1$-$\mathcal{O}_2$	$\mathcal{O}_1'$	% $\mathcal{O}_1$	Recall	$\mathcal{O}_2'$	% $\mathcal{O}_2$	Recall
FMA-NCI	6,512	8%	0.95	12,867	19%	0.97
FMA-SNOMED	20,278	26%	0.92	50,656	17%	0.94
SNOMED-NCI	70,705	23%	0.86	33,829	51%	0.96
Mouse-NCI$_{Anat}$	1,864	68%	0.93	1,894	57%	0.93

Note that LogMap could find in all cases more than 94% of the gold standard mappings having ISUB similarity above 0.95. However, only 21% of the gold standard FMA-SNOMED mappings exceeded this value (in contrast to 88% between FMA and NCI), showing that these ontologies use very different naming conventions. To achieve a high recall for FMA-SNOMED mappings, LogMap would need to use a mapping threshold of 0.5, which would introduce an unmanageable amount of 'noisy' mappings, thus damaging both precision and scalability.

3.3 Overlapping Estimation

Our results concerning overlapping are summarised in Table 9, where $\mathcal{O}_1'$ and $\mathcal{O}_2'$ are the fragments of the input ontologies computed by LogMap.

We can see that the output fragments are relatively small (e.g., only 8% of FMA and 19% of NCI for FMA-NCI and only 26% of FMA and 17% of SNOMED for FMA-SNOMED). Our results also confirm the hypothesis that 'correct' mappings involving an entity outside these fragments are rare. As shown in the table, a minimum of 86% and a maximum of 97% of Gold Standard UMLS mappings involve *only* classes in the computed fragments. Thus, these results confirm our hypothesis even for FMA-SNOMED and SNOMED-NCI, where LogMap could only compute a relatively small fraction of the Gold Standard mappings.

4   Conclusion and Future Work

In this paper, we have presented LogMap—a highly scalable ontology matching tool with built-in reasoning and diagnosis capabilities. LogMap's features and scalability behaviour make it well-suited for matching large-scale ontologies. LogMap, however, is still an early-stage prototype and there is plenty of room for improvement. We are currently working on further optimisations, and in the near future we are planning to integrate LogMap with a Protege-based front-end, such as the one implemented in our tool ContentMap [12].

Acknowledgements. We would like to acknowledge the funding support of the Royal Society and the EPSRC project *LogMap*, and also thank V. Nebot and R. Berlanga for their support in our first experiments with structural indexation.

References

1. Agrawal, R., Borgida, A., Jagadish, H.V.: Efficient management of transitive relationships in large data and knowledge bases. SIGMOD Rec. 18, 253–262 (1989)

2. Bodenreider, O.: The Unified Medical Language System (UMLS): integrating biomedical terminology. Nucleic Acids Research 32 (2004)
3. Bodenreider, O., Hayamizu, T.F., et al.: Of mice and men: Aligning mouse and human anatomies. In: AMIA Annu. Symp. Proc., pp. 61–65 (2005)
4. Christophides, V., Plexousakis, D., Scholl, M., Tourtounis, S.: On labeling schemes for the Semantic Web. In: Proc. of WWW, pp. 544–555. ACM (2003)
5. Cimino, J.J., Min, H., Perl, Y.: Consistency across the hierarchies of the UMLS semantic network and metathesaurus. J. of Biomedical Informatics 36(6) (2003)
6. Cuenca Grau, B., Horrocks, I., Kazakov, Y., Sattler, U.: Just the right amount: extracting modules from ontologies. In: Proc. of WWW, pp. 717–726 (2007)
7. Dowling, W.F., Gallier, J.H.: Linear-time algorithms for testing the satisfiability of propositional Horn formulae. J. Log. Program., 267–284 (1984)
8. Euzenat, J., Meilicke, C., Stuckenschmidt, H., Shvaiko, P., Trojahn, C.: Ontology Alignment Evaluation Initiative: six years of experience. J. Data Semantics (2011)
9. Geller, J., Perl, Y., Halper, M., Cornet, R.: Special issue on auditing of terminologies. Journal of Biomedical Informatics 42(3), 407–411 (2009)
10. Giunchiglia, F., Shvaiko, P., Yatskevich, M.: S-Match: an Algorithm and an Implementation of Semantic Matching. In: Bussler, C.J., Davies, J., Fensel, D., Studer, R. (eds.) ESWS 2004. LNCS, vol. 3053, pp. 61–75. Springer, Heidelberg (2004)
11. Jean-Mary, Y.R., Shironoshita, E.P., Kabuka, M.R.: Ontology matching with semantic verification. J. of Web Semantics 7(3), 235–251 (2009)
12. Jiménez-Ruiz, E., Cuenca Grau, B., Horrocks, I., Berlanga, R.: Ontology Integration Using Mappings: Towards Getting the Right Logical Consequences. In: Aroyo, L., Traverso, P., Ciravegna, F., Cimiano, P., Heath, T., Hyvönen, E., Mizoguchi, R., Oren, E., Sabou, M., Simperl, E. (eds.) ESWC 2009. LNCS, vol. 5554, pp. 173–187. Springer, Heidelberg (2009)
13. Jiménez-Ruiz, E., Cuenca Grau, B., Horrocks, I., Berlanga, R.: Logic-based assessment of the compatibility of UMLS ontology sources. J. Biomed. Sem. 2 (2011)
14. Meilicke, C., Stuckenschmidt, H.: An efficient method for computing alignment diagnoses. In: Proc. of Web Reasoning and Rule Systems, RR, pp. 182–196 (2009)
15. Meilicke, C., Stuckenschmidt, H., Tamilin, A.: Reasoning Support for Mapping Revision. J. Logic Computation 19(5), 807–829 (2009)
16. Morrey, C.P., Geller, J., Halper, M., Perl, Y.: The Neighborhood Auditing Tool: A hybrid interface for auditing the UMLS. J. of Biomedical Informatics 42(3) (2009)
17. Motik, B., Shearer, R., Horrocks, I.: Hypertableau Reasoning for Description Logics. Journal of Artificial Intelligence Research 36, 165–228 (2009)
18. Nebot, V., Berlanga, R.: Efficient retrieval of ontology fragments using an interval labeling scheme. Inf. Sci. 179(24), 4151–4173 (2009)
19. Niepert, M., Meilicke, C., Stuckenschmidt, H.: A probabilistic-logical framework for ontology matching. In: Proc. of AAAI (2010)
20. Noessner, J., Niepert, M.: CODI: Combinatorial optimization for data integration results for OAEI 2010. In: Proc. of OM Workshop (2010)
21. Reul, Q., Pan, J.Z.: KOSIMap: Use of description logic reasoning to align heterogeneous ontologies. In: Proc. of DL Workshop (2010)
22. Simancik, F., Kazakov, Y., Horrocks, I.: Consequence-based reasoning beyond Horn ontologies. In: IJCAI (2011)
23. Stoilos, G., Stamou, G.B., Kollias, S.D.: A String Metric for Ontology Alignment. In: Gil, Y., Motta, E., Benjamins, V.R., Musen, M.A. (eds.) ISWC 2005. LNCS, vol. 3729, pp. 624–637. Springer, Heidelberg (2005)

Generating Resource Profiles by Exploiting the Context of Social Annotations

Ricardo Kawase[1], George Papadakis[1,2], and Fabian Abel[3]

[1] L3S Research Center, Leibniz University Hannover, Germany
kawase@L3S.de
[2] ICCS, National Technical University of Athens, Greece
gpapadis@mail.ntua.gr
[3] Web Information Systems, TU Delft, The Netherlands
f.abel@tudelft.nl

Abstract. Typical tagging systems merely capture that part of the tagging interactions that enrich the semantics of tag assignments according to the system's purposes. The common practice is to build tag-based resource or user profiles on the basis of statistics about tags, disregarding the additional evidence that pertain to the resource, the user or the tag assignment itself. Thus, the main bulk of this valuable information is ignored when generating user or resource profiles.

In this work, we formalize the notion of tag-based and context-based resource profiles and introduce a generic strategy for building such profiles by incorporating available context information from all parts involved in a tag assignment. Our method takes into account not only the contextual information attached to the tag, the user and the resource, but also the metadata attached to the tag assignment itself. We demonstrate and evaluate our approach on two different social tagging systems and analyze the impact of several context-based resource modeling strategies within the scope of tag recommendations. The outcomes of our study suggest a significant improvement over other methods typically employed for this task.

1   Introduction

One of the most popular innovations conveyed by Web 2.0 technologies is the introduction of *tagging*, a novel method of annotating resources with relevant keywords or terms in order to describe and enrich them with useful metadata. In *resource sharing systems* like Flickr[1], users mainly attach tags to their own resources, while *social tagging systems* like Delicious[2] enable users to create *tag assignments*[3] for any resource shared with the community (i.e., free-for-all tagging [12]). Hence, there are two categories of tags: the *personalized* and the

---

[1] See http://www.flickr.com
[2] See http://www.delicious.com
[3] A tag assignment is a user-tag-resource triple that describes which user assigned which tag to which resource.

L. Aroyo et al. (Eds.): ISWC 2011, Part I, LNCS 7031, pp. 289–304, 2011.

collective ones [8]. Similarly, the benefits for the users are twofold: tags of the former category facilitate the organization and management of the resources, making search and retrieval more effective [11,20]; collective tags, on the other hand, enhance the visibility of community content, associating relevant items with the same annotation(s) [8,3].

Tag assignments are typically marked with subjectivity: different authors can interpret the same tag in different ways. Although this conveys significant benefits in the case of personalized tags, it also poses significant obstacles to the usefulness of the collective ones: the purpose of a tag assignment is not always clear to users other than its creator. For example, a tag associated with an image may describe it with respect to different aspects: the place and the persons depicted, the owner, an opinion or even its usage context (i.e., associated task). Thus, tags can be valid solely from a user-specific point of view [7]. This also explains why not all tags are suitable for search [4]; even those tags that mainly aim at describing the content of an item might characterize just a small part of the resource, without being representative of the entire resource. Some systems like LabelMe [17] and TagMe![4] [1] offer solutions to this problem by providing tags of finer granularity to their users.

In addition, tag assignments suffer from the ambiguity, inherent in any natural language: multiple meanings can be associated with the same tag (*polysemy*), while a specific tag can have multiple interpretations (*synonymy*). To disambiguate the meaning of tags for specific tag assignments, frameworks like MOAT [16] enable their users to associate each assignment with a URI specifying its meaning. This procedure is also incorporated in Faviki[5], which uses Wikipedia as the source for URIs that clarify the meaning of an annotation (i.e., *semantic tagging*). A more flexible social tagging model is maintained in TagMe!, where users can enrich tag assignments with additional facets: semantic categories, URIs and spatial information. These facets represent contextual information that contribute to the disambiguation of the tag assignments, thus facilitating the search and the recommendation of tags or resources to a great extent.

In this paper, we argue that the aforementioned shortcomings of social annotations can be ameliorated by considering their context. In a previous work, we have already demonstrated the benefits of context for recommendation strategies [2]. However, the methods presented there were tailored to a particular system and, thus, were not generalizable to other social tagging systems. Instead, this work introduces a general, versatile modeling approach that builds comprehensive resource profiles, easily adapted to any folksonomy. It exploits the contextual information that is available in tagging systems rich in metadata, which are usually neglected.

At the core of our approach lies the idea of encapsulating not only the information that exclusively pertain to tags, but also additional contextual facets that refer to the other components of a tag assignment: the user, the resource and the

---

[4] See http://tagme.groupme.org
[5] See http://faviki.com.

tag assignment itself. Merging these facets appropriately, we can derive weighted tag lists that form comprehensive contextual profiles, which are compatible and easily combined with typical tag-based profiles. These profiles can be employed in a diversity of common tasks that rely on tags, such as personalization, search and tag recommendation. We further describe how context-based profiles can be transformed into semantic URI-based profiles. We also put our generic resource modeling approaches into practice, demonstrating its applicability in two different social tagging systems: TagMe! and BibSonomy[6]. In both cases, we evaluate the impact of context-based profiles on the task of tag recommendations. The outcomes of our experimental study verify our premise that contextual profiles convey significant improvements in the performance of a social tagging system.

On the whole, the main contributions of our paper can be summarized as follows:

- we introduce the notion of tag-based and context-based resource profiles and present a generic context model for social tagging systems,
- we propose a generic strategy for exploiting context information embodied in social annotations, exemplifying it with a variety of resource modeling strategies, and
- we evaluate our strategies in two different tagging systems, verifying that the incorporation of contextual information clearly outperforms typical methods for generating resource profiles.

The remainder of the paper is structured as follows: in Section 2, we first elaborate on traditional folksonomies and recommenders proposed in the literature and then introduce our generic modeling strategy. Section 3 analyzes the potential benefits of exploiting context to model resources and describes the methodology of our experiments. In Section 4 we present and discuss the results of our evaluation, while in Section 5 we conclude the paper together with plans for future work.

2   Generating Resource Profiles

In the following, we elaborate on existing and novel strategies for generating generic resource profiles, which rely not only on the social annotations, but also on their context.

2.1   Modeling Social Annotations and Context

The structure that emerges from social annotations is called *folksonomy* [13]; it basically constitutes a set of user-tag-resource bindings, optionally coupled with the time each of them was performed [19]. In the context of our work, we consider the folksonomy model that was formally defined by Hotho et al. in [11].

---

[6] See http://www.bibsonomy.org.

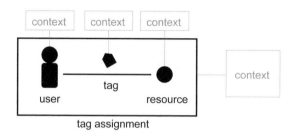

Fig. 1. Contextual information of social annotations can refer to the *user* that performed the tag assignment, to the *tag* that was designated by the user, to the *resource* that was annotated, or to the entire *tag assignment* itself.

Definition 1 (Folksonomy). *A* folksonomy *is a quadruple* $\mathbb{F} := (U, T, R, Y)$, *where* U, T, R *are finite sets of instances of* users, tags, *and* resources, *respectively.* Y *defines the* tag assignment, *which is a relation between these sets (i.e.,* $Y \subseteq U \times T \times R$) *that is potentially enriched with a timestamp indicating* when *it was performed.*

The above definition abstracts from the tagging activities and does not incorporate contextual information. The latter refers either to the entities involved in a tag assignment (i.e., the user, the tag, and the resource), or to the tag assignment itself. This is clearly illustrated in Figure 1.

In the following, we consider all possible dimensions of contextual information: the meta-data attached to the tags, to the resources and to the users, as well as the usage context attached to tag assignments, as a whole. To cover the last case, we need to accommodate the attachment of any kind of context to a tag assignment. We employ an extension of Definition 1, namely the *context folksonomy model* [1].

Definition 2 (Context Folksonomy). *A* context folksonomy *is a tuple* $\mathbb{F} := (U, T, R, Y, C, Z)$, *where:*

- *U, T, R, C are finite sets of instances of* users, tags, resources, *and* context information, *respectively,*
- *Y defines the* tag assignment, *which is a relation between U, T, and R (i.e.,* $Y \subseteq U \times T \times R$), *and*
- *Z defines the* context assignment, *which is a relation between Y and C (i.e.,* $Z \subseteq Y \times C$).

2.2   Tag-Based Profiles

At the core of this work lies the notion of folksonomy structures from the perspective of resources. Similar to a *personomy* (i.e., the user-specific part of a folksonomy, coined by Hotho et al. in [11]), we formally define the resource-specific fraction of a context folksonomy, called *personomy of a resource* from now on, as follows:

Definition 3 (Resource Personomy). *Personomy* $\mathbb{P}_r = (U_r, T_r, Y_r, C_r, Z_r)$ *of a given resource* $r \in R$ *is the restriction of* $\mathbb{F}$ *to* r, *where* U_r *and* T_r *are the finite sets of* users *and* tags, *respectively, that are referenced from the tag assignments* Y_r *that are attached to* r. C_r *comprises the contextual information that are associated with the tag assignments in* Y_r, *and* Z_r *are the corresponding context assignments.*

In essence, a resource personomy encompasses the tag assignments that refer to a specific item along with their context. In a more abstract level, the *tag-based resource profile* $P(r)$ represents a resource as a set of weighted tags.

Definition 4 (Tag-based Resource Profile). *The* tag-based profile $P(r)$ *of a resource* $r \in R$ *is a set of weighted tags, where the weight of a tag* t *is computed by a certain strategy* w *with respect to the given resource* r:

$$P(r) = \{(t, w(r, t)) | t \in T, r \in R\}, \tag{1}$$

where $w(r, t)$ *is the weight that is associated with tag* t *for a given resource* r.

$P(r)@k$ denotes the subset of a tag-based profile $P(r)$ that contains the k tag-weight pairs with the highest weights. $\bar{P}(r)$ represents a tag-based profile whose weights are normalized, so that their sum is equal to 1, while $|P(r)|$ expresses the number of distinct tags contained in $P(r)$. It is worth clarifying at this point that the tags contained in $P(r)$ are not restricted to the tags that are explicitly associated with r (i.e., the tags included in the resource's personomy $\mathbb{P}_r$). Instead, $P(r)$ may also specify the weight for a tag t_i that is associated to the resource r indirectly, through another element of its context. We illustrate this situation in Section 2.4 and Section 2.5 where we present our strategies for weighting tags.

In line with Definition 4, tag-based profiles can be built for a given user $u \in U$ and for a particular context $c \in C$, as well. For instance, tag-based user profiles (i.e., $P(u)$) have been studied by Firan et al. [6] and Michlmayr and Cayzer [14]. A straightforward approach to create a **tag-based context profile** $P(c)$ is to consider the tag assignments that pertain to c and to weight each of them according to the number of annotations that are contextualized with c and mention it. More formally: $w(c, t) = |\{(u, t, r) \in Y : (c, (u, t, r) \in Z)\}|$ (cf. Definition 2). In Section 2.5, we introduce more advanced strategies that exploit the characteristics of the respective type of context and show how these context profiles can be employed to enhance tag-based resource profiles.

2.3 Baseline Strategies for Tag-Based Resource Profiles

The main challenge in generating tag-based profiles for resources is the definition of a strategy w that appropriately assigns weights to the involved tags. In the following, we present two weighting approaches that are typically used in the literature, but do not exploit all aspects of the context of tag assignments.

Tag Frequency. The rationale behind this approach is the assumption that the more users annotate a resource r with a tag t, the more salient is t for the description of r. Given the personomy of a resource $\mathbb{P}_r$, the corresponding tag-based resource profile $P(r)$ can be formed by counting the number of distinct users that assigned at least one tag $t \in T_r$ to the resource r. Hence, the weight $w(r,t)$ attached to a specific tag t in $P(r)$ is equal to: $w(r,t) = |\{u \in U_r : (u,t,r) \in Y_r\}|$. This approach was essentially employed by Cai and Li in [5] with the aim of improving tag-based personalized search.

Tag-based Co-Occurrence. In tagging systems like Flickr, resources are typically annotated with a limited number of distinct tags [18]. For this reason, Sigurbjörnsson and Zwol suggested in [18] to enrich the profile of a resource r with those tags that frequently co-occur with the tags assigned to r (i.e., T_r). The weight of those additional tags is equal to the frequency of their co-occurrence in the folksonomy:

$$w(r,t) = |\{(u,t_i,r_j) \in Y : \exists t_i \in T_r \wedge t \in T_{r_j}\}|.$$

The second method is typically employed in the context of tag recommendation techniques, which rely on association rules to capture the co-occurrence patterns (see, for instance, a recent, state-of-the-art method, introduced by Heymann et al. in [9]). For this reason, we employ it as the baseline method in our experimental study that examines the applicability of our algorithms in the tag recommendation task.

2.4 Generic Strategy for Generating Context-Based Resource Profiles

Context-based resource profiling strategies rely on the contextual information available in folksonomies, and in resource personomies in particular: they build the profile of a resource r by merging (some of) the tag-based context profiles $P(c)$ associated with r. Moreover, one can also consider contextual information attached to the tag assignments referring to r (cf. Figure 1). The process of generating context-based resource profiles is outlined in the form of a generic approach in Definition 5.

Definition 5 (Context-based Resource Profile). *Given a tag-based profile $P(r)$ of a resource r and the set of tag-based context profiles $P(c_1),..., P(c_n)$, where $c_1, .., c_n \in C_r$ form the context information available in the resource personomy $\mathbb{P}_r$, the context-based resource profile $P_c(r)$ is computed by aggregating the tag-weight pairs (t_j, w_j) of the given profiles according to the following algorithm. Note that the parameter α_i allows for (de-)emphasizing the weights originating from profile $P(c_i)$.*

Input:
$P(r), ContextProfiles = \{(P(c_1), \alpha_1), ..., (P(c_n), \alpha_n)\}$
Initialize: $P_c(r) = P(r)$

for $(P(c_i), \alpha_i) \in ContextProfiles:$
  $P(c_i) = \bar{P}(c_i)$
  for $(t_j, w_j) \in P(c_i):$
    if $(t_j, w_{P_c(r)}) \in P_c(r):$
      replace $(t_j, w_{P_c(r)})$ in $P_c(r)$ with $(t_j, w_{P_c(r)} + \alpha_i \cdot w_j)$
    else:
      add $(t_j, \alpha_i \cdot w_j)$ to $P_c(r)$
    end
  end
end
Output: $\bar{P}_c(r)$

The above algorithm is independent from the type of context information that is exploited to construct the context-based profiles and is, thus, generalizable to any tagging system. The construction of context-based resource profiles $P_c(r)$ depends, however, on the type of context that is considered. In the following, we present several weighting strategies for building them in systems rich in metadata, like TagMe! and BibSonomy.

2.5 Domain- and Application-Specific Strategies for Generating Context-Based Resource Profiles

TagMe!. We begin with describing the strategies used to build contexts for resources in TagMe!. This system offers spatial tag assignments, enabling users to draw a rectangle that specifies the part of the image that is relevant to the corresponding tag. The resulting rectangular areas carry implicit information, which add more value to a tag assignment. Consider, for instance, the size and the distance of the tag's area from the center of the resource; the former represents the portion of the visual space that is covered by the tag, with larger areas denoting tags that are more representative of the whole resource (i.e., tags with small area pertain to a particular object depicted in the picture, whereas large areas correspond to tags describing the picture in its entirety) [1]. Similarly, the latter expresses the relevance of tag assignments to the resource: tags closer to its center might be more important than tags placed at the margin of a resource [1].

In addition to this spatial facet, TagMe! provides two additional dimensions that are suitable for building context-based resource profiles: the categories and the semantic-meaning of tags. Categories can be freely entered by users via the tagging interface, in order to provide a more general description that disambiguates and describes tags more clearly. For instance, the tag "Brandenburger Tor" can be assigned to the category "Building". In addition, TagMe! automatically enriches tags and categories assignments with DBpedia URIs to further disambiguate the meaning of a tag. In the following, we introduce strategies for building context-based profiles with the help of the tagging facets of TagMe!. Although the choice of these facets may seem rather intuitive, they have all been empirically evaluated in [1].

User-based Co-occurrence. The rationale behind this weighting method is that an individual typically annotates similar resources, thus employing relevant tags in her tag assignments. This strategy considers, therefore, all users that assigned a tag to a given resource r and aggregates all the tags that they used (even for annotating other resources) into the context-based resource profile $P(r)$. The weight $w(r,t)$ is calculated by accumulating the frequencies of the tags available in the tag-based profiles of these users:

$$w(r,t) = \sum_{u \in U_r} |\{r_k \in R : (u,t,r_k) \in Y, r_k \neq r\}|.$$

Semantic Category Frequency. This strategy considers as evidence for the significance of a tag, the popularity of the category(ies) associated with the respective tag assignment(s). The premise here is that a tag associated with a category is more important than a tag without a category and, thus, more relevant to the annotated resource. In fact, the more frequent its category is, the more relevant it is. Thus, the weight of each tag is equal to the frequency of its category. In case a tag is associated with multiple categories, its weight amounts to the sum of the respective frequencies:

$$w(r,t) = \sum_i |\{(c_i, (u_j, t_k, r_l)) \in Z : \exists(c_i, (u,t,r)) \in Z_r\}|.$$

Co-occurring Semantic Category Frequency. The incentive for this strategy is the idea that tags described by the same categories are semantically relevant to each other. Consequently, when one of them is assigned to a particular resource r, the rest are also representative of r. Given a resource r, this weighting method retrieves all categories associated with r and places all tags associated with them (even through another resource) in the profile of r, $P(r)$. In line with the previous strategy, the value of each tag is set equal to the (sum of) frequency(ies) of the related category(ies):

$$w(r,t) = \sum_i |\{(c_i, (u_i, t_j, r_k)) \in Z : c_i \in C_r \wedge \exists(c_i, (u,t,r)) \in Z_r\}|.$$

Semantic Meaning. The rationale behind this approach is the assumption that semantically annotated tags constitute the more carefully selected annotations of a resource, thus being more representative of it and the basis for a more comprehensive description. Depending on whether a tag has been linked to a URI that uniquely identifies its meaning, this strategy defines two levels of importance. In other words, it assigns a binary value to each tag, with those tags that satisfy this condition receiving the value of 1, while the rest take the value of 0. More formally:

$$w(r,t) = 1 \text{ if } \exists(URI, (u,r,t)) \in Z_r.$$

Co-occurring Semantic Meaning. At the core of this strategy lies the idea that tags that are semantically equal to, but more popular than the tags directly associated with r, are more representative of its content. Thus, given a resource r, this strategy aggregates all the URIs involved in the tag assignments of r and builds the resource profile $P(r)$ by aggregating all tags that were associated with these URIs, independently of the respective resource. Tags are weighted according to the frequency(ies) of the URI(s) assigned to them: $w(r,t) = \sum_{URI_i \in C_r} |\{(URI_i, (u_j, r_k, t_l)) \in Z : \exists(URI_i, (u,r,t)) \in Z_r\}|.$

Area Size. The intuition behind this method is that the importance of tags is proportional to their size: the larger the area occupied by a tag, the more relevant the tag is to the annotated resource. On the other hand, tags that have been associated with a particular part of a resource, are considered more specific, and thus less significant. Thus, this strategy assigns to each tag a weight proportional to its area. More formally: $w(r,t) = |x_1 - x_2| \cdot |y_1 - y_2|$, where (x_1, y_1) and (x_2, y_2) are the Cartesian coordinates of the lower left and the upper right edge of the tag's rectangle $(x_1, x_2, y_1, y_2 \in [0,1])$.

Distance From Center. This strategy is based on the assumption that the closer a tag is to the center of a resource (e.g., image), the more relevant it is. Hence, it weights tags according to their distance from the resource's central point, with smaller distances corresponding to higher values. Expressed mathematically, we have: $w(r,t) = \frac{1}{\sqrt{(x_{t_c} - x_{r_c})^2 + (y_{t_c} - y_{r_c})^2}}$, where (x_{r_c}, y_{r_c}) and (x_{t_c}, y_{t_c}) are the coordinates of the center of the resource and the center of the tag, respectively $(x_{r_c}, x_{t_c}, y_{r_c}, y_{t_c} \in [0,1])$. Note that, with respect to the annotations of the above strategy, we have $x_{t_c} = \frac{x_1 + x_2}{2}$ and $y_{t_c} = \frac{y_1 + y_2}{2}$.

It should be stressed at this point that the above strategies rely on different facets of the context folksonomy of TagMe!. Thus, instead of being competitive to each other, they are complementary and can be arbitrarily combined. In total, we can have $(2^7 - 1 =)127$ distinct strategies, either *atomic* (i.e., composed of a single weighting method) or *composite* ones (i.e., derived from the combination of multiple weighting techniques).

BibSonomy. We now further demonstrate the adaptability and generality of our approach by proposing concrete context modeling strategies for the folksonomy of BibSonomy.

Co-occurring Journal Frequency. BibSonomy resources (i.e., publications) are typically associated with the journals or conferences, where they were published. This strategy exploits these metadata information, assuming that each specific journal is focused on a particular subject that represents the aggregation of similar resources. Thus, its publications are highly relevant to each other, and the tags assigned to one of them are probably applicable to the rest, as well. Given a resource r, this weighting method retrieves the *Journal* metadata associated with r and aggregates in $P(r)$ the tags of all the resources that were published by the same journal. The value of each tag is equal to its frequency:
$w(r,t) = |\{(c_j, (u_j, t, r_l)) \in Z : \exists (c_j, (u, t, r)) \in Z_r\}|$, where c_j stands for the journal metadata of the given resource r.

Co-occurring Journal-Year Frequency. The rationale behind this strategy is the assumption that the topics of the papers published in a specific journal drift with the passage of time. As a result, the papers published in the same journal in a particular year are more relevant in with each other than with the papers published at a different point in time. In this context, this

weighting method retrieves for every resource r the *Journal* and *Year* metadata associated with it; then, it generates a list of the tags of all resources that were also published within the same journal in the same year. Tag weights are set equal to the frequency of the tags:
$w(r, t) = |\{(c_{j,y}, (u_j, t, r_l)) \in Z : \exists (c_{j,y}, (u, t, r)) \in Z_r\}|$, where $c_{j,y}$ stands for the journal and year metadata of the given resource r.

2.6   Transforming Tag-Based and Context-Based Profiles into Semantic Profiles

The aforementioned context-based modeling strategies form the basis for the creation of **semantic profiles**; these are profiles that explicitly specify the semantics of a tag by means of URIs. For social tagging systems that assign meaningful URIs to tag assignments (e.g., TagMe!) or systems that make use of the MOAT framework [16] (e.g., LODr [15]), we propose the transformation of tag-based profiles into semantic profiles that, instead of a list of tags, consist of a weighted list of URIs.

It is worth noting at this point that the semantic meaning of tags depends on the context of their use. For example, the tag "paris" most likely refers to the city, but for some tag assignments it could also refer to a person. It is not possible, therefore, to have a global mapping of tags to URIs. Instead, it is necessary to map each individual tag assignment to a particular URI. Thus, we propose to transform the personomy $\mathbb{P}_r$ (see Definition 3) and its tag assignments as follows:

Definition 6 (URI-based Resource Personomy). *Given the tag-based personomy $\mathbb{P}_r = (U_r, T_r, Y_r, C_r, Z_r)$ of a specific resource r and its URI assignments $Z_{r,uri} \subseteq Y \times C_{uri} \subseteq Z_r$, where C_{uri} is the set of URIs, the URI-based resource personomy, $\mathbb{P}_{r,uri} = (U_r, T_{r,uri}, Y_{r,uri}, C_r, Z_r)$, can be constructed by iterating over the tag assignments and replacing the tags with URIs of the corresponding URI assignments according to the following algorithm:*

$T_{r,uri} = T_r \cup C_{uri}$
$Y_{r,uri} = \{\}$
for $(u, t, r) \in Y_r$:
  for $((u, t, r), uri) \in Z_{r,uri}$:
    $Y_{r,uri} = Y_{r,uri} \cup (u, uri, r)$
  end
end
$\mathbb{P}_{r,uri} = (U_r, T_{r,uri}, Y_{r,uri}, C_r, Z_r)$

Given the URI-based Resource Personomy and a URI-based Context Folksonomy (which can be constructed in a similar manner as the semantic personomy), we can apply the resource modeling strategies presented in Sections 2.3 and 2.5 in order to generate semantic resource profiles. In this way, the resource modeling framework presented above supports tag-based tasks in both the social tagging and the Semantic Web systems.

Table 1. Technical characteristics of the TagMe! data set

Tag Assignments (TAs)	1,288
TAs with Spatial Information	671
TAs with Category Information	917
TAs with URI Information	1,050
TAs with all information	432

3   Experimental Setup

To measure the quality of the above, context-based resource modeling strategies, we apply them to the *tag recommendation* task: given a set of resources annotated with tags and metadata, the goal is to predict other tags that are also relevant to a specific resource, but have not yet been assigned to it. In the subsequent paragraphs, we describe the setup of the thorough, experimental evaluation we conducted in this context.

3.1   Social Tagging Data Sets

In the course of our experiments, we employed two real-world data sets that stem from the aforementioned social tagging applications: TagMe! and BibSonomy. A detailed description of the technical characteristics of the data sets is presented below.

TagMe! This web application constitutes a multifaceted social tagging system that allows users to associate their annotations with a variety of (optional) metadata, which are suitable for building context-based resource profiles. The data we collected comprise the whole activity of the first three weeks after the launch of the system in June, 2009. In total, its user base comprises 30 users; half of them had a Flickr account and, thus, were able to tag their own pictures, while the rest assigned tags to random pictures and pictures of their own interest. A summary of the technical characteristics of this data set is presented in Table 1.

BibSonomy. BibSonomy [10] is a social bookmarking and publication-sharing system that has been running for over four years. The resources in Bibsonomy are publications, stored in BibTeX format. Each resource has several additional metadata, such as the corresponding journal, volume, year, as well as the author names. We employed Bibsonomy's public data set that is available on-line from the 1st July 2010. It consists of 566,939 resources, described and annotated by 6,569 users. In total, there are 2,622,423 tag assignments and 189,664 unique tags. For our experimental study, we considered those resources that had the *journal* information and were tagged with at least five distinct tags. We randomly selected 500 of those resources and derived their context-based profiles from the entire data set.

3.2   Leave-one-out Cross-Validation

To evaluate the effect of context-based resource profiles on tag recommendations, we employed the leave-one-out cross-validation methodology in the following way: at each step, we hid one of the tag assignments and, then, we built the profile of the corresponding resource according to the selected strategy, based on the remaining assignments. The resulting profile encompasses a ranked list of tags, whose value is estimated according to the facets of the folksonomy that the current strategy considers. The goal is to predict the hidden tag by placing it in the top positions of the ranking.

To estimate the performance of the algorithms, we considered the following *metrics*:

Success Rate at 1 (S@1) denotes the percentage of tag predictions that had the missing tag at the first position of the ranking. It takes values in the interval $[0, 1]$, with higher values corresponding to higher performance.

Success Rate at 10 (S@10) stands for the percentage of tag predictions that had the missing tag in one of the top 10 positions of the ranking. Similar to $S@1$, it takes values in the interval $[0, 1]$, and the higher the value, the better the performance of the corresponding method.

As *baseline* strategies, we consider the approaches described in Section 2.3, which exclusively rely on the information encapsulated in tag assignments (i.e., user, tag, and resource). Note that the *tag frequency* strategy adds to a resource profile $P(r)$ only tags that have already been assigned to the resource. Consequently, it cannot be applied to the tag prediction problem without any further extension. Thus, we employ *tag-based co-occurrence* as the main baseline strategy and compare it to the context-based strategies of Section 2.5. These strategies enrich the traditional tag frequency with context-based profiles, following the process described in Definition 5.

4   Results

4.1   TagMe!

As mentioned above, the large number of facets of the TagMe! data leads to a total of 127 distinct context-based strategies. For the sake of readability and due to space limitations, we provide the results only for the atomic ones (see Definition 5) together with the best performing composite methods. It is worth noting at this point that our methods are employed as extensions to the baseline one, merging them with a weight α as described in Definition 5.

A summary of the performance of the baseline method and the atomic weighting strategies is presented in Table 2. It is evident that all context-based methods improve over the baseline, to a varying, but statistically significant extent ($p < 0.01$). The Semantic Description brings about a minor increase in $S@1$ of 2.6%, whereas the Spatial Annotation Distance and the Category-based Co-Occurrence account for an improvement well above 30%. Equally significant is

Table 2. S@1 and S@10 results for the atomic context-based strategies combined with the baseline in TagMe! data set

Context ID	Context	S@1	S@10	Context Weight
0	Baseline	0.076	0.331	–
1	User-based Co-Ocurrence	0.087	0.407	0.8
2	Spatial Annotation Size	0.089	0.408	0.4
3	Spatial Annotation Distance	0.094	0.377	0.9
4	Categorized Tag Frequency	0.085	0.352	0.5
5	Category-based Co-Occurrence	0.102	0.401	0.7
6	Semantic Description	0.078	0.407	0.8
7	Semantic-based Co-Ocurrence	0.083	0.406	0.1

the improvement with respect to the S@10 metric that varies from 6.3% for Categorized Tag Frequency context up to 23.3% for Spatial Annotation Distance. The latter indicates that annotations attached closer to the center of a resource are more valuable than those tags assigned to the margin (cf. Section 2.5).

The fourth column of Table 2 contains the optimal value of the weight used to merge the corresponding individual strategy with the baseline method. This value was determined through an exhaustive search of all values in the interval [0,2] with a step of 0.1. The actual effect of this parameter is demonstrated in Figure 2, where the performance for weight 1 (i.e., merging the baseline and the contextual strategy on an equal basis) is compared with the best performing weight. With the exception of the Semantic Description, we can notice that the calibration of this parameter conveys significant improvement, ranging from 2% for User-based Co-Occurrence to 12% for Categorized Tag Frequency.

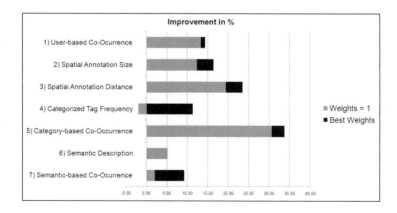

Fig. 2. S@1 improvement (in percentage) of each context over the baseline in the TagMe! data set. Gray bars show the results when the Context-Weight is set to 1, while black bars correspond to the performance of the best performing Context-Weight of each context.

Table 3. S@1 results for the composite context-based methods that have the optimal performance on the TagMe! data set. ContextIDs refer to the methods of Table 2.

ContextID (Context-Weight)	S@1	Improvement(%)
2(0.4) & 5(0.7) & 7(0.1)	0.106	38.8
2(0.4) & 3(0.1) & 5(0.7) & 7(0.1)	0.105	37.7
2(0.4) & 5(0.7)	0.105	37.7
5(0.7) & 7(0.1)	0.104	36.7

Table 4. S@1 and S@10 results for the baseline and the contextualized strategies (strategy-weight one to one) on the Bibsonomy data set

Context	S@1	Improvement(%)
Baseline	0.00712	-
Co-occurring Journal Frequency	0.00991	39.02
Co-occurring Journal-Year Frequency	0.01425	100.00
Context	**S@10**	**Improvement(%)**
Baseline	0.0701	-
Co-occurring Journal Frequency	0.0770	10.42
Co-occurring Journal-Year Frequency	0.1045	49.13

Additionally, we experimented with all possible composite strategies (i.e., combinations of the atomic ones), employing again a variety of context-weights for each of them (i.e., $w_i \in [0, 2]$ for each method i). The best performing ones are presented in Table 3, along with the respective weight and the improvement they convey with respect to $S@1$. We can see that all of them perform significantly better than the individual methods comprising them. Note, though, that they all involve the atomic strategy with the highest value for $S@1$ (i.e., Category-based Co-Occurrence) and assign to it the highest weight. However, they improve its performance by just 2%. This clearly means that merging different contexts does not result in a cumulative improvement, because their combination leads to noise in the form of contradictory evidence: a tag rated high by a specific weighting strategy can be rated lower by another one.

In summary, we can conclude that contextualized strategies that rely on the spatial features, the categories and the semantics produce the best results in the case of TagMe!. They perform individually well enough and can be slightly improved when combined with the appropriate weights. Our semantic resource profiling strategies can be applied to any other semantic tagging system, thus being reusable and appropriate for other applications, as well.

4.2 BibSonomy

The use case of BibSonomy demonstrates how our model of context-based resource profiles can be beneficially applied to any folksonomy, and how we can derive contextual information from the relations between the tag assignments. The outcomes of our evaluation are summarized in Table 4. We can see that both context-based methods substantially improve over the baseline, with the Co-occurring Journal-Year Frequency doubling its precision. Nevertheless, the overall success rate remains very low ($\sim 1\%$) in all cases. Note that the combination of the contextual weighting strategies with the baseline was done on an equal basis ($Context - Weight = 1$).

5   Conclusions

In this paper, we proposed novel approaches to generating and enriching resource profiles that exploit the multiple types of contextual information, available in most social tagging systems. We demonstrated that context can be derived from almost any metadata of the components of a tag assignment (i.e., user, tag, and resource) as well as from the tag assignment as a whole. We formalized the approach for modeling context-based profiles and described various, versatile strategies for combining them.

To verify the benefits of context-based resource profiles, we considered the task of tag recommendation, which typically relies on naive resource profiles that are derived from tag co-occurrences. We applied our strategies on two real-world datasets, with the outcomes indicating a considerable improvement over the baseline recommendation method. This verifies our premise that items sharing similar metadata (with respect to the same part of their tag assignments) are highly likely to be annotated with the same tags. We also demonstrated that contextual information pertaining to entire tag assignments provide significant evidence for modeling the resource profiles. This was proven to be particularly true for the cases where tag assignments are categorized, and spatially or semantically annotated.

Finally, we validated that merging different contexts does not result in a cumulative gain, since their arbitrary combination may lead to contradictory results. This issue actually lies at the core of our future work: we intend to develop techniques that identify complementary contexts, distinguishing them from the competitive ones. So far, there is no relevant work on this field, although techniques that optimally combine context models are expected to enhance the performance of many other tasks, as well, like personalization.

References

1. Abel, F., Henze, N., Kawase, R., Krause, D.: The Impact of Multifaceted Tagging on Learning Tag Relations and Search. In: Aroyo, L., Antoniou, G., Hyvönen, E., ten Teije, A., Stuckenschmidt, H., Cabral, L., Tudorache, T. (eds.) ESWC 2010. LNCS, vol. 6089, pp. 90–105. Springer, Heidelberg (2010)
2. Abel, F., Henze, N., Krause, D.: Exploiting additional Context for Graph-based Tag Recommendations in Folksonomy Systems. In: WI 2008, pp. 148–154. IEEE (2008)
3. Ames, M., Naaman, M.: Why we tag: motivations for annotation in mobile and online media. In: Proceedings of the SIGCHI Conference on Human Factors in Computing Systems (CHI 2007), pp. 971–980. ACM, New York (2007)

4. Bischoff, K., Firan, C., Paiu, R., Nejdl, W.: Can All Tags Be Used for Search? In: Proc. of Conf. on Information and Knowledge Management 2008, ACM (2008)

5. Cai, Y., Li, Q.: Personalized search by tag-based user profile and resource profile in collaborative tagging systems. In: CIKM, pp. 969–978 (2010)

6. Firan, C.S., Nejdl, W., Paiu, R.: The benefit of using tag-based profiles. In: Almeida, V.A.F., Baeza-Yates, R.A. (eds.) LA-WEB, pp. 32–41. IEEE Computer Society (2007)

7. Golder, S., Huberman, B.A.: The structure of collaborative tagging systems. Journal of Information Sciences 32(2), 198–208 (2006)

8. Golder, S.A., Huberman, B.A.: Usage patterns of collaborative tagging systems. J. Inf. Sci. 32(2), 198–208 (2006)

9. Heymann, P., Ramage, D., Garcia-Molina, H.: Social tag prediction. In: Myaeng, S.H., Oard, D.W., Sebastiani, F., Chua, T.S., Leong, M.K. (eds.) SIGIR, pp. 531–538. ACM (2008)

10. Hotho, A., Jäschke, R., Schmitz, C., Stumme, G.: BibSonomy: A Social Bookmark and Publication Sharing System. In: Proc. First Conceptual Structures Tool Interoperability Workshop, pp. 87–102. Aalborg (2006)

11. Hotho, A., Jäschke, R., Schmitz, C., Stumme, G.: Information Retrieval in Folksonomies: Search and Ranking. In: Sure, Y., Domingue, J. (eds.) ESWC 2006. LNCS, vol. 4011, pp. 411–426. Springer, Heidelberg (2006)

12. Marlow, C., Naaman, M., Boyd, D., Davis, M.: HT06, tagging paper, taxonomy, flickr, academic article, to read. In: Proc. of the 17th Conf. on Hypertext and Hypermedia, pp. 31–40. ACM Press (2006)

13. Vander Wal, T.: Folksonomy (2007), http://vanderwal.net/folksonomy.html

14. Michlmayr, E., Cayzer, S.: Learning User Profiles from Tagging Data and Leveraging them for Personal(ized) Information Access. In: Proc. of the Workshop on Tagging and Metadata for Social Information Organization, 16th Int. World Wide Web Conference (WWW 2007) (2007)

15. Passant, A.: LODr – A Linking Open Data Tagging System. In: First Workshop on Social Data on the Web (SDoW 2008). CEUR Workshop Proceedings, vol. 405 (2008)

16. Passant, A., Laublet, P.: Meaning Of A Tag: A collaborative approach to bridge the gap between tagging and Linked Data. In: Proceedings of the WWW 2008 Workshop Linked Data on the Web (LDOW 2008), Beijing, China (2008)

17. Russell, B.C., Torralba, A.B., Murphy, K.P., Freeman, W.T.: LabelMe: A Database and Web-based tool for Image Annotation. International Journal of Computer Vision 77(1-3), 157–173 (2008)

18. Sigurbjörnsson, B., van Zwol, R.: Flickr tag recommendation based on collective knowledge. In: Proc. of 17th Int. World Wide Web Conference (WWW 2008), pp. 327–336. ACM Press (2008)

19. Wu, X., Zhang, L., Yu, Y.: Exploring social annotations for the semantic web. In: Carr, L., Roure, D.D., Iyengar, A., Goble, C.A., Dahlin, M. (eds.) WWW, pp. 417–426. ACM (2006)

20. Xu, S., Bao, S., Fei, B., Su, Z., Yu, Y.: Exploring folksonomy for personalized search. In: SIGIR 2008, pp. 155–162. ACM, New York (2008)

Concurrent Classification of $\mathcal{EL}$ Ontologies

Yevgeny Kazakov, Markus Krötzsch, and František Simančík

Department of Computer Science, University of Oxford, UK

Abstract. We describe an optimised consequence-based procedure for classification of ontologies expressed in a polynomial fragment $\mathcal{ELH}_{\mathcal{R}+}$ of the OWL 2 EL profile. A distinguishing property of our procedure is that it can take advantage of multiple processors/cores, which increasingly prevail in computer systems. Our solution is based on a variant of the 'given clause' saturation algorithm for first-order theorem proving, where we assign derived axioms to 'contexts' within which they can be used and which can be processed independently. We describe an implementation of our procedure within the Java-based reasoner ELK. Our implementation is light-weight in the sense that an overhead of managing concurrent computations is minimal. This is achieved by employing lock-free data structures and operations such as 'compare-and-swap'. We report on preliminary experimental results demonstrating a substantial speedup of ontology classification on multi-core systems. In particular, one of the largest and widely-used medical ontologies SNOMED CT can be classified in as little as 5 seconds.

1 Introduction

Ontology classification is one of the key reasoning services used in the development of OWL ontologies. The goal of classification is to compute the hierarchical representation of the subclass (a.k.a. 'is-a') relations between the classes in the ontology based on their semantic definitions. Ontology classification, and ontology reasoning in general, is a computationally intensive task which can introduce a considerable delay into the ontology development cycle.

Many works have focused on the development of techniques to reduce classification times by optimizing the underlying (mostly tableaux-based) procedures so that they produce fewer inferences. In this paper we study another way of reducing the classification time, which is achieved by performing several inferences in parallel, i.e., concurrently. Concurrent algorithms and data structures have gained substantial practical importance due to the widespread availability of multi-core and multi-processor systems.

Nonetheless, concurrent classification of ontologies is challenging and only few works cover this subject. Approaches range from generic 'divide-and-conquer' strategies when the ontology is divided into several independent components [18] to more specific strategies involving parallel construction of taxonomies [1], concurrent execution of tableau rules [11,12], distributed resolution procedures [15], and MapReduce-based distribution approaches [14]. The practical improvements offered by these strategies, however, remain yet to be demonstrated, as

L. Aroyo et al. (Eds.): ISWC 2011, Part I, LNCS 7031, pp. 305–320, 2011.

empirical evaluation of the proposed approaches is rather limited. As of today, none of the commonly used ontology reasoners, including CEL, FaCT++, HermiT, jCEL, Pellet, and RACER, can make use of multiple processors or cores.

In this paper, we consider a 'consequence-based' classification procedure, which works by deriving logical consequences from the axioms in the ontology. Such procedures were first introduced for the $\mathcal{EL}$ family of tractable description logics (DLs) [2], which became the basis of the OWL 2 EL profile [13]. Later the $\mathcal{EL}$-style classification procedures have been formulated for more expressive languages, such as Horn-$\mathcal{SHIQ}$ [7] and $\mathcal{ALC}$ [16]. Consequence-based procedures have several distinguished properties, such as optimal worst-case complexity, 'pay-as-you-go' behaviour, and lack of non-determinism (even for expressive DLs such as $\mathcal{ALC}$). In this paper we will demonstrate that such procedures are also particularly suitable for concurrent classification of ontologies.

The contributions of this paper can be summarised as follows:

(i) We formulate a consequence-based procedure for the fragment $\mathcal{ELH}_{\mathcal{R}+}$ of an OWL 2 EL profile. The procedure does not require the usual axiom normalisation preprocessing [2] but works directly with the input ontology. Although normalisation is usually fast and its effect on the overall running time is negligible, the removal of this preprocessing step simplifies the presentation of the algorithm and reduces the implementation efforts.

(ii) We describe a concurrent strategy for saturation of the input axioms under inference rules. The strategy works by assigning axioms to 'contexts' in which the inferences can be performed independently, and can be used with arbitrary deterministic inference systems.

(iii) We describe an implementation of our concurrent saturation strategy for $\mathcal{ELH}_{\mathcal{R}+}$ within a Java-based reasoner ELK. We demonstrate empirically that the concurrent implementation can offer a significant speedup in ontology classification (e.g., a factor of 2.6 for SNOMED CT on a 4-core machine) and can outperform all existing highly-optimised (sequential) reasoners on the commonly used $\mathcal{EL}$ ontologies. The improved performance is achieved by minimising the overheads for managing concurrency and is comparable to the 'embarrassingly parallel' algorithm on the pre-partitioned input.

2   Preliminaries

The vocabulary of $\mathcal{ELH}_{\mathcal{R}+}$ consists of countably infinite sets N_R of *(atomic) roles* and N_C of *atomic concepts*. Complex *concepts* and *axioms* are defined recursively using the constructors in Table 1. We use the letters R, S, T for roles, C, D, E for concepts and A, B for atomic concepts. A *concept equivalence* $C \equiv D$ stands for the two inclusions $C \sqsubseteq D$ and $D \sqsubseteq C$. An *ontology* is a finite set of axioms. Given an ontology $\mathcal{O}$, we write $\sqsubseteq^*_{\mathcal{O}}$ for the smallest reflexive transitive binary relation over roles such that $R \sqsubseteq^*_{\mathcal{O}} S$ holds for all $R \sqsubseteq S \in \mathcal{O}$.

$\mathcal{ELH}_{\mathcal{R}+}$ has Tarski-style semantics. An *interpretation* $\mathcal{I}$ consists of a non-empty set $\Delta^{\mathcal{I}}$ called the domain of $\mathcal{I}$ and an interpretation function $\cdot^{\mathcal{I}}$ that

Table 1. Syntax and semantics of $\mathcal{ELH}_{\mathcal{R}^+}$

	Syntax	Semantics
Roles:		
atomic role	R	$R^{\mathcal{I}}$
Concepts:		
atomic concept	A	$A^{\mathcal{I}}$
top	$\top$	$\Delta^{\mathcal{I}}$
conjunction	$C \sqcap D$	$C^{\mathcal{I}} \cap D^{\mathcal{I}}$
existential restriction	$\exists R.C$	$\{x \mid \exists y : \langle x,y\rangle \in R^{\mathcal{I}} \wedge y \in C^{\mathcal{I}}\}$
Axioms:		
concept inclusion	$C \sqsubseteq D$	$C^{\mathcal{I}} \subseteq D^{\mathcal{I}}$
role inclusion	$R \sqsubseteq S$	$R^{\mathcal{I}} \subseteq S^{\mathcal{I}}$
transitive role	$\mathrm{Trans}(T)$	$T^{\mathcal{I}}$ is transitive

Table 2. Inference rules for $\mathcal{ELH}_{\mathcal{R}^+}$

$$\mathbf{R}_{\sqsubseteq} \; \frac{C \sqsubseteq D}{C \sqsubseteq E} : D \sqsubseteq E \in \mathcal{O} \qquad \mathbf{R}_{\sqcap}^+ \; \frac{C \sqsubseteq D_1 \quad C \sqsubseteq D_2}{C \sqsubseteq D_1 \sqcap D_2} : D_1 \sqcap D_2 \text{ occurs in } \mathcal{O}$$

$$\mathbf{R}_{\sqcap}^- \; \frac{C \sqsubseteq D_1 \sqcap D_2}{\begin{array}{c} C \sqsubseteq D_1 \\ C \sqsubseteq D_2 \end{array}} \qquad \mathbf{R}_{\exists}^+ \; \frac{C \sqsubseteq D}{\exists S.C \to \exists S.D} : \exists S.D \text{ occurs in } \mathcal{O}$$

$$\mathbf{R}_{\exists}^- \; \frac{C \sqsubseteq \exists R.D}{D \sqsubseteq D} \qquad \mathbf{R}_{\mathcal{H}} \; \frac{D \sqsubseteq \exists R.C \quad \exists S.C \to E}{D \sqsubseteq E} : R \sqsubseteq_{\mathcal{O}}^* S$$

$$\mathbf{R}_{\top}^+ \; \frac{C \sqsubseteq C}{C \sqsubseteq \top} : \top \text{ occurs in } \mathcal{O} \qquad \mathbf{R}_{\mathcal{T}} \; \frac{D \sqsubseteq \exists R.C \quad \exists S.C \to E \quad R \sqsubseteq_{\mathcal{O}}^* T \sqsubseteq_{\mathcal{O}}^* S}{\exists T.D \to E} : \mathrm{Trans}(T) \in \mathcal{O}$$

assigns to each R a binary relation $R^{\mathcal{I}} \subseteq \Delta^{\mathcal{I}} \times \Delta^{\mathcal{I}}$ and to each A a set $A^{\mathcal{I}} \subseteq \Delta^{\mathcal{I}}$. The interpretation function is extended to complex concepts as shown in Table 1.

An interpretation $\mathcal{I}$ *satisfies* an axiom α (written $\mathcal{I} \models \alpha$) if the corresponding condition in Table 1 holds. If an interpretation $\mathcal{I}$ satisfies all axioms in an ontology $\mathcal{O}$, then $\mathcal{I}$ is a *model* of $\mathcal{O}$ (written $\mathcal{I} \models \mathcal{O}$). An axiom α is a *consequence* of an ontology $\mathcal{O}$ (written $\mathcal{O} \models \alpha$) if every model of $\mathcal{O}$ satisfies α. A concept C is *subsumed* by D w.r.t. $\mathcal{O}$ if $\mathcal{O} \models C \sqsubseteq D$. *Classification* is the task of computing all subsumptions $A \sqsubseteq B$ between atomic concepts such that $\mathcal{O} \models A \sqsubseteq B$.

3   A Classification Procedure for $\mathcal{ELH}_{\mathcal{R}^+}$

Table 2 lists the inference rules of our classification procedure, which are closely related to the original completion rules for $\mathcal{EL}^{++}$ [2]. The rules operate with two types of axioms: (*i*) *subsumptions* $C \sqsubseteq D$ and (*ii*) *(existential) implications* $\exists R.C \to \exists S.D$, where C and D are concepts, and R and S roles. The implications $\exists R.C \to \exists S.D$ have the same semantic meaning as $\exists R.C \sqsubseteq \exists S.D$; we use the symbol $\to$ just to distinguish the two types of axioms in the inference rules.

$$\text{KneeJoint} \equiv \text{Joint} \sqcap \exists \text{isPartOf.Knee} \tag{1}$$
$$\text{LegStructure} \equiv \text{Structure} \sqcap \exists \text{isPartOf.Leg} \tag{2}$$
$$\text{Joint} \sqsubseteq \text{Structure} \tag{3}$$
$$\text{Knee} \sqsubseteq \exists \text{hasLocation.Leg} \tag{4}$$
$$\text{hasLocation} \sqsubseteq \text{isPartOf} \tag{5}$$
$$\text{Trans(isPartOf)} \tag{6}$$

Fig. 1. A simple medical ontology describing some anatomical relations

Note that we make a distinction between the premises of a rule (appearing above the horizontal line) and its side conditions (appearing after the colon), and use axioms from $\mathcal{O}$ as the side conditions, not as the premises.

It is easy to see that the inference rules are sound in the sense that every conclusion is always a logical consequence of the premises and the ontology, assuming that the side conditions are satisfied. For all rules except the last one this is rather straightforward. The last rule works similarly to the propagation of universal restrictions along transitive roles if we view axioms $\exists S.C \rightarrow E$ and $\exists T.D \rightarrow E$ as $C \sqsubseteq \forall S^-.E$ and $D \sqsubseteq \forall T^-.E$ respectively. Before we formulate a suitable form of completeness for our system, let us first consider an example.

Example 1. Consider the ontology $\mathcal{O}$ in Fig. 1 expressing that a knee joint is a joint that is a part of the knee (1), a leg structure is a structure that is a part of the leg (2), a joint is a structure (3), a knee has location in the leg (4), has-location is more specific than part-of (5), and part-of is transitive (6).

Below we demonstrate how the inference rules in Table 2 can be used to prove that a knee joint is a leg structure. We start with a tautological axiom saying that knee joint is a knee joint and then repeatedly apply the inference rules:

$$\text{KneeJoint} \sqsubseteq \text{KneeJoint} \qquad \text{input axiom,} \tag{7}$$
$$\text{KneeJoint} \sqsubseteq \text{Joint} \sqcap \exists \text{isPartOf.Knee} \qquad \text{by } \mathbf{R}_{\sqsubseteq} \ (7):(1), \tag{8}$$
$$\text{KneeJoint} \sqsubseteq \text{Joint} \qquad \text{by } \mathbf{R}_{\sqcap}^- \ (8), \tag{9}$$
$$\text{KneeJoint} \sqsubseteq \exists \text{isPartOf.Knee} \qquad \text{by } \mathbf{R}_{\sqcap}^- \ (8). \tag{10}$$

In the last axiom, we have obtained an existential restriction on knee, which now allows us to start deriving subsumption relations for this concept thanks to $\mathbf{R}_{\exists}^-$:

$$\text{Knee} \sqsubseteq \text{Knee} \qquad \text{by } \mathbf{R}_{\exists}^- \ (10), \tag{11}$$
$$\text{Knee} \sqsubseteq \exists \text{hasLocation.Leg} \qquad \text{by } \mathbf{R}_{\sqsubseteq} \ (11):(4). \tag{12}$$

Similarly, the last axiom lets us start deriving subsumptions for leg:

$$\text{Leg} \sqsubseteq \text{Leg} \qquad \text{by } \mathbf{R}_{\exists}^- \ (12). \tag{13}$$

This time, we can use (13) to derive existential implications using the fact that $\exists$hasLocation.Leg and $\exists$isPartOf.Leg occur in $\mathcal{O}$:

$$\exists\mathsf{hasLocation.Leg} \rightarrow \exists\mathsf{hasLocation.Leg} \qquad \text{by } \mathbf{R}_\exists^+ \text{ (13)}, \qquad (14)$$

$$\exists\mathsf{isPartOf.Leg} \rightarrow \exists\mathsf{isPartOf.Leg} \qquad \text{by } \mathbf{R}_\exists^+ \text{ (13)}. \qquad (15)$$

The last implication, in particular, can be used to replace the existential restriction in (12) using hasLocation $\sqsubseteq_{\mathcal{O}}^*$ isPartOf, which is a consequence of (5):

$$\mathsf{Knee} \sqsubseteq \exists\mathsf{isPartOf.Leg} \qquad \text{by } \mathbf{R}_{\mathcal{H}} \text{ (12), (15)}. \qquad (16)$$

Similarly, we can derive a new implication using hasLocation $\sqsubseteq_{\mathcal{O}}^*$ isPartOf $\sqsubseteq_{\mathcal{O}}^*$ isPartOf and transitivity of isPartOf (6):

$$\exists\mathsf{isPartOf.Knee} \rightarrow \exists\mathsf{isPartOf.Leg} \qquad \text{by } \mathbf{R}_{\mathcal{T}} \text{ (12), (15)}. \qquad (17)$$

This implication can now be used to replace the existential restriction in (10):

$$\mathsf{KneeJoint} \sqsubseteq \exists\mathsf{isPartOf.Leg} \qquad \text{by } \mathbf{R}_{\mathcal{H}} \text{ (10), (17)}. \qquad (18)$$

Finally, we "construct" the definition of leg structure (2) using (3) and the fact that the concept Structure $\sqcap \exists$isPartOf.Leg occurs in $\mathcal{O}$:

$$\mathsf{KneeJoint} \sqsubseteq \mathsf{Structure} \qquad \text{by } \mathbf{R}_{\sqsubseteq} \text{ (9)} : \text{(3)}, \qquad (19)$$

$$\mathsf{KneeJoint} \sqsubseteq \mathsf{Structure} \sqcap \exists\mathsf{isPartOf.Leg} \qquad \text{by } \mathbf{R}_{\sqcap}^+ \text{ (18), (19)}, \qquad (20)$$

$$\mathsf{KneeJoint} \sqsubseteq \mathsf{LegStructure} \qquad \text{by } \mathbf{R}_{\sqsubseteq} \text{ (20)} : \text{(2)}. \qquad (21)$$

We have thus proved that a knee joint is a leg structure.

In the above example we have demonstrated that a consequence subsumption KneeJoint $\sqsubseteq$ LegStructure can be derived using the inference rules in Table 2 once we start with a tautology KneeJoint $\sqsubseteq$ KneeJoint. It turns out that all implied subsumption axioms with KneeJoint on the left-hand-side and a concept occurring in the ontology on the right-hand-side can be derived in this way:

Theorem 1. *Let* $\mathbf{S}$ *be any set of axioms closed under the inference rules in Table 2. For all concepts C and D such that $C \sqsubseteq C \in \mathbf{S}$ and D occurs in $\mathcal{O}$ we have $\mathcal{O} \models C \sqsubseteq D$ implies $C \sqsubseteq D \in \mathbf{S}$.*

Proof (Sketch). Due to lack of space, we can only present a proof sketch. A more detailed proof can be found in the accompanying technical report [8].

The proof of Theorem 1 is by canonical model construction. We construct an interpretation $\mathcal{I}$ whose domain consists of distinct elements x_C, one element x_C for each axiom $C \sqsubseteq C$ in $\mathbf{S}$. Atomic concepts are interpreted so that $x_C \in A^{\mathcal{I}}$ iff $C \sqsubseteq A \in \mathbf{S}$. Roles are interpreted by minimal relations satisfying all role inclusion and transitivity axioms in $\mathcal{O}$ such that $\langle x_C, x_D \rangle \in R^{\mathcal{I}}$ for all $C \sqsubseteq \exists R.D \in \mathbf{S}$.

Using rules $\mathbf{R}_{\sqcap}^{-}$ and $\mathbf{R}_{\exists}^{-}$ one can prove that for all elements x_C and all concepts D we have:

$$C \sqsubseteq D \in \mathbf{S} \text{ implies } x_C \in D^{\mathcal{I}}. \tag{22}$$

Conversely, using rules $\mathbf{R}_{\top}^{+}$, $\mathbf{R}_{\sqcap}^{+}$, $\mathbf{R}_{\exists}^{+}$, $\mathbf{R}_{\mathcal{H}}$ and $\mathbf{R}_{\mathcal{T}}$ one can prove that for all concepts D occurring in $\mathcal{O}$ and all concepts C we have:

$$x_C \in D^{\mathcal{I}} \text{ implies } C \sqsubseteq D \in \mathbf{S}. \tag{23}$$

Properties (22) and (23) guarantee that $\mathcal{I}$ is a model of $\mathcal{O}$. To see this, let $D \sqsubseteq E \in \mathcal{O}$ and x_C be an arbitrary element of $\mathcal{I}$. If $x_C \in D^{\mathcal{I}}$, then $C \sqsubseteq D \in \mathbf{S}$ by (23), then $C \sqsubseteq E \in \mathbf{S}$ by rule $\mathbf{R}_{\sqsubseteq}$, so $x_C \in E^{\mathcal{I}}$ by (22). Thus $D^{\mathcal{I}} \subseteq E^{\mathcal{I}}$.

Finally, to complete the proof of Theorem 1, let C and D be such that $C \sqsubseteq C \in \mathbf{S}$ and D occurs in $\mathcal{O}$. We will show that $C \sqsubseteq D \notin \mathbf{S}$ implies $\mathcal{O} \not\models C \sqsubseteq D$. Suppose $C \sqsubseteq D \notin \mathbf{S}$. Then $x_C \notin D^{\mathcal{I}}$ by (23). Since $C \sqsubseteq C \in \mathbf{S}$, we have $x_C \in C^{\mathcal{I}}$ by (22). Hence $C^{\mathcal{I}} \not\subseteq D^{\mathcal{I}}$ and, since $\mathcal{I}$ is a model of $\mathcal{O}$, $\mathcal{O} \not\models C \sqsubseteq D$. $\qquad\square$

It follows from Theorem 1 that one can classify $\mathcal{O}$ by exhaustively applying the inference rules to the initial set of tautologies input $= \{A \sqsubseteq A \mid A$ is an atomic concept occurring in $\mathcal{O}\}$.

Corollary 1. *Let* $\mathbf{S}$ *be the closure of* input *under the inference rules in Table 2. Then for all atomic concepts A and B occurring in $\mathcal{O}$ we have $\mathcal{O} \models A \sqsubseteq B$ if and only if $A \sqsubseteq B \in \mathbf{S}$.*

Finally, we note that if all initial axioms $C \sqsubseteq D$ are such that both C and D occur in $\mathcal{O}$ (as is the case for the set input defined above), then the inference rules derive only axioms of the form (*i*) $C \sqsubseteq D$ and (*ii*) $\exists R.C \rightarrow E$ with C, D, E and R occurring in $\mathcal{O}$. There is only a polynomial number of such axioms, and all of them can be computed in polynomial time.

4   Concurrent Saturation under Inference Rules

In this section we describe a general approach for saturating a set of axioms under inference rules. We first describe a high level procedure and then introduce a refined version which facilitates concurrent execution of the inference rules.

4.1   The Basic Saturation Strategy

The basic strategy for computing a saturation of the input axioms under inference rules can be described by Algorithm 1. The algorithm operates with two collections of axioms: the queue of 'scheduled' axioms for which the rules have not been yet applied, initialized with the input axioms, and the set of 'processed' axioms for which the rules are already applied, initially empty. The queue of scheduled axioms is repeatedly processed in the while loop (lines 3–8): if the next scheduled axiom has not been yet processed (line 5), it is moved to the set of processed axioms (line 6), and every conclusion of inferences involving this

Algorithm 1. saturate(input): saturation of axioms under inference rules

Data: input: set of input axioms
Result: the saturation of input is computed in processed

```
1  scheduled ← input;
2  processed ← ∅;
3  while scheduled ≠ ∅ do
4      axiom ← scheduled.poll();
5      if not processed.contains(axiom) then
6          processed.add(axiom);
7          for conclusion ∈ deriveConclusions(processed, axiom) do
8              scheduled.put(conclusion);
```

axiom and the processed axioms is added to the queue of scheduled axioms. This strategy is closely related to the 'given clause algorithm' used in saturation-based theorem proving for first-order logic (see, e.g., [3]).

Soundness, completeness, and termination of Algorithm 1 is a consequence of the following (semi-) invariants that can be proved by induction:

(i) Every scheduled and processed axiom is either an input axiom, or is obtained by an inference rule from the previously processed axioms (soundness).

(ii) Every input axiom and every conclusion of inferences between processed axioms occurs either in the processed or scheduled axioms (completeness).

(iii) In every iteration of the while loop (lines 3–8) either the set of processed axiom increases, or, otherwise, it remains the same, but the queue of scheduled axioms becomes shorter (termination).

Therefore, when the algorithm terminates, the saturation of the input axioms under the inference rules is computed in the set of processed axioms.

The basic saturation strategy described in Algorithm 1 can already be used to compute the saturation concurrently. Indeed, the while loop (lines 3–8) can be executed from several independent workers, which repeatedly take the next axiom from the shared queue of scheduled axiom and perform inferences with the shared set of processed axioms. To remain correct with multiple workers, it is essential that Algorithm 1 adds the axiom to the set of processed axioms in line 6 before deriving conclusions with this axiom, not after that. Otherwise, it may happen that two workers simultaneously process two axioms between which an inference is possible, but will not be able to perform this inference because neither of these axioms is in the processed set.

4.2 The Refined Saturation Strategy

In order to implement Algorithm 1 in a concurrent way, one first has to ensure that the shared collections of processed and scheduled axioms can be safely accessed and modified from different workers. In particular, one worker should

---

Algorithm 2. saturate(input): saturation of axioms under inference rules

---

Data: input: set of input axioms
Result: the saturation of input is computed in context.processed

```
 1  activeContexts ← ∅;
 2  for axiom ∈ input do
 3  │   for context ∈ getContexts(axiom) do
 4  │   │   context.scheduled.add(axiom);
 5  │   └   activeContexts.activate(context);

 6  loop
 7  │   context ← activeContexts.poll();
 8  │   if context = null then break;
 9  │   loop
10  │   │   axiom ← context.scheduled.poll();
11  │   │   if axiom = null then break;
12  │   │   if not context.processed.contains(axiom) then
13  │   │   │   context.processed.add(axiom);
14  │   │   │   for conclusion ∈ deriveConclusions(context.processed, axiom) do
15  │   │   │   │   for conclusionContext ∈ getContexts(conclusion) do
16  │   │   │   │   │   conclusionContext.scheduled.add(conclusion);
17  │   │   │   │   └   activeContexts.activate(conclusionContext);

18  │   activeContexts.deactivate (context);
```

---

be able to derive conclusions in line 7 at the same time when another worker is inserting an axiom into the set of processed axioms. The easiest way to address this problem is to guard every access to the shared collection using locks. But this will largely defeat the purpose of concurrent computation, since the workers will have to wait for each other in order to proceed.

Below we describe a lock-free solution to this problem. The main idea is to distribute the axioms according to 'contexts' in which the axioms can be used as premises of inference rules and which can be processed independently by the workers. Formally, let $\mathcal{C}$ be a finite set of *contexts*, and getContexts(axiom) a function assigning a non-empty subset of contexts for every axiom such that, whenever an inference between several axioms is possible, the axioms will have at least one common context assigned to them. Furthermore, assume that every context has its own queue of scheduled axiom and a set of processed axioms (both initially empty), which we will denote by context.scheduled and context.processed.

The refined saturation strategy is described in Algorithm 2. The key idea of the algorithm is based on the notion of an active context. We say that a context is *active* if the scheduled queue for this context is not empty. The algorithm maintains the queue of active contexts to preserve this invariant. For every input axiom, the algorithm takes every context assigned to this axiom and adds this axiom to the queue of the scheduled axioms for this context (lines 2–4). Because the queue of scheduled axiom becomes non-empty, the context is *activated* by adding it to the queue of active contexts (line 5).

Afterwards, each active context is repeatedly processed in the loop (lines 6–17) by essentially performing similar operations as in Algorithm 1 but with the context-local collections of scheduled and processed axioms. The only difference is that the conclusions of inferences computed in line 14 are inserted into (possibly several) sets of scheduled axioms for the contexts assigned to each conclusion, in a similar way as it is done for the input axiom. Once the context is processed, i.e., the queue of the scheduled axioms becomes empty and the loop quits at line 8, the context is deactivated (line 18).

Similar to Algorithm 1, the main loop in Algorithm 2 (lines 6–17) can be processed concurrently by several workers. The advantage of the refined algorithm, however, is that it is possible to perform inferences in line 14 without locking the (context-local) set of processed axioms provided we can guarantee that no context is processed by more than one worker at a time. For the latter, it is sufficient to ensure that a context is never inserted into the queue of active contexts if it is already there or it is being processed by a worker. It seems that this can be easily achieved using a flag, which is set to true when a context is activated and set to false when a context is deactivated—a context is added to the queue only the first time the flag is set to true:

activeContexts.activate (context):
| if not context.isActive then
| context.isActive ← **true**;
| activeContexts.put (context);

activeContexts.deactivate (context):
| context.isActive ← **false**;

Unfortunately, this strategy does not work correctly with multiple workers: it can well be the case that two workers try to activate the same context at the same time, both see that the flag is false, set it to true, and insert the context into the queue two times. To solve this problem we would need to ensure that when two workers are trying to change the value of the flag from false to true, only one of them succeeds. This can be achieved without locking by using an atomic operation 'compare-and-swap' which tests and updates the value of the flag in one instruction. Algorithm 3 presents a safe way of activating contexts.

Algorithm 3. activeContexts.activate(context)

1 if context.isActive.compareAndSwap(**false**, **true**) then
2 | activeContexts.put (context);

Deactivation of contexts in the presence of multiple workers is also not as easy as just setting the value of the flag to false. The problem is that during the time after quitting the loop in line 8 and before deactivation of context in line 18, some other worker could insert an axiom into the queue of scheduled axioms for this context. Because the flag was set to true at that time, the context will not be inserted into the queue of active contexts, thus we end up with a context which is active in the sense of having a non-empty scheduled queue, but not activated according to the flag. To solve this problem, we perform an additional emptiness test for the scheduled axioms as shown in Algorithm 4.

Algorithm 4. activeContexts.deactivate(context)

---

1 context.isActive ← **false**;
2 **if** context.scheduled $\neq \emptyset$ **then** activeContexts.activate(context);

---

5 Implementation

In this section we describe an implementation of the inference rules for $\mathcal{ELH}_{\mathcal{R}^+}$ in Table 2 using the refined concurrent saturation strategy presented in Section 4.2. There are two functions in Algorithm 2 whose implementation we need to explain, namely getContexts(axiom) and deriveConclusions(processed, axiom).

Recall that the function getContexts(axiom) is required to assign a set of contexts to every axiom such that, whenever an inference between several axioms is possible, the premises will have at least one context in common. This is necessary in order to ensure that no inference between axioms gets lost because the inferences are applied only locally within contexts. A simple solution would be to use the inference rules themselves as contexts and assign to every axiom the set of inference rules in which the axiom can participate. Unfortunately, this strategy can provide only as many contexts as there are inference rules—not that much. To come up with a better solution, note that all premises of the inference rules in Table 2 always have a common concept denoted as C. Instead of assigning axioms with inference rules, we can assign them with the set of concepts that match the respective position of C in the rule applications. This idea leads to the implementation described in Algorithm 5.

Algorithm 5. getContexts(axiom)

---

1 result ← $\emptyset$;
  // matching the premises of the rules $\mathsf{R}_\sqsubseteq$, $\mathsf{R}_\sqcap^-$, $\mathsf{R}_\exists^-$, $\mathsf{R}_\top^+$, $\mathsf{R}_\sqcap^+$, $\mathsf{R}_\exists^+$
2 **if** axiom **match** $C \sqsubseteq D$ **then** result.add(C);
  // matching the left premise of the rules $\mathsf{R}_\mathcal{H}$ and $\mathsf{R}_\mathcal{T}$
3 **if** axiom **match** $D \sqsubseteq \exists R.C$ **then** result.add(C);
  // matching the right premise of the rules $\mathsf{R}_\mathcal{H}$ and $\mathsf{R}_\mathcal{T}$
4 **if** axiom **match** $\exists S.C \rightarrow E$ **then** result.add(C);
5 **return** result;

To implement the other function deriveConclusions(processed, axiom), we need to compute the conclusions of all inference rules in which one premise is axiom and remaining premises come from processed. A naïve implementation would iterate over all possible subsets of such premises, and try to match them to the inference rules. To avoid unnecessary enumerations, we use index data structures to quickly find applicable inference rules. For example, for checking the side conditions of the rule $\mathsf{R}_\sqcap^+$, for every concept D occurring in the ontology

---

Algorithm 6. deriveConclusions(processed, axiom) for context C

---

1 result $\leftarrow \emptyset$;
2 **if** axiom **match** $C \sqsubseteq D$ **then**
3   **for** $D_1 \in (C.\text{subsumptions} \cap D.\text{leftConjuncts})$ **do**
4   result.add$(C \sqsubseteq D_1 \sqcap D)$;   `// rule `$\mathbf{R}_{\sqcap}^{+}$`, right premise`
5   **for** $D_2 \in (C.\text{subsumptions} \cap D.\text{rightConjuncts})$ **do**
6   result.add$(C \sqsubseteq D \sqcap D_2)$;   `// rule `$\mathbf{R}_{\sqcap}^{+}$`, left premise`
       `// similarly for rules `$\mathbf{R}_{\sqsubseteq}$`, `$\mathbf{R}_{\sqcap}^{-}$`, `$\mathbf{R}_{\exists}^{-}$`, `$\mathbf{R}_{\top}^{+}$`, `$\mathbf{R}_{\exists}^{+}$
7 **if** axiom **match** $D \sqsubseteq \exists R.C$ **then**
8   **for** $S \in (C.\text{implications.keySet}() \cap R.\text{superRoles})$ **do**
9   **for** $E \in C.\text{implications.get}(S)$ **do**
10   result.add$(D \sqsubseteq E)$;   `// rule `$\mathbf{R}_{\mathcal{H}}$`, left premise`
       `// similarly for rule `$\mathbf{R}_{\mathcal{T}}$`, left premise`
11 **if** axiom **match** $\exists S.C \to E$ **then**
12   **for** $R \in (C.\text{predecessors.keySet}() \cap S.\text{subRoles})$ **do**
13   **for** $D \in C.\text{predecessors.get}(R)$ **do**
14   result.add$(D \sqsubseteq E)$;   `// rule `$\mathbf{R}_{\mathcal{H}}$`, right premise`
       `// similarly for rule `$\mathbf{R}_{\mathcal{T}}$`, right premise`
15 **return** result;

$\mathcal{O}$ we store a set of concepts with which D co-occur in conjunctions:

$$D.\text{rightConjuncts} = \{D' \mid D \sqcap D' \text{ occurs in } \mathcal{O}\},$$
$$D.\text{leftConjuncts} = \{D' \mid D' \sqcap D \text{ occurs in } \mathcal{O}\}.$$

Similarly, for checking the side condition of the rule $\mathbf{R}_{\mathcal{H}}$, for each role R occurring in $\mathcal{O}$, we precompute the sets of its sub-roles and super-roles:

$$R.\text{superRoles} = \{S \mid R \sqsubseteq_{\mathcal{O}}^{*} S\},$$
$$R.\text{subRoles} = \{S \mid S \sqsubseteq_{\mathcal{O}}^{*} R\}.$$

The index computation is usually quick and can be also done concurrently.

To identify relevant premises from the set of processed axioms for the context associated with C, we store the processed axioms for this context in three records according to the cases by which these axioms were assigned to C in Algorithm 5:

$$C.\text{subsumptions} = \{D \mid C \sqsubseteq D \in \text{processed}\},$$
$$C.\text{predecessors} = \{\langle R, D \rangle \mid D \sqsubseteq \exists R.C \in \text{processed}\},$$
$$C.\text{implications} = \{\langle R, E \rangle \mid \exists R.C \to E \in \text{processed}\}.$$

The pairs in the last two records are indexed by the key R, so that for every role R one can quickly find all concepts D and, respectively, E that occur in the

Table 3. Statistics for studied ontologies and classification times on 'laptop' for commonly used $\mathcal{EL}$ reasoners and ELK; times are in seconds; timeout was 1 hour

	SNOMED CT	GALEN	FMA	GO
#concepts	315,489	23,136	78,977	19,468
#roles	58	950	7	1
#axioms	430,844	36,547	121,712	28,897
CB	13.9	1.4	0.7	0.1
FaCT++	387.1	timeout	5.4	7.5
jCEL	661.6	32.5	12.4	2.9
Pellet	509.3	stack overflow	88.5	2.5
Snorocket	24.5	2.3	1.6	0.3
ELK (1 worker)	13.15	1.33	0.44	0.20
ELK (2 workers)	7.65	0.90	0.38	0.18
ELK (3 workers)	5.66	0.80	0.39	0.19
ELK (4 workers)	5.02	0.77	0.39	0.19

pairs with R. Given this index data structure, the function deriving conclusions within a context C can be described by Algorithm 6. Note that the algorithm extensively uses iterations over intersections of two sets; optimizing such iterations is essential for achieving efficiency of the algorithm.

6   Experimental Evaluation

To evaluate our approach, we have implemented the procedure in the Java-based reasoner ELK (version 0.1.0) using lock-free data structures, such as ConcurrentLinkedQueue, and objects such as AtomicBoolean, which allow for 'compare-and-swap' operations, provided by the java.util.concurrent package.[1] The goal of this section is (a) to compare the performance of ELK in practical situations with other popular reasoners, and (b) to study the extent in which concurrent processing contributes to the improved classification performance. To evaluate (a) we have used a notebook with Intel Core i7-2630QM 2GHz quad core CPU and 6GB of RAM, running Microsoft Windows 7 (experimental configuration 'laptop'). To evaluate (b) we have additionally used a server-type computer with an Intel Xeon E5540 2.53GHz with two quad core CPUs and 24GB of RAM running Linux 2.6.16 (experimental configuration 'server'). In both configurations we ran Java 1.6 with 4GB of heap space. All figures reported here were obtained as the average over 10 runs of the respective experiments. More detailed information for all experiments can be found in the technical report [8].

Our test ontology suite includes SNOMED CT, GO, FMA-lite, and an OWL EL version of GALEN.[2] The first part of Table 3 provides some general

---

[1] The reasoner is available open source from http://elk-reasoner.googlecode.com/
[2] SNOMED CT from http://ihtsdo.org/ (needs registration); GO from http://lat.inf.tu-dresden.de/~meng/toyont.html, FMA-lite from http://www.bioontology.org/wiki/index.php/FMAInOwl, GALEN from http://condor-reasoner.googlecode.com/

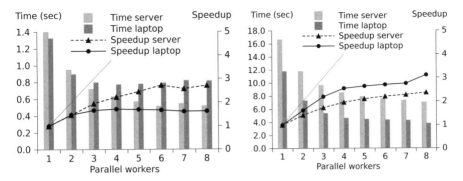

Fig. 2. Classification time and speedup for n workers on GALEN (left) and SNOMED CT (right)

statistics about the sizes of these ontologies. We have measured performance on these ontologies for the reasoners CB r.12 [7], FaCT++ 1.5.2 [19], jCEL 0.15.0,[3] Pellet 2.2.2 [17] and Snorocket 1.3.4 [10]. We selected these tools as they provide specific support for $\mathcal{EL}$-type ontologies. FaCT++ and jCEL, and Pellet were accessed through OWL API; CB, ELK, and Snorocket were accessed through their command line interface. The second part of Table 3 shows the time needed by the tested reasoners to classify the given ontologies in the 'laptop' scenario. The times are as the reasoners report for the classification stage, which does not include the loading times.[4] The last part of Table 3 presents the result for ELK tested under the same conditions for a varying number of workers. As can be seen from these results, ELK demonstrates a highly competitive performance already for 1 worker, and adding more workers can substantially improve the classification times for SNOMED CT and GALEN.

The results in Table 3 confirm that concurrent processing can offer improvements for ontology classification on common computing hardware. On the other hand, the experiments demonstrate that the improvement factor degrades with the number of workers. There can be many causes for this effect, such as dynamic CPU clocking, shared Java memory management and garbage collection, hardware bottlenecks in CPU caches or data transfer, and low-level mechanisms like Hyper-Threading. To check to what extent the overhead of managing the workers can contribute to this effect, we have performed a series of further experiments.

First, we have investigated classification performance for varying numbers of parallel workers. Figure 2 shows the results for GALEN and SNOMED CT obtained for $1 - 8$ workers on 'laptop' and 'server' configurations. The reported data is classification time (left axis) and speedup, i.e., the quotient of single worker classification time by measured multi-worker classification time (right axis). The ideal linear speedup is indicated by a dotted line. On the laptop (4

---

[3] http://jcel.sourceforge.net/

[4] Loading times can vary depending on the syntax/parser/API used and are roughly proportional to the size of the ontology; it takes about 8 seconds to load the largest tested ontology SNOMED CT in functional-style syntax by ELK.

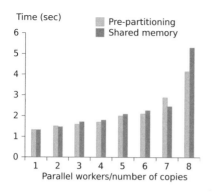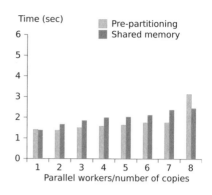

Fig. 3. Classification time for n workers on n copies of GALEN on laptop (left) and server (right)

cores), there is no significant change in the classification times for more than 4 workers, while the classification times for the server (2×4 cores) can slightly improve for up to 8 workers.

Next, we measured how our procedure would score against the 'embarrassingly parallel' algorithm in the situation when an ontology consists of n disjoint and equal components, which can be classified completely independently. We have created ontologies consisting of n disjoint copies of our test ontologies, and ran n independent ELK reasoners with 1 worker on these n copies. We compared the average classification times of this pre-partitioning approach with the times needed by ELK when using n workers on the union of these partitions. Both approaches compute exactly the same results.

The results of this experiment for copies of GALEN are presented in Fig. 3. As can be seen from these results, the classification time for both scenarios grows with n (ideally, it should remain constant up to the number of available cores, but in practice it is not the case because, e.g., the CPU clock speed drops with the load to compensate for the heat). More importantly, the difference between the two approaches is not considerable. This proves that, the performance impact of managing multiple workers is relatively small and interaction between unrelated components is avoided due to the indexing strategies discussed in Section 5. The fact that pre-partitioning requires additional time for initial analysis and rarely leads to perfect partitioning [18] suggests that our approach is more suitable for the (single machine, shared memory) scenario that we target.

7   Related Works and Conclusion

Our work is not the first one to address the problem of parallel/concurrent OWL reasoning. Notable earlier works include an approach for parallelization of (incomplete) structural reasoning algorithms [4], a tableaux procedure that concurrently explores non-deterministic choices [11], a resolution calculus for $\mathcal{ALCHIQ}$ where inferences are exchanged between distributed workers [15], and a distributed classification algorithm that can be used to concurrently invoke (serial) OWL reasoners for checking relevant subsumptions [1]. Experimental evaluations in each case indicated potential on selected examples, but further implementation and evaluation will be needed to demonstrate a clear performance advantage of these systems over state-of-the-art systems.

Some other works have studied concurrency in light-weight ontology languages. Closest to our approach is a distributed MapReduce-based algorithm for $\mathcal{EL}^+$ [14]. However, this idea has not been empericaly evaluated. Works dedicated to OWL RL [13] include an approach for pre-partitioning inputs that inspired the evaluation in Section 6 [18], and recent MapReduce approaches [6,20].

Many further works focus on the distribution of reasoning with assertional data using weaker schema-level modelling languages pD^* (a.k.a. OWL-Horst) and (fragments of) RDFS [5,21,22,9]. These works are distinguished from our approach by their goal to manage large-scale data (in the range of billions of axioms), which is beyond the memory capacity of a single machine. Accordingly, computation is distributed to many servers without memory sharing. Yet, we can find similarities in term-based distribution strategies [5,21,22,6,20,14] and indexing of rules [6] with our strategy of assigning contexts to axioms.

In conclusion, we can say that this work does indeed appear to be the first to demonstrate a compelling performance advantage for terminological reasoning in OWL through exploiting shared-memory parallelism on modern multi-core systems. We hope that these encouraging results will inspire further works in this area, by exploiting existing general techniques for parallel computing, such as the MapReduce framework, as well as new approaches for parallelization specific to OWL reasoning, such as consequence-based reasoning procedures.

Acknowledgements. Yevgeny Kazakov and František Simančík are sponsored by EPSRC grant EP/G02085X/1. Markus Krötzsch is sponsored by EPSRC grant EP/F065841/1. Some experimental results were obtained using computing resources provided by the Oxford Supercomputing Centre.

References

1. Aslani, M., Haarslev, V.: Parallel TBox classification in description logics – first experimental results. In: Proc. 19th European Conf. on Artificial Intelligence (ECAI 2010), pp. 485–490. IOS Press (2010)
2. Baader, F., Brandt, S., Lutz, C.: Pushing the $\mathcal{EL}$ envelope. In: Proc. 19th Int. Joint Conf. on Artificial Intelligence (IJCAI 2005), pp. 364–369. Professional Book Center (2005)
3. Bachmair, L., Ganzinger, H.: Resolution theorem proving. In: Handbook of Automated Reasoning, pp. 19–99. Elsevier and MIT Press (2001)
4. Bergmann, F.W., Quantz, J.: Parallelizing Description Logics. In: Wachsmuth, I., Brauer, W., Rollinger, C.-R. (eds.) KI 1995. LNCS, vol. 981, pp. 137–148. Springer, Heidelberg (1995)
5. Hogan, A., Harth, A., Polleres, A.: Scalable authoritative OWL reasoning for the Web. Int. J. of Semantic Web Inf. Syst. 5(2), 49–90 (2009)
6. Hogan, A., Pan, J.Z., Polleres, A., Decker, S.: SAOR: Template Rule Optimisations for Distributed Reasoning Over 1 Billion Linked Data Triples. In: Patel-Schneider, P.F., Pan, Y., Hitzler, P., Mika, P., Zhang, L., Pan, J.Z., Horrocks, I., Glimm, B. (eds.) ISWC 2010, Part I. LNCS, vol. 6496, pp. 337–353. Springer, Heidelberg (2010)

7. Kazakov, Y.: Consequence-driven reasoning for Horn $\mathcal{SHIQ}$ ontologies. In: Proc. 21st Int. Conf. on Artificial Intelligence (IJCAI 2009), pp. 2040–2045. IJCAI (2009)
8. Kazakov, Y., Krötzsch, M., Simančík, F.: Concurrent classification of $\mathcal{EL}$ ontologies. Tech. rep., University of Oxford (2011), http://code.google.com/p/elk-reasoner/wiki/Publications
9. Kotoulas, S., Oren, E., van Harmelen, F.: Mind the data skew: distributed inferencing by speeddating in elastic regions. In: Proc. 19th Int. Conf. on World Wide Web (WWW 2010), pp. 531–540. ACM (2010)
10. Lawley, M.J., Bousquet, C.: Fast classification in Protégé: Snorocket as an OWL 2 EL reasoner. In: Proc. 6th Australasian Ontology Workshop (IAOA 2010). Conferences in Research and Practice in Information Technology, vol. 122, pp. 45–49. Australian Computer Society Inc. (2010)
11. Liebig, T., Müller, F.: Parallelizing Tableaux-Based Description Logic Reasoning. In: Chung, S., Herrero, P. (eds.) OTM-WS 2007, Part II. LNCS, vol. 4806, pp. 1135–1144. Springer, Heidelberg (2007)
12. Meissner, A.: Experimental analysis of some computation rules in a simple parallel reasoning system for the $\mathcal{ALC}$ description logic. Int. J. of Applied Mathematics and Computer Science 21(1), 83–95 (2011)
13. Motik, B., Cuenca Grau, B., Horrocks, I., Wu, Z., Fokoue, A., Lutz, C. (eds.): OWL 2 Web Ontology Language: Profiles. W3C Recommendation (October 27, 2009), http://www.w3.org/TR/owl2-profiles/
14. Mutharaju, R., Maier, F., Hitzler, P.: A MapReduce algorithm for $\mathcal{EL}^{+}$. In: Proc. 23rd Int. Workshop on Description Logics (DL 2010). CEUR Workshop Proceedings, vol. 573, pp. 464–474. CEUR-WS.org (2010)
15. Schlicht, A., Stuckenschmidt, H.: Distributed Resolution for Expressive Ontology Retworks. In: Polleres, A., Swift, T. (eds.) RR 2009. LNCS, vol. 5837, pp. 87–101. Springer, Heidelberg (2009)
16. Simančík, F., Kazakov, Y., Horrocks, I.: Consequence-based reasoning beyond Horn ontologies. In: Proc. 22nd Int. Conf. on Artificial Intelligence (IJCAI 2011), pp. 1093–1098. AAAI Press/IJCAI (2011)
17. Sirin, E., Parsia, B., Grau, B.C., Kalyanpur, A., Katz, Y.: Pellet: A practical OWL-DL reasoner. J. of Web Semantics 5(2), 51–53 (2007)
18. Soma, R., Prasanna, V.K.: Parallel inferencing for OWL knowledge bases. In: Proc. Int. Conf. on Parallel Processing (ICPP 2008), pp. 75–82. IEEE Computer Society (2008)
19. Tsarkov, D., Horrocks, I.: FaCT++ Description Logic Reasoner: System Description. In: Furbach, U., Shankar, N. (eds.) IJCAR 2006. LNCS (LNAI), vol. 4130, pp. 292–297. Springer, Heidelberg (2006)
20. Urbani, J., Kotoulas, S., Maassen, J., van Harmelen, F., Bal, H.: WebPIE: a Webscale parallel inference engine using MapReduce. J. of Web Semantics (in press, 2011), (accepted manuscript, preprint), http://www.cs.vu.nl/~frankh/postscript/JWS11.pdf
21. Urbani, J., Kotoulas, S., Oren, E., van Harmelen, F.: Scalable Distributed Reasoning Using MapReduce. In: Bernstein, A., Karger, D.R., Heath, T., Feigenbaum, L., Maynard, D., Motta, E., Thirunarayan, K. (eds.) ISWC 2009. LNCS, vol. 5823, pp. 634–649. Springer, Heidelberg (2009)
22. Weaver, J., Hendler, J.A.: Parallel Materialization of the Finite RDFS Closure for Hundreds of Millions of Triples. In: Bernstein, A., Karger, D.R., Heath, T., Feigenbaum, L., Maynard, D., Motta, E., Thirunarayan, K. (eds.) ISWC 2009. LNCS, vol. 5823, pp. 682–697. Springer, Heidelberg (2009)

Capturing Instance Level Ontology Evolution
for DL-Lite

Evgeny Kharlamov and Dmitriy Zheleznyakov

KRDB Research Centre, Free University of Bozen-Bolzano, Italy
last_name@inf.unibz.it

Abstract. Evolution of Knowledge Bases (KBs) expressed in Description Logics (DLs) proved its importance. Recent studies of the topic mostly focussed on model-based approaches (MBAs), where an evolution (of a KB) results in a set of models. For KBs expressed in tractable DLs, such as *DL-Lite*, it was shown that the evolution suffers from inexpressibility, i.e., the result of evolution cannot be expressed in *DL-Lite*. What is missing in these studies is *understanding*: in which *DL-Lite* fragments evolution can be captured, what causes the inexpressibility, which logics is sufficient to express evolution, whether and how one can approximate it in *DL-Lite*. This work provides some understanding of these issues for eight of MBAs which cover the case of both update and revision. We found what causes inexpressibility and isolated a fragment of *DL-Lite* where evolution is expressible. For this fragment we provided polynomial-time algorithms to compute evolution results. For the general case we proposed techniques (based on what we called prototypes) to capture *DL-Lite* evolution corresponding to a well-known Winslett's approach in a DL $\mathcal{SHOIQ}$ (which is subsumed by OWL 2 DL). We also showed how to approximate this evolution in *DL-Lite*.

1 Introduction

Description Logics (DLs) provide excellent mechanisms for representing structured knowledge by means of Knowledge Bases (KBs) $\mathcal{K}$ that are composed of two components: TBox (describes intensional or general knowledge about an application domain) and ABox (describes facts about individual objects). DLs constitute the foundations for various dialects of OWL, the Semantic Web ontology language.

Traditionally DLs have been used for modeling *static* and structural aspects of application domains [1]. Recently, however, the scope of KBs has broadened, and they are now used also for providing support in the maintenance and *evolution* phase of information systems. This makes it necessary to study *evolution of Knowledge Bases* [2], where the goal is to incorporate new knowledge $\mathcal{N}$ into an existing KB $\mathcal{K}$ so as to take into account changes that occur in the underlying application domain. In general, $\mathcal{N}$ is represented by a set of formulas denoting properties that should be true after $\mathcal{K}$ has evolved, and the result of evolution, denoted $\mathcal{K} \diamond \mathcal{N}$, is also intended to be a set of formulas. In the case where $\mathcal{N}$ interacts with $\mathcal{K}$ in an undesirable way, e.g., by causing the KB or relevant parts of it to become unsatisfiable, $\mathcal{N}$ cannot be simply added to the KB. Instead, suitable changes need to be made in $\mathcal{K}$ so as to avoid this undesirable interaction, e.g., by deleting parts of $\mathcal{K}$ conflicting with $\mathcal{N}$. Different choices for changes are possible, corresponding to different approaches to semantics for KB evolution [3,4,5].

L. Aroyo et al. (Eds.): ISWC 2011, Part I, LNCS 7031, pp. 321–337, 2011.

An important group of approaches to evolution semantics, that we focus in this paper, is called *model-based* (MBAs). Under MBAs the result of evolution $\mathcal{K} \diamond \mathcal{N}$ is a *set of models* of $\mathcal{N}$ that are minimally distanced from models of $\mathcal{K}$. Depending on what the distance between models is and how to measure it, eight different MBAs were introduced (see Section 2.2 for details). Since $\mathcal{K} \diamond \mathcal{N}$ is a set of models, while $\mathcal{K}$ and $\mathcal{N}$ are logical theories, it is desirable to represent $\mathcal{K} \diamond \mathcal{N}$ as a logical theory using the same language as for $\mathcal{K}$ and $\mathcal{N}$. Thus, looking for representations of $\mathcal{K} \diamond \mathcal{N}$ is the main challenge in studies of evolution under MBAs. When $\mathcal{K}$ and $\mathcal{N}$ are propositional theories, representing $\mathcal{K} \diamond \mathcal{N}$ is well understood [5], while it becomes dramatically more complicated as soon as $\mathcal{K}$ and $\mathcal{N}$ are first-order, e.g., DL KBs [6].

Model based evolution of KBs where $\mathcal{K}$ and $\mathcal{N}$ are written in a language of the *DL-Lite* family [7] has been recently extensively studied [6,8,9]. The focus on *DL-Lite* is not surprising since it is the basis of OWL 2 QL, a tractable OWL 2 profile. It has been shown that for every of the eight MBAs one can find *DL-Lite* $\mathcal{K}$ and $\mathcal{N}$ such that $\mathcal{K} \diamond \mathcal{N}$ cannot be expressed in *DL-Lite* [10,11], i.e., *DL-Lite* is *not closed* under MBA evolution. This phenomenon was also noted in [6,10] for some of the eight semantics. What is missing in all these studies of evolution for *DL-Lite* is *understanding* of

(1) *DL-Lite wrt evolution*: What *DL-Lite* fragments are closed under MBAs? What *DL-Lite* formulas are in charge of inexpressibility?

(2) *Evolution wrt DL-Lite* : Is it possible and how to capture evolution of *DL-Lite* KBs in richer logics? What are these logics?

(3) *Approximation of evolution results:* For *DL-Lite* KB $\mathcal{K}$ and an ABox $\mathcal{N}$, is it possible and how to do "good" approximations of $\mathcal{K} \diamond \mathcal{N}$ in *DL-Lite*?

In this paper we study the problems *(1)-(3)* for so-called *ABox evolution*, i.e., $\mathcal{N}$ is a new ABox and the TBox of $\mathcal{K}$ should remain the same after the evolution. ABox evolution is important for areas, e.g., artifact-centered service interoperation (http://www.acsi-project.eu/), where the structural knowledge (TBox) is well crafted and stable, while (ABox) facts about individuals may get changed. These ABox changes should be reflected in KBs in a way that the TBox is not affected. Our study covers both the case of ABox updates and ABox revision [4].

The contributions of the paper are: We provide relationships between MBAs for *DL-Lite$_{\mathcal{R}}$* by showing which approaches subsume each other (Section 3). We introduce *DL-Lite$_{\mathcal{R}}^{pr}$*, a restriction on *DL-Lite$_{\mathcal{R}}$* where disjointness of concepts with role projections is forbidden. We show that *DL-Lite$_{\mathcal{R}}^{pr}$* is closed under most of MBA evolutions and provide polynomial-time algorithms to compute (representations of) $\mathcal{K} \diamond \mathcal{N}$ (Section 4). For *DL-Lite$_{\mathcal{R}}$* we focus on an important MBA corresponding to a well accepted Winslett's semantics and show how to capture $\mathcal{K} \diamond \mathcal{N}$ for this semantics in a DL $\mathcal{SHOIQ}$ (Section 5). We show what combination of assertions in $\mathcal{T}$ together with $\mathcal{N}$ can lead to inexpressibility of $(\mathcal{T}, \mathcal{A}) \diamond \mathcal{N}$ in *DL-Lite$_{\mathcal{R}}$* (Section 5.1). For the case when $\mathcal{K} \diamond \mathcal{N}$ is not expressible in *DL-Lite$_{\mathcal{R}}$* we study how to approximate it in *DL-Lite$_{\mathcal{R}}$* (Section 5.4).

2   Preliminaries

2.1   DL-Lite$_\mathcal{R}$

We introduce some basic notions of DLs, (see [1] for more details). We consider a logic *DL-Lite$_\mathcal{R}$* of *DL-Lite* family of DLs [7,12]. *DL-Lite$_\mathcal{R}$* has the following constructs for (complex) *concepts* and *roles*: *(i)* $B ::= A \mid \exists R$, *(ii)* $C ::= B \mid \neg B$, *(iii)* $R ::= P \mid P^-$, where A and P stand for an *atomic concept* and *role*, respectively, which are just names. A *knowledge base* (KB) $\mathcal{K} = (\mathcal{T}, \mathcal{A})$ is compounded of two sets of *assertions*: TBox $\mathcal{T}$, and ABox $\mathcal{A}$. *DL-Lite$_\mathcal{R}$* TBox assertions are *concept inclusion assertions* of the form $B \sqsubseteq C$ and *role inclusion assertions* $R_1 \sqsubseteq R_2$, while ABox assertions are *membership assertions* of the form $A(a)$, $\neg A(a)$, and $R(a, b)$. The *active domain* of $\mathcal{K}$, denoted $adom(\mathcal{K})$, is the set of all constants occurring in $\mathcal{K}$. In Section 5 we will also talk about a DL $\mathcal{SHOIQ}$ [1] while we do not define it here due to space limit.

The semantics of DL-Lite KBs is given in the standard way, using first order interpretations, all over the same countable domain Δ. An *interpretation* $\mathcal{I}$ is a function $\cdot^\mathcal{I}$ that assigns to each C a subset $C^\mathcal{I}$ of Δ, and to R a binary relation $R^\mathcal{I}$ over Δ in a way that $(\neg B)^\mathcal{I} = \Delta \setminus B^\mathcal{I}$, $(\exists R)^\mathcal{I} = \{a \mid \exists a'.(a, a') \in R^\mathcal{I}\}$, and $(P^-)^\mathcal{I} = \{(a_2, a_1) \mid (a_1, a_2) \in P^\mathcal{I}\}$. We assume that Δ contains the constants and that $c^\mathcal{I} = c$ (we adopt *standard names*). Alternatively, we view interpretations as sets of atoms and say that $A(a) \in \mathcal{I}$ iff $a \in A^\mathcal{I}$ and $P(a, b) \in \mathcal{I}$ iff $(a, b) \in P^\mathcal{I}$. An interpretation $\mathcal{I}$ is a *model* of a membership assertion $A(a)$ (resp., $\neg A(a)$) if $a \in A^\mathcal{I}$ (resp., $a \notin A^\mathcal{I}$), of $P(a, b)$ if $(a, b) \in P^\mathcal{I}$, and of an assertion $D_1 \sqsubseteq D_2$ if $D_1^\mathcal{I} \subseteq D_2^\mathcal{I}$.

As usual, we use $\mathcal{I} \models F$ to denote that $\mathcal{I}$ is a model of an assertion F, and $\mathcal{I} \models \mathcal{K}$ denotes that $\mathcal{I} \models F$ for each F in $\mathcal{K}$. We use $Mod(\mathcal{K})$ to denote the set of all models of $\mathcal{K}$. A KB is *satisfiable* if it has at least one model. The *DL-Lite* family has nice computational properties, for example, KB satisfiability has polynomial-time complexity in the size of the TBox and logarithmic-space in the size of the ABox [12,13]. We use entailment on KBs $\mathcal{K} \models \mathcal{K}'$ in the standard sense. An ABox $\mathcal{A}$ $\mathcal{T}$-*entails* an ABox $\mathcal{A}'$, denoted $\mathcal{A} \models_\mathcal{T} \mathcal{A}'$, if $\mathcal{T} \cup \mathcal{A} \models \mathcal{A}'$, and $\mathcal{A}$ is $\mathcal{T}$-*equivalent* to $\mathcal{A}'$, denoted $\mathcal{A} \equiv_\mathcal{T} \mathcal{A}'$, if $\mathcal{A} \models_\mathcal{T} \mathcal{A}'$ and $\mathcal{A}' \models_\mathcal{T} \mathcal{A}$.

The deductive *closure of a TBox* $\mathcal{T}$, denoted $cl(\mathcal{T})$, is the set of all TBox assertions F such that $\mathcal{T} \models F$. For satisfiable KBs $\mathcal{K} = (\mathcal{T}, \mathcal{A})$, a *full closure of $\mathcal{A}$* (wrt $\mathcal{T}$), denoted $fcl_\mathcal{T}(\mathcal{A})$, is the set of all membership assertions f (both positive and negative) over $adom(\mathcal{K})$ such that $\mathcal{A} \models_\mathcal{T} f$. In *DL-Lite$_\mathcal{R}$* both $cl(\mathcal{T})$ and $fcl_\mathcal{T}(\mathcal{A})$ are computable in time quadratic in, respectively, $|\mathcal{T}|$, i.e., the number of assertions of $\mathcal{T}$, and $|\mathcal{T} \cup \mathcal{A}|$. In our work we assume that all TBoxes and ABoxes are closed, while results are extendable to arbitrarily KBs.

A *homomorphism* h from a model $\mathcal{I}$ to a model $\mathcal{J}$ is a mapping from Δ to Δ satisfying: *(i)* $h(a) = a$ for every constant a; *(ii)* if $\alpha \in A^\mathcal{I}$ (resp., $(\alpha, \beta) \in P^\mathcal{I}$), then $h(\alpha) \in A^\mathcal{J}$ (resp., $(h(\alpha), h(\beta)) \in P^\mathcal{J}$) for every A (resp., P). A canonical model of $\mathcal{K}$ is a model which can be homomorphically embedded in every model of $\mathcal{K}$, denoted $\mathcal{I}_\mathcal{K}^{can}$ or just $\mathcal{I}^{can}$ when $\mathcal{K}$ is clear from the context.

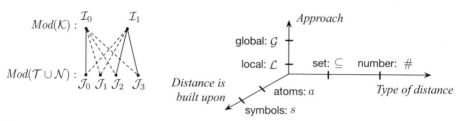

Fig. 1. Left: measuring distances between models and finding local minimums.
Right: three-dimensional space of approaches to model-based evolution semantics.

2.2 Evolution of Knowledge Bases

This section is based on [10]. Let $\mathcal{K} = (\mathcal{T}, \mathcal{A})$ be a *DL-Lite*$_{\mathcal{R}}$ KB and $\mathcal{N}$ a "new" ABox.
We study how to incorporate $\mathcal{N}$'s assertions into $\mathcal{K}$, that is, how $\mathcal{K}$ *evolves* under $\mathcal{N}$ [2].
More practically, we study *evolution operators* that take $\mathcal{K}$ and $\mathcal{N}$ as input and return,
possibly in *polynomial time*, a *DL-Lite*$_{\mathcal{R}}$ $\mathcal{K}' = (\mathcal{T}, \mathcal{A}')$ (with the same TBox as $\mathcal{K}$) that
captures the evolution, and which we call *the (ABox) evolution of $\mathcal{K}$ under $\mathcal{N}$*. Based
on the evolution principles of [10], we require $\mathcal{K}$ and $\mathcal{K}'$ to be satisfiable. A *DL-Lite*$_{\mathcal{R}}$
KB $\mathcal{K} = (\mathcal{T}, \mathcal{A})$ and an ABox $\mathcal{N}$ is a *evolution setting* if $\mathcal{K}$ and $(\mathcal{T}, \mathcal{N})$ are satisfiable.

Model-Based Semantics of Evolution. In model-based approaches (MBAs), the result
of evolution of a KB $\mathcal{K}$ wrt new knowledge $\mathcal{N}$ is a set $\mathcal{K} \diamond \mathcal{N}$ of models. The idea of
MBAs is to choose as $\mathcal{K} \diamond \mathcal{N}$ some models of $(\mathcal{T}, \mathcal{N})$ depending on their distance to $\mathcal{K}$'s
models. Katsuno and Mendelzon [4] considered two ways, so called *local* and *global*,
of choosing these models of $(\mathcal{T}, \mathcal{N})$, where the first choice corresponds to *knowledge
update* and the second one to *knowledge revision*.
  The idea of the local approaches is to consider all models $\mathcal{I}$ of $\mathcal{K}$ one by one and for
each $\mathcal{I}$ to take those models $\mathcal{J}$ of $(\mathcal{T}, \mathcal{N})$ that are minimally distant from $\mathcal{I}$. Formally,

$$\mathcal{K} \diamond \mathcal{N} = \bigcup_{\mathcal{I} \in Mod(\mathcal{K})} \mathcal{I} \diamond \mathcal{N}, \text{ where } \mathcal{I} \diamond \mathcal{N} = \underset{\mathcal{J} \in Mod(\mathcal{T} \cup \mathcal{N})}{\arg\min} dist(\mathcal{I}, \mathcal{J}).$$

where $dist(\cdot, \cdot)$ is a function whose range is a partially ordered domain and $\arg\min$
stands for the *argument of the minimum,* that is, in our case, the set of models $\mathcal{J}$
for which the value of $dist(\mathcal{I}, \mathcal{J})$ reaches its minimum value, given $\mathcal{I}$. The distance
function $dist$ varies from approach to approach and commonly takes as values either
numbers or subsets of some fixed set. To get a better intuition of the local semantics,
consider Figure 1, left, where we present two model $\mathcal{I}_0$ and $\mathcal{I}_1$ of a KB $\mathcal{K}$ and four
models $\mathcal{J}_0, \dots, \mathcal{J}_3$ of $(\mathcal{T}, \mathcal{N})$. We represent the distance between a model of $\mathcal{K}$ and
a model of $\mathcal{T} \cup \mathcal{N}$ by the length of a line connecting them. Solid lines correspond to
minimal distances, dashed lines to distances that are not minimal. In this figure $\{\mathcal{J}_0\} =$
$\arg\min_{\mathcal{J} \in \{\mathcal{J}_0, \dots, \mathcal{J}_3\}} dist(\mathcal{I}_0, \mathcal{J})$ and $\{\mathcal{J}_2, \mathcal{J}_3\} = \arg\min_{\mathcal{J} \in \{\mathcal{J}_0, \dots, \mathcal{J}_3\}} dist(\mathcal{I}_1, \mathcal{J})$.
  In the global approach one choses models of $\mathcal{T} \cup \mathcal{N}$ that are minimally distant from
$\mathcal{K}$:

$$\mathcal{K} \diamond \mathcal{N} = \underset{\mathcal{J} \in Mod(\mathcal{T} \cup \mathcal{N})}{\arg\min} dist(Mod(\mathcal{K}), \mathcal{J}), \tag{1}$$

where $dist(Mod(\mathcal{K}), \mathcal{J}) = \min_{\mathcal{I} \in Mod(\mathcal{K})} dist(\mathcal{I}, \mathcal{J})$. Consider again Figure 1, left, and assume that the distance between $\mathcal{I}_0$ and $\mathcal{J}_0$ is the global minimum, hence, $\{\mathcal{J}_0\} = \arg\min_{\mathcal{J} \in \{\mathcal{J}_0, \ldots, \mathcal{J}_3\}} dist(\{\mathcal{I}_0, \mathcal{I}_1\}, \mathcal{J})$.

Measuring Distance Between Interpretations. The classical MBAs were developed for propositional theories [5], where interpretation were sets of propositional atoms, two distance functions were introduced, respectively based on symmetric difference "$\ominus$" and on the cardinality of symmetric difference:

$$dist_{\subseteq}(\mathcal{I}, \mathcal{J}) = \mathcal{I} \ominus \mathcal{J} \qquad \text{and} \qquad dist_{\#}(\mathcal{I}, \mathcal{J}) = |\mathcal{I} \ominus \mathcal{J}|, \qquad (2)$$

where $\mathcal{I} \ominus \mathcal{J} = (\mathcal{I} \setminus \mathcal{J}) \cup (\mathcal{J} \setminus \mathcal{I})$. Distances under $dist_{\subseteq}$ are sets and are compared by set inclusion, that is, $dist_{\subseteq}(\mathcal{I}_1, \mathcal{J}_1) \leq dist_{\subseteq}(\mathcal{I}_2, \mathcal{J}_2)$ iff $dist_{\subseteq}(\mathcal{I}_1, \mathcal{J}_1) \subseteq dist_{\subseteq}(\mathcal{I}_2, \mathcal{J}_2)$. Finite distances under $dist_{\#}$ are natural numbers and are compared in the standard way.

One can extend these distances to DL interpretations in two different ways. One way is to consider interpretations $\mathcal{I}$, $\mathcal{J}$ as sets of *atoms*. Then $\mathcal{I} \ominus \mathcal{J}$ is again a set of atoms and we can define distances as in Equation (2). We denote these distances as $dist_{\subseteq}^a(\mathcal{I}, \mathcal{J})$ and $dist_{\#}^a(\mathcal{I}, \mathcal{J})$. Another way is to define distances at the level of the concept and role *symbols* in the signature Σ underlying the interpretations:

$$dist_{\subseteq}^s(\mathcal{I}, \mathcal{J}) = \{S \in \Sigma \mid S^{\mathcal{I}} \neq S^{\mathcal{J}}\}, \quad \text{and} \quad dist_{\#}^s(\mathcal{I}, \mathcal{J}) = |\{S \in \Sigma \mid S^{\mathcal{I}} \neq S^{\mathcal{J}}\}|.$$

Summing up across the different possibilities, we have three dimensions, which give eight semantics of evolution according to MBAs by choosing: (1) the *local* or the *global* approach, (2) *atoms* or *symbols* for defining distances, and (3) *set inclusion* or *cardinality* to compare symmetric differences. In Figure 1, right, we depict these three dimensions. We denote each of these eight possibilities by a combination of three symbols, indicating the choice in each dimension, e.g., $\mathcal{L}_{\#}^a$ denotes the local semantics where the distances are expressed in terms of cardinality of sets of atoms.

Closure Under Evolution. Let $\mathcal{D}$ be a DL and M one of the eight MBAs introduced above. We say $\mathcal{D}$ is *closed under evolution wrt M* (or evolution wrt M is *expressible* in $\mathcal{D}$) if for every evolution setting $\mathcal{K}$ and $\mathcal{N}$ written in $\mathcal{D}$, there is a KB $\mathcal{K}'$ written in $\mathcal{D}$ such that $Mod(\mathcal{K}') = \mathcal{K} \diamond \mathcal{N}$, where $\mathcal{K} \diamond \mathcal{N}$ is the evolution result under semantics M.

We showed in [10,11] that *DL-Lite* is not closed under any of the eight model based semantics. The observation underlying these results is that on the one hand, the minimality of change principle intrinsically introduces implicit disjunction in the evolved KB. On the other hand, since *DL-Lite* is a slight extension of Horn logic [14], it does not allow one to express genuine disjunction (see Lemma 1 in [10] for details).

Let M be a set of models that resulted from the evolution of $(\mathcal{T}, \mathcal{A})$ with $\mathcal{N}$. A KB $(\mathcal{T}, \mathcal{A}')$ is a *sound approximation* of M if $M \subseteq Mod(\mathcal{T}, \mathcal{A}')$. A sound approximation $(\mathcal{T}, \mathcal{A}')$ is *minimal* if for every sound approximation $(\mathcal{T}, \mathcal{A}'')$ inequivalent to $(\mathcal{T}, \mathcal{A}')$, it holds that $Mod(\mathcal{T}, \mathcal{A}'') \not\subseteq Mod(\mathcal{T}, \mathcal{A}')$, i.e., $(\mathcal{T}, \mathcal{A}')$ is minimal wrt "$\subseteq$".

3   Relationships between Model-Based Semantics

Let $\mathcal{S}_1$ and $\mathcal{S}_2$ be two evolution semantics and $\mathcal{D}$ a logic language. Then $\mathcal{S}_1$ *is subsumed by $\mathcal{S}_2$ wrt $\mathcal{D}$*, denoted $(\mathcal{S}_1 \preccurlyeq_{sem} \mathcal{S}_2)(\mathcal{D})$, or just $\mathcal{S}_1 \preccurlyeq_{sem} \mathcal{S}_2$ when $\mathcal{D}$ is clear from the

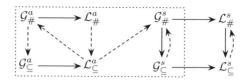

Fig. 2. Subsumptions for evolution semantics.
" $\longrightarrow$ ": for *DL-Lite$_\mathcal{R}$* (Theorem 1). " $\dashrightarrow$ ": for *DL-Lite$_\mathcal{R}^{pr}$* (Theorems 4, 5).
Dashed frame surrounds semantics under which *DL-Lite$_\mathcal{R}^{pr}$* is closed.

context, if $\mathcal{K} \diamond_{\mathcal{S}_1} \mathcal{N} \subseteq \mathcal{K} \diamond_{\mathcal{S}_2} \mathcal{N}$ for all satisfiable KBs $\mathcal{K}$ and $\mathcal{N}$ written in $\mathcal{D}$, where $\mathcal{K} \diamond_{\mathcal{S}_i} \mathcal{N}$ denotes evolution under $\mathcal{S}_i$. Two semantics $\mathcal{S}_1$ and $\mathcal{S}_2$ are *equivalent* (wrt $\mathcal{D}$), denoted $(\mathcal{S}_1 \equiv_{sem} \mathcal{S}_2)(\mathcal{D})$, if $(\mathcal{S}_1 \preccurlyeq_{sem} \mathcal{S}_2)(\mathcal{D})$ and $(\mathcal{S}_2 \preccurlyeq_{sem} \mathcal{S}_1)(\mathcal{D})$. Further in this section we will consider $\mathcal{K}$ and $\mathcal{N}$ written in *DL-Lite$_\mathcal{R}$*. The following theorem shows the subsumption relation between different semantics. We depict these relations in Figure 2 using solid arrows. The figure is complete in the following sense: there is a solid path (a sequence of solid arrows) between any two semantics $\mathcal{S}_1$ and $\mathcal{S}_2$ iff there is a subsumption $\mathcal{S}_1 \preccurlyeq_{sem} \mathcal{S}_2$.

Theorem 1. *Let $\beta \in \{a, s\}$ and $\alpha \in \{\subseteq, \#\}$. Then for DL-Lite$_\mathcal{R}$ it holds that*

$$\mathcal{G}_\alpha^\beta \preccurlyeq_{sem} \mathcal{L}_\alpha^\beta, \quad \mathcal{G}_\#^s \preccurlyeq_{sem} \mathcal{G}_\subseteq^s, \quad and \quad \mathcal{L}_\#^s \preccurlyeq_{sem} \mathcal{L}_\subseteq^s.$$

Proof. Let $dist_\alpha^\beta$ be a distance function, $\mathcal{E}_\mathcal{G} = \mathcal{K} \diamond \mathcal{N}$ wrt $\mathcal{G}_\alpha^\beta$ and $\mathcal{E}_\mathcal{L} = \mathcal{K} \diamond \mathcal{N}$ wrt $\mathcal{L}_\alpha^\beta$ be corresponding global and local semantics based on $dist_\alpha^\beta$. For an evolution setting $\mathcal{K}$ and $\mathcal{N}$, let $\mathcal{J}' \in \mathcal{E}_\mathcal{G}$. Then, there is $\mathcal{I}' \models \mathcal{K}$ such that for every $\mathcal{I}'' \models \mathcal{K}$ and $\mathcal{J}'' \models \mathcal{T} \cup \mathcal{N}$ it *does not* hold that $dist_\alpha^\beta(\mathcal{I}'', \mathcal{J}'') \preccurlyeq dist_\alpha^\beta(\mathcal{I}', \mathcal{J}')$. In particular, when $\mathcal{I}'' = \mathcal{I}'$, there is no $\mathcal{J}'' \models \mathcal{T} \cup \mathcal{N}$ such that $dist_\alpha^\beta(\mathcal{I}', \mathcal{J}'') \preccurlyeq dist_\alpha^\beta(\mathcal{I}', \mathcal{J}')$, which yields that $\mathcal{J}' \in \arg\min_{\mathcal{J} \in Mod(\mathcal{T} \cup \mathcal{N})} dist_\alpha^\beta(\mathcal{I}', \mathcal{J})$, and $\mathcal{J}' \in \mathcal{E}_\mathcal{L}$. We conclude that: $\mathcal{G}_\#^a \preccurlyeq_{sem} \mathcal{L}_\#^a$, $\mathcal{G}_\subseteq^a \preccurlyeq_{sem} \mathcal{L}_\subseteq^a$, $\mathcal{G}_\#^s \preccurlyeq_{sem} \mathcal{L}_\#^s$, $\mathcal{G}_\subseteq^s \preccurlyeq_{sem} \mathcal{L}_\subseteq^s$.

Consider $\mathcal{E}_\# = \mathcal{K} \diamond \mathcal{N}$ wrt $\mathcal{L}_\#^\beta$, which is based on $dist_\#$, and $\mathcal{E}_\subseteq = \mathcal{K} \diamond \mathcal{N}$ wrt $\mathcal{L}_\subseteq^\beta$, which is based on $dist_\subseteq$. We now check whether $\mathcal{E}_\# \preccurlyeq_{sem} \mathcal{E}_\subseteq$ holds. Assume $\mathcal{J}' \in \mathcal{E}_\#$ and $\mathcal{J}' \notin \mathcal{E}_\subseteq$. Then, from the former assumption we conclude existence of $\mathcal{I}' \models \mathcal{K}$ such that $\mathcal{J}' \in \arg\min_{\mathcal{J} \in Mod(\mathcal{T} \cup \mathcal{N})} dist_\#(\mathcal{I}', \mathcal{J})$. From the latter assumption, $\mathcal{J}' \notin \mathcal{E}_\subseteq$, we conclude existence of a model $\mathcal{J}''$ such that $dist_\subseteq(\mathcal{I}', \mathcal{J}'') \subsetneq dist_\subseteq(\mathcal{I}', \mathcal{J}')$. This yields that $dist_\#(\mathcal{I}', \mathcal{J}'') \preccurlyeq dist_\#(\mathcal{I}', \mathcal{J}')$, which contradicts the fact that $\mathcal{J}' \in \mathcal{E}_\#$, assuming that $dist_\subseteq(\mathcal{I}', \mathcal{J}')$ is finite. Thus, $\mathcal{E}_\# \preccurlyeq_{sem} \mathcal{E}_\subseteq$ as soon as $dist_\subseteq(\mathcal{I}, \mathcal{J})$ is finite. This finiteness condition always holds for when $\beta = s$ since the signature of $\mathcal{K} \cup \mathcal{N}$ is finite. It is easy to check that $dist_\subseteq(\mathcal{I}, \mathcal{J})$ may not be finite when $\beta = a$, hence, $\mathcal{L}_\#^a \npreccurlyeq_{sem} \mathcal{L}_\subseteq^a$.

Similarly, one can show that $\mathcal{G}_\#^s \preccurlyeq_{sem} \mathcal{G}_\subseteq^s$ and $\mathcal{G}_\#^a \npreccurlyeq_{sem} \mathcal{G}_\subseteq^a$ in *DL-Lite$_\mathcal{R}$*. $\square$

4   Evolution of *DL-Lite$_\mathcal{R}^{pr}$* KBs

Consider a restriction of *DL-Lite$_\mathcal{R}$*, which we call *DL-Lite$_\mathcal{R}^{pr}$* (*pr* stands for *positive role interaction*), where disjointness that involves roles is forbidden (only positive inclusions

---

INPUT   : satisfiable *DL-Lite*$_\mathcal{R}^{pr}$ KB $(\mathcal{T}, \mathcal{A})$ and *DL-Lite*$_\mathcal{R}$ ABox $\mathcal{N}$
OUTPUT: a set $\mathcal{A}' \subseteq fcl_\mathcal{T}(\mathcal{A})$ of ABox assertions

1 $\mathcal{A}' := \emptyset; \mathcal{S} := fcl_\mathcal{T}(\mathcal{A});$
2 **repeat**
3 | **choose some** $\phi \in \mathcal{S}; \ \mathcal{S} := \mathcal{S} \setminus \{\phi\};$
4 | **if** $\{\phi\} \cup fcl_\mathcal{T}(\mathcal{N})$ *is satisfiable* **then** $\mathcal{A}' := \mathcal{A}' \cup \{\phi\}$
5 **until** $\mathcal{S} = \emptyset$;

Algorithm 1. Algorithm $AlignAlg((\mathcal{T}, \mathcal{A}), \mathcal{N})$ for $\mathcal{A}'$ deterministic computation

involving roles are permitted). Formally, $\mathcal{T}$ is in *DL-Lite*$_\mathcal{R}^{pr}$ if it is in *DL-Lite*$_\mathcal{R}$ and $\mathcal{T} \not\models \exists R \sqsubseteq \neg B$ for any role R and any concept B. *DL-Lite*$_\mathcal{R}^{pr}$ is defined semantically, while one can syntactically check (in quadratic time), given a *DL-Lite*$_\mathcal{R}$ TBox $\mathcal{T}$ whether it is in *DL-Lite*$_\mathcal{R}^{pr}$: compute a closure of $\mathcal{T}$, check that no assertion of the form $\exists R \sqsubseteq \neg B$ is in the closure and if it is the case, then $\mathcal{K}$ is in *DL-Lite*$_\mathcal{R}^{pr}$. Note that *DL-Lite*$_\mathcal{R}^{pr}$ is an extension of RDFS ontology language (of its first-order logics fragment). *DL-Lite*$_\mathcal{R}^{pr}$ adds to RDFS the ability of expressing disjointness of atomic concepts ($A_1 \sqsubseteq \neg A_2$) and mandatory participation ($A \sqsubseteq \exists R$). In the rest of the section we investigate whether and how to capture $\mathcal{K} \diamond \mathcal{N}$ in *DL-Lite*$_\mathcal{R}^{pr}$ for *DL-Lite*$_\mathcal{R}^{pr}$ KBs $\mathcal{K}$ under all the eight MBAs.

4.1   Capturing Atom-Based Evolution

We first study evolution under $\mathcal{L}_\subseteq^a$. Let $\mathcal{I}$ be an interpretation. An *alignment* $Align(\mathcal{I}, \mathcal{N})$ of $\mathcal{I}$ with an ABox $\mathcal{N}$, is the interpretation $\{f \mid f \in \mathcal{I}$ and f is satisfiable with $\mathcal{N}\}$.

Consider an algorithm AlignAlg (see Algorithm 1) that inputs an evolution setting $\mathcal{K}, \mathcal{N}$, and returns the alignment $Align(\mathcal{I}^{can}, \mathcal{N})$ of a canonical model $\mathcal{I}^{can}$ of $\mathcal{K}$: it drops all the assertions of $fcl_\mathcal{T}(\mathcal{A})$ contradicting $\mathcal{N}$ and keeps the rest. Using AlignAlg we can compute representation of $\mathcal{K} \diamond \mathcal{N}$ in *DL-Lite*$_\mathcal{R}^{pr}$:

Theorem 2. *Let* $\mathcal{K} = (\mathcal{T}, \mathcal{A})$ *and* $\mathcal{N}$ *be an evolution setting, and* $\mathcal{T}$ *be in* *DL-Lite*$_\mathcal{R}^{pr}$. *Then there exists a DL-Lite*$_\mathcal{R}^{pr}$ *representation of* $\mathcal{K} \diamond \mathcal{N}$ *under* $\mathcal{L}_\subseteq^a$, *and it can be computed in time polynomial in* $|\mathcal{K} \cup \mathcal{N}|$ *as follows:*

$$\mathcal{K} \diamond \mathcal{N} = Mod(\mathcal{T}, AlignAlg(\mathcal{K}, \mathcal{N}) \cup \mathcal{N}).$$

Example 3. Consider $\mathcal{T} = \{B_0 \sqsubseteq B, \ B \sqsubseteq \neg C\}$, $\mathcal{A} = \{C(a)\}$, and $\mathcal{N} = B(a)$. Then, $fcl_\mathcal{T}(\mathcal{A}) = \{C(a), \neg B_0(a), \neg B(a)\}$ and $AlignAlg((\mathcal{T}, \mathcal{A}), \mathcal{N}) = \{\neg B_0(a)\}$. Hence, the result of evolution $(\mathcal{T}, \mathcal{A}) \diamond \mathcal{N}$ under $\mathcal{L}_\subseteq^a$ is $(\mathcal{T}, \{B(a), \neg B_0(a)\})$.   ∎

Relationships between Atom-Based Semantics. Next theorem shows that in *DL-Lite*$_\mathcal{R}^{pr}$ all four atom-based MBAs coincide We depict these relations between semantics in Figure 2 using dashed arrows, e.g., as between $\mathcal{L}_\subseteq^a$ and $\mathcal{G}_\#^a$. Note that there is a path with solid or dashed arrows (a sequence of such arrows) between any two semantics if and only if in *DL-Lite*$_\mathcal{R}^{pr}$ there is a subsumption between them.

Theorem 4. *For DL-Lite$_{\mathcal{R}}^{pr}$: $\mathcal{L}_{\#}^{a} \equiv_{sem} \mathcal{L}_{\subseteq}^{a} \equiv_{sem} \mathcal{G}_{\#}^{a} \equiv_{sem} \mathcal{G}_{\subseteq}^{a}$.*

Theorems 2 and 4 imply that in *DL-Lite$_{\mathcal{R}}^{pr}$* one can use AlignAlg to compute (a representation of) evolution under all MBAs on atoms.

4.2   Capturing Symbol-Based Evolution

Observe that symbol-based semantics behave differently from atom-based ones: two local semantics (on set inclusion and cardinality) coincide, as well as two global semantics, while there is no subsumption between local and global ones, as depicted in Figure 2:

Theorem 5. *The following relations on symbols-based MBAs hold for DL-Lite$_{\mathcal{R}}^{pr}$:*
   (i) $\mathcal{L}_{\subseteq}^{a} \preccurlyeq_{sem} \mathcal{G}_{\#}^{s}$, while $\mathcal{G}_{\#}^{s} \npreccurlyeq_{sem} \mathcal{L}_{\subseteq}^{s}$;
   (ii) $\mathcal{L}_{\subseteq}^{s} \equiv_{sem} \mathcal{L}_{\#}^{s}$, and $\mathcal{G}_{\subseteq}^{s} \equiv_{sem} \mathcal{G}_{\#}^{s}$, while $\mathcal{L}_{\subseteq}^{s} \npreccurlyeq_{sem} \mathcal{G}_{\#}^{s}$.

As a corollary of Theorem 5, in general the approach presented in Theorem 2 does not work for computing $\mathcal{K} \diamond \mathcal{N}$ under any of the symbol-based MBAs. At the same time, as follows from the following Theorems 6 and 8, this approach gives complete approximations of all symbol-based semantics, while it approximates global semantics better than the local ones.

   Consider the algorithm SymAlg in Algorithm 2 that will be used for evolutions on symbols. It works as follows: it inputs an evolution setting $(\mathcal{T}, \mathcal{A})$, $\mathcal{N}$ and a unary property Π of assertions. Then for every atom ϕ in $\mathcal{N}$ it checks whether ϕ satisfies Π (Line 4). If it the case, SymAlg deletes from $AlignAlg((\mathcal{T}, \mathcal{A}), \mathcal{N})$ all literals ϕ' that share concept name with ϕ. Both local and global semantics have their own Π: Π_G and $\Pi_{\mathcal{L}}$.

Capturing Global Semantics. $\Pi_G(\phi)$ checks whether ϕ of $\mathcal{N}$ $\mathcal{T}$-contradicts $\mathcal{A}$: $\Pi_G(\phi)$ is true iff $\neg\phi \in fcl_{\mathcal{T}}(\mathcal{A}) \setminus AlignAlg((\mathcal{T}, \mathcal{A}), \mathcal{N})$. Intuitively, SymAlg for global semantics works as follows: having contradiction between $\mathcal{N}$ and $\mathcal{A}$ on $\phi = B(c)$, the change of B's interpretation is inevitable. Since the semantics traces changes on symbols only, and B is already changed, one can drop from $\mathcal{A}$ all the assertions of the form $B(d)$. Clearly, $SymAlg(\mathcal{K}, \mathcal{N}, \Pi_G)$ can be computed in time polynomial in $|\mathcal{K} \cup \mathcal{N}|$. The following theorem shows correctness of this algorithm.

Theorem 6. *Let $\mathcal{K} = (\mathcal{T}, \mathcal{A})$ and $\mathcal{N}$ be an evolution setting, and $\mathcal{T}$ be in DL-Lite$_{\mathcal{R}}^{pr}$. Then a DL-Lite$_{\mathcal{R}}^{pr}$ representation of $\mathcal{K} \diamond \mathcal{N}$ under both $\mathcal{G}_{\subseteq}^{s}$ and $\mathcal{G}_{\#}^{s}$ exists and can be computed in time polynomial in $|\mathcal{K} \cup \mathcal{N}|$ as follows:*

$$\mathcal{K} \diamond \mathcal{N} = Mod(\mathcal{T}, SymAlg(\mathcal{K}, \mathcal{N}, \Pi_G)).$$

Capturing Local Semantics. Observe that $\mathcal{L}_{\subseteq}^{s}$ and $\mathcal{L}_{\#}^{s}$ are not expressible in *DL-Lite$_{\mathcal{R}}^{pr}$* because they require for a disjunction which is not available in *DL-Lite$_{\mathcal{R}}$* (we omit details due to space limit).

Theorem 7. *DL-Lite$_{\mathcal{R}}^{pr}$ is not closed under $\mathcal{L}_{\subseteq}^{s}$ and $\mathcal{L}_{\#}^{s}$ semantics.*

INPUT   : satisfiable $DL\text{-}Lite_{\mathcal{R}}^{pr}$ KB $(\mathcal{T}, \mathcal{A})$ and ABox $\mathcal{N}$, a property Π of assertions
OUTPUT: a set $\mathcal{A}' \subseteq fcl_{\mathcal{T}}(\mathcal{A}) \cup fcl_{\mathcal{T}}(\mathcal{N})$ of ABox assertions

1 $\mathcal{A}' := \emptyset; \mathcal{S}_1 := AlignAlg((\mathcal{T}, \mathcal{A}), \mathcal{N}); \mathcal{S}_2 := fcl_{\mathcal{T}}(\mathcal{N});$
2 **repeat**
3   **choose some** $\phi \in \mathcal{S}_2; \ \mathcal{S}_2 := \mathcal{S}_2 \setminus \{\phi\};$
4   **if** $\Pi(\phi) = TRUE$ **then** $\mathcal{S}_1 := \mathcal{S}_1 \setminus \{\phi' \mid \phi$ and ϕ' have the same concept name$\}$
5 **until** $\mathcal{S}_2 = \emptyset$;
6 $\mathcal{A}' := \mathcal{S}_1 \cup fcl_{\mathcal{T}}(\mathcal{N})$

Algorithm 2. Algorithm $SymAlg((\mathcal{T}, \mathcal{A}), \mathcal{N}, \Pi)$ for deterministic computation of $\mathcal{K} \diamond \mathcal{N}$ under $\mathcal{G}_{\subseteq}^s$ and $\mathcal{G}_{\#}^s$ semantics and minimal sound approximation under $\mathcal{L}_{\subseteq}^s$ and $\mathcal{L}_{\#}^s$ semantics

To compute a minimal sound approximations under local semantics on symbols, we use SymAlg with the following $\Pi_{\mathcal{L}}$: $\Pi_{\mathcal{L}}(\phi)$ is true iff $\phi \notin \mathcal{S}_1$. That is, $\Pi_{\mathcal{L}}$ checks whether the ABox $\mathcal{A}$ $\mathcal{T}$-entails $A(c) \in fcl_{\mathcal{T}}(\mathcal{N})$, and if it does not, then the algorithm deletes all the assertions from $fcl_{\mathcal{T}}(\mathcal{A})$ that share the concept name with $A(c)$. This property is weaker than the one for global semantics, since it is easier to get changes in interpretation of A by choosing a model of $\mathcal{K}$ which does not include $A(c)$. The following theorem shows correctness and complexity of the algorithm.

Theorem 8. *Let* $\mathcal{K} = (\mathcal{T}, \mathcal{A})$ *and* $\mathcal{N}$ *be an evolution setting, and* $\mathcal{T}$ *be in* $DL\text{-}Lite_{\mathcal{R}}^{pr}$. *Then a* $DL\text{-}Lite_{\mathcal{R}}^{pr}$ *minimal sound approximation* $\mathcal{K}''$ *of* $\mathcal{K} \diamond \mathcal{N}$ *under both* $\mathcal{L}_{\subseteq}^s$ *and* $\mathcal{L}_{\#}^s$ *exists and can be computed in time polynomial in* $|\mathcal{K} \cup \mathcal{N}|$ *as follows:*

$$\mathcal{K}'' = (\mathcal{T}, SymAlg(\mathcal{K}, \mathcal{N}, \Pi_{\mathcal{L}})).$$

Example 9. Consider the following $DL\text{-}Lite_{\mathcal{R}}^{pr}$ KB $\mathcal{K} = (\emptyset, \mathcal{A})$ and $\mathcal{N}$:

$$\mathcal{A} = \{A(a), A(b), B(c), B(d)\}; \quad \mathcal{N} = \{\neg A(a), B(e)\}.$$

It is easy to see that $\mathcal{A}' = SymAlg(\mathcal{K}, \mathcal{N}, \Pi_{\mathcal{G}})$ is $\{\neg A(a), B(c), B(d), B(e)\}$, and $\mathcal{A}'' = SymAlg(\mathcal{K}, \mathcal{N}, \Pi_{\mathcal{L}})$ is $\{\neg A(a), B(e)\}$. That is, $\mathcal{K} \diamond \mathcal{N}$ under $\mathcal{G}_{\beta}^s$ is equal to $Mod(\emptyset, \mathcal{A}')$, and under $\mathcal{L}_{\beta}^s$ is approximated by $Mod(\emptyset, \mathcal{A}'')$, where $\beta \in \{\subseteq, \#\}$. A closer look at $\mathcal{A}'$ shows that the behaviour of the evolution under $\mathcal{G}_{\beta}^s$ is very counter-intuitive: as soon as we declare that the object a is not in A, all the information about another objects in A is erased. Local semantics $\mathcal{L}_{\beta}^s$ are even worse: the evolution under them erases information about B as soon as we just add information about a new object e in B.   ∎

To sum up on $DL\text{-}Lite_{\mathcal{R}}^{pr}$: atom-based approaches (which all coincide) can be captured using a polynomial-time time algorithm based on alignment. Moreover, the evolution results produced under these MBAs are intuitive and expected, e.g., see Example 3, while symbol-based approaches produce quite unexpected and counterintuitive results (these semantics delete too much data). Furthermore, two out of four of the latter approaches cannot be captured in $DL\text{-}Lite_{\mathcal{R}}^{pr}$. Based on these results we conclude that using atom-based approaches for applications seem to be more practical. In Figure 2 we framed in a dashed rectangle six out of eight MBAs under which $DL\text{-}Lite_{\mathcal{R}}^{pr}$ is closed.

5 $\mathcal{L}_{\subseteq}^a$ Evolution of *DL-Lite*$_\mathcal{R}$ KBs

In the previous section we showed that atom-based MBAs behave well for *DL-Lite*$_\mathcal{R}^{pr}$ evolution settings, while symbol-based ones do not. This suggests to investigate atom-based MBAs for the entire *DL-Lite*$_\mathcal{R}$. Moreover, one of the atom-based semantics $\mathcal{L}_{\subseteq}^a$ which is essentially the same as a so-called *Winslett's semantics* [15] (WS) was widely studied in the literature [6,8]. Liu, Lutz, Milicic, and Wolter studied WS for expressive DLs [6], and KBs with empty TBoxes. Most of the DLs they considered are not closed under WS. Poggi, Lembo, De Giacomo, Lenzerini, and Rosati applied WS to the same setting as we have in this work: to what they called instance level (ABox) update for *DL-Lite* [8]. They proposed an algorithm to compute the result of updates, which has technical issues, i.e., it is neither sound, nor complete [10]. They further use this algorithm to compute approximations of ABox updates in sublogics of *DL-Lite*, which inherits these technical issues. Actually, ABox update algorithm cannot exist since Calvanese, Kharlamov, Nutt, and Zheleznyakov showed that *DL-Lite* is not closed under $\mathcal{L}_{\subseteq}^a$ [11]. We now investigate $\mathcal{L}_{\subseteq}^a$ evolution for *DL-Lite*$_\mathcal{R}$ and firstly explain *why DL-Lite*$_\mathcal{R}$ is not closed under $\mathcal{L}_{\subseteq}^a$.

5.1   Understanding Inexpressibility of Evolution in *DL-Lite*$_\mathcal{R}$

Recall that for every *DL-Lite* KB $\mathcal{K}$, the set *Mod*($\mathcal{K}$) has a canonical model. The following example illustrates the lack of canonical models for $\mathcal{K} \diamond \mathcal{N}$ under $\mathcal{L}_{\subseteq}^a$, which yields inexpressibility of $\mathcal{K} \diamond \mathcal{N}$ in *DL-Lite*.

Example 10. Consider the following *DL-Lite* KB $\mathcal{K}_1 = (\mathcal{T}_1, \mathcal{A}_1)$ and $\mathcal{N}_1 = \{C(b)\}$:

$$\mathcal{T}_1 = \{A \sqsubseteq \exists R, \exists R^- \sqsubseteq \neg C\}; \quad \mathcal{A}_1 = \{A(a), C(e), C(d), R(a,b)\}.$$

Consider a model $\mathcal{I}$ of $\mathcal{K}_1$: $A^\mathcal{I} = \{a, x\}$, $C^\mathcal{I} = \{d, e\}$, and $R^\mathcal{I} = \{(a,b), (x,b)\}$, where $x \in \Delta \setminus adom(\mathcal{K}_1 \cup \mathcal{N}_1)$. The following models belong to $\mathcal{I} \diamond \mathcal{N}_1$:

$$
\begin{array}{llll}
\mathcal{J}_0: & A^\mathcal{I} = \emptyset, & C^\mathcal{I} = \{d, e, b\}, & R^\mathcal{I} = \emptyset, \\
\mathcal{J}_1: & A^\mathcal{I} = \{x\}, & C^\mathcal{I} = \{e, b\}, & R^\mathcal{I} = \{(x,d)\}, \\
\mathcal{J}_2: & A^\mathcal{I} = \{x\}, & C^\mathcal{I} = \{d, b\}, & R^\mathcal{I} = \{(x,e)\}.
\end{array}
$$

Indeed, all the models satisfy $\mathcal{N}_1$ and $\mathcal{T}_1$. To see that they are in $\mathcal{I} \diamond \mathcal{N}_1$ observe that every model $\mathcal{J}(I) \in (\mathcal{I} \diamond \mathcal{N}_1)$ can be obtained from $\mathcal{I}$ by making modifications that guarantee that $\mathcal{J}(I) \models (\mathcal{N}_1 \cup \mathcal{T}_1)$ and that the distance between $\mathcal{I}$ and $\mathcal{J}(I)$ is minimal. What are these modifications? Since in every $\mathcal{J}(I)$ the new assertion $C(b)$ holds and $(\exists R^- \sqsubseteq \neg C) \in \mathcal{T}_1$, there should be no R-atoms with b-fillers (at the second coordinate) in $\mathcal{J}(\mathcal{I})$. Hence, the necessary modifications of $\mathcal{I}$ are either to drop (some of) the R-atoms $R(a,b)$ and $R(x,b)$, or to modify (some of) them, by substituting the b-fillers with another ones, while keeping the elements a and x on the first coordinate. The model $\mathcal{J}_0$ corresponds to the case when both R-atoms are dropped, while in $\mathcal{J}_1$ and $\mathcal{J}_2$ only $R(a,b)$ is dropped and $R(x,b)$ is modified to $R(x,d)$ and $R(x,e)$, respectively. Note that the modification in $R(x,b)$ leads to a further change in the interpretation of C in both $\mathcal{J}_1$ and $\mathcal{J}_2$, namely, $C(d)$ and $C(e)$ should be dropped, respectively.

One can verify that any model $\mathcal{J}_{can}$ that can be homomorphically embedded into $\mathcal{J}_0$, $\mathcal{J}_1$, and $\mathcal{J}_2$ is such that $A^{\mathcal{J}_{can}} = R^{\mathcal{J}_{can}} = \emptyset$, and $e, d \notin C^{\mathcal{J}_{can}}$. It is easy to check that such a model does not belong to $\mathcal{K}_1 \diamond \mathcal{N}_1$. Hence, there is no canonical model in $\mathcal{K}_1 \diamond \mathcal{N}_1$ and it is inexpressible in *DL-Lite*. ∎

We now give an intuition *why* in $\mathcal{K} \diamond \mathcal{N}$ under $\mathcal{L}^a_{\sqsubseteq}$ canonical models may be missing. Observe that in Example 10, the role R is affected by the old TBox $\mathcal{T}_1$ as follows:

(i) $\mathcal{T}_1$ *places* (i.e., enforces the *existence* of) R-atoms in the evolution result, and on *one* of coordinates of these R-atoms, there are constants from specific sets, e.g., $A \sqsubseteq \exists R$ of $\mathcal{T}_1$ enforces R-atoms with constants from A on the first coordinate, and

(ii) $\mathcal{T}_1$ *forbids* R-atoms in $\mathcal{K}_1 \diamond \mathcal{N}_1$ with specific constants on the *other* coordinate, e.g., $\exists R^- \sqsubseteq \neg C$ forbids R-atoms with C-constants on the second coordinate.

Due to this *dual-affection* (both positive and negative) of the role R in $\mathcal{T}_1$, we were able to provide ABoxes $\mathcal{A}_1$ and $\mathcal{N}_1$, which together triggered the case analyses of modifications on the model $\mathcal{I}$, that is, $\mathcal{A}_1$ and $\mathcal{N}_1$ were *triggers* for R. Existence of such an affected R and triggers $\mathcal{A}_1$ and $\mathcal{N}_1$ made $\mathcal{K}_1 \diamond \mathcal{N}_1$ inexpressible in *DL-Lite*$_\mathcal{R}$. Therefore, we now formally define and then learn how to detect dually-affected roles in TBoxes $\mathcal{T}$ and how to understand whether these roles are triggered by $\mathcal{A}$ and $\mathcal{N}$.

Definition 11. *Let $\mathcal{T}$ be a DL-Lite$_\mathcal{R}$ TBox. Then a role R is* dually-affected *in $\mathcal{T}$ if for some concepts A and B it holds that $\mathcal{T} \models A \sqsubseteq \exists R$ and $\mathcal{T} \models \exists R^- \sqsubseteq \neg B$. A dually-affected role R is* triggered *by $\mathcal{N}$ if there is a concept C such that $\mathcal{T} \models \exists R^- \sqsubseteq \neg C$ and $\mathcal{N} \models_\mathcal{T} C(b)$ for some constant b.*

As we saw in Example 10, even one dually-affected role in a TBox can cause inexpressibility of evolution. Moreover, if there is a dually affected role, we can always find $\mathcal{A}$ and $\mathcal{N}$ to trigger it. We generalize this observation as follows:

Theorem 12. *Let $\mathcal{T}$ be a DL-Lite$_\mathcal{R}$ TBox and R be a role dually affected in $\mathcal{T}$. Then there exist ABoxes $\mathcal{A}$ and $\mathcal{N}$ s.t. $(\mathcal{T}, \mathcal{A}) \diamond \mathcal{N}$ is inexpressible in DL-Lite$_\mathcal{R}$ under $\mathcal{L}^a_{\sqsubseteq}$.*

5.2 Prototypes

Closer look at the sets of models $\mathcal{K} \diamond \mathcal{N}$ for *DL-Lite*$_\mathcal{R}$ KBs $\mathcal{K}$ gives a surprising result:

Theorem 13. *The set of models $\mathcal{K} \diamond \mathcal{N}$ under $\mathcal{L}^a_{\sqsubseteq}$ can be divided (but in general not partitioned) into worst-case exponentially many in $|\mathcal{K} \cup \mathcal{N}|$ subsets $\mathcal{S}_0, \dots, \mathcal{S}_n$, where each $\mathcal{S}_i$ has a canonical model $\mathcal{J}_i$, which is a minimal element in $\mathcal{K} \diamond \mathcal{N}$ wrt homomorphisms.*

We call these $\mathcal{J}_i$s *prototypes*. Thus, capturing $\mathcal{K} \diamond \mathcal{N}$ in some logics boils down to (i) capturing each $\mathcal{S}_i$ with some theory $\mathcal{K}_{\mathcal{S}_i}$ and (ii) taking the disjunction across all $\mathcal{K}_{\mathcal{S}_i}$. This will give the desired theory $\mathcal{K}' = \mathcal{K}_{\mathcal{S}_1} \vee \cdots \vee \mathcal{K}_{\mathcal{S}_n}$ that captures $\mathcal{K} \diamond \mathcal{N}$. As we will see some of $\mathcal{K}_{\mathcal{S}_i}$s are not *DL-Lite* theories (while they are $\mathcal{SHOIQ}$ theories, see Section 5.4 for details). We construct each $\mathcal{K}_{\mathcal{S}_i}$ in two steps. First, we construct a *DL-Lite*$_\mathcal{R}$ KB $\mathcal{K}(\mathcal{J}_i)$ which is a sound approximations of $\mathcal{S}_i$, i.e., $\mathcal{S}_i \subseteq Mod(\mathcal{K}(\mathcal{J}_i))$. Second, based on $\mathcal{K}$ and $\mathcal{N}$, we construct a $\mathcal{SHOIQ}$ formula Ψ, which cancels out all the models in $Mod(\mathcal{K}(\mathcal{J}_i)) \setminus \mathcal{S}_i$, i.e., $\mathcal{K}_{\mathcal{S}_i} = \Psi \wedge \mathcal{K}(\mathcal{J}_i)$. Finally,

$$\mathcal{K}_{\mathcal{S}_0} \vee \cdots \vee \mathcal{K}_{\mathcal{S}_n} = (\Psi \wedge \mathcal{K}(\mathcal{J}_0)) \vee \cdots \vee (\Psi \wedge \mathcal{K}(\mathcal{J}_n)) = \Psi \wedge (\mathcal{K}(\mathcal{J}_0) \vee \cdots \vee \mathcal{K}(\mathcal{J}_n)).$$

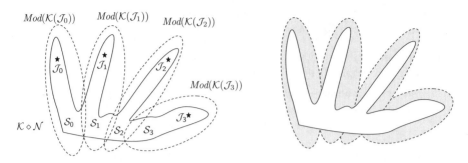

Fig. 3. Graphical representation of our approach to capture the result of evolution under $\mathcal{L}^a_\subseteq$

To get a better intuition on our approach consider Figure 3, where the result of evolution $\mathcal{K} \diamond \mathcal{N}$ is depicted as the figure with solid-line borders (each point within the figure is a model of $\mathcal{K} \diamond \mathcal{N}$). For the sake of example, let $\mathcal{K} \diamond \mathcal{N}$ under $\mathcal{L}^a_\subseteq$ can be divided in four subsets $\mathcal{S}_0, \ldots, \mathcal{S}_3$. To emphasize this fact, $\mathcal{K} \diamond \mathcal{N}$ looks similar to a hand with four fingers, where each finger represents an $\mathcal{S}_i$. Consider the left part of Figure 3, where the canonical $\mathcal{J}_i$ model of each $\mathcal{S}_i$ is depicted as a star. Using $DL\text{-}Lite_\mathcal{R}$, we can provide KBs $\mathcal{K}(\mathcal{J}_i)$s that are sound approximation of corresponding $\mathcal{S}_i$s. We depict the models $Mod(\mathcal{K}(\mathcal{J}_i))$ as ovals with dashed-line boarders. In the right part of Figure 3 we depict in grey the models $Mod(\mathcal{K}(\mathcal{J}_i)) \setminus \mathcal{S}_i$ that are cut off by Ψ.

We now define prototypes formally and proceed to procedures discussed above.

Definition 14. *Let $\mathcal{K}$ and $\mathcal{N}$ be an evolution setting. A* prototypal set *for $\mathcal{K} \diamond \mathcal{N}$ under $\mathcal{L}^a_\subseteq$ is a minimal subset $= \{\mathcal{J}_0, \ldots, \mathcal{J}_n\}$ of $\mathcal{K} \diamond \mathcal{N}$ satisfying the following property:*

for every $\mathcal{J} \in \mathcal{K} \diamond \mathcal{N}$ there exists $\mathcal{J}_i \in$ homomorphically embeddable in $\mathcal{J}$.

We call every $\mathcal{J}_i \in$ a *prototype* for $\mathcal{K} \diamond \mathcal{N}$. Note that prototypes generalize canonical models in the sense that every set of models with a canonical one, say $Mod(\mathcal{K})$ for a $DL\text{-}Lite_\mathcal{R}$ KB $\mathcal{K}$, has a prototype, which is exactly this canonical model.

5.3   Computing $\mathcal{L}^a_\subseteq$ Evolution for *DL-Lite$_\mathcal{R}$*

For the ease of exhibition of our procedure that computes evolution $\mathcal{K} \diamond \mathcal{N}$ under $\mathcal{L}^a_\subseteq$ semantics we restrict $DL\text{-}Lite_\mathcal{R}$ by assuming that TBoxes $\mathcal{T}$ should satisfy: for any two roles R and R', $\mathcal{T} \not\models \exists R \sqsubseteq \exists R'$ and $\mathcal{T} \not\models \exists R \sqsubseteq \neg \exists R'$. That is, we forbid direct interaction (subsumption and disjoint) between role projections and call such $\mathcal{T}$ as *without direct role interactions*. Some interaction between R and R' is still possible, e.g., role projections may contain the same concept. This restriction allows us to analyze evolution that affects roles independently for every role. We will further comment on how the following techniques can be extended to the case when roles interact in an arbitrary way.

Components for Computation. We now introduce several notions and notations that we further use in the description of our procedure. The notion of *alignment* was introduced

$$BZP(\mathcal{K},\mathcal{N})$$

1. $\mathcal{J}_0 := Align(\mathcal{I}^{can},\mathcal{N}) \cup \mathcal{N}$, where $\mathcal{I}^{can}$ is the canonical model of $\mathcal{K}$.
2. For each $R(a,b) \in AA(\mathcal{K},\mathcal{N})$, do $\mathcal{J}_0 := \mathcal{J}_0 \setminus \{R(a,b)\}$,
    if there is no $R(a,\beta) \in \mathcal{I}^{can} \setminus AA(\mathcal{K},\mathcal{N})$ do $\mathcal{J}_0 := \mathcal{J}_0 \setminus root_{\mathcal{T}}^{at}(\exists R(a))$.
3. Return $\mathcal{J}_0$.

Fig. 4. The procedure of building zero-prototype

in Section 4.1. An auxiliary set of atoms *AA* (*Auxiliary Atoms*) that, due to evolution, should be deleted from the original KB and have some extra condition on the first coordinate is:

$$AA(\mathcal{T},\mathcal{A},\mathcal{N}) = \{R(a,b) \in fcl_{\mathcal{T}}(\mathcal{A}) \mid \mathcal{T} \models A \sqsubseteq \exists R, \mathcal{A} \models_{\mathcal{T}} A(a), \mathcal{N} \models_{\mathcal{T}} \neg\exists R^-(b)\}.$$

If R_i is a dually-affected role of $\mathcal{T}$ triggered by $\mathcal{A}$ and $\mathcal{N}$, then the set of *forbidden atoms* (of the original ABox) $FA[\mathcal{T},\mathcal{A},\mathcal{N}](R_i)$ for R_i is:

$$\{D(c) \in fcl_{\mathcal{T}}(\mathcal{A}) \mid \exists R_i^-(c) \wedge D(c) \models_{\mathcal{T}} \perp \text{ and } \mathcal{N} \not\models_{\mathcal{T}} D(c), \text{ and } \mathcal{N} \not\models_{\mathcal{T}} \neg D(c)\}.$$

Consequently, the set of forbidden atoms for the entire KB $(\mathcal{T},\mathcal{A})$ and $\mathcal{N}$ is

$$FA(\mathcal{T},\mathcal{A},\mathcal{N}) = \bigcup_{R_i \in TR} FA(\mathcal{T},\mathcal{A},\mathcal{N})(R_i),$$

where $TR(\mathcal{T},\mathcal{N})$ (or simply *TR*, which stands for *triggered roles*) is the set of all roles dually-affected in $\mathcal{T}$ that are triggered by $\mathcal{N}$. In the following we omit the arguments $(\mathcal{T},\mathcal{A},\mathcal{N})$ in FA when they are clear from the context. For a role R, the set $SC(R)$, where *SC* stands for *sub-concepts*, is a set of concepts that are *immediately* under $\exists R$ in the concept hierarchy generated by $\mathcal{T}$:

$$SC(R) = \{A \mid \mathcal{T} \models A \sqsubseteq \exists R \text{ and there is no } A' \text{ s.t. } \mathcal{T} \models A \sqsubseteq A' \text{ and } \mathcal{T} \models A' \sqsubseteq \exists R\}.$$

If f is an ABox assertion, then $root_{\mathcal{T}}^{at}(f)$ is a set of all the atoms that $\mathcal{T}$-entail f. For example, if $\mathcal{T} \models A \sqsubseteq \exists R$, then $A(x) \in root_{\mathcal{T}}^{at}(\exists R(x))$.

We are ready to proceed to construction of prototypes.

Constructing Zero-Prototype. The procedure $BZP(\mathcal{K},\mathcal{N})$ (*Build Zero Prototype*) in Figure 4 constructs the main prototype $\mathcal{J}_0$ for $\mathcal{K}$ and $\mathcal{N}$, which we call *zero-prototype*. Based on $\mathcal{J}_0$ we will construct all the other prototypes. To build $\mathcal{J}_0$ one should align the canonical model $\mathcal{I}^{can}$ of $\mathcal{K}$ with $\mathcal{N}$, and then delete from the resulting set of atoms all the auxiliary atoms $R(a,b)$ of $AA(\mathcal{K},\mathcal{N})$. If $\mathcal{I}^{can}$ contains *no* atoms $R(a,\beta) \in AA(\mathcal{K},\mathcal{N})$ for some β, then we further delete atoms $root_{\mathcal{T}}^{at}(\exists R(a))$ from $\mathcal{J}_0$, otherwise would we get a contradiction with the TBox. Note that $\mathcal{J}_0$ can be infinite.

Constructing Other Prototypes. The procedure $BP(\mathcal{K},\mathcal{N},\mathcal{J}_0)$ (*Build Prototypes*) of constructing   takes $\mathcal{J}_0$ and, based on it, builds the other prototypes by *(i)* dropping FA-atoms from $\mathcal{J}_0$ and then *(ii)* adding atoms necessary to obtain a model of $\mathcal{K} \diamond \mathcal{N}$. This procedure can be found in Figure 5.

We conclude the discussion on the procedures with a theorem:

$$BP(\mathcal{K}, \mathcal{N}, \mathcal{J}_0)$$

1. $:= \{\mathcal{J}_0\}.$
2.   For each subset $\mathcal{D} = \{D_1(c_1), \ldots, D_k(c_k)\} \subseteq \mathsf{FA}$ do
      for each $\mathcal{R} = (R_{i_1}, \ldots, R_{i_k})$ such that $D_j(c_j) \in \mathsf{FA}(R_{i_j})$ for $j = 1, \ldots, k$ do
      for each $\mathcal{B} = (A_{i_1}, \ldots, A_{i_k})$ such that $A_j \in \mathsf{SC}(R_j)$ do
      $\mathcal{J}[\mathcal{D}, \mathcal{R}, \mathcal{B}] := \left[\mathcal{J}_0 \setminus \bigcup_{i=1}^k root_{\mathcal{T}}(D_i(c_i)) \right] \cup \bigcup_{j=1}^k \left[fcl_{\mathcal{T}}(R_{i_j}(x_j, c_j)) \cup \{A_{R_{i_j}}(x_j)\} \right],$
      where all x_i's are different constants from $\Delta \setminus adom(\mathcal{K})$, fresh for $\mathcal{I}^{can}$.
      $:= \cup \{\mathcal{J}[\mathcal{D}, \mathcal{R}, \mathcal{B}]\}.$
3.   Return .

Fig. 5. The procedure of building prototypes in *DL-Lite$_\mathcal{R}$* without direct role interactions based on the zero prototype $\mathcal{J}_0$

Theorem 15. *Let $\mathcal{K} = (\mathcal{T}, \mathcal{A})$, $\mathcal{N}$ be an evolution setting and $\mathcal{T}$ without direct role interactions. Then the set $BP(\mathcal{K}, \mathcal{N}, BZP(\mathcal{K}, \mathcal{N}))$ is prototypal for $\mathcal{K} \diamond \mathcal{N}$ under $\mathcal{L}^a_{\sqsubseteq}$.*

Continuing with Example 10, one can check that the prototypal set for $\mathcal{K}_1$ and $\mathcal{N}_1$ is $\{\mathcal{J}_0, \mathcal{J}_1, \mathcal{J}_2, \mathcal{J}_3\}$, where $\mathcal{J}_0$, $\mathcal{J}_1$, and $\mathcal{J}_2$ are as in the example and $A^{\mathcal{J}_3} = \{x, y\}$, $C^{\mathcal{J}_3} = \{b\}$, and $R^{\mathcal{J}_3} = \{(x, d), (y, e)\}$.

We proceed to correctness of BP in capturing evolution in *DL-Lite$_\mathcal{R}$*, where we use the following set $\mathsf{FC}[\mathcal{T}, \mathcal{A}, \mathcal{N}](R_i) = \{c \mid D(c) \in \mathsf{FA}[\mathcal{T}, \mathcal{A}, \mathcal{N}](R_i)\}$, that collects all the constants that participate in the forbidden atoms.

Theorem 16. *Let $\mathcal{K} = (\mathcal{T}, \mathcal{A})$, $\mathcal{N}$ be an evolution setting, $\mathcal{T}$ without direct role interactions, and $BP(\mathcal{K}, \mathcal{N}, BZP(\mathcal{K}, \mathcal{N})) = \{\mathcal{J}_0, \ldots, \mathcal{J}_n\}$ a prototypal set for $\mathcal{K} \diamond \mathcal{N}$. Then $\mathcal{K} \diamond \mathcal{N}$ under $\mathcal{L}^a_{\sqsubseteq}$ is expressible in $\mathcal{SHOIQ}$ and moreover*

$$\mathcal{K} \diamond \mathcal{N} = Mod\big(\Psi \wedge ((\mathcal{T}, \mathcal{A}_0) \vee \cdots \vee (\mathcal{T}, \mathcal{A}_n))\big),$$

where A_i is a DL-Lite$_\mathcal{R}$ ABox such that $\mathcal{J}_i$ is a canonical model for $(\mathcal{T}, \mathcal{A}_i)$, $\Psi = \forall R_{\top}.(\Phi_1 \sqcap \Phi_2)$, where $R_{\top}$ is the top role (which is present in OWL 2) and

$$\Phi_1 \equiv \prod_{R_i \in TR} \prod_{c_j \in FC[R_i]} \left[\big(\exists R_i.\{c_j\} \sqsubseteq (\leq 1R.\top)\big) \sqcap \big(\exists R_i.\{c_j\} \sqsubseteq \bigsqcup_{B(x) \in root^{at}_{\mathcal{T}}(\exists R_i(x))} \neg\{x\} \sqcup B\big) \right],$$

$$\Phi_2 \equiv \prod_{R(a,b) \in AA} \left[\neg(\{a\} \sqsubseteq \exists R.\top) \sqcup \prod_{C(a) \in root^{at}_{\mathcal{T}}(\exists R(a)) \sqcap fcl_{\mathcal{T}}(\mathcal{A})} \{a\} \sqsubseteq C \right].$$

What is missing in the theorem above is *how* to compute the ABoxes $\mathcal{A}_i$s. One can do it using a similar procedure to the one of constructing $\mathcal{J}_i$s, with the difference that one has to take the original ABox $\mathcal{A}$ instead of $\mathcal{I}^{can}$ as the input. Note that $\mathcal{A}$ may include negative atoms, like $\neg B(c)$, which should be treated in the same way as positive ones.

Continuing with Example 10, the ABoxes $\mathcal{A}_0$ and $\mathcal{A}_1$ are as follows:

$$\mathcal{A}_0 = \{C(d), C(e), C(b)\}; \quad \mathcal{A}_1 = \{A(x), C(e), C(b), R(x, d)\}.$$

$\mathcal{A}_2$ and $\mathcal{A}_3$ can be built in the similar way. Note that only $\mathcal{A}_0$ is in *DL-Lite$_\mathcal{R}$*, while writing $\mathcal{A}_1, \ldots, \mathcal{A}_3$ requires variables in ABoxes. Variables, also known as *soft constants*, are not allowed in *DL-Lite$_\mathcal{R}$* ABoxes, while present in *DL-Lite$_{\mathcal{RS}}$* ABoxes. Soft constants x are constants not constrained by the Unique Name Assumption: it is not necessary that $x^\mathcal{I} = x$. Since *DL-Lite$_{\mathcal{RS}}$* is tractable and first-order rewritable [12], expressing $\mathcal{A}_1$ in *DL-Lite$_{\mathcal{RS}}$* instead of *DL-Lite$_\mathcal{R}$* does not affect tractability.

Note that the number of prototypes is exponential in the number of constants, and therefore the size of the $\mathcal{SHOIQ}$ theory described in Theorem 16 is also exponential in the number of constants.

Capturing $\mathcal{L}^a_\sqsubseteq$ Semantics for DL-Lite$_\mathcal{R}$ KBs with Direct Role Interactions. In this general case the BP procedure does return prototypes but not all of them. To capture the $\mathcal{L}^a_\sqsubseteq$ for such KBs one should iterate BP over (already constructed) prototypes until no new prototypes can be constructed. Intuitively the reason is that BP deletes forbidden atoms (atoms of FA) and add new atoms of the form $R(a, b)$ for some triggered dually-affected role R which may in turn trigger another dually-affected role, say P, and such triggering may require further modifications, already for P. This further modification require a new run of BP. For example, if we have $\exists R^- \sqsubseteq \neg \exists P^-$ in the TBox and we set $R(a, b)$ in a prototype, say $\mathcal{J}_k$, this modification triggers role P and we should run BP recursively with the prototype $\mathcal{J}_k$ as if it was the zero prototype. We shall not discuss the general procedures in more details due to space limit.

5.4   Practical Considerations on $\mathcal{L}^a_\sqsubseteq$ Evolution

As a summary of Sections 4 and 5, we now discuss how one can compute $\mathcal{L}^a_\sqsubseteq$-evolution of *DL-Lite$_\mathcal{R}$* KBs in practice. For an evolution setting $\mathcal{K}$ and $\mathcal{N}$ consider the following procedure of computing $\mathcal{K}'$ such that $Mod(\mathcal{K}') = \mathcal{K} \diamond \mathcal{N}$:

1. Check whether $\mathcal{K}$ is in *DL-Lite$^{pr}_\mathcal{R}$*. This test can be done in polynomial time, see Section 4.
2. If $\mathcal{K}$ is in *DL-Lite$^{pr}_\mathcal{R}$*, then $\mathcal{K}'$ is in *DL-Lite$^{pr}_\mathcal{R}$* and can be computed in polynomial time using the algorithm AlignAlg as described in Theorem 2 of Section 4.
3. If $\mathcal{K}$ is *not* in *DL-Lite$^{pr}_\mathcal{R}$*, then check whether any dually-affected role of $\mathcal{K}$ is triggered by $\mathcal{N}$. This test can be done in polynomial time, see Section 5.1.
4. If the test of Case 3 fails, then $\mathcal{K}'$ is in *DL-Lite$_\mathcal{R}$* and can be computed as in Case 2.
5. If the test of Case 3 succeeds, then $\mathcal{K}'$ is in $\mathcal{SHOIQ}$, but *not* in *DL-Lite$_\mathcal{R}$*, and can be computed using prototype-based techniques as described in Theorem 16. The size of this $\mathcal{K}'$ is polynomial in the number of prototypes for $\mathcal{K} \diamond \mathcal{N}$. Since the number of prototypes is worst-case exponential in the number of constants in $\mathcal{A}$, the size of $\mathcal{K}'$ could be exponential in $|\mathcal{K} \cup \mathcal{N}|$.

The case when computation of $\mathcal{K}'$ can be intractable is of $\mathcal{K}$ with dually-affected roles triggered by $\mathcal{N}$. It is unclear how often this case may occur in practice. While the tractable case of *DL-Lite$^{pr}_\mathcal{R}$* where we disallow assertions of the form $\exists R \sqsubseteq \neg A$ seems to be practical, since it extends the (first-order fragment) of RDFS.

We now discuss a way to approximate $\mathcal{K} \diamond \mathcal{N}$ with a *DL-Lite$_\mathcal{R}$* KB when this set is not expressible in *DL-Lite$_\mathcal{R}$*. Let $\mathcal{K} = (\mathcal{T}, \mathcal{A})$ and $\mathcal{N}$ be an evolution setting, then a *DL-Lite$_\mathcal{R}$* KB $\mathcal{K}^c = (\mathcal{T}, \mathcal{A}^c)$ is a *certain $\mathcal{L}^a_\sqsubseteq$-approximation* of $\mathcal{K} \diamond \mathcal{N}$ if $\mathcal{A}^c = \{F \mid \mathcal{K} \diamond \mathcal{N} \models F\}$. We called $\mathcal{K}^c$ certain since it resembles certain answers for queries over KBs.

Proposition 17. *Let $\mathcal{K}, \mathcal{N}$ be an evolution setting. Then the certain $\mathcal{L}_{\subseteq}^{a}$-approximation of $\mathcal{K} \diamond \mathcal{N}$ exists, unique, and can be computed in non-deterministic exponential time.*

Proof. Clearly all ABox assertions of $\mathcal{A}^c$ are over concepts, roles, and constants of $\mathcal{K}$, thus, there are at most a quadratic many (in $|\mathcal{K} \cup \mathcal{N}|$) of them, and we can simply test whether $F \in \mathcal{A}^c$ for each such assertion F. Since $\mathcal{K} \diamond \mathcal{N}$ is representable in $\mathcal{SHOIQ}$, this test can be reduced to the subsumption problem for $\mathcal{SHOIQ}$ (checking whether $\mathcal{K}' \models C(a)$ is equivalent to checking whether $\mathcal{K}' \models \{a\} \sqsubseteq C$). Subsumption for $\mathcal{SHOIQ}$ is **NExpTime-complete** and can be tested using the algorithms of [16].

The proposition above gives the upper bound for $\mathcal{K}^c$ computations. We do not know the lower bound, but conjecture it to be in polynomial time. Note that **NExpTime** lower bound for $\mathcal{SHOIQ}$ subsumption checking holds for arbitrary $\mathcal{SHOIQ}$ concepts, while Theorem 16 gives us $\mathcal{K}'$ with concepts of a specific kind. Moreover, the authors of [16] argue that despite the high complexity of subsumption checking their algorithms should behave well in many typically encountered cases. Note also that for $DL\text{-}Lite_{\mathcal{R}}^{pr}$ KBs certain approximations in fact capture the evolution result, that is $Mod(\mathcal{K}^c) = \mathcal{K} \diamond \mathcal{N}$.

6   Conclusion

We studied model-based approaches to ABox evolution (update and revision) over $DL\text{-}Lite_{\mathcal{R}}$ and its fragment $DL\text{-}Lite_{\mathcal{R}}^{pr}$, which both extend (first-order fragment of) RDFS. $DL\text{-}Lite_{\mathcal{R}}^{pr}$ is closed under most of the MBAs, while $DL\text{-}Lite_{\mathcal{R}}$ is *not* closed under any of them. We showed that if the TBox of $\mathcal{K}$ entails a pair of assertions of the form $A \sqsubseteq \exists R$ and $\exists R^- \sqsubseteq \neg C$, then an interplay of $\mathcal{N}$ and $\mathcal{A}$ may lead to inexpressibility of $\mathcal{K} \diamond \mathcal{N}$. For $DL\text{-}Lite_{\mathcal{R}}^{pr}$ we provided algorithms how to compute evolution results for six model-based approaches and approximate for the remaining two. For $DL\text{-}Lite_{\mathcal{R}}$ we capture evolution of KBs under a local model-based approach with $\mathcal{SHOIQ}$ using novel techniques based on what we called prototypes. We believe that prototypes are important since they can be used to study evolution for ontology languages other than $DL\text{-}Lite_{\mathcal{R}}$. Finally, we showed how to approximate evolution when it is not expressible in $DL\text{-}Lite_{\mathcal{R}}$ using what we called certain approximations.

It is the first attempt to provide an understanding of inexpressibility of MBAs for *DL-Lite* evolution. Without this understanding it is unclear how to proceed with the study of evolution in more expressive DLs and what to expect from MBAs in such logics. We also believe that our techniques of capturing semantics based on prototypes give a better understanding of how MBAs behave.

Acknowledgements. We are thankful to Diego Calvanese, Balder ten Cate, and Werner Nutt for insightful discussions. We thank anonymous reviewers for constructive comments. The authors are supported by EU projects ACSI (FP7-ICT-257593) and Ontorule (FP7-ICT-231875); the first author is supported by ERC FP7 grant Webdam (agreement n. 226513).

References

1. Baader, F., Calvanese, D., McGuinness, D., Nardi, D., Patel-Schneider, P.F. (eds.): The Description Logic Handbook: Theory, Implementation and Applications. Cambridge University Press (2003)

2. Flouris, G., Manakanatas, D., Kondylakis, H., Plexousakis, D., Antoniou, G.: Ontology change: Classification and survey. Knowledge Engineering Review 23(2), 117–152 (2008)
3. Abiteboul, S., Grahne, G.: Update semantics for incomplete databases. In: VLDB (1985)
4. Katsuno, H., Mendelzon, A.: On the difference between updating a knowledge base and revising it. In: Proc. of KR 1991, pp. 387–394 (1991)
5. Eiter, T., Gottlob, G.: On the complexity of propositional knowledge base revision, updates and counterfactuals. Artificial Intelligence 57, 227–270 (1992)
6. Liu, H., Lutz, C., Milicic, M., Wolter, F.: Updating description logic ABoxes. In: KR (2006)
7. Calvanese, D., Giacomo, G.D., Lembo, D., Lenzerini, M., Rosati, R.: Tractable reasoning and efficient query answering in description logics: The DL-Lite family. J. of Automated Reasoning 39(3), 385–429 (2007)
8. Giacomo, G.D., Lenzerini, M., Poggi, A., Rosati, R.: On instance-level update and erasure in description logic ontologies. J. of Logic and Computation 19(5), 745–770 (2009)
9. Wang, Z., Wang, K., Topor, R.W.: A new approach to knowledge base revision in DL-Lite. In: AAAI (2010)
10. Calvanese, D., Kharlamov, E., Nutt, W., Zheleznyakov, D.: Evolution of DL-lite Knowledge Bases. In: Patel-Schneider, P.F., Pan, Y., Hitzler, P., Mika, P., Zhang, L., Pan, J.Z., Horrocks, I., Glimm, B. (eds.) ISWC 2010, Part I. LNCS, vol. 6496, pp. 112–128. Springer, Heidelberg (2010)
11. Calvanese, D., Kharlamov, E., Nutt, W., Zheleznyakov, D.: Evolution of DL-Lite knowledge bases (extended version). Technical report, KRDB Research Centre, Free Univ. of Bolzano (2011)
12. Artale, A., Calvanese, D., Kontchakov, R., Zakharyaschev, M.: The DL-Lite family and relations. J. of Artificial Intelligence Research 36, 1–69 (2009)
13. Poggi, A., Lembo, D., Calvanese, D., Giacomo, G.D., Lenzerini, M., Rosati, R.: Linking data to ontologies. J. on Data Semantics, 133–173 (2008)
14. Calvanese, D., Kharlamov, E., Nutt, W.: A proof theory for DL-Lite. In: DL (2007)
15. Winslett, M.: Updating Logical Databases. Cambridge University Press (1990)
16. Horrocks, I., Sattler, U.: A tableau decision procedure for SHOIQ. J. Autom. Reasoning 39(3), 249–276 (2007)

Querying OWL 2 QL and Non-monotonic Rules

Matthias Knorr and José Júlio Alferes

CENTRIA, FCT, Universidade Nova de Lisboa

Abstract. Answering (conjunctive) queries is an important reasoning task in Description Logics (DL), hence also in highly expressive ontology languages, such as OWL. Extending such ontology languages with rules, such as those expressible in RIF-Core, and further with non-monotonic rules, integrating default negation as described in the RIF-FLD, yields an even more expressive language that allows for modeling defaults, exceptions, and integrity constraints.

Here, we present a top-down procedure for querying knowledge bases (KB) that combine non-monotonic rules with an ontology in $DL\text{-}Lite_\mathcal{R}$ – the DL underlying the OWL 2 profile OWL 2 QL. This profile aims particularly at answering queries in an efficient way for KB with large ABoxes. Our procedure extends the query-answering facility to KB that also include non-monotonic rules, while maintaining tractability of reasoning (w.r.t. data complexity). We show that the answers are sound and complete w.r.t. the well-founded MKNF model for hybrid MKNF KB $\mathcal{K}$.

1 Introduction

Combining highly expressive ontology languages, such as OWL [9], (or their underlying DL) and rules, such as in RIF-Core [3], is an important task in the on-going standardization driven by the World Wide Web Consortium[1] (W3C). Both languages are quite different in terms of expressiveness and how decidability is achieved, and providing a joint formalism is non-trivial, all the more if the rules include mechanisms for non-monotonic reasoning, such as the NAF operator described in the RIF-FLD [4].

Non-monotonic rules provide expressive features, such as the ability to model defaults, exceptions, and integrity constraints, and its usefulness is frequently being voiced. E.g., in [14], an ontology language is used for matching clinical trials criteria with patient records, but one open problem is that medication of patients, which is fully known, should be modeled by defaults.

Several approaches that combine rules and DL have been defined (see, e.g., [11,13] for an overview) but, among them, the approach of hybrid MKNF knowledge bases [13], which is based on the logics of minimal knowledge and negation as failure (MKNF) [12], is one of the most advanced. The integration of monotonic and non-monotonic reasoning is seamless yet allows for a modular re-use of reasoning algorithms of each of its components. Thus, hybrid MKNF is more expressive than comparable approaches but at least as competitive in terms of

---

[1] http://www.w3.org/

L. Aroyo et al. (Eds.): ISWC 2011, Part I, LNCS 7031, pp. 338–353, 2011.

computation. At the cost of having a weaker form of reasoning, the well-founded MKNF semantics for hybrid MKNF knowledge bases [11] achieves an even lower computational (data) complexity in general. For example, if reasoning in the DL is in PTIME, then the computation of the well-founded MKNF model remains polynomial, while [13] is in NP. This is clearly preferable for large applications, such as the one described in [14], which uses data of over 240,000 patients.

A further improvement in efficiency can be achieved if we query for information in a top-down manner: instead of computing the entire model of a knowledge base we could, e.g., just query for the medication of one patient ignoring all the others. Queries are considered in the W3C with SPARQL [15] and answering (conjunctive) queries is an important reasoning task in DL [7].

In [1], **SLG**($\mathcal{O}$) is introduced, which allows us to pose queries to a hybrid MKNF KB and whose semantics is shown to correspond to that of [11]. The work is based on an extension of SLG – a procedure for query-answering in normal logic programs under the well-founded semantics resorting to tabling techniques – that besides the operations for resolution in the rules, also incorporates calls to a parametric oracle that deals with the reasoning task in the DL part of the KB. It is shown that, if the number of answers of the oracle is appropriately restricted, the favorable computational complexity of [11] is maintained. However, it is not spelled out how these conditions are achieved for a concrete DL.

In this paper, we present a top-down procedure based on **SLG**($\mathcal{O}$) for querying KB that combine non-monotonic rules with an ontology in $DL\text{-}Lite_{\mathcal{R}}$ – the DL underlying the OWL 2 profile OWL 2 QL. This profile aims particularly at answering queries in an efficient way for KB with large ABoxes. It is thus a natural choice as DL language for a procedure for query answering in KB that, besides the ontology with large ABoxes, also includes the features of expressing defaults, constraints and exceptions, provided by non-monotonic rules. Our procedure achieves that, while maintaining tractability of reasoning (w.r.t. data complexity) on hybrid KB and reasoning in LOGSPACE on $DL\text{-}Lite_{\mathcal{R}}$ alone. In particular, query-answering is obtained by a combination of techniques from top-down procedures in logic programs, and reasoning in relational databases as done in $DL\text{-}Lite_{\mathcal{R}}$. We show that the answers are sound and complete w.r.t. the well-founded MKNF model for hybrid MKNF knowledge bases $\mathcal{K}$ if $\mathcal{K}$ is MKNF-consistent, and a paraconsistent approximation of that model otherwise. As such, our work provides a way for querying KB consisting of an ontology in the profile OWL 2 QL and a RIF dialect that allows for non-monotonic rules, and, together with [10], such a query procedure for each tractable OWL profile.

2   Preliminaries

2.1 $DL\text{-}Lite_{\mathcal{R}}$

The description logic underlying OWL 2 QL is $DL\text{-}Lite_{\mathcal{R}}$, one language of the $DL\text{-}Lite$ family [5], which we recall in the following.

The syntax of $DL\text{-}Lite_{\mathcal{R}}$ is based on three disjoint sets of individual names $\mathsf{N_I}$, concept names $\mathsf{N_C}$, and role names $\mathsf{N_R}$. Complex concepts and roles can be formed according to the following syntax, where $A \in \mathsf{N_C}$ is a concept name, $P \in \mathsf{N_R}$ a role name, and P^- its inverse.

$$C \longrightarrow A \mid \exists R \qquad R \longrightarrow P \mid P^- \qquad D \longrightarrow C \mid \neg C \qquad E \longrightarrow R \mid \neg R$$

A *DL-Lite$_\mathcal{R}$* knowledge base $\mathcal{O} = (\mathcal{T}, \mathcal{A})$ consists of a TBox $\mathcal{T}$ and an ABox $\mathcal{A}$. The TBox contains general inclusion axioms (GCI) of the form $C \sqsubseteq D$, where C and D are defined as above. Thus, the left and right hand sides of GCI are of different expressiveness in *DL-Lite$_\mathcal{R}$*. Additionally, *DL-Lite$_\mathcal{R}$* TBoxes contain role inclusion axioms (RI) of the form $R \sqsubseteq E$ where R and E are formed as introduced above. Such axioms permit to express properties, such as symmetry.

The ABox contains assertions of the form $A(a)$ and $P(a,b)$ where $A \in \mathsf{N_C}$, $P \in \mathsf{N_R}$, and $a, b \in \mathsf{N_I}$. Assertions $C(a)$ for general concepts C are included by adding $A \sqsubseteq C$ to the TBox and $A(a)$ to the ABox for a new concept name A.

The semantics of *DL-Lite$_\mathcal{R}$* is based on interpretations $\mathcal{I} = (\Delta^\mathcal{I}, \cdot^\mathcal{I})$ consisting of a nonempty interpretation domain $\Delta^\mathcal{I}$ and an interpretation function $\cdot^\mathcal{I}$ that assigns to each individual a a distinct[2] element $a^\mathcal{I}$ of $\Delta^\mathcal{I}$, to each concept C a subset $C^\mathcal{I}$, and to each role name R a binary relation $R^\mathcal{I}$ over $\mathcal{I}$. This can be generalized to complex expressions as usual:

$$\begin{aligned}
(P^-)^\mathcal{I} &= \{(i_2, i_1) \mid (i_1, i_2) \in P^\mathcal{I}\} & (\neg C)^\mathcal{I} &= \Delta^\mathcal{I} \setminus C^\mathcal{I} \\
(\exists R)^\mathcal{I} &= \{i \mid (i, i') \in R^\mathcal{I}\} & (\neg R)^\mathcal{I} &= \Delta \times \Delta \setminus R^\mathcal{I}
\end{aligned}$$

An interpretation $\mathcal{I}$ is a model of the GCI $C \sqsubseteq D$ if $C^\mathcal{I} \subseteq D^\mathcal{I}$. It is a model of an RI $R \sqsubseteq E$ if $R^\mathcal{I} \subseteq E^\mathcal{I}$. $\mathcal{I}$ is also a model of an assertion $A(a)$ $(P(a,b))$ if $a^\mathcal{I} \in A^\mathcal{I}$ $((a^\mathcal{I}, b^\mathcal{I}) \in P^\mathcal{I})$. Given an axiom/assertion α we denote by $\mathcal{I} \models \alpha$ that $\mathcal{I}$ is a model of α. A model of a *DL-Lite$_\mathcal{R}$* KB $\mathcal{O} = (\mathcal{T}, \mathcal{A})$ is an interpretation $\mathcal{I}$ such that $\mathcal{I} \models \alpha$ holds for all $\alpha \in \mathcal{T}$ and all $\alpha \in \mathcal{A}$. A KB $\mathcal{O}$ is satisfiable if it has at least one model.

Standard reasoning tasks in *DL-Lite$_\mathcal{R}$* are polynomial in the size of the TBox, and in LOGSPACE in the size of the ABox, i.e., in data complexity. The same holds for answering conjunctive queries, but if we consider the combined complexity (including the query), then answering conjunctive queries is NP-complete [5].

2.2 Well-Founded Semantics for Hybrid MKNF

Hybrid MKNF knowledge bases are introduced in [13] as a combination of non-monotonic rules and a DL that is translatable into first-order logic, and in which standard reasoning tasks, namely satisfiability and instance checking, are decidable. Here, we recall only the version with rules without disjunction in heads and the computation of the complete well-founded MKNF model for such knowledge bases [11], for which we define a top-down procedure for *DL-Lite$_\mathcal{R}$*. Recalling the computation of the complete well-founded MKNF model is not strictly needed for the definition of the top-down procedure itself, but we present it here since

---

[2] Hence, the unique name assumption is applied and, as shown in [2], dropping it would increase significantly the computational complexity of *DL-Lite$_\mathcal{R}$*.

it provides valuable insights into the combined model for which we query. More-over, the operator $D_{\mathcal{K}_G}$, defined in this section, provides exactly the counterpart of the oracle to the DL in the top-down procedure we present in this paper.

Definition 1. *Let $\mathcal{O}$ be a DL knowledge base. A function-free first-order atom $P(t_1, \ldots, t_n)$ such that P is $\approx$ or occurs in $\mathcal{O}$ is called a DL-atom; otherwise it is called non-DL-atom. An MKNF rule r has the following form where H, A_i, and B_i are function-free first-order atoms:*

$$\mathbf{K}\, H \leftarrow \mathbf{K}\, A_1, \ldots, \mathbf{K}\, A_n, \mathbf{not}\, B_1, \ldots, \mathbf{not}\, B_m \tag{1}$$

K-*atoms (resp.* **not**-*atoms) are atoms with a leading* **K** *(resp.* **not***). A program is a finite set of MKNF rules, and a* hybrid MKNF knowledge base $\mathcal{K}$ *is a pair $(\mathcal{O}, \mathcal{P})$ and* positive *if $m = 0$ holds for all MKNF rules in $\mathcal{K}$. The* ground instantiation *of $\mathcal{K}$ is the KB $\mathcal{K}_G = (\mathcal{O}, \mathcal{P}_G)$ where $\mathcal{P}_G$ is obtained from $\mathcal{P}$ by replacing each rule r of $\mathcal{P}$ with a set of rules substituting each variable in r with constants from $\mathcal{K}$ in all possible ways.*

There is no restriction on the interaction between $\mathcal{O}$ and $\mathcal{P}$, i.e., DL-atoms may appear anywhere in the rules.

The semantics of $\mathcal{K}$ is based on a transformation of $\mathcal{K}$ into an MKNF formula to which the MKNF semantics can be applied (see [11,12,13] for details). De-cidability is achieved by applying the well-known notion of DL-safety, in which each variable in a rule appears in at least one non-DL **K**-atom [13]. This es-sentially restricts the application of rules to individuals explicitly appearing in the knowledge base in consideration [13]. Instead of spelling out the technical details of the original MKNF semantics [13] or its three-valued counterpart [11], we focus on a compact representation of models for which the computation of the well-founded MKNF model is defined[3]. This representation is based on a set of **K**-atoms and $\pi(\mathcal{O})$, the translation of $\mathcal{O}$ into first-order logic.

Definition 2. *Let $\mathcal{K}_G = (\mathcal{O}, \mathcal{P}_G)$ be a ground hybrid MKNF knowledge base. The set of* **K**-*atoms of $\mathcal{K}_G$, written $\mathsf{KA}(\mathcal{K}_G)$, is the smallest set that contains (i) all ground* **K**-*atoms occurring in $\mathcal{P}_G$, and (ii) a* **K**-*atom* **K**ξ *for each ground* **not**-*atom* **not** ξ *occurring in $\mathcal{P}_G$. For a subset S of $\mathsf{KA}(\mathcal{K}_G)$, the* objective knowledge *of S w.r.t. $\mathcal{K}_G$ is the set of first-order formulas $\mathsf{OB}_{\mathcal{O},S} = \{\pi(\mathcal{O})\} \cup \{\xi \mid \mathbf{K}\xi \in S\}$.*

The set $\mathsf{KA}(\mathcal{K}_G)$ contains all atoms occurring in $\mathcal{K}_G$, only with **not**-atoms substituted by corresponding modal **K**-atoms, while $\mathsf{OB}_{\mathcal{O},S}$ provides a first-order representation of $\mathcal{O}$ together with a set of known/derived facts.

In the three-valued MKNF semantics, this set of **K**-atoms can be divided into true, undefined and false modal atoms. Next, we recall operators from [11] that derive consequences based on $\mathcal{K}_G$ and a set of **K**-atoms that is considered to hold. To further simplify notation, in the remainder of the paper we abuse notation and consider all operators **K** implicit.

---

[3] Strictly speaking, this computation yields the so-called well-founded partition from which the well-founded MKNF model is defined (see [11] for details).

Definition 3. *Let* $\mathcal{K}_G = (\mathcal{O}, \mathcal{P}_G)$ *be a positive, ground hybrid MKNF knowledge base. The operators* $R_{\mathcal{K}_G}$, $D_{\mathcal{K}_G}$, *and* $T_{\mathcal{K}_G}$ *are defined on subsets of* $\mathsf{KA}(\mathcal{K}_G)$:

$$R_{\mathcal{K}_G}(S) = \{H \mid \mathcal{P}_G \text{ contains a rule of the form } H \leftarrow A_1, \dots A_n$$
$$\text{such that, for all } i, 1 \leq i \leq n, A_i \in S\}$$
$$D_{\mathcal{K}_G}(S) = \{\xi \mid \xi \in \mathsf{KA}(\mathcal{K}_G) \text{ and } \mathsf{OB}_{\mathcal{O},S} \models \xi\}$$
$$T_{\mathcal{K}_G}(S) = R_{\mathcal{K}_G}(S) \cup D_{\mathcal{K}_G}(S)$$

The operator $T_{\mathcal{K}_G}$ is monotonic, and thus has a least fixpoint $T_{\mathcal{K}_G} \uparrow \omega$. Transformations can be defined that turn an arbitrary hybrid MKNF KB $\mathcal{K}_G$ into a positive one (respecting the given set S) to which $T_{\mathcal{K}_G}$ can be applied. To ensure coherence, i.e., that classical negation in the DL enforces default negation in the rules, two slightly different transformations are defined (see [11] for details).

Definition 4. *Let* $\mathcal{K}_G = (\mathcal{O}, \mathcal{P}_G)$ *be a ground hybrid MKNF knowledge base and* $S \subseteq \mathsf{KA}(\mathcal{K}_G)$. *The MKNF transform* $\mathcal{K}_G/S$ *is defined as* $\mathcal{K}_G/S = (\mathcal{O}, \mathcal{P}_G/S)$, *where* $\mathcal{P}_G/S$ *contains all rules* $H \leftarrow A_1, \dots, A_n$ *for which there exists a rule* $H \leftarrow A_1, \dots, A_n, \mathbf{not}\, B_1, \dots, \mathbf{not}\, B_m$ *in* $\mathcal{P}_G$ *with* $B_j \notin S$ *for all* $1 \leq j \leq m$. *The MKNF-coherent transform* $\mathcal{K}_G//S$ *is defined as* $\mathcal{K}_G//S = (\mathcal{O}, \mathcal{P}_G//S)$, *where* $\mathcal{P}_G//S$ *contains all rules* $H \leftarrow A_1, \dots, A_n$ *for which there exists a rule* $H \leftarrow A_1, \dots, A_n, \mathbf{not}\, B_1, \dots, \mathbf{not}\, B_m$ *in* $\mathcal{P}_G$ *with* $B_j \notin S$ *for all* $1 \leq j \leq m$ *and* $\mathsf{OB}_{\mathcal{O},S} \not\models \neg H$. *We define* $\Gamma_{\mathcal{K}_G}(S) = T_{\mathcal{K}_G/S} \uparrow \omega$ *and* $\Gamma'_{\mathcal{K}_G}(S) = T_{\mathcal{K}_G//S} \uparrow \omega$.

Based on these two antitonic operators [11], two sequences $\mathbf{P}_i$ and $\mathbf{N}_i$ are defined, which correspond to the true and non-false derivations.

$$\mathbf{P}_0 = \emptyset \qquad\qquad \mathbf{N}_0 = \mathsf{KA}(\mathcal{K}_G)$$
$$\mathbf{P}_{n+1} = \Gamma_{\mathcal{K}_G}(\mathbf{N}_n) \qquad\qquad \mathbf{N}_{n+1} = \Gamma'_{\mathcal{K}_G}(\mathbf{P}_n)$$
$$\mathbf{P}_\omega = \bigcup \mathbf{P}_i \qquad\qquad \mathbf{N}_\omega = \bigcap \mathbf{N}_i$$

The fixpoints, which are reached after finitely many iterations, yield the well-founded MKNF model [11].

Definition 5. *The well-founded MKNF model of an MKNF-consistent ground hybrid MKNF knowledge base* $\mathcal{K}_G = (\mathcal{O}, \mathcal{P}_G)$ *is defined as* $(\mathbf{P}_\omega, \mathsf{KA}(\mathcal{K}_G) \setminus \mathbf{N}_\omega)$.

If $\mathcal{K}_G$ is MKNF-inconsistent, then there is no MKNF model, hence no well-founded MKNF model.

We use a simple example adapted from [5] to illustrate this computation.

Example 1. Consider the hybrid MKNF KB $\mathcal{K}^4$ consisting of $\mathcal{O}$:

Professor $\sqsubseteq \exists$TeachesTo	Student $\sqsubseteq \exists$HasTutor
$\exists$TeachesTo$^-$ $\sqsubseteq$ Student	$\exists$HasTutor$^-$ $\sqsubseteq$ Professor
Professor $\sqsubseteq \neg$Student	HasTutor$^-$ $\sqsubseteq$ TeachesTo

and the rules (including the facts):

---

[4] We use capital letters for DL-atoms and individuals. Note that $\mathcal{K}$ is not DL-safe, but we assume that each rule contains implicitly, for all variables x appearing in the rule, an additional atom o(x), and that $\mathcal{K}$ contains facts o(x), for every object x.

$$\text{hasKnownTutor}(x) \leftarrow \text{Student}(x), \text{HasTutor}(x, y) \tag{2}$$

$$\text{hasUnknownTutor}(x) \leftarrow \text{Student}(x), \textbf{not } \text{HasTutor}(x, y) \tag{3}$$

$$\text{Student}(\text{Paul}) \leftarrow \quad \text{HasTutor}(\text{Jane}, \text{Mary}) \leftarrow \quad \text{TeachesTo}(\text{Mary}, \text{Bill}) \leftarrow$$

We only consider the set $\text{KA}(\mathcal{K}_G)$ for the computation, i.e., only the atoms that actually appear in the ground rules and we abbreviate the names in this example appropriately. Starting with $\mathbf{P}_0 = \emptyset$ and $\mathbf{N}_0 = \text{KA}(\mathcal{K}_G)$, we compute $\mathbf{P}_1 = \{\text{S(P)}, \text{HT(J, M)}, \text{TT(M, B)}, \text{S(J)}, \text{S(B)}, \text{HT(B, M)}, \text{hKT(J)}, \text{hKT(B)}\}$ and $\mathbf{N}_1 = \mathbf{P}_1 \cup \{\text{hUT(P)}, \text{hUT(J)}, \text{hUT(B)}\}$. We continue with $\mathbf{P}_2 = \mathbf{P}_1 \cup \{\text{hUT(P)}\}$ and $\mathbf{N}_2 = \mathbf{P}_2$, and these are already the fixpoints. We obtain that $\texttt{Jane}$ and $\texttt{Bill}$ have a known tutor, while $\texttt{Paul}$ has not. Other derivations, such as Professor(Mary) can be obtained from the fixpoints and $\mathcal{O}$. Note that rule (3) can be understood as a default: it states that by default students have unknown tutors.

3   Queries in Hybrid MKNF

In [1], a procedure, called $\mathbf{SLG}(\mathcal{O})$, is defined for querying arbitrary hybrid MKNF knowledge bases. This procedure extends SLG resolution with tabling [6] with an oracle to $\mathcal{O}$ that handles ground queries to the DL-part of $\mathcal{K}_G$: given the already derived information and $\mathcal{O}$, the oracle returns a (possibly empty) set of atoms that allows us, together with $\mathcal{O}$ and the already derived information, to derive the queried atom. Recalling the full procedure would be beyond the scope of this paper, so we just give an intuitive overview of the general procedure and only point out the specific technical details that are required for the concrete oracle to $DL\text{-}Lite_{\mathcal{R}}$ in terms of an interface. All the details can be found at [1].

The general idea is that $\mathbf{SLG}(\mathcal{O})$ creates a forest of derivation trees starting with the initial query. If the initial query is unifiable with a rule head, then the rule body is added as a child node to the respective tree. In each such new node, we consider each atom according to a given selection order, create new trees, and try to resolve these so-called goals. Due to potential circularities in the derivations, some goals may be delayed, possibly giving rise to conditional answers. If we achieve an unconditional answer for such a goal, i.e., a leaf node without any remaining goals, then the (instantiated) atom in the root is true, and the goal is resolved. If no further operation is applicable, but no leaf node is empty, then the root(s) corresponding to such a (set of) tree(s) is (are) considered false. If a (default) negated atom $\textbf{not } A$ is selected, then a tree for A is created: if A succeeds then $\textbf{not } A$ fails; otherwise $\textbf{not } A$ is resolved.

If the queried/selected atom is a DL-atom, then a query is posed to an oracle, which encapsulates reasoning in the DL part of the KB, to check whether the DL-atom can be derived from the DL-part. For that purpose, all DL-atoms that are already known to be true in the derivation forest are added to the ontology before calling the oracle. It may happen that only a subsequent derivation step contains the necessary information to derive the queried atom, so we would have to call the oracle over and over again. To avoid this problem, rather than answering whether the queried atom is derivable, an oracle in **SLG($\mathcal{O}$)** returns a set (conjunction) of atoms that, if proven true, would allow us to derive the queried atom. This set is added as a child to the respective tree and treated like the result of an operation on rules.

In the end, if we achieve an unconditional answer for the initial query, the (instantiated) query is true; if all answers are conditional, then the query is undefined; otherwise it is false. The following example shall clarify the idea.

Example 2. Consider again the hybrid MKNF knowledge base $\mathcal{K}$ from Example 1 and the query Student(Bill). A root Student(Bill) is created and since no rule head is unifiable, we call the oracle. If TeachesTo(x, Bill) holds for some x, then Student(Bill) would hold. Thus a child with the goal TeachesTo(x, Bill) is added to the tree. Then, a new tree for TeachesTo(x, Bill) is created whose root unifies with TeachesTo(Mary, Bill). We obtain an unconditional answer, and, after resolving TeachesTo(x, Bill) in the tree with root Student(Bill), an unconditional answer for Student(Bill), thus finishing the derivation.

Alternatively, consider the query hUT(P) (with abbreviations as in Example 1). Since the root of the corresponding tree is a non-DL-atom, only rules can be considered, and, in fact, rule (3) is applicable. The resulting child contains two goals, namely S(P) and **not** HT(P, y). Note that implicitly o(y) also occurs so that y is ground when querying for **not** HT(P, y). A separate tree is created for S(P) and easily resolved with the given fact. Then we consider any of the meanwhile grounded **not** HT(P, y). A new tree is created for HT(P, y) (with ground y) but neither the rules nor the oracle allow us to derive an unconditional answer and eventually each such tree is failed, since there is no known tutor for P. If the tree for HT(P, y) fails, then **not** HT(P, y) holds and can be resolved, which yields an unconditional answer for hUT(P).

As already said, we only want to consider specifically the mechanism that provides the oracle for *DL-Lite$_\mathcal{R}$*, and here we only recall the relevant notions for that task. We start by defining the reasoning task in consideration, namely DL-safe conjunctive queries.

Definition 6. *A (DL-safe) conjunctive query q is a non-empty set, i.e. conjunction, of literals where each variable in q occurs in at least one non-DL atom in q. We also write q as a rule $q(X_i) \leftarrow A_1, \ldots, A_n, \textbf{not } B_1, \ldots, \textbf{not } B_m$ where X_i is the (possibly empty) set of variables, appearing in the body, which are requested.*

This guarantees that **SLG($\mathcal{O}$)** does always pose ground queries to the oracle, avoiding problems with DL where conjunctive query answering is undecidable in general, and, in particular, with inconsistent *DL-Lite$_\mathcal{R}$* KB that would simply return arbitrary solutions [5] in opposite to our intentions (see Definition 8).

Now, we recall the definition of a complete oracle that provides the relation for the intended derivation. We point out that $\mathcal{F}_n$ is the current derivation forest, where n is increased with each applied **SLG($\mathcal{O}$)** operation. In it, $I_{\mathcal{F}_n}$ corresponds to all already derived (true or false) atoms in a concrete derivation forest.

Definition 7. *Let* $\mathcal{K} = (\mathcal{O}, \mathcal{P})$ *be a hybrid MKNF knowledge base, S a ground goal, L a set of ground atoms that appear in at least one rule head in $\mathcal{P}_G$, and $I^+_{\mathcal{F}_n} = I_{\mathcal{F}_n} \setminus \{\mathbf{not}\, A \mid \mathbf{not}\, A \in I_{\mathcal{F}_n}\}$. The* complete oracle *for $\mathcal{O}$, denoted $compT_{\mathcal{O}}$, is defined by $compT_{\mathcal{O}}(I_{\mathcal{F}_n}, S, L)$ iff $\mathcal{O} \cup I^+_{\mathcal{F}_n} \cup L \models S$.*

This notion of a complete oracle is used to define the $\mathbf{SLG}(\mathcal{O})$ operation that handles the oracle calls to the DL (see [1] for notation of trees in $\mathbf{SLG}(\mathcal{O})$).

–   ORACLE RESOLUTION: Let $\mathcal{F}_n$ contain a tree with root node $N = S :- |S$, where S is ground. Assume that $compT_{\mathcal{O}}(I_{\mathcal{F}_n}, S, Goals)$. If N does not have a child $N_{child} = S :- |Goals$ in $\mathcal{F}_n$ then add N_{child} as a child of N.

It is shown (in Theorem 5.3 of [1]) that answers to queries in $\mathbf{SLG}(\mathcal{O})$ correspond to the hybrid MKNF well-founded model as in [11].

The complete oracle is unfortunately not efficient, in that it potentially creates a lot of superfluous answers, such as supersets of correct minimal answers. This problem is tackled with the introduction of a partial oracle [1].

Definition 8. *Let $\mathcal{K}_G = (\mathcal{O}, \mathcal{P}_G)$ be a hybrid MKNF knowledge base, S a ground goal, and L a set of atoms that are unifiable with at least one rule head in $\mathcal{P}_G$. A* partial oracle *for $\mathcal{K}_G$, denoted $pT_{\mathcal{O}}$, is a relation $pT_{\mathcal{O}}(I_{\mathcal{F}_n}, S, L)$ such that if $pT_{\mathcal{O}}(I_{\mathcal{F}_n}, S, L)$, then $\mathcal{O} \cup I^+_{\mathcal{F}_n} \cup L \models S$ and $\mathcal{O} \cup I^+_{\mathcal{F}_n} \cup L$ consistent. A partial oracle $pT_{\mathcal{O}}$ is* correct *w.r.t. $compT_{\mathcal{O}}$ iff, for all MKNF-consistent $\mathcal{K}_G$, replacing $compT_{\mathcal{O}}$ in $\mathbf{SLG}(\mathcal{O})$ with $pT_{\mathcal{O}}$ succeeds for exactly the same set of queries.*

There are three main differences between complete and partial oracles. First, in the latter we do not have to consider all possible answers, thus restricting the number of returned answers for a given query. This is important, because, as pointed out in [1], the favorable computational properties of the well-founded MKNF model are only maintained if the number of returned answers is appropriately bounded. Second, only derivations based on a consistent $\mathcal{O} \cup I^+_{\mathcal{F}_n} \cup L$ are considered. This removes all pointless attempts to derive an inconsistency for an MKNF-consistent KB $\mathcal{K}$. It also removes derivations based on explosive behavior w.r.t. inconsistencies, which is why the correctness result of the partial oracle is limited to MKNF-consistent KB. This may not be a problem, since it has already been conjectured in [1] that the resulting paraconsistent behavior should be beneficial in practice. Finally, instead of requiring ground sets L, we admit sets of atoms that are unifiable with heads. This simplifies notation and postpones the necessary grounding to be handled in $\mathbf{SLG}(\mathcal{O})$ w.r.t. the rules.

How such a partial oracle is defined for a concrete DL is not specified in [1] and, in Section 5, we present a concrete oracle for $DL\text{-}Lite_{\mathcal{R}}$.

4   Satisfiability in $DL\text{-}Lite_{\mathcal{R}}$

For defining an oracle for $DL\text{-}Lite_{\mathcal{R}}$ we rely on the reasoning algorithms provided for $DL\text{-}Lite_{\mathcal{R}}$ in [5]. The basic reasoning service to which all others are reduced is satisfiability. Satisfiability of a $DL\text{-}Lite_{\mathcal{R}}$ KB is checked by evaluating a suitable

Boolean first-order logic query w.r.t. a canonical model of its ABox. We recall the construction and evaluation of such a formula here specifically, since our work is based on an adaptation of that algorithm. First, we recall the definition of a canonical interpretation of the ABox from [5].

Definition 9. *Let $\mathcal{A}$ be a DL-Lite$_\mathcal{R}$ ABox. By $db(\mathcal{A}) = (\Delta^{db(\mathcal{A})}, \cdot^{db(\mathcal{A})})$ we denote the interpretation defined as follows:*

- *$\Delta^{db(\mathcal{A})}$ is the nonempty set consisting of all constants occurring in $\mathcal{A}$;*
- *$a^{db(\mathcal{A})} = a$ for each constant a;*
- *$A^{db(\mathcal{A})} = \{a \mid A(a) \in \mathcal{A}\}$ for each atomic concept A; and*
- *$R^{db(\mathcal{A})} = \{(a,b) \mid R(a,b) \in \mathcal{A}\}$ for each atomic role R.*

It is easy to see that $db(\mathcal{A})$ is in fact a model of $\mathcal{A}$ and a minimal one [5].

This forms the basis for checking satisfiability in *DL-Lite$_\mathcal{R}$*, i.e., we are able to reduce satisfiability to evaluating a Boolean FOL query.

Definition 10. *Satisfiability in DL-Lite$_\mathcal{R}$ is FOL-reducible if, for every TBox $\mathcal{T}$ expressed in DL-Lite$_\mathcal{R}$, there exists a Boolean FOL query q, over the alphabet of $\mathcal{T}$, such that, for every nonempty ABox $\mathcal{A}$, $(\mathcal{T}, \mathcal{A})$ is satisfiable if and only if q evaluates to false in $db(\mathcal{A})$.*

Now, a query is constructed, by first splitting all GCI into those with $\neg$ on the right hand side (called *negative inclusions* (NI)) and those without (called *positive inclusions* (PI)) and by considering the negative inclusions as a starting point to compute all derivable negative inclusions.

Definition 11. *Let $\mathcal{T}$ be a DL-Lite$_\mathcal{R}$ TBox. We call NI-closure of $\mathcal{T}$, denoted by $cln(\mathcal{T})$, the set defined inductively as follows:*

1. *All negative inclusion assertions in $\mathcal{T}$ are also in $cln(\mathcal{T})$.*
2. *If $B_1 \sqsubseteq B_2$ is in $\mathcal{T}$ and $B_2 \sqsubseteq \neg B_3$ or $B_3 \sqsubseteq \neg B_2$ is in $cln(\mathcal{T})$, then also $B_1 \sqsubseteq \neg B_3$ is in $cln(\mathcal{T})$.*
3. *If $R_1 \sqsubseteq R_2$ is in $\mathcal{T}$ and $\exists R_2 \sqsubseteq \neg B$ or $B \sqsubseteq \neg\exists R_2$ is in $cln(\mathcal{T})$, then also $\exists R_1 \sqsubseteq \neg B$ is in $cln(\mathcal{T})$.*
4. *If $R_1 \sqsubseteq R_2$ is in $\mathcal{T}$ and $\exists R_2^- \sqsubseteq \neg B$ or $B \sqsubseteq \neg\exists R_2^-$ is in $cln(\mathcal{T})$, then also $\exists R_1^- \sqsubseteq \neg B$ is in $cln(\mathcal{T})$.*
5. *If $R_1 \sqsubseteq R_2$ is in $\mathcal{T}$ and $R_2 \sqsubseteq \neg R_3$ or $R_3 \sqsubseteq \neg R_2$ is in $cln(\mathcal{T})$, then also $R_1 \sqsubseteq \neg R_3$ is in $cln(\mathcal{T})$.*
6. *If one of the assertions $\exists R \sqsubseteq \neg\exists R$, $\exists R^- \sqsubseteq \neg\exists R^-$, or $R \sqsubseteq \neg R$ is in $cln(\mathcal{T})$, then all three such assertions are in $cln(\mathcal{T})$.*

A translation function from assertions in $cln(\mathcal{T})$ to FOL formulas is defined [5].

Definition 12. *Let $\mathcal{O}$ be a DL-Lite$_\mathcal{R}$ KB and $cln(\mathcal{T})$ the NI-closure of $\mathcal{T}$. The translation function δ from axioms in $cln(\mathcal{T})$ to first-order formulas is:*

$$\delta(B_1 \sqsubseteq \neg B_2) = \exists x. \gamma_1(x) \wedge \gamma_2(x)$$

$$\delta(R_1 \sqsubseteq \neg R_2) = \exists x, y. \rho_1(x,y) \wedge \rho_2(x,y)$$

where $\gamma_i(x) = A_i(x)$ if $B_i = A_i$, $\gamma_i(x) = \exists y_i. R_i(x, y_i)$ if $B_i = \exists R_i$, and $\gamma_i(x) = \exists y_i. R_i(y_i, x)$ if $B_i = \exists R^-$; and $\rho_i(x,y) = P_i(x,y)$ if $R_i = P_i$, and $\rho_i(x,y) = P_i(y,x)$ if $R_i = P_i^-$.

Require: DL-$Lite_{\mathcal{R}}$ KB $\mathcal{O} = (\mathcal{T}, \mathcal{A})$
Ensure: true if $\mathcal{O}$ is satisfiable, false otherwise

> $q_{unsat} = \bot$
> **for all** $\alpha \in cln(\mathcal{T})$ **do**
> > $q_{unsat} = q_{unsat} \vee \delta(\alpha)$
> **end for**
> **if** $q_{unsat}^{db(\mathcal{A})} = \emptyset$ **then**
> > return **true**
> **else**
> > return **false**
> **end if**

Fig. 1. Algorithm Consistent

The algorithm in Fig. 1 checks satisfiability of a DL-$Lite_{\mathcal{R}}$ knowledge base $\mathcal{O}$ by testing if q_{unsat} is not satisfied if evaluated over $db(\mathcal{A})$. Of course, if $\mathcal{O}$ is free of negative inclusions, then the algorithm succeeds automatically.

With this, instance checking is straightforwardly obtained in [5] as a reduction to satisfiability checking.

Theorem 1. *Let $\mathcal{O}$ be DL-$Lite_{\mathcal{R}}$ KB, and H either a general concept (with ground argument t_i) appearing in $\mathcal{O}$ where $\hat{A}$ an atomic concept not appearing in $\mathcal{O}$ or a role name or its inverse (with ground arguments t_i) appearing in $\mathcal{O}$ and $\hat{A}$ an atomic role not appearing in $\mathcal{O}$. Then $\mathcal{O} \models H(t_i)$ iff $\mathcal{O} \cup \{\hat{A} \sqsubseteq \neg H, \hat{A}(t_i)\}$ is unsatisfiable.*

Note that this theorem is a generalization of two separate theorems for concepts and roles in [5] joined here for reasons of notation.

5   An Oracle for DL-$Lite_{\mathcal{R}}$

The material presented in Section 4 suffices to handle the bottom-up computation of the well-founded MKNF model w.r.t. DL-$Lite_{\mathcal{R}}$. In the subsequent example, which we recall from [5], and which is a modification of Example 1 now without rules and with the facts turned into ABox assertions, we not only present how satisfiability and instance checking work, but also intuitively sketch the solution for defining an oracle.

Example 3. Consider the DL-$Lite_{\mathcal{R}}$ KB $\mathcal{O}$ consisting of the axioms in the TBox:

$$\text{Professor} \sqsubseteq \exists\text{TeachesTo} \qquad (4) \qquad \exists\text{HasTutor}^- \sqsubseteq \text{Professor} \qquad (7)$$

$$\text{Student} \sqsubseteq \exists\text{HasTutor} \qquad (5) \qquad \text{Professor} \sqsubseteq \neg\text{Student} \qquad (8)$$

$$\exists\text{TeachesTo}^- \sqsubseteq \text{Student} \qquad (6) \qquad \text{HasTutor}^- \sqsubseteq \text{TeachesTo} \qquad (9)$$

and the simple ABox:

$$\text{Student}(\text{Paul}) \qquad \text{HasTutor}(\text{Jane}, \text{Mary}) \qquad \text{TeachesTo}(\text{Mary}, \text{Bill}). \qquad (10)$$

For checking satisfiability, we consider $db(\mathcal{A})$ with $\Delta^{db(\mathcal{A})} = \{\texttt{Paul}, \texttt{Jane}, \texttt{Mary},$ $\texttt{Bill}\}$ whose elements are all mapped to themselves, and the interpretation of all concept and role names according to $\mathcal{A}$, e.g., $\texttt{Student}^{db(\mathcal{A})} = \{\texttt{Paul}\}$ and $\texttt{Professor}^{db(\mathcal{A})} = \emptyset$.

Then, we compute $cln(\mathcal{T})$ as follows.

$$\texttt{Professor} \sqsubseteq \neg\texttt{Student} \qquad (11)$$
$$\exists\texttt{HasTutor}^- \sqsubseteq \neg\texttt{Student} \qquad (12)$$
$$\exists\texttt{TeachesTo}^- \sqsubseteq \neg\texttt{Professor} \qquad (13)$$
$$\exists\texttt{TeachesTo} \sqsubseteq \neg\texttt{Student} \qquad (14)$$
$$\exists\texttt{HasTutor} \sqsubseteq \neg\texttt{Professor} \qquad (15)$$

Axiom 11 occurs in $\mathcal{T}$, Axiom 12 follows from (11) and (7), Axiom 13 follows from (11) and (6), and (14) and (15) follow from (9) and (12), respectively (13).

The translation function δ can be applied to each negative inclusion in $cln(\mathcal{T})$.

$$\delta(\texttt{Professor} \sqsubseteq \neg\texttt{Student}) = \exists x.\texttt{Professor}(x) \wedge \texttt{Student}(x) \qquad (16)$$
$$\delta(\texttt{HasTutor}^- \sqsubseteq \neg\texttt{Student}) = \exists x.(\exists y \texttt{HasTutor}(y, x)) \wedge \texttt{Student}(x) \qquad (17)$$
$$\delta(\texttt{TeachesTo}^- \sqsubseteq \neg\texttt{Professor}) = \exists x.(\exists y \texttt{TeachesTo}(y, x)) \wedge \texttt{Professor}(x) \qquad (18)$$
$$\delta(\texttt{TeachesTo} \sqsubseteq \neg\texttt{Student}) = \exists x.(\exists y \texttt{TeachesTo}(x, y)) \wedge \texttt{Student}(x) \qquad (19)$$
$$\delta(\texttt{HasTutor} \sqsubseteq \neg\texttt{Professor}) = \exists x.(\exists y \texttt{HasTutor}(x, y)) \wedge \texttt{Professor}(x) \qquad (20)$$

Considering $db(\mathcal{A})$ and the disjunction of first-order formulas resulting from the translation yields a successful test for satisfiability.

If we want to verify, e.g., $\texttt{Student}(\texttt{Paul})$, then we extend $\mathcal{O}$ with $\hat{\texttt{A}}(\texttt{Paul})$ and $\hat{\texttt{A}} \sqsubseteq \neg\texttt{Student}$ resulting in $\mathcal{O}'$, update $db(\mathcal{A}')$ appropriately, and add three more negative inclusions to $cln(\mathcal{T})$ resulting in $cln(\mathcal{T}')$:

$$\hat{\texttt{A}} \sqsubseteq \neg\texttt{Student} \qquad \exists\texttt{TeachesTo}^- \sqsubseteq \neg\hat{\texttt{A}} \qquad \exists\texttt{HasTutor} \sqsubseteq \neg\hat{\texttt{A}}$$

These axioms can again be translated, and it can be verified that the resulting check yields unsatisfiability. From this, we derive that $\texttt{Student}(\texttt{Paul})$ holds.

$$\delta(\hat{\texttt{A}} \sqsubseteq \neg\texttt{Student}) = \exists x.\hat{\texttt{A}}(x) \wedge \texttt{Student}(x) \qquad (21)$$
$$\delta(\texttt{TeachesTo}^- \sqsubseteq \neg\hat{\texttt{A}}) = \exists x.(\exists y \texttt{TeachesTo}(y, x)) \wedge \hat{\texttt{A}}(x) \qquad (22)$$
$$\delta(\texttt{HasTutor} \sqsubseteq \neg\hat{\texttt{A}}) = \exists x.(\exists y \texttt{HasTutor}(x, y)) \wedge \hat{\texttt{A}}(x) \qquad (23)$$

If we want to incorporate this into the top-down query procedure $\mathbf{SLG}(\mathcal{O})$, then there are two possible ways for the concrete example. First, we may, e.g., query for $\texttt{Student}(\texttt{Paul})$ and the previously presented steps would derive this from $\mathcal{O}$ alone, so that we would expect the empty answer for $\mathbf{SLG}(\mathcal{O})$. I.e., nothing needs to be added to $\mathcal{O}$ to derive the queried atom from $\mathcal{O}$ and an unconditional answer is created in the tree for $\texttt{Student}(\texttt{Paul})$.

Alternatively, consider that the ABox is not present, but that, for simplicity, the corresponding statements occur as rule facts as in Example 1. In this case, we want the oracle to return a set of atoms, which if resolved prove the original query. Clearly, we can derive Student(Paul) if the satisfiability test for $\mathcal{O}$ fails. This is the case if one of the disjuncts in $q_{unsat}^{db(\mathcal{A})}$ is satisfiable, e.g., if there is an x such that Professor(x) ∧ Student(x). Of course, it is counterintuitive to prove that Paul is a student by showing that there is some other individual that is a professor and a student, i.e., by deriving some inconsistency in the interaction of $\mathcal{O}$ and the rules. Thus, all the disjuncts resulting from (16)–(20), do not yield meaningful derivations. Instead they yield derivations based on some general MKNF-inconsistency, which is not possible in a partial oracle (cf. Definition 8).

However, if we resolve the disjuncts resulting from (21)–(23) with Â(Paul), then we obtain more meaningful answers that can be used in the derivation tree for Student(Paul). Namely, Student(Paul) itself is obtained, which is immediately discarded in **SLG($\mathcal{O}$)** since Student(Paul) is already a child in this tree, and (∃yTeachesTo(y, Paul)) and (∃yHasTutor(Paul, y)) are also obtained as possible children. Both do not contribute to the derivation of Student(Paul) itself, which is in fact obtained from the rule fact, but if we query for Student(Jane) or Student(Bill), then in each case one of the two goals unifies with a fact in the given rules.

The insights gained with this example can be formalized in the algorithm (Fig. 2) that provides an oracle for *DL-Lite$_\mathcal{R}$*. We only have to formalize the resolution step of the newly introduced query atom with each of the results of applications of δ. The result of such a resolution step is either a ground (unary or binary) atom or a binary atom with one existentially quantified variable. To check whether adding this atom to $\mathcal{O}$ and the already derived information remains consistent, we additionally introduce a uniform notion that turns the new atom into DL notation.

Definition 13. *Let $\mathcal{O}$ be a DL-Lite$_\mathcal{R}$ KB, α an axiom in $cln(\mathcal{T})$, and $\delta(\alpha) = \exists \boldsymbol{x}.(C_1 \wedge C_2)$ such that H is unifiable with mgu[5] θ with C_i, for some i, in $\delta(\alpha)$. Then $res(\delta(\alpha), H)$ is defined as $(C_2)\theta$ if $i = 1$, and $(C_1)\theta$ otherwise. The DL representation $res_{DL}(\delta(\alpha), H)$ of $res(\delta(\alpha), H)$ is defined depending on the form of $res(\delta(\alpha), H)$:*

$$res_{DL}(\delta(\alpha), H) = \begin{cases} res(\delta(\alpha), H) & if\ res(\delta(\alpha), H)\ is\ a\ ground\ atom \\ (\exists R)(a) & if\ res(\delta(\alpha), H) = \exists y.R(a, y)\ for\ ground\ a \\ (\exists R^-)(a) & if\ res(\delta(\alpha), H) = \exists y.R(y, a)\ for\ ground\ a \end{cases}$$

We recall that assertions for complex concepts such as $(\exists R)(a)$ are represented by $A(a)$ and $A \sqsubseteq \exists R$ for a new concept name A. This encoding may also affect atoms appearing in $I_{\mathcal{F}_n}^+$ which is why we directly incorporate $I_{\mathcal{F}_n}^+$ into $\mathcal{O}$ in Fig. 2 to avoid a more complicated notation.

The algorithm itself proceeds as outlined in the example. It checks first whether $\mathcal{O}$ together with the already proven true knowledge yields a satisfiable knowledge base. If not, the algorithm stops and returns the empty set; thus, $\mathcal{O}$ is not

---
[5] most general unifier

Require: $DL\text{-}Lite_{\mathcal{R}}$ KB $\mathcal{O} = (\mathcal{T}, \mathcal{A})$, which already contains $I^+_{\mathcal{F}_n}$, and a ground atomic
    query $q = H(t_i)$
Ensure: a set $\mathcal{L}$ of L_i such that $\mathcal{O} \cup L_i \models H(t_i)$ with $\mathcal{O} \cup L_i$ consistent
    $\mathcal{L} = \emptyset$
    $q_{unsat} = \bot$
    for all $\alpha \in cln(\mathcal{T})$ **do**
        $q_{unsat} = q_{unsat} \vee \delta(\alpha)$
    end for
    if $q^{db(\mathcal{A})}_{unsat} \neq \emptyset$ **then**
        $\mathcal{L} = \emptyset$
    else
        $q_{inst} = q_{unsat}$
        $\mathcal{T}' = \mathcal{T} \cup \{\hat{A} \sqsubseteq \neg H\}$
        $\mathcal{A}' = \mathcal{A} \cup \{\hat{A}(t_i)\}$
        $\mathcal{O}' = (\mathcal{T}', \mathcal{A}')$
        for all $\alpha \in cln(\mathcal{T}') \setminus cln(\mathcal{T})$ **do**
            $q_{inst} = q_{inst} \vee \delta(\alpha)$
        end for
        if $q^{db(\mathcal{A}')}_{inst} \neq \emptyset$ **then**
            $\mathcal{L} = \{\emptyset\}$
        else
            for all $\alpha \in cln(\mathcal{T}') \setminus cln(\mathcal{T})$ **do**
                $\mathcal{O}'' = \mathcal{O} \cup \{res_{DL}(\delta(\alpha), \hat{A}(t_i))\}$
                $q_{uns} = q_{unsat}$
                for all $\beta \in cln(\mathcal{T}'') \setminus cln(\mathcal{T})$ **do**
                    $q_{uns} = q_{uns} \vee \delta(\beta)$
                end for
                if $q^{db(\mathcal{A}'')}_{uns} = \emptyset$ **then**
                    $\mathcal{L} = \mathcal{L} \cup \{res(\delta(\alpha), \hat{A}(t_i))\}$
                end if
            end for
        end if
        return $\mathcal{L}$
    end if

Fig. 2. Algorithm $DL\text{-}Lite_{\mathcal{R}}$ Oracle

used for further derivations. Otherwise, it proceeds with an instance check for
the query, i.e., by checking for unsatisfiability of the extended knowledge base,
and, in the case of success, returns a set containing only the empty answer,
hence, an unconditional answer in the respective tree of $\mathbf{SLG}(\mathcal{O})$. If the instance check fails, then, for all newly introduced axioms in $cln(\mathcal{T}')$, it is verified
whether $res_{DL}(\delta(\alpha), \hat{A}(t_i))$ does not cause an inconsistency if added to $\mathcal{O}$ and
the already derived knowledge. If this is successful then the corresponding atom
$res(\delta(\alpha), \hat{A}(t_i))$ is included in the set of returned answers, which if proven true,
allow us to derive the considered query.

   We show that this algorithm provides a correct partial oracle for $DL\text{-}Lite_{\mathcal{R}}$
w.r.t. $\mathbf{SLG}(\mathcal{O})$.

Theorem 2. *The algorithm DL-Lite$_\mathcal{R}$ Oracle is sound and complete, i.e., the returned answers in L correspond to the definition of a partial oracle for DL-Lite$_\mathcal{R}$ and the algorithm allows the computation of all the minimal sets L according to the partial oracle for DL-Lite$_\mathcal{R}$.*

Proof. We consider $\mathcal{O} \cup I_{\mathcal{F}_n}^+ \cup L \models q$ where $\mathcal{O} \cup I_{\mathcal{F}_n}^+ \cup L$ is consistent (Definition 8) and q is the queried atom.

We show soundness, i.e., we show that the returned answers in L correspond to the definition of a partial oracle for DL-Lite$_\mathcal{R}$.

If the algorithm returns $\mathcal{L} = \emptyset$, then $\mathcal{O} \cup I_{\mathcal{F}_n}^+$ is not consistent, which means that the result from the algorithm is sound for this case. Otherwise, if the algorithm returns $\mathcal{L} = \{\emptyset\}$, then the instance check for the query succeeds and in this case (for consistent $\mathcal{O} \cup I_{\mathcal{F}_n}^+$ as checked) $\mathcal{O} \cup I_{\mathcal{F}_n}^+ \models q$ holds, which is also a sound answer. Finally, if the algorithm returns $\mathcal{L} = \{L_1, \ldots, L_n\}$, then the direct instance check failed, but $\mathcal{O} \cup I_{\mathcal{F}_n}^+ \cup L_i$ is consistent and $\mathcal{O} \cup I_{\mathcal{F}_n}^+ \cup L_i \models q$ holds because the addition of L to $I_{\mathcal{F}_n}^+$ would exactly enable the instance check to succeed (see Theorem 1).

To show completeness, we have to show that the algorithm enables us to compute all the minimal sets L according to the partial oracle for DL-Lite$_\mathcal{R}$. First, if $\mathcal{O} \cup I_{\mathcal{F}_n}^+$ is not consistent, then the partial oracle does not return any answer and this is covered by the returned empty set $\mathcal{L}$ in the algorithm. Then, if $\mathcal{O} \cup I_{\mathcal{F}_n}^+ \cup L \models q$ holds for empty L, then the only minimal answer for the partial oracle is the empty set. The algorithm DL-Lite$_\mathcal{R}$ Oracle returns exactly only the empty set. It remains to be shown that the correctness result holds for nonempty L as well. So suppose that L' is a nonempty minimal set such that $\mathcal{O} \cup I_{\mathcal{F}_n}^+ \cup L' \models q$ and $\mathcal{O} \cup I_{\mathcal{F}_n}^+ \cup L'$ is consistent. First, any minimal set can only consist of one atom due to the restricted syntax of GCI in DL-Lite$_\mathcal{R}$. Furthermore, joining $\mathcal{O} \cup I_{\mathcal{F}_n}^+ \cup L'$ and $q = H(t_i)$ together with $\hat{A} \sqsubseteq \neg H$ and $\hat{A}(t_i)$ yields an inconsistent DL-Lite$_\mathcal{R}$ KB $\mathcal{O}_1$, hence a successful instance check for q (Theorem 1). If we remove L', then the KB is consistent but the check for consistency would still compute all axioms such that the boolean disjunctive query q_{uns} w.r.t. $\mathcal{O}_1$ without L' would be unsatisfiable as such, but satisfiable upon addition of L', i.e., indicate a successful instance check. This is exactly what the algorithm DL-Lite$_\mathcal{R}$ Oracle computes and, for consistent $\mathcal{O} \cup I_{\mathcal{F}_n}^+ \cup L'$, L' is returned as one of the answers. Note that none of the $\alpha \in cln(\mathcal{T})$ is considered since if one of these succeeds, the entire knowledge base is inconsistent. Thus, considering only $\alpha \in cln(\mathcal{T}') \setminus cln(\mathcal{T})$ suffices to find all possible sets L, since *res* is applicable by construction in each such case.   □

Building on the results on computational complexity in DL-Lite$_\mathcal{R}$ ([5]), we can show that the algorithm ensures that the oracle is polynomial.

Theorem 3. *Let $\mathcal{K} = (\mathcal{O}, \mathcal{P})$ be a hybrid MKNF knowledge base with $\mathcal{O}$ in DL-Lite$_\mathcal{R}$. An $\mathbf{SLG}(\mathcal{O})$ evaluation of a query q in the algorithm DL-Lite$_\mathcal{R}$ Oracle is decidable with combined complexity PTime and with data complexity LogSpace.*

Proof. We know from the proof of Theorem 43 in [5] that the combined complexity for computing the disjunctive formula using δ on α in $cln(\mathcal{T})$ is polynomial, while the evaluation w.r.t. $db(\mathcal{A})$ is in LogSpace. Consequently, instance checking and checking satisfiability for $DL\text{-}Lite_{\mathcal{R}}$ is in PTime and LogSpace respectively. The algorithm $DL\text{-}Lite_{\mathcal{R}}$ Oracle applies one such satisfiability check for $\mathcal{O} \cup I_{\mathcal{F}_n}$ and conditionally a further one for the instance check. Then, conditionally a set of (polynomially many in combined complexity) $\delta(\alpha)$ is processed (each in the worst case containing a further satisfiability check, which is a slight extension of the first). We conclude that the combined complexity of $DL\text{-}Lite_{\mathcal{R}}$ Oracle is in PTime and the data complexity LogSpace.   □

Intuitively, this result is achieved because GCI and RI are of a particular restricted form, so that the oracle only returns single atoms, and does not need to compute minimal subsets of arbitrary size in the power set of all atoms.

Consequently, we obtain the computational complexity of answering DL-safe conjunctive queries in hybrid MKNF knowledge bases with a $DL\text{-}Lite_{\mathcal{R}}$ DL part.

Theorem 4. *Let $\mathcal{K} = (\mathcal{O}, \mathcal{P})$ be a hybrid MKNF knowledge base with $\mathcal{O}$ in DL-$Lite_{\mathcal{R}}$. Answering a DL-safe conjunctive query q in $\mathbf{SLG}(\mathcal{O})$ is decidable with data complexity PTime and LogSpace if $\mathcal{P}$ is empty.*

Proof. This is a direct consequence of Theorem 5.4 in [1] and Theorem 3 and the fact that, for nonempty $\mathcal{P}$, PTime from the rules includes LogSpace.   □

Hence, reasoning can still partially be done with relational databases.

Theorem 5. *Let $\mathcal{K} = (\mathcal{O}, \mathcal{P})$ be a consistent hybrid MKNF knowledge base with $\mathcal{O}$ in DL-$Lite_{\mathcal{R}}$. The answer to a DL-safe conjunctive query q in $\mathbf{SLG}(\mathcal{O})$ corresponds to the well-founded MKNF model.*

Proof. The result follows from Theorem 5.3 in [1] and Theorem 2.   □

If $\mathcal{K}$ is MKNF-inconsistent, then there is no well-founded MKNF model, but we obtain a paraconsistent approximation. Consider that the KB from Example 1 is part of a larger KB that is MKNF-inconsistent, but the inconsistency is not related to the predicates shown in the example, i.e., no rule or GCI links predicates from the example to those causing the inconsistency. Then, querying for atoms from Example 1 yields the same results as if $\mathcal{K}$ was MKNF-consistent.

6   Conclusions

In [10] we provided a concrete procedure for KB with non-monotonic rules and an ontology in the DL $\mathcal{REL}$, a fragment of the DL underlying OWL 2 EL. This slightly easier to obtain procedure relies, after preprocessing, on translating the DL to rules rather than on defining an oracle in the true sense of [1] as done here for $DL\text{-}Lite_{\mathcal{R}}$. We note that the resulting data complexity is identical, but higher than the one for $DL\text{-}Lite_{\mathcal{R}}$, simply because it has a lower complexity than $\mathcal{REL}$ in [10]. By translating the DL into rules, one can also easily obtain a procedure for DLP [8] – the DL underlying OWL 2 RL.

With the results in this paper, query-answering procedures that do not jeopardize tractability are now available for MKNF KB with rules and ontologies for the DL underlying all the three OWL2 profiles defined by W3C. As the next step, we want to provide an implementation of our work, building on XSB[6] and QuOnto/Mastro[7]. Moreover, OWL 2 QL has some expressive features not contained in $DL\text{-}Lite_{\mathcal{R}}$ and an extension is considered for future work.

References

1. Alferes, J.J., Knorr, M., Swift, T.: Queries to Hybrid MKNF Knowledge Bases Through Oracular Tabling. In: Bernstein, A., Karger, D.R., Heath, T., Feigenbaum, L., Maynard, D., Motta, E., Thirunarayan, K. (eds.) ISWC 2009. LNCS, vol. 5823, pp. 1–16. Springer, Heidelberg (2009)
2. Artale, A., Calvanese, D., Kontchakov, R., Zakharyaschev, M.: The DL-Lite family and relations. J. Artif. Intell. Res. 36, 1–69 (2009)
3. Boley, H., Hallmark, G., Kifer, M., Paschke, A., Polleres, A., Reynolds, D. (eds.): RIF Core Dialect. W3C Candidate Recommendation (June 22, 2010)
4. Boley, H., Kifer, M. (eds.): RIF Framework for Logic Dialects. W3C Candidate Recommendation (June 22, 2010), http://www.w3.org/TR/rif-fld/
5. Calvanese, D., de Giacomo, G., Lembo, D., Lenzerini, M., Rosati, R.: Tractable reasoning and efficient query answering in description logics: The DL-Lite family. Journal of Automated Reasoning 39(3), 385–429 (2007)
6. Chen, W., Warren, D.S.: Tabled Evaluation with Delaying for General Logic Programs. J. ACM 43(1), 20–74 (1996)
7. Glimm, B., Lutz, C., Horrocks, I., Sattler, U.: Answering conjunctive queries in the $\mathcal{SHIQ}$ description logic. J. Artif. Intell. Res. 31, 150–197 (2008)
8. Grosof, B.N., Horrocks, I., Volz, R., Decker, S.: Description logic programs: Combining logic programs with description logics. In: Proc. of the World Wide Web Conference (WWW 2003), pp. 48–57. ACM (2003)
9. Hitzler, P., Krötzsch, M., Parsia, B., Patel-Schneider, P.F., Rudolph, S. (eds.): OWL 2 Web Ontology Language: Primer. W3C Recommendation (October 2009)
10. Knorr, M., Alferes, J.J.: Querying in $\mathcal{EL}^+$ with nonmonotonic rules. In: Coelho, H., Studer, R., Wooldridge, M. (eds.) 19th European Conf. on Artificial Intelligence, ECAI 2010, pp. 1079–1080. IOS Press (2010)
11. Knorr, M., Alferes, J.J., Hitzler, P.: Local closed world reasoning with description logics under the well-founded semantics. Artificial Intelligence 175(9-10), 1528–1554 (2011)
12. Lifschitz, V.: Nonmonotonic databases and epistemic queries. In: Mylopoulos, J., Reiter, R. (eds.) 12th Int. Joint Conf. on AI, IJCAI 1991, pp. 381–386 (1991)
13. Motik, B., Rosati, R.: Reconciling Description Logics and Rules. Journal of the ACM 57(5), 93–154 (2010)
14. Patel, C., Cimino, J., Dolby, J., Fokoue, A., Kalyanpur, A., Kershenbaum, A., Ma, L., Schonberg, E., Srinivas, K.: Matching Patient Records to Clinical Trials Using Ontologies. In: Aberer, K., Choi, K.-S., Noy, N., Allemang, D., Lee, K.-I., Nixon, L.J.B., Golbeck, J., Mika, P., Maynard, D., Mizoguchi, R., Schreiber, G., Cudré-Mauroux, P. (eds.) ASWC 2007 and ISWC 2007. LNCS, vol. 4825, pp. 816–829. Springer, Heidelberg (2007)
15. Prud'hommeaux, E., Seaborne, A. (eds.): SPARQL Query Language for RDF. W3C Candidate Recommendation (January 2008)

---

[6] http://xsb.sourceforge.net/
[7] http://www.dis.uniroma1.it/quonto/

ShareAlike Your Data: Self-referential Usage Policies for the Semantic Web

Markus Krötzsch[1] and Sebastian Speiser[2]

[1] Department of Computer Science, University of Oxford, UK
`markus.kroetzsch@cs.ox.ac.uk`
[2] Institute AIFB, Karlsruhe Institute of Technology, DE
`speiser@kit.edu`

Abstract. Numerous forms of policies, licensing terms, and related conditions are associated with Web data and services. A natural goal for facilitating the re-use and re-combination of such content is to model usage policies as part of the data so as to enable their exchange and automated processing. This paper thus proposes a concrete policy modelling language. A particular difficulty are *self-referential* policies such as *Creative Commons ShareAlike*, that mandate that derived content is published under some license with the same permissions and requirements. We present a general semantic framework for evaluating such recursive statements, show that it has desirable formal properties, and explain how it can be evaluated using existing tools. We then show that our approach is compatible with both OWL DL and Datalog, and illustrate how one can concretely model self-referential policies in these languages to obtain desired conclusions.

1 Introduction

Semantic technologies facilitate the sharing and re-use of data and associated services, but in practice such uses are often governed by a plethora of policies, licensing terms, and related conditions. Most data and service providers reserve certain rights, but an increasing number of providers also choose usage terms that encourage the re-use of content, e.g. by using a Creative Commons[1] license. Even such policies still impose restrictions, and it has been estimated that 70% – 90% of re-uses of Flickr images with Creative Commons Attribution license actually violate the license terms [29]. A possible reason for frequent violations is that checking license compliance is a tedious manual task that is often simply omitted in the process of re-using data.

A natural goal therefore is to accurately model usage policies as part of the data so as to enable their easy exchange and automated processing. This resonates with multiple topical issues in Semantic Web research. On the one hand, it is increasingly acknowledged that the distribution of semantic data and services may also require transparent licensing for such content [33,10]. This closely relates to the wider goal of semantically representing *provenance* information about the origin and context of data items. Not surprisingly, the W3C Incubator Group on Provenance also lists support for usage policies and licenses of artefacts in their requirements report [9].

---

[1] `http://creativecommons.org/`

L. Aroyo et al. (Eds.): ISWC 2011, Part I, LNCS 7031, pp. 354–369, 2011.

On the other hand, modelling of policy information is also promising as an application area for semantic technologies [17,7]. Capturing the variety of relevant conditions involves domain-specific concepts such as "non-commercial" or "fair use" but also (when thinking about distribution policies that are internal to an organisation) levels of confidentiality, and personal access permissions. Semantic technologies offer powerful tools and methodologies for developing shared conceptualisations for such complex modelling problems.

This paper presents a new policy modelling language to address the specific challenges of this domain. A primary task is to enable the computation of policy containment, i.e. the automatic decision whether all uses that are allowed by one policy are also allowed by another [8]. But some policies go a step further and require such containments to hold *as part of their condition*. A well-known example are the Creative Commons ShareAlike licenses which mandate that content is published under some license that involves the same permissions and requirements – including the requirement to share under such licenses only. Such self-referential policies introduce recursive dependencies and a form of meta-modelling not found in ontology languages like OWL.

Our main contributions to solving this problem are as follows.

(1) We develop the syntax and semantics of a general policy modelling language. Our formalisation is guided by an analysis of the requirements for a policy (meta) model that supports self-referential policies as given by the Creative Commons licenses.
(2) We show that this policy language has desirable formal properties under reasonable syntactic restrictions on policy conditions and background theories. In particular we establish how to utilise standard first-order reasoning in a non-trivial way for computing conclusions under our new semantics.
(3) Using this connection to first-order logic, we instantiate this general policy language for the Web Ontology Language OWL and for the basic rule language Datalog. Both cases lead to expressive policy representation languages that can readily be used in practice by taking advantage of existing tools. Concretely, we show how to express the well-known Creative Commons licenses and verify that the expected relationships are derived.

Section 2 introduces our main use case and Section 3 presents a basic vocabulary to model policies. In Section 4 we discuss challenges in modelling self-referential policies formally. We introduce a formal policy semantics in Section 5 and apply it to our use case in Section 6. Related work is discussed in Section 7. The technical results at the core of this paper are not obvious and require a notable amount of formal argumentation. However, the focus of this presentation is to motivate and explain the rationale behind our proposal. Formal proofs and further details are found in an extended report [20].

2   Use Case: Creative Commons ShareAlike

To motivate our formalisation of policies we discuss some common requirements based on the popular Creative Commons (CC) licenses. CC provides a family of license models for publishing creative works on the Web, which share the common goal of enabling re-use as an alternative to the "forbidden by default" approach of traditional copyright law. Each license specifies how the licensed work may be used by stating, e.g., in which cases it can be further distributed (shared) and if derivative works are allowed.

The most permissive CC license is *Creative Commons Attribution* (CC BY), which allows all types of uses (sharing and derivation) provided that the original creator of the work is attributed. Various restrictions can be added to CC BY:

- NoDerivs (ND): the work can be used and redistributed, but it must remain unchanged, i.e., no derivations can be created.
- NonCommercial (NC): re-use is restricted to non-commercial purposes.
- ShareAlike (SA): derived works have to be licensed under the identical terms.

The CC ShareAlike restriction is particularly interesting, as it does not only restrict processes using the protected data artefact, but the policy of artefacts generated by those processes. ShareAlike is formulated in legal code as follows:

> "You may Distribute or Publicly Perform an Adaptation only under: (i) the terms of this License; (ii) a later version of this License [...]; (iii) a Creative Commons jurisdiction license [...] that contains the same License Elements as this License [...]"[2]

Thus derived artefacts can only be published under some version of the exact same CC license. This could easily be formalised by simply providing an exhaustive list of all licenses that are currently admissible for derived works. In this case, policies would be identified by their name, not by the permissions and restrictions that they impose.

This effect can be desired, e.g. for the GPL which thus ensures its "viral" distribution. However, the name-based restriction is not intended for Creative Commons, as noted by Lessig who originally created CC: rather, it would be desirable to allow the combination of licenses that share the same intentions but that have a different name, e.g. to specify that an artefact must be published under a license that allows only non-commercial uses instead of providing a list of all (known) licenses to which this characterisation applies [21]. To overcome this incompatibility problem, we propose *content-based* policy restrictions that are based on the allowed usages of a policy.

3   Schema for Modelling Policies

Before we can formally specify the semantics of a policy language that can formalise the "intention" of a policy like CC, we need some basic conceptual understanding of the modelling task, and also some shared vocabulary that enables the comparison of different licenses. In this section, we provide a high-level schema that we use for modelling policies in this paper.

In general, we understand a policy as a specification that defines what one is allowed to *do* with an artefact that has this policy. Thus, a policy can be viewed as a collection of admissible usages. In order to align with the terminology of the Open Provenance Model OPM [23] below we prefer to speak of admissible "processes" as the most general type of use. The admissible processes can be viewed as "desired states"

---

[2] Section 4(b) in http://creativecommons.org/licenses/by-nc-sa/3.0/legalcode

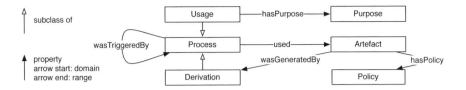

Fig. 1. Informal view of a simple provenance model

(in the sense of "states of the world" such as when an artefact has been published with suitable attribution), which corresponds to the notion of goal-based policies as defined by Kephart and Walsh [19].

To specify the conditions of a policy, we need a model for further describing such usage processes and their relationships to concrete artefacts. This model in particular must represent the origin of the artefact, and the context in which it has been published. Such *provenance* information can be described in various ways, e.g. with a provenance graph that specifies the dependencies between processes and the artefacts they use and generate. Here we use the very simple provenance model illustrated informally in Fig. 1. This base model can of course be further specialised for specific applications and other use cases; we just require a minimal setup for our examples.

The provenance model re-uses the vocabulary elements *artefact*, *process*, *used*, *was-GeneratedBy*, and *wasTriggeredBy* from the Open Provenance Model. For our particular application, we further split processes into *derivations* (processes that generate a new artefact) and other *usages* that only use artefacts without change. To cover the CC use case, we introduce the *hasPurpose* property relating a usage to its *purpose*, e.g., stating that a usage was non-commercial. The *hasPolicy* property assigns to an artefact a *policy*, which means that all processes using the artefact are (legally) required to comply to its policy.

According to OPM, a process p_1 *wasTriggeredBy* another process p_2, if p_1 can only have started after p_2 started. So, somewhat contrary to intuition, the "triggering" is rather a precondition but not a necessary cause of the triggered one. A usage restriction that requires attribution would thus be formalised as a policy requiring that the usage process *wasTriggeredBy* an attribution process, and not the other way around.

The provenance model provides a basic vocabulary for specifying information about artefacts and policies. To realise content-based restrictions we further want to talk about the relationships of policies. For example, ShareAlike requires the value of *hasPolicy* to refer to a policy which allows exactly the same uses as the given CC SA license. This subsumption between policies is called *policy containment*, and we introduce a predicate containedIn to express it. Informally speaking, the fact containedIn(p, q) can also be read as: any process that complies with policy p also complies with policy q. When allowing policy conditions to use containedIn, the question whether or not a process complies to a policy in turn depends on the evaluation of containedIn. Our goal therefore is to propose a formal semantics that resolves this recursive dependency in a way that corresponds to our intuitive understanding of the policies that occur in practice.

4   Challenges of Defining a Semantics for Policies

For formalising our above understanding of policies, we use the syntax of first-order logic as a general framework. Thus, our earlier 'classes' and 'properties' become predicates of arity 1 and 2, respectively. A policy that represents a set of allowed processes then corresponds to a formula $\varphi[x]$ with one free variable x, representing the set of individuals that make $\varphi[x]$ true when assigned as values to x.[3] For example, a policy p that allows no uses other than derivations that generate artefacts with policy p can be described as:

$$p : \text{Derivation}(x) \wedge \exists y.(\text{wasGeneratedBy}(y, x) \wedge \text{hasPolicy}(y, p)). \tag{1}$$

More generally, we can use containedIn to allow derived artefacts to use any policy that is at least as restrictive as p:

$$p : \text{Derivation}(x) \wedge \exists y.(\text{wasGeneratedBy}(y, x) \wedge \\ \exists z.(\text{hasPolicy}(y, z) \wedge \text{containedIn}(z, p))). \tag{2}$$

A collection of such policy definitions $p : \varphi_p[x]$ will be called a *policy system*. Given a policy system with definitions $p : \varphi_p$ for all policy names $p \in N_P$, we can formalise some general restrictions that conform to our intuition:

$$\forall x.\text{conformsTo}(x, p) \leftrightarrow \varphi_p[x] \qquad \text{for all } p \in N_P, \tag{3}$$

$$\forall x, y.\text{containedIn}(x, y) \leftrightarrow \forall z.(\text{conformsTo}(z, x) \rightarrow \text{conformsTo}(z, y)). \tag{4}$$

Formula (3) defines conformsTo to relate processes to the policies they conform to. Please note the difference between conformsTo (actual semantic conformance) and hasPolicy (legally required conformance). Formula (4) ensures that containedIn relates two policies exactly if fewer (or at most the same) processes conform to the first, i.e. if the first policy is at least as restrictive as the second. The set of these two types of sentences (for a given set of policy names N_P) is denoted by T_{ct}.

Unfortunately, these formulae under first-order semantics do not lead to the intended interpretation of policies. Consider the policy (2), and a second policy q that is defined by exactly the same formula, but with p replaced by q. Intuitively, p and q have the same conditions but merely different names, so they should be in a mutual containedIn relationship. Indeed, there are first-order models of T_{ct} where this is the case: if containedIn(p, q) holds, then $\forall x.\varphi_p[x] \rightarrow \varphi_q[x]$ is also true. However, this is not the only possible interpretation: if containedIn(p, q) does not hold, then $\forall x.\varphi_p[x] \rightarrow \varphi_q[x]$ is not true either. First-order logic does not prefer one of these interpretations, so in consequence we can conclude neither containedIn(p, q) nor $\neg$containedIn(p, q).

Working with first-order interpretations still has many advantages for defining a semantics, in particular since first-order logic is widely known and since many tools and knowledge representation languages are using it. This also enables us to specify additional background knowledge using first-order formalisms of our choice, e.g. the OWL DL ontology language. However, we would like to restrict attention to first-order

---

[3] We assume basic familiarity with first-order logic. Formal definitions are given in [20].

models that conform to our preferred reading of containedIn. Logical consequences can still be defined as the statements that are true under all of the preferred interpretations, but undesired interpretations will be ignored for this definition. Our goal of defining the semantics of self-referential policies thus boils down to defining the "desired" interpretations of a given first-order theory that uses containedIn. To do this, we propose a semantics for policy containment that, intuitively speaking, always prefers containedIn(p, q) to hold if this is possible without making additional unjustified assumptions. For illustration, consider the following policy q that further restricts p from (2) to non-commercial uses:

$$q : \text{Derivation}(x) \land \forall w.(\text{hasPurpose}(x, w) \rightarrow \text{NonCommercial}(w)) \land \tag{5}$$
$$\exists y.(\text{wasGeneratedBy}(y, x) \land \exists z.(\text{hasPolicy}(y, z) \land \text{containedIn}(z, q))).$$

Though the policy q is clearly more restrictive than p, there still is a first-order interpretation that satisfies containedIn(p, q) by simply assuming that all things that conform to p happen to have non-commercial uses only. Nothing states that this is not the case, yet we do not want to make such assumptions to obtain more containedIn conclusions.

We thus distinguish *basic predicates* such as NonCommercial and hasPolicy from the two "special" predicates containedIn and conformsTo. Basic predicates are given by the data, and represent the available information, and their interpretation should not be considered a matter of choice. Special predicates in turn should be interpreted to reflect our intended understanding of policy containment, and as shown in the above example it is often desirable to maximise containedIn entailments. In other words, we would like to ensure that the consideration of a policy system does not lead to new logical consequences over basic predicates – merely defining license conditions should not increase our knowledge of the world. More formally: the policy semantics should be *conservative* over first-order semantics w.r.t. sentences that use only basic predicates.

Unfortunately, this is not easy to accomplish, and indeed Theorem 1 only achieves a limited version of this. One reason is that even T_{ct} may entail undesired consequences. Consider policies as follows (we use abstract examples to highlight technical aspects):

$$p : A(x) \land \text{containedIn}(p, q) \qquad q : B(x). \tag{6}$$

This policy system entails containedIn(p, q). Indeed, if containedIn(p, q) would not hold, then nothing would conform to p by (3). But the empty set is clearly a subset of every other set, hence containedIn(p, q) would follow by (4). Thus all interpretations that satisfy T_{ct} must satisfy $\forall x.A(x) \land \text{containedIn}(p, q) \rightarrow B(x)$, and thus $\forall x.A(x) \rightarrow B(x)$ is a consequence over basic predicates. Clearly, the mere definition of licenses should not entail that some otherwise unrelated class A is a subclass of B.

5   A Formal Language for Policy Definitions

In order to address the challenges discussed in the previous section, we now formally define a policy language. More precisely, we define a language for policies *and* a first-order language that is to be used for background theories. These definitions are intended to be very general to impose only those restrictions that we found necessary to obtain a well-behaved semantics. Section 6 shows how this general framework can be instantiated in various well-known modelling languages.

The basic restriction that we impose on the logic is *connectedness*. Intuitively, this ensures that a formula can only refer to a connected relational structure of individuals. In our setting the conformance of a process to a policy thus only depends on the characteristics of individuals directly or indirectly reachable from the process. We argue that this is a small restriction. It might even be a best practice for "controlled" modelling in an open environment like the Web, as it ensures that the classification of any object is based only on its "environment" and not on completely unrelated individuals.

Our formal definition is reminiscent of the *Guarded Fragment* (GF) of first-order logic [4] and indeed it can be considered as a generalization of GF, though without the favourable formal properties that motivated GF. We first define open connected formulae (with free variables) and then closed ones. We write $\varphi[x]$ to indicate that φ has at most the free variables that occur in x (or possibly less). For technical reasons, our first definition distinguishes "guard predicates" that must not use constant symbols from "non-guard predicates" where constants are allowed:

Definition 1. *Consider a first-order signature Σ where each predicate in Σ is marked as a* guard predicate *or as a* non-guard predicate. *The* connected open fragment *COF of first-order logic over Σ is the smallest set of formulae over Σ that satisfies the following properties:*

1. *Every atomic formula $p(t)$ with t a vector of terms that contain at least one variable belongs to COF, provided that t contains only variables if p is a guard predicate.*
2. *If φ_1 and φ_2 are in COF then so are $\neg\varphi_1$, $\varphi_1 \wedge \varphi_2$, $\varphi_1 \vee \varphi_2$, and $\varphi_1 \rightarrow \varphi_2$.*
3. *Consider a formula $\varphi[x, y]$ in COF, and a conjunction $\alpha[x, y] = \alpha_1[x, y] \wedge \ldots \wedge \alpha_n[x, y]$ of atomic formulae α_i that contain only guard predicates and variables, such that x, y are both non-empty and do not share variables. Then the formulae*

$$\exists y.\alpha[x, y] \wedge \varphi[x, y] \qquad \forall y.\alpha[x, y] \rightarrow \varphi[x, y],$$

   are in COF provided that for each variable y in y, there is some variable x in x and some atom $\alpha_i[x, y]$ where both x and y occur.

The distinction of guard and non-guard predicates is important, but a suitable choice of guard predicates can be easily made for a given formula set of formulae in COF by simply using exactly those predicates as guards that do not occur in atomic formulae with constants. The only predicate that we really need to be a non-guard is containedIn. Therefore, we will omit the explicit reference to the signature Σ in the following and simply assume that one signature has been fixed.

Definition 2. *The* connected fragment *CF of first-order logic consists of the following sentences:*

– *Every formula without variables is in CF.*
– *If $\varphi[x]$ is a COF formula with one free variable x, then $\forall x.\varphi[x]$ and $\exists x.\varphi[x]$ are in CF.*

We will generally restrict to background theories that belong to CF. As discussed in Section 6 below, large parts of OWL DL and Datalog fall into this fragment. A typical example for a non-CF sentence is the formula $\neg\exists x.A(x) \vee \neg\exists x.B(x)$. Also note that the formulae (3) and (4) of T_{ct} are not in CF – we consider them individually in all our formal arguments. On the other hand, the policy conditions (1), (2), (5), and (6) all are in COF. Using the terminology of connected formulae, we can define policy conditions, policy descriptions, and policy systems that we already introduced informally above:

Definition 3. *Let N_P be a set of* policy names. *A* policy condition φ *for N_P is a formula that may use an additional binary predicate* containedIn *that cannot occur in background theories, and where:*

- *φ is a COF formula with one free variable,*
- *φ contains at most one constant symbol $p \in N_P$ that occurs only in atoms of the form* containedIn(y, p) *or* containedIn(p, y),
- *every occurrence of* containedIn *in φ is positive (i.e. not in the scope of a negation) and has the form* containedIn(y, p) *or* containedIn(p, y).

A policy description *for a policy $p \in N_P$ is a pair $\langle p, \varphi \rangle$ where φ is a policy condition. A* policy system *P for N_P is a set of policy descriptions that contains exactly one description for every policy $p \in N_P$.*

This definition excludes the problematic policy p in (6) above while allowing (1), (2), and (5). Moreover, it generally requires containedIn to be a non-guard predicate.

We define the semantics of policy containment as the greatest fixed point of an operator introduced next. Intuitively, this computation works by starting with the assumption that all named policies are contained in each other. It then refers to the policy definitions to compute the actual containments that these assumptions yield, and removes all assumptions that cannot be confirmed. This computation is monotone since the assumptions are reduced in each step, so it also has a greatest fixed point.

Definition 4. *Consider a set of CF sentences T (background theory), a set of policy names N_P that includes the* top policy $p_\top$ *and the* bottom policy $p_\bot$, *and a policy system P for N_P such that $\langle p_\top, \top(x) \rangle, \langle p_\bot, \bot(x) \rangle \in P$.[4] Let T_{ci} be the following theory:*

$$T_{ci} = \{\forall x, y, z.\text{containedIn}(x, y) \wedge \text{containedIn}(y, z) \rightarrow \text{containedIn}(x, z),$$
$$\forall x.\text{containedIn}(x, p_\top), \forall x.\text{containedIn}(p_\bot, x)\}.$$

For a set $C \subseteq N_P^2$, define $\text{Cl}(C) := \{\text{containedIn}(p, q) \mid \langle p, q \rangle \in C\}$. *An operator $P_T : \mathcal{P}(N_P^2) \rightarrow \mathcal{P}(N_P^2)$, where $\mathcal{P}(N_P^2)$ is the powerset of N_P^2, is defined as follows:*

$$P_T(C) = \{\langle p, q \rangle \mid \langle p, \varphi_p \rangle, \langle q, \varphi_q \rangle \in P \text{ and } T \cup T_{ci} \cup \text{Cl}(C) \models \forall x.\varphi_p[x] \rightarrow \varphi_q[x]\}.$$

Proposition 1. *The operator P_T has a greatest fixed point $\text{gfp}(P_T)$ that can be obtained by iteratively applying P_T to N_P^2 until a fixed point is reached. More concretely, the greatest fixed point is of the form $P_T^n(N_P^2)$ for some natural number $n \leq |N_P|^2$ where P_T^n denotes n-fold application of P_T.*

---

[4] As usual, we consider $\top/\bot$ as unary predicates that are true/false for all individuals.

The fact that P_T requires the existence of policies $p_\top$ and $p_\bot$ is not restricting the applicability of our approach since the according standard policy declarations can always be added. Using the greatest fixed point of P_T, we now define what our "preferred" models for a policy system and background theory are.

Definition 5. *Given a policy system P, a P-model for a theory T is a first-order interpretation $\mathcal{I}$ that satisfies the following theory:*

$$\mathcal{I} \models T \cup T_{ci} \cup Cl(gfp(P_T)) \cup T_{ct}, \tag{7}$$

where T_{ci} and $Cl(gfp(P_T))$ are as in Definition 4, and where T_{ct} is the collection of all sentences of the form (3) and (4). In this case, we say that $\mathcal{I}$ P-satisfies T. A sentence φ is a P-consequence of T, written $T \models_P \varphi$, if $\mathcal{I} \models \varphi$ for all P-models $\mathcal{I}$ of T.

It is essential to note that the previous definition uses a fixed point computation only to obtain a minimal set of containments among named policies that must be satisfied by all P-models. It is not clear if and how the semantics of P-models could be captured by traditional fixed point logics (cf. Section 7). At the core of this problem is that policy conformance is inherently non-monotonic in some policies that we want to express. A policy p might, e.g., require that the policy of all derived artefacts admits *at least* all uses that are allowed by p. Then the fewer uses are allowed under the p, the more policies allow these uses too, and the more uses conform to p. This non-monotonic relationship might even preclude the existence of a model.

The policy semantics that we defined above is formal and well-defined for all policy systems and background theories, even without the additional restrictions of Definition 2 and 3. However, three vital questions have to be answered to confirm that it is appropriate for our purpose: (1) How can we compute the entailments under this new semantics? (2) Does this semantics avoid the undesired conclusions discussed in Section 4? (3) Does the semantics yield the intended entailments for our use cases? The last of these questions will be discussed in Section 6. Questions (1) and (2) in turn are answered by the following central theorem of this paper:

Theorem 1. *Consider a theory T and a policy system P. For every φ that is a CF formula over the base signature, or a variable-free atom (fact) over the predicates* containedIn *or* conformsTo *we have:*

$$T, T_{ci}, Cl(gfp(P_T)), T_{ct}^- \models \varphi \qquad iff \qquad T \models_P \varphi, \tag{8}$$

where T_{ci} and $Cl(gfp(P_T))$ are defined as in Definition 4, and where T_{ct}^- is the collection of all sentences of the form (3).

Let us first discuss how Theorem 1 answers the above questions.

(1) The theorem reduces P-entailment to standard first-order logic entailment. Since $gfp(P_T)$ can be computed under this semantics as well, this means that reasoning under our semantics is possible by re-using existing tools given that one restricts to fragments of (CF) first-order logic for which suitable tools exist. We pursue this idea in Section 6.

(2) The theorem asserts that all CF formulae that are P-entailments are entailed by the first-order theory $T \cup T_{ci} \cup Cl(gfp(P_T))$. It is easy to see that T_{ci} and $Cl(gfp(P_T))$ only affect the interpretation of formulae that use containedIn. All other CF formulae are P-entailments of T if and only if they are first-order entailments of T. Thus, new entailments over base predicates or even inconsistencies are not caused by considering a policy system.

The proof of Theorem 1 is not straightforward. At its core, it hinges on the fact that every model $\mathcal{I}$ of $T \cup T_{ci} \cup Cl(gfp(P_T))$ can be extended into a P-model $\hat{\mathcal{I}}$ of T that satisfies no containedIn or conformsTo facts that have not already been satisfied by $\mathcal{I}$. Constructing this P-model requires a number of auxiliary constructions centred around the idea that, for every policy containment not in $Cl(gfp(P_T))$, one can find a witness (a process conforming to the one policy but not to the other) in some model of $T \cup T_{ci} \cup Cl(gfp(P_T))$. This witness (and all of its environment) is then copied into the P-model that we want to construct. This is only feasible since the CF formulae in T are inherently "local" and will not change their truth value when extending the model by new (disjoint) individuals. After enough witnesses have been included to refute all non-entailed containedIn facts, the construction of $\hat{\mathcal{I}}$ is completed by defining suitable extensions for conformsTo where care is needed to do this for "unnamed" policies so that T_{ct} is satisfied. A full formal argument is found in the technical report [20].

6 Practical Policy Languages

In this section, we provide concrete instantiations of the general formalism introduced above. The CF fragment still is overly general for practical use, in particular since the computation of entailments in this logic is undecidable which precludes many desired applications where policy containment would be checked automatically without any user interaction.[5] However, Theorem 1 asserts that we can generally evaluate formal models under the semantics of first-order logic which is used in many practical knowledge representation languages. By identifying the CF fragments of popular modelling formalisms, we can therefore obtain concrete policy modelling languages that are suitable for specific applications.

There are various possible candidates for knowledge representation languages that can be considered under a first-order semantics and for which good practical tool support is available. Obvious choices include the Web Ontology Language OWL under its Direct Semantics [32], and the rule language Datalog under first-order semantics [3] which we will discuss in more detail below.

As we will explain for the case of Datalog, one can also model policy conditions as (conjunctive/disjunctive) queries with a single result, given that the query language uses a first-order semantics. Query evaluation is known to be difficult for expressive modelling languages, but can be very efficient when restricting to a light-weight background theory. A possible example is the combination of SPARQL for OWL [11] with

---

[5] This is easy to see in many ways, for example since (as noted below) CF allows us to express description logics like $\mathcal{SRIQ}$, whereas CF does not impose the regularity or acyclicity conditions that are essential for obtaining decidability of reasoning in these logics [15].

the lightweight OWL QL or OWL RL languages [32]. The below cases thus can only serve as an illustration of the versatility of our approach, not as a comprehensive listing.

6.1 Modelling Policies in OWL DL

The Direct Semantics of OWL 2 is based on description logics which in turn are based on the semantics of first-order logic [32]. The ontology language OWL 2 DL for which this semantics is defined can therefore be viewed as a fragment of first-order logic to which we can apply the restrictions of Section 5. The standard translation to first-order logic (see, e.g., [14]) produces formulae that are already very close to the syntactic form of CF sentences described above. Moreover, OWL class expressions are naturally translated to first-order formulae with one free variable, and are thus suitable candidates for expressing policies. Policy containment then corresponds to class subsumption checking – a standard inferencing task for OWL reasoners. The binary predicates of our simple provenance model, as well as the special predicates containedIn and conformsTo can be represented by OWL properties, whereas unary predicates from the provenance model correspond to primitive OWL classes.

Some restrictions must be taken into account to ensure that we consider only ontologies that are CF theories, and only classes that are valid policy conditions. Nominals (enumerated classes as provided by ObjectOneOf in OWL) are expressed in first-order logic using constant symbols, and must therefore be excluded from background ontologies. On the other hand nominals must be used in containedIn in policy descriptions (in OWL this particular case can conveniently be expressed with ObjectHasValue). Besides nominals, the only non-connected feature of OWL 2 that must be disallowed is the universal role (owl:topObjectProperty). On the other hand, cardinality restrictions are unproblematic even though they are usually translated using a special built-in equality predicate $\approx$ that we did not allow in first-order logic in Section 5. The reason is that $\approx$ can easily be emulated in first-order logic using a standard equality theory [20], so that all of our earlier results carry over to this extension.

To apply Theorem 1 for reasoning, we still must be able to express T_{ci} of Definition 4 in OWL. Transitivity of containedIn is directly expressible, and the remaining axioms can be written as follows:[6]

$$\top \sqsubseteq \exists \mathsf{containedIn}.\{p_\top\} \qquad\qquad \top \sqsubseteq \exists \mathsf{containedIn}^-.\{p_\bot\}$$

Note that the represented axioms are not in CF, and likewise the restriction to nominal-free OWL is not relevant here.

Concrete policies are now easily modelled. The public domain (PD) policy that allows every type of usage and derivation is expressed as:

$$\mathsf{PD}: \mathsf{Usage} \sqcup \mathsf{Derivation}\,.$$

Processes compliant to CC BY are either usages that were triggered by some attribution, or derivations for which all generated artefacts have only policies that also require

---

[6] Throughout this section we use the usual DL notation for concisely writing OWL axioms and class expressions; see [14] for an extended introduction to the relationship with OWL 2 syntax.

attributions, i.e., which are contained in BY:

$$BY: (Usage \sqcap \exists\, wasTriggeredBy.Attribution) \sqcup$$
$$(Derivation \sqcap \forall\, wasGeneratedBy^{-1}.\forall\, hasPolicy.\exists\, containedIn.\{BY\}).$$

To account for the modular nature of CC licenses, it is convenient to re-use class expressions as the one for BY. Thus, we will generally write C_{BY} to refer to the class expression for BY, and similarly for the other policies we define. To define NoDerivs (ND) licenses that allow all processes that are not derivations, we introduce C_{ND} as an abbreviation for Process $\sqcap$ ¬Derivation. We can thus express CC BY-ND as

$$BY\text{-}ND: C_{BY} \sqcap C_{ND}.$$

The ShareAlike (SA) condition cannot be modelled as an independent building block, as it refers directly to the policy in which it is used. As an example, we model the condition for the CC BY-SA policy as a requirement that all policies of all generated artefacts are equivalent to BY-SA, i.e., they are contained in BY-SA and BY-SA is contained in them:

$$BY\text{-}SA: C_{BY} \sqcap \forall\, wasGeneratedBy^{-1}.\forall\, hasPolicy.(\exists\, containedIn.\{BY\text{-}SA\} \sqcap$$
$$\exists\, containedIn^{-1}.\{BY\text{-}SA\}).$$

To validate the basic practicability of this modelling approach, we used the OWL reasoner HermiT[7] to compute the fixed point semantics of the policy system. We then conducted some basic tests with the formalised CC policies.[8] Not surprisingly, it can be observed that the fixed point of P_T is reached after just 2 iterations, which is significantly less than the rough upper bound of $|N_P|^2$ which was 49 in case of the 7 CC licenses. In general, one may presume that even big numbers of policies do rarely expose a linear dependency that would lead to long iterations for reaching a fixed point.

As a basic example of how to apply automated conformance checking, we modelled for every combination (p_{orig}, p_{deriv}) of Creative Commons licenses a derivation which uses an artefact with policy p_{orig} and generates a new artefact with policy p_{deriv}. If such a derivation is compliant to p_{orig}, we know that p_{deriv} is a valid license for derivations of p_{orig} licensed artefacts. The results (as expected) agree with the official Creative Commons compatibility chart.[9]

It can be noted that, besides its use for conformance checking, the computation of containedIn can also assist in modelling policies. For example, one can readily infer that any ShareAlike (SA) requirement is redundant when a NoDerivs (ND) requirement is present as well: adding SA to any ND license results in an equivalent license, i.e. one finds that the licenses are mutually contained in each other.

---

[7] http://www.hermit-reasoner.com/

[8] For reasons of space, we did not include all formalisations for all CC licenses here; the complete set of example policies for OWL and Datalog is available at http://people.aifb.kit.edu/ssp/creativecommons_policies.zip

[9] see Point 2.16 in http://wiki.creativecommons.org/FAQ, accessed 15th June 2011

6.2   Modelling Policies in Datalog

Datalog is the rule language of function-free definite Horn clauses, i.e., implications with only positive atoms and a single head atom. It can be interpreted under first-order semantics [3]. The syntax corresponds to first-order logic with the only variation that quantifiers are omitted since all variables are understood to be quantified universally. Datalog rules can thus be used to express a background theory. Policies can be expressed by conjunctive or disjunctive queries, i.e., by disjunctions and conjunctions of atomic formulae where one designated variable represents the free variable that refers to the conforming processes, while the other variables are existentially quantified.

Again we have to respect syntactic restrictions of Section 5. Thus we can only use rules that are either free of variables, or that contain no constants. In the latter case, all variables in the rule head must occur in its body (this is known as *safety* in Datalog), and the variables in the rule body must be connected via the atoms in which they co-occur. For policy queries, we also require this form of connection, and we allow constants in containedIn. The (non-CF) theory T_{ci} of Definition 4 is readily expressed in Datalog.

Containment of conjunctive and disjunctive queries is decidable, and can be reduced to query answering [2]. Namely, to check containment of a query q_1 in a query q_2, we first create for every conjunction in q_1 (which is a disjunction of conjunctive queries) a grounded version, i.e., we state every body atom in the conjunction as a fact by uniformly replacing variables with new constants. If, for each conjunction in q_1, these new facts provide an answer to the query q_2, then q_1 is contained in q_2. Note that Datalog systems that do not support disjunctive query answering directly can still be used for this purpose by expressing disjunctive conditions with multiple auxiliary rules that use the same head predicate, and querying for the instances of this head.

As above, the simplest policy is the public domain (PD) license:

$$PD : \mathsf{Usage}(x) \vee \mathsf{Derivation}(x).$$

Here and below, we always use x as the variable that represents the corresponding process in a policy description. CC BY can now be defined as follows:

$$BY : (\mathsf{Usage}(x) \wedge \mathsf{wasTriggeredBy}(x, y) \wedge \mathsf{Attribution}(y)) \vee$$
$$(\mathsf{Derivation}(x) \wedge \mathsf{wasGeneratedBy}(z, x) \wedge$$
$$\mathsf{hasPolicy}(z, v) \wedge \mathsf{containedIn}(v, BY)) .$$

This formalisation alone would leave room for derivations that are falsely classified as compliant, since the condition only requires that there exists one artefact that has one contained policy. Further artefacts or policies that violate these terms might then exist. We can prevent this by requiring hasPolicy to be *functional* and wasGeneratedBy to be *inverse functional* (as before, we assume that $\approx$ has been suitably axiomatised, which is possible in Datalog; see [20] for details):

$$v_1 \approx v_2 \leftarrow \mathsf{hasPolicy}(x, v_1) \wedge \mathsf{hasPolicy}(x, v_2),$$
$$z_1 \approx z_2 \leftarrow \mathsf{wasGeneratedBy}(z_1, x) \wedge \mathsf{wasGeneratedBy}(z_2, x) .$$

Using this auxiliary modelling, we can easily express BY-ND and BY-SA as well [20].

7   Related Work

The formalisation of policies and similar restrictions has been considered in many works, but the relationship to our approach is often limited. For example, restrictions in Digital Rights Management (DRM) systems can be specified in a rights expression language such as ODRL [16]. Policy containment or self-referentiality is not considered there. Similarly, ccREL offers an RDF representation for Creative Commons licenses but uses a static name-based encoding that cannot capture the content-based relationships that we model [1]. Using rules in the policy language AIR [18], the meaning of ccREL terms has been further formalised but without attempting to overcome the restrictions of name-based modelling [30].

Bonatti and Mogavero consider policy containment as a formal reasoning task, and restrict the *Protune* policy language so that this task is decidable [8]. Reasoning about policy conformance and containment also motivated earlier studies by the second author, where policies have been formalised as conjunctive queries [31]. Our present work can be viewed as a generalisation of this approach.

Other related works have focussed on different aspects of increasing the expressiveness of policy modelling. Ringelstein and Staab present the history-aware PAPEL policy language that can be processed by means of a translation to Datalog [27]. The data-purpose algebra by Hanson et al. allows the modelling of usage restrictions of data and the transformation of the restrictions when data is processed [13].

Many knowledge representation formalisms have been proposed to accomplish non-classical semantics (e.g. fixed point semantics) and meta-modelling (as present in our expression of containment as an object-level predicate). However, both aspects are usually not integrated, or come with technical restrictions that do not suit our application.

Fixed point operators exist in a number of flavours. Most closely related to our setting are works on fixed point based evaluation of terminological cycles in description logic ontologies [5,25]. Later works have been based on the relationship to the μ-calculus, see [6, Section 5.6] for an overview of the related literature. As is typical for such constructions, the required monotonicity is ensured on a logical level by restricting negation. This is not possible in our scenario where we focus on the entailment of implications (policy containments). Another approach of defining preferred models where certain predicate extensions have been minimised/maximised is Circumscription [22]. This might provide an alternative way to define a semantics that can capture desired policy containments, but it is not clear if and how entailments could then be computed.

Meta-modelling is possible with first- and higher-order approaches (see, e.g., [24] for an OWL-related discussion) yet we are not aware of any approaches that provide the semantics we intend. Glimm et al. [12], e.g., show how some schema entailments of OWL 2 DL can be represented with ontological individuals and properties, but the classical semantics of OWL would not yield the desired policy containments.

For relational algebra, it has been proposed to store relation names as individuals, and to use an expansion operator to access the extensions of these relations [28]. This allows for queries that check relational containment, but based on a fixed database (closed world) rather than on all possible interpretations (open world) as in our case.

8   Conclusions and Future Work

To the best of our knowledge, we have presented the first formal language for modelling self-referential policies. A particular advantage of our approach is that it can be instantiated in more specific knowledge representation formalisms, such as rule or ontology languages, to take advantage of existing tools for automated reasoning.

This opens up a number of directions for practical studies and exploitations. Refined provenance models, better tool support, and best practices for publishing policies are still required. On the conceptual side it would also be interesting to ask if our CF-based syntactic restrictions could be further relaxed without giving up the positive properties of the semantics.

Acknowledgements. We would like to thank Piero Bonatti, Clemens Kupke and the anonymous reviewers for their comments. Markus Krötzsch is sponsored by EPSRC grant EP/F065841/1. Sebastian Speiser is sponsored by the EU FP7 grant 257641.

References

1. Abelson, H., Adida, B., Linksvayer, M., Yergler, N.: ccREL: The Creative Commons Rights Expression Language. Tech. rep., Creative Commons (2008),
   http://creativecommons.org/projects/ccREL
2. Abiteboul, S., Duschka, O.M.: Complexity of answering queries using materialized views. In: Proc. 17th ACM SIGACT-SIGMOD-SIGART Symp. on Principles of Database Systems (PODS 1998), pp. 254–263. ACM (1998)
3. Abiteboul, S., Hull, R., Vianu, V.: Foundations of Databases. Addison Wesley (1994)
4. Andréka, H., van Benthem, J., Németi, I.: Back and forth between modal logic and classical logic. Logic Journal of the IGPL 3(5), 685–720 (1995)
5. Baader, F.: Terminological cycles in KL-ONE-based knowledge representation languages. In: 8th National Conf. on Artificial Intelligence (AAAI 1990), pp. 621–626. AAAI Press (1990)
6. Baader, F., Calvanese, D., McGuinness, D., Nardi, D., Patel-Schneider, P. (eds.): The Description Logic Handbook, 2nd edn. Cambridge University Press (2007)
7. Bonatti, P.A., De Coi, J.L., Olmedilla, D., Sauro, L.: A rule-based trust negotiation system. IEEE Transactions on Knowledge and Data Engineering 22(11), 1507–1520 (2010)
8. Bonatti, P.A., Mogavero, F.: Comparing rule-based policies. In: 9th IEEE Int. Workshop on Policies for Distributed Systems and Networks (POLICY 2008), pp. 11–18 (2008)
9. Cheney, J., Gil, Y., Groth, P., Miles, S.: Requirements for Provenance on the Web. W3C Provenance Incubator Group (2010),
   http://www.w3.org/2005/Incubator/prov/wiki/User_Requirements
10. Dodds, L.: Rights statements on the Web of Data. Nodalities Magazine, 13–14 (2010)
11. Glimm, B., Krötzsch, M.: SPARQL beyond subgraph matching. In: Patel-Schneider, et al. [26], pp. 241–256
12. Glimm, B., Rudolph, S., Völker, J.: Integrated metamodeling and diagnosis in OWL 2. In: Patel-Schneider, et al. [26], pp. 257–272
13. Hanson, C., Berners-Lee, T., Kagal, L., Sussman, G.J., Weitzner, D.: Data-purpose algebra: Modeling data usage policies. In: 8th IEEE Int. Workshop on Policies for Distributed Systems and Networks (POLICY 2007), pp. 173–177 (2007)
14. Hitzler, P., Krötzsch, M., Rudolph, S.: Foundations of Semantic Web Technologies. Chapman & Hall/CRC (2009)

15. Horrocks, I., Sattler, U.: Decidability of $\mathcal{SHIQ}$ with complex role inclusion axioms. Artificial Intelligence 160(1), 79–104 (2004)
16. Iannella, R.: Open Digital Rights Language (ODRL) Version 1.1. W3C Note (September 19, 2002), http://www.w3.org/TR/odrl/
17. Kagal, L., Finin, T., Joshi, A.: A policy language for a pervasive computing environment. In: 4th IEEE Int. Workshop on Policies for Distributed Systems and Networks (POLICY 2003), pp. 63–74 (2003)
18. Kagal, L., Hanson, C., Weitzner, D.: Using dependency tracking to provide explanations for policy management. In: 9th IEEE Int. Workshop on Policies for Distributed Systems and Networks (POLICY 2008), pp. 54–61 (2008)
19. Kephart, J.O., Walsh, W.E.: An artificial intelligence perspective on autonomic computing policies. In: 5th IEEE Int. Workshop on Policies for Distributed Systems and Networks (POLICY 2004), pp. 3–12 (2004)
20. Krötzsch, M., Speiser, S.: Expressing self-referential usage policies for the Semantic Web. Tech. Rep. 3014, Institute AIFB, Karlsruhe Institute of Technology (2011), http://www.aifb.kit.edu/web/Techreport3014
21. Lessig, L.: CC in Review: Lawrence Lessig on Compatibility (2005), http://creativecommons.org/weblog/entry/5709 (accessed July 1, 2011)
22. Lifshitz, V.: Circumscriptive theories: A logic-based framework for knowledge representation. Journal of Philosophical Logic 17, 391–441 (1988)
23. Moreau, L., Clifford, B., Freire, J., Futrelle, J., Gil, Y., Groth, P., Kwasnikowska, N., Miles, S., Missier, P., Myers, J., Plale, B., Simmhan, Y., Stephan, E., Van den Bussche, J.: The Open Provenance Model core specification (v1.1). Future Generation Computer Systems 27, 743–756 (2011)
24. Motik, B.: On the properties of metamodeling in OWL. J. of Logic and Computation 17(4), 617–637 (2007)
25. Nebel, B.: Terminological cycles: Semantics and computational properties. In: Sowa, J.F. (ed.) Principles of Semantic Networks: Explorations in the Representation of Knowledge, pp. 331–361. Kaufmann (1991)
26. Patel-Schneider, P.F., Pan, Y., Glimm, B., Hitzler, P., Mika, P., Pan, J., Horrocks, I.: ISWC 2010, Part I. LNCS, vol. 6496. Springer, Heidelberg (2010)
27. Ringelstein, C., Staab, S.: PAPEL: A Language and Model for Provenance-Aware Policy Definition and Execution. In: Hull, R., Mendling, J., Tai, S. (eds.) BPM 2010. LNCS, vol. 6336, pp. 195–210. Springer, Heidelberg (2010)
28. Ross, K.A.: Relations with relation names as arguments: algebra and calculus. In: Proc. 11th ACM SIGACT-SIGMOD-SIGART Symp. on Principles of Database Systems (PODS 1992), pp. 346–353. ACM (1992)
29. Seneviratne, O., Kagal, L., Berners-Lee, T.: Policy-Aware Content Reuse on the Web. In: Bernstein, A., Karger, D.R., Heath, T., Feigenbaum, L., Maynard, D., Motta, E., Thirunarayan, K. (eds.) ISWC 2009. LNCS, vol. 5823, pp. 553–568. Springer, Heidelberg (2009)
30. Seneviratne, O.W.: Framework for Policy Aware Reuse of Content on the WWW. Master thesis. Massachusetts Institute of Technology (2009)
31. Speiser, S., Studer, R.: A self-policing policy language. In: Patel-Schneider, et al. [26], pp. 730–746
32. W3C OWL Working Group: OWL 2 Web Ontology Language: Document Overview. W3C Recommendation (October 27, 2009), http://www.w3.org/TR/owl2-overview/
33. Weitzner, D.J., Hendler, J., Berners-Lee, T., Connolly, D.: Creating a policy-aware Web: Discretionary, rule-based access for the World Wide Web. In: Web and Information Security, ch. I, pp. 1–31. IRM Press (2006)

A Native and Adaptive Approach for Unified Processing of Linked Streams and Linked Data*

Danh Le-Phuoc[1], Minh Dao-Tran[2], Josiane Xavier Parreira[1],
and Manfred Hauswirth[1]

[1] Digital Enterprise Research Institute, National University of Ireland, Galway
{danh.lephuoc,josiane.parreira,manfred.hauswirth}@deri.org
[2] Institut für Informationssysteme, Technische Universität Wien
dao@kr.tuwien.ac.at

Abstract. In this paper we address the problem of scalable, native and adaptive query processing over Linked Stream Data integrated with Linked Data. Linked Stream Data consists of data generated by stream sources, e.g., sensors, enriched with semantic descriptions, following the standards proposed for Linked Data. This enables the integration of stream data with Linked Data collections and facilitates a wide range of novel applications. Currently available systems use a "black box" approach which delegates the processing to other engines such as stream/event processing engines and SPARQL query processors by translating to their provided languages. As the experimental results described in this paper show, the need for query translation and data transformation, as well as the lack of full control over the query execution, pose major drawbacks in terms of efficiency. To remedy these drawbacks, we present CQELS (*Continuous Query Evaluation over Linked Streams*), a native and adaptive query processor for unified query processing over Linked Stream Data and Linked Data. In contrast to the existing systems, CQELS uses a "white box" approach and implements the required query operators natively to avoid the overhead and limitations of closed system regimes. CQELS provides a flexible query execution framework with the query processor dynamically adapting to the changes in the input data. During query execution, it continuously reorders operators according to some heuristics to achieve improved query execution in terms of delay and complexity. Moreover, external disk access on large Linked Data collections is reduced with the use of data encoding and caching of intermediate query results. To demonstrate the efficiency of our approach, we present extensive experimental performance evaluations in terms of query execution time, under varied query types, dataset sizes, and number of parallel queries. These results show that CQELS outperforms related approaches by orders of magnitude.

Keywords: Linked Streams, RDF Streams, Linked Data, stream processing, dynamic query planning, query optimisation.

* This research has been supported by Science Foundation Ireland under Grant No. SFI/08/CE/I1380 (Lion-II), by the Irish Research Council for Science, Engineering and Technology (IRCSET), by the European Commission under contract number FP7-2007-2-224053 (CONET), by Marie Curie action IRSES under Grant No. 24761 (Net2), and by the Austrian Science Fund (FWF) project P20841.

L. Aroyo et al. (Eds.): ISWC 2011, Part I, LNCS 7031, pp. 370–388, 2011.

1 Introduction

Sensing devices have become ubiquitous. Mobile phones (accelerometer, compass, GPS, camera, etc.), weather observation stations (temperature, humidity, etc.), patient monitoring systems (heart rate, blood pressure, etc.), location tracking systems (GPS, RFID, etc.), buildings management systems (energy consumption, environmental conditions, etc.), and cars (engine monitoring, driver monitoring, etc.) continuously produce information streams. Also on the Web, services like Twitter, Facebook and blogs, deliver streams of (typically unstructured) real-time data on various topics. The heterogeneous nature of such diverse streams makes their use and integration with other data sources a difficult and labor-intensive task, which currently requires a lot of "hand-crafting."

To address some of the problems, there have been efforts to lift stream data to a semantic level, e.g., by the W3C Semantic Sensor Network Incubator Group[1] and [12,32,37]. The goal is to make stream data available according to the Linked Data principles [10] – a concept that is known as *Linked Stream Data* [31]. This would allow an easy and seamless integration, not only among heterogenous sensor data, but also between sensor and Linked Data collections, enabling a new range of "real-time" applications.

However, one distinguishing aspect of streams that the Linked Data principles do not consider is their temporal nature. Usually, Linked Data is considered to change infrequently. Data is first crawled and stored in a centralised repository before further processing. Updates on a dataset are usually limited to a small fraction of the dataset and occur infrequently, or the whole dataset is replaced by a new version entirely. Query processing, as in traditional relational databases, is *pull* based and *one-time*, i.e., the data is read from the disk, the query is executed against it once, and the output is a set of results for that point in time. In contrast, in Linked Stream Data, new data items are produced continuously, the data is often valid only during a time window, and it is continually *pushed* to the query processor. Queries are continuous, i.e., they are registered once and then are evaluated continuously over time against the changing dataset. The results of a continuous query are updated as new data appears. Therefore, current Linked Data query processing engines are not suitable for handling Linked Stream Data. It is interesting to notice that in recent years, there has been work that points out the dynamics of Linked Data collections [35]. Although at a much slower pace compared to streams, it has been observed that centralised approaches will not be suitable if *freshness* of the results is important, i.e., the query results are consistent with the actual "live" data under certain guarantees, and thus an element of "live" query execution will be needed [34]. Though this differs from stream data, some of our findings may also be applicable to this area.

Despite its increasing relevance, there is currently no native query engine that supports unified query processing over Linked Stream and Linked Data inputs. Available systems, such as C-SPARQL [9], SPARQLstream [14] and EP-SPARQL [3], use a "black box" approach which delegates the processing to other engines such as stream/event processing engines and SPARQL query processors by translating to their provided languages. This dependency introduces the overhead of query translation and data transformation. Queries first need to be translated to the language used in the un-

---

[1] http://www.w3.org/2005/Incubator/ssn/

derlying systems. The data also needs to be transformed to feed into the system. For instance, in C-SPARQL and SPARQLstream, the data is stored in relational tables and relational streams before any further processing, and EP-SPARQL uses logic facts. This strategy also does not allow full control over the execution plan nor over the implementation of the query engine's elements. Consequently, the possibilities for query optimisations are very limited.

To remedy these drawbacks, we present CQELS (*Continuous Query Evaluation over Linked Streams*), a native and adaptive query processing engine for querying over unified Linked Stream Data and Linked Data. In contrast to the existing systems, CQELS uses a "white box" approach. It defines its own native processing model, which is implemented in the query engine. CQELS provides a flexible query execution framework with the query processor dynamically adapting to changes in the input data. During query execution, it continuously reorders operators according to some heuristics to achieve improved query execution in terms of delay and complexity. External disk access on large Linked Data collections is reduced with the use of data encoding and caching of intermediate query results, and faster data access is obtained with indexing techniques. To demonstrate the efficiency of our approach, we present extensive experimental performance evaluations in terms of query execution time, under varied query types, dataset sizes, and number of parallel queries. Results show that CQELS performs consistently well, and in most cases outperforms related approaches by orders of magnitude.

The remainder of this paper is organised as follows: Section 2 discusses our contribution in relation to relational database management systems, data stream management systems, Linked Data processing, and Linked Stream Data processing. Our processing model is described in Section 3, and the query engine is discussed in Section 4. Section 5 presents an experimental evaluation of our approach, and Section 6 provides our conclusions and a brief discussion about ongoing work and next steps.

2   Related Work

RDF stores. A fair amount of work on storage and query processing for Linked Data is available, including Sesame [13], Jena [38], RISC-3X [28], YARS2 [23], and Oracle Database Semantic Technologies [16]. Most of them focus on scalability in dataset size and query complexity. Based on traditional database management systems (DBMSs), they typically assume that data changes infrequently, and efficiency and scalability are achieved by carefully choosing appropriate data storage and indexing optimised for read access, whereas stream data is characterised by high numbers and frequencies of updates. The Berlin SPARQL benchmark[2] shows that the throughput of a typical triple store currently is less than 200 queries per second, while in stream applications continuous queries need to be processed every time there is a new update in the data, which can occur at rates up to 100,000 updates per second. Nevertheless, some of the techniques and design principles of triple stores are still useful for scalable processing of Linked Stream Data, for instance some of the physical data organisations [1,13,38] and indexing schemas [16,23,28].

---

[2] http://www4.wiwiss.fu-berlin.de/bizer/BerlinSPARQLBenchmark/

Data stream management. Data stream management systems (DSMSs) such as STREAM [4], Aurora [15], and TelegraphCQ [26] were built to overcome limitations of traditional database management systems in supporting streaming applications [20]. The STREAM system proposes CQL [4] (Continuous Query Language) which extends standard SQL syntax with new constructs for temporal semantics and defines a mapping between streams and relations. The query engine consists of three components: operators, that handle the input and output streams, queues, that connect input operators to output operators, and synopses, that store the intermediate states needed by continuous query plans. In the Aurora/Borealis project [15] users can compose stream relationships and construct queries in a graphical representation which is then used as input for the query planner. TelegraphCQ introduces StreaQuel as a language, which follows a different path and tries to isolate temporal semantics from the query language through external definitions in a C-like syntax. TelegraphCQ also uses a technique called *Eddies* [6], which continuously reorders operators in a query plan as it runs, adapting to changes in the input data. DSMSs perform better compared to traditional DBMSs in the context of high volumes of updates. Even though DSMSs can not directly process Linked Stream Data, such processing is still possible by translating the queries and mapping the data to fit into the data storage. This is currently done by available systems that process Linked Stream Data. The CQELS query engine, on the other hand, can directly process Linked Stream Data, yielding consistently better performance, as we will demonstrate later on in the paper.

Streams and semantics. Semantic Streams [37] was among the first systems to propose semantic processing of streams. It uses Prolog-based logic rules to allow users to pose declarative queries over semantic interpretations of sensor data. Semantic System S [12] proposes the use of the Web Ontology Language (OWL) to represent sensor data streams, as well as processing elements for composing applications from input data streams. The Semantic Sensor Web project [8,32] also focuses on interoperability between different sensor sources, as well as providing contextual information about the data. It does so by annotating sensor data with spatial, temporal, and thematic semantic metadata. Research like the one carried by W3C Semantic Sensor Network Incubator Group[3] aims at the integration of stream data with Linked Data sources by following the Linked Data principles for representing the data. In parallel, the concept of *Linked Stream Data* was introduced [31], in which URIs were suggested for identifying sensors and stream data.

In contrast to these approaches, our work focuses on the efficient processing of Linked Stream Data integrated with other Linked Data sources. Existing work with this focus comprises Streaming SPARQL [11], C-SPARQL [9], SPARQLstream [14], and EP-SPARQL [3] as the main approaches. They all extend SPARQL with sliding window operators for RDF stream processing. Streaming SPARQL simply extends SPARQL to support window operators without taking into account performance issues regarding the choice of the data structures and the sharing of computing states for continuous execution. Continuous SPARQL (C-SPARQL) proposes an execution framework built of top of existing stream data management systems and triple stores. These systems are used independently as "black boxes." In C-SPARQL, continuous queries are divided

---

[3] http://www.w3.org/2005/Incubator/ssn/

into static and dynamic parts. The framework orchestrator loads bindings of the static parts into relations, and the continuous queries are executed by processing the stream data against these relations. C-SPARQL is not designed for large static data sets, which can degrade the performance of the stream processing considerably.

Along the same lines, SPARQLstream also translates its SPARQLstream language to another relational stream language based on mapping rules. Event Processing SPARQL (EP-SPARQL), a language to describe event processing and stream reasoning, can be translated to ETALIS [3], a Prolog-based complex event processing framework. First, RDF-based data elements are transformed into logic facts, and then EP-SPARQL queries are translated into Prolog rules. In contrast to these systems, CQELS is based on a unified "white box" approach which implements the required query operators for the triple-based data model natively, both for streams and static data. This native approach enables better performance and can dynamically adapt to changes in the input data.

3   Processing Model

The adaptive processing model of CQELS captures all the aspects of both data modelling and query processing over Linked Stream Data and Linked Data in one single theoretical framework. It defines two types of data sources, RDF streams and RDF datasets, and three classes of operators for processing these types of data sources. Operators used in a query are organised in a data flow according to defined query semantics, and the adaptive processing model provides functions to reorder the query operators to create equivalent, more efficient data flows. The details of the processing model are described in the following.

3.1   Definitions

In continuous query processing over dynamic data, the temporal nature of the data is crucial and needs to be captured in the data representation. This applies to both types of data sources, since updates in Linked Data collections are also possible. We define RDF streams to represent Linked Stream Data, and we model Linked Data by generalising the standard definition of RDF datasets to include the temporal aspect.

Similar to RDF temporal [22], C-SPARQL, and SPARQLstream, we represent temporal aspects of the data as a timestamp label. We use $t \in \mathbb{N}$ to indicate a *logical* timestamp to facilitate ordered logical clocks for local and distributed data sources as done by classic time-synchronisation approaches [24]. The issues of distributed time synchronization and flexible time management are beyond the scope of this paper. We refer the reader to [19,27,33] for more details.

Let I, B, and L be *RDF nodes* which are pair-wise disjoint infinite sets of Information Resource Identifiers (IRIs), blank nodes and literals, and $IL = I \cup L$, $IB = I \cup B$ and $IBL = I \cup B \cup L$ be the respective unions. Thereby,

1. A triple $(s, p, o) \in IB \times I \times IBL$ is an *RDF triple*.
2. An *RDF dataset at timestamp t*, denoted by $G(t)$, is a set of *RDF triples* valid at time t. An *RDF dataset* is a sequence $G = [G(t)], t \in \mathbb{N}$, ordered by t. When it holds that $G(t) = G(t+1)$ for all $t \geq 0$, we call G a *static RDF dataset* and denote $G^s = G(t)$.

3. An *RDF stream* S is a bag of elements $\langle(s,p,o):[t]\rangle$, where (s,p,o) is an *RDF triple* and t is a timestamp. $S^{\leq t}$ denotes the bag of elements in S with timestamps $\leq t$, i.e., $\{\langle(s,p,o):[t']\rangle \in S \mid t' \leq t\}$.

Let V be an infinite set of variables disjoint from IBL. A *mapping* is a partial function $\mu\colon V \to IBL$. The domain of μ, $dom(\mu)$, is the subset of V where μ is defined. Two mappings μ_1 and μ_2 are *compatible* if $\forall x \in dom(\mu_1) \cap dom(\mu_2), \mu_1(x) = \mu_2(x)$.

A tuple from $(IB \cup V) \times (I \cup V) \times (IBL \cup V)$ is a *triple pattern*. For a given triple pattern P, the set of variables occurring in P is denoted as $var(P)$ and the triple obtained by replacing elements in $var(P)$ according to μ is denoted as $\mu(P)$. A *graph template* $\mathbb{T}$ is a set of triple patterns.

3.2 Operators

Our processing model takes as input RDF datasets and RDF streams containing possibly infinite numbers of RDF triples, applies a query Q and continuously produces outputs.

In processing Q, snapshots of the input at discrete times t, i.e., finite amounts of data, are used in the evaluation of the query. This requires dedicated operators to (i) take snapshots of the input and filter its valid part w.r.t. some condition, (ii) operate on the finite, intermediate data, and (iii) convert the final results back into a stream. The required operators are called *window*, *relational*, and *streaming operators*.

Window Operators. These operators extract triples from an RDF stream or dataset that match a given triple pattern and are valid within a given time window. Similar to SPARQL, we define a triple matching pattern operator on an RDF dataset at timestamp t as

$$[\![P,t]\!]_G = \{\mu \mid dom(\mu) = var(P) \wedge \mu(P) \in G(t)\}.$$

A window operator $[\![P,t]\!]_S^\omega$ is then defined by extending the operator above as follows.

$$[\![P,t]\!]_S^\omega = \{\mu \mid dom(\mu) = var(P) \wedge \langle\mu(P):[t']\rangle \in S \wedge t' \in \omega(t)\}.$$

where $\omega(t)\colon \mathbb{N} \to 2^{\mathbb{N}}$ is a function mapping a timestamp to a (possibly infinite) set of timestamps. This gives us the flexibility to choose between different window modes [5]. For example, a time-based sliding window of size T can be expressed as $\omega_{RANGE}(t) = \{t' \mid t' \leq t \wedge t' \geq \max(0, t-T)\}$, and a window that extracts only events happening at the current time corresponds to $\omega_{NOW}(t) = \{t\}$. Moreover, we can similarly define *triple-based windows* that return the latest N triples ordered by the timestamps.

We define a *result set* Γ as a function from $\mathbb{N} \cup \{-1\}$ to finite but unbounded bags of mappings, where $\Gamma(-1) = \emptyset$. A *discrete result set* $\Omega = \Gamma(t)$, $t \geq 0$, denotes the bag of mappings at time t. Discrete result sets are the input of relational operators described below.

Relational Operators. Our processing model supports the operators found in traditional relational database management systems [18]. Similar to the semantics of SPARQL [29], the operators work on the mappings from discrete result sets. As an example, given two discrete result sets, Ω_1 and Ω_2, the join and union operators are defined as

$\Omega_1 \bowtie \Omega_2 = \{\mu_1 \cup \mu_2 \mid \mu_1 \in \Omega_1, \mu_2 \in \Omega_2 \text{ are compatible } \}$

$\Omega_1 \cup \Omega_2 = \{\mu \mid \mu \in \Omega_1 \vee \mu \in \Omega_2\}.$

Streaming Operators. Similarly to the relation-to-stream operator of CQL [5], we define an operator, based on some patterns, to generate RDF streams from result sets. From a graph template $\mathbb{T}$, that provides a set of triple patterns, and a result set Γ, a streaming operator $\mathbb{C}$ is defined as

$$\mathbb{C}(\mathbb{T}, \Gamma) = \bigcup_{t \geq 0} \{\langle \mu(P) : [f(t)]\rangle \mid \mu \in \Gamma(t) \setminus \Gamma(t-1) \wedge P \in \mathbb{T}\},$$

where $f : \mathbb{N} \to \mathbb{N}$ is a function mapping t to a new timestamp to indicate when we want to stream out the result. In the simplest case, f is the identity function, indicating that triples are streamed out immediately.

Query Semantics. Operators of a query are organised in a *data flow*. A data flow D is a directed tree of operators, whose root node is either a relational or a streaming operator, while leaves and intermediate nodes are window and relational operators, respectively.

Suppose the inputs to the leaves of D are RDF streams $S_1, \ldots, S_n$ ($n \geq 1$) and RDF datasets $G_1, \ldots, G_m$ ($m \geq 0$). The *query semantics* of D is then defined as follows: If the root of D is a streaming (resp., relational) operator, producing a stream S (resp., result set Γ), then the result of D at time t is $S^{\leq t}$ (resp., $\Gamma(t)$), which is produced by recursively applying the operators comprising D to $S_1^{\leq t}, \ldots, S_n^{\leq t}$ and $G_1, \ldots, G_m$. Next we introduce the "localisation scenario" to illustrate the query semantics of our processing model. This scenario will also be used in following sections of the paper.

Localisation scenario: Consider a group of people wearing devices that constantly stream their locations in a building, i.e., in which room they currently are, and assume we have information about the direct connectivity between the rooms, given by a static RDF dataset G with triples of the form $P_3 = (?loc_1, conn, ?loc_2)$, where $G^S = \{(r_1, conn, r_2), (r_1, conn, r_3), (r_2, conn, r_1), (r_3, conn, r_1)\}$. Also assume that people's locations are provided in a single stream S with triples of form $(?person, detectedAt, ?loc)$. We are interested in answering the following continuous query: "Notify two people when they can reach each other from two different and directly connected rooms."

Figure 1a depicts a possible data flow D_1 for the query in the localisation scenario. It suggests to extract two windows from stream S using the functions $\omega_1 = \omega_{NOW}$ and $\omega_2 = \omega_{RANGE}$. The former looks at the latest detected person, and the latter monitors people during the last T logical clock ticks by which we can assume that they are still in the same room. For the example, we assume $T = 2$. Let Γ_1 and Γ_2 be the outputs of the window operators. We use the triple patterns $P_i = (?person_i, detectedAt, ?loc_i)$ for $i = 1, 2$ at the window operators; hence, mappings in Γ_i are of the form $\{?person_i \mapsto pid, ?loc_i \mapsto lid\}$.

The join $\bowtie_{12}$ of discrete result sets from Γ_1 and Γ_2 in Figure 1a gives us the output result set in Γ_3 to check the reachability based on the latest detected person. After joining elements of Γ_3 with those of Γ_4 (the direct connectivity between locations

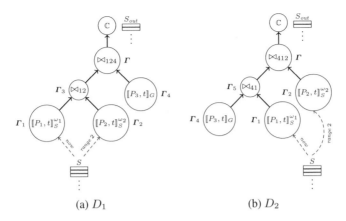

(a) D_1 (b) D_2

Fig. 1. Possible data flows for the query in the localisation scenario

Table 1. Input and output of D_1 as time progresses

t	S	Γ_1	Γ_2	S_{out}
0	$\langle(m_0, dA, r_1) : [0]\rangle$	$\{?p_1 \mapsto m_0, ?\ell_1 \mapsto r_1\}$	$\{?p_2 \mapsto m_0, ?\ell_2 \mapsto r_1\}$	$\emptyset$
1	$\langle(m_0, dA, r_1) : [0]\rangle$ $\langle(m_1, dA, r_2) : [1]\rangle$	$\{?p_1 \mapsto m_1, ?\ell_1 \mapsto r_2\}$	$\{?p_2 \mapsto m_0, ?\ell_2 \mapsto r_1\}$ $\{?p_2 \mapsto m_1, ?\ell_2 \mapsto r_2\}$	$\langle(m_0, reaches, m_1) : [1]\rangle$
2	$\langle(m_0, dA, r_1) : [0]\rangle$ $\langle(m_1, dA, r_2) : [1]\rangle$ $\langle(m_2, dA, r_1) : [2]\rangle$	$\{?p_1 \mapsto m_2, ?\ell_1 \mapsto r_1\}$	$\{?p_2 \mapsto m_1, ?\ell_2 \mapsto r_2\}$ $\{?p_2 \mapsto m_2, ?\ell_2 \mapsto r_1\}$	$\langle(m_0, reaches, m_1) : [1]\rangle$ $\langle(m_1, reaches, m_2) : [2]\rangle$
3	$\langle(m_0, dA, r_1) : [0]\rangle$ $\langle(m_1, dA, r_2) : [1]\rangle$ $\langle(m_2, dA, r_1) : [2]\rangle$ $\langle(m_3, dA, r_2) : [3]\rangle$	$\{?p_1 \mapsto m_3, ?\ell_1 \mapsto r_2\}$	$\{?p_2 \mapsto m_2, ?\ell_2 \mapsto r_1\}$ $\{?p_2 \mapsto m_3, ?\ell_2 \mapsto r_2\}$	$\langle(m_0, reaches, m_1) : [1]\rangle$ $\langle(m_1, reaches, m_2) : [2]\rangle$ $\langle(m_2, reaches, m_3) : [3]\rangle$
$\vdots$	$\vdots$	$\vdots$	$\vdots$	$\vdots$

provided by G) via $\bowtie_{124}$, we have the result set Γ to answer the query. To return this result in terms of a stream S_{out}, the operator $\mathbb{C}$ is used at the root of D_1.

Table 1 shows the input/output of D_1 as time progresses. To reduce space consumption, we use abbreviations as follows: dA for *detectedAt*, $?p$ for $?person$, and $?\ell$ for $?loc$.

3.3 Adaptation Strategies

A data flow contains inner relational operators which can be reordered to create new equivalent data flows. For instance, Figures 1a and 1b show two equivalent data flows for the query in the localisation scenario. With respect to each alternative, an operator might have a different next/parent operator. For example, $[\![P_1, t]\!]_S^{\omega_1}$ has $\bowtie_{12}$ as its parent in D_1 while in D_2, its parent is $\bowtie_{41}$.

In a stream processing environment, due to updates in the input data, during the query lifetime the engine constantly attempts to determine the data flow that currently provides the most efficient query execution. We propose an adaptive query processing

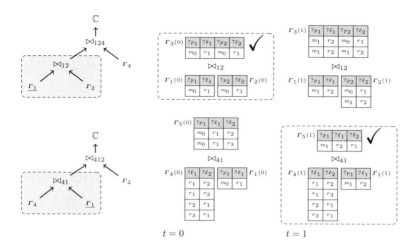

Fig. 2. Dynamically choose the next operator after Γ_1 at timestamps 0 and 1

mechanism similar to Eddies [6], which continuously routes the outputs of an operator to the next operator on the data flow. The routing policy will dynamically tell the operator what is the next operator it should forward data to, as shown in Algorithm 1.

Algorithm 1. route($routingEntry, \mathbb{O}, t$)

Input: $routingEntry$: timestamped triple/mapping, $\mathbb{O}$: operator, t : timestamp
$\Omega := $ compute($routingEntry, \mathbb{O}, t$)
if $\mathbb{O}$ *is not root* **then**
$\quad nextOp := $ findNextOp($\mathbb{O}, t$)
$\quad$ **for** $\mu \in \Omega$ **do** route($\mu, nextOp, t$)
else deliver Ω

Function route($routingEntry, \mathbb{O}, t$) is used to recursively apply the operator $\mathbb{O}$ on a mapping or timestamped triple $routingEntry$ and to route the output mappings to the next operator. It uses the following primitives:

- compute($routingEntry, \mathbb{O}, t$): apply $\mathbb{O}$, a window, relational, or streaming operator, to $routingEntry$, a timestamped triple or a mapping, at timestamp t, and return a discrete result set.
- findNextOp($\mathbb{O}, t$): find the next operator to route the output mapping to, at timestamp t, based on a given routing policy.

The routing policy decides the order in which the operators are executed at runtime. There are many ways to implement a routing policy. However, choosing the optimal order on every execution is not trivial. We are investigating mechanisms for dynamic cost-based optimisation. Preliminary findings are reported in [25]. A possible solution, common to DBMSs, is a cost-based strategy: the routing policy computes an estimated "cost" to each possible data flow, and chooses the one with the smallest cost. While the definition of cost is not fixed, it is usually measured by estimating the number of output mappings the operator will produce.

The following example illustrates how the adaptation strategies work as a whole.

Example 1. Consider again the query in the localisation scenario at timestamps 0 and 1, and assume the routing policy implemented is the cost-based strategy mentioned above. Figure 2 illustrates the decision of which operator to choose next after extracting the latest triple at Γ_1. In this figure, two simplified versions of D_1 and D_2 are on the left. On the right hand side, we show the input/output of the join operators $\bowtie_{12}$ and $\bowtie_{41}$. At timestamp 0, $|\Gamma_1(0)| = |\Gamma_2(0)| = 1$ as the first triple is streamed into the system. It is preferable at this point to use D_1, i.e., to join $\Gamma_1(0)$ with $\Gamma_2(0)$ using $\bowtie_{12}$ because the intermediate result $\Gamma_3(0)$ has size 1. If we follow D_2 then joining $\Gamma_1(0)$ with $\Gamma_4(0)$ using $\bowtie_{41}$ yields $\Gamma_5(0)$ with size 2. However, at $t = 1$, D_2 is preferred because $|\Gamma_3(1)| = 2$ and $|\Gamma_5(1)| = 1$.

4   CQELS's Query Engine

The CQELS query engine implements the model introduced in Section 3. Continuous queries can be registered using our CQELS language, an extension of the declarative SPARQL 1.1 language, which is described next. We then explain the details of the engine. We show how data is encoded for memory savings, how caching and indexing are used for faster data access, and how operators and the routing policy are implemented. Before moving onto the query language, we first need to introduce our second scenario, the "conference scenario," which is also used in the evaluation section.

Conference scenario: This scenario is based on the Live Social Semantics experiment presented in [2]. We extend the localisation scenario by considering that people are now authors of research papers and they are attending a conference. These authors have their publication information stored in a DBLP dataset. To enhance the conference experience, each participant would have access to the following services, which can all be modelled as continuous queries:

(Q1) Inform a participant about the name and description of the location he just entered,

(Q2) Notify two people when they can reach each other from two different and directly connected (from now on called *nearby*) locations,

(Q3) Notify an author of his co-authors who have been in his current location during the last 5 seconds,

(Q4) Notify an author of the editors that edit a paper of his and have been in a nearby location in the last 15 seconds,

(Q5) Count the number of co-authors appearing in nearby locations in the last 30 seconds, grouped by location.

4.1 CQELS Language

Based on our query semantics, we introduce a declarative query language called CQELS by extending the SPARQL 1.1 grammar[4] using the EBNF notation. We add a query pattern to apply window operators on RDF Streams into the *GraphPatternNotTriples* pattern.

$GraphPatternNotTriples ::= GroupOrUnionGraphPattern \mid OptionalGraphPattern$

$\mid MinusGraphPattern \mid GraphGraphPattern \mid$ **StreamGraphPattern**
$\mid ServiceGraphPattern \mid Filter \mid Bind$

Assuming that each stream is identified by an IRI as identification, the **Stream-GraphPattern** pattern is defined as follows.

$\textbf{StreamGraphPattern} ::= \text{'STREAM' '[' } Window \text{ ']' } VarOrIRIref \text{ '\{' } TriplesTemplate \text{ '\}'}$

$Window ::= Range \mid Triple \mid \text{'NOW'} \mid \text{'ALL'}$

$Range ::= \text{'RANGE' } Duration \text{ ('SLIDE' } Duration)?$

$Triple ::= \text{'TRIPLES' INTEGER}$

$Duration ::= (\text{INTEGER 'd'} \mid \text{'h'} \mid \text{'m'} \mid \text{'s'} \mid \text{'ms'} \mid \text{'ns'})^+$

where *VarOrIRIRef* and *TripleTemplate* are patterns for the *variable/IRI* and *triple template* of SPARQL 1.1, respectively. *Range* corresponds to a time-based window while *Triple* corresponds to a triple-based window. The keyword *SLIDE* is used for specifying the sliding parameter of a time-based window, whose time interval is specified by *Duration*. More details of the syntax are available at http://code.google.com/p/cqels/.

Given the CQELS language defined above, we can represent the five queries from the conference scenario as follows, where $Name$ is replaced by a constant when instantiating the query.[5]

```
SELECT ?locName ?locDesc
FROM NAMED <http://deri.org/floorplan/>
WHERE {
  STREAM<http://deri.org/streams/rfid> [NOW] {?person lv:detectedAt ?loc}
  GRAPH <http://deri.org/floorplan/>{?loc lv:name ?locName. ?loc lv:desc ?locDesc}
  ?person foaf:name ''$Name$''. }
```

Query Q1

```
CONSTRUCT {?person1 lv:reachable ?person2}
FROM NAMED <http://deri.org/floorplan/>
WHERE {
  STREAM<http://deri.org/streams/rfid>[NOW]        {?person1 lv:detectedAt ?loc1}
  STREAM<http://deri.org/streams/rfid>[RANGE 3s]{?person2 lv:detectedAt ?loc2}
  GRAPH <http://deri.org/floorplan/>              {?loc1    lv:connected ?loc2} }
```

Query Q2

```
SELECT ?coAuthName
FROM NAMED <http://deri.org/floorplan/>
WHERE {
  STREAM <http://deri.org/streams/rfid> [TRIPLES 1] {?auth    lv:detectedAt ?loc}
  STREAM <http://deri.org/streams/rfid> [RANGE 5s]  {?coAuth lv:detectedAt ?loc}
  { ?paper dc:creator ?auth.      ?paper  dc:creator ?coAuth.
    ?auth  foaf:name ''$Name$''. ?coAuth foaf:name  ?coAuthorName}
  FILTER (?auth != ?coAuth) }
```

Query Q3

---

[4] http://www.w3.org/TR/sparql11-query/#grammar

[5] For the sake of space we omit the PREFIX declarations of lv, dc, foaf, dcterms and swrc.

```
SELECT ?editorName
WHERE {
  STREAM <http://deri.org/streams/rfid> [TRIPLES 1] {?auth    lv:detectedAt ?loc1}
  STREAM <http://deri.org/streams/rfid> [RANGE 15s] {?editor lv:detectedAt ?loc2}
  GRAPH <http://deri.org/floorplan/> {?loc1    lv:connected   ?loc2}
  ?paper       dc:creator  ?auth.  ?paper   dcterms:partOf ?proceeding.
  ?proceeding swrc:editor ?editor. ?editor foaf:name        ?editorName.
  ?auth        foaf:name   ''$Name$'' }
```

Query Q4

```
SELECT ?loc2 ?locName count(distinct ?coAuth) as ?noCoAuths
FROM NAMED <http://deri.org/floorplan/>
WHERE {
  STREAM<http://deri.org/streams/rfid>[TRIPLES 1]{?auth    lv:detectedAt ?loc1}
  STREAM<http://deri.org/streams/rfid>[RANGE 30s]{?coAuth lv:detectedAt ?loc2}
  GRAPH <http://deri.org/floorplan/>{?loc2 lv:name?locName.loc2 lv:connected?loc1}
  {?paper dc:creator ?auth. ?paper dc:creator ?coAuth. ?auth foaf:name ''$Name$''}
  FILTER (?auth != ?coAuth) }
  GROUP BY ?loc2 ?locName
```

Query Q5

4.2   Data Encoding

When dealing with large data collections, it is very likely that data will not fit into the machine's main memory for processing, and parts of it will have to be temporarily stored on disk. In the particular case of RDF data, with IRIs or literals stored as strings, a simple join operation on strings could generate enough data to trigger a large number of disk reads/writes. However, these are among the most expensive operations in query processing and should be avoided whenever possible. While we cannot entirely avoid disk access, we try to reduce it by encoding the data such that more triples can fit into main memory.

We apply *dictionary encoding*, a method commonly used by triple stores [1,16,13]. An RDF node, i.e., literal, IRI or blank node, is mapped to an integer identifier. The encoded version of an RDF node is considerably smaller than the original, allowing more data to fit into memory. Moreover, since data comparison is now done on integers rather than strings, operations like pattern matching, perhaps the most common operator in RDF streams and datasets, are considerably improved.

However, in context of RDF streams, data is often fed into the system at a high rate, and there are cases when the cost of updating a dictionary and decoding the data might significantly hinder the performance. Therefore, our engine does not encode the RDF nodes into dictionary if they can be represented in 63 bits. As such, a node identifier is presented as a 64-bit integer. The first bit is used to indicate whether the RDF node is encoded or not. If the RDF nodes does not have to be encoded, the next 5 bits represent the data type of the RDF node (e.g. integer, double or float) and the last 58 bits store its value. Otherwise, the RDF node is stored in the dictionary and its identifier is stored in the remaining 63 bits.

4.3   Caching and Indexing

While data encoding allows a smaller data representation, caching and indexing aim at providing faster access to the data. Caching is used to store intermediate results of sub-queries over RDF data sets. Indexing is applying on top of caches, as well as on output mapping sets from window operators, for faster data look-ups. Similar to data warehouses, cached data is initially kept on disk with indexes and only brought to memory when needed.

In continuous query processing, RDF datasets are expected to have a much slower update rate than RDF streams. Therefore, the output of a sub-query over an RDF dataset rarely changes during a series of updates of RDF streams. Based on this observation, as soon as a query is registered, we materialise the output of its sub-queries over the RDF datasets and store them in a cache that is available to the remaining query operators. Thereby, a possibly large portion of the query does not need to be re-executed when new stream triples arrive.

To keep the cache updated, we use triggers to notify changes in the RDF datasets. The CQELS engine has a triple store that allows the engine to load and update RDF datasets as named graphs. This triple store provides triggers that will notify the engine to update the respective cached data. For the RDF datasets that are not loaded, we manually set a timer to trigger an update. At the moment, a cache update is done by recomputing the full sub-query as a background process and replacing the old cached data by the new results as soon as they are ready. We are investigating adaptive caching [7] and materialised view maintenance [21] techniques to create more efficient cache updating mechanisms.

For faster lookups on the cache, indexes are built on the variables shared among the materialised sub-queries and other operator's inputs. We use similar indexing schemas as in popular triple stores [13,16,23,28,38]. Vigals et al. [36] showed that, in stream processing, building hash tables for multi-way joins can accelerate the join operation. Therefore, we also index data coming from window operators, which are the input to the relational operators. Similar to caching, there is an update overheard attached to indexes. In CQELS, the decision to create an index is as follows: cache data is always indexed. For data coming from window operators, an index is maintained as long as it can be updated faster than the window's stream rate. If this threshold is reached, the index is dropped, and the relational operators that depend on this index will be replaced by equivalent ones that can work without indexes.

4.4   Operators and Routing Policy

To recap, the CQELS processing model contains three groups of operators: window, relational and streaming operators. In the current implementation, we support two types of window operators: *triple-based window* and *sliding window*. We implement all relational operators needed to support the CQELS language. In particular, one of the join operators is a binary index join that uses indexing for faster processing. The implementation of the streaming operator is rather simple: as soon as a mapping arrives at the streaming operator, it simply binds the mapping to the graph template, then sends the output triples, tagged with the time they were created, to the output stream.

To allow adaptive query execution, our engine currently support a "cardinality-based" routing policy, based on some heuristics. For a given query, the engine keeps all possible left-deep data flows that start with a window operator. For instance, Figure 3 shows the four data flows that are maintained for the query in the localisation scenario from Section 3.

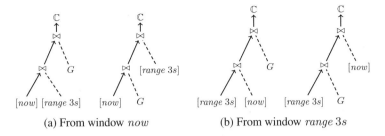

(a) From window *now*   (b) From window *range 3s*

Fig. 3. Left-deep data flows for the query in the localisation scenario

---

Algorithm 2. findNextOp($\mathbb{O}, t$)

Input: $\mathbb{O}$: operator, t : timestamp
$nextOp := null$
for $unaryOp \in nextUnaryOp(\mathbb{O})$ **do**
    $\quad$ **if** $unaryOp$ *is a filter operator* **then** **return** $unaryOp$ **else** $nextOp := unaryOp$
$mincard := +\infty$
for $binaryOp \in nextBinaryOpOnLeftDeepTree(\mathbb{O})$ **do**
    $\quad$ **if** $mincard > card(binaryOp.rightChildOp, t)$ **then**
        $\qquad$ $mincard := card(binaryOp.rightChildOp, t)$
        $\qquad$ $nextOp := binaryOp$

return $nextOp$

Algorithm 2 shows the findNextOp function used in the current routing policy (see Algorithm 1). It applies two simple heuristics: the first one, common in DBMSs, pushes operators like filters closer to the data sources. The rationale here is that the earlier we prune the triples that will not make it to the final output, the better, since operators will then process fewer triples. The second looks at the cardinality of the operators' output and sorts them in increasing order of this value, which also helps in reducing the number of mappings to process.

Function $nextUnaryOp(\mathbb{O})$ returns the set of possible next unary operators that $\mathbb{O}$ can route data to, while $nextBinaryOpOnLeftDeepTree(\mathbb{O})$ returns the binary ones. Examples of unary operators are filters and projections, and they can be directly executed on the output produced by $\mathbb{O}$. Binary operators, such as joins and unions, have two inputs, called left and right child, due to the tree shape of the data flows. $\mathbb{O}$ will be the left child, since the data flows are all left-deep. The right child is given by the $rightChildOp$ attribute. For each binary operator, we obtain the cardinality of the right child at time t from $card(binaryOp.rightChildOp, t)$. We then route the output of $\mathbb{O}$ to the one whose cardinality function returns the smallest value.

5 Experimental Evaluation

To evaluate the performance of CQELS, we compare it against two existing systems that also offer integrated processing of Linked Streams and Linked Data – C-SPARQL [9] and ETALIS [3].[6] Note that EP-SPARQL is implemented on top of ETALIS. We first planned to express our queries in EP-SPARQL, which would then be translated into the language used in ETALIS. However, the translation from EP-SPARQL to ETALIS is currently not mature enough to handle all queries in our setup, so we decided to represent the queries directly in the ETALIS language. We also considered comparing our system against SPARQLstream [14], but its current implementation does not support querying on both RDF streams and RDF dataset. Next, we describe our experimental setup, and then report and discuss the results obtained. All experiments presented in this paper are reproducible. Both systems and datasets used are available at `http://code.google.com/p/cqels/`.

5.1 Experimental Setup

We use the conference scenario introduced in Section 4. For the stream data, we use the RFID-based tracking data streams provided by the Open Beacon community.[7] The data is generated from active RFID tags, the same hardware used in the Live Social Semantics deployment [2]. The data generator from SP^2Bench [30] is used to create simulated DBLP datasets. We have also created a small RDF dataset, 172 triples, to represent the connectivity between the locations given in the Open Beacon dataset.

The experiments were executed on a standard workstation with 1 x Quad Core Intel Xeon E5410 2.33 GHz, 8GB memory, 2 x 500GB Enterprise SATA disks, running Ubuntu 11.04/x86_64, Java version "1.6", Java HotSpot(TM) 64-Bit Server VM, and SWI-Prolog 5.10.4. The maximum heap size on JVM instances when running CQELS and C-SPARQL was set to 4GB. For ETALIS, the global stack size is also 4GB.

We evaluate performance in terms of average query execution time. At each run, after registering the query, we stream a number of triples into the system and every time the query is re-executed we measure its processing time. We then average these values over multiple runs.

The queries used follow the templates specified in Section 4.1. They were selected in a way that cover many operators with different levels of complexity, for instance joins, filters and aggregations. One query instance is formed by replacing $Name$ in the template with a particular author's name from the DBLP dataset. We have performed the following three types of experiments:

Exp.(1) *Single query*: For each of the Q1, Q3, Q4 and Q5 templates we generate 10 different query instances. For query template Q2, since it has no constants, we create one instance only. Then we run each instance at a time and compute the average query execution time.

Exp.(2) *Varying size of the DBLP dataset*: We do the same experiment as in (1) but varying the numbers of triples of the DBLP dataset, ranging from 10^4 to 10^7 triples. We do not include Q2 in this experiment, since it does not involve DBLP dataset.

---

[6] We would like to thank the C-SPARQL, ETALIS, and SPARQLstream teams for their support in providing their implementations and helping us to understand and correctly use their systems.

[7] `http://www.openbeacon.org/`

Table 2. Average query execution time for single queries (in milliseconds)

	Q_1	Q_2	Q_3	Q_4	Q_5
CQELS	0.47	3.90	0.51	0.53	21.83
C-SPARQL	332.46	99.84	331.68	395.18	322.64
ETALIS	0.06	27.47	79.95	469.23	160.83

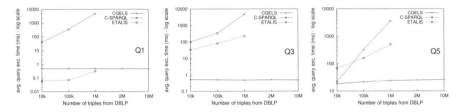

Fig. 4. Average query execution time for varying sizes of simulated DBLP dataset

Exp.(3) *Multiple queries*: For query templates Q1, Q3 and Q4, we register 2^M query instances at the same time, with $0 \leq M \leq 10$, and execute them in parallel.

In experiments Exp.(1) and Exp.(3), the numbers of triples from DBLP is fixed to 10^5.

5.2   Results and Analysis

Table 2 shows the results for Exp.(1). We can see that, for most of the cases, CQELS outperforms the other approaches by orders of magnitude; sometimes it is over 700 times faster. The only exception is query Q1, where ETALIS is considerably faster. The reason is that ETALIS supports three consumption policies, namely *recent*, *chronological*, and *unrestricted*, where *recent* is very efficient for queries containing only simple filters on the stream data. For more complex queries, the performance of ETALIS drops significantly. C-SPARQL is currently not designed to handle large datasets, which explains its poor performance in our setup. CQELS, on the other hand, is able to constantly deliver great performance, due to its combination of pre-processing and adaptive routing policy.

The results from Exp.2 are shown in Figure 4, for query templates Q1, Q3 and Q5. The results for query template Q4 are very similar to those from query template Q3, so we omit them for the sake of space.

We can see how the performance is affected when the size of the RDF dataset increases. For both ETALIS and C-SPARQL, not only does the average execution time increase with the size of the RDF dataset, but they are only able to run up to a certain number of triples. They can execute queries with a RDF dataset of 1 million triples, but at 2 million ETALIS crashes and C-SPARQL does not respond. CQELS' performance is only marginally affected by the RDF dataset's size, even for values as high as 10 million triples, and the performance gains sometimes were three orders of magnitude. This is mainly due to the cache and indexes used for storing and accessing pre-computed

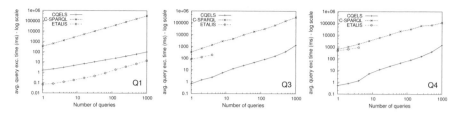

Fig. 5. Average query execution time when running multiple query instances

intermediate results. We have observed that the size of the cache, which stores the co-authors and editors of a certain author, does not increase linearly with the size of the dataset. Moreover, by using indexes on this cache, the access time of a mapping increases only logarithmically with the cache size. This behaviour shows the importance of having such cache and index structures for efficient query processing.

As a scalability test, we wanted to analyse how the systems perform with a number of queries running in parallel. Figure 5 presents the results for Exp.(3). Again, ETALIS delivers the best performance when there is no join operator on the stream data (Q1). But, for the other cases, the number of queries it can handle in parallel is very limited (less than 10). Both C-SPARQL and CQELS can scale to a large number of queries, but in C-SPARQL queries face a long execution time that exceeds 100 seconds, while in CQELS, even with 1000 queries running, the average execution time is still around one second. This scalability is mainly due to our encoding technique, which allows more efficient use of main memory, consequently reducing read/write disk operations.

In summary, our experimental evaluation shows the great performance of CQELS, both in terms of efficiency and scalability. Its query engine, with the cache, index, and routing policy, adapts well to different query complexities and it can scale with the size of the RDF datasets. Our encoding technique enhances memory usage, which is crucial when handling multiple queries. Even though ETALIS performed better for simpler queries, CQELS performs consistently well in all the experiments, and in most cases outperforms the other approaches by orders of magnitude.

6   Conclusions

This paper presented CQELS, a native and adaptive approach for integrated processing of Linked Stream Data and Linked Data. While other systems use a "black box" approach which delegates the processing to existing engines, thus suffering major efficiency drawbacks because of lack of full control over the query execution process, CQELS implements the required query operators natively, enabling improved query execution. Our query engine can adapt to changes in the input data, by applying heuristics to reorder the operators in the data flows of a query. Moreover, external disk access on large Linked Data collections is reduced with the use of data encoding, and caching/indexing enables significantly faster data access. Our experimental evaluation shows the good performance of CQELS, in terms of efficiency, latency and scalability. CQELS performs consistently well in experiments over a wide range of test cases, outperforming other approaches by orders of magnitude.

Our promising results indicate that an integrated and native approach is in fact necessary to achieve the required query execution efficiency. For future work, we plan to improve the performance of CQELS further. Query optimisation in adaptive query processing is still an open problem under active research [17]. We have already started investigating cost-based query optimisation policies [25] and we plan to look into adaptive caching [7] and materialised view maintenance [21] to enhance the efficiency of our query execution algorithms.

References

1. Abadi, D.J., Marcus, A., Madden, S.R., Hollenbach, K.: Scalable semantic web data management using vertical partitioning. In: VLDB 2007, pp. 411–422 (2007)
2. Alani, H., Szomszor, M., Cattuto, C., Van den Broeck, W., Correndo, G., Barrat, A.: Live Social Semantics. In: Bernstein, A., Karger, D.R., Heath, T., Feigenbaum, L., Maynard, D., Motta, E., Thirunarayan, K. (eds.) ISWC 2009. LNCS, vol. 5823, pp. 698–714. Springer, Heidelberg (2009)
3. Anicic, D., Fodor, P., Rudolph, S., Stojanovic, N.: Ep-sparql: a unified language for event processing and stream reasoning. In: WWW 2011, pp. 635–644 (2011)
4. Arasu, A., Babu, S., Widom, J.: The CQL continuous query language: semantic foundations and query execution. The VLDB Journal 15(2), 121–142 (2006)
5. Arasu, A., Widom, J.: A denotational semantics for continuous queries over streams and relations. SIGMOD Record 33(3), 6–12 (2004)
6. Avnur, R., Hellerstein, J.M.: Eddies: continuously adaptive query processing. SIGMOD Rec. 29(2), 261–272 (2000)
7. Babu, S., Munagala, K., Widom, J., Motwani, R.: Adaptive Caching for Continuous Queries. In: ICDE 2005, pp. 118–129 (2005)
8. Balazinska, M., Deshpande, A., Franklin, M.J., Gibbons, P.B., Gray, J., Hansen, M., Liebhold, M., Nath, S., Szalay, A., Tao, V.: Data Management in the Worldwide Sensor Web. IEEE Pervasive Computing 6(2), 30–40 (2007)
9. Barbieri, D.F., Braga, D., Ceri, S., Grossniklaus, M.: An execution environment for C-SPARQL queries. In: EDBT 2010, pp. 441–452 (2010)
10. Bizer, C., Heath, T., Berners-Lee, T.: Linked Data - The Story So Far. International Journal on Semantic Web and Information Systems 5(3), 1–22 (2009)
11. Bolles, A., Grawunder, M., Jacobi, J.: Streaming SPARQL - Extending SPARQL to Process Data Streams. In: Bechhofer, S., Hauswirth, M., Hoffmann, J., Koubarakis, M. (eds.) ESWC 2008. LNCS, vol. 5021, pp. 448–462. Springer, Heidelberg (2008)
12. Bouillet, E., Feblowitz, M., Liu, Z., Ranganathan, A., Riabov, A., Ye, F.: A Semantics-Based Middleware for Utilizing Heterogeneous Sensor Networks. In: Aspnes, J., Scheideler, C., Arora, A., Madden, S. (eds.) DCOSS 2007. LNCS, vol. 4549, pp. 174–188. Springer, Heidelberg (2007)
13. Broekstra, J., Kampman, A., van Harmelen, F.: Sesame: An architecture for storing and querying rdf data and schema information (2003)
14. Calbimonte, J.P., Corcho, O., Gray, A.J.G.: Enabling Ontology-Based Access to Streaming Data Sources. In: Patel-Schneider, P.F., Pan, Y., Hitzler, P., Mika, P., Zhang, L., Pan, J.Z., Horrocks, I., Glimm, B. (eds.) ISWC 2010, Part I. LNCS, vol. 6496, pp. 96–111. Springer, Heidelberg (2010)
15. Carney, D., Çetintemel, U., Cherniack, M., Convey, C., Lee, S., Seidman, G., Stonebraker, M., Tatbul, N., Zdonik, S.: Monitoring streams: a new class of data management applications. In: VLDB 2002, pp. 215–226 (2002)

16. Chong, E.I., Das, S., Eadon, G., Srinivasan, J.: An efficient SQL-based RDF querying scheme. In: VLDB 2005, pp. 1216–1227 (2005)
17. Deshpande, A., Ives, Z., Raman, V.: Adaptive query processing. Found. Trends Databases (January 2007)
18. Elmasri, R., Navathe, S.B.: Fundamentals of Database Systems, 5th edn. Addison-Wesley Longman Publishing Co., Inc., Boston (2006)
19. Fidge, C.J.: Logical time in distributed computing systems. IEEE Computer 24(8), 28–33 (1991)
20. Golab, L., Özsu, M.T.: Issues in data stream management. SIGMOD Rec. 32(2), 5–14 (2003)
21. Gupta, A., Mumick, I.S.: Maintenance of materialized views: problems, techniques, and applications. In: Materialized Views, pp. 145–157 (1999)
22. Gutierrez, C., Hurtado, C.A., Vaisman, A.: Introducing Time into RDF. IEEE Transactions on Knowledge and Data Engineering 19, 207–218 (2007)
23. Harth, A., Umbrich, J., Hogan, A., Decker, S.: YARS2: A Federated Repository for Querying Graph Structured Data from the Web. In: Aberer, K., Choi, K.-S., Noy, N., Allemang, D., Lee, K.-I., Nixon, L.J.B., Golbeck, J., Mika, P., Maynard, D., Mizoguchi, R., Schreiber, G., Cudré-Mauroux, P. (eds.) ASWC 2007 and ISWC 2007. LNCS, vol. 4825, pp. 211–224. Springer, Heidelberg (2007)
24. Lamport, L.: Time, clocks, and the ordering of events in a distributed system. Commun. ACM 21(7), 558–565 (1978)
25. Le-Phuoc, D., Parreira, J.X., Hausenblas, M., Hauswirth, M.: Continuous query optimization and evaluation over unified linked stream data and linked open data. Technical report, DERI, 9 (2010)
26. Madden, S., Shah, M., Hellerstein, J.M., Raman, V.: Continuously adaptive continuous queries over streams. In: 2002 ACM SIGMOD International Conference on Management of Data, pp. 49–60 (2002)
27. Mattern, F.: Virtual time and global states of distributed systems. In: Parallel and Distributed Algorithms, pp. 215–226. North-Holland (1989)
28. Neumann, T., Weikum, G.: The RDF-3X engine for scalable management of RDF data. The VLDB Journal 19(1), 91–113 (2010)
29. Pérez, J., Arenas, M., Gutierrez, C.: Semantics and complexity of SPARQL. ACM Trans. Database Syst. 34(3), 1–45 (2009)
30. Schmidt, M., Hornung, T., Lausen, G., Pinkel, C.: Sp2bench: A sparql performance benchmark. In: ICDE 2009, pp. 222–233 (2009)
31. Sequeda, J.F., Corcho, O.: Linked stream data: A position paper. In: SSN 2009 (2009)
32. Sheth, A.P., Henson, C.A., Sahoo, S.S.: Semantic Sensor Web. IEEE Internet Computing 12(4), 78–83 (2008)
33. Srivastava, U., Widom, J.: Flexible time management in data stream systems. In: PODS 2004, pp. 263–274 (2004)
34. Stuckenschmidt, H., Vdovjak, R., Houben, G.-J., Broekstra, J.: Index structures and algorithms for querying distributed rdf repositories. In: WWW, pp. 631–639 (2004)
35. Umbrich, J., Karnstedt, M., Land, S.: Towards understanding the changing web: Mining the dynamics of linked-data sources and entities. In: KDML, Workshop (2010)
36. Viglas, S.D., Naughton, J.F., Burger, J.: Maximizing the output rate of multi-way join queries over streaming information sources. In: VLDB 2003 (2003)
37. Whitehouse, K., Zhao, F., Liu, J.: Semantic Streams: A Framework for Composable Semantic Interpretation of Sensor Data. In: Römer, K., Karl, H., Mattern, F. (eds.) EWSN 2006. LNCS, vol. 3868, pp. 5–20. Springer, Heidelberg (2006)
38. Wilkinson, K., Sayers, C., Kuno, H.A., Reynolds, D.: Efficient RDF storage and retrieval in Jena2, pp. 35–43 (2003)

Learning Relational Bayesian Classifiers from RDF Data

Harris T. Lin*, Neeraj Koul**, and Vasant Honavar

Department of Computer Science
Iowa State University
Ames, IA 50011 USA
{htlin,neeraj,honavar}@iastate.edu

Abstract. The increasing availability of large RDF datasets offers an exciting opportunity to use such data to build predictive models using machine learning algorithms. However, the massive size and distributed nature of RDF data calls for approaches to learning from RDF data in a setting where the data can be accessed only through a query interface, e.g., the SPARQL endpoint of the RDF store. In applications where the data are subject to frequent updates, there is a need for algorithms that allow the predictive model to be incrementally updated in response to changes in the data. Furthermore, in some applications, the attributes that are relevant for specific prediction tasks are not known a priori and hence need to be discovered by the algorithm. We present an approach to learning Relational Bayesian Classifiers (RBCs) from RDF data that addresses such scenarios. Specifically, we show how to build RBCs from RDF data using statistical queries through the SPARQL endpoint of the RDF store. We compare the communication complexity of our algorithm with one that requires direct centralized access to the data and hence has to retrieve the entire RDF dataset from the remote location for processing. We establish the conditions under which the RBC models can be incrementally updated in response to addition or deletion of RDF data. We show how our approach can be extended to the setting where the attributes that are relevant for prediction are not known a priori, by selectively crawling the RDF data for attributes of interest. We provide open source implementation and evaluate the proposed approach on several large RDF datasets.

1 Introduction

The *Semantic Web* as envisioned by Berners-Lee, Hendler, and others [14,3] aims to describe the semantics of Web content in a form that can be processed by computers [5,2]. A key step in realizing this vision is to cast knowledge and data on the Web in a form that is conducive to processing by computers [15]. Resource Description Framework (RDF) ([23] for a primer) offers a formal language for

---

* Primary author.
** Primary author, contributed equally.

L. Aroyo et al. (Eds.): ISWC 2011, Part I, LNCS 7031, pp. 389–404, 2011.
© Springer-Verlag Berlin Heidelberg 2011

describing structured information on the Web. RDF represents data in the form of subject-predicate-object triples, also called RDF triples, which describe a directed graph where the directed labeled edges encode binary relations between labeled nodes (also called resources). RDF stores or triple stores and associated query languages such as SPARQL [26] offer the means to store and query large amounts of RDF data. Over the past decade, RDF has emerged as a basic representation format for the Semantic Web [15]. Cyganiak [8] estimated in 2010 that there are 207 RDF datasets containing over 28 billion triples published in the Linked Open Data cloud.

The increasing availability of large RDF datasets on the web offers unprecedented opportunities for extracting useful knowledge or predictive models from RDF data, and using the resulting models to guide decisions in a broad range of application domains. Hence, it is natural to consider the use of machine learning approaches, and in particular, statistical relational learning algorithms [11], to extract knowledge from RDF data [17,4,28]. However, existing approaches to learning predictive models from RDF data have significant shortcomings that limit their applicability in practice. Specifically, existing approaches rely on the learning algorithm having direct access to RDF data. However, in many settings, it may not be feasible to transfer data a massive RDF dataset from a remote location for local processing by the learning algorithm. Even in settings where it is feasible to provide the learning algorithm direct access to a local copy of an RDF dataset, algorithms that assume in-memory access to data cannot cope with RDF datasets that are too large to fit in memory. Hence, there is an urgent need for approaches to learning from RDF data in a setting where the data can be accessed only through a query interface, e.g., the SPARQL endpoint for the RDF store. In applications where the data are subject to frequent updates, there is a need for algorithms that allow the predictive model to be incrementally updated in response to changes in the data. Furthermore, in some applications, the attributes that are relevant for specific prediction tasks are not known a priori and hence need to be discovered by the algorithm. We present an approach to learning Relational Bayesian Classifiers from RDF data that addresses such scenarios.

Our approach to learning Relational Bayesian Classifiers (RBCs) from RDF data adopts the general framework introduced by Caragea et al. [6] for transforming a broad class of standard learning algorithms that assume in memory access to a dataset into algorithms that interact with the data source(s) only through statistical queries or procedures that can be executed on the remote data sources. This involves decomposing the learning algorithm into two parts: (i) a component that poses the relevant statistical queries to a data source to acquire the information needed by the learner; and (ii) a component that uses the resulting statistics to update or refine a partial model (and if necessary, further invoke the statistical query component). This approach has been previously used to learn a variety of classifiers from relational databases [20] using SQL queries and from biomolecular sequence data [19]. It has recently become feasible to use a similar approach to learning RBCs from RDF data due to the incorporation of support for aggregate queries in SPARQL. (SPARQL 1.1 supports aggregate queries whereas SPARQL 1.0 does not).

We show how to learn RBCs from RDF data using only aggregate queries through the SPARQL endpoint of the RDF store. This approach does not require in-memory access to RDF data to be processed by the learning algorithm, and hence can scale up to very large data sets. Because the predictive model is built using aggregate queries against a SPARQL endpoint, it can be used to learn RBCs from large remote RDF stores without having to transfer the data to a local RDF store for processing (in general, the cost of retrieving the statistics needed for learning is much lower than the cost of retrieving the entire dataset). Under certain conditions which we identify in the paper, we show how the RBC models can be incrementally updated in response to changes (addition or deletion of triples) from the RDF store. We further show how our approach can be extended to the setting where the attributes that are relevant for prediction are not known a priori, by selectively crawling the RDF data for attributes of interest. We have implemented the proposed approach into INDUS [21], an open source suite of learning algorithms, that learn from massive data sets only using statistical queries. We describe results of experiments on several large RDF datasets that demonstrate the feasibility of the proposed approach to learning RBCs from RDF stores.

The rest of the paper is organized as follows: Section 2 introduces a precise formulation of the problem of learning RBCs from RDF data. Section 3 describes how to build RBCs from RDF data using only aggregate queries. Section 4 identifies the conditions under which it is possible to incrementally update an RBC learned model from an RDF store in response to updates to the underlying RDF store. Section 5 presents an analysis of the communication complexity of learning RBCs from RDF stores. Section 6 describes how to extend to the setting where the attributes that are relevant for prediction are not known a priori, by selectively crawling the RDF data for attributes of interest. Section 7 describes results of experiments with several RDF datasets that demonstrate the feasibility proposed approach. Finally Sec. 8 concludes with a summary and a brief discussion of related work.

2   Problem Formulation

In this section we formulate the problem of learning predictive models from RDF data. Assume there are pairwise disjoint infinite sets I, B, L and V (IRIs, Blank nodes, Literals and Variables respectively). A triple $(s, p, o) \in (I \cup B) \times I \times (I \cup B \cup L)$ is called an RDF triple. In this triple, s is the subject, p the predicate, and o the object. An RDF graph is a set of RDF triples.

As a running example for the following definitions, we consider the RDF schema for the movie domain as shown in Fig. 1. We wish to predict whether a movie receives more than \$2M in its opening week.

Definition 1 (Target Class). *Given an RDF graph $\mathcal{G}$, a target class is a distinguished IRI of type* rdfs:Class *in $\mathcal{G}$. For example, Movie.*

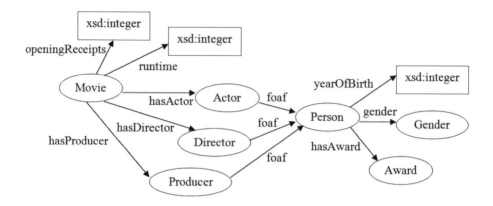

Fig. 1. RDF schema for the movie domain

Definition 2 (Instances). *Given an RDF graph $\mathcal{G}$ and a target class $\mathcal{T}$, the instances of $\mathcal{T}$, denoted $\mathcal{T}(\mathcal{G})$ is the set $\{x : (x, rdf{:}type, \mathcal{T}) \in \mathcal{G}\}$.*

Definition 3 (Attribute). *Given an RDF graph $\mathcal{G}$ and a target class $\mathcal{T}$, an attribute A (of a target class $\mathcal{T}$) is a tuple of IRIs $(p_1, \ldots, p_n)$ such that the domain of p_1 is $\mathcal{T}$, the range of p_i is the domain of p_{i+1}, and the range of p_n is a literal. For example, (hasActor, foaf, yearOfBirth). We also refer the range of the attribute A as the range of p_n.*

Definition 4 (Attribute Graph). *Given an instance x of the target class $\mathcal{T}$ in the RDF graph $\mathcal{G}$ and an attribute $A = (p_1, \ldots, p_n)$, the attribute graph of the instance x, denoted by $A(x)$, is the union of the sets of triples that match the Basic Graph Pattern [26]*

$$((x, p_1, ?v_1) \; AND \; (?v_1, p_2, ?v_2) \; AND \; \ldots \; AND \; (?v_{n-1}, p_n, ?v_n)) \qquad (1)$$

where $v_i \in V$ are variables.

Given an additional literal value a, we also define a filtered attributed graph, denoted $A(x, a)$, which includes the filter constraint $FILTER(?v_n = a)$ in the graph pattern (1). Further, if $\mathcal{A}$ is a tuple of attributes $(A_1, \ldots, A_n)$, then we define $\mathcal{A}(x)$ to be $(A_1(x), \ldots, A_n(x))$

Definition 5 (Target Attribute). *Given an RDF graph $\mathcal{G}$ and a target class $\mathcal{T}$, a target attribute is a distinguished attribute denoted by C. For example, (openingReceipts).*

$C(x)$ is intended to describe the *class label* of the instance x, hence we assume that each instance has exactly one class label, i.e., $|C(x)| = 1$ for every $x \in \mathcal{T}(\mathcal{G})$. Given a target attribute $C = (p_1, \ldots, p_n)$, we define $v(C, x)$ to be the value of $?v_n$ matched by the graph pattern (1).

Definition 6 (Class Label). *Given a target attribute $C = (p_1, \ldots, p_n)$, the set of class labels is the the range of p_n. For brevity we denote this set by $\mathcal{C}$.*

Definition 7 (RDF Dataset). *An* RDF *dataset* D *is a tuple* $(\mathcal{G}, \mathcal{T}, \mathcal{A}, C)$ *where* $\mathcal{G}$ *is an* RDF *graph,* $\mathcal{T}$ *a target class in* $\mathcal{G}$, $\mathcal{A}$ *a tuple of attributes, and* C *is a target attribute. We also denote the tuple* $(\mathcal{T}, \mathcal{A}, C)$ *as* $Desc(D)$ *corresponding to the descriptor of the dataset.*

Definition 8 (Induced Attribute Graph Dataset). *Given an* RDF *dataset* $D = (\mathcal{G}, \mathcal{T}, \mathcal{A}, C)$, *its induced attribute graph dataset, denoted* $I(D)$, *is defined as* $\{(\mathcal{A}(x), v(C, x)) : x \in \mathcal{T}(\mathcal{G})\}$.

We now formalize the the problem of learning from RDF data.

Problem 1. Given an RDF dataset $D = (\mathcal{G}, \mathcal{T}, \mathcal{A}, C)$ and its induced attribute graph dataset $I(D)$, a hypothesis class H, and a performance criterion P, the learning algorithm L outputs a classifier $h \in H$ that optimizes P. The input to the classifier h is $\mathcal{A}(x)$ where x is an instance of a target class $\mathcal{T}$, and the output $h(x) \in C$ is a class label.

3   Learning from RDF Data

We reduce the problem of learning from RDF data to the problem of learning from multiset attribute data which is defined below. This reduction allows for application of algorithms for learning from multiset attribute data (e.g. Relational Bayesian Classifier [25]) to this setting. Given an RDF dataset $D = (\mathcal{G}, \mathcal{T}, \mathcal{A}, C)$ and its induced attribute graph dataset $I(D)$, consider an attribute A and the attribute graph $A(x)$ of an instance $x \in \mathcal{T}(\mathcal{G})$. The attribute graph $A(x)$ can be viewed as a directed acyclic graph (DAG) rooted in x, and here we are interested in only the leaves of this DAG. The following definition captures this notion.

Definition 9 (Leaf). *Given an attribute* A_i, *we define the leaf function* $\mathcal{L}(A_i(x))$ *that returns the multiset of leaves of* $A_i(x)$, *such that each leaf* $a \in A_i(x)$ *is replaced with* n *copies of* a *where* n *is the number of unique paths from* x *to* a. *For brevity we write* $\mathcal{L}(A_i(x))$ *as* $\mathcal{L}_i(x)$ *and* $\mathcal{L}(A_i(x, a))$ *as* $\mathcal{L}_i(x, a)$.
    Also, we overload the leaf function on a tuple of attributes $\mathcal{A} = (A_1, \dots, A_n)$ *by* $\mathcal{L}(\mathcal{A}(x)) = (\mathcal{L}_1(x), \dots, \mathcal{L}_n(x))$.

Using the leaf function, we reduce $I(D)$ into a multiset attributed dataset $\mathcal{M}(D) = \{(\mathcal{L}(\mathcal{A}(x)), v(C, x)) : x \in \mathcal{T}(\mathcal{G})\}$. To learn from $\mathcal{M}(D)$ we focus our attention on Relational Bayesian Classifiers (RBC) motivated from modeling relational data [25]. RBC assumes that attribute multisets are independent given the class, and the most probable class of an instance is given by:

$$h_{RBC}(x) = \underset{c \in \mathcal{C}}{\operatorname{argmax}} \, p(c) \prod_i p(\mathcal{L}_i(x) : c) \tag{2}$$

Several methods to estimate the probabilities $p(\mathcal{L}_i(x) : c)$ are described in [25]:

- Aggregation: $\hat{p}_{agg}(\mathcal{L}_i(x) : c) = \hat{p}(agg(\mathcal{L}_i(x)) : c)$, where agg is an aggregation function such as min, max, average for continuous attributes; and mode for discrete attributes.
- Independent Value: $\hat{p}_{ind}(\mathcal{L}_i(x) : c) = \prod_{a \in \mathcal{L}_i(x)} \hat{p}(a : c)$, which assumes each value in the multiset is independently drawn from the same distribution (attribute value independence).
- Average Probability: $\hat{p}_{avg}(\mathcal{L}_i(x) : c) = \frac{\sum_{a \in \mathcal{L}_i(x)} \hat{p}(a: c)}{|\mathcal{L}_i(x)|}$, which also assumes attribute value independence as in *Independent Value*, however during inference the probabilities are averaged instead of multiplied.

For estimating the parameters in (2), we assume that the learner does not have access to the RDF graph $\mathcal{G}$ but instead only has knowledge $\mathcal{T}, \mathcal{A}$, and C. In addition, we assume that the RDF store answers statistical queries over the RDF graph $\mathcal{G}$ which in our setting correspond to aggregate SPARQL queries submitted to a SPARQL endpoint. Given a descriptor $Desc(D) = (\mathcal{T}, \mathcal{A}, C)$ where $C = (c_1, \ldots, c_m)$ we assume that the RDF store supports the following type of primitive queries:

(Q1) $S(\mathcal{G}, \mathcal{T}) = |\mathcal{T}(\mathcal{G})|$, the number of instances of target type $\mathcal{T}$ in $\mathcal{G}$. This corresponds to the SPARQL query:

```
SELECT COUNT(*) WHERE { ?x rdf:type <T> . }
```

(Q2) $S(\mathcal{G}, \mathcal{T}, C = c) = |\{x \in \mathcal{T}(\mathcal{G}) : v(C, x) = c\}|$, the number of instances of target type $\mathcal{T}$ in which the target attribute takes the class label c. This corresponds to the SPARQL query:

```
SELECT COUNT(*) WHERE {
    ?x rdf:type <T> .
    ?x <c1> ?c1 .  ...  ?cm-1 <cm> c .
}
```

(Q3) $S(\mathcal{G}, \mathcal{T}, C = c, A_i) = \sum_{x \in \mathcal{T}(\mathcal{G}) \text{ and } v(C,x)=c} |\mathcal{L}_i(x)|$. Assuming the attribute $A_i = (p_1, \ldots, p_j)$ this corresponds to the SPARQL query:

```
SELECT COUNT(*) WHERE {
    ?x rdf:type <T> .
    ?x <c1> ?c1 .  ...  ?cm-1 <cm> c .
    ?x <p1> ?v1 .  ...  ?vj-1 <pj> ?vj .
}
```

(Q4) $S(\mathcal{G}, \mathcal{T}, C = c, A_i = a) = \sum_{x \in \mathcal{T}(\mathcal{G}) \text{ and } v(C,x)=c} |\mathcal{L}_i(x, a)|$. Assuming the attribute $A_i = (p_1, \ldots, p_j)$ this corresponds to the SPARQL query:

```
SELECT COUNT(*) WHERE {
    ?x rdf:type <C> .
    ?x <c1> ?c1 .  ...  ?cm-1 <cm> c .
    ?x <p1> ?v1 .  ...  ?vj-1 <pj> a .
}
```

(Q5) $S(\mathcal{G}, \mathcal{T}, C = c, A_i, agg, [v_l, v_h])$. Given a range $[v_l, v_h]$ this corresponds to the SPARQL query:

```
SELECT COUNT(*) WHERE {
  { SELECT (agg(?vj) AS ?aggvalue) WHERE {
      ?x rdf:type <T> .
      ?x <c1> ?c1 .  ...  ?cm-1 <cm> c .
      OPTIONAL { ?x <p1> ?v1 .  ...  ?vj-1 <pj> ?vj . }
    } GROUP BY ?x
  } FILTER(?aggvalue >= vl && ?aggvalue <= vh)
}
```

We now proceed to describe how an RBC can be built using the supported SPARQL queries without requiring access to the underlying dataset. The RBC estimates the following probabilities from training data:

1. $\hat{p}(c)$
2. $\hat{p}(\mathrm{agg}(A_i) : c)$ for each attribute A_i where aggregation is used to estimate probabilities. For simplicity, we discretize the aggregated values and predetermine the bins prior to learning. Hence, we estimate $\hat{p}(\mathrm{agg}(A_i) \in [v_l, v_h] : c)$ for each bin $[v_l, v_h]$
3. $\hat{p}(a : c)$ where a is in the range of A_i, for each attribute A_i where independent value or average probability is used to estimate the probabilities.

The above three probabilities can be estimated (using Laplace correction for smoothing) as follows:

1. $\hat{p}(c) = \frac{S(\mathcal{G},\mathcal{T},c)+1}{S(\mathcal{G},\mathcal{T})+m}$ where m is the number of class labels
2. $\hat{p}(\mathrm{agg}(A_i) \in [v_l, v_h] : c) = \frac{S(\mathcal{G},\mathcal{T},C=c,A_i,\mathrm{agg},[v_l,v_h])+1}{S(\mathcal{G},\mathcal{T},c)+m}$ where m is the number of bins (ranges)
3. $\hat{p}(a : c) = \frac{S(\mathcal{G},\mathcal{T},C=c,A_i=a)+1}{S(\mathcal{G},\mathcal{T},C=c,A_i)+m}$ where a is in the range of A_i and m is the size of range of A_i

Hence, it is possible to learn RBCs from an RDF graph by interacting with the RDF store only through SPARQL queries. This approach does not require access to the underlying dataset and in most practical settings requires much less bandwidth as compared to transferring the data to a local store for processing (see Sec. 5).

4   Updatable Models

In many settings, the RDF store undergoes frequent updates i.e., addition or deletion of sets of RDF triples. In such settings, it is necessary to update the predictive model to reflect the changes in the RDF store used to build the model. While in principle, the algorithm introduced in Sec. 3 can be re-executed each time there is an update to the RDF store, it is of interest to explore more efficient solutions for incrementally updating the RBC model by updating only the relevant statistics.

Given a dataset D and a learning algorithm L, let $L(D)$ be a predictive model built from the dataset D. Let θ be a primitive query required over the dataset D to build $L(D)$.

Definition 10 (Updatable Model [19]). *Given datasets D_1 and D_2 such that $D_1 \subseteq D_2$, we say that a primitive query θ is updatable iff we can specify functions f and g such that:*

1. $\theta(D_2) = f(\theta(D_2 - D_1), \theta(D_1))$
2. $\theta(D_1) = g(\theta(D_2), \theta(D_2 - D_1))$

We say that the predictive model constructed using L is updatable iff all primitive queries required over the dataset D to build $L(D)$ are updatable.

The following propositions show that the primitive query $(Q1)$ of the RBC model is updatable, whereas the rest of the queries are not updatable. Hence, in general, the RBC model is not updatable.

Proposition 1. *The primitive query $S(\mathcal{G}, \mathcal{T})$ is updatable.*

Proof. This query counts the number of instances of target type $\mathcal{T}$ in $\mathcal{G}$, which is the cardinality of $\{x : (x, \text{rdf:type}, \mathcal{T}) \in \mathcal{G}\}$. Since $\mathcal{G}_1 \subseteq \mathcal{G}_2$ we have $S(\mathcal{G}_2, \mathcal{T}) = S(\mathcal{G}_2 - \mathcal{G}_1, \mathcal{T}) + S(\mathcal{G}_1, \mathcal{T})$, and also $S(\mathcal{G}_1, \mathcal{T}) = S(\mathcal{G}_2, \mathcal{T}) - S(\mathcal{G}_2 - \mathcal{G}_1, \mathcal{T})$.

Proposition 2. *The primitive query $S(\mathcal{G}, \mathcal{T}, C = c, A = a)$ is not updatable.*

Proof. We prove by showing a counter example. Let the target class be $\mathcal{T}$, the target attribute be $C = (c_1)$, an attribute $A = (p_1, \ldots, p_i, \ldots, p_n)$, and suppose we have the following RDF graphs: $S_1 = \{(x, \text{rdf:type}, \mathcal{T}), (x, c_1, c)\}$, $S_2 = \{(x, p_1, o_1), \ldots, (o_{i-1}, p_i, o_i)\}$, and $S_3 = \{(o_i, p_{i+1}, o_{i+1}), \ldots, (o_{n-1}, p_n, a)\}$. Suppose the graph before update is $\mathcal{G}_1 = S_1 \cup S_2$ and after an insertion of S_3 the graph becomes $\mathcal{G}_2 = S_1 \cup S_2 \cup S_3$. For brevity let $\theta(\mathcal{G}) = S(\mathcal{G}, \mathcal{T}, C = c, A = a)$. We will show that there exists no functions f for the query $\theta(\mathcal{G}_2)$, which counts the total number of leaves of $A(x, a)$ such that x has the class label c. In $\mathcal{G}_2$ the attribute graph $A(x)$ is $S_2 \cup S_3$, and hence $\theta(\mathcal{G}_2) = 1$. However, $A(x)$ is partitioned among S_2 and S_3, so $\theta(\mathcal{G}_1) = 0$ and $\theta(\mathcal{G}_2 - \mathcal{G}_1) = 0$, therefore in this case $f(\theta(\mathcal{G}_2 - \mathcal{G}_1), \theta(\mathcal{G}_1)) = f(0, 0) = 1 = \theta(\mathcal{G}_2)$. Now consider another case where initially the graph is $\mathcal{G}_3 = S_1$ and after insertion of S_3 the graph becomes $\mathcal{G}_4 = S_1 \cup S_3$. In this case we have $\theta(\mathcal{G}_4) = 0$, $\theta(\mathcal{G}_3) = 0$, and $\theta(\mathcal{G}_4 - \mathcal{G}_3) = 0$, and so $f(0, 0) = 0$. Since a function can not map an input to more than one output, this shows that there exists no function f to maintain the query result.

Similarly, can show that the primitive queries $S(\mathcal{G}, \mathcal{T}, C = c)$, $S(\mathcal{G}, \mathcal{T}, C = c, A)$, and $S(\mathcal{G}, \mathcal{T}, C = c, A, \text{agg}, [v_l, v_h])$ are not updatable.

Corollary 1. *RBC model is not updatable.*

The proof of Prop. 2 shows that when an attribute graph is partitioned across multiple updates, there exists no function to update the required counts. This raises the question as to whether we can ensure updatability by requiring that each update involves only complete attribute graphs. However, this requirement is not sufficient for the query to be updatable. To see why, consider $\mathcal{G}_1 = \{(x, \text{rdf:type}, \mathcal{T}), (x, c_1, c), (x, p_1, o), (o, p_2, a)\}$ and $\mathcal{G}_2 - \mathcal{G}_1 =$

$\{(y, \text{rdf:type}, \mathcal{T}), (y, c_1, c), (y, p_1, o), (o, p_2, b)\}$, then $f(0, 1) = 2$. The extra count from $\theta(\mathcal{G}_2)$ is due to o being shared between two datasets despite the fact that each attribute graph is complete. This motivates the restriction of not allowing the update to *reuse* certain subjects or objects. We formalize this notion as follows.

Definition 11 (Clean Update). *Assume* $\mathcal{G}_1 \subseteq \mathcal{G}_2$, *and let* $V(\mathcal{G}) = \{s : (s, p, o) \in \mathcal{G}\} \cup \{o : (s, p, o) \in \mathcal{G}\}$ *denote the set of all subjects and objects of an RDF graph* $\mathcal{G}$. *An update (from* $\mathcal{G}_1$ *to* $\mathcal{G}_2$ *by insertion, or from* $\mathcal{G}_2$ *to* $\mathcal{G}_1$ *by deletion) is said to be clean if* $[\forall (s, p, o) \in \mathcal{G}_2][s \notin V(\mathcal{G}_1) \cap V(\mathcal{G}_2 - \mathcal{G}_1)]$. *That is, triples in* $\mathcal{G}_2 - \mathcal{G}_1$ *share objects with only the leaves of attribute graphs in* $\mathcal{G}_1$.

Proposition 3. *RBC models are updatable if every update is clean.*

Proof. Let D_1 and D_2 be two RDF datasets such that $D_1 \subseteq D_2$. We first consider the primitive query $\theta(\mathcal{G}) = S(\mathcal{G}, \mathcal{T}, C = c, A = a)$. Since every update is clean, the attribute graphs $A(x)$ for all attributes in A, and all instances $x \in \mathcal{T}(\mathcal{G}_1)$ and $x \in \mathcal{T}(\mathcal{G}_2 - \mathcal{G}_1)$ remain the same after insertion (or deletion). Hence, $\mathcal{M}(D_2) = \mathcal{M}(D_1) \cup \mathcal{M}(D_2 - D_1)$ and similarly $\mathcal{M}(D_1) = \mathcal{M}(D_2) - \mathcal{M}(D_2 - D_1)$ for the multiset attributed dataset reductions. It follows that $\theta(\mathcal{G}_2) = \theta(\mathcal{G}_1) + \theta(\mathcal{G}_2 - \mathcal{G}_1)$ and $\theta(\mathcal{G}_1) = \theta(\mathcal{G}_2) - \theta(\mathcal{G}_2 - \mathcal{G}_1)$. Similar argument also holds true for the other queries used for learning a RBC.

Thus, RBC model can be updated incrementally in a restricted setting where every update is clean in the sense defined above. When clean updates are not available, RBC models can still be incrementally updated if we are willing to sacrifice some accuracy; and rebuild the model periodically by querying the entire RDF store, with the frequency of rebuild chosen based on the desired tradeoff between computational efficiency and model accuracy. Regardless of whether the RBC model is updatable or not, answering of aggregate queries from RDF stores answering can be optimized using an aggregate view maintenance algorithm [16]. Since we assume that the data descriptor does not change as frequently as the data, the aggregate queries needed by the RBC model can be set up and maintained as views on the RDF store.

5   Communication Complexity

In this section, we analyze the communication complexity, i.e., the amount of data transfer needed to build an RBC model. We compare the communication complexity of building an RBC model from RDF data in the following two scenarios: (i) posing statistical queries needed for learning the model against a remote RDF store which is the approach proposed in this paper; and (ii) retrieving the entire RDF dataset from a remote RDF store for local processing.

Given an RDF dataset $D = (\mathcal{G}, \mathcal{T}, \mathcal{A}, C)$ where $\mathcal{A} = (A_1, \dots, A_n)$. Suppose the RDF store holds the RDF graph $\mathcal{G}$, and let $|\mathcal{G}|$ denotes the size of this graph.

The communication complexity in scenario (ii) is simply $O(|\mathcal{G}|)$. We now analyze the communication complexity in scenario (i). Let l_C denotes the length of tuple C, let r_C denotes the size of range of C, and let l_A denotes the maximum length of an attribute tuple. Also let r_A^1 denotes the maximum number of bins of those attributes estimated by aggregation, let r_A^2 denotes the maximum size of range of the remaining attributes, and we define r_A to be $\max(r_A^1, r_A^2)$.

The size of query expressed in SPARQL, is $O(1)$ for $(Q1)$, $O(l_C)$ for $(Q2)$, and $O(l_C + l_A)$ for $(Q3)$, $(Q4)$, and $(Q5)$. Further, to estimate the probabilities to build an RBC, the following number of calls for each query described in Sec. 3 are required:

(Q1) one.
(Q2) r_C, once for each class label.
(Q3) $r_C \cdot n$, once for each class label and each attribute.
(Q4) $O(r_C \cdot n \cdot r_A)$, once for each class label, each attribute, and each value of the attribute.
(Q5) $O(r_C \cdot n \cdot r_A)$, same as $(Q4)$.

Therefore, the total complexity is $O(1) + O(l_C r_C) + O((l_C + l_A)r_C n) + O((l_C + l_A)r_C \cdot n \cdot r_A) + O((l_C + l_A)r_C \cdot n \cdot r_A) = O((l_C + l_A)r_C \cdot n \cdot r_A)$. In Sec. 7.1 we provide results of experiments which show that $O((l_C + l_A)r_C \cdot n \cdot r_A)$ is usually less than $O(|\mathcal{G}|)$ in practice.

6   Selective Attribute Crawling

In previous sections we have considered the problem of learning RBCs given an RDF dataset $D = (\mathcal{G}, \mathcal{T}, \mathcal{A}, C)$ in the setting where the learner has direct access to $\mathcal{T}, \mathcal{A}$, and C, but not $\mathcal{G}$. Here we consider a more general problem where the learner does not have a priori knowledge of $\mathcal{A}$. This requires the learner to interact with the RDF store containing $\mathcal{G}$ in order to determine $\mathcal{A}$ (e.g. by crawling and selecting attributes) that best optimizes a predetermined performance criterion P. Since the number of attributes in an RDF store can be arbitrarily large we specify an additional constraint Z to guarantee termination (e.g. number of attributes crawled, number of queries posed, time spent, etc.).

Problem 2. Given an RDF dataset without attributes, $D = (\mathcal{G}, \mathcal{T}, C)$, a hypothesis class H, a performance criterion P, and constraint Z, the learning algorithm L outputs the following while respecting Z: (i) The selected tuple of attributes $\mathcal{A}$, and (ii) a classifier $h \in H$ that optimizes P.

For simplicity, we focus the setting where the constraint Z specifies the maximum the number of attributes crawled. We consider the problem of identifying $\mathcal{A}$ of cardinality at most Z so as to optimize P. This problem is a variant of the well-studied feature subset selection problem [22,12], albeit in a setting where the set of features is a priori unknown. Identifying attributes one at a time to optimize P can be seen as a search over a tree rooted at $\mathcal{T}$, where the edges are IRIs of properties and the nodes are the domain/range of properties, and an attribute corresponds to a path from the root to an RDF literal (a leaf in this tree). To complete the specification of the search problem, we need to specify operations for expanding a node to generate its successors and define the scoring function for evaluating nodes. Expanding a node consists of querying (i) the set of distinct properties outgoing from a node, (ii) the range of each property, and (iii) the type of each range (e.g. numeric, string, non-literal), each of which can be expressed as SPARQL queries. We define the score of a node based on the degree of correlation of the node with the target attribute C. Specifically, for each attribute (represented by a leaf), we compute mutual information [7] between it and the the target attribute C. The score of an internal node is defined (recursively) as a function of its descendants, e.g. average of the scores of its children.

Formally, the score of an attribute A is:

$$Score(A) = \sum_{C=c, A=a} p(A = a, C = c) \log_2 \frac{p(A = a, C = c)}{p(A = a)p(C = c)} \qquad (3)$$

These probabilities can also be estimated based on the queries described in Sec. 3. Given this framework, a variety of alternative search strategies can be considered, along with several alternative scoring functions.

7   Experiments

We conduct three experiments each with a different goal. The first measures the communication complexity using the LinkedMDB [13] dataset. The second experiment combines the US Census dataset with a government dataset to evaluate the accuracy of models using different attribute crawling strategies. Finally we demonstrate learning of RBC from another government dataset through a live SPARQL endpoint.

7.1   Communication Complexity Experiment

The goal of this experiment is to measure the communication complexity under two different approaches described in Sec. 5.

Dataset and Experiment Setup. The IMDB dataset is a standard benchmark that has been used to evaluate probabilistic relational models including RBCs [25]. The task is to predict whether a movie receives more than \$2M in its opening week. We used LinkedMDB [13], which is an RDF store extracted from IMDB, with links to other datasets on the Linked Open Data cloud [8]. We used links to Freebase[1] which includes *foaf* property to the *Person* class and three properties of class *Person*. Fig. 1 shows the RDF schema of the extracted dataset. Since LinkedMDB does not have openingReceipts, we add

---

[1] http://www.freebase.com

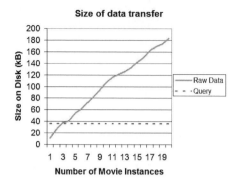

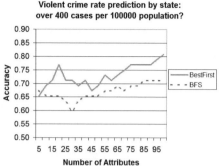

Fig. 2. Comparison of size of data transfer for Experiment 7.1

Fig. 3. Comparison of two crawling strategies for Experiment 7.2

them by crawling the IMDB website[2]; also for the Freebase data, we parse the yearOfBirth property for each Person from the dateOfBirth property. We extract 20 movies which are released after 2006 such that each movie has at least one actor, one director, and one producer. The target class is *Movie* and the target attribute is (*openingReceipts*). We consider a total of 10 attributes: (*runtime*), and (*h, foaf, a*) where $h \in \{hasActor, hasDirector, hasProducer\}$ and $a \in \{yearOfBirth, gender, hasAward\}$.

To show the growth of data transfer, we prepared 20 subsets of the dataset by corresponding to 1 to 20 movies. A movie instance consists of the URI of the movie and all reachable linked data for it. For communication complexity of learning RBC from RDF stores using statistical queries, we used the proposed approach to build an RBC for each subset and logged the SPARQL queries sent, saved the log in a plain text format, and measured the size of the logs. We compared the results with the communication complexity of learning RBC by first retrieving the data from a remote store for local processing as measured by the size of the corresponding dataset in RDF/XML format on disk.

Results. Figure 2 shows that the size of the raw data exceeds that of the query when there are more than three movie instances in the dataset. We also considered the case where the RDF store compresses the raw data before transfer, and in this case the size of the compressed raw data exceeds that of the query when there are more than 90 movie instances.

7.2 Selective Attribute Crawling Experiment

The goal of this experiment is to evaluate the accuracy of RBC models built using different attribute crawling strategies. Recall that in this setting the learner is only given a SPARQL endpoint of the RDF store, the target class, and the target attribute.

---
[2] http://www.imdb.com

Dataset and Experiment Setup. In this experiment we use datasets from Data.gov and US census 2000. The target class is 52 US states and we wish to predict whether a state's violent crime rate is over 400 per 100,000 population, which is from dataset 311 of the Data-gov project [9]. We link this with the US Census 2000 dataset for the corresponding states. This dataset was converted to over 1 billion RDF triples by [27]. Part of its RDF schema is shown in [27]. It uses a property as a way to sub-divide the population, and a number at a leaf represents the population that satisfies the conditions (properties) on the path from root. In our experiment we normalize by dividing every number of a state by the state's total population. We vary the maximum number of attributes to be crawled. We set the constraint to be the number of attributes the learner is allowed to crawl. We apply two different attribute crawling strategies (described below) separately and build RBC models using the crawled attributes. To measure the accuracy of the built models, we randomly partition 52 states into 13 groups (of 4 states each) and perform cross validation. That is, for each group, the 4 states in the group are held out and used for prediction, and the remaining are used for training the model. The overall accuracy is the total number of correct predictions divided by the number of states (52).

We experiment with two crawling strategies: BreadthFirst (BFS) and Best-First. BFS chooses the node with the least depth to expand and BestFirst chooses the node that has the highest score as defined in Sec. 6.

Results. As shown in Fig. 3, BestFirst outperforms BFS with the exception of the case where the number of attributes is 5. We examined the crawled attributes for BestFirst from for choices of Z from 20 to 45, and found that the strategy focused on expanding the households property. This is because attribute selection is guided by mutual information between a candidate attribute and the target class. The sub-divisions of this property may provide very minimal additional information compared to the first one crawled in this group, and hence they may not contribute to the predictive accuracy. One way to circumvent this problem is to use a scoring function to that measures the amount of *information gain* resulting from a candidate attribute given all the attributes that have already been chosen. Another approach is to penalize the attributes based on the depth of search. A third approach is to use the marginal improvement in the accuracy of the RBC classifier resulting from inclusion of the attribute to decide whether to retain it. Other alternatives worth exploring include different search strategies such as Iterative Deepening Search (IDS) [18].

7.3 Live Demonstration

The goal of this experiment is to demonstrate learning of RBC from a government dataset through a live SPARQL endpoint[3] hosted on Rensselaer Polytechnic Institute [10]. This endpoint supports aggregate and nested queries proposed in SPARQL 1.1.

---

[3] http://logd.tw.rpi.edu/sparql

Dataset and Experiment Setup. We used a Health Information National Trends Survey (HINTS) [24] from NCI which has been converted into RDF as part of the Data-gov project [9]. The survey represents a cross-sectional study of health media use and cancer-related knowledge among adults in the United States, and it has been used by [1] to study associations of covariates with different smoking statuses. There are 12080 participants across two years (2003 and 2005), represented by 623544 total number of RDF triples, and the raw RDF data (as TTL dump) has a size 35.9MB on disk. The task in our setting is to predict the smoking status (never, former, or current) of a participant from 16 other attributes such as race, sex, household income, and education. The dataset is propositional in nature although represented in RDF format; that is, every attribute has exactly one value in terms of the reduced multiset attribute data, hence the task reduces to learning of a conventional Naive Bayes classifier. Nevertheless, the experiment demonstrates learning of RBC from large and remote RDF store by querying its SPARQL endpoint.

Results. A total of 159 queries were posed to the live SPARQL endpoint, and the model was learned in approximately 30 secs, using 2.8 GHz processor with 4 GB memory, and the network download and upload speed is approximately 3 Mbps.

8   Summary and Related Work

Summary. The emergence of RDF as a basic data representation format for Semantic Web has led to increasing availability of all kinds of data in RDF. Transforming this data into knowledge calls for approaches to learning predictive models from massive RDF stores in settings where (i) the learning algorithm can interact with the data store only through a SPARQL endpoint; (ii) the model needs to be updated in response to updates to the underlying RDF store; and (iii) the attributes that can be used to build the predictive models are not known a priori and hence need to be identified by crawling the RDF store. We have introduced an approach to learning predictive models from RDF stores in such settings using Relational Bayesian Classifiers (RBCs) as an example. We have implemented our solutions in an open source system available as part of the INDUS toolkit for learning predictive models from massive data sets [21] and demonstrated the its feasibility using experiments with several RDF datasets.

Related Work. The work on SPARQL-ML [17] extends SPARQL with data mining support to build classifiers, including statistical relational models such as RBC from RDF data. Other works on learning predictive models from RDF data include [4] and [28]. In [4] kernel machines are defined over RDF data where features are constructed by ILP-based dynamic propositionalization. In [28] RDF triples are represented as entries in a Boolean matrix, and matrix completion methods are used to train the model and predict unknown triples off-line. However, all the approaches assume that the learner has direct access to RDF data.

In contrast, our approach does not require the learning algorithm to have direct access to RDF data, and relies only on the ability of the RDF store to answer aggregate SPARQL queries.

Acknowledgments. The authors would like to thank our anonymous reviewers who have provided valuable comments. This work is supported in part by a grant (IIS 0711356) from the National Science Foundation and in part by the Iowa State University Center for Computational Intelligence, Learning, and Discovery. The work of Vasant Honavar was supported by the National Science Foundation, while working at the Foundation. Any opinion, finding, and conclusions contained in this article are those of the authors and do not necessarily reflect the views of the National Science Foundation.

References

1. Ackerson, L.K., Viswanath, K.: Communication inequalities, social determinants, and intermittent smoking in the 2003 health information national trends survey. Prev. Chronic. Dis. 6(2) (2009)
2. Antoniou, G., van Harmelen, F.: A Semantic Web Primer, 2nd edn. MIT Press, Cambridge (2008)
3. Berners-Lee, T., Hendler, J., Lassila, O.: The semantic web. Scientific American (2001)
4. Bicer, V., Tran, T., Gossen, A.: Relational Kernel Machines for Learning from Graph-Structured RDF Data. In: Antoniou, G., Grobelnik, M., Simperl, E., Parsia, B., Plexousakis, D., De Leenheer, P., Pan, J. (eds.) ESWC 2011, Part I. LNCS, vol. 6643, pp. 47–62. Springer, Heidelberg (2011)
5. Breitmann, K., Casanova, M., Truszkowski, W.: Semantic Web: Concepts, Technologies and Applications. Springer (2007)
6. Caragea, D., Zhang, J., Bao, J., Pathak, J., Honavar, V.: Algorithms and Software for Collaborative Discovery from Autonomous, Semantically Heterogeneous, Distributed Information Sources. In: Hoffmann, A., Motoda, H., Scheffer, T. (eds.) DS 2005. LNCS (LNAI), vol. 3735, p. 14. Springer, Heidelberg (2005)
7. Cover, T.M., Thomas, J.A.: Elements of information theory. Wiley Interscience, New York (1991)
8. Cyganiak, R., Jentzsch, A.: Linking open data cloud diagram, http://lod-cloud.net/ (accessed 2011)
9. Ding, L., DiFranzo, D., Graves, A., Michaelis, J.R., Li, X., McGuinness, D.L., Hendler, J.: Data-gov wiki: Towards linking government data. In: AAAI Spring Symposium on Linked Data Meets Artificial Intelligence (2010)
10. Ding, L., DiFranzo, D., Graves, A., Michaelis, J.R., Li, X., McGuinness, D.L., Hendler, J.A.: TWC data-gov corpus: incrementally generating linked government data from data. gov. In: Proceedings of the 19th International Conference on World Wide Web, pp. 1383–1386 (2010)
11. Getoor, L., Taskar, B.: Introduction to Statistical Relational Learning. The MIT Press (2007)
12. Guyon, I., Elisseeff, A.: An introduction to variable and feature selection. J. Mach. Learn. Res. 3, 1157–1182 (2003)
13. Hassanzadeh, O., Consens, M.: Linked movie data base. In: WWW 2009 LDOW Workshop (2009)

14. Hendler, J.: Science and the semantic web. Science 299, 520–521 (2003)
15. Hitzler, P., Krötzsch, M., Rudolph, S.: Foundations of Semantic Web Technologies. Chapman & Hall/CRC (2009)
16. Hung, E., Deng, Y., Subrahmanian, V.S.: RDF aggregate queries and views. In: 21st International Conference on Data Engineering, pp. 717–728 (2005)
17. Kiefer, C., Bernstein, A., Locher, A.: Adding Data Mining Support to SPARQL Via Statistical Relational Learning Methods. In: Bechhofer, S., Hauswirth, M., Hoffmann, J., Koubarakis, M. (eds.) ESWC 2008. LNCS, vol. 5021, pp. 478–492. Springer, Heidelberg (2008)
18. Korf, R.E.: Depth-first iterative-deepening: an optimal admissible tree search. Artif. Intell. 27, 97–109 (1985)
19. Koul, N., Bui, N., Honavar, V.: Scalable, updatable predictive models for sequence data. In: BIBM, pp. 681–685 (2010)
20. Koul, N., Caragea, C., Honavar, V., Bahirwani, V., Caragea, D.: Learning classifiers from large databases using statistical queries. In: Web Intelligence, pp. 923–926 (2008)
21. Koul, N., Lin, H.T.: Indus learning framework. Google Code (2011), http://code.google.com/p/induslearningframework/
22. Liu, H., Motoda, H.: Feature Extraction, Construction and Selection: A Data Mining Perspective. Kluwer Academic Publishers, Norwell (1998)
23. Manola, F., Miller, E. (eds.): RDF Primer. W3C Recommendation. World Wide Web Consortium (February 2004)
24. Nelson, D., Kreps, G., Hesse, B., Croyle, R., Willis, G., Arora, N., Rimer, B., Viswanath, K.V., Weinstein, N., Alden, S.: The health information national trends survey (HINTS): Development, design, and dissemination. Journal of Health Communication: International Perspectives 9(5), 443–460 (2004)
25. Neville, J., Jensen, D., Gallagher, B.: Simple estimators for relational bayesian classifiers. In: Proceedings of the Third IEEE International Conference on Data Mining, pp. 609–612 (2003)
26. Prud'ommeaux, E., Seaborne, A.: SPARQL query language for RDF, http://www.w3.org/TR/2008/REC-rdf-sparql-query-20080115/ (accessed 2011)
27. Tauberer, J.: The 2000, U.S. census: 1 billion RDF triples, http://www.rdfabout.com/demo/census/ (accessed 2011)
28. Tresp, V., Huang, Y., Bundschus, M., Rettinger, A.: Materializing and querying learned knowledge. In: Proceedings of the ESWC 2009 IRMLeS Workshop (2009)

Large Scale Fuzzy pD^* Reasoning Using MapReduce

Chang Liu[1], Guilin Qi[2], Haofen Wang[1], and Yong Yu[1]

[1] Shanghai Jiaotong University, China
{liuchang,whfcarter,yyu}@apex.sjtu.edu.cn
[2] Southeast University, China
gqi@seu.edu.cn

Abstract. The MapReduce framework has proved to be very efficient for data-intensive tasks. Earlier work has tried to use MapReduce for large scale reasoning for pD^* semantics and has shown promising results. In this paper, we move a step forward to consider scalable reasoning on top of semantic data under fuzzy pD^* semantics (i.e., an extension of OWL pD^* semantics with fuzzy vagueness). To the best of our knowledge, this is the first work to investigate how MapReduce can help to solve the scalability issue of fuzzy OWL reasoning. While most of the optimizations used by the existing MapReduce framework for pD^* semantics are also applicable for fuzzy pD^* semantics, unique challenges arise when we handle the fuzzy information. We identify these key challenges, and propose a solution for tackling each of them. Furthermore, we implement a prototype system for the evaluation purpose. The experimental results show that the running time of our system is comparable with that of WebPIE, the state-of-the-art inference engine for scalable reasoning in pD^* semantics.

1 Introduction

The Resource Description Framework (RDF) is one of the major representation standards for the Semantic Web. RDF Schema (RDFS) is used to describe vocabularies used in RDF descriptions. However, RDF and RDFS only provide a very limited expressiveness. In [3], a subset of Ontology Web Language (OWL) vocabulary (e.g., owl:sameAs) was introduced, which extends the RDFS semantics to the pD^* fragment of OWL. Unlike the standard OWL (DL or Full) semantics which provides the full if and only if semantics, the OWL pD^* fragment follows RDF(S)'s if semantics. That is, the OWL pD^* fragment provides a complete set of entailment rules, which guarantees that the entailment relationship can be determined within polynomial time under a non-trivial condition (if the target graph is ground). It has become a very promising ontology language for the Semantic Web as it trades off the high computational complexity of OWL Full and the limited expressiveness of RDFS.

Recently, there is an increasing interest in extending RDF to represent vague information on Web. Fuzzy RDF allows us to state to a certain degree, a triple

L. Aroyo et al. (Eds.): ISWC 2011, Part I, LNCS 7031, pp. 405–420, 2011.

is true. For example, $(Tom, eat, pizza)$ is true with degree at least 0.8. However, Fuzzy RDF (or fuzzy RDFS) has limited expressive power to represent information in some real life applications of ontologies, such as biomedicine and multimedia. In [4], we extended the OWL pD^* fragment with fuzzy semantics to provide more expressive power than fuzzy RDF(S). In that work, we focused on some theoretical problems, such as the complexity issues, without providing an efficient reasoning algorithm for the new semantics. Since fuzzy pD^* semantics is targeted to handle large scale semantic data, it is critical to provide a scalable reasoning algorithm for it.

Earlier works (e.g. [12]) have proved that MapReduce is a very efficient framework to handle the computation of the closure containing up to 100 billion triples under pD^* semantics. One may wonder if it is helpful to scalable reasoning in fuzzy pD^* semantics. It turns out that this is a non-trivial problem as the computation of the closure under fuzzy pD^* semantics requires the computation of the *Best Degree Bound (BDB)* of each triple. The BDB of a triple is the greatest lower bound of the fuzzy degrees of this triple. Although most of the optimizations for reasoning with MapReduce in pD^* semantics are also applicable for the fuzzy pD^* semantics, unique challenges arise when we handle the fuzzy information.

In this paper, we first identify some challenges to apply the MapReduce framework to deal with reasoning in fuzzy pD^* semantics. We then propose an algorithm for fuzzy $pD*$ reasoning by separately considering fuzzy D rules and fuzzy p rules. After that, we propose the *map* function and the *reduce* function for several fuzzy pD^* rules that may cause difficulties. Finally, we implement a prototype system to evaluate these optimizations. The experimental results show that the running time of our system is comparable with that of WebPIE [12] which is the state-of-the-art inference engine for OWL pD^* fragment.

2   Background Knowledge

In this section, we first introduce the fuzzy pD^* entailment rule set in Section 2.1, then explain the MapReduce framework for reasoning in OWL pD^* fragment in Section 2.2.

2.1   Fuzzy RDF and Fuzzy pD^* Reasoning

A fuzzy RDF graph is a set of fuzzy triples which are in form of $t[n]$. Here t is a triple, and $n \in (0, 1]$ is the fuzzy degree of t.

Fuzzy pD^* semantics, given in [4], extends pD^* semantics with fuzzy semantics so that there is a complete and sound entailment rule set in fuzzy OWL pD^* fragment. We list part of them in Table 1 by excluding some naive rules. The notion of a (partial) closure can be easily extended to the fuzzy case.

The key notion in fuzzy pD^* semantics is called the *Best Degree Bound (BDB)* of a triple. The BDB n of an arbitrary triple t from a fuzzy RDF graph G is defined to be the largest fuzzy degree n such that $t[n]$ can be derived from G by applying fuzzy pD^*-entailment rules, or 0 if no such fuzzy triple can be derived.

Table 1. Difficult part of fuzzy pD^*-entailment rules

Condition		
	Conclusion	
f-rdfs2	$(p,\texttt{domain},u)[n]\ (v,p,w)[m]$	$(v,\texttt{type},u)[n\otimes m]$
f-rdfs3	$(p,\texttt{range},w)[n]\ (v,p,w)[m]$	$(v,\texttt{type},w)[n\otimes m]$
f-rdfs5	$(v,\texttt{subPropertyOf},w)[n]\ (w,\texttt{subPropertyOf},u)[m]$	$(v,\texttt{subPropertyOf},u)[n\otimes m]$
f-rdfs7x	$(p,\texttt{subPropertyOf},q)[n]\ (v,p,w)[m]$	$(v,q,w)[n\otimes m]$
f-rdfs9	$(v,\texttt{subClassOf},w)[n]\ (u,\texttt{type},v)[m]$	$(u,\texttt{type},w)[n\otimes m]$
f-rdfs11	$(v,\texttt{subClassOf},w)[n]\ (w,\texttt{subClassOf},u)[m]$	$(v,\texttt{subClassOf},u)[n\otimes m]$
f-rdfs12	$(v,\texttt{type},\texttt{ContainerMembershipProperty})[n]$	$(v,\texttt{subPropertyOf},\texttt{member})[n]$
f-rdfs13	$(v,\texttt{type},\texttt{Datatype})[n]$	$(v,\texttt{subClassOf},\texttt{Literal})[1]$
f-rdfp1	$(p,\texttt{type},\texttt{FunctionalProperty})[n]\ (u,p,v)[m]\ (u,p,w)[l]$	$(v,\texttt{sameAs},w)[l\otimes m\otimes n]$
f-rdfp2	$(p,\texttt{type},\texttt{InverseFunctionalProperty})[n]$	
	$(u,p,w)[m]\ (v,p,w)[l]$	$(u,\texttt{sameA},v)[l\otimes m\otimes n]$
f-rdfp3	$(p,\texttt{type},\texttt{SymmetricProperty})[n]\ (v,p,w)[m]$	$(w,p,v)[n\otimes m]$
f-rdfp4	$(p,\texttt{type},\texttt{TransitiveProperty})[n]\ (u,p,v)[m]\ (v,p,w)[l]$	$(u,p,w)[n\otimes m\otimes l]$
f-rdfp5(ab)	$(v,p,w)[n]$	$(v,\texttt{sameAs},v)[1],\ (w,\texttt{sameAs},w)[1]$
f-rdfp6	$(v,\texttt{sameAs},w)[n]$	$(w,\texttt{sameAs},v)[n]$
f-rdfp7	$(u,\texttt{sameAs},v)[n]\ (v,\texttt{sameAs},w)[m]$	$(u,\texttt{sameAs},w)[n\otimes m]$
f-rdfp8ax	$(p,\texttt{inverseOf},q)[n]\ (v,p,w)[m]$	$(w,q,v)[n\otimes m]$
f-rdfp8bx	$(p,\texttt{inverseOf},q)[n]\ (v,q,w)[m]$	$(w,p,v)[n\otimes m]$
f-rdfp9	$(v,\texttt{type},\texttt{Class})[n]\ (v,\texttt{sameAs},w)[m]$	$(v,\texttt{subClassOf},w)[m]$
f-rdfp10	$(p,\texttt{type},\texttt{Property})[1]\ (p,\texttt{sameAs},q)[m]$	$(p,\texttt{subPropertyOf},q)[m]$
f-rdfp11	$(u,p,v)[n]\ (u,\texttt{sameAs},u')[m]\ (v,\texttt{sameAs},v')[l]$	$(u',p,v')[n\otimes m\otimes l]$
f-rdfp12(ab)	$(v,\texttt{equivalentClass},w)[n]\Rightarrow(v,\texttt{subClassOf},w)[n],(w,\texttt{subClassOf},w)[n]$	
f-rdfp12c	$(v,\texttt{subClassOf},w)[n]\ (w,\texttt{subClassOf},v)[m]$	$(v,\texttt{equivalentClass},w)[\min(n,m)]$
f-rdfp13(ab)	$(v,\texttt{equivalentProperty},w)[n]\Rightarrow(v,\texttt{subPropertyOf},w)[n],(w,\texttt{subPropertyOf},w)[n]$	
f-rdfp13c	$(v,\texttt{subPropertyOf},w)[n]\ (w,\texttt{subPropertyOf},v)[m]$	$(v,\texttt{equivalentClass},w)[\min(n,m)]$
f-rdfp14a	$(v,\texttt{hasValueOf},w)[n]\ (v,\texttt{onProperty},p)[m]\ (u,p,w)[l]$	$(u,\texttt{type},v)[n\otimes m\otimes l]$
f-rdfp14bx	$(v,\texttt{hasValueOf},w)[n]\ (v,\texttt{onProperty},p)[m]\ (u,\texttt{type},v)[l]$	$(u,p,w)[n\otimes m\otimes l]$
f-rdfp15	$(v,\texttt{someValueFrom},w)[n]\ (v,\texttt{onProperty},p)[m]$	
	$(u,p,x)[l]\ (x,\texttt{type},w)[k]$	$(u,\texttt{type},v)[n\otimes m\otimes l\otimes k]$
f-rdfp16	$(v,\texttt{allValuesfrom},w)[m]\ (v,\texttt{onProperty},p)[n]$	
	$(u,\texttt{type},v)[l]\ (u,p,x)[k]$	$(x,\texttt{type},w)[n\otimes m\otimes l\otimes k]$

2.2 MapReduce Algorithm for pD^* Reasoning

MapReduce is a programming model introduced by Google for large scale data processing [1]. A MapReduce program is composed of two user-specified functions, *map* and *reduce*. When the input data is appointed, the *map* function scans the input data and generates intermediate key/value pairs. Then all pairs of key and value are partitioned according to the key and each partition is processed by a *reduce* function.

We use an example to illustrate how to use a MapReduce program to apply a rule. Here, we consider Rule rdfs2 which reads:

$$(p,\texttt{domain},u),(v,p,w)\Rightarrow(v,\texttt{type},u)$$

The *map* function scans the data set, and checks every triple if it has the form $(p,\texttt{domain},u)$ or (v,p,w). If a triple has the form $(p,\texttt{domain},u)$, then the *map* function generates an output (key=p, value={flag='L', u}). While a triple in the form of (v,p,w) is scanned, the *map* function generates an output (key=p, value={flag='R', v}). The *reduce* function gets all outputs of the *map* function that share the same key together. Then it enumerates all values with flag 'L' to get all u and enumerates all values with flag 'R' to get all v. For each pair of u and v, the *reduce* function generates a new triple $(v,\texttt{type},u)$ as output.

There are several key factors to make a MapReduce program efficient. Firstly, since the *map* function operates on single pieces of data without dependencies, partitions can be created arbitrarily and can be scheduled in parallel across many nodes. Secondly, the *reduce* function operates on an iterator of values since the set of values is typically too large to fit in memory. So the reducer should treat the values as a stream instead of a set. Finally, all the outputs of mappers sharing the same key will be processed by the same *reduce* function. Therefore, the reducer that processes one popular key will run very slowly. The mappers' output keys should be carefully designed to ensure that the sizes of all partitions should be balanced.

Due to these reasons, the naive implementations of rules may be very inefficient. In [12] and [13], several optimizations are proposed that improve the performance of inference in OWL pD^* fragment significantly. We list them as follows.

- **Loading schema triples in memory.** Since the set of schema triples is generally small enough to fit in memory, when performing a join over schema triples, we can load them into memory. Then, the join can be performed directly between the loaded data and the in-memory schema triples.
- **Data grouping to avoid duplicates.** Some RDFS rules may generate duplicates. However, using carefully designed algorithms, such duplicates can be avoided.
- **Ordering the RDFS rule applications.** Arbitrarily applying the rules will result in a fixpoint iteration. For RDFS rules, such a fixpoint iteration can be avoided by applying rules in a specific order.
- **Transitive algorithm.** An efficient algorithm to calculate the transitive closure is designed, which will produce a minimal amount of duplicates and minimize the number of iterations.
- **Sameas algorithm.** For OWL pD^* fragment, [12] uses the canonical representation to deal with the **sameas** rules. This method greatly reduces both the computation time and the space required.
- **someValuesFrom and allValuesFrom algorithm.** In both rules involve someValuesFrom and allValuesFrom, three joins among four triples are needed. However, two of the four triples are schema triples so that they can be loaded into memory. Furthermore by choosing the output key, the *map* function will generate balanced partitions for the *reduce* function.

3   MapReduce Algorithm for Fuzzy pD^* Reasoning

In this section, we first illustrate how to use a MapReduce program to apply a fuzzy rule. Then, we give an overview of the challenges in fuzzy pD^* reasoning when applying the MapReduce framework. Finally, we present our solutions to handle these challenges.

3.1   Naive MapReduce Algorithm for Fuzzy Rules

We consider rule f-rdfs2 to illustrate our naive MapReduce algorithms:

Algorithm 1. map function for rule f-rdfs2

Input: key, triple
 1: **if** triple.predicate == 'domain' **then**
 2:     emit({p=triple.subject}, {flag='L', u=triple.object, n=triple.degree});
 3: **end if**
 4: emit({p=triple.predicate}, {flag='R', v=triple.subject, m=triple.degree});

Algorithm 2. reduce function for rule f-rdfs2

Input: key, iterator values
 1: unSet.clear();
 2: vmSet.clear();
 3: **for** value $\in$ values **do**
 4:     **if** value.flag == 'L' **then**
 5:         unSet.update(value.u, value.n);
 6:     **else**
 7:         vmSet.update(value.v, value.m);
 8:     **end if**
 9: **end for**
10: **for** $i \in$ unSet **do**
11:     **for** $j \in$ vmSet **do**
12:         emit(null, triple(i.u, 'type', j.v, i.n$\otimes$j.m));
13:     **end for**
14: **end for**

$$(p, \text{domain}, u)[n], (v, p, w)[m] \Rightarrow (v, \text{type}, u)[n \otimes m]$$

In this rule, we should find all fuzzy triples that are either in the form of $(p, \text{domain}, u)[n]$ or in the form of $(v, p, w)[m]$. A join should be performed over the variable p. The *map* and *reduce* functions are given in Algorithms 1 and 2 respectively. In the *map* function, when a fuzzy triple is in the form of $(p, \text{domain}, u)[n]$ (or $(v, p, w)[m]$), the mapper emits p as the key and u (or v) along with the degree n (or m) as the value. The reducer can use the flag in the mapper's output value to identify the content of the value. If the flag is 'L' (or 'R'), the content of the value is the pair (u, n) (or the pair (v, m)). The reducer uses two sets to collect all the u, n pairs and the v, m pairs. After all pairs are collected, the reducer enumerates pairs (u, n) and (v, m) to generate $(u, \text{type}, v)[n \otimes m]$ as output.

3.2 Challenges in Fuzzy pD^* Reasoning

Even though the fuzzy pD^* entailment rules are quite similar to the pD^* rules, several difficulties arise when we calculate the BDB for each triple by applying the MapReduce framework. We summarize these challenges as follows:

Ordering the rule applications. In fuzzy pD^* semantics, the reasoner might produce the duplicated triple with different fuzzy degrees before the fuzzy BDB triple is derived. For example, suppose the data set contains a fuzzy triple $t[m]$, when we derive a fuzzy triple $t[n]$ with $n > m$, a duplicate is generated. When duplicates are generated, we should employ a duplicate deleting program to reproduce the data set to ensure that only the fuzzy triples with maximal degrees are in the data set. Different rule applications' order will result in different number of such duplicates. To achieve the best performance, we should choose a proper order to reduce the number of duplicates. For instance, the subproperty rule (f-rdfs7x) should be applied before the domain rule (f-rdfs2); and the equivalent class rule (f-rdfp12(abc)) should be considered together with the subclass rule (f-rdfs9, f-rdfs10). The solution for this problem will be discussed in Section 3.3.

Shortest path calculation. In OWL pD^* fragment, the three rules, rdfs5 (subproperty), rdfs11 (subclass) and rdfp4 (transitive property) are essentially used to calculate the transitive closure over a subgraph of the RDF graph. In fuzzy OWL pD^*, when we treat each fuzzy triple as a weighted edge in the RDF graph, then calculating the closure by applying these three rules is essentially a variation of the all-pairs shortest path calculation problem. We have to find out efficient algorithms for this problem. In Section 3.4, we will discuss the solutions for rules f-rdfs5 and f-rdfs11, while discuss rule f-rdfp4 in Section 3.5.

Sameas rule. For OWL pD^* fragment, the traditional technique to handle the semantics of sameas is called canonical representation. Rules rdfp6 and rdfp7 enforce that sameas is a symmetric and transitive property, thus the sameas closure obtained by applying these two rules is composed of several complete subgraphs. The instances in the same subgraph are all synonyms, so we can assign a unique key, which we call the canonical representation, to all of them. Replacing all the instances with its unique key results in a more compact representation of the RDF graph without loss of completeness for inference.

However, in the fuzzy pD^* semantics, we cannot choose such a canonical representation as illustrated by the following example. Suppose we use the *min* as the t-norm function. Given a fuzzy RDF graph G containing seven triples:

$(a, \text{sameas}, b)[0.8]$ $(b, \text{sameas}, c)[0.1]$ $(c, \text{sameas}, d)[0.8]$
$(a, \text{range}, r)[0.9]$ $(u, b, v)[0.9]$ $(c, \text{domaine})[1]$ $(u', d, v')[0.9]$

From this graph, we can derive $(v, \text{type}, r)[0.8]$. Indeed, we can derive $(b, \text{range}, r)[0.8]$ by applying rule f-rdfp11 over $(a, \text{sameas}, b)[0.8]$, $(a, \text{range}, r)[0.9]$ and $(r, \text{sameas}, r)[1.0]$. Then we can apply rule f-rdfs3 over $(b, \text{range}, r)[0.8]$ and $(u, b, v)[0.9]$ to derive $(v, \text{type}, r)[0.8]$.

In this graph, four instances, a, b, c and d are considered as synonyms in the classical pD^* semantics. Suppose we choose c as the canonical representation, then the fuzzy RDF graph is converted into the following graph G' containing four fuzzy triples:

$(c, \text{range}, r)[0.1]$ $(u, c, v)[0.1]$ $(c, \text{domaine})[1]$ $(u', c, v')[0.8]$

From this graph, we can derive the fuzzy triple $(v, \text{type}, r)[0.1]$, and this is a fuzzy BDB triple from G', which means we cannot derive the fuzzy triple, $(v, \text{type}, r)[0.8]$. The reason is that after replacing a and b with c, the fuzzy information between a and b, e.g. the fuzzy triple $(a, \mathtt{sameas}, b)[0.8]$, is missing. Furthermore, no matter how we choose the canonical representation, some information will inevitably get lost during the replacement. We will discuss the solution for this problem in Section 3.6.

3.3   Overview of the Reasoning Algorithm

Our main reasoning algorithm is Algorithm 3, which can be separated into two phases: the first phase (line 3) applies the fuzzy D rules (from f-rdfs1 to f-rdfs13), and the second phase (lines 7 to line 9) applies the fuzzy p rules (from rdfp1 to rdfp16). However, since some fuzzy p rules may generate some fuzzy triples having effect on fuzzy D rules, we execute these two phases iteratively (line 2 to line 11) until a fix point is reached (line 4 to line 6).

In the first phase, we consider the following order of rule applications such that we can avoid a fix point iteration. Firstly, we apply the property inheritance rules (f-rdfs5 and f-rdfs7), so that domain rule (f-rdfs2) and range rule (f-rdfs3) can be applied consecutively without loosing any important fuzzy triples. Then the class inheritance rules (f-rdfs9 and f-rdfs11) along with the rest rules are applied together. Similar techniques have been used in [13] for RDFS. Compared with [13], our algorithm relies on the computation of rules f-rdfs5, f-rdfs7, f-rdfs9 and f-rdfs11. We will discuss this point in the next section.

For the fuzzy p rules, there is no way to avoid a fixpoint iteration. So we employ an iterative algorithm to calculate the p closure. In each iteration, the program can be separated into five steps. In the first step, all non-iterative rules (rules f-rdfp1, 2, 3, 8) are applied. The second step processes the transitive property (rule f-rdfp4) while the $\mathtt{sameas}$ rules (rule f-rdfp6, 7, 10, 11) are applied in the third step. The rules related to $\mathtt{equivalentClass}$, $\mathtt{equivalentProperty}$ and $\mathtt{hasValue}$ are treated in the fourth step, because we can use the optimizations for reasoning in OWL pD^* fragment to compute the closure of these rules in a non-iterative manner. The $\mathtt{someValuesFrom}$ and $\mathtt{allValuesFrom}$ rules are applied in the fifth step which needs a fixpoint iteration. The first step and the last two steps can employ the same optimization discussed in [12]. We will discuss the solution to deal with transitive property in Section 3.5. Finally the solution to tackle $\mathtt{sameas}$ rules will be discussed in Section 3.6.

3.4   Calculating $\mathtt{subClassOf}$ and $\mathtt{subPropertyOf}$ Closure

Rules f-rdfs5, 6, 7, f-rdfp13(abc) process the semantics of $\mathtt{subPropertyOf}$ property while rules f-rdfs9, 10, 11 and f-rdfp12(abc) mainly concern the semantics of $\mathtt{subClassOf}$ property. Since they can be disposed similarly, we only discuss the rules that are relevant to $\mathtt{subClassOf}$. Since f-rdfs10 only derives a triple $(v, \mathtt{subClassOf}, v)[1]$ which will have no affect on other rules, we only consider rules f-rdfs5, 7 and f-rdfp12(abc).

We call the triples in the form of $(u, \mathtt{subClassOf}, v)[n]$ to be $\mathtt{subClassOf}$ triples. The key task is to calculate the closure of $\mathtt{subClassOf}$ triples by applying rule f-rdfs11. Since the set of $\mathtt{subClassOf}$ triples is relatively small, we can

Algorithm 3. Fuzzy pD^* reasoning

1: first_time = true;
2: **while** true **do**
3:     derived = apply_fD_rules();
4:     **if** derived == 0 and not first_time **then**
5:         break;
6:     **end if;**
7:     **repeat**
8:         derived = apply_fp_rules();
9:     **until** derived == 0;
10:     first_time = false;
11: **end while**

Algorithm 4. Calculate the `subClassOf` closure

1: **for** $k \in I$ **do**
2:     **for** $i \in I$ and $w(i, k) > 0$ **do**
3:         **for** $j \in I$ and $w(k, j) > 0$ **do**
4:             **if** $w(i, k) \otimes w(k, j) > w(i, j)$ **then**
5:                 $w(i, j) = w(i, k) \otimes w(k, j);$
6:             **end if**
7:         **end for**
8:     **end for**
9: **end for**

load them into memory. We can see that calculating the `subClassOf` closure by applying rule f-rdfs11 is indeed a variation of the all-pairs shortest path calculation problem, according to the following property:

Property 1. *For any fuzzy triple in the form of (u, `subClassOf`, v)[n] that can be derived from the original fuzzy RDF graph by only applying rule f-rdfs11, there must be a chain of classes $w_0 = u, w_1, ..., w_k = v$ and a list of fuzzy degrees $d_1, ..., d_k$ where for every $i = 1, 2, ..., k$, $(w_{i-1}, \text{subClassOf}, w_i)[d_k]$ is in the original fuzzy graph and $n = d_1 \otimes d_2 \otimes ... \otimes d_k$.*

So we can use the FloydCWarshall style algorithm given in Algorithm 4 to calculate the closure. In the algorithm, I is the set of all the classes, and $w(i, j)$ is the fuzzy degree of triple $(i, \text{subClassOf}, j)$. The algorithm iteratively update the matrix w. When it stops, the subgraph represented by the matrix $w(i, j)$ is indeed the `subClassOf` closure.

The worst-case running complexity of the algorithm is $O(|I|^3)$, and the algorithm uses $O(|I|^2)$ space to store the matrix w. When $|I|$ goes large, this is unacceptable. However, we can use nested hash map instead of 2-dimension arrays to only store the positive matrix items. Furthermore, since $0 \otimes n = n \otimes 0 = 0$, in line 2 and line 3, we only enumerate those i and j where $w(k, i) > 0$ and $w(k, j) > 0$. In this case, the running time of the algorithm will be greatly reduced.

After the `subClassOf` closure is computed, rules f-rdfs9 and f-rdfs11 can be applied only once to derive all fuzzy triples: for rule f-rdfs9 (or f-rdfs11), when

---

Algorithm 5. map function for rule f-rdfp4

Input: length, triple=(subject, predicate, object)[degree], n
  if getTransitiveDegree(predicate) == 0 **then**
    return;
  end if
  if length ==2^{n-2} or length == 2^{n-1} **then**
    emit({predicate, object}, {flag=L, length, subject, degree};
  end if
  if length > 2^{n-2} and length $\leq 2^{n-1}$ **then**
    emit({predicate, subject}, {flag=R, length, object, degree};
  end if

---

we find a fuzzy triple $(i, \texttt{type}, v)[n]$ (or $(i, \texttt{subClassOf}, v)[n]$), we enumerate all classes j so that $w(v,j) > 0$ and output a fuzzy triple $(i, \texttt{type}, j)[n \otimes w(v,j)]$ (or $(i, \texttt{subClassOf}, j)[n \otimes w(v,j)]$).

For rule f-rdfp12(abc), since `equivalentClass` triples are also schema triples, we load them into memory and combine them into the `subClassOf` graph. Specifically, when we load a triple $(i, \texttt{equivalentClass}, j)[n]$ into memory, if $n > w(i,j)$ (or $n > w(j,i)$), we update $w(i,j)$ (or $w(j,i)$) to be n . After the closure is calculated, two fuzzy triples $(i, \texttt{equivalentClass}, j)[n]$ and $(j, \texttt{equivalentClass}, i)[n]$ are output for each pair of classes $i, j \in I$, if $n = \min(w(i,j), w(j,i)) > 0$.

3.5 Transitive Closure for `TransitiveProperty`

The computation of the transitive closure by applying rule f-rdfp4 is essentially calculating the all-pairs shortest path on the instance graph. To see this point, we consider the following property:

Property 2. *Suppose there is a fuzzy triple* $(p, \texttt{Type}, \texttt{TransitiveProperty})[n]$ *in the fuzzy RDF graph G, and $(a, p, b)[m]$ is a fuzzy triple that can be derived from G using only rule f-rdfp4. Then there must be a chain of instances $u_0 = a, u_1, ..., u_k = b$ and a list of fuzzy degree $d_1, ..., d_k$ such that $m = d_1 \otimes n \otimes d_2 \otimes ... \otimes n \otimes d_k$, and for every $i = 1, 2, ..., k$, $(u_{i-1}, p, u_p)[d_i]$ is in the original fuzzy RDF graph. Furthermore, in one of such chains, $u_i \neq u_j$, if $i \neq j$ and $i, j \geq 1$.*

We use an iterative algorithm to calculate this transitive closure. In each iteration, we execute a MapReduce program using Algorithm 5 as the *map* function and Algorithm 6 as the *reduce* function.

We use `getTransitiveDegree(p)` function to get the maximal n such that $(p, \texttt{type}, \texttt{TransitiveProperty})[n]$ is in the graph. Since these triples are schema triples, we can load them into memory before the mappers and reducers are executed. In the *map* function, n is the number of iterations that the transitive closure calculation algorithm already executes. Since for any fuzzy triple $(a, p, b)[m]$, there is at least one chain $u_0 = a, u_1, ..., u_k = b$ according to Property 2, we use variable *length* to indicate the length of the shortest chain of $(a, p, b)[m]$. At the

---
Algorithm 6. reduce function for rule f-rdfp4

Input: key, iterator values
  left.clear();
  right.clear();
  for value $\in$ values **do**
    if value.flag == 'L' **then**
      left.update(value.subject, {value.degree, value.length});
    else
      right.update(value.object, {value.degree, value.length});
    end if
  end for
  for $i \in$ left **do**
    for $j \in$ right **do**
      newLength = i.length + j.length;
      emit(newLength, triple(i.subject, key.predicate, j.object,
          $i.degree \otimes j.degree\otimes$ getTransitiveDegree($key.predicate$)));
    end for
  end for
---

beginning of the algorithm, for every triple $(a, p, b)[n]$ in the fuzzy RDF graph, *length* is assigned to be one.

If $(a, p, b)[m]$ has a chain $u_0, u_1, ..., u_k$ with length k, and it can be derived from $(a, p, t)[m_1]$ and $(t, p, b)[m_2]$ in the n-th iteration, then we have $m_1 + m_2 = m$, $m_1 = 2^{n-2}$ or 2^{n-1}, and $2^{n-2} < m_2 \leq 2^{n-1}$. We can prove the integer equation $m_1 + m_2 = m$ has a unique solution satisfying $m_1 = 2^{n-2}$ or $m_1 = 2^{n-1}$, and $2^{n-2} < m_2 \leq 2^{n-1}$. Thus for such a chain, the triple $(a, p, b)[m]$ will be generated only once. As a consequence, a fuzzy triple will be generated at most as many times as the number of chains it has. In most circumstances, every fuzzy triple will be generated only once.

Furthermore, based on the above discussion, if a fuzzy triple $(a, p, b)[m]$ has a chain with length $2^{n-1} < l \leq 2^n$, it will be derived within n iterations. As a consequence, the algorithm will terminate within $\log N$ iterations where N is the number of all instances in the graph.

3.6   Handling SameAs Closure

The rules related to sameas are f-rdfp5(ab), 6, 7, 9, 10 and 11. Rules f-rdfp5(ab) are naive rules which can be implemented directly. The conclusion of Rule f-rdfp9 can be derived by applying rules f-rdfs10 and f-rdfp11. Rule f-rdfp10 allows replacing the predicate with its synonyms. Thus we only consider the rules f-rdfp6 and f-rdfp7, and the following variation of rule f-rdfp11, called f-rdfp11x:

$$(u, p, v)[n], (u, \mathsf{sameas}, u')[m], (v, \mathsf{sameas}, v')[l], (p, \mathsf{sameas}, p')[k]$$
$$\Rightarrow (u', p', v')[n \otimes m \otimes l \otimes k]$$

The first two rules only affect the computation of sameas closure, and the rule f-rdfp11x influences the other rules' computation.

For convenience, we call a fuzzy triple in the form of $(i, \text{sameas}, j)[n]$ a sameas triple. We further call the sameas triples with fuzzy degree 1 the *certain* sameas *triples*, and the others with fuzzy degree less than 1 the *vague* sameas *triples*. The sameas problem is caused by those vague sameas triples. Thus for certain sameas triples, the canonical representation technique is still applicable. In real applications, such as Linking Open Data project, most of the sameas triples are certain in order to link different URIs across different datasets. Thus the traditional technique will be helpful to solve the problem.

However, the fuzzy pD^* semantics allows using sameas triples to represent the similarity information. For these triples, we must store all of them and calculate the sameas closure using rules f-rdfp6 and f-rdfp7 to ensure the inference to be complete.

Materializing the result by applying the rule f-rdfp11x will greatly expand the dataset which may cause fatal efficiency problems. To accelerate the computation, we do not apply rule f-rdfp11x directly. Instead, we modify the algorithms for other rules to consider the effect of rule f-rdfp11x.

In the following, we use rule f-rdfs2 mentioned in 3.1 as an example to illustrate the modification. In rule f-rdfs2, two fuzzy triples join on p. Considering rule f-rdfp11x, if the dataset contains a fuzzy triple $(p, \text{sameas}, p')[n]$, then we can make the following inference by applying f-rdfp11x and f-rdfs2:

$$(p, \text{domain}, u)[m], (v, p', w)[k], (p, \text{sameas}, p')[n] \Rightarrow (v, \text{type}, u)[n \otimes m \otimes k]$$

We use Algorithm 7 to replace Algorithm 1 as the *map* function. The difference is that Algorithm 7 uses a loop between line 2 and line 5 instead of line 2 in Algorithm 1. In practice, vague sameas triples are relatively few. Thus we can load them into memory and compute the sameas closure before the mappers are launched. When the mapper scans a triple in the form of $(p, \text{domain}, u)[m]$, the mapper looks up the sameas closure to find the set of fuzzy triples in the form of $(p, \text{sameas}, p')[n]$. For each pair (p', n), the mapper outputs a key p' along with a value {flag='L', u=triple.object, $m \otimes n$}. While processing key p', the reducer will receive all the values of u and $m \otimes n$. Furthermore, the reducer will receive all values of v and k outputted by the mapper in line 7. Thus the reducer will generate fuzzy triples in the form of $(v, \text{type}, u)[n \otimes m \otimes k]$ as desired. Similarly we can modify the algorithms for other rules to consider the effect of rule f-rdfp11x.

Finally, we discuss the sameas problem while processing the rules f-rdfp1 and f-rdfp2, since they generate sameas triples. We only discuss the rule f-rdfp1 since the other is similar. Consider a fuzzy graph G containing the following $n + 1$ fuzzy triples:

$(a, p, b_1)[m_1]\ (a, p, b_2)[m_2] \ ... \ (a, p, b_n)[m_n]$
$(p, \text{type}, \text{FunctionalProperty})[k]$

By applying the rule f-rdfp1, we can derive $n(n-1)/2$ fuzzy triples in the form of $(b_i, \text{sameas}, b_j)[k \otimes m_i \otimes m_j]$.

Algorithm 7. map function for rules f-rdfs2 and f-rdfp11

Input: key, triple
```
1: if triple.predicate == 'domain' then
2:     for (subject, sameas, p')[n] is in the sameas closure do
3:         m = triple.degree;
4:         emit({p=p'}, {flag='L', u=triple.object, m ⊗ n});
5:     end for
6: end if
7: emit({p=triple.predicate}, {flag='R', v=triple.subject, k=triple.degree});
```

---

4   Experiment

We implemented a prototype system based on the Hadoop framework[1], which is an open-source Java implementation of MapReduce. Hadoop uses a distributed file system, called HDFS[2] to manage executions details such as data transfer, job scheduling, and error management.

Since there is no system supporting fuzzy pD^* reasoning, we run our system over the standard LUBM data, and validate it against the WebPIE reasoner which supports inference of Horst fragment to check the correctness of our algorithms. Our system can produce the same results as WebPIE. Furthermore, we build some small fuzzy pD^* ontologies, and a naive inference system for validation purpose. Our system can produce the same results on all these ontologies.

The experiment was conducted in a Hadoop cluster containing 25 nodes. Each node is a PC machine with a 4-core, 2.66GHz, Q8400 CPU, 8GB main-memory, 3TB hard disk. In the cluster, each node is assigned three processes to run *map* tasks, and three process to run *reduce* tasks. So the cluster allows running at most 75 mappers or 75 reducers simultaneously. Each mapper and each reducer can use at most 2GB main-memory.

4.1   Datasets

Since there is no real fuzzy RDF data available, we generate synthesis fuzzy ontology, called fpdLUBM[3], for experimental purpose. Our system is based on a fuzzy extension of LUBM [2], called fLUBM, which is used for testing querying ability under fuzzy DL-Lite semantics in [7]. The fLUBM dataset adds two fuzzy classes, called Busy and Famous. The fuzzy degrees of how an individual belongs to these classes are generated according to the number of courses taught or taken by the individual, and the publications of the individual respectively.

However, since there is no hierarchy among these fuzzy classes, we cannot use fLUBM to test our reasoning algorithm. To tackle this problem, we further added six fuzzy classes, VeryBusy, NormalBusy, LessBusy, VeryFamous, NormalFamous and LessFamous. Given an individual i, suppose its membership degree w.r.t.

---

[1] http://hadoop.apache.org/

[2] http://hadoop.apache.org/hdfs/

[3] Available at http://apex.sjtu.edu.cn/apex_wiki/fuzzypd

class Busy (the fuzzy degree how i belongs to class Busy) is b_i. If $b_i < 0.5$, we add a fuzzy triple $(i,$ type, LessBusy$)$ $[b_i/0.5]$ into the dataset; if $0.5 \leq b_i < 0.7$, we generated a fuzzy triple $(i,$ type, NormalBusy$)$ $[b_i/0.7]$; otherwise, we generate a fuzzy triple $(i,$ type, VeryBusy$)$ $[b]$. We added two fuzzy triples, (LessBusy, subClassOf, Busy) [0.5] and (VeryBusy, subClassOf, Busy) [1.0] to the TBox. Similarly, we can generate the fuzzy triples related to Famous.

We further added a transitive property call youngerThan to test calculation of the transitive closure. In each university ontology, we assigned a randomly generated age to each student. Then we generated n youngerThan triples. For each triple, we randomly chose two different students i and j satisfying $age_i < age_j$, and added a fuzzy triple $(i,$ youngerThan, $j)$ $[age_i/age_j]$ into the data set.

Finally, we added a TBox triple to assert that emailAddress is an inverse functional property. In fact, e-mail is usually used for identifying a person online. Furthermore, for each faculty f, since we know the university from which he got his bachelor degree, we picked one email address e belonging to an undergraduate student in that university, and added a triple $(f,$ emailAddress, $e)$ $[d]$ into the data set. Here we assigned the fuzzy degrees d to be either 1.0 or 0.9. Then sameas triples were derived using the semantics of inverseFunctionalProperty. We set the probability when $d = 0.9$ to be 1%, so that a small set of vague sameas triples can be generated. Similarly, we can generate other emailAddress triples according to the master and doctoral information similarly.

4.2   Experimental Results

Comparison with WebPIE. We compared the performance of our system with that of the baseline system WebPIE[4]. We run both systems over the same dataset fpdLUBM8000. The results are shown in Table 2. Notice that the dataset is a fuzzy dataset, for WebPIE, we simply omit the fuzzy degree, and submit all crisp triples to the system. So our system (FuzzyPD) output a little more triples than WebPIE, because our system also updates the fuzzy degrees. The running time difference between our system and WebPIE is from -5 to 20 minutes. However, since a Hadoop job's execution time is affected by the statuses of the machines in the cluster, several minutes' difference between the two systems is within a rational range. Thus we conclude that our system is comparable with the state-of-the-art inference system.

Scalability. To test the scalability of our algorithms, we run two experiments. In the first experiment, we tested the inference time of our system over datasets with different sizes to see the relation between the data volume and the throughput. In the second experiment, we run our system over fpdLUBM1000 dataset with different number of units (mappers and reducers) to see the relation between the processing units and the throughput. Furthermore, in the second experiment, we set the number of mappers to be the same as the number of reducers. Thus a total number of 128 units means launching 64 mappers and 64 reducers.

---

[4] We fix some bugs in the source code which will cause performance problem.

Table 2. Experimental results of our system and WebPIE

Number of Universities	Time of FuzzyPD	Time of WebPIE
1000	38.8	41.32
2000	66.97	74.57
4000	110.40	130.87
8000	215.48	210.01

Table 3. Scalability over number of mappers

Number of units	Time (minutes)	Speedup
128	38.80	4.01
64	53.15	2.93
32	91.58	1.70
16	155.47	1.00

Table 4. Scalability over data volume

Number of universities	Input (MTriples)	Output (MTriples)	Time (minutes)	Throughput (KTriples/second)
1000	155.51	92.01	38.8	39.52
2000	310.71	185.97	66.97	46.28
4000	621.46	380.06	110.40	57.37
8000	1243.20	792.54	215.50	61.29

The results for the first experiment can be found in table 4. From the table, we can see that the throughput increases significantly while the volume increases. The throughput while processing fpdLUBM8000 dataset is 50% higher than the throughput while processing dataset containing 1000 universities. We attribute this performance gain to the platform startup overhead which is amortized over a larger processing time for large datasets. The platform overhead is also responsible for the non-linear speedup in Table 3 which contains the results of the second test. Figure 1 gives a direct illustration of the overhead effect. In Figure 1, if we subtract a constant from the time dimension of each data point, then the time is inversely proportional to the number of units. Since the running time should be inversely proportional to the speed, after eliminating the effect of the platform overhead, the system's performance speeds up linearly to the increase of number of units.

5   Related Work

[10] is the first work to extend RDFS with fuzzy vagueness. In [4], we further propose the fuzzy pD^* semantics which allows some useful OWL vocabularies, such as TransitiveProperty and SameAs. In [11] and [6], a more general framework called annotated RDF to represent annotation for RDF data and a query language called AnQL were proposed.

As far as we know, this is the first work on applying the MapReduce framework to tackle large scale reasoning in fuzzy OWL. Pan et al. in [7] propose a framework of fuzzy query languages for fuzzy ontologies. However, they mainly concerns the query answering algorithms for these query languages over fuzzy DL-Lite ontologies. Our work concerns the inference problem over large scale fuzzy pD^* ontologies.

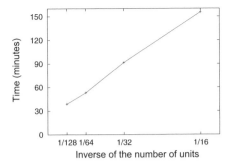

Fig. 1. Time versus inverse of number of mappers

We briefly discuss some related work on scalable reasoning in OWL and RDF. None of them takes into account of fuzzy information.

Schlicht and Stuckenschmidt [8] show peer-to-peer reasoning for the expressive ALC logic but focusing on distribution rather than performance. Soma and Prasanna [9] present a technique for parallel OWL inferencing through data partitioning. Experimental results show good speedup but only on very small datasets (1M triples) and runtime is not reported.

In Weaver and Hendler [14], straightforward parallel RDFS reasoning on a cluster is presented. But this approach splits the input to independent partitions. Thus this approach is only applicable for simple logics, e.g. RDFS without extending the RDFS schema, where the input is independent.

Newman et al. [5] decompose and merge RDF molecules using MapReduce and Hadoop. They perform SPARQL queries on the data but performance is reported over a dataset of limited size (70,000 triples).

Urbani et al. [13] develop the MapReduce algorithms for materializing RDFS inference results. In [12], they further extend their methods to handle OWL pD^* fragment, and conduct experiment over a dataset containing 100 billion triples.

6   Conclusion

In this paper, we proposed MapReduce algorithms to process forward inference over large scale data using fuzzy pD^* semantics (i.e. an extension of OWL Horst semantics with fuzzy vagueness). We first identified the major challenges to handle the fuzzy information using the MapReduce framework, and proposed a solution for tackling each of them. Furthermore, we implemented a prototype system for the evaluation purpose. The experimental results show that the running time of our system is comparable with that of WebPIE, the state-of-the-art inference engine for large scale OWL ontologies in pD^* fragment. As a future work, we will apply our system to some applications, such as Genomics and multimedia data management.

420   C. Liu et al.

Acknowledgement. Guilin Qi is partially supported by NSFC (61003157), Jiangsu Science Foundation (BK2010412), and the Key Laboratory of Computer Network and Information Integration (Southeast University).

References

1. Dean, J., Ghemawat, S.: MapReduce: Simplified Data Processing on Large Clusters. In: Proc. of OSDI 2004, pp. 137–147 (2004)
2. Guo, Y., Pan, Z., Heflin, J.: Lubm: A benchmark for owl knowledge base systems. Journal of Web Semantics 3(2), 158–182 (2005)
3. Horst, H.J.: Completeness, decidability and complexity of entailment for rdf schema and a semantic extension involving the owl vocabulary. Journal of Web Semantics 3(2-3), 79–115 (2005)
4. Liu, C., Qi, G., Wang, H., Yu, Y.: Fuzzy Reasoning over RDF Data using OWL Vocabulary. In: Proc. of WI 2011 (2011)
5. Newman, A., Li, Y.-F., Hunter, J.: Scalable semantics - the silver lining of cloud computing. In: Proc. of ESCIENCE 2008 (2008)
6. Lopes, N., Polleres, A., Straccia, U., Zimmermann, A.: AnQL: SPARQLing up Annotated RDFS. In: Patel-Schneider, P.F., Pan, Y., Hitzler, P., Mika, P., Zhang, L., Pan, J.Z., Horrocks, I., Glimm, B. (eds.) ISWC 2010, Part I. LNCS, vol. 6496, pp. 518–533. Springer, Heidelberg (2010)
7. Pan, J.Z., Stamou, G., Stoilos, G., Taylor, S., Thomas, E.: Scalable querying services over fuzzy ontologies. In: Proc. of WWW 2008, pp. 575–584 (2008)
8. Schlicht, A., Stuckenschmidt, H.: Peer-to-peer reasoning for interlinked ontologies, vol. 4, pp. 27–58 (2010)
9. Soma, R., Prasanna, V.: Parallel inferencing for owl knowledge bases. In: Proc. of ICPP 2008, pp. 75–82 (2008)
10. Straccia, U.: A minimal deductive system for general fuzzy RDF. In: Proc. of RR 2009, pp. 166–181 (2009)
11. Straccia, U., Lopes, N., Lukacsy, G., Polleres, A.: A general framework for representing and reasoning with annotated semantic web data. In: Proc. of AAAI 2010, pp. 1437–1442. AAAI Press (2010)
12. Urbani, J., Kotoulas, S., Maassen, J., van Harmelen, F., Bal, H.: OWL Reasoning with WebPIE: Calculating the Closure of 100 Billion Triples. In: Aroyo, L., Antoniou, G., Hyvönen, E., ten Teije, A., Stuckenschmidt, H., Cabral, L., Tudorache, T. (eds.) ESWC 2010. LNCS, vol. 6088, pp. 213–227. Springer, Heidelberg (2010)
13. Urbani, J., Kotoulas, S., Oren, E., van Harmelen, F.: Scalable Distributed Reasoning Using MapReduce. In: Bernstein, A., Karger, D.R., Heath, T., Feigenbaum, L., Maynard, D., Motta, E., Thirunarayan, K. (eds.) ISWC 2009. LNCS, vol. 5823, pp. 634–649. Springer, Heidelberg (2009)
14. Weaver, J., Hendler, J.A.: Parallel materialization of the finite RDFS closure for hundreds of millions of triples. In: Bernstein, A., Karger, D.R., Heath, T., Feigenbaum, L., Maynard, D., Motta, E., Thirunarayan, K. (eds.) ISWC 2009. LNCS, vol. 5823, pp. 682–697. Springer, Heidelberg (2009)

On Blank Nodes*

Alejandro Mallea[1], Marcelo Arenas[1], Aidan Hogan[2], and Axel Polleres[2,3]

[1] Department of Computer Science, Pontificia Universidad Católica de Chile, Chile
{aemallea,marenas}@ing.puc.cl
[2] Digital Enterprise Research Institute, National University of Ireland Galway, Ireland
{aidan.hogan,axel.polleres}@deri.org
[3] Siemens AG Österreich, Siemensstrasse 90, 1210 Vienna, Austria

Abstract. Blank nodes are defined in RDF as 'existential variables' in the same way that has been used before in mathematical logic. However, evidence suggests that actual usage of RDF does not follow this definition. In this paper we thoroughly cover the issue of blank nodes, from incomplete information in database theory, over different treatments of blank nodes across the W3C stack of RDF-related standards, to empirical analysis of RDF data publicly available on the Web. We then summarize alternative approaches to the problem, weighing up advantages and disadvantages, also discussing proposals for Skolemization.

1 Introduction

The Resource Description Framework (RDF) is a W3C standard for representing information on the Web using a common data model [18]. Although adoption of RDF is growing (quite) fast [4, § 3], one of its core features—blank nodes—has been sometimes misunderstood, sometimes misinterpreted, and sometimes ignored by implementers, other standards, and the general Semantic Web community. This lack of consistency between the standard and its actual uses calls for attention.

The standard semantics for blank nodes interprets them as existential variables, denoting the existence of some unnamed resource. These semantics make even *simple entailment checking* intractable. RDF and RDFS entailment are based on simple entailment, and are also intractable due to blank nodes [14].

However, in the documentation for the RDF standard (*e.g.*, RDF/XML [3], RDF Primer [19]), the existentiality of blank nodes is not directly treated; ambiguous phrasing such as "blank node identifiers" is used, and examples for blank nodes focus on representing resources which do not have a natural URI. Furthermore, the standards built upon RDF sometimes have different treatment and requirements for blank nodes. As we will see, standards and tools are often, to varying degrees, ambivalent to the existential semantics of blank nodes, where, *e.g.*, the standard query language SPARQL can return different results for two graphs considered equivalent by the RDF semantics [1].

Being part of the RDF specification, blank nodes are a core aspect of Semantic Web technology: they are featured in several W3C standards, a wide range of tools, and

---

* The work presented in this report has been funded in part by Science Foundation Ireland under Grant No. SFI/08/CE/I1380 (Líon-2), by an IRCSET postgraduate grant, by Marie Curie action IRSES under Grant No. 24761 (Net2), and by FONDECYT grant No. 1090565.

L. Aroyo et al. (Eds.): ISWC 2011, Part I, LNCS 7031, pp. 421–437, 2011.

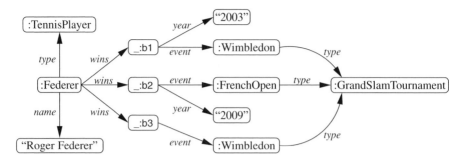

Fig. 1. An RDF graph for our running example. In this graph, URIs are preceded by ':' blank nodes by '_:' and literals are enclosed in quotation marks.

hundreds of datasets across the Web, but not always with the same meaning. Dealing with the issue of blank nodes is thus not only important, but also inherently complex and potentially costly: before weighing up alternatives for blank nodes, their interpretation and adoption—across legacy standards, tools, and published data—must be considered.

In this paper, we first look at blank nodes from a background theoretical perspective, additionally introducing *Skolemization* (§ 3). We then look at how blank nodes are used in the Semantic Web standards: we discuss how they are supported, what features rely on them, issues surrounding blank nodes, and remarks on adoption (§ 4). Next, we look at blank nodes in publishing, their prevalence for Linked Data, and what blank node graph-structures exist in the wild (§ 5). Finally, in light of the needs of the various stakeholders already introduced, we discuss proposals for handling blank nodes (§ 6).

Throughout this document, we use the the RDF graph given in Figure 1 to illustrate our discussion. The graph states that the tennis player :Federer won an event at the :FrenchOpen in 2009; it also states twice that he won :Wimbledon, once in 2003.

2   Preliminaries

We follow the abstract representation [13,20] of the formal RDF model [14,18]. This means we will consider standard notation for the sets of all URIs (**U**), all literals (**L**) and all blank nodes (**B**), all being pairwise disjoint. For convenience of notation, we write **UBL** for the union of **U**, **B** and **L**, and likewise for other combinations. In general, we write (s, p, o) for an RDF triple, and assume that $(s, p, o) \in \mathbf{UB} \times \mathbf{U} \times \mathbf{UBL}$.

For the purposes of our study, we define an *interpretation* of a graph as per [20], but without considering literals or classes since they are irrelevant for the study of blank nodes. Also, we do not consider the use of vocabularies with predefined semantics (*e.g.*, RDFS or OWL). Graphs that do not use such vocabularies are called *simple*. More precisely, define a *vocabulary* as a subset of **UL**. Given an RDF graph G, denote by terms(G) the set of elements of **UBL** that appear in G, and denote by voc(G) the set terms(G) ∩ **UL**. Then an *interpretation* $\mathcal{I}$ over a vocabulary V is a tuple $\mathcal{I} = (Res, Prop, Ext, Int)$ such that: (1) *Res* is a non-empty set of *resources*, called the *domain* or *universe* of $\mathcal{I}$; (2) *Prop* is a set of property names (not necessarily disjoint from or a subset of *Res*); (3) *Ext* : *Prop* → $2^{Res \times Res}$, a mapping that assigns an *extension* to each property name; and (4) *Int* : $V \rightarrow$ *Res* ∪ *Prop*, the *interpretation mapping*, a mapping that assigns a resource or a property name to each element of V, and such that *Int* is the identity for literals. Given an interpretation mapping *Int* and a function $A : \mathbf{B} \rightarrow$ *Res*, we define the *extension* function $Int_A : V \cup \mathbf{B} \rightarrow$ *Res* ∪ *Prop* as the extension of *Int* by A, that is, $Int_A(x) = Int(x)$ if $x \in V$, and $Int_A(x) = A(x)$ if $x \in \mathbf{B}$. Based on interpretations, we have the fundamental notion of *model*:

Definition 1. *An interpretation* $\mathcal{I} = (Res, Prop, Ext, Int)$ *is a* model *of* G *if* $\mathcal{I}$ *is an interpretation over* voc(G) *and there exists a function* $A : \mathbf{B} \rightarrow$ *Res such that for each* $(s, p, o) \in G$, *it holds that* $Int(p) \in Prop$ *and* $(Int_A(s), Int_A(o)) \in Ext(Int(p))$.

The existentiality of blank nodes is given by the extension A above. We say that a graph is *satisfiable* if it has a model. Simple RDF graphs are trivially satisfiable thanks to Herbrand interpretations [14], in which URIs and literals are interpreted as their (unique) syntactic forms instead of "real world" resources. This will be important in distinguishing Skolemization in first-order logic from Skolemization in RDF.

Recall that in the context of RDF a subgraph is a subset of a graph. Then a graph G is *lean* if there is no map $h : \mathbf{UBL} \rightarrow \mathbf{UBL}$ that preserves URIs and literals ($h(u) = u$ for all $u \in \mathbf{UL}$) such that the RDF graph obtained from G by replacing every element u mentioned in G by $h(u)$, denoted by $h(G)$, is a proper subgraph of G; otherwise, the graph is *non-lean* and (formally speaking) contains redundancy. For instance, the graph G in Figure 1 is *non-lean* as the triple (_:b3, *event*, :Wimbledon) is redundant: if h maps _:b3 to _:b1 and is the identity elsewhere, then $h(G)$ is a proper subgraph of G.

3   Theoretic Perspective

In this section we look at theoretic aspects of blank nodes, focusing on background theory with respect to existentials (§ 3.1) and Skolemization (§ 3.2) in first-order logic.

3.1   Existential Variables

As per Section 2, the existentiality of blank nodes is given by the extension function A for an interpretation mapping *Int*. We now show that this interpretation of blank nodes can be precisely characterized in terms of existential variables in first-order logic.

Let G be an RDF graph. We define $Th(G)$ to be a first-order sentence with a ternary predicate triple as follows. Let $\mathbf{V}$ be an infinite set of variables disjoint with $\mathbf{U}$, $\mathbf{L}$ and $\mathbf{B}$, and assume that $\rho : \mathbf{ULB} \rightarrow \mathbf{ULV}$ is a one-to-one function that is the identity on $\mathbf{UL}$. Now, for every triple $t = (s, p, o)$ in G, define $\rho(t)$ as the fact triple$(\rho(s), \rho(p), \rho(o))$, and define $Th(G)$ as $\exists x_1 \cdots \exists x_n (\bigwedge_{t \in G} \rho(t))$, where $x_1, \ldots, x_n$ are the variables from $\mathbf{V}$ mentioned in $\bigwedge_{t \in G} \rho(t)$. Then we have the following equivalence between the notion of entailment for RDF graphs and the notion of logical consequence for first-order logic.

Theorem 1 ([9]). *Given RDF graphs* G *and* H, *it holds that* $G \models H$ *if and only if* $Th(G) \models Th(H)$. □

This theorem reveals that the implication problem for RDF can be reduced to implication for existential first-order formulae without negation and disjunction.

RDF implication in the presence of existentials is NP-Complete [14,26]. However, Pichler *et al.* [22] note that for common blank node morphologies, entailment becomes tractable. Let G be a simple RDF graph, and consider the (non-RDF) graph $\mathsf{blank}(G) = (V, E)$ as follows: $V = \mathbf{B} \cap \mathsf{terms}(G)$ and $E = \{(b, c) \in V \times V \mid b \neq c$ and there exists $P \in \mathsf{terms}(G)$ such that $(b, P, c) \in G$ or $(c, P, b) \in G\}$. Thus, $\mathsf{blank}(G)$ gives an undirected graph connecting blank nodes appearing in the same triple in G. Let G and H denote two RDF graphs with m and n triples respectively. Now, performing the entailment check $G \models H$ has the upper bound $O(n^2 + mn^{2k})$, where $k = \mathsf{tw}(\mathsf{blank}(H)) + 1$ for $\mathsf{tw}(\mathsf{blank}(H))$ the *treewidth* of $\mathsf{blank}(H)$ [22]. Note that we will survey the treewidth of blank node structures in published data in Section 5.1.2, which gives insights into the expense of RDF entailment checks in practice.

3.2 Skolemization

In first-order logic (FO), Skolemization is a way of removing existential quantifiers from a formula in prenex normal form (a chain of quantifiers followed by a quantifier-free formula). The process was originally defined and used by Thoralf Skolem to generalize a theorem by Jacques Herbrand about models of universal theories [5].

The central idea of Skolemization is to replace existentially quantified variables for "fresh" constants that are not used in the original formula. For example, $\exists x \forall y \, R(x, y)$ can be replaced by $\forall y \, R(c, y)$ where c is a fresh constant, as this new formula also states that there exists a value for the variable x (in fact, $x = c$) such that $R(x, y)$ holds for every possible value of variable y. Similarly, $\forall x \exists y \, (P(x) \rightarrow Q(y))$ can be replaced by $\forall x (P(x) \rightarrow Q(f(x)))$, where the unary function f does not belong to the underlying vocabulary, as in this case we know that for every possible value of variable x, there exists a value of variable y that depends on x and such that $P(x) \rightarrow Q(y)$ holds. When the original formula does not have universal quantifiers, only constants are needed in the Skolemization process; since only existential quantifiers are found in simple RDF graphs (see Definition 1), we need only talk about Skolem constants. However, if Skolemization was used to study the satisfiability of logical formulae in more expressive languages (*e.g.*, OWL), Skolem functions would be needed.

The most important property of Skolemization in first-order logic is that it preserves satisfiability of the formula being Skolemized. In other words, if ψ is a Skolemization of a formula φ, then φ and ψ are equisatisfiable, meaning that φ is satisfiable (in the original vocabulary) if and only if ψ is satisfiable (in the extended vocabulary, with the new Skolem functions and constants). Nevertheless, this property is of little value when Skolemizing RDF graphs since we recall that all simple RDF graphs are satisfiable.

4 Blank Nodes in the Standards

We now look at the treatment of blank nodes in the RDF-related standards, *viz.* RDF syntaxes, RDFS, OWL, RIF and SPARQL; we also cover RDB2RDF and SPARQL 1.1.

4.1 RDF Syntaxes

We first give general discussion on the role of blank nodes in RDF syntaxes.

Support for blank nodes. All RDF syntaxes allow blank nodes to be explicitly labeled; in N-Triples, explicit labeling is necessary. Explicit labels allow blank nodes to be referenced outside of nested elements and thus to be used in arbitrary graph-based data even though the underlying syntaxes (*e.g.*, XML) are inherently tree-based. Note that we will study cyclic blank node structures in published data later in Section 5.1.2.

Features requiring blank nodes. Blank nodes play two major (and related) roles in all but the N-Triples syntax. First, aside from N-Triples, all syntaxes passively default to representing resources as blank nodes when optional URIs are omitted. Second, blank nodes are also used in shortcuts for n-ary predicates and RDF lists (aka. containers) [14, § 3.3.1] in Turtle and RDF/XML, as well as containers [14, § 3.3.2] and reification [14, § 3.3.1] in RDF/XML. For example, the Turtle shortcut:

   :GrandSlam :order (:AustralianOpen :FrenchOpen :Wimbledon :USOpen) .

represents an RDF list. This would be equivalently representable in Turtle's square-bracket syntax (*left*), and full verbose forms (*right*) as follows:

```
:GrandSlam :order                          :GrandSlam :order _:b1 .
[ rdf:first :AustralianOpen ; rdf:rest     _:b1 rdf:first :AustralianOpen .   _:b1 rdf:rest _:b2 .
[ rdf:first :FrenchOpen ; rdf:rest         _:b2 rdf:first :FrenchOpen .       _:b2 rdf:rest _:b3 .
[ rdf:first :Wimbledon ; rdf:rest          _:b3 rdf:first :Wimbledon .        _:b3 rdf:rest _:b4 .
[ rdf:first :USOpen ; rdf:rest rdf:nil ]]]] _:b4 rdf:first :USOpen .          _:b4 rdf:rest rdf:nil .
```

In the two shortcut notations, the labels of the "structural" blank nodes are left implicit. Similar shortcuts using unlabeled blank nodes hold for n-ary predicates, reification and containers in RDF/XML. Note that such shortcuts can only induce "trees" of blank nodes; *e.g.*, _:b1 :p _:b2 . _:b2 :p _:b1 . cannot be expressed without explicit labels.

Issues with blank nodes. Given a fixed, serialized RDF graph (*i.e.*, a document), labeling of blank nodes can vary across parsers and across time. Checking if two representations originate from the same data thus often requires an isomorphism check, for which in general, no polynomial algorithms are known (*cf. e.g.* [6] in the RDF context). Further, consider a use-case tracking the changes of a document over time; given that parsers can assign arbitrary labels to blank nodes, a simple syntactic change to the document may cause a dramatic change in blank node labels, making precise change detection difficult (other than on a purely syntactic level). We note that isomorphism checking is polynomial for "blank node trees" (*e.g.*, as generated for documents without explicit blank node labels) [17].

In practice. Parsers typically feature a systematic means of labeling blank nodes based on the explicit blank node labels and the order of appearance of implicit blank nodes. The popular Jena Framework[1] offers methods for checking the RDF-equivalence of two graphs. We will further discuss blank nodes in publishing in Section 5.

4.2   RDF Schema (RDFS)

RDF Schema (RDFS) is a lightweight language for describing RDF vocabularies, which allows for defining sub-class, sub-property, domain and range relations between class

---

[1] http://jena.sourceforge.net/

and property terms (as appropriate). The RDFS vocabulary—incl., *e.g.*, rdfs:domain, rdfs:range, rdfs:subClassOf and rdfs:subPropertyOf—is well-defined by means of a (normative) model-theoretic semantics, accompanied by a (non-normative) set of entailment rules to support inferencing [14,20].

Support for blank nodes. RDFS entailment is built on top of simple entailment, and thus supports an existential semantics for blank nodes.

Features requiring blank nodes. Blank nodes are used as "surrogates" for literals through entailment rules LG/GL in [14]. This is necessary for completeness of the entailment rules w.r.t. the formal semantics such that literal terms can "travel" to the subject position of triples by means of their surrogates [14]; for instance, the triples

:Federer atp:name "Roger Federer" . atp:name rdfs:range atp:PlayerName .

entail the triple _:b$_{Federer}$ rdf:type atp:PlayerName by the RDFS rules LG and RDFS2 [14]. Here, the "surrogate" blank node _:b$_{Federer}$ represents the actual literal, whereas the direct application of rule RDFS2 (without LG) would result in a non-valid RDF triple having a literal in the subject position; *viz.* "Roger Federer" rdf:type atp:PlayerName .

Issues with blank nodes. Existential semantics for blank nodes makes RDFS entailment NP-Complete [14,13,26,20]. Further, ter Horst [26] showed that the RDFS entailment lemma in the non-normative section of the RDF Semantics is incorrect: blank node surrogates are still not enough for the completeness of rules in [14, § 7], where blank nodes would further need to be allowed in the predicate position. For example, consider the three triples (1) :Federer :wins _:b1, (2) :wins rdfs:subPropertyOf _:p, and (3) _:p rdfs:domain :Competitor, we still cannot infer the triple :Federer rdf:type :Competitor, since the required intermediate triple :Federer _:p _:b1 is not valid in RDF.

In practice. To overcome the NP-completeness of simple entailment, RDFS rule-based reasoners often apply Herbrand interpretations over blank nodes such that they denote their own syntactic form: this "ground RDFS entailment" is equisatisfiable and tractable [26]. Avoiding the need for (or supplementing) literal surrogates, reasoners often allow various forms of "generalized triples" in intermediate inferencing steps, with relaxed restrictions on where blank nodes and literals can appear [26,11].

4.3 Web Ontology Language (OWL)

The Web Ontology Language (OWL) is a more expressive language than RDFS (but which partially re-uses RDFS vocabulary). With the advent of OWL 2, there are now eight standard profiles of OWL. OWL Full and OWL 2 Full are given an RDF-Based Semantics [24] and so are compatible with arbitrary RDF graphs, but are undecidable [11]. OWL Lite, OWL DL, OWL 2 EL, OWL 2 QL and OWL 2 RL are given a Direct Semantics based on Description Logics (DL) formalisms and are decidable [11], but are not compatible with arbitrary RDF graphs. Further, OWL 2 RL features the OWL 2 RL/RDF entailment ruleset: a partial axiomatization of the RDF-Based Semantics.

Support for blank nodes. The RDF-Based Semantics is built on top of simple entailment [24], and considers blank nodes as existentials. The OWL Direct Semantics does not treat blank nodes in assertions, where special DL-based existentials are instead supported; *e.g.*, the implicit assertion that :Federer won "something" can be expressed in the DL axiom Federer $\in \exists$wins.$\top$ (*i.e.*, on a level above blank nodes).

Features requiring blank nodes. For the DL-based (sub)languages of OWL, blank nodes are used to map between DL-based structural syntaxes and RDF representations: the structural syntaxes feature special n-ary predicates represented with blank nodes in RDF. For example, the DL concept $\exists$wins.$\top$ is expressed structurally as ObjectSomeValuesFrom(OPE(:wins) CE(owl:Thing)), which maps to the following three RDF triples:

   _:x a owl:Restriction . _:x owl:someValuesFrom owl:Thing . _:x owl:onProperty :wins .

Blank nodes are also required to represent RDF lists used in the mapping, *e.g.*, of OWL union classes, intersection classes, enumerations, property chains, complex keys, *etc*. An important aspect here is the locality of blank nodes: if the above RDF representation is valid in a given graph, it is still valid in an Open World since, *e.g.*, an external document cannot add another value for owl:onProperty to _:x.

Issues with blank nodes. Once RDF representations are parsed, DL-based tools are agnostic to blank nodes; existentials are handled on a higher level. RDF-based tools encounter similar issues as for RDFS; *e.g.*, OWL 2 RL/RDF supports generalized triples [11] (but otherwise gives no special treatment to blank nodes).

In practice. Rule-based reasoners—supporting various partial-axiomatizations of the RDF-Based Semantics such as DLP [12], pD* [26] or OWL 2 RL/RDF [11]—again often apply Herbrand interpretations over blank nodes. ter Horst proposed pD*sv [26] which contains an entailment rule with an existential blank node in the head to support owl:someValuesFrom, but we know of no system supporting this rule.

4.4   SPARQL Protocol and RDF Query Language (SPARQL)

SPARQL [23] is the standard query language for RDF, and includes an expressive set of query features. An important aspect of SPARQL is the notion of *Named Graphs*: SPARQL querying is defined over a *dataset* given as $\{G_0, (u_1, G_1), \ldots, (u_n, G_n)\}$ such that $u_1, \ldots, u_n \in \mathbf{U}$, and $G_0, \ldots, G_n$ are RDF graphs; each pair (u_i, G_i) is called a *named graph* and G_0 is called the *default graph*.

Support for blank nodes. With respect to querying over *blank nodes in the dataset*, SPARQL considers blank nodes as constants scoped to the graph they appear in [23, § 12.3.2]. Thus, for example, the query:

   SELECT DISTINCT ?X WHERE { :Federer :wins ?X . ?X :event :Wimbledon . }

issued over the graph depicted in Figure 1 would return $\{\{(?X, \_:b1)\}, \{(?X, \_:b3)\}\}$ as distinct solution mappings, here effectively considering blank nodes as constants. Interestingly, SPARQL 1.1—currently a Working Draft—introduces a COUNT aggregate, which, with an analogue of the above query, would answer that :Federer won an event at :Wimbledon *twice*. Posing the same COUNT query over a lean (and thus RDF equivalent [14]) version of Figure 1 would return *once*.

Features requiring blank nodes. SPARQL uses *blank nodes in the query* to represent *non-distinguished variables*, *i.e.*, variables which can be arbitrarily bound, but which cannot be returned in a solution mapping. For example:

SELECT ?p WHERE { ?p :wins _:t . _:t :event :Wimbledon . _:t :year _:y . }

requests players who have won *some* Wimbledon event in *some* year (*viz.*, :Federer).[2]
We note that such queries can be expressed by replacing blank nodes with fresh query
variables. A special case holds for CONSTRUCT templates which generate RDF from
solution mappings: a blank node appearing in a query's CONSTRUCT clause is replaced
by a fresh blank node for each solution mapping in the resulting RDF.

Issues with blank nodes. A practical problem posed by blank nodes refers to "round-
tripping", where a blank node returned in a solution mapping cannot be referenced in a
further query. Consider receiving the result binding (?X, _:b1) for the previous DISTINCT
query; one cannot ask a subsequent query for what year the tournament _:b1 took place
since the _:b1 in the solution mapping no longer has any relation to that in the originating
graph; thus, the labels need not correspond to the original data. Further, SPARQL's
handling of blank nodes can cause different behavior for RDF-equivalent graphs, where,
e.g., leaning a graph will affect results for COUNT.

In practice. Where round-tripping is important, SPARQL engines often offer a special
syntax for effectively Skolemizing blank nodes. For example, ARQ[3] is a commonly
(re)used SPARQL processor which supports a non-standard <_:b1> style syntax for
terms in queries, indicating that the term can only be bound by a blank node labeled
"b1" in the data. Virtuoso[4] [8] supports the <nodeID://b1> syntax with similar purpose,
but where blank nodes are only externalized in this syntax and (interestingly) where
isBlank(<nodeID://b1>) evaluates as true.

4.5 RDB2RDF

In the last 10 years, we have witnessed an increasing interest in publishing relational
data as RDF. This has resulted in the creation of the RDB2RDF W3C Working Group,
whose goal is to standardize a language for mapping relational data into RDF [7,2].
Next we show the current proposal of the Working Group about the use of blank nodes
in the mapping language.

Support for blank nodes. The input of the mapping language being developed by
the RDB2RDF Working Group is a relational database, including the schema of the
relations being translated and the set of keys and foreign keys defined over them. The
output of this language is an RDF graph that may contain blank nodes.

Features requiring blank nodes. The RDF graph generated in the translation process
identifies each tuple in the source relational database by means of a URI. If the tuple
contains a primary key, then this URI is based on the value of the primary key. If the
tuple does not contain such a constraint, then a blank node is used to identify it in the
generated RDF graph [2].

Issues with blank nodes. In the mapping process, blank nodes are used as identifiers
of tuples without primary keys [2], and as such, two of these blank nodes should not be

---

[2] Further note that blank nodes are scoped to Basic Graph Patterns (BGPs) of queries.
[3] http://jena.sourceforge.net/ARQ/
[4] http://virtuoso.openlinksw.com/

considered as having the same value. Thus, the existential semantics of blank nodes in RDF is not appropriate for this use.

5 Blank Nodes in Publishing

In this section, we survey the use of blank nodes in RDF data published on the Web. The recent growth in RDF Web data is thanks largely to the pragmatic influence of the Linked Data community [4,15], whose guidelines are unequivocal on the subject of blank node usage; in a recent book, Bizer *et al.* [15] only mention blank nodes in the section entitled "RDF Features Best Avoided in the Linked Data Context", as follows:

> *"The scope of blank nodes is limited to the document in which they appear,* [...] *reducing the potential for interlinking between different Linked Data sources.* [...] *it becomes much more difficult to merge data from different sources when blank nodes are used,* [...] *Therefore, all resources in a data set should be named using URI references."*   —[15, § 2.4.1]

With this (recent) guideline discouraging blank nodes in mind, we now provide an empirical study of the prevalence of blank nodes in published data (§ 5.1.1), and of the morphology of blank nodes in such data (§ 5.1.2). Finally, we briefly discuss the results of a poll conducted on public Semantic Web mailing lists (§ 5.2).

5.1 Empirical Survey of Blank Nodes in Linked Data

With the previous guidelines in mind, we now present an empirical survey of the prevalence and nature of blank nodes in Linked Data published on the Web. Our survey is conducted over a corpus of 1.118 g quadruples (965 m unique triples) extracted from 3.985 m RDF/XML documents through an open-domain crawl conducted in May 2010. The corpus consists of data from 783 different pay-level domains, which are direct subdomains of either top-level domains (such as dbpedia.org), or country code second-level domains (such as bbc.co.uk). We performed a domain-balanced crawl: we assign a queue to each domain, and in each round, poll a URI to crawl from each queue in a round-robin fashion. This strategy led to 53.2% of our raw data coming from the hi5.com FOAF exporter, which publishes documents with an average of 2,327 triples per document: an order of magnitude greater than the 140 triple/doc average from all other domains [16]. Note that this corpus represents a domain-agnostic *sample* of RDF published on the Web. The bias of sampling given by the dominance of hi5.com is important to note; thus, along with measures from the monolithic dataset, we also present per-domain statistics. Details of the crawl and the corpus are available (in significant depth) in [16, § 4].

5.1.1 Prevalence of Blank Nodes in Published Data
We looked at terms in the data-level position of triples in our corpus: *i.e.*, positions other than the predicate or object of rdf:type triples which are occupied by property and class terms respectively. We found 286.3 m unique terms in such positions, of which

Table 1. Top publishers of blank nodes in our corpus

# domain	bnodes	%bnodes	LOD?
1 hi5.com	148,409,536	87.5	✗
2 livejournal.com	8,892,569	58.0	✗
3 ontologycentral.com	2,882,803	86.0	✗
4 opiumfield.com	1,979,915	17.4	✗
5 freebase.com	1,109,485	15.6	✓
6 vox.com	843,503	58.0	✗
7 rdfabout.com	464,797	41.7	✓
8 opencalais.com	160,441	44.9	✓
9 soton.ac.uk	117,390	19.1	✓
10 bbc.co.uk	101,899	7.4	✓

165.4 m (57.8%) were blank nodes, 92.1 m (32.2%) were URIs, and 28.9 m (10%) were literals. Each blank node had on average 5.233 data-level occurrences (the analogous figure for URIs was 9.41 data-level occurrences: 1.8× that for blank nodes). Each blank node occurred, on average, 0.995 times in the object position of a non-rdf:type triple, with 3.1 m blank nodes (1.9% of all blank nodes) not occurring in the object position; conversely, each blank node occurred on average 4.239 times in the subject position of a triple, with 69 k (0.04%) not occurring in the subject position.[5] Thus, we surmise that (i) blank nodes are prevalent on the Web; (ii) most blank nodes appear in both the subject and object position, but occur most prevalently in the former, possibly due to the tree-based RDF/XML syntax.

Again, much of our corpus consists of data crawled from high-volume exporters of FOAF profiles—however, such datasets are often not considered as Linked Data, where, *e.g.*, they are omitted from the Linked Open Data (LOD) cloud diagram due to lack of links to external domains.[6] Table 1 lists the top ten domains in terms of publishing unique blank nodes found in our corpus; "**%bnodes**" refers to the percentage of all unique data-level terms appearing in the domain's corpus which are blank nodes; "**LOD?**" indicates whether the domain features in the LOD cloud diagram. Of the 783 domains contributing to our corpus, 345 (44.1%) did not publish any blank nodes. The average percentage of unique terms which were blank nodes for each domain—*i.e.*, the average of **%bnodes** for all domains—was 7.5%, indicating that although a small number of high-volume domains publish many blank nodes, many other domains publish blank nodes more infrequently. The analogous figure including only those domains appearing in the LOD cloud diagram was 6.1%.

5.1.2 Structure of Blank Nodes in Published Data

As per Section 3.1, checking $G \models H$ has the upper bound $\mathcal{O}(n^2 + mn^{2k})$, where k is one plus the *treewidth* of the blank node structure blank(H) [22]. Intuitively, treewidth

---

[5] We note that in RDF/XML syntax—essentially a tree-based syntax—blank nodes can only ever occur once in the object position of a triple unless rdf:nodeID is used, but can occur multiple times in the subject position.

[6] See http://lod-cloud.net/

provides a measure of how close a given graph is to being a tree; for example, the treewidth of trees and forests is 1, the treewidth of cycles is 2 (a cycle is "almost" a tree: take out an edge from a cycle and you have a tree) and the treewidth of a complete graph on n vertexes is $n - 1$. A detailed treatment is out of scope, but we refer the interested reader to [22]. However, we note that all graphs whose treewidth is greater than 1 are cyclic. From this, it follows that simple entailment checking only becomes difficult when blank nodes form cycles:

> "[...] in practice, an RDF graph contains rarely blank nodes, and even less blank triples.[7] Hence, most of the real RDF graphs are acyclic or have low treewidth such as 2, and the entailment can be tested efficiently [...]."   —[22, § 4]

To cross-check this claim, we ran the following analysis over all documents (RDF graphs) in our corpus. For each document G, we extracted blank(G) and separated out the connected components thereof using a UNION-FIND algorithm [25]. We found 918 k documents (23% of all documents) containing blank nodes, with 376 k (9% of all documents) containing "non-reflexive" blank triples. We found a total of 527 k non-singleton connected components, an average of 1.4 components per document with some blank triple. We then checked the treewidth of all 527 k components using the QUICKBB algorithm [10], where the distribution of values is given in Table 2. Notably, 98.4% of the components are acyclical, but a significant number are cyclical (treewidth greater than 1). One document[8] contained a single component C with 451 blank nodes and 887 (undirected) edges and a treewidth of 7. Figure 2 renders this graph, where vertexes are blank nodes and edges are based on blank triples; for clarity, we collapse groups of n disconnected vertexes with the same neighbors into single nodes labeled n (note that these are not n-cliques).

We conclude that the majority of documents surveyed contain tree-based blank node structures. However, a small fraction contain complex blank node structures for which entailment is potentially very expensive to compute.

Table 2. tw distribution

treewidth	# components
1	518,831
2	8,134
3	208
4	99
5	23
6	–
7	1

Fig. 2. C where $\text{tw}(C) = 7$

---

[7] In the terminology of [22], a *blank triple* is an element of $\mathbf{B} \times \mathbf{U} \times \mathbf{B}$.

[8] http://www.rdfabout.com/rdf/usgov/congress/people/B000084

5.2   Survey of Publishing

To further understand how blank nodes are used, we made a simple poll[9] asking users what is their intended meaning when they publish triples with blank nodes. Herein, we briefly discuss the results, and we refer the interested reader to the web page for more details. We sent the poll to two W3C's public mailing lists: Semantic Web and Linked Open Data, and got 88 responses. In order to identify active publishers, we asked participants to indicate which datasets appearing in the LOD cloud (if any) they have contributed to; 10 publishers claimed contributions to a current LOD dataset.

At the top of the web page, before the questions, we explicitly stated that *"...the poll is trying to determine what you intend when you publish blank nodes. It is not a quiz on RDF Semantics. There is no correct answer"*.

In the first question, we asked participants in which scenarios they would publish a graph containing (only) the following triple: :John :telephone _:b. We chose the :telephone predicate as an abstract example which could be read as having a literal or URI value. Participants were told to select multiple options which would cover their reason(s) for publishing such a triple. The options were: *(1.a) John has a tel. number whose value is unknown*; *(1.b) John has a tel. number but its value is hidden, e.g., for privacy*; *(1.c) John has no tel. number*; *(1.d) John may or may not have a tel. number*; *(1.e) John's number should not be externally referenced*; *(1.f) I do not want to mint a URI for the tel. number*; and *(1.g) I would not publish such a triple*. The results were as follows:

	1.a	1.b	1.c	1.d	1.e	1.f	1.g
all (88)	46.6%	23.9%	0%	2.3%	18.2%	37.5%	41.0%
lod (10)	20%	0%	0%	0%	0%	30%	70%

In the second, we asked participants to select zero or more scenarios in which they would publish a graph containing (only) the following triples: :John :telephone _:b1, _:b2. The options were *(2.a) John does not have a tel. number*; *(2.b) John may not have a tel. number*; *(2.c) John has at least one tel. number*; *(2.d) John has two different tel. numbers*; *(2.e) John has at least two different tel. numbers*; and *(2.f) I would not publish such triples*. The results were as follows:

	2.a	2.b	2.c	2.d	2.e	2.f
all (88)	0%	0%	23.9%	23.9%	35.2%	50.0%
lod (10)	0%	0%	0%	10%	40%	70%

The poll had an optional section for comments; a number of criticisms (∼12) were raised about the :telephone example used and the restriction of having only one or two triples in the graph. This leaves ambiguity as to whether the participant would publish blank nodes at all (*intended*) or would not publish that specific example (*unintended*). Thus, we note that answers *1.g* and *2.f* might be over-represented. Also, one concern was raised about the "right" semantics of blank nodes in RDF (namely, that John has a telephone number, without saying anything about our knowledge of the number) not being an alternative; this was a deliberate choice.

---

[9] http://db.ing.puc.cl/amallea/blank-nodes-poll

Table 2. Implications of existential semantics, local constants semantics and total absence of blank nodes in theoretical aspects, standards and publishing

	standard	issue	Existential		Local constants		No Blank Nodes	
THEORY	RDF	*entailment*	NP-Complete	X	NP-Complete	X	PTime	✓
		equivalence	NP-Complete	X	NP-Complete	X	PTime	✓
	RDFS	*entailment*	NP-Complete	X	NP-Complete	X	PTime	✓
STANDARDS	Syntaxes	*shortcuts*	no change	✓	no change	✓	Sk. scheme	~
	OWL	*RDF mapping*	no change	✓	no change	✓	needs attention	X
	RIF	—	—	—	—	—	—	—
	SPARQL	*semantics*	mismatch	X	aligns	✓	aligns	✓
		query syntax	no change	✓	may need attention	~	may need attention	~
		round tripping	Sk. scheme	~	Sk. scheme	~	no change	✓
	RDB2RDF	*no primary key*	mismatch	X	aligns	✓	Sk. scheme	~
PUBL.	RDF	*unknown values*	no change	✓	no change	✓	Sk. scheme	~
		legacy data	ambiguous	~	unambiguous	✓	unambiguous	✓

Despite the limitations of the poll, we can see that blank nodes are not typically published with the intent of a non-existent/non-applicable semantics (*1c,1d,2b*). Obvious as it might be, the most conclusive result is that blank nodes are a divisive issue.

6   Alternative Treatments of Blank Nodes

Having covered the various desiderata for blank nodes across the several stakeholders, we now consider some high-level alternatives in light of the discussion thus far. Table 2 summarizes the implications of three conceptual paradigms for blank nodes in different aspects of theory and practice. In the column "Existential", we consider blank nodes with the current semantics of RDF (existential variables). In the column "Local constants" we consider blank nodes as constants with local scope. In the last column, we assume that blank nodes are eliminated from the standard.

The approaches in the second and third columns often require an agreed Skolemization scheme or at least the definition of frame conditions/best practices that guarantee non-conflicting Skolemization across the Web (when considering to eliminate blank nodes in published datasets) or within implementations (when merging datasets with "local constants"). Wherever we state "Sk. scheme", we refer to issues which may be solved by such agreed mechanisms used to generate globally unique URIs from syntax with implicit or locally scoped labels. We discuss the practicalities of (and proposals for) such schemes in Section 6.1. The core principle here is that locally-scoped artifacts can not "interfere" with each other across documents. For example, to enable syntactic shortcuts in the absence of blank nodes (*i.e.*, when the RDF universe consists of **UL**), the URIs generated must be globally unique to ensure that legacy resources are not unintentionally referenced and redefined when labels are not given.

In terms of the SPARQL *query syntax*, the use of blank nodes is purely syntactic and is decoupled from how RDF handles them. Even if blank nodes were discontinued, the blank node syntax in queries could still be supported (though a fresh syntax may be preferable). In terms of SPARQL *round tripping*, we refer to enabling blank nodes (either local or existential) to persist across scopes, where one possible solution is, again, Skolemization schemes. Note that SPARQL 1.1 will allow to "mint" custom URIs in results as an alternative to blank nodes.

For RDB2RDF mappings where *no primary key* is available, when removing blank nodes completely, the problem is again essentially one of implicit labeling, where a (possibly specialized) Skolemization scheme would be needed. We note that, unlike for existentials, a constant semantics for blank nodes would align better with the underlying semantics of the database.

For RDF *legacy data*, our intention is to note that for a consumer, how ad-hoc data should be interpreted remains ambiguous in the presence of non-leanness. Again, taking our original example and assuming it has been found on the Web, with the existential blank node semantics it is not clear whether leaning the data would honor the intent of the original publisher. Deciding whether to lean or not may then, *e.g.*, affect SPARQL query answers. For the other alternatives, leanness is not possible and so the ambiguity disappears. Conversely, we note that in a practical sense, publishers do not lose the ability to state *unknown values* in the absence of existential variables; such values can be expressed as *unique* constants which have no further information attached. Finally, we note that Table 2 does not quite cover all concrete alternatives. One other possibility we considered was not allowing blank nodes to ever be explicitly labeled, such that they form trees in the syntaxes, essentially enforcing all graphs to have a blank-treewidth of 1. Note that some syntaxes (like Turtle and RDF/XML without nodeID) guarantee a tree-structure for blank nodes. As discussed in Sections 3.1 & 4.1, this would make the isomorphism-checks and simple and RDF(S) entailment-checks tractable, although still with an implementational cost.

6.1   Skolemization Schemes

The following ideas have been proposed as recommended treatment of blank nodes in RDF. They do not necessarily require changes in the standards—or at least not to the semantics of RDF—but are intended as guidelines for publishers (*i.e.*, "best practices"). We will consider a set S of Skolem constants, such that every time Skolemization occurs, blank nodes are replaced with elements of S. Different alternatives will consider different behaviors and nature for this set.

6.1.1 $S \subseteq U$, Centralized

The idea of this alternative is to offer a centralized service that "gives out" fresh URIs upon request, ensuring uniqueness of the generated constants on a global scale. Formally, there would be a distinguished subset of the URIs, $S \subseteq U$, such that all Skolem constants belong to S. Every time the service gets a request, it returns an element $s \in S$ such that s has not been used before. This is very similar to what URL shorteners do[10]. Since the returned constants are also URIs, they can be used in published documents.

It is not clear who should pay and take responsibility for such a service (the obvious candidate being the W3C). The costs of bandwidth and maintenance can be too high.

---

[10] See, for example, http://bit.ly/ or http://is.gd/.

For example, the system should be able to cope with huge individual requests (see table 1). Also, the very idea of centralizing a service like this seems to go against the spirit of the Semantic Web community, though this might not be relevant for everyone. Further, the mere existence of such a system will not guarantee that users will only use this option when Skolemizing. Publishers will still have the freedom of using other methods to replace blank nodes with constants of their own choice.

6.1.2   S ⊆ U, Decentralized

Similar to the previous case, but with no central service, so each publisher will generate their constants locally. This has been discussed already by the RDF Working Group[11] and by the general community through the public mailing list of the Semantic Web Interest Group[12]. In both cases, the proposal is to establish a standard (but voluntary) process for generating globally unique URIs to replace blank nodes for querying, publishing and performing other operations with RDF graphs. A requirement of this process is that it contains a best practice that avoids naming conflicts between documents. Naming conflicts on the Web are typically avoided by pay level domains, since they guarantee a certain level of "authority"[16]. Along these lines, at the time of writing, the proposal of the RDF Working Group is to add a small section on how to replace blank nodes with URIs to the document "RDF Concepts" [18, §6.6]. The idea would be that blank nodes are replaced with well-known URIs [21] with a registered name that is to be decided (probably "genid" or "bnode") and a locally-unique identifier, which would make a globally-unique URI, since publishers would only be supposed to use their own domains. For example, the authority responsible for the domain example.com could mint the following URI for a blank node:

http://example.com/.well-known/bnode/zpHvSwfgDjU7kXTsrc0R

This URI can be recognized as the product of Skolemization. If desired, a user could replace it with a blank node. Moreover, publishers can encode information about the original blank node in the identifier, such as the name of the source graph, the label of the blank node in that graph, and the date-time of Skolemization.

It should be noted though that, although no central control authority is needed for such decentralized Skolemization, this proposal would allow third parties malicious, non-authoritative use of bnode-URIs which are not in their control (*i.e.* outside their domain): the often overlooked "feature" of blank nodes as local constants not modifiable/redefinable outside the graph in which they are published would be lost.

7   Conclusions

In this paper, we have provided detailed discussion on the controversial and divisive issue of blank nodes. Starting with formal considerations, we covered treatment of blank nodes in the W3C standards, how they are supported, and what they are needed for. The main use-case for blank nodes is as locally-scoped artifacts which need not be explicitly

---

[11] See http://www.w3.org/2011/rdf-wg/wiki/Skolemisation.
[12] See http://www.w3.org/wiki/BnodeSkolemization.

labeled. We also looked at how blank nodes are being published in Linked Data, where they are perhaps more prevalent than the best-practices would suggest; we also note that although rare, complex structures of blank nodes are present on the Web.

Informed by our earlier discussion, we proposed and compared three conceptual paradigms for viewing blank nodes. The first one is rather radical and involves eliminating blank nodes from the standard, with the additional consequences of having to deal with legacy data and with possible changes in other standards that rely on blank nodes in one way or another. The second one consists in standardizing the already widespread interpretation of blank nodes as local constants; most standards (like OWL and SPARQL) would not need to change at all. The third one is twofold: keeping the existential nature of blank nodes in RDF, and ensuring this is the meaning that other standards follow, for example, by means of a best-practices document; in this case, even if SPARQL were to follow the notions of leanness and entailment, query answering would not be expensive in most cases due to a good structure for blank nodes in currently published data. In all these alternatives, a Skolemization scheme would be handy as an aid for publishers to update their would-be obsolete data. Finally, we note that no alternative stands out as "the one solution to all issues with blank nodes". Discussion is still open and proposals are welcome, but as the amount of published data grows rapidly, a consensus is very much needed. However, in the absence of an undisputed solution, the community may need to take an alternative which might not be the most beneficial, but the least damaging for current and future users.

References

1. Arenas, M., Consens, M., Mallea, A.: Revisiting Blank Nodes in RDF to Avoid the Semantic Mismatch with SPARQL. In: RDF Next Steps Workshop (June 2010)
2. Arenas, M., Prud'hommeaux, E., Sequeda, J.: Direct mapping of relational data to RDF. W3C Working Draft (March 24, 2011), http://www.w3.org/TR/rdb-direct-mapping/
3. Beckett, D., McBride, B.: RDF/XML Syntax Specification (Revised). W3C Recommendation (February 2004), http://www.w3.org/TR/rdf-syntax-grammar/
4. Bizer, C., Heath, T., Berners-Lee, T.: Linked Data – The Story So Far. Int. J. Semantic Web Inf. Syst. 5(3), 1–22 (2009)
5. Buss, S.R.: On Herbrand's Theorem. In: Leivant, D. (ed.) LCC 1994. LNCS, vol. 960, pp. 195–209. Springer, Heidelberg (1995)
6. Carroll, J.J.: Signing RDF graphs. In: Fensel, D., Sycara, K., Mylopoulos, J. (eds.) ISWC 2003. LNCS, vol. 2870, pp. 369–384. Springer, Heidelberg (2003)
7. Das, S., Sundara, S., Cyganiak, R.: R2RML: RDB to RDF mapping language. W3C Working Draft (March 24, 2011), http://www.w3.org/TR/r2rml/
8. Erling, O., Mikhailov, I.: RDF Support in the Virtuoso DBMS. In: Pellegrini, T., Auer, S., Tochtermann, K., Schaffert, S. (eds.) Networked Knowledge - Networked Media. SCI, vol. 221, pp. 7–24. Springer, Heidelberg (2009)
9. Franconi, E., de Bruijn, J., Tessaris, S.: Logical reconstruction of normative RDF. In: OWLED (2005)
10. Gogate, V., Dechter, R.: A Complete Anytime Algorithm for Treewidth. In: UAI, pp. 201–208 (2004)

11. Grau, B.C., Motik, B., Wu, Z., Fokoue, A., Lutz, C.: OWL 2 Web Ontology Language: Profiles. W3C Recommendation (October 2009), http://www.w3.org/TR/owl2-profiles/
12. Grosof, B., Horrocks, I., Volz, R., Decker, S.: Description Logic Programs: Combining Logic Programs with Description Logic. In: WWW (2004)
13. Gutierrez, C., Hurtado, C., Mendelzon, A.O.: Foundations of Semantic Web Databases. In: 23rd ACM SIGACT-SIGMOD-SIGART (June 2004)
14. Hayes, P.: RDF Semantics. W3C Recommendation (February 2004)
15. Heath, T., Bizer, C.: Linked Data: Evolving the Web into a Global Data Space, vol. 1. Morgan & Claypool (2011)
16. Hogan, A.: Exploiting RDFS and OWL for Integrating Heterogeneous, Large-Scale, Linked Data Corpora. PhD thesis, DERI Galway (2011)
17. Kelly, P.J.: A congruence theorem for trees. Pacific Journal of Mathematics 7(1) (1957)
18. Klyne, G., Carroll, J.J.: Resource Description Framework (RDF): Concepts and Abstract Syntax. W3C Recommendation (February 2004), http://www.w3.org/TR/rdf-concepts/
19. Manola, F., Miller, E., McBride, B.: RDF Primer. W3C Recommendation (February 2004)
20. Muñoz, S., Pérez, J., Gutierrez, C.: Simple and Efficient Minimal RDFS. J. Web Sem. 7(3), 220–234 (2009)
21. Nottingham, M., Hammer-Lahav, E.: Defining Well-Known Uniform Resource Identifiers (URIs). RFC 5785 (April 2010), http://www.ietf.org/rfc/rfc5785.txt
22. Pichler, R., Polleres, A., Wei, F., Woltran, S.: dRDF: Entailment for Domain-Restricted RDF. In: Bechhofer, S., Hauswirth, M., Hoffmann, J., Koubarakis, M. (eds.) ESWC 2008. LNCS, vol. 5021, pp. 200–214. Springer, Heidelberg (2008)
23. Prud'hommeaux, E., Seaborne, A.: SPARQL Query Language for RDF. W3C Recommendation (January 2008), http://www.w3.org/TR/rdf-sparql-query/
24. Schneider, M.: OWL 2 Web Ontology Language RDF-Based Semantics. W3C Recommendation (October 2009), http://www.w3.org/TR/owl2-rdf-based-semantics/
25. Tarjan, R.E., van Leeuwen, J.: Worst-case analysis of set union algorithms. J. ACM 31(2), 245–281 (1984)
26. ter Horst, H.J.: Completeness, decidability and complexity of entailment for RDF Schema and a semantic extension involving the OWL vocabulary. J. of Web Sem. 3, 79–115 (2005)

Inspecting Regularities in Ontology Design Using Clustering

Eleni Mikroyannidi, Luigi Iannone, Robert Stevens, and Alan Rector

The University of Manchester
Oxford Road, Manchester, M13 9PL
{mikroyannidi,iannone,stevens,rector}@cs.manchester.ac.uk

Abstract. We propose a novel application of clustering analysis to identify regularities in the usage of entities in axioms within an ontology.
We argue that such regularities will be able to help to identify parts of
the schemas and guidelines upon which ontologies are often built, especially in the absence of explicit documentation. Such analysis can also
isolate irregular entities, thus highlighting possible deviations from the
initial design. The clusters we obtain can be fully described in terms of
generalised axioms that offer a synthetic representation of the detected
regularity. In this paper we discuss the results of the application of our
analysis to different ontologies and we discuss the potential advantages
of incorporating it into future authoring tools.

1 Introduction

Ontologies are often built according to guidelines or schemas that give rise to
repeating regularities in the use of entities in axioms. Recognising those regularities is important in understanding the ontology and assuring that it conforms
to the schemas. A regular ontology shows organisation of the knowledge and
thus its coherence. For example, in the wine ontology[1], all the wines at the individual level are described in similar ways; they have a particular type of wine
and there are property assertions defining their body, origin, maker, flavour etc.
The inspection of these regularities can give an insight into the construction of
the ontology. By looking at the description of some wines, the user can have an
insight about the template that was used to describe them. Deviations from this
regular design might either be legitimate exceptions in the regularities or defects
in modelling. For example, the individual TaylorPort does not have any property
assertions referring to its origin, as this is defined on its type (Port SubClassOf
locatedIn **value** PortugalRegion). Having a mechanism that can help to isolate such
deviations could be a useful tool for quality assurance.

It is, however, difficult to trace irregularities and regularities by eye, especially
in big ontologies. The reasoner can check the satisfiability of concepts in the
ontology, and there are tools [9,5] that provide explanations for unsatisfiabilities

---

[1] http://www.w3.org/TR/owl-guide/wine

L. Aroyo et al. (Eds.): ISWC 2011, Part I, LNCS 7031, pp. 438–453, 2011.
© Springer-Verlag Berlin Heidelberg 2011

and other entailments. However, they cannot trace irregularities in the design as these are not logical errors.

Ontology environments such as Protégé-4, Swoop, NeOn Toolkit and Top Braid Composer provide some structural information about the ontology through visualisations, such as hierarchies and panels showing the usage of the entities. The task of spotting regularities in the usage of axioms in entity descriptions is, however, not supported beyond this basic exposure of usage.

Ontology engineering has adopted the idea of design patterns [4] to capture accepted modelling solutions to common issues [2,3]. Patterns of axioms, however, can exist throughout an ontology without being an accepted design pattern. In this paper we focus on the general notion of patterns. In the remainder of the paper we will refer to them as *regularities*. Such a regularity is illustrated in the following axioms in the wine ontology:

$\alpha_1 =$ PinotBlanc *EquivalentTo* wine **and**

(madeFromGrape **value** PinotBlancGrape) **and** (madeFromGrape **max** 1 Thing)

$\alpha_2 =$ CabernetSauvignon *EquivalentTo* wine **and** (madeFromGrape **value**

CabernetSauvignonGrape) **and** (madeFromGrape **max** 1 Thing)

The regularity can be expressed with the abstract axiom:

$\alpha =$?xWine *EquivalentTo* wine **and**

(madeFromGrape **some** ?xGrape) **and** (madeFromGrape **max** 1 Thing)

where ?xWine = {PinotBlanc, CabernetSauvignon }, ?xGrape ={PinotBlancGrape, CabernetSauvignonGrape} are variables holding similar entities.

We present a novel framework for inspecting such *regularities* and *clustering* entities in the ontology according to these kinds of regularities. Clustering is a common scientific approach for identifying similarities [13]. The proposed method is based on capturing similar usage of the entities; such similar entities are expected to result in the same cluster. The description of the clusters is provided in abstract forms; axioms containing meaningful *placeholders* (e.g. ?xWine *EquivalentTo* wine **and** (madeFromGrape **some** ?xGrape) **and** (madeFromGrape **max** 1 Thing)). With such a framework, tools could help the inspection of regularities when attempting to comprehend ontologies, where information on the design style and schemas is often poorly documented and not represented explicitly in the ontology itself.

Efforts for developing or transforming ontologies using patterns are reported in [12], [2]. In [7], the use of the Ontology Pre-Processor Language (OPPL) as a means of embedding knowledge patterns in OWL ontologies is presented. However, little work is known in the field of regularity detection in ontologies. In [11] an analysis of collections of OWL ontologies with the aim of determining the frequency of several combined name and graph patterns is described. However, this work mainly focuses on lexical patterns, with some additional examination on how these apply in the taxonomy hierarchy. It does not explicitly involve identification of regularities in axiom usage. There appears to be no framework that can identify regularities in the axioms of an ontology and highlight entities that have been designed according to these regularities.

2   Clustering Framework

The purpose of cluster analysis is to divide data into groups (clusters) that are meaningful, useful or both [13]. Our particular problem is to partition the set of entities in an ontology according to their usage, i.e.: entities in the same cluster occur with *similar* axioms playing *similar roles*. Cluster analysis relies on the notion of distance to quantify how similar (or dissimilar) and, therefore, how close or far apart two entities are in the clustering space. In the literature there are several clustering techniques; we chose agglomerative clustering and we report an informal description of the algorithm we used in Algorithm 1. In our algorithm, the implementation of step 1, and in particular, the distance adopted to compute the proximity matrix is the most important aspect of the implementation and the main focus of the rest of this section; steps 2–7 are typical steps for agglomerative cluster analysis and are not explained further. However, in our implementation, the algorithm will continue agglomerating until the distance between all possible pairs of elements in the two closest clusters is less than 1.

Algorithm 1. Clustering algorithm

Require: A set of entities.
Ensure: A set of clusters.

1: Compute the proximity matrix {matrix containing the values of the distance between all entities}
2: Assign to the current set of clusters the set of singleton clusters, each representing an input entity.
3: **repeat**
4:    Merge the closest two clusters, replace them with the result of the merge in the current set of clusters.
5:    Update the proximity matrix, with the new distance between the newly created cluster and the original ones.
6: **until** The stopping criterion is met
7: **return** The current set of clusters.

In the approach taken here, the calculation of the distance is based on the similarity of the structure of the axioms in the ontology. For axiom comparison, we transform them into a more abstract form following a *place-holder replacement policy*. We will define the notion of similar structure of axioms more formally in the following, and show how this leads to the particular distance we adopted for our cluster analysis.

Comparing axioms: Let us consider the Pizza Ontology[2]. Its main scope is pizzas and their toppings along with other information such as the country of

---

[2] http://www.co-ode.org/ontologies/pizza/

origin for each pizza, the spiciness of some toppings, or their classification as vegetables, fish, or meat. Among other things, in this ontology, we observe that all the topping classes are used as fillers in axioms like:

> *aPizza* SubClassOf hasTopping **some** *aTopping*
>
> *aPizza* SubClassOf hasTopping **only** (*aTopping* **or** *anotherTopping* or ...)

In other words, classes like MozzarellaTopping and TomatoTopping seem similar because they appear as fillers of an existential restriction on the property hasTopping within a sub-class axiom where the left-hand side is a pizza. They also appear as disjuncts in a universal restriction on the same property in another sub-class axiom whose left-hand side is, again, a pizza. Likewise, pizzas in this ontology tend to appear on the left-hand side of sub-class axioms describing their toppings, base, and country of origin. Therefore, our cluster analysis should, in the case of the pizza ontology, put together all toppings in a single cluster, pizzas in another, and countries of origin in a third one and so on. More formally, we need to introduce a distance that quantifies the difference between the usage of two entities in a set of axioms (ontology).

Definition 1. *[Place-holder replacement] Let $\mathcal{O}$ be an ontology and let $\Phi = \{owlClass, owlObjectProperty, owlDataProperty, owlAnnotationProperty, owlIndividual, *\}$ be a set of six symbols that do not appear in the signature[3] of $\mathcal{O}$ - $sig(\mathcal{O})$. A place-holder replacement is a function $\phi : sig(\mathcal{O}) \rightarrow sig(\mathcal{O}) \cup \Phi$ satisfying the following constraints: Consider $e \in \mathcal{O}$ then $\phi(e) =$*

- *e or * or owlClass if e is a class name;*
- *e or * or owlObjectProperty if e is a object property name;*
- *e or * or owlDataProperty if e is a data property name;*
- *e or * or owlAnnotationProperty if e is a annotation property name;*
- *e or * or owlIndividual if e is an individual property name.*

We define the particular placeholder replacement ϕ^S as $\phi^S(e) =$

- *owlClass if e is a class name;*
- *owlObjectProperty if e is an object property name;*
- *owlDataProperty if e is a data property name;*
- *owlAnnotationProperty if e is an annotation property name;*
- *owlIndividual if e is an individual property name.*

Definition 2. *[Place-holder replacement in axioms] Let $\mathcal{O}$ be an ontology, $\alpha \in \mathcal{O}$ one of its axioms and ϕ a place-holder replacement function. We define ϕ_{Ax} as a function that for the input α returns a new axiom by applying ϕ to all the entities $e \in sig(\alpha)$.*

Example 1. Let our $\mathcal{O}$ be the Pizza ontology mentioned above and let us define ϕ as follows. $\forall e \in \mathcal{O}, \phi(e) =$

---

[3] For signature here we mean the set of class names, data/object/annotation property names, individuals referenced in the axioms of an ontology $\mathcal{O}$.

- * if $e \in \{$Margherita, Capricciosa$\}^4$;
- $\phi^S(e)$ otherwise;

Let us now compute the values of $\phi_{Ax}(\alpha)$ for some of the axioms in $\mathcal{O}$

- $\alpha =$ Margherita *DisjointWith* Cajun, $\phi_{Ax}(\alpha)=$ * *DisjointWith* owlClass;
- $\alpha =$ Capricciosa *DisjointWith* Cajun, $\phi_{Ax}(\alpha)=$ * *DisjointWith* owlClass;
- $\alpha =$ Margherita *SubClassOf* hasTopping **some** TomatoTopping,
  $\phi_{Ax}(\alpha)=$ * *SubClassOf* owlObjectProperty **some** owlClass;
- $\alpha =$ Capricciosa *SubClassOf* hasTopping **some** TomatoTopping,
  $\phi_{Ax}(\alpha)=$ * *SubClassOf* owlObjectProperty **some** owlClass;

We have defined the replacement function that transforms the axioms into abstractions and we can proceed with the measure of the distance.

Distance measure: We define the distance function as follows:

Definition 3. *[Distance] Let $\mathcal{O}$ be an ontology, e_1 and e_2 be two entities from $sig(\mathcal{O})$ and ϕ a place-holder replacement function. We denote $Ax_\phi(e)$ the set $\{\phi_{Ax}(\alpha), \alpha \in \mathcal{O}, e \in sig(\alpha)\}$, i.e: the set of pace-holder replacements for the axioms in $\mathcal{O}$ that reference e.*

We define the distance between the two entities, $d_\phi(e_1, e_2)$ as:

$$d(e_1, e_2) = \frac{|(Ax_\phi(e_1) \cup Ax_\phi(e_2)| - |Ax_\phi(e_1) \cap Ax_\phi(e_2))|}{|(Ax_\phi(e_1) \cup Ax_\phi(e_2)|}$$

From this we can observe that $\forall \mathcal{O}, \phi, e_1, e_2 : 0 \leq d_\phi(e_1, e_2) \leq 1$. The place-holder replacement function ϕ is a way to control the granularity of our distance.

Example 2. Let our $\mathcal{O}$ be the Pizza ontology again and let us define ϕ_1 as follows. $\forall e \in \mathcal{O}, \phi_1(e)=$

- * if $e \in \{$TomatoTopping, PizzaBase$\}$;
- $\phi^S(e)$ otherwise;

Let us now compute the values of $\phi_{1\,Ax}(\alpha)$ for a pair of axioms in $\mathcal{O}$

- $\alpha =$ Margherita *SubClassOf* hasTopping **some** TomatoTopping,
  $\phi_{1\,Ax}(\alpha) =$ owlClass *SubClassOf* owlObjectProperty **some** *;
- $\alpha =$ Pizza *SubClassOf* hasBase **some** PizzaBase,
  $\phi_{1\,Ax}(\alpha) =$ owlClass *SubClassOf* owlObjectProperty **some** *.

This means that $d_{\phi 1}($TomatoTopping, PizzaBase$) < 1$ as $|Ax_{\phi_1}(e_1) \cap Ax_{\phi_1}(e_2)| > 0$ and, therefore, $|(Ax_{\phi_1}(e_1) \cup Ax_{\phi_1}(e_2)| - |Ax_{\phi_1}(e_1) \cap Ax_{\phi_1}(e_2)| < |(Ax_{\phi_1}(e_1) \cup Ax_{\phi_1}(e_2)|$.

The consequence would be that our distance $d_{\phi 1}$ does not separate as cleanly as possible TomatoTopping (and likewise several sub-classes of PizzaTopping) from PizzaBase. Let us now compare it with another place-holder replacement function, ϕ_2, defined as follows:

---

[4] The * placeholder represents entities, which distance is computed. We denote this placeholder for keeping track of the position of the entities in the referencing axioms.

- * if $e \in \{$TomatoTopping, PizzaBase$\}$;
- e if e is a object property name;
- $\phi^S(e)$ otherwise;

Then our ϕ_{2Ax} for the same values for α will be:

- $\alpha =$ Margherita *SubClassOf* hasTopping **some** TomatoTopping,
  $\phi_{2Ax}(\alpha) =$ owlClass *SubClassOf* hasTopping **some** *;
- $\alpha =$ Pizza *SubClassOf* hasBase **some** PizzaBase,
  $\phi_{2Ax}(\alpha) =$ owlClass *SubClassOf* hasBase **some** *

This will keep $d_{\phi 2}($TomatoTopping, PizzaBase$) = 1$

Changing the granularity of the place-holder replacement function produces more or less sensitive distance functions. The two extremes are replacing every entity with a place-holder or not replacing any of them. Whilst the former produces a distance that is far too tolerant and puts together entities that seem unrelated, the latter will most likely result in a distance that scores 1 (maximal distance) for most entity pairs. In this work we propose a tradeoff where we delegate the decision of whether to replace an entity in an axiom to a measure of its **popularity** with respect to the other entities in the same kind of axiom within the ontology. More formally:

Definition 4 (Popularity). *Let $\mathcal{O}$ be an ontology, $e \in sig(\mathcal{O})$ an entity. The place-holder replacement function $\phi^S{}_{Ax}$ for the axioms of $\mathcal{O}$ will extract the* **structure** *of each axiom.*

Given an axiom $\alpha \in \mathcal{O}$, let us define the set $Ax^\alpha = \{\beta \in \mathcal{O}, \phi^S{}_{Ax}(\beta) = \phi^S{}_{Ax}(\alpha)\}$, that is, the set of axioms in $\mathcal{O}$ that have the same **structure** *as α. We can, finally, define popularity π_{Ax^α} of an entity $f \in sig(\mathcal{O})$ as*

$$\pi_{Ax^\alpha}(f) = \frac{|\{\beta \in Ax^\alpha, f \in sig(\beta)\}|}{|Ax^\alpha|}$$

that is, the number of axioms in Ax^α that reference f over the size of Ax^α itself.

We can plug-in popularity as defined above into a place-holder replacement function and therefore in our distance as follows: When computing a distance between two entities, namely e_1 and e_2, for each axiom α where either occurs, the function replaces e_1 or e_2 with * and decides whether to replace the other entities with a place-holder depending on their popularity across all the axioms that have the same structure as α.

Definition 5 (Popularity based place-holder replacement). *Let $\mathcal{O}$ be an ontology, $e \in sig(\mathcal{O})$ an entity, and $\alpha \in \mathcal{O}$ an axiom. Let Ax^α and π_{Ax^α} be respectively the set of axioms sharing the same structure as α and the popularity metric defined in Definition 4. Finally, let σ be a function that we call* **popularity criterion** *and maps a popularity value into the set $\{$true, false$\}$.*

$\forall f \in sig(\mathcal{O})$, we define our function as follows: $\phi_e^\alpha(f)$

- * *if* $f = e$;
- f *if* $\sigma(\pi_{Ax^\alpha}(f)) = $ *true;*
- $\phi^S(f)$ *otherwise.*

We can now use the popularity based place-holder replacement defined above in our distance (Definition 3). Given two entities e_1 and e_2 according to the formula we need to compute $\mathsf{Ax}_\phi(e_1)$ and $\mathsf{Ax}_\phi(e_2)$. For every axiom α in the ontology $\mathcal{O}$ that references e_1 (resp. e_2), we compute $\phi_{Ax}(\alpha) = \phi^\alpha_{e_1\,Ax}(\alpha)$ (resp. $\phi_{Ax}(\alpha) = \phi^\alpha_{e_2\,Ax}(\alpha)$). Informally, for each axiom, we compute our replacement function based on the popularity of the entities across the set of axioms sharing the same structure as the axiom we are currently considering. In the definition above we deliberately parameterised the decision criterion to make our distance framework independent from any particular implementation. In this work, however, we compute a confidence interval $[l, u]$ for the mean value of π_{Ax^α}. (95% confidence). We assume the variance is unknown; therefore in order to compute the area under the distribution function (z), we use the values for the t *distribution*, rather than the *normal* one in the formulas:

$$l = M - z \cdot \frac{\mathsf{sd}}{\sqrt{N}}, \qquad u = M + z \cdot \frac{\mathsf{sd}}{\sqrt{N}}$$

where with sd we denote the standard deviation and with M the mean computed on the set of entities (whose size is N) in the ontology. If the popularity of a given entity is greater than u then we assign **true** to our σ (see Definition 5), false otherwise.

Example 3. Once again, let our ontology be the Pizza ontology and let us use as our place-holder replacement function ϕ, the one in Definition 5 (based on popularity). Let us compute the replacements for the same axioms as in Example 2. We omit the calculations but the confidence interval for the popularity when applied to such axioms are such that the only entities which will not be replaced are: hasTopping and TomatoTopping, therefore:

- $\alpha = $ Margherita *SubClassOf* hasTopping **some** TomatoTopping,
  $\phi_{1\,Ax}(\alpha) = $ owlClass *SubClassOf* hasTopping **some** *;
- $\alpha = $ Pizza *SubClassOf* hasBase **some** PizzaBase,
  $\phi_{1\,Ax}(\alpha) = $ owlClass *SubClassOf* owlObjectProperty **some** *.

The extensive usage of object property hasTopping in this particular kind of axiom is the reason why our place-holder replacement function deems it as important and preserves it in the replacement result.

We observe, however, that deciding replacements based on confidence intervals is strongly dependant on the quality of the sample data. TomatoTopping, for instance, in the example above, is judged *popular* too. The reason is that all pizzas in the ontology have TomatoTopping (and MozzarellaTopping) among their toppings. Conversely, the formula correctly spots that several other entities (Margherita, Pizza, hasBase, ...) are not *relevant* when dealing with axioms presenting a particular structure (owlClass *SubClassOf* owlObjectProperty **some** owlClass). We claim that this is preferable w.r.t. making an *a priori* decision, maybe based on users' intuitions, on what should be replaced and when.

Agglomerative hierarchical clustering: To complete the discussion of our implementation for Algorithm 1, we need to illustrate how we update the distances in our proximity matrix at every agglomeration and what we use as our stopping criterion. For the former we use the Lance-Williams formula (see Section 8.3.3 in [13] - page 524). This formula computes the distance between cluster Q and R, where R is the result of a merger between clusters A and B, as a function of the distances between Q, A, and B. The distance between two sets (clusters) is a function of the distance between their single elements. There are several approaches to compute this, each corresponds to a different value configuration of the coefficients in the general Lance-Williams formula. In the experiments described in the following sections, we used the so-called *centroid* configuration[5].

As its stopping criterion, our implementation uses a heuristic decision:

Definition 6 (Agglomerate decision function). *Let $\mathcal{O}$ be an ontology and d a distance function. We define the function* $\mathsf{agg}_d : 2^{sig(\mathcal{O})} \times 2^{sig(\mathcal{O})} \rightarrow \{\textsf{true}, \textsf{false}\}$ *as follows: Given $E = \{e_1, e_2, \ldots, e_n\}$ and $F = \{f_1, f_2, \ldots, f_m\}$ be two clusters,* $\mathsf{agg}_d(E, F) =$

  – *false if* $\exists 1 \leq i \leq n (\exists 1 \leq j \leq m : d(e_i, f_j) = 1)$;
  – *true otherwise.*

The algorithm terminates when no pair in the current set of clusters returns true for the Agglomeration decision function agg_d, defined above. When clustering the Pizza ontology, our implementation returns 17 clusters containing over 110 entities in total; these include: a cluster for the toppings that are used in pizzas; one for the named pizzas (pizza with a name and a description of their toppings); and one for the country of origin of the toppings.

As intuitive these groups may seem, given the average familiarity people have with the pizza domain, this represents a cluster analysis based on the actual usage of the entities in the ontology. In this example clusters seem to follow the taxonomy quite well, however, as we shall see in the next section this may not be the case. Performing this kind of analysis can indeed reveal common use between entities that are far apart in the taxonomical hierarchy.

Description of the clusters: Once the clusters are available, the axioms that reference entities in the same cluster can be *generalised* and provide a more abstract view on the entire cluster. We can define a generalisation as a simple substitution of an entity with a variable within an axiom. More formally

Definition 7 (Generalisation). *Let $\mathcal{O}$ be an ontology, $E = \{e \in \mathsf{sig}(\mathcal{O})\}$ a set of entities, and $\alpha \in \mathcal{O}$ an axiom. Let us now choose a symbol (variable name),*

---

[5] Although vaguely related, not to be confused with a *centroid* in the K-MEANS cluster analysis - see Chapter 8 in [13].

say ?x. We generalise over E with ?x in α ($g(\alpha, E, ?x)$) when we replace every element of E in α with ?x.

In the definition above, as well as in the remainder of the paper, we will borrow the syntax for variables from OPPL[6], a declarative language for manipulating OWL ontologies [6]. However, for the purpose of this paper, it is sufficient to say that an OPPL variable can be: **Input or Non generated**, i.e.: they can replace entities in axioms of the corresponding type (there are OPPL variables for each type of entity in a signature); **Generated**, i.e: their value is the result of an expression depending on other variables.

Example 4 (Generalised Pizzas). Let $\mathcal{O}$ be our Pizza ontology and let cluster$_1$ be the cluster of all the toppings used in pizzas obtained using our cluster analysis above, and cluster$_2$ be the cluster of all pizzas. Given α = Margherita *SubClassOf* hasTopping **some** TomatoTopping:

- $g(\alpha, \text{cluster}_1, ?\text{cluster}_1)$ = Margherita *SubClassOf* hasTopping **some** ?cluster$_1$;
- $g(\alpha, \text{cluster}_2, ?\text{cluster}_2)$ = ?cluster$_2$ *SubClassOf* hasTopping **some** TomatoTopping; or composing the two
- $g(g(\alpha, \text{cluster}_2, ?\text{cluster}_2), \text{cluster}_1, ?\text{cluster}_1)$ = ?cluster$_2$ *SubClassOf* hasTopping **some** ?cluster$_1$

where ?cluster$_1$ and ?cluster$_2$ are two variables of type class. (In OPPL: ?cluster$_1$:CLASS, ?cluster$_2$:CLASS).

Generalisations provide a synthetic view of all the axioms that contribute to generate a cluster of entities. Each of these axioms can indeed be regarded as an instantiation of a generalisation, as they can be obtained by replacing each variable in g with entities in the signature of the ontology.

3 Results and Evaluation

Four ontologies (AminoAcid [7], OBI[8], a module of the SNOMED-CT[9] containing axioms about hypertension and KUPO[10]) were selected for testing the clustering framework; Table 1 summarises some results.

The AminoAcid ontology has been developed internally, allowing comments on the clusters from the ontology's authors. The remaining ontologies are documented, enabling further analysis and evaluation of the clusters and regularities. All of the selected ontologies preexisted the clustering framework. A quantitive analysis was also performed on 85 ontologies from the BioPortal[11] repository for

---

[6] http://oppl2.sourceforge.net
[7] http://www2.cs.man.ac.uk/ mikroyae/2011/ iswc/files/amino-acid-original.owl
[8] http://purl.obolibrary.org/obo/obi.owl
[9] http://www2.cs.man.ac.uk/ mikroyae/2011/iswc/files/ sct-20100731-stated_Hypertension-subs_module.owl
[10] http://www.e-lico.eu/public/kupkb
[11] http://bioportal.bioontology.org/

Table 1. Clustering results on the four selected ontologies

Ontology name	No of Clusters	No of Clustered entities	Cluster coverage per generalisation (%)
AminoAcid	16	77 (84%)	78
OBI	445	2832 (70%)	70
KUPKB	26	470 (44%)	51
SNOMED-CT	77	420 (80%)	62

which clusters were computed in less than 3 minutes. The framework was tested with the asserted information of the ontologies, for analysing the regularities of their construction.

The number of clusters in Table 1 shows that in most cases more than 50% of the entities in the ontologies were clustered. In principal, the union of the generalisations describes the cluster, thus a single generalisation might not be necessarily applicable for all the values in the cluster. However, in all four ontologies the majority of the values in a cluster is covered by a single generalisation (cluster coverage percentiles in Table 1).

The results in all of the ontologies are of similar impact and can be found in detail online[12]. In this section, however, we will highlight some cases from each ontology.

Inspecting regularities: We detected regularities in all four ontologies and we evaluate them by referring to their documentation or ontology authors. In OBI ontology, ?$cluster_7$ includes 47 T cell epitopes of specific type that are equivalent classes. The structure of the definition of these terms is similar. To demonstrate this regularity we will consider two classes from ?$cluster_7$, the "epitope specific killing by T cells" and the "epitope specific T cell activation". These classes appear on the left hand side of the axioms:

α = 'epitope specific killing by T cells' *EquivalentTo* 'T cell mediated cytotoxicity'
    and ('process is result of' **some** 'MHC:epitope complex binding to TCR')

α = 'epitope specific T cell activation' *EquivalentTo* 'T cell activation'
    and ('process is result of' **some** 'MHC:epitope complex binding to TCR')

The generalisation for these axioms is:

$g(\alpha)$ = cluster$_7$ *EquivalentTo* ?cluster$_8$ **and** ('process is result of' **some** cluster$_{91}$)
where ?cluster$_8$ = {'T cell mediated cytotoxicity', 'T cell activation},
?cluster$_{91}$ = {'MHC:epitope complex binding to TCR'}

?$cluster_8$, ?$cluster_{91}$ are placeholders for the corresponding classes, while the object property 'process is result of' does not belong to any cluster, thus it is not represented by a placeholder.

In [10], a methodology for developing 'analyte assay' terms in OBI using a template based on spreadsheets is described. The clustering framework, gave such a

---

[12] http://www2.cs.man.ac.uk/~mikroyae/2011/iswc/

cluster of classes (cluster$_{35}$) and their descriptions in the form of generalisations highlighting their commonalities. For example, there are 13 axioms in cluster$_{35}$ covered by the following generalisation, which describes the analyte assays that are used to *"achieve a planned objective"*:

$g(\alpha)$ = ?cluster$_{35}$ *SubClassOf* ?cluster$_{117}$ **some** ?cluster$_{16}$

Example instantiation: α = 'genotyping assay' *SubClassOf* achieves_planned_objective **some** 'sequence feature identification objective'

In [8] the design process of the KUP ontology is explained and two main patterns are described for generating the cell types in the ontology. The results of the clustering framework showed such clusters of cells and clusters of classes used as fillers of the properties describing the cells (e.g. participates_in, part_of). Two example generalisations capturing these regularities are:

1. $g(\alpha)$ = ?cluster$_{13}$ *EquivalentTo* ?cluster$_{27}$ **and** (part_of **some** ?cluster$_2$),
   where ?cluster$_{13}$, ?cluster$_{27}$, ?cluster$_2$: CLASS, ?cluster$_{27}$ = {cell}
Example Instantiation:
α = 'bladder cell' *EquivalentTo* cell **and** (part_of **some** 'bladder')
2. $g(\alpha)$ = ?cluster$_1$ *SubClassOf* (participates_in **some** ?cluster$_{16}$) **and** (participates_in **some** ?cluster$_{19}$), *where* ?cluster$_1$, ?cluster$_{16}$, ?cluster$_{19}$: CLASS
Example Instantiation:
α = 'kidney interstitial fibroblast' *SubClassOf* (participates_in **some** 'cytokine production') **and** (participates_in **some** 'extracellular matrix constituent secretion')

Each one of these generalisations corresponds to a different cluster in the ontology. Also, these regularities were described in [8]. The first regularity is encapsulated in the description of the first pattern and the second regularity is encapsulated in the description of the second pattern. Additional regularities were also detected that refer to longer conjunctions of the previous generalisations (e.g. a conjuction of part_of relationships on the right hand side of the axiom).

In addition, the clustering framework gave an alternative view based on the similar usage of the entities in the ontology. It could selectively reveal repeating structures in the ontology that were more difficult to inspect manually. For example, in SNOMED the highlighted classes of Figure 1(a) are grouped in the same cluster and their similar definition is given by the generalisation of Figure 1(b). However, the inspection of regularities through navigation in the class hierarchy is not an easy task because of the high level of nesting and complexity of the hierarchy. The form of regularity of Figure 1(b) is also described in the technical guide of the ontology [1](section 17.2.2., page 180).

Fully covered generalisations: As it has been shown in Table 1 a single generalisation does not necessarily apply to all of the values in the cluster.

However, there are cases that an individual generalisation can be applicable to all values of its cluster. Such an example is shown in Figure 2, taken from

▼ ⊖'Procedure by site (procedure)'
  ⊖'Intracranial procedure (procedure)'
  ▼ ⊖'Procedure on abdomen (procedure)'
    ⊞'Abdominal cavity operation (procedure)'
  ⊖'Procedure on body system (procedure)'
  ▼ ⊖'Procedure on cardiovascular system (procedure)'
    ⊞'Cardiovascular operative procedure (procedure)'
   ▶ ⊖'Procedure on artery (procedure)'
   ▶ ⊖'Procedure on blood vessel (procedure)'
  ▼ ⊖'Procedure on endocrine system (procedure)'
    ⊞'Operation on endocrine system (procedure)'
   ▶ ⊖'Procedure on genitourinary system (procedure)'
   ▶ ⊖'Procedure on head AND/OR neck (procedure)'
   ▶ ⊖'Procedure on nervous system (procedure)'
  ⊖'Procedure on organ (procedure)'
   ▶ ⊖'Procedure on soft tissue (procedure)'
  ▼ ⊖'Procedure on thorax (procedure)'
    ⊞'Surgical procedure on thorax (procedure)'
    ⊖'Thorax implantation (procedure)'
  ▼ ⊖'Procedure on trunk (procedure)'
    ⊖'Operation on trunk (procedure)'
   ▼ ⊖'Procedure on mediastinum (procedure)'
     ⊞'Operation on mediastinum (procedure)'
    ▼ ⊖'Procedure on heart (procedure)'
     ▼ ⊖'Operation on heart (procedure)'
      ⊖'Ventricular operation (procedure)'
   ▼ ⊖'Procedure on pelvis (procedure)'

(a) Classes of $?cluster_{18}$ as shown in Protégé class hierarchy view. The similar entities do not appear to be siblings in the class hierarchy.

Generalisation:

$g(\alpha) = ?cluster_{18}$ *EquivalentTo* $?cluster_4$
    and $?cluster_6$
      and (RoleGroup **some**
        $((?cluster_{10}$ **some** $?cluster_2)$
        and $(?cluster_{74}$ **some** $?cluster_{16})))$

where

$?cluster_6, ?cluster_{18}, ?cluster_4, ?cluster_2,$
$?cluster_{16}$: CLASS
$?cluster_{10}, ?cluster_{74}$: OBJECTPROPERTY.

Example Instantiation:

$\alpha =$ SCT_79537002 *EquivalentTo* SCT_118681009
    and SCT_387713003
      and (RoleGroup **some**
        ((SCT_260686004 **some** SCT_129284003))
        and (SCT_363704007 **some** SCT_113331007)))

SCT_79537002: 'Operation on endocrine
    system (procedure)'
SCT_118681009: 'Procedure on
    endocrine system (procedure)'
SCT_387713003: 'Surgical procedure (procedure)'
SCT_260686004: 'Method (attribute)'
SCT_129284003: 'Surgical action (qualifier value)'
SCT_113331007: 'Procedure site (attribute)'
SCT_113331007: 'Structure of endocrine
    system (body structure)'

(b) Generalisation and example instantiation of $?cluster_{18}$

Fig. 1. View and description of entities of $cluster_{18}$ in SNOMED hierarchy

the AminoAcid ontology. The first generalisation covers more than one axiom corresponding to a single entity in $?cluster_1$ (an example instantiation is shown in Figure 2). All the amino acids in the ontology are described following the same template, expressed by the set of 4 generalisations of Figure 2, abstracting 202 axioms. This is a clear indication of the impact the abstraction can achieve when trying to comprehend the ontology. The information that describes these entities is gathered in one place and expressed in a synthetic and meaningful way. That is, because each variable represents a cluster of entities. For example, the ontology engineer by looking the instantiation of the first generalisation understands that $cluster_5$ holds all the physicochemical properties of the amino acids and $cluster_2$ holds all the fillers of these properties.

An analysis of the generalisation coverage and axiom coverage for 85 ontologies from BioPortal is presented in Figure 3(a) and Figure 3(b) respectively. The results of Figure 3 show that there are ontologies, which have cluster coverage higher than 30% and in many cases the average number of instantiations per generalisation exceeds 20. It should be marked that most of the ontologies that have

Values:

?$cluster_1$: CLASS

?$cluster_1$ = {Alanine, Arginine, Aspartate, Cysteine, Glutamate, Glutamine, Histidine, Isoleucine, Leucine, Lysine, Methionine, Phenylalanine, Proline, Serine, Threonine, TinyAromaticAminoAcid, Tryptophan, Tyrosine, Valine, Glycine}

Generalisations:

1. ?$cluster_1$ *SubClassOf* ?$cluster_5$ **some** ?$cluster_3$
2. ?$cluster_1$.IRI?$cluster_7$"constant"
3. ?$cluster_1$ *SubClassOf* AminoAcid
4. *DisjointClasses*:'set(?$cluster_1$.VALUES)'

where ?$cluster_3$: CLASS,
?$cluster_7$: ANNOTATIONPROPERTY,
?$cluster_5$: OBJECTPROPERTY

Example Instantiations:

for the value ?$cluster_1$ = {Alanine}

1. Alanine *SubClassOf* hasSize **some** Tiny
   Alanine *SubClassOf* hasSideChainStructure
      some Aliphatic
   Alanine *SubClassOf* hasCharge **some** Neutral
   Alanine *SubClassOf* hasPolarity **some** Non-Polar
   Alanine *SubClassOf* hasHydrophobicity
      some Hydrophobic
2. Alanine *label* "Alanine"
3. Alanine *SubClassOf* AminoAcid
4. *DisjointClasses*: Alanine, Cysteine, Aspartate,
   Glutamate, Phenylalanine, Glycine, Histidine,
   Isoleucine, Lysine, Leucine, Methionine,
   Asparagine, Proline, Glutamine, Arginine, Serine,
   Threonine, Valine, Tryptophan, Tyrosine

Fig. 2. Values, generalisations and example instantiation of ?$cluster_1$ in the AminoAcid Ontology

a high cluster coverage percentile they also have a high average number of instantiations per generalisation. For example, in ontology 34 the cluster coverage per generalisation is 87% and the average number of axioms per generalisation is 560. These cases show that very few generalisations can summarise big number of axioms and can give an inclusive description of the clusters in the ontology.

Inspecting irregularities: In the example of Figure 2, we notice that a possible value of ?$cluster_1$ is the TinyAromaticAminoAcid. This value is covered only by the third generalisation. The axioms describing this class are:

1. TinyAromaticAminoAcid *EquivalentTo* AminoAcid **and** hasSize **some** Tiny
2. TinyAromaticAminoAcid *SubClassOf* AminoAcid

The second axiom is redundant causing the TinyAromaticAminoAcid class to result in the same cluster with the amino acids. By removing this axiom, the TinyAromaticAminoAcid no longer is a value of ?$cluster_1$. This irregularity is a design defect.

However, there were cases that entities were not included in a cluster because their description was a deliberate exception in the regularity. E.g., in SNOMED, the 'Surgical insertion - action (qualifier value)' is not included in $cluster_{16}$ as it is not used in a regular axiom like the 'Transplantation - action (qualifier value)':

α = 'Transplantation (procedure)' *EquivalentTo* 'Procedure by method (procedure)' **and** (RoleGroup **some** ('Method (attribute)' **some** 'Transplantation-action (qualifier value)'))),

$g(\alpha)$ = ?$cluster_{30}$ *EquivalentTo* 'Procedure by method (procedure) **and** (RoleGroup **some** (?$cluster_{74}$ **some** ?$cluster_{16}$)))

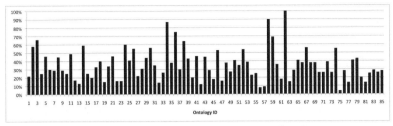

(a) average cluster coverage per generalisation

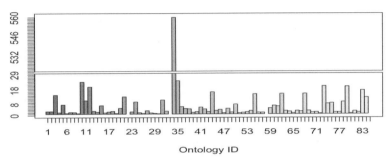

(b) average instantiations per generalisation

Fig. 3. Graph showing selected clustering results for 85 ontologies in BioPortal

Values:

?cluster$_6$: CLASS

?cluster$_6$ = {SmallHydrophilicAminoAcid,
  SmallHydrophobicAminoAcid,
  SmallNonPolarAminoAcid, SmallPolarAminoAcid,
  SmallPositiveAminoAcid},

Generalisation :

$g(\alpha)$ = ?cluster$_6$ *EquivalentTo* ?cluster$_2$ **and** ?cluster$_9$

where cluster$_2$, cluster$_9$: CLASS

Example instantiation:

α = SmallPositiveAminoAcid *EquivalentTo*
  PositiveChargedAminoAcid **and** SmallAminoAcid

Fig. 4. Values, generalisations and
example instantiation of ?cluster$_6$ in
the AminoAcid Ontology

Values:

?cluster$_8$ = {TinyHydrophobicAminoAcid,
TinyNonPolarAminoAcid, TinyPolarAminoAcid}

Generalisation :

$g(\alpha)$ = ?cluster$_8$ *EquivalentTo* AminoAcid
  and (?cluster$_5$ **some** ?cluster$_{10}$)
  and (?cluster$_5$ **some** ?cluster$_3$)

where ?cluster$_3$ = {Tiny (CLASS)}

?cluster$_5$: OBJECTPROPERTY,

?cluster$_{10}$: CLASS.

Example instantiation:

α_2 = TinyPolarAminoAcid *EquivalentTo*
  AminoAcid **and** hasSize **some** Tiny
  and hasPolarity **some** Polar

Fig. 5. Values, generalisations and
example instantiation of ?cluster$_8$ in
the AminoAcid Ontology

To evaluate the sensitivity of the framework, we edited regular axioms of the
ontologies to check if these can be identified by the algorithm. In the AminoAcid
ontology these irregularities already existed. In particular, cluster$_6$ and cluster$_8$
include equivalent classes in the ontology, which are used to categorise the amino
acids according to their chemical properties. The equivalent classes are grouped
into different clusters according to their regular usage and description. In Figures

4, 5 the description of cluster$_6$ and cluster$_8$ is presented respectively. Cluster$_6$ includes small amino acids with an additional chemical property. On the other hand, cluster$_8$ has a different design from this of cluster$_6$ (e.g. it would be expected to be α =TinyPolarAminoAcid *EquivalentTo* TinyAminoAcid **and** PolarAminoAcid). For preserving a regular design in the ontology, the terms of cluster$_8$ are transformed similar to the design of terms of cluster$_6$. This change has an impact on the clustering; the number of clusters is decreased, but these are more homogenous.

For the rest of the three ontologies, we removed 2 axioms that appear to have a regularity and check if the clustering framework corresponded to these changes. All the entities of which regular axioms were modified were discarded from their initial cluster because the regularity no longer existed.

4   Conclusions and Future Work

We presented a framework for identifying regularities and clustering the entities in an ontology according to these regularities. The application of the approach to 4 ontologies detected regularities that were expected to be found. These findings we confirm and evaluate through access to the ontology authors or the published documentation from the ontology's authors. The framework also provided an alternative presentation of an ontology based on its regularities, giving an insight about the major components of its construction. The generalisations provided a meaningful abstract form of axioms, in which each variable was representing a cluster of similar entities. This abstraction has potential as a tool for comprehension of an ontology, as it manages to summarise axioms in a coherent way. The method tended to give good coverage of an ontology's axioms within clusters, suggesting that it could show the significant portions of the ontology. The analysis on the BioPortal ontologies also highlighted such cases. That not all axioms are clustered is also meaningful; not all axioms can be part of regularities and those that do not cluster can indicate deviations from a style or simply deliberate authoring techniques. Either reason could be informative to a person comprehending an ontology.

The evaluation showed that changing regularities affected clustering results. Altering referencing axioms of entities that belonged to a cluster either caused them to be regrouped in a different cluster or their exclusion from any cluster. This shows the method to be sensitive to changes and a potential tool for helping authors to " tidy up" their ontology. Future work will include the development of plugins for the Protégé editor for supporting the development process dynamically through access to these features.

The inspection of regularities in known ontologies helped us to derive knowledge patterns when they existed. We expect that the further exploitation of the generalisation forms and the examination of possible combinations of these can lead to the induction of patterns in the ontology (e.g. like the pattern describing the amino acids in Figure 2). Future work will also involve alternative decisions of a transformation policy and of a clustering algorithm. Although the current implementation, which is based on a popularity transformation, produced adequate results, it will be worth examining other techniques that can generate homogenous and well defined clusters. The inspection of regularities in known ontologies helped us to derive knowledge patterns when they existed.

Our use of a basic clustering approach has a demonstrable use in finding regularities and irregularities in an ontology. It has potential for offering authors a means to gain generalisation of the major portions of an ontology; to detect deviations from a given style of representation and to facilitate the comprehension of what can be large and complex logical artefacts. As ontologies using OWL can be large and complex, the provision of techniques to manage this complexity, especially when attempting to understand an ontology's construction, should be an important addition to an ontology author's toolbox.

References

1. Snomed technical reference guide (January 2011), http://www.ihtsdo.org/fileadmin/user_upload/Docs_01/Publications/SNOMED_CT
2. Egaña, M., Rector, A., Stevens, R., Antezana, E.: Applying ontology design patterns in bio-ontologies (2008)
3. Gangemi, A.: Ontology Design Patterns for Semantic Web Content. In: Gil, Y., Motta, E., Benjamins, V.R., Musen, M.A. (eds.) ISWC 2005. LNCS, vol. 3729, pp. 262–276. Springer, Heidelberg (2005)
4. Gangemi, A., Presutti, V.: Ontology design patterns. In: Handbook on Ontologies, pp. 221–243. Springer, Heidelberg (2009)
5. Horridge, M., Parsia, B., Sattler, U.: Laconic and Precise Justifications in OWL. In: Sheth, A.P., Staab, S., Dean, M., Paolucci, M., Maynard, D., Finin, T., Thirunarayan, K. (eds.) ISWC 2008. LNCS, vol. 5318, pp. 323–338. Springer, Heidelberg (2008)
6. Iannone, L., Aranguren, M.E., Rector, A.L., Stevens, R.: Augmenting the expressivity of the ontology pre-processor language. In: OWLED. CEUR Workshop Proceedings, vol. 432. CEUR-WS.org (2008)
7. Iannone, L., Rector, A., Stevens, R.: Embedding Knowledge Patterns into OWL. In: Aroyo, L., Traverso, P., Ciravegna, F., Cimiano, P., Heath, T., Hyvönen, E., Mizoguchi, R., Oren, E., Sabou, M., Simperl, E. (eds.) ESWC 2009. LNCS, vol. 5554, pp. 218–232. Springer, Heidelberg (2009)
8. Jupp, S., Horridge, M., Iannone, L., Klein, J., Owen, S., Schanstra, J., Stevens, R., Wolstencroft, K.: Populous: A tool for populating ontology templates. Arxiv preprint arXiv:1012.1745 (2010)
9. Parsia, B., Sirin, E., Kalyanpur, A.: Debugging OWL ontologies. In: Proceedings of the 14th International Conference on World Wide Web, pp. 633–640. ACM (2005)
10. Peters, B., Ruttenberg, A., Greenbaum, J., Courtot, M., Brinkman, R., Whetzel, P., Schober, D., Sansone, S., Scheuerman, R., Rocca-Serra, P.: Overcoming the ontology enrichment bottleneck with quick term templates (2009)
11. Šváb-Zamazal, O., Svátek, V.: Analysing Ontological Structures Through Name Pattern Tracking. In: Gangemi, A., Euzenat, J. (eds.) EKAW 2008. LNCS (LNAI), vol. 5268, pp. 213–228. Springer, Heidelberg (2008)
12. Šváb-Zamazal, O., Svátek, V., Iannone, L.: Pattern-Based Ontology Transformation Service Exploiting OPPL and OWL-API. In: Cimiano, P., Pinto, H.S. (eds.) EKAW 2010. LNCS, vol. 6317, pp. 105–119. Springer, Heidelberg (2010)
13. Tan, P.-N., Steinbach, M., Kumar, V.: Introduction to Data Mining. Addison-Wesley (2005)

Dbpedia SPARQL Benchmark – Performance Assessment with Real Queries on Real Data*

Mohamed Morsey, Jens Lehmann, Sören Auer, and Axel-Cyrille Ngonga Ngomo

Department of Computer Science, University of Leipzig
Johannisgasse 26, 04103 Leipzig, Germany.
{lastname}@informatik.uni-leipzig.de

Abstract. Triple stores are the backbone of increasingly many Data Web applications. It is thus evident that the performance of those stores is mission critical for individual projects as well as for data integration on the Data Web in general. Consequently, it is of central importance during the implementation of any of these applications to have a clear picture of the weaknesses and strengths of current triple store implementations. In this paper, we propose a generic SPARQL benchmark creation procedure, which we apply to the DBpedia knowledge base. Previous approaches often compared relational and triple stores and, thus, settled on measuring performance against a relational database which had been converted to RDF by using SQL-like queries. In contrast to those approaches, our benchmark is based on queries that were actually issued by humans and applications against existing RDF data not resembling a relational schema. Our generic procedure for benchmark creation is based on query-log mining, clustering and SPARQL feature analysis. We argue that a pure SPARQL benchmark is more useful to compare existing triple stores and provide results for the popular triple store implementations Virtuoso, Sesame, Jena-TDB, and BigOWLIM. The subsequent comparison of our results with other benchmark results indicates that the performance of triple stores is by far less homogeneous than suggested by previous benchmarks.

1 Introduction

Triple stores, which use IRIs for entity identification and store information adhering to the RDF data model [9] are the backbone of increasingly many Data Web applications. The RDF data model resembles directed labeled graphs, in which each labeled edge (called *predicate*) connects a *subject* to an *object*. The intended semantics is that the *object* denotes the value of the *subject's* property *predicate*. With the W3C SPARQL standard [17] a vendor-independent query language for the RDF triple data model exists. SPARQL is based on powerful graph matching allowing to bind variables to fragments in the input RDF graph. In addition, operators akin to the relational joins, unions, left outer joins, selections and projections can be used to build more expressive queries [18]. It is evident that the performance of triple stores offering a SPARQL query interface is mission critical for individual projects as well as for data integration on the Web in

---

* This work was supported by a grant from the European Union's 7th Framework Programme provided for the project LOD2 (GA no. 257943).

L. Aroyo et al. (Eds.): ISWC 2011, Part I, LNCS 7031, pp. 454–469, 2011

general. It is consequently of central importance during the implementation of any Data Web application to have a clear picture of the weaknesses and strengths of current triple store implementations.

Existing SPARQL benchmark efforts such as LUBM [16], BSBM [4] and SP^2 [18] resemble relational database benchmarks. Especially the data structures underlying these benchmarks are basically relational data structures, with relatively few and homogeneously structured classes. However, RDF knowledge bases are increasingly heterogeneous. Thus, they do not resemble relational structures and are not easily representable as such. Examples of such knowledge bases are curated bio-medical ontologies such as those contained in Bio2RDF [2] as well as knowledge bases extracted from unstructured or semi-structured sources such as DBpedia [10] or LinkedGeoData [1]. DBpedia (version 3.6) for example contains 289,016 classes of which 275 classes belong to the DBpedia ontology. Moreover, it contains 42,016 properties, of which 1335 are DBpedia-specific. Also, various datatypes and object references of different types are used in property values. Such knowledge bases can *not* be easily represented according to the relational data model and hence performance characteristics for loading, querying and updating these knowledge bases might potentially be fundamentally different from knowledge bases resembling relational data structures.

In this article, we propose a generic SPARQL benchmark creation methodology. This methodology is based on a flexible data generation mimicking an input data source, query-log mining, clustering and SPARQL feature analysis. We apply the proposed methodology to datasets of various sizes derived from the DBpedia knowledge base. In contrast to previous benchmarks, we perform measurements on *real* queries that were issued by humans or Data Web applications against existing RDF data. We evaluate two different methods *data generation* approaches and show how a representative set of resources that preserves important dataset characteristics such as indegree and outdegree can be obtained by sampling across classes in the dataset. In order to obtain a representative set of *prototypical queries* reflecting the typical workload of a SPARQL endpoint, we perform a query analysis and clustering on queries that were sent to the official DBpedia SPARQL endpoint. From the highest-ranked query clusters (in terms of aggregated query frequency), we derive a set of 25 SPARQL query templates, which cover most commonly used SPARQL features and are used to generate the actual benchmark queries by parametrization. We call the benchmark resulting from this dataset and query generation methodology *DBPSB* (i.e. DBpedia SPARQL Benchmark). The benchmark methodology and results are also available online[1]. Although we apply this methodology to the DBpedia dataset and its SPARQL query log in this case, the same methodology can be used to obtain application-specific benchmarks for other knowledge bases and query workloads. Since the DBPSB can change with the data and queries in DBpedia, we envision to update it in yearly increments and publish results on the above website. In general, our methodology follows the four key requirements for domain specific benchmarks are postulated in the Benchmark Handbook [8], i.e. it is (1) relevant, thus testing typical operations within the specific domain, (2) portable, i.e. executable on different platforms, (3) scalable, e.g. it is possible to run the benchmark on both small and very large data sets, and (4) it is understandable.

---

[1] http://aksw.org/Projects/DBPSB

We apply the DBPSB to assess the performance and scalability of the popular triple stores *Virtuoso* [7], *Sesame* [5], *Jena-TDB* [15], and *BigOWLIM* [3] and compare our results with those obtained with previous benchmarks. Our experiments reveal that the performance and scalability is by far less homogeneous than other benchmarks indicate. As we explain in more detail later, we believe this is due to the different nature of DBPSB compared to the previous approaches resembling relational databases benchmarks. For example, we observed query performance differences of several orders of magnitude much more often than with other RDF benchmarks when looking at the runtimes of individual queries. The main observation in our benchmark is that previously observed differences in performance between different triple stores amplify when they are confronted with actually asked SPARQL queries, i.e. there is now a wider gap in performance compared to essentially relational benchmarks.

The remainder of the paper is organized as follows. In Section 2, we describe the dataset generation process in detail. We show the process of query analysis and clustering in detail in Section 3. In Section 4, we present our approach to selecting SPARQL features and to query variability. The assessment of four triple stores via the DBPSB is described in Sections 5 and 6. The results of the experiment are discussed in Section 7. We present related work in Section 8 and conclude our paper in Section 9.

2   Dataset Generation

A crucial step in each benchmark is the generation of suitable datasets. Although we describe the dataset generation here with the example of DBpedia, the methodology we pursue is dataset-agnostic.

The data generation for DBPSB is guided by the following requirements:

- The DBPSB data should resemble the original data (i.e., DBpedia data in our case) as much as possible, in particular the large number of classes, properties, the heterogeneous property value spaces as well as the large taxonomic structures of the category system should be preserved.
- The data generation process should allow to generate knowledge bases of various sizes ranging from a few million to several hundred million or even billion triples.
- Basic network characteristics of different sizes of the network should be similar, in particular the in- and outdegree.
- The data generation process should be easily repeatable with new versions of the considered dataset.

The proposed dataset creation process starts with an input dataset. For the case of DBpedia, it consists of the datasets loaded into the official SPARQL endpoint[2]. Datasets of multiple size of the original data are created by duplicating all triples and changing their namespaces. This procedure can be applied for any scale factors. While simple, this procedure is efficient to execute and fulfills the above requirements.

For generating smaller datasets, we investigated two different methods. The first method (called "rand") consists of selecting an appropriate fraction of all triples of

---

[2] Endpoint: http://dbpedia.org/sparql, Loaded datasets: http://wiki.dbpedia.org/DatasetsLoaded

Table 1. Statistical analysis of DBPSB datasets

Dataset	Indegree w/ literals	Outdegree w/ literals	Indegree w/o literals	Outdegree w/o literals	No. of nodes	No. of triples
Full DBpedia	5.45	30.52	3.09	15.57	27,665,352	153,737,776
10% dataset (seed)	6.54	45.53	3.98	23.05	2,090,714	15,267,418
10% dataset (rand)	3.82	6.76	2.04	3.41	5,260,753	16,739,055
50% dataset (seed)	6.79	38.08	3.82	18.64	11,317,362	74,889,154
50% dataset (rand)	7.09	26.79	3.33	10.73	9,581,470	78,336,781

the original dataset randomly. If RDF graphs are considered as small world graphs, removing edges in such graphs should preserve the properties of the original graph. The second method (called "seed") is based on the assumption that a representative set of resources can be obtained by sampling across classes in the dataset. Let x be the desired scale factor in percent, e.g. $x = 10$. The method first selects $x\%$ of the classes in the dataset. For each selected class, 10% of its instances are retrieved and added to a queue. For each element of the queue, its concise bound description (CBD) [19] is retrieved. This can lead to new resources, which are appended at the end of the queue. This process is iterated until the target dataset size, measured in number of triples, is reached.

Since the selection of the appropriate method for generating small datasets is an important issue, we performed a statistical analysis on the generated datasets for DBpedia. The statistical parameters used to judge the datasets are the average indegree, the average outdegree, and the number of nodes, i.e. number of distinct IRIs in the graph. We calculated both the in- and the outdegree for datasets once with literals ignored, and another time with literals taken into consideration, as it gives more insight on the degree of similarity between the dataset of interest and the full DBpedia dataset. The statistics of those datasets are given in Table 1. According to this analysis, the seed method fits our purpose of maintaining basic network characteristics better, as the average in- and outdegree of nodes are closer to the original dataset. For this reason, we selected this method for generating the DBPSB.

3   Query Analysis and Clustering

The goal of the query analysis and clustering is to detect prototypical queries that were sent to the official DBpedia SPARQL endpoint based on a query-similarity graph. Note that two types of similarity measures can been used on queries, i.e. string similarities and graph similarities. Yet, since graph similarities are very time-consuming and do not bear the specific mathematical characteristics necessary to compute similarity scores efficiently, we picked string similarities for our experiments. In the query analysis and clustering step, we follow a four-step approach. First, we select queries that were executed frequently on the input data source. Second, we strip common syntactic constructs (e.g., namespace prefix definitions) from these query strings in order to increase the conciseness of the query strings. Then, we compute a query similarity graph from the stripped queries. Finally, we use a soft graph clustering algorithm for computing clusters on this graph. These clusters are subsequently used to devise the query generation patterns used in the benchmark. In the following, we describe each of the four steps in more detail.

Query Selection. For the DBPSB, we use the DBpedia SPARQL query-log which contains all queries posed to the official DBpedia SPARQL endpoint for a three-month period in 2010[3]. For the generation of the current benchmark, we used the log for the period from April to July 2010. Overall, 31.5 million queries were posed to the endpoint within this period. In order to obtain a small number of distinctive queries for benchmarking triple stores, we reduce those queries in the following two ways:

– *Query variations.* Often, the same or slight variations of the same query are posed to the endpoint frequently. A particular cause of this is the renaming of query variables. We solve this issue by renaming all query variables in a consecutive sequence as they appear in the query, i.e., *var0*, *var1*, *var2*, and so on. As a result, distinguishing query constructs such as REGEX or DISTINCT are a higher influence on the clustering.
– *Query frequency.* We discard queries with a low frequency (below 10) because they do not contribute much to the overall query performance.

The application of both methods to the query log data set at hand reduced the number of queries from 31.5 million to just 35,965. This reduction allows our benchmark to capture the essence of the queries posed to DBpedia within the timespan covered by the query log and reduces the runtime of the subsequent steps substantially.

String Stripping. Every SPARQL query contains substrings that segment it into different clauses. Although these strings are essential during the evaluation of the query, they are a major source of noise when computing query similarity, as they boost the similarity score without the query patterns being similar per se. Therefore, we remove all SPARQL syntax keywords such as PREFIX, SELECT, FROM and WHERE. In addition, common prefixes (such as http://www.w3.org/2000/01/rdf-schema# for RDF-Schema) are removed as they appear in most queries.

Similarity Computation. The goal of the third step is to compute the similarity of the stripped queries. Computing the Cartesian product of the queries would lead to a quadratic runtime, i.e., almost 1.3 billion similarity computations. To reduce the runtime of the benchmark compilation, we use the LIMES framework [13][4]. The LIMES approach makes use of the interchangeability of similarities and distances. It presupposes a metric space in which the queries are expressed as single points. Instead of aiming to find all pairs of queries such that $sim(q, p) \geq \theta$, LIMES aims to find all pairs of queries such that $d(q, p) \leq \tau$, where *sim* is a similarity measure and *d* is the corresponding metric. To achieve this goal, when given a set of *n* queries, it first computes

---

[3] The DBpedia SPARQL endpoint is available at: http://dbpedia.org/sparql/ and the query log excerpt at: ftp://download.openlinksw.com/support/dbpedia/
[4] Available online at: http://limes.sf.net

$\sqrt{n}$ so-called *exemplars*, which are prototypical points in the affine space that subdivide it into regions of high heterogeneity. Then, each query is mapped to the exemplar it is least distant to. The characteristics of metrics spaces (especially the triangle inequality) ensures that the distances from each query q to any other query p obeys the following inequality

$$d(q, e) - d(e, p) \leq d(q, p) \leq d(q, e) + d(e, p), \tag{1}$$

where e is an exemplar and d is a metric. Consequently,

$$d(q, e) - d(e, p) > \tau \Rightarrow d(q, p) > \tau. \tag{2}$$

Given that $d(q, e)$ is constant, q must only be compared to the elements of the list of queries mapped to e that fulfill the inequality above. By these means, the number of similarity computation can be reduced significantly. In this particular use case, we cut down the number of computations to only 16.6% of the Cartesian product without any loss in recall. For the current version of the benchmark, we used the *Levenshtein* string similarity measure and a threshold of 0.9.

Clustering. The final step of our approach is to apply graph clustering to the query similarity graph computed above. The goal of this step is to discover very similar groups queries out of which prototypical queries can be generated. As a given query can obey the patterns of more than one prototypical query, we opt for using the soft clustering approach implemented by the BorderFlow algorithm[5].

BorderFlow [12] implements a seed-based approach to graph clustering. The default setting for the seeds consists of taking all nodes in the input graph as seeds. For each seed v, the algorithm begins with an initial cluster X containing only v. Then, it expands X iteratively by adding nodes from the direct neighborhood of X to X until X is node-maximal with respect to a function called the border flow ratio. The same procedure is repeated over all seeds. As different seeds can lead to the same cluster, identical clusters (i.e., clusters containing exactly the same nodes) that resulted from different seeds are subsequently collapsed to one cluster. The set of collapsed clusters and the mapping between each cluster and its seeds are returned as result. Applying BorderFlow to the input queries led to 12272 clusters, of which 24% contained only one node, hinting towards a long-tail distribution of query types. To generate the patterns used in the benchmark, we only considered clusters of size 5 and above.

4 SPARQL Feature Selection and Query Variability

After the completion of the detection of similar queries and their clustering, our aim is now to select a number of frequently executed queries that cover most SPARQL features and allow us to assess the performance of queries with single as well as combinations of features. The SPARQL features we consider are:

– the overall number of triple patterns contained in the query ($|GP|$),
– the graph pattern constructors UNION (*UON*), OPTIONAL (*OPT*),

---

[5] An implementation of the algorithm can be found at http://borderflow.sf.net

```
1  SELECT * WHERE {
2    { ?v2    a            dbp-owl:Settlement ;
3                          rdfs:label  %%v%% .
4       ?v6  a            dbp-owl:Airport . }
5    { ?v6  dbp-owl:city   ?v2 . }
6    UNION
7    { ?v6  dbp-owl:location   ?v2 . }
8    { ?v6  dbp-prop:iata      ?v5 . }
9    UNION
10   { ?v6  dbp-owl:iataLocationIdentifier  ?v5 . }
11   OPTIONAL { ?v6  foaf:homepage  ?v7 . }
12   OPTIONAL { ?v6  dbp-prop:nativename  ?v8 . }
13 }
```

Fig. 1. Sample query with placeholder

- the solution sequences and modifiers DISTINCT (*DST*),
- as well as the filter conditions and operators FILTER (*FLT*), LANG (*LNG*), REGEX (*REG*) and STR (*STR*).

We pick different numbers of triple patterns in order to include the efficiency of JOIN operations in triple stores. The other features were selected because they frequently occurred in the query log. We rank the clusters by the sum of the frequency of all queries they contain. Thereafter, we select 25 queries as follows: For each of the features, we choose the highest ranked cluster containing queries having this feature. From that particular cluster we select the query with the highest frequency.

In order to convert the selected queries into query templates, we manually select a part of the query to be varied. This is usually an IRI, a literal or a filter condition. In Figure 1 those varying parts are indicated by %%v%% or in the case of multiple varying parts %%vn%%. We exemplify our approach to replacing varying parts of queries by using Query 9, which results in the query shown in Figure 1. This query selects a specific settlement along with the airport belonging to that settlement as indicated in Figure 1. The variability of this query template was determined by getting a list of all settlements using the query shown in Figure 2. By selecting suitable placeholders, we ensured that the variability is sufficiently high ($\geq$ 1000 per query template). Note that the triple store used for computing the variability was different from the triple store that we later benchmarked in order to avoid potential caching effects.

For the benchmarking we then used the list of thus retrieved concrete values to replace the %%v%% placeholders within the query template. This method ensures, that (a) the actually executed queries during the benchmarking differ, but (b) always return results. This change imposed on the original query avoids the effect of simple caching.

5 Experimental Setup

This section presents the setup we used when applying the DBPSB on four triple stores commonly used in Data Web applications. We first describe the triple stores and their configuration, followed by our experimental strategy and finally the obtained results. All experiments were conducted on a typical server machine with an AMD Opteron

```
1    SELECT DISTINCT ?v WHERE {
2        { ?v2   a              dbp-owl:Settlement ;
3                rdfs:label  ?v .
4            ?v6  a              dbp-owl:Airport . }
5        { ?v6  dbp-owl:city  ?v2 . }
6        UNION
7            { ?v6  dbp-owl:location  ?v2 . }
8            { ?v6  dbp-prop:iata      ?v5 . }
9        UNION
10           { ?v6  dbp-owl:iataLocationIdentifier ?v5 . }
11       OPTIONAL { ?v6  foaf:homepage    ?v7 . }
12       OPTIONAL { ?v6  dbp-prop:nativename  ?v8 . }
13   } LIMIT 1000
```

Fig. 2. Sample auxiliary query returning potential values a placeholder can assume

6 Core CPU with 2.8 GHz, 32 GB RAM, 3 TB RAID-5 HDD running Linux Kernel 2.6.35-23-server and Java 1.6 installed. The benchmark program and the triple store were run on the same machine to avoid network latency.

Triple Stores Setup. We carried out our experiments by using the triple stores *Virtuoso* [7], *Sesame* [5], *Jena-TDB* [15], and *BigOWLIM* [3]. The configuration and the version of each triple store were as follows:

1. **Virtuoso** Open-Source Edition version 6.1.2: We set the following memory-related parameters: NumberOfBuffers = 1048576, MaxDirtyBuffers = 786432.
2. **Sesame** Version 2.3.2 with Tomcat 6.0 as HTTP interface: We used the native storage layout and set the spoc, posc, opsc indices in the native storage configuration. We set the Java heap size to 8GB.
3. **Jena-TDB** Version 0.8.7 with Joseki 3.4.3 as HTTP interface: We configured the TDB optimizer to use statistics. This mode is most commonly employed for the TDB optimizer, whereas the other modes are mainly used for investigating the optimizer strategy. We also set the Java heap size to 8GB.
4. **BigOWLIM** Version 3.4, with Tomcat 6.0 as HTTP interface: We set the entity index size to 45,000,000 and enabled the predicate list. The rule set was empty. We set the Java heap size to 8GB.

In summary, we configured all triple stores to use 8GB of memory and used default values otherwise. This strategy aims on the one hand at benchmarking each triple store in a real context, as in real environment a triple store cannot dispose of the whole memory up. On the other hand it ensures that the whole dataset cannot fit into memory, in order to avoid caching.

Benchmark Execution. Once the triple stores loaded the DBpedia datasets with different scale factors, i.e. 10%, 50%, 100%, and 200%, the benchmark execution phase began. It comprised the following stages:

1. **System Restart:** Before running the experiment, the triple store and its associated programs were restarted in order to clear memory caches.

2. **Warm-up Phase:** In order to measure the performance of a triple store under normal operational conditions, a warm-up phase was used. In the warm-up phase, query mixes were posed to the triple store. The queries posed during the warm-up phase were disjoint with the queries posed in the hot-run phase. For DBPSB, we used a warm-up period of 20 minutes.
3. **Hot-run Phase:** During this phase, the benchmark query mixes were sent to the tested store. We kept track of the average execution time of each query as well as the number of query mixes per hour (QMpH). The duration of the hot-run phase in DBPSB was 60 minutes.

Since some benchmark queries did not respond within reasonable time, we specified a 180 second timeout after which a query was aborted and the 180 second maximum query time was used as the runtime for the given query even though no results were returned. The benchmarking code along with the DBPSB queries is freely available[6].

6   Results

We evaluated the performance of the triple stores with respect to two main metrics: their overall performance on the benchmark and their query-based performance.

The overall performance of any triple store was measured by computing its query mixes per hour (QMpH) as shown in Figure 4. Please note that we used a logarithmic scale in this figure due to the high performance differences we observed. In general, Virtuoso was clearly the fastest triple store, followed by BigOWLIM, Sesame and Jena-TDB. The highest observed ratio in QMpH between the fastest and slowest triple store was 63.5 and it reached more than 10 000 for single queries. The scalability of stores did not vary as much as the overall performance. There was on average a linear decline in query performance with increasing dataset size. Details will be discussed in Section 7.

We tested the queries that each triple store failed to executed withing the 180s timeout and noticed that even much larger timeouts would not have been sufficient most of those queries. We did not exclude the queries completely from the overall assessment, since this would have affected a large number of the queries and adversely penalized stores, which complete queries within the time frame. We penalized failure queries with 180s, similar to what was done in the SP2-Benchmark [18]. Virtuoso was the only store, which completed all queries in time. For Sesame and OWLIM only rarely a few particular queries timed out. Jena-TDB had always severe problems with queries 7, 10 and 20 as well as 3, 9, 12 for the larger two datasets.

The metric used for query-based performance evaluation is Queries per Second (QpS). QpS is computed by summing up the runtime of each query in each iteration, dividing it by the QMpH value and scaling it to seconds. The QpS results for all triple stores and for the 10%, 50%, 100%, and 200% datasets are depicted in Figure 3.

The outliers, i.e. queries with very low QpS, will significantly affect the mean value of QpS for each store. So, we additionally calculated the geometric mean of all the QpS timings of queries for each store. The geometric mean for all triple stores is also depicted in Figure 4. By reducing the effect of outliers, we obtained additional information from this figure as we will describe in the subsequent section.

---

[6] https://akswbenchmark.svn.sourceforge.net/svnroot/akswbenchmark/

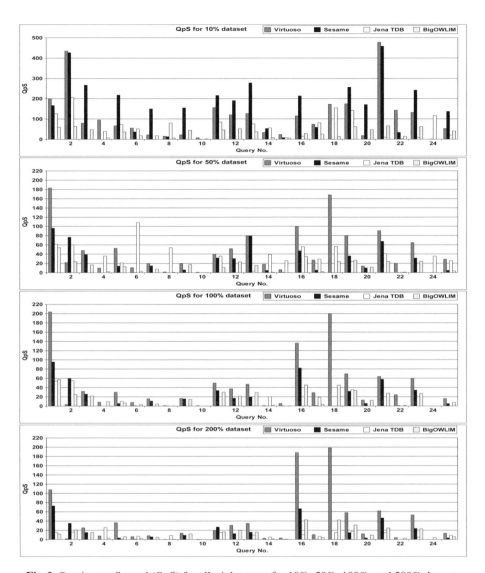

Fig. 3. Queries per Second (QpS) for all triple stores for 10%, 50%, 100%, and 200% datasets

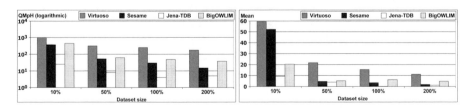

Fig. 4. QMpH for all triple stores (left). Geometric mean of QpS (right).

7   Discussion

This section consists of three parts: First, we compare the general performance of the systems under test. Then we look individual queries and the SPARQL features used within those queries in more detail to observe particular strengths and weaknesses of stores. Thereafter, we compare our results with those obtained with previous benchmarks and elucidate some of the main differences between them.

General Performance. Figure 4 depicts the benchmark results for query mixes per hour for the four systems and dataset sizes. Virtuoso leads the field with a substantial head start of double the performance for the 10% dataset (and even quadruple for other dataset sizes) compared to the second best system (BigOWLIM). While Sesame is able to keep up with BigOWLIM for the smaller two datasets it considerably looses ground for the larger datasets. Jena-TDB can in general not deliver competitive performance with being by a factor 30-50 slower than the fastest system.

If we look at the geometric mean of all QpS results in Figure 4, we observe similar insights. The spreading effect is weakened, since the geometric mean reduces the effect of outliers. Still Virtuoso is the fastest system, although Sesame manages to get pretty close for the 10% dataset. This shows that most, but not all, queries are fast in Sesame for low dataset sizes. For the larger datasets, BigOWLIM is the second best system and shows promising scalability, but it is still by a factor of two slower than Virtuoso.

Scalability, Individual Queries and SPARQL Features. Our first observation with respect to individual performance of the triple stores is that Virtuoso demonstrates a good scaling factor on the DBPSB. When dataset size changes by factor 5 (from 10% to 50%), the performance of the triple store only degrades by factor 3.12. Further dataset increases (i.e. the doubling to the 100% and 200% datasets) result in only relatively small performance decreases by 20% and respectively 30%.

Virtuoso outperforms Sesame for all datasets. In addition, Sesame does not scale as well as Virtuoso for small dataset sizes, as its performance degrades sevenfold when the dataset size changes from 10% to 50%. However, when the dataset size doubles from the 50% to the 100% dataset and from 100% to 200% the performance degrades by just half.

The performance of Jena-TDB is the lowest of all triple stores and for all dataset sizes. The performance degradation factor of Jena-TDB is not as pronounced as that of Sesame and almost equal to that of Virtuoso when changing from the 10% to the 50% dataset. However, the performance of Jena-TDB only degrades by a factor of 2 for the transition between the 50% and 100% dataset, and reaches 0.8 between the 100% and 200% dataset, leading to a slight increase of its QMpH.

BigOWLIM is the second fastest triple store for all dataset sizes, after Virtuoso. BigOWLIM degrades with a factor of 7.2 in transition from 10% to 50% datasets, but it decreases dramatically to 1.29 with dataset size 100%, and eventually reaches 1.26 with dataset size 200%.

Due to the high diversity in the performance of different SPARQL queries, we also computed the geometric mean of the QpS values of all queries as described in the previous section and illustrated in Figure 4. By using the geometric mean, the resulting values are less prone to be dominated by a few outliers (slow queries) compared to standard QMpH values. This allows for some interesting observations in DBPSB by comparing Figure 4 and 4. For instance, it is evident that Virtuoso has the best QpS values for all dataset sizes.

With respect to Virtuoso, query 10 performs quite poorly. This query involves the features FILTER, DISTINCT, as well as OPTIONAL. Also, the well performing query 1 involves the DISTINCT feature. Query 3 involves a OPTIONAL resulting in worse performance. Query 2 involving a FILTER condition results in the worst performance of all of them. This indicates that using complex FILTER in conjunction with additional OPTIONAL, and DISTINCT adversely affects the overall runtime of the query.

Regarding Sesame, queries 4 and 18 are the slowest queries. Query 4 includes UNION along with several free variables, which indicates that using UNION with several free variables causes problems for Sesame. Query 18 involves the features UNION, FILTER, STR and LANG. Query 15 involves the features UNION, FILTER, and LANG, and its performance is also pretty slow, which leads to the conclusion that introducing this combination of features is difficult for Sesame. Adding the STR feature to that feature combination affects the performance dramatically and prevents the query from being successfully executed.

For Jena-TDB, there are several queries that timeout with large dataset sizes, but queries 10 and 20 always timeout. The problem with query 10 is already discussed with Virtuoso. Query 20 contains FILTER, OPTIONAL, UNION, and LANG. Query 2 contains FILTER only, query 3 contains OPTIONAL, and query 4 contains UNION only. All of those queries run smoothly with Jena-TDB, which indicates that using the LANG feature, along with those features affects the runtime dramatically.

For BigOWLIM, queries 10, and 15 are slow queries. Query 10 was already problematic for Virtuoso, as was query 15 for Sesame.

Query 24 is slow on Virtuoso, Sesame, and BigOWLIM, whereas it is faster on Jena-TDB. This is due to the fact that most of the time this query returns many results. Virtuoso, and BigOWLIM return a bulk of results at once, which takes long time. Jena-TDB just returns the first result as a starting point, and iteratively returns the remaining results via a buffer.

It is interesting to note that BigOWLIM shows in general good performance, but almost never manages to outperform any of the other stores. Queries 11, 13, 19, 21 and 25 were performed with relatively similar results across triple stores thus indicating that the features of these queries (i.e. UON, REG, FLT) are already relatively well supported. With queries 3, 4, 7, 9, 12, 18, 20 we observed dramatic differences between the different implementations with factors between slowest and fastest store being higher than 1000. It seems that a reason for this could be the poor support for OPT (in queries 3, 7, 9, 20) as well as certain filter conditions such as LNG in some implementations, which demonstrates the need for further optimizations.

Comparison with Previous Benchmarks. In order to visualize the performance improvement or degradation of a certain triple store compared to its competitors, we calculated the relative performance for each store compared to the average and depicted it for each dataset size in Figure 5. We also performed this calculation for BSBM version 2 and version 3. Overall, the benchmarking results with DBPSB were less homogeneous than the

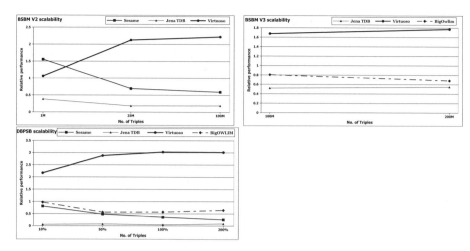

Fig. 5. Comparison of triple store scalability between BSBM V2, BSBM V3, DBPSB

results of previous benchmarks. While with other benchmarks the ratio between fastest and slowest query rarely exceeds a factor of 50, the factor for the DBPSB queries (derived from real DBpedia SPARQL endpoint queries) reaches more than 1 000 in some cases.

As with the other benchmarks, Virtuoso was also fastest in our measurements. However, the performance difference is even higher than reported previously: Virtuoso reaches a factor of 3 in our benchmark compared to 1.8 in BSBM V3. BSBM V2 and our benchmark both show that Sesame is more suited to smaller datasets and does not scale as well as other stores. Jena-TDB is the slowest store in BSBM V3 and DBPSB, but in our case they fall much further behind to the point that Jena-TDB can hardly be used for some of the queries, which are asked to DBpedia. The main observation in our benchmark is that previously observed differences in performance between different triple stores amplify when they are confronted with actually asked SPARQL queries, i.e. there is now a wider gap in performance compared to essentially relational benchmarks.

8   Related Work

Several RDF benchmarks were previously developed. The *Lehigh University Benchmark* (LUBM) [16] was one of the first RDF benchmarks. LUBM uses an artificial data generator, which generates synthetic data for universities, their departments, their professors, employees, courses and publications. This small number of classes limits the variability of data and makes LUMB inherent structure more repetitive. Moreover, the SPARQL queries used for benchmarking in LUBM are all plain queries, i.e. they contain only triple patterns with no other SPARQL features (e.g. FILTER, or REGEX). LUBM performs each query 10 consecutive times, and then it calculates the average response time of that query. Executing the same query several times without introducing any variation enables query caching, which affects the overall average query times.

Table 2. Comparison of different RDF benchmarks

	LUBM	SP²Bench	BSBM V2	BSBM V3	DBPSB
RDF stores tested	DLDB-OWL, Sesame, OWL-JessKB	ARQ, Redland, SDB, Sesame, Virtuoso	Virtuoso, Sesame, Jena-TDB, Jena-SDB	Virtuoso, 4store, BigData Jena-TDB BigOwlim	Virtuoso, Jena-TDB, BigOWLIM Sesame
Test data	Synthetic	Synthetic	Synthetic	Synthetic	Real
Test queries	Synthetic	Synthetic	Synthetic	Synthetic	Real
Size of tested datasets	0.1M, 0.6M, 1.3M, 2.8M, 6.9M	10k, 50k, 250k, 1M,	1M, 25M, 100M, 5M, 25M	100M, 200M	14M, 75M, 150M, 300M
Dist. queries	14	12	12	12	25
Multi-client	–	–	x	x	–
Use case	Universities	DBLP	E-commerce	E-commerce	DBpedia
Classes	43	8	8	8	239 (internal) +300K(YAGO)
Properties	32	22	51	51	1200

SP²Bench [18] is another more recent benchmark for RDF stores. Its RDF data is based on the Digital Bibliography & Library Project (DBLP) and includes information about publications and their authors. It uses the SP²Bench Generator to generate its synthetic test data, which is in its schema heterogeneity even more limited than LUMB. The main advantage of SP²Bench over LUBM is that its test queries include a variety of SPARQL features (such as FILTER, and OPTIONAL). The main difference between the DBpedia benchmark and SP²Bench is that both test data and queries are synthetic in SP²Bench. In addition, SP²Bench only published results for up to 25M triples, which is relatively small with regard to datasets such as DBpedia and LinkedGeoData.

Another benchmark described in [14] compares the performance of BigOWLIM and AllegroGraph. The size of its underlying synthetic dataset is 235 million triples, which is sufficiently large. The benchmark measures the performance of a variety of SPARQL constructs for both stores when running in single and in multi-threaded modes. It also measures the performance of adding data, both using bulk-adding and partitioned-adding. The downside of that benchmark is that it compares the performance of only two triple stores. Also the performance of each triple store is not assessed for different dataset sizes, which prevents scalability comparisons.

The Berlin SPARQL Benchmark (BSBM) [4] is a benchmark for RDF stores, which is applied to various triple stores, such as Sesame, Virtuoso, and Jena-TDB. It is based on an e-commerce use case in which a set of products is provided by a set of vendors and consumers post reviews regarding those products. It tests various SPARQL features on those triple stores. It tries to mimic a real user operation, i.e. it orders the query in a manner that resembles a real sequence of operations performed by a human user. This is an effective testing strategy. However, BSBM data and queries are artificial and the data schema is very homogeneous and resembles a relational database. This is reasonable for comparing the performance of triple stores with RDBMS, but does not give many insights regarding the specifics of RDF data management.

A comparison between benchmarks is shown in Table 2. In addition to general purpose RDF benchmarks it is reasonable to develop benchmarks for specific RDF data management aspects. One particular important feature in practical RDF triple store usage scenarios (as was also confirmed by DBPSB) is full-text search on RDF literals. In [11] the LUBM benchmark is extended with synthetic scalable fulltext data and corresponding queries for fulltext-related query performance evaluation. RDF stores are benchmarked for basic fulltext queries (classic IR queries) as well as hybrid queries (structured and fulltext queries).

9   Conclusions and Future Work

We proposed the DBPSB benchmark for evaluating the performance of triple stores based on non-artificial data and queries. Our solution was implemented for the DBpedia dataset and tested with 4 different triple stores, namely Virtuoso, Sesame, Jena-TDB, and BigOWLIM. The main advantage of our benchmark over previous work is that it uses real RDF data with typical graph characteristics including a large and heterogeneous schema part. Furthermore, by basing the benchmark on queries asked to DBpedia, we intend to spur innovation in triple store performance optimisation towards scenarios, which are actually important for end users and applications. We applied query analysis and clustering techniques to obtain a diverse set of queries corresponding to feature combinations of SPARQL queries. Query variability was introduced to render simple caching techniques of triple stores ineffective.

The benchmarking results we obtained reveal that real-world usage scenarios can have substantially different characteristics than the scenarios assumed by prior RDF benchmarks. Our results are more diverse and indicate less homogeneity than what is suggested by other benchmarks. The creativity and inaptness of real users while constructing SPARQL queries is reflected by DBPSB and unveils for a certain triple store and dataset size the most costly SPARQL feature combinations.

Several improvements can be envisioned in future work to cover a wider spectrum of features in DBPSB:

- Coverage of more SPARQL 1.1 features, e.g. reasoning and subqueries.
- Inclusion of further triple stores and continuous usage of the most recent DBpedia query logs.
- Testing of SPARQL update performance via DBpedia Live, which is modified several thousand times each day. In particular, an analysis of the dependency of query performance on the dataset update rate could be performed.

In addition, we will further investigate the data generation process in future work, in particular based on recent work such as [6].

References

1. Auer, S., Lehmann, J., Hellmann, S.: LinkedGeoData: Adding a Spatial Dimension to the Web of Data. In: Bernstein, A., Karger, D.R., Heath, T., Feigenbaum, L., Maynard, D., Motta, E., Thirunarayan, K. (eds.) ISWC 2009. LNCS, vol. 5823, pp. 731–746. Springer, Heidelberg (2009)

2. Belleau, F., Nolin, M.-A., Tourigny, N., Rigault, P., Morissette, J.: Bio2rdf: Towards a mashup to build bioinformatics knowledge systems. Journal of Biomedical Informatics 41(5), 706–716 (2008)
3. Bishop, B., Kiryakov, A., Ognyanoff, D., Peikov, I., Tashev, Z., Velkov, R.: Owlim: A family of scalable semantic repositories. Semantic Web 2(1), 1–10 (2011)
4. Bizer, C., Schultz, A.: The Berlin SPARQL Benchmark. Int. J. Semantic Web Inf. Syst. 5(2), 1–24 (2009)
5. Broekstra, J., Kampman, A., van Harmelen, F.: Sesame: A generic architecture for storing and querying RDF and RDF schema. In: Horrocks, I., Hendler, J. (eds.) ISWC 2002. LNCS, vol. 2342, pp. 54–68. Springer, Heidelberg (2002)
6. Duan, S., Kementsietsidis, A., Srinivas, K., Udrea, O.: Apples and oranges: a comparison of RDF benchmarks and real RDF datasets. In: Proceedings of the ACM SIGMOD International Conference on Management of Data, pp. 145–156. ACM (2011)
7. Erling, O., Mikhailov, I.: RDF support in the virtuoso DBMS. In: Auer, S., Bizer, C., Müller, C., Zhdanova, A.V. (eds.) CSSW. LNI, vol. 113, pp. 59–68. GI (2007)
8. Gray, J. (ed.): The Benchmark Handbook for Database and Transaction Systems, 1st edn. Morgan Kaufmann (1991)
9. Klyne, G., Carroll, J.J.: Resource description framework (RDF): Concepts and abstract syntax. W3C Recommendation (February 2004)
10. Lehmann, J., Bizer, C., Kobilarov, G., Auer, S., Becker, C., Cyganiak, R., Hellmann, S.: DBpedia - a crystallization point for the web of data. Journal of Web Semantics 7(3), 154–165 (2009)
11. Minack, E., Siberski, W., Nejdl, W.: Benchmarking Fulltext Search Performance of RDF Stores. In: Aroyo, L., Traverso, P., Ciravegna, F., Cimiano, P., Heath, T., Hyvönen, E., Mizoguchi, R., Oren, E., Sabou, M., Simperl, E. (eds.) ESWC 2009. LNCS, vol. 5554, pp. 81–95. Springer, Heidelberg (2009)
12. Ngonga Ngomo, A.-C., Schumacher, F.: BorderFlow: A local graph clustering algorithm for natural language processing. In: Gelbukh, A. (ed.) CICLing 2009. LNCS, vol. 5449, pp. 547–558. Springer, Heidelberg (2009)
13. Ngonga Ngomo, A.-C., Auer, S.: Limes - a time-efficient approach for large-scale link discovery on the web of data. In: Proceedings of IJCAI (2011)
14. Owens, A., Gibbins, N., Schraefel, m.c.: Effective benchmarking for rdf stores using synthetic data (May 2008)
15. Owens, A., Seaborne, A., Gibbins, N., Schraefel, m.c.: Clustered TDB: A clustered triple store for jena. Technical report, Electronics and Computer Science, University of Southampton (2008)
16. Pan, Z., Guo, Y., Heflin, J.: LUBM: A benchmark for OWL knowledge base systems. Journal of Web Semantics 3, 158–182 (2005)
17. Prud'hommeaux, E., Seaborne, A.: SPARQL Query Language for RDF. W3C Recommendation (2008)
18. Schmidt, M., Hornung, T., Lausen, G., Pinkel, C.: SP2Bench: A SPARQL performance benchmark. In: ICDE, pp. 222–233. IEEE (2009)
19. Stickler, P.: CBD - concise bounded description (2005), http://www.w3.org/Submission/CBD/ (retrieved February 15, 2011)

A Novel Approach to
Visualizing and Navigating Ontologies

Enrico Motta[1], Paul Mulholland[1], Silvio Peroni[2], Mathieu d'Aquin[1],
Jose Manuel Gomez-Perez[3], Victor Mendez[3], and Fouad Zablith[1]

[1] Knowledge Media Institute, The Open University, MK7 6AA, Milton Keynes, UK
{e.motta,p.mulholland,m.daquin,f.zablith}@open.ac.uk
[2] Dept. of Computer Science, University of Bologna, 40127 Bologna, Italy
speroni@cs.unibo.it
[3] Intelligent Software Components (iSOCO) S.A., 28042 Madrid Spain
{jmgomez,vmendez}@isoco.com

Abstract. Observational studies in the literature have highlighted low levels of
user satisfaction in relation to the support for ontology visualization and
exploration provided by current ontology engineering tools. These issues are
particularly problematic for non-expert users, who rely on effective tool support
to abstract from representational details and to be able to make sense of the
contents and the structure of ontologies. To address these issues, we have
developed a novel solution for visualizing and navigating ontologies, *KC-Viz*,
which exploits an empirically-validated ontology summarization method, both
to provide concise views of large ontologies, and also to support a 'middle-out'
ontology navigation approach, starting from the most information-rich nodes
(*key concepts*). In this paper we present the main features of KC-Viz and also
discuss the encouraging results derived from a preliminary empirical
evaluation, which suggest that the use of KC-Viz provides performance
advantages to users tackling realistic browsing and visualization tasks.
Supplementary data gathered through questionnaires also convey additional
interesting findings, including evidence that prior experience in ontology
engineering affects not just objective performance in ontology engineering tasks
but also subjective views on the usability of ontology engineering tools.

Keywords: Ontology Visualization, Key Concepts, Ontology Summarization,
Ontology Navigation, Ontology Engineering Tools, Empirical Evaluation.

1 Introduction

Browsing ontologies to make sense of their contents and organization is an essential
activity in ontology engineering. This is particularly the case today, as the significant
increase in the number of ontologies available online means that ontology engineering
projects often include a reuse activity, where people first locate ontologies which may
be relevant to their project – e.g., by using ontology search engines, such as Sindice
[1] or Watson [2], and then examine them to understand to what extent they provide
solutions to their modelling needs.

L. Aroyo et al. (Eds.): ISWC 2011, Part I, LNCS 7031, pp. 470–486, 2011.

In addition, ontologies are no longer developed and used exclusively by specialized researchers and practitioners. On the contrary, as ontologies are increasingly used in a variety of scenarios, such as research, healthcare, and business, more and more domain experts and other relatively inexperienced users are involved in the ontology engineering process, especially in the context of community-wide ontology development activities [3].

However, evidence gathered through observational studies [4] indicates low levels of user satisfaction with the tool support currently available to users for visualizing and navigating ontologies, in particular in relation to the lack of effective mechanisms for 'content-level visualization', including support for selective visualization of ontology parts, summaries, and overviews [4]. Needless to say, these problems affect in particular inexperienced users, who rely on effective tool support to abstract from representational details and make sense of the contents and the structure of ontologies.

Attempting to address these issues, we have developed a novel solution for visualizing and navigating ontologies, *KC-Viz,* which builds on our earlier work on *key concepts extraction* [5], both as a way to provide concise overviews of large ontologies, and also to support a 'middle-out' ontology navigation approach, starting from the most information-rich nodes[1] (*key concepts*). Building on its ability to abstract out from large ontologies through key concept extraction, KC-Viz provides a rich set of navigation and visualization mechanisms, including flexible *zooming* into and *hiding* of specific parts of an ontology, *history* browsing, saving and loading of customized *ontology views,* as well as essential interface customization support, such as graphical zooming, font manipulation, tree layout customization, and other functionalities. KC-Viz is a core plugin of the NeOn Toolkit and can be downloaded from http://neon-toolkit.org.

In this paper we introduce KC-Viz and we present the results from a preliminary empirical evaluation, which suggest that the use of KC-Viz provides performance advantages to users tackling realistic browsing and visualization tasks. Moreover, we also report on additional findings gathered through questionnaires, which offer a number of other insights, including evidence that prior experience in ontology engineering affects not just objective performance in ontology engineering tasks but also subjective views on the usability of ontology engineering tools.

2 Approaches to Visualizing and Navigating Ontologies

2.1 Literature Review

The issue of how best to support visualization and navigation of ontologies has attracted much attention in the research community. As Wang and Parsia emphasize [6], "effective presentation of the hierarchies can be a big win for the users", in particular, but not exclusively, during the early stages of a *sensemaking*[2] process,

---

[1] In the paper we will use the terms 'node', 'concept', and 'class' interchangeably to refer to classes in an ontology.

[2] In the rest of the paper we will use the term 'sensemaking' to refer to a specific ontology engineering task, where the user is primarily concerned with understanding the contents and overall structure of the ontology, i.e., acquiring an overview of the concepts covered by the ontology and the way they are organized in a taxonomy.

when a user is trying to build an initial mental model of an ontology, focusing less on specific representational details than on understanding the overall organization of the ontology. In particular, as discussed in [7], there are a number of functionalities that an effective visualization system needs to support, including (but not limited to) the ability to provide *high level overviews* of the data, *to zoom in* effectively on specific parts of the data, and *to filter out* irrelevant details and/or irrelevant parts of the data.

An approach to addressing the issue of providing high level overviews of hierarchical structures focuses on maximizing the amount of information on display, through *space-filling* solutions, such as those provided by *treemaps* [8]. Treemaps have proved to be a very successful and influential visualization method, used not just to represent conceptual hierarchies but also to visualize information in several mainstream sectors, including news, politics, stock market, sport, etc. However, while treemaps define a clever way to provide concise overviews of very large hierarchical spaces, they are primarily effective when the focus is on leaf nodes and on a particular dimension of visualization, in particular if colour-coding can be used to express different values for the dimension in question. However, as pointed out in [6], treemaps are not necessarily effective in supporting an understanding of topological structures, which is what is primarily needed in the ontology sensemaking context highlighted earlier.

State of the art ontology engineering toolkits, such as Protégé[3] and TopBraid Composer[4], include visualization systems which use the familiar *node-link diagram* paradigm to represent entities in an ontology and their taxonomic or domain relationships. In particular, both the OwlViz visualizer in Protégé and the 'Graph View' in TopBraid make it possible for users to navigate the ontology hierarchy by selecting, expanding and hiding nodes. However OwlViz arguably provides more flexibility, allowing the user to customize the expansion radius and supporting different modalities of use, including the option of automatically visualizing in OwlViz the current selection shown in the Protégé Class Browser.

SpaceTree [9], which also follows the *node-link diagram* paradigm, is able to maximize the number of nodes on display, by assessing how much empty space is available. At the same time it also avoids clutter by utilizing informative preview icons. These include miniatures of a branch, which are able to give the user an idea of the size and shape of an un-expanded subtree at a very high level of abstraction, while minimizing the use of real estate.

Like treemaps, CropCircles [6] also uses geometric containment as an alternative to classic node-link displays. However, it tries to address the key weakness of treemaps, by sacrificing space in order to make it easier for users to understand the topological relations in an ontology, including both parent-child and sibling relations. An empirical evaluation comparing the performance of users on topological tasks using treemaps, CropCircles and SpaceTree showed that, at least for some tasks, users of CropCircles performed significantly better than those using treemaps [6]. However, SpaceTree appears to perform significantly better than either treemaps or CropCircles on node finding tasks.

---

[3] http://protege.stanford.edu/

[4] http://www.topquadrant.com/products/TB_Composer.html

A number of 'hybrid' solutions also exist, such as Jambalaya [10] and Knoocks [11], which attempt to combine the different strengths of containment-based and node-link approaches in an integrated framework, by providing both alternative visualizations as well as hybrid, integrated views of the two paradigms.

The group of techniques categorized in [12] as *"context + focus and distortion"* are based on "the notion of distorting the view of the presented graph in order to combine context and focus. The node on focus is usually the central one and the rest of the nodes are presented around it, reduced in size until they reach a point that they are no longer visible" [12]. These techniques are normally based on hyperbolic views of the data and offer a good trade-off – a part of the ontology is shown in detailed view, while the rest is depicted around. A good exemplar of this class of approaches is *HyperTree* [13].

Finally, we should also consider in this short survey the most ubiquitous and least visual class of tools, exemplified by plugins such as the Class Browser in Protégé and the Ontology Navigator in the NeOn Toolkit. These follow the classic file system navigation metaphor, where clicking on a folder opens up its sub-folders. This approach is ubiquitous in both file system interfaces and ontology engineering tools and, in the case of ontologies, it allows the user to navigate the ontology hierarchy simply by clicking on the identifier of a class, to display its subclasses, and so on. While superficially a rather basic solution, especially when compared to some of the sophisticated visual metaphors that can be found in the literature, this approach can be surprisingly effective for two reasons: i) it is very familiar to users and ii) it makes it possible to display quite a lot of information in a rather small amount of space, in contrast with node-link displays, which can be space-hungry. As a result it is not surprising that these interfaces often perform better in evaluation scenarios than the graphical alternatives. For instance, the evaluation reported in [14] shows that subjects using the Protégé Class Browser fared better than those using alternative visualization plugins in a number of ontology engineering tasks.

2.2 Discussion

It is clear from the review in the previous section that different approaches exhibit different strengths and weaknesses and that in general the effectiveness of a particular solution depends on the specific task it is being used for. For example, the evaluation presented in [6] suggests that CropCircles may perform well in 'abstract' topological tasks, such as "Find the class with the most direct subclasses", but SpaceTree appears to be better in locating a specific class. As already mentioned, here we are primarily concerned with the ontology sensemaking task, so what we are looking for is effective support for the user in quickly understanding what are the main areas covered by the ontology, how is the main hierarchy structured, etc.

The problem is a particularly tricky one because, once an ontology is large enough, it is not possible to show its entire structure in the limited space provided by a computer screen and therefore a difficult trade-off needs to be addressed. On the one hand the information on display needs to be coarse-grained enough to provide an overview of the ontology, thus ensuring the user can maintain an overall mental model of the ontology. On the other hand, an exploration process needs to be supported, where the user can effectively home in on parts of the ontology, thus changing the level of analysis, while at the same time not losing track of the overall organization of the ontology. In sum, we can say that the main (although obviously not the only) issue is one of reconciling abstraction with focus.

However, a problem affecting all the approaches discussed in the review is that all of them essentially use geometric techniques to providing abstraction, whether it is the use of a hyperbolic graph, geometric containment, or the miniature subtrees provided by SpaceTree.

In contrast with these approaches, human experts are able to provide effective overviews of an ontology, simply by highlighting the key areas covered by the ontology and the classes that best describe these areas. In particular, the work reported in [5] provides empirical evidence that there is a significant degree of agreement among experts in identifying the main concepts in an ontology, and it also shows that our algorithm for key concept extraction (*KCE*) is also able to do so, while maintaining the same level of agreement with the experts, as they have among themselves [5]. Hence, the main hypothesis underlying our work on KC-Viz is that effective abstraction mechanisms for ontology visualization and navigation can be developed by building on the KCE algorithm, thus going beyond purely geometric approaches and focusing instead on displaying the concepts which are identified as the most useful for making sense of an ontology.

3   Overview of KC-Viz

3.1   Key Concept Extraction

Our algorithm for key concept extraction [5] considers a number of criteria, drawn from psychology, linguistics, and formal knowledge representation, to compute an 'importance score' for each class in an ontology. In particular, we use the notion of *natural category* [15], which is drawn from cognitive psychology, to identify concepts that are information-rich in a psycho-linguistic sense. Two other criteria are drawn from the topology of an ontology: the notion of *density* highlights concepts which are information-rich in a formal knowledge representation sense, i.e., they have been richly characterized with properties and taxonomic relationships, while the notion of *coverage* states that the set of key concepts identified by our algorithm should maximize the coverage of the ontology with respect to its is-a hierarchy[5]. Finally, the notion of *popularity*, drawn from lexical statistics, is introduced as a criterion to identify concepts that are likely to be most familiar to users.

The *density* and *popularity* criteria are both decomposed in two sub-criteria, *global* and *local density*, and *global* and *local popularity* respectively. While the global measures are normalized with respect to all the concepts in the ontology, the local ones consider the relative density or popularity of a concept with respect to its surrounding concepts in the is-a hierarchy. The aim here is to ensure that 'locally significant' concepts get a high score, even though they may not rank too highly with respect to global measures.

Each of the seven aforementioned criteria produces a score for each concept in the ontology and the final score assigned to a concept is a weighted sum of the scores

---

[5] By 'is-a hierarchy' here, we refer to the hierarchy defined by rdfs:subClassOf relations.

resulting from individual criteria. As described in [5], which provides a detailed account of our approach to key concept extraction, the KCE algorithm has been shown to produce ontology summaries that correlate significantly with those produced by human experts.

3.2 Exploring Ontologies with KC-Viz

Normally, a KC-Viz session begins by generating an initial summary of an ontology, to get an initial 'gestalt' impression of the ontology. This can be achieved in a number of different ways, most obviously by i) selecting the ontology in question in the 'Ontology Navigator' tab of the NeOn Toolkit, ii) opening up a menu of options by right clicking on the selected ontology, and then iii) choosing Visualize Ontology → Visualize Key Concepts, through a sequence of menus. Figure 1[6] shows the result obtained after performing this operation on the SUMO ontology, a large upper level ontology, which comprises about 4500 classes. The version used in these examples can be downloaded from http://www.ontologyportal.org/SUMO.owl.

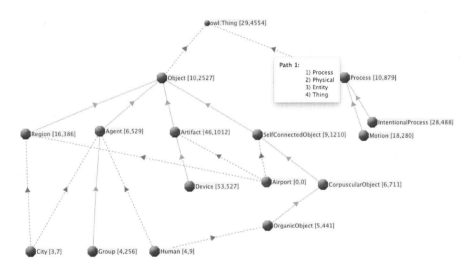

Fig. 1. Initial visualization of the SUMO ontology

The summary shown in Figure 1, which has been generated by the KCE algorithm, includes 16 concepts because we have set the size of our ontology summary to 15 and the algorithm has automatically added the most generic concept, owl:Thing, to ensure that the visualization displays a connected graph. If we wish to display more or less

---

[6] As shown in Figure 1, KC-Viz is based on the *node-link diagram* paradigm. However, as correctly pointed out by an anonymous reviewer, the KCE algorithm can in principle be used with alternative visualization styles, and indeed this is something we plan to explore in the future. The rationale for adopting the *node-link diagram* paradigm in the first instance is that this is a familiar representation for users and we wish to test our hypothesis that the use of key concepts can succeed in equipping this approach with effective abstraction mechanisms.

succinct graphs, we can do so by changing the size of the ontology summary. The solid grey arrows in the figure indicate *direct* rdfs:subClassOf links, while the dotted green arrows indicate *indirect* rdfs:subClassOf links. As shown in the figure, by hovering the mouse over an indirect rdfs:subClassOf links, we can see the chain of rdfs:subClassOf relations, which have been summarized by the indirect link. In this case, we can see that an indirect rdfs:subClassOf link in the display summarizes the chain of direct rdfs:subClassOf relations, [Process -> Physical -> Entity -> owl:Thing].

In order to help users to quickly get an idea of the size of a particular part of the ontology, for each node displayed, KC-Viz shows two numbers, indicating the number of direct and indirect subclasses. We refer to these as *subtree summaries*. For instance, Figure 1 tells us that class Process has 10 direct subclasses and 879 indirect ones. More information about a class can be found by hovering over the node in question, as shown in Figure 2. Alternatively, if a more thorough analysis of the definition is required, the user can right-click on the node and then select the Inspect menu item, to open up the definition of the class in the Entity Properties View of the NeOn Toolkit.

Once an initial visualization is produced, it is possible to use it as the starting point for a more in-depth exploration of the various parts of the ontology. To this purpose, KC-Viz provides a flexible set of options, allowing the user to control at a rather fine-grained level the extent to which she wishes to open up a particular part of the ontology. For example, let's assume we wish to explore the subtree of class Process in more detail, to get a better understanding of the type of processes covered by the ontology. Figure 3 shows the menu which is displayed, when right-clicking on class Process and selecting Expand. In particular, the following four options (corresponding to the four panes of the window shown in Figure 3) for customizing node expansion are available:

Fig. 2. Tooltips provide additional information about a class

- Whether to open up the node using taxonomic relations, other relations (through domain and range), or any combination of these. That is, while we primarily use KC-Viz to support sensemaking, focusing on taxonomic relations, KC-Viz can also be used to visualize domain (i.e., non taxonomic) relations.
- Whether or not to make use of the ontology summarization algorithm, which in this case will be applied only to the selected subtree of class Process. As in the case of generating a summary for the entire ontology, the user is given the option to specify the size of the generated summary. Here it is important to emphasize that this option makes it possible to use KC-Viz in a 'traditional' way, by

expanding a tree in a piecemeal way, without recourse to key concept extraction. This is especially useful when dealing with small ontologies, or when the user is aware that only a few nodes will be added by the expansion operation, even without recourse to the KCE algorithm.

- Whether or not to limit the range of the expansion – e.g., by expanding only to 1, 2, or 3 levels.
- Whether to display the resulting visualization in a new window ('Hide'), or whether to add the resulting nodes to the current display. In the latter case, some degree of control is given to the user with respect to the redrawing algorithm, by allowing her to decide whether she wants the system to redraw all the nodes in the resulting display (Redraw), or whether to limit the freedom of the graph layout algorithm to rearrange existing nodes (Block Soft, Block Hard). The latter options are particularly useful in those situations where expansion only aims to add a few nodes, and the user does not want the layout to be unnecessarily modified – e.g., because she has already manually rearranged the nodes according to her own preferences. In our view, this feature is especially important to avoid the problems experienced by users with some 'dynamic' visualization systems, where each node selection/expansion/hiding operation causes the system to rearrange the entire layout, thus making it very difficult for a user to retain a consistent mental map of the model.

Fig. 3. Options for further exploration starting from class Process

The result of expanding the subtree under class Process, using key concepts with the size of the summary set to 15, with no limit to the expansion level, while hiding all other concepts, is shown in Figure 4.

While the flexible expansion mechanism is the key facility provided by KC-Viz to support exploration of ontology trees under close user control, a number of other functionalities are also provided, to ensure a comprehensive visualization and navigation support. These include:

- A flexible set of options for hiding nodes from the display.
- Integration with the core components of the NeOn Toolkit, including the Entity Properties View and Ontology Navigator. This means that it is possible to click on nodes in KC-Viz and highlight them in these components, as well as clicking on items shown in the Ontology Navigator and adding them to the visualization in KC-Viz.
- A dashboard, shown in Figure 5, which allows the user to move back and forth through the history of KC-Viz operations, to modify the formatting of the layout, and to save the current display to a file, among other things.
- A preferences panel, which allows the user to set defaults for the most common operations and also enables her to switch to a more efficient (but sub-optimal) algorithm when dealing with very large ontologies.

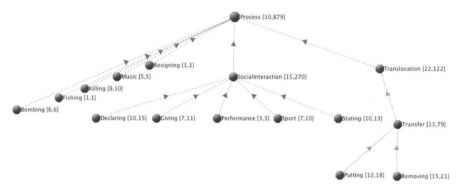

Fig. 4. Expanding class Process by key concepts

Fig. 5. The KC-Viz dashboard

4   Empirical Evaluation

4.1 Experimental Setup

4.1.1 Tool Configurations
In order to gather initial data about the performance of KC-Viz, we have carried out a preliminary empirical evaluation, which required 21 subjects to perform four ontology engineering tasks, involving ontology exploration. The 21 subjects were drawn from the members of the Knowledge Media Institute, the Computer Science Department at

the University of Bologna, and Isoco iLab and were randomly allocated to three different groups, labeled A, B, and C, where each group used a particular configuration of ontology engineering tools.

In particular members of group A carried out the tasks using the NeOn Toolkit v2.5, without any visualization support. More precisely, they were only allowed to use the search functionality, the Ontology Navigator and the Entity Properties View. The role of this group was to provide a baseline to the experiment, providing us with some data on how effectively people can tackle ontology exploration tasks, without any visualization support. The members of Group C were asked to solve the tasks using KC-Viz[7], together with the search functionality provided by the NeOn Toolkit. To ensure a separation between groups A and C, members of the latter group were explicitly forbidden from using the Ontology Navigator for exploration, although they were allowed to use it as an interface between the search facility in the NeOn Toolkit and KC-Viz[8]. Finally, the members of Group B carried out the tasks using the Protégé 4 environment, v4.1.0, in particular using the search functionality, the class browser and the OwlViz plugin. This configuration was chosen for three reasons: i) we wanted to compare KC-Viz to a robust tool, widely used in concrete projects by members of the ontology engineering community[9], to maximize the value of the experiment to the community; ii) while OwlViz uses the same node-link paradigm as KC-Viz, its design is rather different from KC-Viz; and iii) having considered the visualizers available in other state of the art ontology engineering tools, such as the NeOn Toolkit (Kaon Visualizer) and TopBraid (Graph View), we subjectively concluded that OwlViz appears to provide a more user friendly and flexible functionality, than the comparable ones available in TopBraid and the NeOn Toolkit.

4.1.2 Exploration Tasks

For the tasks we used a smaller version of the SUMO ontology, compared to the one referred to in section 4, which comprises 630 classes[10]. SUMO was chosen because, as an upper-level ontology, it is reasonable to expect most people to have familiarity with the notions it covers, in contrast with highly specialized ontologies in technical domains. This particular version of SUMO was chosen for a number of reasons:

- After running a couple of pilots, it became obvious that the ontology provided enough complexity to challenge the subjects and to potentially provide data about the effectiveness of different tool configurations.
- An ontology with thousands, rather than hundreds, of nodes would have required more time for the experiment, potentially reducing the number of subjects willing to take part.

---

[7] The members of Group C used version 1.3.0 of the KC-Viz plugin, which is part of the core set of plugins included with version 2.5 of the NeOn Toolkit.

[8] The search facility in the NeOn Toolkit locates an entity in the Ontology Navigator and then the user can use the "Visualize in KC-Viz" menu item, to add the located entity to the current KC-Viz display.

[9] To our knowledge Protégé is the most widely used ontology engineering environment currently available.

[10] This can be found at http://www.ontologyportal.org/translations/SUMO.owl

- The more complex the ontology, the higher the risk that a high number of subjects (many of whom could not be considered as experienced ontology engineers) would not complete the task, thus potentially reducing the number of useful data points.

The tasks given to the subjects are shown in Table 1. This set of tasks was designed to ensure coverage of different exploration strategies, which are typically required in the context of a sensemaking activity[11]. Task 1 can be seen as a 'pure' topological task, along the lines of the tasks used in the evaluation described in [6], in the sense that it asks the user to locate a node with a specific topological property. Task 2 is similar to Task 1, however it also requires the user to examine, as a minimum, the labels of the classes, rather than considering them only as abstract nodes in a node-link diagram. Tasks 3 and 4 require a mix of top-down and bottom-up exploration of the ontology and in addition Task 4 requires the user to understand part of the ontology at a deeper level than mere topological structure. Moreover, Task 4 also allowed us to test to what extent tools are able to help when the ontology has a non-standard conceptualization, which may easily confuse users, whether experts or novices. In particular, the SUMO ontology models class CurrencyCoin as a subclass of class Text, which is something many people could find surprising.

Table 1. Ontology Engineering Tasks

T1. Which class has the highest number of direct subclasses in the ontology?
T2. What is the most developed (i.e., has the biggest subtree) subclass of class Quantity found in the ontology at <u>a concrete level of granularity</u> (i.e., do not consider abstract classes which have the term 'quantity' in their id)?
T3. Find three subclasses of Agent, at the most abstract level possible (under Agent of course), which are situated at the same level in the hierarchy as each other, and are also subclasses of CorpuscularObject.
T4. We have two individual entities (a particular copy of the book War&Peace and a particular 5p coin). Find the most specific classes in the ontology, to which they belong, say P1 and P2, and then identify the most specific class in the ontology, say C1, which is a superclass of both P1 and P2 – i.e., the lowest common superclass of both P1 and P2.

For each task, the subjects were given a 15 minutes time slot. If they were not able to solve a particular task within 15 minutes, that task would be recorded as 'fail'. Before the experiment, every subject filled a questionnaire, answering questions about his/her expertise in ontology engineering, knowledge representation languages, and with various ontology engineering tools, including (but not limited to) NeOn and Protégé. None of the subjects had much direct experience with the SUMO ontology.

---

[11] It is important to emphasize that there is no direct mapping between KC-Viz features and the evaluation tasks chosen for this study. This is not accidental, as we are not interested in performing experiments on tasks which are artificially manufactured for KC-Viz. In addition, to ensure repeatability, the evaluation tasks used in this study are rather fine-grained and are associated to precise performance criteria.

4.1.3 Session Setup
At the beginning of the session a subject would be briefed about the purpose of the experiment. To avoid biases in favour or against a particular tool, subjects were simply told that the purpose of the experiment was "to test different configurations of ontology engineering tools". The subject would then be given a tutorial (max 10 minutes) about the specific set of tools he/she would be using. The tutorial was given by the person in charge of the experiment (the 'administrator'). In total four administrators were used. To minimize differences between the tutorials given by different administrators, these were given a precise list of the features that ought to be shown to each specific group. For the tutorial we used the Pizza ontology v1.5, which can be found at http://www.co-ode.org/ontologies/pizza/2007/02/12/.

After the tutorial, the subjects were asked to do a 'warm-up task'. This was exactly the same as T1, however it was carried out on a rather small ontology, AKTLite, a subset of the AKT reference ontology[12], which has been used in a variety of projects and applications for representing data about academic organizations. While the AKT ontology contains 170 classes, the AKTLite ontology only contains 85 classes. The AKTLite ontology consists of two sub-ontologies, AKT Support and AKT Portal, and can be found at http://technologies.kmi.open.ac.uk/KC-Viz/evaluation/AKTLite.zip[13]. The subjects were given 10 minutes to solve the warm-up task.

All the tasks, including the warm-up task, were recorded using screen capture software. After completing the task, the subjects were asked to fill a SUS usability questionnaire[14] and to provide qualitative data about their experience with the particular tool configuration they used, including overall impression of the tool, strengths and weaknesses, etc. Finally, the subjects in groups A and B were given a demo of KC-Viz and asked to provide feedback about the tool. This allowed us to get feedback about KC-Viz from all 21 participants in the evaluation.

4.2 Results

4.2.1 Task Performance
Out of 84 tasks in total (4 * 21), 71 were completed within the 15 minutes time limit, while 13 tasks were not completed, a 15.47% percentage failure. The 13 failures were distributed as follows: 5 in group A (NTK), 6 in group B (OwlViz), and 2 in group C (KC-Viz). Table 2 shows the average time taken by each group in each task, as well as the total averages across groups and tasks[15]. As shown in the table, on each of the four tasks the fastest mean performance was with KC-Viz, whose overall mean performance was about 13 minutes faster than OWLViz, which in turn was about two

[12] http://www.aktors.org/publications/ontology/
[13] For the sake of repeatability all tools and all ontologies used in the evaluation are publicly available online, while the tasks carried out in the evaluation are described in this paper.
[14] http://www.usabilitynet.org/trump/documents/Suschapt.doc
[15] For tasks not completed within the time limit, we consider a 15 minutes performance. This could be modified to consider 'penalties', such as a 5 minutes penalty for a non-completed task. However, adding such penalties does not lead to meaningful changes in the interpretation of the data, other than increasing the performance gap between the KC-Viz group and the others.

minutes faster than NTK. The mean performance time for OwlViz was faster than NTK for task 3, slower for the others. Although not significant, the difference in total time taken across the four tasks with the three different tools appeared to be approaching significance, $F(2, 20) = 2.655$, $p = 0.098$.

The difference in performance across the three tools on Task 1, was statistically significant $F(2, 20) = 9.568$, $p < 0.01$. A Tukey HSD pairwise comparison revealed a significant difference between both KC-Viz and NTK ($p < 0.01$) and KC-Viz and OwlViz ($p < 0.01$), however not between NTK and OwlViz. Although mean performance was faster for KC-Viz across the board, performance differences on the other three tasks did not reach statistical significance. By some margin, the least significant result was found for Task 4 ($p = 0.755$). As discussed earlier, Task 4 involved more problem solving steps than the other tasks (i.e., finding direct parent classes for suggested instances and then their common parent) and an answer that was counter-intuitive to many of the subjects (i.e., a CurrencyCoin being a subclass of Text). Due to the more complex nature of the problem, we hypothesize that other factors, beyond the features provided by a particular ontology engineering tool, influenced performance on this task.

Nevertheless these results suggest advantages for KC-Viz in supporting users in such realistic browsing and visualization tasks. In particular it is reasonable to assume that increasing the sample size beyond the seven per condition in the current study could be expected to lead to statistical significance for overall performance and possibly also for other individual tasks.

Table 2. Experimental results (in min:secs)

	NTK		OWLViz		KCViz		Overall	
	mean	s.d.	mean	s.d.	mean	s.d.	mean	s.d.
Task 1	12:03	02:51	12:19	04:16	05:10	03:07	09:50	04:44
Task 2	06:43	04:45	07:20	03:55	04:03	02:15	06:02	03:52
Task 3	11:00	05:31	07:24	04:27	06:25	05:06	08:16	05:12
Task 4	08:01	05:56	08:23	05:28	06:17	05:15	07:34	05:21
Total	37:47	15:02	35:26	15:36	21:55	10:32	31:43	15:01

It is interesting to note that it is the first task that most clearly distinguished the performance of KC-Viz relative to the other tools. This was the first of the four tasks that used the SUMO ontology. Performance on this task would therefore have involved developing some initial overall conceptualization of the ontology, its structure, size and scope, as well as finding ways to navigate it. It is possible therefore that the use of KC-Viz is particularly effective, when users are confronted with large and unfamiliar ontologies.

4.2.2   Other Quantitative Findings

Usability scores were calculated using the SUS formula for each of the three conditions – see Table 3. The mean usability score was slightly higher for KC-Viz, though very similar across the three tools and not statistically significant.

Table 3. Usability scores

	NKT		OwlViz		KC-Viz	
	mean	s.d.	mean	s.d.	mean	s.d.
Usability score	26.9	5.1	25.7	4.3	27.1	5.8

However, for each subject, two sub-scores were calculated from the experience questionnaire. The first seven questions in the questionnaire are related to experience with ontologies and ontology languages. The scores on these questions were summed to give a measure of ontology experience. The scores on the final six questions were summed to give a score related to experience with ontology engineering tools.

A positive correlation was found between the ontology experience score and the usability score, $r = 0.546$, $p < 0.05$, while a significant correlation was not found between the score for experience with tools and the usability score. This appears to indicate that perceived usability probably reflects the greater ability of subjects, who are more experienced in the use of ontologies, to adapt to the features, and compensate for the shortcomings, of whatever tool provided for the task. These findings also suggest that the results of usability questionnaires in this kind of evaluations should be treated with much caution and ideally triangulated with other sources of data.

The ontology experience score also had a significant negative correlation with the total time spent across the four tasks (i.e. the higher the ontology experience, the lower the time to complete the task), $r = -0.476$, $p < 0.05$, as well as on task 3, $r = -0.511$, $p < 0.05$. Correlation between experience of ontology engineering tools and task performance was statistically significant for task 1 ($r = -0.469$, $p < 0.5$) and task 3 ($r = -0.452$), and was close to significance on overall performance ($r = -0.410$, $p = 0.065$).

These findings suggest that prior experience with both ontologies and associated tools increases task performance regardless of the toolset used. The deeper understanding that the expert has of the underlying ontological constructs and the heuristics and techniques developed through experience allows the expert to more easily interpret and adapt to whatever tool is provided. Therefore, both differences in performance and usability judgements can be expected to be harder to find when testing with experts than when testing with users with lower levels of experience.

Given the effect of experience on usability judgements and performance, an analysis was conducted to verify that performance differences across the three tools were not due to a skewed distribution of experience across the three conditions. However, experience scores were similar across the subjects in the three conditions and were not statistically significant. This demonstrates that variation in performance across the tools was not due to a bias in the distribution of experience across the three conditions.

4.2.3 Qualitative Results

As already mentioned, the free text questions on the post-task questionnaire elicited views on the perceived strengths and weaknesses of the tool used by each subject. Additionally, subjects who did not use KC-Viz provided feedback following a demo.

A *grounded theory* approach [16] was used to build categories of comments that either expressed positive feedback, offered criticism, or suggested improvements. Categories were discarded when they only contained comments from a single subject. Because of the page limit constraint on this paper, we do not have enough space here to discuss this analysis in detail, hence we only highlight the main findings.

The three main categories of positive comments concerned the flexible support provided by KC-Viz to manipulate the visual displays; the abstraction power enabled by the KCE algorithm; and the value of the subtree summaries provided by KC-Viz. These results are encouraging in the sense that they provide some initial indication that there is probably a direct causal link between the use of key concepts as an abstraction mechanism and the good performance of KC-Viz on the evaluation tasks, even though these were not designed specifically to map directly to KC-Viz features.

The three main categories of negative comments included: criticism of the tree layout algorithm used by KC-Viz, which does not allow display rotation and at times generates overlapping labels; the lack of transparency of the KCE algorithm, which does not allow the user to configure it, or to clarify why a node is considered more important than others; and the lack of integration between KC-Viz and reasoning/query support in the NeOn Toolkit.

5   Conclusions

Exploring a large ontology, particularly when it is unfamiliar to the user, can be characterized as *information foraging* [17]. Information foraging theory, drawing on ecological models of how animals hunt for food, proposes the notion of *information scent*. An animal foraging for food will follow a scent in order to locate a promising patch rich in sources of food. Analogously, an information forager will follow an information scent in order to locate rich information sources. In a hypertext environment, the name of a link, a preview of the information source, or even the source URL may give the information forager clues as to the potential profitability of following the link. This helps the forager to choose between alternative paths in the search for information.

The support provided by KC-Viz for displaying key concepts and using these as the basis for further exploration can be seen as assisting information foraging from an ontology. In particular, the flexible set of options provided by KC-Viz for manipulating the visualization enables the user to construct a view on the ontology that allows them to compare the information scent of different paths through the ontology and control how they pursue these paths. Key concepts use a number of factors to estimate the importance of a particular class and therefore provide means for estimating the potential profitability of an information scent. In addition, subtree summaries provide a topological clue as to the potential profitability of a class.

This perspective might help explain why KC-Viz was found to be particularly advantageous when exploring a large ontology for the first time and indeed both key concepts and subtree summaries were highlighted as strengths of the approach, while the lack of these kinds of abstraction/summarization mechanisms were identified by users as a deficit of both the NTK and OwlViz configurations.

The empirical study also offers lessons learned for future evaluations of ontology browsing and visualization tools. In particular it showed that significant prior experience in ontology engineering enables users to adapt well to different tool configurations and perform well regardless of the specific configuration they are using. In addition, it also showed that experts can have relatively positive usability judgments of interfaces. Both of these observations suggest that users having a broad range of expertise should be gathered for usability testing and that the results from experiments that fail to triangulate multiple sources of data, including usability scores, task performance and qualitative feedback, should be treated with caution.

Our future work has two broad strands. First of all, we intend to further develop KC-Viz, taking on board the feedback gathered during the evaluation. In particular, improving the layout algorithm, opening up the KCE algorithm to users, and integrating KC-Viz with ontology reasoning and querying are all priorities for development. In addition, we also intend to capitalize on the 21 videos collected during the evaluation, both to undertake a fine-grained analysis of the navigational strategies employed by users, and also to uncover possible misconceptions revealed in the use of the various ontology tools. It is hoped that from these analyses we will then be able to generate additional design recommendations for future versions of KC-Viz.

References

1. Oren, E., Delbru, R., Catasta, M., Cyganiak, R., Stenzhorn, H., Tummarello, G.: Sindice.com: a document-oriented lookup index for open linked data. Int. J. Metadata, Semantics and Ontologies 3(1), 37–52 (2008)
2. d'Aquin, M., Motta, E.: Watson, more than a Semantic Web search engine. Semantic Web 2(1) (2011)
3. Noy, N.F., Tudorache, T.: de Coronado, Sh., and Musen, M. A. Developing Biomedical Ontologies Collaboratively. In: Proceedings of the AMIA Annual Symposium, pp. 520–524 (2008)
4. Dzbor, M., Motta, E., Buil Aranda, C., Gomez-Perez, J.M., Goerlitz, O., Lewen, H.: Developing ontologies in OWL: An observational study. In: OWL: An observational study. Workshop on OWL: Experiences and Directions, Georgia, US (November 2006)
5. Peroni, S., Motta, E., d'Aquin, M.: Identifying key concepts in an ontology through the integration of cognitive principles with statistical and topological measures. In: Third Asian Semantic Web Conference, Bangkok, Thailand (2008)
6. Wang, T.D., Parsia, B.: Cropcircles: Topology Sensitive Visualization of Owl Class Hierarchies. In: Proceedings of the 5th International Semantic Web Conference, Georgia, US (2006)
7. Shneiderman, B.: The Eyes Have It: A Task by Data Type Taxonomy for Information Visualizations. In: Proceedings of the 1996 IEEE Symposium on Visual Languages (VL 1996), IEEE Computer Society Press, Washington, DC, USA (1996)
8. Shneiderman, B.: Tree Visualization with Tree-Maps: A 2d Space-Filling Approach. ACM Trans. Graph. 11(1), 92–99 (1992)
9. Plaisant, C., Grosjean, J., Bederson, B.B.: Spacetree: Supporting Exploration in Large Node Link Tree, Design Evolution and Empirical Evaluation. In: Proc. of the Intl. Symposium on Information Visualization (2002)

10. Storey, M.A., Musen, M.A., Silva, J., Best, C., Ernst, N., Fergerson, R., Noy, N.F.: Jambalaya: Interactive visualization to enhance ontology authoring and knowledge acquisition in Protege. In: Workshop on Interactive Tools for Knowledge Capture, K-CAP-2001, Victoria, B.C., Canada (2001)
11. Kriglstein, S., Motschnig-Pitrik, R.: Knoocks: A New Visualization Approach for Ontologies. In: Proceedings of the 12th International Conference on Information Visualisation (IV 2008), IEEE Computer Society, Washington, DC, USA (2008)
12. Katifori, A., Halatsis, C., Lepouras, G., Vassilakis, C., Giannopoulou, E.: Ontology Visualization Methods - a Survey. ACM Computing Surveys 39(4) (2007)
13. Souza, K., Dos Santos, A., et al.: Visualization of Ontologies through Hypertrees. In: Proc. of the Latin American Conference on Human-Computer Interaction, pp. 251–255 (2003)
14. Katifori, A., Torou, E., Halatsis, C., Lepouras, G., Vassilakis, C.: A Comparative Study of Four Ontology Visualization Techniques in Protege: Experiment Setup and Preliminary Results. In: Proceedings of the 10th Int. Conference on Information Visualisation (IV 2006), London, UK, pp. 417–423 (2006)
15. Rosch, E.: Principles of Categorization. Cognition and Categorization. Lawrence Erlbaum, Hillsdale (1978)
16. Birks, M., Mills, J.: Grounded Theory: A Practical Guide. SAGE Publications Ltd. (2011)
17. Pirolli, P., Card, S.K.: Information Foraging. Psychological Review 106(4), 643–675 (1999)

Wheat and Chaff – Practically Feasible Interactive Ontology Revision

Nadeschda Nikitina[1], Birte Glimm[2], and Sebastian Rudolph[1]

[1] Institute AIFB, Karlsruhe Institute of Technology, DE
[2] Ulm University, Institute of Artificial Intelligence, DE

Abstract. When ontological knowledge is acquired automatically, quality control is essential. We consider the tightest possible approach – an exhaustive manual inspection of the acquired data. By using automated reasoning, we partially automate the process: after each expert decision, axioms that are entailed by the already approved statements are automatically approved, whereas axioms that would lead to an inconsistency are declined. Adequate axiom ranking strategies are essential in this setting to minimize the amount of expert decisions.

In this paper, we present a generalization of the previously proposed ranking techniques which works well for arbitrary validity ratios – the proportion of valid statements within a dataset – whereas the previously described ranking functions were either tailored towards validity ratios of exactly 100% and 0% or were optimizing the worst case. The validity ratio – generally not known *a priori* – is continuously estimated over the course of the inspection process. We further employ partitioning techniques to significantly reduce the computational effort. We provide an implementation supporting all these optimizations as well as featuring a user front-end for successive axiom evaluation, thereby making our proposed strategy applicable to practical scenarios. This is witnessed by our evaluation showing that the novel parameterized ranking function almost achieves the maximum possible automation and that the computation time needed for each reasoning-based, automatic decision is reduced to less than one second on average for our test dataset of over 25,000 statements.

1 Introduction

Many real-world applications in the Semantic Web make use of ontologies in order to enrich the semantics of the data on which the application is based. As a popular example, consider DBpedia, which consists of structured information from Wikipedia. DBpedia uses a background ontology, which defines the meaning of and relationships between terms. For example, if two terms are related via the property *river*, the first one can be inferred to be an instance of the class *Place* and the latter one of the class *River*.

In order to guarantee very high quality standards, the DBpedia background ontology has been created manually. For many applications, however, the time requirements of a completely manual knowledge acquisition process are too high. An additional application of (semi-) automatic knowledge acquisition methods such as ontology learning or matching is, therefore, often considered to be a reasonable way to reduce the

L. Aroyo et al. (Eds.): ISWC 2011, Part I, LNCS 7031, pp. 487–503, 2011.

expenses of ontology development. The results produced by such automatic methods usually need to be manually inspected either partially, to estimate the overall quality of the resulting data, or to the full extent, to keep the quality of the developed ontology under control.

So far, the knowledge representation community has been focusing on restoring the consistency of ontologies enriched with new axioms as done in various belief revision and repair approaches, see, e.g., [1,10]. Thereby, new axioms not causing inconsistency are treated as valid facts, which do not require further inspection. Our goal is to support a more restrictive quality control process in which a domain expert inspects a set of candidate axioms and decides for each of them whether it is a desired logical consequence. Based on this decision, we automatically discard or include yet unevaluated axioms depending on their logical relationships with the already evaluated axioms. In the following, we call this interactive process *ontology revision*.

Throughout the paper, we use the following running example, which we write in OWL's functional-style syntax using an imaginary prefix ex to abbreviate IRIs:

Example 1. Let us assume that we have already confirmed that the axioms, which state subclass relations between classes, belong to the desired consequences:

> SubClassOf(ex:AluminiumNitrideNanotube ex:AluminiumNitride)
> SubClassOf(ex:AluminiumNitride ex:NonOxideCeramics)
> SubClassOf(ex:NonOxideCeramics ex:Ceramics)
> SubClassOf(ex:Ceramics ex:MaterialByMaterialClass)
> SubClassOf(ex:MaterialByMaterialClass ex:Material)
> SubClassOf(ex:Material ex:PortionOfMaterial)
> SubClassOf(ex:Material ex:TangibleObject)

We further assume that the following axioms, which define several different types for the individual ex:nanotube1, are still to be evaluated:

ClassAssertion(ex:AluminiumNitrideNanotube ex:nanotube1)	(1)
ClassAssertion(ex:AluminiumNitride ex:nanotube1)	(2)
ClassAssertion(ex:NonOxideCeramics ex:nanotube1)	(3)
ClassAssertion(ex:Ceramics ex:nanotube1)	(4)
ClassAssertion(ex:MaterialByMaterialClass ex:nanotube1)	(5)
ClassAssertion(ex:Material ex:nanotube1)	(6)
ClassAssertion(ex:PortionOfMaterial ex:nanotube1)	(7)
ClassAssertion(ex:TangibleObject ex:nanotube1)	(8)

If Axiom (8) is declined, we can immediately also decline Axioms (1) to (6) assuming OWL or RDFS reasoning since accepting the axioms would implicitly lead to the undesired consequence (8). Note that no automatic decision is possible for Axiom (7) since it is not a consequence of Axiom (8) and the already approved subsumption axioms. Similarly, if Axiom (1) is approved, Axioms (2) to (8) are implicit consequences, which can be approved automatically. If we start, however, with declining Axiom (1), no automatic evaluation can be performed. It can, therefore, be observed that

- a high grade of automation requires a good evaluation order, and
- approval and decline of an axiom has a different impact.

Which axioms have the highest impact on decline or approval and which axioms can be automatically evaluated once a particular decision has been made can be determined with the help of algorithms for automated reasoning, e.g., for RDFS or OWL reasoning. One of the difficulties is, however, that it is not known in advance, which of the two decisions the user makes. In our previous work [8], we tackle this problem by showing that, if the quality of the acquired axioms is known, a prediction about the decision of the user can be made: if the quality is high, the user is likely to approve an axiom. Hence, axioms that have a high impact on approval should be evaluated with higher priority. For low quality data, the situation is reversed. We measure the quality by means of the *validity ratio*, i.e., the percentage of accepted axioms, and show in [8] that, depending on the validity ratio of a dataset, different impact measures used for axiom ranking are beneficial. In this paper, we extend the previous results in several directions:

- First, we generalize the ranking functions proposed in [8], which are tailored towards validity ratios of 100% and 0% by parametrizing the ranking function by an estimated validity ratio. In our evaluation, we show that the revision based on the novel ranking function almost achieves the maximum possible automation. The gain is particularly important for datasets with a validity ratio close to 50%, since the currently existing ranking function for those datasets only optimizes the worst case and does not fully exploit the potential of automation.
- Second, since the expected validity ratio is not necessarily known in advance, we suggest a ranking function where the validity ratio is learned on-the-fly during the revision. We show that, even for small datasets (50-100 axioms), it is worthwhile to rank axioms based on this learned validity ratio instead of evaluating them in a random order. Furthermore, we show that, in case of larger datasets (e.g., 5,000 axioms and more) with an unknown validity ratio, learning the validity ratio is particularly effective due to the law of large numbers, thereby making the assumption of a known or expected validity ratio unnecessary. For such datasets, our experiments show that the proportion of automatically evaluated axioms when learning the validity ratio is nearly the same (difference of 0.3%) as in case where the validity ratio is known in advance.

Even for not very expressive knowledge representation formalisms, reasoning is an expensive task and, in an interactive setting as described above, a crucial challenge is to minimize the number of expensive reasoning tasks while maximizing the number of automated decisions. In our previous work [8], we have developed *decision spaces* – data structures, which exploit the characteristics of the logical entailment relation between axioms to maximize the amount of information gained by reasoning. Decision spaces further allow for reading off the impact that an axiom will have in case of an approval or decline. In this paper, we extend the latter work by combining decision spaces with a partitioning technique in order to further improve the efficiency of the revision process. It is interesting to observe that partitioning intensifies the effectiveness of decision spaces, since it increases the relative density of dependencies between axioms considered together during the revision.

We further present *revision helper*: an interactive application supporting ontology revision. We evaluate the proposed techniques and demonstrate that even for expressive OWL reasoning, an interactive revision process is feasible with on average 0.84 seconds (7.4 reasoning calls) per expert decision, where the automatic evaluation significantly reduces the number of expert decisions.[1]

The remainder of this paper is organized as follows. Next, we describe relevant preliminaries. Section 3 describes the proposed new ranking function and how the validity ratio can be learned during the revision. Section 4 introduces partitioning as a way of optimizing the efficiency of the revision process. We then evaluate the approach in Section 5 and present the user front-end of *revision helper* in Section 6. In Section 7, we discuss the existing related approaches and then conclude in Section 8.

2   Preliminaries

In this section, we introduce the basic notions that are relevant to the revision of an ontology. The ontologies that are to be revised can be written in standard semantic web languages such as RDFS or OWL. We focus, however, on OWL 2 DL ontologies.

The revision of an ontology O aims at a separation of its axioms (i.e., logical statements) into two disjoint sets: the set of intended consequences $O^{\vDash}$ and the set of unintended consequences $O^{\nvDash}$. This motivates the following definitions.

Definition 1 (Revision State). *A revision state is defined as a tuple $(O, O^{\vDash}, O^{\nvDash})$ of ontologies with $O^{\vDash} \subseteq O$, $O^{\nvDash} \subseteq O$, and $O^{\vDash} \cap O^{\nvDash} = \emptyset$. Given two revision states $(O, O_1^{\vDash}, O_1^{\nvDash})$ and $(O, O_2^{\vDash}, O_2^{\nvDash})$, we call $(O, O_2^{\vDash}, O_2^{\nvDash})$ a refinement of $(O, O_1^{\vDash}, O_1^{\nvDash})$, if $O_1^{\vDash} \subseteq O_2^{\vDash}$ and $O_1^{\nvDash} \subseteq O_2^{\nvDash}$. A revision state is* complete, *if $O = O^{\vDash} \cup O^{\nvDash}$, and* incomplete *otherwise. An incomplete revision state $(O, O^{\vDash}, O^{\nvDash})$ can be refined by evaluating a further axiom $\alpha \in O \setminus (O^{\vDash} \cup O^{\nvDash})$, obtaining $(O, O^{\vDash} \cup \{\alpha\}, O^{\nvDash})$ or $(O, O^{\vDash}, O^{\nvDash} \cup \{\alpha\})$. We call the resulting revision state an* elementary refinement *of $(O, O^{\vDash}, O^{\nvDash})$.*

Since we expect that the deductive closure of the intended consequences in $O^{\vDash}$ must not contain unintended consequences, we introduce the notion of *consistency* for revision states. If we want to maintain consistency, a single evaluation decision can predetermine the decision for several yet unevaluated axioms. These implicit consequences of a refinement are captured in the *revision closure*.

Definition 2 (Revision State Consistency and Closure). *A (complete or incomplete) revision state $(O, O^{\vDash}, O^{\nvDash})$ is* consistent *if there is no $\alpha \in O^{\nvDash}$ such that $O^{\vDash} \models \alpha$. The*

---

[1] Anonymized versions of the used ontologies and the *revision helper* tool can be downloaded from http://people.aifb.kit.edu/nni/or2010/Interactive_Ontology_Revision/

Algorithm 1. Interactive Ontology Revision

Data: $(O, O_0^\vDash, O_0^\#)$ a consistent revision state
Result: $(O, O^\vDash, O^\#)$ a complete and consistent revision state
1 $(O, O^\vDash, O^\#) \leftarrow \mathrm{clos}(O, O_0^\vDash, O_0^\#)$;
2 **while** $O^\vDash \cup O^\# \neq O$ **do**
3 $\quad$ choose $\alpha \in O \setminus (O^\vDash \cup O^\#)$;
4 $\quad$ **if** *expert confirms* α **then**
5 $\quad\quad | \quad (O, O^\vDash, O^\#) \leftarrow \mathrm{clos}(O, O^\vDash \cup \{\alpha\}, O^\#)$;
6 $\quad$ **else**
7 $\quad\quad \lfloor \quad (O, O^\vDash, O^\#) \leftarrow \mathrm{clos}(O, O^\vDash, O^\# \cup \{\alpha\})$;

---

revision closure $\mathrm{clos}(O, O^\vDash, O^\#)$ *of* $(O, O^\vDash, O^\#)$ *is* $(O, O_c^\vDash, O_c^\#)$ *with* $O_c^\vDash := \{\alpha \in O \mid O^\vDash \vDash \alpha\}$ *and* $O_c^\# := \{\alpha \in O \mid O^\vDash \cup \{\alpha\} \vDash \beta$ *for some* $\beta \in O^\#\}$.

We observe that, for a consistent revision state $(O, O^\vDash, O^\#)$, the closure $\mathrm{clos}(O, O^\vDash, O^\#)$ is again consistent and that every further elementary refinement of $\mathrm{clos}(O, O^\vDash, O^\#)$ is also consistent; furthermore, any consistent and complete refinement of $(O, O^\vDash, O^\#)$ is a refinement of $\mathrm{clos}(O, O^\vDash, O^\#)$ [8, Lemma 1]. Algorithm 1 employs these properties to implement a general methodology for interactive ontology revision. Instead of initializing $O_0^\vDash$ and $O_0^\#$ with the empty set, one can initialize $O_0^\vDash$ with already approved axioms, e.g., from a previous revision, and $O_0^\#$ with declined axioms from a previous revision and with axioms that express inconsistency and unsatisfiability of classes (or properties), which we assume to be unintended consequences.

In line 3, an axiom is chosen that is evaluated next. As motivated in the introduction, a random decision can have a detrimental effect on the amount of manual decisions. Ideally, we want to rank the axioms and choose one that allows for a high number of consequential automatic decisions. The notion of *axiom impact* captures how many axioms can be automatically evaluated when the user approves or declines an axiom. Note that after an approval, the closure might extend both $O^\vDash$ and $O^\#$, whereas after a decline only $O^\#$ can be extended. We further define $?(O, O^\vDash, O^\#)$ as the number of yet unevaluated axioms and write $|S|$ to denote the cardinality of a set S:

Definition 3 (Impact). *Let* $(O, O^\vDash, O^\#)$ *be a consistent revision state with* $\alpha \in O$ *and let* $?(O, O^\vDash, O^\#) := |O \setminus (O^\vDash \cup O^\#)|$. *For an axiom* α, *we define*

> *the* approval impact *$impact^+(\alpha) = ?(O, O^\vDash, O^\#) - ?(\mathrm{clos}(O, O^\vDash \cup \{\alpha\}, O^\#))$,*
>
> *the* decline impact *$impact^-(\alpha) = ?(O, O^\vDash, O^\#) - ?(\mathrm{clos}(O, O^\vDash, O^\# \cup \{\alpha\}))$,*
>
> *the* guaranteed impact *$guaranteed(\alpha) = \min(impact^+(\alpha), impact^-(\alpha))$.*

We further separate $impact^+(\alpha)$ into the number of automatic approvals, $impact^{+a}(\alpha)$, and the number of automatic declines, $impact^{+d}(\alpha)$:

$$impact^{+a}(\alpha) = |\{\beta \in O \mid O^\vDash \cup \{\alpha\} \vDash \beta\}|,$$
$$impact^{+d}(\alpha) = |\{\beta \in O \mid O^\vDash \cup \{\alpha, \beta\} \vDash \gamma, \gamma \in O^\#\}|.$$

Table 1. Example axiom dependency graph and the corresponding ranking values

$impact^+ \rightarrow$ ⬤ (1)
  ◐ (2)
  ◐ (3)
$guaranteed \rightarrow$ ◯ (4)
$guaranteed \rightarrow$ ◯ (5)
  ◐ (6)
$impact^- \rightarrow$ ◯ (7) ◯ (8)

Axiom	$impact^{+a}$	$impact^{+d}$	$impact^-$	guaranteed
(1)	7	0	0	0
(2)	6	0	1	1
(3)	5	0	2	2
(4)	4	0	3	3
(5)	3	0	4	3
(6)	2	0	5	2
(7)	0	0	6	0
(8)	0	0	6	0

Note that $impact^+(\alpha) = impact^{+a}(\alpha) + impact^{+d}(\alpha)$. The function $impact^+$ privileges axioms, for which the number of automatically evaluated axioms in case of an accept is high. Going back to our running example, Axiom (1), which yields 7 automatically accepted axioms in case it is accepted, will be ranked highest. The situation is the opposite for $impact^-$. It privileges axioms, for which the number of automatically evaluated axioms in case of a decline is high (Axioms (7) and (8)). The function $guaranteed$ privileges axioms with the highest guaranteed impact, i.e., axioms with the highest number of automatically evaluated axioms in the worst-case (Axioms (4) and (5)). Table 1 lists the values for all ranking functions for the axioms from Example 1.

Which ranking function should be chosen for an ontology revision in order to maximize the amount of automatic decisions depends on the expected validity ratio within the axiom set under revision. For a validity ratio of 100% the function $impact^+$ is the most effective, whereas for a validity ratio of 0%, $impact^-$ clearly performs best. In cases when the expected validity ratio is close to 50%, the guaranteed impact can be used to get a reasonable compromise between $impact^+$ and $impact^-$. The ranking functions do, however, not adapt to validity ratios that divert from these extremes. We address this in the next section, by introducing a parametrized ranking function.

Since computing such an impact as well as computing the closure after each evaluation (lines 1, 5, and 7) can be considered very expensive due to the high worst-case complexity of reasoning, we developed *decision spaces* [8] as auxiliary data structures which significantly reduce the cost of computing the closure upon elementary revisions and provide an elegant way of determining high impact axioms. Intuitively, a decision space keeps track of the dependencies between the axioms, i.e., if an axiom β is entailed by the approved axioms together with an unevaluated axiom α, then an "entails" relationship is added linking α to β. Similarly, if adding β to the approved axioms together with an unevaluated axiom α would yield an inconsistency, then a "conflicts" relationship is established between α to β. We show a simplified graph capturing the entails relation for our running example on the left-hand side of Table 1 (the conflicts relation for the example is empty). Note that the entails relation is transitive and reflexive, but for a clearer presentation, we show a transitively reduced version of the graph. From this graph we can see, for example, that if we approve Axiom (5), then we can automatically approve Axioms (6) to (8) as indicated by the (entails) edges in the graph. Thus, decision spaces allow for simply reading-off the consequences of revision state refinements upon an approval or a decline of an axiom, thereby reducing the required reasoning operations. Furthermore, updating a decision space after an approval or a decline can be performed more efficiently compared to a recomputation of all dependencies.

3 Parametrized Ranking

In order to motivate the introduction of the parametrized ranking and to clarify its difference to the three previously proposed ranking functions, we now consider so-called *key axioms* for a path in the *entails*-graph of the decision space: an axiom α on a path p is a key axiom for p if

1. any axiom β on p such that α entails β is correct and
2. any axiom γ on p such that γ entails α is incorrect.

It can be observed that each path is such that there is at least one and there are at most two key axioms (one axiom in case all axioms on the path are correct or incorrect). Intuitively, by making a decision about the key axioms of a path first, we can automatically make a decision for all remaining axioms on the path. While these decisions are made, we might also automatically find conflicts and perform further automatic declines. Conflicts allow, however, for fewer automatic decisions. Hence we focus on the entails paths in the decision space. From this perspective, the behavior of *impact*$^+$ in a tree-shaped structure corresponds to the search for such a key axiom starting from the source axioms with only outgoing entails edges, while the behavior of *impact*$^-$ corresponds to the search from the sink axioms with only incoming entails edges. On the other hand, the behavior of *guaranteed* corresponds to binary search. For instance, if we assume that in our example, Axioms (1) and (2) are incorrect and we choose Axiom (4) among the two highest ranked axioms under the *guaranteed* ranking function, then Axioms (5) to (8) will be automatically evaluated, leaving us with Axioms (1) to (3). This time, Axiom (2) will receive the highest ranking value (1 in contrast to 0 for Axioms (1) and (3)). After another expert decision, Axiom (1) will remain, which is ranked with 0. Therefore, after each expert decision, the remaining axioms are again divided into more or less equally large sets until the set of unevaluated axioms is empty. The improvement achieved by *guaranteed* in comparison to a revision in random order becomes more and more visible with the growing size of the set of unevaluated axioms forming a connected decision space graph, since, in this scenario, the probability of incidentally choosing an axiom with the above specified property becomes lower.

Under the assumption that the dataset in the example has a validity ratio of 75%, the ranking technique *guaranteed* will (theoretically) require 2.8 expert decisions. This is an average for the different possible choices among the highest ranked axioms assuming that these have the same probability of being chosen. In contrast to that, *impact*$^+$ will require 3 expert decisions, while *impact*$^-$ will require even 7 decisions. It is obvious that if the expected validity ratio would have been taken into account, the corresponding ranking strategy would choose Axioms (2) and (3) and require only two expert decisions. In the following, we generalize the ranking techniques *impact⁺* and *impact⁻*, which assume the expected validity ratio to be 100% and 0%, respectively, to a ranking technique, which is parametrized by the actual expected validity ratio. The new ranking technique then chooses axioms based on the expected validity ratio for the dataset.

The goal of the parametrized ranking is to privilege axioms that are most probably key axioms under the assumption that the validity ratio is R. While in Example 1, Axioms (2) and (3) would be the clear choice, in an arbitrary graph, more than two axioms can have such a property. Interestingly, the examination of decision space structures computed within our experiments indicates that the number of possible axioms with such a property is close to two within the connected components of such graphs.

3.1 The Ranking Function *norm*

We now define the ranking function $norm_R$ according to the above set goals. We first normalize the number of automatic approvals and declines to values between 0 and 1. Since in the case of an approval we can possibly accept and decline axioms, we split the approval impact accordingly. We can then normalize the obtained values with respect to the expected validity ratio which allows for choosing an axiom that behaves best according to our expectation.

Definition 4. *Let $O^?$ be a connected component of the decision space and R the expected validity ratio. The* normalized impact functions *are:*

$$impact_N^{+a} = \frac{1 + impact^{+a}}{|O^?|}, \quad impact_N^{+d} = \frac{impact^{+d}}{|O^?|}, \quad impact_N^{-} = \frac{1 + impact^{-}}{|O^?|}.$$

The ranking functions $norm_R^{+a}$, $norm_R^{+d}$ and $norm_R^{-}$ are then defined by

$$norm_R^{+a} = -|R - impact_N^{+a}|, \quad norm_R^{+d} = -|1 - R - impact_N^{+d}|, \quad norm_R^{-} = -|1 - R - impact_N^{-}|.$$

Finally, the ranking function $norm_R$ is:

$$norm_R = max(norm_R^{+a}, norm_R^{+d}, norm_R^{-}).$$

Note that we do not add 1 for $impact_N^{+d}$ since the axiom itself is not declined, i.e., we capture just the "side-effect" of accepting another axiom. Table 2 shows the computation of $norm_{0.75}$ for Example 1. The function $norm_R^{+a}$ captures how the fraction of automatically accepted axioms deviates from the expected overall ratio of wanted consequences, e.g., accepting Axiom (2) or (4) deviates by 12.5%: for the former axiom we have automatically accepted too many axioms, while for the latter we do not yet have accepted enough under the premise that the validity ratio is indeed 75%. Since Example 1 does not allow for automatic declines after an approval, the function $norm_R^{+d}$ shows that for each accept, we still deviate 25% from the expected ratio of *invalid*

Table 2. The values for $norm_{0.75}$ and the intermediate functions (shown in percentage)

Axiom	$impact_N^{+a}$	$impact_N^{+d}$	$impact_N^-$	$norm_{0.75}^{+a}$	$norm_{0.75}^{+d}$	$norm_{0.75}^-$	$norm_{0.75}$
(1)	100.0%	0.0%	12.5%	-25.0%	-25.0%	-12.5%	-12.5%
(2)	87.5%	0.0%	25.0%	-12.5%	-25.0%	0.0%	0.0%
(3)	75.0%	0.0%	37.5%	0.0%	-25.0%	-12.5%	0.0%
(4)	62.5%	0.0%	50.0%	-12.5%	-25.0%	-25.0%	-12.5%
(5)	50.0%	0.0%	62.5%	-25.0%	-25.0%	-37.5%	-25.0%
(6)	37.5%	0.0%	75.0%	-37.5%	-25.0%	-50.0%	-25.0%
(7)	12.5%	0.0%	87.5%	-62.5%	-25.0%	-62.5%	-25.0%
(8)	12.5%	0.0%	87.5%	-62.5%	-25.0%	-62.5%	-25.0%

axioms, which is $1 - R$. The function $norm_R^-$ works analogously for declines. Hence, $norm_R$ is defined in a way that it takes the greatest value if the chance that all wanted (unwanted) axioms are accepted (declined) at once becomes maximal.

Note that the expected validity ratio needs to be adjusted after each expert decision, to reflect the expected validity ratio of the remaining axioms. For instance, after Axiom (2) has been declined, $norm_{1.00}$ needs to be applied to rank the remaining axioms. If, however, Axiom (3) has been accepted, $norm_{0.00}$ is required. Note that employing $norm_{0.00}$ for ranking yields the same behavior as $impact^-$. On the other hand, $norm_{1.00}$ corresponds to $impact^+$ in case no conflicting axioms are involved.

3.2 Learning the Validity Ratio

Users might only have a rough idea of the validity ratio of a dataset in advance of the revision or the validity ratio might not be known at all. Hence, it might be difficult or impossible to decide upfront which R should be used for $norm_R$. To address this problem, we investigate how efficient we can "learn" the validity ratio on-the-fly. In this setting, the user gives an estimate for R (or we use 50% as default) and with each revision step, R is adjusted based on the number of accepted and declined axioms. Thus, the algorithm tunes itself towards an optimal ranking function, which relieves the user from choosing a validity ratio. We call the according ranking function *dynnorm* as it dynamically adapts the estimated validity ratio over the course of the revision.

In our experiments, we show that, already for small datasets, *dynnorm* outperforms random ordering and, in case of sufficiently large datasets, the estimate converges towards the actual validity ratio, thereby making the assumption of a known validity ratio unnecessary.

4 Partitioning

Since reasoning operations are very expensive (the reasoner methods take 99.2% of the computation time in our experiments according to our profiling measurements), we combine the optimization using decision spaces with a straight-forward partitioning approach for ABox axioms (i.e., class and property assertions):

Definition 5. *Let $\mathcal{A}$ be a set of ABox axioms, ind($\mathcal{A}$) the set of individual names used in $\mathcal{A}$, then $\mathcal{A}$ is* connected *if, for all pairs of individuals $a, a' \in ind(\mathcal{A})$, there exists a sequence $a_1, \ldots, a_n$ such that $a = a_1$, $a' = a_n$, and, for all $1 \le i < n$, there exists a property assertion in A containing a_i and a_{i+1}. A collection of ABoxes $\mathcal{A}_1, \ldots, \mathcal{A}_k$ is a partitioning of $\mathcal{A}$ if $\mathcal{A} = \mathcal{A}_1 \cup \ldots \cup \mathcal{A}_k$, $ind(\mathcal{A}_i) \cap ind(\mathcal{A}_j) = \emptyset$ for $1 \le i < j \le k$, and each $\mathcal{A}_i$ is connected.*

In the absence of nominals (OWL's oneOf constructor), the above described partitions or clusters of an ABox are indeed independent. Thus, we take each partition separately, join the partition with the TBox/schema axioms and perform the revision. In order to also partition TBox axioms or to properly take axioms with nominals into account, the signature decomposition approach by Konev et al.[6] could be applied. This approach partitions the signature of an ontology (i.e., the set of occurring class, property, and individual names) into subsets that are independent regarding their meaning. The resulting independent subsets of the ontology can then be reviewed independently from each other analogously to the clusters of ABox axioms used in our evaluation. We show in our experiments that:

- In particular in case of large datasets containing several partitions, the additional optimization based on partitioning significantly reduces the computational effort.
- Partitioning intensifies the effectiveness of decision spaces, since the density of entailment and contradiction relations are significantly higher within each partition than the density within a set of independent partitions.

5   Experimental Results

We evaluate our revision support methodology within the project *NanOn*[2] aiming at ontology-supported literature search. During this project, a hand-crafted ontology modeling the scientific domain of nano technology has been developed, including substances, structures, and procedures used in that domain. The ontology, denoted here with O, is specified in the Web Ontology Language OWL 2 DL [11] and comprises 2,289 logical axioms. This ontology is used as the core resource to automatically analyze scientific documents for the occurrence of NanOn classes and properties by the means of lexical patterns. When such classes and properties are found, the document is automatically annotated with those classes and properties to facilitate topic-specific information retrieval on a fine-grained level. In this way, one of the project outputs is a large amount of class and property assertions associated with the *NanOn* ontology. In order to estimate the accuracy of such automatically added annotations, they need to be inspected by human experts, which provides a natural application scenario for our approach. The manual inspection of annotations provided us with sets of valid and invalid annotation assertions (denoted by $\mathcal{A}^+$ and $\mathcal{A}^-$, respectively). To investigate how the quality and the size of each axiom set influences the results, we created several distinct annotation sets with different *validity ratios* $|\mathcal{A}^+|/(|\mathcal{A}^+| + |\mathcal{A}^-|)$. As the annotation tools provided rather reliable data, we manually created additional frequently occurring

---

[2] http://www.aifb.kit.edu/web/NanOn

Table 3. Revision results of *norm* in comparison with other ranking functions for the sets L_1-L_5

	validity ratio	*optimal*	*norm*	*best previous*	*random*
L_1	90%	65.6%	65.4%	(*impact*$^+$) 65.4%	41.7%
L_2	76%	59.8%	55.8%	(*impact*$^+$) 59.9%	35.8%
L_3	50%	47.8%	47.6%	(*guaranteed*) 36.5%	24.4%
L_4	25%	59.9%	59.8%	(*impact*$^-$) 54.9%	37.6%
L_5	10%	63.9%	63.9%	(*impact*$^-$) 63.9%	40.3%

wrong patterns and applied them for annotating texts to obtain datasets with a lower validity ratio.

For each set, we applied our methodology starting from the revision state $(O \cup O^- \cup \mathcal{A}^+ \cup \mathcal{A}^-, O, O^-)$ with O containing the axioms of the NanOn ontology and with O^- containing axioms expressing inconsistency and class unsatisfiability. We obtained a complete revision state $(O \cup O^- \cup \mathcal{A}^+ \cup \mathcal{A}^-, O \cup \mathcal{A}^+, O^- \cup \mathcal{A}^-)$ where on-the-fly expert decisions about approval or decline were simulated according to the membership in $\mathcal{A}^+$ or $\mathcal{A}^-$. For computing the entailments, we used the OWL reasoner HermiT.[3]

For each set, our baseline is the reduction of expert decisions when axioms are evaluated in random order, i.e., no ranking is applied and only the revision closure is used to automatically evaluate axioms. For this purpose, we repeat the experiments 10 times and compute the average values of effort reduction. The upper bound for the in principle possible reduction of expert decisions is obtained by applying the *optimal* ranking as suggested by the "impact oracle" for each axiom α that is to be evaluated:

$$\text{KnownImpact}(\alpha) = \begin{cases} impact^+(\alpha) & \text{if } \alpha \in \mathcal{A}^+, \\ impact^-(\alpha) & \text{if } \alpha \in \mathcal{A}^-. \end{cases}$$

5.1 Evaluation of *norm*

To compare the effectiveness of the three previously proposed impact measures and the new impact measure, we created five sets of annotations L_1 to L_5, each comprising 5,000 axioms and validity ratios varying from 10% to 90%.

Table 3 shows the results for the different ranking techniques: the column *optimal* shows the upper bound achieved by using the impact oracle, *norm* shows the reduction for our novel ranking parametrized with the actual validity ratio, *best previous* shows the best possible value achievable with the previously introduced ranking functions *impact*$^+$, *guaranteed* and *impact*$^-$, and, finally, the column *random* states the effort reduction already achieved by presenting the axioms in random order.

The results show that *norm* consistently achieves almost the maximum effort reduction with an average difference of 0.1%. The previously introduced ranking functions only work well for the high and low quality datasets, as expected. For the dataset with the validity ratio of 50%, *norm* achieves an additional 11.1% of automation by using the parametrized ranking.

---

[3] http://www.hermit-reasoner.com

Table 4. Revision results for datasets S_1 to S_5, M_1 to M_5, and L_1 to L_5

	validity ratio	optimal	norm	$dynnorm_{0.50}$	$dynnorm_{1.00}$	$dynnorm_{0.00}$	random
S_1	90%	72.4%	72.4%	58.6%	72.4%	65.5%	40.8%
S_2	77%	68.6%	65.7%	57.1%	62.9%	48.6%	38.2%
S_3	48%	65.1%	65.1%	65.1%	60.3%	61.9%	22.0%
S_4	25%	68.3%	68.3%	64.6%	63.4%	67.1%	37.6%
S_5	10%	72.5%	72.5%	71.6%	67.6%	72.5%	29.2%
M_1	91%	66.4%	66.0%	66.2%	66.4%	65.6%	40.8%
M_2	77%	60.0%	60.0%	59.6%	59.8%	59.2%	38.2%
M_3	44%	40.8%	40.6%	40.4%	40.6%	40.4%	22.0%
M_4	25%	60.0%	60.0%	59.6%	59.2%	59.8%	37.6%
M_5	10%	53.2%	53.0%	52.8%	52.8%	53.2%	29.2%
L_1	90%	65.6%	65.4%	65.4%	65.4%	65.3%	41.7%
L_2	76%	59.8%	59.8%	59.8%	59.8%	59.9%	35.8%
L_3	50%	47.8%	47.6%	47.4%	47.2%	47.3%	24.4%
L_4	25%	59.9%	59.8%	59.8%	59.8%	59.8%	37.6%
L_5	10%	63.9%	63.9%	63.9%	63.8%	63.9%	40.3%

In general, the actual difference in performance achieved by the more precise parametrized ranking increases with the increasing average maximum path length within connected decision space graphs. To see this, consider again the decision space shown in Table 1 and 2. It is clear that the distance between the highest ranked axioms for different ranking functions increases with the increasing height of the presented tree.

5.2   Evaluation of *dynnorm*

In order to evaluate our solution for situations where the validity ratio is unknown or only very rough estimates can be given upfront, we now analyze the effectiveness of the dynamically learning ranking function *dynnorm*. For this, we created the following annotation sets in addition to the datasets $L_1 - L_5$:

- small datasets S_1 to S_5 with the size constantly growing from 29 to 102 axioms and validity ratios varying from 10% to 90%,
- medium-sized datasets M_1 to M_5 with 500 axioms each and validity ratios varying from 10% to 91%.

Table 4 shows the results of the revision: the columns *optimal* and *random* are as described above, the column *norm* shows the results that we would obtain if we were to assume that the validity ratio is known and given as parameter to the *norm* ranking function, the columns $dynnorm_{0.50}$, $dynnorm_{1.00}$ and $dynnorm_{0.00}$ show the results for starting the revision with a validity ratio of 50%, 100%, and 0%, respectively, where over the course of the revision, we update the validity ratio estimate.

We observe that, in case of small datasets (S_i), the deviation from *norm* (on average 5%) as well as the dependency of the results on the initial value of the validity ratio are clearly visible. However, the results of *dynnorm* are significantly better (45.0%) than those of a revision in random order. It is also interesting to observe that the average deviation from *norm* decreases with the size of a dataset (6.9%, 10.5%, 2.7%, 3.3%,

1.9% for S_1 to S_5, respectively) and that the probability of a strong deviation is lower for datasets with an extreme validity ratio (close to 100% or 0%).

For medium-sized and large datasets (M_i and L_i), the deviation from *norm* (on average 0.3% for both) as well as the dependency on the initial value of the validity ratio are significantly lower. We conclude that

- ranking based on learning validity ratio is already useful for small datasets (30-100 axioms), and improves significantly with the growing size of the dataset under revision;
- in case of large datasets, the performance difference between the results with a validity ratio known in advance and a learned validity ratio almost disappears, thereby making the assumption of a known average validity ratio not necessary for axiom ranking.

5.3   Computational Effort

During our experiments, we measured the average number of seconds after each expert decision required for the automatic evaluation and ranking as well as the average number of reasoning calls. If we compute the average values for the revision based on *dynnorm* ranking for all 15 datasets, the revision takes on average 0.84 seconds (7.4 reasoning calls) after each expert decision. In the case of small datasets, partitioning yields additionally an improvement by an order of magnitude in terms of reasoning calls. For medium-sized datasets, the first step out of on average 153 evaluation steps took already 101,101 reasoning calls (ca. 3 hours) even when using decision spaces. Without the decision spaces, the required number of reasoning calls would be more than 500,000 judging from the required reasoning calls to build the corresponding decision space in the worst case. For this reason, we did not try to run the experiment for large datasets, which would require more than 50 million reasoning calls without decision spaces. In contrast to that, the average number of required reasoning calls for a complete revision of the sets M_1 to M_5 amounts to 3,380. The revision of datasets L_1 to L_5 required overall on average 16,175 reasoning calls, which corresponds to between 6 and 7 reasoning calls per evaluation decision. We can summarize the evaluation results as follows:

- The proposed reasoning-based support performs well in an interactive revision process with on average 0.84 seconds per expert decision.
- In particular in case of large datasets containing several partitions, the additional optimization based on partitioning significantly reduces the computational effort.
- Decision spaces save in our experiments on average 75% of reasoner calls. As measured in case of small datasets, partitioning further intensifies the effect of decision spaces and we save even 80% of reasoner calls.

6   User Front-End

Figure 1 shows the user front-end of the *revision helper* tool. It allows the user to load the set O of axioms under revision and save or load an evaluation state for the currently loaded set O. Thereby, the user can interrupt the revision at any time and proceed later

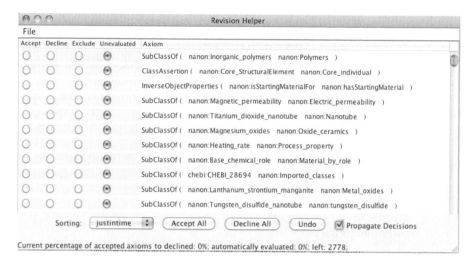

Fig. 1. Revision Helper GUI

on. If partitioning is activated, revision helper shows the partitions one after another and the revision of each partition is independent from the revision of all other partitions.

By default, revision helper initializes the set $O^{\not\models}$ of undesired statements with the minimal set of statements expressing the inconsistency of the ontology or unsatisfiability of its classes. The set of desired statements $O^{\models}$ can be initialized by loading an arbitrary ontology. A statement can be evaluated by choosing one of the values *Accept* or *Decline*, and it can be excluded from the revision process by choosing *Exclude*. The latter option should be used if the meaning of a statement is not clear and the user cannot decide whether to accept or to decline it. After the statement has been evaluated, it disappears from the revision list as well as all statements that could be evaluated automatically, unless the checkbox *Propagate Decisions* is deactivated. The ranking strategy used for sorting the statements can be selected or deactivated at any time and is taken into account after the next evaluation decision. At any stage of the revision, it is possible to export the current set $O^{\models}$ of accepted statements as an ontology. For the export, we exclude, however, axioms with which $O^{\models}$ has been initialized at the beginning of the revision.

7   Related Work

We are aware of two approaches for supporting the revision of ontological data based on logical appropriateness: an approach by Meilicke et al.[7] and another one called *ContentMap* by Jiménez-Ruiz et al.[5]. Both approaches are applied in the context of mapping revision. An extension of *ContentMap* called *ContentCVS* [4] supports an integration of changes into an evolving ontology.

In all of these approaches, dependencies between evaluation decisions are determined based on a set of logical criteria each of which is a subset of the criteria that can be derived from the notion of revision state consistency introduced in Def. 1.

In contrast to our approach, the focus of ContentMap and ContentCVS lies within the visualization of consequences and user guidance in case of difficult evaluation decisions. These approaches selectively materialize and visualize the logical consequences caused by the axioms under investigation and support the revision of those consequences. Subsequently, the approved and declined axioms are determined in correspondence with the revision of the consequences. The minimization of the manual and computational effort required for the revision is out of scope. In contrast to our approach, which requires at most a polynomial number of entailment checks, ContentMap and ContentCVS require an exponential number of reasoning operations compared to the size of the ontology under revision. The reason for this is that ContentMap is based on the computation of *justifications*, i.e., sets of axioms causing an entailment, and, in the worst-case, there can be exponentially many justifications for a particular statement.

Similarly to our approach, Meilicke et al. aim at reducing the manual effort of mapping revision. However, their results are difficult to generalize to the revision of ontologies, since the notion of impact is defined based on specific properties of mapping axioms. For every mapping axiom possible between the entities of the two mapped ontologies O_1 and O_2, they define the impact as the corresponding number of possible entailed and contradicting mapping axioms. The assumption is that the set of possible mapping axioms and the set of possible axioms in O_1 and O_2 are mostly disjoint, since axioms in O_1 and O_2 usually refer only to entities from the same ontology, while mapping axioms are assumed to map only entities from different ontologies. In case of ontology revision in general, no such natural distinction criteria for axioms under revision can be defined. Moreover, in contrast to our approach, Meilicke et al. abstract from the interactions between more than one mapping axiom.

Another strand of work is related to the overall motivation of enriching ontologies with additional expert-curated knowledge in a way that minimizes the workload of the human expert: based on the *attribute exploration* algorithm from formal concept analysis (FCA) [3], several works have proposed structured interactive enumeration strategies of inclusion dependencies or axioms of certain fragments of description logics which then are to be evaluated by the expert [9,2]. While similar in terms of the workflow, the major difference of these approaches to ours is that the axioms are not pre-specified but created on the fly and therefore, the exploration may require (in the worst case exponentially) many human decisions.

8   Summary

In our previous work [8], we established the theoretical ground for partially automated interactive revision of ontologies. In this paper, we present the implementation of the approach including an optimization based on partitioning, which significantly reduces the required computational effort. We further define a generalization of the previously proposed ranking techniques, called *norm*, which is parametrized by the expected validity ratio. The ranking function *norm* works well for any validity ratio, whereas the previous functions were tailored towards validity ratios of 100% or 0%. We define a variant of *norm*, called *dynnorm*, which can be used without knowing the validity ratio beforehand: starting with an initial estimate, e.g., 50%, the estimate is more and more refined over the course of the revision. We evaluate our implementation in a revision of ontology-based annotations of scientific publications comprising over 25,000 statements and show that

- All claims made in [8] hold also in case of large datasets under revision; on average, we were able to reduce the number of required evaluation decisions by 36% when the statements were reviewed in an arbitrary order, and by 55.4% when the ranking techniques suggested in [8] were used.
- The proposed reasoning-based support is feasible for an interactive revision process requiring on average less than one second after each expert decision in our evaluation.
- The parametrized ranking technique proposed in this paper almost achieved the maximum possible automation (59.4% of evaluation decisions) thereby reducing the manual effort of revision by 59.3%. The gain is particularly important for datasets with a validity ratio close to 50%, since for those datasets the potential of automation was not fully exploited by the other ranking techniques. In our experiments, we managed to achieve an additional 11.1% of automation for the dataset with the validity ratio of 50% by using the parametrized ranking.
- In case of large datasets with an unknown validity ratio, learning the validity ratio is particularly effective due to the law of large numbers. In our experiments, the proportion of automatically evaluated statements is nearly the same as in case where the validity ratio is known *a priori* and is used as a fixed parameter of *norm*, thereby making the assumption of known average validity ratio not necessary for axiom ranking.

As part of our future work, we intend to study more general partitioning methods, e.g., [6], to increase the applicability of the partitioning optimization. Another interesting approach in this direction would also be to study the effects of separating the ontology into parts that are not logically independent. In such a case, we might miss automatic decisions, but the potential performance gain, due to the reasoning with smaller subsets of the ontology, might compensate for this drawback.

Acknowledgments. This work is supported by the German Federal Ministry of Education and Research (BMBF) under the SAW-project NanOn.

References

1. Alchourrón, C.E., Gärdenfors, P., Makinson, D.: On the logic of theory change: Partial meet contraction and revision functions. Journal of Symbolic Logic 50, 510–530 (1985)
2. Baader, F., Ganter, B., Sertkaya, B., Sattler, U.: Completing description logic knowledge bases using formal concept analysis. In: Proceedings of the 20th International Joint Conference on Artificial Intelligence (IJCAI 2007), pp. 230–235 (2007)
3. Ganter, B., Wille, R.: Formal Concept Analysis: Mathematical Foundations. Springer, Heidelberg (1997)
4. Jiménez-Ruiz, E., Cuenca Grau, B., Horrocks, I., Llavori, R.B.: Building ontologies collaboratively using ContentCVS. In: Proceedings of the 22nd International Workshop on Description Logics (DL 2009). CEUR Workshop Proceedings, vol. 477. CEUR-WS.org (2009)

5. Jiménez-Ruiz, E., Cuenca Grau, B., Horrocks, I., Berlanga, R.: Ontology integration using mappings: Towards getting the right logical consequences. In: Aroyo, L., Traverso, P., Ciravegna, F., Cimiano, P., Heath, T., Hyvönen, E., Mizoguchi, R., Oren, E., Sabou, M., Simperl, E. (eds.) ESWC 2009. LNCS, vol. 5554, pp. 173–187. Springer, Heidelberg (2009)
6. Konev, B., Lutz, C., Ponomaryov, D., Wolter, F.: Decomposing description logic ontologies. In: Proceedings of the 12th International Confonference on Principles of Knowledge Representation and Reasoning, KR 2010 (2010)
7. Meilicke, C., Stuckenschmidt, H., Tamilin, A.: Supporting manual mapping revision using logical reasoning. In: Proceedings of the 23rd Conference on Artificial Intelligence (AAAI 2008), pp. 1213–1218. AAAI Press (2008)
8. Nikitina, N., Rudolph, S., Glimm, B.: Reasoning-supported interactive revision of knowledge bases. In: Proceedings of the 22nd International Joint Conference on Artificial Intelligence, IJCAI 2011 (2011)
9. Rudolph, S.: Exploring Relational Structures Via FLE. In: Wolff, K.E., Pfeiffer, H.D., Delugach, H.S. (eds.) ICCS 2004. LNCS (LNAI), vol. 3127, pp. 196–212. Springer, Heidelberg (2004)
10. Schlobach, S., Cornet, R.: Non-standard reasoning services for the debugging of description logic terminologies. In: Proceedings of the 18th International Joint Conference on Artificial Intelligence (IJCAI 2003), pp. 355–362. Morgan Kaufmann (2003)
11. W3C OWL Working Group: OWL 2 Web Ontology Language: Document Overview. W3C Recommendation (October 2009), http://www.w3.org/TR/owl2-overview/

Getting the Meaning Right: A Complementary Distributional Layer for the Web Semantics*

Vít Nováček, Siegfried Handschuh, and Stefan Decker

Digital Enterprise Research Institute (DERI)
National University of Ireland Galway (NUIG)
IDA Business Park, Lower Dangan, Galway, Ireland
vit.novacek@deri.org

Abstract. We aim at providing a complementary layer for the web semantics, catering for bottom-up phenomena that are empirically *observable* on the Semantic Web rather than being merely *asserted* by it. We focus on meaning that is not associated with particular semantic descriptions, but emerges from the multitude of explicit and implicit links on the web of data. We claim that the current approaches are mostly top-down and thus lack a proper mechanisms for capturing the emergent aspects of the web meaning. To fill this gap, we have proposed a framework based on distributional semantics (a successful bottom-up approach to meaning representation in computational linguistics) that is, however, still compatible with the top-down Semantic Web principles due to inherent support of rules. We evaluated our solution in a knowledge consolidation experiment, which confirmed the promising potential of our approach.

1   Introduction

The Semantic Web has been designed for asserting meaning of things mostly in a top-down manner (via explicit specifications of RDF descriptions or ontologies). We conjecture that there is also another, bottom-up meaning of the web (both the 'semantic' and 'human' one). Similarly to the meaning of natural languages arising from the complex system of interactions between their individual speakers [1], we conceive the bottom-up web semantics as consisting of implicit patterns. In the Semantic Web case, though, the complex patterns of meaning emerge from a simple language of countless triple statements, which may come from the evolving Linked Open Data cloud, but also from the human web (mediated to machines by methods like data or text mining).

The proposed alternative way of looking at the Semantic Web can bring better solutions to problems in areas like knowledge consolidation (by which we basically mean clustering of related entities and properties). For instance, in our CORAAL prototype (see http://coraal.deri.ie), users can search for properties linking particular life science entities (like genes or diseases). CORAAL

---

* This work has been supported by the 'Líon II' project funded by SFI under Grant No. SFI/08/CE/I1380. We are indebted to Ed Hovy (ISI, USC), who had an indirect, yet substantial influence on the presented research. Also, we would like to thank to Václav Belák (DERI) for discussions about possible interpretations of our results.

L. Aroyo et al. (Eds.): ISWC 2011, Part I, LNCS 7031, pp. 504–519, 2011.
© Springer-Verlag Berlin Heidelberg 2011

extracts all the underlying statements automatically from text, which leads to thousands of properties occurring only in very small number of triples. This may result in too specific query answers and user frustration, as they have to struggle to figure out how to get more general information. Imagine one wants to know more about organs involved in the production of a hormone H. A query for that can look like *H secreted_in ?x*. However, such a query may retrieve only a single result. More results could be retrieved via related properties like *excreted_in* or *produced_in*, but it is rather tedious to try all such possibilities without knowing precisely how exactly one should ask. A solution grouping extracted content into more general inter-related clusters would significantly improve the user satisfaction and efficiency, as hitting a single property would also reveal all the related ones. Yet for achieving such a consolidation, one needs to know not (only) what is meant by the statements at the level of the particular documents (which is covered by the current approaches). What is more important (and less explored) are the minuscule contextual features distributed across the whole data set (e.g., properties and ⟨ subject, object ⟩ tuples that tend to co-occur at a larger scale with sufficient significance). This is what constitutes the global evidence of what is actually *meant* by the data set at large (and not just *asserted* at the level of local semantic descriptions). By capturing these aspects, one can consolidate the little scattered chunks of related knowledge in an empirically valid manner. As detailed in Section 2, we lack a comprehensive solution for this, though.

Therefore we have proposed (in Section 3) a framework stemming from recent advances in distributional semantics. This sub-field of computational linguistics is based on a hypothesis that "a word is characterized by the company it keeps" [2]. In our case, we can rephrase this to characterise the meaning of a thing on the web by the company of things linked to it. In order for such meaning to be representative, though, we have to analyse the 'company' across as much content as possible. To do so, we employ an approach utilising simple, yet universal and powerful tensor-based representation of distributional semantics proposed in [3]. We adapt it to the Semantic Web specifics and show how one can execute rules on the top of it, which effectively leads to a smooth combination of the bottom-up (distributional) and top-down (symbolic) approaches to the representation of meaning. Apart of that, we dedicate a substantial a part of the paper (Section 4) to an experimental application of our approach to automated consolidation of knowledge in life sciences. We conclude the paper in Section 5.

2   Related Work

We define emergent meaning of Semantic Web expressions using tensors to elucidate various distributional effects, which stems from the comprehensive approach in [3]. However, we extended this approach with a symbolic (rule-based) layer in order to combine it with the top-down Semantic Web principles. A tensor-based representation of the Semantic Web data was presented for instance in [4], which, however, focuses mostly on ranking and decomposition, not on providing generic means for an analysis of various bottom-up semantics phenomena. Approaches to induction of implicit patterns or schemata from data are also related to our work (particular examples include [5] and [6] in the fields of databases and Semantic Web, respectively). Yet these approaches are usually focused on rather limited sets of problems (e.g., query optimisation or concept induction) and thus are not as comprehensive and theoretically uniform as our framework. The NELL project [7] aims at incremental and continuous induction of triple patterns from the web data, which is a goal very similar to ours. The main differences are that NELL needs manually provided seeds of knowledge and a slight supervision in the form of pruning. Also, the type of extracted patterns is limited to instances of fixed generic relations in NELL, whereas we allow for bottom-up inference of rather varied and dynamic set of phenomena. Works like [8] or [9] deal with emergent semantics, but they mostly investigate how to gather the semantics (from ontologies or simple peer-to-peer interactions in heterogeneous data systems). Less attention is paid to how the semantics can be uniformly represented and utilised later on. Finally, our experiment in knowledge consolidation is closely related to ontology matching [10] and life science data integration [11]. Most of the ontology matching algorithms are designed to operate at the schema level, though, and not at the data level that is most pertinent to our work. Regarding extant methods for knowledge integration in life sciences, majority of them uses quite a limited set of specifically tuned lexical or structural similarities. Thus our approach can provide for a more adaptive and empirically driven data-based integration in this context.

3   Distributional Web Semantics

The proposed distributional web semantics framework has two major layers – the bottom-up and top-down one. The former caters for the implicit meaning, while the latter allows for adding more value to the bottom-up analysis by utilising the current Semantic Web resources (e.g., RDF Schema or ontologies). A general way of using the framework follows this pipeline: (1) convert a set of simple RDF documents into the internal distributional representation; (2) extract interesting patterns from it; (3) make use of extant top-down semantic resources to materialise more implicit knowledge by means of inference (optional); (4) utilise the results to improve the quality of the initial RDF data set. The last step can consist of exporting the distributional patterns as RDF statements to be added to the input data (e.g., as links between the entities or properties found to be similar). Alternatively, one can present the patterns directly to users along the original data set to facilitate its machine-aided augmentation.

3.1   Bottom-Up Layer

Source Representation. The basic structure of the bottom-up layer is a so called source (or graph) representation $\mathbf{G}$, which captures the co-occurrence of things (i.e., subjects and objects) within relations (i.e., predicates) across a set of documents (i.e., RDF graphs). Let A_l, A_r be sets representing left and right arguments of binary co-occurrence relationships (i.e., statements), and L the types of the relationships. A_l, A_r, L correspond to sets of RDF subjects, objects and predicates, respectively. Furthermore, let P be a set representing provenances of particular relationships (i.e., graph names). We define the source representation as a 4-ary labeled tensor $\mathbf{G} \in \mathbb{R}^{|A_l| \times |L| \times |A_r| \times |P|}$. It is a four-dimensional array structure indexed by subjects, predicates, objects and provenances, with values reflecting a frequency or weight of statements in the context of particular provenance sources (0 if a statement does not occur in a source). For instance, if a statement (a_l, l, a_r) occurs k-times in a data source d (a single graph or a set of graphs in general), then the element $g_{a_l,l,a_r,d}$ of $\mathbf{G}$ will be set to k to reflect it. More details are illustrated in the following example.

Example 1. Let us consider 7 statements (acquired from biomedical texts):

> (*protein domain*, *different*, *protein*), (*protein domain*, *type*, *domain*), (*gene*, *different*, *protein*), (*internal tandem duplications*, *type*, *mutations*), (*internal tandem duplications*, *in*, *juxtamembrane*), (*internal tandem duplications*, *in*, *extracelullar domains*), (*protein domain*, *type*, *domain*)

with provenances $D_1, D_1, D_2, D_3, D_3, D_3, D_4$, respectively. The source representation (using statement occurrence frequencies as values) is:

$s \in A_l$	$p \in L$	$o \in A_r$	$d \in P$	$g_{s,p,o,d}$
protein domain	different	protein	D_1	1
protein domain	type	domain	D_1	1
gene	different	protein	D_2	1
internal tandem duplications	type	mutations	D_3	1
internal tandem duplications	in	juxtamembrane	D_3	1
internal tandem duplications	in	extracelullar domains	D_3	1
protein domain	type	domain	D_4	1

We omit all zero values and use the tabular notation as a convenient and concise representation of a 4-dimensional tensor, with the three first columns for indices and the fourth one for the corresponding value.

Corpus Representation. The source tensor is merely a low-level data representation preserving the association of statements with their provenance contexts. Before allowing for actual distributional analysis, the data have to be transformed into a more compact structure $\mathbf{C}$ called corpus representation. $\mathbf{C} \in \mathbb{R}^{|A_l| \times |L| \times |A_r|}$ is a ternary (three-dimensional) labeled tensor, devised according to [3] in order to provide for a universal and compact distributional representation for the proposed bottom-up web semantics framework. A corpus $\mathbf{C}$ can be constructed from a source representation $\mathbf{G}$ using functions $a : \mathbb{R} \times \mathbb{R} \to \mathbb{R}, w : P \to \mathbb{R}, f : A_l \times L \times A_r \to \mathbb{R}$. For each $\mathbf{C}$ element $c_{s,p,o}$, $c_{s,p,o} = a(\sum_{d \in P} w(d) g_{s,p,o,d}, h(s,p,o))$, where $g_{s,p,o,d}$ is an element of the source tensor $\mathbf{G}$ and the a, f, w functions act as follows: (1) w assigns a relevance degree to each source; (2) f reflects the relevance of the statement elements (e.g., a mutual information score of the subject and object within the sources); (3) a aggregates the result of the w, f functions' application. This way of constructing the elements of the corpus tensor from the low-level source representation essentially aggregates the occurrences of statements within the input data, reflecting also two important things – the relevance of $\mathbf{G}$ particular sources (via the w function), and the relevance of the statements themselves (via the f function). The specific implementation of the functions is left to applications – possible examples include (but are not limited to) ranking (both at the statement and document level) or statistical analysis of the statements within the input data. In Section 4.1, we provide a detailed description of a particular source-to-corpus conversion we used in the evaluation experiment.

Example 2. A corpus corresponding to the source tensor from Example 1 can be represented (again in a tabular notation) as given below. The w values were 1 for all sources and a, f aggregated the source values using relative frequency (in a data set containing 7 statements).

$s \in A_l$	$p \in L$	$o \in A_r$	$c_{s,p,o}$
protein domain	different	protein	1/7
protein domain	type	domain	2/7
gene	different	protein	1/7
internal tandem duplications	type	mutations	1/7
internal tandem duplications	in	juxtamembrane	1/7
internal tandem duplications	in	extracelullar domains	1/7

Corpus Perspectives. The elegance and power of the corpus representation lays in its compactness and universality that, however, yields for many diverse possibilities of the underlying data analysis. The analysis are performed using a process of so called matricisation of the corpus tensor **C**. Essentially, matricisation is a process of representing a higher-order tensor using a 2-dimensional matrix perspective. This is done by fixing one tensor index as one matrix dimension and generating all possible combinations of the other tensor indices within the remaining matrix dimension. In the following we illustrate the process on the simple corpus tensor from Example 2. Detailed description of matricisation and related tensor algebra references can be found in [3].

Example 3. When fixing the subjects (A_l set members) of the corpus tensor from Example 2, one will get the following matricised perspective (the rows and columns with all values equal to zero are omitted here and in the following examples):

$s/\langle p, o \rangle$	$\langle d, p \rangle$	$\langle t, dm \rangle$	$\langle t, m \rangle$	$\langle i, j \rangle$	$\langle i, e \rangle$
protein domain	1/7	2/7	0	0	0
gene	1/7	0	0	0	0
internal tandem duplications	0	0	1/7	1/7	1/7

The abbreviations d, p, t, dm, m, i, j, e stand for *different, protein, type, domain, mutations, in, juxtamembrane, extracellular domains*. One can clearly see that the transformation is lossless, as the original tensor can be easily reconstructed from the matrix by appropriate re-grouping of the indices.

The corpus tensor matricisations correspond to vector spaces consisting of elements defined by particular rows of the matrix perspectives. Each of the vectors has a name (the corresponding matrix row index) and a set of features (the matrix column indices). The features represent the distributional attributes of the entity associated with the vector's name – the contexts aggregated across the whole corpus. Thus by comparing the vectors, one essentially compares the meaning of the corresponding entities emergently defined by the underlying data. For exploring the matricised perspectives, one can uniformly use the linear algebra methods that have been successfully applied to vector space analysis tasks for the last couple of decades. Large feature spaces can be reliably reduced to a couple of hundreds of the most significant indices by techniques like singular value decomposition or random indexing (see http://en.wikipedia.org/wiki/Dimension_reduction for details). Vectors can be compared in a well-founded manner by various metrics or by the cosine similarity (see http://en.wikipedia.org/wiki/Distance or http://en.wikipedia.org/wiki/Cosine_similarity, respectively). This way matrix perspectives can be combined with vector space analysis techniques in order to investigate a wide range of semantic phenomena related to synonymy, clustering, ambiguity resolution, taxonomy detection or analogy discovery. In this introductory paper, we focus only on clustering of similar entities (subjects and/or objects) and properties. The following example explains how to perform these particular types of analysis.

Example 4. Let us add two more matrix perspectives to the $s/\langle p, o \rangle$ one provided in Example 3. The first one represents the distributional features of objects (based on the contexts of predicates and subjects they tend to co-occur with in the corpus):

$o/\langle p, s \rangle$	$\langle d, pd \rangle$	$\langle t, pd \rangle$	$\langle d, g \rangle$	$\langle t, itd \rangle$	$\langle i, itd \rangle$
protein	1/7	0	1/7	0	0
domain	0	2/7	0	0	0
mutations	0	0	0	1/7	0
juxtamembrane	0	0	0	0	1/7
extracellular domains	0	0	0	0	1/7

d, pd, t, g, itd, i stand for *different, protein domain, type, gene, internal tandem duplications, in.* Similarly, the second perspective represents the distributional features of properties:

$p/\langle s, o \rangle$	$\langle pd, p \rangle$	$\langle pd, d \rangle$	$\langle g, p \rangle$	$\langle itd, m \rangle$	$\langle itd, j \rangle$	$\langle itd, ed \rangle$
different	1/7	0	1/7	0	0	0
type	0	2/7	0	1/7	0	0
in	0	0	0	0	1/7	1/7

$itd, pd, p, d, g, m, j, ed$ stand for *internal tandem duplications, protein domain, protein, domain, gene, mutations, juxtamembrane, extracellular domains.*

The vector spaces induced by the matrix perspectives $s/\langle p, o \rangle$ and $o/\langle p, s \rangle$ can be used for finding similar entities by comparing their corresponding vectors. Using the cosine vector similarity, one finds that $sim_{s/\langle p,o \rangle}(\textit{protein domain, gene}) =$ $\frac{\frac{1}{7}\frac{1}{7}}{\sqrt{(\frac{1}{7})^2 + (\frac{2}{7})^2}\sqrt{(\frac{1}{7})^2}} \doteq 0.2972$ and $sim_{o/\langle p,s \rangle}(\textit{juxtamembrane, extracel-lular domains})$ $= \frac{\frac{1}{7}\frac{1}{7}}{\sqrt{(\frac{1}{7})^2}\sqrt{(\frac{1}{7})^2}} = 1.$ These are the only non-zero similarities among the subject and object entities present in the corpus. As for the predicates, all of them have a zero similarity. This quite directly corresponds to the intuition a human observer can get from the data represented by the initial statements from Example 1. Protein domains and genes seem to be different from proteins, yet protein domain is a type of domain and gene is not, therefore they share some similarities but are not completely equal according to the data. Juxtamembranes and extracellular domains are both places where internal tandem duplications can occur, and no other information is available, so they can be deemed equal until more data comes. Among the particular predicates, no patterns as clear as for the entities can be observed, therefore they can be considered rather dissimilar given the current data.

3.2   Top-Down Layer

A significant portion of the expressive Semantic Web standards (RDFS, OWL) and widely used extensions (such as N3, cf. `http://www.w3.org/DesignIssues/Notation3.html`) can be expressed by conjunctive rules (see `http://www.w3.org/TR/rdf-mt/`), `http://www.w3.org/TR/owl2-profiles/` or [12]). To allow for a seamless combination of this top-down layer of the Semantic Web with the bottom-up principles introduced in the previous section, we propose a straightforward adaptation of state of the art rule-based reasoning methods.

Conjunctive rules can be described as follows in the 'language' of the bottom-up semantics. Let $\mathcal{S} = \mathbb{R}^{|A_l \cup V| \times |L \cup V| \times |A_r \cup V|}$ be a set of corpus tensors with their index domains (A_l, L, A_r) augmented by a set of variables V. Then $(\mathbf{L}, \mathbf{R}, w)$, where $\mathbf{L}, \mathbf{R} \in \mathcal{S}, w \in \mathbb{R}$, is a rule with an antecedent $\mathbf{L}$, a consequent $\mathbf{R}$ and a weight w. The values of the rule tensors are intended to represent the structure of the rule statements – a non-zero value reflects the presence of a statement consisting of the corresponding indices in the rule. However, the antecedent tensor values can also specify the weights of the relationship instances to be matched and thus facilitate uncertain rule pattern matching. The weights can be used to set relative importance of rules. This is especially useful when combining rules from rule sets of variable relevance – one can assign higher weights to rule coming from more reliable resources and the other way around. We assume the weights to be set externally – if this is not the case, they are assumed to be 1 by default.

Example 5. An RDFS entailment rule for transitivity can be stated in N3 as: `{?x rdfs:subClassOf ?y . ?y rdfs:subClassOf ?z } => {?x rdfs:subClassOf ?z }`. The rule is transformed to the tensor form as:

$s \in A_l \cup V$	$p \in L \cup V$	$o \in A_r \cup V$	$l_{s,p,o}$
$?x$	$rdfs : subClassOf$	$?y$	1
$?y$	$rdfs : subClassOf$	$?z$	1

$s \in A_l \cup V$	$p \in L \cup V$	$o \in A_r \cup V$	$r_{s,p,o}$
$?x$	$rdfs : subClassOf$	$?z$	1

$, 1)$.

Rules can be applied to a corpus by means of Algorithm 1. The particular rule-based reasoning method we currently use is a modified version of the efficient RETE algorithm for binary predicates [13]. The *conditionTrees()* function in Algorithm 1 generates a set of trees of antecedent conditions from a rule set $\mathcal{R}$.

Algorithm 1. Rule Evaluation

1: $RESULTS \leftarrow \emptyset$
2: $FOREST \leftarrow conditionTrees(\mathcal{R})$
3: **for** $T \in FOREST$ **do**
4:     **for** $(I, \mathbf{R}, w) \in matches(T)$ **do**
5:         $\mathbf{R}' \leftarrow w \cdot materialise(I, \mathbf{R})$
6:         $RESULTS \leftarrow RESULTS \cup \mathbf{R}'$
7:     **end for**
8: **end for**
9: **return** $\sum_{\mathbf{X} \in RESULTS} \mathbf{X}$

Example 6. For instance, let us imagine the following rule set (described in N3 again): R_1 : {?x rdfs:subClassOf ?y . ?y rdfs:subClassOf ?z } => {?x rdfs:subClas-sOf ?z}. R_2 :{?x rdfs:subClassOf ?y . ?z rdf:type ?x} => {?z rdf:type ?y}. For simplicity, we assume the weights of the rules R_1, R_2 to be 1.0. Given this rule set, the *conditionTrees()* function returns a single tree with a root condition ?x rdfs:subClassOf ?y and the ?y rdfs:subClassOf ?z, ?z rdf:type ?x conditions as the root's children. The tree leafs (i.e., children of the root's children) then point to the consequents and weights of the rules R_1, R_2, respectively.

The rule condition forest allows for optimised incremental generation of all possible corpus instance assignments to the variables in the rule conditions – each condition is being evaluated only once even if it occurs in multiple rules. The generation of instance assignments for particular condition variables is realised by the function *matches()* in Algorithm 1. It produces tuples $(I, \mathbf{R}, w)$, where I is an assignment of instances to the antecedent variables along a particular root-leaf path in the given tree T. $\mathbf{R}, w$ are then the rule consequent and weight in the leaf of the corresponding instance assignment path.

The function *materialise()* takes the computed instance assignment I and applies it to the consequent $\mathbf{R}$. The values of the materialised consequent tensor $\mathbf{R}'$ are computed as $r_{s,p,o} = \top\{c_{i_1,i_2,i_3}|(i_1,i_2,i_3) \in I\}$ for each (s,p,o) element of the consequent that has a non-zero value in the original $\mathbf{R}$ tensor. The c_{i_1,i_2,i_3} elements of the tensor $\mathbf{C}$ (the corpus representation, i.e., knowledge base) correspond to all statements in the instantiated rule conditions along the assignment path I. Finally, the $\top$ operation is an application of a fuzzy conjunction (t-norm, cf. http://en.wikipedia.org/wiki/T-norm) to a set of values[1]. The result of Algorithm 1 is a sum of all the tensors resulting from the particular consequent materialisations weighted by the corresponding rule weights.

Example 7. To exemplify an iterative rule materialisation (knowledge base closure), let us add two more elements to the corpus from Example 2 (the weights are purely illustrative):

A_l	L	A_r	value
domain	$rdfs : subClassOf$	molecular structure	2/9
molecular structure	$rdfs : subClassOf$	building block	1/9

If we assume that the **type** relation from the previous examples is equivalent to the rdf:type relation from the rule R_2 in Example 6, we can apply the R_1, R_2 rules to the extended corpus representation with the following results. After assigning instances to the antecedent variables, the only instance path leading towards R_1 in the condition tree consists of the statements domain rdfs:subClassOf molecular structure and molecular structure rdfs:subClassOf building block. The R_2 branch generates four possible instance paths. The root can have two values: domain rdfs:subClassOf molecular structure, molecular structure rdfs:subClassOf building block. Similarly for the child – there are two statements in the

---

[1] Note that although the minimum t-norm, $t(a,b) = min(a,b)$, can be applied to any positive values in the corpus representation tensors with the intuitively expected (fuzzy-conjunctive) semantics, any other t-norm, such as the product one, $t(a,b) = ab$, would lead to rather meaningless results if the tensor values were not normalised to the $[0,1]$ interval first.

corpus that fit the corresponding condition: `protein domain rdf:type domain` and `internal tandem duplications rdf:type mutations`. When using the minimum t-norm we can enrich the knowledge base by the following materialised consequents:

$s \in A_l$	$p \in L$	$o \in A_r$	$r_{s,p,o}$
domain	*rdfs : subClassOf*	*building block*	1/9
protein domain	*rdf : type*	*molecular structure*	2/9

If we apply Algorithm 1 again, we get one more new statement:

$s \in A_l$	$p \in L$	$o \in A_r$	$r_{s,p,o}$
protein domain	*rdf : type*	*building block*	2/9

After that the corpus representation already remains stable (its closure has been computed), as no further application of the rules produces new results.

4   Evaluation

In the evaluation, we addressed life sciences, a domain where the information overload is now more painful than ever and where efficient data/knowledge integration can bring a lot of benefit [11]. Specifically, we looked into knowledge consolidation, by which we mean—at the abstract level—grouping of possibly isolated, yet related simple facts into more general chunks of knowledge with similar meaning. Drilling down to a more concrete level of the actual experiments, we applied the framework proposed in this paper to clustering of entities (i.e., subjects and objects) and relations based on their distributional features within a corpus of input resources. We considered two types of inputs – existing linked data sets and statements extracted from texts associated with the linked data sets. Details on the data, experiments, evaluation methods and results are provided in the corresponding sections below.

4.1   Method

Data. The first type of data we used were four RDF documents (parts of the Linked Open Data cloud) that were converted into RDF from manually curated life science databases and served on the D2R web site (`http://www4.wiwiss.fu-berlin.de/`). To keep the data set focused, we chose resources dealing with drugs and diseases: Dailymed, Diseasome, Drugbank and Sider (see `http://goo.gl/cODqo`, `http://goo.gl/sbq8E`, `http://goo.gl/ydMSD` and `http://goo.gl/LgmlF`, respectively). This data set is referred to by the LD identifier in the rest of the paper. We pre-processed the data as follows. Most importantly, we converted the identifiers of entities to their human-readable names to facilitate the evaluation. Also, we added new statements for each explicitly defined synonym in the LD data set by "mirroring" the statements of the descriptions associated with the corresponding preferred term. More technical details are available at `http://goo.gl/38bGK` (an archive containing all the data, source code and additional descriptions relevant to the paper).

The second data set we used was generated from the textual content of the LD documents, which contain many properties with string literal objects representing natural language (English) definitions and detailed descriptions of the entries (e.g., `drugbank:pharmacology` describes the biochemical mechanism of drug functions). We extracted the text from all such properties, cleaned it up (removing spurious HTML mark-up and irregularities in the sentence segmentation) and applied a simple NLP relation extraction pipeline on it, producing a data set of extracted statements (XD in the following text). In the extraction pipeline we first split the text into sentences and then associated each word in a sentence with a corresponding part-of-speech tag. The tagged sentences were shallow-parsed into a tree structure with annotated NPs (noun phrases). These trees were then used to generate statements particular statements as follows. From any $NP_1 [verb|preposition]^+ NP_2$ sequence in the parsed tree, we created subject from NP_1, object from NP_2 and predicate from the intermediate verb or prepositional phrase. Additional statements were generated by decomposing compound noun phrases. More details and examples are out of scope here, but we made them available for interested readers as a part of the data package provided at `http://goo.gl/38bGK`.

Concerning the size of the experimental data, the linked data sets contained ca. 630 thousand triples, 126 properties and around 270 thousands of simple entities (i.e., either subjects or objects) corresponding to almost 150 thousands of unique identifiers (i.e., preferred labels). The size of the extracted data set was around 3/4 of the linked data one, however, the number of extracted properties was much higher – almost 35 thousand. Apart of the LD, XD data sets, we also prepared their LD⁻, XD⁻ alternatives, where we just 'flattened' all the different properties to uniform links. We did so to investigate the influence the multiple property types have on the distributional web semantics features within the experiments.

Knowledge Consolidation. Before performing the knowledge consolidation, we had to incorporate the RDF data (the LD, XD sets) into the framework introduced in Section 3, i.e., to populate the graph and source representation tensors $\mathbf{G}, \mathbf{C}$ (separate tensors for each of the LD, XD, LD⁻, XD⁻ data sets). The $\mathbf{G}$ indices were filled by the lexical elements of triples and by the corresponding source graph identifiers (there were five provenance graphs – one for each of the four linked data documents and one for the big graph of extracted statements). The $\mathbf{G}$ values were set to 1 for all elements $g_{s,p,o,d}$ such that the statement (s, p, o) occurred in the graph d; all other values were 0. To get the $\mathbf{C}$ tensor values $c_{s,p,o}$, we multiplied the frequency of the (s, p, o) triples (i.e., $\sum_{d \in P} g_{s,p,o,d}$) by the point-wise mutual information score of the (s, o) tuple (see `http://en.wikipedia.org/wiki/Pointwise_mutual_information` for details on the mutual information score theory and applications). This method is widely used for assigning empirical weights to distributional semantics representations [3], we only slightly adapted it to the case of our "triple corpora" by using the frequencies of triple elements and triples themselves. As we were incorporating triples from documents with equal relevance, we did not use any specific provenance weights in the $\mathbf{C}$ tensor computation. After the population of the corpus tensor, we used its $s/\langle p, o \rangle, o/\langle p, s \rangle$ perspectives for generating similar entities and the $p/\langle s, o \rangle$ perspective for similar properties, proceeding exactly as described in Example 4. A cluster of size x related to a vector $\mathbf{u}$ in a perspective π was generated as a set of up to x most similar vectors $\mathbf{v}$ such that $sim_\pi(\mathbf{u}, \mathbf{v}) > 0$. Our implementation of the large scale tensor/matrix representation and analysis is open source and available as a part of the data package provided at http://goo.gl/38bGK. Note that one might also employ rules in the knowledge consolidation experiments, for instance to materialise more implicit statements providing additional features for the distributional analysis. However, a comprehensive evaluation of such a combined approach does not fit into the scope of this paper, therefore we will elaborate on it in a separate technical report.

To evaluate the entity consolidation, we employed a gold standard – MeSH (see http://www.nlm.nih.gov/mesh/), a freely available controlled vocabulary and thesaurus for life sciences. MeSH is manually designed and covers a lot of disease, gene and drug terms, therefore the groups of related things within its taxonomical structure are a good reference comparison for artificially generated clusters of entities from the same domain[2]. To the best of our knowledge, no similar applicable gold standard that would cover our property consolidation data sets exists, thus we had to resort to manual assessment of the corresponding results. As a baseline, we used randomly generated clusters of entities and properties. Other baseline methods are possible, such as various ontology matching techniques [10]. However, these methods are designed rather for 'schema-level' matching between two semantic resources. Their application to the 'data-level' consolidation of many statements possibly contained in a single resource is a research question in its own right, which we leave for future work.

Evaluation Metrics. For the entity clustering evaluation, we first need a 'gold standard' similarity between two terms, based on their paths in the MeSH taxonomy[3]. Every MeSH entry (and its synonyms) are associated with one or more tree codes. These take form of alphanumeric tree level identifiers divided by dots that determine the position of particular entries in the MeSH trees. The path from the root (most generic term in a category) to a given entry corresponds to its tree code read from left to right. For instance, *abdominal wall* and *groin* have MeSH tree codes *A01.047.050* and *A01.047.365*, which means they have a path of length 2 in com-

---

[2] We considered, e.g., GO, as well as vaccine and disease ontologies from the OBO dataset for the gold standard. Unfortunately, either the coverage of the ontologies w.r.t. the experimental dataset was worse than for MeSH, or the definition of 'gold standard' similarities was trickier (requiring possibly substantial further research) due to additional relations and/or rather complex (OWL) semantics. Thus, for the time being, we chose to focus on something simpler, but representative and already convincing.

[3] This essentially edge-based approach is motivated by the similarity measures commonly used in the context of life science knowledge bases [14]. A possible alternative would be a node-based similarity utilising the information content measure (also discussed in detail in [14]). However, the computation of the information content depends on a clear distinction between classes (or concepts) and instances (or terms) in the data set. In our case, this is rather difficult – an instance term can become a class term as soon as it becomes a type of another term in an extracted statement. Therefore an application of a node-based (or a hybrid node-edge) similarity would require additional investigations that unfortunately do not fit the rather limited scope of this paper.

mon from their tree's root (going from *Body Regions* through *Abdomen* with the respective tree codes *A01, A01.047*). The length of the path shared by two terms can be used as a simple and naturally defined similarity measure – the bigger the relative portion of the MeSH root path shared by two terms, the closer—i.e., more similar—they are in the MeSH hierarchy.

More formally, we can define a MeSH similarity measure s_M between two terms s, t as follows. If s and t do not share any node in their tree paths, $s_M(s, t) = 0$. If any of the s, t tree paths subsumes the other one, $s_M(s, t) = 1$. In the remaining cases, $s_M(s, t) = k^{max(|p(s)|, |p(t)|) - |mcp(s,t)|}$, where $k \in (0, 1)$ is a coefficient, $p(x)$ refers to the tree path of a term x and $mcp(s, t)$ is a maximum common path shared between the tree paths of s, t. For our experiment, we chose coefficient $k = 0.9$, which provides for clearly visible but not too abrupt changes in the similarity values. Note that the particular choice of k is rather cosmetic as it does not influence the descriptive power of the results, it only changes the absolute values of the similarity.

To assess the clusters computed in the entity consolidation experiments, we defined the overlap (o) and quality (q) evaluation measures: $o(C) = \frac{|\{t | t \in M \wedge t \in C\}|}{|C|}$, $q(C) = \frac{\sum_{(s,t) \in comb_2(C_M)} s_M(s,t)}{|comb_2(C_M)|}$, where C is the cluster (a set of terms) being measured, M is a set of all MeSH terms, $C_M = \{t | t \in M \wedge t \in C\}$ are all terms from C that are in MeSH as well, $comb_2(C_M)$ selects all combinations of two different terms from C_M, and, finally, s_M is the MeSH term similarity. If $C_M = \emptyset$, $q(C) = 0$ by definition. The overlap is an indication of how many terms from MeSH are contained in a cluster, while the quality is the actual evaluation measure that tells us how good the part of the cluster covered by the gold standard is (i.e., how close it is to the structure of the manually designed MeSH thesaurus). The quality is computed as an average similarity of all possible combinations of term pairs in a cluster that are contained in MeSH. Such a measure may seem to be a little restrictive when the MeSH-cluster overlap is low. However, the low overlap is not as much caused by the noise in the clusters as by insufficient coverage of the gold standard itself, which is quite a common problem of gold standards in as dynamic and large domains as life sciences [11].

Since we lack a proper gold standard for the property consolidation experiment, the corresponding evaluation measures will inevitably be a slightly less solid than the ones for entity clustering. We base them on human assessment of two factors – an adequacy (aq) and accuracy (ac) of property clusters. Given a property cluster C, $ac = \frac{|C| - |N|}{|C|}$, $aq = \frac{|R|}{|C| - |N|}$, where N is a set of properties deemed as noise by a human evaluator, and R is a set of properties considered to be relevant to the seed property the cluster was generated from. For the manual evaluation of the property consolidation experiment, we used two human experts (one bioinformatician and one clinical researcher). They both had to agree on determining whether properties are not noise and whether they are relevant. For the opposite decisions, a single vote only was enough to mark the corresponding property as a true negative result (thus making the evaluation rather pessimistic in order to reduce the likelihood of bias).

In both entity and property consolidation experiments, we randomly selected 10 batches of 10 terms that served as cluster seeds for each evaluated cluster size, and computed the arithmetic means of the metrics of all the clusters. Thus we made sure that the results closely approximated the whole data set (for the automatic evaluation with gold standard, at least 75% of the particular results in all the selected batches were differing from the mean value by less than 5%, although some fluctuations were present in the manual evaluation). We tested several different sizes of the generated clusters – $10^<, 10, 25, 50, 100, 250, 500$. The $10^<$ clusters contained terms related to the seed term with a similarity of at least 0.75 (no such cluster was larger than 10 when abstracting from synonyms). Most interesting clusters were of size up to 50, bigger sizes already converged to the random baseline.

4.2 Results and Discussion

The evaluation results are summarised in Table 1. The first column represents labels of the experimental data sets. The EC, PC parts of the labels indicate the entity and property consolidation experiments. XD and LD refer to the usage of the extracted and linked data sets. The $^-$ superscripts refer to flattened, "property-less" data sets. Finally, EC-BL, PC-BL-LD, PC-BL-XD refer to the particular random baseline experiments. The columns of Table 1 represent the evaluation measures per each tested cluster size.

Firstly we will discuss the results of property clustering. To reduce the already quite substantial workload of the human evaluators, we considered only clusters of size up to 50. The accuracy of the LD batch was obviously 1, since the properties were manually defined there. The adequacy of clustering was best (0.875) for small, crisp LD clusters (with a rather strict similarity threshold of 0.75), while for bigger clusters without a similarity threshold restriction, it was decreasing, yet still significantly better than the baseline. For the extracted data set (XD), roughly one half of extracted properties was deemed to be accurate. Out of these, around 46.4% in average were adequate members of the analysed property clusters, which is quite promising, as it would allow for reduction of the space of about 35,000 extracted properties to several hundreds with an error rate around 50%. This may not be enough for a truly industry-strength solution, but it could already be useful in prototype applications if

Table 1. Results summary

cl. size	$10^<$		10		25		50		100		250		500	
metric	o	q	o	q	o	q	o	q	o	q	o	q	o	q
EC-LD	0.021	0.071	0.048	0.103	0.024	0.070	0.017	0.064	0.019	0.052	0.031	0.071	0.019	0.062
EC-LD$^-$	0.075	0.248	0.055	0.176	0.060	0.285	0.066	0.258	0.079	0.242	0.086	0.237	0.084	0.199
EC-XD	0.021	0.045	0.016	0.047	0.022	0.042	0.030	0.081	0.029	0.064	0.023	0.050	0.037	0.073
EC-XD$^-$	0.053	0.127	0.038	0.109	0.049	0.121	0.046	0.109	0.067	0.127	0.072	0.130	0.091	0.092
EC-BL	0.011	0.000	0.020	0.000	0.040	0.000	0.044	0.110	0.064	0.130	0.067	0.118	0.066	0.119
metric	aq	ac	aq	ac	aq	ac	aq	ac	aq	ac	aq	ac	aq	ac
PC-LD	0.875	1.000	0.603	1.000	0.578	1.000	0.596	1.000	N/A	N/A	N/A	N/A	N/A	N/A
PC-BL-LD	0.134	1.000	0.140	1.000	0.048	1.000	0.027	1.000	N/A	N/A	N/A	N/A	N/A	N/A
PC-XD	0.417	0.448	0.593	0.550	0.395	0.429	0.450	0.589	N/A	N/A	N/A	N/A	N/A	N/A
PC-BL-XD	0.016	0.497	0.027	0.450	0.017	0.523	0.024	0.511	N/A	N/A	N/A	N/A	N/A	N/A

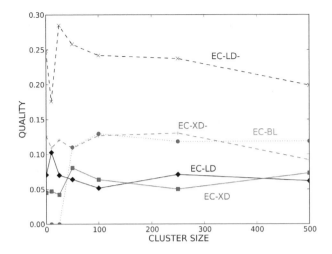

Fig. 1. Dependency of the cluster quality on their sizes

extended by result filtering (e.g., ranking). Examples of interesting property clusters we generated are: $C_1 = \{$ *secreted_in, excreted_in, appear_in, detected_in, accounted_for_in, produced_in, eliminated_in*$\}$, $C_2 = \{$ *from, following_from, gathered_from*$\}$, $C_3 = \{$ *increase, increased_by, diminish*$\}$. C_1 appears to be related to production/consumption of substances in organs, C_2 to origin of substances in location and C_3 to quantity change. More examples are available in the data package at `http://goo.gl/38bGK`.

To discuss the entity clustering results, let us have a look at Figure 1, which shows the dependency of the quality (q) measure on the cluster sizes for all the evaluated data sets. The dotted green line (circle markers) represents the baseline (EC-BL), while the red and black lines (square/plus and diamond/cross markers) are for the extracted and linked open data sets, respectively (EC-XD/EX-XD⁻, EC-LD/EC-LD⁻). The solid lines are for original data sets (with properties), whereas the dashed lines indicate resources with properties reduced to mere links. One can immediately see that the results of entity consolidation are significantly better in terms of quality than the baseline for clusters of size up to 25. This holds for all evaluated data sets and thus demonstrates a clear contribution of our approach. This is, we believe, not the most interesting thing, though. Quite surprisingly, the data sets flattened to mere links between entities (the dotted lines) produce much better results than the original resources with multiple property types. This is especially the case of flattened linked data resources (the dashed black line), which perform better than the baseline for all cluster sizes. Another counterintuitive finding is that the flattened automatically extracted resources (red dashes) perform better than the manually created linked data sets (without flattened properties). For clusters of size 50 and bigger, the flattened extracted batch oscillates around the random baseline, while both batches with actual properties are consistently worse.

It would be easy to say that the reason for such results is that the properties in triples do not have any semantics that can be empirically useful. This would be quite a controversial claim in the Semantic Web context, though, and we believe that it is most likely false, as the properties have been proven useful in countless other applications already. Alternative explanation could be that particular authors of the original linked data resources used the properties in a rather haphazard way when contributing to the knowledge base, thus introducing noise at a larger scale. This might partly be a culprit of the observed results, but there is yet another, perhaps more intriguing and plausible hypothesis: what if the empirical similarities induced by the full and flattened data are actually different? The gold standard similarity imposed by the MeSH thesaurus may be directly related to its taxonomy structure. The flattened resources may produce a related, rather simple 'subsumption' type of distributional similarity, while the resources with multiple property types can give rise to a more complex 'structural' similarity. This could be the reason for a better fit of the flattened data to the gold standard. Also, it could explain the poor (i.e., worse than random) performance of the full-fledged data sets for larger cluster sizes. In these cases, the flattened resources may be producing bigger clusters of more general and more specific (but still related) terms, whereas the other type of similarity just increases the noise by adding more and more specific and mutually unrelated structural sub-clusters. Experimenting with alternative similarity measures for the comparison of the results with the gold standard should help to clarify these rather surprising results. Apart of the hypothetical explanations mentioned so far, also the actual method of computing the corpus representation values (see Section 3.1) may play a significant role here. Whatever the actual reasons for the obtained results are, we believe that a further investigation of the suggested hypotheses could lead to interesting findings about more fundamental principles of the web semantics than investigated in this introductory paper.

5   Conclusion and Future Work

We have proposed to complement the currently prevalent top-down approaches to web semantics by an additional distributional layer. This layer allows for so far unexplored representation and analysis of bottom-up phenomena emerging from the Semantic Web resources. We demonstrated the usefulness of our framework by applying it to an experiment in consolidation of life science knowledge. The results showed promising potential of our approach and, in addition, revealed unexpected findings that are inspiring for futher research.

Our next plans include an experiment with a full-fledged combination of the top-down and bottom-up semantics (i.e., with rule-based materialisations in the loop). We will also explore a far wider range of the distributional semantics phenomena within various practical applications (e.g., automated thesaurus construction, rule learning or discovery of analogies). Then we want to engage in a continuous collaboration with sample users to assess qualitative aspects and practical value of tools based on our work. Finally, we want to ensure web-scale applicability of our approach by its distributed and parallel implementations.

References

1. de Saussure, F.: Course in General Linguistics, Open Court, Illinois (1983)
2. Firth, J.: A synopsis of linguistic theory 1930-1955. Studies in Ling. Anal. (1957)
3. Baroni, M., Lenci, A.: Distributional memory: A general framework for corpus-based semantics. Computational Linguistics (2010)
4. Franz, T., Schultz, A., Sizov, S., Staab, S.: TripleRank: Ranking Semantic Web Data by Tensor Decomposition. In: Bernstein, A., Karger, D.R., Heath, T., Feigenbaum, L., Maynard, D., Motta, E., Thirunarayan, K. (eds.) ISWC 2009. LNCS, vol. 5823, pp. 213–228. Springer, Heidelberg (2009)
5. Goldman, R., Widom, J.: DataGuides: Enabling query formulation and optimization in semistructured databases. In: VLDB. Morgan Kaufmann (1997)
6. Chemudugunta, C., Holloway, A., Smyth, P., Steyvers, M.: Modeling Documents by Combining Semantic Concepts with Unsupervised Statistical Learning. In: Sheth, A.P., Staab, S., Dean, M., Paolucci, M., Maynard, D., Finin, T., Thirunarayan, K. (eds.) ISWC 2008. LNCS, vol. 5318, pp. 229–244. Springer, Heidelberg (2008)
7. Carlson, A., Betteridge, J., Kisiel, B., Settles, B., Hruschka Jr., E.R., Mitchell, T.M.: Toward an architecture for never-ending language learning. In: AAAI 2010. (2010)
8. Maedche, A.: Emergent semantics for ontologies. In: Emergent Semantics. IEEE Intelligent Systems, pp. 85–86. IEEE Press (2002)
9. Aberer, K., Cudré-Mauroux, P., Catarci, A.M.O(e.) T., Hacid, M.-S., Illarramendi, A., Kashyap, V., Mecella, M., Mena, E., Neuhold, E.J., De Troyer, O., Risse, T., Scannapieco, M., Saltor, F., Santis, L.d., Spaccapietra, S., Staab, S., Studer, R.: Emergent Semantics Principles and Issues. In: Lee, Y., Li, J., Whang, K.-Y., Lee, D. (eds.) DASFAA 2004. LNCS, vol. 2973, pp. 25–38. Springer, Heidelberg (2004)
10. Euzenat, J., Shvaiko, P.: Ontology matching. Springer, Heidelberg (2007)
11. Taubert, J., Hindle, M., Lysenko, A., Weile, J., Köhler, J., Rawlings, C.J.: Linking Life Sciences Data Using Graph-Based Mapping. In: Paton, N.W., Missier, P., Hedeler, C. (eds.) DILS 2009. LNCS, vol. 5647, pp. 16–30. Springer, Heidelberg (2009)
12. ter Horst, H.J.: Completeness, decidability and complexity of entailment for RDF schema and a semantic extension involving the OWL vocabulary. Journal of Web Semantics, 79–115 (2005)
13. Doorenbos, R.B.: Production Matching for Large Learning Systems. PhD thesis (1995)
14. Pesquita, C., Faria, D., Falco, A.O., Lord, P., Couto, F.M.: Semantic similarity in biomedical ontologies. PLoS Computational Biololgy 5(7) (2009)

Encyclopedic Knowledge Patterns from Wikipedia Links

Andrea Giovanni Nuzzolese[1,2], Aldo Gangemi[1],
Valentina Presutti[1], and Paolo Ciancarini[1,2]

[1] STLab-ISTC Consiglio Nazionale delle Ricerche, Rome, Italy
[2] Dipartimento di Scienze dell'Informazione, Università di Bologna, Italy

Abstract. What is the most intuitive way of organizing concepts for describing things? What are the most relevant types of things that people use for describing other things? Wikipedia and Linked Data offer knowledge engineering researchers a chance to empirically identifying invariances in conceptual organization of knowledge i.e. knowledge patterns. In this paper, we present a resource of Encyclopedic Knowledge Patterns that have been discovered by analyizing the Wikipedia page links dataset, describe their evaluation with a user study, and discuss why it enables a number of research directions contributing to the realization of a meaningful Semantic Web.

1 Introduction

The realization of the Web of Data (aka Semantic Web) partly depends on the ability to make meaningful knowledge representation and reasoning. Elsewhere [5] we have introduced a vision of a pattern science for the Semantic Web as the means for achieving this goal. Such a science envisions the study of, and experimentation with, *knowledge patterns* (KP): small, well connected units of meaning which are 1) task-based, 2) well-grounded, and 3) cognitively sound. The first requirement comes from the ability of associating ontology vocabularies or schemas with explicit tasks, often called *competency questions* [7]: if a schema is able to answer a typical question an expert or user would like to make, it is a useful schema. The second requirement is related to the ability of ontologies to enable access to large data (which typically makes them successful) as well as being grounded in textual documents so as to support semantic technology applications that hybridize RDF data and textual documents. The third requirement comes from the expectation that schemas that more closely mirror the human ways of organizing knowledge are better. Unfortunately, evidence for this expectation is only episodic until now for RDF or OWL vocabularies [5].

Linked data and social web sources such as Wikipedia give us the chance to empirically study what are the patterns in organizing and representing knowledge i.e. knowledge patterns. KPs can be used for evaluating existing methods and models that were traditionally developed with a top-down approach, and open new research directions towards new reasoning procedures that better fit

L. Aroyo et al. (Eds.): ISWC 2011, Part I, LNCS 7031, pp. 520–536, 2011.

the actual Semantic Web applications need. In this study, we identify a set of invariances from a practical, crowd-sourced repository of knowledge: Wikipedia page links (wikilinks), which satisfy those three requirements, hence constituting good candidates as KPs. We call them Encyclopedic Knowledge Patterns (EKP) for emphasizing that they are grounded in encyclopedic knowledge expressed as linked data, i.e. DBpedia[1], and as natural language text, i.e. Wikipedia[2]. We have collected such set of EKPs in an open repository[3]. EKPs are able to answer the following (generic) competency question:

What are the most relevant entity types that provide an effective and intuitive description of entities of a certain type?

For example, when describing "Italy" (a country), we typically indicate its neighbor countries, cities, administrative regions, spoken languages, etc. The EKP for describing countries should then include such a set of entity types: the most relevant for describing a country. We assume EKPs as cognitively sound because they emerge from the largest existing multi-domain knowledge source, collaboratively built by humans with an encyclopedic task in mind. This assumption is bound to our working hypothesis about the process of knowledge construction realized by the Wikipedia crowds: each article is linked to other articles when *explaining* or *describing* the entity referred to by the article. Therefore, the article's main subject can be said to be soundly and centrally related to the linked articles' main subjects. DBpedia, accordingly with this intuition, has rdf-ized a) the subjects referred to by articles as *resources*, b) the wikilinks as relations between those resources, and c) the types of the resources as OWL classes.

Hypotheses. Assuming that the articles linked from a Wikipedia page constitute a major source of descriptive knowledge for the subject of that page, we hypothesize that (i) the types of linked resources that occur most often for a certain type of resource constitute its EKP (i.e., the most relevant concepts to be used for describing resources of that type), and (ii) since we expect that any cognitive invariance in explaining/describing things is reflected in the wikilink graph, discovered EKPs are cognitively sound.

Contribution. The contribution of this paper is twofold: (i) we define an EKP discovery procedure, extract 184 EKPs, and publish them in OWL2 (ii) we support our hypotheses through a user-based evaluation, and discuss a number of research directions opened by our findings.

The paper is organized as follows: Section 2 discusses related work, Section 3 describes the resources we have used and the basic assumptions we have made, Section 4 focuses on the results we have gathered, Section 5 presents a user study for the evaluation and fine-tuning of EKPs, and Section 6 draws conclusions and gives an overview of research directions we are concentrating upon.

---

[1] http://dbpedia.org

[2] http://en.wikipedia.org

[3] The EKP repository is available at http://stlab.istc.cnr.it/stlab/WikiLinkPatterns

2   Related Work

To the best of our knowledge this work is the first attempt to extract knowledge patterns (KPs) from linked data. Nevertheless, there is valuable research on exploiting Wikipedia as a knowledge resource as well as on creating knowledge patterns.

Knowledge patterns. [5] argues that KPs are basic elements of the Semantic Web as an empirical science, which is the vision motivating our work. [4,16] present experimental studies on KPs, focusing on their creation and usage for supporting ontology design with shared good practices. Such KPs are usually stored in online repositories[4]. Contrary to what we present in this work, KPs are typically defined with a top-down approach, from practical experience in knowledge engineering projects, or extracted from existing, e.g. foundational, ontologies. These KPs are close to EKPs, but although some user-study proved that their use is beneficial in ontology design [3], yet they miss some of the aspects that we study here: evaluation of their cognitive soundness, and adequacy to provide access to large-scale linked data. [14] presents a resource of KPs derived from a lexical resource i.e., FrameNet [2]. In future work, we plan a compared analysis between EKPs and other KPs.

Building the web of data. Research focusing on feeding the Web of Data is typically centered on extracting knowledge from structured sources and transforming it into linked data. Notably, [8] describes how DBpedia is extracted from Wikipedia, and its linking to other Web datasets.

Another perspective is to apply knowledge engineering principles to linked data in order to improve its quality. [18] presents YAGO, an ontology extracted from Wikipedia categories and infoboxes that has been combined with taxonomic relations from WordNet. Here the approach can be described as a reengineering task for transforming a thesaurus, i.e. Wikipedia category taxonomy, to an ontology, which required accurate ontological analysis.

Extracting knowledge from wikipedia. Wikipedia is now largely used as a reference source of knowledge for empirical research. Research work from the NLP community, e.g., [20,9,15], exploits it as background knowledge for increasing the performance of algorithms addressing specific tasks. Two approaches are close to ours. [6] presents a method for inducing thesauri from Wikipedia by exploiting the structure of incoming wikilinks. The graph of wikilinks is used for identifying meaningful terms in the linked pages. In contrast, in our case we exploit outgoing wikilinks, as well as the full potential of the linked data semantic graph for identifying semantic entities as opposed to terms. [19] presents a statistical approach for the induction of expressive schemas for RDF data. Similarly to our study, the result is an OWL ontology, while in our experiment we extract novel schemas from wikilink structures that are encoded in RDF. Some studies have produced reusable results for improving the quality of the Web of

---

[4] E.g. the ontology design patterns semantic portal,
  http://www.ontologydesignpatterns.org

Data. We mention two notable examples: [13,1], which address the extraction of relations between Wikipedia entities, and [12] that presents a multi-lingual network of inter-connected concepts obtained by mining Wikipedia.

3   Materials and Methods

Our work grounds on the assumption that wikilink relations in DBpedia, i.e. instances of the dbpo:wikiPageWikiLink property[5], convey a rich encyclopedic knowledge that can be formalized as EKPs, which are good candidates as KPs [5].

Informally, an EKP is a small ontology that contains a concept S and its relations to the most relevant concepts C_j that can be used to describe S.

Representing invariances from wikipedia links. For representing wikilink invariances, we define *path* (type) as an extension of the notion of *property path*[6]:

Definition 1 (Path). *A path (type) is a property path (limited to length 1 in this work, i.e. a triple pattern), whose occurrences have (i) the same* rdf:type *for their subject nodes, and (ii) the same* rdf:type *for their object nodes. It is denoted here as:*

$$P_{i,j} = [S_i, p, O_j]$$

where S_i is a subject type, p is a property, and O_j is an object type of a triple. In this work, we only extract paths where p=dbpo:wikiPageWikiLink. Sometimes we use a simplified notation $[S_i, O_j]$, assuming $p = $ dbpo:wikiPageWikiLink.

We extract EKPs from paths (see Definition 2), however in order to formalize them, we perform a heuristic procedure to reduce multi-typing, to avoid redundancies, and to replace dbpo:wikiPageWikiLink with a contextualized object property. In practice, given a triple s dbpo:wikiPageWikiLink o, we construct its path as follows:

- the subject type S_i is set to the most specific type(s) of s
- the object type O_j is set to the most specific type(s) of o
- the property p of the path is set to the most general type of o

For example, the triple:

dbpedia:Andre_Agassi dbpo:wikiPageWikiLink dbpedia:Davis_Cup

would count as an occurrence of the following path:

$$Path_{Agassi,Davis} = [\text{dbpo:TennisPlayer}, \text{dbpo:Organisation}, \text{dbpo:TennisLeague}]$$

Figure 1 depicts such procedure for the path $Path_{Agassi,Davis}$:

---

[5] Prefixes dbpo:, dbpedia:, and ka: stand for
http:dbpedia.org/ontology/, http:dbpedia.org/resource/
and http://www.ontologydesignpatterns.org/ont/lod-analysis-path.owl,
respectively.

[6] In SPARQL1.1 (http://www.w3.org/TR/sparql11-property-paths/) property paths can have length n, given by their route through the RDF graph.

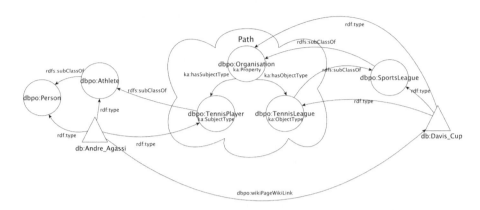

Fig. 1. Path discovered from the triple `dbpedia:Andre_Agassi dbpo:wikiPageWikiLink dbpedia:Davis_Cup`

- `dbpo:TennisPlayer` is the subject type because it is the most specific type of `dbpedia:Andre_Agassi`, i.e., `dbpo:TennisPlayer ⊑ dbpo:Person`;
- `dbpo:TennisLeague` is the object type because it is the most specific type of `dbpedia:Davis_Cup`, i.e., `dbpo:TennisLeague ⊑ dbpo:SportsLeague ⊑ dbpo:Organisation`
- `dbpo:Organisation` is the property of the path because it is the most general type of `dbpedia:Davis_Cup`.

Indicators. We use a set of indicators that are described in Table 1. Their application and related interpretation in this work are discussed in the following sections.

Table 1. Indicators used for empirical analysis of wikilink paths

Indicator	Description
$nRes(C)$	number of resources typed with a certain class C, $|\{r_i \; \texttt{rdf:type} \; C\}|$
$nSubjectRes(P_{i,j})$	number of distinct resources that participate in a path as subjects, $|\{(s_i \; \texttt{rdf:type} \; S_i) \in P_{i,j} = [S_i, p, O_j]\}|$
$pathPopularity(P_{i,j}, S_i)$	The ratio of how many distinct resources of a certain type participate as subject in a path to the total number of resources of that type. Intuitively, it indicates the popularity of a path for a certain subject type, $nSubjectRes(P_{i,j} = [S_i, p, O_j])$ divided by $nRes(S_i)$
$nPathOcc(P_{i,j})$	number of occurrences of a path $P_{i,j} = [S_i, p, O_j]$
$nPath(S_i)$	number of distinct paths having a same subject type S_i, e.g. the number of paths having `dbpo:TennisPlayer` as subject type
$AvPathOcc(S_i)$	sum of all $nPathOcc(P_{i,j})$ having a subject type S_i divided by $nPath(S_i)$ e.g. the avarage number of occurrences of paths having `dbpo:Philosopher` as subject type

Boundaries of Encyclopedic Knowledge Patterns. We choose the boundaries of an EKP by defining a threshold t for $pathPopularity(P_{i,j}, S_i)$. Accordingly, we give the following definition of $EKP(S_i)$ for a DBpedia type S_i.

Definition 2 (Encyclopedic Knowledge Patterns). *Let S_i be a DBpedia type, O_j $(j = 1, .., n)$ a list of DBpedia types, $P_{i,j} = [S_i, p, O_j]$ and t a threshold value.*
Given the triples:

```
dbpedia:s dbpedia-ont:wikiPediaWikiLink dbpedia:o
            dbpedia:s rdf:type dbpedia:Si
            dbpedia:o rdf:type dbpedia:Oj
```

we state that $EKP(S_i)$ is a set of paths, such that

$$P_{i,j} \in EKP(S_i) \iff pathPopularity(P_{i,j}, S_i) \geq t \qquad (1)$$

We hypothesize values for t in Section 4, and evaluate them in Section 5.

OWL2 formalization of EKPs. We have stored paths and their associated indicators in a dataset, according to an OWL vocabulary called *knowledge architecture*[7]. Then, we have generated the Encyclopedic Knowledge Patterns (EKPs) repository[8] by performing a refactoring of the knowledge architecture data into OWL2 ontologies). Given the namespace ekp: and an $EKP(S_i) = [S_i, p_1, O_1], \dots, [S_i, p_n, O_n]$, we formalize it in OWL2 by applying the following translation procedure:

- the name of the OWL file is ekp:[9] followed by the local name of S e.g., ekp:TennisPlayer.owl. Below we refer to the namespace of a specific EKP through the generic prefix ekpS:;
- S_i and O_j $j = 1, \dots, n$ are refactored as owl:Class entities (they keep their original URI);
- p_j keep their original URI and are refactored as owl:ObjectProperty entities;
- for each O_j we create a sub-property of p_{i+n}, $ekpS:O_j$ that has the same local name as O_j and the ekpS: namespace; e.g. ekp:TennisPlayer.owl#TennisLeague.
- for each $ekpS:O_j$ we add an owl:allVauesFrom restriction to S_i on $ekpS:O_j$, with range O_j.

For example, if $Path_{Agassi,Davis}$ (cf. Figure 1) is part of an EKP, it gets formalized as follows:

```
Prefix: dbpo: <http://dbpedia.org/ontology/>
Prefix:
    ekptp: <http://www.ontologydesignpatterns.org/ekp/TennisPlayer.owl#>
Ontology: <http://www.ontologydesignpatterns.org/ekp/TennisPlayer.owl>
Class: dbpo:TennisPlayer
    SubClassOf:
```

---

[7] http://www.ontologydesignpatterns.org/ont/lod-analysis-path.owl
[8] The EKP repository is available at http://stlab.istc.cnr.it/stlab/WikiLinkPatterns.
[9] The prefix ekp: stands for the namespace
http://www.ontologydesignpatterns.org/ekp/.

Table 2. Dataset used and associated figures

Dataset	Description	Indicator	Value
DBPO	DBpedia ontology	Number of classes	272
dbpedia_instance_types_en	Resource types i.e. rdf:type triples	Number of resources having a DBPO type	1,668,503
		rdf:type triples	6,173,940
dbpedia_page_links_en	Wikilinks triples	Number of resources used in wikilinks	15,944,381
		Number of wikilinks	107,892,317
DBPOwikilinks	Wikilinks involving only resources typed with DBPO classes	Number of resources used in wikilinks	1,668,503
		Number of wikilinks	16,745,830

```
        ekptp:TennisLeague only dbpo:TennisLeague
Class: dbpo:TennisLeague
ObjectProperty: ekptp:TennisLeague
   SubPropertyOf: dbpo:Organisation
...
```

Materials. We have extracted EKPs from a subset of the DBpe-dia wikilink dataset (*dbpedia_page_links_en*), and have created a new dataset (*DBPOwikilinks*) including only links between resources that are typed by DBpedia ontology version 3.6 (DBPO) classes (15.52% of the total wikilinks in *dbpedia_page_links_en*). *DBPOwikilinks* excludes a lot of links that would create semantic interpretation issues, e.g. images (e.g. dbpedia:Image:Twitter_2010_logo.svg), Wikipedia categories (e.g. dbpedia:CAT:Vampires_in_comics), untyped resources (e.g. dbpedia:%23Drogo), etc.

DBPO currently includes 272 classes, which are used to type 10.46% of the resources involved in *dbpedia_page_links_en*. We also use *dbpedia_instance_types_en*, which contains type axioms, i.e. rdf:type triples. This dataset contains the materialization of all inherited types (cf. Section 4). Table 2 summarizes the figures described above.

4   Results

We have extracted 33,052 paths from the English wikilink datasets, however many of them are not relevant either because they have a limited number of occurrences, or because their subject type is rarely used. In order to select the paths useful for EKP discovery (our goal) we have considered the following criteria:

- *Usage in the wikilink dataset.* The resources involved in *dbpedia_page_links_en* are typed with any of 250 DBPO classes (out of 272). Though, we are inter-ested in *direct types*[10] of resources in order to avoid redundancies when count-ing path occurrences. For example, the resource dbpedia:Ludwik_Fleck has three types dbpo:Scientist;dbpo:Person;owl:Thing because type asser-tions in DBpedia are materialized along the hirerachy of DBPO. Hence, only

---

[10] In current work, we are also investigating indirectly typed resource count, which might lead to different EKPs, and to empirically studying EKP ordering.

`dbpo:Scientist` is relevant to our study. Based on this criterion, we keep only 228 DBPO classes and the number of paths decreases to 25,407.

- *Number of resources typed by a class C (i.e., nRes(C))*. Looking at the distribution of resource types, we have noticed that 99.98% of DBPO classes have at least 30 resource instances. Therefore we have decided to keep paths whose subject type C has at least $nRes(C) = 30$.
- *Number of path occurrences having a same subject type (i.e., $nPathOcc(P_{i,j})$)*. The average number of outgoing wikilinks per resource in *dbpedia_page_links_en* is 10. Based on this observation and on the previous criterion, we have decided to keep paths having at least $nPathOcc(P) = 30*10 = 300$.

After applying these two criteria, only 184 classes and 21,503 paths are retained. For example, the path [Album,Drug] has 226 occurrences, and the type `dbpo:AustralianFootballLeague` has 3 instances, hence they have been discarded.

EKP discovery. At this point, we had each of the 184 classes used as subject types associated with a set of paths, each set with a cardinality ranging between 2 and 191 (with 86.29% of subjects bearing at least 20 paths). Our definition of EKP requires that its backbone be constituted of a small number of object types, typically below 10, considering the existing resources of models that can be considered as KPs (see later in this section for details). In order to generate EKPs from the extracted paths, we need to decide what threshold should be used for selecting them, which eventually creates appropriate *boundaries* for EKPs. In order to establish some meaningful threshold, we have computed the ranked distributions of $pathPopularity(P_{i,j}, S_i)$ for each selected subject type, and measured the correlations between them. Then, we have fine-tuned these findings by means of a user study (cf. Section 5), which had the dual function of both evaluating our results, and suggesting relevance criteria for generating the EKP resource. Our aim is to build a *prototypical* ranking of the $pathPopularity(P_{i,j}, S_i)$ of the selected 184 subject types, called $pathPopularity_{DBpedia}$, which should show how relevant paths for subject types are typically distributed according to the Wikipedia crowds, hence allowing us to propose a threshold criterion for any subject type. We have proceeded as follows.

1. We have chosen the top-ranked 40 paths $(P_{i,j})$ for each subject type (S_i), each constituting a $pathPopularity(P_{i,j}, S_i)$. Some subject types have less than 40 paths: in such cases, we have added 0 values until filling the gap. The number 40 has been chosen so that it is large enough to include not only paths covering at least 1% of the resources, but also much rarer ones, belonging to the long tail.
2. In order to assess if a prototypical ranking $pathPopularity_{DBpedia}$ would make sense, we have performed a multiple correlation between the different $pathPopularity(P_{i,j}, S_i)$. In case of low correlation, the prototypical ranking would create odd effects when applied to heterogeneous rank distributions across different S_i. In case of high correlation, the prototype would make

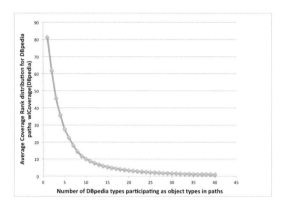

Fig. 2. Distribution of $pathPopularity_{DBpedia}$: the average values of popularity rank i.e., $pathPopularity(P_{i,j}, S_i)$, for DBpedia paths. The x-axis indicates how many paths (on average) are above a certain value t of $pathPopularity(P, S)$.

sense, and we can get reassured that the taxonomy we have used (DBPO in this experiment) nicely fits the way wikilinks are created by the Wikipedia crowds.

3. We have created a prototypical distribution $pathPopularity_{DBpedia}$ that is representative for all S_i distributions. Such a distribution is then used to hypothesize some thresholds for the relevance of $P_{i,j}$ when creating boundaries for EKPs. The thresholds are used in Section 5 to evaluate the proposed EKPs with respect to the rankings produced during the user study.

In order to measure the distribution from step 2, we have used the Pearson correlation measure ρ, ranging from -1 (no agreement) to +1 (complete agreement), between two variables X and Y i.e. for two different S_i in our case. The correlation has been generalized to all 16,836 pairs of the 184 $pathPopularity(P_{i,j}, S_i)$ ranking sets $(184 * 183/2)$, in order to gather a multiple correlation. The value of such multiple correlation is 0.906, hence excellent.

Once reassured on the stability of $pathPopularity(P_{i,j}, S_i)$ across the different S_i, we have derived (step 3) $pathPopularity_{DBpedia}$, depicted in Figure 2.

In order to establish some reasonable relevance thresholds, $pathPopularity_{DBpedia}$ has been submitted to K-Means Clustering, which generates 3 small clusters with popularity ranks above 22.67%, and 1 large cluster (85% of the 40 ranks) with popularity ranks below 18.18%. The three small clusters includes seven paths: this feature supports the buzz in cognitive science about a supposed amount of 7 ± 2 objects that are typically manipulated by the cognitive systems of humans in their recognition tasks [11,10]. While the 7 ± 2 conjecture is highly debated, and possibly too generic to be defended, this observation has been used to hypothesize a first threshold criterion: since the seventh rank is at 18.18% in $pathPopularity_{DBpedia}$, this value of $pathPopularity(P_{i,j}, S_i)$ will be our first guess for including a path in an EKP. We propose a second threshold based on FrameNet [2], a lexical database, grounded in a textual corpus, of situation types called *frames*. FrameNet is

Table 3. Sample paths for the subject type `Album`: number of path occurrences, distinct subject resources, and popularity percentage value

Path	$nPathOcc(P_{i,j})$	$nSubjectRes(P_{i,j})$	$pathPopularity(P_{i,j}, S_i)$ (%)
[Album,Album]	170,227	78,137	78.89
[Album,MusicGenre]	108,928	68,944	69.61
[Album,MusicalArtist]	308,619	68,930	69.59
[Album,Band]	125,919	62,762	63.37
[Album,Website]	62,772	49,264	49.74
[Album,RecordLabel]	56,285	47,058	47.51
[Album,Single]	114,181	29,051	29.33
[Album,Country]	40,296	25,430	25.67

currently the only cognitively-based resource of potential knowledge patterns (the frames, cf. [14]). The second threshold (11%) is provided by the average number of frame elements in FrameNet frames (frame elements roughly correspond to paths for EKPs), which is 9 (the ninth rank in $pathPopularity_{DBpedia}$ is at 11%). The mode value of frame elements associated with a frame is 7, which further supports our proposal for the first threshold. An example of the paths selected for a subject type according to the first threshold is depicted in Tab. 3, where some paths for the type `Album` are ranked according to their $pathPopularity(P_{i,j}, S_i)$. In Section 5 we describe an evaluation of these threshold criteria by means of a user study.

Threshold criteria are also used to enrich the formal interpretation of EKPs. Our proposal, implemented in the OWL2 EKP repository, considers the first threshold as an indicator for an existential quantification over an OWL restriction representing a certain path. For example, `[Album,MusicGenre]` is a highly-popular path in the `Album` EKP. We interpret high-popularity as a feature for generating an existential interpretation, i.e.: Album $\sqsubseteq$ ($\exists$MusicGenre.MusicGenre). This interpretation suggests that each resource typed as an `Album` has at least one `MusicGenre`, which is intuitively correct. Notice that even if all paths have a $pathPopularity(P_{i,j}, S_i)$ of less that 100%, we should keep in mind that semantic interpretation over the Web is made in open-world, therefore we feel free to assume that such incompleteness is a necessary feature of Web-based knowledge (and possibly of any crowd-sourced knowledge).

5   Evaluation

Although our empirical observations on DBpedia could give us means for defining a value for the threshold t (see Definition 2 and Section 4), we still have to prove that emerging EKPs provide an intuitive schema for organizing knowledge. Therefore, we have conducted a user study for making users identify the EKPs associated with a sample set of DBPO classes, and for comparing them with those emerging from our empirical observations.

User study. We have selected a sample of 12 DBPO classes that span social, media, commercial, science, technology, geographical, and governmental

Table 4. DBPO classes used in the user-study and their related figures

DBPO class type	nRes(S)	$nPath(S_i)$	$AvPathOcc(S_i)$
Language	3,246	99	29.27
Philosopher	1,009	112	18.29
Writer	10,102	172	15.30
Ambassador	286	85	15.58
Legislature	453	83	25.11
Album	99,047	172	11.71
Radio Station	16,310	151	7.31
Administrative Region	31,386	185	11.30
Country	2,234	169	35.16
Insect	37,742	98	9.16
Disease	5,215	153	12.10
Aircraft	6,420	126	10.32

domains. They are listed in Table 4. For each class, we indicate the number of its resources, the number of paths it participates in as subject type, and the average number of occurrences of its associated paths. We have asked the users to express their judgement on how relevant were a number of (object) types (i.e., paths) for describing things of a certain (subject) type. The following sentence has been used for describing the user study task to the users:

> We want to study the best way to describe things by linking them to other things. For example, if you want to describe a person, you might want to link it to other persons, organizations, places, etc. In other words, what are the most relevant types of things that can be used to describe a certain type of things?

We asked the users to fill a number of tables, each addressing a class in the sample described in Table 4. Each table has three columns:

– *Type 1* indicating the class of things (subjects) to be described e.g. `Country`;
– A second column to be filled with a relevance value for each row based on a scale of five relevance values, Table 5 shows the scale of relevance values and their interpretations as they have been provided to the users. Relevance values had to be associated with each element of Type 2;
– *Type 2* indicating a list of classes of the paths (i.e. the object types) in which *Type 1* participates as subject type. These were the suggested types of things that can be linked for describing entities of *Type 1* e.g. `Administrative Region`, `Airport`, `Book`, etc.

By observing the figures of DBPO classes (cf. Table 4) we realized that the entire list of paths associated with a subject type would have been too long to be proposed to the users. For example, if *Type 1* was `Country`, the users would have been submitted 169 rows for *Type 2*. Hence, we decided a criterion for selecting a representative set of such paths. We have set a value for t to 18% and have included, in the sample set, all $P_{i,j}$ such that $pathPopularity(P_{i,j}, S_i) \geq 18\%$. Furthermore, we have also included an additional random set of 14 $P_{i,j}$ such that $pathPopularity(P_{i,j}, S_i) < 18\%$.

We have divided the sample set of classes into two groups of 6. We had ten users evaluating one group, and seven users evaluating the other group. Notice

Table 5. Ordinal (Likert) scale of relevance scores

Relevance score	Interpretation
1	The type is irrelevant;
2	The type is slightly irrelevant;
3	I am undecided between 2 and 4;
4	The type is relevant but can be optional;
5	The type is relevant and should be used for the description.

Table 6. Average coefficient of concordance for ranks (Kendall's W) for the two groups of users

User group	Average inter-rater agreement
Group 1	0.700
Group 2	0.665

that the users come from different cultures (Italy, Germany, France, Japan, Serbia, Sweden, Tunisia, and Netherlands), and speak different mother tongues. In practice, we wanted to avoid focusing on one specific language or culture, at the risk of reducing consensus.

In order to use the EKPs resulting from the user study as a reference for next steps in our evaluation task, we needed to check the inter-rater agreement. We have computed the Kendall's coefficient of concordance for ranks (W), for all analyzed DBPO classes, which calculates agreements between 3 or more rankers as they rank a number of subjects according to a particular characteristic. Kendall's W ranges from 0 (no agreement) to 1 (complete agreement). Table 6 reports such values for the two groups of users, which show that we have reached a good consensus in both cases. Additionally, Table 7 reports W values for each class in the evaluation sample.

Table 7. Inter-rater agreement computed with Kendall's W (for all values $p < 0.0001$) and reliability test computed with Cronbach's alpha

DBPO class	Agreement	Reliability	DBPO class	Agreement	Reliability
Language	0.836	0.976	Philosopher	0.551	0.865
Writer	0.749	0.958	Ambassador	0.543	0.915
Legislature	0.612	0.888	Album	0.800	0.969
Radio Station	0.680	0.912	Administrative Region	0.692	0.946
Country	0.645	0.896	Insect	0.583	0.929
Disease	0.823	0.957	Aircraft	0.677	0.931

Evaluation of emerging DBpedia EKPs through correlation with user-study results: how good is DBpedia as a source of EKPs? The second step towards deciding t for the generation of EKPs has been to compare DBpedia EKPs to those emerging from the users' choices. DBpedia $EKP(S_i)$ would result from a selection of paths having S_i as subject type, based on their associated $pathPopularity(P_{i,j}, S_i)$ values (to be $\geq t$). We had to compare the $pathPopularity(P_{i,j}, S_i)$ of the paths associated with the DBPO sample classes (cf. Table 4), to the relevance scores assigned by the users. Therefore, we needed to define a mapping function between $pathPopularity(P_{i,j}, S_i)$ values and the 5-level scale of relevance scores (Table 5).

We have defined the mapping by splitting the $pathPopularity_{DBpedia}$ distribution (cf. Figure 2) into 5 intervals, each corresponding to the 5 relevance scores of the Likert scale used in the user-study. Table 8 shows our hypothesis

Table 8. Mapping between $wlCoverage_{DBpedia}$ intervals and the relevance score scale

$pathPopularity_{DBpedia}$ **interval**	**Relevance score**
[18, 100]	5
[11, 18[	4
]2, 11[	3
]1, 2]	2
[0, 1]	1

Table 9. Average multiple correlation (Spearman ρ) between users' assigned scores, and $pathPopularity_{DBpedia}$ based scores

User group	Correl. with DBpedia
Group 1	0.777
Group 2	0.717

Table 10. Multiple correlation coefficient (ρ) between users's assigned score, and $pathPopularity_{DBpedia}$ based score

DBPO class	Correl. users / DBpedia	DBPO class	Correl. users / DBpedia
Language	0.893	Philosopher	0.661
Writer	0.748	Ambassador	0.655
Legislature	0.716	Album	0.871
Radio Station	0.772	Administrative Region	0.874
Country	0.665	Insect	0.624
Disease	0.824	Aircraft	0.664

of such mapping. The hypothesis is based on the thresholds defined in Section 4. The mapping function serves our purpose of performing the comparison and identifying the best value for t, which is our ultimate goal. In case of scarce correlation, we expected to fine-tune the intervals for finding a better correlation and identifying the best t. Based on the mapping function, we have computed the relevance scores that DBpedia would assign to the 12 sample types, and calculated the Spearman correlation value (ρ) wich ranges from -1 (no agreement) to $+1$ (complete agreement) by using the *means* of relevance scores assigned by the users. This measure gives us an indication on how precisely DBpedia wikilinks allow us to identify EKPs as compared to those drawn by the users. As shown in Table 9, there is a good correlation between the two distributions. Analogously, Table 10 shows the multiple correlation values computed for each class, which are significantly high. Hence, they indicate a satisfactory precision.

We can conclude that our hypothesis (cf. Section 1) is supported by these findings, and that Wikipedia wikilinks are a good source for EKPs. We have tested alternative values for t, and we have found that our hypothesized mapping (cf. Table 8) provides the best correlation values among them. Consequently, we have set the threshold value for EKP boundaries (cf. Definition 2) as $t = 11\%$.

6   Discussion and Conclusions

We have presented a study for discovering Encyclopedic Knowledge Patterns (EKP) from Wikipedia page links. In this study we have used the DBPO classes to create a wikilink-based partition of crowd-sourced encyclopedic knowledge expressed as paths of length 1, and applied several measures to create a *boundary* around the most relevant object types for a same subject type out of wikilink triples.

Data have been processed and evaluated by means of both statistical analysis over the paths, and a user study that created a reference ranking for a subset of subject types and their associated paths. Results are very good: *stable criteria for boundary creation* (high correlation of path popularity distributions across subject types), *large consensus* among (multicultural) users, and *good precision* (high correlation between users' and EKP rankings).

The 184 EKP so generated have been formalized in OWL2 and published, and can be used either as lenses for the exploration of DBpedia, or for designing new ontologies that inherit the data and textual grounding provided by DBpedia and Wikipedia. Also data linking can take advantage of EKPs, by modularizing the datasets to be linked.

There are many directions that the kind of research we have done opens up: some are presented in the rest of this section.

Applying EKPs to resource concept maps. An application of EKPs is the creation of synthetic concept maps out of the wikilinks of a resource. For example, a concept map of all wikilinks for the resource about the scientist dbpr:Ludwik_Fleck contains 44 unordered resources, while a concept map created with a lens provided by the Scientist EKP provides the 13 most typical resources with explicit relations. We should remark that EKPs typically (and intentionally) exclude the "long tail" features of a resource, which sometimes are important. Investigating how to make these relevant long tail features emerge for specific resources and requirements is one of the research directions we want to explore.

Wikilink relation semantics. An obvious elaboration of EKP discovery is to infer the object properties that are implicit in a wikilink. This task is called *relation discovery*. Several approaches have been used for discovering relations in Wikipedia, (cf. Section 2, [9] is an extensive overview), and are being investigated. Other approaches are based on the existing semantic knowledge from DBpedia: three of them are exemplified here because their results have already been implemented in the EKP resource.

Induction from infobox properties. For example, the path [Album,MusicalArtist] features a distribution of properties partly reported in Table 11. There is a clear majority for the producer property, but other properties are also present, and some are even clear anomalies (e.g. *[Album,dbprop:nextAlbum,MusicalArtist][11]). In general, there are two typical situations: the first is exemplified by [Album,MusicalArtist], where the most frequent property covers only part of the possible semantics of the wikilink paths. The second situation is when the most frequent property is maximally general, and repeats the name of the object type, e.g. [Actor,dbprop:film,Film]. In our EKP resource, we add the most frequent

---

[11] * indicates a probably wrong path.

534   A.G. Nuzzolese et al.

Table 11. Sample paths for the subject type `Album` from the infobox DBpedia dataset, with their frequency. Some paths are clear mistakes.

Path	$nPathOcc(P_{i,j})$
[Album,dbprop:producer,MusicalArtist]	3,413
[Album,dbprop:artist,MusicalArtist]	236
[Album,dbprop:writer,MusicalArtist]	46
[Album,dbprop:lastAlbum,MusicalArtist]	35
*[Album,dbprop:nextAlbum,MusicalArtist]	33
[Album,dbprop:thisAlbum,MusicalArtist]	27
[Album,dbprop:starring,MusicalArtist]	20

properties from the infobox dataset as annotations, accompanied by a frequency attribute.

Induction from top superclasses. For example, the path [Album,MusicalArtist] can be enhanced by inducing the top superclass of MusicalArtist, i.e. Person, as its property. This is possible either in RDFS, or in OWL2 (via punning). The path would be in this case [Album,Person,MusicalArtist]. This solution has not precision problems, but is also quite generic on the semantics of a wikilink.

Punning of the object type. For example, the path [Album,MusicalArtist] can be enriched as [Album,MusicalArtist,MusicalArtist]. This solution is pretty uninformative at the schema level, but can be handy when an EKP is used to visualize knowledge from wikilinks, for example in the application described above of a resource concept map, where resources would be linked with the name of the object type: this results to be very informative for a concept map user. In our EKP resource, we always reuse the object type as a (locally defined) object property as well.

Additional approaches we have conceived would exploit existing resources created by means of NLP techniques (e.g. WikiNet, [12]), or by game-based crowdsourcing (e.g. OpenMind [17]).

Intercultural issues. Given the multilingual and multicultural nature of Wikipedia, comparison between EKPs extracted from different versions of Wikipedia is very interesting. We have extracted EKPs from English and Italian versions, and we have measured the correlation between some English- and Italian-based EKPs. The results are encouraging; e.g. for the subject type Album the Spearman correlation between the top 20 paths for Italian resources and those for English ones is 0.882%, while for Language is 0.657%. This despite the fact that Italian paths have lower popularity values than English ones, and much fewer wikilinks (3.25 wikilinks per resource on average).

Schema issues. DBPO has been generated from Wikipedia infoboxes. The DBpedia infobox dataset contains 1,177,925 object property assertions, and their objects are also wikilinks. This means that 7.01% of the 16,745,830 wikilink triples that we have considered overlap with infobox triples. This is a potential bias on our results, which are based on DBPO; however, such bias is very limited:

removing 7% of the wikilinks is not enough to significantly decrease the high correlations we have found.

Finally, one might wonder if our good results could be obtained by using other ontologies instead of DBPO. We are experimenting with wikilink paths typed by Yago [18], which has more than 288,000 classes, and a broader coverage of resources (82% vs. 51.9% of DBPO). Working with Yago is very interesting, but also more difficult, since it applies multityping extensively, and the combinatorics of its paths is orders of magnitude more complex than with paths typed by DBPO. Sample Yago paths include e.g.: [Coca-ColaBrands,BoycottsOfOrganizations], [DietSodas,LivingPeople]. Those paths are domain-oriented, which is a good thing, but they also share a low popularity (ranging around 3% in top ranks) in comparison to DBPO classes. In other words, the skewness of Yago $pathPopularity(P_{i,j}, S_i)$ is much higher than that of DBPO, with a very long tail. However, the clustering factor is not so different: a Yago EKP can be created e.g. for the class yago:AmericanBeerBrands, and its possible thresholds provided by K-Means Clustering appear very similar to the ones found for DBPO EKPs: we should only scale down the thresholds, e.g. from 18% to 1%.

Acknowledgements. This work has been part-funded by the European Commission under grant agreement FP7-ICT-2007-3/ No. 231527 (IKS - Interactive Knowledge Stack). We would like to thank Milan Stankovic for his precious advise.

References

1. Akbik, A., Broß, J.: Wanderlust: Extracting Semantic Relations from Natural Language Text Using Dependency Grammar Patterns. In: Proc. of the Workshop on Semantic Search (SemSearch 2009) at the 18th International World Wide Web Conference (WWW 2009), Madrid, Spain, pp. 6–15 (2009)
2. Baker, C.F., Fillmore, C.J., Lowe, J.B.: The Berkeley FrameNet Project. In: Proc. of the 17th International Conference on Computational Linguistics, Morristown, NJ, USA, pp. 86–90 (1998)
3. Blomqvist, E., Presutti, V., Gangemi, A.: Experiments on pattern-based ontology design. In: Proceeding of K-CAP 2009, Redondo Beach, California, USA, pp. 41–48. ACM (2009)
4. Blomqvist, E., Sandkuhl, K., Scharffe, F., Svatek, V.: Proc. of the Workshop on Ontology Patterns (WOP, collocated with the 8th International Semantic Web Conference (ISWC-2009), Washington D.C., USA. CEUR Workshop Proceedings, vol. 516 (October 25 (2009)
5. Gangemi, A., Presutti, V.: Towards a Pattern Science for the Semantic Web. Semantic Web 1(1-2), 61–68 (2010)
6. Giuliano, C., Gliozzo, A.M., Gangemi, A., Tymoshenko, K.: Acquiring Thesauri from Wikis by Exploiting Domain Models and Lexical Substitution. In: Aroyo, L., Antoniou, G., Hyvönen, E., ten Teije, A., Stuckenschmidt, H., Cabral, L., Tudorache, T. (eds.) ESWC 2010. LNCS, vol. 6089, pp. 121–135. Springer, Heidelberg (2010)

7. Gruninger, M., Fox, M.S.: The role of competency questions in enterprise engineering. In: Proc. of the IFIP WG5.7 Workshop on Benchmarking - Theory and Practice, Trondheim, Norway, pp. 83–95 (1994)

8. Lehmann, J., Bizer, C., Kobilarov, G., Auer, S., Becker, C., Cyganiak, R., Hellmann, S.: DBpedia - A Crystallization Point for the Web of Data. Journal of Web Semantics 7(3), 154–165 (2009)

9. Medelyan, O., Milne, D., Legg, C., Witten, I.H.: Mining meaning from wikipedia. Int. J. Hum.-Comput. Stud. 67(9), 716–754 (2009)

10. Migliore, M., Novara, G., Tegolo, D.: Single neuron binding properties and the magical number 7. Hippocampus 18(11), 1122–1130 (2008)

11. Miller, G.A.: The magical number seven, plus or minus two: some limits on our capacity for processing information. Psychological Review 63(2), 81–97 (1956)

12. Nastase, V., Strube, M., Boerschinger, B., Zirn, C., Elghafari, A.: WikiNet: A Very Large Scale Multi-Lingual Concept Network. In: Calzolari, N., Choukri, K., Maegaard, B., Mariani, J., Odijk, J., Piperidis, S., Rosner, M., Tapias, D. (eds.) Proc. of the Seventh International Conference on Language Resources and Evaluation (LREC), pp. 1015–1022. European Language Resources Association (2010)

13. Nguyen, D.P.T., Matsuo, Y., Ishizuka, M.: Relation extraction from wikipedia using subtree mining. In: Proc. of the 22nd National Conference on Artificial Intelligence, vol. 2, pp. 1414–1420. AAAI Press (2007)

14. Nuzzolese, A.G., Gangemi, A., Presutti, V.: Gathering Lexical Linked Data and Knowledge Patterns from FrameNet. In: Proc. of the 6th International Conference on Knowledge Capture (K-CAP), Banff, Alberta, Canada, pp. 41–48 (2011)

15. Ponzetto, S.P., Navigli, R.: Large-Scale Taxonomy Mapping for Restructuring and Integrating Wikipedia.. In: Boutilier, C. (ed.) IJCAI, Pasadena, USA, pp. 2083–2088 (2009)

16. Presutti, V., Chaudhri, V.K., Blomqvist, E., Corcho, O., Sandkuhl, K.: Proc. of the Workshop on Ontology Patterns (WOP 2010) at ISWC-2010, Shangai, China. CEUR Workshop Proceedings (November 8, 2010)

17. Singh, P.: The Open Mind Common Sense project. Technical report, MIT Media Lab (2002)

18. Suchanek, F., Kasneci, G., Weikum, G.: Yago - A Large Ontology from Wikipedia and WordNet. Elsevier Journal of Web Semantics 6(3), 203–217 (2008)

19. Völker, J., Niepert, M.: Statistical Schema Induction. In: Antoniou, G., Grobelnik, M., Simperl, E., Parsia, B., Plexousakis, D., De Leenheer, P., Pan, J. (eds.) ESWC 2011, Part I. LNCS, vol. 6643, pp. 124–138. Springer, Heidelberg (2011)

20. Zesch, T., Müller, C., Gurevych, I.: Extracting Lexical Semantic Knowledge from Wikipedia and Wiktionary.. In: Proc. of the Sixth International Conference on Language Resources and Evaluation (LREC), Marrakech, Morocco, pp. 1646–1652. European Language Resources Association (2008)

An Ontology Design Pattern for Referential Qualities

Jens Ortmann and Desiree Daniel

Institute for Geoinformatics, University of Muenster, Germany
jens.ortmann@uni-muenster.de,
desiree.daniel@yahoo.com

Abstract. Referential qualities are qualities of an entity taken with reference to another entity. For example the vulnerability of a coast to sea level rise. In contrast to most non-relational qualities which only depend on their host, referential qualities require a referent additional to their host, i.e. a quality Q of an entity X taken with reference to another entity R. These qualities occur frequently in ecological systems, which make concepts from these systems challenging to model in formal ontology. In this paper, we discuss exemplary resilience, vulnerability and affordance as qualities of an entity taken with reference to an external factor. We suggest an ontology design pattern for referential qualities. The design pattern is anchored in the foundational ontology DOLCE and evaluated using implementations for the notions affordance, resilience and vulnerability.

1 Introduction

Environmental problems are major challenges of the 21st century and occur in all parts of the world on local, regional and global scales, especially in face of climate variability. The rate of global environmental change is surpassing response and without action to mitigate its drivers and enhance societal resilience, these changes will retard economic, social, environmental and developmental goals [16]. According to the International Council for Science [16], to meet this challenge, what is required is a robust information infrastructure that can combine data and knowledge both past and present with new observations and modeling techniques to provide integrated, interdisciplinary datasets and other information products. One key functionality within this infrastructure is to assess vulnerability and resilience.

Before such an infrastructure can be realized, there is a need to sort out the bottleneck that occurs on the conceptual level with notions such as vulnerability and resilience. In an attempt to introduce ecological concepts to formal ontology, the problem of modeling qualities arises. In addition to their host, qualities in the ecological domain often depend on external entities. A recurring pattern is *a quality Q of an entity X taken with reference to another entity R*. These qualities have one host and cannot exist without an external factor. To name just one example, the vulnerability of a coast cannot be assessed per se, but only with respect to an external threat, like the vulnerability to sea level rise. So we have

L. Aroyo et al. (Eds.): ISWC 2011, Part I, LNCS 7031, pp. 537–552, 2011.

the vulnerability of a coast with reference to sea level rise. Yet, vulnerability is not considered a relational quality, but is attributed to the coast. Furthermore, the same coast has different vulnerabilities with reference to different external factors (e.g., a hurricane or an oil-spill).

Modeling qualities that inhere in more than one host represents a conceptual quandary that applies to ecological notions but it appears in many other fields that deal with systems (e.g. economic systems, social systems, communication networks). We take a look at the notions of resilience, vulnerability and affordance, which lack formal definitions. An approach that can model referential qualities in ontologies can lead to these concepts being implementable and thereby usable in semantic web applications.

This paper therefore brings into focus the research question *"how can referential qualities be modeled in ontologies?"*. To solve the quandary, an Ontology Design Pattern (ODP) [8] is suggested as a template to model different referential qualities. This work draws from Kuhn's [20,21] Semantic Reference Systems. Semantic Reference Systems provide a theory of how categories in an ontology can be described with reference to other, already established categories. At the same time the theory of Semantic Reference Systems paves the road to make these ecological qualities operational. A high level formalization of the Ontology Design Pattern as well as an OWL-DL[1] implementation is carried out by the authors. Both are aligned to the Descriptive Ontology for Linguistic and Cognitive Engineering (DOLCE) [23]. The ecological qualities resilience, vulnerability and affordance serve to illustrate applications of the suggested ODP.

This paper proceeds by outlining the background on non-relational qualities, relational qualities and Ontology Design Patterns in Sect. 2. Subsequently, Section 3 motivates this research by discussing three examples of referential qualities. Section 4 describes the Ontology Design Pattern for referential qualities and its implementation in OWL-DL. The implementation of the examples are discussed in Sect. 5. In Sect. 6 the conducted research is evaluated against the use cases as well as analytically. Finally, Sect. 7 concludes the paper.

2   Background

This section introduces the notions of non-relational qualities and relational qualities with respect to DOLCE as well as the idea of the ODP approach.

2.1   Non-relational and Relational Qualities

In philosophy there exist several dichotomies of qualities (or properties), they are labeled for example "intrinsic-extrinsic", "relational-non-relational", "unary-n-ary" (cf. eg., [38]). The differences between these distinctions are sometimes hard to grasp for non-experts, as for instance many standard examples of non-relational qualities are intrinsic and unary (e.g. color of a rose), and many

---

[1] The description logic (DL) variant of the web ontology language (OWL). See http://www.w3.org/TR/owl-guide/ for more information. All websites referred to in this paper were last accessed on 21.06.2011.

standard examples of relational qualities are extrinsic and binary (e.g. gravitational attractions). However, there exist intrinsic relational qualities and extrinsic non-relational qualities. In the following we will introduce the relational non-relational distinction, as we perceive this one as the most intuitive to non-experts in metaphysics. We only mention the intrinsic-extrinsic distinction as additional criterion when necessary.

Non-Relational Qualities. Non-relational qualities can be modeled in DOLCE. In essence, this can be seen as an ODP in itself, even though to our knowledge it is never officially declared as such. The design pattern for modeling qualities in DOLCE follows the EQQS (Endurant, Quality, Quale, Spaces) pattern described in [22]. The pattern comprises four categories. An `endurant`[2] (E) that serves as host for the `quality` (Q), with a `quale` (Q) that is located in a `quality-space` (S). In DOLCE all entities are particulars [23]. Every particular endurant has its own particular qualities. For example, we can say that two boys have the same height, however, in DOLCE each boy has his own distinct height quality. This allows qualities to change, for instance, if the boys grow, especially when they grow at different rates and do not have the same heights as adults any more. To make the distinct qualities operational, a quale is needed. The quale is a representation of the quality in a quality space. For example, in a meter space the height quality of the two boys is reflected by the same quale. The quale is modeled as a `region` in a `quality-space`, which allows certain computations, depending on the structure of the space. An illustration of the DOLCE quality ODP is shown in Fig. 1. The formal relation and implementation of the quality ODP can be found in [23].

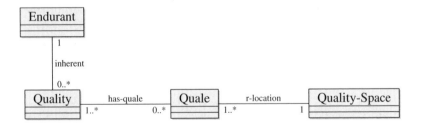

Fig. 1. UML class diagram of the quality ODP in DOLCE. A quality inheres in an endurant. The quality has a quale that is located in a quality space.

Relational Qualities. Relational qualities inhere in more than one host. This is not allowed in DOLCE, where the quality can only inhere in one host. For example the solubility of a substance in a liquid cannot be expressed in the EQQS pattern in DOLCE, because the solubility inheres in both, the substance and the

---

[2] Throughout the paper we use truetype fonts like `quality` to refer to categories in a formal ontology.

liquid. DOLCE falls short modelling qualities like solubility, distance, direction [29]. Qualities that depend on two entities are often extrinsic and relational. They usually qualify a relation between two entities. The formal relation that expresses a relational quality has more than two entities, namely the quality and the hosts. Therefore, a direct implementation in Description Logics is not possible. However, there exists a logical ontology design pattern to transform n-ary relations expressed in a more expressive logic into OWL [26]. This pattern can be used to encode relational qualities as well. Nonetheless, this model diverts from the DOLCE EQQS pattern for qualities, and it violates DOLCE's constraint that a quality can only inhere in one host.

2.2 Ontology Design Patterns

Ontology Design Patterns are recurring structures in ontologies. Similar to design patterns in software engineering they provide templates to specific problems in engineering ontologies. The first patterns were not designed from scratch but emerged in ontologies. A first account of this phenomenon in ontology engineering along with a framework to document and classify ontology design patterns was given by Gangemi [8]. Gangemi and Presutti [10] list different types of ontology design patterns, depending on their intended use. They distinguish between Structural ODPs, Reasoning ODPs, Presentation ODPs, Correspondence ODPs, Lexico-Syntactic ODPs and Content ODPs for example. The details on these different ODPs can be found in [10] or looked up on the website ontology-designpatterns.org. In this paper we are only interested in Content ODPs, and will refer to them as "ODP" for short. They can be defined as:

> [Content ODPs] are distinguished ontologies. They address a specific set of competency questions, which represent the problem they provide a solution for. Furthermore, [Content ODPs] show certain characteristics, i.e. they are: computational, small, autonomous, hierarchical, cognitively relevant, linguistically relevant, and best practises. [10, p. 231]

Most important in this definition is that Content ODPs are small ontologies. Gangemi and Presutti [10] also suggest a new paradigm for ontology engineering, based on small modular Content ODPs, which form that backbone of a new ontology. The website ontologydesignpatterns.org is a community portal intended to serve as hub for finding and contributing ODPs of all types.

3 Examples

In this section, we discuss the salience of affordance, resilience and vulnerability as well as the problems that occur on the conceptual level when modeling these notions.. A major problem is still, that these qualities cannot be modelled in formal ontology. This hinders modelling ecological systems in information infrastructures.

3.1 Resilience

The notion of resilience originated within the Ecology domain through the work of Holling [15]. From its beginning in ecology, resilience has transcended across various disciplines such as Sustainable Science, Disaster Management, Climate Change and Psychology [1,36,18,13] as it is deemed a salient concept pertaining to systems in face of adversity. Holling [15] ascertained that ecological resilience is a characteristic of ecosystems to maintain themselves in face of disturbance. In other words, resilience can be viewed as a quality of a system.

After four decades of the resilience debate, the concept still remains on the abstraction level. Instead of domains moving into an operational direction, there is a constant re-invention of the wheel that hinders research. This impediment can be attributed to the lack of a common lexicon across domains [24]. Walker et al. [35] state that in order for resilience to be operational we have to consider resilience in a specific context 'from what to what', for example the resilience of corn in South Africa to the impacts of drought. This specificity brings into focus the problem of having a quality that is dependent on a referent and the limitations of DOLCE in this regard.

Since the inception of resilience in Ecology, several definitions across disciplines have emerged. Holling initially described resilience in contrast to stability:

> Resilience determines the persistence of relationships within a system and is a measure of the ability of these systems to absorb changes of state variable, driving variables, and parameters and still persist. [15, p. 17]

In an attempt to disambiguate resilience, we [6] suggested a generalized definition of resilience for the ecological and socio-ecological domain:

> The ability of a system to cope with an external factor that undermines it, with the system bouncing back. [6, p. 121]

This definition was proven to be compatible with definitions given by the Resilience Alliance (c.f. [5]) and by the United Nations International Strategy for Disaster Reduction (c.f. [34, Annex1 p. 6]). To our knowledge, no attempts have been made to model resilience as quality in DOLCE.

Resilience has been linked to vulnerability as both concepts aim to understand socio-ecological systems in face of disturbance.

3.2 Vulnerability

Vulnerability can be described according to [27] as the extent to which a natural or social system is susceptible to sustaining damage, for example from climate change. As societies aim to enhance resilience in face of global climate change, the assessment of vulnerability solely, provides a one sided approach to foster adaptation strategies. Consequently, there is a need for the convergence of both theoretical and practical approaches of both concepts [24].

The MONITOR[3] project embarked on the challenge of developing a risk ontology to act as a reference framework for risk management. The concept of vulnerability was modeled and defined as the quality of the objects of an environment, which determines damage, given a defined hazardous event [19].

3.3   Affordances

The notion of affordance was introduced by Gibson [11,12]. Gibson defined affordances as follows:

> The affordance of anything is a specific combination of its substance and its surfaces taken with reference to an animal [11, p. 67]

> [A]ffordances are properties of things *taken with reference to an observer* [12, p. 137]

For example, a chair offers a sitting affordance to a human and a hollow tree has a shelter affordance to a bat or owl. Yet, there exists no definition of affordances that is commonly agreed-upon. Affordances are often seen as either relational qualities, that exist as relation between the environment and the agent (e.g. [32]), or as qualities of (an object in) the environment taken with reference to the agent. The latter view would benefit from an ODP for referential qualities. Definitions supporting the latter view have been given for example by Heft [14, p. 3] and Turvey [33, p. 174].

Today affordances play an important role, not only in their original field of psychology, but they are applied for example in design [25], robotics [31], navigation and way-finding [30] and similarity analysis [17]. However, the potential of affordances is far from exploited. One reason for this is that affordances are challenging to model in formal theories, especially ontologies for the semantic web. The first author has recently suggested to model affordances as `qualities` in DOLCE [28]. However, in the cases where affordances were modeled in ontology, the reference to the observing agent was not made explicit. This takes away the key idea of an affordance.

4   An Ontology Design Pattern for Referential Qualities

In the cases addressed in this paper, we do not need to violate the constraint of a unique host, but we need to extend the quality pattern by a referent entity that is required for the quality. Therefore, qualities like affordance, resilience and vulnerability are not relational and they are not of the kind of non-relational qualities that fit the DOLCE EQQS pattern. They have one host entity, but an additional referent that is necessary for the quality to exist. However, Ellis suggested distinct (extrinsic) qualities that objects have in virtue of outside forces [7, as discussed in [38]]. The notion of force comes very close to the terminology of definitions for vulnerability and resilience, and it also fits the dynamic

---

[3] see http://www.monitor-cadses.org/

and potential nature of affordances. A referential quality could then be called a non-relational extrinsic quality.

The resulting pattern can be described as EQRQS pattern with an endurant (E), a quality (Q), a referent entity (R), a quale (Q) and a quality-space (S). The referent is characterized as playing a special referent-role.

The remainder of the section presents the Ontology Design Pattern for referential qualities. In the classification of the Ontology Pattern Initiative[4] this pattern is a Content Ontology Design Pattern. We do not suggest referential qualities as distinct type of quality to be considered in metaphysics, but intend to give knowledge engineers and domain scientists a practical tool to model the described qualities and to be able to account for qualities that do not solely depend on one or more host entities. For practical reasons the scope of this discussion is mostly limited to DOLCE [23] and its commitments.

In the following the ODP for referential qualities is introduced according to the Content ODP Frame suggested in [8]. This frame provides eleven slots to introduce a new Content ODP.

UML Diagram. Figure 2 shows the UML diagram of the referential quality design pattern.

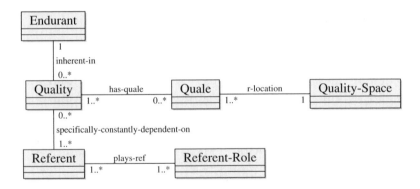

Fig. 2. UML class diagram of the referential quality ODP. The relations are to be read starting from the category quality or from left to right

Generic Use Case (GUC). The referential quality pattern's overall aim is to enable modeling of qualities that have one major host but existentially depend on some external factor.This is a recurring pattern of concepts in Ecology, Economics, Natural Disaster Management, Climate Change, Computer Science, just to name a few. The generic pattern can be phrased as *"a quality Q of an entity X taken with reference to another entity R"*. These qualities occur in particular when following system-theoretic approaches to model a domain. A system is

---

[4] see http://ontologydesignpatterns.org

described not only by its elements but also by relations and interdependencies between these elements [2]. In such systems there often exist qualities that an entity has by virtue of an external factor. For this paper, attention is brought to ecological systems, however, the pattern is not restricted to the domain of ecology, but is kept generic enough to be used in the other aforementioned domains as well.

Local Use Case(s). The local uses cases are equivalent to the examples described in Sect. 3. The referential quality pattern should be capable to account for:

- an affordance as quality of an object taken with reference to an observer.
- resilience as the capacity of a sysem to cope with an external factor.
- vulnerability as the quality of an object with respect to a potential damage.

Affordance, resilience and vulnerability lack a formal ontological definition that makes the concepts operational and allows a consideration of these concepts in and across information systems. Even though conceptual definitions exist, there is a dearth in formal ontological solutions that will allow for semantic integration of information sources and models.

Logic Addressed. The Ontology Design Pattern is modeled in the Unified Modeling Language (UML), it is then encoded in Description Logic and implemented in the Web Ontology Language OWL-DL.

Reference Ontologies. The Ontology Design Pattern uses DOLCE Lite[5] as upper level ontology. As this pattern provides a template to model certain types of qualities, this paper uses especially the DOLCE category `qualities`. In addition to DOLCE Lite we use the Extended Descriptions and Situations module[6] to specify a `role`.

Specialized Content ODPs. The ODP for referential qualities is a specialization of the quality ODP in DOLCE. Masolo and Borgo have discussed several approaches to model qualities in formal ontology [22]. In DOLCE, a `quality` depends on one host `entity`. The quality invokes a quale, i.e. a magnitude of the quality. The quale is located in an abstract quality space. DOLCE uses a framework of `entity`, `quality`, `quale` and `quality-space` [23]. This ODP extends the DOLCE EQQS pattern with a referent.

Composed Content ODPs. The ODP for referential qualities does not formally compose of other ODPs in its most general form. Nonetheless, the pattern can be seen to compose of the quality pattern that is specialized with a role pattern. Unfortunately, there is no role pattern that is general enough to allow endurants, perdurants or qualities to play the referent role. There exist patterns for special roles, for example the Objectrole[7], that can be used to model the role part of the referential quality pattern. Hence, in case the `referent-role`

---

[5] http://www.loa-cnr.it/ontologies/DOLCE-Lite.owl

[6] http://www.loa-cnr.it/ontologies/ExtendedDnS.owl

[7] see http://ontologydesignpatterns.org/wiki/Submissions:Objectrole

is played by a `physical-object`, the pattern is implemented as composition of the DOLCE quality pattern and the Objectrole pattern.

A special case occurs when the referent is a `referential-quality` itself, then the pattern can be recursively applied to realize a composition of various referential quality patterns.

Formal Relation. $Referential\_Quality(e, q, ql, qs, r, rr)$, where e is an endurant, perdurant or quality, q is a referential quality that inheres in e, ql is a quale of q in the quality space qs, r is an endurant, perdurant or quality that plays a referent-role rr for the referential quality q.

Sensitive Axioms.
$$Referential\_Quality(e, q, r) =_{df} QU(q)$$
$$\wedge \, (ED(e) \vee PD(e) \vee QU(e)) \wedge inheres(q, e)$$
$$\wedge \, (ED(r) \vee PD(r) \vee QU(r)) \wedge \quad \wedge (\exists \Phi, s : subsumes(\Theta, \Phi) \wedge \Phi(s))$$
$$\wedge \, specifically - constantly - dependent - on(q, r)$$
$$\wedge \, plays - ref(r, s)$$
where Θ is the category `Role` in DOLCE. Note that according to the DOLCE definition of constant specific dependence, q and r can not be equivalent [23, Dd(70), p. 31].

Explanation. Many existing approaches to model a quality that depends on an external factor lead to unsatisfactory results. The referential quality design pattern suggests a practical approach to model such qualities that is compatible with DOLCE. The DOLCE pattern for qualities comes with qualia and quality-spaces. This already anticipates a future operationalization of these qualities.

OWL(DL) encoding (Manchester syntax, excerpt[8])
Class: referent-role
   SubClassOf:
      edns:role

Class: referential-quality
   SubClassOf:
      dol:quality
      and (dol:specifically-constantly-dependent-on
         some (rq:plays-ref **some** rq:referent-role))

5 Implementation of Affordance, Resilience and Vulnerability

This section presents the exemplary implementations of vulnerability taken with reference to a hazardous event, resilience taken with reference to an external factor, and an affordances taken with reference to an agent. These examples

---

[8] the complete implementation along with the examples presented in this paper encoded in the OWL Manchester Syntax are available online at `http://www.jensortmann.de/ontologies/odp4refprop.html`

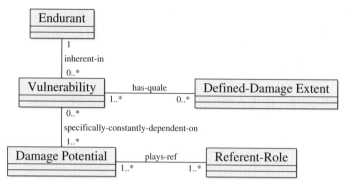

Fig. 3. UML class diagram of vulnerability in the MONITOR project, modeled according to the referential quality ODP

demonstrate potential applications. However, the ontological choices made to restrict the host and referent of the quality do not necessarily reflect the full scope of the modeled concepts.

The implementation of vulnerability is based on results from the MONITOR project. Hence, it comprises of the `referential-quality vulnerability`, that is specifically constantly dependent on a `damage-potential`[9]. The MONITOR project describes damage potential as a quality of the environment. In our implementation we only model the damage potential as `quality`, but do not model an environment explicitly. From the MONITOR account of vulnerability we concluded that the quale of vulnerability is a defined damage extent, which resides in a special quality space for damage extents. `damage-extent` is a subcategory of `quale`. Our implementation does not make the quality space explicit. Figure 3 shows the UML diagram of the vulnerability implementation.

As an example of how to apply vulnerability as a referential quality, we implemented a small set of instances that reflect the vulnerability of a society to an earthquake for the case of Haiti. We introduced a subcategory `society` of `agent`, because we treat the society as one whole, abstracting from its constituting institutions, government and people. The category `society` has one individual called `haitian-society`. The `vulnerability` individual is called `haitian-vulnerability-to-earthquake` and is member of the newly introduced subcategory `vulnerability-to-earthquake`. The `referent-role` is played by an individual `haitian-damage-potential` that we simply introduce as member of `quality`. The quale of `vulnerability` is a member of the `defined-damage-extent`, which is labeled as `earthquake-damage-extent`. The quale requires further characterization. It could stand for a damage assessed in monetary value or number of injured and dead. The quality space would be modeled accordingly, for example as metric space for money, or as space for counts in the case of victim numbers.

---

[9] Note that MONITOR uses generic dependence here.

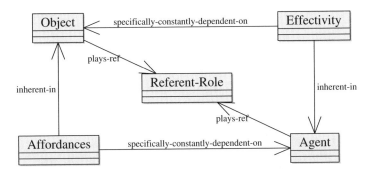

Fig. 4. UML class diagram of an affordance modeled according to the referential quality ODP. For better clarity the cardinalities are not depicted here, but are the same as in Fig. 2. Instead, the associations are directed in this diagram.

The implementation of resilience is based on [6]. It classifies `resilience` as `referential-quality`. Aiming for a general account, we leave the external factor as well as the host for resilience unspecified and resort to the category `spatio-temporal-particular` in DOLCE. We have implemented a specific example and made it available online, further details are left out here.

The implementation of an affordance is based on Turvey's [33] definition of an affordance contrasted with an effectivity. Therefore the implementation comprises two `referential-qualities`: `affordance` and `effectivity`. Both are implemented as direct subcategories of `referential-quality`. The implementation enforces the referent of an `affordance` to be the host of an `effectivitiy`, which in turn needs to be a DOLCE `agent`. To do justice to Turvey's dispositional nature of affordance, the referent of the `effectivity` must host the `affordance`. Figure 4 illustrates the implementation of affordance as UML diagram.

An additional exemplary implementation with individuals of a concrete affordance serves to make the affordances example more graspable and easier to evaluate. One of the most cited investigation of an affordance is Warren's [37] account of the stair-climbing affordance that is provided by a step with a riser height to a person with matching leg-length. The implementation of stair-climbability introduces two new categories for the stair-climbing affordances and for the stair-climbing effectivity. Individuals exist for the `physical-object` `step1` that hosts the `referential-quality` `step1-climbability-affordance` and depends on the `agent` Susan with the `effectivity` leg-length[10]. Two `referent-role` individuals fill the respective object-property slots for the individuals of `affordance` and `effectivity`. Qualia and quality spaces are not modeled in the stair-climbing example.

---

[10] The leg length as effectivity is an abstraction that Warren [37] made in his model.

6 Evaluation

There is no strict methodology to evaluate ODPs as yet but only some criteria, such as reusability, to assess the success of an ODP [10]. Therefore, a full evaluation of the suggested ODP cannot be given at this stage. However, this section will shed some light on how well the ODP solved the local use-cases and asses some analytical criteria taken form information systems evaluation.

6.1 Evaluation of the ODP against the Local Use Cases

The ontological account of vulnerability is taken from the MONITOR project. It fits the referential quality pattern nearly perfectly. The only difference is that MONITOR suggests a generic dependence between vulnerability and its host, whereas the referential quality pattern requires a constant specific dependence. We chose the specific dependence to express that a particular referential quality depends on a particular referent, but not on a category of referents. Additionally, generic dependence cannot be implemented in OWL-DL, the target language of this ODP. Furthermore, the choice of specific or generic dependence seems to boil down to the philosophical stance on the environment as being shared or being individual. This issue is out of scope of this paper, but both views are supported in the literature. Especially, Turvey's [33] use of singular forms suggests that his objects of discourse are particulars, not universals.

The example of vulnerability to an earthquake could be implemented in DOLCE. Nonetheless, in a real scenario this example has to be integrated with domain ontologies that account for earthquakes. However, the evaluation should rather prioritize the potential for application in a real project like MONITOR. Here we find our pattern is compatible with the ontology suggested in MONITOR, and can lend support to their modeling decisions.

The implementation of resilience as category in DOLCE was straight forward and directly captures the general definition of resilience [6]. A small example showed that it is possible to create individuals that reflect the resilience of a society to sea-level rise. However, the implementations of resilience and also vulnerability do not make these concepts fully operational as yet. The definitions of structured quality spaces are still missing. As soon as qualia and structures are identified in the domain qualia and quality spaces can be defined. Probst [29] suggested a classification of structured quality spaces and an approach to model them in DOLCE.

The challenge in the affordance use case was to model an affordance as property of a thing taken with reference to an observer [12]. The ODP gives a direct implementation of this in modeling the affordance as quality of an object that is specifically constantly dependent on an agent. The effectivity is modeled as quality of the agent who plays the referent role. An actual realization of one example of a stair-climbing affordance showed that it is possible and feasible to implement affordances with this ODP in DOLCE and that this implementation complies with a well-established theory of affordances. The affordance example also demonstrates how two referential quality patterns can be composed.

6.2   Analytical Evaluation

In [4], the authors suggested five criteria to evaluate information systems: Significance, internal validity, external validity, confirmability/objectivity and reliability/dependability/auditability. Information system engineering is close enough to ontology engineering and the suggested categories are defined broad enough to use them to evaluate the ODP for referential qualities.

Significance. Referential qualities can be deemed as important building blocks of domain ontologies that account for systems or networks. The whole idea of an ontology design pattern is to turn the theoretical significance of a concept such as referential quality, into a practical significance. The pattern provides a small modular extension to DOLCE that allows to account for these qualities. The authors make clear that the theory of referential qualities is not a contribution to metaphysics, and that the aim is rather to give domain experts a handy template to model common structures in their domain. The practical significance can only be evaluated in terms of the adoption and use of this pattern in the future. The MONITOR project already employs this pattern, without making the pattern itself explicit. The authors own ongoing work on resilience and affordances will also benefit from this pattern.

Internal Validity. The implementation of the three local use cases demonstrates that the ODP works and can be applied in practice. A comparison with relational and non-relational qualities and their respective modeling approaches has revealed the shortcomings of modelling these qualities in formal ontology with respect to requirements stated in the literature on affordances, resilience and vulnerability. The ODP for referential qualities overcomes these shortcomings.

External Validity. The ODP extends an existing pattern for non-relational qualities and uses DOLCE. The implementation of the use-cases conforms with theory in the domain, but rival theories exist that might require a different pattern. The findings in this paper are compatible with for example the specification suggested in the MONITOR project [19] and in Warren's stair climbing experiment [37].

In general, the idea of ODPs and modularity in ontologies is to increase the reusability of ontologies. This pattern is intentionally kept simple to achieve this goal. The nature of ODPs is to make the theory transferable. However, boundaries for the use have not been suggested. The authors see a huge potential for application in a variety of domains that deal with systems and networks in any form. Both authors are currently employing the pattern in their ongoing thesis work.

Confirmability/Objectivity. The design pattern is introduced in a formal way according to [8]. An OWL-DL implementation of the ODP and of the examples is available online. The theory of referential qualities is inspired by ecological systems and the language used by domain scientist in ecology. An ecological bias cannot be ruled out. Furthermore, the commitment to DOLCE and its way of modeling qualities and qualia implies a cognitive bias of the theory.

Reliability/Dependability/Auditability. The research objective has been made clear in natural language. A detailed description of the OPD is available in formal and natural language. The basic constructs can be easily retrieved from the ODP frame given in Sect. 4, while the implementations are available online.

7   Conclusion and Future Work

This paper formally introduced an Ontology Design Pattern for referential qualities. We have discussed the idea of referential qualities and introduced the pattern according to a suggested frame. Three use cases demonstrated the practicability of the pattern. Finally, the paper evaluated the patterns with respect to the use cases together with its general engineering grounds. The pattern is kept general and simple to ensure a flexible and easy application. The pattern has the potential to facilitate ontological modeling of ecological systems. It thereby paves the way for a stronger integration of ecological models and datasets.

Future work will be at first a further application and exploration of the ODP. A promising aspect of referential qualities is their compatibility with Kuhn's Semantic Reference Systems [20,21] and Probst's semantic reference system for observations and measurements of qualities [29]. This opens the door for the operationalization that the presented concepts still lack. Operationalisation entails the physical and mental measurement of a concept relative to a reference framework [3]. The process of operationalization is an important step to a clear understanding of a concept. Finally, one of the reviewers suggested to consider the Description and Situations Ontology [9] as alternative to the comparably strong commitment to DOLCE.

Acknowledgements. The authors gratefully acknowledge the reviewers comments and suggestions. This work has been partly supported through the International Research Training Group on Semantic Integration of Geospatial Information by the DFG (German Research Foundation), GRK 1498.

References

1. Adger, W.: Social and ecological resilience: are they related? Progress in Human Geography 24(3), 347–363 (2000)
2. von Bertalanffy, L.: An outline of general system theory. The British Journal for the Philosophy of Science 1(2), 134–165 (1950)
3. Bridgeman, P.: The logics of modern physics. Beaufort Books (1927)
4. Burstein, F., Gregor, S.: The systems development or engineering approach to research in information systems: An action research perspective. In: Hope, B., Yoong, P. (eds.) Proceedings of the 10th Australasian Conference on Information Systems, pp. 122–134. Victoria University of Wellington, New Zealand (1999)
5. Carpenter, S., Walker, B., Anderies, J.M., Abel, N.: From Metaphor to Measurement: Resilience of What to What? Ecosystems 4(8), 765–781 (2001)
6. Daniel, D., Ortmann, J.: Disambiguating Resilience. In: Schwering, A., Pebesma, E., Behnke, K. (eds.) Geoinformatik 2011 - Geochange. ifgiprints, vol. 41, pp. 117–125. AKA Verlag, Heidelberg (2011)

7. Ellis, B.: Scientific Essentialism. Cambridge Studies in Philosophy. Cambridge University Press, Cambridge (2001)
8. Gangemi, A.: Ontology Design Patterns for Semantic Web Content. In: Gil, Y., Motta, E., Benjamins, V., Musen, M. (eds.) ISWC 2005. LNCS, vol. 3729, pp. 262–276. Springer, Heidelberg (2005)
9. Gangemi, A., Mika, P.: Understanding the Semantic web Through Descriptions and Situations. In: Chung, S., Schmidt, D.C. (eds.) CoopIS 2003, DOA 2003, and ODBASE 2003. LNCS, vol. 2888, pp. 689–706. Springer, Heidelberg (2003)
10. Gangemi, A., Presutti, V.: Ontology design patterns. In: Handbook on Ontologies, 2nd edn. International Handbooks on Information Systems, pp. 221–243. Springer, Heidelberg (2009)
11. Gibson, J.: The Theory of Affordances. In: Shaw, R.E., Bransford, J. (eds.) Perceiving, Acting, and Knowing: Toward an Ecological Psychology, pp. 67–82. Lawrence Erlbaum Associates Inc., Hillsdale (1977)
12. Gibson, J.: The ecological approach to visual perception. Lawrence Erlbaum, Hillsdale (1979)
13. Handmer, J.W., Dovers, S.R.: A typology of resilience: Rethinking institutions for sustainable development. Organization & Environment 9(4), 482–511 (1996)
14. Heft, H.: Affordances and the body: An intentional analysis of Gibson's ecological approach to visual perception. Journal for the Theory of Social Behaviour 19(1), 1–30 (1989)
15. Holling, C.: Resilience and stability of ecological systems. Annual Review of Ecology and Systematics 4, 1–23 (1973)
16. ICSU: Earth system science for global sustainability: The grand challenges. Tech. rep., International Council for Science, Paris (2010)
17. Janowicz, K., Raubal, M.: Affordance-Based Similarity Measurement for Entity Types. In: Winter, S., Duckham, M., Kulik, L., Kuipers, B. (eds.) COSIT 2007. LNCS, vol. 4736, pp. 133–151. Springer, Heidelberg (2007)
18. Klein, R., Nicholls, J., Thomalla, F.: Resilience to natural hazards: How useful is this concept? Global Environmental Change Part B: Environmental Hazards 5(1-2), 35–45 (2003)
19. Kollartis, S., Wergels, N.: MONITOR – an ontological basis for risk managment. Tech. rep., MONITOR (January 2009)
20. Kuhn, W.: Semantic reference systems. International Journal of Geographical Information Science 17(5), 405–409 (2003)
21. Kuhn, W., Raubal, M.: Implementing Semantic Reference Systems. In: Gould, M., Laurini, R., Coulondre, S. (eds.) Proceedings of the AGILE 2003 - 6th AGILE Conference on Geographic Information Science, Lyon, France, pp. 63–72. Presses polytechniques et universitaires romandes, Lausanne (2003)
22. Masolo, C., Borgo, S.: Qualities in formal ontology. In: Foundational Aspects of Ontologies (FOnt 2005) Workshop at KI, pp. 2–16 (2005)
23. Masolo, C., Borgo, S., Gangemi, A., Guarino, N., Oltramari, A.: Wonderweb deliverable D18 ontology library (final). ICT Project 33052 (2003)
24. Miller, F., Osbahr, H., Boyd, E., Thomalla, F., Bharwani, S., Ziervogel, G., Walker, B., Birkmann, J., van der Leeuw, S., Rockström, J.: Resilience and vulnerability: Complementary or conflicting concepts? Ecology and Society 15(3) (2010)
25. Norman, D.: The Design of Everyday Things (Originally published: The psychology of everyday things). Basic Books, New York (1988)
26. Noy, N., Rector, A.: Defining N-ary Relations on the Semantic Web. Working group note, W3C (April 2006), http://www.w3.org/TR/swbp-n-aryRelations/

27. Olmos, S.: Vulnerability and adaptation to climate change: Concepts, issues, assessment methods. Tech. rep., Climate Change Knowledge Network, CCKN (2001)
28. Ortmann, J., Kuhn, W.: Affordances as Qualities. In: Galton, A., Mizoguchi, R. (eds.) Formal Ontology in Information Systems Proceedings of the Sixth International Conference (FOIS 2010). Frontiers in Artificial Intelligence and Applications, vol. 209, pp. 117–130. IOS Press, Amsterdam (2010)
29. Probst, F.: Observations, measurements and semantic reference spaces. Applied Ontology 3(1-2), 63–89 (2008)
30. Raubal, M.: Ontology and epistemology for agent-based wayfinding simulation. International Journal of Geographical Information Science 15(7), 653–665 (2001)
31. Rome, E., Hertzberg, J., Dorffner, G.: Towards Affordance-Based Robot Control: International Seminar Dagstuhl Castle, Germany. LNCS. Springer, Heidelberg (June 2006) (revised papers)
32. Stoffregen, T.A.: Affordances as properties of the animal-environment system. Ecological Psychology 15(2), 115–134 (2003)
33. Turvey, M.: Affordances and prospective control: An outline of the ontology. Ecological Psychology 4(3), 173–187 (1992)
34. UN/ISDR: Living with risk: A global review of disaster reduction initiatives. Tech. rep., Inter-Agency Secretariat of the International Strategy for Disaster Reduction (2004)
35. Walker, B., Carpenter, S., Anderies, J., Abel, N., Cumming, G., Janssen, M., Lebel, L., Norberg, J., Peterson, G., Prichard, R.: Resilience Management in Social-ecological System: a Working Hypothesis for a Participatory Approach. Conservation Ecology 6(1) (2002)
36. Waller, M.A.: Resilience in ecosystemic context: evolution of the concept. Am. J. Orthopsychiatry 71(3), 290–297 (2001)
37. Warren, W.: Perceiving affordances: Visual guidance of stair climbing. J. of Experimental Psychology: Human Perception and Performance 10(5), 683–703 (1984)
38. Weatherson, B.: Intrinsic vs. Extrinsic Properties. In: The Stanford Encyclopedia of Philosophy, Fall 2006 edn. (2006),
http://plato.stanford.edu/archives/fall2008/entries/
intrinsic-extrinsic/

Connecting the Dots:
A Multi-pivot Approach to Data Exploration

Igor O. Popov, M.C. Schraefel, Wendy Hall, and Nigel Shadbolt

School of Electronics and Computer Science,
University of Southampton,
SO17 1BJ, Southampton, UK
{ip2g09,mc,wh,nrs}@ecs.soton.ac.uk
http://users.ecs.soton.ac.uk/{ip2g09,mc,wh,nrs}

Abstract. The purpose of data browsers is to help users identify and
query data effectively without being overwhelmed by large complex graphs
of data. A proposed solution to identify and query data in graph-based
datasets is *Pivoting* (or *set-oriented browsing*), a many-to-many graph
browsing technique that allows users to navigate the graph by starting
from a set of instances followed by navigation through common links.
Relying solely on navigation, however, makes it difficult for users to find
paths or even see if the element of interest is in the graph when the points
of interest may be many vertices apart. Further challenges include finding
paths which require combinations of forward and backward links in order
to make the necessary connections which further adds to the complexity
of pivoting. In order to mitigate the effects of these problems and enhance the strengths of pivoting we present a *multi-pivot* approach which
we embodied in tool called Visor. Visor allows users to explore from multiple points in the graph, helping users connect key points of interest in
the graph on the conceptual level, visually occluding the remainder parts
of the graph, thus helping create a road-map for navigation. We carried
out an user study to demonstrate the viability of our approach.

Keywords: Data browsing, graph-data, pivoting, interaction.

1 Introduction

Challenges in browsing large graphs of data are rich: large numbers of ontology
concepts, high entropy and diversity in links between individual data instances,
often makes it hard for users to understand both the overall content of a dataset,
as well as understand and find the particular bits of the data that might be of
interest. Such problems can often overshadow the benefits of interacting over
large highly inter-connected data. The goal of data browsers has been, in part, to
tackle the problem of making sense of such rich and complex domains. A common
technique that has been adopted by a number of data browsers for exploring large
graphs of data is *pivoting* (otherwise known as *set-oriented browsing*). Pivoting
leverages the rich semantic descriptions within the data to extend the commonly
used *one-to-one* browsing paradigm to a *many-to-many* navigation for data.

L. Aroyo et al. (Eds.): ISWC 2011, Part I, LNCS 7031, pp. 553–568, 2011.

In this paper we focus on several commonly observed design patterns found in pivot-based data browsers: (1) exploration is often restricted to starting from a single point in the data, (2) navigation is typically supported in a single direction, and (3) immediate instance level exploration is regularly preferred without gaining familiarity with the domain or setting the exploration context first. We argue that these characteristics impose a number of limitations: in the case of (1) they reduce flexibility and therefore the ability to quickly find data that are related to the initial set multiple hops away, in the case of (2) they reduce the expressivity of the browser, and in the case of (3) the absence of an overview of the domain to be explored can often lead to difficulties in retracing exploration steps as well as make potential alternative exploration paths difficult to recognise. In this paper we introduce a novel approach we call *multi-pivot* which extends the traditional pivoting techniques to mitigate the aforementioned limitations. We approach we designed a demonstrator tool named *Visor* and carried out a user study to test the viability of our approach.

The outline of the paper is as follows. In the following section we examine related work in the area of pivot-based data browsers, and discuss these limitations in more detail. In Section 3 we discuss our approach, and lay out key design requirements in our approach. Section 4 describes Visor, a tool which we developed to test the multi-pivot technique. In Section 5 we carry out an evaluation study to test the viability of our approach and discusses the implications for design. The paper concludes with a summary of our work and planned future work.

2   Related Work

Pivoting, as an interaction method, has been adopted by a number of data browsers. Tabulator [3] can be considered an early pioneer of pivoting. Users browse data by starting from a single resource following links to other resources. Tabulator allows users to select patterns by selecting fields in the explored context and tabulate any results that are following the same pattern. Explorator [2] uses pivoting as a metaphor for querying, where users select subjects, objects and predicates to create sets of things, subsequently combining them with unions and intersections operations. The Humboldt browser [12] provides a list of instances and faceted filters from which the user can either choose to pivot or refocus. Parallax [10] shows the set of instances, accompanied by a list of facets for filtering and a list of connections showing the available properties that can be used in a pivoting operation. In VisiNav [7] users can drag and drop properties and instances in order to pivot and filter through results. A common characteristic of these interfaces is the notion that pivoting never occurs in branching i.e. a user cannot pivot with two different properties from the current focus and keep the context of both trails of exploration. Parallax, however, supports branching to some extent in the tabular view where generating a table allows this feature. gFacet [8] also mitigates the problem of branching; exploration starts from a set of instances and multiple properties can be selected to surface related sets of instances. The lists of instances, generated through successive pivoting operations, are used as facets and spatially arranged in a graph visualisation. Finding relationships between remote it typically unsupported by these browsers. RelFinder [9] allows finding relationships on the instance level, but not on the conceptual level. Fusion [1] offers discovery of relationships however the framework is designed more for programmers rather than non-technical end users.

2.1   Limitations of Pivot-Based Browsers

In the following we expand our discussion of the aforementioned limitations. To better illustrate these, we consider the following example: a user exploring the DBpedia [4] dataset to find basketball players, their affiliation to radio stations, the radio frequency of these stations, and the cities from which they operate. Figure 1a depicts the subset of the domain that is needed to answer the given query. Figure 1b, on the other hand, depicts the entire domain of the dataset[1] from which the subset domain needs to be surfaced in order to answer the given query in our scenario. We highlight the specific problems through describing a hypothetical exploration process provided by a typical pivot-based data browser.

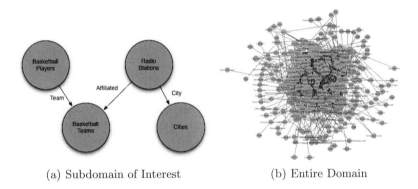

(a) Subdomain of Interest   (b) Entire Domain

Fig. 1. (a) A portion of the DBPedia ontology showing how instances of the classes Basketball Players, Basketball Teams, Radio Stations and Cites are connected. (b) A graph visualisation of the DBPedia ontology showing which concepts are related. Each node represents a class and each arc means that there is at least one property connecting instances between two classes.

Exploration starts for a single point. In a standard pivot-based browser, exploration of a dataset begins with a particular set of instances. The initial set of instances usually pertains to instances from a certain class that is typically found through a keyword search. In our example, we can start our exploration from either "Basketball players" or "Radio stations". Once the initial instances are shown, users are presented with a number of properties which can be selected to pivot and simultaneously get all the instances which are related to the instances in the first set through that property. In such a way navigation

---

[1] http://wiki.dbpedia.org/Ontology

through the graph is facilitated. Let us suppose that "Basketball players" or "Radio stations" are commonly used keywords for our query, which would surface the corresponding two classes of instances. At this point we can choose to start with the instances of either class as our initial set. Suppose we choose, for example, "Basketball players". As can be seen from Figure 1a we need to perform two pivoting operations to get to "Radio stations". Since the two are not directly linked we need to do a little bit exploring to find out the ways "Basketball players" are related to "Radio stations". Unfortunately, no cues are given to guide us in which direction to start exploring so we can connect them with "Radio stations". In a situation where the domain is unfamiliar this presents a problem. Property labels, which are used to show to what is being navigated, do not hold any information about the path two or three arcs way of the current set of instances. The problem is further exacerbated when a high number of possible choices for pivoting is present and the number of choices increases exponentially if the relating instances of interest are multiple arcs way.

Navigation is uni-directional. The direction of pivoting in pivot-based browsers is often uni-directional i.e. navigation is enabled only from outgoing links from the instances in the current focus. The restriction can sometimes limit the query expressivity of the interface. In our example (refer to Figure 1a) we notice that whatever set of instances we start from ("Basketball players" or "Radio Stations" alike) we cannot pivot in a single direction to all the sets of instances we need, since the direction of the links we require for pivoting in "Basketball teams" are all incoming links.

Exploration and domain overview absence. Current pivoting practices are predominantly instance-centric. As such overexposing instance data, filtering and repeated pivoting operations, can often times result in lack of overview about the sub-domain being explored as part of the exploration. The general lack of overview that can often lead to unseen relations in the data and therefore contribute to a lack of understanding about the domain being explored. Research from the HCI community suggest that when confronted with complex information spaces, information-seeking interfaces should follow the Visual Information Seeking Mantra [14]: overview first, zoom and filter, then details-on-demand. The complexity and size of large datasets suggests that using such a paradigm might suitable for data browsers.

3   Multi-pivot Approach

In the previous section we pointed out several challenges to standard pivoting as an interaction technique for exploring graph-based data. The aim of our research was to ascertain whether we can fashion an interaction model that mitigates these limitations and test if the solution we propose will introduce any major usability problems for end users. We integrated our ideas into a tool that followed four design requirements (R1 - R4):

R1. Exploration can be initiated by selecting multiple items of interest. Rather than being limited in starting from a single point, we wanted users to be able to start from multiple points of interest, and discover how the selected points of interest are connected with each other. As an analogy, we considered a puzzle solving example. When solving a puzzle, the solvers can start piecing the puzzle from multiple points: they can select several different pieces, find pieces that match, create several greater pieces and then piece these together to slowly gain an understanding of the overall picture. Similarly, we wanted users to grab different portions of the domain simultaneously, navigate either back or forth using either normal links or back-links, build their own subset of the domain related to their interest. Since there was no central point where the exploration starts and users would be able to pivot freely from anywhere in any direction we named this approach a *multi-pivot*.

R2. Overview first, instance data on demand. We didn't want to overburden users by immediately exposing instance data during exploration. Rather we wanted them to always have an overview of their exploration path and be able to quickly access the individual instance data if required.

R3. Allow navigation to be bi-directional. In addition to being able to start from multiple points, we wanted to support navigation in both directions. Additionally, we wanted to enable users to execute queries which paths include both forward and backward links.

R4. Creating custom spreadsheets as a way to query data. Once users have created and explored a sub-domain of the dataset they can query the sub-domain for instance data. Querying is done by assembling custom created spreadsheets from instance data by choosing among the concepts in the sub-domain and specifying the relationships between them. We choose spreadsheets as an representation, because its relative familiarity among non-technical data users. An additional motivator was that these spreadsheets can than be exported in a variety of formats which can be picked up and reused in different applications. For example, they can be published as an visualisation using ManyEyes [15] or published on the Web as a standalone dataset using an Exhibit [11].

4 User Interface

In order to test the approach outlined in Section 3 we developed a demonstrator tool called *Visor*[2]. Visor is a generic data explorer that can be configured on any SPARQL endpoint. For the purposes of testing and evaluating Visor we made an initial deployment on the DBPedia[3] SPARQL endpoint. The following sections describe the user experience in Visor. Throughout the section we refer to specific areas where the description of the UI meets the design characteristics outlined in Section 3.

---

[2] A demo version of Visor is available online at `http://visor.psi.enakting.org/`
[3] `http://dbpedia.org`

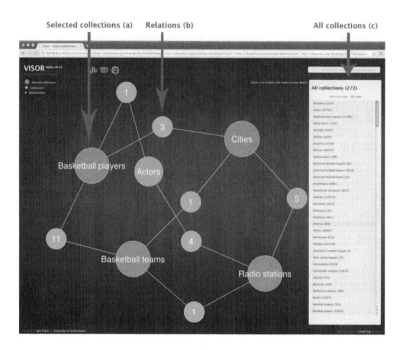

Selected collections (a)   Relations (b)   All collections (c)

Fig. 2. Generating a subset of the DBPedia ontology generated by selecting concepts in the ontology in Visor. Selected concepts (e.g. "Basketball Players") are coloured in green, while suggested concepts are coloured in gray. Arcs between two collection with a blue node in the middle indicates links between items (e.g. "Basketball players" have a direct relationships with "Basketball teams".

4.1 Data and Ontology Exploration

In Visor, exploration starts by selecting ontological classes of interest (named *collections* in Visor). Users can choose from collections either by viewing an entire list of all the known collections or browse in the hierarchical view of the collections. The collections are listed in a panel on the right hand side of the user interface (Figure 2c). Alternatively a search bar is provided where the user can execute a keyword search to get results to both individual instances and collections in the data.

Instead of choosing a single collection as a starting point for exploration, Visor allows users to select multiple collections simultaneously. The UI represents a canvas where a graph rendering of selected collections takes place. The graph rendering consists of the following nodes:

- **Selected collections.** Collections selected through the collections menu or searched are rendered with the title of the collection on top (Figure 2a).
- **Relations.** If there are properties linking the instances between two selected collections we indicate to the user that items from these collections are inter-related by displaying an arc with a blue node in the middle (Figure 2b).

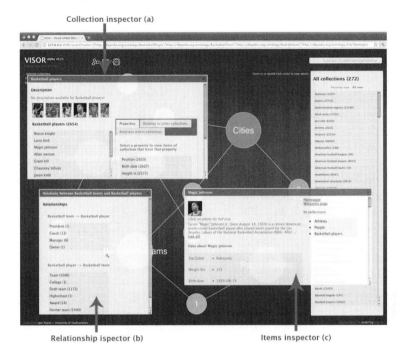

Collection inspector (a)

Relationship ispector (b)   Items inspector (c)

Fig. 3. Inspectors showing various information about the concepts in the ontology. Information about "Basketball Players" is shown in (a), and information about a particular basketball player (in this case Magic Johnson) is shown in (b). The various links that exist between "Basketball players" and "Basketball teams" is shown in (c).

The number in middle of the blue node is a indicator of the total number of properties (named *relations* in Visor) that link instances between the two collections in either direction. We adopted this approach to mitigate generating a large and incomprehensible graph [13]. This is true in cases where datasets include classes with a large number of links between them.

– **Intermediary collections.** In some cases there are no properties linking two collections. In such a circumstance, Visor tries to find the shortest path in the ontology by seeking an intermediate collection to which both selected collections can be linked from. If there is none, a path with two intermediary collections is looked up. The process is repeated until a path is found. Currently, Visor finds the first shortest path it can find and suggests it to the user by adding it to the current graph. While multiple shortest paths might exist, Visor recommends only the first one it finds. In cases where users want to find another path in the ontology they can simply select another collection, and the interface will attempt to link the last selected collection to all other selected collections. In such a way we ensure that whatever collections are selected the resulting sub-ontology is always connected and thus query-able. To distinguish selected and intermediary collections the latter are coloured in grey and are smaller in size.

The graph representation is rendered using a force directed layout and can be zoomed and dragged to improve visibility. Each node can be double clicked which opens up different inspector windows. These allow to view details about the sub-domain. In the following we describe the different kind of inspectors in Visor.

Collection inspector. Double clicking on the collections brings up a *collection inspector* (Figure 3a). The inspector shows the individual instances which are part of the collection, a description of the collection if one is available, and a list showing the possible properties that items from that collection can have. In Visor, object and datatype properties are listed separately. Object properties (or *relations*) are shown together with a corresponding collection to which they link (Figure 4). Furthermore object properties linking to and from other collection are show in separate lists (Figure 4). Users can than add these classes to the canvas. In such a way

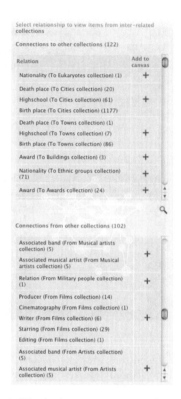

Fig. 4. Displaying properties in the collection ispector that link from or to items of a collection

we support bi-directional set-oriented navigation, however in Visor we do so on the ontology level which serves as a potential roadmap for querying (Design requirement R3). Users can also view filters of items in the inspector by selecting any property (object or datatype). This shows the instances that only have that property and show the corresponding property value.

Item inspector. Clicking on any of the items in the collection inspector opens up an *item inspector* where all the data pertaining to the individual instance is shown (Figure 3c). A predefined lens is used to render the individual resource, including rendering images if any exist, a description of the item and the data associated with that item. If geographic coordinates are found for the item for example, a map is presented to the user. Additionally we show the collections that include the item. In the data panel, links to other resources opens the item inspector associated with that item. In such a way browsing from one item to another item is also supported in Visor.

Relations inspector. The relation nodes (the blue nodes in the visualisation) can also be inspected to quickly access properties that interlink items from two collections (Figure 3b). Clicking on any of the relations will display the items from both collections that are linked with that property.

With selecting collections users can create a subset of the ontology which is composed of concepts of their interest without restricting them to selecting a single collection and use navigation (Design requirement R1). With the inspector windows users can surface up the data on demand (Design requirement R2) to explore how collections are related, what are their individual instances, and if required inspect the instances themselves.

4.2 Spreadsheet Creation

Once a subset of the ontology is selected the user can query this information space by creating custom spreadsheets based on the selecting concepts and relations from the ontology subset (Design requirement R4). After the spreadsheet has been created users can export their custom made data collection in a format of their particular liking. Currently we support exporting data in CSV and JSON formats, however the system is extendable and multiple formats may be supported. In the following we describe the query interface and procedure for creating custom spreadsheets.

Main collection. The "Create a table" button located in the top menu of the UI opens up a query interface which guides users in selecting things from the previously explored domain (Figure 5).

The first step in creating a spreadsheet is selecting the *main collection* i.e. the collection that will be the focus of the spreadsheet (Figure 5a). This will instantiate a spreadsheet with a single column (the *main* column) composed of the instances from the main collection. All subsequent columns added to the table will be facets of the first column each created by specifying a path showing how the items of the newly created column are related to the items in the first column.

Adding columns. Once the main collection is selected, adding additional columns is the next step. The first choice of columns are the datatype properties of the main collection shown in Figure 5b. Users can select a property and click on the "Add column" button to add the column to the table. By default when a column is added, Visor queries and tries to find a corresponding value for all the instances in the main collection. If such a value does not exist a "No value" cell informs users that the item in the main column does not have that property. The default option corresponds to generating a SPARQL query with an OPTIONAL statement. To filter for non-empty values users can check the "Show only" option before adding the column. In such a way user have the flexibility of selecting which columns are optional and which are mandatory to have a value in each cell. Additionally users can also choose the "Count" option to count the values in a cell in the corresponding to the item in the main column. Similarly, selecting the "Count" option corresponds to having a COUNT query in the SPARQL query.

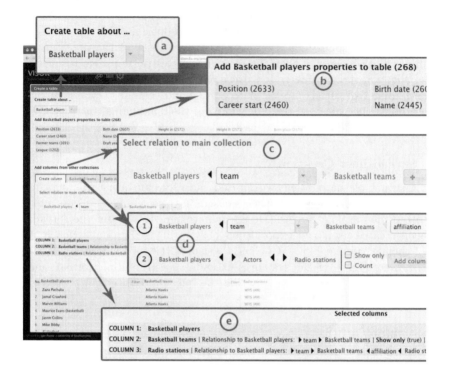

Fig. 5. Spreadsheet creating/query interface in Visor. Users start from selecting the main collection (a), datatype properties (b), columns other collections in the sub-ontology by specifying relations to the main collection (c) and (d). A preview of the columns in shown in (e).

Defining column paths. Users can also select to add columns based on other collection in the current sub-domain. The query interface allows users to specify a path that connects items from the main collection to items in the newly added column. This can be implemented in two ways:

1. The first way is by using a path creation tool (Figure 5c). The path creation tool starts a path with the first element in the path being the main collection. Users can than select a collection that is related to the main collection using a drop-down list of available choices (Figure 6). Once a relating collection is selected, a property that links them is selected again from a selection of choices in a drop-down list. Then another collection can be chained to the previous one and again a property between them is specified and so on. When a column is added based on the specified path the column pertains to instances from the last collection in the path. We note that in the path creation tool enables users to connect the collections by properties going in both direction (the left and right arrows shown in Figure 5c,d).

2. To help speed up the process an alternative way of adding columns is supported in the UI. Each tab relates to a concept in the sub-ontology. Each

Fig. 6. Specifying a path from main collection to an arbitrary collection in Visor

tab panel contains suggested paths for reaching that node (Figure 5d). It basically list all the paths from the main collection to the collection specified in the tab. Then users need only to specify the properties in-between the collections. This will save users time, as well as give cues into all the different ways items from two collections can be related.

At any time users can update the current spreadsheet to monitor their progress. An overview of selected columns is shown to the user (Figure 5e) which allows to backtrack on choices made as well as rearrange the ordering. The spreadsheet also supports filtering for specific values in a column. Once users are satisfied with their custom spreadsheet they can choose to export it in a number of different formats.

4.3 Implementation

The implementation in Visor is composed of a front end (UI) and back-end system. The UI is based on HTML5 and Javascript together with the jQuery library[4]. For visualising the ontology we relied on the Protovis visualisation toolkit [5]. The Visor back-end server is a Python/Django application that serves data in a JSON format to the front-end by exposing a RESTful interface. Thus the UI side of the application does not rely on any raw SPARQL query generation or parsing SPARQL results.

5 User Study

In order to ascertain whether people will be able to learn and use Visor we conducted a user study. The purpose of our study was two fold: (1) we wanted to test if there was any major issues in the ability of users to comprehend and use our UI and (2) identify specific usability problems and areas where interaction can be improved. Thus, our goal was to test if our approach was viable.

5.1 Study Design and Procedure

For our study we recruited ten participants through an email advertisement among graduate students at the University of Southampton. Seven of the participants were male and three female and their ages ranged between 21-41. We wanted to have a diverse group of users with respect to knowledge of Semantic

---

[4] http://jquery.com/

Web/Linked Data technologies and see if there were any particular difficulties among users with different skill levels. We asked them to rate their knowledge of Semantic Web/Linked Data technologies on a scale of one to three, one being *"very basic understanding or no knowledge"*, two being *"some knowledge and understanding"*, and three being *"high or expert knowledge"*. To gain further insight in their skills, we also asked them to rate their knowledge on the same scale to several specific areas: Linked Data application development, (2) Use of SPARQL, and (3) Ontology Engineering and/or data authoring. Calculating the averages of the responses by participants 50.25% or roughly half or the participants had no or very little understanding of Semantic/Linked data technologies, 30% or about 3 participants had intermediary knowledge and 17.5% or about 2 participants had expert knowledge.

For our study we relied on a cooperative protocol analysis or "think aloud" method [6]. We choose this method because we wanted to pinpoint any potential usability issues introduced by the design requirements R1-R4 and get the users insight into what were the problems.

Each participant went through a study session that took approximately one hour to complete. A session was structured in the following way. First, the participant was shown a 6 minute video[5] tutorial of Visor. The tutorial explained the terminology of the UI and showed a complete example worked out in Visor. Second, the participants were handed 3 written tasks to complete. During this time the "think aloud" protocol was observed, and we recorded the users screen and audio. Finally, participants were required to fill in a questionnaire, in order to reflect and give feedback based on all the entire session with questions targeting specific portions of the UI.

Two of the tasks were structured tasks i.e. the users we given a concrete task with a clear result. The tasks were given with increasing difficulty: the first task required a three column table with generated with one-hop links in a single direction, while the second required more column, specifying a loop pattern, and setting paths with bi-directional patterns. One example task was the following:

> In a history course you are required to find which royals from which countries have intermarried. You decide that you will need a table showing: All the royals you can find, which country these royals were born, the royals who are spouses of the royals, and the country the spouses were born.

The third task was unstructured i.e. we gave users a general area to browse and explore and come up with some data of their particular interest. The task was to find some data pertaining to *Scientists*, *Universities*, and *Awards*.

5.2   Results

During the task we focused primarily on three things: (1) Observing user behaviour during data finding tasks, (2) observing when users chose to view the

---

[5] The video is available at: http://vimeo.com/24174055

actual instance data and for what reasons, and (3) observing what problems participants experienced when attempting to create their spreadsheets.

Data finding. When searching for collections to build up their sub-domain for answering their tasks most participants (nine out of ten) choose to use multiple collection selection rather than use navigation after selecting their first collection. Only when the the resulting connections contained intermediary nodes that did not meet the requirements of the sub-domain did they resort to navigating to other potentially useful collections. This was particularly the case during the exploratory task. Most users used a keyword search option to search for collections. Beyond using it for finding collections participants suggested additional ways search can be useful for finding additional data. For example, one user commented that the use of synonyms would be helpful to find collections. Another user, for example, wanted to search for a particular instance during the exploratory task because the user wasn't sure in which collection that particular item can be found. Beyond searching for instances and collection we observed that some users tried looking up things that we currently did not support e.g. searching for relations. For example, one user thought that there might be a collection named "Spouses" before realising that it might be a relation instead.

Showing instance data. We also observed users to see how much would they need to view the underlying instance data during various stages and across different tasks. We concentrated our observations on two things. One was to observe if users needed to examine the underlying data during the initial exploration phase when the user was building a sub-domain of the data. Second we wanted to test if they can specify relationships that were three or more hops away without seeing the intermediary data that relates them. During the structured tasks we found that users did spend very little time or no time exploring the generated sub-domains with the inspector tools. Six participants chose to directly open the spreadsheet creation tool mentioning that they felt confident they had everything they needed to answer the query. Three others noted that they just wanted to open up the inspectors to explore and but mentioned no particular reason except for just casual exploration. While we observed a slight increase during the exploratory tasks we did observe that the spreadsheet creation tool was used as an exploratory tool as well. When creating their spreadsheet more than half of the participants chose to view their progression with each added column. At the start of the sessions, novice and intermediary users reported difficulties in grasping how paths worked, but once explained they felt confident in generating paths two or more hops away without viewing the intermediary data.

Spreadsheet creation. As expected, users found the spreadsheet creation tool was the most difficult part of the interface the user to learn and use. Some suggested better integration with the visualisation by either selecting or being able to drag and drop directly from the graph into the header row of the spreadsheet. While for the two expert participants and one non-expert specifying direction of the relationship made sense, for most other participants, specifying the direction of the relationship seamed irrelevant. When asked what they would prefer, most of them responded that they would like a single list of how to relate two collections instead of two separate lists. However we did not observe that users had any difficulties in specifying paths where bi-directional patterns occurred. When faced with the choice of using the template paths or create the paths manually, the preference of participants were split among the two choices. Participants recommended, however, that rather than specifying a new path every time they would have been able to reuse paths from existing columns which were sub-paths of the new path.

Task completion and survey. Task completion was generally high: eight out of ten participants were able to complete the all the tasks and create spreadsheets to the specified requirements of each task. Overall we found that the users were able to easily learn and create their spreadsheets after the one hour session. After going through the tasks, participants were asked to submit a survey rate the overall difficulty of using the tool on a Likert scale of one to five. Two participants reported that the found the tool very easy (1) to use, six reported it easy (2), one user reported it average (3) and one user reported it difficult (4) to use. When asked to rate specific components of the UI most participants (8 of 10) reported that the graph visualisation useful and easy to use while they gave the spreadsheet creation tool an average (3.2) score.

5.3   Implications for Design

Based on the results we compiled a set of recommendations which we encourage future designers to consider when designing data-centred exploration interfaces.

Integrate keyword search with direct manipulation techniques. The approach in this paper gave more flexibility than standard browsers by enabling users to select multiple points of interest. We noticed that users not only took advantage of this flexibility, they even wanted more freedom when trying to find the portion of the data domain that is of their interest. We suggest that rather than being able to just add multiple collections, users should be able to search more freely for thing such as properties, instance data and view how they are relevant in the already explored data. Therefore we suggest integrating keyword search techniques with direct manipulation techniques as a possible way of providing flexibility in finding data during initial exploration. So far direct manipulation techniques have been mostly focused on supporting only direct manipulation techniques for navigating graph data, while interfaces supporting keyword search have focused on entity retrieval or question answering. We encourage future designers of data browsers to consider closely integrating keyword search with the direct manipulation features of the data browser.

Support bidirectional navigation. In our related work section we noted that a lot of interfaces support navigation in a single direction i.e. only form outgoing properties which limits the expressivity of the queries that can be answered by the data browsing UI. While we realise that from an implementation point supporting finding back-links on the open Web of Data is much harder task than when the data is contained in a single store, our study showed that from an interaction point of view supporting both does not have any significant impact on users when browsing.

Show data on demand. One of the thing our study showed was that users can browse and query for data without relying too much on always viewing the data. We recommend that when future UI designers of data browser develop their tools they use less screen real estate on showing too much instance data at once, or at least give several views to users. Retaining context while exploring or combining querying with visual aids can be utilised to give overview of the exploration path and make querying easier.Users should, however, always retain the option of viewing the instance data of current result of the exploration at any time.

6   Conclusion

In this paper, we examined some of the interaction challenges associated with pivoting as a exploration technique for data browsers. We also presented Visor, a tool for users to create custom spreadsheets by exploring a dataset using a combination of multiple selections as well as link navigation. With Visor we have shown a flexible way of finding data in large graphs and presented an overview-first data-on-demand approach to browsing data.

Our future work includes extending Visor based on the recommendations we laid out in this paper. We plan on adding additional features to further increase the flexibility for understanding complex data domains. Our plan is to support multiple potentially complementary ways of finding data including better integration with keyword search as well as more visual aids. We then intend to extend our evaluation with comparing and measuring support of different data-seeking tasks in UIs supporting different navigation and exploration techniques to better understand for what kind of task what exploration model works best.

Acknowledgments. This work was supported by the EnAKTing project, funded by EPSRC project number EI/G008493/1. Many thanks to Manuel Salvadores and Christos Koumenides for providing valuable feedback on this work.

References

1. Araujo, S., Houben, G.-J., Schwabe, D., Hidders, J.: Fusion – Visually Exploring and Eliciting Relationships in Linked Data. In: Patel-Schneider, P.F., Pan, Y., Hitzler, P., Mika, P., Zhang, L., Pan, J.Z., Horrocks, I., Glimm, B. (eds.) ISWC 2010, Part I. LNCS, vol. 6496, pp. 1–15. Springer, Heidelberg (2010)
2. Araujo, S., Schwabe, D., Barbosa, S.: Experimenting with explorator: a direct manipulation generic rdf browser and querying tool. In: VISSW 2009 (February 2009)

3. Berners-lee, T., Chen, Y., Chilton, L., Connolly, D., Dhanaraj, R., Hollenbach, J., Lerer, A., Sheets, D.: Tabulator: Exploring and analyzing linked data on the semantic web. In: Procedings of the 3rd International Semantic Web User Interaction Workshop (SWUI 2006), p. 06 (2006)
4. Bizer, C., Lehmann, J., Kobilarov, G., Auer, S., Becker, C., Cyganiak, R., Hellmann, S.: Dbpedia - a crystallization point for the web of data. Web Semantics: Science, Services and Agents on the World Wide Web 7(3), 154–165 (2009); The Web of Data
5. Bostock, M., Heer, J.: Protovis: A graphical toolkit for visualization. IEEE Transactions on Visualization and Computer Graphics 15(6), 1121–1128 (2009)
6. Dix, A., Finlay, J., Abowd, G.D., Beale, R.: Human Computer Interaction, 3rd edn. Pearson, Harlow (2003)
7. Harth, A.: VisiNav: Visual Web Data Search and Navigation. In: Bhowmick, S.S., Küng, J., Wagner, R. (eds.) DEXA 2009. LNCS, vol. 5690, pp. 214–228. Springer, Heidelberg (2009)
8. Heim, P., Ertl, T., Ziegler, J.: Facet Graphs: Complex Semantic Querying Made Easy. In: Aroyo, L., Antoniou, G., Hyvönen, E., ten Teije, A., Stuckenschmidt, H., Cabral, L., Tudorache, T. (eds.) ESWC 2010. LNCS, vol. 6088, pp. 288–302. Springer, Heidelberg (2010)
9. Heim, P., Hellmann, S., Lehmann, J., Lohmann, S., Stegemann, T.: RelFinder: Revealing Relationships in RDF Knowledge Bases. In: Chua, T.-S., Kompatsiaris, Y., Mérialdo, B., Haas, W., Thallinger, G., Bailer, W. (eds.) SAMT 2009. LNCS, vol. 5887, pp. 182–187. Springer, Heidelberg (2009)
10. Huynh, D.F., Karger, D.R.: Parallax and companion: Set-based browsing for the data web. In: WWW Conference. ACM (2009)
11. Huynh, D.F., Karger, D.R., Miller, R.C.: Exhibit: lightweight structured data publishing. In: Proceedings of the 16th International Conference on World Wide Web, WWW 2007, pp. 737–746. ACM, New York (2007)
12. Kobilarov, G., Dickinson, I.: Humboldt: Exploring linked data (2008)
13. Schraefel, m.c., Karger, D.: The pathetic fallacy of rdf. In: International Workshop on the Semantic Web and User Interaction, SWUI 2006 (2006)
14. Shneiderman, B.: The eyes have it: a task by data type taxonomy for information visualizations. In: Proceedings of IEEE Symposium on Visual Languages, pp. 336–343 (September 1996)
15. Viegas, F.B., Wattenberg, M., van Ham, F., Kriss, J., McKeon, M.: Manyeyes: a site for visualization at internet scale. IEEE Transactions on Visualization and Computer Graphics 13, 1121–1128 (2007)

strukt—A Pattern System for Integrating Individual and Organizational Knowledge Work

Ansgar Scherp, Daniel Eißing, and Steffen Staab

University of Koblenz-Landau, Germany
{scherp,eissing,staab}@uni-koblenz.de

Abstract. Expert-driven business process management is an established means for improving efficiency of organizational knowledge work. Implicit procedural knowledge in the organization is made explicit by defining processes. This approach is not applicable to individual knowledge work due to its high complexity and variability. However, without explicitly described processes there is no analysis and efficient communication of best practices of individual knowledge work within the organization. In addition, the activities of the individual knowledge work cannot be synchronized with the activities in the organizational knowledge work.

Solution to this problem is the semantic integration of individual knowledge work and organizational knowledge work by means of the pattern-based core ontology *strukt*. The ontology allows for defining and managing the dynamic tasks of individual knowledge work in a formal way and to synchronize them with organizational business processes. Using the strukt ontology, we have implemented a prototype application for knowledge workers and have evaluated it at the use case of an architectural office conducting construction projects.

1 Introduction

There is an increasing interest in investigating means for improving quality and efficiency of knowledge work [6]. An established means for improving efficiency of organizational knowledge work is expert-driven business process management [1]. The implicit procedural knowledge found within the organization is made explicit by defining and orchestrating corresponding business processes (cf. [15]). By this, procedural knowledge of the organization is explicitly captured and made accessible for analysis, planning, and optimization. Important supplement to organizational knowledge work is individual knowledge work. It is present in domains in which acquiring and applying new knowledge plays a central role such as research, finance, and design [14]. Due to its complexity and variability, individual knowledge work is typically not amenable to planning. In addition, activities of individual knowledge work that occur only rarely do not justify the effort of business process modeling. Nevertheless, it seems to be worthwhile to consider individual knowledge work from the perspective of business process optimization. Even if the activities of individual knowledge work are not entirely accessible to planning, they are often embedded in organizational business processes defining,

L. Aroyo et al. (Eds.): ISWC 2011, Part I, LNCS 7031, pp. 569–584, 2011.

e.g., some constraints on the activities, deadlines, communication partners, and others [20]. Individual knowledge work often contains some sub-activities that actually provide a fixed structure and thus can be explicitly planned, e.g., to obtain approval for some activities in a large construction project [20]. These activities of individual knowledge work need to be synchronized with the organizational knowledge work. However, today's models for business processes and weakly structured workflows do not allow for representing such an integration of the activities.

Solution to this problem is the semantic integration of individual knowledge work and organizational knowledge work based on the pattern-based core ontology *strukt*[1]. The *strukt* ontology allows for modeling weakly structured workflows of individual knowledge work in a formal and precise way. It allows for decomposing the tasks of the individual knowledge work into sub-tasks, which again can be structured along a specific order of execution and dependencies between the tasks. The tasks can be semantically connected with any kinds of documents, information, and tools of a particular domain. In addition, the strukt ontology provides for modeling structured workflows of organizational knowledge work and combing the weakly structured workflows with the structured ones. The ontology is used in the strukt application that allows knowledge workers to collaboratively create, modify, and execute the dynamic tasks of individual knowledge work and to synchronize them with organizational business processes.

The need for integrating individual and organizational knowledge work is motivated by a scenario of an architectural office in Section 2. Based on the scenario, the requirements on the strukt ontology are derived in Section 3. In Section 4, existing models and languages for business process modeling and weakly structured workflows are compared to the requirements. The pattern-based design of the core ontology strukt is described in Section 5. An example application of the ontology design patterns defined in strukt is provided in Section 6. The strukt application for collaboratively executing tasks of individual knowledge work and synchronizing it with business processes is presented in Section 7, before we conclude the paper.

2   Scenario

The scenario is based on a real architectural office. The work in the architectural office is highly knowledge oriented as the acquisition and application of knowledge plays a crucial role in planning and conducting construction projects. One finds some organizational business processes in the architectural office that are repeated with each project. Figure 1 depicts an excerpt of typical steps in the process of planing an apartment construction in Business Process Modeling Notation (BPMN) [17]. Subsequent to the activity Initiate construction project (a) are the activities Prepare building application (b) and File building application (c). The activities are strictly separated from each other and executed in a determined, sequential order. The resource (d) defines the input and output

---

[1] strukt comes from the German word Struktur and means structure in English.

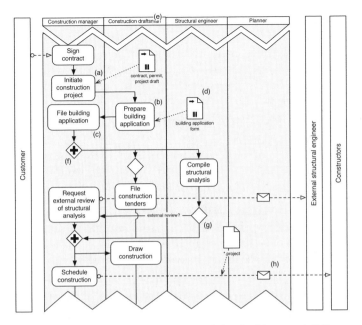

Fig. 1. Example Business Process of the Architectural Office

documents of an activity. In the case of the activity Prepare building application these documents are, e.g., the building application form and all required
attachments. The activity Prepare building application is associated with the role
Construction draftsman (e), whereas the other activities are conducted by roles
like Construction manager, Structural engineer, or Planner. Branches are used to
represent parallel activities (f) and conditions (g). Besides the processes within
the company also the communication with external project partners is explicitly
captured (h).

The organizational knowledge work is already well described on the level of
business processes. However, the core area of the architectural office's activities
is insufficiently captured. For example, activities such as Prepare building application or Draw construction consist of a large number of sub-activities and are
usually collaboratively executed by multiple persons. These activities of individual knowledge work are characterized by high complexity and variability when
executing the tasks. As an example, we consider the business process Prepare
building application of Figure 1 in more detail: For preparing a building application one has to fill a corresponding application form. This form requires some
attachments such as ground plan, site plan, and others that are used to prepare
the administrative permit for the construction project. Depending on the type
of building construction, however, different attachments are needed. In addition,
the construction projects may have specific requirements to be considered like
terrestrial heat, timber construction, accessibility, and others. In some cases a
complete structural engineering calculation has to be conducted at application
time whereas this is not required in other cases.

3   Requirements to strukt Ontology

We have derived the requirements to the strukt ontology from the scenario in Section 2 as well as from related work in information systems research such as [12,11,20,15]. We briefly discuss each requirement and provide a reference number *REQ-<number>*.

Weakly Structured Workflows (*REQ-1*): Individual knowledge work is characterized by a high complexity and variability [12]. Resources and activities for conducting tasks are often not known a priori (see Section 2). A support for representing weakly structured workflows is needed that can be adapted during execution time without violating the consistency of other running processes.

Support for Structured Workflows (*REQ-2*): Despite the high flexibility of indidivual knowledge work, there are also some organizational requirements and framework directives that need to be strictly followed (see scenario in Section 2). Thus, support is needed to represent structured workflows in the sense of traditional business process management [11].

Integrating Weakly Structured and Structured Workflows (*REQ-3*): Within an organization there is typically a need to represent both weakly structured workflows and structured workflows (see Section 2). Today's models and systems, however, lack in formally integrating weakly structured workflows and structured workflows and thus cannot benefit from this integration. In order to leverage the strength of both weakly structured and structured workflows, an appropriate model must be able to formally integrate and synchronize them into a common workflow.

Workflow Models and Instances (*REQ-4*): Distinguishing workflow models and workflow instances is a common feature of traditional business process models [20]. In individual knowledge work, however, such a distinction is often not made as the individual knowledge work is high in complexity and variability. However, also from the execution of weakly structured workflows one can learn some generic procedural knowledge. Thus, also for weakly structured workflows the distinction between instance and model should be made. In addition, it shall be possible to modify a workflow instance without affecting its workflow model or other workflow instances. In addition, it shall be possible to derive workflow models from executed workflow instances.

Descriptive Workflow Information (*REQ-5*): Structured workflows and weakly structured workflows are characterized by the resources involved. A core ontology for integrating individual and organizational knowledge work should therefore support describing the necessary information for the workflow execution, like resources used, processed, or created (which is a central aspect in particular for individual knowledge work [15]), the tools applied, the status of the workflow execution, as well as scheduling information (cf. Section 2).

4   Comparing Models for Knowledge Work

We analyze and evaluate existing models for structured workflows and weakly structured workflows with respect to the requirements introduced in Section 3. The traditional business process models like BPMN [17] and extended Event-driven Process Chain (EPC) [18] are available as semantic models in form of the sBPMN [13] and sEPC [13] ontologies. However, they still lack support for representing weakly structured workflows and thus are less applicable to our problem. Also OWL-S [22] shares these characteristics of traditional business process models. Ad-hoc and weakly structured models like the Process Meta-Model (PMM) [2] and the Task-Concept-Ontology (TCO) [19] do not require a strictly determined process flow like the traditional business process models and may be automatically extracted from natural language descriptions [10]. Such models are suitable to represent individual knowledge work. However, the lack of formal precision and missing integration with traditional business processes hinder their reuse.

The DOLCE+DnS Plan Ontology (DDPO) [8] provides a rich axiomatization and formal precision. It obtains its high level of formal precision from the foundational ontology DOLCE [3] and specializes the ontology design pattern Descriptions and Situations (DnS). The central concepts defined in the DDPO are Plan, Goal, Task, and PlanExecution [8]. A Plan is a description of at least one Task and one agentive role participating in the task. In addition, a Plan has at least one Goal as a part. A Goal is a desire that shall be achieved. Tasks are activities within plans. They are used to organize the order of courses. Finally, PlanExecutions are actual executions of a plan, i.e., they are real-world situations that satisfy a Plan. It is in principle possible to represent both traditional workflows as well as weakly structured workflows using the DDPO. However, DDPO does not distinguish structured and weakly structured workflows (*REQ-3*) and does not support descriptive workflow information (*REQ-5*). *REQ-4* is present in DDPO but not explicitly specified. Nevertheless, due to its high formality and using the foundational ontology DUL as basis, the DDPO is well suited for extensions and serves as basis for our work.

In conclusion, one can say that none of the existing models fulfill all requirements stated to strukt. Traditional business process models miss representing weakly structured workflows of individual knowledge work. On contrary, weakly structured workflows are in principle enabled to represent the activities of individual knowledge work. However, they lack the formal precision required and do not allow for an integration with traditional business process models. The DDPO model differs from the other models insofar as it in principle allows for modeling both organizational business processes and activities of individual knowledge work. In addition, it enables integration with other systems due to its formal nature. Thus, it is used as basis in our work and will be adapted and extended towards the requirements stated in Section 3.

5   Pattern-Based Core Ontology strukt

The foundational ontology DOLCE+DnS Ultralight (DUL) [3] serves as basis for the core ontology strukt. Foundational ontologies like DUL provide a highly axiomatized representation of the very basic and general concepts and relations that make up the world [16]. As such, foundational ontologies are applicable to a wide variety of different fields. Foundational ontologies like DUL follow a pattern-oriented design. Ontology design patterns [9] are similar to design patterns in software engineering [7]. Adapted from software engineering, an ontology design pattern provides (i) a description of a specific, recurring modeling problem that appears in a specific modeling context and (ii) presents a proven, generic solution to it [4,7]. The solution consists of a description of the required concepts, their relationships and responsibilities, and the possible collaboration between these concepts [4]. An ontology design pattern is independent of a concrete application domain [7] and can be used in a variety of different application contexts.

In the following, we briefly introduce the patterns of DUL that are of particular interest in this work:[2] The Descriptions and Situations Pattern provides a formal specification of context [16]. The Description concept formalizes the description of a context by using roles, parameters, and other concepts. The Situation represents an observable excerpt of the real world that satisfies the Description. By using the Descriptions and Situations Pattern, different views onto the same entities can be formally described. The patterns of the core ontology strukt are based on the Descriptions and Situations Pattern. This means that they reuse or specialize concepts or relations defined in the pattern. The foundational ontology DOLCE+DnS Ultralight provides a specialization of the DDPO [8] (see Section 4) for planning activities, called the Workflow Pattern. Central entity of the Workflow Pattern is the Workflow concept that formalizes the planning of processes. The Workflow concept is specialized from DDPO's Plan, which itself is derived from Description of the Descriptions and Situations Pattern. The WorkflowExecution concept represents the concrete execution of a workflow instance. It is derived from DDPO's PlanExecution, which is a specialization of Situation. The Task Execution Pattern formalizes the processing of tasks in activities. The Role Task Pattern enables association of roles to tasks. The Part-of Pattern represents the (de-)composition of entities into wholes and parts [21]. The Sequence Pattern describes the order of entities through the relations precedes, follows, directlyPrecedes, and directlyFollows.

A core ontology refines a foundational ontology towards a particular field by adding detailed concepts and relations [16]. However, core ontologies are still applicable in a large variety of different domains. The core ontology strukt reuses and specializes different ontology design patterns that DUL offers. Central patterns of the core ontology strukt are the Weakly Structured Workflow Pattern (*REQ-1*), the Structured Workflow Pattern in combination with the Transition Pattern (*REQ-2*), the Workflow Integration Pattern to integrate weakly structured workflows and structured workflows (*REQ-3*), and the Workflow Model

---

[2] For a detailed description we refer to http://ontologydesignpatterns.org/

Pattern for differentiating workflow models and workflow instances (*REQ-4*). Weakly structured workflows and structured workflows can be further described by applying strukt's Condition Pattern, Resource Pattern, Status Pattern, and Scheduling Pattern (*REQ-5*). Each pattern of the core ontology strukt solves a specific modeling problem that distinguishes it from the other patterns. However, strukt is not just a collection of some otherwise independent ontology design patterns. Rather, the set of ontology design patterns strukt defines relate to each other and are designed to be applied together. Such a set of related patterns is called a pattern system [4]. The core ontology strukt can be applied in various domains that need to represent knowledge work and workflows, respectively. Finally, strukt can be extended by domain ontologies such as an architectural ontology or financial administration ontology. In the following, we describe the patterns of the strukt ontology.

5.1   Weakly Structured Workflow Pattern

The Weakly Structured Workflow Pattern depicted in Figure 2 refines the generic Workflow Pattern of DUL. The concept WeaklyStructuredWorkflow specializes the Workflow concept of DUL's Workflow Pattern. Using the defines relation, different Roles and Tasks are defined. Roles abstract from the characteristics, skills, or procedures relevant for the execution of a specific task and allows for differentiating Agents and Objects participating in activities (see Role Task Pattern of DUL). The classifies relation determines the Role of an Object in the context of a specific workflow. The concept Agent is a specialization of the Object concept and describes the entity acting such as a person. Objects and Agents are defined as participants of an Action by using the hasParticipant relation. Tasks are used to sequence activities [8]. They structure a workflow into different sub-tasks relevant for the workflow execution and can be hierarchically ordered (see Task Execution Pattern of DUL). Tasks are associated to Actions using the relation isExecutedIn. Action is a specialization of DUL's Event and describes the actual processing of a task. Tasks can be ordered using the precedes relation. The order of tasks may be underspecified, i.e., the actual sequence of processing may only be determined on a short-term basis and day-to-day requirements when executing the workflow. Thus, a strict order of processing the tasks is not enforced and the order may even change during execution time.

In knowledge-intensive activities, it may not be possible to define *a priori* all details of a complex task. Thus, the Weakly Structured Workflow Pattern allows for defining additional (sub-)tasks during the execution of the workflow using the hasPart relation. A Task is associated with a Role using the isTaskOf relation (see Part-of Pattern of DUL). Also Actions can be decomposed using the relation hasPart. Typically, the decomposition of an Action is bound with the decomposition of the corresponding Task.

The goal that is to be reached by executing a workflow is represented using the Goal concept and associated to WeaklyStructuredWorkflow using the hasComponent relation. It can be further decomposed into sub-goals using the hasPart relation. Goals are explicitly associated to corresponding sub-tasks

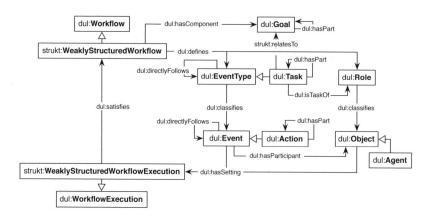

Fig. 2. Weakly Structured Workflow Pattern

using the relatesTo relation. The Goal concept is central to the weakly structured workflow pattern and is used by the Workflow Integration Pattern described in Section 5.4 to link the Weakly Structured Workflow Pattern with the Structured Workflow Pattern.

5.2   Structured Workflow Pattern and Transition Pattern

The Structured Workflow Pattern provides a formal specification of traditional business processes (see Section 4). It is applied in combination with the Transition Pattern that defines the transitions between processes, i.e., the Events. Thus, the Structured Workflow Pattern is an abstraction from the concepts of traditional business process models.

Figure 3 depicts the Structured Workflow Pattern. It specifies the concepts StructuredWorkflow and StructuredWorkflowExecution as specialization of DUL's Workflow and WorkflowExecution concepts. This eases the integration with the Weakly Structured Workflow Pattern that specializes the same concepts and thereby supports the integration of individual and organizational knowledge work. The distinction between StructuredWorkflow and StructuredWorkflowExecution reflects the two phases of traditional business process management, namely the definition phase and execution phase [23]. In the definition phase, existing business processes are captured and orchestrated into a (semi-)formal business process model. In the execution phase, the previously created process model is implemented.

Using the defines relation, the StructuredWorkflow specifies the Roles, Tasks, EventTypes, and TransitionTypes of the workflow as in the definition phase. Roles determine the roles played by Objects such as Agents participating in processes of the workflow. The roles are associated with some concrete Tasks using the isTaskOf relation. The TransitionType is part of the Transition Pattern and allows for formally defining the transition between two concepts, which classify processes represented by DUL's Event concept. The concepts TransitionAction, Event, and Object constitute the entities of the workflow execution phase. Like

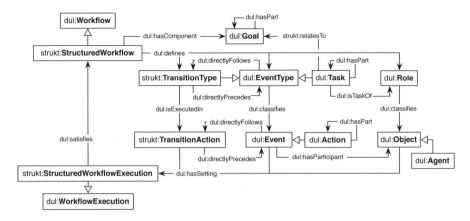

Fig. 3. Structured Workflow Pattern

the WeaklyStructuredWorkflow concept, also the StructuredWorkflow concept defines a Goal concept, which captures the goal of the workflow.

The transitions between business processes are defined using the Transition Pattern. It provides the four basic transition types [18,17,5] *sequence, condition, fork,* and *join,* defined as specializations of the generic Transition Pattern. The Sequence Transition Pattern specifies a strict sequence of process execution as depicted in Figure 4(a). The corresponding operator in BPMN is shown in Figure 4(b). The Sequence Transition Pattern defines a SequenceTransitionType as specialization of the generic TransitionType. It determines a strict sequential order of execution of two EventTypes. Thus, the SequenceTransitionAction connects exactly two concrete business processes represented as Events. The Condition-based Transition Pattern models process executions that are bound to some process conditions. The Fork Transition Pattern is used to model fork/join transitions.

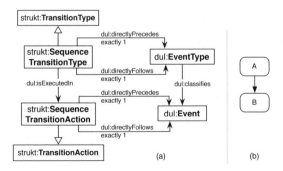

Fig. 4. Sequence Transition Pattern

5.3   Condition Pattern, Resource Pattern, Status Pattern, and Scheduling Pattern

The workflows specified using the Weakly Structured Workflow Pattern and Structured Workflow Pattern can be further described with information about the conditions, resources, status, and scheduling of activities. Information about conditions for executing an activity are added by combining the Weakly Structured Workflow Pattern or Structured Workflow Pattern with the Condition Pattern. The Condition Pattern allows for defining some preconditions and post-conditions such as that a document needs to be signed. Using the Resource Pattern, one can define if an activity produces a resource (create), uses a resource (without exactly knowing if the resource is modified or not), views a resource (without modifying it, i.e., read), edits a resource (update), consumes a resource (delete), or locks a resource. The status of activities and processes can be set to active, inactive, or finished using the Status Pattern. The pattern can be extended to domain specific requirements such as initiated, suspended, and failed. Activities may have to be executed at a specific time and/or place. This can be represented using the Scheduling Pattern.

5.4   Workflow Integration Pattern

The integration of individual knowledge work and organizational knowledge work is conducted using the Workflow Integration Pattern specialized from DUL's Workflow Pattern and is depicted in Figure 5. The alignment of the concepts defined in the Weakly Structured Workflow Pattern and the Structured Workflow Pattern is supported by using the Workflow Pattern of DUL as common modeling basis. As described in Section 5.1 and Section 5.2, it is possible to associate a Goal to each Task using the relatesTo relation. The Goal concept is connected to the workflow via the hasComponent relation. Using the Goal concept, a formal mapping of weakly structured workflows and structured workflows can be conducted. It is based on the assumption that if some individual knowledge work is carried out in the context of an organizational business process or vice versa, they share a common Goal. Finally, the association between concrete activities carried out in the individual knowledge work and organizational knowledge work is established through the relatesTo relation that connects the Goals with Tasks in the Weakly Structured Workflow Pattern and the Structured Workflow Pattern.

5.5   Workflow Model Pattern

The Workflow Model Pattern allows for explicitly distinguishing workflow models and workflow instances for both, weakly structured workflows and structured workflows. To create a workflow model, the Workflow Model Pattern is able to represent on a *generic*, i.e., conceptual level, the flow of Tasks, their dependencies, and the resources required. In contrast to the traditional business process modeling (see Section 4), however, the workflow instances created from a workflow model do not need to be strictly in accordance with the model. This is in

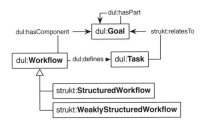

Fig. 5. Workflow Integration Pattern

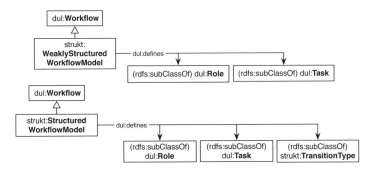

Fig. 6. Workflow Model Pattern

particular important for weakly structured workflows that can be adapted to the requirements of a concrete execution situation.

Figure 6 depicts the Workflow Model Pattern. It consists of two parts, one for the WeaklyStructuredWorkflowModel and one for the StructuredWorkflowModel. In the case of the StructuredWorkflowModel, subclasses of Role, Task, and TransitionType are defined as valid components of the workflow model definition. For weakly structured workflow models, only Roles and Tasks can be defined.

6   Example Application of the strukt Core Ontology

The application of the strukt core ontology is shown at the example of an apartment construction by the architectural office introduced in Section 2. Figure 7 (bottom part) depicts the application of the Weakly Structured Workflow Pattern wsw-prepare-building-application-1 for preparing a building application. It defines the Tasks t-compute-statics-1 and t-create-groundplan-1. The relation isExecutedIn classifies the individuals a-compute-statics-1 and a-create-groundplan-1 as Actions, executing the tasks. The isTaskOf relation specifies that the task t-compute-statics-1 has to be conducted by an agent playing the role of a StructuralEngineer r-structural-engineer-1, here the NaturalPerson tmueller-1. The NaturalPerson tmueller-1 is specified as participant of the Action a-compute-statics-1. The participant of the Action a-create-groundplan-1 is not specified.

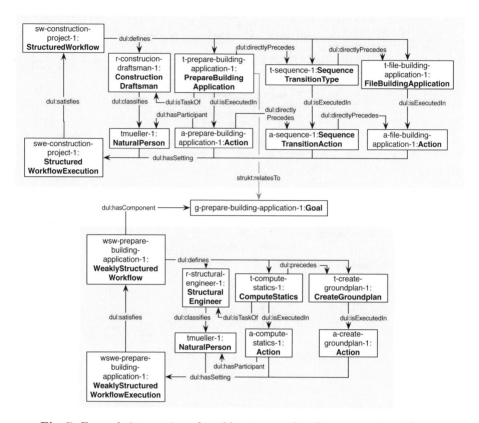

Fig. 7. Example integration of weakly structured and structured workflow

The weakly structured workflow belongs to the organizational business process depicted in Figure 7 (top part) using the Structured Workflow Pattern. It models an excerpt of the business process shown in Figure 1 of the scenario in Section 2. The StructuredWorkflow sw-residential-object-1 defines the Tasks t-prepare-building-application-1 and t-file-building-application-1, the SequenceTransitionType tt-sequence-1, and the role r-draftsman-1. The tasks t-prepare-building-application-1 and t-submit-application are connected in a sequence using the relations directlyPrecedes and directlyFollows of t-sequence-1. The Role r-draftsman-1 is connected using the isTaskOf relation with the Task t-prepare-building-application-1. In the context of this workflow, the NaturalPerson tmueller-1 acts as r-draftsman-1. The Actions a-prepare-building-application-1, a-sequence-1, and a-submit-application-1 constitute the execution of the Tasks and SequenceTransitionType, respectively. The integration of the weakly structured workflow wsw-prepare-building-application-1 and structured workflow sw-construction-project-1 is conducted by defining g-prepare-building-application-1 as Goal of the t-prepare-building-application-1 task. The Goal g-prepare-building-application-1 is then connected with the Weakly Structured Workflow wsw-prepare-building-application-1 using the has-Component relation. As described above, the wsw-prepare-building-application-1

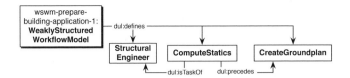

Fig. 8. Application of the Workflow Model Pattern

captures the individual activities, concrete sub-tasks, and roles involved in actually writing the building application.

An instance of a workflow such as the example of the weakly structured workflow wsw-prepare-building-application-1 in Figure 7 (bottom part) can be abstracted to a workflow model using the Workflow Model Pattern. As shown in Figure 8, the abstraction from a workflow instance to a model is basically the upper part of the Descriptions and Situations Pattern of DUL. In our example, the WeaklyStructuredWorkflowModel wswm-prepare-building-application-1 consists of the domain-specific concepts of the role StructuralEngineer, the two tasks ComputeStatics and CreateGroundplan, and the relations.

As shown in Figure 7, using the patterns of struct allows for modeling and integrating structured workflows and weakly structured workflows. Using the ontology design pattern Descriptions and Situations as design principle for representing workflows in strukt allows for modifying workflow instances without affecting the original workflow model. This is achieved by contextualizing the workflow instances using the individuals sw-construction-project-1 and wsw-prepare-building-application-1. Other instances of the same WeaklyStructuredWorkflow like a wsw-prepare-building-application-2 can have different roles and tasks defined for the actual execution and the tasks can be executed in different order.

7   Prototype Application

The prototype application supports individual and organizational knowledge work and their combination. It instantiates the pattern of the strukt ontology. A domain-specific construction ontology aligned to DUL is used to describe the roles such as manager, draftsman, and engineer. The user interface for the individual knowledge worker is depicted in Figure 9. It consists of a *task space* for managing the weakly structured workflows with their tasks and sub-tasks. The task space allows for receiving details of a task, create new tasks, modify tasks, save a workflow instance as workflow model, instantiating a workflow model, and deleting tasks and workflows, respectively.

The left hand side of the screenshot depicted in Figure 9 shows example weakly structured workflows from the architecture scenario presented in Section 2. The tasks and subtasks of a weakly structured workflow can be shown by clicking on the small triangle symbol next to the task like the Building application Mornhinweg Inc example. Important details of a task are shown on the right hand side of the screenshot such as deadlines, appointments, and others. Tasks can be marked as finished by clicking on the checkbox on the left to the task name. When

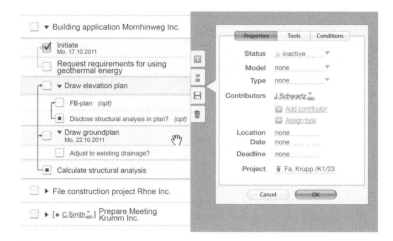

Fig. 9. Task Space for the Individual Knowledge Worker

there is a lock symbol in the checkbox (indicated as small box), the task cannot be accomplished due to unfulfilled dependencies (indicated by the arrows). For example, the task Calculate structural analysis cannot be processed as the tasks Draw elevation plan and Draw ground plan are not completed. Optional tasks are indicated with the keyword *(opt)*. The order of tasks in a weakly structured workflow can be changed by the knowledge worker using simple drag and drop interaction. The right hand side of the screenshot in Figure 9 provides details of a task such as its status and the responsible agent. Additional agents can be added as well as the responsibility of tasks can be forwarded. Thus, the strukt prototype enables a collaborative execution of a weakly structured workflow by multiple knowledge workers. Further details can be investigated using the tab Tools showing the tools used to process a task and the tab Conditions showing detailed information about the conditions associated with the task, e.g., when a specific role needs to sign a specific document.

In order to abstract a workflow model from a workflow instance, the strukt application provides the *workflow transformation menu* depicted in Figure 10. It allows for defining the components of the workflow model. To this end, all components of the workflow instance to be transformed are depicted in a table. Each row of the table represents a task of the weakly structured workflow. Subtasks are indicated by indentions. The columns Task, Conditions, Optional, Role conditions, Documents, and Tools show the details of the tasks relevant for creating a workflow model. Tasks can be removed from the workflow at this point from the transformation process. In addition, the order of tasks can be changed by drag and drop interaction and new tasks can be added.

We have implemented a simple workflow management system in our strukt prototype. It provides a test environment for synchronizing the activities in the weakly structured workflows and some pre-defined business processes of the architectural office. A user interface is not provided as it is assumed that strukt is integrated in an existing business process engine with its own interface.

Fig. 10. Transformation Menu for Creating a Workflow Model from an Instance

8   Conclusions

We have presented an approach for integrating individual knowledge work and organizational knowledge work by means of the pattern-based core ontology *strukt*. The core ontology strukt defines several ontology design patterns for capturing weakly structured workflows of individual knowledge work and structured workflows in the context of organizational knowledge work. A formal alignment and synchronization of the activities in individual knowledge work and organizational knowledge work is conducted by basing on the DOLCE+DnS Plan Ontology [8]. Concrete instances of weakly structured workflows can be transformed into generic workflow models, enabling reuse of procedural knowledge. On basis of the strukt ontology, we have developed a prototypical software system for the collaborative planning and execution of weakly structured workflows and applied it to the use case of an architectural office. The strukt prototype connects with a simple workflow management system to synchronize the flexible, individual knowledge work with the strict execution of business processes. This work has been co-funded by the EU in FP7 in the ROBUST project (257859).

References

1. Becker, J., Weiss, B., Winkelman, A.: Developing a Business Process Modeling Language for the Banking Sector - A Design Science Approach. In: Americas Conference on Information Systems, pp. 1–12 (2009)
2. Bolinger, J., Horvath, G., Ramanathan, J., Ramnath, R.: Collaborative workflow assistant for organizational effectiveness. In: Applied Computing. ACM (2009)
3. Borgo, S., Masolo, C.: Foundational choices in DOLCE. In: Handbook on Ontologies, 2nd edn. Springer, Heidelberg (2009)

4. Buschmann, F., Meunier, R., Rohnert, H., Sommerlad, P., Stal, M.: Pattern-Oriented Software Architecture: A System of Patterns, vol. 1. Wiley (1996)

5. Carlsen, S.: Action port model: A mixed paradigm conceptual workflow modeling language. In: Cooperative Information Systems, pp. 300–309. IEEE (1998)

6. Drucker, P.: Knowledge-Worker Productivity: The Biggest Challenge. California Management Review 41(2), 79–94 (1999)

7. Gamma, E., Helm, R., Johnson, R., Vlissides, J.: Design patterns: elements of reusable object-oriented software. Addison-Wesley (July 2004)

8. Gangemi, A., Borgo, S., Catenacci, C., Lehmann, J.: Task Taxonomies for Knowledge Content. In: METOKIS Deliverable, vol. D7, pp. 20–42 (2004), http://www.loa-cnr.it/Papers/D07_v21a.pdf

9. Gangemi, A., Presutti, V.: Ontology Design Patterns. In: Handbook on Ontologies. Springer, Heidelberg (2009)

10. Groth, P.T., Gil, Y.: A scientific workflow construction command line. In: Intelligent User Interfaces. ACM (2009)

11. Hammer, M., Champy, J.: Reengineering the Corporation: A Manifesto for Business Revolution. Harper Paperbacks (2003)

12. Hart-Davidson, W., Spinuzzi, C., Zachry, M.: Capturing & Visualizing Knowledge Work: Results & Implications of a Pilot Study of Proposal Writing Activity. In: Design of Communication, pp. 113–119. ACM (2007)

13. Hepp, M., Belecheanu, R., Domingue, J., FilipowskaG, A., Kaczmarek, M., Kaczmarek, T., Nitzsche, J., Norton, B., Pedrinaci, C., Roman, D., et al: Business Process Modelling Ontology and Mapping to WSMO. Technical report, SUPER Project IST-026850 (2006)

14. Kidd, A.: The marks are on the knowledge worker. In: Human Factors in Computing Systems, pp. 186–191. ACM (1994)

15. Nonaka, I.: The Knowledge-Creating Company. Harvard Business Review 69(6), 96–104 (1991)

16. Oberle, D.: Semantic Management of Middleware. Springer, Heidelberg (2006)

17. OMG. Business process model and notation (BPMN), version 2.0 beta 2 (2010), http://www.omg.org/cgi-bin/doc?dtc/10-06-04.pdf

18. Scheer, A.: Aris-Business Process Frameworks. Springer, Heidelberg (1998)

19. Schwarz, S.: Task-Konzepte: Struktur und Semantik für Workflows. In: Professionelles Wissesmanagement, Luzern, Switzerland. GI e.V. (2003)

20. Schwarz, S., Abecker, A., Maus, H., Sintek, M.: Anforderungen an die Workflow Unterstützung für Wissensintensive Geschäftsprozesse. In: Professionelles Wissensmanagement, Baden-Baden, Germany (2001)

21. Varzi, A.: Parts, Wholes, and Part-Whole Relations: The Prospects of Mereotopology. Data & Knowledge Engineering 20(3), 259–286 (1996)

22. W3C. OWL-S (2004), http://www.w3.org/Submission/OWL-S/

23. WFMC. Terminology & Glossary. Technical Report WFMC-TC-1011, Version 3.0 (1999), http://www.wfmc.org/standards/docs/TC-1011_term_glossary_v3.pdf

FedBench: A Benchmark Suite for Federated Semantic Data Query Processing

Michael Schmidt[1], Olaf Görlitz[2], Peter Haase[1], Günter Ladwig[3],
Andreas Schwarte[1], and Thanh Tran[3]

[1] Fluid Operations AG, Walldorf, Germany
[2] Institute for Web Science and Technology, University of Koblenz-Landau, Germany
[3] Institute AIFB, Karlsruhe Institute of Technology, Germany

Abstract. In this paper we present *FedBench*, a comprehensive benchmark suite for testing and analyzing the performance of federated query processing strategies on semantic data. The major challenge lies in the *heterogeneity of semantic data use cases*, where applications may face different settings at both the data and query level, such as varying data access interfaces, incomplete knowledge about data sources, availability of different statistics, and varying degrees of query expressiveness. Accounting for this heterogeneity, we present a highly *flexible benchmark suite*, which can be customized to accommodate a variety of use cases and compare competing approaches. We discuss design decisions, highlight the flexibility in customization, and elaborate on the choice of data and query sets. The practicability of our benchmark is demonstrated by a rigorous evaluation of various application scenarios, where we indicate both the benefits as well as limitations of the state-of-the-art federated query processing strategies for semantic data.

1 Introduction

Driven by the success of the Linking Open Data initiative, the amount of semantic data that is published on the Web in the form of RDF is increasing at a tremendous pace. While offering great potentials for innovative applications that integrate heterogeneous data from different sources, on the data management side, this development comes along with a variety of new challenges, last but not least due to the sheer amount of data that may be utilized by such applications. Most research contributions in the context of RDF data processing have focused on the problem of query evaluation over local, centralized repositories (see e.g. [2, 24, 19]) – and for these scenarios different benchmarks have been proposed [11, 6, 21]. Accounting for the decentralized nature of the Semantic Web, though, one can observe an ongoing shift from localized to federated semantic data processing, where independent endpoints provide data, and semantic data applications utilize both local repositories and remote data sources at the same time to satisfy their information needs. In response to this paradigm shift, different federated RDF processing strategies – targeted at different use cases and application scenarios – have been proposed [15, 14, 16, 9, 17].

L. Aroyo et al. (Eds.): ISWC 2011, Part I, LNCS 7031, pp. 585–600, 2011.

With distributed semantic data processing becoming increasingly important, we identify a clear need for *a benchmark tailored to the problem of federated semantic data query processing*. We employ a broad definition of *semantic data*, which includes Linked Data sources, datasets, and ontologies represented in RDF. The main challenge here lies in the diversity and heterogeneity of semantic data use cases, and the demands they pose to a benchmark: First, we can observe *heterogeneity at data level* along several dimensions: applications are facing different physical distribution of datasets, different interfaces for data access, incomplete knowledge about the existence of entry points into the Web of data, and different types of metadata and statistics. Apart from the challenges at data level, applications may also exhibit *different demands w.r.t. query evaluation*, including aspects such as query languages, expressiveness, and ranking.

The overall setting in which a concrete semantic data application is settled may have severe impact on query processing strategies. Ultimately, there cannot exist a single "one-size-fits-all" benchmark to measure each and every aspect of an application – or to compare the performance of orthogonal federated query processing strategies. Hence, taking an existing benchmark and distributing its data across several endpoints may cover some, but not all challenges that arise in the context of federated semantic data processing. What is needed instead is a *collection* of *customizable* benchmark scenarios that *accommodate a multitude of dimensions* as well as *essential challenges* – and from which one can choose data sets, queries, and settings that fit the specific needs.

Contributions. (1) Based on a review of federated query processing scenarios our community has dealt with so far, we discuss orthogonal *dimensions at data and query level* that can be used to classify existing approaches. (2) Accounting for the heterogeneity of these dimensions, we present a rich collection of *queries, data*, and *data statistics*, which can be flexibly combined and customized. This makes our benchmark generic enough to cover a broad range of use cases, such as testing the performance of the underlying federation approach, data access mechanisms, static optimization based on metadata and statistics, queries with varying characteristics, and many more. All queries and datasets were carefully chosen to reflect a variety of domains, query patterns, and typical challenges in query processing (in particular in distributed settings). While some of the queries were specifically designed to test and vary in these aspects, others were taken from prototypical, domain-specific use cases (e.g. in the Life Science domain) built by participants in other projects. (3) In order to *show the flexibility* and illustrate *the broad range of scenarios covered by FedBench*, we provide results for selected scenarios and implementations, identifying areas where ongoing work is required. Our results are published in a Wiki, where we also maintain data, statistics, queries, and scenarios. We invite researchers and benchmark users to customize and extend the benchmark suite according to their own needs.

We point out that the resulting benchmark suite is available online[1], including a flexible and extensible Open Source Java-based evaluation framework.

---

[1] See `http://fbench.googlecode.com/`

Related Work. Apart from benchmarks that target structural properties of RDF schemas (such as [18]), several benchmarks for RDF, RDFS, and OWL data processing have been proposed. The Lehigh University Benchmark [11], for instance, has been designed to test the reasoning capabilities of systems over a single ontology. The SPARQL-specific, use-case driven Berlin SPARQL Benchmark [6] comes with a set of queries implementing meaningful requests on top of an eCommerce scenario modeled in RDF. Complementary, the SPARQL Performance Benchmark (SP^{2}Bench) [21] puts a stronger focus on language-specific features of the SPARQL query language, addressing optimization in complex scenarios. None of the above benchmarks considers federation at data level, nor does provide data collections consisting of multiple interlinked datasets.

To the best of our knowledge, the only work addressing the latter issue was our previous work in [13], which – focusing on selected federation approaches – served as a starting point for designing FedBench. Going far beyond this initial work, in this paper we present a holistic benchmark suite, including a variety of new data and query sets, new scenarios such as Linked Data access (i.e., via HTTP requests), an automated evaluation framework (with support for various metrics like counting the number of requests, automated evaluation, interfaces for connecting new systems etc.), a discussion and classification of state-of-the-art approaches, a discussion of statistics, as well as novel experiments and findings.

Our benchmark has been designed to be compatible with the current state-of-the-art in Linked Data query processing. In particular, in our benchmark we use queries defined in SPARQL 1.0, without requiring specific extensions for explicit specification of the endpoint services as proposed by the SPARQL 1.1 federation extensions. To our knowledge, there is currently no public implementation of the SPARQL 1.1 federation extension [1]. Examples of systems expected to support SPARQL 1.1 federation in future releases include SPARQL DQP [5], Sesame [7], and FedX [23]. While in this paper we focus on SPARQL 1.0 features, on our project page we also provide SPARQL 1.1 versions of the benchmark queries, ready to be used as soon as implementations become available.

An analysis and comparison of the structure of different semantic data sets and benchmarks from the Linked Data domain is presented in [8]. It shows that artificial datasets used in benchmarks are typically highly structured, while Linked Data are less structured. They conclude that benchmarks should not solely rely on artificial data but also consider real world datasets. Resuming this discussion, in Section 3.1 we will present an analysis and comparison of our data sets using the methods proposed in [8].

Outline. In the next section, we identify and discuss essential dimensions of federated semantic data processing, which form the groundwork for the benchmark suite presented in this paper. Next, in Section 3 we motivate benchmark design goals and present the suite in detail, namely benchmark datasets (Section 3.1), covering aspects such as their properties and associated statistics, benchmark queries and their properties (Section 3.2), as well as the benchmark driver (Section 3.3). We turn towards a comprehensive evaluation of concrete application scenarios in Section 4 and wrap up with some concluding remarks in Section 5.

2   Heterogeneity of Semantic Data Use Cases

To date, several approaches to semantic data query processing have been studied, comprising both centralized and decentralized settings. Centralized approaches, where all data is periodically crawled and stored locally, come with the merits of high controllability and reliability. While Google and the likes have shown that this is a successful approach for dealing with Web documents that primarily comprise text, it has also been adopted for semantic data, where projects like Factforge[2] collect large amounts of RDF(S) data in centralized stores.

If federation is set up dynamically or the underlying sources exhibit high update rates, though, it may not be affordable to keep imported semantic data up-to-date. Then, decentralized, federated query processing strategies are required, typically implemented on top of public SPARQL endpoints or directly on Linked Data accessible via HTTP requests. In the best case, not only the existence of data sources but also detailed statistics can be assumed and exploited for optimization [14]. A stronger relaxation is to assume only partial knowledge about the sources, e.g. past information that is stored and employed for future runs, thereby obtaining more entry points and hints to explore unknown sources in a goal-directed fashion. In fact, it has been shown that already little knowledge obtained at runtime can be used to adaptively correct the query processing strategy chosen at compile time, to improve performance [16]. In the worst case, engines have to deal with federated scenarios where no data and knowledge are available, so queries have to be processed based on iterative URI lookups [15].

In summary, previous approaches reveal several dimensions along the data and query level, which we use to characterize approaches. They determine the challenges, including the major issues of centralized vs. decentralized processing and knowledge about datasets discussed above. At data level, we identify heterogeneity along the following dimensions:

(D1) Physical Distribution: Federated query processing systems may either access and process global data from the Web, process locally stored data sources, or mix up both paradigms connecting *local* with *global data.*

(D2) Data Access Interface: Semantic data may be accessible through different interfaces. There may be native repositories, SPARQL endpoints, and Linked Data accessible through HTTP requests. These interfaces provide different access paths to the data, ranging from iterators at data level, URI lookups, to expressive queries in different languages.

(D3) Data Source Existence: In particular in Linked Data scenarios, not all sources may be known a priori. Hence, applications may have only few entry points into the data graph, which can be used to iteratively deep-dive by exploring links (see for instance the setting described in [16]).

(D4) Data Statistics: In the best case, advanced statistical information about properties, counts, and distributions in the form of histograms for all data sets are available; in the worst case – in particular if data is not stored locally – only few or no information about the data sources may be given.

---

[2] http://ontotext.com/factforge/

The concrete setting an application faces at data level – i.e., the classification within dimension (D1)–(D4) – implies challenges in data processing and imposes an upper bound on the efficiency in query processing: applications built on top of local repositories exploiting detailed statistical knowledge for query optimization, for instance, are generally faster than applications that rely on federated Linked Data accessible via HTTP lookups, where network delay and incomplete knowledge about data sets impose hard limits on query efficiency.

Apart from the challenges at data level, applications may also face different challenges at query level. Like the dimensions at data level, also those at query level drive the challenges behind semantic data applications and should be covered in a benchmark. In particular, we identify the following dimensions.

(Q1) Query Language: The expressiveness of the query language needed by applications may vary from case to case: while some applications get around with simple conjunctive queries, others may rely on the full expressive power of RDF query languages, such as the de facto standard SPARQL [20, 22].

(Q2) Result Completeness: Certain applications may rely on complete results, while others cannot afford it when responsiveness is first priority. In particular in Linked Data scenarios where complete knowledge cannot be assumed (s.t., beginning from some entry points, further sources have to be discovered via online link traversal) not all data sources may be found [15, 16].

(Q3): Ranking: Applications may be interested in queries that enable ranking according to some predefined metrics, or maybe only in top-k results.

3   FedBench: Benchmark Description

In order to support benchmarking of the different scenarios that emerge along all the dimensions, FedBench consists of three components, all of which can be *customized and extended* to fit the desired scenario: (i) multiple datasets, (ii) multiple query sets, and (iii) a comprehensive evaluation framework. We first elaborate on the datasets and statistics (addressing dimension (D4)), then present the queries (addressing dimensions (Q1)–(Q3)), and conclude with a discussion of our evaluation framework, which addresses dimensions (D1)–(D3).

3.1   Benchmark Data

Accounting for the heterogeneity of semantic data use cases, we provide three *data collections*, each consisting of a number of interlinked datasets. The data collections have been selected to represent both real-world and artificial data federations over multiple representative semantic datasets. The collections differ in size, coverage, and types of interlinkage. Two of them are subsets of the Linked Open Data cloud: The first spans different domains of general interest, representing typical scenarios of combining cross-domain data with heterogeneous types of interlinkage; the second contains datasets from the Life Science area, representing a federation scenario in a very domain-specific setting. Additionally, we use a partitioned synthetic data collection, whose advantage lies in the ability to simulate federations of varying size with well-defined characteristics of the data.

General Linked Open Data Collection. This first data collection consists of datasets from different domains: **DBpedia** is a central hub in the Linked Data world, containing structured data extracted from Wikipedia. Many datasets are linked to DBpedia instances. **GeoNames** provides information about geographic entities like countries and cities. **Jamendo** is a music database containing information about artists, records, tracks, publishers and publication dates. **Linked-MDB** exhibits details about films, genres, actors, producers, etc. and connects its instances to the corresponding DBpedia entities. The **New York Times** dataset contains about 10,000 subject headings covering different topics, which are linked with with people, organizations and locations. Finally, the **Semantic Web Dog Food** dataset provides information about Semantic Web conferences of the past years, including paper descriptions, authors, and so on.

Life Science Data Collection. In this collection, we again included the DBpedia subset from the General Linked Open Data dataset as a central hub. **KEGG** (Kyoto Encyclopedia of Genes and Genomes) contains data about chemical compounds and reactions, with a focus on information relevant for geneticists. It is published in a number of separate modules; in the dataset we included the modules KEGG Drug, Enzyme, Reaction and Compound. Next, **ChEBI** (Chemical Entities of Biological Interest) is a dictionary of molecular entities focused on "small" chemical compounds, describing constitutionally or isotopically distinct atoms, molecules, ions, ion pairs, radicals, radical ions, complexes, conformers, etc. **DrugBank** is a bioinformatics and cheminformatics resource that combines detailed drug (i.e. chemical, pharmacological, and pharmaceutical) data with comprehensive drug target (i.e. sequence, structure, and pathway) information. The datasets are linked in different ways: Drugbank is linked with DBpedia via `owl:sameAs` statements, and other datasets are linked via special properties, e.g. Drugbank links to KEGG via the property `keggCompoundId`. KEGG and Drugbank can be joined via identifiers of the CAS database (Chemical Abstract Service). Some links are implicit by the use of common identifiers in literal values, e.g. the `genericName` in Drugbank corresponds to the `title` in ChEBI.

SP^{2}Bench is a synthetic dataset generated by the SP^{2}Bench data generator [21], which mirrors vital characteristics (such as power law distributions or Gaussian curves) encountered in the DBLP bibliographic database. The data generator provides a single dataset, from which we created a collection by clustering by the types occurring in the dataset, finally obtaining 16 sub-datasets (for persons, inproceedings, articles, etc.) that can be deployed independently in a distributed scenario. The data collection consists of 10M triples in total.

Metadata and Statistics about data sources are important for identifying suitable data sources for answering a given query, as well as for query optimization. They can be used to parametrize the benchmark along dimension (D4).

Table 1. Basic Statistics of Datasets[1]

Collection	Dataset	version	#triples	#subj.	#pred.	#obj.	#types	#links	strct.
Cross Domain	DBpedia subset[2]	3.5.1	43.6M	9.50M	1063	13.6M	248	61.5k	0.19
	NY Times	2010-01-13	335k	21.7k	36	192k	2	31.7k	0.73
	LinkedMDB	2010-01-19	6.15M	694k	222	2.05M	53	63.1k	0.73
	Jamendo	2010-11-25	1.05M	336k	26	441k	11	1.7k	0.96
	GeoNames	2010-10-06	108M	7.48M	26	35.8M	1	118k	0.52
	SW Dog Food	2010-11-25	104k	12.0k	118	37.5k	103	1.6k	0.43
Life Sciences	DBpedia subset[2]	3.5.1	43.6M	9.50M	1063	13.6M	248	61.5k	0.19
	KEGG	2010-11-25	1.09M	34.3k	21	939k	4	30k	0.92
	Drugbank	2010-11-25	767k	19.7k	119	276k	8	9.5k	0.72
	ChEBI	2010-11-25	7.33M	50.5k	28	772k	1	-	0.34
SP²Bench	SP²Bench 10M	v1.01	10M	1.7M	77	5.4M	12	-	0.76

[1] All datasets are available at http://code.google.com/p/fbench/.
[2] Includes the ontology, infobox types plus mapped properties, titles, article categories with labels, Geo coordinates, images, SKOS categories, and links to New York Times and Linked Geo Data.

Some Linked Data sources provide basic VoiD [3] statistics such as *number of triples, distinct subjects, predicates, objects*, and information about the vocabulary and links to other sources. Table 1 surveys such basic statistics for our datasets. These and other statistics (such as *predicate and type frequency, histograms, full pattern indexes* obtained by counting all combinations of values in triple patterns, *full join indexes* obtained by counting all join combinations, and *link statistics*) can be exploited by engines in the optimization process.

Duan et al. [8] introduced the notion of structuredness, which indicates whether the instances in a dataset have only a few or all attributes of their types set. They show that artificial datasets are typically highly structured and "real" datasets are less structured. As shown in the last column of Table 1, the structuredness (range $[0, 1]$) varies for our datasets, e.g. DBpedia has a low structuredness value whereas Jamendo and KEGG are highly structured.

3.2 Benchmark Queries

There are two reasonable options for the design of benchmark queries [10]: language-specific vs. use case driven design. The query sets we propose cover both dimensions. We choose SPARQL as a query language: It is known to be relationally complete, allowing us to encode a broad range of queries with varying complexity, from simple conjunctive queries to complex requests involving e.g. negation [4, 20, 22]. We restrict ourselves on general characteristics, pointing to the FedBench project page for a complete listing and description.

Life Science (LS) and Cross Domain Queries (CD). These two query sets implement realistic, real-life use cases on top of the cross-domain and life science data collection, respectively. Their focus is on federation-specific aspects, in particular (1) number of data sources involved, (2) join complexity, (3) types

Example, Life Science Query 4: For all drugs in DBpedia, find all drugs they interact with, along with an explanation of the interaction.

```
SELECT ?Drug ?IntDrug ?IntEffect WHERE {
    ?Drug rdf:type dbpedia-owl:Drug .
    ?y owl:sameAs ?Drug .
    ?Int drugbank:interactionDrug1 ?y .
    ?Int drugbank:interactionDrug2 ?IntDrug .
    ?Int drugbank:text ?IntEffect . }
```

This query includes a star-shaped sub pattern of drugs which is connected via `owl:sameAs` link to DBpedia drug entities.

Example, Cross Domain Query 5: Find the director and the genre of movies directed by Italians.

```
SELECT ?film ?director ?genre WHERE {
    ?film dbpedia-owl:director ?director.
    ?director dbpedia-owl:nationality dbpedia:Italy .
    ?x owl:sameAs ?film .
    ?x linkedMDB:genre ?genre . }
```

A chain-like query for finding film entities (in LinkedMDB and in DBpedia) linked via `owl:sameAs` and restricted on genre and director.

Example, Linked Data Query 4: Find authors of papers at the ESWC 2010 conference who were also involved in the conference organization.

```
SELECT * WHERE {
    ?role swc:isRoleAt <http://data.semanticweb.org/conference/eswc/2010> .
    ?role swc:heldBy ?p .
    ?paper swrc:author> ?p .
    ?paper swc:isPartOf ?proceedings .
    ?proceedings swc:relatedToEvent <http://data.semanticweb.org/conference/eswc/2010> }
```

Fig. 1. Selected Benchmark Queries

of links used to join sources, and (4) varying query (and intermediate) result size. Figure 1 exemplarily discusses three queries taken from these query sets.

SP^{2}Bench Queries (SP). Next, we reuse the queries from the SP^{2}Bench SPARQL performance benchmark, which were designed to test a variety of SPARQL constructs and operator constellations, but also cover characteristics like data access patterns, result size, and different join selectivities. Some of the SP^{2}Bench queries have high complexity, implementing advanced language constructs such as negation and double negation. They are intended to be run on top of the distributed SP^{2}Bench dataset described in Section 3.1. A thorough discussion of the queries and their properties can be found in [21].

Table 2. Query Characteristics. Operators: **A**nd ("."), **U**nion, **F**ilter, **O**ptional; Modifiers: **D**istinct, **L**imit, **O**ffset, **O**rderBy; Structure: **S**tar, **C**hain, **H**ybrid

Life Science (LS)						SP^{2}Bench (SP)						Linked Data (LD)			
	Op.	Mod.	Struct.	#Res.	#Src		Op.	Mod.	Struct.	#Res.	#Src		Op.	Mod.	Struct.	#Res.
1	U	-	-	1159	2	1	A	-	S	1	11	1	A	-	C	309
2	AU	-	-	333	4	2	AO	Or	S	>500k	12	2	A	-	C	185
3	A	-	H	9054	2	3a	AF	-	S	>300k	16	3	A	-	C	162
4	A	-	H	3	2	3b	AF	-	S	2209	16	4	A	-	C	50
5	A	-	H	393	3	3c	AF	-	S	0	16	5	A	-	S	10
6	A	-	H	28	3	4	AF	D	C	>40M	14	6	A	-	H	11
7	AFO	-	H	144	3	5a	AF	D	C	>300k	14	7	A	-	S	1024
Cross Domain (CD)						5b	AF	D	C	>300k	14	8	A	-	H	22
1	AU	-	S	90	2	6	AFO	-	H	>700k	16	9	A	-	C	1
2	A	-	S	1	2	7	AFO	D	H	>2k	14	10	A	-	C	3
3	A	-	H	2	5	8	AFU	D	H	493	16	11	A	-	S	239
4	A	-	C	1	5	9	AU	D	-	4	16					
5	A	-	C	2	5	10	-	-	-	656	12					
6	A	-	C	11	4	11	-	LOfOr	-	10	8					
7	A	-	C	1	5											

Linked Data Queries (LD). Today's Linked Data engines typically focus on basic graph patterns (Conjunctive Queries). Therefore, all LD queries are basic graph pattern queries, designed to deliver results when processing Linked Data in an exploration-based way (cf. the bottom-up strategy described in Section 4.1).

Queries LD1–LD4 use the SW Dog Food dataset to extract information about conference and associated people. LD1–LD3 all contain a single URI to be used as a starting-point for exploration-based query processing, whereas LD4 has two URIs that could be used to speed up processing by starting the exploration from multiple points in the Linked Data graph (cf. Figure 1). The other queries operate on DBpedia, LinkedMDB, NewYork Times, and the Life Science collection. In summary, the Linked Data queries vary in a broad range of characteristics, such as number of sources involved, number of query results, and query structure.

Query Characteristics. Table 2 surveys the query properties, showing that they vastly vary in their characteristics. We indicate the SPARQL operators used inside the query (*Op.*), the solution modifiers that were used additionally (*Sol.*), categorize the query structure (*Struct.*), roughly distinguishing different join combinations – like subject-subject or subject-object joins – leading to different query structures commonly referred to as star-shaped, chain, or hybrid queries, and indicate the number of results (*#Res.*) on the associated datasets (we provide an estimated lower bound when the precise number is unknown). In addition, we denote the number of datasets that potentially contribute to the result (*#Src*) , i.e. those that match at least one triple pattern in the query. Note that the number of data sets used for evaluation depends on the evaluation strategy (e.g., an engine may substitute variable bindings at runtime and, in turn, some endpoints would no longer yield results for it), or intermediate results delivered by endpoints may be irrelevant for the final outcome. We observe that the (LS) queries typically address 2–3 sources, the (CD) queries up to 5 sources, while the (SP) queries have intermediate results in up to 16 sources (where, however, typically only few sources contribute to the result).

3.3   Benchmark Evaluation Framework

To help users executing FedBench in a standardized way and support parametrization along the dimensions from Section 2, we have developed a Java benchmark driver, which is available in Open Source. It provides an integrated execution engine for the different scenarios and is highly configurable. Using the Sesame[3] API as a mediator, it offers support for querying local repositories, SPARQL endpoints, and Linked Data in a federated setting. Systems that are not built upon Sesame can easily be integrated by implementing the generic Sesame interfaces.

The driver comes with predefined configurations for the benchmark scenarios that will be discussed in our experimental results. Custom scenarios can be created intuitively by writing config files that define properties for data configuration and other benchmark settings (query sets, number of runs, timeout, output mediator, etc). In particular, one can specify the types of repositories

---

[3] http://www.openrdf.org/

(e.g., native vs. SPARQL endpoints), automatically load datasets into repositories (while measuring loading time), and execute arbitrary queries while simulating real-world conditions like execution via HTTP with a customizable network delay. Combining this flexibility with the predefined data and query sets thus allows the user to customize the benchmark along the dimensions relevant for the setting under consideration. Designed with the goal to position the benchmark as an ongoing community effort, the underlying evaluation framework is Open Source and has been designed with extensibility in mind at different levels. In particular, it is easy to specify complex evaluation settings by means of simple configuration files (i.e., without code modifications), plug in new systems, implement new metrics, evaluate and visualize results, etc.

To standardize the output format, the driver ships two default mediators for writing results in CSV and RDF; for the latter we have implemented an Information Workbench [12] module to visualize benchmark results automatically.

4   Evaluation

The central goal of our experimental evaluation is to demonstrate the usefulness of our benchmark suite. Thus, in order to show that our framework is a useful tool to assess strengths and weaknesses in a variety of semantic data use cases, we investigate different scenarios that vary in the dimensions sketched in Section 2, in particular in data distribution, access interfaces, and query complexity. For space limitations, aspects (Q2) *Result Completeness* and (Q3) *Ranking* are not covered; further, there are currently no systems that improve their behavior when an increasing amount of statistics are provided, so (D4) *Data Statistics* could only be assessed by comparing systems that make use of different statistics. All experiments described in the following are supported by our benchmark driver out-of-the-box and were realized by setting up simple benchmark driver config files specifying data and query sets, setup information, etc. We refer the interested reader to [23] for additional results on other systems, such as DARQ and FedX. We start with a description of the scenarios, then discuss the benchmark environment, and conclude with a discussion of the evaluation results.

4.1   Description of Scenarios

(A) RDF Databases vs. SPARQL Endpoints. This first set of scenarios was chosen to demonstrate the capabilities of FedBench to compare federation approaches for data stored in local RDF databases or accessible via SPARQL endpoints. In particular, they were designed to test how dimensions (D1) *Physical Distribution* of data and (D2) *Data Access* Interfaces affect query evaluation while the remaining dimensions are fixed across the scenarios, namely

(A1) centralized processing, where all data is held in a local, central store, vs.
(A2) local federation, where we use a federation of local repository endpoints, all of which are linked to each other in a federation layer, vs.

(A3) a federation of SPARQL endpoints, also linked to each other in a common local federation layer, pursuing the goal to test the overhead that is imposed by the SPARQL requests exchanged over the associated HTTP layer.

For all scenarios, we carried out experiments with the Sesame 2.3.2 engine using AliBaba version 2.0 beta 3, a federation layer for the Sesame framework which links integrated federation members together. In addition, we carried out experiments with the SPLENDID federation system from [9]. In contrast to the AliBaba federation, the latter uses statistical data about predicates and types to select appropriate data sources for answering triple patterns, which offers a wide range of optimization opportunities. Patterns that need to be sent to the same source are grouped and the join order between them is optimized with a dynamic programming approach, using e.g. the number of distinct subjects and objects per predicate to estimate the cardinality of intermediate results. The evaluation strategy relies on hash joins to allow for parallel execution and to reduce the number of HTTP requests, instead of sending individual result bindings to endpoints in a nested-loop join.

(B) Linked Data. Complementing the previous scenarios, we also evaluated a Linked Data scenario, where the data is distributed among a large number of sources that can only be retrieved using URI lookup and queries are evaluated on the combined graph formed by the union of these sources. Hence, in this setting the focus is on the knowledge about (D3) *Data Source Existence*. When all relevant sources are known, all of them are retrieved using their URIs before executing the query on the retrieved data (*top-down*) [14]. Another scheme that does not require a priori knowledge about data sources is an exploration-based approach [15]. Here, the query is assumed to contain at least one constant that is a URI. This URI is then used for retrieving the first source, and new sources are iteratively discovered in a bottom-up fashion starting with links found in that source (*bottom-up*). The (*mixed*) approach in [16] combines bottom-up and top-down to discover new sources at runtime as well as leverage known sources.

All three approaches in scenario (B) were evaluated based on the Linked Data query set (LD) using the prototype system from [16], which implements a stream-based query processing engine based on symmetric hash join operators. Note that all three approaches yield complete results (by design of the queries).

4.2   Setup and Evaluation Metrics

All experiments were carried out on an Integrated Lights-Out 2 (ILO2) HP server ProLiant DL360 G6 with 2000MHz 4Core CPU and 128KB L1 Cache, 1024KB L2 Cache, 4096KB L3 Cache, 32GB 1333MHz RAM, and a 160GB SCSI hard drive. They were run on top of a 64bit Windows 2008 Server operating system and executed with a 64bit Java VM 1.6.0_22 (all tested systems were Java-based). In the centralized setting (A1) we reserved a maximum of $28\,GB$ RAM to the VM, while in the distributed settings we assigned $20\,GB$ to the server process and $1\,GB$ to the client processes (i.e., the individual endpoints). Note that we run

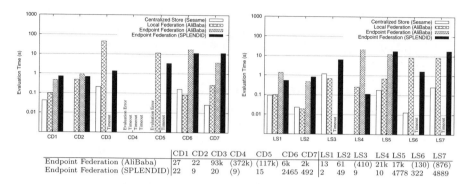

Fig. 2. Scenario (A): Evaluation Time (top) and Number of Requests to Endpoints (bottom) for the Cross Domain and Life Science Queries

all tests against local SPARQL endpoints, one for each federation member, to avoid unpredictable effects like network delay and high load of remote sources.

All query mixes in each setup have been run three times with a timeout of ten minutes per query. We report on the average total time over all runs. Queries that failed (e.g. with a system-internal exception) or delivered an unsound or incomplete result are indicated by "Evaluation Error". To exclude the influence of cold start effects, for each setup we ran a "ramp-up" query mix prior to query evaluation. We executed each query set separately and in order, counting the number of results (but not materializing them on disk). In addition to the evaluation time, we counted the number of requests sent to the individual endpoints (which is supported by our benchmark driver out-of-the-box).

For the Linked Data scenario, a CumulusRDF[4] Linked Data server was used to deploy the dataset on the local network. Both the server process and the Linked Data server were started with a maximum of 10GB RAM reserved. To simulate internet conditions, an artificial delay of 750ms was introduced, which resembles typical response times of current Linked Data servers.

4.3 Experimental Results

Figure 2 summarizes our results for the cross domain (CD) and life science (LS) queries in scenarios (A1)–(A3). The two plots visualize evaluation time, while the table at the bottom shows the number of requests sent by the systems to the federation members (numbers in parentheses are lower bounds for queries that failed due to timeout). Comparing Sesame and AliBaba first, we observe that the centralized approach in almost all cases outperforms the local federation approach, for 5 out of 14 queries even by an order of one magnitude or more (the time for the centralized store for queries CD2 and LS4 was about 1ms, which is not visible in the diagram). We observe an additional performance overhead for the AliBaba SPARQL Endpoint federation approach, which delivered results

---

[4] http://code.google.com/p/cumulusrdf/

only for 8 out of the 14 queries. Upon closer investigation, we could identify several reasons for the poor performance of the AliBaba federation approaches:

- Due to lack of statistics, effective join order optimization is often impossible, resulting in a high number of triples being exchange in the federation.
- Also caused by the lack of statistics, the AliBaba federation layer iteratively sends triple patterns to all federation members, to obtain candidate bindings for free variables. These bindings are then instantiated in the remaining part of the query, subsequently sending instantiated triple patterns to the federation members in a nested-loop fashion. This often results in a very high number of requests to the federation members (cf. the table at the bottom), which cause the high evaluation time. Given that AliBaba's strategy is identical in the local and SPARQL Endpoint federation scenario, the results indicate an enormous overhead imposed by the HTTP layer in the SPARQL Endpoint federation, typically in the order of one magnitude or more.
- Sesame's ability to deal with parallelization is limited. In the SPARQL Endpoint scenario, where our driver simulates endpoints by servlets that process incoming HTTP requests, we experimented with different degrees of parallelization. When instantiating more than 5–10 worker threads for answering the queries (each having its own repository connection), we could observe a performance drop down, manifesting in high waiting times for the worker threads, probably caused by Sesame's internal locking concept.

For the SPLENDID federation, we can observe that the number of HTTP requests is significantly lower: in contrast to AliBaba, which evaluates the query starting with a single triple pattern and iteratively substitutes results in subsequent patterns, SPLENDID also generates execution plans which send the patterns independently to relevant endpoints and join them together locally, at the server. Therefore, SPLENDID still returns results where AliBaba's naive nested-loop join strategy times out. For queries CD3, CD5, LS4 and LS6 it even beats the local federation.

Figure 3 summarizes our results in the SP^2Bench scenario. We observe that even the centralized Sesame store has severe problems answering the more complex queries, which is in line with previous investigations from [21].[5] Except for the outlier query SP1, the trends are quite similar to those observed in the (LS) and (CD) scenario, with the centralized scenario being superior to the local federation, which is again superior to the SPARQL Endpoint federation. Query SP1 is a simple query that asks for a specific journal in the data set. It can be answered efficiently on top of the local federation, because the federation is split up by type and the journals are distributed across only two federation members, so the system benefits from the parallelization of requests. In summary, the SP^2Bench experiments show that, for more complex queries and federation among a larger number of federation members (as indicated in Table 2), current federation approaches are still far from being applicable in practice.

---

[5] In the experiments from [21] Sesame was provided with all possible index combinations, whereas in these experiments we use only the standard indices. This explains why in our setting Sesame behaves slightly worse than in the experiments from [21].

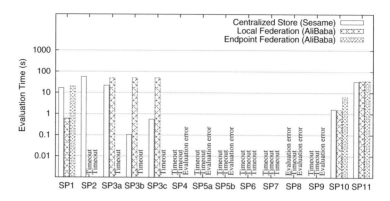

Fig. 3. Scenario (A): Results of SP^{2}Bench Queries

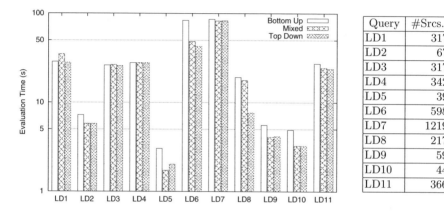

Query	#Srcs.
LD1	317
LD2	67
LD3	317
LD4	342
LD5	39
LD6	598
LD7	1219
LD8	217
LD9	59
LD10	44
LD11	366

Fig. 4. Scenario (B): Query Evaluation Times (left) and Number of Sources Dereferenced for Answering the Query (right)

Figure 4 visualizes our results for the Linked Data scenario (B). Regarding overall query time, the bottom-up and mixed approaches behave similarly: both perform run-time discovery and the mixed strategy cannot use its partial knowledge to restrict sources, but only to load relevant sources earlier. This leads to earlier result reporting, but is not reflected in the overall query time. In some cases the mixed approach is even slightly worse than bottom-up, due to the overhead imposed by using local source indices. The top-down approach, though, is able to restrict the number of sources to be retrieved, leading to better query times in many cases. For example, for query LD8 the query time for bottom up evaluation is 19.1s, while top-down requires 4.2s, an improvement of 75%, made possible by the lower number of sources retrieved in the top-down scenario.

Overall, the top-down approach uses its centralized-complete knowledge to identify relevant sources and exclude non-relevant sources. In a dynamic scenario, though, such as Linked Data, it may be infeasible to keep local indexes up-to-date, so exploration-based approaches like bottom-up or the mixed approach, which do not rely on complete knowledge, may be more suitable.

5   Conclusion

As witnessed by the evaluation, our benchmark is flexible enough to cover a wide range of semantic data application processing strategies and use cases, ranging from centralized processing over federation to pure Linked Data processing. Clearly, our experimental analysis is not (and does not intend to be) complete with respect to covering all existing systems and solutions – yet we have provided a flexible benchmark suite that can be used and extended by others to evaluate alternative approaches, systems, and scenarios.

Extensions we are planning to address in future work particularly include queries targeted at the new SPARQL features that will be published in the coming SPARQL 1.1 release such as aggregation, nested subqueries, and built-in support for federation. Further, given that the current data sets and queries focus on instance data and query answering over ground RDF graphs, extensions for testing reasoning capabilities (e.g., over RDFS and OWL data) in a distributed setting are left as future work. With our flexible framework, though, it is straightforward to incorporate such scenarios with little effort, and we invite the community to contribute to FedBench with own data and query sets.

Finally, our evaluation has revealed severe deficiencies of today's federation approaches, which underline the practicability of FedBench. As one of our major findings, data statistics – which are explicitly included in our benchmark suite – play a central role in efficient federated data processing: as indicated by our results, they are crucial in optimization to minimize the size of results shipped across federation members and requests exchanged within the federation.

Acknowledgments. Research reported in this paper was partially supported by the German BMBF in the project CollabCloud. http://collabcloud.de/

References

1. SPARQL 1.1 Federation Extensions. W3C Working Draft (June 1, 2010), http://www.w3.org/TR/sparql11-federated-query/
2. Abadi, D.J., Marcus, A., Madden, S.R., Hollenbach, K.: Scalable Semantic Web Data Management Using Vertical Partitioning. In: VLDB (2007)
3. Alexander, K., Hausenblas, M.: Describing Linked Datasets – On the Design and Usage of voiD. In: Linked Data on the Web Workshop (2009)
4. Angles, R., Gutierrez, C.: The Expressive Power of SPARQL. In: Sheth, A.P., Staab, S., Dean, M., Paolucci, M., Maynard, D., Finin, T., Thirunarayan, K. (eds.) ISWC 2008. LNCS, vol. 5318, pp. 114–129. Springer, Heidelberg (2008)
5. Buil-Aranda, C., Arenas, M., Corcho, O.: Semantics and Optimization of the SPARQL 1.1 Federation Extension. In: Antoniou, G., Grobelnik, M., Simperl, E., Parsia, B., Plexousakis, D., De Leenheer, P., Pan, J. (eds.) ESWC 201. LNCS, vol. 6644, pp. 1–15. Springer, Heidelberg (2011)

6. Bizer, C., Schultz, A.: The Berlin SPARQL Benchmark. Int. J. Semantic Web Inf. Syst. 5(2), 1–24 (2009)
7. Broekstra, J., Kampman, A., van Harmelen, F.: Sesame: A Generic Architecture for Storing and Querying RDF and RDF Schema. In: Horrocks, I., Hendler, J. (eds.) ISWC 2002. LNCS, vol. 2342, pp. 54–68. Springer, Heidelberg (2002)
8. Duan, S., Kementsietsidis, A., Srinivas, K., Udrea, O.: Apples and Oranges: A Comparison of RDF Benchmarks and Real RDF Datasets. In: SIGMOD (2011)
9. Görlitz, O., Staab, S.: Federated Data Management and Query Optimization for Linked Open Data. In: New Directions in Web Data Management (2011)
10. Gray, J.: Database and Transaction Processing Performance Handbook. In: The Benchmark Handbook (1993)
11. Guo, Y., Pan, Z., Heflin, J.: LUBM: A benchmark for OWL knowledge base systems. J. Web Sem. 3(2-3), 158–182 (2005)
12. Haase, P., Eberhart, A., Godelet, S., Mathäß, T., Tran, T., Ladwig, G., Wagner, A.: The Information Workbench - Interacting with the Web of Data. Technical report, fluid Operations & AIFB Karlsruhe (2009)
13. Haase, P., Mathäß, T., Ziller, M.: An Evaluation of Approaches to Federated Query Processing over Linked Data. In: I-SEMANTICS (2010)
14. Harth, A., Hose, K., Karnstedt, M., Polleres, A., Sattler, K.-U., Umbrich, J.: Data Summaries for On-Demand Queries over Linked Data. In: WWW (2010)
15. Hartig, O., Bizer, C., Freytag, J.C.: Executing SPARQL Queries Over the Web of Linked Data. In: Bernstein, A., Karger, D.R., Heath, T., Feigenbaum, L., Maynard, D., Motta, E., Thirunarayan, K. (eds.) ISWC 2009. LNCS, vol. 5823, pp. 293–309. Springer, Heidelberg (2009)
16. Ladwig, G., Tran, T.: Linked Data Query Processing Strategies. In: Patel-Schneider, P.F., Pan, Y., Hitzler, P., Mika, P., Zhang, L., Pan, J.Z., Horrocks, I., Glimm, B. (eds.) ISWC 2010, Part I. LNCS, vol. 6496, pp. 453–469. Springer, Heidelberg (2010)
17. Ladwig, G., Tran, T.: SIHJoin: Querying Remote and Local Linked Data. In: Antoniou, G., Grobelnik, M., Simperl, E., Parsia, B., Plexousakis, D., De Leenheer, P., Pan, J. (eds.) ESWC 2011, Part I. LNCS, vol. 6643, pp. 139–153. Springer, Heidelberg (2011)
18. Magkanaraki, A., Alexaki, S., Christophides, V., Plexousakis, D.: Benchmarking RDF Schemas for the Semantic Web. In: Horrocks, I., Hendler, J. (eds.) ISWC 2002. LNCS, vol. 2342, p. 132. Springer, Heidelberg (2002)
19. Neumann, T., Weikum, G.: Rdf-3X: a RISC-style engine for RDF. PVLDB 1(1) (2008)
20. Pérez, J., Arenas, M., Gutierrez, C.: Semantics and Complexity of SPARQL. ACM Trans. Database Syst. 34(3) (2009)
21. Schmidt, M., Hornung, T., Lausen, G., Pinkel, C.: SP2Bench: A SPARQL Performance Benchmark. In: ICDE, pp. 222–233 (2009)
22. Schmidt, M., Meier, M., Lausen, G.: Foundations of SPARQL Query Optimization. In: ICDT, pp. 4–33 (2010)
23. Schwarte, A., Haase, P., Hose, K., Schenkel, R., Schmidt, M.: FedX: Optimization Techniques for Federated Query Processing on Linked Data. In: Aroyo, L., et al. (eds.) ISWC 2011, Part I. LNCS, vol. 7031, pp. 601–616. Springer, Heidelberg (2011)
24. Weiss, C., Karras, P., Bernstein, A.: Hexastore: Sextuple Indexing for Semantic Web Data Management. PVLDB 1(1), 1008–1019 (2008)

FedX: Optimization Techniques for Federated Query Processing on Linked Data

Andreas Schwarte[1], Peter Haase[1], Katja Hose[2],
Ralf Schenkel[2], and Michael Schmidt[1]

[1] Fluid Operations AG, Walldorf, Germany
[2] Max-Planck Institute for Informatics, Saarbrücken, Germany

Abstract. Motivated by the ongoing success of Linked Data and the growing amount of semantic data sources available on the Web, new challenges to query processing are emerging. Especially in distributed settings that require joining data provided by multiple sources, sophisticated optimization techniques are necessary for efficient query processing. We propose novel join processing and grouping techniques to minimize the number of remote requests, and develop an effective solution for source selection in the absence of preprocessed metadata. We present FedX, a practical framework that enables efficient SPARQL query processing on heterogeneous, *virtually integrated* Linked Data sources. In experiments, we demonstrate the practicability and efficiency of our framework on a set of real-world queries and data sources from the Linked Open Data cloud. With FedX we achieve a significant improvement in query performance over state-of-the-art federated query engines.

1 Introduction

In recent years, the Web more and more evolved from a Web of Documents to a Web of Data. This development started a few years ago, when the Linked Data principles[3] were formulated with the vision to create a globally connected data space. The goal to integrate semantically similar data by establishing links between related resources is especially pursued in the Linking Open Data initiative, a project that aims at connecting distributed RDF data on the Web. Currently, the Linked Open Data cloud comprises more than 200 datasets that are interlinked by RDF links, spanning various domains ranging from Life Sciences over Media to Cross Domain data.

Following the idea of Linked Data, there is an enormous potential for integrated querying over multiple distributed data sources. In order to join information provided by these different sources, efficient query processing strategies are required, the major challenge lying in the natural distribution of the data. A commonly used approach for query processing in this context is to integrate relevant data sets into a local, centralized warehouse. However, accounting for the decentralized structure of the Semantic Web, recently one can observe a paradigm shift towards federated approaches over the distributed data sources with the ultimate goal of *virtual integration*[8, 13, 14, 16]. From the user perspective this means that data of multiple heterogeneous sources can be queried transparently as if residing in the same database.

L. Aroyo et al. (Eds.): ISWC 2011, Part I, LNCS 7031, pp. 601–616, 2011.

While there are efficient solutions to query processing in the context of RDF for local, centralized repositories [5, 15, 23], research contributions and frameworks for federated query processing are still in the early stages. Available systems offer poor performance, do not support the full SPARQL standard, and/or require local preprocessed metadata and statistics. The problem we deal with in this work is to find optimization techniques that allow for efficient SPARQL query processing on federated Linked Data. Our goal is to provide optimizations that do not require any preprocessing – thus allowing for on-demand federation setup – and that are realizable using SPARQL 1.0 language features. Given that in a distributed setting communication costs induced by network latency and transfer of data are a considerable factor, we claim that reducing the number of (remote) requests that are necessary to answer a query must be minimized. Thus, join processing strategies as well as other sophisticated optimization approaches are needed to find an appropriate solution.

In summary, our contributions are:

- We propose novel optimization techniques for federated RDF query processing (Section 3), including new join processing strategies for query processing targeted at minimizing the number of requests sent to federation members, mechanisms to group triple patterns that can be exclusively evaluated at single endpoints, and an effective approach for source selection without the need of preprocessed metadata.

- We present FedX (Section 4), a practical framework allowing for *virtual integration* of heterogeneous Linked Open Data sources into a federation. Our novel sophisticated optimization techniques combined with effective variants of existing approaches constitute the FedX query processing engine and allow for efficient SPARQL query processing. Linked Data sources can be integrated into the federation on-demand without preprocessing.

- We evaluate our system (Section 5) using experiments with a set of real-world queries and data sources. We demonstrate the practicability and efficiency of our framework on the basis of real data from the Linked Open Data cloud and compare our performance to other competitive systems.

2 Related Work

Driven by the success of Linked Data, recently various solutions for federated query processing of heterogeneous RDF data sources have been discussed in the literature. A recent overview and analysis of federated data management and query optimization techniques is presented in [6]. [9, 10] discuss the consumption of Linked Data from a database perspective. Federated query processing from a relational point of view has been studied in database research for a long time [11, 21]. Although the architectures and optimization approaches required in the context of RDF query processing have the same foundation, several problems arise due to differences in the data models.

Generally, in the context of Linked Data query processing, we can distinguish (a) bottom-up strategies that discover sources during query processing by following links between sources, and (b) top-down strategies that rely on upfront knowledge about relevant sources [7, 12]. Several bottom-up techniques including active discovery of new sources based on Linked Data HTTP lookups have been proposed in the literature [8, 13]. New relevant sources are discovered at runtime by following URIs of intermediate results using an iterator-based pipelining approach [8] or using the novel *Symmetric Index Hash Join* operator [13].

In our work, we focus on top-down strategies, where the relevant sources are known, hence guaranteeing sound and complete results over a *virtually integrated* data graph. In the research community various systems implementing these strategies have been proposed. *DARQ* [16] is a query engine allowing for SPARQL query processing on a number of (distributed) SPARQL endpoints. DARQ uses so-called *service descriptions* to summarize capabilities and statistics of data providers. This information is used in the optimization steps for source selection, i.e. sources for a triple pattern are determined based on predicate index lookups. Consequently, DARQ restricts query processing to queries in which all predicates are bound. A similar approach is employed in *SemWIQ* [14]. SemWIQ uses a concept-based approach and performs source selection based on type information of RDF entities available in a local dynamic catalog. SemWIQ requires that all subjects in a SPARQL query are variables. In addition, the type of each subject must be explicitly or implicitly known. In contrast to previous systems, our solution does not need any local preprocessed metadata since a different technique is employed for source selection. This makes it suitable for on-demand federation setup and practical query processing. Moreover, there is no limitation with respect to the SPARQL query language.

The W3C's SPARQL Working Group started to work on language extensions targeting the requirements and challenges arising in the context of distributed SPARQL processing. In a recent working draft[1], they propose the SERVICE operator, which allows for providing source information directly within the SPARQL query. In addition, BINDING clauses are introduced, which make it possible to efficiently communicate constraints to SPARQL endpoints. [2] provides a formal semantics for these features and presents a system called *SPARQL DQP*, which is capable of interpreting the new SERVICE keyword. SPARQL DQP does not need any preprocessed metadata, however, requires the endpoint to interpret SPARQL 1.1, which is typically not implemented in existing endpoints, as SPARQL 1.1 is currently available as a W3C working draft only.

In contrast to SPARQL DQP, FedX does not require any SPARQL 1.1 extensions and achieves automatic source selection over a set of defined sources (which can be dynamically extended) without additional input from the user. Thus, query formulation is more intuitive for the user, while query processing in most cases is as efficient as with manual specification of service providers. In fact, this is not a restriction: when implementing the SPARQL 1.1 federation extensions in a future release, FedX can exploit the SERVICE keyword for improved source selection and use the BINDING clauses to further optimize queries.

---

[1] http://www.w3.org/TR/sparql11-federated-query/

Statistics can influence performance tremendously in a distributed setting. The VoID vocabulary (Vocabulary of Interlinked Datasets) [1] allows to specify various statistics and features of datasets in a uniform way at the endpoint. In addition, the SPARQL Working Group proposes the SPARQL 1.1 service descriptions[2], which allow discovery of basic information about the SPARQL service. Although these (remote) statistics are a good foundation for various optimizations, the expressiveness is limited to basic statistics, such as the number of triples or distinct subjects. Currently, we focus on optimizations without these statistics, yet we are planning to incorporate them in a future release.

3   Optimization Techniques for Federated Linked Data

In a federated setting with distributed data sources it is important to optimize the query in such a way that the number of intermediate requests is minimized, while still guaranteeing fast execution of the individual requests. While we support full SPARQL 1.0, our optimization techniques focus on conjunctive queries, namely basic graph patterns (BGPs). A BGP is a set of triple patterns, a triple pattern being a triple (subject, predicate, object) with variables in zero or more positions.

Given that the SPARQL semantics is compositional, our strategy is to apply the optimizations to all conjunctive subqueries independently (including, e.g., BGPs nested inside OPTIONAL clauses) to compute the intermediate result sets. Since we aim at a practical federation framework capable of on-demand configuration, we additionally focus on optimizations that do not require pre-processed metadata and that are realizable using SPARQL 1.0.

In practice, there are two basic options to evaluate a SPARQL query in a federated setting: either (1) all triple patterns are individually and completely evaluated against every endpoint in the federation and the query result is constructed locally at the server or (2) an engine evaluates the query iteratively pattern by pattern, i.e., starting with a single triple pattern and substituting mappings from the pattern in the subsequent evaluation step, thus evaluating the query in a nested loop join fashion (NLJ). The problem with (1) is that, in particular when evaluating queries containing non-selective triple patterns (such as e.g. (?a,sameAs,?b)), a large amount of potentially irrelevant data needs to be shipped from the endpoints to the server. Therefore, we opt for the second approach. The problem with (2), though, is that the NLJ approach causes many remote requests, in principle one for each join step. We show that, with careful optimization, we can minimize the number of join steps (e.g., by grouping triple patterns) and minimize the number of requests sent in the NLJ approach.

3.1   Federated Query Processing Model

In our work, we focus on top-down strategies, where a set of user-configured sources is known at query time, hence guaranteeing sound and complete results over a virtually integrated data graph. Figure 1 depicts our federated query

---

[2] http://www.w3.org/TR/sparql11-service-description/

processing model, which closely follows the common workflow for general distributed query processing [11]. First, the SPARQL query is parsed and transformed into an internal representation (cf. Figure 2). Next, the relevant sources for each triple pattern are determined from the configured federation members using SPARQL ASK requests in conjunction with a local cache (Section 3.2). The remaining optimization steps include join order optimization (Section 3.3) as well as forming *exclusive groups* (Section 3.4). The outcome of the optimization step is the actual query execution plan. During query execution, subqueries are generated and evaluated at the relevant endpoints. The retrieved partial results are aggregated locally and used as input for the remaining operators. For iterative join processing the *bound joins* technique (Section 3.5) is applied to reduce the number of remote requests. Once all operators are executed, the final query result is returned to the client.

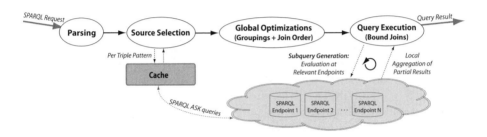

Fig. 1. Federated Query Processing Model

As a running example, Figure 2 depicts Life Science query 6 from our benchmark collections (Section 5) and illustrates the corresponding unoptimized query plan. The query computes all drugs in Drugbank[3] belonging to the category "Micronutrient" and joins computed information with corresponding drug names from the KEGG dataset[4]. A standard SPARQL query processing engine implementing the NLJ technique evaluates the first triple pattern in a single request, while the consecutive joins are performed in a nested loop fashion meaning that intermediate mappings of the left join argument are fed into the right join pattern one by one. Thus, the number of requests directly correlates with the number of intermediate results. In a federation, it must additionally be ensured that the endpoints appear *virtually integrated* in a combined RDF graph. This can in practice be achieved by sending each triple pattern to all federation members, using the union of partial results as input to the next operator.

3.2   Source Selection

Triple patterns of a SPARQL query need to be evaluated only at those data sources that can contribute results. In order to identify these *relevant sources*, we

---

[3] http://www4.wiwiss.fu-berlin.de/drugbank/
[4] http://kegg.bio2rdf.org/sparql

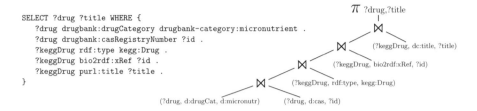

Fig. 2. Life Science Query 6 and the Corresponding Unoptimized Query Plan

use an effective technique, which does not require preprocessed metadata: before optimizing the query, we send SPARQL `ASK` queries for each triple pattern to the federation members and, based on the results, annotate each pattern in the query with its relevant source(s). Although this technique possibly overestimates the set of relevant data sources (e.g., for (`?s, rdf:type, ?o`) any data source will likely match during source selection, however, during join evaluation with actual mappings substituted for `?s` and `?o` there might not be results), in practical queries many triple patterns are specific to a single data source. Note also that FedX uses a cache to remember binary provenance information (i.e., whether source S is relevant/irrelevant for a triple pattern) in order to minimize the number of remote `ASK` queries.

Source selection has been discussed in previous works, e.g., [7, 14, 16]. However, existing approaches either require extensive local metadata or are too restrictive with respect to the SPARQL query language. In DARQ [16], for instance, relevant sources are determined using predicate lookups in so-called preprocessed *service descriptions*, hence requiring all predicates to be bound in a SPARQL query. The SPARQL 1.1 federation extension requires to specify sources in the query using the `SERVICE` keyword. In our approach we do not oblige the user to specify sources, while still offering efficient query computation.

3.3   Join Ordering

The join order determines the number of intermediate results and is thus a highly influential factor for query performance. For the federated setup, we propose a rule-based join optimizer, which orders a list of join arguments (i.e., triple patterns or groups of triple patterns) according to a heuristics-based cost estimation. Our algorithm uses a variation of the variable counting technique proposed in [22] and is depicted in Algorithm 1. Following an iterative approach it determines the argument with lowest cost from the remaining items (line 5-10) and appends it to the result list (line 13). For cost estimation (line 6) the number of free variables is counted considering already bound variables, i.e., the variables that are bound through a join argument that is already ordered in the result list. Additionally, we apply a heuristic that prefers *exclusive groups* (c.f. Section 3.4) since these in many cases can be evaluated with the highest selectivity.

Algorithm 1. Join Order Optimization

```
order(joinargs: list of n join arguments) {
 1: left ← joinargs
 2: joinvars ← ∅
 3: for i = 1 to n do
 4:     mincost ← MAX_VALUE
 5:     for all j ∈ left do
 6:         cost ← estimateCost(j, joinvars)
 7:         if cost < mincost then
 8:             arg ← j
 9:             mincost ← cost
10:        end if
11:    end for
12:    joinvars ← joinvars ∪ vars(arg))
13:    result[i] ← arg
14:    left ← left − arg
15: end for
16: return  result }
```

3.4   Exclusive Groups

High cost in federated query processing results from the local execution of joins at the server, in particular when joins are processed in a nested loop fashion. To minimize these costs, we introduce so-called *exclusive groups*, which play a central role in the FedX optimizer:

Definition 1. *Let $t_1 \ldots t_n$ be a set of triple patterns (corresponding to a conjunctive query), $S_1 \ldots S_n$ be distinct data sources, and S_t the set of relevant sources for triple pattern t. For $s \in \{S_1, \ldots, S_n\}$ we define $E_s := \{t \mid t \in \{t_1..t_n\} \text{ s.t. } S_t = \{S\} \}$ as the exclusive groups for source S, i.e. the triple patterns whose single relevant source is S.*

Exclusive groups with size ≥ 2 can be exploited for query optimization in a federated setting: instead of sending the triple patterns of such a group sequentially to the (single) relevant source, we can send them together (as a conjunctive query), thus executing them in a single subquery at the respective endpoint. Hence, for such groups only a single remote request is necessary, which typically leads to a considerably better performance because the amount of data to be transferred through the network and the number of requests often can be minimized by evaluating the subquery at the endpoint. This is because in many cases triple patterns that are not relevant for the final result are filtered directly at the endpoint, and on the other hand because the communication overhead of sending subqueries resulting from a nested loop join is avoided entirely. Correctness is guaranteed as no other data source can contribute to the group of triple patterns with further information.

In Figure 3, we illustrate the optimized query execution plan for our running example. During source selection, we annotate each triple pattern with its relevant sources and identify two exclusive groups, denoted as $\sum_{excl}$. For this query, we can reduce the number of local joins from four to just two.

3.5   Bound Joins

By computing the joins in a block nested loop fashion, i.e., as a distributed semijoin, it is possible to reduce the number of requests by a factor equivalent to

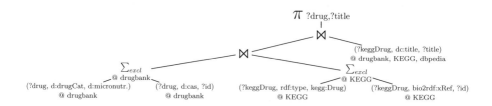

Fig. 3. Execution Plan of Life Science Query 6 (Including Optimizations)

the size of a *block*, in the following referred to as an *input sequence*. The overall idea of this optimization is to group a set of mappings in a single subquery using SPARQL UNION constructs. This grouped subquery is then sent to the relevant data sources in a single remote request. Finally, some post-processing is applied locally to retain correctness. We propose the *bound join* technique and discuss the technical insights below.

In the following, we illustrate bound join processing for the triple pattern (?S, name, ?O). For the example, assume that values for ?S have been computed yielding the input sequence $I:=$[?S=Person1,?S=Person2,?S=Person3]. Further, let us assume that the database (where we evaluate the triple pattern) contains the RDF triples $t_1=$(Person1, name, 'Peter') and $t_2=$(Person3, name, 'Andreas'). When evaluating the query sequentially for the bindings in the input sequence I, we obtain the result depicted in Figure 4 a). While the naive NLJ approach requires distinct subqueries for each input mapping substituted into the triple pattern (e.g., Person1, name, ?O), our bound join solution allows to evaluate the complete input sequence in a single grouped subquery. The concrete subquery for this example is depicted in Figure 4 b).

To guarantee correctness of the final result, we have to address three issues within the subquery: (1) we need to keep track of the original mappings, (2) possibly not all triple patterns yield results, and (3) the results of the subquery may be in arbitrary order. Our solution to this is an effective renaming technique: we annotate variable names in the subquery with the index of the respective mapping from the input sequence, e.g., for the first input mapping the constructed bound triple pattern is (Person1, name, ?O_1). This renaming technique allows to implicitly identify correspondences between partial subquery results and input mappings in a post-processing step. Figure 4 c) depicts the results of this subquery evaluated against our sample database. In the post-processing step the final result is reconstructed by matching the retrieved partial results to the

a) Expected Result

?S	?O
Person1	Peter
Person3	Andreas

b) SPARQL subquery

```
SELECT ?O_1 ?O_2 ?O_3 WHERE {
  { Person1 name ?O_1 } UNION
  { Person2 name ?O_2 } UNION
  { Person3 name ?O_3} }
```

c) Subquery result

?O_1	?O_2	?O_3
Peter		
		Andreas

Fig. 4. Sample execution for bound join processing of (?S, name, ?O)

corresponding input mapping using the index annotation in the variable name, and then performing the inverse renaming. In our running example, for instance, variable ?0_1 is linked to the first position in the input sequence; therefore, the binding from ?0_1 to 'Peter' is combined with the first binding for ?S in the input sequence, yielding the first result in Figure 4 a). Bound join processing can be trivially generalized to an input sequence of N mappings. For a detailed formalization and a technical discussion we refer the interested reader to [19].

A similar technique is discussed in [6, 24]. The authors propose to use a distributed semijoin sending the buffered mappings as additional conditions in a SPARQL FILTER expression. Although the theory behind this technique is similar to *bound joins*, in practice it is far less efficient than using UNIONs. We observed that for many available SPARQL endpoints the whole extension for a triple pattern is evaluated prior to applying the FILTER expressions. In the working draft for SPARQL 1.1 federation extensions, the W3C proposes the BINDINGS keyword to efficiently communicate constraints in the form of mappings to SPARQL endpoints, allowing to process sets of mappings corresponding to a block in a single subquery. We achieve a distributed semijoin without requiring this feature, using only SPARQL 1.0 language constructs. Clearly, our approach can easily be extended to utilize SPARQL 1.1 BINDINGS in the future.

4 FedX - Implementation

Having introduced various optimizations for distributed federated query processing on Linked Data, in this section we present FedX[5], a solution implementing the previously discussed techniques. FedX represents a practical solution for efficient federated query processing on heterogeneous, *virtually integrated* Linked Data sources. The practicability of FedX has been demonstrated in various scenarios in the Information Workbench[6] [20].

4.1 Architecture and Design

FedX has been developed to provide an efficient solution for distributed query processing on Linked Data. It is implemented in Java and extends the Sesame framework with a federation layer. FedX is incorporated into Sesame as SAIL (Storage and Inference Layer), which is Sesame's mechanism for allowing seamless integration of standard and customized RDF repositories. The underlying Sesame infrastructure enables heterogeneous data sources to be used as endpoints within the federation. On top of Sesame, FedX implements the logics for efficient query execution in the distributed setting utilizing the basic Sesame infrastructure (i.e., query parsing, Java mappings, I/O components) and adding the necessary functionality for data source management, endpoint communication and – most importantly – optimizations for distributed query processing.

---

[5] http://www.fluidops.com/FedX

[6] http://www.fluidops.com/information-workbench/

In FedX, data sources can be added to a federation in the form of any implementation of a Sesame repository. Standard implementations are provided for local, native Sesame repositories as well as for remote SPARQL endpoints. Furthermore, customized mediators can be integrated by implementing the appropriate Sesame interface. With these mediators different types of federations are possible: SPARQL federations integrating (remote) SPARQL endpoints, local federations consisting of native, local Sesame repositories, or hybrid forms. In the SPARQL federation, communication with the endpoints is done using HTTP-based SPARQL requests, while in the local case the native Java interfaces are employed. In the remainder of this paper, we focus on SPARQL federations.

4.2   Parallelization

Query processing in a federated, distributed environment is highly parallelizable meaning that different subqueries can be executed at the data sources concurrently. FedX incorporates a sophisticated parallelization infrastructure, which uses a multithreaded worker pool to execute the joins, i.e., *bound joins* (Section 3.5), and union operators in a highly parallelized fashion. In addition, we employ a pipelining approach such that intermediate results can be processed in the next operator as soon as they are ready – yielding higher throughput.

The parallelization architecture in FedX is realized by means of a `Scheduler` implementation managing a set of `ParallelTask`s and a pool of `WorkerThread`s. A `ParallelTask` refers to a prepared subquery to be executed at a particular data source. As an example, consider a task representing a single step of a nested loop join. In the scheduler, all tasks are maintained in a FIFO queue that the workers *pull* new tasks from. To reduce synchronization costs, worker threads are *paused* when they are idle, and notified when there are new tasks available. Note that only a single worker thread is notified if a new task arrives to avoid unnecessary synchronization overhead. Moreover, worker threads only go to sleep when there are no further tasks in the queue to avoid context switches. After experimenting with different configurations, we defined 25 worker threads for the scheduler as default.

4.3   Physical Join and Union Operators

For the physical `JOIN` operator, we tested with two variants: (1) parallel execution using a simple nested loop join and (2) our *bound join* technique, which we call *controlled worker join* (CJ) and *controlled bound worker join* (CBJ), respectively. Both variants generate tasks for each (block) nested loop iteration and submit them to the scheduler (Section 4.2). The scheduler then takes care of the controlled parallel execution of the tasks. For both, the CJ and CBJ implementation, synchronization is needed because the partial results of tasks belonging to the same join are merged, i.e., all partial results of a particular join are added to the same result set. In SPARQL federations, where (remote) requests cause a certain base cost, the CBJ operator improves performance significantly (see Section 5.2 for details) because the number of requests can be reduced tremendously. This is also the default implementation used in a SPARQL federation.

Note that in local federations with native Sesame stores the first approach, i.e., the CJ operator, outperforms bound joins because simple subqueries with a single triple pattern only, can be evaluated faster. This is because the data source can be accessed through native Java interfaces using Sesame's `getStatements` method, i.e., without prior SPARQL query construction.

Similarly, we provide two implementations for the `UNION` operator: a *synchronous union* (SU) and a *controlled worker union* (CU). The *synchronous union* executes its operands in a synchronous fashion, i.e., one union task after the other, thus avoiding synchronization overhead. In contrast, the *controlled worker union* executes the particular operands using the above described parallelization infrastructure (Section 4.2). The decision which implementation to use in a particular setup depends on the tradeoff between synchronization overhead and execution cost of an operand. In a remote setup, for instance, FedX benefits from parallel execution of a union since network latency and HTTP overhead typically outweigh synchronization costs. Note that union in this context does not solely refer to the SPARQL `UNION` operator but also to subqueries, which have to be evaluated at several relevant data sources resulting in a union of intermediate results. Consequently, for a SPARQL federation the *controlled worker union* is the implementation of choice, and for a local federation unions are evaluated using the *synchronous union* implementation. Note that SPARQL `UNION`s are always executed in the parallelization architecture described above.

5   Evaluation

In this section, we evaluate FedX and analyze the performance of our optimization techniques. With the goal of assessing the practicability of our system, we run various benchmarks and compare the results to state-of-the-art federated query processing engines. In our benchmark, we compare the performance of FedX with the competitive systems DARQ and AliBaba[7] since these are comparable to FedX in terms of functionality and the implemented query processing approach. Unfortunately, we were not able to obtain a prototype of the system presented in [2] for comparison.

5.1   Benchmark Setup

As a basis for our evaluation we use FedBench[8] [17], a comprehensive benchmark suite, which in contrast to other SPARQL benchmarks[4, 18] focuses on analyzing the efficiency and effectiveness of *federated* query processing strategies over semantic data. FedBench covers a broad range of scenarios and provides a benchmark driver to perform the benchmark in an integrative manner.

We select the Cross Domain (CD) as well as the Life Science (LS) data collections from the FedBench benchmark. The reason for our choice lies in the nature of the queries and data sets: both query sets implement realistic queries on top of real-world data from the Linked Open Data cloud. The queries focus on aspects

---

[7] http://www.openrdf.org/
[8] FedBench project page: http://code.google.com/p/fbench/

Table 1. Query characteristics of benchmark queries (a) and datasets used (b): Number
of triple patterns (#Tp.), data sources (#Src) and results (#Res); number of triples
included in datasets (#Triples) and preprocessing time for DARQ Service Description
(SD) in hh:mm:ss

a) **Query Characteristics**

Cross Domain (CD)				Life Science (LS)			
	#Tp.	#Src	#Res		#Tp.	#Src	#Res
1	3	2	90	1	2	2	1159
2	3	2	1	2	3	4	333
3	5	5	2	3	5	3	9054
4	5	5	1	4	7	2	3
5	4	5	2	5	6	3	393
6	4	4	11	6	5	3	28
7	4	5	1	7	5	3	144

b) **Datasets**

	#Triples	DARQ SD
DBpedia	43.6M	01:05:46
NYTimes	335k	00:00:09
LinkedMDB	6.15M	01:07:39
Jamendo	1.05M	00:00:20
GeoNames	108M	n/a
KEGG	1.09M	00:00:18
Drugbank	767k	00:00:12
ChEBI	7.33M	00:01:16

relevant for query processing over multiple sources and vary in join complexity,
query result size, the number of data sources involved, and structure (i.e., star
shaped, chain, or hybrid). Figure 2 in Section 3.1 depicts Life Science query 6
as an example; for space reasons we refer the interested reader to the FedBench
project page for the complete query set. To give a better understanding of the
queries, we summarize some characteristics in Table 1 a). In particular, we depict
the number of triple patterns, reference the number of results on the domain's
data sets, and an estimate of the relevant data sources (possibly overestimated).

The used data sources in the two scenarios are part of the Linked Open
Data cloud. Table 1 b) summarizes the included data collections. Details to the
datasets and various advanced statistics are provided at the FedBench project
page. To ensure reproducibility and reliability of the service, we conducted our
experiments on local copies of the SPARQL endpoints using the infrastructure
provided by FedBench, i.e. for each data source a local process is started publishing the respective data as individual SPARQL endpoint; we did not introduce
an additional delay to simulate network latency. All federation engines access
the data sources via the SPARQL protocol.

For the respective scenarios, we specify the relevant data sources as federation members upfront (Cross Domain: DBpedia, NYTimes, LinkedMDB, Jamendo, GeoNames; Life Sciences: KEGG, Drugbank, ChEBI, DBpedia). Note
that DARQ required additional preprocessing of the *service descriptions*, which
are needed for their source selection approach. The duration of this preprocessing
is depicted in Table 1 b). Even with 32GB RAM provided, a service description
for GeoNames could not be generated with DARQ's tools. Hence, we had to
omit the evaluation of DARQ for queries CD6 and CD7 (which require data
from GeoNames). Thus, the federation for the Cross Domain scenario had one
member less for DARQ.

All experiments are carried out on an HP Proliant DL360 G6 with 2GHz
4Core CPU with 128KB L1 Cache, 1024KB L2 Cache, 4096KB L3 Cache, 32GB
1333MHz RAM, and a 160 GB SCSI hard drive. A 64bit Windows 2008 Server
operating system and the 64bit Java VM 1.6.0_22 constitute the software environment. Sesame is integrated in version 2.3.2 and AliBaba's 2.0b3 build was
used. In all scenarios we assigned 20GB RAM to the process executing the query,

i.e. the query processing engine that is wrapped in the FedBench architecture. In the SPARQL federation we additionally assign 1GB RAM to each individual SPARQL endpoint process. For all experiments we defined a timeout of 10 minutes and all queries are executed 5 times, following a single warm-up run. All systems are run in their standard configurations.

5.2   Experimental Results

Figure 5 summarizes our experimental results of the Cross Domain and Life Science scenarios in a SPARQL federation. We depict the average query runtimes for AliBaba, DARQ, and FedX in Figure 5 a). As an overall observation, we find that FedX improves query performance significantly for most queries. Only in Query CD2 DARQ outperforms FedX. The reason is that FedX' exclusive group optimization in this query is more expensive than using simple triple patterns because the used SPARQL endpoint is more efficient for simple triple patterns for very small intermediate result sets (which is the case in this query as each triple pattern yields only a single result). For many queries the total runtime is improved by more than an order of magnitude. Moreover, timeouts and evaluation errors for this set of realistic queries are removed entirely. The improvement is best explained by the reduction in the number of requests, for which we provide a detailed analysis below. With our optimization techniques, we are able to reduce the number of requests significantly, e.g., from 170,579 (DARQ) and 93,248 (AliBaba) to just 23 (FedX) for query CD3. Such a reduction is made possible by the combined use of our optimization approaches, in particular source selection, exclusive groups, join reordering, and bound joins. Note that query CD2 and LS2 are not supported in DARQ since the query contains an unbound predicate, and that CD6 and CD7 are omitted since we were not able to generate the service description with 32GB RAM.

To measure the influence of caching the results of ASK requests during source selection, we performed a benchmark with activated and deactivated cache. The results are summarized in Figure 5 b). We observe that there is a slight overhead due to the additional communication. However, even with these ASK requests FedX significantly outperforms the state-of-the art systems for most queries. Our source selection technique – which in contrast to DARQ does not need preprocessed metadata – thus is effective in the federated setting.

Figure 5 c) summarizes the total number of requests sent to the data sources during query evaluation in the SPARQL federation. In particular, we indicate the results for AliBaba and DARQ, as well as for FedX with a nested loop implementation of the *controlled worker join* (CJ) and in the bound join variant using the *controlled worker bound join* (CBJ). These numbers immediately explain the improvements in query performance of FedX. With our optimization techniques, FedX is able to minimize the number of subqueries necessary to process the queries. Consider as an example query CD5: FedX is able to answer this query in just 18 requests, while DARQ needs 247,343. This is obviously immediately reflected in query runtime, which is just 0.097s in the case of FedX and 294.890s for DARQ. Note that the timeouts and the long runtimes of AliBaba and DARQ are easily explained with the number of requests sent to the endpoints.

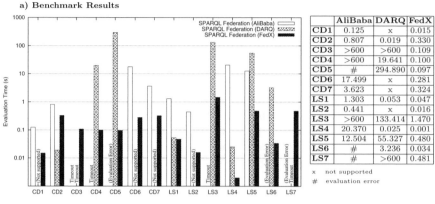

a) Benchmark Results

	AliBaba	DARQ	FedX
CD1	0.125	x	0.015
CD2	0.807	0.019	0.330
CD3	>600	>600	0.109
CD4	>600	19.641	0.100
CD5	#	294.890	0.097
CD6	17.499	x	0.281
CD7	3.623	x	0.324
LS1	1.303	0.053	0.047
LS2	0.441	x	0.016
LS3	>600	133.414	1.470
LS4	20.370	0.025	0.001
LS5	12.504	55.327	0.480
LS6	#	3.236	0.034
LS7	#	>600	0.481

x   not supported
evaluation error

b) Caching in FedX

	No Caching	Caching
CD1	0.044	0.015
CD2	0.374	0.330
CD3	0.219	0.109
CD4	0.134	0.100
CD5	0.131	0.097
CD6	0.508	0.281
CD7	0.449	0.324
LS1	0.062	0.047
LS2	0.038	0.016
LS3	2.202	1.470
LS4	0.018	0.001
LS5	0.633	0.480
LS6	0.063	0.034
LS7	0.686	0.481

c) Number of Requests

	AliBaba	DARQ	FedX CJ	FedX CBJ
CD1	27	x	7	7
CD2	22	5	2	2
CD3	(93,248)	(170,579)	63	23
CD4	(372,339)	22,331	69	38
CD5	(117,047)	247,343	35	18
CD6	6,183	x	2,457	185
CD7	1,883	x	1,508	138
LS1	13	1	1	1
LS2	61	x	38	18
LS3	(410)	101,386	14,221	2059
LS4	21,281	3	3	3
LS5	16,621	2,666	6,537	458
LS6	(130)	98	315	45
LS7	(876)	(576,089)	5,027	485

d) Join Operators

	CJ	CBJ
CD1	0.016	0.015
CD2	0.349	0.330
CD3	0.203	0.109
CD4	0.134	0.100
CD5	0.115	0.097
CD6	1.560	0.281
CD7	1.336	0.324
LS1	0.053	0.047
LS2	0.025	0.016
LS3	5.435	1.470
LS4	0.001	0.001
LS5	2.146	0.480
LS6	0.103	0.034
LS7	1.763	0.481

Fig. 5. Experimental Results of Cross Domain (CD) and Life Science (LS) Queries in SPARQL Federation: a) Benchmark Results of AliBaba, DARQ, and FedX. b) Influence of Caching ASK Requests for Source Selection. c) Total Number of Requests sent to Endpoints; Parentheses Indicate Timeouts after 10min or Evaluation Errors. d) Comparison of Join Operator Implementations in the SPARQL Federation: *Controlled Worker Join* (CJ), *Controlled Worker Bound Join* (CBJ). All Runtimes in Seconds.

In Figure 5 d) we compare the physical join operators of the *controlled worker join* (CJ) and *controlled worker bound join* (CBJ), which use the NLJ and bound joins (BNLJ) technique, respectively. We observe that the CBJ implementation significantly improves performance over the simple CJ variant since in a SPARQL federation we tremendously benefit from the reduction in the number of requests due to bound joins.

6   Conclusion and Outlook

In this paper, we proposed novel optimization techniques for efficient SPARQL query processing in the federated setting. As revealed by our benchmarks, bound joins combined with our grouping and source selection approaches are effective in terms of performance. By minimizing the number of intermediate requests, we are able to improve query performance significantly compared to state-of-the-art systems. We presented FedX, a practical solution that allows for querying multiple distributed Linked Data sources as if the data resides in a *virtually integrated* RDF graph. Compatible with the SPARQL 1.0 query language, our framework allows clients to integrate available SPARQL endpoints on-demand into a federation without any local preprocessing. While we focused on optimization techniques for conjunctive queries, namely basic graph patterns (BGPs), there is additional potential in developing novel, operator-specific optimization techniques for distributed settings (in particular for OPTIONAL queries), which we are planning to address in future work. As our experiments confirm, the optimization of BGPs alone (combined with common equivalence rewritings) already yields significant performance gains.

Important features for federated query processing are the federation extensions proposed for the upcoming SPARQL 1.1 language definition. These allow to specify data sources directly within the query using the SERVICE operator, and moreover to attach mappings to the query as data using the BINDINGS operator. When implementing the SPARQL 1.1 federation extensions for our next release, FedX can exploit these language features to further improve performance. In fact, the SPARQL 1.1 SERVICE keyword is a trivial extension, which enhances our source selection approach with possibilities for manual specification of new sources and gives the query designer more control.

Statistics can influence performance tremendously in a distributed setting. Currently, FedX does not use any local statistics since we follow the design goal of on-demand federation setup. We aim at providing a federation framework, in which data sources can be integrated ad-hoc, and used immediately for query processing. In a future release, (remote) statistics (e.g., using VoID [1]) can be incorporated for source selection and to further improve our join order algorithm.

Acknowledgments. Research reported in this paper was partially supported by the German BMBF in the project CollabCloud. http://collabcloud.de/

References

1. Alexander, K., Cyganiak, R., Hausenblas, M., Zhao, J.: Describing linked datasets - on the design and usage of void. In: Linked Data on the Web Workshop (LDOW 2009), in Conjunction with WWW 2009 (2009)
2. Buil-Aranda, C., Corcho, O., Arenas, M.: Semantics and Optimization of the SPARQL 1.1 Federation Extension. In: Antoniou, G., Grobelnik, M., Simperl, E., Parsia, B., Plexousakis, D., De Leenheer, P., Pan, J. (eds.) ESWC 2011. LNCS, vol. 6644, pp. 1–15. Springer, Heidelberg (2011)
3. Berners-Lee, T.: Linked data - design issues (2006), http://www.w3.org/DesignIssues/LinkedData.html (retrieved August 25, 2011)
4. Bizer, C., Schultz, A.: The Berlin SPARQL Benchmark. Int. J. Semantic Web Inf. Syst. 5(2), 1–24 (2009)
5. Erling, O., Mikhailov, I.: RDF support in the virtuoso DBMS. In: CSSW. LNI, vol. 113, pp. 59–68. GI (2007)
6. Görlitz, O., Staab, S.: Federated Data Management and Query Optimization for Linked Open Data. In: Vakali, A., Jain, L.C. (eds.) New Directions in Web Data Management 1. SCI, vol. 331, pp. 109–137. Springer, Heidelberg (2011)

7. Harth, A., Hose, K., Karnstedt, M., Polleres, A., Sattler, K.-U., Umbrich, J.: Data summaries for on-demand queries over linked data. In: WWW (2010)
8. Hartig, O., Bizer, C., Freytag, J.-C.: Executing SPARQL Queries over the Web of Linked Data. In: Bernstein, A., Karger, D.R., Heath, T., Feigenbaum, L., Maynard, D., Motta, E., Thirunarayan, K. (eds.) ISWC 2009. LNCS, vol. 5823, pp. 293–309. Springer, Heidelberg (2009)
9. Hartig, O., Langegger, A.: A database perspective on consuming linked data on the web. Datenbank-Spektrum 10, 57–66 (2010)
10. Hose, K., Schenkel, R., Theobald, M., Weikum, G.: Database Foundations for Scalable RDF Processing. In: Polleres, A., d'Amato, C., Arenas, M., Handschuh, S., Kroner, P., Ossowski, S., Patel-Schneider, P. (eds.) Reasoning Web 2011. LNCS, vol. 6848, pp. 202–249. Springer, Heidelberg (2011)
11. Kossmann, D.: The state of the art in distributed query processing. ACM Computing Surveys 32(4), 422–469 (2000)
12. Ladwig, G., Tran, T.: Linked Data Query Processing Strategies. In: Patel-Schneider, P.F., Pan, Y., Hitzler, P., Mika, P., Zhang, L., Pan, J.Z., Horrocks, I., Glimm, B. (eds.) ISWC 2010, Part I. LNCS, vol. 6496, pp. 453–469. Springer, Heidelberg (2010)
13. Ladwig, G., Tran, T.: SIHJoin: Querying remote and local linked data. In: Antoniou, G., Grobelnik, M., Simperl, E., Parsia, B., Plexousakis, D., De Leenheer, P., Pan, J. (eds.) ESWC 2011, Part I. LNCS, vol. 6643, pp. 139–153. Springer, Heidelberg (2011)
14. Langegger, A., Wöß, W., Blöchl, M.: A Semantic Web Middleware for Virtual Data Integration on the Web. In: Bechhofer, S., Hauswirth, M., Hoffmann, J., Koubarakis, M. (eds.) ESWC 2008. LNCS, vol. 5021, pp. 493–507. Springer, Heidelberg (2008)
15. Neumann, T., Weikum, G.: The RDF-3X engine for scalable management of RDF data. The VLDB Journal 19, 91–113 (2010)
16. Quilitz, B., Leser, U.: Querying Distributed RDF Data Sources with SPARQL. In: Bechhofer, S., Hauswirth, M., Hoffmann, J., Koubarakis, M. (eds.) ESWC 2008. LNCS, vol. 5021, pp. 524–538. Springer, Heidelberg (2008)
17. Schmidt, M., Görlitz, O., Haase, P., Ladwig, G., Schwarte, A., Tran, T.: FedBench: A Benchmark Suite for Federated Semantic Data Query Processing. In: Aroyo, L., et al. (eds.) ISWC 2011, Part I. LNCS, vol. 7031, pp. 585–600. Springer, Heidelberg (2011)
18. Schmidt, M., Hornung, T., Lausen, G., Pinkel, C.: SP2Bench: A SPARQL Performance Benchmark. In: ICDE, pp. 222–233 (2009)
19. Schwarte, A.: FedX: Optimization Techniques for Federated Query Processing on Linked Data. Master's thesis, Saarland University, Germany (2011)
20. Schwarte, A., Haase, P., Hose, K., Schenkel, R., Schmidt, M.: FedX: A Federation Layer for Distributed Query Processing on Linked Open Data. In: Antoniou, G., Grobelnik, M., Simperl, E., Parsia, B., Plexousakis, D., De Leenheer, P., Pan, J. (eds.) ESWC 2011. LNCS, vol. 6644, pp. 481–486. Springer, Heidelberg (2011)
21. Sheth, A.P.: Federated Database Systems for Managing Distributed, Heterogeneous, and Autonomous Databases. In: VLDB 1991, p. 489 (1991)
22. Stocker, M., Seaborne, A., Bernstein, A., Kiefer, C., Reynolds, D.: SPARQL basic graph pattern optimization using selectivity estimation. In: WWW, pp. 595–604. ACM (2008)
23. Weiss, C., Karras, P., Bernstein, A.: Hexastore: sextuple indexing for semantic web data management. PVLDB 1(1), 1008–1019 (2008)
24. Zemanek, J., Schenk, S., Svatek, V.: Optimizing SPARQL Queries over Disparate RDF Data Sources through Distributed Semi-Joins. In: ISWC 2008 Poster and Demo Session Proceedings. CEUR-WS (2008)

Local Closed World Semantics: Grounded Circumscription for OWL

Kunal Sengupta, Adila Alfa Krisnadhi, and Pascal Hitzler

Wright State University, Dayton OH 45435, USA
{kunal,adila,pascal}@knoesis.org

Abstract. We present a new approach to adding closed world reasoning to the Web Ontology Language OWL. It transcends previous work on circumscriptive description logics which had the drawback of yielding an undecidable logic unless severe restrictions were imposed. In particular, it was not possible, in general, to apply local closure to roles.

In this paper, we provide a new approach, called *grounded circumscription,* which is applicable to $\mathcal{SROIQ}$ and other description logics around OWL without these restrictions. We show that the resulting language is decidable, and we derive an upper complexity bound. We also provide a decision procedure in the form of a tableaux algorithm.

1 Introduction

The semantics of the Web Ontology Language OWL [8] (which is based on the description logic $\mathcal{SROIQ}$ [9]) adheres to the Open World Assumption (OWA): statements which are *not* logical consequences of a given knowledge base are not necessarily considered false. The OWA is a reasonable assumption to make in the World Wide Web context (and thus for Semantic Web applications). However, situations naturally arise where it would be preferable to use the Closed World Assumption (CWA), that is, statements which are *not* logical consequences of a given knowledge base are considered false. Such situations include, for example, when data is being retrieved from a database, or when data can be considered *complete* with respect to the application at hand (see, e.g., [6,23]).

As a consequence, efforts have been made to combine OWA and CWA modeling for the Semantic Web, and knowledge representation languages which have both OWA and CWA modeling features are said to adhere to the *Local Closed World Assumption* (LCWA). Most of these combinations are derived from non-monotonic logics which have been studied in logic programming [10] or on first-order predicate logic [19,20,24]. Furthermore, many of them are of a *hybrid* nature, meaning that they achieve the LCWA by combining, e.g., description logics with (logic programming) rules. Please see [14, Section 4].

On the other hand, there are not that many approaches which provide a seamless (non-hybrid) integration of OWA and CWA, and each of them has its drawbacks. This is despite the fact that the modeling task, from the perspective of the application developer, seems rather simple: Users would want to specify,

L. Aroyo et al. (Eds.): ISWC 2011, Part I, LNCS 7031, pp. 617–632, 2011.

simply, that individuals in the extension of a predicate should be exactly those which are *necessarily required* to be in it, i.e., extensions should be *minimized*. Thus, what is needed for applications is a simple, intuitive approach to closed world modeling which caters for the above intuition, and is also sound, complete and computationally feasible.

Among the primary approaches to non-monotonic reasoning, there is one approach which employs the minimization idea in a very straightforward and intuitively simple manner, namely *circumscription* [19]. However, a naive transfer of the circumscription approach to description logics, which was done in [2,3,6,7], appears to have three primary drawbacks.

1. The approach is undedicable for expressive description logics (e.g., for the description logic $\mathcal{SROIQ}$) unless awkward restrictions are put in place. More precisely, it is not possible to have non-empty TBoxes plus minimization of roles if decidability is to be retained.
2. Extensions of minimized predicates can still contain elements which are not named individuals (or pairs of such, for roles) in the knowledge base, which is not intuitive for modeling (see also [6]).
3. Complexity of the approach is very high.

The undecidability issue (point 1) hinges, in a sense, also on point 2 above. In this paper, we provide a modified approach to circumscription for description logics, which we call *grounded circumscription*, that remedies both points 1 and 2.[1] Our idea is simple yet effective: we modify the circumscription approach from [2,3,6,7] by adding the additional requirement that extensions of minimized predicates may only contain named individuals (or pairs of such, for roles). In a sense, this can be understood as porting a desirable feature from (hybrid) MNKF description logics [5,12,13,21] to the circumscription approach. In another (but related) sense, it can also be understood as employing the idea of DL-safety [22], respectively of DL-safe variables [17] or nominal schemas [4,15,16].

The paper is a substantial extension of the workshop paper [14] and will be structured as follows. In Section 2, we introduce the semantics of grounded circumscription. In Section 3, we show that the resulting language is decidable. Next, we provide a tableaux calculus in Section 4 to reason with grounded circumscription. We conclude with a discussion of further work in Section 5.

2   Local Closed World Reasoning with Grounded Circumscription

In this section we describe LCW reasoning with grounded circumscription (GC) and also revisit the syntax and semantics of the Description Logic $\mathcal{ALC}$ and extend it with GC. Some results in this paper also apply to many other description logics besides $\mathcal{ALC}$, and we will point this out in each case.

---

[1] We are not yet addressing the complexity issue; this will be done in future work.

2.1   The Description Logic $\mathcal{ALC}$

Let $\mathsf{N_C}, \mathsf{N_R}$ and $\mathsf{N_I}$ be countably infinite sets of concept names, role names and individual names, respectively. The set of $\mathcal{ALC}$ *concepts* is the smallest set that is created using the following grammar where $A \in \mathsf{N_C}$ denotes an atomic concept, $R \in \mathsf{N_R}$ is a role name and C, D are concepts.

$$C \longrightarrow \top \mid \bot \mid A \mid \neg C \mid C \sqcap D \mid C \sqcup D \mid \exists R.C \mid \forall R.C$$

An $\mathcal{ALC}$ *TBox* is a finite set of axioms of the form $C \sqsubseteq D$, called *general concept inclusion* (GCI) *axioms*, where C and D are concepts. An $\mathcal{ALC}$ *ABox* is a finite set of axioms of the form $C(a)$ and $R(a, b)$, which are called *concept* and *role assertion axioms*, where C is a concept, R is a role and a, b are individual names. An $\mathcal{ALC}$ *knowledge base* is a union of an $\mathcal{ALC}$ ABox and an $\mathcal{ALC}$ TBox

The semantics is defined in terms of interpretations $\mathcal{I} = (\Delta^{\mathcal{I}}, \cdot^{\mathcal{I}})$, where $\Delta^{\mathcal{I}}$ is a non-empty set called the *domain* of interpretation and $\cdot^{\mathcal{I}}$ is an interpretation function which maps each individual name to an element of the domain $\Delta^{\mathcal{I}}$ and interprets concepts and roles as follows.

$$\top^{\mathcal{I}} = \Delta^{\mathcal{I}}, \quad \bot^{\mathcal{I}} = \emptyset, \quad A^{\mathcal{I}} \subseteq \Delta^{\mathcal{I}}, \quad R^{\mathcal{I}} \subseteq \Delta^{\mathcal{I}} \times \Delta^{\mathcal{I}}$$
$$(\neg C)^{\mathcal{I}} = \Delta^{\mathcal{I}} \setminus C^{\mathcal{I}}, \quad (C_1 \sqcap C_2)^{\mathcal{I}} = C_1^{\mathcal{I}} \cap C_2^{\mathcal{I}}, \quad (C_1 \sqcup C_2)^{\mathcal{I}} = C_1^{\mathcal{I}} \cup C_2^{\mathcal{I}}$$
$$(\forall r.C)^{\mathcal{I}} = \{x \in \Delta^{\mathcal{I}} \mid (x, y) \in r^{\mathcal{I}} \text{ implies } y \in C^{\mathcal{I}}\}$$
$$(\exists r.C)^{\mathcal{I}} = \{x \in \Delta^{\mathcal{I}} \mid \text{there is some } y \text{ with } (x, y) \in r^{\mathcal{I}} \text{ and } y \in C^{\mathcal{I}}\}$$

An interpretation $\mathcal{I}$ *satisfies* (is a *model* of) a GCI $C \sqsubseteq D$ if $C^{\mathcal{I}} \subseteq D^{\mathcal{I}}$, a concept assertion $C(a)$ if $a^{\mathcal{I}} \in C^{\mathcal{I}}$, a role assertion $R(a, b)$ if $(a^{\mathcal{I}}, b^{\mathcal{I}}) \in R^{\mathcal{I}}$. We say $\mathcal{I}$ *satisfies* (is a *model* of) a knowledge base K if it satisfies every axiom in K. K is *satisfiable* if such a model $\mathcal{I}$ exists.

The negation normal form of a concept C, denoted by $\mathsf{NNF}(C)$, is obtained by pushing the negation symbols inward, as usual, such that negation appears only in front of atomic concepts, e.g., $\mathsf{NNF}(\neg(C \sqcup D)) = \neg C \sqcap \neg D$.

Throughout the paper, we will often talk about $\mathcal{L}$ knowledge bases ($\mathcal{L}$-KBs for short), where $\mathcal{L}$ is some decidable description logic. When we do this, then this indicates that the result does not only hold for $\mathcal{L}$ being $\mathcal{ALC}$, but rather for many decidable description logics around OWL. We will point out restrictions in each case. For general background on various description logics, as well as for established names (like $\mathcal{ALC}$) for different description logics, see [1,9].

Besides widely known DL constructors, we will also make use of Boolean role constructors (in limited form), which can be added to many description logics without loss of decidability or even of complexity [25,28]. We also make limited use of the *concept product*, written $C \times D$ with C, D concepts in $\mathcal{L}$, which allows a role to be constructed from the Cartesian product of two concepts, and which can actually be eliminated in the presence of Boolean role constructors [16,26]. In terms of interpretations $\mathcal{I}$, concept products are characterized by the equation $(C \times D)^{\mathcal{I}} = \{(x, y) \mid x \in C^{\mathcal{I}}, y \in D^{\mathcal{I}}\}$.

2.2 Grounded Circumscription

We now describe a very simple way for ontology engineers to model local closed world aspects in their ontologies: simply use a description logic (DL) knowledge base (KB) as usual, and augment it with *meta*-information which states that some predicates (concept names or role names) are *closed*. Semantically, those predicates are considered minimized, i.e., their extensions contain only what is absolutely required, and furthermore only contain *known* (or *named*) individuals, i.e., individuals which are explicitly mentioned in the KB. In the case of concept names, the idea of restricting their extensions only to known individuals is similar to the notion of nominal schema [4,16] (and thus, DL-safe rules [17,22]) and also the notion of DBox [27], while the minimization idea is borrowed from circumscription [19], one of the primary approaches to non-monotonic reasoning.

In the earlier efforts to carry over circumscription to DLs [2,3,6,7], circumscription is realized by the notion of *circumscription pattern*. A circumscription pattern consists of three disjoint sets of predicates (i.e., concept names and role names) which are called *minimized, fixed* and *varying* predicates, and a preference relation on interpretations.[2] The preference relation allows us to pick *minimal* models as the *preferred* models with respect to set inclusion of the extensions of the minimized predicates.

Our formalism here is inspired by one of the approaches described by Makinson in [18], namely restricting the set of valuations to get more logical consequences than what we can get as classical consequences. Intuitively, this approach is a simpler version of the circumscription formalism for DLs as presented in [3,7] in the sense that we restrict our attention only to models in which the extension of minimized predicates may only contain known individuals from the KB. Furthermore, the predicates (concept names and role names) in KB are partitioned into two disjoint sets of minimized and non-minimized predicates, i.e., no predicate is considered fixed.[3] The non-minimized predicates would be viewed as varying in the more general circumscription formalism mentioned above.

The non-monotonic feature of the formalism is given by restricting models of an $\mathcal{L}$-KB such that the extension of closed predicates may only contain individuals (or pairs of them) which are explicitly occurring in the KB, plus a minimization of the extensions of these predicates. We define a function Ind that maps each $\mathcal{L}$-KB to the set of individual names it contains, i.e., given an $\mathcal{L}$-KB K, $\mathsf{Ind}(K) = \{b \in \mathsf{N}_I \mid b \text{ occurs in } K\}$. Among all possible models of K that are obtained by the aforementioned restriction to $\mathsf{Ind}(K)$, we then select a model that is minimal w.r.t. concept inclusion or role inclusion, in accordance with the following definition.

---

[2] There is also a notion of *prioritization* which we will not use, mainly because we are not convinced yet that it is a desirable modeling feature for local closed world reasoning for the Semantic Web.

[3] Fixed predicates can be simulated in the original circumscriptive DL approach if negation is available, i.e., for fixed concept names, concept negation is required, while for fixed role names, role negation is required. The latter can be added to expressive DLs without jeopardizing decidability [16,28].

Definition 1. *A* GC-$\mathcal{L}$-KB *is a pair* (K, M) *where* K *is an* $\mathcal{L}$-KB *and* $M \subseteq$ $N_C \cup N_r$. *For every concept name and role name* $W \in M$, *we say that* W *is* closed *with respect to* K. *For any two models* $\mathcal{I}$ *and* $\mathcal{J}$ *of* K, *we furthermore say that* $\mathcal{I}$ *is* smaller than *(or* preferred over*)* $\mathcal{J}$ *w.r.t.* M, *written* $\mathcal{I} \prec_M \mathcal{J}$, *iff all of the following hold: (i)* $\Delta^{\mathcal{I}} = \Delta^{\mathcal{J}}$ *and* $a^{\mathcal{I}} = a^{\mathcal{J}}$ *for every* $a \in N_I$; *(ii)* $W^{\mathcal{I}} \subseteq W^{\mathcal{J}}$ *for every* $W \in M$; *and (iii) there exists a* $W \in M$ *such that* $W^{\mathcal{I}} \subset W^{\mathcal{J}}$

The following notion will be helpful.

Definition 2 (grounded model). *Given a GC-$\mathcal{L}$-KB* (K, M), *a model* $\mathcal{I}$ *of* K *is called a* grounded model *w.r.t* M *if all of the following hold:*
(1) $C^{\mathcal{I}} \subseteq \{b^{\mathcal{I}} \mid b \in \mathsf{Ind}(K)\}$ *for each concept* $C \in M$; *and*
(2) $R^{\mathcal{I}} \subseteq \{(a^{\mathcal{I}}, b^{\mathcal{I}}) \mid a, b \in \mathsf{Ind}(K)\}$ *for each role* $R \in M$

We now define models and logical consequence of GC-$\mathcal{L}$-KBs as follows.

Definition 3. *Let* (K, M) *be a GC-$\mathcal{L}$-KB. An interpretation* $\mathcal{I}$ *is a* GC-model *of* (K, M) *if it is a grounded model of* K *w.r.t* M *and* $\mathcal{I}$ *is minimal w.r.t.* M, *i.e., there is no model* $\mathcal{J}$ *of* K *with* $\mathcal{J} \prec_M \mathcal{I}$. *A statement (GCI, concept assertion, or role assertion)* α *is a* logical consequence *(a* GC-inference*) of* (K, M) *if every GC-model of* (K, M) *satisfies* α. *Finally, a GC-$\mathcal{L}$-KB is said to be* GC-satisfiable *if it has a GC-model.*

Note that every GC-model is also a grounded model. Moreover, in comparison with the more general circumscription formalism for DLs as presented in [3,7], every GC-model of a KB is also a circumscriptive model,[4] hence every circumscriptive inference is also a valid GC-inference.

To give an example, consider the knowledge base K consisting of the axioms

$$\mathsf{hasAuthor}(\mathsf{paper1}, \mathsf{author1}) \qquad \mathsf{hasAuthor}(\mathsf{paper1}, \mathsf{author2})$$

$$\mathsf{hasAuthor}(\mathsf{paper2}, \mathsf{author3}) \qquad \top \sqsubseteq \forall \mathsf{hasAuthor}.\mathsf{Author}$$

Consider the following (ABox) statements: $\neg\mathsf{hasAuthor}(\mathsf{paper1}, \mathsf{author3})$ and $(\leq 2\ \mathsf{hasAuthor}.\mathsf{Author})(\mathsf{paper1})$.[5] Neither of them is a logical consequence of K under classical DL semantics. However, if we assume that we have complete information on authorship relevant to the application under consideration, then it would be reasonable to *close* parts of the knowledge base in the sense of the LCWA. In the original approach to circumscriptive DLs, we could close the concept name Author, but to no avail. But if we close $\mathsf{hasAuthor}$, we obtain $(\leq 2\ \mathsf{hasAuthor}.\mathsf{Author})(\mathsf{paper1})$ as a logical consequence. In addition, if we adopt the Unique Name Assumption (UNA), $\neg\mathsf{hasAuthor}(\mathsf{paper1}, \mathsf{author3})$ is also a logical consequence of K. Even without UNA, we can still obtain this as a logical consequence if we add the following axioms to K, which essentially

---

[4] This can be seen, e.g., by a straightforward proof by contradiction.
[5] The semantics is $(\leq n\ R.C)^{\mathcal{I}} = \{x \in \Delta^{\mathcal{I}} \mid |\{y \mid (x,y) \in R^{\mathcal{I}} \text{ and } y \in C^{\mathcal{I}}\}| \leq n\}$; this *qualified number restriction* is not part of $\mathcal{ALC}$, though it makes a very good example without depending on the UNA.

forces the UNA:[6] $A_1(\texttt{author1}); A_2(\texttt{author2}); A_3(\texttt{author3}); A_i \sqcap A_j \sqsubseteq \bot$ for all $i \neq j$. With regard to this example, note that the closure of roles in the original circumscriptive DL approach leads to undecidability [3]. The GC-semantics, in contrast, is decidable even under role closure (see Section 3 below), and also yields the desired inferences.

3 Decidability of Grounded Circumscription

As noted earlier, circumscription in many expressive DLs is undecidable [3]. Undecidability even extends to the basic DL $\mathcal{ALC}$ when non-empty TBoxes are considered and roles are allowed as minimized predicates. Such a bleak outlook would greatly discourage useful application of circumscription, despite the fact that there is a clear need of such a formalism to model LCWA.

Our formalism aims to fill this gap by offering a simpler approach to circumscription in DLs that is decidable provided that the underlying DL is also decidable. The decidability result is obtained due to the imposed restriction of minimized predicates to known individuals in the KB as specified in Definition 3. Let $\mathcal{L}$ be any standard DL. We consider the reasoning task of *GC-KB satisfiability*: "given a GC-$\mathcal{L}$-KB (K, M), does (K, M) have a GC-model?" and show in the following that this is decidable.

Assume that $\mathcal{L}$ is any DL featuring nominals, concept disjunction, concept products, role hierarchies and role disjunctions. We show that GC-KB satisfiability in $\mathcal{L}$ is decidable if satisfiability in $\mathcal{L}$ is decidable.

Let (K, M) be a GC-$\mathcal{L}$-KB. We assume that $M = M_A \cup M_r$ where $M_A = \{A_1, \ldots, A_n\}$ is the set of minimized concept names and $M_r = \{r_1, \ldots, r_m\}$ is the set of minimized role names. Now define a family of $(n + m)$-tuples as

$$\mathcal{G}_{(K,M)} = \{(X_1, \ldots, X_n, Y_1, \ldots, Y_m) \mid X_i \subseteq \mathsf{Ind}(K), Y_j \subseteq \mathsf{Ind}(K) \times \mathsf{Ind}(K)\}$$

with $1 \leq i \leq n, 1 \leq j \leq m$. Note that there are

$$\left(2^{|\mathsf{Ind}(K)|}\right)^n \cdot \left(2^{|\mathsf{Ind}(K)|^2}\right)^m = 2^{n \cdot |\mathsf{Ind}(K)| + m \cdot |\mathsf{Ind}(K)|^2} \tag{1}$$

of such tuples; in particular note that $\mathcal{G}_{(K,M)}$ is a finite set.

Now, given (K, M) and some $G = (X_1, \ldots, X_n, Y_1, \ldots, Y_m) \in \mathcal{G}_{(K,M)}$, let K_G be the $\mathcal{L}$-KB consisting of all axioms in K together with all of the following axioms, where the A_i and r_j are all the predicates in M—note that we require role disjunction and concept products for this.

$$A_i \equiv \bigsqcup \{a\} \quad \text{for every } a \in X_i \text{ and } i = 1, \ldots, n$$

$$r_j \equiv \bigsqcup (\{a\} \times \{b\}) \quad \text{for every pair } (a, b) \in Y_j \text{ and } j = 1, \ldots, m$$

Then the following result clearly holds.

---

[6] The UNA can be enforced in an $\mathcal{ALC}$ KB by adding ABox statements $A_i(a_i)$, where a_i are all individuals and A_i are new concept names, to the knowledge base, together with all disjointness axioms of the form $A_i \sqcap A_j \sqsubseteq \bot$ for all $i \neq j$.

Lemma 1. *Let (K, M) be a GC-$\mathcal{L}$-KB. If K has a grounded model $\mathcal{I}$ w.r.t. M, then there exists $G \in \mathcal{G}_{(K,M)}$ such that K_G has a (classical) model $\mathcal{J}$ which coincides with $\mathcal{I}$ on all minimized predicates. Likewise, if there exists $G \in \mathcal{G}_{(K,M)}$ such that K_G has a (classical) model $\mathcal{J}$, then K has a grounded model $\mathcal{I}$ which coincides with $\mathcal{J}$ on all minimized predicates.*

Now consider the set

$$\mathcal{G}'_{(K,M)} = \{G \in \mathcal{G}_{(K,M)} \mid K_G \text{ has a (classical) model}\},$$

and note that this set is finite and computable in finite time since $\mathcal{G}_{(K,M)}$ is finite and $\mathcal{L}$ is decidable. Furthermore, consider $\mathcal{G}'_{(K,M)}$ to be ordered by the pointwise ordering $\prec$ induced by $\subseteq$. Note that the pointwise ordering of the finite set $\mathcal{G}'_{(K,M)}$ is also computable in finite time.

Lemma 2. *Let (K, M) be a GC-$\mathcal{L}$-KB and let*

$$\mathcal{G}''_{(K,M)} = \{G \in \mathcal{G}'_{(K,M)} \mid G \text{ is minimal in } (\mathcal{G}'_{(K,M)}, \prec)\}.$$

Then (K, M) has a GC-model if and only if $\mathcal{G}''_{(K,M)}$ is non-empty.

Proof. This follows immediately from Lemma 1 together with the following observation: Whenever K has two grounded models $\mathcal{I}$ and $\mathcal{J}$ such that $\mathcal{I}$ is smaller than $\mathcal{J}$, then there exist $G_{\mathcal{I}}, G_{\mathcal{J}} \in \mathcal{G}'_{(K,M)}$ with $G_{\mathcal{I}} \prec G_{\mathcal{J}}$ such that $K_{G_{\mathcal{I}}}$ and $K_{G_{\mathcal{J}}}$ have (classical) models $\mathcal{I}'$ and $\mathcal{J}'$, respectively, which coincide with $\mathcal{I}$, respectively, $\mathcal{J}$, on the minimized predicates.

Theorem 1. *GC-KB-satisfiability is decidable.*

Proof. This follows from Lemma 2 since the set $\mathcal{G}''_{(K,M)}$, for any given GC-KB (K, M), can be computed in finite time, i.e., it can be decided in finite time whether $\mathcal{G}''_{(K,M)}$ is empty.

Some remarks on complexity are as follows. Assume that the problem of deciding KB satisfiability in $\mathcal{L}$ is in the complexity class C. Observe from equation (1) that there are exponentially many possible choices of the $(n + m)$-tuples in $\mathcal{G}_{(K,M)}$ (in the size of the input knowledge base). Computation of $\mathcal{G}'_{(K,M)}$ is thus in Exp^C, and subsequent computation of $\mathcal{G}''_{(K,M)}$ is also in Exp. We thus obtain the following upper bound.

Proposition 1. *The problem of finding a GC-model (if one exists) of a given GC-$\mathcal{L}$-KB is in Exp^C, where C is the complexity class of $\mathcal{L}$. Likewise, GC-$\mathcal{L}$-KB satisfiability is in Exp^C.*

4   Algorithms for Grounded Circumscriptive Reasoning

We now present algorithms for reasoning with grounded circumscription. We start with a tableaux algorithm to decide knowledge base GC-satisfiability and then discuss how to extend it to other reasoning tasks. For simplicity of presentation, we only consider GC-KB-satisfiability in $\mathcal{ALC}$, but the procedure should be adaptable to other DLs. Inspiration for the algorithm comes from [7,11].

4.1 Decision Procedure for GC-Satisfiability in $\mathcal{ALC}$

The algorithm is a tableaux procedure as usual where the expansion rules are defined to be compatible with the semantics of the language, and for easier reference, we call the resulting algorithm *Tableau1*. It starts with an initial graph F_i constructed using the ABox of a given GC-$\mathcal{ALC}$-KB (K, M), such that all known individuals are represented as nodes along with their labels that consist of the concepts that contain them in the ABox. Additionally, links are added for all role assertions using labels that consist of the roles in the ABox assertion axioms. We call this set of nodes and labels the *initial graph*. The creation of the initial graph F_i is described in terms of the following steps called the initialization process:

- create a node a, for each individual a that appears in at least one assertion of the form $C(a)$ in K (we call these nodes *nominal nodes*),
- add C to $\mathcal{L}(a)$, for each assertion of the form $C(a)$ or $R(a, b)$ in K,
- add R to $\mathcal{L}(a, b)$, for each assertion of the form $R(a, b)$ in K,
- initialize a set $T := \{\mathsf{NNF}(\neg C \sqcup D) \mid C \sqsubseteq D \in K\}$.

The algorithm begins with the initial graph F_i along with the sets T and M, and proceeds by non-deterministically applying the rules defined in Table 1, a process which can be understood as creating a candidate model for the knowledge base. The $\longrightarrow_{TBox}$, $\longrightarrow_{\sqcap}$, $\longrightarrow_{\exists}$ and $\longrightarrow_{\forall}$ rules are deterministic rules, whereas the $\longrightarrow_{\sqcup}$, $\longrightarrow_{GC_C}$ and $\longrightarrow_{GC_R}$ rules are non-deterministic rules, as they provide a choice, with each choice leading to possibly a different graph. The algorithm differs from the usual tableaux algorithm for $\mathcal{ALC}$, as it provides extra $\longrightarrow_{GC_C}$ and $\longrightarrow_{GC_R}$ non-deterministic rules, such that the candidate models are in fact *grounded* candidate models as defined in Definition 2. The rules are applied until a clash is detected or until none of the rules is applicable. A graph is said to contain an *inconsistency clash* when one of the node labels contains both C and $\neg C$, or it contains $\bot$, and it is called *inconsistency-clash-free* if it does not contain an inconsistency clash. The algorithm by application of the rules upon termination generates a so-called *completion graph*. A notion of blocking is required to ensure termination, and we define it as follows.

Definition 4 (Blocking). *A non-nominal node x is* blocked
1. *if it has a* blocked *ancestor; or*
2. *if it has a non-nominal ancestor x' such that $\mathcal{L}(x) \subseteq \mathcal{L}(x')$ and the path between x' and x consists only of non-nominal nodes.*
In the second case, we say that x is directly blocked *by the node x'. Note that any non-nominal successor node of x is also blocked.*

For a GC-$\mathcal{ALC}$-KB (K, M), the tableau expansion rules when applied exhaus-tively, generate a completion graph which consists of nodes, edges and their labels, each node x of the graph is labeled with a set of (complex or atomic) concepts and each edge (x, y) is labeled with a set of roles.

Lemma 3 (termination). *Given any GC-$\mathcal{ALC}$-KB (K, M), the tableaux procedure for (K, M) terminates.*

Table 1. Tableau1 expansion rules for GC-$\mathcal{ALC}$-KBs (K, M). The first five rules are taken directly from the $\mathcal{ALC}$ tableaux algorithm. Input: F_i, T and M.

$\longrightarrow_{TBox}$:	**if** $C \in T$ and $C \notin \mathcal{L}(x)$
	then $\mathcal{L}(x) := \mathcal{L}(x) \cup \{C\}$
$\longrightarrow_\sqcap$:	**if** $C_1 \sqcap C_2 \in \mathcal{L}(x), x$ is not blocked, and $\{C_1, C_2\} \not\subseteq \mathcal{L}(x)$
	then $\mathcal{L}(x) := \mathcal{L}(x) \cup \{C_1, C_2\}$
$\longrightarrow_\sqcup$:	**if** $C_1 \sqcup C_2 \in \mathcal{L}(x), x$ is not blocked, and $\{C_1, C_2\} \cap \mathcal{L}(x) = \emptyset$
	then $\mathcal{L}(x) := \mathcal{L}(x) \cup \{C_1\}$ or $\mathcal{L}(x) := \mathcal{L}(x) \cup \{C_2\}$
$\longrightarrow_\exists$:	**if** $\exists R.C \in \mathcal{L}(x), x$ is not blocked, and x has no R-successor y
	with $C \in \mathcal{L}(y)$
	then add a new node y with $\mathcal{L}(y) := \{C\}$ and $\mathcal{L}(x, y) := \{R\}$
$\longrightarrow_\forall$:	**if** $\forall R.C \in \mathcal{L}(x), x$ is not blocked, and x has an R-successor y
	with $C \notin \mathcal{L}(y)$
	then $\mathcal{L}(y) := \mathcal{L}(y) \cup \{C\}$
$\longrightarrow_{GC_C}$:	**if** $C \in \mathcal{L}(x), C \in M, x \notin \mathsf{Ind}(K)$ and x is not blocked
	then for some $a \in \mathsf{Ind}(K)$ do
	1. $\mathcal{L}(a) := \mathcal{L}(a) \cup \mathcal{L}(x)$,
	2. if x has a predecessor y, then $\mathcal{L}(y, a) := \mathcal{L}(y, a) \cup \mathcal{L}(y, x)$,
	3. remove x and all incoming edges to x in the completion graph
$\longrightarrow_{GC_R}$:	**if** $R \in \mathcal{L}(x, y), R \in M$ and y is not blocked.
	then initialize variables $x' := x$ and $y' := y$, and do
	1. if $x \notin \mathsf{Ind}(K)$ then for some $a \in \mathsf{Ind}(K), \mathcal{L}(a) := \mathcal{L}(a) \cup \mathcal{L}(x)$,
	$x' := a$.
	2. if $y \notin \mathsf{Ind}(K)$ for some $b \in \mathsf{Ind}(K), \mathcal{L}(b) := \mathcal{L}(b) \cup \mathcal{L}(y)$ and
	$y' := b$
	3. if $x' = a$ and x has a predecessor z,
	then $\mathcal{L}(z, a) := \mathcal{L}(z, a) \cup \mathcal{L}(z, x)$.
	4. $\mathcal{L}(x', y') := \mathcal{L}(x', y') \cup \{R\}$
	5. if $x' = a$ remove x and all incoming edges to x and
	if $y' = b$ remove y and all incoming edges to y
	from the completion graph.

Proof. First note that node labels can only consist of axioms from K in NNF or of subconcepts of axioms from K in NNF. Thus, there is only a finite set of possible node labels, and thus there is a global bound, say $m \in \mathbb{N}$, on the cardinality of node labels.

Now note the following. (1) The number of times any rule can be applied to a node is finite, since the labels trigger the rules and the size of labels is bounded by m. (2) The outdegree of each node is bounded by the number of possible elements of node labels of the form $\exists R.C$, since only the $\longrightarrow_\exists$ rule generates new nodes. Thus the outdegree is also bounded by m. Further, infinite non-looping paths cannot occur since there are at most 2^m possible different labels, and so the blocking condition from Definition 4 implies that some node along such a path would be blocked, contradicting the assumption that the path would be infinite. (3) While the $\longrightarrow_{GC_C}$ rule and the $\longrightarrow_{GC_R}$ rule delete nodes, they can only change labels of nominal nodes by possibly adding elements to nominal node labels. Since the number of possible elements of node labels is bounded by m, at some stage application of the $\longrightarrow_{GC_C}$ rule or the $\longrightarrow_{GC_R}$ rule will no longer add anything to nominal node labels, and then no new applications of rules can be enabled by this process.

From (1), (2) we obtain a global bound on the size of the completion graphs which can be generated by the algorithm, and from (3) we see that infinite loops due to deletion and recreation of nodes cannot occur. Thus, the algorithm necessarily terminates.

Before we show that the tableaux calculus is sound and complete, we define a function called read function which will be needed for clarity of the proof and verification of minimality of the models.

Definition 5 (read function). *Given an inconsistency-clash-free completion graph F, we define a read function* r *which maps the graph to an interpretation* $\mathsf{r}(F) = \mathcal{I}$ *in the following manner. The interpretation domain $\Delta^{\mathcal{I}}$ contains all the non-blocked nodes in the completion graph. Further, for each atomic concept A, we set $A^{\mathcal{I}}$ to be the set of all non-blocked nodes x for which $A \in \mathcal{L}(x)$. For each role name R, we set $R^{\mathcal{I}}$ to be the set of pairs (x, y) which satisfy any of the following conditions:*
  – $R \in \mathcal{L}(x, y)$ and y is not blocked; or
  – x is an immediate R-predecessor of some node z, and y directly blocks z
The mapping just defined is then lifted to complex concept descriptions as usual.

The second condition is due to the well-known technique of unraveling (see, e.g., [11]): while disregarding blocked nodes, an incoming edge from an immediate R-predecessor x of the blocked node z is considered to be replaced by an edge from the predecessor to the node y which directly blocks z. This accounts for the intuition that a path ending in a blocked node stands for an infinite but repetitive path in the model.

Lemma 4 (soundness). *If the expansion rules are applied to a GC-$\mathcal{ALC}$-KB (K, M), such that they result in an inconsistency-clash-free completion graph F, then K has a grounded model $\mathcal{I} = \mathsf{r}(F)$. Furthermore, the extension $A^{\mathcal{I}}$ of each concept $A \in M$ under $\mathcal{I}$ coincides with the set $\{x \mid x \in A^{\mathsf{r}(F)}\}$, the extension $R^{\mathcal{I}}$ of each role $R \in M$ under $\mathcal{I}$ coincides with the set $\{(x, y) \mid (x, y) \in R^{\mathsf{r}(F)}\}$, and both these sets can be read off directly from the labels of the completion graph.*

Proof. From the inconsistency-clash-free completion graph F, we create an interpretation $\mathcal{I} = \mathsf{r}(F)$ where r is the read function defined in Definition 5. Since the completion graph is free of inconsistency clashes, and the first five expansion rules from Table 1 follow the definition of a model from Section 2, the resulting interpretation is indeed a model of K.[7] Moreover, the $\longrightarrow_{GC_C}$ and $\longrightarrow_{GC_R}$ rules ensure that the extensions of minimized predicates contain only (pairs of) known individuals. Hence, $\mathsf{r}(F) = \mathcal{I}$ is a grounded model of K w.r.t M, and Definition 5 shows how the desired extensions can be read off from the completion graph.

---

[7] This can be proven formally by structural induction on formulas as in [11].

Lemma 5 (completeness). *If a GC-$\mathcal{ALC}$-KB (K, M) has a grounded model $\mathcal{I}$, then the expansion rules can be applied to the initial graph F_i of (K, M) in such a way that they lead to an inconsistency-clash-free completion graph F, and such that the following hold.*

- $\Delta^{r(F)} \subseteq \Delta^{\mathcal{I}}$
- $a^{r(F)} = a^{\mathcal{I}}$ *for every nominal node a*
- $W^{r(F)} \subseteq W^{\mathcal{I}}$ *for every $W \in M$*
- *the extensions, under $r(F)$, of the closed concept and role names can be read off from F as in the statement of Lemma 4.*

Proof. Given a grounded model $\mathcal{I}$ for K w.r.t M, we can apply the completion rules to F_i in such a way that they result in an inconsistency-clash-free completion graph F. To do this we only have to ascertain that, for any nodes x and y in the graph, the conditions $\mathcal{L}(x) \subseteq \{C \mid \pi(x) \in C^{\mathcal{I}}\}$ and $\mathcal{L}(x, y) \subseteq \{R \mid (\pi(x), \pi(y)) \in R^{\mathcal{I}}\}$ are satisfied, where π is mapping from nodes to $\Delta^{\mathcal{I}}$. This construction is very similar to the one in [11, Lemma 6], to which we refer for details of the argument.

The remainder of the statement follows from the fact that the two conditions just given are satisfied, and from the reading-off process specified in Lemma 4.

We have provided an algorithm that generates a set of completion graphs and each inconsistency-clash-free completion graph represents a grounded model. In fact (K, M) is GC-satisfiable if at least one of the completion graphs is inconsistency-clash-free.

Theorem 2. *Let (K, M) be a GC-$\mathcal{ALC}$-KB. Then (K, M) has a grounded model if and only if it is GC-satisfiable.*

Proof. The *if* part of the proof is trivial.

We prove the *only if* part. For any grounded model $\mathcal{I}$, let $|M_{\mathcal{I}}|$ denote the sum of the cardinalities of all extensions of all the minimized predicates in M, and note that, for any two grounded models $\mathcal{I}$ and $\mathcal{J}$ of K w.r.t. M, we have $|M_{\mathcal{J}}| < |M_{\mathcal{I}}|$ whenever $\mathcal{J} \prec_M \mathcal{I}$. Hence, for any grounded model $\mathcal{I}$ of K w.r.t. M which is not a GC-model of (K, M), there is a grounded model $\mathcal{J}$ of K w.r.t. M with $\mathcal{J} \prec_M \mathcal{I}$ and $|M_{\mathcal{J}}| < |M_{\mathcal{I}}|$. Since $|M_{\mathcal{I}}| > 0$ for all grounded models $\mathcal{I}$ (and because $\prec_M$ is transitive), we obtain that, given some grounded model $\mathcal{I}$, it is not possible that there is an infinite descending chain of grounded models preferred over $\mathcal{I}$. Consequently, there must be some grounded model $\mathcal{J}$ of K w.r.t. M which is minimal w.r.t. $|M_{\mathcal{J}}|$ among all models which are preferred over $\mathcal{I}$. This model $\mathcal{J}$ must be a GC-model, since otherwise it would not be minimal.

The following is a direct consequence of Lemmas 3, 4, 5, and Theorem 2.

Theorem 3. *The tableaux algorithm Tableau1 presented above is a decision procedure to determine GC-satisfiability of GC-$\mathcal{ALC}$-KBs.*

4.2   Inference Problems beyond GC-Satisfiability

Unlike in other description logics, common reasoning tasks such as concept satisfiability or instance checking cannot be readily reduced to GC-satisfiability
checking.[8] To cover other inference tasks, we need to extend the previously described algorithm. To do this, we first describe a tableaux algorithm *Tableau2*
which is a modification of Tableau1, as follows. All computations are done with
respect to an input GC-$\mathcal{ALC}$-KB (K, M).

(i) Initialization of Tableau2 is done on the basis of a inconsistency-clash-free
    completion graph F, as follows. We create a finite set of nodes which is
    exactly the domain $\Delta^{\mathcal{I}}$ of a grounded model $\mathcal{I} = \mathsf{r}(F)$. We distinguish be-
    tween two different kinds of nodes, the $\mathcal{I}$-*nominal nodes*, which are nodes
    corresponding to some $a^{\mathcal{I}} \in \Delta^{\mathcal{I}}$ where a is an individual name, and the
    remaining nodes which we call *variable nodes*. For initialization, we fur-
    thermore add all information from the ABox of K to the graph and create
    the set T from K, as in the initialization of Tableau1.
(ii) We modify the $\longrightarrow_\exists$ rule as follows.

> $\longrightarrow_\exists$: **if** $\exists R.C \in \mathcal{L}(x)$, and x has no R-successor y with $C \in \mathcal{L}(y)$
>   **then**   select an existing node y and
>   set $\mathcal{L}(y) := \{C\}$ and $\mathcal{L}(x, y) := \{R\}$

The above change in the $\longrightarrow_\exists$-rule enables us to restrict the graph to
contain only the nodes it was initialized with, which means new nodes are
not created.
(iii) We retain all other completion rules, however we dispose of blocking.
(iv) We retain the notion of inconsistency clash, and add a new notion of *pref-
     erence clash* as follows. A graph F' obtained during the graph construction
     performed by Tableau2 is said to *contain a preference clash with $\mathcal{I}$* if at
     least one of the following holds.
     - $W^{\mathsf{r}(F')} = W^{\mathcal{I}}$ for each predicate $W \in M$
     - $W^{\mathsf{r}(F')} \cap \{a^{\mathcal{I}} \mid a$ an individual $\} \not\subseteq W^{\mathcal{I}}$ for some concept name $W \in M$
     - $W^{\mathsf{r}(F')} \cap \{(a^{\mathcal{I}}, b^{\mathcal{I}}) \mid a, b$ individuals $\} \not\subseteq W^{\mathcal{I}}$ for some role name $W \in M$

Proposition 2. *Tableau2 always terminates. If it terminates by constructing
an inconsistency- and preference-clash-free completion graph F', then $\mathsf{r}(F')$ is
preferred over $\mathcal{I}$, i.e., it shows that $\mathcal{I}$ is not a GC-model. If no such graph F' is
found, then $\mathcal{I}$ has been verified to be a GC-model.*

Proof. Termination is obvious due to the fact that no new nodes are created, i.e.,
the algorithm will eventually run out of choices for applying completion rules.

---

[8] E.g., say we want to decide whether (K, M) GC-entails $C(a)$. We cannot do this,
in general, by using the GC-satisfiability algorithm in the usual way, i.e., by adding
$\neg C(a)$ to K with subsequent checking of its GC-satisfiability. This is because in
general it does not hold that (K, M) does not GC-entail $C(a)$ if $(K \cup \neg C(a), M)$ is
GC-satisfiable. This is due to the non-monotonic nature of circumscription.

Now assume that the algorithm terminates by finding an inconsistency- and preference-clash-free completion graph F'. We have to show that $r(F')$ is preferred over $\mathcal{I}$, i.e., we need to verify the properties listed in Definition 1. $\Delta^{\mathcal{I}} = \Delta^{r(F')}$ holds because we initiate the algorithm with nodes being elements from $\Delta^{\mathcal{I}}$ and no new nodes are created. In case nodes are lost due to the grounding rules of Tableau2, we can simply extend $\Delta^{r(F')}$ with some additional elements which are not otherwise of relevance for the model. The condition $a^{\mathcal{I}} = a^{r(F')}$ for every $a^{\mathcal{I}} \in \Delta^{r(F')}$ holds because this is how the algorithm is initialized. The remaining two conditions hold due to the absence of a preference clash.

For the last statement of the proposition, note that Tableau2 will non-deterministically find an inconsistency- and preference-clash-free completion graph if such a graph exists. This can be seen in a similar way as done in the proof of Lemma 5.

We next use Tableau1 and Tableau2 together to create an algorithm which finds GC-models for (K, M) if they exists. We call this algorithm *GC-model finder*. The algorithm is specified as follows, on input (K, M).

1. Initialize and run Tableau1 on (K, M). If no inconsistency-clash-free completion graph is found, then (K, M) has no GC-model and the algorithm terminates. Otherwise let F be the resulting completion graph.
2. Initialize Tableau2 from F and run it. If no inconsistency- and preference-clash-free completion graph is found, then $r(F)$ is a GC-model of (K, M) and the algorithm terminates with output $r(F)$. Otherwise let F' be the resulting completion graph.
3. Set $F = F'$ and go to step 2.

The loop in steps 2 and 3 necessarily terminates, because whenever step 2 finds a completion graph F' as specified, then $r(F')$ is preferred over $r(F)$. As argued in the proof of Theorem 2, there are no infinite descending chains of grounded models w.r.t. the *preferred over* relation, so the loop necessarily terminates. The output $r(F)$ of the GC-model finder is a GC-model of (K, M), and we call F a *GC-model graph* of (K, M) in this case.

Theorem 4. *On input a GC-$\mathcal{ALC}$-KB (K, M), the GC-model finder creates a GC-model $\mathcal{I}$ of (K, M) if such a model exists. Conversely, for every GC-model $\mathcal{J}$ of (K, M), there exist non-deterministic choices of rule applications in the GC-model finder such that they result in a model $\mathcal{I}$ which coincides with $\mathcal{J}$ on all extensions of minimized predicates.*

Proof. The first statement follows from Propositon 2 together with the explanations already given. The second statement follows due to Lemma 5, since Tableau1 can already create the sought GC-model $\mathcal{I}$.

We now consider the reasoning tasks usually known as instance checking, concept satisfiability and concept subsumption. We provide a convenient way to utilize the GC-model finder algorithm to solve these problems by use of another notion of clash called entailment clash. The following definition describes the inference tasks and provides the notion of entailment clash for each of them as well.

Definition 6. *For a GC-$\mathcal{ALC}$-KB (K, M).*

- Instance checking: *Given an atomic concept C and an individual a in (K, M), $(K, M) \models_{GC} C(a)$ if and only if $a^{\mathcal{I}} \in C^{\mathcal{I}}$ for all GC-models $\mathcal{I}$ of (K, M). For instance checking of $C(a)$, a GC-model graph F is said to contain an entailment clash if $C \in \mathcal{L}(a)$ in F.*
- Concept satisfiability: *Given an atomic concept C in (K, M), C is GC-satisfiable if and only if $C^{\mathcal{I}} \neq \emptyset$ for some GC-model of (K, M). For checking satisfiability of C, a GC-model graph F is said to contain an entailment clash if $C \in \mathcal{L}(x)$ for any node x in F.*
- Concept subsumption: *Given concepts C and D in (K, M), $(K, M) \models_{GC} C \sqsubseteq D$ if and only if $C^{\mathcal{I}} \subseteq D^{\mathcal{I}}$ for all models $\mathcal{I}$ in (K, M). Subsumption can be reduced to concept satisfiability: GC-$\mathcal{ALC}$-KB$(K, M) \models_{GC} C \sqsubseteq D$ if and only if $C \sqcap \neg D$ is not GC-satisfiable.*

We use the following process to solve these inference problems:

To determine if $C(a)$ is entailed by a GC-$\mathcal{ALC}$-KB (K, M), we invoke the GC-model finder until we find a GC-model. If this non-deterministic procedure results in a GC-model graph which does not contain an entailment clash, then $(K, M) \not\models_{GC} C(a)$. If no such GC-model graph can be generated this way, then $(K, M) \models_{GC} C(a)$.

To determine if C is GC-satisfiable, we invoke the GC-model finder until we find a GC-model. If this non-deterministic procedure results in a GC-model graph which contains an entailment clash, then C is satisfiable. If no such GC-model graph can be generated this way, then C is unsatisfiable.

5   Conclusion

We have provided a new approach for incorporating the LCWA into description logics. Our approach, grounded circumscription, is a variant of circumscriptive description logics which avoids two major issues of the original approach: Extensions of minimized predicates can only contain named individuals, and we retain decidability even for very expressive description logics while we can allow for the minimization of roles. We have also provided a tableaux algorithm for reasoning with grounded circumscription.

While the contributions in this paper provide a novel and, in our opinion, very reasonable perspective on LCWA reasoning with description logics, there are obviously also many open questions. A primary theoretical task is to investigate the complexity of our approach. Of more practical relevance would be an implementation of our algorithm with a substantial evaluation to investigate its efficiency empirically. More work also needs to be done in carrying over the concrete algorithm to description logics which are more expressive than $\mathcal{ALC}$.

It also remains to investigate the added value and limitations in practice of modeling with grounded circumscription. This will also shed light onto the question whether fixed predicates and prioritization are required for applications.

Acknowledgements. This work was supported by the National Science Foundation under award 1017225 "III: Small: TROn—Tractable Reasoning with Ontologies," and by State of Ohio Research Incentive funding in the Kno.e.CoM project. Adila Krisnadhi acknowledges support by a Fulbright Indonesia Presidential Scholarship PhD Grant 2010.

References

1. Baader, F., Calvanese, D., McGuinness, D., Nardi, D., Patel-Schneider, P.: The Description Logic Handbook: Theory, Implementation, and Applications, 2nd edn. Cambridge University Press (2007)
2. Bonatti, P.A., Lutz, C., Wolter, F.: Expressive Non-Monotonic Description Logics Based on Circumscription. In: Proc. of the 10th Int. Conf. on Principles of Knowledge Representation and Reasoning (KR 2006), pp. 400–410. AAAI Press (2009)
3. Bonatti, P.A., Lutz, C., Wolter, F.: The Complexity of Circumscription in Description Logic. Journal of Artificial Intelligence Research 35, 717–773 (2009)
4. Carral Martínez, D., Krisnadhi, A., Maier, F., Sengupta, K., Hitzler, P.: Reconciling OWL and rules. Tech. rep., Kno.e.sis Center, Wright State University, Dayton, Ohio, USA (2011), http://www.pascal-hitzler.de/
5. Donini, F.M., Nardi, D., Rosati, R.: Description Logics of Minimal Knowledge and Negation as Failure. ACM Trans. on Computational Logic (TOCL) 3(2), 177–225 (2002)
6. Grimm, S., Hitzler, P.: Semantic Matchmaking of Web Resources with Local Closed-World Reasoning. International Journal of Electronic Commerce 12(2), 89–126 (2007)
7. Grimm, S., Hitzler, P.: A Preferential Tableaux Calculus for Circumscriptive $\mathcal{ALCO}$. In: Polleres, A., Swift, T. (eds.) RR 2009. LNCS, vol. 5837, pp. 40–54. Springer, Heidelberg (2009)
8. Hitzler, P., Krötzsch, M., Parsia, B., Patel-Schneider, P.F., Rudolph, S. (eds.): OWL 2 Web Ontology Language: Primer. W3C Recommendation (October 27, 2009), http://www.w3.org/TR/owl2-primer/
9. Hitzler, P., Krötzsch, M., Rudolph, S.: Foundations of Semantic Web Technologies. Chapman & Hall/CRC (2009)
10. Hitzler, P., Seda, A.K.: Mathematical Aspects of Logic Programming Semantics. CRC Press (2010)
11. Horrocks, I., Sattler, U.: Ontology reasoning in the SHOQ(D) description logic. In: Proc. IJCAI-2001, pp. 199–204 (2001)
12. Knorr, M., Alferes, J.J., Hitzler, P.: Local Closed-World Reasoning with Description Logics under the Well-founded Semantics. Artificial Intelligence 175(9-10), 1528–1554 (2011)
13. Knorr, M., Alferes, J.J., Hitzler, P.: A Coherent Well-founded Model for Hybrid MKNF Knowledge Bases. In: Ghallab, M., Spyropoulos, C.D., Fakotakis, N., Avouris, N.M. (eds.) Proceedings of the 18th European Conference on Artificial Intelligence. Frontiers in Artificial Intelligence and Applications, Patras, Greece, July 21-25, vol. 178, pp. 99–103. IOS Press, Amsterdam (2008)
14. Krisnadhi, A., Sengupta, K., Hitzler, P.: Local closed world semantics: Keep it simple, stupid! In: Rosati, R., Rudolph, S., Zakharyaschev, M. (eds.) 2011 International Workshop on Description Logics. CEUR Workshop Proceedings, vol. 745. CEUR-WS.org (2011)

15. Krisnadhi, A., Maier, F., Hitzler, P.: OWL and Rules. In: Polleres, A., d'Amato, C., Arenas, M., Handschuh, S., Kroner, P., Ossowski, S., Patel-Schneider, P. (eds.) Reasoning Web 2011. LNCS, vol. 6848, pp. 382–415. Springer, Heidelberg (2011)
16. Krötzsch, M., Maier, F., Krisnadhi, A.A., Hitzler, P.: A better uncle for OWL: Nominal schemas for integrating rules and ontologies. In: Sadagopan, S., Ramamritham, K., Kumar, A., Ravindra, M.P., Bertino, E., Kumar, R. (eds.) Proceedings of the 20th International World Wide Web Conference, WWW 2011, Hyderabad, India, pp. 645–654. ACM, New York (2011)
17. Krötzsch, M., Rudolph, S., Hitzler, P.: ELP: Tractable Rules for OWL 2. In: Sheth, A.P., Staab, S., Dean, M., Paolucci, M., Maynard, D., Finin, T., Thirunarayan, K. (eds.) ISWC 2008. LNCS, vol. 5318, pp. 649–664. Springer, Heidelberg (2008)
18. Makinson, D.: Bridges from Classical to Nonmonotonic Logic, Texts in Computing, vol. 5. King's College Publications (2005)
19. McCarthy, J.: Circumscription – A Form of Non-Monotonic Reasoning. Artificial Intelligence 13(1-2), 27–39 (1980)
20. Moore, R.: Possible-worlds Semantics for Autoepistemic Logic. In: Proceedings of the 1984 Non-monotonic Reasoning Workshop. AAAI, Menlo Park (1984)
21. Motik, B., Rosati, R.: Reconciling Description Logics and Rules. Journal of the ACM 57(5), 1–62 (2010)
22. Motik, B., Sattler, U., Studer, R.: Query Answering for OWL-DL with Rules. Journal of Web Semantics 3, 41–60 (2005)
23. Patel, C., Cimino, J.J., Dolby, J., Fokoue, A., Kalyanpur, A., Kershenbaum, A., Ma, L., Schonberg, E., Srinivas, K.: Matching Patient Records to Clinical Trials Using Ontologies. In: Aberer, K., Choi, K.S., Noy, N.F., Allemang, D., Lee, K.I., Nixon, L.J.B., Golbeck, J., Mika, P., Maynard, D., Mizoguchi, R., Schreiber, G., Cudré-Mauroux, P. (eds.) ASWC 2007 and ISWC 2007. LNCS, vol. 4825, pp. 816–829. Springer, Heidelberg (2007)
24. Reiter, R.: A Logic for Default Reasoning. Artificial Intelligence 13, 81–132 (1980)
25. Rudolph, S., Krötzsch, M., Hitzler, P.: Cheap Boolean Role Constructors for Description Logics. In: Hölldobler, S., Lutz, C., Wansing, H. (eds.) JELIA 2008. LNCS (LNAI), vol. 5293, pp. 362–374. Springer, Heidelberg (2008)
26. Rudolph, S., Krötzsch, M., Hitzler, P.: All Elephants are Bigger than All Mice. In: Baader, F., Lutz, C., Motik, B. (eds.) Proceedings of the 21st International Workshop on Description Logics (DL 2008), Dresden, Germany, May 13-16. CEUR Workshop Proceedings, vol. 353 (2008)
27. Seylan, I., Franconi, E., de Bruijn, J.: Effective query rewriting with ontologies over DBoxes. In: Boutilier, C. (ed.) IJCAI 2009, Proceedings of the 21st International Joint Conference on Artificial Intelligence, Pasadena, California, USA, July 11-17, pp. 923–925 (2009)
28. Tobies, S.: Complexity Results and Practical Algorithms for Logics in Knowledge Representation. Ph.D. thesis, RWTH Aachen, Germany (2001)

Extending Logic Programs with Description Logic Expressions for the Semantic Web

Yi-Dong Shen[1] and Kewen Wang[2]

[1] State Key Laboratory of Computer Science, Institute of Software
Chinese Academy of Sciences, Beijing 100190, China
ydshen@ios.ac.cn
[2] School of Computing and Information Technology, Griffith University
Brisbane, QLD 4111, Australia
k.wang@griffith.edu.au

Abstract. Recently much attention has been directed to extending logic programming with description logic (DL) expressions, so that logic programs have access to DL knowledge bases and thus are able to reason with ontologies in the Semantic Web. In this paper, we propose a new extension of logic programs with DL expressions, called normal DL logic programs. In a normal DL logic program arbitrary DL expressions are allowed to appear in rule bodies and atomic DL expressions (i.e., atomic concepts and atomic roles) allowed in rule heads. We extend the key condition of well-supportedness for normal logic programs under the standard answer set semantics to normal DL logic programs and define an answer set semantics for DL logic programs which satisfies the extended well-supportedness condition. We show that the answer set semantics for normal DL logic programs is decidable if the underlying description logic is decidable (e.g. $\mathcal{SHOIN}$ or $\mathcal{SROIQ}$).

1 Introduction

In the development of Semantic Web languages we are concerned with two major components: ontologies and rules. Ontologies describe terminological knowledge and rules model constraints and exceptions over the ontologies. Since the two components provide complementary descriptions of the same problem domain, they are supposed to be integrated in some ways (e.g., [2,4,16,18,19]; see [3] for a survey). The core of the Web ontology language OWL (more recently, OWL 2) [13,9] is description logics (DLs) [1] and thus in this paper we assume an ontology is represented as a knowledge base in DLs.

Logic programming under the (standard) answer set semantics [8] is currently a widely used declarative language paradigm for knowledge representation and reasoning. A normal logic program Π consists of rules of the form $H \leftarrow A_1, \cdots, A_m, not\ B_1, \cdots, not\ B_n$, where H and each A_i and B_i are atoms. Such a rule states that if the body $A_1, \cdots, A_m, not\ B_1, \cdots, not\ B_n$ holds, then the head H holds. The semantics of Π is defined by *answer sets*, which are Herbrand models of Π satisfying the *well-supportedness* condition [8,7]. Informally,

L. Aroyo et al. (Eds.): ISWC 2011, Part I, LNCS 7031, pp. 633–648, 2011.
© Springer-Verlag Berlin Heidelberg 2011

a Herbrand model I is well-supported if for any $H \in I$, there is a rule as above from Π such that I satisfies the rule body and for no A_i the evidence of the truth of A_i is circularly dependent on H in I. It is this well-supportedness condition that lets rules in a logic program differ from formulas (implications) in classical logic and guarantees that answer sets are free of circular justifications.

Recently, much attention has been directed to using logic programs to express rules in the Semantic Web by extending logic programming under the answer set semantics with DL expressions [4,16,19]. By allowing DL expressions to appear in rules, logic programs have access to DL knowledge bases and thus are able to reason with ontologies in the Semantic Web. Major current extensions of logic programs with DL expressions include *description logic programs* (or dl-programs) [4], $\mathcal{DL}$+log [19] and *disjunctive dl-programs* [16].

Given an external DL knowledge base L, a dl-program extends a normal logic program Π by adding *dl-atoms* to rule bodies as an interface to access to L [4]. A dl-atom is of the form $DL[S_1 op_1 P_1, \cdots, S_m op_m P_m; Q](\mathbf{t})$, where each $S_i op_i P_i$ semantically maps a predicate symbol P_i in Π to a concept or role S_i in L via a special interface operator $op_i \in \{\uplus, \cup, \cap\}$, and $Q(\mathbf{t})$ is a DL expression which will be evaluated against L after the predicate mapping. For instance, $p(a) \leftarrow DL[c \uplus p, b \cap q; c \sqcap \neg b](a)$ is a rule, where the dl-atom queries L if a is in the concept c but not in the concept b, given the mapping that for any x, if $p(x)$ is true then x is in c and if $q(x)$ is false then x is not in b. Note that predicate symbols in Π must be disjoint from predicate symbols (i.e., atomic concepts and atomic roles) in L. Moreover, DL expressions are not allowed to appear in the head of a rule, thus no conclusion about L can be inferred from Π.

It is necessary to allow DL expressions to occur in rule heads because DL knowledge bases (ontologies) define only general terminological knowledge, while additional constraints and exceptions over some DL concepts/roles must be defined by rules. To avoid predicate mappings between L and Π and allow DL expressions to appear in rule heads, another extension, called $\mathcal{DL}$+log, is introduced [19]. $\mathcal{DL}$+log lets Π and L share some predicate symbols and allows atomic DL expressions (i.e. atomic concepts and atomic roles) to appear either in bodies or heads of rules without using any predicate mapping operators. One restriction of this extension is that DL expressions are not allowed to appear behind the negation operator *not*.

Disjunctive dl-programs [16] are a third extension of logic programs with DL expressions. This extension allows atomic DL expressions to appear anywhere in a rule, and has a semantics substantially different from that of $\mathcal{DL}$+log.

For dl-programs, three answer set semantics are introduced in [4,5], called the *weak, strong,* and *FLP-reduct* based semantics, respectively. These semantics are proper extensions of the standard answer set semantics, but their answer sets do not satisfy the well-supportedness condition and thus may incur circular justifications by self-supporting loops. For $\mathcal{DL}$+log, a semantics is defined with a class of first-order models, called *NM-models* [19]. Such NM-models are not well-supported models. For disjunctive dl-programs, an FLP-reduct based answer set semantics is defined [16]. It is a proper extension of the standard answer set semantics, but its answer sets do not satisfy the well-supportedness condition either and thus may also incur circular justifications.

Observe that the three major extensions of logic programs with DL expressions have complementary features. In syntax, dl-programs allow arbitrary DL expressions in rule bodies, while $\mathcal{DL}$+log and disjunctive dl-programs allow atomic DL expressions in rule heads. In semantics, in dl-programs and $\mathcal{DL}$+log, DL concepts and roles occurring in Π are all interpreted against the external DL knowledge base L under the first-order semantics, while in disjunctive dl-programs, these DL concepts and roles are all included in the Herbrand base of Π and interpreted under the answer set semantics. These observations suggest that it is desirable to have a new extension of logic programs with DL expressions, which allows arbitrary DL expressions in rule bodies and atomic DL expressions in rule heads, and interprets DL concepts and roles occurring in Π flexibly in either of the above ways. Moreover, as we mentioned earlier, well-supportedness is a key condition of logic programming under the standard answer set semantics, so it is desirable to extend this condition to logic programs with DL expressions. In fact, the well-supportedness condition has recently been extended to dl-programs and a new answer set semantics for dl-programs has been developed which satisfies the extended well-supportedness condition [20].

Therefore, in this paper we advance one step further by introducing a fourth extension of logic programs with DL expressions as follows: (1) Given an external DL knowledge base L, we extend a normal logic program Π with DL expressions relative to L by introducing rules of the form $H \leftarrow A_1, \cdots, A_m, not\ B_1, \cdots, not\ B_n$, where H is an atom or an atomic DL expression, and each A_i and B_i are either atoms or arbitrary DL expressions. We call this extension Π relative to L a *normal DL logic program*. (2) We allow DL concepts and roles occurring in Π to flexibly choose between first-order interpretations and Herbrand interpretations, as described above. (3) We extend the well-supportedness condition of the standard answer set semantics from normal logic programs to normal DL logic programs, and define an answer set semantics which satisfies the extended well-supportedness condition and thus whose answer sets are free of circular justifications. (4) We show that the answer set semantics for normal DL logic programs is decidable if the underlying description logic is decidable (e.g. $\mathcal{SHOIN}$ or $\mathcal{SROIQ}$ [13,12]).

The paper is arranged as follows. Section 2 briefly reviews logic programs and DL knowledge bases. Section 3 defines normal DL logic programs. Section 4 mentions related approaches, and Section 5 concludes with future work.

2   Preliminaries

2.1   Logic Programs

Consider a vocabulary $\Phi = (\mathbf{P}, \mathbf{C})$, where $\mathbf{P}$ is a finite set of predicate symbols and $\mathbf{C}$ a nonempty finite set of constants. A *term* is either a constant from $\mathbf{C}$ or a variable. Predicate symbols begin with a capital letter, and constants with a lower case letter. We use strings starting with X, Y or Z to denote variables.

An *atom* is of the form $P(t_1, ..., t_m)$, where P is a predicate symbol from $\mathbf{P}$, and t_i is a term. A *rule* r is of the form

$$H \leftarrow A_1, \cdots, A_m, not\ B_1, \cdots, not\ B_n \qquad (1)$$

where H and each A_i and B_i are atoms. Each A_i is called a *positive literal*, and each $not\ B_i$ called a *negative literal*. We use $head(r)$ and $body(r)$ to denote the head H and the body $A_1, \cdots, A_m, not\ B_1, \cdots, not\ B_n$, respectively. We also use $pos(r)$ to denote the positive literals $A_1, \cdots, A_m$, and $neg(r)$ to denote the negative literals $not\ B_1, \cdots, not\ B_n$. Therefore, a rule r can simply be written as $head(r) \leftarrow body(r)$ or $head(r) \leftarrow pos(r), neg(r)$.

A *normal logic program* Π consists of a finite set of rules. A *ground instance* of a rule r is obtained by replacing every variable in r with a constant from $\mathbf{C}$. We use $ground(\Pi)$ to denote the set of all ground instances of rules in Π. The *Herbrand base* of Π, denoted HB_Π, is the set of all ground atoms $P(t_1, ..., t_m)$, where $P \in \mathbf{P}$ occurs in Π and t_i is in $\mathbf{C}$. Any subset of HB_Π is a *Herbrand interpretation* (or *interpretation* for short) of Π. For an interpretation I, let $I^- = HB_\Pi \setminus I$ and $\neg I^- = \{\neg A \mid A \in I^-\}$.

An interpretation I *satisfies* a ground atom $A \in HB_\Pi$ if $A \in I$, and I satisfies $not\ A$ if $A \notin I$. For a rule r in $ground(\Pi)$, I satisfies $body(r)$ if for each (positive or negative) literal l in $body(r)$, I satisfies l; I satisfies r if I does not satisfy $body(r)$ or I satisfies $head(r)$. I is a *model* of Π if I satisfies all $r \in ground(\Pi)$. A *minimal model* is a model that is minimal in terms of set inclusion.

Let $\Pi^I = \{A \leftarrow pos(r) \mid A \leftarrow pos(r), neg(r) \in ground(\Pi)$ and I satisfies $neg(r)\}$. Since Π^I has no negative literals in rule bodies, it has a least model. The *standard answer set semantics* defines I to be an *answer set* of Π if I is the least model of Π^I [8].

2.2   DL Knowledge Bases

We assume familiarity with the basics of description logics (DLs) [1], and for simplicity consider $\mathcal{SHOIN}$, a DL underlying the Web ontology language OWL DL [13]. The approach presented in this paper can easily be extended to other more expressive DLs such as $\mathcal{SROIQ}$ (a logical underpinning for OWL 2) [12,9], and to DLs with *datatypes* such as $\mathcal{SHOIN}(\mathbf{D})$ and $\mathcal{SROIQ}(\mathbf{D})$.

Consider a vocabulary $\Psi = (\mathbf{A} \cup \mathbf{R}, \mathbf{I})$, where $\mathbf{A}$, $\mathbf{R}$ and $\mathbf{I}$ are pairwise disjoint (denumerable) sets of *atomic concepts*, *atomic roles* and *individuals*, respectively. A *role* is either an atomic role R from $\mathbf{R}$ or its inverse, denoted R^-. General *concepts* C are formed from atomic concepts, roles and individuals, according to the following syntax:

$$C ::= \top \mid \bot \mid A \mid \{a\} \mid C \sqcap C_1 \mid C \sqcup C_1 \mid \neg C \mid \exists R.C \mid \forall R.C \mid\ \geq_n R \mid\ \leq_n R$$

where A is an atomic concept from $\mathbf{A}$, R is a role, a is an individual from $\mathbf{I}$, C and C_1 are concepts, and n is a non-negative integer. An *axiom* is of the form $C \sqsubseteq D$ (*concept inclusion axiom*), $R \sqsubseteq R_1$ (*role inclusion axiom*), Trans(R) (*transitivity axiom*), $C(a)$ (*concept membership axiom*), $R(a, b)$ (*role membership axiom*), $=(a, b)$ (*equality axiom*), or $\neq(a, b)$ (*inequality axiom*), where C, D are concepts, R, R_1 are atomic roles in $\mathbf{R}$, and a, b are individuals in $\mathbf{I}$. We use $C \equiv D$ to denote $C \sqsubseteq D$ and $D \sqsubseteq C$.

A *DL knowledge base* L is a finite set of axioms. Since DLs are fragments of first-order logic with equality, where atomic concepts (resp. roles) are unary (resp. binary) predicate symbols, and individuals are constants, L has the first-order semantics. When we say predicate symbols in L, we refer to atomic concepts or atomic roles in L. L is *consistent* (or *satisfiable*) if L has a first-order model. For an axiom F, the *entailment* relation $L \models F$ is defined as in first-order logic, i.e., L entails F if all models of L are models of F. Note that if L is inconsistent, L entails everything.

A *DL expression*, also called a *DL query* in [4], which is allowed to appear in rules of a logic program, is either (i) a concept inclusion axiom F or its negation $\neg F$; or (ii) of the form $C(t)$ or $\neg C(t)$, where C is a concept, and t is a term (i.e., a variable or a constant); or (iii) of the form $R(t_1, t_2)$ or $\neg R(t_1, t_2)$, where R is a role, and t_1 and t_2 are terms; or (iv) of the form $=(t_1, t_2)$ or $\neq(t_1, t_2)$, where t_1 and t_2 are terms. An *atomic DL expression* is either $C(t)$ or $R(t_1, t_2)$, where C is an atomic concept and R an atomic role. For convenience, we denote a DL expression by $Q(\mathbf{t})$, where $\mathbf{t}$ denotes all terms occurring in the expression (e.g., t_1 and t_2 in (iii)), and Q denotes the remaining part of the expression (e.g., R or $\neg R$ in (iii)).

3 Normal DL Logic Programs

Let L be a DL knowledge base built over a vocabulary $\Psi = (\mathbf{A} \cup \mathbf{R}, \mathbf{I})$, and Π be a normal logic program built over $\Phi = (\mathbf{P}, \mathbf{C})$. To extend Π with DL expressions relative to L, we first extend Φ such that: (i) all constants in $\mathbf{C}$ are individuals in $\mathbf{I}$ (i.e., $\mathbf{C} \subseteq \mathbf{I}$), so that constants occurring in DL expressions are individuals, and (ii) some atomic concepts and roles in $\mathbf{A} \cup \mathbf{R}$ are included in $\mathbf{P}$ (as unary and binary predicate symbols, respectively), so that we can make conclusions about them in the same way as other predicate symbols in $\mathbf{P}$. To ensure decidability, we require that $\mathbf{P}$ and $\mathbf{C}$ be finite. Let $\Omega = \mathbf{P} \cap (\mathbf{A} \cup \mathbf{R})$ denote the set of predicate symbols shared by Π and L.

Definition 1. Let L be a DL knowledge base. A *normal DL logic program* Π with DL expressions relative to L consists of a finite set of rules of form (1), where H is an atom, and each A_i and B_i are either atoms or DL expressions.

Note that when the predicate symbol of an atom in Π is in Ω, the atom is also an atomic DL expression.

A *ground instance* of a rule (resp. a DL expression) in Π is obtained by replacing all variables with constants in $\mathbf{C}$. Let $ground(\Pi)$ denote the set of ground instances of all rules in Π. The *Herbrand base* HB_Π of Π relative to L is the set of all ground atoms $P(t_1, ..., t_m)$, where $P \in \mathbf{P}$ occurs either in Π or in L and each t_i is in $\mathbf{C}$. Any subset of HB_Π is a *Herbrand interpretation* (or simply *interpretation*) of Π relative to L. When the context is clear, we omit the phrase "relative to L."

For an interpretation I, let $I|_\Omega = \{A \in I \mid$ the predicate symbol of A is in $\Omega\}$ and $I^-|_\Omega = \{A \in I^- \mid$ the predicate symbol of A is in $\Omega\}$. We say that I is *consistent* with L if $L \cup I|_\Omega \cup \neg I^-|_\Omega$ is consistent. Note that when I is consistent with L, L must be consistent.

Since DL expressions must be evaluated against L, the satisfaction relation for normal logic programs needs to be extended to normal DL logic programs. In the sequel, by a *literal* we refer to A or *not* A, where A is an atom or a DL expression.

Definition 2. Let Π be a normal DL logic program with DL expressions relative to a DL knowledge base L, I an interpretation, and l a ground literal. We use $I \models_L l$ to denote that I *satisfies* l *under* L, which is defined as follows:

1. For a ground atom $A \in HB_\Pi$, which is not an atomic DL expression, $I \models_L A$ if $A \in I$.
2. For a ground DL expression A, $I \models_L A$ if $L \cup I|_\Omega \cup \neg I^-|_\Omega \models A$.
3. For a ground atom or a ground DL expression A, $I \models_L$ *not* A if $I \not\models_L A$.

For a rule r in $ground(\Pi)$, $I \models_L body(r)$ if for each (positive or negative) literal l in $body(r)$, $I \models_L l$; $I \models_L r$ if $I \not\models_L body(r)$ or $I \models_L head(r)$. I is a *model* of Π relative to L if I is consistent with L and $I \models_L r$ for all $r \in ground(\Pi)$. Note that when L is inconsistent, Π has no model relative to L.

Example 1. Let $L = \{\neg B(a)\}$ and $\Pi = \{A(X) \leftarrow not \ \neg(A \sqcup B)(X)\}$. Let $\mathbf{P} = \{A\}, \mathbf{C} = \{a\}$ and $\Omega = \{A\}$. Note that $\neg(A \sqcup B)(X)$ is a DL expression, and $A(X)$ is both an atom and an atomic DL expression. We have $HB_\Pi = \{A(a)\}$ and $ground(\Pi) = \{A(a) \leftarrow not \ \neg(A \sqcup B)(a)\}$. Π has two models relative to L: $I_1 = \emptyset$ and $I_2 = \{A(a)\}$. For the rule r in $ground(\Pi)$, $I_1 \not\models_L body(r)$, $I_2 \models_L body(r)$, and $I_2 \models_L head(r)$.

3.1   Well-Supported Models

The notion of well-supportedness in logic programming is defined by Fages in [7] as a key characterization of the standard answer set semantics. For a normal logic program Π, an interpretation I is *well-supported* if there exists a strict well-founded partial order $\prec$ on I such that for any $A \in I$, there is a rule $A \leftarrow body(r)$ in $ground(\Pi)$ such that I satisfies $body(r)$ and for every positive literal B in $body(r)$, $B \prec A$. A binary relation $\leq$ is *well-founded* if there is no infinite decreasing chain $A_0 \geq A_1 \geq \cdots$. A well-supported interpretation I guarantees that every $A \in I$ is free of circular justifications in I.

To extend Fages' well-supportedness condition to normal DL logic programs with DL expressions, we introduce a notion of *up to satisfaction*.

Definition 3. Let Π be a normal DL logic program with DL expressions relative to a DL knowledge base L, I an interpretation consistent with L, and l a ground literal. For any $E \subseteq I$, we use $(E, I) \models_L l$ to denote that E *up to* I *satisfies* l *under* L, which is defined as follows: For any ground atom or ground DL expression A, $(E, I) \models_L A$ if for every F with $E \subseteq F \subseteq I$, $F \models_L A$; $(E, I) \models_L$ *not* A if for no F with $E \subseteq F \subseteq I$, $F \models_L A$. For a rule r in $ground(\Pi)$, $(E, I) \models_L body(r)$ if for every literal l in $body(r)$, $(E, I) \models_L l$.

As the phrase "up to" suggests, for any ground (positive or negative) literal l, $(E, I) \models_L l$ means that for all interpretations F between E and I, $F \models_L l$. This implies that the truth of l depends on E and $\neg I^-$ and is independent of atoms in $I \setminus E$, since for any $A \in I \setminus E$ and any interpretation F with $E \subseteq F \subseteq I$, whether or not A is in F, $F \models_L l$.

Theorem 1. *Let l be a ground literal. For any $E_1 \subseteq E_2 \subseteq I$, if $(E_1, I) \models_L l$ then $(E_2, I) \models_L l$.*

Proof: Straightforward from Definition 3. □

Theorem 1 shows that the up to satisfaction is monotonic. In addition, it has the following two properties.

Proposition 1. *For any ground DL expression A, $(E, I) \models_L A$ iff $L \cup E|_{\Omega} \cup \neg I^-|_{\Omega} \models A$.*

Proof: $(E, I) \models_L A$ means that $E \subseteq I$ and for every F with $E \subseteq F \subseteq I$, $F \models_L A$. Then, by Definition 2, $(E, I) \models_L A$ means that for every F with $E \subseteq F \subseteq I$, $L \cup F|_{\Omega} \cup \neg F^-|_{\Omega} \models A$. Note that $F = E \cup (F \setminus E)$ and $F^- = I^- \cup (I \setminus F)$. So $(E, I) \models_L A$ means that for every F with $E \subseteq F \subseteq I$,

$$L \cup E|_{\Omega} \cup \neg I^-|_{\Omega} \cup (F \setminus E)|_{\Omega} \cup \neg (I \setminus F)|_{\Omega} \models A \qquad (2)$$

Then, to prove this proposition it suffices to prove that the entailment (2) holds for every F with $E \subseteq F \subseteq I$ iff the following entailment holds:

$$L \cup E|_{\Omega} \cup \neg I^-|_{\Omega} \models A \qquad (3)$$

Note that for any model M of the left side of the entailment (2) or (3), we have $E \subseteq M \subseteq I$.

Assume that the entailment (2) holds for every F with $E \subseteq F \subseteq I$. Let M be a model of the left side of the entailment (3). Since $E \subseteq M \subseteq I$, M is a model of the left side of the entailment (2), where $F = M$. Then, M is a model of A (the right side of the entailment (2)). This means the entailment (3) holds.

Conversely, assume the entailment (3) holds. Let M be a model of the left side of the entailment (2). M is also a model of the left side of the entailment (3) and thus M is a model of A (the right side of the entailment (3)). This means the entailment (2) holds. □

Proposition 2. *For any ground atom $A \in HB_{\Pi}$, which is not an atomic DL expression, $(E, I) \models_L A$ iff $A \in E$; $(E, I) \models_L$ not A iff $A \notin I$.*

Proof: Straightforward from Definitions 3 and 2. □

Next we extend the well-supportedness condition for normal logic programs to normal DL logic programs by means of the up to satisfaction.

Definition 4. Let Π be a normal DL logic program with DL expressions relative to a DL knowledge base L, and I an interpretation consistent with L. I is *well-supported* if there exists a strict well-founded partial order $\prec$ on I such that for any $A \in I$, there exists $E \subset I$, where for every $B \in E$, $B \prec A$, such that either (i) $L \cup E|_\Omega \cup \neg I^- |_\Omega \models A$, or (ii) there is a rule $A \leftarrow body(r)$ in $ground(\Pi)$ such that $(E, I) \models_L body(r)$.

The above conditions (i) and (ii) imply that the truth of $A \in I$ is determined by E and $\neg I^-$. Since for every $B \in E$, $B \prec A$, the truth of A is not circularly dependent on itself. As a result, a well-supported interpretation I of Π guarantees that every $A \in I$ is free of circular justifications in I.

Observe in Definition 4 that due to the occurrence of DL expressions, some $A \in I$ may be supported by no rule $A \leftarrow body(r)$ in $ground(\Pi)$ such that $I \models_L body(r)$. Instead, A is supported by L such that $L \cup I|_\Omega \cup \neg I^- |_\Omega \models A$. This is a special property of the well-supportedness condition for normal DL logic programs. The next example further illustrates this property.

Example 2. Let $L = \{B(a), B \sqsubseteq A\}$ and $\Pi = \{A(X) \leftarrow C(X)\}$. Let $\mathbf{P} = \{A, C\}$, $\mathbf{C} = \{a\}$ and $\Omega = \{A\}$. We have $HB_\Pi = \{A(a), C(a)\}$ and $ground(\Pi) = \{A(a) \leftarrow C(a)\}$. Π has two models relative to L: $I_1 = \{A(a)\}$ and $I_2 = \{A(a), C(a)\}$. Only I_1 is a well-supported model, where for $A(a) \in I_1$, we have $E = \emptyset$ and condition (i) of Definition 4 holds. Note that there is no rule of the form $A(a) \leftarrow body(r)$ in $ground(\Pi)$ such that $I_1 \models_L body(r)$.

The following result shows that Definition 4 is a proper extension to Fages' well-supportedness condition.

Theorem 2. *Let $L = \emptyset$ and Π be a normal logic program without DL expressions. An interpretation I is a well-supported model of Π relative to L iff I is a well-supported model of Π under Fages' definition.*

Proof: Let I be an interpretation of Π relative to L. I is also an interpretation of Π. Since $L = \emptyset$, for any $A \in I$ the condition (i) of Definition 4 does not hold. Since Π is a normal logic program without DL expressions, each A_i and B_i occurring in the body of each rule r of the form $A \leftarrow A_1, \cdots, A_m, not\ B_1, \cdots, not\ B_n$ in $ground(\Pi)$ are ground atoms. For such rules, $I \models_L body(r)$ iff every A_i is in I and no B_i is in I.

Assume that I is a well-supported model of Π relative to L. By Definition 4, there exists a strict well-founded partial order $\prec$ on I such that for any $A \in I$, there exists $E \subset I$, where for every $B \in E$, $B \prec A$, and there is a rule r as above in $ground(\Pi)$ such that $(E, I) \models_L body(r)$. Note that $(E, I) \models_L body(r)$ implies $E \models_L body(r)$ and $I \models_L body(r)$, which implies that both I and E satisfy $body(r)$. This means that for every positive literal A_i in $body(r)$, $A_i \in E$ and thus $A_i \prec A$. As a result, for any $A \in I$, there is a rule r as above in $ground(\Pi)$ such that I satisfies $body(r)$ and for every positive literal A_i in $body(r)$, $A_i \prec A$. This shows that I is a well-supported model of Π under Fages' definition.

Assume I is a well-supported model of Π under Fages' definition. There exists a strict well-founded partial order $\prec$ on I such that for any $A \in I$, there is a rule r

as above in $ground(\Pi)$ such that I satisfies $body(r)$ and for every positive literal A_i in $body(r)$, $A_i \prec A$. Let $E \subset I$ and for every $A_i \in body(r)$, $A_i \in E$. Then, E contains no B_i in $body(r)$, since no B_i is in I. For any F with $E \subseteq F \subseteq I$, F satisfies $body(r)$ and thus $F \models_L body(r)$. That means $(E, I) \models_L body(r)$. As a result, for any $A \in I$, there exists $E \subset I$, where for every $B \in E$, $B \prec A$, and there is a rule r as above in $ground(\Pi)$ such that $(E, I) \models_L body(r)$. This shows that I is a well-supported model of Π relative to L. $\square$

3.2 Well-Supported Answer Set Semantics

We define an answer set semantics for normal DL logic programs whose answer sets are well-supported models. We first define an immediate consequence operator.

Definition 5. Let Π be a normal DL logic program relative to a DL knowledge base L, and I an interpretation consistent with L. For $E \subseteq I$, define

$$\mathcal{T}_\Pi(E, I) = \{A \mid A \leftarrow body(r) \in ground(\Pi) \text{ and } (E, I) \models_L body(r)\}.$$

By Theorem 1, when the second argument I is a model of Π, $\mathcal{T}_\Pi$ is monotone w.r.t. its first argument E.

Theorem 3. *Let Π be a normal DL logic program with DL expressions relative to a DL knowledge base L, and I a model of Π relative to L. For any $E_1 \subseteq E_2 \subseteq I$, $\mathcal{T}_\Pi(E_1, I) \subseteq \mathcal{T}_\Pi(E_2, I) \subseteq I$.*

Proof: For any $A \in \mathcal{T}_\Pi(E_1, I)$, there is a rule $A \leftarrow body(r)$ in $ground(\Pi)$ such that $(E_1, I) \models_L body(r)$. Since $E_1 \subseteq E_2$, by Theorem 1, $(E_2, I) \models_L body(r)$, and thus $A \in \mathcal{T}_\Pi(E_2, I)$. This shows $\mathcal{T}_\Pi(E_1, I) \subseteq \mathcal{T}_\Pi(E_2, I)$. Since $E_2 \subseteq I$, it follows $(I, I) \models_L body(r)$ and $A \in \mathcal{T}_\Pi(I, I)$. Therefore, $\mathcal{T}_\Pi(E_1, I) \subseteq \mathcal{T}_\Pi(E_2, I) \subseteq \mathcal{T}_\Pi(I, I)$. Note that $(I, I) \models_L body(r)$ means $I \models_L body(r)$. Since I is a model of Π relative to L, $I \models_L body(r)$ implies $A \in I$. This shows that when I is a model of Π relative to L, every $A \in \mathcal{T}_\Pi(I, I)$ is in I. Hence, $\mathcal{T}_\Pi(E_1, I) \subseteq \mathcal{T}_\Pi(E_2, I) \subseteq \mathcal{T}_\Pi(I, I) \subseteq I$. $\square$

Therefore, for any model I of Π relative to L, the sequence $\langle \mathcal{T}_\Pi^i(\emptyset, I)\rangle_{i=0}^\infty$, where $\mathcal{T}_\Pi^0(\emptyset, I) = \emptyset$ and $\mathcal{T}_\Pi^{i+1}(\emptyset, I) = \mathcal{T}_\Pi(\mathcal{T}_\Pi^i(\emptyset, I), I)$, converges to a fixpoint, denoted $\mathcal{T}_\Pi^\alpha(\emptyset, I)$. This fixpoint has the following properties.

Theorem 4. *Let I be a model of Π relative to L. (1) $\mathcal{T}_\Pi^\alpha(\emptyset, I) \subseteq I$. (2) $L \cup \mathcal{T}_\Pi^\alpha(\emptyset, I)|_\Omega \cup \neg I^-|_\Omega$ is consistent. (3) For any model J of Π relative to L with $J \subset I$, $\mathcal{T}_\Pi^\alpha(\emptyset, I) \subseteq \mathcal{T}_\Pi^\alpha(\emptyset, J) \subseteq J$.*

Proof: (1) It suffices to prove that for any $i \geq 0$, $\mathcal{T}_\Pi^i(\emptyset, I) \subseteq I$. It is obvious for $i = 0$. Assume $\mathcal{T}_\Pi^k(\emptyset, I) \subseteq I$ for $k \geq 0$. For $i = k + 1$, by Theorem 3, $\mathcal{T}_\Pi^{k+1}(\emptyset, I) = \mathcal{T}_\Pi(\mathcal{T}_\Pi^k(\emptyset, I), I) \subseteq I$. Therefore, $\mathcal{T}_\Pi^\alpha(\emptyset, I) \subseteq I$.

(2) Since I is a model of Π relative to L, $L \cup I|_\Omega \cup \neg I^-|_\Omega$ is consistent. Since $\mathcal{T}_\Pi^\alpha(\emptyset, I) \subseteq I$, $L \cup \mathcal{T}_\Pi^\alpha(\emptyset, I)|_\Omega \cup \neg I^-|_\Omega$ is also consistent.

(3) By Definition 3, for any rule body $body(r)$ in $ground(\Pi)$ and any E_1 and E_2 with $E_1 \subseteq E_2 \subseteq J$, if $(E_1, I) \models_L body(r)$ then $(E_2, J) \models_L body(r)$. Then, by Definition 5, $\mathcal{T}_\Pi(E_1, I) \subseteq \mathcal{T}_\Pi(E_2, J)$. Next we prove that for any $i \geq 0$, $\mathcal{T}_\Pi^i(\emptyset, I) \subseteq \mathcal{T}_\Pi^i(\emptyset, J) \subseteq J$. It is obvious for $i = 0$. Assume $\mathcal{T}_\Pi^i(\emptyset, I) \subseteq \mathcal{T}_\Pi^i(\emptyset, J) \subseteq J$ for any $i \leq k \geq 0$. For $i = k + 1$, by Theorem 3, $\mathcal{T}_\Pi^{k+1}(\emptyset, I) = \mathcal{T}_\Pi(\mathcal{T}_\Pi^k(\emptyset, I), I) \subseteq \mathcal{T}_\Pi^{k+1}(\emptyset, J) = \mathcal{T}_\Pi(\mathcal{T}_\Pi^k(\emptyset, J), J) \subset J$. Therefore, $\mathcal{T}_\Pi^\alpha(\emptyset, I) \subseteq \mathcal{T}_\Pi^\alpha(\emptyset, J) \subseteq J$.   □

We define answer sets for normal DL logic programs using the above fixpoint.

Definition 6. Let Π be a normal DL logic program relative to a DL knowledge base L, and I a model of Π relative to L. I is an *answer set* of Π relative to L if for every $A \in I$, either $A \in \mathcal{T}_\Pi^\alpha(\emptyset, I)$ or $L \cup \mathcal{T}_\Pi^\alpha(\emptyset, I)|_\Omega \cup \neg I^-|_\Omega \models A$.

It is immediate that when $L = \emptyset$, a model I is an answer set of Π relative to L iff $I = \mathcal{T}_\Pi^\alpha(\emptyset, I)$.

The answer set semantics for Π relative to L is then defined by answer sets of Π. That is, a ground literal l is *credulously* (resp. *skeptically*) true in Π relative to L if $I \models_L l$ for some (resp. every) answer set I of Π relative to L.

Example 3. Consider Example 1. For $I_1 = \emptyset$, $\mathcal{T}_\Pi^\alpha(\emptyset, I_1) = \emptyset$, so I_1 is an answer set of Π relative to L. For $I_2 = \{A(a)\}$, $\mathcal{T}_\Pi^0(\emptyset, I_2) = \emptyset$ and $\mathcal{T}_\Pi^1(\emptyset, I_2) = \mathcal{T}_\Pi(\emptyset, I_2) = \emptyset$, so $\mathcal{T}_\Pi^\alpha(\emptyset, I_2) = \emptyset$. For $A(a) \in I_2$, $A(a) \notin \mathcal{T}_\Pi^\alpha(\emptyset, I_2)$ and $L \cup \mathcal{T}_\Pi^\alpha(\emptyset, I_2)|_\Omega \cup \neg I_2^-|_\Omega \not\models A(a)$. Thus I_2 is not an answer set of Π relative to L.

Consider Example 2. For $I_1 = \{A(a)\}$, $\mathcal{T}_\Pi^0(\emptyset, I_1) = \emptyset$ and $\mathcal{T}_\Pi^1(\emptyset, I_1) = \mathcal{T}_\Pi(\emptyset, I_1) = \emptyset$, so $\mathcal{T}_\Pi^\alpha(\emptyset, I_1) = \emptyset$. For $A(a) \in I_1$, $A(a) \notin \mathcal{T}_\Pi^\alpha(\emptyset, I_1)$, but $L \cup \mathcal{T}_\Pi^\alpha(\emptyset, I_1)|_\Omega \cup \neg I_1^-|_\Omega \models A(a)$, so I_1 is an answer set of Π relative to L. It is easy to verify that $I_2 = \{A(a), C(a)\}$ is not an answer set of Π relative to L.

The following result shows that answer sets must be minimal models.

Theorem 5. *Let Π be a normal DL logic program with DL expressions relative to a DL knowledge base L. If I is an answer set of Π relative to L, then I is a minimal model of Π relative to L.*

Proof: Assume, on the contrary, that I is not a minimal model relative to L. Let $J \subset I$ be a minimal model relative to L. By Theorem 4, $\mathcal{T}_\Pi^\alpha(\emptyset, I) \subseteq \mathcal{T}_\Pi^\alpha(\emptyset, J) \subseteq J$. Let $S = I \setminus J$. Note that S is not empty and for any $A \in S$, $A \notin \mathcal{T}_\Pi^\alpha(\emptyset, I)$. Since I is an answer set of Π, for any $A \in S$, $L \cup \mathcal{T}_\Pi^\alpha(\emptyset, I)|_\Omega \cup \neg I^-|_\Omega \models A$. Since $J^- \supset I^-$, $L \cup J|_\Omega \cup \neg J^-|_\Omega \models A$. Since every $A \in S$ is in J^-, $L \cup J|_\Omega \cup \neg J^-|_\Omega \models \neg A$. This means that $L \cup J|_\Omega \cup \neg J^-|_\Omega$ is not consistent, and thus J is not a model of Π relative to L. We then have a contradiction. Therefore, I is a minimal model of Π relative to L.   □

The next result shows that answer sets are exactly well-supported models.

Theorem 6. *Let Π be a normal DL logic program with DL expressions relative to a DL knowledge base L, and I a model of Π relative to L. I is an answer set of Π relative to L iff I is a well-supported model of Π relative to L.*

Proof: Assume that I is an answer set relative to L. We can construct a level mapping $f : I \to N$, where N is an integer, as follows: For each $A \in I$, we assign $f(A) = i$, where $i \geq 0$ is the smallest number such that either $L \cup T_\Pi^i(\emptyset, I)|_\Omega \cup \neg I^-|_\Omega \models A$ or there is a rule $A \leftarrow body(r)$ in $ground(\Pi)$ such that $(T_\Pi^i(\emptyset, I), I) \models_L body(r)$.

We then define a strict well-founded partial order $\prec$ on I such that for any $A, B \in I$, $B \prec A$ iff $f(B) < f(A)$. For each $A \in I$ with $f(A) = i$, we always have $E = T_\Pi^i(\emptyset, I) \subset I$, where for every $B \in E$, $B \prec A$, such that either $L \cup E|_\Omega \cup \neg I^-|_\Omega \models A$ or there is a rule $A \leftarrow body(r)$ in $ground(\Pi)$ such that $(E, I) \models_L body(r)$. By Definition 4, I is a well-supported model relative to L.

Conversely, assume that I is a well-supported model relative to L. Then, there exists a strict well-founded partial order $\prec$ on I such that for any $A \in I$, there exists $E \subset I$, where for every $B \in E$, $B \prec A$, such that either $L \cup E|_\Omega \cup \neg I^-|_\Omega \models A$ or there is a rule $A \leftarrow body(r)$ in $ground(\Pi)$ such that $(E, I) \models_L body(r)$. Such a partial order establishes a level mapping $f : I \to N$ so that for any $A \in I$, A can be derived from some $E \subset I$ at lower levels in the way as above. Next, we show that for every $A \in I$ at level $i \geq 0$ we have $E = T_\Pi^i(\emptyset, I)$ satisfying the above conditions.

First, each $A \in I$ at the lowest level ($i = 0$) does not depend on any other atom $B \in I$, i.e., there is no $B \in I$ with $B \prec A$. By the assumption that there exists $E \subset I$, where for every $B \in E$, $B \prec A$, such that either $L \cup E|_\Omega \cup \neg I^-|_\Omega \models A$ or there is a rule $A \leftarrow body(r)$ in $ground(\Pi)$ such that $(E, I) \models_L body(r)$, we have $E = \emptyset$. Therefore, for each $A \in I$ at level 0, we have $E = T_\Pi^0(\emptyset, I)$ which satisfies the above conditions.

As the induction hypothesis, assume that for any $i \leq n$ and any $A \in I$ at level i, we have $E = T_\Pi^i(\emptyset, I)$ such that either $L \cup E|_\Omega \cup \neg I^-|_\Omega \models A$ or there is a rule $A \leftarrow body(r)$ in $ground(\Pi)$ such that $(E, I) \models_L body(r)$. Then, by Theorem 3, for each $A \in I$ at level $i \leq n$, we have $E = T_\Pi^n(\emptyset, I)$ which satisfies the above conditions.

Consider $A \in I$ at level $n + 1$. Then, there exists $E \subset I$, where for every $B \in E$, $B \prec A$, such that either (1) $L \cup E|_\Omega \cup \neg I^-|_\Omega \models A$, or (2) there is a rule $A \leftarrow body(r)$ in $ground(\Pi)$ such that $(E, I) \models_L body(r)$. Next, we show that when using $T_\Pi^{n+1}(\emptyset, I)$ to replace E, the conditions (1) and (2) still hold for every $A \in I$ at level $n + 1$.

For every $B \in E$, since B is at a level below $n+1$, by the induction hypothesis, either (a) $L \cup T_\Pi^n(\emptyset, I)|_\Omega \cup \neg I^-|_\Omega \models B$, or (b) there is a rule $B \leftarrow body(r)$ in $ground(\Pi)$ such that $(T_\Pi^n(\emptyset, I), I) \models_L body(r)$. For case (a), we distinguish between two cases: (i) $B \in T_\Pi^n(\emptyset, I)$. In this case, if we replace B in E by $T_\Pi^{n+1}(\emptyset, I)$, the conditions (1) and (2) above still hold for each $A \in I$ at level $n + 1$. (ii) $B \notin T_\Pi^n(\emptyset, I)$. Then, for no $i \leq n$, $B \in T_\Pi^i(\emptyset, I)$; thus B is an atomic DL expression. In this case, if we replace B in E by $T_\Pi^{n+1}(\emptyset, I)$, the condition (1) above still holds, since $L \cup T_\Pi^n(\emptyset, I)|_\Omega \cup \neg I^-|_\Omega \models B$. Consider the condition (2). $(E, I) \models_L body(r)$ means that for every F with $E \subseteq F \subseteq I$, $F \models_L body(r)$. Let us replace B in E by $T_\Pi^{n+1}(\emptyset, I)$. Since B is a ground atomic DL expression, by Proposition 1, $(E \setminus \{B\} \cup T_\Pi^{n+1}(\emptyset, I), I) \models_L B$, because

$L \cup (E \setminus \{B\} \cup \mathcal{T}_\Pi^{n+1}(\emptyset, I))|_\Omega \cup \neg I^-|_\Omega \models B$. This shows that for any $body(r)$, $(E \setminus \{B\} \cup \mathcal{T}_\Pi^{n+1}(\emptyset, I), I) \models_L body(r)$ iff $(E \cup \mathcal{T}_\Pi^{n+1}(\emptyset, I), I) \models_L body(r)$. Then, when $(E, I) \models_L body(r)$, $(E \cup \mathcal{T}_\Pi^{n+1}(\emptyset, I), I) \models_L body(r)$ and thus $(E \setminus \{B\} \cup \mathcal{T}_\Pi^{n+1}(\emptyset, I), I) \models_L body(r)$. This shows that after replacing B in E by $\mathcal{T}_\Pi^{n+1}(\emptyset, I)$, the condition (2) above still holds, Therefore, if we replace B in E by $\mathcal{T}_\Pi^{n+1}(\emptyset, I)$, the conditions (1) and (2) above still hold. For case (b), $B \in \mathcal{T}_\Pi^{n+1}(\emptyset, I)$. So if we replace B in E by $\mathcal{T}_\Pi^{n+1}(\emptyset, I)$, the conditions (1) and (2) above still hold.

As a result, if we replace all $B \in E$ by $\mathcal{T}_\Pi^{n+1}(\emptyset, I)$, the conditions (1) and (2) above still hold. Therefore, for every $A \in I$ at level $n + 1$. we have $E = \mathcal{T}_\Pi^{n+1}(\emptyset, I)$ such that either $L \cup E|_\Omega \cup \neg I^-|_\Omega \models A$ or there is a rule $A \leftarrow body(r)$ in $ground(\Pi)$ such that $(E, I) \models_L body(r)$.

Consequently, for every $A \in I$, either $L \cup \mathcal{T}_\Pi^\alpha(\emptyset, I)|_\Omega \cup \neg I^-|_\Omega \models A$ or $A \in \mathcal{T}_\Pi^\alpha(\emptyset, I)$. This shows that I is an answer set of Π relative to L.   □

Example 4. Let $L = \emptyset$ and

$$\Pi: \quad A(g). \quad B(g) \leftarrow C(g). \quad C(g) \leftarrow ((A \sqcap \neg C) \sqcup B)(g).$$

Let $\mathbf{P} = \{A, B, C\}$, $\mathbf{C} = \{g\}$ and $\Omega = \{A, B, C\}$. $HB_\Pi = \{A(g), B(g), C(g)\}$ and $ground(\Pi) = \Pi$. Π has only one model relative to L, $I = \{A(g), B(g), C(g)\}$. This model is not an answer set, since it is not a well-supported model of Π relative to L.

Note that we can use a fresh DL concept D to replace the DL expression $(A \sqcap \neg C) \sqcup B$ and add to L an axiom $D \equiv (A \sqcap \neg C) \sqcup B$. This yields

$$\Pi': \quad A(g). \quad B(g) \leftarrow C(g). \quad C(g) \leftarrow D(g).$$
$$L': \quad D \equiv (A \sqcap \neg C) \sqcup B.$$

Using the same $\mathbf{P}$, $\mathbf{C}$ and Ω as above, Π' has the same answer sets relative to L' as Π relative to L.

The following result shows that this answer set semantics is a proper extension to the standard answer set semantics for normal logic programs.

Theorem 7. *Let $L = \emptyset$ and Π be a normal logic program without DL expressions. An interpretation I is an answer set of Π relative to L iff I is an answer set of Π under the standard answer set semantics.*

Proof: By Theorem 6, I is an answer set of Π relative to L iff I is a well-supported model of Π relative to L. By Theorem 2, I is a well-supported model of Π relative to L iff I is a well-supported model of Π under Fages' definition. Then as shown in [7], the well-supported models of Π under Fages' definition are exactly the answer sets of Π under the standard answer set semantics.   □

3.3   Decidability Property

For a normal DL logic program Π with DL expressions relative to a DL knowledge base L, the decidability of computing answer sets of Π relative to L depends on the decidability of satisfiability of L. Since DLs are fragments of first-order logic, the satisfiability of L is undecidable in general cases. However, if L is built from the description logic $\mathcal{SHOIN}$ or $\mathcal{SROIQ}$, its satisfiability is decidable [4,13,12].

Let L be a DL knowledge base built from a decidable description logic such as $\mathcal{SHOIN}$ or $\mathcal{SROIQ}$. Since HB_Π and $ground(\Pi)$ are finite, it is decidable to determine if an interpretation I is a model of Π relative to L. For any $E \subseteq I$ and any ground atom or DL expression A in $ground(\Pi)$, it is decidable to determine if $(E, I) \models_L A$ (resp. $(E, I) \models_L not\ A$) holds, and thus it is decidable to determine if $(E, I) \models_L body(r)$ holds for each rule r in $ground(\Pi)$. Since $ground(\Pi)$ consists of a finite set of rules, it takes finite time to compute the fixpoint $T_\Pi^\alpha(\emptyset, I)$. As a result, it is decidable to determine if an interpretation I is an answer set of Π relative to L. Since Π has only a finite set of interpretations, it is decidable to compute all answer sets of Π relative to L.

4   Related Work

Although many approaches to integrating rules and DLs have been proposed in the literature [2,4,10,11,14,16,18,19], to the best of our knowledge dl-programs [4] are the first framework which extends normal logic programs under the standard answer set semantics to logic programs with arbitrary DL expressions relative to an external DL knowledge base. Four different answer set semantics have been defined for dl-programs. The first one, called *weak* answer set semantics [4], easily incurs circular justifications by self-supporting loops, so a second one, called *strong* answer set semantics, was introduced [4]. Answer sets under the strong answer set semantics are not minimal models of a dl-program, then a third one, called *FLP-reduct* based answer set semantics, was proposed [5]. This semantics is based on the concept of FLP-reduct from [6]. It turns out, however, that none of the three answer set semantics extends the key well-supportedness condition of the standard answer set semantics to dl-programs, so that their answer sets may incur circular justifications by self-supporting loops. To resolve this problem, a fourth semantics, called *well-supported* answer set semantics, was recently introduced [20], which extends the well-supportedness condition to dl-programs. Dl-programs differ in fundamental ways from normal DL logic programs. First, in a dl-program, Π and L share no predicate symbols, so DL expressions $Q(\mathbf{t})$ must occur together with predicate mapping operations $S_i op_i P_i$. Note that in dl-programs one cannot use only dl-atoms of the form $DL[Q](\mathbf{t})$ to express all DL expressions $Q(\mathbf{t})$ because that would cut the knowledge flow from Π to L. Second, in a dl-program, DL expressions (dl-atoms) are not allowed to occur in a rule head, so no conclusions about L can be inferred from Π. Third, in this paper we extend the well-supportedness condition to normal DL logic programs. The extension process is similar to that in [20] by introducing an up to satisfaction relation (Definition 3), but the formalization of the well-supportedness condition is significantly different. For dl-programs, since Π and L share no predicate symbols, a model I of Π is well-supported if and only if for each $A \in I$ there is a rule $A \leftarrow body(r)$ in $ground(\Pi)$ such that I satisfies $body(r)$ and the evidence of the truth of $body(r)$ is not circularly dependent on A in I. For normal DL logic programs, however, the situation is much more complicated. As illustrated in Example 2, since Π and L share some predicate symbols, a model I of Π relative to L would be well-supported even if some $A \in I$ is not supported by any rule $A \leftarrow body(r)$ in $ground(\Pi)$ such that I satisfies $body(r)$. This presents additional difficulties in formalizing the well-supportedness condition for normal DL logic programs.

$\mathcal{DL}+\log$ [19] (and its variant such as *guarded hybrid knowledge bases* [11]) is closely related to but differs significantly from normal DL logic programs. Syntactically, it divides predicate symbols into *Datalog* predicates and *DL* predicates. The former type can only occur in Π, while the latter is not allowed to occur (as DL expressions) behind the negation operator *not*. Semantically, it considers first-order interpretations, instead of Herbrand interpretations, and defines a semantics with a class of first-order models, called *NM-models*. In an NM-model, DL predicates can take arbitrary truth values (as in first-order logic), but Datalog predicates take truth values that must be minimal (as in logic programming) when the truth values of all DL predicates are fixed. Consider a $\mathcal{DL}+\log$ program with $\Pi = \{B(g) \leftarrow A(g)\}$ and $L = \{A \sqcup C\}$. A, C must be DL predicates. Let B be a Datalog predicate. This program has at least three NM-models: $I_1 = \{A(g), B(g)\}$, $I_2 = \{A(g), B(g), C(g)\}$ and $I_3 = \{C(g)\}$. In contrast, if we take Π as a normal DL logic program relative to L, where $\mathbf{P} = \{B\}$, $\mathbf{C} = \{g\}$ and $\Omega = \emptyset$, Π has a unique well-supported model/answer set $\emptyset$ relative to L.

Disjunctive dl-programs [16] are closely related to normal DL logic programs, but differ significantly at least in three ways. Let Π be a disjunctive dl-program relative to L, where Π is built over a vocabulary $\Phi = (\mathbf{P}, \mathbf{C})$, $\mathbf{P}$ is a finite set of predicate symbols, and $\mathbf{C}$ is a nonempty finite set of constants. (1) All concepts and roles occurring in Π are required to be included in $\mathbf{P}$, so that all of them are interpreted over the Herbrand base HB_Π of Π. This strict requirement does not seem to be intuitive in some cases. For Example 4, since D is a fresh concept of L' introduced to represent $(A \sqcap \neg C) \sqcup B$, D is expected to be interpreted against L' in first-order logic. But in a disjunctive dl-program, D must be included in $\mathbf{P}$ and thus be interpreted over the Herbrand base $HB_{\Pi'}$. (2) The semantics of disjunctive dl-programs is based on FLP-reduct. Like the FLP-reduct based semantics for dl-programs [5], this FLP-reduct based semantics for disjunctive dl-programs yields answer sets that are minimal but not necessarily well-supported models. For Example 4, let Π' be a disjunctive dl-program. $I = \{A(g), B(g), C(g), D(g)\}$ is an answer set of Π' relative to L' under the FLP-reduct based semantics. Observe that the evidence of the truth of $B(g), C(g), D(g)$ in the answer set can only be inferred via a self-supporting loop $B(g) \Leftarrow C(g) \Leftarrow D(g) \Leftarrow B(g)$. (3) Disjunctive dl-programs allow only atomic DL expressions in rule bodies. We cannot have a disjunctive dl-program with $\Pi = \{A \leftarrow \neg A\}$ and $L = \emptyset$, since $\neg A$ is not an atomic DL expression. One might think that this issue could be handled by introducing a fresh concept B to represent the DL expression $\neg A$, which yields a disjunctive dl-program with $\Pi' = \{A \leftarrow B\}$ and $L' = \{B \equiv \neg A\}$.

However, this would produce an answer set $I = \{A\}$, which is not a well-supported model of Π.

Extensions of logic programs with DL expressions, such as dl-programs, $\mathcal{DL}$+log, disjunctive dl-programs, and normal DL logic programs, are different in fundamental ways from embeddings of rules and DLs into some unifying logic formalisms, such as the embedding [2] to first-order autoepistemic logic [17] and the embedding [18] to the logic of Minimal Knowledge and Negation as Failure (MKNF) [15]. The two embeddings employ modal logics and transform rules Π and DL axioms L to autoepistemic (resp. MKNF) logic formulas Π' and L' with modal operators. Then, the semantics of Π and L is defined by the semantics of $\Pi' \cup L'$ under autoepistemic (resp. MKNF) modal logic.

5 Summary

We have introduced a new extension, called normal DL logic programs, of normal logic programs with DL expressions relative to an external DL knowledge base. In normal DL logic programs, arbitrary DL expressions are allowed to appear in rule bodies and atomic DL expressions allowed in rule heads. We extended the key condition of well-supportedness for normal logic programs under the standard answer set semantics to normal DL logic programs and defined an answer set semantics which satisfies the extended condition of well-supportedness. As a result, answer sets under the well-supported semantics are free of circular justifications. We show that the answer set semantics for normal DL logic programs is decidable if the underlying description logic is decidable (e.g. $\mathcal{SHOIN}$ or $\mathcal{SROIQ}$).

As future work, we will study computational properties of normal DL logic programs w.r.t. different DLs, and extend normal DL logic programs to disjunctive DL logic programs, where the head of a rule is a disjunction of atoms.

Acknowledgments. We would like to thank all anonymous reviewers for their helpful comments. This work is supported in part by the National Natural Science Foundation of China (NSFC) grants 60970045 and 60833001, and by the Australia Research Council (ARC) Discovery Projects DP110101042 and DP1093652.

References

1. Baader, F., Calvanese, D., McGuinness, D., Nardi, D., Patel-Schneider, P.F. (eds.): The Description Logic Handbook: Theory, Implementation and Applications. Cambridge University Press (2003)
2. de Bruijn, J., Eiter, T., Tompits, H.: Embedding approaches to combining rules and ontologies into autoepistemic logic. In: Proceedings of the Eleventh International Conference on Principles of Knowledge Representation and Reasoning (KR 2008), pp. 485–495 (2008)
3. Eiter, T., Ianni, G., Krennwallner, T., Polleres, A.: Rules and ontologies for the semantic web. In: Proceedings of the 2nd International Conference on Web Reasoning and Rule Systems (RR 2008), pp. 1–53 (2008)

4. Eiter, T., Ianni, G., Lukasiewicz, T., Schindlauer, R., Tompits, H.: Combining answer set programming with description logics for the semantic web. Artificial Intelligence 172(12-13), 1495–1539 (2008)
5. Eiter, T., Ianni, G., Schindlauer, R., Tompits, H.: A uniform integration of higher-order reasoning and external evaluations in answer-set programming. In: IJCAI 2005, pp. 90–96 (2005)
6. Faber, W., Leone, N., Pfeifer, G.: Recursive aggregates in disjunctive logic programs: Semantics and complexity. In: Logics in Artificial Intelligence: European Workshop (JELIA 2004), pp. 200–212 (2004)
7. Fages, F.: Consistency of clark's completion and existence of stable models. Journal of Methods of Logic in Computer Science 1, 51–60 (1994)
8. Gelfond, M., Lifschitz, V.: The stable model semantics for logic programming. In: Proceedings of the 5th International Conference on Logic Programming (ICLP-1988), pp. 1070–1080 (1988)
9. Grau, B.C., Horrocks, I., Motik, B., Parsia, B., Patel-Schneider, P.F., Sattler, U.: OWL 2: The next step for OWL. Journal of Web Semantics 6(4), 309–322 (2008)
10. Grosof, B., Horrocks, I., R. Volz, S.D.: Description logic programs: Combining logic programs with description logics. In: WWW 2003, pp. 48–57 (2003)
11. Heymans, S., Bruijn, J.D., Predoiu, L., Feier, C., Niewenborgh, D.V.: Guarded hybrid knowledge bases. Theory and Practice of Logic Programming 8, 411–429 (2008)
12. Horrocks, I., Kutz, O., Sattler, U.: The even more irresistible SROIQ. In: Proceedings of the Tenth International Conference on Principles of Knowledge Representation and Reasoning (KR-2006), pp. 57–67 (2006)
13. Horrocks, I., Patel-Schneider, P.F., van Harmelen, F.: From SHIQ and RDF to OWL: the making of a web ontology language. Journal of Web Semantics 1(1), 7–26 (2003)
14. Horrocks, I., Patel-Schneider, P.F., Boley, H., Tabet, S., Grosof, B., Dean, M.: SWRL: A Semantic Web rule language combining OWL and RuleML. In: W3C Member Submission (2004), http://www.w3.org/Submission/SWRL
15. Lifschitz, V.: Nonmonotonic databases and epistemic queries. In: IJCAI-1991, pp. 381–386 (1991)
16. Lukasiewicz, T.: A novel combination of answer set programming with description logics for the semantic web. IEEE Transactions on Knowledge and Data Engineering 22(11), 1577–1592 (2010)
17. Moore, R.C.: Semantical considerations on nonmonotonic logic. Artificial Intelligence 25(1), 75–94 (1985)
18. Motik, B., Rosati, R.: Reconciling description logics and rules. J. ACM 57(5) (2010)
19. Rosati, R.: DL+log: Tight integration of description logics and disjunctive datalog. In: Proceedings of the Tenth International Conference on Principles of Knowledge Representation and Reasoning (KR-2006), pp. 68–78 (2006)
20. Shen, Y.D.: Well-supported semantics for description logic programs. In: IJCAI 2011, pp. 1081–1086 (2011)

Automatically Generating Data Linkages Using a Domain-Independent Candidate Selection Approach

Dezhao Song and Jeff Heflin

Department of Computer Science and Engineering, Lehigh University,
19 Memorial Drive West, Bethlehem, PA 18015, USA
{des308,heflin}@cse.lehigh.edu

Abstract. One challenge for Linked Data is scalably establishing high-quality *owl:sameAs* links between instances (e.g., people, geographical locations, publications, etc.) in different data sources. Traditional approaches to this entity coreference problem do not scale because they exhaustively compare every pair of instances. In this paper, we propose a candidate selection algorithm for pruning the search space for entity coreference. We select candidate instance pairs by computing a character-level similarity on discriminating literal values that are chosen using domain-independent unsupervised learning. We index the instances on the chosen predicates' literal values to efficiently look up similar instances. We evaluate our approach on two RDF and three structured datasets. We show that the traditional metrics don't always accurately reflect the relative benefits of candidate selection, and propose additional metrics. We show that our algorithm frequently outperforms alternatives and is able to process 1 million instances in under one hour on a single Sun Workstation. Furthermore, on the RDF datasets, we show that the entire entity coreference process scales well by applying our technique. Surprisingly, this high recall, low precision filtering mechanism frequently leads to higher F-scores in the overall system.

Keywords: Linked Data, Entity Coreference, Scalability, Candidate Selection, Domain-Independence.

1 Introduction

One challenge for the Linked Data [4] is to scalably establish high quality *owl:sameAs* links between instances in different data sources. According to the latest statistics[1], there are currently 256 datasets (from various domains, e.g., Media, Geographic, Publications, etc.) in the Linked Open Data (LOD) Cloud with more than 30 billion triples and about 471 million links across different datasets. This large volume of data requires automatic approaches be adopted for detecting *owl:sameAs* links. Prior research to this entity coreference problem[2]

---

[1] http://www4.wiwiss.fu-berlin.de/lodcloud/state/

[2] Entity Coreference is also referred to as Entity Resolution, Disambiguation, etc.

L. Aroyo et al. (Eds.): ISWC 2011, Part I, LNCS 7031, pp. 649–664, 2011.

[1,10,15] has focused on how to precisely and comprehensively detect coreferent instances and good results were achieved. However, one common problem with previous algorithms is that they were only applied to a small number of instances because they exhaustively compare every pair of instances in a given dataset. Therefore, such algorithms are unlikely to be of practical use at the scale of Linked Data. Although Sleeman and Finin [14] adopted a filtering mechanism to select potentially matching pairs, their filter checks every pair of instances by potentially having to consider all associated properties of an instance; this is unlikely to scale for datasets with many properties.

To scale entity coreference systems, one solution would be to efficiently determine if an instance pair could be coreferent by only comparing part of the pair's context, i.e., candidate selection. Other researchers have used the term *blocking* [12] but with two different meanings: finding non-overlapping blocks of instances such that all instances in a block will be compared to each other or simply locating similar instance pairs. This second usage is what we refer to as *candidate selection* in this paper. For an instance, we select other instances that it could be coreferent with, i.e., selecting a candidate set of instance pairs. Several interesting questions then arise. First, manually choosing the information to compare might not work for all domains due to insufficient domain expertise. Also, candidate selection should cover as many true matches as possible and reduce many true negatives. Finally, the candidate selection algorithm itself should scale to very large datasets.

In this paper, we propose a candidate selection algorithm with the properties discussed above. Although our algorithm is designed for RDF data, it generalizes to any structured dataset. Given an RDF graph and the types of instances to do entity coreference on, through unsupervised learning, we learn a set of datatype properties as the candidate selection key that both discriminates and covers the instances well in a domain-independent manner. We then utilize the object values of such predicates for candidate selection. In order to support efficient look-up for similar instances, we index the instances on the learned predicates' object values and adopt a character level n-gram based string similarity measure to select candidate pairs. We evaluate our algorithm on 3 instance categories from 2 RDF datasets and on another 3 well adopted structured datasets for evaluating entity coreference systems. Instead of only using traditional metrics (to be described in Section 2) for evaluating candidate selection results, we propose to apply an actual entity coreference system to the selected candidate pairs to measure the overall runtime and the final F1-score of the coreference results. We show that our proposed algorithm frequently outperforms alternatives in terms of the overall runtime and the F1-score of the coreference results; it also commonly achieved the best or comparably good results on the non-RDF datasets.

We organize the rest of the paper as following. We discuss the related work in Section 2. Section 3 presents the process of learning the predicates for candidate selection and Section 4 describes how to efficiently look up and select candidate instance pairs by comparing the object values of the learned predicates. We evaluate our algorithm in Section 5 and conclude in Section 6.

2   Related Work

Several candidate selection algorithms have been proposed. Best Five [18] is a set of manually identified rules for matching census data. However, developing such rules can be expensive, and domain expertise may not be available for various domains. ASN [21] learns dynamically sized blocks for each record with a manually determined key. The authors claim that changing to different keys didn't affect the results but no data was reported. Marlin [3] uses an unnormalized Jaccard similarity on the tokens between attributes by setting a threshold to 1, which is to find an identical token between the attributes. Although it was able to cover all true matches on some datasets, it only reduced the pairs to consider by 55.35%.

BSL [12] adopted supervised learning to learn a blocking scheme: a disjunction of conjunctions of (method, attribute) pairs. It learns one conjunction each time to reduce as many pairs as possible; by running iteratively, more conjunctions would be learned to increase coverage on true matches. However, supervised approaches require sufficient training data that may not always be available. As reported by Michelson and Knoblock [12], compared to using 50% of the groundtruth for training, 4.68% fewer true matches were covered on some dataset by training on only 10% of the groundtruth. In order to reduce the needs of training data, Cao et. al. [5] proposed a similar algorithm that utilizes both labeled and unlabeled data for learning the blocking scheme.

Adaptive Filtering (AF) [9] is unsupervised and is similar to our approach in that it filters record pairs by computing their character level bigram similarity. All-Pairs [2], PP-Join(+) [20] and Ed-Join [19] are all inverted index based approaches. All-Pairs is a simple index based algorithm with certain optimization strategies. PP-Join(+) proposed a positional filtering principle that exploits the ordering of tokens in a record. Ed-Join proposed filtering methods that explore the locations and contents of mismatching n-grams. Silk [17] indexes ontology instances on the values of manually specified properties to efficiently retrieve similar instance pairs. Customized rules are then used to detect coreferent pairs.

Compared to Best Five and ASN, our approach automatically learns the candidate selection key for various domains. Unlike Marlin, our system can both effectively reduce candidate set size and achieve good coverage on true matches. Although BSL achieved good results on various domains, its drawbacks are that it requires sufficient training data and is not able to scale to large datasets [13]. Cao et. al. [5] used unlabeled data for learning. However the supervised nature of their method still requires a certain amount of available groundtruth; while our algorithm is totally unsupervised. Similar to AF and Ed-Join, we also exploit using n-grams. However, later we show that our method covers 5.06% more groundtruth than AF on a census dataset and it generally selects one order of magnitude (or even more) fewer pairs than Ed-Join. All-Pairs and PP-Join(+) treat each token in a record as a feature and select features by only considering their frequency in the entire document collection; while we select the information for candidate selection on a predicate-basis and consider both if a predicate discriminates well and if it is used by a sufficient number of instances.

The Ontology Alignment Evaluation Initiative (OAEI) [7] includes an instance matching track that provides several benchmark datasets to evaluate entity coreference systems for detecting equivalent ontology instances; however, some of the datasets are of small scale and thus cannot sufficiently demonstrate the scalability of a candidate selection algorithm. Three metrics have been well adopted for evaluating candidate selection (Eq. 1): Pairwise Completeness (PC), Reduction Ratio (RR) and F-score (F_{cs}) [6,21]. PC and RR evaluate how many true positives are returned by the algorithm and the degree to which it reduces the number of comparisons needed respectively; F_{cs} is their F1-score, giving a comprehensive view of how well a system performs.

$$PC = \frac{|true\ matches\ in\ candidate\ set|}{|true\ matches|}, RR = 1 - \frac{|candidate\ set|}{N * M} \qquad (1)$$

where N and M are the sizes of two instance sets that are matched to one another. As we will show in Section 5.3, when applied to large datasets (with tens of thousands of instances), a large change in the size of the candidate set may only be reflected by a small change in RR due to its large denominator.

3   Learning the Candidate Selection Key

As discussed, candidate selection is the process of efficiently selecting possibly coreferent instance pairs by only comparing part of their context information. Therefore, the information we will compare needs to be useful in disambiguating the instances. For example, a person instance may have the following triples:

person#100   *has-last-name*   "Henderson"
person#100   *has-first-name*   "James"
person#100   *lives-in*   "United States"

Intuitively, we might say that last name could disambiguate this instance from others better than first name which is better than the place where he lives in. The reason could be that the last name *Henderson* is less common than the first name *James*; and a lot more people live in the United States than those using *James* as first name. Therefore, for person instances, we might choose the object values of *has-last-name* for candidate selection. However, we need to be able to automatically learn such disambiguating predicate(s) in a domain-independent manner. Furthermore, the object values of a single predicate may not be sufficiently disambiguating to the instances. Take the above example again, it could be more disambiguating if we use both last name and first name.

Algorithm 1 presents the process for learning the candidate selection key, a set of datatype predicates, whose object values are then utilized for candidate selection. Triples with datatype predicates use literal values as objects. The goal is to iteratively discover a predicate set (the candidate selection key) whose values are sufficiently discriminating (discriminability) such that the vast majority of instances in a given dataset use at least one of the learned predicates (coverage). The algorithm starts with an RDF graph G (a set of triples, <i,p,o>)

Algorithm 1. Learn_Key(G, C), G is an RDF graph, consisting a set of triples, C is a set of instance types

---
1. $key\_set \leftarrow a\ set\ of\ datatype\ properties\ in\ G$
2. $I_C \leftarrow \{i| < i, \text{rdf:type}, c > \in G \wedge c \in C\}$
3. $satisfied \leftarrow$ **false**
4. **while not** $satisfied$ **and** $key\_set \neq \emptyset$ **do**
5.   **for** $key \in key\_set$ **do**
6. $discriminability \leftarrow dis(key, I_C, G)$
7.   **if** $discriminability < \beta$ **then**
8. $key\_set \leftarrow key\_set - key$
9.   **else**
10. $coverage \leftarrow cov(key, I_C, G)$
11. $F_L \leftarrow \frac{2*discriminability*coverage}{discriminability+coverage}$
12. $score[key] \leftarrow F_L$
13.   **if** $F_L > \alpha$ **then**
14. $satisfied \leftarrow$ **true**
15.   **if not** $satisfied$ **then**
16. $dis\_key \leftarrow \arg\max_{key \in key\_set} dis(key, I_C, G)$
17. $key\_set \leftarrow combine\ dis\_key\ with\ all\ other\ keys$
18. $G \leftarrow update(I_C, key\_set, G)$
19. **return** $\arg\max_{key \in key\_set} score[key]$

---

and it extracts all the datatype predicates (key_set) and the instances (I_C) of certain categories (C) (e.g., person, publication, etc.) from G. Then, for each predicate $key \in key\_set$, the algorithm retrieves all the object values of the key for instances in I_C. Next, it computes three metrics: discriminability, coverage as shown in Equations 2 and 3 respectively and a F1-score (F_L) on them.

$$dis(key, I_C, G) = \frac{|\{o|t =< i, key, o > \in G \wedge i \in I_C\}|}{|\{t|t =< i, key, o > \in G \wedge i \in I_C\}|} \tag{2}$$

$$cov(key, I_C, G) = \frac{|\{i|t =< i, key, o > \in G \wedge i \in I_C\}|}{|I_C|} \tag{3}$$

Note, i and o represent the subject and object of a triple respectively. In the learning process, we remove low-discriminability predicates. Because the discriminability of a predicate is computed based upon the diversity of its object values, having low-discriminability means that many instances have the same object values on this predicate; therefore, when utilizing such object values to look up similar instances, we will not get a suitable reduction ratio.

If any predicate has an F_L (line 11) higher than the given threshold α, the predicate with the highest F_L will be chosen to be the candidate selection key. If none of the keys have an F_L above the threshold α, the algorithm combines the predicate that has the highest discriminability with every other predicate to form $|key\_set|$-1 virtual predicates, add them to key_set and remove the old ones. Furthermore, via the function $update(I_C, key\_set, G)$, for a new key, we concatenate the object values of different predicates in the key for the same instance to form new triples that use the combined virtual predicate as their predicate and the concatenated object values as their objects. These new triples and predicates are added to G. The same procedure can then be applied iteratively.

Worst case Algorithm 1 is exponential in the number of candidate keys because of its two loops; but typically only a few passes are made through the outer loop before the termination criteria is met in our current evaluations. For future work, we will explore how to prune the initial list of candidate keys and reduce the data complexity of functions dis and cov with sampling techniques.

4 Index Based Candidate Selection

With the learned predicates, for each instance, we present how to efficiently look up similar instances and compute their similarity based on the objects of such predicates. One simple approach is to compare the object values of the learned predicates for all pairs of instances, e.g., comparing names for people instances. However, this simple method itself might not even scale for large scale datasets. So, we need a technique that enables efficient look-up for similar instances.

4.1 Indexing Ontology Instances

We adopt a traditional technique in Information Retrieval (IR) research, the inverted index, to speed up the look-up process. Many modern IR systems allow us to build separate indexes for different fields. Given an RDF graph G and the datatype properties PR learned by Algorithm 1, we use this feature to build indexes for the learned predicates, each of which has posting lists of instances for each token in that field. For a learned predicate $p \in PR$, we extract tokens from the object values of triples using p; for each such token tk, we collect all instances that are subjects of at least one triple with predicate p and token tk contained in its object value. With the learned predicates, each instance is associated with tuple(s) in the form of (instance, predicate, value) by using the learned predicates individually. We define a function search(Idx, q, $pred$) that returns the set of instances for which the $pred$ field matches the boolean query q using inverted index Idx.

4.2 Building Candidate Set

With the index, Algorithm 2 presents our candidate selection process where t is a tuple and $t.v$, $t.p$ and $t.i$ return the value, predicate and instance of t respectively. For each tuple t, we issue a Boolean query, the disjunction of its tokenized values, to the index to search for tuples ($results$) with similar values on all predicates comparable to that of t. The search process performs an exact match on each query token. $is\_sim(t, t')$ returns true if the similarity between two tuple values is higher than a threshold.

First of all, we look up instances on comparable fields. For example, in one of our datasets used for evaluation, we try to match person instances of both the *citeseer:Person* and the *dblp:Person* classes where the key is the combination

Algorithm 2. Candidate_Selection(T,Idx), T is a set of tuples using predicates in the learned key; Idx is an inverted index

---

1. *candidates* ← ∅
2. **for all** $t \in T$ **do**
3.   *query* ← *the disjunction of tokens of t.v*
4.   *results* ← $\bigcup_{p \in Comparable(t.p)} search(Idx, query, p)$
5.   **for all** $t' \in results$ **do**
6.   **if** *is_sim(t, t')* **then**
7.   *candidates* ← *candidates* ∪ $(t.i, t'.i)$
8. **return** *candidates*

---

of *citeseer:Name* and *foaf:Name*. So, for a tuple, we need to search for similar tuples on both predicates. Assuming we have an alignment ontology where mappings between classes and predicates are provided, two predicates p and q are comparable if the ontology entails $p \sqsubseteq q$ (or vice versa).

To further reduce the size of the candidate set, it would be necessary to adopt a second level similarity measure between a given instance (i) and its returned similar instances from the Boolean query. Otherwise, any instance that shares a token with i will be returned. In this paper, we compare three different definitions of the function *is_sim*. The first one is to directly compare (*direct_comp*) two tuple values (e.g., person names) as shown in Equation 4.

$$String\_Matching(t.v, t'.v) > \delta \tag{4}$$

where t and t' are two tuples; *String_Matching* computes the similarity between two strings. If the score is higher than the threshold δ, this pair of instances will be added to the candidate set. Although this might give a good pairwise completeness by setting δ to be low, it could select a lot of non-coreference pairs. One example is person names. Person names can be expressed in different forms: first name + last name; first initial + last name, etc.; thus, adopting a low δ could help to give a very good coverage on true matches; however, it may also match people with the same family name and similar given names.

Another choice is to check the percentage of their shared highly similar tokens (*token_sim*) as shown in Equation 5:

$$\frac{|sim\_token(t.v, t'.v)|}{\min(|token\_set(t.v)|, |token\_set(t'.v)|)} > \theta \tag{5}$$

where *token_set* returns the tokens of a string; *sim_token* is defined in Eq. 6:

$$sim\_token(s_i, s_j) =$$
$$\{token_i \in token\_set(s_i) | \exists token_j \in token\_set(s_j),$$
$$String\_Matching(token_i, token_j) > \delta)\} \tag{6}$$

where $s_{i/j}$ is a string and $token_{i/j}$ is a token from it. Without loss of generality, we assume that the number of tokens of s_i is no greater than that of s_j. The intuition is that two coreferent instances could share many similar tokens, though the entire strings may not be sufficiently similar on their entirety. One potential problem is that it may take longer to calculate because the selected literal values could be long for some instances types (e.g., publication titles).

Instead of computing token level similarity, we can check how many character level n-grams are shared between two tuple values as computed in Equation 7:

$$\frac{|gram\_set(n, t.v) \cap gram\_set(n, t'.v)|}{\min(|gram\_set(n, t.v)|, |gram\_set(n, t'.v)|)} > \theta \tag{7}$$

where $gram\_set(n, t.v)$ extracts the character level n-grams from a string. We hypothesize that the n-gram based similarity measure is the best choice. The intuition is that we can achieve a good coverage on true matches to the Boolean query by examining the n-grams (which are finer grained than both tokens and entire strings) while at the same time effectively reducing the candidate set size by setting an appropriate threshold. We use min in the denominator for Equations 5 and 7 to reduce the chance of missing true matches due to missing tokens, spelling variations or misspellings (e.g., some tokens of people names can be missing or spelled differently). When building/querying the index and comparing the literal values, we filter stopwords, use lowercase for all characters and ignore the ordering of the tokens and n-grams.

5   Evaluation

Our system is implemented in Java and we conducted experiments on a Sun Workstation with an 8-core Intel Xeon 2.93GHz processor and 6GB memory.

5.1   Datasets

We evaluate our n-gram based approach on 2 RDF datasets: RKB[3] [8] and SWAT[4]. For RKB, we use 8 subsets of it: ACM, DBLP, CiteSeer, EPrints, IEEE, LAAS-CNRS, Newcastle and ECS. The SWAT dataset consists of RDF data parsed from downloaded XML files of CiteSeer and DBLP. Both datasets describe publications and share some information; but they use different ontologies, and thus different predicates are involved. Their coverage of publications is also different. We compare on 3 instance categories: RKB Person, RKB Publication and SWAT Person. The groundtruth was provided as *owl:sameAs* statements that can be crawled from RKB and downloaded from SWAT as an RDF dump respectively. Since the provided groundtruth was automatically derived and was incomplete and erroneous, we randomly chose 100K instances for each category, applied our algorithm with different thresholds to get candidate selection results, and manually checked the false positives/negatives to verify and augment the groundtruth to improve their quality. We are in the process of completing SWAT Publication groundtruth and will conduct relevant experiments for future work.

---

[3] http://www.rkbexplorer.com/data/
[4] http://swat.cse.lehigh.edu/resources/data/

We also evaluate on 3 other structured datasets frequently used for evaluating entity coreference systems. Each dataset has a pre-defined schema with several attributes: name, date, etc. We convert them into RDF by treating each attribute as a datatype property. The first one is the Restaurant dataset [16], matching segmented online posts (records) from Fodors (331 records) to Zagat (533 records) with 112 duplicates. It has 4 attributes: name, address, type and city. Another dataset is the Hotel dataset [13] that has 5 attributes: name, rating, area, price and date, matching 1,125 online hotel bidding posts from the Bidding For Travel website[5] to another 132 hotel information records from the Bidding For Travel hotel guides with 1,028 coreferent pairs. The last one is *dataset4* [9], a synthetic census dataset, with 10K records and 5K duplicates within themselves. We remove the Social Security Number from it as was done in BSL [12] to perform a fair comparison and match the 10K records to themselves.

5.2   Evaluation Methods and Metrics

In this paper, we adopt a two-phase approach for evaluating our proposed candidate selection algorithm. In phase one, we use the 3 well adopted metrics PC, RR and F_{cs} from previous works [21,6] as discussed in Section 2. For phase two, we adopt an actual entity coreference algorithm for detecting *owl:sameAs* links between ontology instances [15] that measures the similarity of two instances by utilizing the triples in an RDF graph as context information. Not only does this context include the direct triples but also triples two steps away from an instance. We apply our candidate selection technique on the RDF datasets discussed in the previous section to select candidate pairs and run this algorithm on the candidate sets to get the F-score of the coreference phase and the runtime of the entire process, including indexing, candidate selection and coreference.

As for parameter settings, for the learning process (Algorithm 1), there are two parameters α, determining if a key could be used for candidate selection and β, determining if a key should be removed. To show the domain independence of our algorithm, we set them to be 0.9 and 0.3 respectively for all experiments. We tested our algorithm on different α and β values and it is relatively insensitive to β, but requires high values for α for good performance. When β is low, only a few predicates are removed for not being discriminating enough; when α is high, then we only select keys that discriminate well and are used by most of the instances. For Algorithm 2, different similarity measures may use different parameters. For Equations 4, 5, 6 and 7, we set θ to be 0.8; for *direct_comp* and *token_sim*, we varied δ from 0.1 to 0.9 and report the best results. We extract bigrams and compute Jaccard similarity for string matching in all experiments.

5.3   Evaluation Results on RDF Datasets

From Algorithm 1, we learned the key for each RDF dataset as following:

RKB Person: full-name, job, email, web-addr and phone
RKB Publication: title
SWAT Person: citeseer:name and foaf:name

For RKB Person, *full-name* has good coverage but is not sufficiently discriminating; while the other selected predicates have good discriminability but poor coverage. So, they were combined to be the key. For SWAT Person, neither of the two selected predicates has sufficient coverage; thus both were selected.

We compare our method *bigram* (Eq. 7) to *direct_comp* (Eq. 4) and *token_sim* (Eq. 5) that use different string similarity measures; we also compare to *All-Pairs* [2], *PP-Join(+)* [20] and *Ed-Join* [19]; lastly, we compare to *Naive* [15] that detects *owl:sameAs* links without candidate selection. Since *Ed-Join* is not compatible with our Sun machine, we run it on a Linux machine (dual-core 2GHz processor and 4GB memory), and estimate its runtime on the Sun machine by examining runtime difference of *bigram* on the two machines. For coreference results, we report a system's best F-Score from threshold 0.1-0.9. We split each 100K dataset into 10 non-overlapping and equal-sized subsets, index each subset, run all algorithms on the same input and report the average. We conduct a two-tailed t-test to test the statistical significance on the results of the 10 subsets from two systems. On average, there are 6,096, 4,743 and 684 coreferent pairs for each subset of RKB Person, RKB Publication and SWAT Person respectively.

The results are shown in Table 1. Comparing within our own alternatives, for

Table 1. Candidate Selection Results on RDF Datasets $|Pairs|$: candidate set size; RR: Reduction Ratio; PC: Pairwise Completeness; F_{cs}: the F1-score for RR and PC; F-Score: the F1-Score of Precision and Recall for the coreference results; Total: the runtime for the entire entity coreference process

Dataset	System	Candidate Selection					Coref	Total (s)
		$\|Pairs\|$	$RR(\%)$	$PC(\%)$	$F_{cs}(\%)$	$Time(s)$	F-score (%)	
RKB Per	bigram (Eq. 7)	14,024	**99.97**	99.33	99.65	13.32	**93.48**	**25.45**
	direct_comp (Eq. 4)	104,755	99.79	**99.82**	**99.80**	14.00	92.55	51.04
	token_sim (Eq. 5)	**13,156**	**99.97**	98.52	99.24	15.72	93.37	27.13
	All-Pairs [2]	680,403	98.64	99.76	99.20	**1.34**	92.04	195.37
	PP-Join [20]	680,403	98.64	99.76	99.20	1.36	92.04	195.38
	PP-Join+ [20]	680,403	98.64	99.76	99.20	1.39	92.04	195.42
	Ed-Join [19]	150,074	99.70	99.72	99.71	1.73	92.38	72.79
	Naive [15]	N/A	N/A	N/A	N/A	N/A	91.64	4,765.46
RKB Pub	bigram (Eq. 7)	6,831	**99.99**	**99.97**	**99.98**	18.26	**99.74**	**31.73**
	direct_comp (Eq. 4)	7,880	99.98	**99.97**	99.97	22.23	99.68	36.74
	token_sim (Eq. 5)	**5,028**	**99.99**	99.80	99.89	79.91	99.70	88.96
	All-Pairs [2]	1,527,656	96.94	97.95	97.44	3.93	98.59	877.80
	PP-Join [20]	1,527,656	96.94	97.95	97.44	**3.79**	98.59	877.66
	PP-Join+ [20]	1,527,656	96.94	97.95	97.44	4.00	98.59	877.87
	Ed-Join [19]	2,579,333	94.84	98.57	96.67	409.08	99.04	1,473.47
	Naive [15]	N/A	N/A	N/A	N/A	N/A	99.55	34,566.73
SWAT Per	bigram (Eq. 7)	7,129	**99.99**	98.72	99.35	13.46	**95.07**	**21.21**
	direct_comp (Eq. 4)	90,032	99.82	99.86	**99.84**	14.30	95.06	51.33
	token_sim (Eq. 5)	**6,266**	**99.99**	96.81	98.37	16.58	**95.07**	23.70
	All-Pairs [2]	508,505	98.98	**99.91**	99.44	**1.00**	95.06	108.89
	PP-Join [20]	508,505	98.98	**99.91**	99.44	1.01	95.06	108.90
	PP-Join+ [20]	508,505	98.98	**99.91**	99.44	1.04	95.06	108.92
	Ed-Join [19]	228,830	99.54	99.79	99.66	1.48	95.01	51.66
	Naive [15]	N/A	N/A	N/A	N/A	N/A	95.02	12,139.60

all datasets, both *bigram* and *token_sim* have the best *RR* while *direct_comp* commonly has better *PC*. *bigram* selected almost as few pairs as *token_sim*, and always has better *PC*.

On RKB Person, *bigram*'s F_{cs} was not as good as that of *direct_comp* and *Ed-Join*; statistically, the difference between *bigram* and *direct_comp* is significant with a P value of 0.0106; the difference between *bigram* and *token_sim* and *All-Pairs/PP-Join*(+) is statistically significant with P values of 0.0004 and 0.0001 respectively. Also, by applying our entity coreference system to the selected pairs, *bigram* has the best F-score that is statistically significant compared to that of *All-Pairs/PP-Join*(+) with a P value of 0.0093.

We observed similar results on SWAT Person. For F_{cs}, the difference between *bigram* to *direct_comp*, *token_sim* and *Ed-Join* is statistically significant with P values of 0.0011, 0.0001 and 0.0263 respectively but not to *All-Pairs/PP-Join*(+). Similarly, all other systems took longer to finish the entire process than *bigram*. As for the *F-score* of the coreference results, we didn't observe any significant difference among the different systems.

On RKB Publication, *bigram* dominates the others in all aspects except for |*pairs*|. For F_{cs}, except for *direct_comp*, the difference between *bigram* and others is statistically significant with P values of 0.0001. As for the coreference results, although no statistical significance was observed between *bigram* and *direct_comp/token_sim*, statistically, *bigram* achieved a better F-score than *All-Pairs/PP-Join*(+)/*Ed-Join* with P values of 0.0001. Similarly, adopting *bigram* gave the best runtime.

Note that *token_sim* took longer to finish than *bigram* even with fewer selected pairs because it took longer to select candidate pairs. It would potentially have to compare every pair of tokens from two strings, which was time-consuming. This was even more apparent on RKB Publication where titles generally have more tokens than people names do.

Finally, we ran our coreference algorithm (*Naive*) on the subsets of each RDF dataset. Although our proposed candidate selection algorithms were typically slower than their competitors, they filtered out many more pairs, which led to faster times for the complete system. Table 1 shows that using *bigram* was the fastest of all options; it was 169.08, 529.65 and 938.30 times faster than *Naive* on RKB Person, SWAT Person and RKB Publication; and by applying candidate selection, the F-score of the coreference results doesn't drop and even noticeably better performance was achieved. For RKB Person and RKB Publication, the improvement on the F-score is statistically significant with P values of 0.0020 and 0.0005. Such improvement comes from better precision: by only comparing the disambiguating information selected by Algorithm 1, candidate selection filtered out some false positives that could have been returned as coreferent by *Naive*. E.g., *Naive* might produce a false positive for RKB Person for two frequent co-authors, because the titles and venues of their papers are often the same; however, by only considering their most disambiguating information, they could be filtered out. In this case, candidate selection doesn't only help to scale the entire entity coreference process but also improves its overall F-score.

To further demonstrate the capability of our technique (*bigram*) in scaling entity coreference systems, we run *Naive* with and without it on up to 20K instances from each of the RDF datasets respectively and measure the speedup factor, computed as the runtime without *bigram* divided by the runtime with it, as shown in Figure 1. The runtime includes both the time for candidate selection

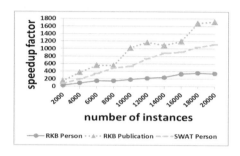

Fig. 1. Runtime Speedup by Applying Candidate Selection

and entity coreference. The entire coreference process was speeded up by 2 to 3 orders of magnitude. RKB Person shows less speedup than the others: first, candidate selection found more pairs for RKB Person; second, RKB Person has fewer paths than the other datasets, so there is less to prune.

5.4   Evaluation Results Using Standard Coreference Datasets

To show the generality of our proposed algorithm, we also evaluate it on three non-RDF but structured datasets frequently used for evaluating entity coreference algorithms: Restaurant, Hotel and Census as described earlier. We learned the candidate selection key for each dataset as following:

Restaurant: name
Hotel: name
Census: date-of-birth, surname and address_1

Here, we compare to five more systems: BSL [12], ASN [21], Marlin [3], AF [9] and Best Five [18] by referencing their published results. We were unable to obtain executables for these systems.

Here, we apply candidate selection on each of the three full datasets. First, the scale of the datasets and their groundtruth is small. Also, each of the Restaurant and the Hotel datasets is actually composed of two subsets and the entity coreference task is to map records from one to the other; while for other datasets, we detect coreferent instances within all the instances of a dataset itself. So, it is difficult to split such datasets. We didn't apply any actual coreference systems to the candidate set here due to the small scale and the fact that we couldn't run some of the systems to collect the needed candidate sets. Instead, in order to accurately reflect the impact of RR, we suggest a new metric RR_{log} computed as $1 - \frac{\log|candidate\ set|}{\log(N*M)}$. In Table 1, on RKB Person, an order of magnitude difference in detected pairs between *bigram* and *Ed-Join* is only represented by less than 1 point in RR; however, a more significant difference in the total runtime was observed. With this new metric, *bigram* and *Ed-Join* have an RR_{log} of 46.13% and 32.77% respectively where the difference is now better represented by 13.36%. We also compute a corresponding F_{cs_log} using RR_{log}. For systems where we reused reported results, we calculated $|pairs|$ from their reported RR; because BSL is supervised (thus the blocking was not done on the full dataset), we assumed the same RR as if it was done on the full dataset.

Table 2 shows the results. Since not all systems reported results on all datasets, we only report the available results here. Comparing within our own alternatives,

Table 2. Candidate Selection Results on Standard Coreference Datasets

Dataset	System	Candidate Selection					
		$|Pairs|$	$RR(\%)$	$PC(\%)$	$F_{cs}(\%)$	$RR\_log(\%)$	$F_{cs\_log}(\%)$
Restaurant	bigram (Eq. 7)	182	**99.90**	98.21	99.05	**56.92**	**72.07**
	direct_comp (Eq. 4)	2,405	98.64	100.00	99.31	35.56	52.46
	token_sim (Eq. 5)	184	**99.90**	95.54	97.67	56.83	71.27
	All-Pairs [2]	1,967	98.89	99.11	99.00	37.22	54.12
	PP-Join [20]	1,967	98.89	99.11	99.00	37.22	54.12
	PP-Join+ [20]	1,967	98.89	99.11	99.00	37.22	54.12
	Ed-Join [19]	6,715	96.19	96.43	96.31	27.06	42.26
	BSL [12]	1,306	99.26	98.16	98.71	40.61	57.45
	ASN [21]	N/A	N/A	<96	<98	N/A	N/A
	Marlin [3]	78,773	55.35	**100.00**	71.26	6.67	12.51
Hotel	bigram (Eq. 7)	**4,142**	**97.21**	94.26	**95.71**	30.06	45.58
	direct_comp (Eq. 4)	10,036	93.24	96.69	94.94	22.63	36.67
	token_sim (Eq. 5)	4,149	**97.21**	90.56	93.77	30.04	45.12
	All-Pairs [2]	6,953	95.32	95.91	95.62	25.71	40.55
	PP-Join [20]	6,953	95.32	95.91	95.62	25.71	40.55
	PP-Join+ [20]	6,953	95.32	95.91	95.62	25.71	40.55
	Ed-Join [19]	17,623	88.13	98.93	93.22	17.90	30.31
	BSL [12]	27,383	81.56	**99.79**	89.76	14.20	24.86
Census	bigram (Eq. 7)	166,844	99.67	97.76	98.70	32.17	48.41
	direct_comp (Eq. 4)	738,945	98.52	98.08	98.30	23.77	38.27
	token_sim (Eq. 5)	163,207	99.67	96.36	97.99	32.30	48.38
	All-Pairs [2]	**5,231**	**99.99**	100.00	**99.99**	**51.70**	**68.16**
	PP-Join [20]	**5,231**	**99.99**	100.00	**99.99**	**51.70**	**68.16**
	PP-Join+ [20]	**5,231**	**99.99**	100.00	**99.99**	**51.70**	**68.16**
	Ed-Join [19]	11,010	99.98	99.50	99.74	47.50	64.30
	AF [9]	49,995	99.9	92.7	96.17	38.97	54.87
	BSL [12]	939,906	98.12	99.85	98.98	22.42	36.62
	Best Five [18]	239,976	99.52	99.16	99.34	30.12	46.21

for all datasets, *direct_comp* has the best PC; *bigram* and *token_sim* have identical RR, but *bigram* always has better PC. Furthermore, *bigram* always has the best F_{cs_log} and has better RR_{log} on Restaurant and Hotel but only slightly worse on Census than *token_sim*.

Compared to other systems, on both Restaurant and Hotel, *bigram* has the best RR, F_{cs}, RR_{log} and F_{cs_log}, though its F_{cs_log} was only slightly better than that of *All-Pairs/PP-Join(+)*. Also, with better RR, it only has slightly worse PC than *All-Pairs/PP-Join(+)/Marlin* on Restaurant. Particularly, *bigram* has significantly better RR (15.65% and 9.08% higher) than *BSL* and *Ed-Join* on Hotel; however it was not able to achieve a PC as good as these two systems did. If we consider larger datasets, such a significant difference in RR may save a great amount of runtime. Note that with the two new metrics, the impact of the number of selected pairs becomes more apparent, which we believe more accurately reflects its impact. On Census, $All\text{-}Pairs/PP\text{-}Join(+)$ achieved the best F_{cs} and F_{cs_log}; while $bigram$ still achieved better RR than BSL and $BestFive$ with slightly worse PC. $bigram$ only has a PC of 97.76% because our method only performs exact look-ups into the index; however, in this synthetic dataset, coreferent records were generated by modifying the original records, including adding misspellings, removing white spaces, etc. Therefore, some of the coreferent records couldn't even be looked up. In future work, we will explore techniques for efficient fuzzy retrieval to overcome this problem.

5.5   Scalability of Candidate Selection

Figure 2 presents the runtime by applying $bigram$ on up to 1 million instances of RKB Person, RKB Publication, SWAT Person and Census, showing that it scales well on large scale datasets. Due to limited availability of high quality groundtruth, we only measured the runtime. For SWAT Person, there are only 500K instances in the dataset. $bigram$ scales better on RKB Person since few instances actually use the selected predicates other than $full\text{-}name$. Note that

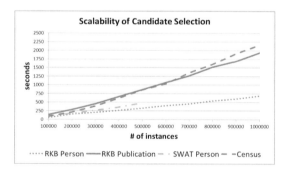

Fig. 2. Scalability of the Proposed Candidate Selection Algorithm

$All\text{-}Pairs/PP\text{-}Join(+)$ couldn't scale to 200K instances on any of these datasets due to insufficient memory, though they fared significantly better F_{cs} on Census.

5.6   Discussion

One limitation of our algorithm is that it currently targets datasets that are primarily composed of strings, and we adopt the same string similarity measure for numeric values, e.g., telephone number. Given that a lot of telephone numbers could be very similar to each other, counting the shared bigrams between two such numbers might greatly increase candidate set size, particularly when the data is primarily describing instances in the same geographic area.

Another problem is that we currently perform exact match on each query token when looking up similar instances with the index. This should work well on datasets with decent data quality; however, when there are a lot of errors (e.g., misspellings, missing characters or tokens, etc.), our algorithm may not even be able to retrieve all coreferent instances for a given instance. One possible solution to this problem is to adopt fuzzy matching. We could compute the Soundex code for each token and tokens with the same code are treated *similar*. For a given token, we query the index with all its similar tokens.

Finally, although *bigram* was only tested on 1 million instances (which is relatively small compared to the entire Linked data), it is larger than the number of instances in many Linked Data sets. Also, the number of instances is much smaller than the number of triples (e.g., DBPedia has 672 million triples but only 3.5 million instances), and we perform an initial filtering that instances must be of *comparable types*. Assuming around 100 million instances exist in Linked Data, they could be conservatively grouped into at least 10 sets of *comparable types* with no more than 10 million instances each. Extrapolating from Figure 2, our candidate selection could be computed in about 5.5 hours for each.

6 Conclusion

In this paper, we present an index based domain-independent candidate selection algorithm for scalably detecting *owl:sameAs* links. We learn a set of predicates for candidate selection through unsupervised learning. By indexing the instances on the learned predicates' object values, our algorithm is able to efficiently look up similar instances. In the author, publication, restaurant, hotel and census domains, using a bigram-based similarity measure, our algorithm almost always had a better RR than all alternatives, and when a full entity coreference algorithm was applied to the results, it led to the best F-score. As a result of its high RR, it frequently runs the fastest. Interestingly, our technique enables the overall system to produce coreference results with better F1-score by filtering out possible false positives when comparing only on the most disambiguating information. In the future, we will apply our technique to other entity coreference systems (e.g., [11]) to verify its capability of scaling those systems and improving their overall performances. Also, instead of doing exact lookup into the index, we are interested in exploring methods for efficient fuzzy retrieval.

References

1. Aswani, N., Bontcheva, K., Cunningham, H.: Mining Information for Instance Unification. In: Cruz, I., Decker, S., Allemang, D., Preist, C., Schwabe, D., Mika, P., Uschold, M., Aroyo, L.M. (eds.) ISWC 2006. LNCS, vol. 4273, pp. 329–342. Springer, Heidelberg (2006)
2. Bayardo, R.J., Ma, Y., Srikant, R.: Scaling up all pairs similarity search. In: Proceedings of the 16th International Conference on World Wide Web, pp. 131–140 (2007)
3. Bilenko, M., Mooney, R.J.: Adaptive duplicate detection using learnable string similarity measures. In: Proceedings of the Ninth ACM SIGKDD International Conference on Knowledge Discovery and Data Mining, pp. 39–48 (2003)

4. Bizer, C., Heath, T., Berners-Lee, T.: Linked data - the story so far. Int. J. Semantic Web Inf. Syst. 5(3), 1–22 (2009)
5. Cao, Y., Chen, Z., Zhu, J., Yue, P., Lin, C.Y., Yu, Y.: Leveraging unlabeled data to scale blocking for record linkage. In: Proceedings of the 22nd International Joint Conference on Artificial Intelligence, IJCAI (2011)
6. Elfeky, M.G., Elmagarmid, A.K., Verykios, V.S.: Tailor: A record linkage tool box. In: Proceedings of the 18th International Conference on Data Engineering (ICDE), pp. 17–28 (2002)
7. Euzenat, J., Ferrara, A., Meilicke, C., Nikolov, A., Pane, J., Scharffe, F., Shvaiko, P., Stuckenschmidt, H., Svb-Zamazal, O., Svtek, V., Trojahn dos Santos, C.: Results of the ontology alignment evaluation initiative 2010. In: Proceedings of the 4th International Workshop on Ontology Matching (2010)
8. Glaser, H., Millard, I., Jaffri, A.: RKBExplorer.com: A Knowledge Driven Infrastructure for Linked Data Providers. In: Bechhofer, S., Hauswirth, M., Hoffmann, J., Koubarakis, M. (eds.) ESWC 2008. LNCS, vol. 5021, pp. 797–801. Springer, Heidelberg (2008)
9. Gu, L., Baxter, R.A.: Adaptive filtering for efficient record linkage. In: Proceedings of the Fourth SIAM International Conference on Data Mining (2004)
10. Hassell, J., Aleman-Meza, B., Arpinar, I.B.: Ontology-Driven Automatic Entity Disambiguation in Unstructured Text. In: Cruz, I., Decker, S., Allemang, D., Preist, C., Schwabe, D., Mika, P., Uschold, M., Aroyo, L.M. (eds.) ISWC 2006. LNCS, vol. 4273, pp. 44–57. Springer, Heidelberg (2006)
11. Hu, W., Chen, J., Qu, Y.: A self-training approach for resolving object coreference on the semantic web. In: Proceedings of the 20th International Conference on World Wide Web (WWW), pp. 87–96 (2011)
12. Michelson, M., Knoblock, C.A.: Learning blocking schemes for record linkage. In: The Twenty-First National Conference on Artificial Intelligence, AAAI (2006)
13. Michelson, M., Knoblock, C.A.: Creating relational data from unstructured and ungrammatical data sources. J. Artif. Intell. Res. 31, 543–590 (2008)
14. Sleeman, J., Finin, T.: Computing FOAF co-reference relations with rules and machine learning. In: Third International Workshop on Social Data on the Web (2010)
15. Song, D., Heflin, J.: Domain-independent entity coreference in RDF graphs. In: Proceedings of the 19th ACM Conference on Information and Knowledge Management (CIKM), pp. 1821–1824 (2010)
16. Tejada, S., Knoblock, C.A., Minton, S.: Learning domain-independent string transformation weights for high accuracy object identification. In: Proceedings of the 8th ACM SIGKDD International Conference on Knowledge Discovery and Data Mining, pp. 350–359 (2002)
17. Volz, J., Bizer, C., Gaedke, M., Kobilarov, G.: Discovering and Maintaining Links on the Web of Data. In: Bernstein, A., Karger, D.R., Heath, T., Feigenbaum, L., Maynard, D., Motta, E., Thirunarayan, K. (eds.) ISWC 2009. LNCS, vol. 5823, pp. 650–665. Springer, Heidelberg (2009)
18. Winkler, W.E.: Approximate string comparator search strategies for very large administrative lists. Tech. rep., Statistical Research Division, U.S. Census Bureau (2005)
19. Xiao, C., Wang, W., Lin, X.: Ed-join: an efficient algorithm for similarity joins with edit distance constraints. Proc. VLDB Endow. 1(1), 933–944 (2008)
20. Xiao, C., Wang, W., Lin, X., Yu, J.X.: Efficient similarity joins for near duplicate detection. In: Proceedings of the 17th International Conference on World Wide Web (WWW), pp. 131–140 (2008)
21. Yan, S., Lee, D., Kan, M.Y., Giles, C.L.: Adaptive sorted neighborhood methods for efficient record linkage. In: ACM/IEEE Joint Conference on Digital Libraries (JCDL), pp. 185–194 (2007)

A Machine Learning Approach to Multilingual and Cross-Lingual Ontology Matching

Dennis Spohr[1], Laura Hollink[2], and Philipp Cimiano[1]

[1] Semantic Computing Group, CITEC, University of Bielefeld
[2] Web Information Systems Group, Delft University of Technology

Abstract. Ontology matching is a task that has attracted considerable attention in recent years. With very few exceptions, however, research in ontology matching has focused primarily on the development of monolingual matching algorithms. As more and more resources become available in more than one language, novel algorithms are required which are capable of matching ontologies which share more than one language, or ontologies which are multilingual but do not share any languages. In this paper, we discuss several approaches to learning a matching function between two ontologies using a small set of manually aligned concepts, and evaluate them on different pairs of financial accounting standards, showing that multilingual information can indeed improve the matching quality, even in cross-lingual scenarios. In addition to this, as current research on ontology matching does not make a satisfactory distinction between multilingual and cross-lingual ontology matching, we provide precise definitions of these terms in relation to monolingual ontology matching, and quantify their effects on different matching algorithms.

Keywords: Multilingual and cross-lingual ontology matching, machine learning, interoperability of financial information.

1 Introduction

Ontology matching is a discipline that has matured considerably over the last years, which is shown by the fact that many ontology matching algorithms have been successfully implemented and evaluated.[1] However, while these algorithms generally focus on using information in a single language that is shared by two ontologies, *multilingual* and *cross-lingual* aspects of ontology matching are – with the notable exception of [3,10] – still largely understudied. Assuming that more and more ontological resources will become available on the Semantic Web in more than one language, it is necessary to develop novel algorithms which are able to leverage multilingual information in case it is available, as well as capable of bridging the gap in case two ontological resources which do not share any languages need to be matched.

---

[1] See e.g. http://oaei.ontologymatching.org/

L. Aroyo et al. (Eds.): ISWC 2011, Part I, LNCS 7031, pp. 665–680, 2011.

In this paper, we present a new approach that relies on machine learning techniques in order to match concepts in two ontologies using both multilingual and cross-lingual information. As a very challenging use case, we have chosen financial accounting standards (FAS) such as the *United States Generally Accepted Accounting Principles* (US-GAAP) or the German *Grundsätze ordnungsmäßiger Buchführung* (GoB) – not only because they represent a type of taxonomic resource that raises a number of methodological issues, but also because the lack of interoperability of financial information across jurisdictional barriers is one of the most central problems in the business domain.[2] On the one hand, this is because companies from different countries are required to report their financial statements against different FAS. As these use different financial concepts with different interpretations, financial data reported against e.g. US-GAAP cannot be compared to financial data based on GoB, as the concepts need to be matched before any meaningful data integration can take place. On the other hand, FAS are frequently multilingual (i.e. annotated with more than one language) and have more than one label per language, which raises the question as to how to match financial concepts from ontologies sharing more than one language, or concepts from ontologies which are multilingual but do not share any languages. FAS thus represent one of the primary obstacles to achieving interoperability, and the impact of an approach that helps to solve this problem is expected to be considerably high.

This paper addresses the afore-mentioned issues in several respects. Firstly, we will give precise definitions of multilingual and cross-lingual ontology matching in relation to monolingual ontology matching, and discuss general research questions arising in such settings (Section 2). Moreover, we discuss several approaches which leverage multilingual and cross-lingual information in order to learn a matching function between two ontologies, using a small set of manually aligned financial concepts as training data. In contrast to the predominant view of ontology matching as a *classification problem* (as e.g. in [4,9]), we understand it as a *ranking problem*, similar to relevance ranking in information retrieval. In particular, we describe a novel approach that trains a *ranking support vector machine* (see [5]) on relative preference constraints between a concept in a source ontology and all possible concepts in a target ontology, with the goal of ranking good matches higher than bad matches. The approach is described in detail in Section 3 and evaluated on different pairs of FAS in different scenarios, in order to quantify the impact of multilingual and cross-lingual information on the performance of ontology matching algorithms (see Section 4).

2   Background and Preliminaries

2.1   Monolingual, Multilingual and Cross-Lingual Ontology Matching

Current research on ontology matching does not make a consistent distinction between *multilingual* and *cross-lingual* ontology matching (see e.g. [3,6,10]). In

---

[2] See http://www.xbrl.org/2010TechDiscussion/2010TechDiscussion.pdf

the following, we will define these notions, based on the general definition of ontology matching as "the process of finding relationships or correspondences between entities of different ontologies" (cf. [2]). As the definitions focus on those aspects of ontology matching which are relevant to multilingual and cross-lingual scenarios, specific strategies – such as structure-based or instance-based matching – will be ignored. Given a source ontology S and a target ontology T, the sets $S(l)$ and $T(l)$ of labels of S and T in a language l, and the sets L_S and L_T of languages in S and T respectively, these notions can be defined as follows.

Definition 1. *Monolingual ontology matching is the process of matching entities in S and T by comparing the labels in $S(l)$ and $T(l)$ in a single language $l \in L_S \cap L_T$.*

Definition 2. *Multilingual ontology matching is the process of matching entities in S and T by comparing the labels in $S(l_i)$ and $T(l_i)$ in at least two languages $l_i \in L_S \cap L_T$, with $|L_S \cap L_T| \geq 2$.*

Definition 3. *Cross-lingual ontology matching is the process of matching entities in S and T either*

  a. *by translating the labels in $S(l)$ to at least one language $l' \in L_T$ and comparing the labels in $S(l')$ with those in $T(l')$, or*
  b. *by translating the labels in $T(l)$ to at least one language $l' \in L_S$ and comparing the labels in $S(l')$ with those in $T(l')$, or*
  c. *by translating the labels $S(l)$ and the labels $T(l')$ to at least one language $l'' \notin L_S \cup L_T$ and comparing the labels in $S(l'')$ with those in $T(l'')$.*

For example, given a source ontology S with labels in English, German and Italian, monolingual ontology matching is a process that matches entities in S to entities in a target ontology T_1 with English labels by comparing the English labels in S with those of T_1. Multilingual ontology matching is a process that matches entities in S with entities in a target ontology T_2 with English and German labels by considering the labels in English and German. Cross-lingual ontology matching is a process that matches entities in S to entities in a target ontology T_3 with French labels either by translating the labels of S to French (Definition 3a.), by translating the labels of the T_3 to one of the languages in L_S (Definition 3b.), or by translating the labels of S and T to a third language (Definition 3c.). We believe that e.g. what Fu et al. [3] refer to as "multilingual ontologies" can thus be described more accurately as a cross-lingual matching scenario involving two (or more) monolingual ontologies.

2.2   Financial Accounting Standards and XBRL

As was mentioned in the introduction, financial accounting standards (FAS) differ between countries, and thus inhibit interoperability of financial information. However, there have been important developments towards solving this problem in recent years. From a technological perspective, the *eXtensible Business Reporting Language* (XBRL; [11]) solves the syntactic aspects of this interoperability issue by providing a common XML-based framework for expressing financial information. In XBRL, a FAS is commonly referred to as a *taxonomy*, as it specifies – among others – a hierachical structure according to which financial concepts appear in a financial statement (called *presentation hierarchy* in XBRL). In addition to this, for financial concepts which determine monetary values, such taxonomies specify how the value in question is to be calculated (e.g. "Assets" is the sum of "Current assets" and "Non-current assets"). Similar to the hierarchical presentation structure, these calculations are recursive (e.g. "Non-current assets" is, in turn, the sum of "Property, plant and equipment", "Investment property" etc.). This means that a monetary concept has a number of calculation items, each of which may itself be calculated on the basis of further calculation items. As such, we can distinguish between the *direct calculation items* of a monetary concept ("Current assets" and "Non-current assets" for "Assets"), as well as the *elementary calculation items* (e.g. "Investment property"), and likewise between *direct* and *elementary children* in the case of the presentation hierarchy. Finally, a concept can have more than one calculation, and more than one presentation.

The move towards XBRL has been a crucial development towards achieving interoperability. However, it does not solve the conceptual aspects of the problem, as companies from different countries still use different vocabularies to file their financial reports. Hence, the semantics of individual pieces of information is still not interchangeable. The *XBRL Europe Business Registers Working Group* (XEBR WG) has approached this problem from a conceptual perspective, by defining a set of core financial concepts which are believed to be shared by most FAS. The main idea behind this is that if the taxonomy of core financial concepts defines exact and close matches to the different national FAS, financial information reported against each individual FAS becomes interoperable through these mappings. While the work of the XEBR WG is still ongoing, first manual matches between the XEBR core taxonomy and several national FAS have already been produced, and can thus be leveraged for the approach described in this paper. Moreover, a very beneficial side-effect of the resource created by the XEBR WG is that it is possible to define matches between the individual taxonomies as well, based on the manual matches to the core taxonomy. Since – as was mentioned in the introduction – these taxonomies are frequently annotated in more than one language, the XEBR WG has created a valuable resource for the investigation and evaluation of different multilingual and cross-lingual matching strategies.

2.3 Open Research Questions in Ontology Matching

Trojahn et al. [10] mention multilingual and cross-lingual ontology matching as an open research issue. Below, we to explicate some of the research questions arising in such scenarios, with a particular focus on the machine learning aspect.

Impact of machine translation in cross-lingual scenarios. Fu et al. [3] have argued that the matching quality in cross-lingual scenarios strongly depends on the translation quality of label translations generated by machine translation (MT) tools. This certainly holds for the present study as well, since the choice of the MT system determines e.g. whether the Italian term "Conto economico" is translated as "Income statement"[3] or as "I count economical"[4]. The conclusion of Fu et al. [3] is that good translation quality is a prerequisite for achieving good quality cross-lingual ontology matches. While we do believe that this is true when comparing cross-lingual ontology matching to monolingual ontology matching with high-quality labels, it is worth investigating the possibility of translating the labels of both ontologies to a third language, as it may be the case that the quality difference between the labels can thus be reduced. In addition to this, in a machine learning scenario, a learning algorithm may weight structure-based similarity features higher in case string-based ones are found to be less predictive or even unpredictive, thus reducing the importance of high-quality translations.

Impact of structural information in ontology matching. Previous work has already shown the importance of structural information in ontology matching (see e.g. [2]). However, while it seems to be intuitively the case that algorithms capable of leveraging structural information should perform better than those which do not have this kind of information available, the question in a machine learning scenario is whether a learning algorithm which does not have access to structural information can still learn a reasonably predictive matching function. Therefore, a direct comparison between an algorithm using structural information with one not using structural information is necessary in order to answer this question.

Aggregation of scores in multilingual scenarios. Matching concepts with annotations in several languages, as well as several annotations within a single language, raises a number of further questions. One of these is the question how the similarity scores across different annotations should be aggregated within a single language (*intralingual aggregation*) as well as between languages (*interlingual aggregation*; cf. "composition" in [6]). For example, should the fact that one label of a concept C_S in a language l is very similar to one label of a concept C_T in language l suffice to say that C_S and C_T are good matching candidates? Or is the average over all labels within a language – averaged over all languages – a better indicator? To illustrate the importance of the treatment of multilingual information in ontology matching, consider the following example. The XBRL taxonomy of the *International Financial Reporting Standards* 2009 specifies a label "Total property, plant and equipment, gross" for the respective concept PropertyPlantAndEquipment. In the Italian GAAP, the corresponding concept is called "Total tangible fixed assets". Comparing only these two labels in a single language would yield a very low similarity score, as the only overlap consists in the word "total". However, both taxonomies specify labels in Italian and French. While the overlap in the Italian labels is still only marginally higher than in the English ones ("Immobili, impianti e macchinari" vs. "Immobilizzazioni materiali"), the French set of labels assures that the two concepts are in fact equivalent ("Immobilisations corporelles" vs. "II Immobilisations corporelles"). This example not only shows how vital multilingual information is for ontology matching,

---

[3] Using Microsoft's MT system Bing; http://www.microsofttranslator.com/

[4] Using SDL FreeTranslation; http://www.freetranslation.com/

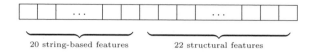

Fig. 1. Vector containing 42 features measuring the similarity between two concepts

but also that different strategies for intralingual and interlingual matching need to be defined. In the following section, we present the main ideas of our approach, as well as the features implementing the different strategies.

3   Machine Learning Approach to Ontology Matching

As was mentioned above, the general idea of our approach is to apply machine learning techniques to ontology matching, based on the notion of *ranking* SVM*s* as defined by Joachims [5]. In order to be able to apply this methodology, we are in need of a set of manually matched concepts to train on, as well as features representing the characteristics of each possible match. For the first issue, we can resort to the work of the *XBRL Europe Business Registers Working Group* (XEBR WG), which is currently in the process of matching different FAS to a set of core concepts. For the second issue, we define an appropriate set of features such that each combination of a source concept C_S with a target concept C_T can be represented as a feature vector, as in Figure 1. As each of these specifies the similarity between C_S and C_T, the value of each feature is between 0 and 1.

In total, we have thus defined 42 different features, comprising 20 string-based features, as well as 22 structure-based ones. These will be discussed in more detail below, before describing how the algorithm can be applied to the resulting similarity vectors.

3.1   Definition of Feature Set

String-based features. Similar to most other approaches in the field, we make use of a number of different string-based comparisons in order to measure the similarity between to concepts. In particular, we use five different measures, each of which represents a feature in the vector in Figure 1. Two similarity features are based on the Levenshtein edit distance measure [2,8], where one is applied to the labels of C_S and C_T, and the other one to the labels after their tokens have been sorted. This is to cover cases like "Current assets" vs. "Assets, current", where the plain (unsorted) Levenshtein distance would be very high although the labels are in fact very similar. Two further features use a bag-of-words cosine similarity measure, one on the original labels and the other one after punctuation has been removed. The fifth string-based measure uses the following substring distance as implemented by Euzenat and Shvaiko [2].

$$sim_{substr}(label_1, label_2) = \frac{2 * |longest\ common\ sequence|}{|label_1| + |label_2|} \qquad (1)$$

As mentioned in the previous section, we distinguish between different ways of aggregating the scores within a language, as well as between languages. In particular, we consider the best and average scores of all labels within a language, as well as the best and average scores across all shared languages. Thus, in order to cover these different intra- and interlingual matching strategies, we have implemented four different aggregations for each of the above measures. For example, the intralingual string similarity between the sets of labels of C_S and C_T in a language k (i.e. $C_S(k)$ and $C_T(k)$) using the Levenshtein measure is calculated on the basis of both (2) for the average score $sim_{intra\sim}^k$ over all labels and (3) for the best score sim_{intra+}^k of all labels.

$$sim_{intra\sim}^k(C_S(k), C_T(k)) = \frac{1}{n * m} \sum_{i=1}^{n} \sum_{j=1}^{m} sim_{lev}(l_{C_S}^i, l_{C_T}^j) \qquad (2)$$

$$sim_{intra+}^k(C_S(k), C_T(k)) = max(sim_{lev}(l_{C_S}^1, l_{C_T}^1), \dots, sim_{lev}(l_{C_S}^n, l_{C_T}^m)) \qquad (3)$$

Each of these is then aggregated to yield the interlingual similarity scores. For example, the interlingual score $sim_{inter\sim/\sim}$ between two concepts C_S and C_T is calculated by taking the average similarity of the average intralingual similarity scores over all n languages shared by S and T (i.e. $|L_S \cap L_T| = n$), as in (4), while the interlingual score $sim_{inter+/\sim}$ takes the best similarity score of all average intralingual scores. The scores $sim_{inter\sim/+}$ and $sim_{inter+/+}$ are analogous.

$$sim_{inter\sim/\sim}(C_S, C_T) = \frac{1}{n} \sum_{i=1}^{n} sim_{intra\sim}^i(C_S(i), C_T(i)) \qquad (4)$$

$$sim_{inter+/\sim}(C_S, C_T) = max(sim_{intra\sim}^1(C_S(1), C_T(1)), \qquad (5)$$
$$\dots, sim_{intra\sim}^n(C_S(n), C_T(n)))$$

We thus define string-based features, i.e. five measures with four aggregations.

Structural features. In addition to the string-based features, we have defined a set of 22 features representing the structural similarity between two concepts. As we cannot describe all of these features in depth, we limit ourselves to describing the calculation of two types of features to which most of the other structural features belong. In particular, we discuss the use of calculation information in S and T by considering the sets of direct and elementary items in calculations of C_S and C_T (i.e. $Cal_{C_S}^{dir} \times Cal_{C_T}^{dir}$ and $Cal_{C_S}^{ele} \times Cal_{C_T}^{ele}$). The scores using presentation information by considering direct and elementary children in the presentation hierarchy are calculated analogously.

Table 1. Matrix of string similarities between items of calculations of C_S and C_T

	$item^1_{C_S}$	$item^2_{C_S}$	$item^3_{C_S}$	$item^4_{C_S}$
$item^1_{C_T}$	0.1	0.3	0.4	0.2
$item^2_{C_T}$	0.4	0.7	0.3	0.5

The scores for the first type of features are rather straightforward to calculate, in that the average number[5] of direct (or elementary) items in calculations of a concept C_S is compared with the corresponding number in calculations of C_T. For example, if C_S has five direct calculation items and C_T has three, then $sim^{dir}_{cal\#}(C_S, C_T) = 1 - \frac{5-3}{5} = 0.6$. Similar calculations are done for the minimal and maximal number of elementary and direct calculation items. The second type of structural feature combines structural information with string-based similarity measures. In particular, we compare not the number of direct or elementary items of the calculations of C_S and C_T, but their similarities in all languages under consideration[6]. The motivation for this is that concepts whose components have similar labels are expected to be similar, even if e.g. their own labels are very different. Consider the similarity matrix of calculation items of C_S and C_T in Table 1, which shows the pairwise string similarities between all calculation items of C_S and C_T. In order to calculate a similarity value between C_S and C_T, we apply a best-first algorithm to the matrix, which yields that $item^2_{C_S}$ is aligned with $item^2_{C_T}$ (0.7) and $item^3_{C_S}$ with $item^1_{C_T}$ (0.4). As two calculation items have not been aligned, we consider both as having 0.0 similarity with items in C_T. In other words, we divide the sum of the scores by $max(|Cal^{dir}_{C_S}|, |Cal^{dir}_{C_T}|)$. This results in an overall similarity of $sim^{dir}_{cal\$}(C_S, C_T) = \frac{0.7+0.4}{4} = 0.275$ for this example. The scores for elementary calculation items are calculated similarly, as well as the scores for children in the presentation hierarchy. As was mentioned above, we have thus defined 22 structural features, arriving at a total of 42 features on which the ranking SVM algorithm can be trained.

3.2   Learning the Matching Function

On the basis of the features just discussed, we can now represent all combinations of concepts in S with concepts in T in terms of their similarity scores. Moreover, since we want to apply the ranking SVM algorithm as developed by [5] in order to learn the matching function, we need to specify relative relevance preferences between the possible matches. As the XEBR WG has defined exact matches as well as broad and narrow matches, we can state preferences such that exact matches of a concept C_S should be ranked higher than broad and narrow matches, which are in turn to be ranked higher than all other possible combinations of C_S with concepts in T. Therefore, we assign a target value of 3 to exact matches, 2 to broad

---

[5] Recall from Section 2.2 that a concept can have more than one calculation and presentation.

[6] We have used the Levenshtein score of the pair of labels which matched best overall.

and narrow matches, and 1 to all other combinations.[7] Given the features and similarity vectors as presented above, we can apply the ranking SVM algorithm described by Joachims [5], which produces an SVM model predicting scores for each input similarity vector. The matches can then be ranked such that those for which the model predicts higher scores are ranked above those with lower scores. In the experiments described in this paper, we have limited ourselves to learning an SVM with a linear kernel, which produces a single support vector.[8]

4   Evaluation

In order to make the impact of multilingual and cross-lingual information evident, we have defined several scenarios differing as to the type of language information they have available. As defined in Section 2, we differentiate between monolingual matching (using one overlapping language), multilingual matching (using at least two overlapping languages), and cross-lingual matching (involving the translation of at least one of the ontologies into at least one additional language). In the latter case, we further distinguish between monolingual and multilingual cross-lingual ontology matching. In addition to this, we investigate a cross-lingual transfer scenario in which the algorithm is trained on two pairs of taxonomies and evaluated on a third pair. The different scenarios are described in Section 4.2. In order to be able to quantify the contribution of other types of information to the matching process, such as the impact of structural features and the availability of close matches, we have further defined different evaluation settings differing with respect to the amount of information they can leverage. These settings are described in 4.3. Sections 4.4 and 4.5 present and analyse the results of the evaluation.

4.1   Data Set

Thanks to the fact that the XBRL Europe Business Registers Working Group (XEBR WG) has begun to manually match national financial accounting standards to their taxonomy of core financial concepts, we are able to evaluate our matching algorithms on the data produced by the XEBR WG. In particular, we have used version 5 of the XEBR core taxonomy (XEBR), the Italian *Tassonomia relativa ai Principi Contabili Italiani* of 2011 (also called *ITaliaCodiceCivile*; ITCC), and the GAAP taxonomy of the German *Handelsgesetzbuch* of 2011 (HGB). Table 2 lists the sizes of the taxonomies and the languages available, as well as the matches between them as defined by the XEBR WG.

---

[7] This means that we consider both broad and narrow matches as "close" matches, since it did not seem reasonable to assume that narrow matches should generally be ranked higher than broad matches or vice versa. See http://www.cs.cornell.edu/people/tj/svm_light/svm_rank.html for implementational details of the tool used to train the ranking SVM.

[8] First experiments involving a radial basis function kernel have yielded significantly worse results. For this reason, and due to the efficiency drawbacks compared to linear kernels, they have been discarded so far.

Table 2. Taxonomies used in the evaluation, with number of exact (and close) matches

Taxonomy	Languages	Concepts		XEBR	ITCC	HGB
XEBR	EN	269	XEBR	×	61 (77)	64 (475)
ITCC	EN, FR, DE, IT	444	ITCC	61 (77)	×	29 (38)
HGB	EN, DE	3,146	HGB	64 (475)	29 (38)	×

Between XEBR and ITCC, the XEBR WG has defined 61 exact and 77 close matches, with 70 XEBR concepts having at least one exact or close ITCC match. For XEBR and HGB, 64 exact and 475 close matches were defined, with 67 XEBR concepts having at least one exact or close HGB match. Matches between ITCC and HGB were not explicitly created by the XEBR WG. However, given that there are matches from each of these taxonomies to XEBR, we have applied a simple heuristic in order to arrive at a mapping between the two. If an ITCC concept and a HGB concept were marked as an exact match of the same XEBR concept, we defined them as exact matches. If an ITCC concept was marked as a close match of an XEBR concept, and this concept had an exact HGB concept match, we marked them as close matches, and vice versa. After applying this heuristic, we arrived at 29 exact and 38 close matches between ITCC and HGB, all of which were manually inspected and verified.

Regarding the structural content of the taxonomies, it needs to be said that the XEBR taxonomy does not define calculation information, but only provides presentation information. This means that structural information in the XEBR / ITCC and XEBR / HGB pairs is limited to leveraging presentation information, while in the ITCC / HGB both presentation and calculation information is available. Finally, we have used an RDF conversion of the XBRL format in which financial concepts are represented as RDF classes.

4.2   Matching Scenarios

Monolingual scenario. In the monolingual matching scenario, matching is done on the basis of one overlapping language. As English is the only language present in all taxonomies, we have used it for the monolingual matching scenario.

Multilingual scenario. As ITCC and HGB are the only taxonomies in the data set sharing more than one language, the multilingual scenario could only be applied to this pair of taxonomies, using English and German labels.

Cross-lingual scenario, S translated (monolingual). As was mentioned above, we distinguish different cross-lingual scenarios. In this first scenario, we have removed the labels in S and T such that each contained labels only in one non-overlapping language, in order to simulate a cross-lingual matching problem. For the XEBR to ITCC pair, we isolated the Italian labels in T and translated the English labels in S to Italian.[9] For the pair XEBR and HGB, we isolated the German labels in T and translated the English labels in S to German, and for the pair ITCC to HGB, we translated the Italian labels in S to English.

---

[9] All translations were done with the Microsoft MT system Bing.

Cross-lingual scenario, S and T translated (monolingual). This second cross-lingual scenario is similar to the previous one, except for the fact that we do not translate the labels in S to a language in T, but instead translate the labels of both S and T to a pivot language. The motivation behind this is to find out whether the translation quality issues in cross-lingual scenarios can be mitigated to some extent if the quality between both sets of labels is (at least assumed to be) more similar. In this scenario, we have translated the English labels in XEBR as well as the Italian labels in ITCC to German, and performed monolingual matching. Similarly, for XEBR to HGB, we have translated the English source labels as well as the German target labels to Italian, and for the ITCC to HGB case the Italian source labels as well as the German target labels to English.

Cross-lingual scenario, S translated (multilingual). In the third cross-lingual scenario, we have translated the labels in S to at least one other language existing in T, and performed multilingual matching. In the XEBR to ITCC case, we have translated the English labels in S to German and Italian, and performed multilingual matching on English, German and Italian[10]. For XEBR to HGB, we have translated the English source labels to German and matched in English and German. For the third pair, we first removed the English, French and German labels from ITCC, translated the remaining Italian labels to English and German, and performed multilingual matching with the English and German labels in HGB.

Cross-lingual transfer learning scenario, S translated (multilingual). In the final cross-lingual scenario, we wanted to investigate whether the matching function learned on two pairs of taxonomies can be transferred to a third pair of taxonomies, using the similarity scores calculated in the previous scenario (cross-lingual multilingual). In particular, for the XEBR to ITCC pair we trained the ranking SVM on the similarity scores between concepts in XEBR and HGB, as well as between ITCC and HGB, and tested it on the scores for XEBR and ITCC. Similarly, the XEBR to HGB pair was trained on XEBR to ITCC and ITCC to HGB, and the ITCC to HGB pair was trained on XEBR to ITCC and XEBR to HGB respectively.

4.3 Evaluation Settings

In each of the scenarios just described, we have evaluated three different learning settings as well as one baseline setting.[11]

AllInfo. In this setting, a matching function is trained and tested on all available information. In particular, it uses the similarity scores of exact and close matches, as well as all structural features (i.e. similarity scores based on presentation and calculation similarity). The matches in the test set are then ranked according to the score assigned by the learned matching function.

---

[10] Note that five entities in ITCC needed to be translated as well, as they lacked either an English or a German label. We assume this does not distort the results too much.

[11] As the baseline setting does not involve learning, the baseline for the transfer scenario is given by the score it yields in the cross-lingual multilingual setting.

NoClose. This setting is similar to *AllInfo* except for the fact that it does not use close matches for training and testing.

NoStruc. This setting is similar to *AllInfo* except for the fact that it does not use structural information for training and testing. As was mentioned in Section 4.1, in the case of XEBR/ITCC and XEBR/HGB this means that presentation information is ignored, and in the case of ITCC/HGB that presentation and calculation information is ignored.

EqWeights. This is a baseline setting where the matching function is not learned, but instead all features are assigned the same weight. In other words, the matches are ranked simply according to their average score across all features.

4.4   Evaluation Results

The results presented in this section are based on the following configuration. For each of the matching settings *AllInfo*, *NoClose* and *NoStruc*, we have carried out a four-fold cross-validation (i.e. training on three folds and testing on one fold). Each source concept in the training folds contained 20 similarity vectors representing exact, close and random bad matches with concepts in the target ontology, and they are the same 20 matches for all scenarios within one pair of ontologies. In contrast to this, for each source concept in the test folds the similarity vectors for all combinations with concepts in the target ontology T is given. This means, for example, that for the taxonomy pair XEBR and HGB – with matches defined to 67 of 3,146 HGB concepts –, each validation iteration is based on roughly 51 * 20 training examples, and evaluated on roughly 16 * 3,146 test examples (cf. Section 4.1 above).

As was mentioned in Section 3, we have used the algorithm developed by Joachims [5] to train a ranking SVM with a linear kernel. The developer of the corresponding tool SVM^{rank} gives a default value for the regularization parameter C (i.e. the trade-off between training error and margin) of 0.01 for "normal" SVMs, and defines $C_{rank} = C * n$ (where n is the number of queries, i.e. concepts in S) for ranking SVMs. Due to this dependence on the number of concepts in the source ontology, and as the number of matches provided by the XEBR WG – and thus the number of source concepts that can be used for evaluation – differs for each pair of taxonomies, C_{rank} is different for each pair of taxonomies. However, it should be noted that we have neither tried to optimize C for a given pair of taxonomies, nor tried to find the optimal set of training samples. The results are thus all based on the default value 0.01 for C, using a simple uniform random sampling method that produces acceptable results for all taxonomy pairs. We believe that this should make the results comparable.

Table 3 shows the results for the different matching settings. The column entitled "1" indicates the cases in which the matcher has ranked the exact match at rank 1 (i.e. precision), "5" indicates that the times in which the exact match was among the first 5 ranks, and analogously for column "10". As was mentioned above, the baseline for each scenario is given by a matcher that uses the average score over all similarity features (*EqWeights*). In addition to this, we

Table 3. Results for monolingual, multilingual and cross-lingual matching scenarios

Scenario	Setting	XEBR / ITCC			XEBR / HGB			ITCC / HGB		
		1	5	10	1	5	10	1	5	10
Monolingual	$AllInfo_1$	**51.67**	**76.67**	**81.67**	45.90	**73.77**	**78.69**	**44.83**	65.52	68.97
	$NoClose_1$	**51.67**	73.33	80.00	**52.46**	**73.77**	**78.69**	41.38	65.52	68.97
	$NoStruc_1$	46.67	66.67	76.67	50.82	72.13	**78.69**	41.38	55.17	58.62
	$EqWeights_1$	41.67	63.33	78.33	**52.46**	68.85	**78.69**	41.38	**68.97**	**72.41**
Multilingual	$AllInfo_n$	–	–	–	–	–	–	**51.72**	68.97	68.97
	$NoClose_n$	–	–	–	–	–	–	**51.72**	68.97	72.41
	$NoStruc_n$	–	–	–	–	–	–	44.83	55.17	65.52
	$EqWeights_n$	–	–	–	–	–	–	44.83	**72.41**	**75.86**
Cross-lingual, S and T translated	$AllInfo_1^{ST}$	**38.33**	**63.33**	75.00	29.51	45.90	52.46	37.93	**62.07**	65.52
	$NoClose_1^{ST}$	35.00	56.67	75.00	**32.79**	45.90	52.46	**41.38**	55.17	65.52
	$NoStruc_1^{ST}$	20.00	48.33	56.67	29.51	45.90	52.46	34.48	44.83	48.28
	$EqWeights_1^{ST}$	30.00	48.33	66.67	27.87	45.90	52.46	34.48	55.17	62.07
Cross-lingual, S translated to one language	$AllInfo_1^{S}$	35.00	**56.67**	63.33	27.87	**42.62**	47.54	34.48	65.52	**68.97**
	$NoClose_1^{S}$	**38.33**	53.33	**66.67**	**29.51**	39.34	44.26	34.48	**68.97**	**68.97**
	$NoStruc_1^{S}$	28.33	53.33	53.33	**29.51**	40.98	**47.54**	**41.38**	51.72	55.17
	$EqWeights_1^{S}$	30.00	53.33	58.33	24.59	39.34	42.62	**41.38**	65.52	**68.97**
Cross-lingual, S translated to several languages	$AllInfo_n^{S}$	56.67	**78.33**	**86.67**	49.18	**70.49**	75.41	44.83	65.52	**72.41**
	$NoClose_n^{S}$	**58.33**	76.67	83.33	**54.10**	68.85	**77.05**	**48.28**	**68.97**	68.97
	$NoStruc_n^{S}$	46.67	70.00	81.67	44.26	68.85	75.41	**48.28**	58.62	65.52
	$EqWeights_n^{S}$	40.00	71.67	81.67	47.54	**70.49**	73.77	**48.28**	65.52	68.97
Cross-lingual, transfer	$AllInfo_n^{Str}$	45.67	76.67	85.00	39.34	67.21	72.13	41.38	62.07	**75.86**
	$NoClose_n^{Str}$	**53.33**	**80.00**	**88.33**	**47.54**	**70.49**	**75.41**	**51.72**	**72.41**	72.41
	$NoStruc_n^{Str}$	23.33	60.00	73.33	37.70	57.38	68.85	31.03	58.62	65.52
	$EqWeights_n^{S}$	40.00	71.67	81.67	**47.54**	**70.49**	73.77	**51.72**	65.52	65.52
	$AROMA$	38.33	–	–	4.92	–	–	3.45	–	–

have aligned each pair of taxonomies with the state-of-the-art ontology aligner $AROMA$ [1], as it has been among the participants of the OAEI workshop series in the past years, and as it was available for download. In order to provide a level playing field for comparing the results, we have transformed all statements using custom label types to `rdfs:label` statements, and the hierarchical presentation information to `rdfs:subClassOf` (i.e. if x should appear above y in a financial statement, then `y rdfs:subClassOf x`) before applying $AROMA$.

4.5 Discussion of Results

Impact of multilingual and cross-lingual labels. The results clearly indicate that the performance goes up for almost all matchers if multilingual information is available. This means that matching algorithms should be capable of leveraging information in all overlapping languages in S and T. Interestingly, this also seems to hold in cross-lingual scenarios, as the best results have been obtained in cross-lingual multilingual scenarios. This is further supported by the very high scores obtained in the cross-lingual transfer scenario.

Impact of structural features. In almost all scenarios, the setting which did not contain any structural information (*NoStruc*) performed considerably worse than the settings which used structural information. This is as such an expected result, as previous research has already shown the importance of structural information in ontology matching (see chapter 4.3 of Euzenat and Shvaiko [2]).

Impact of close matches. The settings ignoring close matches (*NoClose*) have performed consistently better than the *AllInfo* setting in almost all scenarios. While this may seem counter-intuitive at first sight, there is a reasonable explanation for this. As close matches were also excluded from the test sets, the probability of assigning rank 1 to the exact match seems to be higher, as there are fewer candidates with high scores in the test set. As such, it seems reasonable to assume that the close matches occupy some of the higher ranks in the *AllInfo* settings, which would need to be verified in future experiments. Moreover, manual analysis has revealed that close matches may even be less similar to a source concept than bad matches. For example, combinations of the ITCC source concept `TotaleAttivoCircolante` ("total current assets") with HGB target concepts show that the similarity scores for the bad match `bs.ass.fixAss` ("Fixed assets") are higher than those for the close match `bs.ass.other.comment` ("Other assets, disclosures"). While this may in principle be true for some combinations of concepts, it may as well be an undesired side-effect of the fact that the mapping work of the XEBR WG has not been completed yet, and that some of the bad matches are in fact close (or even exact) matches which have not yet been classified as such. On the one hand, this assumption can be verified by comparing the results of future experiments with the ones presented here. More importantly, however, the (supposedly) bad matches which have been ranked higher than exact or close matches can be used to speed up the work of the XEBR WG, by suggesting potential candidates not considered until now.

Comparison to baselines. Table 3 shows that the best setting in each taxonomy pair clearly outperforms *AROMA*. In the cases of XEBR/HGB and ITCC/HGB, *AROMA* has performed extraordinarily low, which is surprising given the fact that the comparably reasonable score of 38.33% for XEBR/ITCC was obtained using exactly the same default configuration. A possible explanation for this is that the association rules approach followed by *AROMA* does not work well for repetitive labels such as the ones found in FAS. The results of the naive baseline *EqWeights* are surprising as well, though in the opposite respect. While the best setting outperforms the baseline in most scenarios, it is in some scenarios as good as the best setting or better. A possible explanation is that we have not attempted to optimize the learning parameters nor the training samples, and in fact, the baseline can be outperformed by adjusting the parameters.

Summing up these findings, it seems best to translate the labels in the source ontology to all languages available in the target ontology when trying to match a monolingual source ontology to a multilingual target ontology. Moreover, in most cases the settings in which both S and T have been translated perform better than the settings in which only S has been translated. This suggests that issues with translation quality as mentioned by Fu et al. [3] can to some extent be mitigated by translating to a pivot language. However, this claim still needs to be supported by further evidence from several language pairs, as the translation quality of MT systems varies greatly depending on the pair of languages considered (cf. [7]).

5 Related Work

There have been a number of machine learning approaches to ontology matching, such as the ones by Ichise [4] and Nezhadi et al. [9]. In particular, Ichise also follows an SVM-based approach, and Nezhadi et al. evaluate different learned classifiers. In contrast to this, we approach the matching problem by assuming that good matches should be ranked higher than bad ones, instead of attempting to classify a specific pair of concepts as being either a match or no match.

Concerning multilingual and cross-lingual ontology matching, the SOCOM framework (Semantic-oriented cross-lingual ontology mapping; [3]) has presented an approach to cross-lingual ontology matching. Similar to what has been discussed in this paper, they first translate the source ontology to the language of the target ontology, and then apply monolingual matching strategies, which corresponds to cross-lingual ontology matching as defined in Definition 3a. above. However, we are not aware of any attempts to combine this with multilingual matching strategies (in the sense of Definitions 2 and 3c.), nor of an evaluation of different cross-lingual matching scenarios at a scale presented in this paper.

6 Conclusion and Future Work

In this paper, we have presented a novel approach to ontology matching that uses a ranking SVM to learn a matching function that ranks good matches between two ontologies higher than bad matches. In addition to this, we have provided a precise definition of multilingual and cross-lingual ontology matching in relation to monolingual matching, and tried to quantify their effects on the performance of different matching strategies. Our approach was evaluated on different pairs of financial accounting standards in different languages, simulating both monolingual, multilingual and cross-lingual scenarios. The results have shown that multilingual information can indeed improve the performance of ontology matching algorithms, even in cross-lingual scenarios.

As was mentioned above, further work should go into optimising the learning parameters of the ranking SVM, in order to arrive at an estimate for the optimal performance of the SVM-based approach. This optimsation could then be attempted for non-linear kernels as well, in order to be able to compare the results obtained with each kernel. In addition to this, we have tried to use a uniform sampling approach for all pairs of financial standards, although we have observed that the performance of some pairs of taxonomies can be improved when choosing different sampling strategies. As such, it seems reasonable to try to identify the characteristics of the set of training samples that produces optimal results for a given pair of ontologies, in order to improve the composition of the training set. Finally, we have so far neglected deeper linguistic information in the set of features. Here, it should be interesting to investigate the effects of including similarity measures which leverage e.g. the terminological or morphological structure of the labels.

Acknowledgements. This work has been funded by the European Commission under Grant No. 248458 for the Monnet project. We would like to thank Fabian Abel, Jorge Gracia, Gilles Maguet, Susan Marie Thomas, Thomas Verdin, and the anonymous reviewers.

References

1. David, J., Guillet, F., Briand, H.: Matching directories and OWL ontologies with AROMA. In: Proceedings of the 15th ACM International Conference on Information and Knowledge Management, pp. 830–831. ACM, New York (2006)
2. Euzenat, J., Shvaiko, P.: Ontology matching. Springer, Heidelberg (2007)
3. Fu, B., Brennan, R., O'Sullivan, D.: Cross-Lingual Ontology Mapping – An Investigation of the Impact of Machine Translation. In: Gómez-Pérez, A., Yu, Y., Ding, Y. (eds.) ASWC 2009. LNCS, vol. 5926, pp. 1–15. Springer, Heidelberg (2009)
4. Ichise, R.: Machine learning approach for ontology mapping using multiple concept similarity measures. In: Seventh IEEE/ACIS International Conference on Computer and Information Science, ICIS 2008, Portland, Oregon (2008)
5. Joachims, T.: Optimizing search engines using clickthrough data. In: Proceedings of the ACM Conference on Knowledge Discovery and Data Mining (2002)
6. Jung, J.J., Håkansson, A., Hartung, R.: Indirect Alignment between Multilingual Ontologies: A Case Study of Korean and Swedish Ontologies. In: Håkansson, A., Nguyen, N.T., Hartung, R.L., Howlett, R.J., Jain, L.C. (eds.) KES-AMSTA 2009. LNCS(LNAI), vol. 5559, pp. 233–241. Springer, Heidelberg (2009)
7. Koehn, P.: Europarl: A Parallel Corpus for Statistical Machine Translation. In: Proceedings of the Tenth Machine Translation Summit, Phuket, Thailand, pp. 79–86 (2005)
8. Levenshtein, V.I.: Binary Codes capable of Correcting Deletions, Insertions, and Reversals. Soviet Physics Doklady 10(8), 707–710 (1966)
9. Nezhadi, A.H., Shadgar, B., Osareh, A.: Ontology alignment using machine learning techniques. International Journal of Computer Science & Information Technology 3(2) (2011)
10. Trojahn, C., Quaresma, P., Vieira, R.: An API for multi-lingual ontology matching. In: Calzolari, N., Choukri, K., Maegaard, B., Mariani, J., Odijk, J., Piperidis, S., Rosner, M., Tapias, D. (eds.) Proceedings of the Seventh International Conference on Language Resources and Evaluation (LREC 2010). European Language Resources Association (ELRA), Valletta (2010)
11. XBRL International Specification Working Group: Extensible Business Reporting Language 2.1. XBRL Recommendation (2008),
http://www.xbrl.org/SpecRecommendations/

Repairing Ontologies for Incomplete Reasoners

Giorgos Stoilos, Bernardo Cuenca Grau, Boris Motik, and Ian Horrocks

Department of Computer Science, University of Oxford
Wolfson Building, Parks Road, OX1 3QD, Oxford

Abstract. The need for scalable query answering often forces Semantic Web applications to use *incomplete* OWL 2 reasoners, which in some cases fail to derive all answers to a query. This is clearly undesirable, and in some applications may even be unacceptable. To address this problem, we investigate the problem of 'repairing' an ontology $\mathcal{T}$—that is, computing an ontology $\mathcal{R}$ such that a reasoner that is incomplete for $\mathcal{T}$ becomes complete when used with $\mathcal{T} \cup \mathcal{R}$. We identify conditions on $\mathcal{T}$ and the reasoner that make this possible, present a practical algorithm for computing $\mathcal{R}$, and present a preliminary evaluation which shows that, in some realistic cases, repairs are feasible to compute, reasonable in size, and do not significantly affect reasoner performance.

1 Introduction

Answering SPARQL queries over RDF data sets structured using an OWL 2 ontology provides the basis for a large number of Semantic Web applications. Such data sets can, however, be extremely large, and reasoning with OWL 2 DL ontologies is known to be of high computational complexity. As a consequence, complete reasoners—that is, reasoners such as Pellet, HermiT, and RACER that are capable (modulo bugs) of correctly computing all answers to all queries for all ontologies and datasets—often fail to deliver the required level of scalability. Application developers thus often use scalable but *incomplete* reasoners—that is, reasoners that, for *some* query, ontology, and dataset, fail to compute all answers to the query. Examples of such incomplete reasoners include state of the art RDF management systems, such as Jena [8], OWLim [6], DLE-Jena [9], and Oracle's Semantic Store [17], which typically provide completeness guarantees only for ontologies expressed in the OWL 2 RL [11] profile of OWL 2 DL.

The lack of a completeness guarantee may be unacceptable for applications in areas such as healthcare and defence, where missing answers may have serious consequences. Furthermore, even if an application can tolerate some level of incompleteness, it is desirable to provide the highest level of completeness that is compatible with the required scalability. Hence, techniques for improving the completeness of incomplete reasoners have recently been investigated. A common approach is to *materialise* certain kinds of ontology consequences before computing query answers. Such a solution does not require modifying the internals of the reasoner since the relevant consequences can be added as ontology axioms in a preprocessing step. In fact, systems such as DLE-Jena [9],

L. Aroyo et al. (Eds.): ISWC 2011, Part I, LNCS 7031, pp. 681–696, 2011.

PelletDB,[1] TrOWL [12], Minerva [7], and DLDB [4] internally use a complete OWL 2 DL reasoner to transparently materialise certain axioms. Furthermore, materialisation is used in approximation frameworks [12,2], where an OWL 2 DL ontology is projected into OWL 2 QL to allow for scalable reasoning.

Existing materialisation approaches, however, exhibit several important limitations. First, materialisation is commonly performed without taking into account the capabilities of the incomplete reasoner and may thus introduce redundant axioms. Second, to avoid a blowup in the ontology size, typically only subsumptions between (named) classes are materialised. Third, the extent to which materialisation improves a reasoner's completeness is often unclear, particularly if the data set is large, frequently changing, or unknown in advance.

In this paper, we present a novel approach to materialisation that addresses these limitations. Given an OWL 2 DL ontology $\mathcal{T}$ and a reasoner complete for OWL 2 RL, we show how to compute a *repair* $\mathcal{R}$ of $\mathcal{T}$ for the given reasoner. Intuitively, $\mathcal{R}$ is a set of OWL 2 RL consequences of $\mathcal{T}$ that, if added to $\mathcal{T}$, allow the reasoner to become *complete for* $\mathcal{T}$—that is, by using $\mathcal{T} \cup \mathcal{R}$ as input, the reasoner can correctly answer all queries w.r.t. $\mathcal{T}$ for all data sets. We focus on achieving completeness w.r.t. *ground certain answers* (i.e., answers obtained by matching query variables to named individuals). This is consistent with the semantics of SPARQL, and it allows us to ensure the existence of a repair whenever $\mathcal{T}$ can be rewritten into an OWL 2 RL ontology. Our technique is 'guided' by both the input ontology and the reasoner, which limits the size of $\mathcal{R}$ and ensures that adding $\mathcal{R}$ to $\mathcal{T}$ has minimal impact on the reasoner's scalability. Towards this goal, we proceed as follows.

In Section 3, similarly to our previous work [16,15], we devise a way of abstracting concrete reasoners using a notion of a *reasoning algorithm*, and we formalise the notion of an ontology repair for a reasoning algorithm.

In Section 4 we present a practical, two-step technique for computing a repair of an OWL 2 DL ontology $\mathcal{T}$ for a reasoner complete for OWL 2 RL. We first rewrite $\mathcal{T}$ into an OWL 2 RL ontology $\mathcal{T}'$ that is entailed by $\mathcal{T}$ and that preserves all ground answers to arbitrary queries over $\mathcal{T}$, regardless of the data. Based on this rewriting, we subject the incomplete reasoner to a series of tests, whose results identify the subset of the rewriting that constitutes a repair.

In Section 5 we demonstrate empirically that repairs can be computed in practice for well-known ontologies and reasoners. Our experiments show that the size of repairs is typically quite small, and that extending the original ontology with a repair typically has a negligible impact on reasoner performance.

2   Preliminaries

In this paper we use the standard notions of *constants, variables,* (function-free) *atoms, sentences, substitutions, satisfiability, unsatisfiability,* and *entailment* (written $\models$) from first-order logic. An application of a substitution σ to a term, atom, or formula α is written as $\sigma(\alpha)$. The *falsum* symbol (i.e., the symbol that

---

[1] http://clarkparsia.com/files/pdf/pelletdb-whitepaper.pdf

is false in all interpretations) is written as $\bot$. A *datalog rule* r is an expression of the form $B_1 \wedge \ldots \wedge B_n \to H$ where H is either $\bot$ or an atom, each B_i is an atom, and each variable occurring in H occurs in some B_i as well. The *body* of r is the set $\mathsf{body}(r) = \{B_1, \ldots, B_n\}$, and the *head* of r is $\mathsf{head}(r) = H$. Both head and body atoms can contain the *equality predicate* $\approx$, and head atoms can also contain the *inequality predicate* $\not\approx$. A datalog rule is interpreted as a universally-quantified first-order implication. It is well known that checking whether a first-order theory entails a datalog rule can be realised as follows.

Proposition 1. *Let $\mathcal{F}$ be a set of first-order sentences and let r be a datalog rule such that $\mathsf{body}(r) = \{B_1, \ldots, B_n\}$ and $\mathsf{head}(r) = H$. Then, for each substitution σ mapping the variables of r to distinct constants not occurring in $\mathcal{F}$ or r, we have $\mathcal{F} \models r$ if and only if $\mathcal{F} \cup \{\sigma(B_1), \ldots, \sigma(B_n)\} \models \sigma(H)$.*

2.1   OWL 2 DL and OWL 2 RL

We assume the reader to be familiar with the OWL 2 DL ontology language [10]. For succinctness, we use the Description Logics (DL) notation to write down OWL 2 DL axioms; please refer to [1] for an overview of the relationship between DLs and OWL. As is common in the literature, we partition an OWL 2 DL ontology into a *TBox* (i.e., a finite set of axioms describing the classes and properties in a domain of discourse) and an *ABox* (i.e., a finite set of facts). For simplicity, we assume that all ABox assertions refer to classes and properties only (i.e., that they do not contain complex class and property expressions); an ABox is thus allowed to contain class and property assertions, equalities, and inequalities, all of which can involve named and/or unnamed individuals.

OWL 2 RL [11] is a prominent profile of OWL 2 DL. Each OWL 2 RL ontology can be translated into an equivalent datalog program using (a straightforward extension of) the transformation presented in [3]. This close connection with datalog makes OWL 2 RL a popular implementation target since OWL 2 RL reasoners can be implemented by extending RDF triple stores with deductive features. For simplicity, in this paper we assume that each OWL 2 RL axiom α can be translated into a *single* datalog rule $\pi(\alpha)$; this can be ensured by transforming axioms using de Morgan identities to eliminate disjunctions and conjunctions in subclass and superclass positions, respectively.

Example 1. Consider the following OWL 2 DL ontology that describes the organisation of a typical university.

$$\exists \mathsf{take}.\mathsf{Co} \sqsubseteq \mathsf{Student} \quad \rightsquigarrow \quad \mathsf{take}(x,y) \wedge \mathsf{Co}(y) \to \mathsf{Student}(x) \quad (1)$$
$$\mathsf{GradCo} \sqsubseteq \mathsf{Co} \quad \rightsquigarrow \quad \mathsf{GradCo}(x) \to \mathsf{Co}(x) \quad (2)$$
$$\mathsf{PhDSt} \sqsubseteq \mathsf{GradSt} \quad \rightsquigarrow \quad \mathsf{PhDSt}(x) \to \mathsf{GradSt}(x) \quad (3)$$
$$\mathsf{Student} \sqcap \mathsf{Co} \sqsubseteq \bot \quad \rightsquigarrow \quad \mathsf{Student}(x) \wedge \mathsf{Co}(x) \to \bot \quad (4)$$
$$\exists \mathsf{teach}.\top \sqsubseteq \mathsf{Employee} \quad \rightsquigarrow \quad \mathsf{teach}(x,y) \to \mathsf{Employee}(x) \quad (5)$$
$$\mathsf{GradSt} \sqsubseteq \exists \mathsf{take}.\mathsf{GradCo} \quad (6)$$
$$\mathsf{ResAsst} \sqcap \mathsf{PhDSt} \sqsubseteq \exists \mathsf{teach}.\mathsf{LabPrac} \quad (7)$$

According to the definition of OWL 2 RL [11], axioms (1)–(5) are OWL 2 RL axioms, and so each axiom can be transformed into an equivalent datalog rule shown on the righthand side. In contrast, axioms (6) and (7) contain an existential quantifier (someValuesFrom in OWL 2 jargon) in the superclass position, so they cannot be translated into an equivalent datalog rule. The OWL 2 RL profile therefore disallows axioms such as (6) and (7).   ◇

2.2   Queries

A *union of conjunctive queries* (UCQ) $\mathcal{Q}$ with a *query predicate Q* is a datalog program in which each rule contains Q in the head but not in the body. We assume that query predicates do not occur in TBoxes and ABoxes.

Let $\mathcal{Q}$ be a UCQ with query predicate Q; let $\mathcal{F}$ be a set of first-order sentences; let $\mathcal{A}$ be an ABox; let G be a class not occurring in $\mathcal{F}$, $\mathcal{A}$, and $\mathcal{Q}$; let $\mathcal{A}_G$ be the ABox containing the class assertion $G(a)$ for each individual a occurring in $\mathcal{A}$; and let $\mathcal{Q}_G$ be the UCQ obtained from $\mathcal{Q}$ by adding to the body of each rule $r \in \mathcal{Q}$ the atom $G(x)$ for each variable x occurring in r. A tuple of constants $\vec{a}$ is a *certain answer* to $\mathcal{Q}$ w.r.t. $\mathcal{F}$ and $\mathcal{A}$ if the arity of $\vec{a}$ agrees with the arity of Q and $\mathcal{T} \cup \mathcal{A} \cup \mathcal{Q} \models Q(\vec{a})$. The set of all certain answers of $\mathcal{Q}$ w.r.t. $\mathcal{F}$ and $\mathcal{A}$ is written as $\mathsf{cert}(\mathcal{Q}, \mathcal{F}, \mathcal{A})$. If Q is propositional (i.e., if the query is *Boolean*), then $\mathsf{cert}(\mathcal{Q}, \mathcal{F}, \mathcal{A})$ is either empty or it contains the tuple of zero length; in such cases, we commonly write $\mathsf{cert}(\mathcal{Q}, \mathcal{F}, \mathcal{A}) = \mathsf{f}$ and $\mathsf{cert}(\mathcal{Q}, \mathcal{F}, \mathcal{A}) = \mathsf{t}$, respectively. Furthermore, $\vec{a}$ is a *ground certain answer* to $\mathcal{Q}$ w.r.t. $\mathcal{F}$ and $\mathcal{A}$ if the arity of $\vec{a}$ agrees with the arity of Q and $\mathcal{F} \cup \mathcal{A} \cup \mathcal{Q}_G \cup \mathcal{A}_G \models Q(\vec{a})$. The set of all ground certain answers of $\mathcal{Q}$ w.r.t. $\mathcal{F}$ and $\mathcal{A}$ is written as $\mathsf{cert}_G(\mathcal{Q}, \mathcal{F}, \mathcal{A})$.

3   A Framework for Repairing OWL Ontologies

We now introduce the technical framework that the rest of this paper depends on. In particular, in Section 3.1, we formalise the notion of a reasoning algorithm, and in Section 3.2 we formalise the notion of an ontology repair.

3.1   Reasoning Algorithms

As in [16,15], we abstract concrete reasoners using a notion of a *reasoning algorithm*. This has several benefits: it allows us to precisely specify the assumptions that a reasoner must satisfy for our results to be applicable, it allows us to precisely define the notions of completeness and repair, and it allows us to prove that our algorithm for repairing ontologies indeed guarantees completeness.

Definition 1. *A reasoning algorithm* ans *is a computable function that takes as input an arbitrary OWL 2 DL TBox $\mathcal{T}$, an arbitrary ABox $\mathcal{A}$, and either a* special *unsatisfiability query $*$ or an arbitrary UCQ $\mathcal{Q}$. The return value of* ans *is defined as follows:*

 – ans$(*, \mathcal{T}, \mathcal{A})$ *is either* t *or* f*; and*

– $\mathsf{ans}(\mathcal{Q}, \mathcal{T}, \mathcal{A})$ *is defined only if* $\mathsf{ans}(*, \mathcal{T}, \mathcal{A}) = \mathsf{f}$, *in which case the result is a set of tuples each having the same arity as the query predicate of* $\mathcal{Q}$.

Intuitively, $\mathsf{ans}(*, \mathcal{T}, \mathcal{A})$ asks the reasoner to check whether $\mathcal{T} \cup \mathcal{A}$ is unsatisfiable, and $\mathsf{ans}(\mathcal{Q}, \mathcal{T}, \mathcal{A})$ asks the reasoner to evaluate $\mathcal{Q}$ w.r.t. $\mathcal{T}$ and $\mathcal{A}$. If $\mathcal{T} \cup \mathcal{A}$ is unsatisfiable, then each tuple of the same arity as the query predicate of $\mathcal{Q}$ is trivially an answer to $\mathcal{Q}$; therefore, the result of $\mathsf{ans}(\mathcal{Q}, \mathcal{T}, \mathcal{A})$ is of interest only if $\mathsf{ans}(*, \mathcal{T}, \mathcal{A}) = \mathsf{f}$—that is, if ans identifies $\mathcal{T} \cup \mathcal{A}$ as satisfiable.

Example 2. Let rdf, rdfs, rl, and classify be reasoning algorithms that, given a UCQ $\mathcal{Q}$, an OWL 2 TBox $\mathcal{T}$, and an ABox $\mathcal{A}$, proceed as described next.

The algorithm rdf ignores $\mathcal{T}$ and evaluates $\mathcal{Q}$ w.r.t. $\mathcal{A}$; more precisely, we have $\mathsf{rdf}(*, \mathcal{T}, \mathcal{A}) = \mathsf{f}$ and $\mathsf{rdf}(\mathcal{Q}, \mathcal{T}, \mathcal{A}) = \mathsf{cert}(\mathcal{Q}, \emptyset, \mathcal{A})$. Thus, rdf captures the behaviour of RDF reasoners.

The algorithm rdfs constructs a datalog program $\mathcal{P}_{\mathsf{rdfs}}$ by translating each RDFS axiom α in $\mathcal{T}$ into an equivalent datalog rule; then, $\mathsf{rdfs}(*, \mathcal{T}, \mathcal{A})$ is always answered as f; furthermore, $\mathcal{Q}$ is evaluated w.r.t. $\mathcal{T}$ and $\mathcal{A}$ by evaluating $\mathcal{P}_{\mathsf{rdfs}}$ over $\mathcal{A}$—that is, $\mathsf{rdfs}(\mathcal{Q}, \mathcal{T}, \mathcal{A}) = \mathsf{cert}(\mathcal{Q}, \mathcal{P}_{\mathsf{rdfs}}, \mathcal{A})$. Thus, rdfs captures the behaviour of RDFS reasoners such as Sesame.

The algorithm rl constructs a datalog program $\mathcal{P}_{\mathsf{rl}}$ by translating each OWL 2 RL axiom α in $\mathcal{T}$ into an equivalent datalog rule; then, $\mathsf{rl}(*, \mathcal{T}, \mathcal{A})$ is answered by checking whether $\mathcal{P}_{\mathsf{rl}} \cup \mathcal{A}$ is satisfiable—that is, $\mathsf{rl}(*, \mathcal{T}, \mathcal{A}) = \mathsf{t}$ if and only if $\mathcal{P}_{\mathsf{rl}} \cup \mathcal{A} \models \bot$; furthermore, $\mathcal{Q}$ is evaluated w.r.t. $\mathcal{T}$ and $\mathcal{A}$ by evaluating $\mathcal{P}_{\mathsf{rl}}$ over $\mathcal{A}$—that is, $\mathsf{rl}(\mathcal{Q}, \mathcal{T}, \mathcal{A}) = \mathsf{cert}(\mathcal{Q}, \mathcal{P}_{\mathsf{rl}}, \mathcal{A})$. Thus, rl captures the behaviour of OWL 2 RL reasoners such as Jena and Oracle's Semantic Data Store.

The algorithm classify first classifies $\mathcal{T}$ using a complete OWL 2 DL reasoner; that is, it computes a TBox $\mathcal{T}'$ containing each subclass axiom $A \sqsubseteq B$ such that $\mathcal{T} \models A \sqsubseteq B$, and A and B are (named) classes occurring in $\mathcal{T}$. The algorithm then proceeds as rl, but considers $\mathcal{T} \cup \mathcal{T}'$ instead of $\mathcal{T}$; more precisely, $\mathsf{classify}(*, \mathcal{T}, \mathcal{A}) = \mathsf{rl}(*, \mathcal{T} \cup \mathcal{T}', \mathcal{A}_{in})$ and $\mathsf{classify}(\mathcal{Q}, \mathcal{T}, \mathcal{A}) = \mathsf{rl}(\mathcal{Q}, \mathcal{T} \cup \mathcal{T}', \mathcal{A})$. In this way, classify captures the behaviour of OWL 2 RL reasoners such as DLDB and DLE-Jena that try to be 'more complete' by materialising certain consequences of $\mathcal{T}$. ◇

Reasoning algorithms such as the ones specified in Example 2 are incomplete for OWL 2 DL—that is, there exist inputs for which they fail to compute all ground certain answers. These algorithms, however, are complete for a fragment of OWL 2 DL: algorithms rl and classify are complete for OWL 2 RL inputs, and algorithms rdf and rdfs are complete for RDF and RDFS, respectively. We next formally define the notion of an algorithm being complete for a fragment of OWL 2 DL (w.r.t. ground certain answers). Intuitively, for each UCQ, such an algorithm computes at least all ground certain answers for the UCQ and the part of the TBox that fits into the fragment in question.

Definition 2. *Given an OWL 2 DL TBox* $\mathcal{T}$ *and a fragment* $\mathcal{L}$ *of OWL 2 DL,* $\mathcal{T}|_{\mathcal{L}}$ *is the set of all* $\mathcal{L}$-*axioms in* $\mathcal{T}$.

Let ans *be a reasoning algorithm, and let* $\mathcal{L}$ *be a fragment of OWL 2 DL. We say that* ans *is complete for* $\mathcal{L}$ *if the following conditions hold for each OWL 2 DL TBox* $\mathcal{T}$, *each UCQ* $\mathcal{Q}$, *and each ABox* $\mathcal{A}$:

- $\mathcal{T}|_{\mathcal{L}} \cup \mathcal{A} \models \bot$ *implies* $\mathsf{ans}(*, \mathcal{T}, \mathcal{A}) = \mathsf{t}$; *and*
- $\mathsf{ans}(*, \mathcal{T}|_{\mathcal{L}}, \mathcal{A}) = \mathsf{f}$ *implies* $\mathsf{cert}_G(\mathcal{Q}, \mathcal{T}|_{\mathcal{L}}, \mathcal{A}) \subseteq \mathsf{ans}(\mathcal{Q}, \mathcal{T}, \mathcal{A})$.

Note that an $\mathcal{L}$-complete reasoning algorithm need not be sound (i.e., it may compute answers that are not certain answers). Although virtually all existing concrete reasoners are based on *sound* algorithms, their implementation may be unsound due to bugs. The results presented in this paper, however, do not require reasoning algorithms to be sound, so we can repair ontologies for concrete reasoners even if they are unsound. This is important in practice since testing reasoners for soundness is currently infeasible.

3.2   The Notion of a Repair

Intuitively, a repair of an OWL 2 DL TBox $\mathcal{T}$ for an algorithm ans is a TBox $\mathcal{R}$ such that adding $\mathcal{R}$ to $\mathcal{T}$ allows ans to correctly compute all ground certain answers for all UCQs and all ABoxes. For a repair to be useful, $\mathcal{R}$ should not introduce new consequences—that is, $\mathcal{R}$ should be a logical consequence of $\mathcal{T}$. This intuition is captured by the following definition.

Definition 3. *Let* $\mathcal{T}$ *be an OWL 2 DL TBox and let* ans *be a reasoning algorithm. A repair of* $\mathcal{T}$ *for* ans *is an OWL 2 DL TBox* $\mathcal{R}$ *such that* $\mathcal{T} \models \mathcal{R}$, *and the following conditions hold for each UCQ* $\mathcal{Q}$ *and each ABox* $\mathcal{A}$:

- $\mathcal{T} \cup \mathcal{A} \models \bot$ *implies* $\mathsf{ans}(*, \mathcal{T} \cup \mathcal{R}, \mathcal{A}) = \mathsf{t}$; *and*
- $\mathsf{ans}(*, \mathcal{T} \cup \mathcal{R}, \mathcal{A}) = \mathsf{f}$ *implies* $\mathsf{cert}_G(\mathcal{Q}, \mathcal{T}, \mathcal{A}) \subseteq \mathsf{ans}(\mathcal{Q}, \mathcal{T} \cup \mathcal{R}, \mathcal{A})$.

Example 3. Let $\mathcal{T}$ be the TBox containing axioms (1)–(7) from Example 1, let $\mathcal{A} = \{\mathsf{PhDSt}(a), \mathsf{ResAsst}(a)\}$, and let $\mathcal{Q}_1$ and $\mathcal{Q}_2$ be the following UCQs:

$$\mathcal{Q}_1 = \{\mathsf{Student}(x) \rightarrow Q(x)\} \tag{8}$$
$$\mathcal{Q}_2 = \{\mathsf{Employee}(x) \rightarrow Q(x)\} \tag{9}$$

One can check that $\mathsf{cert}_G(\mathcal{Q}_1, \mathcal{T}, \mathcal{A}) = \mathsf{cert}_G(\mathcal{Q}_2, \mathcal{T}, \mathcal{A}) = \{a\}$.

Consider now the algorithm rl from Example 2. Since axioms (6) and (7) are not in OWL 2 RL, the axioms are ignored by the algorithm. Consequently, $\mathcal{P}_{\mathsf{rl}}$ contains only the datalog rules corresponding to axioms (1)–(5), and so $\mathsf{rl}(\mathcal{Q}_1, \mathcal{T}, \mathcal{A}) = \mathsf{rl}(\mathcal{Q}_2, \mathcal{T}, \mathcal{A}) = \emptyset$—that is, rl is not complete for $\mathcal{T}$. One can, however, simulate the relevant consequences of axioms (6) and (7) using the OWL 2 RL TBox $\mathcal{R}_1$ containing the following axioms:

$$\mathsf{GradSt} \sqsubseteq \mathsf{Student} \rightsquigarrow \mathsf{GradSt}(x) \rightarrow \mathsf{Student}(x) \tag{10}$$
$$\mathsf{ResAsst} \sqcap \mathsf{PhDSt} \sqsubseteq \mathsf{Employee} \rightsquigarrow \mathsf{ResAsst}(x) \wedge \mathsf{PhDSt}(x) \rightarrow \mathsf{Employee}(x) \tag{11}$$

Clearly, $\mathcal{T} \models \mathcal{R}_1$; hence, extending $\mathcal{T}$ with $\mathcal{R}_1$ does not change the consequences of $\mathcal{T}$. The addition of axioms (10) and (11) to $\mathcal{T}$, however, changes the behaviour of algorithm rl; indeed, $\mathsf{rl}(Q_1, \mathcal{T} \cup \mathcal{R}_1, \mathcal{A}) = \mathsf{rl}(Q_2, \mathcal{T} \cup \mathcal{R}_1, \mathcal{A}) = \{a\}$. We show in the following section that $\mathcal{R}_1$ is a repair of $\mathcal{T}$ for rl; that is, for an arbitrary UCQ Q and ABox $\mathcal{A}$, running algorithm rl on Q, $\mathcal{T} \cup \mathcal{R}_1$, and $\mathcal{A}$ computes all ground certain answers of Q w.r.t. $\mathcal{T}$ and $\mathcal{A}$.

Next, consider the algorithm classify from Example 2. One can see that $\mathsf{classify}(Q_1, \mathcal{T}, \mathcal{A}) = \{a\}$ but $\mathsf{classify}(Q_2, \mathcal{T}, \mathcal{A}) = \emptyset$—that is, classify is also not complete for $\mathcal{T}$. Moreover, since classify is complete for OWL 2 RL, TBox $\mathcal{R}_1$ is a repair of $\mathcal{T}$ for classify. Note, however, that the classification of $\mathcal{T}$ takes care of axiom (10). Let $\mathcal{R}_2$ be the TBox containing only axiom (11). One can easily see that $\mathsf{rl}(Q_2, \mathcal{T} \cup \mathcal{R}_2, \mathcal{A}) = \{a\}$; in fact, we show in the following section that $\mathcal{R}_2$ is a repair of $\mathcal{T}$ for classify.

Finally, consider the algorithm rdfs form Example 2. In spite of the fact that $\mathsf{rdfs}(Q_1, \mathcal{T} \cup \mathcal{R}_1, \mathcal{A}) = \{a\}$, TBox $\mathcal{R}_1$ is not a repair of $\mathcal{T}$ for rdfs: since (11) is not an RDFS axiom, it is ignored by algorithm rdfs and so $\mathsf{rdfs}(Q_2, \mathcal{T} \cup \mathcal{R}_1, \mathcal{A}) = \emptyset$. In fact, even if we take $\mathcal{R}'$ to be the maximal set of RDFS axioms that logically follow from $\mathcal{T}$ (which is finite for RDFS), we can see that $\mathsf{rdfs}(Q_2, \mathcal{T} \cup \mathcal{R}', \mathcal{A}) = \emptyset$; consequently, no repair of $\mathcal{T}$ for rdfs exists. $\diamondsuit$

4 Repairing OWL 2 RL Reasoners

We now turn our attention to the problem of computing a repair for an OWL 2 DL TBox and a reasoning algorithm. In Section 4.1 we present a straightforward way of repairing via so-called *TBox rewritings*, and in Section 4.2 we show how to optimise repairs for reasoning algorithms that are complete for OWL 2 RL.

4.1 TBox Rewritings as Repairs

We next show that a repair of an OWL 2 TBox $\mathcal{T}$ for an algorithm ans can be obtained by *rewriting* $\mathcal{T}$ into the fragment of OWL 2 DL that ans can handle. Before proceeding, we first recapitulate the formal definition of a TBox rewriting.

Definition 4. *Let $\mathcal{T}$ be an OWL 2 DL TBox and let $\mathcal{L}$ be a fragment of OWL 2 DL. An $\mathcal{L}$-rewriting of $\mathcal{T}$ is a TBox $\mathcal{T}'$ in fragment $\mathcal{L}$ such that $\mathcal{T} \models \mathcal{T}'$ and the following conditions hold for each ABox $\mathcal{A}$ and each UCQ Q:*

- *$\mathcal{T} \cup \mathcal{A} \models \bot$ implies $\mathcal{T}' \cup \mathcal{A} \models \bot$; and*
- *$\mathsf{cert}_G(Q, \mathcal{T}, \mathcal{A}) = \mathsf{cert}_G(Q, \mathcal{T}', \mathcal{A})$.*

Note that, unlike $\mathcal{T}|_{\mathcal{L}}$, an $\mathcal{L}$-rewriting of $\mathcal{T}$ may not be a subset of $\mathcal{T}$, and may even be disjoint from $\mathcal{T}$. Rewritings were introduced mainly to facilitate reasoning in a complex ontology language by reasoning in a simpler language: instead of reasoning directly with an OWL 2 DL TBox $\mathcal{T}$, we compute a TBox $\mathcal{T}'$ in a simpler fragment $\mathcal{L}$ such that, for an arbitrary UCQ Q and an arbitrary ABox $\mathcal{A}$, the ground certain answers of $\mathcal{T}$ and $\mathcal{T}'$ coincide; we can then answer queries over $\mathcal{T}$ by applying to $\mathcal{T}'$ a reasoning algorithm complete for $\mathcal{L}$.

Example 4. Let $\mathcal{T}$ be the TBox consisting of axioms (1)–(7). The OWL 2 RL TBox $\mathcal{T}'$ consisting of axioms (1)–(5) and (10)–(11) is an OWL 2 RL rewriting of $\mathcal{T}$. Thus, instead of answering a query over $\mathcal{T}$ using an OWL 2 DL reasoner, we can answer the query over $\mathcal{T}'$ using an OWL 2 RL reasoner.

Note, however, that no RDFS rewriting of $\mathcal{T}$ exists: even if we take $\mathcal{T}''$ to be the maximal set of RDFS axioms that logically follow from $\mathcal{T}$, for $\mathcal{Q}_2$ and $\mathcal{A}$ as defined in Example 3 we have $\mathsf{cert}_G(\mathcal{Q}_2, \mathcal{T}'', A) = \emptyset$.   ◇

The following proposition, the proof of which is straightforward, establishes the connection between TBox rewritings and repairs. According to this proposition, the TBox $\mathcal{T}'$ in Example 4 is a repair of $\mathcal{T}$ for algorithm rl.

Proposition 2. *Let $\mathcal{T}$ be an OWL 2 TBox, let $\mathcal{L}$ be a fragment of OWL 2, and let* ans *be a reasoning algorithm complete for $\mathcal{L}$. If $\mathcal{R}$ is an $\mathcal{L}$-rewriting of $\mathcal{T}$, then $\mathcal{R}$ is a repair of $\mathcal{T}$ for* ans.

Although this simple result provides us with a straightforward way of repairing certain OWL 2 DL ontologies, as we discuss in the following section, repairs obtained in this way can be unnecessarily large. Therefore, we develop a technique that optimises a repair for the reasoner at hand.

4.2   Repairing a Class of Algorithms Complete for OWL 2 RL

Reasoners based on RDF triple stores and databases, such as Jena, OWLim, Oracle's Semantic Datastore and DLE-Jena, are typically complete at least for OWL 2 RL. Therefore, in the rest of this section we focus on repairing an OWL 2 DL TBox $\mathcal{T}$ for a reasoner ans that is complete for OWL 2 RL.

By Proposition 2, we can solve the aforementioned problem by computing an OWL 2 RL rewriting of $\mathcal{T}$. Depending on the language that $\mathcal{T}$ is expressed in, systems such as REQUIEM [13] and KAON2 [5] can compute a (possibly disjunctive) datalog rewriting; now whenever the rewriting is a datalog program, each datalog rule in the rewriting can always be 'rolled-up' into an OWL 2 RL axiom. Therefore, in order to simplify the presentation, we consider such rewritings to be OWL 2 RL TBoxes rather than datalog programs.

Note, however, that, if an OWL 2 RL rewriting $\mathcal{T}'$ of $\mathcal{T}$ exists, it must capture *all* OWL 2 RL consequences of $\mathcal{T}$ and can thus be very large; in fact, the size of $\mathcal{T}'$ can in the worst case even be exponential in the size of $\mathcal{T}$. Thus, to make our approach practicable, it is desirable to reduce the size of a repair as much as possible. This can be achieved in (at least) two ways.

First, rewritings often contain redundant axioms, so we can try to *minimise* them—that is, we can identify a smallest subset of $\mathcal{T}'$ that is also a rewriting of $\mathcal{T}$. While minimisation can be computationally very expensive, as a bare minimum we can eliminate from $\mathcal{T}'$ each axiom α for which $\mathcal{T}|_{\mathsf{rl}} \models \alpha$ holds; this can be straightforwardly checked using a sound and complete OWL 2 DL reasoner. A repair obtained in this way does not contain axioms whose consequences can be derived from $\mathcal{T}|_{\mathsf{rl}}$ by OWL 2 RL complete reasoning algorithms.

Second, we can exploit the fact that, while a reasoning algorithm might be complete only for OWL 2 RL, the algorithm may actually take into account some consequences of the axioms in $\mathcal{T} \setminus \mathcal{T}|_{\mathsf{rl}}$. Consider again algorithm classify from Example 2 and a TBox consisting of axioms (1)–(7). As shown in Example 4, a rewriting of this TBox consists of axioms (1)–(5) and (10)–(11); however, as discussed in Example 3, only axiom (11) is needed to repair the TBox for classify. Based on this observation, in the rest of this section we show how to reduce the size of a repair beyond what is possible via minimisation of a rewriting.

In order to achieve this goal, we first introduce the notion of a *datalog-reproducible* algorithm, which captures the class of reasoners to which our approach is applicable. This notion was inspired by an observation that many state of the art reasoners that can handle (a fragment of) OWL 2 DL are based on deductive database technologies: given a UCQ Q, a TBox $\mathcal{T}$, and an ABox $\mathcal{A}$, these reasoners first 'saturate' $\mathcal{A}$ by adding all assertions that are entailed by $\mathcal{T} \cup \mathcal{A}$; next, they answer Q by simply evaluating it over the saturated ABox. The ABox saturation process depends only on $\mathcal{T}$ and $\mathcal{A}$, and it can be characterised at an abstract level as evaluating over $\mathcal{A}$ a datalog program that depends only on $\mathcal{T}$. This intuition is formalised by the following definition.

Definition 5. *A reasoning algorithm* ans *is* datalog-reproducible *if, for each OWL 2 DL TBox $\mathcal{T}$, a datalog program $\mathcal{P}_\mathcal{T}$ exists such that the following holds:*

- *for each ABox $\mathcal{A}$ and each UCQ Q,*
  - ans$(*, \mathcal{T}, \mathcal{A}) = \mathsf{t}$ *if and only if* $\mathcal{P}_\mathcal{T} \cup \mathcal{A} \models \bot$, *and*
  - ans$(*, \mathcal{T}, \mathcal{A}) = \mathsf{f}$ *implies* ans$(Q, \mathcal{T}, \mathcal{A}) = \mathsf{cert}(Q, \mathcal{P}_\mathcal{T}, \mathcal{A})$; *and*
- *algorithm* ans *is* monotonic—*that is, for all OWL 2 DL TBoxes $\mathcal{T}$ and $\mathcal{T}'$, we have $\mathcal{P}_{\mathcal{T} \cup \mathcal{T}'} \models \mathcal{P}_\mathcal{T}$.*

If program $\mathcal{P}_\mathcal{T}$ contains predicates or individuals that do not occur in $\mathcal{T}$, these are considered to be 'private' to ans *and are not accessible elsewhere (e.g., in queries, TBoxes, and ABoxes).*

Note that a datalog-reproducible reasoning algorithm does not need to construct $\mathcal{P}_\mathcal{T}$; what matters is that *some* datalog program $\mathcal{P}_\mathcal{T}$ exists that characterises the behaviour of the algorithm.

Example 5. Algorithms rdf, rdfs and rl from Example 2 explicitly construct a datalog program $\mathcal{P}_\mathcal{T}$, so they are clearly datalog-reproducible. Note, however, that algorithm classify is also datalog-reproducible even through it does not directly construct a datalog program: the algorithm's behaviour can be characterised by a program $\mathcal{P}_\mathcal{T}$ containing all rules corresponding to the axioms in $\mathcal{T}|_{\mathsf{rl}}$ extended with the rule $A(x) \to B(x)$ for each pair of classes A and B occurring in $\mathcal{T}$ such that $\mathcal{T} \models A \sqsubseteq B$. $\diamond$

Note also that, even if a reasoner uses a particular datalog program as part of its implementation, the actual rules of the program may not be available to the users of the reasoner. For example, the rules used for reasoning by Oracle's Semantic Data Store are not publicly available; however, the reasoner can still be considered datalog-reproducible as its external behaviour can be captured using a datalog program. As we show next, our approach does not need to know the actual rules in order to repair an ontology: it suffices to know that a suitable datalog program exists.

As we discuss next, not all reasoning algorithms are datalog-reproducible.

Example 6. Reasoning algorithms based on query rewriting (e.g., algorithms underpinning the QuONTO reasoner) are not datalog-reproducible: although they answer queries by first constructing a datalog program, this program depends on *both* on the query and the TBox, and not on the TBox alone.

As another example, consider a reasoning algorithm that behaves as algorithm rdf from Example 2, but that first removes from the input ABox each assertion involving an individual whose IRI belongs to a certain predefined namespace. (This could be done, e.g., for efficiency or trust reasons.) Fact 'removal' cannot be represented using a monotonic theory, so this algorithm is clearly not datalog-reproducible.   ◇

We now show how to compute a repair of a TBox $\mathcal{T}$ for an algorithm ans that is datalog-reproducible and complete for OWL 2 RL. Intuitively, the behaviour of ans on $\mathcal{T}$ is characterised by a datalog program $\mathcal{P}_\mathcal{T}$ so, given an OWL 2 RL rewriting $\mathcal{T}'$ of $\mathcal{T}$, we can safely disregard each axiom in $\mathcal{T}'$ that is logically entailed by $\mathcal{P}_\mathcal{T}$. In other words, a repair of $\mathcal{T}$ for ans needs to contain only the *essential* axioms of $\mathcal{T}'$—that is, the axioms that are not entailed by $\mathcal{P}_\mathcal{T}$. Furthermore, the rewriting $\mathcal{T}'$ is an OWL 2 RL TBox, so each axiom $\alpha \in \mathcal{T}'$ corresponds to an equivalent datalog rule $\pi(\alpha)$; but then, by Proposition 1 we can construct from $\pi(\alpha)$ an ABox $\mathcal{A}_\alpha$ and a query $\mathcal{Q}_\alpha$ such that $\mathsf{ans}(\mathcal{Q}_\alpha, \mathcal{T}, \mathcal{A}_\alpha) = \mathsf{t}$ if and only if $\mathcal{P}_\mathcal{T} \models \alpha$. A repair $\mathcal{R}$ of $\mathcal{T}$ for ans can thus be obtained as a TBox that contains each axiom $\alpha \in \mathcal{T}'$ such that $\mathsf{ans}(\mathcal{Q}_\alpha, \mathcal{T}, \mathcal{A}_\alpha) = \mathsf{f}$. Since ans is complete for OWL 2 RL and $\mathcal{R}$ is an OWL 2 RL TBox, extending $\mathcal{T}$ with $\mathcal{R}$ will allow ans to recover the missing consequences of $\mathcal{T}$ and thus become complete.

Definition 6. *Let $\mathcal{T}$ be an OWL 2 TBox, let $\mathcal{T}'$ be an OWL 2 RL rewriting of $\mathcal{T}$, let* ans *be a datalog-reproducible reasoning algorithm, and let λ be a substitution that maps each variable in the signature to a fresh individual. The* essential subset *of $\mathcal{T}'$ for* ans *is the TBox $\mathcal{R}$ that contains each axiom $\alpha \in \mathcal{T}'$ satisfying the following conditions, where $r = \pi(\alpha)$ and $\mathcal{A}_\lambda^r = \{\lambda(B) \mid B \in \mathsf{body}(r)\}$.[2]*

1. $\mathsf{head}(r) = \bot$ *and* $\mathsf{ans}(*, \mathcal{T}, \mathcal{A}_\lambda^r) = \mathsf{f}$; *or*
2. $\mathsf{head}(r) = H$ *with* $H \neq \bot$ *and* $\mathsf{ans}(\{\lambda(H) \to Q\}, \mathcal{T}, \mathcal{A}_\lambda^r) = \mathsf{f}$, *for Q a propositional query predicate.*

Example 7. Let $\mathcal{T}$ contain axioms (1)–(7), and let $\mathcal{T}'$ be a rewriting of $\mathcal{T}$ that contains axioms (1)–(5) and (10)–(11).

The essential subset of $\mathcal{T}'$ for algorithm rl from Example 2 contains (10) and (11). For example, let α be axiom (10), so $\pi(\alpha) = r = \mathsf{GradSt}(x) \to \mathsf{Student}(x)$. Then for $\mathcal{A}_\lambda^r = \{\mathsf{GradSt}(a)\}$ and $\mathcal{Q} = \{\mathsf{St}(a) \to Q\}$ we have $\mathsf{rl}(\mathcal{Q}, \mathcal{T}, \mathcal{A}_\lambda^r) = \mathsf{f}$, so α must be included into the essential subset of $\mathcal{T}'$. Analogous reasoning applies to axiom (11).

In contrast, the essential subset of $\mathcal{T}'$ for algorithm classify contains only (11) since, for $\mathcal{Q}$ and $\mathcal{A}_\lambda^r$ as defined above, we have $\mathsf{classify}(\mathcal{Q}, \mathcal{T}, \mathcal{A}_\lambda^r) = \mathsf{t}$.   ◇

---

[2] Note that $\pi(\alpha)$ is the translation of α into a datalog rule from Section 2.

We next present the main result of this paper, which shows that essential subsets can be used as repairs.

Theorem 8. *Let $\mathcal{T}$ be an OWL 2 TBox, let $\mathcal{T}'$ be an OWL 2 RL rewriting of $\mathcal{T}$, let* ans *be a datalog-reproducible algorithm complete for OWL 2 RL, and let $\mathcal{R}$ be the essential subset of $\mathcal{T}'$ for* ans. *Then, $\mathcal{R}$ is a repair of $\mathcal{T}$ for* ans.

Proof. Assume that ans is complete for OWL 2 RL and let λ be a substitution that maps each variable in the signature to a fresh individual.

We first show that $\mathcal{P_R} \models \mathcal{R}$. To this end, let $\mathcal{R}_1$ be the subset of all rules $r \in \mathcal{R}$ such that $\mathcal{P_R} \cup \mathcal{A}_\lambda^r \models \bot$, and let $\mathcal{R}_2 = \mathcal{R} \setminus \mathcal{R}_1$. Furthermore, let $\overline{\mathcal{R}}_1$ be the set of rules obtained by replacing the head atom in each rule in $\mathcal{R}_1$ with $\bot$. Since clearly $\overline{\mathcal{R}}_1 \cup \mathcal{R}_2 \models \mathcal{R}$, it suffices to show that $\mathcal{P_R} \models \overline{\mathcal{R}}_1 \cup \mathcal{R}_2$. So, let r be an arbitrary rule in $\overline{\mathcal{R}}_1 \cup \mathcal{R}_2$.

- Assume that $r \in \overline{\mathcal{R}}_1$. Then, by the definition of $\overline{\mathcal{R}}_1$, we have $\mathcal{P_R} \cup \mathcal{A}_\lambda^r \models \bot$. But then, since $\mathsf{head}(r) = \bot$, by Proposition 1 we have $\mathcal{P_R} \models r$, as required.
- Assume that $r \in \mathcal{R}_2$. Then, $\mathcal{P_R} \cup \mathcal{A}_\lambda^r \not\models \bot$. By the definition of datalog-reproducible algorithms, then $\mathsf{ans}(*, \mathcal{R}, \mathcal{A}_\lambda^r) = \mathsf{f}$. Furthermore, we clearly have $\mathcal{R} \cup \mathcal{A}_\lambda^r \models \lambda(H)$, where $H = \mathsf{head}(r)$. Hence, $\mathsf{cert}_G(\mathcal{Q}, \mathcal{R}, \mathcal{A}_\lambda^r) = \mathsf{t}$ for $\mathcal{Q} = \{\lambda(H) \to Q\}$. But then, since ans is complete for OWL 2 RL and $\mathcal{R}$ is an OWL 2 RL TBox, we have that $\mathsf{cert}_G(\mathcal{Q}, \mathcal{R}, \mathcal{A}_\lambda^r) \subseteq \mathsf{ans}(\mathcal{Q}, \mathcal{R}, \mathcal{A}_\lambda^r)$ for each UCQ $\mathcal{Q}$. Therefore, $\mathsf{ans}(\mathcal{Q}, \mathcal{R}, \mathcal{A}_\lambda^r) = \mathsf{t}$, so by the definition of datalog-reproducible algorithms we also have $\mathcal{P_R} \cup \mathcal{A}_\lambda^r \models \lambda(H)$. But then, by Proposition 1, we have $\mathcal{P_R} \models r$, as required.

We next show that, since $\mathcal{R}$ is an essential subset of $\mathcal{T}'$ for ans, we have $\mathcal{P_T} \models \mathcal{T}' \setminus \mathcal{R}$. To this end, consider an arbitrary rule in $r \in \mathcal{T}' \setminus \mathcal{R}$. We have the following possibilities:

- $\mathsf{head}(r) = \bot$. In this case, by the definition of essential subset we have that $\mathsf{ans}(*, \mathcal{T}, \mathcal{A}_\lambda^r) = \mathsf{t}$ and hence $\mathcal{P_T} \cup \mathcal{A}_\lambda^r \models \bot$. But then, by Proposition 1 we have $\mathcal{P_T} \models r$.
- $\mathsf{head}(r) = H$ where $H \neq \bot$. By the definition of essential subset, we have $\mathsf{ans}(\{\lambda(H) \to Q\}, \mathcal{T}, \mathcal{A}_\lambda^r) = \mathsf{t}$. But then, by Proposition 1 we have $\mathcal{P_T} \models r$.

We now show that $\mathcal{P_T} \models \mathcal{T}' \setminus \mathcal{R}$ and $\mathcal{P_R} \models \mathcal{R}$ imply $\mathcal{P_{T \cup R}} \models \mathcal{T}'$. Since we have $\mathcal{P_T} \models \mathcal{T}' \setminus \mathcal{R}$, we also clearly have $\mathcal{P_T} \cup \mathcal{R} \models \mathcal{T}'$; since $\mathcal{P_R} \models \mathcal{R}$, we have $\mathcal{P_T} \cup \mathcal{P_R} \models \mathcal{T}'$ as well. Since ans satisfies the monotonicity property from Definition 5, we have $\mathcal{P_{T \cup R}} \models \mathcal{P_T}$ and $\mathcal{P_{T \cup R}} \models \mathcal{P_R}$; thus, $\mathcal{P_{T \cup R}} \models \mathcal{P_T} \cup \mathcal{P_R}$. But then, $\mathcal{P_{T \cup R}} \models \mathcal{T}'$, as required.

We finally use the fact that $\mathcal{P_{T \cup R}} \models \mathcal{T}'$ to show that the following properties hold for each UCQ $\mathcal{Q}$ and each ABox $\mathcal{A}$.

1. $\mathcal{T} \cup \mathcal{A} \models \bot$ implies $\mathsf{ans}(*, \mathcal{T} \cup \mathcal{R}, \mathcal{A}) = \mathsf{t}$; and
2. $\mathsf{ans}(*, \mathcal{T} \cup \mathcal{R}, \mathcal{A}) = \mathsf{f}$ implies then $\mathsf{cert}_G(\mathcal{Q}, \mathcal{T}, \mathcal{A}) \subseteq \mathsf{ans}(\mathcal{Q}, \mathcal{T} \cup \mathcal{R}, \mathcal{A})$.

(Property 1). Assume that $\mathcal{T} \cup \mathcal{A} \models \perp$. Since $\mathcal{T}'$ is an OWL 2 RL rewriting of $\mathcal{T}$, we have $\mathcal{T}' \cup \mathcal{A} \models \perp$. But then, since $\mathcal{P}_{\mathcal{T} \cup \mathcal{R}} \models \mathcal{T}'$, we also have $\mathcal{P}_{\mathcal{T} \cup \mathcal{R}} \cup \mathcal{A} \models \perp$, as required.

(Property 2). Assume that $\mathsf{ans}(*, \mathcal{T} \cup \mathcal{R}, \mathcal{A}) = \mathsf{f}$ and consider an arbitrary tuple $\vec{a} \in \mathsf{cert}_G(\mathcal{Q}, \mathcal{T}, \mathcal{A})$. Since $\mathcal{T}'$ is an OWL 2 RL rewriting of $\mathcal{T}$, then we have $\vec{a} \in \mathsf{cert}_G(\mathcal{Q}, \mathcal{T}', \mathcal{A})$, so $\mathcal{T}' \cup \mathcal{A} \cup \mathcal{Q}_G \cup \mathcal{A}_G \models Q(\vec{a})$. But then, since $\mathcal{P}_{\mathcal{T} \cup \mathcal{R}} \models \mathcal{T}'$, we also have that $\mathcal{P}_{\mathcal{T} \cup \mathcal{R}} \cup \mathcal{A} \cup \mathcal{Q}_G \cup \mathcal{A}_G \models Q(\vec{a})$. Furthermore, since $\mathcal{P}_{\mathcal{T} \cup \mathcal{R}}$ is a datalog program, we have that $\mathcal{P}_{\mathcal{T} \cup \mathcal{R}} \cup \mathcal{A} \cup \mathcal{Q} \models Q(\vec{a})$ and hence we have $\vec{a} \in \mathsf{cert}(\mathcal{Q}, \mathcal{P}_{\mathcal{T} \cup \mathcal{R}}, \mathcal{A})$; consequently, $\vec{a} \in \mathsf{ans}(Q, \mathcal{T} \cup \mathcal{R}, \mathcal{A})$, as required.

We finally show that $\mathcal{R}$ is a repair of $\mathcal{T}$ for ans. Since $\mathcal{T}'$ is an OWL 2 RL rewriting of $\mathcal{T}$, we have $\mathcal{T} \models \mathcal{T}'$; since $\mathcal{R} \subseteq \mathcal{T}'$, we also have $\mathcal{T} \models \mathcal{R}$. This together with Properties 1 and 2 implies our claim.   □

Theorem 8 and the observations made in Example 7 thus confirm our claims from Example 3: axioms (10) and (11) constitute a repair of $\mathcal{T}$ for algorithm rl, and axiom (11) alone constitutes a repair for algorithm classify.

5   Evaluation

We developed a prototype tool for computing repairs. Our implementation uses the system REQUIEM [13] for the computation of TBox rewritings. We evaluated our approach using the following two well-known ontologies.

First, we used the well-known Lehigh University Benchmark (LUBM) [4]—an ontology extensively used for evaluating performance of ontology-based systems. We used LUBM's generator of large datasets and the supplied 14 test queries.

Second, we used a small subset of the GALEN ontology [14]—a complex medical ontology. We used a subset of GALEN because REQUIEM was unable to handle the full version of GALEN. Since we are not aware of a large ABox or a data generator for GALEN, we created synthetic data by extending the techniques for ABox generation from [16,15]; we thus obtained ABoxes with (approximately) 2000, 4000, 8000, 16000 and 32000 assertions. Furthermore, we tested the systems using four atomic queries presented in [15].

We evaluated the following reasoning systems: OWLim v2.9.1,[3] Jena v2.6.3[4] and DLE-Jena v2.0.[5]

For each test ontology $\mathcal{T}$ and each reasoning system ans mentioned above, we performed the following tasks.

1. We computed an OWL 2 RL rewriting $\mathcal{T}_{\mathsf{rew}}$ of the input TBox $\mathcal{T}$ and recorded the time needed to complete this step.
2. As mentioned in Section 4.2, $\mathcal{T}_{\mathsf{rew}}$ can contain many axioms, so we minimised $\mathcal{T}_{\mathsf{rew}}$ as follows. First, we eliminated each axiom α such that $\mathcal{T}|_{\mathsf{rl}} \models \alpha$. Second,

---

[3] http://www.ontotext.com/owlim/
[4] http://jena.sourceforge.net/
[5] http://lpis.csd.auth.gr/systems/DLEJena/

Table 1. Repairing the LUBM ontology for OWLim, Jena and DLE-Jena

Ontology	$\mathcal{T}_{rew}$	$\mathcal{T}_{min}$	$\mathcal{R}_{owlim}$	$\mathcal{R}_{jena}$	$\mathcal{R}_{dle\text{-}jena}$
Ontology size	331	7	3	3	0
Time to compute ontology (in s)	4.7	7.9	3.3	6.3	14

for all pairs of distinct remaining axioms α_1 and α_2, we eliminated α_2 if $\mathcal{T}|_{rl} \cup \{\alpha_1\} \models \alpha_2$. Let $\mathcal{T}_{min}$ be the resulting set of axioms; clearly, $\mathcal{T}|_{rl} \cup \mathcal{T}_{min}$ is still a rewriting of $\mathcal{T}$. Note that $\mathcal{T}_{min}$ can depend on the order in which we select α_1 and α_2 in the second step; however, we did not notice significant variance in our tests. We conducted all entailment checks using HermiT—a sound and complete OWL 2 DL reasoner—and we recorded the time needed to complete this step.

3. We extracted from $\mathcal{T}_{min}$ the essential subset $\mathcal{R}$ for ans as described in Definition 6, and we recorded the time needed to complete this step. By Theorem 8, $\mathcal{R}$ is a repair of $\mathcal{T}$ for ans.

4. To estimate the effect that repairing $\mathcal{T}$ has on the performance of ans, we proceeded as follows. We first applied ans to $\mathcal{T}$ and each corresponding data set and query, and we recorded the load time, the query evaluation time, and the number of certain answers returned. Next, we repeated the experiment by applying ans to $\mathcal{T} \cup \mathcal{R}$. The results obtained using $\mathcal{T} \cup \mathcal{R}$ are compared against Pellet—a sound and complete OWL 2 DL reasoner.

The results of repairing LUBM are shown in Table 1. Although the initial rewriting is quite large, our procedure computes repairs for OWLim and Jena that consist of only the following three axioms:

GradStudent $\sqsubseteq$ Student   Director $\sqsubseteq$ Employee   ResearchAssist $\sqsubseteq$ Employee

The repair for DLE-Jena is empty—that is, the system is already complete for LUBM. This is due to the fact that the repair for OWLim and Jena consists only of simple subclass axioms, all of which are derived by DLE-Jena's preprocessing phase (DLE-Jena is similar to the classify algorithm from Example 2). In all cases computing the repair took less than 15 seconds.

For OWLim and Jena, $\mathcal{T}_{min} \setminus \mathcal{R}$ is non-empty, which suggests that these systems can process 'more' than just OWL 2 RL. We observed that, for many axioms in $\mathcal{T}_{min} \setminus \mathcal{R}$ of the form $A \sqsubseteq B$, TBox $\mathcal{T}$ contains an axiom $A \sqsubseteq B \sqcap \exists R.C$. The latter is not an OWL 2 RL axiom, so $\mathcal{T}|_{rl} \not\models A \sqsubseteq B$ and $A \sqsubseteq B$ is not removed from $\mathcal{T}_{min}$. The OWL 2 RL/RDF rules from [11], however, correctly handle the conjunction in the superclass position—that is, given an assertion $A(a)$, they derive $B(a)$ and $\exists R.C(a)$. This effectively allows OWLim and Jena to use the $A \sqsubseteq B$ 'part' of $A \sqsubseteq B \sqcap \exists R.C$, so the repair does not need to contain $A \sqsubseteq B$. Thus, tailoring the repair to a particular reasoner can exploit the reasoning capabilities of the reasoner at hand and thus produce smaller repairs.

We observed no measurable performance changes for OWLim and Jena after repairing, which is not surprising since the repairs contained only a few simple

Table 2. Repairing GALEN for OWLim, Jena and DLE-Jena

Ontology	$\mathcal{T}_{rew}$	$\mathcal{T}_{min}$	$\mathcal{R}_{owlim}$	$\mathcal{R}_{jena}$	$\mathcal{R}_{dle\text{-}jena}$
Ontology size	1666	291	11	10	5
Time to compute ontology (in s)	380	126	426	1586	-

Table 3. Number of certain answers for GALEN (without repairs)

	2000				4000				8000		16000		32000	
Queries	J	D	O	P	J	D	O	P	O	P	O	P	O	P
$Q1$	73	65	65	92	226	206	206	280	501	672	1281	1587	1379	1727
$Q2$	49	43	43	72	158	135	135	217	368	532	1008	1353	1148	1502
$Q3$	81	74	74	97	234	212	212	283	515	678	1301	1588	1355	1714
$Q4$	112	226	78	260	334	656	232	756	547	1687	1272	3463	1445	3706
J=Jena, D=DLE-Jena, O=OWLim, P=Pellet														

axioms. Moreover, both OWLim and Jena are complete for the LUBM dataset already when using the original TBox, so no change in the number of answers produced was observed either.

The results of repairing GALEN are shown in Table 2. Again, the repair is quite small, despite the fact that GALEN heavily uses features outside OWL 2 RL such as existential quantification. The repair, however, is more complex than in the case of LUBM, containing axioms such as subsumptions between complex class expressions with several nested existential quantifiers. As with LUBM, the repair for DLE-Jena is smaller than the repairs for OWLim and Jena. Note, however, that DLE-Jena ran out of memory while computing the repair from $\mathcal{T}_{min}$; thus, since DLE-Jena is 'more complete' than Jena, we produced $\mathcal{R}_{dle\text{-}jena}$ by applying Definition 6 to $\mathcal{R}_{jena}$.

Table 3 shows the number of certain answers computed by each system for each dataset, using the original GALEN TBox; please note that we could only load datasets 2000 and 4000 into Jena and DLE-Jena. As expected, all systems returned fewer answers than Pellet. Using the repaired TBoxes, however, *all* systems returned the *same* number of certain answers as Pellet. Thus, repairing an ontology can significantly improve the quality of answers that a reasoner produces for a given ontology.

Table 4 shows the loading times for both the original and the repaired ontologies. As one can see, repairing the ontology leads to an increase in loading times of about 20% on average. The times for the repaired ontology, however, are of the same order of magnitude as the original times; hence, the increase in loading time may be acceptable given that the systems then return complete answers. Furthermore, OWLim is much faster than Pellet even on the repaired ontology, which suggests that using an incomplete reasoner with a repaired ontology might be more appropriate in practice than using a complete reasoner.

Table 5 shows the query answering times for the repaired ontology. Since all systems perform reasoning during loading (i.e., they saturate the input ABox), repairing the ontology produced no noticeable difference on the query answering

Table 4. Loading times for original and repaired TBoxes (in ms)

		2000	4000	8000	16000	32000
OWLim	$\mathcal{T}$	1411	2328	3611	6000	6871
	$\mathcal{T} \cup \mathcal{R}$	1768	2807	4279	7815	8696
Pellet	$\mathcal{T}$	2598	4623	9596	11275	12086
Jena	$\mathcal{T}$	25524	117524	-	-	-
	$\mathcal{T} \cup \mathcal{R}$	34000	139839	-	-	-
DLE-Jena	$\mathcal{T}$	23198	138075	-	-	-
	$\mathcal{T} \cup \mathcal{R}$	24844	139129	-	-	-

Table 5. Query answering times for repaired GALEN (in ms)

	2000				4000				8000		16000		32000	
Queries	J	D	O	P	J	D	O	P	O	P	O	P	O	P
Q1	15	156	39	734	5	155	46	1765	41	5503	53	7460	50	8122
Q2	2	3	1	425	5	11	3	1226	6	2635	15	4271	17	4620
Q3	4	3	1	926	8	8	4	2711	100	6473	18	10188	20	11866
Q4	8	9	1	37	21	21	11	47	20	88	55	127	42	136
J=Jena, D=DLE-Jena, O=OWLim, P=Pellet														

times. Note that all systems are much faster than Pellet, which again suggests that using an incomplete reasoner with a repaired ontology might offer significant advantages compared to using a complete reasoner.

6   Conclusions

In this paper, we studied the problem of *repairing* an ontology for a given incomplete reasoner in a way that guarantees completeness. Our repairs guarantee completeness w.r.t. ground certain answers independently of data and queries: once an ontology has been repaired for a given system, the system will, for any given query and data set, compute all ground certain answers that follow from the original ontology. Our approach tries to limit the size of the repair as much as possible. Our experiments suggest that repairs may indeed be very small, and that their effect on system performance may be negligible. This allows application designers to use highly scalable incomplete reasoners, but with the guarantee that they produce the same answers as provably complete reasoners, thus having the 'the best of both worlds'.

We leave the extension of our techniques to more expressive DLs, as well as a more extensive evaluation, for future work.

Acknowledgments. Research supported by project SEALS (FP7-ICT-238975). B. Cuenca Grau is supported by a Royal Society University Research Fellowship.

References

1. Baader, F., McGuinness, D., Nardi, D., Patel-Schneider, P.F.: The Description Logic Handbook: Theory, implementation and applications. Cambridge University Press (2002)
2. Botoeva, E., Calvanese, D., Rodriguez-Muro, M.: Expressive Approximations in *DL-lite* Ontologies. In: Dicheva, D., Dochev, D. (eds.) AIMSA 2010. LNCS, vol. 6304, pp. 21–31. Springer, Heidelberg (2010)
3. Grosof, B.N., Horrocks, I., Volz, R., Decker, S.: Description logic programs: Combining logic programs with description logic. In: Proc. of WWW, pp. 48–57 (2003)
4. Guo, Y., Pan, Z., Heflin, J.: LUBM: A Benchmark for OWL Knowledge Base Systems. Journal of Web Semantics 3(2), 158–182 (2005)
5. Hustadt, U., Motik, B., Sattler, U.: Deciding Expressive Description Logics in the Framework of Resolution. Information & Computation 206(5), 579–601 (2008)
6. Kiryakov, A., Ognyanov, D., Manov, D.: Owlim-a pragmatic semantic repository for owl. In: Dean, M., Guo, Y., Jun, W., Kaschek, R., Krishnaswamy, S., Pan, Z., Sheng, Q.Z. (eds.) WISE Workshops, pp. 182–192 (2005)
7. Ma, L., Yang, Y., Qiu, Z., Xie, G.T., Pan, Y., Liu, S.: Towards a Complete OWL Ontology Benchmark. In: Sure, Y., Domingue, J. (eds.) ESWC 2006. LNCS, vol. 4011, pp. 125–139. Springer, Heidelberg (2006)
8. McBride, B.: Jena: Implementing the RDF Model and Syntax Specification. In: International Workshop on the Semantic Web 2001 (2001)
9. Meditskos, G., Bassiliades, N.: Combining a DL Reasoner and a Rule Engine for Improving Entailment-Based OWL Reasoning. In: Sheth, A.P., Staab, S., Dean, M., Paolucci, M., Maynard, D., Finin, T., Thirunarayan, K. (eds.) ISWC 2008. LNCS, vol. 5318, pp. 277–292. Springer, Heidelberg (2008)
10. Motik, B., Patel-Schneider, P.F., Parsia, B.: OWL 2 Web Ontology Language: Structural Specification and Functional-Style Syntax, W3C Recommendation (October 27, 2009), http://www.w3.org/TR/owl2-syntax/
11. Motik, B., Cuenca Grau, B., Horrocks, I., Wu, Z., Fokoue, A., Lutz, C. (eds.): OWL 2 Web Ontology Language Profiles. W3C Recommendation (2009)
12. Pan, J.Z., Thomas, E.: Approximating OWL-DL Ontologies. In: Proceedings of the 22nd AAAI Conference on Artificial Intelligence (AAAI-2007), pp. 1434–1439 (2007)
13. Pérez-Urbina, H., Motik, B., Horrocks, I.: Tractable query answering and rewriting under description logic constraints. Journal of Applied Logic 8(2), 186–209 (2010)
14. Rector, A.L., Rogers, J.D.: Ontological and Practical Issues in Using a Description Logic to Represent Medical Concept Systems: Experience from GALEN. In: Barahona, P., Bry, F., Franconi, E., Henze, N., Sattler, U. (eds.) Reasoning Web 2006. LNCS, vol. 4126, pp. 197–231. Springer, Heidelberg (2006)
15. Stoilos, G., Cuenca Grau, B., Horrocks, I.: Completeness guarantees for incomplete reasoners. In: Proc. of ISWC-10. LNCS, Springer (2010)
16. Stoilos, G., Cuenca Grau, B., Horrocks, I.: How incomplete is your semantic web reasoner? In: Proc. of AAAI-2010, pp. 1431–1436 (2010)
17. Wu, Z., Eadon, G., Das, S., Chong, E.I., Kolovski, V., Annamalai, M., Srinivasan, J.: Implementing an inference engine for RDFS/OWL constructs and user-defined rules in oracle. In: Proc. of ICDE, pp. 1239–1248. IEEE (2008)

Watermarking for Ontologies

Fabian M. Suchanek[1], David Gross-Amblard[1,2], and Serge Abiteboul[1]

[1] INRIA Saclay, Paris
[2] LE2I CNRS, Université de Bourgogne, Dijon

Abstract. In this paper, we study watermarking methods to prove the ownership of an ontology. Different from existing approaches, we propose to watermark not by altering existing statements, but by removing them. Thereby, our approach does not introduce false statements into the ontology. We show how ownership of ontologies can be established with provably tight probability bounds, even if only parts of the ontology are being re-used. We finally demonstrate the viability of our approach on real-world ontologies.

1   Introduction

An ontology is a formal collection of world knowledge. Creating an ontology usually involves a major human effort. In the case of manually constructed ontologies, human effort is needed to collect the knowledge, to formalize it and to maintain it. The same applies to ontologies constructed by a community, such as Freebase or DBpedia. In the case of automatically constructed ontologies, human effort comes in the form of scientific investigation and the development of algorithms. Consequently, the creators of an ontology usually do not give away the ontology for free for arbitrary use. Rather, they request their users to pay for the content, to follow the terms of a specific license, or to give credit to the creators of the ontology. In most cases, it is prohibited to re-publish the data, or allowed only with proper acknowledgment.

This restriction is most obvious in the case of commercially sold ontologies such as [9]: The use of the data is restricted by the sale contract. The contract usually prohibits the re-publication of the data. Any dissemination of the data into other data sets constitutes a breach of contract.

One might think that the picture would be different for the public ontologies of the Semantic Web. The general spirit of the Semantic Web wants data to be shared across application and community boundaries[1]. However, even the ontologies of the Semantic Web are not available for arbitrary re-publication. Table 1 shows some popular ontologies mentioned together with their licenses. None of the ontologies is available in the public domain. All of them require at least an acknowledgment when their data is re-published. It is considered dishonest to sell or re-publish the data from an ontology elsewhere without giving due credit to the original.

---

[1] http://www.w3.org/2001/sw/

L. Aroyo et al. (Eds.): ISWC 2011, Part I, LNCS 7031, pp. 697–713, 2011.

Some data sets are not freely available at all (see Table 1). The Wolfram Alpha data set[2], for example, can be queried through an API, but cannot be downloaded. Its terms of use prohibit the systematic harvesting of the API to re-create the data set. Any re-publication of a substantial portion of such data constitutes a breach of the terms of use. Similar observations apply to trueknowledge[3] or the commercial version of Cyc [9]. In all of these cases, the extraction and systematic dissemination of the data is prohibited.

Table 1. Common licenses for ontologies

License	Conditions	Ontologies
GFDL	attribution, copyleft	DBpedia [2]
GPL	attribution, copyleft	SUMO [10]
CC-BY	attribution	YAGO [15], Freebase, Geonames, OpenCyc [9]
CC-BY-ND	attribution, no derivatives	UniProt
–	access under restrictions	TrueKnowledge, KnowItAll, WolframAlpha, full Cyc [9]

This raises the issue of how we can detect whether an ontology has been illegally re-published. We call a person who re-publishes an ontology (or part of it) in a way that is inconsistent with its license an *attacker*. The attacker could, e.g., re-publish the ontology under his own name or use parts of the ontology in his own ontology without giving due credit. We call the source ontology the *original ontology* and the re-published ontology the *suspect ontology*. We want to solve the problem of ownership proof: How can we prove that the suspect ontology contains part of the original ontology? Obviously, it is not sufficient to state that the suspect ontology contains data from the original ontology. This is because ontologies contain world knowledge, that anybody can collect. Take the example of an ontology about scientists: the attacker could claim that he also collected biographies of scientists and that he happened to produce the same data set as the original ontology. A similar argument applies to ontologies that have been derived from public sources. YAGO [15] and DBpedia [2], e.g., have both been extracted from Wikipedia. An attacker on DBpedia could claim that he also extracted data from Wikipedia and happened to produce a similar output.

One might be tempted to assume that we could simply publish the original ontology with a time stamp (in the style of proof of software ownership). For example, we could upload the ontology to a trusted external server. If someone else publishes the same data later, then we could point to the original copy. However, this does not prove our ownership. The other publisher could have had the data before we published ours. The fact that he did not publish his data cannot be used against him.

Therefore, a more sophisticated approach is needed to enable ownership proofs. This is the goal of the present paper. We present an approach that uses digital watermarking to detect whether a suspect ontology contains data derived from

---

[2] http://www.wolframalpha.com/

[3] http://www.trueknowledge.com/

an original ontology. Watermarking techniques aim at hiding some relevant information in a data set, in an invisible or robust way. Finding such information in a suspect data set acts as the proof of ownership. Several works consider watermarking for relational databases by performing voluntarily alteration of data. These approaches could be extended to ontologies. However, data alteration invariably decreases the precision of the ontology.

Therefore, we develop an alternative method that is specifically adapted to the Semantic Web: We propose to watermark an ontology by removing carefully selected statements before publishing. The suspect absence of these statements in an ontology with a significant overlap will act as the proof of theft. We argue that this does less harm than altering statements, because the Semantic Web operates under the Open World Assumption: Most ontologies are incomplete.

More specifically, our contributions are as follows:

1. A formalization of the problem of ontological data re-publication,
2. An algorithm for watermarking ontologies, which allows detecting malicious re-publication without harming the precision,
3. Extensive experiments that show the validity of our approach.

The rest of this paper is structured as follows: Section 2 summarizes related work. Section 3 formalizes our scenario and lists different attack models. Section 4 presents our watermarking algorithm with a formal analysis. Section 5 details our experiments before Section 6 concludes.

2   Related Work

In [4], the authors introduce named graphs as a way to manage trust and provenance on the Semantic Web. Named graphs, however, cannot guard against the misuse of ontologies that are publicly available.

One of the oldest attempts to prove ownership of factual data is the use of fictitious entries in dictionaries. Since the 19th century, dictionaries, maps, encyclopedias and directories have had occasional fake entries. The New Columbia Encyclopedia, for example, contained an entry about a fictitious photographer called Lillian Virginia Mountweazel. If such an entry ever appeared in another encyclopedia, it was clear that the data was copied. To mark an ontology, however, it is not sufficient to add a single isolated entity, because an attacker can simply remove unconnected entities.

A classical way to achieve ownership proofs is to apply watermarking techniques. Some recent proposals have targeted ontologies [6]. This previous effort uses a purely syntactical rewriting of the RDF XML source layout into an equivalent layout to hide information. This approach can be circumvented by normalizing the XML file. This can be done, e.g., by loading the file into a tool such as Protégé [11] and then saving it again as an XML document.

Quite a number of approaches have targeted semi-structured data [13] and relational databases [1,14,7,8,12]. These works provide one way to prove ownership of ontologies. Most of them are blind, i.e., they do not require the original data set for detection. However, all of these approaches presume that the schema of

the data is known. This is not necessarily the case on the Semantic Web. Some
ontologies (such as DBpedia or YAGO) have hundreds of relationships. An attacker just has to map some of them manually to other relationships to obscure
the theft. We will develop approaches that can still identify the suspect data,
but in a non-blind way. Furthermore, the classical methods work by voluntarily
altering data. Therefore, we call these methods *modification approaches* in the
sequel. Such approaches could, e.g., change the birth year of Elvis Presley from
the year 1935 to the year 1936. While the introduced errors are only gradual
for numerical data, they are substantial for categorical data. Such an approach
could, e.g., change Elvis' nationality from American to Russian. Apart from the
fact that an ontology owner will find it controversial to voluntarily alter clean
data, such an approach will also decrease the precision of the ontology with respect to the real world. Furthermore, the altered facts can lead to contradictions
with other data sets. This is not only annoying to the user, but can also allow
the attacker to detect and remove the altered facts. Therefore, we present an
alternative approach in this paper, which works by deleting carefully selected
facts. Since the Semantic Web operates under the open world assumption, the
absence of true information is less harmful than the presence of false information.

3 Model

3.1 Watermarking

A watermarking protocol is a pair of algorithms $(\mathcal{M}, \mathcal{D})$, where $\mathcal{M}$ stands for
the marker and $\mathcal{D}$ the detector (Figure 1). Given an original ontology O and
a secret key $\mathcal{K}$, the marker algorithm outputs a watermarked ontology $O^* =
\mathcal{M}(O, \mathcal{K})$. Given a suspect ontology O', the original ontology and the secret key,
the detector decides if O' contains a mark, that is if $\mathcal{D}(O', O, \mathcal{K})$ is true.

If O' contains the mark, then it is assumed that O' has indeed been derived
from O^*. However, the watermarking protocol may erroneously say that O' has
been derived from O^*, even though O' just contains the mark by chance. For

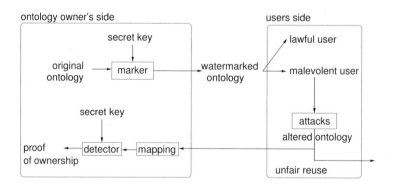

Fig. 1. Watermarking protocol for ontologies

example, O' may be a totally unrelated ontology. In this case, O' is called a *false positive*. Watermarking protocols are designed so that the probability of a false positive is provably below a confidence threshold ξ. The parameter ξ is called the *security parameter* of the protocol and is typically in the order of $\xi = 10^{-6}$.

In an adversarial setting, the attacker can try to evade the detection of the mark in the ontology by various means, including:

- **Subset attack:** The attacker uses only a subset of the stolen ontology. For example, the attacker could choose a certain thematic domain from the original ontology (such as, say, sports), and re-use only this portion. While publishing just a few statements should be allowed, larger stolen parts should be detected by the protocol.
- **Union attack:** The attacker merges the ontology with another one. For example, the attacker could combine two ontologies about gene expressions, thereby hiding the stolen ontology in a larger data set. A naive union attack would keep facts from both ontologies, even if they are inconsistent with each other.
- **Intersection attack:** The attacker keeps only ontology elements that are found in another ontology. For example, an attacker could remove entities from the stolen ontology that do not appear in a reference ontology.
- **Comparison attack:** The attacker compares the ontology to another data source and eliminates inconsistent information. For example, the attacker could cross-check birth dates of one ontology with the birth dates in a reference ontology and remove inconsistent dates.
- **Random alteration attack:** The attacker alters values, relations or entities. This action is limited as the attacker still desires valuable data.

The ability of the watermarking protocol to withstand these attacks is called its *robustness*. Watermarking protocols are typically designed so that the probability of a successful attack is provably below their security parameter ξ.

An attacker may also publish someone else's ontology as part of an offensive or illegal data set. An attacker may, e.g., take an ontology about chemistry and publish it as part of a data set about weapons of mass destruction. If the ontology was marked, it will appear as if the creator of the original ontology authored the offensive data set. Our method can fend off this so-called *copy attack*.

3.2   Ontologies

An RDFS ontology over a set of entities (resources and literals) E and a set of relation names R can be seen as a set of triples $O \subset E \times R \times E$. Each triple is called a *statement*, with its components being called *subject*, *predicate* and *object*. We will write *e.name* to refer to the identifier of the entity in the ontology. This can be the URI (in case of a resource) or the string value (in case of a literal). RDFS specifies certain *semantic rules*. These rules state that if the ontology contains certain statements, certain new statements should be added to the ontology. We call these added statements *redundant* and assume that redundant statements have been removed from the ontology [5]. RDFS rules are strictly positive. Therefore, an RDFS ontology cannot become inconsistent if a statement is removed or altered. This predestines RDFS ontologies for watermarking and we leave more sophisticated ontology models for future work.

To prove that a suspect ontology O' copied data from an original ontology O, we will first have to determine whether O' and O contain similar data. This is a challenging task in itself, because it amounts to finding a mapping from O' to O. This problem has attracted a lot of research in the context of the Web of Linked Data [3]. Finding such a mapping is outside the scope of the present paper. Here, we limit ourselves to 2 assumptions: (1) We know a partial mapping function σ, which maps (some of) the entities of O' to entities of O; (2) there exists a partial mapping function σ_R, which maps the relations of O' to relations of O. Note that we do not need to know σ_R. Our method has been designed so that it is transparent to the "schema" of the ontologies. We need the existence of σ_R just to assure that O and O' have some structural similarity. An investigator needs to find only σ. If the ontology was truly stolen, then the entities will probably be recognizable in some way, because otherwise the stolen ontology would be less useful. In the best case, σ will reflect mainly syntactic variations of the identifiers and values. We note that the modification approaches like [1] also require σ, because they determine the tuples to mark based on the value of the primary key, which has to be available at detection. In addition, the modification approaches also require the mapping of the schema, σ_R, which our approach does not require. In the sequel, we assume that σ has been found and has been applied to O'. We are now concerned with the question that follows: We want to prove that O' and O are similar not just by chance, but because O' copied data from O.

3.3   Ethical Considerations

Ontologies represent world knowledge. Therefore, it is questionable whether one can "own" such data. Can the creator have a copyright on the ontology, given that it is nothing more than a collection of facts about the world? In this paper, we do not deal with the legal implications of owning or copying ontologies. We only provide a method to prove that one data source copied from another source – independently of whether such behavior is considered legal or not.

Our approach will remove facts from the ontology before publishing. Then the question arises whether it is honest to withhold information that one could publish. However, an ontology always represents only part of reality. In most cases, the creator of an ontology is not obliged to make it exhaustive. The Open World Assumption of the Semantic Web makes the absence of information a normal and tolerable circumstance.

In general, the watermarking of an ontology always remains a trade-off between the ability to prove ownership and the truthfulness of the data. One should not willfully alter ontologies that are highly sensitive to even small misrepresentations of reality, such as ontologies in the domain of medicine, security, or aeronautics. However, unlike the modification approaches [1], our approach of deleting facts can be applied even if truthfulness of the data has to be preserved, as long as some incompleteness can be tolerated.

4   Watermarking Ontologies

4.1   Watermarking Basics

Our starting point is a watermarking protocol from Agrawal et al. [1] for relational databases. Their statistical watermarking uses cryptographically secure pseudo-random number generators (CSPRNGs). A CSPRNG is a function which, given an integer seed value, produces a sequence of pseudo-random bits. The bits are pseudo-random in the sense that, without the seed, and given the first k bits of a random sequence, there is no polynomial-time algorithm that can predict the $(k+1)$th bit with probability of success better than 50% [16]. A CSPRNG is a *one-way-function*. This means that, given a CSPRNG and a sequence of random bits it produced, it is close to impossible to determine the seed value that generated it. More formally, given a CSPNG f, for every randomized polynomial time algorithm A, $Pr[f(A(f(x))) = f(x)] < \frac{1}{p(n)}$ for every positive polynomial $p(n)$ and sufficiently large n, assuming that x is chosen from a uniform distribution [17]. In the following, we use a given CSPRNG $\mathcal{G}$.

After $\mathcal{G}$ has been seeded with a value by calling $\mathcal{G}.seed(value)$, one can repeatedly call the function $\mathcal{G}.nextBit()$ to obtain the next bit in the random sequence. By combining multiple calls to this function, we can construct a pseudo-random integer value. The function that delivers a pseudo-random integer number greater or equal to 0 and below a given upper bound k is denoted by $\mathcal{G}.nextInt(k)$.

Our watermarking algorithm makes use of a *secret key*. A secret key is a integer number that is only known to the owner of the original ontology. We will use the key as a seed value for $\mathcal{G}$. We also make use of a *cryptographically secure hash function*. A hash function is a function which, given an object, returns an integer number in a certain range. A cryptographically secure hash function is such that it is infeasible to find two inputs with the same output, or the inverse of the output. We assume a given hash function, such as SHA, denoted *hash*. In order to resist the aforementioned copy attack, we will first compute the secure hash of the original ontology. All our subsequent computations will depend on this hash, so that no attacker can pretend to own the watermarked version by finding another ontology with the same watermarking.

4.1.1   Algorithm

Our watermarking method shall be transparent to relation names, because we do not want to require an investigator to find a mapping of the schema. Therefore, we define the notion of a *fact pair*. A fact pair of an ontology O is a pair of entities $\langle e_1, e_2 \rangle$, such that there exists r such that $\langle e_1, r, e_2 \rangle \in O$. Algorithm 1 marks an ontology by removing fact pairs. It takes as input the original ontology O, a secret key $\mathcal{K}$, and the number of facts *delTotal* to remove. The secret key is an arbitrary chosen number. Section 4.2.1 will discuss how to find a suitable value for *delTotal*. For each fact pair of the ontology, the algorithm seeds $\mathcal{G}$ with the names of the entities, the hash of O, and $\mathcal{K}$. If the next random integer of $\mathcal{G}$ between 0 and the total number of fact pairs in O divided by *delTotal* happens to be 0, the fact pair is removed. This removes *delTotal* facts from the

---

Algorithm 1. subtractiveMark(orig. ontology O, secret key $\mathcal{K}$, integer $delTotal$)

$O^* \leftarrow O$
for all $\langle e_1, e_2 \rangle \in \pi_{subject, object}(O)$ **do**
   $\mathcal{G}.seed(e_1.name \oplus e_2.name \oplus hash(O) \oplus \mathcal{K})$
   if $\mathcal{G}.nextInt(|\pi_{subject, object}(O)|/delTotal) = 0$ **then**
      Remove $\langle e_1, *, e_2 \rangle$ from O^*
   end if
end for
return O^*

---

ontology[4]. Note that our algorithm does not consider the relation names at all. After running Algorithm 1, the marked ontology O^* is published. The original ontology O is kept secret.

To detect whether a suspect ontology stole data from an original ontology, we run Algorithm 2 on the original ontology O (without the mark) and the suspect ontology O' (after having applied the mapping from Section 3.2). The algorithm runs through all fact pairs of O and computes the proportion of published fact pairs that appear in the suspect ontology. It also computes the proportion of removed fact pairs that appear in the suspect ontology. Since we only consider fact pairs and not facts, a mapping of the relation names (the schema) is not necessary. It is possible that the suspect ontology contains some of the fact pairs that we removed from the original before publishing. This can be for two reasons: Either the suspect ontology is innocent and just happens to have a thematic overlap with our original, or the suspect ontology imported data from other sources, thus complementing the facts we removed. The algorithm then compares the ratio of removed fact pairs that appear in O' to the ratio of published fact pairs that appear in O. If O' is innocent, these ratios should be the same. If the ratio of deleted facts is lower than the ratio of published facts, and significantly so, this indicates a theft and the algorithm will return *true*. It seems highly counter-intuitive that the absence of a fact should prove theft of data. Yet, the proof comes from the fact that the removed statements form a pattern of present and absent facts in O. The probability that this pattern appears by chance in another ontology is extremely low.

Let us detail the check of significance. The suspect ontology will cover a certain portion of facts of the original ontology. The central observation is that, if this overlap is by chance, then the suspect ontology should cover the same portion of published facts as it covers of the deleted facts, because the watermarking is randomized. Thus, we have to determine whether any difference between these two ratios is statistically significant. This is the last step in Algorithm 2. This significance is determined by a χ^2 test. Be $pubTotal$ the total number of published fact pairs. Be $pubFound$ the number of published fact pairs that appear in the suspect ontology. Be $delTotal$ the total number of deleted fact pairs. Be $delFound$ the number of deleted fact pairs that appear in the suspect ontology

---

[4] If less than $delTotal$ facts got removed, we rerun the algorithm with a different key.

Algorithm 2. subtractiveDetect(original O, suspect O', key $\mathcal{K}$, integer $delTotal$)

$O^* \leftarrow subtractiveMark(O, \mathcal{K}, delTotal)$
$pubFound \leftarrow 0; delFound \leftarrow 0;$
  for all $\langle e_1, e_2 \rangle \in \pi_{subject,object}(O)$ **do**
    if $\langle e_1, e_2 \rangle \in \pi_{subject,object}(O')$ **then**
      if $\langle e_1, e_2 \rangle \in \pi_{subject,object}(O^*)$ **then** $pubFound + +$
      else $delFound + +$
    end if
  end for
$pubTotal \leftarrow |\pi_{subject,object}(O^*)|$
if $delFound/delTotal \geq pubFound/pubTotal$ **then return false**
return $delFound/delTotal$ significantly different from $pubFound/pubTotal$

and be $N = pubTotal + delTotal$ the total number of fact pairs. We get

$$\chi^2 = \frac{N(delFound \times (pubTotal - pubFound) - pubFound \times (delTotal - delFound))^2}{(pubFound + delFound) \times (N - pubFound - delFound) \times delTotal \times pubTotal}.$$

If $\chi^2 > \chi^2(1, \xi)$, where ξ is the security parameter, then the two ratios are not independent. Since the removal of the fact pairs was purely random, any significant difference between the ratios indicates a dependence on the original ontology. In the extreme case, the suspect ontology reproduces the published facts and omits all (or nearly all) removed facts.

In order to be applicable, the standard χ^2 test requires the total number of samples to be greater than 100 and the expected number of samples for each case to be greater than 5. Therefore, our algorithm returns *true* iff $N > 100$ and $(pubFound + delFound) \times delTotal/N > 5$ and $\chi^2 > \chi^2(1, \xi)$.

4.2 Analysis

4.2.1 Impact
We are interested in how many fact pairs we have to remove in order to achieve significance in the χ^2 test. This number depends on the total number of fact pairs N. It also depends on the types of attacks against which we want to protect. The first property of an attack is the overlap ratio of found facts, $\omega = (pubFound + delFound)/N$. We choose $\omega = 1$ if we want to protect only against a theft of the complete ontology. We choose a smaller value if we want to protect also against a theft of a sub-portion of the ontology. The second property of an attack is the ratio of removed facts $\delta = delFound/delTotal$ that appear in the stolen ontology. If the attacker just republishes our published ontology, $\delta = 0$. If he adds data from other sources, this can complement some of the facts we removed. Ratio δ should be the proportion of removed facts that we expect in the stolen ontology. If δ is larger, and ω is smaller, the protection is safer, but the marking will remove more facts. Abbreviating $delTotal = d$, this yields $pubFound = \omega N - \delta d, delFound = \delta d, pubTotal = N - d$ and thus

$$\chi^2 = \frac{N((\omega N - \delta d)(1 - \delta)d - \delta d((1 - \omega)N - (1 - \delta)d))^2}{\omega N(1 - \omega)Nd(N - d)}.$$

For $\chi^2 > \chi^2(1, \xi)$, this yields

$$d > \frac{N\omega(1 - \omega)\chi^2(1, \xi)}{N(\delta(1 - \omega) - \omega(1 - \delta))^2 + \omega(1 - \omega)\chi^2(1, \xi)}.$$

We have to impose $N > 100$ as a precondition for the χ^2 test. We also have to impose $d > 5/(\omega(1 - \omega))$, i.e., $d > 20$ in the worst case. Finally, $\delta < \omega$, because we cannot prove theft if the ratio of appearing deleted facts is greater than the ratio of appearing published facts. As an example, take a choice of $\xi = 10^{-6}$, which leads to $\chi^2(1, \xi) = 23.9284$. Assuming an overlap ratio of $\omega = \frac{1}{2}$, a fault tolerance of $\delta = 0.2$, and $N = 30 \times 10^6$ fact pairs, we get $d = 67$, i.e., 67 fact pairs have to be deleted from the ontology.

4.2.2   Robustness

In general, the χ^2 test tends to err on the safe side, concluding independence only in the presence of overwhelming evidence. Thus, our algorithm will signal theft only in very clear cases ("in dubio pro reo"). However, our algorithm is also well-protected against attacks. First, a marked ontology is protected against intersection attacks. Intersection attacks can happen, e.g., when an attacker wants to misuse someone else's ontology to clean up his own noisy data set. Since an intersection does not add facts, our marks survive such an attack. The marked ontology is also protected against comparison attacks, because the ontologies we target generally suffer from incompleteness. Thus, a fact that is absent in the original ontology but present in a reference ontology will not raise suspicions. A marked ontology is also safe against subset attacks, if ω is chosen smaller than the proportion of stolen facts. A union attack, in contrast, could add information that fills up some of the removed facts. In this case, the marks will be reduced to those portions of the ontology that do not appear in the other ontology. By choosing $1 - \delta$ equal to the portion that we estimate to be proper to the ontology, and adjusting d accordingly, we can still guard against the union attack. Random alteration attacks fall into the scope of the classical analysis of robustness [1]: an attacker being ignorant on the positions of the deleted facts can only try at random to delete more facts or fill missing ones. For this attack to be successful, a large number of facts have to be altered, a much larger number than the watermark algorithm used. Finally, an attacker can try to modify the fact pairs. Deleted facts are of course not sensitive to this attack, but the number $pubFound$ can be altered. However, as the subset of published facts we are looking for at detection are chosen pseudo-randomly, the attacker has no way to locate them efficiently. The only valid strategy for the attacker is again to alter a huge amount of fact pairs, which reduces drastically the quality of the stolen ontology.

Table 2. Number of facts that have to be marked
$\alpha = 0, \xi = 10^{-6}$ and $N = 30 \times 10^6$

ω	50%	10%	5%	2.5%	0.5%	0.05%
Removing	24	215	456	935	4775	47900
Modifying	96	480	950	1900	9500	95000

4.2.3 Comparison to Modification Watermarking

The modification approach [1] changes a fact from the ontology. In the majority of cases, it will change a correct fact to an incorrect fact. Thereby, precision invariably suffers. In contrast, our approach does not decrease the precision of the ontology at all. To see this, assume that O contains n statements, c of which are correct. If a correct fact is deleted, which will happen in $\frac{c}{n}$ of the cases, the precision drops to $\frac{c-1}{n-1}$. If an incorrect fact is deleted, the precision increases to $\frac{c}{n-1}$. Thus, on average, the precision is $\frac{c}{n} \times \frac{c-1}{n-1} + \frac{n-c}{n} \times \frac{c}{n-1}$, which is $\frac{c}{n}$, just as before. As a comparison, the modification approaches have an average impact of $\frac{c}{n} \times \frac{c-1}{\mathbf{n}} + \frac{n-c}{n} \times \frac{c}{\mathbf{n}}$ (a modified correct fact turns incorrect, and a modified incorrect fact will still be incorrect, while the number of total facts is the same).

Now let us consider the number of facts that have to be modified in the classical approach. Assuming that $\delta = 0, \xi = 10^{-6}$ and $N = 30 \times 10^6$, we used the estimates in [1] to compute the number of tuples (fact pairs, in our setting) that are required to be modified in order to resist a subset attack of parameter ω. This leads to the numbers in Table 2.

The modification method hides a list of 0 and 1 on secretly chosen positions. If such a position is not selected by the subset attack, the marked bit is lost, whatever its value. But for our method, the subset attack has no impact on already deleted facts. Therefore, the modification approach has to modify more fact pairs than we have to delete. Overall, the number of facts that have to be modified is comparable to the number of facts that have to be deleted. Given that removal maintains precision, while modification does not, and that modification yields false facts, the ontology owner might decide to remove instead of to modify.

5 Experiments

5.1 Applicability

5.1.1 Impact

We were interested in how applicable our method is to real world ontologies. For this purpose, we collected 5 ontologies from the Semantic Web that cover a wide range of complexity, size, and topics (Table 3): The core part of YAGO [15], the manually supervised part of DBpedia [2], the Universal Protein Resource[5], an ontology about city finances (provided by the UK government[6]), and a subset of the IMDb[7]. For each ontology, we report the number of facts, fact pairs, relations

---

[5] http://www.uniprot.org/
[6] http://data.gov.uk/
[7] http://imdb.com/

Table 3. Marking different ontologies

	YAGO	DBpedia	UniProt	Finance	IMDb	*Modification*
# relations	83	1,107	4	11	12	
# instances	2,637,878	1,675,741	1,327	1,948	4,657,880	
# facts	18,206,276	19,788,726	6,096	14,926	34,699,697	
dup. objects	3,529,697	450,171	0	0	14,907	
# fact pairs	14,676,579	19,338,555	6,096	14,926	34,685,090	
Facts to remove						
$\delta = 0, \omega = 50\%$	24	24	24	24	24	[97]
$\delta = 0, \omega = 5\%$	456	456	424	442	456	[975]
$\delta = 0, \omega = 0.5\%$	4,775	4,774	2,677	3,618	4,775	[9700]
$\delta = 0, \omega = 0.05\%$	47,820	47,825	-	-	47,909	[97500]
$\delta = 10\%, \omega = 50\%$	37	37	37	37	37	N/A

and instances. We also report the number of entities that are connected to an object by more than one relation. This number is usually small, except for YAGO, where every entity has a label and a (mostly equivalent) preferred name. We compute the number of facts that have to be removed to protect against various subset attacks ($\xi = 10^{-6}$). As a comparison, the last column gives the number of alterations needed for the modification approach [1]. It is roughly independent of the size of the data set. Values for the modification method are not given for $\delta = 10\%$, because the scenario where the attacker irons out the marks has not been considered in [1].

5.1.2   Removing the Marks

We were interested in how an attacker could try to identify the missing facts in order to reinstate them. The attacker could, e.g., compare all instances of a class and see whether some of them lack a relation that all the other instances have. This entity is suspect from an attackers point of view, because he has to assume that we deleted that relation. More precisely, we call an entity e *suspect* in an ontology O, if there exists a class c, an entity e', and a relation r, such that

$$e \in c, e' \in c, |\{e'' : \langle e', r, e'' \rangle\}| > |\{e'' : \langle e, r, e'' \rangle\}|.$$

We call a fact *discreet* if we can remove it without creating a suspect entity. We computed the proportion of discreet facts, their relations and the proportion of instances with at least one discreet fact (Table 4). Roughly half of the facts are discreet, meaning that we can delete them without raising suspicions. Even if the attacker correctly identifies the fact we removed, he cannot simply discard the entity, because this would still keep the mark. Instead, he has to find the correct value for the missing link to plug in the hole. This may be hard or even close to impossible, because the attacker has no access to the original ontology and cannot run the detection algorithm. Also, he does not know the ratio of discreet facts. Furthermore, from the attacker's point of view, nearly all instances are suspect on the original data set already. Thus, nearly every instance could potentially have been marked. Filling up what could be a hole would amount to

Table 4. The ontologies from an attackers' point of view

	YAGO	DBpedia	UniProt	Finance	IMDb
discreet instances	99%	92%	74%	69%	97%
discreet facts	74%	86%	48%	39%	75%
discreet relations	96%	99%	50%	45%	92%
suspect instances	99%	99%	100%	75%	100%

reconstructing the ontology. The only exception to this rule is the finance data set, which contains rather complete information (most entities have the maximal number of links in their class). We note, however, that removing facts might still be preferable to modifying facts in this ontology.

5.1.3 False Positive Detection

A risk with watermarking approaches is the occurrence of false positives. The probability of considering a non-marked ontology suspect has to be bounded. While this poses a problem for blind watermarking methods that do not rely on the original for detection, it is very unlikely that a random ontology is signaled as stolen by our approach, because our approach first checks that the overlap of the suspect ontology with the original ontology is significant. Even in the unlikely situation of a large overlap, the innocent suspect ontology will match with published facts and deleted facts uniformly. Thus the ratio of deleted and found facts will be similar, leading to a correct non-detection. This risk is taken into account and formalized in the significance level of the χ^2 test. We provide next some subset attacks whose overlap is so small that our method does not signal theft.

5.2 Robustness

5.2.1 Attack Simulations

To demonstrate the robustness of our watermarking, we simulated suspect ontologies that overlap only partially with the original ontology. This partial overlap could be due to an incomplete mapping σ or due to the fact that the attacker chose to steal only a subset of the original. We watermarked the Finance ontology with 30 removed facts. This should make the mark resistant to a partial overlap of $\omega = 50\%$ or more. We varied ω and created 10 random overlaps for each value of ω. Figure 2 shows the average rate of successful detection ($\xi = 10^{-6}, \delta = 0$). As predicted, any suspect ontology that overlaps more than half is identified as stolen.

We also simulate suspect ontologies that do contain portions of the deleted facts. This can be the case if the attacker merged the stolen ontology with another data set. We removed 50 facts from the Finance ontology. At an overlap of $\omega = 50\%$, this should protect the ontology up to $\delta = 15\%$. We varied δ and simulated 10 random suspect ontologies for each value of δ. Figure 3 shows that the rate of successful detection. As expected, the rate is 1 for $\delta \leq 15\%$.

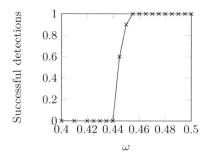

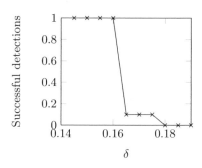

Fig. 2. Rate of successful detection with varying ω, ($\delta = 0$)

Fig. 3. Rate of successful detection with varying δ ($\omega = 0.5$)

5.2.2 Thematic Subset Attacks

Next, we analyze *thematic subset attacks*, i.e., attacks that steal a certain class with all its instances and all their facts. We call such a set facts a *theme*. Table 5 shows populous classes in YAGO together with their number of fact pairs. We computed the ratio ω and the number of fact pairs that would have to be deleted in total (at $\xi = 10^{-6}$). The numbers in brackets show the number of fact pairs that the modification method would consume. Table 6 shows the same characteristics for populous classes in DBpedia. The number of facts to delete fades in comparison to YAGO's 15m fact pairs in total and DBpedia's 19m fact pairs in total. We experimentally verified the subset attacks on DBpedia with 1000 removed facts, achieving significance levels of $\chi^2 = 68, 18, 38, 51, 38$, respectively. This means that, with 1000 removed facts, we could detect all thematic subset attacks except for the one on Television Episodes, because the subset is too small for the chosen number of marks. This experiment confirms our predictions.

Table 5. Protecting different themes in YAGO

Theme	# fact pairs	ω	Fact pairs to remove	
			$\delta = 0$	$\delta = 1\%$
Person	10,594,790	72.19%	20 [67]	20
Village	582,984	3.97%	580 [1,225]	1,037
Album	588,338	4.01 %	574 [1,215]	1,020
Football player	1,200,459	8.18 %	269 [596]	350
Company	400,769	2.73%	854 [1,785]	2,129

5.2.3 Union Attacks

We wanted to evaluate how our method works if an attacker merges the stolen ontology with another, possibly similar ontology. This might fill up some of the removed facts and thus destroy our marks. We simulated attacks that steal a certain theme from DBpedia and merge it into YAGO. We merged by matching

Table 6. Protecting different themes in DBpedia

Theme	# fact pairs	ω	Fact pairs to remove	
			$\delta = 0$	$\delta = 1\%$
Album	120,7561	6.24%	360 [782]	511
TelevisionEpisode	342,277	1.77%	1,331 [2750]	7,035
Village	701,084	3.63%	637 [1345]	1,213
SoccerPlayer	924,299	4.78%	478 [1020]	764
Film	694,731	3.59%	644 [1360]	1,238

resources (DBpedia and YAGO use the same local identifiers with different pre-fixes), matching identical strings and identical numbers, and matching numbers that share the value (ignoring the unit). This yields an overlap of 1.6×10^6 fact pairs between the two ontologies.[8] This overlap is 8% of DBpedia. This means that 8% of the marks that we add to DBpedia can be filled up by merging with YAGO. Hence, we have to choose $\delta > 8\%$ in order to protect against a theft.

Table 7 shows the ontologies obtained from merging YAGO with a certain theme from DBpedia. The table shows the total number of fact pairs of these ontologies as well as the absolute and relative overlap with the original DBpedia. The relative overlap corresponds to ω. The last column shows the number of fact pairs that have to be removed from DBpedia to protect the theme, calculated from ω and $\delta = 8\%$.

Table 7. Merging different themes of DBpedia into YAGO

YAGO + DBpedia Theme	# fact pairs	overlap with DBpedia	overlap as % of DBpedia ($=\omega$)	Fact pairs to remove
Album	15,920,534	2,831,716	15%	624
TelevisionEpisode	15,108,014	2,019,289	10%	5398
Village	15,399,557	2,310,784	12%	1583
SoccerPlayer	15,585,500	2,496,744	13%	1085
Films	15,369,042	2,280,321	12%	1583

We experimentally verified these theoretical results by marking DBpedia with the removal of 2000 facts. Then, we stole different themes from the marked DBpedia and merged them into YAGO. We ran our detection algorithm and obtained significance levels of $\chi^2 = 28, 15, 34, 47$, and 31, respectively. This means that we could successfully detect all thefts, except for the theft of the Television Episode

---

[8] Part of the reason for the small overlap is the rather crude mapping (YAGO normalizes numbers to SI units, while DBpedia does not). However, manual inspection also shows that YAGO knows many types for the entities, many labels, and some facts that DBpedia does not know. DBpedia, in turn, captures many infobox attributes that YAGO does not capture. The ontologies share just 1.4 million instances.

theme. This set is smaller, so that it requires the removal of more facts, as predicted. This shows that our method works even if part of the mark is destroyed.

6 Conclusion

We have presented an alternative approach for the watermarking of ontologies. Instead of altering facts, we remove facts. Thereby, we do not lower the precision of the ontology. We have shown that even on large ontologies, only a few hundred facts have to be removed to guarantee protection from theft. Through experiments, we have shown that our approach is well applicable to real world ontologies. In the future, we intend to explore whether ontologies can also be watermarked by adding artificial facts.

Acknowledgements. This work was supported by the European Research Council under the European Community's Seventh Framework Programme (FP7/2007-2013) / ERC grant Webdam, agreement 226513 (http://webdam.inria.fr/).

References

1. Agrawal, R., Haas, P.J., Kiernan, J.: Watermarking Relational Data: Framework, Algorithms and Analysis. VLDB J. 12(2), 157–169 (2003)
2. Auer, S., Bizer, C., Kobilarov, G., Lehmann, J., Cyganiak, R., Ives, Z.G.: DBpedia: A Nucleus for a Web of Open Data. In: Aberer, K., Choi, K.-S., Noy, N., Allemang, D., Lee, K.-I., Nixon, L.J.B., Golbeck, J., Mika, P., Maynard, D., Mizoguchi, R., Schreiber, G., Cudré-Mauroux, P. (eds.) ASWC 2007 and ISWC 2007. LNCS, vol. 4825, pp. 722–735. Springer, Heidelberg (2007)
3. Bizer, C., Heath, T., Idehen, K., Berners-Lee, T.: Linked data on the Web. In: Proceedings of the International Conference on World Wide Web (WWW), pp. 1265–1266. ACM, New York (2008)
4. Carroll, J.J., Bizer, C., Hayes, P., Stickler, P.: Named graphs, provenance and trust. In: WWW 2005: Proceedings of the 14th International Conference on World Wide Web, pp. 613–622. ACM, New York (2005)
5. Grimm, S., Wissmann, J.: Elimination of Redundancy in Ontologies. In: Antoniou, G., Grobelnik, M., Simperl, E., Parsia, B., Plexousakis, D., De Leenheer, P., Pan, J. (eds.) ESWC 2011, Part I. LNCS, vol. 6643, pp. 260–274. Springer, Heidelberg (2011)
6. Kong, H., Xue, G., Tian, K., Yao, S.: Techniques for owl-based ontology watermarking. In: WRI Global Congress on Intelligent Systems (GCIS), Xiamen, pp. 582–586 (2009)
7. Lafaye, J., Gross-Amblard, D., Constantin, C., Guerrouani, M.: Watermill: An optimized fingerprinting system for databases under constraints. IEEE Trans. Knowl. Data Eng. (TKDE) 20(4), 532–546 (2008)
8. Li, Y., Swarup, V., Jajodia, S.: Fingerprinting relational databases: Schemes and specialties. IEEE Trans. Dependable Sec. Comput. (TDSC) 2(1), 34–45 (2005)

9. Matuszek, C., Cabral, J., Witbrock, M., Deoliveira, J.: An introduction to the syntax and content of cyc. In: Proceedings of the AAAI Spring Symposium on Formalizing and Compiling Background Knowledge and Its Applications to Knowledge Representation and Question Answering, pp. 44–49. AAAI Press, Menlo Park (2006)
10. Niles, I., Pease, A.: Towards a standard upper ontology. In: Proceedings of the international conference on Formal Ontology in Information Systems, pp. 2–9. ACM, New York (2001)
11. Noy, N., Fergerson, R., Musen, M.: The knowledge Model of Protégé-2000: Combining Interoperability and Flexibility. In: Dieng, R., Corby, O. (eds.) EKAW 2000. LNCS (LNAI), vol. 1937, pp. 17–32. Springer, Heidelberg (2000)
12. Shehab, M., Bertino, E., Ghafoor, A.: Watermarking relational databases using optimization-based techniques. IEEE Trans. Knowl. Data Eng. (TKDE) 20(1), 116–129 (2008)
13. Sion, R., Atallah, M., Prabhakar, S.: Resilient Information Hiding for Abstract Semi-Structures. In: Kalker, T., Cox, I., Ro, Y.M. (eds.) IWDW 2003. LNCS, vol. 2939, pp. 141–153. Springer, Heidelberg (2004)
14. Sion, R., Atallah, M., Prabhakar, S.: Protecting rights over relational data using watermarking. IEEE Trans. Knowl. Data Eng. TKDE 16(12), 1509–1525 (2004)
15. Suchanek, F.M., Kasneci, G., Weikum, G.: YAGO: A core of semantic knowledge - unifying WordNet and Wikipedia. In: Williamson, C.L., Zurko, M.E., Patel-Schneider, P.J., Shenoy, P.F. (eds.) Proceedings of the International Conference on World Wide Web (WWW), Banff, Canada, pp. 697–706. ACM (2007)
16. Wikipedia. CSPRNG (2011)
17. Wikipedia. One-way function (2011)

Link Prediction for Annotation Graphs Using Graph Summarization

Andreas Thor[1], Philip Anderson[1], Louiqa Raschid[1], Saket Navlakha[1],
Barna Saha[1], Samir Khuller[1], and Xiao-Ning Zhang[2]

[1] University of Maryland, USA
[2] St. Bonaventure University, USA
thor@umiacs.umd.edu, phand@umd.edu, louiqa@umiacs.umd.edu,
saket@cs.umd.edu, barna@cs.umd.edu, samir@cs.umd.edu, xzhang@sbu.edu

Abstract. Annotation graph datasets are a natural representation of
scientific knowledge. They are common in the life sciences where genes
or proteins are annotated with controlled vocabulary terms (CV terms)
from ontologies. The W3C Linking Open Data (LOD) initiative and
semantic Web technologies are playing a leading role in making such
datasets widely available. Scientists can mine these datasets to discover
patterns of annotation. While ontology alignment and integration across
datasets has been explored in the context of the semantic Web, there is
no current approach to mine such patterns in annotation graph datasets.
In this paper, we propose a novel approach for link prediction; it is a preliminary task when discovering more complex patterns. Our prediction
is based on a complementary methodology of graph summarization (GS)
and dense subgraphs (DSG). GS can exploit and summarize knowledge
captured within the ontologies and in the annotation patterns. DSG uses
the ontology structure, in particular the distance between CV terms, to
filter the graph, and to find promising subgraphs. We develop a scoring
function based on multiple heuristics to rank the predictions. We perform
an extensive evaluation on Arabidopsis thaliana genes.

Keywords: Link prediction, Graph summarization, Dense subgraphs,
Linking Open Data ontology alignment.

1 Introduction

Among the many "killer apps" that could be enabled by the Linking Open Data
(LOD) initiative [2,20] and semantic Web technologies, the ability for scientists
to mine annotation graph datasets and to determine actionable patterns shows
great promise. A majority of the links in LOD datasets are at the instance level
as exemplified by the *owl:sameAs* relationship type. However, there has been a
rapid emergence of biological and biomedical datasets that are typically annotated using controlled vacabulary (CV) terms from ontologies. For example, the
US NIH clinical trial data `ClinicalTrial.gov` has been linked to (1) PubMed
publications and Medical Subject Header (MeSH) terms; (2) drug names and
drug terms from `RxNorm`; (3) disease names and terms from `Diseasome`; etc.

L. Aroyo et al. (Eds.): ISWC 2011, Part I, LNCS 7031, pp. 714–729, 2011.

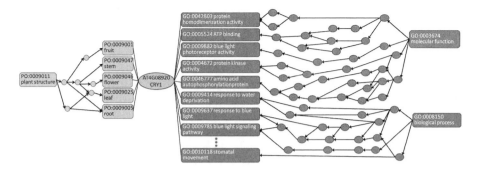

Fig. 1. GO and PO annotations for gene CRY1

This has lead to a rich annotation graph dataset [8]. Semantic Web research has laid the groundwork for research in link prediction and pattern discovery in the context of annotation graph datasets as discussed next.

1.1 Motivating Example

Arabidopsis thaliana is a model organism and TAIR http://www.arabidopsis.org/ is the primary source of annotated data for Arabidopsis genes. Each gene in TAIR is marked up with CV terms from the Gene Ontology and from the Plant Ontology. The resulting tripartite annotation graph (TAG) is illustrated in Figure 1 where we visualize the annotations for gene CRY1; PO terms are on the left and GO terms are on the right of CRY1. The TAG has been enhanced to include relevant fragments of the GO and PO ontologies. As of October 2010 there were 18 GO and 36 PO annotations for CRY1. The figure illustrates partial annotations (due to lack of space). The annotations can be represented using an RDF class gene_GO_PO_TAGtriplet as follows:

    t1: (TAGtripletID rdf:type gene_GO_PO_TAGtriplet)
    t2: (TAGtripletID gene_ID name-of-gene)
    t3: (TAGtripletID GO_ID uri-of-GO-CV-term)
    t4: (TAGtripletID PO_ID uri-of-PO-CV-term)

A scientist is typically interested in a set of genes of interest within a biological context, e.g., flowering time genes or photomorphogenesis genes. Given the resulting large annotation graph dataset, the scientist would like to be presented with interesting patterns. For photomorphogenesis, a pattern may correspond to the following 4 TAG triplets for CRY2 and PHOT1; note that we use a comma separated representation (gene, GO CV term, PO CV term), instead of the RDF triples for ease of presentation and due to space constraints:

$TAGtripletT_1$: (CRY2, GO_5773:vacuole, PO_13:cauline leaf)
$TAGtripletT_2$: (CRY2, GO_5773:vacuole, PO_37:shoot apex)
$TAGtripletT_3$: (PHOT1, GO_5773:vacuole, PO_13:cauline leaf)
$TAGtripletT_4$: (PHOT1, GO_5773:vacuole, PO_37:shoot apex)

Subsequently, she will explore the literature to understand the evidence. PHOT1 and CRY2 belong to two different groups of blue light receptors, namely phototropins (PHOT1) and cryptochromes (CRY2). To date there has been *no evidence reported in the literature that confirm any interactions* between these 2 groups. A literature search identified 2 independent studies of interest [11,24] that provide some background evidence. The set of 4 TAG triplets, in conjunction with the 2 studies, may lead her to design a set of bench experiments to validate the potential interaction in the vacuole between CRY2 and PHOT1.

1.2   Challenges and Contributions

While scientists are interested in complex patterns, in this paper, we examine a simpler task of link prediction. We predict edges between genes and GO CV terms or edges between genes and PO CV terms. We briefly summarize the challenges of link prediction for the annotation graph datasets. First, the TAG is a layered graph. Layered graphs impose restrictions on the link prediction process, e.g., the neighborhoods of two nodes in neighboring layers are disjoint and only edges between neighboring layers should be predicted. This restriction makes many popular prediction approaches ineffective as will be discussed.

The next challenge is the a heterogeneity of biological knowledge. As seen in the previous example, a set of gene_GO_PO_TAGtriplets forms a complex and interesting cross-ontology pattern. The GO ontology is focused on universal biological processes, e.g., DNA binding. It does not capture organism-specific processes, e.g., leaf development. The PO ontology is designed to capture such organism specific knowledge. Thus, a gene_GO_PO_TAGtriplet, or a complex pattern of multiple triplets, may be used to determine when a plant specific biological phenomenon has a relationship with a ubiquitous biological process.

A related challenge is identifying an area or subgraph of the dataset to make predictions or find patterns. Ontologies capture multiple relationship types between CV terms that can be exploited for prediction. GO supports multiple relationship types including *is_a*, *part_of* and *regulates*. From Figure 1, the GO CV term blue light photoreceptor activity is *part_of* blue light signaling pathway which *is_a* cellular response to blue light which *is_a* response to blue light. CRY1 is annotated with blue light photoreceptor activity and response to blue light. PO has relationship types *is_a*, *part_of* and *develops_from*. Our challenge is to restrict the patterns of gene_GO_PO_TAGtriplets so that they favor GO CV terms (or PO CV terms) that are closely related.

Our observation is that the edges of each relationship type are not uniformly distributed across the ontology structure. For GO, the edges of type *is_a* are dominant, and thus all the edges of any path in GO are more likely to be of this type. The edges relevant to regulation are more densely placed in specific areas of the ontology; thus, an edge of this type also has a greater probability that an adjacent edge is of the same type. For PO, while neither *is_a* nor *part_of* dominate, the edge distribution of these types are similarly concentrated so that an edge of one type is more likely to have an adjacent edge of the same type.

Based on these observations, our first attempt at prediction will use the *topological shortest path distance on undirected graphs*, between 2 CV terms, as a proxy for relatedness. We note that this path length metric is affected by both human annotation patterns and the ontology structure representing biological knowledge, e.g., the depth of the tree along any branch. We will consider the impact of the GO (PO) relationship type(s) on the path based distance metric in future research. Our link prediction framework relies on 2 complementary approaches. Graph summarization (GS) is a minimum description length (MDL) encoding that represents a graph with a *signature* and *corrections*. Such a representation is intuitive for both explanation and visualization. Since annotation graph datasets may be large and sparse, high quality predictions must rely on finding good candidate regions or subgraphs. Dense subgraphs (DSG) is a methodology to find such regions that include clique-like structures, i.e., cliques with missing edges. Variations of the dense subgraph whose nodes satisfy some *distance restriction* is also useful to ensure possible relatedness of the CV terms. Our research makes the following contributions:

- We develop a prediction framework that can be used for both unsupervised or supervised learning. We focus on unsupervised learning in this paper. We perform an extensive evaluation on the annotation graph of TAIR.
- Our evaluation illustrates the benefit of the DSG and the *distance restriction* to identify a potential subgraph so as to increase prediction accuracy. We further show that high values of the scoring function, or predicted edges with high confidence, are correlated with increasing prediction accuracy.

Due to space limitations, our examples only involve TAGs; however, our prediction framework is not limited to TAGs. We have applied our framework to a layered graph of 5 layers; beyond 5 layers, we are unclear if the patterns and predictions will be meaningful. We are also studying the clinical trial dataset; this is a star graph with a clinical trial having links to PubMed publications, MeSH terms, (disease) conditions, interventions (drugs or treatments), etc.

1.3 Related Work

Semantic Web research has addressed information integration using ontologies and ontology alignment [9,25]. There are also multiple projects and tools for annotation, e.g., Annotea/Amaya [10] and OntoAnnotate [20].

Graph data mining covers a broad range of methods dealing with the identification of (sub)structures and patterns in graphs. Popular techniques are, amongst others, graph clustering, community detection and finding cliques. Our work builds upon two complementary graph methods: graph summarization [23] and dense subgraphs [27]. To the best of our knowledge, we are the first to consider the synergy of these two approaches.

Link prediction is a subtask of link mining [21]; prediction in bipartite and tripartite graphs is also of interest [15,26]. Prediction methods can be supervised or unsupervised. Supervised link prediction methods (e.g., [1,7,26]) utilize training and test data for the generation and evaluation of a prediction model. Unsupervised link prediction in graphs is a well known problem [18]. There are two types of approaches: methods based on node neighborhoods and methods using the ensemble of all paths between two nodes. We discuss their disadvantages for tripartite graphs in Section 3. Many approaches for predicting annotations in the biological web are available [3,5,17]. The AraNet system [17] predicts GO functional annotations for Arabidopsis using a variety of biological knowledge; details are discussed with our evaluation in Section 4.

2   Problem Definition

A **tripartite annotation graph (TAG)** is an undirected layered tripartite graph $G = ((A, B, C), (X, Y))$ with three pairwise disjoint sets of nodes A, B, and C and two sets of edges $X \subseteq A \times B$ and $Y \subseteq C \times B$. Figure 2 shows an example of a TAG. For example, in the TAIR annotated graph, the node sets A, B, and C correspond to POs, genes, and GOs, respectively. The sets of edges then reflect gene annotations using POs (X) and GOs (Y).

We study the **link prediction** problem for TAGs. Given a TAG G at time t_1 and a future time t_2, we assume that edges will be added during the transition from the original graph G_1 to the new graph G_2, i.e., $G_1 = ((A, B, C), (X, Y))$

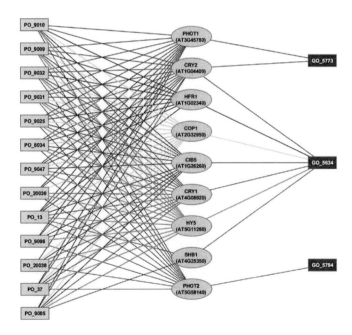

Fig. 2. Example of a TAG $G = ((A, B, C), (X, Y))$ with $|A| = 13$ PO nodes, $|B| = 9$ genes, and $|C| = 3$ GO nodes. The nodes are connected by $|X| = 98$ edges of type $A \times B$ and $|Y| = 10$ edges of type $C \times B$.

and $G_2 = ((A, B, C), (X \cup X_{new}, Y \cup Y_{new}))$. The goal of link prediction is to infer the set of new edges based on the original graph G_1 only. Ideally the **predicted edges** $P_X(G)$ and $P_Y(G)$ are the added edges, i.e., $P_X(G) = X_{new}$ and $P_Y(G) = Y_{new}$.

For a given TAG $G = ((A, B, C), (X, Y))$ we refer to X and Y as the set of observed edges. We call all other possible edges, i.e., $((A \times B) - X) \cup ((C \times B) - Y)$ **potential edges**. Predicted edges $P_X(G)$ and $P_Y(G)$ and new edges are subsets of the corresponding potential edges.

Note that we consider only edge additions and we do not consider node additions for the transition from G_1 to G_2. In biological terms, we plan to use prior annotations to existing PO and GO nodes in G_1 to predict new edges in G_2. We are not attempting to predict new annotations to new PO or GO nodes that do not occur in G_1.

3   Approach

Unsupervised link prediction in graphs is a well known problem, e.g., see [18] for a survey on link prediction approaches in social networks. Basically there are two types of approaches. Neighborhood-based approaches consider the sets of node neighbors $N(a)$ and $N(b)$ for a potential edge (a, b) and determine a prediction probability based on the (relative) overlap of these two sets. Methods based on the ensemble of all paths aggregate all paths from a through b to a combined prediction score. Shorter paths usually have a higher impact than longer paths and the more paths exist the higher the score will be.

Unfortunately, these types of approaches are not suited to TAGs. Neighborhood-based approaches will even fail for TAGs because the sets $N(a)$ and $N(b)$ are disjoint. Given a tripartite graph $G = ((A, B, C), (X, Y))$ and a potential edge (a, b) with $a \in A$ and $b \in B$, the node neighbors of a are in B ($N(a) \subseteq B$) and b's neighbors are in A ($N(b) \subseteq A$) and therefore $N(a) \cap N(b) = \emptyset$. On the other hand, path-based approaches are in general applicable for tripartite graphs but will produce similar prediction scores for many potential edges due to the structure of a tripartite graphs for two reasons. First, the minimal path length for a potential edge (a, b) equals 3 because there are only two possible path types $(a \to b' \to c' \to b)$ or $(a \to b' \to a' \to b)$. Second, most potential

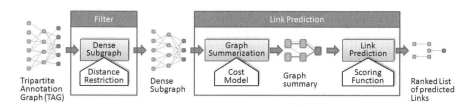

Fig. 3. The proposed link prediction framework combines graph summarization with link prediction functions. The original TAG can be subject to an optional filter step to identify dense subgraphs.

edges will have multiple paths with length 3 because it is very likely in the anno-tated biological web that any two genes b and b' have (at least) one GO (a') or PO (c') in common. Furthermore path-based approaches are not able to benefit from the rich ontology knowledge because they do not distinguish paths between the three layers (GO/genes/PO) and paths within the ontologies (PO, GO).

We therefore propose a different approach that employs graph summarization that transforms a graph into an equivalent compact graph representation using super nodes (groups of nodes) and super edges (edges between super nodes). The summary reflects the basic pattern (structure) of the graph and is accompanied by a list of corrections, i.e., deletions and additions, that express differences between the graph and its simplified pattern. The idea of our link prediction approach is that adding predicted edges reinforces the underlying graph pattern, i.e., predicted edges are the missing building blocks for existing patterns.

Figure 4 illustrates a possible summarization of the graph shown in Figure 2. The utilization of a graph summary has several advantages. First, the summary gives a better understanding of the overall structure of the underlying graph and may itself be used for visualization. Second, the corrections, foremost dele-tions, are intuitive indicators for edge prediction. Third, the summary captures semantic knowledge not only about individual nodes and their connections but also about groups of related nodes.

Figure 3 illustrates the overall scheme of our approach. The input is a TAG G and the output is a ranked list of predicted edges. Our approach consists of three consecutive steps. The first step is optional and deals with the identification of dense subgraphs, i.e., highly connected subgraphs of G like (almost) cliques. The goal is to identify interesting regions of the graph by extracting a relevant subgraph. Next, graph summarization transforms the graph into an equivalent compact graph representation using super nodes (groups of nodes) and super edges (edges between super nodes). The summarized graph is then input to the last step. A prediction function computes prediction scores for potential edges and returns a ranked list. Our approach is not limited to TAGs. A K-partite layered graph can be first converted to a more general (bi-partite) graph before creating a DSG and applying graph summarization.

3.1   Dense Subgraphs

Given an initial tripartite graph, a challenge is to find interesting regions of the graph, i.e., candidate subgraphs, that can lead to accurate predictions. We commence with the premise that an area of the graph that is rich or dense with annotation is an interesting region to identify candidate subgraphs. For example, for a set of genes, if each is annotated with a set of GO terms and/or a set of PO terms, then the set of genes and GO terms, or the set of genes and PO terms, form a clique. We thus exploit cliques, or dense subgraphs (DSG) representing cliques with missing edges.

Density is a measure of the connectedness of a subgraph; it is the ratio of the number of induced edges to the number of vertices in the subgraph. Even though there are an exponential number of subgraphs, a subgraph of maximum density can be found in polynomial time [16,6,4]. In contrast, the maximum clique problem to find the subgraph of largest size having all possible edges is NP-hard; it is even NP hard to obtain any non-trivial approximation. Finding densest subgraphs with additional size constraints is NP hard [13]; yet, they are more amenable to approximation than the maximum clique problem.

Recall that our annotation graph is a tripartite graph $G = ((A, B, C), (X, Y))$. We employ our approach in [27] and thus first transform the tripartite graph G in the form of a bipartite graph $G' = (A, C, E)$ between the two sets A and C of outer nodes in G. The bipartite graph is a weighted graph where each edge $e = (a, c) \in E$ is labeled with the number of nodes $b \in B$ that have links to a and c in the tripartite graph, i.e., $(a, b) \in X$ and $(c, b) \in Y$. We then compute a densest bipartite subgraph $G'_{dense} = (A', C', E)$ by choosing subsets $A' \subset A$ and $C' \subset C$ to maximize the density of the subgraph, which is defined as $\frac{w'(E)}{|A|+|C|}$. Here $w'(E)$ denotes the weight of the edges in the subgraph induced by E. Finally, we build the dense tripartite graph G_{dense} out of the computed dense bipartite graph G'_{dense} by adding all intermediate nodes $b \in B$ that are connected to at least one $a \in A'$ or $c \in C'$.

An interesting variation on the DSG includes a **distance restriction** according to the ontology of nodes. In the annotated biological web (see Figure1) nodes from PO and GO are hierarchically arranged to reflect their relationships (e.g., is-a or part-of). Assume we are given a distance metric d_A (d_C) that specifies distances between pairs of nodes in set A (C). We are also given distance thresholds $\tau_A(\tau_C)$. The goal is to compute a densest subgraph G'_S that ensures that for all node pairs of A (C) are within a given distance. For any pair of vertices $a_1, a_2 \in A_S$ we have $d_A(a_1, a_2) \leq \tau_A$, and the same condition holds for pairs of vertices in C_S, namely that for all $c_1, c_2 \in C_S$ we have $d_C(c_1, c_2) \leq \tau_C$. We will evaluate the influence of a distance restriction in Section 4.

The distance restricted DSG algorithm calls a routine with complexity $O(n^3 \cdot log(n))$, where n is the number of nodes in a valid distance-restricted subgraph; it is called once for each pair of nodes in A, and for each pair in C. We have also implemented a linear time greedy 2-approximation to DSG that greatly outperforms our previous running time results reported in [27]; this solution was previously reported in [4,14].

3.2   Graph Summarization

We start with the intuition that a summary of a tripartite graph is also a graph. The summary must however include a compact representation that can be easily visualized and that can be used for making predictions. While there are many methods to summarize graphs, we focus on the graph summarization (GS) approach of [22,23]. Their graph summary is an aggregate graph comprised of a signature and corrections. It is the first application of minimum description length (MDL) principles to graph summarization and has the added benefit of providing intuitive course-level summaries that are well suited for visualization and link prediction.

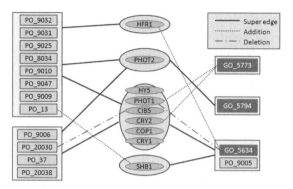

Fig. 4. Possible summary of the graph in Figure 2. The summary has 9 supernodes, 8 superedges, two deletions (PO_20030, CIB5) and (PHOT1, GO_5643) and 6 additions.

A **graph summary** of a graph $G = ((A, B, C), (X, Y))$ consists of a graph **signature** $\Sigma(G)$ and a set of **corrections** $\Delta(G)$. The graph signature is defined as follows: $\Sigma(G) = ((S_{AC}, S_B), S_{XY})$. The sets S_{AC} and S_B are disjoint partitionings of $A \cup C$ and B, respectively, that cover all elements of these sets. Each element of S_{AC} or S_B is a **super node** and consists of one or more nodes of the original graph. Elements of S_{XY} are called **super edges** and they represent edges between super nodes, i.e., $S_{XY} \subseteq S_{AC} \times S_B$. The second part of a summary is the sets of edge additions and deletions $\Delta(G) = (S_{add}, S_{del})$. All edge additions are edges of the original graph G, i.e., $S_{add} \subseteq X \cup Y$. Deletions are edges between nodes of G that do not have an edge in the original graph, i.e., $S_{Del} \subseteq ((A \cup C) \times B) - (X \cup Y)$. Figure 4 depicts a possible summarization of the graph shown in Figure 2.

The summarization algorithms makes sure that $G \equiv (\Sigma(G), \Delta(G))$, i.e., the original graph G can be reconstructed based on the graph summary and the edge corrections $\Delta(G)$. The nodes A, B, and C are "flattened" sets of S_{AC} and S_B, respectively. A super edge between two super nodes $s_{AC} \in S_{AC}$ and $s_B \in S_B$ represents the set of all edges between any node of s_{AC} and any node of s_B. The original edges can therefore be reconstructed by computing the Cartesian product of all super edges with consideration of edge corrections $\Delta(G)$. For example, X is therefore $X = \{(a, b) | a \in A \wedge b \in B \wedge a \in s_{AC} \wedge b \in s_B \wedge (s_{AC}, s_B) \in (S_{XY} \cup S_{add} - S_{del})\}$.

Graph summarization is based on a two-part minimum description length encoding. We use a greedy agglomerative clustering heuristic. At first, each node belongs to its own supernode. Then, in each step, the pair of supernodes are merged that result in the greatest reduction in representation cost. When the cost of merging any pair becomes negative, the algorithm naturally terminates. There are no parameters or thresholds to set. The complexity of the original GS problem is currently unknown. However, if nodes are allowed to belong to more than one super node (i.e., overlapping supernodes), the problem reduces to finding the maximum clique in a graph, which is NP-hard.

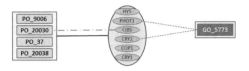

Fig. 5. Detail of Figure 4 for computing prediction scores for potential edges (PO_20030, CIB5) and (GO_5773, CIB5)

The possible summaries of a graph will depend on the **cost model** used for an MDL encoding. In general, the cost model is a triple (α, β, γ) that assigns weights to the number of superedges, deletions, and additions, respectively. Graph summarization looks for a graph summary with a minimal cost of $C(G) = \alpha \cdot |S_{XY}| + \beta \cdot |S_{add}| + \gamma \cdot |S_{del}|$. A simple cost model that gives equal weight to supernodes, superedges and corrections was used in [23] and was used to produce Figure 4.

GS has time complexity $O(d_{av}^3 \cdot (d_{av} + log(n) + log(d_{av})))$, where d_{av} is the average degree of the nodes [23]. The average degree in our datasets is low so average running time is low.

3.3 Prediction Function

A **prediction function** is a function $p : e \mapsto s \in [0, 1]$ that maps each potential edge e of a TAG to a real value between 0 and 1. This value s is called **prediction score**. The function p can be used for ranking all possible edges according to their probability. Consider the graph summary $\Sigma(G)$; let s_{AC} and s_B be the corresponding super nodes of e. Note that this does not imply the existence of an super edge between s_{AC} and s_B. The prediction score for an edge $e \in ((A \cup C) \times B) - (X \cup Y)$ is defined as $p(e) = s(e) \cdot c(e)$ and combines a so-called **supernode factor** $s(e)$ and a **correction factor** $c(e)$. The supernode factor is defined as follows:

$$s(e) = \begin{cases} 1 - \dfrac{|s_{AC} \times s_B \cap S_{del}|}{|s_{AC}| \cdot |s_B|} & \text{if } e \in S_{del} \\[3mm] \dfrac{|s_{AC} \times s_B \cap S_{add}|}{|s_{AC}| \cdot |s_B|} & \text{otherwise} \end{cases}$$

For $e \in S_{del}$ the graph summary contains a super edge between s_{AC} and s_B. The supernode factor determines the fraction of missing edges between the two super nodes. The larger the super nodes and the smaller the number of deletions are, the higher is the supernode factor. On the other hand, $e \notin S_{del}$ implies that there is no super edge between s_{AC} and s_B. The supernode factor then reflects the fraction of additions in all possible edges between these two supernodes. The larger the super nodes and the smaller the number of additions are, the lower is the supernode factor. The correction factor for an edge $e = (a, b)$ is as follows:

$$c(e) = \frac{1}{1 + |S_{corr}(a)|} \cdot \frac{1}{1 + |S_{corr}(b)|}$$

Here $S_{corr}(a)$ and $S_{corr}(b)$ describe the set of corrections involving a and b, respectively, i.e., $S_{corr}(a) = \{b'|b' \neq b \wedge (a, b') \in S_{del} \cup S_{add}\}$ and $S_{corr}(b) = \{a'|a' \neq a \wedge (a', b) \in S_{del} \cup S_{add}\}$. The correction factor accounts for the number of corrections that are relevant to a given edge. The higher the number of corrections, the smaller the correction factor, and thus, the prediction score.

Figure 5 shows the relevant part of the example summarization of Figure 4 for potential edges (PO_20030, CIB5) and (GO_5773, CIB5). The deletion (PO_20030, CIB5) is the only deletion between the two supernodes and the size of the supernodes are 4 and 6, respectively. For (GO_5773, CIB5) there are two additions between the corresponding supernodes of size 1 and 6, respectively. The supernode factors are therefore calculated as follows: $s(\text{PO\_20030}, \text{CIB5}) = 1 - \frac{1}{4 \cdot 6} = \frac{23}{24}$ and $s(\text{GO\_5773}, \text{CIB5}) = \frac{2}{1 \cdot 6} = \frac{1}{3}$. The correction factors for the two example edges are: $c(\text{PO\_20030}, \text{CIB5}) = \frac{1}{1+0} \cdot \frac{1}{1+0} = 1$ and $c(\text{GO\_5773}, \text{CIB5}) = \frac{1}{1+2} \cdot \frac{1}{1+1} = \frac{1}{6}$. Finally, the overall prediction scores are: $p(\text{PO\_20030}, \text{CIB5}) = \frac{23}{24} \cdot 1 \approx 0.96$ and $p(\text{GO\_5773}, \text{CIB5}) = \frac{1}{3} \cdot \frac{1}{6} \approx 0.06$. In other words, the edge (PO_20030, CIB5) seems to be a good prediction whereas edge (GO_5773, CIB5) does not.

4   Experimental Evaluation

4.1   Dataset Preparation

The Arabidopsis Information Resource (TAIR) consists of *Arabidopsis thaliana* genes and their annotations with terms in the Gene Ontology (GO) and Plant Ontology (PO). The entire TAIR dataset includes 34,515 genes, with 201,185 annotations to 4,005 GO terms and 529,722 annotations to 370 PO terms circa October 2010. We created three subsets labeled ds1, ds2 and ds3, respectively. Each dataset was constructed by choosing 10 functionally related genes associated with photomorphogenesis, flowering time and photosynthesis, respectively, and expanding the graph to include all GO and PO terms. The statistics of these 3 dataset are shown in Table 6. Recall that we use the *shortest path distance* between a pair of CV terms as a proxy for relatedness. To test the distance restriction we create subgraphs ds1-DSG, etc. The impact of the distance restriction will be discussed in a later section.

4.2   Evaluation Methodology

We use a simple leave-K-out strategy to evaluate our link prediction approach. Given a dataset, we remove 1 (up to K) edges that are selected at random from the set of all edges. We then predict 1 (up to K) edges.

We report on precision. We consider precision at the Top 1 or P@1 when we predict 1 edge and mean average precision (MAP) when we predict K edges [19].

To further study the quality of our prediction, we report on the scores produced by our scoring function. For those predictions in which we have the *highest confidence*, i.e., those predictions are consistently above a threshold of the scoring function, we report on the true positives (TP) and false positives (FP). A TP is a correct prediction while a FP is an incorrect prediction.

Subset	ds1	ds2	ds3	ds1+ DSG	ds2+ DSG	ds3+ DSG
Genes	10	10	10	7	10	10
GO Terms	68	44	28	14	4	8
PO Terms	44	53	48	22	11	24
Total Nodes	122	108	86	43	25	42
Annotations	395	355	426	159	123	246
Density	3.24	3.29	4.95	3.70	4.92	5.86

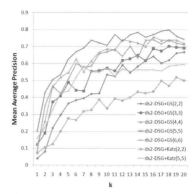

Fig. 6. Statistics of the 3 datasets along with their dense subgraphs.

Fig. 7. MAP of predicting k annotations in ds2 dense subgraphs. Distance restrictions are (GO Distance, PO Distance).

As a baseline, we compute the Katz metric between any 2 pair of nodes [12]. The Katz metric is a path based measure equal to $\sum_{l=1}^{\infty} \beta^\ell \cdot paths_\ell(x, y)$, where $paths_\ell(x, y)$ is the number of paths between nodes x and y of length ℓ. For our experiments, we used $\beta = .005$. All potential edges were ranked and sorted by the value of the Katz metric, creating a ranked list of predictions. This is labeled `dsi-Katz` or `dsi-DSG-Katz` where the prefix `dsi` identifies the dataset.

Three following variations of our prediction approach were considered:

- `dsi-GS`: The prefix represents the dataset and the suffix indicates that there was no DSG created and we only used graph summarization.
- `dsi:DSG+GS`: We created a DSG with no distance restrictions.
- `dsi:DSG+GS(dP,dG)`: We created a DSG with a distance restriction of `dP` for PO and a distance restriction of `dG` for GO.

We note that the DSG with no distance restriction results in the densest subgraph. Imposing a distance restriction may result in a less dense subgraph, but possibly one with greater biological meaning. The cost model for graph summarization is another experimental parameter, but one that we did not vary in our experiments. Equal weights were given to supernodes, superedges and corrections throughout all of our summarizations.

AraNet [17] created an extensive functional gene network for Arabidopsis exploiting pairwise gene-gene linkages from 24 diverse datasets representing > 50 million observations. They report on prediction accuracy of GO *biological process* CV terms for over 27,000 genes. Their prediction method computes a score for each gene and association using its neighborhood and naive Bayes estimation; this is similar in spirit to the Katz metric. Their results demonstrate that for over 55% of gene annotation, their predictions (cumulative likelihood ratio) were more significant compared to random prediction. A direct comparison of our approach with AraNet was not possible since AraNet exploits significant knowledge beyond GO and PO annotations. We note that the mean average

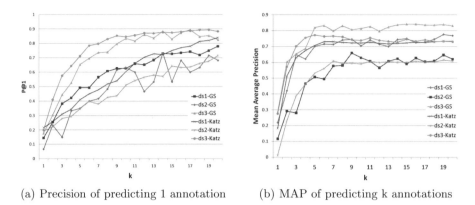

(a) Precision of predicting 1 annotation   (b) MAP of predicting k annotations

Fig. 8. Evaluation using different approaches across all datasets

precision (MAP) for our method and Katz reflect that the prediction accuracies of all three methods appear to occur in a similar range; this is notable since Katz and our method exploit only PO and GO annotation data.

4.3   Summary of Results

Baseline Analysis. Given a ranked list of predictions, precision at one (P@1) provides a useful metric for evaluating the performance of the different approaches. To establish a baseline, Figure 8(a) reports on P@1 for the 3 datasets for `dsi-GS` and `dsi-Katz`. The P@1 values are low for lower K values and increase with higher K. This is expected since larger K provides a larger ground truth and improves prediction accuracy. Both methods perform best on `ds3` and show the worst prediction accuracy on `ds2`. A visual examination of the datasets and the graph summary intuitively illustrates the difference in performance across the 3 datasets. For example, `ds3` is the most dense dataset.

To complete the baseline analysis, we consider the Top K predictions as we leave out K. Figure 8(b) reports on the mean average precision (MAP) of the different approaches as a function of K. As expected MAP for Top K is higher than the values for P@1 since we are making K predictions (and not 1 prediction as before). We note that `ds3-GS` outperforms `ds3-Katz`. Both methods show the least prediction accuracy for `ds2`.

Impact of Varying the Distance Restriction. The average distance between a pair of GO CV terms in ds2 is 5.57. Of the 946 pairs, 402 are within distance 5 of each other; this is the distance restriction used in our previous experiments. 160 pairs are not connected at all, i.e., they are in different parts of the ontology.

Figure   7 reports on MAP for `dsi-DSG+GS` and `dsi-DSG+Katz` for varying PO and GO restrictions on dataset `ds2`. Method `dsi-DSG+GS` dominates `dsi-DSG+Katz` *over all distance restrictions*. This is a very strong validation of the prediction accuracy of our approach. Accuracy initially increases with increasing (PO,GO) distance. The best accuracy was obtained with (5,5) after

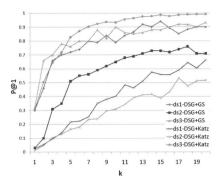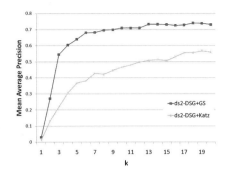

(a) Precision at 1 after removing k edges   (b) MAP of k annotations after removing
from a subgraph   k edges from a subgraph

Fig. 9. Comparison of our graph summarization approach with the Katz metric

which accuracy decreases, e.g., for a (6,6) distance restriction. Figure 9(a) reports on P@1 for `dsi-DSG+GS` and `dsi-DSG+Katz` for the 3 datasets with distance
(5,5). Method `dsi-DSG+GS` dominates `dsi-DSG+Katz` for `ds1` and `ds2`. Surprisingly `ds3-DSG+Katz` outperforms `ds3-DSG+GS` for `ds3`. An examination of the
predictions indicates that `ds3-DSG+GS` makes an incorrect prediction which has
a high prediction score and is therefore ranked high. Since Figure 9(a) reports
on P@1, this has a significant penalty on the accuracy of `ds3-DSG+GS`.

Figure 9(b) reports on the mean average precision (MAP) for `ds2` with distance restriction (5,5). Again, `ds3-DSG+GS` outperforms `ds3-DSG+Katz`, further
confirming the strength of our approach.

Confidence in Predictions. Our final experiment is to validate that high
confidence predictions result in more accurate predictions. High confidence predictions are those that receive a high prediction score. Table 1 reports on the
percentage of true positive (TP) and false positives (FP) for `ds2-Katz`, `ds2-GS`
and `ds2-DSG+GS`, bucketized by the range of prediction score. Note that for Katz,
we normalize the score from 0.0 to 1.0 prior to bucketization. The values on the
left represent the high confidence (high score) prediction buckets and the confidence (score) decreases as we move to the right. As expected, the % of TP values
is greater than the % of FP values for high confidence buckets. The reverse is
true for low confidence buckets. This holds for all the methods. Further, the %
TP values for `ds2-DSG+GS` for the 2 left most buckets, 80% and 89%, dominates
the % TP values of `ds2-GS` (50% and 42%) and `ds2-Katz` (50% and 51%). The
% TP values for `ds2-DSG+GS` is overall higher than the other two methods except for one exception (score between 0.7 to 0.8). These results confirm that
`ds2-DSG+GS` had both higher confidence scores and higher prediction accuracy,
compared to `ds2-GS` and `ds2-Katz`. This held across all 3 datasets and further
validates our prediction approach.

Table 1. Percentage of true positives (TP) and false positives (FP) as a function of the prediction score for predictions made on ds2 with DSG+GS(5,5), GS, and Katz

Score	<1.0	<.90	<.80	<.70	<.60	<.50	<.40	<.30	<.20	<.10
TP ds2-Katz	50	51	77	18	11	4	1	3	6	0
FP ds2-Katz	50	49	23	82	89	96	99	97	94	100
TP ds2-GS	50	42	35	37	38	40	18	3	1	1
FP ds2-GS	50	58	65	63	62	60	82	97	99	99
TP ds2-DSG+GS	80	89	61	61	56	51	18	18	18	25
FP ds2-DSG+GS	20	11	39	39	44	49	82	82	82	75

5   Conclusions and Future Work

We presented a novel approach for link prediction in the layered annotation graph datasets that employs graph summarization for link prediction. Furthermore, the complementary method of identifying dense subgraphs helps find interesting regions for high quality predictions. To the best of our knowledge, we are the first to consider the synergy of these two approaches. Future work includes learning GS cost models using supervised learning.

References

1. Benchettara, N., Kanawati, R., Rouveirol, C.: Supervised machine learning applied to link prediction in bipartite social networks. In: Proc. ASONAM, pp. 326–330 (2010)
2. Bizer, C., Heath, T., Berners-Lee, T.: Linked data - the story so far. International Journal on Semantic Web and Information Systems 5(3), 1–22 (2009)
3. Bogdanov, P., Singh, A.K.: Molecular Function Prediction Using Neighborhood Features. IEEE/ACM Trans. Comput. Biology Bioinform. 7(2), 208–217 (2010)
4. Charikar, M.: Greedy Approximation Algorithms for Finding Dense Components in a Graph. In: Jansen, K., Khuller, S. (eds.) APPROX 2000. LNCS, vol. 1913, pp. 84–95. Springer, Heidelberg (2000)
5. Chua, H.N., Sung, W.-K., Wong, L.: An efficient strategy for extensive integration of diverse biological data for protein function prediction. Bioinformatics 23(24), 3364–3373 (2007)
6. Goldberg, A.V.: Finding a maximum density subgraph. Technical Report UCB/CSD-84-171, EECS Department, University of California, Berkeley (1984)
7. Hasan, M.A., Chaoji, V., Salem, S., Zaki, M.: Link Prediction Using Supervised Learning. In: Proc. on Link Analysis, Counterterrorism and Security (2006)
8. Hassanzadeh, O., et al.: Linkedct: A linked data space for clinical trials. In: Proc. WWW 2009 Workshop on Linked Data on the Web, LDOW 2009 (2009)
9. Jain, P., Yeh, P., Verma, K., Vasquez, R., Damova, M., Hitzler, P., Sheth, A.: Contextual Ontology Alignment of LOD with an Upper Ontology: A Case Study With Proton. In: Antoniou, G., Grobelnik, M., Simperl, E., Parsia, B., Plexousakis, D., De Leenheer, P., Pan, J. (eds.) ESWC 2011, Part I. LNCS, vol. 6643, pp. 80–92. Springer, Heidelberg (2011)
10. Kahan, J., Koivunen, M.: Annotea: an open rdf infrastructure for shared web annotations. In: Proc. of the WWW, pp. 623–632 (2001)

11. Kang, B., Grancher, N., Koyffmann, V., Lardemer, D., Burney, S., Ahmad, M.: Multiple interactions between cryptochrome and phototropin blue-light signalling pathways in arabidopsis thaliana. Planta 227(5), 1091–1099 (2008)
12. Katz, L.: A new status index derived from sociometric analysis. Psychometrika 18, 39–40 (1953)
13. Khuller, S., Saha, B.: On Finding Dense Subgraphs. In: Albers, S., Marchetti-Spaccamela, A., Matias, Y., Nikoletseas, S., Thomas, W. (eds.) ICALP 2009. LNCS, vol. 5555, pp. 597–608. Springer, Heidelberg (2009)
14. Kortsarz, G., Peleg, D.: Generating sparse 2-spanners. J. Algorithms 17(2), 222–236 (1994)
15. Kunegis, J., De Luca, E., Albayrak, S.: The Link Prediction Problem in Bipartite Networks. In: Hüllermeier, E., Kruse, R., Hoffmann, F. (eds.) IPMU 2010. LNCS, vol. 6178, pp. 380–389. Springer, Heidelberg (2010)
16. Lawler, E.: Combinatorial optimization - networks and matroids. Holt, Rinehart and Winston, New York (1976)
17. Lee, I., Ambaru, B., Thakkar, P., Marcotte, E., Rhee, S.: Rational association of genes with traits using a genome-scale gene network for arabidopsis thaliana. Nature Biotechnology (28), 149–156 (2010)
18. Liben-Nowell, D., Kleinberg, J.M.: The link-prediction problem for social networks. Journal of the American Society for Information Science and Technology (JASIST) 58(7), 1019–1031 (2007)
19. Manning, C., Raghavan, P., Schutze, H.: Introduction to Information Retrieval. Cambridge University Press (2008)
20. Mir, S., Staab, S., Rojas, I.: An Unsupervised Approach for Acquiring Ontologies and RDF Data from Online Life Science Databases. In: Aroyo, L., Antoniou, G., Hyvönen, E., ten Teije, A., Stuckenschmidt, H., Cabral, L., Tudorache, T. (eds.) ESWC 2010, Part II. LNCS, vol. 6089, pp. 319–333. Springer, Heidelberg (2010)
21. Namata, G.M., Sharara, H., Getoor, L.: A Survey of Link Mining Tasks for Analyzing Noisy an Incomplete Networks. In: Philip, J.H., Yu, S.S., Faloutsos, C. (eds.) Link Mining: Models, Algorithms, and Applications. Springer, Heidelberg (2010)
22. Navlakha, S., Kingsford, C.: Exploring biological network dynamics with ensembles of graph partitions. In: Proc. 15th Intl. Pacific Symposium on Biocomputing (PSB), vol. 15, pp. 166–177 (2010)
23. Navlakha, S., Rastogi, R., Shrivastava, N.: Graph summarization with bounded error. In: Proc. of Conference on Management of Data, SIGMOD (2008)
24. Ohgishi, M., Saji, K., Okada, K., Sakai, T.: Functional analysis of each blue light receptor, cry1, cry2, phot1, and phot2, by using combinatorial multiple mutants in arabidopsis. Proc. of the National Academy of Sciences 1010(8), 2223–2228 (2004)
25. Parundekar, R., Knoblock, C., Ambite, J.: Linking and Building Ontologies of Linked Data. In: Patel-Schneider, P.F., Pan, Y., Hitzler, P., Mika, P., Zhang, L., Pan, J.Z., Horrocks, I., Glimm, B. (eds.) ISWC 2010, Part I. LNCS, vol. 6496, pp. 598–614. Springer, Heidelberg (2010)
26. Pujari, M., Kanawati, R.: A supervised machine learning link prediction approach for tag recommendation. In: Proc. of HCI (2011)
27. Saha, B., Hoch, A., Khuller, S., Raschid, L., Zhang, X.-N.: Dense Subgraphs with Restrictions and Applications to Gene Annotation Graphs. In: Berger, B. (ed.) RECOMB 2010. LNCS, vol. 6044, pp. 456–472. Springer, Heidelberg (2010)

QueryPIE: Backward Reasoning for OWL Horst over Very Large Knowledge Bases

Jacopo Urbani, Frank van Harmelen, Stefan Schlobach, and Henri Bal

Department of Computer Science, Vrije Universiteit Amsterdam
{j.urbani,frank.van.harmelen,schlobac,he.bal}@few.vu.nl

Abstract. Both materialization and backward-chaining as different modes of performing inference have complementary advantages and disadvantages.

Materialization enables very efficient responses at query time, but at the cost of an expensive up front closure computation, which needs to be redone every time the knowledge base changes. Backward-chaining does not need such an expensive and change-sensitive precomputation, and is therefore suitable for more frequently changing knowledge bases, but has to perform more computation at query time.

Materialization has been studied extensively in the recent semantic web literature, and is now available in industrial-strength systems. In this work, we focus instead on backward-chaining, and we present an hybrid algorithm to perform efficient backward-chaining reasoning on very large datasets expressed in the OWL Horst ($pD*$) fragment.

As a proof of concept, we have implemented a prototype called QueryPIE (Query Parallel Inference Engine), and we have tested its performance on different datasets of up to 1 billion triples. Our parallel implementation greatly reduces the reasoning complexity of a naive backward-chaining approach and returns results for single query-patterns in the order of milliseconds when running on a modest 8 machine cluster.

To the best of our knowledge, QueryPIE is the first reported backward-chaining reasoner for OWL Horst that efficiently scales to a billion triples.

1 Introduction

We are witnessing an exponential growth of semantically annotated data available on the Web. While a few years ago a large RDF dataset would consist of a few hundred thousand triples, now a large dataset is in the order of billions of triples. This growth calls for knowledge-base systems that are able to efficiently process large amounts of data.

The community has provided tools to perform efficient materialization (i.e. calculate the forward closure) using distributed techniques that can scale up to hundreds of billion statements over reasonably expressive logics [14] but there are use cases in which this technique is neither desirable nor possible. In particular, when datasets are frequently updated, materialization is not efficient.

L. Aroyo et al. (Eds.): ISWC 2011, Part I, LNCS 7031, pp. 730–745, 2011.

Currently, there is no alternative to materialization that scales to relatively complex logics and very large data sizes. Backward-chaining reasoning, which does not require materialization, suffers from more complex query evaluation that adversely affects performance and scalability. Thus, it has until now been limited to either small datasets (usually in the context of expressive DL reasoners) or weak logics (RDFS inference).

To overcome this problem, we propose an *hybrid* method to perform backward-chaining reasoning that calculates some derivations in a forward fashion while the majority are computed on-the-fly during query time as necessary. This method strikes a balance between the large pre-processing costs of materialization and the complexity of pure backward-chaining reasoning. Thus, it allows us to do more complex reasoning with competitive performance. Furthermore, our algorithms have been designed to exploit the computational power of a compute cluster.

The costs of reasoning depend on the logic we consider. In this paper, we will consider the OWL Horst fragment [13], also known as the $pD*$ ruleset. OWL Horst is the most widely used complex fragment in Web-scale data to date, as witnessed by our datasets which combine some of the most important parts of the Linked Data Cloud.

Our method abstracts from the actual query language by describing and evaluating the reasoning system in terms of retrieving triples that match a given pattern. As a proof of concept, we have implemented a prototype called QueryPIE and tested its performance. QueryPIE has been built on top of the Ibis framework [1] and it was launched on the DAS-4 cluster with up to 8 machines. As we will describe later in the paper, the results indicate that our algorithms manage to keep the query response time in the order of a few milliseconds over triple stores of up to a billion statements.

The rest of the paper is organized as follows: In Section 2 we formalize our problem while in Section 3 we give a brief overview of rule-based reasoning, positioning our approach within this field. Next, in Section 4, we describe our algorithms for performing efficient backward-chaining reasoning introducing key optimizations. In Section 5 we evaluate the performance of the QueryPIE prototype on real and benchmark data. Finally, in Sections 6 and 7 we report on related work and we draw our conclusions.

2   Querying Complex Web-Scale Data

In this paper, we consider a scenario where a user queries a potentially huge and rapidly changing knowledge base with information modeled in some expressive ontology language. Even for a simple SPARQL such as

```
SELECT ?s WHERE { ?s :lives 'Amsterdam' . ?s rdf:type Person . }
```

additional implicit information can be derived according to the formal semantics of the underlying representation language and those consequences are commonly retrieved from the knowledge base through some form of reasoning. In our case, we will study the specific problem when reasoning is invoked at query-time, i.e. at the moment when the system searches for information in the knowledge base for all triples that match (?s :lives 'Amsterdam') and (?s rdf:type Person). In this paper, we only consider simple conjunctions of triples in a query, which means that we can assume that the input of the reasoning process is a single triple pattern.

We will use the following simple definitions: as usual, a *triple* is a sequence of three RDF terms, and a *triple pattern* is a sequence of three elements where each of them is either a variable (in this case it is preceded by a '?') or an RDF term. A *query* is a triple pattern. A *ground triple pattern* is a triple pattern not containing any variables (i.e. a ground triple pattern is a triple). Triple pattern P_1 is *more specific* than triple pattern P_2 (written ($P_1 < P_2$) if P_1 can be constructed from P_2 by replacing all occurrences of at least one variable with an RDF term.

Formally, the problem that we are addressing is the following: given a set of axioms in the language OWL Horst (which we will call the knowledge base KB) and a query Q as input, we want to derive all the ground triples $T < Q$ that are logically entailed by KB (see [13] for the definition of the entailment relation in OWL Horst).

The most common form of reasoning in ontology languages such as OWL Horst or OWL 2 RL is rule-based. It has been shown that all triples that are entailed by a OWL Horst knowledge base are precisely those triples derivable by the repeated application of a restricted set of rules defined by the language. In the following section we will review some of the rule-based reasoning approaches and the drawback with the current techniques with respect to our problem, which lead to the development of the novel approach described in Section 4.

3   Rule-Based Reasoning

For the purpose of this paper, we consider rule-based reasoning as a process that exhaustively applies a set of rules to a set of triples to infer some conclusions. Rules can be applied either in a forward or in a backward way. The first case is referred as materialization (or forward-chaining) while the second is referred as backward-chaining.

With materialization, the rules are applied over the entire KB until all possible triples are derived, irrespective of the input query. The main advantage of this method is that querying is simple and efficient after the closure has been calculated since it does not require further inference. The main disadvantage is that the closure needs to be updated at every change in the KB and this becomes problematic when the KB is updated frequently or when queries are infrequent compared to updates.

With backward-chaining, the rules are applied only over the strictly necessary data that lead to the derivation of ground triples of the input query. Since reasoning is only performed for the given query, updating the knowledge base is cheap because there is no closure that needs to be recomputed. Unfortunately, this flexibility comes at a price: the system has to perform specific computations for every query.

In this paper we focus on backward-chaining since currently there is no valid technique that can scale to a large extent. We define *backward-chaining reasoning* (or simply backward-chaining) over a ruleset R as a process that takes as input a triple pattern Q (the query) and a knowledge base KB and returns as output a set of triples C (the conclusions) such that each $C_i \in C$ can be derived from KB using the rules in R and $C_i < Q$ (conclusions are instantiations of the query). We call C the *answer-set* of the query Q: all triples $C_i < Q$ that are entailed by the KB by using R.

We consider the rules in the RDFS [6] and OWL Horst fragments. We report in Table 1 the rules in these two fragments because we will frequently refer to them. As we can see from this table, all rules have one or more triple patterns as antecedents and exactly one consequent.

Regardless of the set of considered rules, backward-chaining first searches for all rules with a consequent that is either compatible (i.e. contains variables in the same position) or more specific than the query pattern. After this, it will recursively look at the antecedents of these rules, regarding them as new query patterns. In this way the reasoning process builds an *and-or tree* of all the possible rules that might return some derivations.

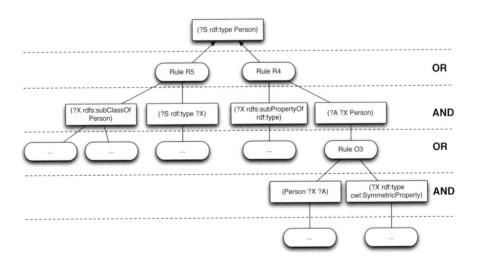

Fig. 1. Example of and-or reasoning tree

In Figure 1 we show an example of an and-or tree. Here, the derived triples are generated by different rules (the OR level) only if all of their antecedents are bound while respecting the shared variables (the AND level). The variable bindings will be propagated to the higher levels until they reach the top of the tree and can be returned as part of the answer-set. The reasoner dynamically

Table 1. *RDFS* and *OWL Horst* rulesets

	Antecedents	Consequent
R1:	p rdfs:domain x, $s\ p\ o$	$\Rightarrow$ s rdf:type x
R2:	p rdfs:range x, $s\ p\ o$	$\Rightarrow$ o rdf:type x
R3:	p rdfs:subPropertyOf q, q rdfs:subPropertyOf r	$\Rightarrow$ p rdfs:subPropertyOf r
R4:	$s\ p\ o$, p rdfs:subPropertyOf q	$\Rightarrow$ $s\ q\ o$
R5:	s rdf:type x, x rdfs:subClassOf y	$\Rightarrow$ s rdf:type y
R6:	x rdfs:subClassOf y, y rdfs:subClassof z	$\Rightarrow$ x rdfs:subClassOf z
O1:	p rdf:type owl:FunctionalProperty, $u\ p\ v$, $u\ p\ w$	$\Rightarrow$ v owl:sameAs w
O2:	p rdf:type owl:InverseFunctionalProperty, $v\ p\ u$, $w\ p\ u$	$\Rightarrow$ v owl:sameAs w
O3:	p rdf:type owl:SymmetricProperty, $v\ p\ u$	$\Rightarrow$ $u\ p\ v$
O4:	p rdf:type owl:TransitiveProperty, $u\ p\ w$, $w\ p\ v$	$\Rightarrow$ $u\ p\ v$
O5:	v owl:sameAs w	$\Rightarrow$ w owl:sameAs v
O6:	v owl:sameAs w, w owl:sameAs u	$\Rightarrow$ v owl:sameAs u
O7a:	p owl:inverseOf q, $v\ p\ w$	$\Rightarrow$ $w\ q\ v$
O7b:	p owl:inverseOf q, $v\ q\ w$	$\Rightarrow$ $w\ p\ v$
O8:	v rdf:type owl:Class, v owl:sameAs w	$\Rightarrow$ v rdfs:subClassOf w
O9:	p rdf:type owl:Property, p owl:sameAs q	$\Rightarrow$ p rdfs:subPropertyOf q
O10:	$u\ p\ v$, u owl:sameAs x, v owl:sameAs y	$\Rightarrow$ $x\ p\ y$
O11a:	v owl:equivalentClass w	$\Rightarrow$ v rdfs:subClassOf w
O11b:	v owl:equivalentClass w	$\Rightarrow$ w rdfs:subClassOf v
O11c:	v rdfs:subClassOf w, w rdfs:subClassOf v	$\Rightarrow$ v rdfs:equivalentClass w
O12a:	v owl:equivalentProperty w	$\Rightarrow$ v rdfs:subPropertyOf w
O12b:	v owl:equivalentProperty w	$\Rightarrow$ w rdfs:subPropertyOf v
O12c:	v rdfs:subPropertyOf w, w rdfs:subPropertyOf v	$\Rightarrow$ v rdfs:equivalentProperty w
O13a:	v owl:hasValue w, v owl:onProperty p, $u\ p\ w$	$\Rightarrow$ u rdf:type v
O13b:	v owl:hasValue w, v owl:onProperty p, u rdf:type v	$\Rightarrow$ $u\ p\ w$
O14:	v owl:someValuesFrom w, v owl:onProperty p, $u\ p\ x$, x rdf:type w	$\Rightarrow$ u rdf:type v
O15:	v owl:allValuesFrom u, v owl:onProperty p, w rdf:type v, $w\ p\ x$	$\Rightarrow$ x rdf:type u

builds such a tree until no rule can be further applied, or when the triple patterns can be read from the knowledge base.

Since such reasoning is executed at query time, it must be efficient, and it is crucial that this unfolding of the and-or tree is limited as much as possible. Therefore, we are required to come up with some optimizations to reduce the size of the and-or tree. In the next section, we will present some algorithms to reduce the size of the tree and hence the execution time.

4   Optimizations for Backward-Chaining Reasoning

In this section, we propose two main optimizations that aim to reduce the and-or tree complexity. These optimizations are:

- Precompute some reasoning branches that will appear often in the tree to avoid their recomputation at query-time.
- Encourage early failure of branches, allowing to prune the and-or tree, by using the precomputed branches.

4.1   Precomputation of Reasoning Branches

If we look at the and-or tree in Figure 1, we notice that the execution of some reasoning branches depends less on the input than others. For example, the pattern (?X rdfs:subPropertyOf rdf:type) is more generic than (?A ?X :Person), because the latter refers to a specific term from the query. Another difference between these two patterns is that the first corresponds to only terminological triples while the second can match any triple. It was already empirically verified [15] that terminological triples are far less than the others on Web-data, therefore, the first pattern will match with many fewer triples than the other.

We make a distinction between these two types of patterns, calling the first terminological triple patterns. A *terminological triple pattern* is a triple pattern which has as predicate or object a term from either the RDFS or the OWL vocabularies.

These terminological patterns are responsible for a notable computational cost that affects many queries. If we precompute all the ground instances of these triple patterns that are entailed by the knowledge-base, then whenever the reasoner needs such patterns it can use the precomputed results avoiding to perform additional computation. This would simplify our task to only having to perform reasoning on the non-terminological patterns. We call this simplified form of reasoning *terminology-independent reasoning* since it can avoid reasoning over terminological patterns.

Algorithm 1. Terminology-independent reasoning algorithm

```
ti-reasoner(Pattern pattern):
//Get rules where pattern is more specific than rule's consequent
Rules applicableRules = ruleSet.applicable(pattern)

Results results = {}
for(Rule rule in applicableRules)
  Patterns antecedents = rule.instantiate_antecedents(pattern)
  for(Pattern antecedent : antecedents) //Perform reasoning to fetch all antecedents
    if (antecedent != terminological)
      antecedents.add(ti-reasoner(antecedent)) //Recursive call to the reasoner
    antecedents.add(KnowledgeBase.read(antecedent))
  results += rule.apply_rule(antecedents) //Apply the rule using the antecedents triples

return results
```

The terminology-independent reasoning algorithm is reported in pseudocode in Algorithm 1. The terminology-independent reasoner will be faster than the standard one, because it can avoid the reasoning on terminological patterns but this algorithm is only complete if all entailed instances of these terminological triple patterns have been added to the knowledge-base.

So now the problem becomes how to calculate all implied terminological triples so that the terminology-independent reasoning is complete. Such precomputation cannot be calculated using traditional forward-chaining techniques because the complexity of the ruleset is such that completeness cannot be reached unless we calculate the entire closure. For this task backward-chaining is more appropriate but we have explained before that a naive approach does not scale for its excessive computation requisites. To solve this issue and improve the performance, we propose an algorithm to calculate the implied terminological triples using the terminology-independent reasoner in an iterative manner. This algorithm is reported in pseudocode in Algorithm 2. The first step in this method consists of listing all the terminological patterns that should be calculated beforehand. Such a list depends on the ruleset and in our case these patterns are reported in Table 2.

Then, the algorithm starts querying the knowledge base with the terminology-independent reasoner using each pattern in the table. If the reasoner will produce some derivation it will be immediately added to the knowledge base. This process is repeated until no new triples can be inferred.

By querying the reasoner using the terminological patterns, we perform the reasoning necessary to calculate the implicit terminological triples. Since the derivation of a terminological triple might require other ground triples of other terminological patterns that might not have been found yet, we need to repeat this operation adding new derivations to the knowledge base until saturation.

In this way, the reasoning tree that leads to the derivation of an implicit terminological triple is built bottom-up. The first time the system will derive only the terminological triples that require only the existence of explicit ground triples while other triples that depend on other implicit terminological triples will be missed. However, at the next iteration the system will be able to use the implicit triples derived before to infer new conclusions and reach completeness when all queries return an empty set of results.

Algorithm 2. Closure of the terminological triple patterns

```
terminological_closure():
    do {
        InferredTriples = {}
        for (Pattern pattern in terminological-patterns)
            InferredTriples += ti-reasoner(pattern)
        KnowledgeBase = KnowledgeBase + InferredTriples
    } while (InferredTriples is not empty)
```

At that point, the answer-set for all terminological queries has been computed. We will call this the *terminological closure*. After the terminological closure is completed, the terminology-independent reasoner will be sound and complete. A proof of these two properties goes beyond the scope of this paper but is available online[1].

4.2 Prune Reasoning Using the Precomputed Branches

The pre-calculation of the terminological closure allows us to implement another optimization that can further reduce the size of the and-or tree by identifying beforehand whether a rule can contribute to derive facts for the parent branch.

---

[1] http://www.few.vu.nl/~jui200/papers/tr-iswc2011.pdf

Table 2. Terminological triple patterns considered for RDFS and OWL Horst fragment

(?X rdfs:subPropertyOf ?Y)	(?X rdfs:subClassOf ?Y)
(?X rdfs:domain ?Y)	(?X rdfs:range ?Y)
(?P rdf:type owl:FunctionalProperty)	(?X owl:sameAs ?Y)
(?P rdf:type owl:InverseFunctionalProperty)	(?X owl:inverseOf ?Y)
(?P rdf:type owl:TransitiveProperty)	(?X rdf:type owl:Class)
(?P rdf:type owl:SymmetricProperty)	(?X rdf:type owl:Property)
(?X owl:equivalentClass ?Y)	(?X owl:onProperty ?Y)
(?X owl:hasValue ?Y)	(?X owl:equivalentProperty ?Y)
(?X owl:someValuesFrom ?Y)	(?X owl:allValuesFrom ?Y)

In this case, the triples in the terminological closure can be used for the purposes of inducing early failures: the truth of these triples is easy to verify since they have been precomputed and no inference is needed. Therefore, when scheduling the derivation of rule-antecedents, we give priority to antecedents that potentially match these precomputed triples so that if these cheap antecedents do not hold, the rule will not apply anyway, and we can avoid the computation of the more expensive antecedents of the rule for which further reasoning would have been required.

To better illustrate this optimization, we proceed with an example. Suppose we have the and-or tree described in Figure 1. In this tree, the reasoner fires rule O3 (concerning symmetric properties in OWL) to be applied on the second antecedent of rule R4.

In this case, Rule O3 will fire only if some of the subjects of the triples part of (?X rdfs:subPropertyOf rdf:type) will also be the subject of triples part of (?X rdf:type owl:SymmetricProperty). Since both patterns are more specific than terminological patterns, we know beforehand all the possible '?X', and therefore we can immediately perform an intersection between the two sets to see whether this is actually the case. If there is an intersection, then the reasoner proceeds executing rule O3, otherwise it can skip its execution since it will never fire.

It is very unlikely that the same property appears in all the terminological patterns, therefore by performing such intersections we are able to further reduce the tree size not considering rules that will derive no conclusion.

5   Evaluation

To evaluate the methods described above, we have implemented a proof-of-concept prototype called *QueryPIE* using the Java language and the Ibis framework [1].

The Ibis framework provides a set of libraries which ease the development of a parallel distributed application. We have used it to develop a distributed reasoner which can work on a variable number of nodes. The data is indexed with 4 indexes (*spo, sop, pos,* and *ops*) and partitioned across the nodes. The nodes load the data in the main memory, and, when the reasoner is invoked with an input pattern, it builds the and-or tree and executes it on the relevant data in the compute nodes. Then, the data is collected to one location and returned to the user. We have used the Hadoop MapReduce framework [4] and WebPIE [14] to create the data indexes and compute the sameAs closure and consolidation, which is a common practice among reasoners [14]. All the code is publicly available[2].

A complete evaluation of our work is beyond the scope of this article. In this section we report the results of a number of experiments that aimed to understand the effectiveness and performance of our algorithms. The evaluation of other aspects like scalability is left as future work.

The evaluation was performed on the DAS4 cluster[3]. Each node in this cluster is equipped with at least 24 GB of main memory, two quad-core processors and 2 TB disk space. The nodes use a 10 Gbit/s ethernet connection. The input consists of triples encoded in N-Triples format. Initially, we have compressed the datasets using the technique presented in [16]. All tests were performed using 8 machines.

We have used three datasets for our experiments, one benchmark tool and two real-world datasets. The artificial benchmark tool that we used is LUBM [5]. LUBM allows the generation of arbitrary numbers of triples. In our experiments we have generated a dataset of about 1 billion triples. The other two real-world datasets that we used are LLD[4] and LDSR[5]. The first dataset is a collection of biomedical data, taken from different sources, and it contains about 700 million triples. LDSR is a collection of generic information, of about 860 million triples. LDSR and LLD are among the largest single collections of triples that are currently available on the Web of Data. All three datasets make use of OWL modeling primitives.

Unfortunately, there is no standard set of queries on real-world datasets to benchmark the reasoner[6]. Therefore, we had to choose a number of input patterns and execute them in our prototype. In Table 3 we report the list of input patterns used in this evaluation and refer to them through this section.

These example query patterns were chosen because:

- they all require inference (i.e. none of them appears in the datasets) with the deliberate exception of pattern nr. 9;
- they are identical or similar[7] to the query-patterns that would be generated from the SPARQL queries that come as examples with LLD and LDSR;
- they differ in the size of the answer-set that they generate;
- they differ in the size of the inference tree that is required to derive them.

---

[2] http://few.vu.nl/~jui200/files/querypie-1.0.0.tar.gz
[3] http://www.cs.vu.nl/das4
[4] LinkedLifeData, available at http://linkedlifedata.com/
[5] Also known as FactForge, available at http://factforge.net/
[6] Standard sets of queries exist for artificial datasets such as LUBM which we use, but not for real-world datasets.
[7] Some queries were slightly changed to evaluate different types of reasoning.

Table 3. List of the input patterns used in the evaluation

Pattern	Dataset	Pattern
1	LUBM	? ? University0
2	LUBM	University0 hasAlumnus ?
3	LUBM	? rdf:type ResearchGroup
4	LUBM	UndergradStudent0 rdf:type ?
5	LDSR	? rdf:type opencyc:Business
6	LDSR	dbpedia:Arnold_Sch...gger ? ?
7	LDSR	? rdf:type umbel:CompactCar
8	LDSR	dbpedia:Lamborghini owl:sameAs ?
9	LLD	? rdf:type gene:Gene
10	LLD	? uniprot:pathway399...145 ? ?
11	LLD	? ? skos:definition
12	LLD	? rdf:type biopax:sequenceFeature

Initially we focused on the effectiveness of our algorithms and we calculated the reduction of the and-or tree size caused by our optimizations on a set of queries. The results and a more complete discussion are reported in subsection 5.1.

After this, we performed some experiments to evaluate whether the performance of our method is competitive when compared to materialization, which is currently the de-facto reasoning method over large data. Such comparison was chosen because to the best of our knowledge there is no other OWL Horst backward-chaining reasoner that works on large amounts of RDF data, and therefore we are unable to perform a comparison of the absolute reasoning performance of our algorithms. The results are discussed in subsection 5.2.

5.1   Effectiveness: Comparison against Naive Backward-Chaining

The main scope of our work was to reduce the size of the and-or tree as described in sections 4.1 and 4.2, in order to decrease the runtime of the reasoning.

The optimizations and algorithms that we described are crucial to perform backward-chaining reasoning on large data. Therefore, in order to evaluate their impact on the performance, we manually calculated how large the tree would be if we did not make any pre-calculation. For this purpose, every time the reasoner had to process a pattern that was already pre-calculated, we added the tree that was needed to calculate it during the initial phase. This method is in fact an underestimate, because we could not deactivate the other optimization, therefore in reality the gain is even higher than the one calculated.

Table 4 reports the actual number of leaves of the and-or tree with and without the pre-calculation. The last column reports the obtained reduction ratio and shows that the number of leaves (= the number of paths in the tree) shrinks by one order of magnitude through our pre-calculation. This shows that our pre-calculation is indeed very effective. For a very small cost in both data space and upfront computation time (see table 5), we substantially reduce the search tree. Apparently, the pre-calculation precisely captures small amounts of inferences

Table 4. Estimated performance gain against naive backward-chaining

Input pattern	#leaves with/out optimiz.	Ratio
Pattern 1	21/174	8.29
Pattern 2	5/58	11.60
Pattern 3	2/3	1.50
Pattern 4	38/291	7.66

that contribute substantially to the reasoning costs because they are being used very often.

As expected, we notice that the reduction ratio is not constant but changes depending on the input pattern. Overall, the optimized reasoning algorithm generated an and-or tree that is between 11.6-1.5 times smaller.

5.2   Performance: Comparison against Full Materialization

With materialization, typically the data provider computes the entire closure of the data beforehand and then loads the input and derived data into a database-like infrastructure where the users can query the data with no reasoning performed on the fly. Here, we can distinguish two phases: the first, where the entire closure is computed, and the second, where the user can query the data.

In this scenario, the first phase takes a lot of time because all the reasoning must be performed while the second is much faster since only a lookup is performed. Instead, in our case the first phase will be much faster, since we do not calculate the entire closure but only a very small part that can be used to speed up the reasoning later, but the second phase will be slower since we do perform some reasoning.

Table 5. Comparison computation terminological closure against full materialization

Input	Terminological closure		Full material.	
	Time (sec.)	# statms.	Time (sec.)	# statms.
LDSR (862M)	89	0.62M	10036	927M
LLD (694M)	332	7.06M	3931	330M
LUBM (1101M)	8	22	4526	495M

In Table 5 we report the reasoning execution time of our pre-calculated closure against the execution time of calculating the entire closure for the datasets that we consider. We have used WebPIE to compute the closure on the same number of machines since it supports the same ruleset and has the best performance on large data [14]. In this table, in the first part we report respectively the runtime and the number of triples derived in the preprocessing stage of our algorithm while in the second we report the same when we compute the entire closure. From the table we observe, as expected, that our method is considerably faster

Table 6. Performance comparison at query-time of our method against full-closure approach

Pattern	#Results	#leaves and-or tree	Time query back. reas. (ms.)	Time query full closure (ms.)	Ratio
Pattern 1	75613	312	55.12	35.75	1.54
Pattern 2	37118	5	38.91	17.47	2.24
Pattern 3	2400836	2	1166.84	1017.85	1.15
Pattern 4	4	38	3.53	1.02	3.46
Pattern 5	26440	411	34.83	13.57	2.57
Pattern 6	4937	60	8.57	2.86	2.99
Pattern 7	182	3	3.38	3.32	1.02
Pattern 8	5	23	3.49	0.92	3.79
Pattern 9	4524379	1	1685.55	1680.87	1.00
Pattern 10	4	134	8.17	1.10	7.43
Pattern 11	0	72	7.00	1.01	6.93
Pattern 12	245831	4	100.89	98.97	1.02

than a traditional forward reasoner performing reasoning in about 5 minutes in the worst case against the almost 3 hours necessary to compute the closure.

Thus, even in the worst case (LLD), the costs of our preprocessing stage are only a fraction of computing the full closure, using the fastest approach for closure computation known in the literature, and using the same hardware setup.

In the second phase (when performing the actual query) our approach will be slower than engines that query a fully computed closure, since we must do reasoning at query time and it is important to evaluate such cost.

A direct comparison of the performance with existing approaches is not appropriate since the majority of the RDF stores has a single-machine architecture and/or consider a different ruleset than ours. Therefore, since our purpose is to evaluate the overhead caused by reasoning while keeping other factors constant, we proceeded as follows. First, we launched a set of queries on our prototype using the datasets with the terminological closure calculated beforehand. After this, we loaded the full closure of the dataset (as derived with WebPIE), we completely disabled reasoning in our engine, and launched the same set of queries. In this way, we kept the infrastructure constant and indirectly made a comparison with a forward reasoning scenario using the pre-calculated derivation on the same infrastructure.

The results of this comparison are presented in Table 6. For every pattern in the input we report the number of returned results, the number of leaves of the and-or tree generated when reasoning was enabled (when querying the full closure, the size of the tree is 1), the execution time when reasoning was enabled and the execution time when reasoning was disabled with the full-closured data used as input. The last column reports the ratio between the two execution times, and represents the reasoning overhead.

We observe that the overhead varies from 1.00 to 7.43. This means that in the best case reasoning does not introduce any overhead while in the worst it slows down the response time almost 7.5 times. However, even in that worst case, the response time of the system is never more than a few milliseconds. The only exceptions are patterns nr. 3 and 9, where the transport of the large number of output triples completely dominates the calculation in both cases. Similarly as before, there is no clear correlation between the input pattern and the response time since it depends on the complexity of the reasoning involved. For example, the pattern 1 generates an and-or tree with 312 leaves and it is only 1.54 times slower whereas pattern 4 generates a tree with only 38 leaves but is 3.46 times slower.

Overall, table 6 shows that the response time of our approach is competitive with querying the forward closure, while table 5 shows that our upfront cost is anywhere from 1 to 2 orders of magnitude smaller than the upfront cost of forward reasoning.

It is interesting to evaluate when our approach becomes more attractive than a full-closure approach. To this purpose, we calculated the average response time of the selected queries for the datasets LDSR and LLD. These are respectively 12.57 and 450.40 milliseconds if reasoning is activated and 5.17 and 445.49 milliseconds when querying the full closure. We used these values to estimate how many queries a system is able to answer in a certain amount of time.

The full-closure approaches start answering the queries later because they need to wait until the closure is computed (an upfront waiting time of close to 1 hour on LLD and more than 3 hours on LDSR), while our approach can start almost immediately to answer queries with an upfront waiting time of a few minutes (see table 5). However, since the response time of our approach is slower (because of reasoning), there will be a point after which the forward approach has a lower total runtime over a large number of queries. For LDSR, it takes about 1.42 million queries in order to gain back the costs of the initial closure computation amounting to about 5 hours of continuous query-load. This means that as soon as the update frequency of the data is lower than once every 5 hours, our method will be more efficient in total runtime, and more convenient because of a much smaller upfront delay.[8]

In the case of LLD, the query times are more or less equal between both approaches, while our method does keep the advantage of having only a 5 minute startup time, instead of 1 hour. This makes our approach much more competitive than before since now the full-closure approach will become convenient only after 733 thousand queries or 91 hours without any update.

6   Related Work

In previous work [15,14], we have shown scalable RDFS and OWL materialization for datasets up to 100 billion triples. There, the MapReduce programming framework was used to encode the logic for the rulesets at hand, and a set of optimizations was introduced to improve load balancing and the efficiency of the

---

[8] This calculation assumes a maximal query-load. As soon as the query load is lower than 100% utilization, the balance shifts even more in favor of backward reasoning.

computation. In this paper, we depart from the full forward closure and take a significant step in the direction of scalable backward-chaining reasoning.

In [17], straightforward parallel RDFS reasoning on a cluster is presented. This approach replicates all schema triples to all processing nodes, partitions instance triples arbitrarily and calculates the closure of each partition. Triples extending the RDFS schema are ignored, thus the reasoning is incomplete.

In [8], a method for distributed reasoning with EL++ using MapReduce is presented, which is applicable to the EL fragment of OWL 2. No experimental results are provided.

The work on Signal/Collect [12], introduces a new programming paradigm, targeted at handling graph data in a distributed manner. Although very promising, it is not comparable to our approach, since current experiments deal with much smaller graphs and are performed on a single machine.

The operation of passing the query bindings to the lower branches of the reasoning tree is likewise applied in the Magic Sets query rewriting technique [2] and it is commonly referred as one type of sideways information passing strategy (SIPS) [11]. However, while in the latter it is used to efficiently rewrite a query, in our case we use it to prune the reasoning branches so that it becomes effective only when combined with the schema closure.

In the context of RDF stores, in [10], backward-chaining reasoning for RDFS on 4Store is presented. The authors show how they perform RDFS reasoning on their architecture but do not report on more complex inferencing than RDFS.

The Jena RDF store [3] uses a hybrid reasoner at its core with a focus on lower expressivity logics. The data store administrator can define so-called hybrid rules which include conditions for firing rules in a backwards fashion. There are no results for using Jena with a more complex ruleset.

The Virtuoso RDF store performs incomplete RDFS and OWL rule-based reasoning. Some results are reported online[9], but no experiments are reported for scaling on the number of nodes or on datasets more complex than LUBM.

7 Conclusion and Future Work

Until now, all inference engines that can handle reasonably expressive logics over very large triple stores have deployed full materialization. In the current paper, we have broken with this mold, showing that it is indeed possible to do efficient backward-chaining over large and reasonably expressive knowledge bases. The key to our approach is two optimizations which substantially reduce the size of the search space that must be navigated by a backward-chaining inference engine.

The first optimization precomputes a small number of inferences which appear very frequently in the derivation trees. By precomputing these inferences upfront instead of during query-time, we reduce the size of the trees by an order of magnitude. This of course re-introduces some amount of preprocessing (making

---

[9] http://virtuoso.openlinksw.com/dataspace/dav/wiki/Main/
VOSArticleLUBMBenchmark

our work strictly speaking a *hybrid* approach), but this computation is measured in terms of minutes, instead of the hours needed for the full closure computation.

The second optimization exploits these precomputed triples, to further reduce computation. It does that by giving priority to the evaluation of antecedents that potentially match these precomputed triples. If there is no match, we can avoid calculating the other more expensive antecedents that would have required additional reasoning.

Performance analysis of our approach on three datasets varying from 0.7 billion to 1.1 billion triples shows that the query response-time for our approach is competitive with that of full materialization, with response times in the low number of milliseconds on our test query patterns, running on only a small cluster of 8 machines. The small loss of response time is offset by the great gain in not having to perform a very expensive computation of many hours before being able to answer the first query.

Obvious next steps in future work would be to investigate how our algorithms scale with the number of machines, and to understand the properties of the knowledge base that influence both the cost of the limited forward computation and the size of the inference tree. Since the proposed approach is not specifically tailored around the OWL Horst ruleset, it would be interesting to extend our prototype to support the OWL 2 RL [7] ruleset and evaluate its performance.

Also, it is worth to explore whether other techniques like memoization, other SIP strategies [11], or ad-hoc query-rewriting techniques [9] can be exploited to further improve the performance.

To the best of our knowledge, this is the first time that logically complete backward-chaining reasoning over realistic OWL Horst knowledge bases of a billion triples has been realized. Our results show that this approach is feasible, opening the door to reasoning over much more dynamically changing datasets than was possible until now.

Acknowledgments. We would like to thank Barry Bishop for reviewing our work. This work was partly supported by the LarKC project (EU FP7-215535) and by the COMMIT project.

References

1. Bal, H.E., Maassen, J., van Nieuwpoort, R., Drost, N., Kemp, R., Palmer, N., Kielmann, T., Seinstra, F.J., Jacobs, C.J.H.: Real-World Distributed Computing with Ibis. IEEE Computer (2010)
2. Bancilhon, F., Maier, D., Sagiv, Y., Ullman, J.: Magic Sets and Other Strange Ways to Implement Logic Programs. In: Proceedings of SIGACT-SIGMOD (1985)
3. Carroll, J.J., Dickinson, I., Dollin, C., Reynolds, D., Seaborne, A., Wilkinson, K.: Jena: Implementing the Semantic Web Recommendations. Technical Report HPL-2003-146, Hewlett Packard Laboratories (2003)
4. Dean, J., Ghemawat, S.: MapReduce: Simplified Data Processing on Large Clusters. In: Proceedings of OSDI (2004)
5. Guo, Y., Pan, Z., Heflin, J.: LUBM: A Benchmark for OWL Knowledge Base Systems. Web Semantics: Science, Services and Agents on the World Wide Web (2005)

6. Hayes, P. (ed.): RDF Semantics. W3C Recommendation (February 2004)
7. Motik, B., Grau, B., Horrocks, I., Wu, Z., Fokoue, A., Lutz, C.: OWL2 Web Ontology Language: Profiles. W3C Working Draft (2008)
8. Mutharaju, R., Maier, F., Hitzler, P.: A MapReduce Algorithm for EL+. In: Proceedings of the 23rd International Workshop on Description Logics, DL 2010 (2010)
9. Pérez-Urbina, H., Horrocks, I., Motik, B.: Efficient Query Answering for OWL 2. In: Bernstein, A., Karger, D.R., Heath, T., Feigenbaum, L., Maynard, D., Motta, E., Thirunarayan, K. (eds.) ISWC 2009. LNCS, vol. 5823, pp. 489–504. Springer, Heidelberg (2009)
10. Salvadores, M., Correndo, G., Harris, S., Gibbins, N., Shadbolt, N.: The Design and Implementation of Minimal RDFS Backward Reasoning in 4store. In: Antoniou, G., Grobelnik, M., Simperl, E., Parsia, B., Plexousakis, D., De Leenheer, P., Pan, J. (eds.) ESWC 2011. LNCS, vol. 6644, pp. 139–153. Springer, Heidelberg (2011)
11. Seshadri, P., Hellerstein, J., Pirahesh, H., Leung, T., Ramakrishnan, R., Srivastava, D., Stuckey, P., Sudarshan, S.: Cost-Based Optimization for Magic: Algebra and Implementation. ACM SIGMOD Record 25, 435–446 (1996)
12. Stutz, P., Bernstein, A., Cohen, W.: Signal/Collect: Graph Algorithms for the (Semantic) Web. In: Patel-Schneider, P.F., Pan, Y., Hitzler, P., Mika, P., Zhang, L., Pan, J.Z., Horrocks, I., Glimm, B. (eds.) ISWC 2010, Part I. LNCS, vol. 6496, pp. 764–780. Springer, Heidelberg (2010)
13. ter Horst, H.J.: Completeness, decidability and complexity of entailment for RDF schema and a semantic extension involving the OWL vocabulary. Journal of Web Semantics 3(2-3), 79–115 (2005)
14. Urbani, J., Kotoulas, S., Maassen, J., van Harmelen, F., Bal, H.: OWL Reasoning with WebPIE: Calculating the Closure of 100 Billion Triples. In: Aroyo, L., Antoniou, G., Hyvönen, E., ten Teije, A., Stuckenschmidt, H., Cabral, L., Tudorache, T. (eds.) ESWC 2010. LNCS, vol. 6088, pp. 213–227. Springer, Heidelberg (2010)
15. Urbani, J., Kotoulas, S., Oren, E., van Harmelen, F.: Scalable Distributed Reasoning Using MapReduce. In: Bernstein, A., Karger, D.R., Heath, T., Feigenbaum, L., Maynard, D., Motta, E., Thirunarayan, K. (eds.) ISWC 2009. LNCS, vol. 5823, pp. 634–649. Springer, Heidelberg (2009)
16. Urbani, J., Maassen, J., Bal, H.: Massive Semantic Web data compression with MapReduce. In: Proceedings of the HPDC (2010)
17. Weaver, J., Hendler, J.: Parallel Materialization of the Finite RDFS Closure for Hundreds of Millions of Triples. In: Bernstein, A., Karger, D.R., Heath, T., Feigenbaum, L., Maynard, D., Motta, E., Thirunarayan, K. (eds.) ISWC 2009. LNCS, vol. 5823, pp. 682–697. Springer, Heidelberg (2009)

Practical RDF Schema Reasoning with Annotated Semantic Web Data

Carlos Viegas Damásio and Filipe Ferreira

CENTRIA, Departamento de Informática Faculdade de Ciências e Tecnologia
Universidade Nova de Lisboa, 2829-516 Caparica, Portugal
cd@di.fct.unl.pt, p110362@fct.unl.pt

Abstract. Semantic Web data with annotations is becoming available, being YAGO knowledge base a prominent example. In this paper we present an approach to perform the closure of large RDF Schema annotated semantic web data using standard database technology. In particular, we exploit several alternatives to address the problem of computing transitive closure with real fuzzy semantic data extracted from YAGO in the PostgreSQL database management system. We benchmark the several alternatives and compare to classical RDF Schema reasoning, providing the first implementation of annotated RDF schema in persistent storage.

Keywords: Annotated Semantic Web Data, Fuzzy RDF Schema, Transitive closure in SQL, Rules.

1 Introduction

The Semantic Web rests on large amounts of data expressed in the form of RDF triples. The need to extend this data with meta-information like trust, provenance and confidence [26,22,3] imposed new requirements and extensions to the Resource Description Framework (Schema) [19] to handle annotations appropriately. Briefly, an annotation v from a suitable mathematical structure is added to the ordinary triples $(s\ p\ o)$ obtaining $(s\ p\ o) : v$, annotating with v the statement that subject s is related via property p to object o. The general semantics of this RDFS extension has been recently addressed [21,22] improving the initial work of [26], but only a memory-based Prolog implementation is available.

The feasibility of large scale classical RDFS reasoning and its extensions has been shown in the literature [27,10,25,13,11]. In this paper we will show that the inclusion of annotated reasoning naturally introduces some overhead but that it is still possible to perform the closure of large annotated RDFS data in reasonable amount of time. The major difficulty is the implementation of transitive closure of RDF Schema `subPropertyOf` and `subClassOf` properties, with the extra problem of maintaining the annotations since a careless implementation might result in worst-case exponential runtime. For this reason, we discuss several alternative approaches for implementing transitive closure of RDFS data exploiting facilities present in modern relational database systems, particularly

L. Aroyo et al. (Eds.): ISWC 2011, Part I, LNCS 7031, pp. 746–761, 2011.

recursive views. We selected PostgreSQL[1] because of the mechanisms it provides as well as ease of integration with other Semantic Web triple stores and reasoners, and use annotated data of the YAGO ontology [24] to evaluate our contribution. The use of existing semantic web data is essential since current "artificial" benchmarks like LUBM [7] do not reflect exactly the patterns of data present in real applications [6]. All the data and code is made publicly available at `http://ardfsql.blogspot.com/`, including instructions for replicating the tests presented in the paper.

An advantage of our approach is that we present an entirely based SQL implementation allowing practitioners to use our technique directly in their preferred standard RDBMS without external imperative code. To the best of our knowledge, we present the first complete database implementation of the fuzzy Annotated Resource Description Framework Schema and assess it with respect to real ontologies. We justify that our approach is competitive and show that semi-naive and a variant of semi-naive (differential) evaluation are not always competitive for performing RDFS closures, especially when annotated data is present.

In the next section, we start by overviewing the Annotated RDF Schema framework. In Section 3 we address the issues storing (annotated) RDFS data in persistent storage, and present our own encoding. Section 4 describes the several closure algorithms for RDFS with annotated data whose benchmarking is performed in Section 5. We proceed with comparisons to relevant work in Section 6 and we finish the paper in Section 7 with conclusions and future work.

2 Annotated RDF Schema

In this section we shortly present the RDFS with Annotations [22] and we assume good knowledge of the classical (or crisp) Resource Description Framework (RDF). We ignore the model-theoretical aspects and focus on the inference rules used to perform the closure. An annotation domain [22] is an algebraic structure $D = \langle L \preceq, \otimes, \top, \bot \rangle$ such that $\langle L, \preceq, \top, \bot \rangle$ is a bounded lattice (i.e. a lattice with a $\top$ top and $\bot$ bottom elements) and where operator $\otimes$ is a t-norm. A t-norm is a generalization of the conjunction operation to the many-valued case, obeying to the natural properties of commutativity, associativity, monotonicity and existence of a neutral element (i.e. $v \otimes \top = \top \otimes v = v$).

The inference rules for annotated RDFS [22] can be found in Fig. 1 and where `sp`, `sc`, `type`, `dom` and `range` are abbreviations for the RDF and RDFS properties `rdfs:subPropertyOf`, `rdfs:subClassOf`, `rdf:type`, `rdfs:domain` and `rdfs:range`, respectively, which is known as the ρdf vocabulary or minimal RDFS [16]. The rules are extensions of the set of crisp rules defined in [16] to handle annotations; the original rules can be obtained by dropping the annotations or equivalently using the algebraic domain $D_{01} = \langle \{0, 1\}, \leq, \min, 0, 1 \rangle$, which corresponds to classical boolean logic. As standard practice, we drop reflexivity rules which provide uninteresting inferences like that any class (property) is a

---

[1] See `http://www.postgresql.org/`

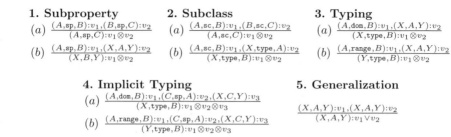

Fig. 1. Inference rules for annotated RDFS [22]

subclass (subproperty) of itself [16]. An important aspect of the inference rules is that, with the exception of rules *1b)* and *5)*, the conclusions of the rules produce triples of the ρdf vocabulary.

In this paper we will restrict mostly to the annotation domain $D_{goedel} = \langle [0,1], \leq, \min, 0, 1 \rangle$ where the t-norm is the minimum operator and the least upper bound is the maximum operator corresponding to Goedel's fuzzy logic [8], obtaining the fuzzy RDFS framework in [21]. Our algorithms will work and terminate with any t-norm in the $[0,1]$ interval. Termination is guaranteed in these circumstances by our own results on fuzzy logic programming [5], and are far from trivial since an infinitely-valued lattice is being used. In fact, one of the distinctions between the work [26] and [22], is that the former is restricted to finite annotation lattices while the latter allows infinite ones. A particularly striking example of the problems that can occur is the use of real-valued product t-norm which can generate "new" values from existing ones, contrasting to the minimum t-norm that can only return existing annotations in the asserted data.

3   Persistent Storing of Annotated RDFS Data

We follow a hybrid approach for representing RDFS data in a relational database with some optimizations. We use a vertically partitioned approach [1] for storing triples (in fact quads) having a table for each of the properties of the ρdf vocabulary, and use a common table to store all the other triples (see Fig. 2). The major differences to a classical relational RDFS representation is that we add an extra column to represent the double valued annotations, and include a column to store graph information in order to be compatible with SPARQL. The schema used is very similar to the one of Sesame's reported in [4] but we do not use property-tables. The `id` column is the key of `Triples` table automatically filled-in with a sequence, and `ref` column has been added to allow traceability support to the reification vocabulary of RDF Schema but it is not used in the current implementation. Moreover, we use a `Resources` table to reduce the size of the database by keeping a single entry for each plain literal (`nodet=1`, `value="`*string*`"`, `type="`*language tag*`"` (or NULL)), typed literal (`nodet=2`, `value="`*string*`"`,`type="`*URI type*`"`), URI (`nodet=3`,`value="`*URI*`"`,`type=NULL`)

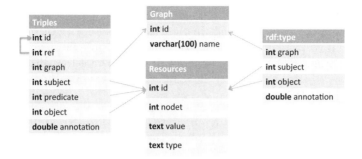

Fig. 2. Annotated RDFS table schema

or blank node (`nodet=4`, `type="`*`identifier`*`"`, `type=NULL`) in any asserted data. The `Resources` table is pre-initialized at database creation time with the ρdf vocabulary in order to have known identifiers for these URIs, reducing joins. Figure 2 shows only the table for `rdf:type`, being the schema of the remaining tables for the properties of the ρdf vocabulary identical. In the common table `Triples`, and ρdf tables `rdf:type`, `rdfs:subPropertyOf`, `rdfs:subClassOf`, `rdfs:domain` and `rdfs:range` we only have an integer foreign key for the subject, property and object of each triple, making each row fixed-length.

Notice that this representation is very similar to the star schema used in data warehouses where fact tables correspond to our triple and property specific tables while dimension tables correspond to our resources table. Moreover as discussed in [1], this representation has several advantages for query answering because it reduces self-joins in tables, which will benefit from when implementing the inference rules presented in Section 3. The choice of a particular schema is not arbitrary and can have significative impact in performance (see [1]). Additionally, we realized that under default database configuration and with the tested data sets, indexing did not bring significant advantages for performing the annotated RDFS closure for the data available, and therefore we do not include any on-disk indexes.

4   Closure of Annotated RDFS Data

In this section we discuss the techniques and algorithms we have developed to perform the closure of Annotated RDFS data. We start by discussing the annotated RDFS specific generalization rule, and afterwards we discuss rule order application. We conclude that the only recursive rules necessary are subproperty and subclass transitivity (rules 1a and 2a), which can be implemented in any current mainstream RDBMSs. The major difficulty is performing the transitive closure with annotated data, for which we specify several algorithms.

4.1   Generalization Rule

The rule which has greater impact in the implementation is the generalization rule which simply states whenever a triple is derived twice with different annotations $(s\ p\ o) : v_1$ and $(s\ p\ o) : v_2$ then their annotations may be combined to obtain a triple with a greater annotation $(s\ p\ o) : v_1 \lor v_2$, and the original annotated triples can be removed. In the case of the $[0, 1]$ interval ordered as usual, this rule corresponds to keeping the annotated triple with maximum value, deleting any other smaller annotated triple. It is important to realize that if subsumed annotated triples are left in the database, then exponential behaviour can be generated for some t-norms.

Example 1. Consider the following annotated RDFS database in the algebraic domain $D_{prod} = \langle [0, 1], \leq, \times, 0, 1 \rangle$ where the t-norm is the usual real-valued multiplication.

$$(a_0, \mathsf{sc}, b_0) : \tfrac{3}{1000} \qquad (a_1, \mathsf{sc}, b_1) : \tfrac{7}{1000} \qquad (a_2, \mathsf{sc}, b_2) : \tfrac{13}{1000}$$

$$(a_0, \mathsf{sc}, c_0) : \tfrac{5}{1000} \qquad (a_1, \mathsf{sc}, c_1) : \tfrac{11}{1000} \qquad (a_2, \mathsf{sc}, c_2) : \tfrac{17}{1000}$$

$$(b_0, \mathsf{sc}, a_1) : 1 \qquad (b_1, \mathsf{sc}, a_2) : 1 \qquad (b_2, \mathsf{sc}, a_3) : 1$$

$$(c_0, \mathsf{sc}, a_1) : 1 \qquad (c_1, \mathsf{sc}, a_2) : 1 \qquad (c_2, \mathsf{sc}, a_3) : 1$$

There are 8 paths between a_0 and a_3 by selecting at each step a path via b_i or a c_i, having assigned a different annotation obtained by multiplying the annotations in the path edges going out of each a_i, originating 8 subclass annotated triples (a_0, sc, a_3) each with a different annotation. It is immediate to see that the construction can be iterated more times with different prime-number based annotations, obtaining an exponential number of subclass relations on the number of a nodes, all with different annotations. By applying the generalization rule one can see that all but one of these annotated triples are redundant.

For this reason, our major concern will always be to never introduce in the tables duplicated triples with different annotations. To achieve this we will extensively resort to a mixture of SQL aggregations using MAX function, and in-built the rule in the other rules. Therefore, we will only take care of domains over the real-valued $[0, 1]$ to achieve better performance, otherwise one would require from the DBMS facilities to implement new aggregation functions[2]. The encoding of inference rule 3a in SQL can be seen below, where i_graph is a parameter with the graph identifier to be closed:

```
CREATE OR REPLACE FUNCTION Rule3a(i_graph integer) RETURNS integer AS $typa$
BEGIN

UPDATE "rdf:type" as r SET annotation=q.annotation FROM
    (SELECT t.graph, t.subject, d.object, MAX(tnorm(d.annotation,t.annotation)) AS annotation
    FROM "rdfs:domain" d INNER JOIN "Triples" t ON (d.subject=t.predicate)
    WHERE d.graph=i_graph AND t.graph=i_graph
    GROUP BY t.subject, d.object, t.graph ) AS q
    WHERE (r.subject,r.object,r.graph)=(q.subject,q.object,q.graph) AND
            r.annotation<q.annotation;

INSERT INTO "rdf:type" (
    SELECT t.graph, t.subject, d.object, MAX(tnorm(d.annotation,t.annotation)) AS annotation
    FROM "rdfs:domain" d INNER JOIN "Triples" t ON (d.subject=t.predicate)
    WHERE d.graph=i_graph AND t.graph=i_graph AND
```

---

[2] These are available in some commercial RDBMSs.

```
        NOT EXISTS (SELECT * FROM "rdf:type" AS old WHERE old.graph=i_graph
                    AND old.subject=t.subject AND old.object=d.object)
    GROUP BY t.subject, d.object, t.graph );

 RETURN 1;
 END
```

First we update the table `rdf:type` table with the better inferred annotations for already existing triples and afterwards we `INSERT` completely new triples accordingly to the `NOT EXISTS` clause. Both statements only generate an annotation for a given triple therefore not introducing redundant information. The user-defined function `tnorm` implements in a stored function the intended t-norm function (in our experiments, minimum).

4.2 Fixpoint Iteration and Rule Ordering

The rules present in Fig. 1 have to be iterated till a fixpoint is reached, i.e. no new annotated triples are generated. A clever implementation does not require the execution of the whole set of rules at a time. It is easy to see that rules depend on each other (see Fig. 3) and rules can be ordered to reduce computation time.

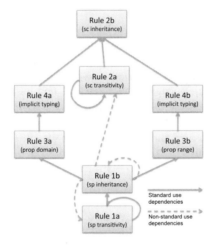

Fig. 3. Dependency graph of annotated RDFS inference rules

The dependency graph of Fig. 3 is very similar to the one presented in [27] with three distinctive features. First, we have the implicit typing rules, and we show dependencies regarding "non-standard" use of ρdf vocabulary (the dashed dependencies). More important is that we do not have to iterate the subclass and subproperty inheritance rules (rules 1*b* and 2*b*) because of the following result:

Lemma 1. *Subclass inheritance rule is idempotent[3], and subproperty inheritance is idempotent with standard use of vocabulary.*

Proof. Consider two chained applications of rule $2b$

$$1.\frac{(A,\mathsf{sc},B):v_1,(X,\mathsf{type},A):v_2}{(X,\mathsf{type},B):v_1\otimes v_2} \qquad 2.\frac{(B,\mathsf{sc},C):v_3,(X,\mathsf{type},B):v_1\otimes v_2}{(X,\mathsf{type},C):v_1\otimes v_2\otimes v_3}$$

However since the subclass relationship is closed with respect to rule $2a$ we conclude that $(A,\mathsf{sc},C):v_1\otimes v_3$ and thus we will get by one application of rule $2b$ that $(X,\mathsf{type},C):v_1\otimes v_2\otimes v_3$, by commutativity and associativity of $\otimes$:

$$\frac{(A,\mathsf{sc},B):v_1,(B,\mathsf{sc},C):v_3}{(A,\mathsf{sc},C):v_1\otimes v_3} \qquad 3.\frac{(A,\mathsf{sc},C):v_1\otimes v_3,(X,\mathsf{type},A):v_2}{(X,\mathsf{type},B):v_1\otimes v_3\otimes v_2}$$

If in the subclass transitivity closure we get an annotation for $(A,\mathsf{sc},C):v_4$ such that $v_4 > v_1\otimes v_3$ the inferred triple by chained application in step 2 will be subsumed by the triple in corresponding step 3.

A similar argument shows the result for the case of subproperty inheritance, but in this situation we have to guarantee that we cannot generate triples where $B = \mathsf{sp}$. However, this can only happens whenever we have a statement $(A,\mathsf{sp},\mathsf{sp}):v$ in the original graph, i.e. with non-standard use of the ρdf vocabulary.

We believe that it has not been realized before in the literature that a single application of rules $1b$ and $2b$ is necessary to generate all the triples, whenever the subclass and subproperty relationships are closed, even though apparently it is assumed to hold in [9] for the classical case. Therefore, we conclude that the only recursive rules necessary to perform annotated RDFS closure are transitive closures. This analysis carries over for the classical RDFS case as well.

Assuming standard use of the ρdf vocabulary we are guaranteed that rule $1b$ inserts data only in table `Triples`, further simplifying its implementation.

4.3 Transitive Closure with Annotated Data

In order to conclude the implementation of annotated RDFS we just have to consider transitivity of the subclass and subproperty relations. The definition of transitive closure and shortest-path algorithms for databases has been addressed in the literature [14,2,18,15].

To formalize our algorithms we will resort to an extension of the notion of fuzzy binary relation [28]. A fuzzy binary relation R is a mapping $R : U \times U \to [0,1]$, associating to each pair of elements of the universe U a membership degree in $[0,1]$. This can be appropriately generalized to our setting by defining annotated relations over annotated domain $D = \langle L, \preceq, \otimes, \top, \bot \rangle$ as mappings $R : U \times U \to L$.

---

[3] By idempotent rule we mean that the rule will always produce the same results whether it is applied once or several times, thus there is no need to apply it more than once.

Definition 1. *Consider binary annotated relations R_1 and R_2 with universe of discourse $U \times U$ and over annotated domain $D = \langle L \preceq, \otimes, \top, \bot \rangle$. Define composition $R_1 \circ R_2$ and union $R_1 \vee R_2$ of annotated relations R_1 and R_2 as:*

$$(R_1 \circ R_2)(u, w) = \bigvee_{v \in U} \{R_1(u, v) \otimes R_2(v, w)\} \quad (R_1 \vee R_2)(u, v) = R_1(u, v) \vee R_2(u, v)$$

In the case of fuzzy annotated domains of the form $D = \langle [0, 1], \leq, \otimes, 1, 0 \rangle$, we can use relational algebra to implement the composition and union of annotated relations. For ease of presentation we overload the symbols of $\vee$ and $\circ$:

Definition 2. *Consider binary fuzzy annotated relations R_1 and R_2, represented in relational algebra by relations $r_1 \subseteq U \times U \times [0, 1]$ and $r_2 \subseteq U \times U \times [0, 1]$ with relational schema (sub, obj, ann) and obeying to functional dependency $sub\ obj \to ann$. Operations $r_1 \circ r_2$ and $r_1 \vee r_2$ are defined as:*

$$r_1 \circ r_2 = {}_{sub,obj}\mathcal{G}_{MAX(ann)} \big($$
$$\Pi_{r1.sub\ as\ sub,\ r2.obj\ as\ obj,\ r1.ann \otimes r2.ann\ as\ ann}\ \sigma_{r1.obj=r2.sub}(r_1 \times r_2))\big)$$
$$r_1 \vee r_2 = {}_{sub,obj}\mathcal{G}_{MAX(ann)}(r_1 \cup r_2)$$

Using standard relational algebra notation, where Π is the projection operator, $\mathcal{G}$ is the aggregation operator, σ the selection operator and $\times$ the Cartesian-Product operation[4]. Finally $\otimes$ is the t-norm operation.

To simplify matters, we have assumed that only non-zero annotated information will be present in the relations, reducing a lot the storage requirements. Since $0 \vee v = v \vee 0 = v$ and $0 \otimes v = v \otimes 0 = 0$ we will be able to respect all the original operations in relational algebra.

The transitived closure of annotated fuzzy binary relations R^+ is defined by the least fixpoint of the equation $R^+ = R \vee (R^+ \circ R)$, corresponding to determining shortest-paths between all nodes in a weighted graph. In Fig. 4 we present five different ways of obtaining the annotated fuzzy relation R^+, corresponding to well-known transitive closure algorithms found in the literature [14,2,18,15]. The naive algorithm is a direct implementation of an iterated computation of the fixpoint of the recursive definition. The popular semi-naive algorithm is an improvement of the first algorithm, by at each step just propagating the changes performed in the previous iteration. The differential semi-naive algorithm is particularly optimized for relational database systems and was proposed and implemented in the DLV^{DB} system [25], which reduces the number of joins necessary with respect to the semi-naive algorithm. The matrix algorithm corresponds to Warshall's algorithm, and the logarithmic algorithm is a variant specially constructed for reducing joins [18]. We show below the translation to SQL of the Matrix method that will prove to be the best and more reliable algorithm:

---

[4] For the exact definition of these operators see any standard database manual, e.g. [20].

Naive algorithm

$R^+ = R$
LOOP
$\quad R^+ := R \vee (R^+ \circ R)$
WHILE R^+ changes

Semi-naive algorithm

$old_R := \emptyset$
$R^+ = R$
$\delta = R$
WHILE $\delta \neq \emptyset$
$\quad old_R := R^+$
$\quad R^+ := R^+ \vee \delta \circ R$
$\quad \delta := R^+ - old_R$
END

Differential semi-naive alg.

$R^+ = R$
$\delta = R$
LOOP
$\quad \Delta := (R^+ \circ \delta) \vee (\delta \circ R^+) \vee (\delta \circ \delta)$
$\quad \Delta := (\Delta - \delta) - R^+$
$\quad R^+ := R^+ \vee \delta$
$\quad \delta = \Delta$
WHILE $\delta \neq \emptyset$

Matrix algorithm

$R^+ = R$
LOOP
$\quad R^+ := R^+ \vee (R^+ \circ R^+)$
WHILE R^+ changes

Logarithmic algorithm

$R^+ = R$
$\Delta = R$
$\delta = R$
LOOP
$\quad \delta := \delta \circ \delta$
$\quad \Delta := R^+ \circ \delta$
$\quad R^+ := R^+ \vee \delta \vee \Delta$
WHILE R^+ changes

Fig. 4. Transitive closure algorithms for annotated binary relations

```
CREATE OR REPLACE FUNCTION MatrixRule2a(i_graph integer) RETURNS integer
DECLARE
    nrow_upd integer;
    nrow_ins integer;
BEGIN
LOOP
  UPDATE "rdfs:subClassOf" as r SET annotation=a.annotation FROM (
  SELECT   q1.graph, q1.subject, q2.object, MAX(tnorm(q1.annotation,q2.annotation)) annotation
  FROM "rdfs:subClassOf" AS q1 INNER JOIN "rdfs:subClassOf" AS q2 ON ( q1.object=q2.subject )
  WHERE q1.graph=i_graph AND q2.graph=i_graph
  GROUP BY q1.subject, q2.object, q1.graph
  ) AS a
  WHERE (r.subject,r.object,r.graph)=(a.subject,a.object,a.graph) AND r.annotation<a.annotation;
  GET DIAGNOSTICS nrow_upd = ROW_COUNT;

  INSERT INTO "rdfs:subClassOf" (
  SELECT   q1.graph, q1.subject, q2.object, MAX(tnorm(q1.annotation,q2.annotation)) annotation
  FROM "rdfs:subClassOf" AS q1 INNER JOIN "rdfs:subClassOf" AS q2 ON ( q1.object=q2.subject )
  WHERE q1.graph=i_graph AND q2.graph=i_graph AND
  NOT EXISTS (SELECT * FROM "rdfs:subClassOf" AS sc
             WHERE sc.subject = q1.subject AND sc.object=q2.object AND sc.graph=q1.graph)
  GROUP BY q1.subject, q2.object, q1.graph );
  GET DIAGNOSTICS nrow_ins = ROW_COUNT;

  IF (nrow_upd+nrow_ins=0) THEN
   EXIT;
  END IF;
 END LOOP;

RETURN 1;
END
```

As before the translation proceeds in two steps. First, we update the annotation of any already existing `rdfs:subClassOf` triple. Afterwards, we insert new `rdfs:subClassOf` triples of newly generated paths between nodes. In the case of the classical transitive closure implementation, it is not necessary to have the `UPDATE` statement, and that aggregation with `GROUP BY` to obtain the `MAX` of t-norm combined annotations, and of course the annotations. Therefore it is expected to have overhead with respect to the classical RDFS closure. We have implemented all the algorithms being the translation of the formal descriptions in Fig. 4 to SQL along the same lines.

5   Evaluation

In this section we present and discuss the results of the evaluation of the algorithms that we have developed. We start presenting the evaluation methods and the tests we have performed. First, we compare the several algorithms for classical RDFS closure and compare to the DLV^{DB} system [25]. We proceed by performing the testing with the annotated version algorithms in order to test effectiveness and obtain the overhead.

The algorithms were evaluated with respect to completeness and performance. We started by guaranteeing that the RDFS closure algorithms produced the correct outputs, namely that the output does not have duplicate or missing triples and, in the case of annotated graphs, that the annotations in each annotated triple are correct. For that matter, we produced several small tests (that we believe are representative of at least most kinds of graphs) to test the correctness of the implementation of each method. Afterwards we compared the number of triples in the output of each method to each other and with the DLV^{DB} system, for the same test. Since every method, and DLV^{DB}, returns the same amount of triples for each test, we have reasons to believe in the correctness of our implementation. The second aspect of evaluation was the amount of time that each method took to compute the closure for each test, which we now detail. The tests were built from the data sets YAGO, YAGO2 and WordNet 2.0. We chose YAGO and YAGO2 as our main sources of testing sets because YAGO contains large amounts of data, annotated with values in the [0,1] interval, so its the ideal dataset for the evaluation of annotated RDFS closure. Since WordNet was used in past works as test data set, in order to be possible to compare with other applications, we also tested the closure of this data set.

We devised six tests[5] to the Recursive, Semi-naive, Matrix-based, Differential, Logarithmic implementations and annotated versions. For the tests 1, 2, 3 and 4 only the code for the rule 2a) was executed since the objective of the tests is to show the way the diferent implementations react to transitive closure. For the remaining tests the full RDFS closure was computed. We also tested the closure time using DLV^{DB}. The description, sizes of the test sets and the output are shown in Table 1.

---

[5] The tested data and all the code needed to perform the tests is available at http://ardfsql.blogspot.com/

Table 1. Test sizes and specification

Test	1*	2*	3*	4*	5+	6+
Input Size	0.066M	0.366M	0.599M	3.617M	0.417M	1.942M
Output Size	0.599M	3.617M	0.599M	3.617M	3.790M	4.947M

* - only transitive closure of rdfs:subClassOf
+ - full RDFS closure of the input data

Test 1: Contains rdfs:subClassOf data from YAGO in the WordNetLinks file.
Test 2: Contains all rdfs:subClassOf data from YAGO2.
Test 3: Contains the output graph from Test 1.
Test 4: Contains the output graph from Test 2.
Test 5: Full RDFS closure of a subset of YAGO, containing the subclass data of the WordNetLinks file, all rdfs:subPropertyOf, rdfs:domain, rdfs:range triples and the triples from relations created, givenNameOf, inTimeZone, isLeaderOf, isPartOf, isSubstanceOf.
Test 6: Full RDFS closure of the WordNet 2.0 Full data set.

The tests were performed using a Laptop with an Intel i5 2.27GHz processor, 4Gb of RAM and running Windows 7 64-bit. The PostgreSQL 9.0 database was installed with default options, and no modification was made to the DB server. The data for each test was stored in a new database with no use of commands for the gathering of statistics. All database constraints were disabled and no indexing was used. We repeated each test three times, and in this paper we present the reasoning average time, excluding as usual data loading time.

Table 2. Results for the classical algorithms, time results presented in seconds

	1 Time	Deviation	2 Time	Deviation	3 Time	Deviation	4 Time	Deviation	5 Time	Deviation	6 Time	Deviation
Matrix	32.33	0.45%	325.18	17.93%	**8.93**	11.00%	69.85	5.24%	85.49	5.24%	103.06	2.24%
Logarithmic	41.99	1.23%	283.01	6.03%	60.50	0.59%	244.30	1.68%	89.02	1.45%	84.78	19.64%
Differential	38.53	0.84%	507.70	43.84%	48.87	2.96%	402.00	5.81%	169.45	6.81%	99.07	25.00%
Semi-Naive	142.10	1.04%	936.91	2.49%	40.83	0.38%	253.07	5.97%	188.66	3.57%	**73.65**	5.72%
Recursive	**12.18**	0.62%	**87.99**	1.29%	*		*		**65.71**	2.35%	96.89	22.25%
DLVdb	39.13	1.55%	327.36	7.72%	20.62	1.93%	133.54	3.37%	161.94	2.35%	179.14	4.35%

* timeout

The results for the classical algorithms for the six tests can be found in Table 2, where the best results for each test are in bold. The recursive implementation, using built-in recursive views of PostgreSQL, is either very well-behaved or extremely bad. In general, the semi-naive method has poor performance when compared to the best algorithm in all tests, except for test 6. The differential semi-naive performs better than semi-naive in tests 1,2 and 5, where the graph closure contains a large number of new triples obtained by transitive closure of the rdf:subClassOf relation. The semi-naive and differential semi-naive methods have worse performance than the other algorithms, justifying also the comparatively bad timings of the DLV^{DB} system which uses the differential one. The logarithmic method can be very good but its behavior oscillates more than the matrix method although they have similar performances.

The results of test 6 are particularly significant since in [27] the same task is performed in a 32 cluster machine in more than 3 minutes. In test 6 the five methods implemented by us have very similar times, in fact the algorithms with worse performance so far (semi-naive and differential) seem to perform much better than in test 5. This happens because the difference in graph closure times between the implementations comes from the time taken evaluating the transitive closure rules (rules 1a and 2a) and in test 6 the number of triples in the `rdfs:subPropertyOf` and `rdfs:subClassOf` relations is not big enough (respectivly 11 and 42) to show efficiency differences among transitive closure algorithms.

We have also performed a partial test with larger data for the classical case for determining the full RDFS closure of a larger subset of YAGO, containing all `rdfs:subClassOf`, `rdfs:subPropertyOf`, `rdfs:domain`, `rdfs:range` triples all `rdf:type` triples except those in "IsAExtractor", plus the triples of the YAGO relations bornOnDate, directed, familyNameOf, graduatedFrom, isMemberOf, isPartOf, isSubstanceOf, locatedIn, and worksAt. This test has 5.505M input triples generating 29.462M output triples. We have run the matrix and logarithmic methods obtaining 1547 seconds and 1947 seconds, respectively.

Table 3. Results for the annotated algorithms, time results presented in seconds

	1			2			3		
	Time	Deviation	Overhead	Time	Deviation	Overhead	Time	Deviation	Overhead
Matrix	**113.06**	3.52%	253%	1484.91	21.70%	356%	**30.48**	2.63%	241%
Logarithmic	126.90	1.27%	202%	**829.24**	10.24%	193%	155.02	0.39%	156%
Differential	163.36	4.86%	323%	1324.84	61.73%	160%	153.33	2.09%	213%
Semi-Naive	230.15	1.94%	61.96%	2761.80	115.54%	195%	149.05	3.05%	265%
Recursive	6679.70	57.74%	5496%	*	*	*	*	*	*

	4			5			6		
	Time	Deviation	Overhead	Time	Deviation	Overhead	Time	Deviation	Overhead
Matrix	**230.14**	0.01	229%	272.59	4.74%	218%	336.92	20.74%	226%
Logarithmic	1033.00	0.37	322%	**223.04**	3.71%	150%	341.63	35.34%	302%
Differential	978.63	0.10	143%	458.25	1.04%	170%	337.37	8.97%	240%
Semi-Naive	2177.60	1.16	760%	336.00	1.26%	78%	**195.18**	6.11%	165%
Recursive	*	*	*	6075.24	30.48%	914%	344.89	6.45%	245%

* timeout

The results for the annotated versions of the algorithm can be found in Table 3. It is clear that the recursive version does not scale with annotations, which we believe is due to the problem identified in Lemma 1. Most of the times the matrix algorithm is better and the logarithmic algorithm suffers from large variance problems. The differential algorithm is well-behaved in the case of the WordNet test data but most of the times it is not competitive.

The overhead introduced by the annotated versions is consistently around 250% for the case of the matrix method. This is expected since more queries and more complex are necessary to obtain the closure.

We have also performed some tests using a different database server configuration and indexes in specific tables. In these tests annotated and non annotated versions of the algorithms have performed better than with default configuration, in some cases three or four times faster. This leads us to believe that server configuration optimization may lead to great improvement of performance of these algorithms. This was expected and future work will be developed in order to provide more definitive conclusions.

6   Comparison and Related Work

Current triple stores like Sesame[6] and Jena[7] apparently do not perform the RDFS closure directly in the RDBMSs over stored data, and first load the data into memory, perform the inferences and store them afterwards. Moreover, their inference reasoning algorithms do not handle annotated data.

An extensive analysis of inferencing with RDF(S) and OWL data can be found in the description of the SOAR system [10]. A first major distinction to our approach is that part of the information is kept in main-memory, basically to what corresponds to our special tables for handling ρdf vocabulary and thus the transitive closure is performed in main memory, and no annotations are available. However, SOAR has more rules and implements a subset of OWL inferences, which we do not address here. It is used a technique call partial-indexing which relies on pre-processing a comparatively small-sized T-box with respect to the assertional data [11]. This terminological data extends the knowledge in our `rdfs:subClassof`, `rdfs:subPropertyOf`, `rdfs:domain`, and `rdfs:range` properties, and is not the major concern of the authors.

A novel extension of the SOAR system has been reported with scalable and distributed algorithms to handle particular annotations for dealing with trust/provenance including blacklisting, authoritativeness and ranking [3]. They cover a subset of the OWL 2 RL/RDF rules in order to guarantee a linear number of inferences on the size of the assertional data. The computation of the transitive closure of subclass and subproperty relations with annotations are performed by semi-naive evaluation with specialized algorithms, and thus it is not comparable to our approach.

Here we are not trying to assess query and storage trade-offs like [23] where just part of the RDFS closure graph is stored and the remaining triples are inferred at query time. However, we believe the same kind of balance still holds for annotated RDFS Semantic Web data.

Our approach also differs from [12] since we do not allow neither SPARQL querying nor change of entailment regime. A quite recent improvement of the annotated Semantic Web framework has been made available at [17,29]. The semantics have been extended with aggregate operators, and AnQL an extension of SPARQL to handle annotations has also been detailed with an available

---

[6] See http://www.openrdf.org/
[7] See http://jena.sourceforge.net/

memory-based implementation in Prolog using constraint logic programming. Since persistent storage is not supported we did not compare with this approach.

7   Conclusions and Future Work

In this paper we present a full relational database implementation of the annotated RDFS closure rules of [22] supporting any t-norm real-valued intervals, namely in the unit interval $[0, 1]$. We have analysed the dependencies of inference rules and proved that recursion is solely necessary for performing transitive closure of the subproperty and subclass relationships. We presented several algorithms for performing the transitive closure with annotated data, and implemented them in SQL. We performed practical evaluation of the several algorithms over existing data of YAGO and Wordnet knowledge bases for the case of minimum t-norm, and concluded that the matrix and logarithmic versions have better average behaviour than the other versions in the case of annotated data, but the logarithmic method is less reliable. The standard semi-naive evaluation shows poor performance, except when the subClassOf instances are in small number. We have shown that our approach for the case of non-annotated data has comparable or better performance than the DLV^{DB} system, and that the overhead imposed by annotations can be significant (from 150% to 350%). The relative behaviour of the compared algorithms carries over from the non-annotated to the annotated versions, except for the recursive algorithm. The recursive annotated transitive closure algorithm is extremely bad behaved, contrary to the non-annotated one.

We plan to extend the experimental evaluation to the large graphs in the Linked Data in order to confirm scalability of the techniques proposed as well as to evaluate the effect of indexing structures. We also would like to evaluate the impact of different t-norms in the running time of our algorithms. Moreover, since our proposal relies on standard database technology we would like to explore vendor specific facilities to improve performance of the developed system as well as increase generality. We also intend to explore alternative memory-based implementations and compare them with the current system.

References

1. Abadi, D.J., Marcus, A., Madden, S., Hollenbach, K.: SW-store: a vertically partitioned dbms for semantic web data management. VLDB 18(2), 385–406 (2009)
2. Biskup, J., Stiefeling, H.: Transitive closure algorithms for very large databases. In: Proceedings of the 14th International Workshop on Graph-Theoretic Concepts in Computer Science, London, UK, pp. 122–147. Springer, Heidelberg (1989)
3. Bonatti, P.A., Hogan, A., Polleres, A., Sauro, L., and: Robust and scalable linked data reasoning incorporating provenance and trust annotations. Journal of Web Semantics (to appear, 2011)
4. Broekstra, J., Kampman, A., van Harmelen, F.: Sesame: A Generic Architecture for Storing and Querying RDF and RDF Schema. In: Horrocks, I., Hendler, J. (eds.) ISWC 2002. LNCS, vol. 2342, pp. 54–68. Springer, Heidelberg (2002)

5. Damásio, C.V., Medina, J., Ojeda-Aciego, M.: Termination of logic programs with imperfect information: applications and query procedure. J. Applied Logic 5(3), 435–458 (2007)

6. Duan, S., Kementsietsidis, A., Srinivas, K., Udrea, O.: Apples and oranges: a comparison of RDF benchmarks and real RDF datasets. In: Proceedings of the 2011 International Conference on Management of Data, SIGMOD 2011, pp. 145–156. ACM, New York (2011)

7. Guo, Y., Pan, Z., Heflin, J.: Lubm: A benchmark for owl knowledge base systems. J. Web Sem. 3(2-3), 158–182 (2005)

8. Hajek, P.: The Metamathematics of Fuzzy Logic. Kluwer (1998)

9. Hogan, A., Harth, A., Polleres, A.: SAOR: Authoritative Reasoning for the Web. In: Domingue, J., Anutariya, C. (eds.) ASWC 2008. LNCS, vol. 5367, pp. 76–90. Springer, Heidelberg (2008)

10. Hogan, A., Harth, A., Polleres, A.: Scalable authoritative OWL reasoning for the web. Int. J. Semantic Web Inf. Syst. 5(2), 49–90 (2009)

11. Hogan, A., Pan, J.Z., Polleres, A., Decker, S.: SAOR: Template Rule Optimisations for Distributed Reasoning Over 1 Billion Linked Data Triples. In: Patel-Schneider, P.F., Pan, Y., Hitzler, P., Mika, P., Zhang, L., Pan, J.Z., Horrocks, I., Glimm, B. (eds.) ISWC 2010, Part I. LNCS, vol. 6496, pp. 337–353. Springer, Heidelberg (2010)

12. Ianni, G., Krennwallner, T., Martello, A., Polleres, A.: Dynamic Querying of Mass-Storage RDF Data with Rule-Based Entailment Regimes. In: Bernstein, A., Karger, D.R., Heath, T., Feigenbaum, L., Maynard, D., Motta, E., Thirunarayan, K. (eds.) ISWC 2009. LNCS, vol. 5823, pp. 310–327. Springer, Heidelberg (2009)

13. Ianni, G., Martello, A., Panetta, C., Terracina, G.: Efficiently querying RDF(S) ontologies with answer set programming. J. Log. Comput. 19(4), 671–695 (2009)

14. Ioannidis, Y.E.: On the computation of the transitive closure of relational operators. In: Proceedings of the 12th International Conference on Very Large Data Bases, VLDB 1986, pp. 403–411. Morgan Kaufmann Publishers Inc., San Francisco (1986)

15. Mohri, M.: Semiring frameworks and algorithms for shortest-distance problems. J. Autom. Lang. Comb. 7, 321–350 (2002)

16. Muñoz, S., Pérez, J., Gutierrez, C.: Minimal Deductive Systems for RDF. In: Franconi, E., Kifer, M., May, W. (eds.) ESWC 2007. LNCS, vol. 4519, pp. 53–67. Springer, Heidelberg (2007)

17. Lopes, N., Polleres, A., Straccia, U., Zimmermann, A.: AnQL: SPARQLing Up Annotated RDFS. In: Patel-Schneider, P.F., Pan, Y., Hitzler, P., Mika, P., Zhang, L., Pan, J.Z., Horrocks, I., Glimm, B. (eds.) ISWC 2010, Part I. LNCS, vol. 6496, pp. 518–533. Springer, Heidelberg (2010)

18. Nuutila, E.: Efficient transitive closure computation in large digraphs. Acta Polytechnica Scandinavia: Math. Comput. Eng. 74, 1–124 (1995)

19. RDF Semantics. W3C Recommendation, Edited by Patrick Hayes (February 10, 2004),
http://www.w3.org/TR/2004/REC-rdf-mt-20040210/

20. Silberschatz, A., Korth, H.F., Sudarshan, S.: Database System Concepts, 6th Edition, 6th edn. McGraw-Hill (2010), Material available at
http://codex.cs.yale.edu/avi/db-book/

21. Straccia, U.: A Minimal Deductive System for General Fuzzy RDF. In: Polleres, A., Swift, T. (eds.) RR 2009. LNCS, vol. 5837, pp. 166–181. Springer, Heidelberg (2009)

22. Straccia, U., Lopes, N., Lukacsy, G., Polleres, A.: A general framework for representing and reasoning with annotated semantic web data. In: Fox, M., Poole, D. (eds.) Procs. of AAAI 2010, AAAI Press (2010)
23. Stuckenschmidt, H., Broekstra, J.: Time – Space Trade-Offs in Scaling up RDF Schema Reasoning. In: Dean, M., Guo, Y., Jun, W., Kaschek, R., Krishnaswamy, S., Pan, Z., Sheng, Q.Z. (eds.) WISE 2005 Workshops. LNCS, vol. 3807, pp. 172–181. Springer, Heidelberg (2005)
24. Suchanek, F.M., Kasneci, G., Weikum, G.: Yago: A Core of Semantic Knowledge. In: 16th international World Wide Web Conference (WWW 2007), ACM Press, New York (2007)
25. Terracina, G., Leone, N., Lio, V., Panetta, C.: Experimenting with recursive queries in database and logic programming systems. TPLP 8(2), 129–165 (2008)
26. Udrea, O., Recupero, D.R., Subrahmanian, V.S.: Annotated rdf. ACM Trans. Comput. Log. 11(2) (2010)
27. Urbani, J., Kotoulas, S., Oren, E., van Harmelen, F.: Scalable Distributed Reasoning Using MapReduce. In: Bernstein, A., Karger, D.R., Heath, T., Feigenbaum, L., Maynard, D., Motta, E., Thirunarayan, K. (eds.) ISWC 2009. LNCS, vol. 5823, pp. 634–649. Springer, Heidelberg (2009)
28. Zadeh, L.A.: The concept of a linguistic variable and its application to approximate reasoning - parts I, II and III. Inf. Sciences 8(3), 199–249, 301–357, 43–80 (1975)
29. Zimmermann, A., Lopes, N., Polleres, A., Straccia, U.: A general framework for representing, reasoning and querying with annotated semantic web data. In: CoRR, abs/1103.1255 (2011)

Enabling Fine-Grained HTTP Caching of SPARQL Query Results

Gregory Todd Williams and Jesse Weaver

Tetherless World Constellation, Rensselaer Polytechnic Institute, Troy, NY, USA
{willig4,weavej3}@cs.rpi.edu

Abstract. As SPARQL endpoints are increasingly used to serve linked data, their ability to scale becomes crucial. Although much work has been done to improve query evaluation, little has been done to take advantage of caching. Effective solutions for caching query results can improve scalability by reducing latency, network IO, and CPU overhead. We show that simple augmentation of the database indexes found in common SPARQL implementations can directly lead to effective caching at the HTTP protocol level. Using tests from the Berlin SPARQL benchmark, we evaluate the potential of such caching to improve overall efficiency of SPARQL query evaluation.

1 Introduction

SPARQL endpoints are increasingly being used to provide access to large amounts of linked data. As use increases, both in frequency and complexity, and as the amount of data being served increases, scaling these systems to handle the increased load is crucial. Much work has been done on improving performance of SPARQL processors through more intelligent query planners, optimized index structures, and parallelization. However, there has been little work on addressing scalability through the use of caching.

Caching of query results can benefit both the SPARQL endpoint and the client. When a client uses a conditional HTTP request to which the server responds with a "Not Modified" message, only the IO for the response header is required. On the server, validating a conditional request is likely to be faster and require fewer resources (both CPU time and working memory) than evaluating the whole query. This allows the server to respond to more and/or more complex queries given fixed resources (or, conversely, response to the same queries with fewer resources). Moreover, if successfully validating a conditional request is faster than evaluating the query, the client benefits not only from reduced IO but also reduced latency and potentially by avoiding the need to parse the response (if a client's local cache is able to store a parsed representation).

The benefits of caching are only realized if both the client and server support the caching protocol and if requests are repeatedly made for already-cached results. Client-side support for caching is already available due to the widespread support in HTTP libraries that are used to implement the SPARQL protocol.

L. Aroyo et al. (Eds.): ISWC 2011, Part I, LNCS 7031, pp. 762–777, 2011.

In the rest of this work, we propose how to enable support for caching in the data structures used on the server. Because of the widespread support for HTTP caching, and the high frequency of repeated queries, caching of SPARQL query results has the potential to significantly improve efficiency. We restrict our work to only consider caching at the HTTP level as the standard SPARQL Protocol is defined in terms of HTTP[1].

The rest of this paper is organized as follows. Section 2 reviews related work. Section 3 reviews the caching features of HTTP. Section 4 defines "relevant" data as it pertains to caching the result of a SPARQL query pattern, and introduces the data structures and algorithms necessary for enabling caching in SPARQL query evaluation. Section 5 presents experimental results showing the effects of caching using the Berlin SPARQL Benchmark (BSBM). Section 6 concludes and discusses possible future work.

2   Related Work

Using caching to increase scalability touches upon several areas research, including statistical distributions of queries affecting their cacheability, indexing structures used in Semantic Web query answering systems, and caching as it relates to both the Semantic Web and to databases. In this section we discuss work related to these areas.

Regarding the repetition of queries, work on analyzing web access logs by Breslau, et al.[1] found that the statistical distribution of requests followed a "Zipf-like distribution" with the distribution exponent varying between different user communities. This finding suggests that caching can have a significant impact on real-world access patterns because a large portion of requests are made for a small set of resources. More recently, and related specifically to SPARQL requests, Gallego, et al.[2] analyzed a set of SPARQL endpoint query logs, and found a high degree of queries duplicated from the same hosts. However, Gallego, et al. only mention the repeated queries in passing, without specific details on the distribution of repeated queries.

There has been a trend in SPARQL systems to use search trees to efficiently index RDF data (following similar use in relational databases) and to use many indexes to support a range of access patterns. The YARS system[3] (and subsequently Hexastore[4] and RDF-3X[5]) demonstrated the effectiveness of maintaining many search tree indexes to provide direct access to RDF data matching a certain triple- or quad-pattern. YARS made use of six B+ tree indexes ($\langle SPOG \rangle$, $\langle POG \rangle$, $\langle OGS \rangle$, $\langle GSP \rangle$, $\langle GP \rangle$, $\langle OS \rangle$) to cover all sixteen possible quad-access patterns. Hexastore and RDF-3X, while only considering triples, both made use of six indexes ($\langle SPO \rangle$, $\langle SOP \rangle$, $\langle PSO \rangle$, $\langle POS \rangle$, $\langle OSP \rangle$, $\langle OPS \rangle$) to provide complete indexing of triples, covering all eight triple access patterns

---

[1] The SPARQL 1.0 Protocol is defined in terms of WSDL with bindings for HTTP and SOAP. However, the SPARQL 1.1 Protocol is defined only for HTTP as there was no widespread support for non-HTTP implementations.

and all possible orderings. The design of 4store[6] makes use of only three indexes ($\langle PS \rangle$, $\langle PO \rangle$, $\langle G \rangle$), using RADIX tries for the $\langle PS \rangle$ and $\langle PO \rangle$ indexes[2]. While all of these systems utilize search trees for performance, their use is restricted only to indexing the RDF data. In our work, we make use of the search trees not only to maintain many indexes over the RDF data but also to store additional metadata about when that data was modified. As described in the following sections, keeping such metadata allows a query processor to validate existing cached query results.

Caching database query results has been studied widely. Goldstein and Larson[7] show the potential of materializing views within a relational system to dramatically improve performance of expensive queries. Both Amiri, et al.[8] and Larson, Goldstein, and Zhou[9] addresses caching relational query results using materialized views. Amiri, et al. perform caching in edge caches separate from the origin database which reduces load on the origin server, but maintaining consistency of cached results requires that the origin database propagate every update, delete, and insert operation to all caches, making it unsuitable for environments with high write throughput. Larson, Goldstein, and Zhou improve upon the approach by Amiri, et al. by allowing more flexible materialization of views, allowing the query optimizer to choose whether to evaluate the query on the origin server even in the presence of cached data, and improving support for parameterized queries. In contrast to these approaches, the fixed structure of RDF data makes supporting caching much easier. No complex logic or knowledge of database schemas is needed to determine which tables or columns might benefit from caching as all data is structured in terms of triples.

Caching as it relates to the Semantic Web is a much more recent field of study. The work on caching SPARQL query results by Martin, Unbehauen, and Auer[10] shares the same goal and many details with our work (we frequently cite this work herin as a source of common groundwork and greater detail) However, they perform caching by coupling a caching layer with an existing SPARQL processor. This has the benefit of portability across SPARQL implementations, but incurs high cache maintenance costs and is suitable only for caches which are tightly integrated with the underlying system and can intercept all write operations (e.g. ISP caches or caches built into a user agent cannot be used). Finally, the work in [10] deals with only a subset of SPARQL; in this work we make specific note of several features of SPARQL that deserve special attention in the context of caching.

Work by Hartig[11] shows performance improvements of using a local cache in evaluating queries by linked data link traversal. This work is complementary to ours in that both use caching to improve performance of query answering, one over static linked data, the other over potentially dynamic data available via structured queries.

---

[2] 4store actually maintains two RADIX tries for each predicate, but for our purposes these may be understood as being equivalent to tries with P prepended to the actual keys.

3   HTTP Caching

In this section we introduce the caching features available in HTTP upon which our system relies.

HTTP supports two primary caching mechanisms, allowing servers to explicitly indicate a caching expiration (with an `Expires` date or a `max-age` duration) or indicating a cache validator (with a `Last-Modified` date or ETag value). Here we concern ourselves only with cache validators – specifically, `Last-Modified` dates – as they are a more natural fit for caching data that may be updated in the database at any time. However, as they relate to our work, both the `Last-Modified` and `ETag` headers may be understood as being effectively equivalent as we do not use the more expressive "weak validation" that ETags allow.

The `Last-Modified` validator works as follows. A client user-agent requests a resource (in this case a SPARQL query) to the server:

```
GET /sparql?query=SELECT... HTTP/1.1
Host: example.org
```

The server sends back a response whose header includes the `Last-Modified` validator with a date indicating when the resource was last modified:

```
HTTP/1.1 200 OK
Last-Modified: Wed, 1 Jun 2011 12:45:15 GMT
Content-Type: application/sparql-results+xml

<?xml version="1.0"?>
<sparql xmlns="http://www.w3.org/2005/sparql-results#">
...
</sparql>
```

At some point in the future, the client requests the resource again and, noting that the response is cached from the last time it was requested, indicates the request as *conditional* by using the `If-Modified-Since` header with the previously returned validator date:

```
GET /sparql?query=SELECT... HTTP/1.1
Host: example.org
If-Modified-Since: Wed, 1 Jun 2011 12:45:15 GMT
```

If the resource has not changed since the validator date, the server sends a response indicating that the already-cached content is still valid:

```
HTTP/1.1 304 Not Modified
```

Otherwise, the server responds as usual with a full response, including the resource content and any applicable cache validators (the updated `Last-Modified` time).

In terms of SPARQL, HTTP caching ought to make query results appear as valid ("fresh") so long as the query results do not change. Since determining precisely if results to a query have changed may require re-evaluating the query (negating one of the benefits of caching), we settle for a less strict condition: caching ought to make the query results appear as valid so long as data "relevant" to the query have not changed. Once a query result has been cached, updates to "irrelevant" data in the SPARQL endpoint should have no affect on the caching – upon re-submitting the query, the server should indicate that the cached results are still valid. "Relevant" data being updated prior to the query being re-submitted should result in the server re-evaluating the query and returning fresh query results. In section 4.2 we define "relevant" and "irrelevant" data.

4   Methodology

We propose modifying the search trees used to index RDF data in a simple way to enable determining the effective modification time of data relevant to a query. In the following sections we show how the modification time data stored in the search trees can be maintained during database updates, and how the data can be retrieved and used at query time. In determining what data is relevant to a specific query, we extend the work done by Martin, Unbehauen, and Auer[10] (what they call "Graph Pattern Solution Invariance") to support the much more expressive queries and graph patterns of SPARQL 1.1[3]. This includes the use of named graphs, property paths, and DESCRIBE queries.

In this work we assume that the SPARQL processor is built using a quad-store, and that the SPARQL "RDF Dataset" is mapped directly to the statements in the quad-store (with a special graph name representing the default graph). This assumption simplifies the following discussion, but is not required by our approach. The algorithms we present can be extended to work with arbitrary mappings between RDF dataset and quad-store, or to work with graph-stores.

4.1   Search Tree Indexing

Search trees are a common data structure used to implement efficient access to RDF data for varying access patterns. To determine the modification-time (mtime) of "relevant" data in a search tree, we propose adding an mtime field to each node in the search tree. During an update operation (insertion or deletion of RDF data), we update the mtime field in each affected search tree node. Moreover, during an update we ensure the mtime of each node in the tree is greater than or equal to the mtime of all of its children. In this way, the mtime of a node in the tree can be used as a conservative proxy value for the mtime of any of its descendant nodes.

For each access pattern in a query, we can now determine an mtime after which we can be assured that no update operation has affected data matching

---
[3] http://www.w3.org/TR/sparql11-query/

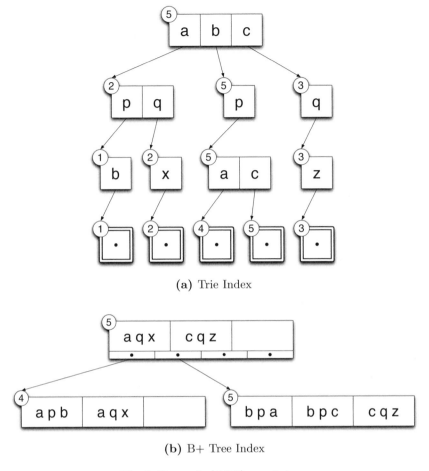

(a) Trie Index

(b) B+ Tree Index

Fig. 1. Example $\langle SPO \rangle$ search trees

the pattern. By calculating the maximum mtime over all the query access patterns, we arrive at a single timestamp which is at least as recent as the actual modification time of the query's "relevant" data.

While different SPARQL systems make use of different types of search trees, and use varying numbers of indexes, we propose a general solution that works with any number of indexes and across a variety of tree types (we discuss specifics of both B+ trees and tries). Although the caching results in our proposed system are complete, soundness is affected by the choice of a specific search tree type and number of available indexes. For example, fewer indexes, or the use of B+ trees versus tries, may cause some query results to appear as if they have changed when in fact they haven't. However, query results that are asserted as being the same as cached results will always in fact be the same.

Figure 1 shows both a B+ tree of order 4 and a trie with example data[4]. The mtime of each tree node appears inside the circle attached to each node, and shows the mtimes that result from this example 5-update sequence (with mtimes starting at 1, and incrementing on subsequent operations):

1. Insert triple { a p b }
2. Insert triple { a q x }
3. Insert triple { c q z }
4. Insert triple { b p a }
5. Insert triple { b p c }

As can be seen, the trie maintains the correct mtime for each leaf node while the mtimes in the B+ tree only indicate that the triples with subject a were not updated by operation 5. We discuss the reasons for this unsoundness in section 4.3.

4.2   Relevant Data and Graph Patterns

We define data "relevant" to a SPARQL query as being data that *may* affect the results of the query. Martin, Unbehauen, and Auer claim: "the solution of a graph pattern stays the same at least until a triple, which matches any of the triple patterns being part of the graph pattern, is added to or deleted from the [graph]." This is true when considering only *triple*-patterns in the default graph, but to support the full expressivity of SPARQL, we must extend this claim: The solution of a query with a graph pattern stays the same at least until:

- a triple, which matches any of the triple patterns being part of the graph pattern, is added to or deleted from the default graph
- a triple, which matches any of the triple patterns being part of a GRAPH <iri> pattern, is added to or deleted from the <iri> named graph
- a triple, which matches any of the triple patterns being part of a GRAPH ?var pattern, is added to or deleted from any named graph
- a triple is added to a new named graph and the graph pattern includes an empty GRAPH ?var pattern
- a triple is removed from a named graph, leaving the graph empty, and the graph pattern includes an empty GRAPH ?var pattern
- a triple, with predicate matching any part of a property path being part of the graph pattern, is added to or deleted from the dataset
- a triple is added or removed from the dataset, and the graph pattern includes a zero-length or negated property path
- a triple is added to or deleted from the dataset, and the query uses the DESCRIBE form

We discuss each of these cases and how they relate to relevant data below.

---

[4] The example data used here is comprised of triples for brevity; the handling of mtimes is identical for quad data.

Named Graph Patterns. As noted in [10], the addition or deletion of a triple (to the default graph) may change the solutions of a graph pattern. To fully support SPARQL datasets (which contain not just the default graph, but also any number of named graphs), we must also consider graph patterns scoped to a named graph. These patterns may either be scoped to a specific named graph (using the `GRAPH <iri> { ... }`) syntax) or be scoped to *any* named graph (using the `GRAPH ?var { ... }` syntax). For a graph pattern scoped to a specific named graph, `iri`, the solutions to the graph pattern may change with the addition or deletion of a triple matching the graph pattern to the named graph `iri`. For a pattern scoped to any named graph, the solutions to the pattern may change with the addition or deletion of a triple matching the graph pattern to any named graph.

Empty Named Graph Patterns. Beyond graph patterns scoped to named graphs, special handling is required for the *empty* named graph pattern:

`SELECT ?g WHERE { GRAPH ?g {} }`

This query returns the set of graph names in the dataset. The query has no triple-patterns which might match triples being added or removed, yet its results may change based on added or removed data. Specifically, adding a triple to a new named graph, or removing the final triple from a named graph[5] may change the solutions to this pattern.

Paths. Property paths greatly increase the expressiveness of SPARQL, but as they relate to relevant data, may be reduced to the matching of triple patterns. We can partition property paths into two categories: fixed-length and variable-length. Fixed-length property paths are those that can be syntactically represented by basic graph patterns (BGPs). Due to their equivalence with BGPs (sets of triples), these paths can be handled the same as triple patterns.

Variable-length paths are those that cannot be reduced to BGPs, and may rely on new algebraic operations to match data. These include zero-length paths (`?s <p>{0} ?o`), zero-or-more paths (`?s <p>* ?o`), one-or-more paths (`?s <p>+ ?o`), and negated paths (`?s !<p> ?o`). Due to their complexity, we discuss only a subset of the expressivity of these path types.

With respect to "relevant" data, one-or-more paths with simple predicates (those in which the + path operator applies to an IRI) can be reduced to triple pattern matching with predicate-bound triple patterns. For example, the path `<s> <p>+ ?o` has the same relevant data as the triple pattern `?s <p> ?o`. Note that while the subject is bound to `<s>` in the path pattern, it must be unbound in the triple pattern equivalent as the path may be affected by triples where the subject is not `<s>`.

Zero-length paths and negated paths require special attention. The zero-length path connects a graph node (any subject or object in the graph) to itself. Therefore, any insertion (deletion) in a graph may affect the results to a zero-length

---

[5] Some SPARQL systems allow empty named graphs to exist. Removing all triples from a named graph would not affect the set of graph names on such systems.

path pattern by adding (removing) a node to the graph that didn't exist before (doesn't exist after) the update. Similarly, a negated path ?s !<p> ?o implies that any insertion or deletion *not* using the <p> predicate may impact the results. While the relevant data for such a negated path is a subset of all the data, we assume that realistic datasets will contain a range of predicates and so the relevant data will be very large (in many cases approximating the size of the dataset itself). Therefore, we conservatively assume that the entire dataset is relevant to a negated path pattern.

DESCRIBE Queries. DESCRIBE queries present a challenge in determining relevant data. These queries involve matching a graph pattern just as SELECT queries do (a DESCRIBE query without a WHERE clause being semantically equivalent to one with an empty WHERE clause). However, the final results of a DESCRIBE query depend on the WHERE clause *and* the algorithm used for enumerating the RDF triples that comprise the description of a resource.

A naïve DESCRIBE algorithm would be to return all the triples in which the resource appeared as the subject. For our purposes, this algorithm would make this query:

```
DESCRIBE ?s
WHERE { ?s a <Class> }
```

roughly equivalent to a SELECT query with an additional triple pattern:

```
SELECT ?s ?p ?o
WHERE { ?s a <Class> .
        ?s ?p ?o }
```

Since most DESCRIBE algorithms will include *at least* these triples, and given the course-grained nature of the triple pattern ?s ?p ?o (matching every triple in the database), we consider the "relevant" data for a DESCRIBE query to be all data in the database. We note that the work in [10] does not (and need not) address this issue as that work is concerned with caching of *graph pattern* results, not *query* results. Since the DESCRIBE query form takes *graph pattern* results (or ground IRIs) as input, and outputs an implementation-dependent set *query* results, the caching of *query* results must respect this process.

Given that the algorithm used for DESCRIBE queries is implementation dependent, our definition of "relevant" data for DESCRIBE queries is intentionally conservative and we do not discuss specific handling of DESCRIBE queries in any further detail.

4.3   Maintaining and Probing Cache Status

In this section we briefly describe the algorithms used during update operations to maintain the mtime field in the search tree. We then describe the probing algorithm used to determine the effective mtime of the relevant data for a specific query.

Cache Maintenance. Maintaining the mtime field in the search tree is a simple process:

1. Before each tree node is written to disk (due to an insertion of deletion), update the node's mtime to the current time.
2. For each node that is written to disk, write its parent to disk (thereby updating its parent's mtime).

This process will ensure our condition that every tree node's mtime is greater or equal to those of its descendants and can be used as the effective mtime of descendant, relevant data.

We distinguish between the effective mtime of data matching an access pattern, and that data's actual mtime. As discussed in section 4.1, the specific data structure used for the search tree affects the granularity (and therefore the expected accuracy) of the effective mtime. Due to their design, tries yield effective mtimes that are exactly the same as the most recent mtime of data matching an access pattern. B+ trees yield effective mtimes of matching data that may be affected by any non-matching data that is co-located on a leaf node with matching data.

During the update process, we note that the parent node(s) may already need to be written to disk (in the case of a node split), so step 2 may already be required on any given update. Moreover, an update at a leaf node in append-only and counted B+ trees cause a cascade of writes up to the (possibly new) root. In these cases, all IO incurred by the cache update algorithm is already required by the update operation, and so the cost of maintaining the cache data is effectively free.

Cache Probing. The algorithm used for probing a database index to retrieve the effective mtime for a query is shown in algorithm 1. Given a query and a set of available search tree indexes, for each access pattern in the query, the algorithm probes the index that will yield the most accurate effective mtime, and returns the most recent of the mtimes. The index that will yield the most accurate effective mtime is the one with a key ordering that will allow descending as deep into the tree as there are bound terms in the access pattern. If no such index exists, a suitable replacement index is chosen that maximizes the possible depth into the tree that some subset of bound terms in the access pattern will allow. In the case of the completely unbound access pattern, the effective mtime is the same as the mtime of the entire dataset and so can be retrieved from the root node of *any* available index.

While this algorithm describes how the effective mtime of a query may be computed, it is worth noting that the specific steps described may be implemented in more or less efficient ways. For example, the algorithm calls for finding the lowest common ancestor (LCA) of data matching the access pattern. For a system using B+ trees, a naïve implementation might traverse the tree to find the leaves with matching data and then walk up the tree to find the LCA. A more efficient implementation could avoid having to find all leaves with matching data by

---

Algorithm 1. Probe database for effective mtime of query results

---

Input: A SPARQL query graph pattern *query*, a set of available database indexes *indexes*

Output: $effectiveMtime$, the effective modification time of relevant data for the query

1 $mtimes = \emptyset$;
2 **foreach** $ap \in query$ **do**
3      $orderedIndexes = \{i | i \in indexes, \exists s \subseteq boundPositions(ap)$ s.t. the key order of i starts with $s\}$;
4   **if** $|orderedIndexes| > 0$ **then**
5          $index = \underset{i \in orderedIndexes}{argmax} |s|$;
6   n = LCA of data matching ap in $index$;
7          $mtimes = mtimes \cup \{mtime(n)\}$;
8   **else**
9   i = any index in $indexes$;
10         $mtimes = mtimes \cup \{mtime(root(i))\}$;
11   **end**
12 **end**
13 $effectiveMtime = Max(mtimes)$;
14 **return** $effectiveMtime$

---

traversing tree edges until finding the LCA by using the bounds data contained in internal nodes.

As discussed above and in section 4.1, the soundness of results is affected by the choice of the search tree data structure used. B+ trees produce less sound results as a result of maintaining less accurate effective mtimes. Tries will result in more sound results as a result of being able to maintain accurate effective mtimes. Even though tries maintain accurate effective mtimes, their use does not guarantee perfectly sound cache validation as updates that affect data relevant to a query may not change the results to that query. This can occur when the relevant updated data does not appear in the query results due to join conditions, filter expressions, or projection. In these cases, cache validation will fail and the query must be evaluated again, despite accurate results already being cached.

One final case that is worth noting is the special case of determining the effective mtime for an *empty* named graph pattern (GRAPH ?g). As discussed in section 4.2, this pattern returns the set of available named graphs. If the set of available indexes are all covering indexes (using key orders that are just permutations of subject, predicate, object, and graph), then there is no way to determine an accurate effective mtime for this pattern. However, if there is an available index over just $\langle G \rangle$, an accurate effective mtime for the set of named graphs is stored in the $\langle G \rangle$ index root node.

5   Evaluation

We evaluated the potential impact on performance of query result caching by implementing a simple SPARQL process in C with B+ tree indexes. To index data, we use the six index orderings $\langle SPOG \rangle$, $\langle SGOP \rangle$, $\langle POGS \rangle$, $\langle OGSP \rangle$, $\langle OSPG \rangle$, and $\langle GPSO \rangle$. We evaluated our system using a slightly modified version of the Berlin SPARQL Benchmark[6] using both the Explore (read-only) and the Explore and Update (read-write) use cases. All BSBM evaluation was performed on a dual Intel Xeon E5504 Quad Core 2.0GHz processor with 24GB of memory, with 5 warmup runs, and 10 timed runs.

5.1 Modified Berlin SPARQL Benchmark

We believe the standard BSBM benchmark fails to account for the skewed distribution of real-world queries and so, following the work in Martin[10], modified the benchmark test driver to use a Pareto distribution for benchmark queries. The evaluation tests were performed with varying query repetition as represented by the α parameter. We also modified the benchmark test driver to support HTTP caching by storing query results when they are returned with caching headers, and validating existing cached query results using conditional requests.

5.2 Explore Use Case

The Explore use case of BSBM consists of a set ("query mix") of read-only queries that simulate a consumer looking for product information in an e-commerce setting. In our evaluation, this use case tests performance gains from caching on a static dataset. No updates (neither relevant nor irrelevant) are performed and so once cached, query results are always valid.

Figure 2 shows the performance improvement of our caching system on the BSBM Explore use case (as a percentage increase over the same tests run without the use of caching). The test was run with α distribution values ranging from 0.1 to 4.0, and shows between 35–650% increase in benchmark performance.

5.3 Explore and Update Use Case

The Explore and Update use case of BSBM consists of the same queries as in the Explore use case, with occasional updates to the dataset representing new products, reviews, and offers being added to, and old offers being removed from the dataset. This use case tests performance gains from caching on both static datasets (intra-query mix) and updated dataset (inter-query mix, after an update set). The updates contain both relevant and irrelevant data to the queries in the Explore use case.

Figure 3 shows the performance improvement of our caching system on the BSBM Explore and Update use case, again with α distribution values ranging from 0.1 to 4.0, and shows between 2–160% increase in benchmark performance.

5.4 Cost of Caching

To evaluate whether implementing our caching system increases overall processing cost, we evaluated the difference in performance between two versions of

---

[6] http://www4.wiwiss.fu-berlin.de/bizer/BerlinSPARQLBenchmark/

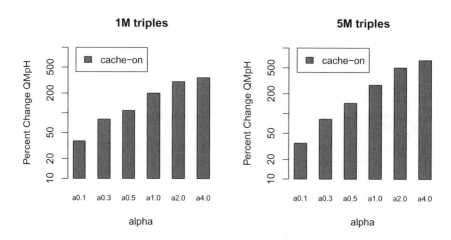

Fig. 2. BSBM Explore Use Case

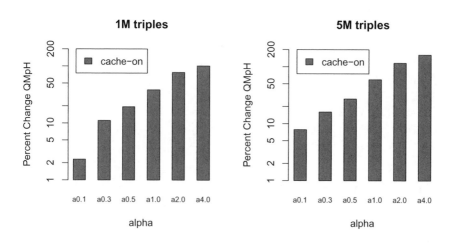

Fig. 3. BSBM Explore and Update Use Case

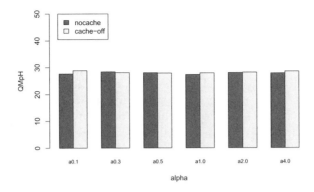

Fig. 4. Cost of Caching, BSBM Explore and Update Use Case

our system. In one version ("nocache"), we compile our system without any cache-supporting code. In the other ("cache-off"), caching support is included. However, both of these versions were tested against the Explore and Update use case with no caching support enabled in the test driver. In the cache-enabled version, this tests both the cost of cache probing to generate the `Last-Modified` response header during the explore phase and the cost of maintaining mtimes during the update phase. As can be seen in figure 4, the cache-enabled system performs roughly the same as the version without caching (executing at times both slightly faster and slower than the baseline "nocache" system).

5.5 Discussion

The performance of this system is not competitive with existing SPARQL stores, which achieve dramatically higher scores on the Berlin SPARQL benchmark. We attribute this to our system being a testing implementation meant to demonstrate our cache-supporting indexes and algorithms, not meant to compete head to head with production systems. Specifically, our system uses a very basic B+ tree implementation with no optimization to reduce disk IO, and lacks any query optimizer or memory management of database pages. We would have liked to evaluate our caching approach using a more efficient implementation, but found that modifying the low-level index structures with mtime fields *and* making those fields available through many layers of API abstractions was very difficult. Overall, we suggest attention should be paid to the large relative improvement of performance with query result caching, not on the specific QMpH figure of our implementation.

We believe our system's lack of database page management may hurt overall caching performance. The ability to cache database pages in memory would not only improve overall performance, but in some cases would specifically improve performance of the cache query (probe) algorithm. Specifically, in cases where the upper levels of index nodes reside entirely in memory, accessing the LCA nodes that provide the mtime of relevant data may require no disk access whatsoever.

Conversely, a highly optimized system tuned for very fast pattern matching and joins would narrow the performance gap between validating a conditional query with cache probing and evaluating the query in full. This situation would seem to provide less benefit from caching. However, this narrowing of performance gap is only one aspect of the benefits from caching. Even on a very efficient implementation, caching would still reduce the network IO required to transfer the query results (which our evaluation does not address) and the memory usage on the server.

6   Conclusion

Caching of SPARQL query results is a promising approach to improving scalability. In this paper we have shown that simple modification of the indexing structures commonly used in SPARQL processors can allow fine-grained caching of query results based on the freshness of data relevant to the query. We evaluated this system using the Berlin SPARQL Benchmark and found that caching can dramatically improve performance in the presence of repeated queries. Moreover, maintaining the data required for caching and using it to service conditional query requests has low cost compared to fully evaluating queries.

In the future we hope to apply the presented caching structures and algorithms to an existing, optimized SPARQL processor and evaluate it using much larger datasets and more expressive queries than those provided by BSBM. We also believe this work could be improved in many ways. The precision (and therefore the soundness) of the cache probing algorithm might be improved by taking into account the ways in which relevant data is combined and modified (e.g. using joins, filters, and projection). In many cases, typical queries use only a subset of available database indexes. Cache-enabled search trees for only the most frequent access patterns could be augmented with non-tree indexes, allowing the system to leverage the benefits of certain non-tree index structures while keeping the precision of caching for frequent queries. Finally, other indexing structures might also be modified to store similar fine-grained caching data, allowing informed indexing structure choices while maintaining the benefits of caching.

Acknowledgements. We thank Timothy Lebo and Lee Feigenbaum for their helpful comments and suggestions about this work.

References

1. Breslau, L., Cao, P., Fan, L., Phillips, G., Shenker, S.: Web caching and Zipf-like distributions: Evidence and implications. In: Proceedings of INFOCOM 1999, Eighteenth Annual Joint Conference of the IEEE Computer and Communications Societies (1999)

2. Gallego, M., Fernández, J., Martínez-Prieto, M., Fuente, P.: An empirical study of real-world SPARQL queries. In: USEWOD 2011 - 1st International Workshop on Usage Analysis and the Web of Data (2011)
3. Harth, A., Decker, S.: Optimized index structures for querying RDF from the web. In: Proceedings of the 3rd Latin American Web Congress (2005)
4. Weiss, C., Karras, P., Bernstein, A.: Hexastore: sextuple indexing for semantic web data management. In: Proceedings of the VLDB Endowment Archive (2008)
5. Neumann, T., Weikum, G.: RDF-3X: a RISC-style engine for rdf. In: Proceedings of the VLDB Endowment Archive (2008)
6. Harris, S., Lamb, N., Shadbolt, N.: 4store: The design and implementation of a clustered rdf store. In: Proceedings of the 5th International Workshop on Scalable Semantic Web Knowledge Base Systems, SSWS 2009 (2009)
7. Goldstein, J., Larson, P.: Optimizing queries using materialized views: a practical, scalable solution. In: Proceedings of the 2001 ACM SIGMOD International Conference on Management of Data (2001)
8. Amiri, K., Park, S., Tewari, R., Padmanabhan, S.: Dbproxy: A dynamic data cache for web applications. In: Proceedings of the 19th International Conference on Data Engineering, ICDE 2003 (2003)
9. Larson, P., Goldstein, J., Zhou, J.: Mtcache: transparent mid-tier database caching in sql server. In: Proceedings of 20th International Conference on Data Engineering, pp. 177–188 (2004)
10. Martin, M., Unbehauen, J., Auer, S.: Improving the Performance of Semantic web Applications with SPARQL Query Caching. In: Aroyo, L., Antoniou, G., Hyvönen, E., ten Teije, A., Stuckenschmidt, H., Cabral, L., Tudorache, T. (eds.) ESWC 2010. LNCS, vol. 6089, pp. 304–318. Springer, Heidelberg (2010)
11. Hartig, O.: How caching improves efficiency and result completeness for querying linked data. In: Proceedings of the 4th Linked Data on the Web (LDOW) Workshop (March 2011)
12. Guéret, C., Groth, P., Oren, E., Schlobach, S.: eRDF: A scalable framework for querying the web of data, pp. 1–17 (October 2010)

dipLODocus[RDF]—Short and Long-Tail RDF Analytics for Massive Webs of Data

Marcin Wylot, Jigé Pont, Mariusz Wisniewski, and Philippe Cudré-Mauroux

eXascale Infolab
University of Fribourg, Switzerland
{firstname.lastname}@unifr.ch

Abstract. The proliferation of semantic data on the Web requires RDF database systems to constantly improve their scalability and transactional efficiency. At the same time, users are increasingly interested in investigating or visualizing large collections of online data by performing complex analytic queries. This paper introduces a novel database system for RDF data management called dipLODocus[RDF] , which supports both transactional and analytical queries efficiently. dipLODocus[RDF] takes advantage of a new hybrid storage model for RDF data based on recurring graph patterns. In this paper, we describe the general architecture of our system and compare its performance to state-of-the-art solutions for both transactional and analytic workloads.

1 Introduction

Despite many recent efforts, the lack of efficient infrastructures to manage RDF data is often cited as one of the key problems hindering the development of the Semantic Web. Last year at ISWC, for instance, the two industrial keynote speakers (from the New York Times and Facebook) pointed out that the lack of an open-source, efficient and scalable alternative to MySql for RDF data was the number one problem of the Semantic Web.

The Semantic Web community is not the only one suffering from a lack of efficient data infrastructures. Researchers and practitioners in many other fields, from business intelligence to life sciences or astronomy, are currently crumbling under gigantic piles of data they cannot manage or process. The current crisis in data management is from our perspective the result of three main factors: i) rapid advances in CPU and sensing technologies resulting in very cheap and efficient processes to create data ii) relatively slow advances in primary, secondary and tertiary storage (PCM memories and SSD disks are still expensive, while modern SATA disks are singularly slow–with seek times between 5ms and 10ms typically) and iii) the emergence of new data models and new query types (e.g., graph reachability queries, analytic queries) that cannot be handled properly by legacy systems. This situation resulted in a variety of novel approaches to solve specific problem, for large-scale batch-processing [10], data warehousing [20], or array processing [8].

L. Aroyo et al. (Eds.): ISWC 2011, Part I, LNCS 7031, pp. 778–793, 2011.

Nonetheless, we believe that the data infrastructure problem is particularly acute for the Semantic Web, because of its peculiar and complex data model (which can be modeled as a constrained graph, as a ternary or n-ary relation, or as an object-oriented model depending on the context) and of the very different types of queries a typical SPARQL end-point must support (from relatively simple transactional queries to elaborate business intelligence queries). The recent emergence of distributed Linked Open Data processing and visualization applications relying on complex analytic and aggregate queries is aggravating the problem even further.

In this paper, we propose dipLODocus[RDF] , a new system for RDF data processing supporting both simple transactional queries and complex analytics efficiently. dipLODocus[RDF] is based on a novel hybrid storage model considering RDF data both from a graph perspective (by storing RDF subgraphs or RDF molecules) and from a "vertical" analytics perspective (by storing compact lists of literal values for a given attribute). dipLODocus[RDF] trades insert complexity for analytics efficiency: isolated inserts and simple look-up are relatively complex in our system due to our hybrid model, which on the other hand enables us to considerably speed-up complex queries.

The rest of this paper is structured as follows: we start by discussing related work in Section 2. Section 3 gives a high-level overview of our system and introduces our hybrid storage scheme. We give a more detailed description of the various data structures in dipLODocus[RDF] in Section 4. We describe how our system handles common operation like bulk inserts, updates, and various types of queries in Section 5. Section 6 is devoted to a performance evaluation study, where we compare the performance of dipLODocus[RDF] to state-of-the-art systems both for a popular Semantic Web benchmark and for various analytic queries. Finally, we conclude in Section 7.

2   Related Work

Approaches for storing RDF data can be broadly categorized in three subcategories: triple-table approaches, property-table approaches, and graph-based approaches. Many approaches have been proposed to optimize RDF query processing; we list below some of the most popular approaches and systems. We refer the reader to recent surveys of the field (such as [15], [13], or [16]) for a more comprehensive coverage.

Triple-Table Storage: since RDF data can be seen as sets of *subject-predicate-object* triples, many early approaches used a giant triple table to store all data. Our GridVine [2,7] system, for instance, uses a triple-table storage approach to distribute RDF data over decentralized P2P networks using the P-Grid [1] distributed hash-table. More recently, Hexastore [21] suggests to index RDF data using six possible indices, one for each permutation of the set of columns in the triple table, leading to shorter response times but also a worst-case five-fold increase in index space. Similarly, RDF-3X [17] creates various indices from a giant triple-table, including indices based on the six possible permutations of the triple columns, and aggregate indices storing only two out of the three columns. All indices are heavily compressed using dictionary encoding and byte-wise compression mechanisms. The query executor of RDF-3X implements a dedicated cost-model to optimize join orderings and determine the cheapest query plan automatically.

Property-Table Storage: various approaches propose to speed-up RDF query processing by considering structures clustering RDF data based on their *properties*. Wilkinson *et al.* [22] propose the use of two types of property tables: one containing clusters of values for properties that are often co-accessed together, and one exploiting the type property of subjects to cluster similar sets of subjects together in the same table. Chong *et al.* [6] also suggest the use of property tables as materialized views, complementing a primary storage using a triple-table. Going one step further, Abadi *et al.* suggest a fully-decomposed storage model for RDF: the triples are in that case rewritten into n two-column tables where n is the number of unique properties in the data. In each of these tables, the first column contains the subjects that define that property and the second column contains the object values for those subjects. The authors then advocate the use of a column-store to compactly store data and efficiently resolve queries.

Graph-Based Storage: a number of further approaches propose to store RDF data by taking advantage of its graph structure. Yan *et al.* [23] suggest to divide the RDF graph into subgraphs and to build secondary indices (e.g., Bloom filters) to quickly detect whether some information can be found inside an RDF subgraph or not. BitMat [4] is an RDF data processing system storing the RDF graph as a compressed bit matrix structure in main-memory. gStore [24] is a recent system storing RDF data as a large, labeled, and directed multi-edge graph; SPARQL queries are then executed by being transformed into subgraph matching queries, that are efficiently matched to the graph using a novel indexing mechanism.

Several of the academic approaches listed above have also been fully implemented, open-sourced, and used in a number of projects (e.g., GridVine[1], Jena[2], and RDF-3X[3]).

A number of more industry-oriented efforts have also been proposed to store and manage RDF data. Virtuoso[4] is an object-relational database system offering bitmap indices to optimize the storage and processing of RDF data. Sesame[5] [5] is an extensible architecture supporting various back-ends (such as PostgreSQL) to store RDF data using an object-relational schema. Garlik's 4Store[6] is a parallel RDF database distributing triples using a round-robin approach. It stores triple in triple-tables (or quadruple-tables more precisely). BigOWLIM[7] is a scalable RDF database taking advantage of ordered indices and data statistics to optimize

---

[1] http://lsirwww.epfl.ch/GridVine/
[2] http://jena.sourceforge.net/
[3] http://www.mpi-inf.mpg.de/neumann/rdf3x/
[4] http://virtuoso.openlinksw.com/
[5] http://www.openrdf.org/
[6] http://4store.org/
[7] http://www.ontotext.com/owlim/

queries. AllegroGraph[8], finally, is a native RDF database engine based on a quadruple storage.

3   System Rationale

Our own storage system in dipLODocus[RDF] can be seen as a hybrid structure extending several of the ideas from above. Our system is built on three main structures: RDF molecule clusters (which can be seen as hybrid structures borrowing both from property tables and RDF subgraphs), template lists (storing literals in compact lists as in a column-oriented database system) and an efficient hash-table indexing URIs and literals based on the clusters they belong to.

Figure 1 gives a simple example of a few molecule clusters—storing information about students—and of a template list—compactly storing lists of student IDs. Molecules can be seen as *horizontal* structures storing information about a given object instance in the database (like rows in relational systems). Template lists, on the other hand, store *vertical* lists of values corresponding to one *type* of object (like columns in a relational system). Hence, we say that dipLODocus[RDF] is a *hybrid* system, following the terminology used for approaches such as Fractured Mirrors [19] or our own recent Hyrise system [12].

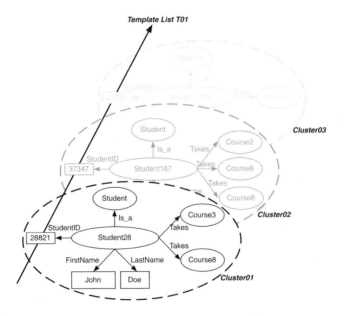

Fig. 1. The two main data structures in dipLODocus[RDF] : molecule clusters, storing in this case RDF subgraphs about students, and a template list, storing a list of literal values corresponding to student IDs

---

[8] http://www.franz.com/agraph/allegrograph/

Molecule clusters are used in two ways in our system: to logically group sets of related URIs and literals in the hash-table (thus, pre-computing joins), and to physically co-locate information relating to a given object on disk and in main-memory to reduce disk and CPU cache latencies. Template lists are mainly used for analytics and aggregate queries, as they allow to process long lists of literals efficiently. We give more detail about both structures below as we introduce the overall architecture of our system.

4   Architecture

Figure 2 gives a simplified architecture of dipLODocus[RDF] . The *Query Processor* receives the query from the client, parses it, optimizes it, and creates a query plan to execute it. The *hash-table* uses a lexicographical tree to assign a unique numeric key to each URI, stores metadata associated to that key, and points to two further data structures: the molecule *clusters*, which are managed by the *Cluster Manager* and store RDF sub-graphs, and the *template lists*, managed by the *Template Manager*. All data structures are stored on disk and are retrieved using a page manager and buffered operations to amortize disk seeks. Those components are described in greater detail below.

4.1   Query Processor

The query processor receives inserts, updates, deletes and queries from the clients. It offers a SPARQL [18] interface and supports the most common features

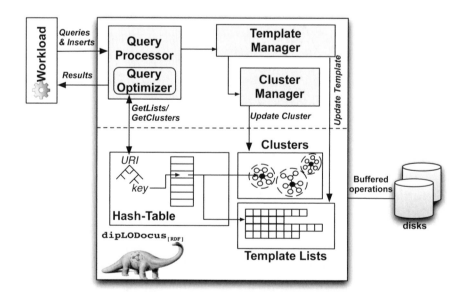

Fig. 2. The architecture of dipLODocus[RDF]

of the SPARQL query language, including conjunctions and disjunctions of triple patterns and aggregate operations. We use the RASQAL RDF Query Library[9] to parse both incoming triples serialized in XML, as well as to parse SPARQL queries. New triples are then handed to the Template and Cluster managers to be inserted into the database. As for incoming queries, after being parsed, they are rewritten as query trees in order to be executed. The query trees are passed to the *Query Optimizer*, which rewrites the queries to optimize their execution plans (cf. below Section 5). Finally, the queries are resolved bottom-up, by executing the leaf-operators first in the query tree. Examples of query processing are given below in Section 5.

4.2   Template Manager

One of the key innovations of dipLODocus[RDF] revolves around the use of *declarative storage patterns* [9] to efficiently co-locate large collections of related values on disk and in main-memory. When setting-up a new database, the database administrator may give dipLODocus[RDF] a few hints as to how to store the data on disk: the administrator can give a list of triple patterns to specify the *root nodes*, both for the template lists and the molecule clusters (see for instance above Figure 1, where "Student" is the root node of the molecule, and "StudentID" is the root node for the template list). Cluster roots are used to determine which clusters to create: a new cluster is created for each instance of a root node in the database. The clusters contain all triples departing from the root node when traversing the graph, until another instance of a root node is crossed (thus, one can join clusters based on the root nodes). Template roots are used to determine which literals to store in template lists.

In case the administrator gives no hint about the root nodes, the system inspects the templates created by the template manager (see below) and takes all classes as molecule roots and all literals as template roots (this is for example the case for the performance evaluation we describe in Section 6). Optimizing the automated selection of root nodes based on samples of the input data and an approximate query workload is a typical automated design problem [3] and is the subject of future work.

Based on the storage patterns, the template manager handles two main operations in our system: i) it maintains a schema of triple templates in main-memory and ii) it manages template lists. Whenever a new triples enters the system, it is passed to the template manager, which associates template IDs corresponding to the triple by considering the type of the subject, the predicate, and the type of the object. Each distinct list of "(subject-type, predicate, object-type)" defines a new triple template. The triple templates play the role of an instance-based RDF schema in our system. We don't rely on the explicit RDF schema to define the templates, since a large proportions of constraints (e.g., domains, ranges) are often omitted in the schema (as it is for example the case for the data we consider in our experiments, see Section 6). In case a new template is detected (e.g., a new predicate is used), then the template manager updates its

---

[9] http://librdf.org/rasqal/

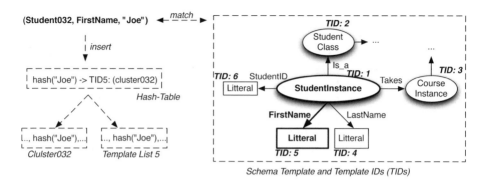

Fig. 3. An insert using templates: an incoming triple (upper left) is matched to the current RDF template of the database (right), and inserted into the hash-table, a cluster, and a template list

in-memory triple template schema and inserts new template IDs to reflect the new pattern it discovered. Figure 3 gives an example of a template. In case of very inhomogeneous data sets containing millions of different triple templates, wildcards can be used to regroup similar templates (e.g., "Student - likes - *"). Note that this is very rare in practice, since all the datasets we encountered so far (even those in the LOD cloud) typically consider a few thousands triple templates at most.

The triple is then passed to the Cluster Manager, which inserts it in one or several molecules. If the triple's object corresponds to a root template list, the object is also inserted into the template list corresponding to its template ID. Templates lists store literal values along with the key of their corresponding cluster root. They are stored compactly and segmented in sublists, both on disk and in main-memory. Template lists are typically sorted by considering a lexical order on their literal values—though other orders can be specified by the database administrator when he declares the template roots. In that sense, template lists are reminiscent of *segments* in a column-oriented database system. Finally, the triple is inserted into the hash-table as well (see Figure 3 for an example).

4.3   Cluster Manager

The Cluster Manager takes care of updating and querying the molecule clusters. When receiving a new triple from the Template Manager, the cluster manager inserts it in the corresponding cluster(s) by interrogating the hash-table (see Figure 3). In case the corresponding cluster does not exist yet, the Cluster Manager creates a new molecule cluster, inserts the triple in the molecule, and inserts the cluster in the list of clusters it maintains.

Similarly to the template lists, the molecule clusters are serialized in a very compact form, both on disk and in main-memory. Each cluster is composed of two parts: a list of offsets, containing for each template ID in the molecule the offset at which the keys corresponding for the template ID are stored, and the list of keys themselves. Thus, the size of a molecule, both on-disk and in main-memory, is $\#TEMPLATES + (KEY\_SIZE * \#TRIPLES)$, where KEY_SIZE is the size of a key (in bytes), $\#TEMPLATES$ is the number of templates IDs in the molecule, and $\#TRIPLES$ is the number of triples in the molecule (we note that this storage structure is much more compact than a standard list of triples). To retrieve a given information in a molecule, the system first determines the position of the template ID corresponding to the information sought in the molecule (e.g., "FirstName" is the sixth template ID for the "Student" molecule above in Figure 3). It then jumps to the offset corresponding to that position (e.g., 6th offset in our example), reads that offset and the offset of the following template ID, and finally retrieves all the keys/values between those two offsets to get all the values corresponding to that template ID in the molecule.

4.4   Hash-Table

The hash-table is the central index in dipLODocus[RDF] ; the hash-table uses a lexicographical tree to parse each incoming URI or literal and assign it a unique numeric key value. The hash-table then stores, for every key and every template ID, an ordered list of all the clusters IDs containing the key (e.g., "key 10011, corresponding to a Course object [template ID 17], appears in clusters 1011, 1100 and 1101"; see also Figure 3 for another example). This may sound like a pretty peculiar way of indexing values, but we show below that this actually allows us to execute many queries very efficiently simply by reading or intersecting such lists in the hash-table directly.

5   Common Operations

Given the main components and data structures described above, we describe below how common operation such as inserts, updates, and triple pattern queries are handled by our system.

5.1   Bulk Inserts

Inserts are relatively complex and costly in dipLODocus[RDF] , but can be executed in a fairly efficient manner when considered in bulk; this is a tradeoff we are willing to make in order to speed-up complex queries using our various data structures (see below), especially in a Semantic Web or LOD context where isolated inserts or updates are from our experience rather infrequent.

Bulk insert is a n-pass algorithm (where n is the deepest level of a molecule) in dipLODocus[RDF] , since we need to construct the RDF molecules in the clusters (i.e., we need to materialize triple joins to form the clusters). In a first pass, we identify all root nodes and their corresponding template IDs, and create all clusters. The subsequent passes are used to join triples to the root nodes (hence, the student clusters depicted above in Figure 1 are built in two phases, one for the Student root node, and one for the triples directly connected to the Student).

During this operation, we also update the template lists and the hash-table incrementally. Bulk inserts have been highly optimized in dipLODocus$_{[RDF]}$, and use an efficient page-manager to execute inserts for large datasets that cannot be kept in main-memory.

5.2   Updates

As for other hybrid or analytic systems, updates can be relatively expensive in dipLODocus$_{[RDF]}$. We distinguish between two kinds of updates: in-place and complex updates. In-place updates are punctual updates on literal values; they can be processed directly in our system by updating the hash-table, the corresponding cluster, and the template lists if necessary. Complex updates are updates modifying object properties in the molecules. They are more complex to handle than in-place updates, since they might require a rewrite of a list of clusters in the hash-table, and a rewrite of a list of keys in the molecule clusters. To allow for efficient operations, complex updates are treated like updates in a column-store (see [20]): the corresponding structures are flagged in the hash-table, and new structures are maintained in write-optimized structures in main-memory. Periodically, the write-optimized structures are merged with the hash-table and the clusters on disk.

5.3   Queries

Query processing in dipLODocus$_{[RDF]}$ is very different from previous approaches to execute queries on RDF data, because of the three peculiar data structures in our system: a hash-table associating URIs and literals to template IDs and cluster lists, clusters storing RDF molecule clusters in a very compact fashion, and template lists storing compact lists of literals. We describe below how a few common queries are handled in dipLODocus$_{[RDF]}$.

Triple Patterns: Triple patterns are relatively simple in dipLODocus$_{[RDF]}$: they are usually resolved by looking for a bound-variable (URI) in the hash-table, retrieving the corresponding cluster numbers, and finally retrieving values from the clusters when necessary. Conjunctions and disjunctions of triples patterns can be resolved very efficiently in our system. Since the RDF nodes are logically grouped by clusters in the hash-table, it is typically sufficient to read the corresponding list of clusters in the hash table (e.g., for "return all students following Course0"), or to intersect or take the union of several lists of clusters in the hash table (e.g., for "return all students following Course0 whose last names are Doe") to answer the queries. In most cases, no join operation is needed since joins are implicitly materialized in the hash-table and in the clusters. When more complex join occurs, dipLODocus$_{[RDF]}$ resolves them using standard hash-join operators.

Molecule Queries: Molecule queries or queries retrieving many values/ instances around a given instance (for example for visualization purposes) are also extremely efficient in our system. In most cases, the hash-table is invoked to find the corresponding cluster, which contains then all the corresponding values. For bigger scopes (such as the ones we consider in our experimental evaluation below), our system can efficiently join clusters based on the various root nodes they contain.

Aggregates and Analytics: Finally, aggregate and analytic queries can also be very efficiently resolved by our system. Many analytic queries can be solved by first intersecting lists of clusters in the hash-table, and then looking up values in the remaining molecule clusters. Large analytic or aggregate queries on literals (such as our third analytic query below, returning the names of all graduate students) can be extremely efficiently resolved by taking advantage of template lists (containing compact and sorted lists of literal values for a given template ID), or by filtering template lists based on lists of cluster IDs retrieved from the hash-table.

6   Performance Evaluation

To evaluate the performance of our system, we compared it to various RDF database systems. The details of the hardware platform, the data sets and the workloads we used are give below.

6.1   Hardware Platform

All experiments were run on a HP ProLiant DL360 G7 server with two Quad-Core Intel Xeon Processor E5640, 6GB of DDR3 RAM and running Linux Ubuntu 10.10 (Maverick Meerkat). All data were stored on recent 1.4 TB Serial ATA disk.

6.2   Data Sets

The benchmark we used is one of the oldest and most popular benchmarks for Semantic Web data called Lehigh University Benchmark (LUBM) [14]. It provides an ontology describing universities together with a data generator and fourteen queries. We used two data sets, the first one consisting of ten LUBM universities (1'272'814 distinct triples, 315'003 distinct strings), and the second regrouping one hundred universities (13'876'209 distinct triples, 3'301'868 distinct strings).

6.3   Workload

We compared the runtime execution for LUBM queries and for three analytic queries inspired by an RDF analytic benchmark we recently proposed (the BowlognaBench benchmark [11]). LUBM queries are criticized by some for their reasoning coverage; this was not an issue in our case, since we focused on RDF DB query processing rather than on reasoning capabilities. We keep an in-memory representation of subsumption trees in dipLODocus[RDF] and rewrite queries automatically to support subclass inference for the LUBM queries. We manually rewrote inference queries for the systems that do not support such functionalities (e.g., RDF-3X).

The three additional analytic/aggregate queries that we considered are as follows: 1) a query returning the professor who supervises the most Ph.D. students 2) a query returning a big molecule containing all triples within a scope of 2 of Student0 and 3) a query returning all graduate students.

6.4   Methodology

As for other benchmarks (e.g., tpc-x^{10}) we include a warm-up phase before measuring the execution time of the queries. We first run all the queries in sequence once to warm-up the systems, and then repeat the process ten times (i.e., we run in total 11 batches containing all the queries in sequence for each system). We report the mean values for each query and each system below as well as a 95% confidence interval on run times. We assumed that the maximum time for each query shouldn't exceed 2 hours (we stopped the tests if one query took more than two hours to be executed). We compared the output of all queries running on all systems to ensure that all results were correct.

We tried to do a reasonable optimization job for each system, by following the recommendations given in the installation guides for each system. We did not try to optimize the systems any further, however. We performed no fine-tuning or optimization for dipLODocus$_{[RDF]}$.

We avoided the artifact of connecting to the server, initializing the DB from files and printing results for all systems; we measured instead the query execution times only.

6.5   Systems

We compared our prototype implementation of dipLODocus$_{[RDF]}$ to five other well-known database systems: Postgres, AllegroGraph, BigOWLIM, Jena, Virtuoso, and RDF 3X. We chose those systems to have different comparison points using well-known systems, and because they were all freely available on the Web. We give a few details about each system below.

Postgres: We used Postgres 8.4 with Redland RDF Library 1.0.13; Postgres is a well-known relational database, but as the numbers below show, it is not optimized for RDF storage. We couldn't run our 100-universities on it because its load time took more than one week. It also had huge difficulties to cope with some of the queries for the 10-universities data set. Since the time of query execution was particularly long for this system, we ran each query five times only and simply report the best run below.

AllegroGraph: We used AllegroGraph RDFStore 4.2.1 AllegroGraph unfortunately poses some limits on the number of triples that can be stored for the free edition, such that we couldnt load the big data set. It also showed difficulty to deal with one query. For AllegroGraph, we prepared a SPARQL Python script using libraries supported by the vendor.

---

[10] http://www.tpc.org/

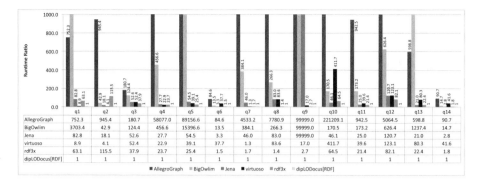

Fig. 4. Runtime ratios for the 10 universities data set

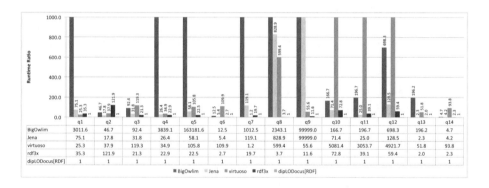

Fig. 5. Runtime ratios for the 100 universities data set

BigOWLIM: We used BigOWLIM 3.5.3436. OWLIM provides us with a java
application to run the LUBM benchmark, so we used it directly for our tests.

Jena: We used Jena-2.6.4 and the TDB-0.8.10 storage component. We created
the database by using the "tdbloader" provided by Jena. We created a Java
application to run and measure the execution time of each query.

Virtuoso: We used Virtuoso Open-Source Edition 6.1.3. Virtuoso supports
ODBC connections, and we prepared a Python script using the PyODBC
library for our queries.

RDF-3X: We used RDF-3X 0.3.5. For this system, we converted our dataset
to NTriples/Turtle. We also hacked the system to measure the execution
time of the queries only, without taking into account the initialization of the
database and turning off the print-outs.

6.6   Results

Relative execution times for all queries and all systems are given below, in
Figure 4 for 10 universities and in Figure 5 for 100 universities. Results are

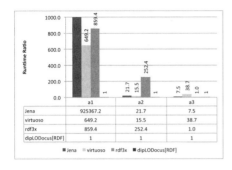
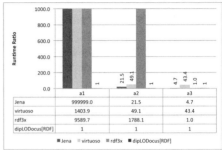

	a1	a2	a3
Jena	925367.2	21.7	7.5
virtuoso	649.2	15.5	38.7
rdf3x	859.4	252.4	1.0
dipLODocus[RDF]	1	1	1

■ Jena ▨ virtuoso ■ rdf3x ■ dipLODocus[RDF]

	a1	a2	a3
Jena	999999.0	21.5	4.7
virtuoso	1403.9	49.1	43.4
rdf3x	9589.7	1788.1	1.0
dipLODocus[RDF]	1	1	1

■ Jena ▨ virtuoso ■ rdf3x ■ dipLODocus[RDF]

Fig. 6. Runtime ratios for 10 (left) and 100 (right) universities for the analytic/aggregate queries

		10 UNI -- Query Execution Time [s]								100 UNI -- Query Execution Time [s]				
		dipLODocus	AllegroGrap	BigOwlim	virtuoso	rdf3x	Jena			dipLODocus	BigOwlim	virtuoso	rdf3x	Jena
q1	AVG	1.45E-05	1.09E-02	5.37E-02	1.29E-04	9.14E-04	1.20E-03	q1	AVG	1.73E-05	5.21E-02	4.38E-04	6.10E-04	1.30E-03
	CI	6.47E-08	4.81E-05	6.27E-05	1.00E-06	2.23E-07	7.93E-06		CI	5.17E-08	5.35E-05	5.88E-07	3.53E-07	9.09E-06
q2	AVG	1.21E-02	1.14E+01	5.19E-01	4.96E-02	1.40E+00	2.19E-01	q2	AVG	1.27E-01	5.94E+00	4.83E+00	1.55E+01	2.27E+00
	CI	5.63E-05	2.96E-03	9.04E-04	4.72E-05	1.85E-05	1.42E-04		CI	6.41E-05	3.79E-03	8.82E-04	1.52E-04	4.09E-04
q3	AVG	2.09E-05	3.78E-03	2.60E-03	1.10E-03	7.91E-04	1.10E-03	q3	AVG	3.14E-05	2.90E-03	3.75E-03	6.68E-04	1.00E-03
	CI	9.57E-08	8.77E-06	1.32E-05	4.00E-06	1.11E-06	5.95E-06		CI	2.54E-08	1.07E-05	6.01E-07	1.56E-07	0.00E+00
q4	AVG	7.95E-05	4.62E+00	3.63E-02	1.82E-03	1.89E-03	2.20E-03	q4	AVG	8.33E-05	3.20E-01	2.91E-03	1.90E-03	2.20E-03
	CI	3.17E-07	8.49E-04	1.18E-04	1.04E-06	2.22E-07	7.93E-06		CI	7.15E-08	5.44E-04	3.32E-06	1.29E-07	7.93E-06
q5	AVG	5.32E-05	4.74E+00	8.19E-01	2.08E-03	1.35E-03	2.90E-03	q5	AVG	5.34E-05	8.71E+00	5.65E-03	1.20E-03	3.10E-03
	CI	2.56E-07	1.11E-03	2.89E-04	1.71E-05	3.89E-08	1.07E-05		CI	6.16E-08	2.28E-03	4.37E-06	1.61E-07	2.07E-05
q6	AVG	1.65E-02	1.40E+00	2.23E-01	6.22E-01	2.51E-02	5.52E-02	q6	AVG	1.42E-01	1.77E+00	1.56E+01	3.77E-01	7.61E-01
	CI	8.65E-05	3.93E-04	8.00E-04	1.58E-03	8.81E-06	2.70E-04		CI	1.64E-05	1.15E-03	1.25E-02	5.24E-05	7.28E-03
q7	AVG	1.22E-03	7.03E+01	5.96E+00	1.55E-03	4.82E-03	7.14E-01	q7	AVG	2.63E-03	6.34E+01	3.11E-03	5.18E-02	7.46E+00
	CI	3.21E-07	1.12E-02	2.18E-03	2.10E-06	5.04E-05	2.25E-04		CI	3.39E-05	1.55E-02	2.09E-06	9.53E-04	1.54E-03
q8	AVG	6.54E-03	5.09E+01	1.74E+00	5.47E-01	8.94E-03	5.43E-01	q8	AVG	6.34E-03	1.49E+01	3.80E+00	2.36E-02	5.26E+00
	CI	2.21E-05	8.28E-03	7.10E-04	1.09E-03	1.57E-06	1.47E-03		CI	1.71E-06	3.93E-03	7.60E-04	6.69E-06	5.19E-03
q9	AVG	6.74E-02	longer than	longer than	1.14E+00	1.83E-01	longer than	q9	AVG	2.61E-01	longer than	1.45E+01	3.01E+00	longer than
	CI	1.98E-06	two hours	two hours	4.38E-03	9.06E-05	two hours		CI	2.29E-05	two hours	1.92E-03	1.07E-03	two hours
q10	AVG	2.17E-05	4.80E+00	3.70E-03	8.93E-03	1.40E-03	1.00E-03	q10	AVG	1.68E-05	2.80E-03	8.54E-02	1.22E-03	1.20E-03
	CI	7.32E-08	1.15E-03	9.09E-06	7.49E-06	1.80E-06	0.00E+00		CI	1.19E-08	7.93E-06	3.59E-06	1.49E-07	7.93E-06
q11	AVG	6.41E-05	6.04E-02	1.11E-02	2.54E-03	3.75E-03	1.60E-03	q11	AVG	6.00E-05	1.18E-02	1.83E-01	2.35E-03	1.50E-03
	CI	2.32E-07	5.08E-04	5.95E-06	1.76E-05	3.25E-07	1.32E-05		CI	1.54E-08	7.93E-06	4.72E-05	7.76E-07	9.91E-06
q12	AVG	1.74E-05	8.81E-02	1.09E-02	2.14E-03	1.43E-03	2.10E-03	q12	AVG	1.79E-05	1.25E-02	8.81E-02	1.06E-03	2.30E-03
	CI	5.19E-08	8.05E-05	5.95E-06	1.93E-06	2.70E-07	5.95E-06		CI	5.95E-09	3.68E-05	8.23E-05	1.06E-06	9.09E-06
q13	AVG	4.76E-05	2.85E-02	5.89E-02	3.82E-03	1.06E-03	1.00E-03	q13	AVG	5.62E-04	1.10E-01	2.91E-02	1.11E-03	1.30E-03
	CI	1.18E-07	8.19E-05	5.71E-05	8.27E-07	7.82E-08	0.00E+00		CI	1.99E-08	2.39E-04	5.56E-05	1.01E-07	9.09E-06
q14	AVG	1.29E-02	1.17E+00	1.90E-01	5.37E-01	2.28E-02	3.62E-02	q14	AVG	1.41E-01	6.68E-01	1.33E+01	3.27E-01	5.98E-01
	CI	6.00E-05	3.79E-04	2.17E-04	1.91E-03	9.93E-06	1.82E-04		CI	2.18E-06	1.21E-03	6.99E-03	1.30E-04	6.59E-03
a1	AVG	1.16E-03	not run	not run	7.50E-01	9.93E-01	1.07E+03	a1	AVG	1.04E-02	not run	1.45E+01	9.93E+01	longer than
	CI	2.24E-06			4.90E-04	3.90E-03	3.42E-02		CI	1.22E-07		6.52E-04	1.05E-03	two hours
a2	AVG	5.07E-05	not run	not run	7.85E-04	1.28E-02	1.10E-03	a2	AVG	6.50E-05	not run	3.19E-03	1.16E-01	1.40E-03
	CI	1.72E-07			4.62E-06	1.25E-05	5.95E-06		CI	1.54E-08		2.35E-06	1.21E-04	9.71E-06
a3	AVG	1.07E-02	not run	not run	4.13E-01	1.10E-02	8.01E-02	a3	AVG	1.55E-01	not run	6.72E+00	1.49E-01	7.25E-01
	CI	1.57E-07			2.30E-03	2.49E-06	7.00E-04		CI	2.65E-07		1.94E-03	3.61E-05	2.88E-03

	dipLODocus	AllegroGrap	BigOwlim	virtuoso	rdf3x	Jena		dipLODocus	BigOwlim	virtuoso	rdf3x	Jena
Load Time	31s	13s	50s	88s	16s	98s	Load Time	427s	748s	914s	214s	1146s
size	87MB	696MB	209MB	140MB	66MB	118MB	size on disk	913MB	2012MB	772MB	694MB	1245MB

Fig. 7. Absolute query execution and load times [s], plus size of the databases on disk for both data sets

given as runtime ratios, with dipLODocus[RDF] taken as a basis for ratio 1.0 (i.e., a bar indicating 752.3 means that the execution time of that query on that system was 752.3 times slower than the dipLODocus[RDF] execution). Figure 6 gives relative execution times for analytics executed on a selection of the fastest systems. Absolute times with confidence intervals at 95%, database sizes on disk and load times are given in Figures 7 for both datasets.

We observe that dipLODocus[RDF] is generally speaking very fast, both for bulk inserts, for LUBM queries and especially for analytic queries. dipLODocus[RDF] is not the fastest system for inserts, and produces slightly larger databases on disk than some other systems (like RDF-3x), but performs overall very-well for all queries. Our system is on average 30 times faster than the fastest RDF data management system we have considered (i.e., RDF-3X) for LUBM queries, and on average 350 times faster than the fastest system (Virtuoso) on analytic queries. Is is also very scalable (both bulk insert and query processing scale gracefully from 10 to 100 universities).

7 Conclusions

In this paper, we have described dipLODocus$_{[RDF]}$, a new RDF management system based on a hybrid storage model and RDF templates to execute various kinds of queries very efficiently. In our performance evaluation, dipLODocus$_{[RDF]}$ is on average 30 times faster than the fastest RDF data management system we have considered (i.e., RDF-3X) on LUBM queries, and on average 350 times faster than the fastest system we have considered on analytic queries. More importantly, dipLODocus$_{[RDF]}$ is the only system to consistently show low processing times for *all* the queries we have considered (i.e., our system is the only system being able to answer any of the queries we considered in less than one second), thus making it an extremely versatile RDF management system capable of efficiently supporting both short and long-tail queries in real deployments.

This impressive performance can be explained by several salient features of our system, including: its extremely compact structures based on molecule templates to store data, its redundant structures to optimize different types of operations, its very efficient ways of coping with disk and memory reads (avoiding seeks and memory jumps as much as possible since they are extremely expensive on modern machines), and its way of materializing various joins in all its data structures. This performance is counterbalanced by relatively complex and expensive updates and inserts, which can however be optimized if considered in bulk.

In the near future, we plan to work on cleaning, proof-testing, and extending our code base to deliver an open-source release of our system as soon as possible[11]. We also have longer-term research plans for dipLODocus$_{[RDF]}$; our next research efforts will revolve around parallelizing many of the operations in the system, to take advantage of multi-core architectures on one hand, and large cluster of commodity machines on the other hand. Also, we plan to work on the automated database design problem in order to automatically suggest sets of optimal root nodes to the database administrator given some sample input data and an approximate query workload.

Acknowledgment. This work is supported by the Swiss National Science Foundation under grant number PP00P2_128459.

References

1. Aberer, K., Cudré-Mauroux, P., Datta, A., Despotovic, Z., Hauswirth, M., Punceva, M., Schmidt, R.: P-grid: A self-organizing structured p2p system. ACM SIGMOD Record 32(3) (2003)

---

[11] visit `http://diuf.unifr.ch/xi/diplodocus` for updates.

2. Aberer, K., Cudré-Mauroux, P., Hauswirth, M., Van Pelt, T.: GridVine: Building Internet-Scale Semantic Overlay Networks. In: McIlraith, S.A., Plexousakis, D., van Harmelen, F. (eds.) ISWC 2004. LNCS, vol. 3298, pp. 107–121. Springer, Heidelberg (2004)

3. Agrawal, S., Chaudhuri, S., Narasayya, V.: Automated selection of materialized views and indexes in SQL databases. In: International Conference on Very Large Data Bases, VLDB (2000)

4. Atre, M., Chaoji, V., Weaver, J., Williamss, G.: Bitmat: An in-core rdf graph store for join query processing. In: Rensselaer Polytechnic Institute Technical Report (2009)

5. Broekstra, J., Kampman, A., Harmelen, F.V.: Sesame: An architecture for storing and querying rdf data and schema information. In: Semantics for the WWW. MIT Press (2001)

6. Chong, E.I., Das, S., Eadon, G., Srinivasan, J.: An efficient sql-based rdf querying scheme. In: Proceedings of the 31st International Conference on Very Large Data Bases, VLDB 2005, pp. 1216–1227. VLDB Endowment (2005)

7. Cudré-Mauroux, P., Agarwal, S., Aberer, K.: Gridvine: An infrastructure for peer information management. IEEE Internet Computing 11(5) (2007)

8. Cudré-Mauroux, P., Lim, K., Simakov, R., Soroush, E., Velikhov, P., Wang, D.L., Balazinska, M., Becla, J., DeWitt, D., Heath, B., Maier, D., Madden, S., Patel, J.M., Stonebraker, M., Zdonik, S.: A Demonstration of SciDB: A Science-Oriented DBMS. Proceedings of the VLDB Endowment (PVLDB) 2(2), 1534–1537 (2009)

9. Cudré-Mauroux, P., Wu, E., Madden, S.: The Case for RodentStore, an Adaptive, Declarative Storage System. In: Biennial Conference on Innovative Data Systems Research, CIDR (2009)

10. Dean, J., Ghemawat, S.: Mapreduce: simplified data processing on large clusters. Commun. ACM 51, 107–113 (2008)

11. Demartini, G., Enchev, I., Gapany, J., Cudré-Maurox, P.: BowlognaBench— Benchmarking RDF Analytics. In: SIMPDA 2011: First International Symposium on Process Data (2011)

12. Grund, M., Krüger, J., Plattner, H., Zeier, A., Cudré-Mauroux, P., Madden, S.: Hyrise - a main memory hybrid storage engine. PVLDB 4(2), 105–116 (2010)

13. Guo, Y., Pan, Z., Heflin, J.: An Evaluation of Knowledge Base Systems for Large OWL Datasets. In: McIlraith, S.A., Plexousakis, D., van Harmelen, F. (eds.) ISWC 2004. LNCS, vol. 3298, pp. 274–288. Springer, Heidelberg (2004)

14. Guo, Y., Pan, Z., Heflin, J.: Lubm: A benchmark for owl knowledge base systems. Web Semant. 3, 158–182 (2005)

15. Haslhofer, B., Roochi, E.M., Schandl, B., Zander, S.: Europeana RDF Store Report. University of Vienna, Technical Report (2011), http://eprints.cs.univie.ac.at/2833/1/europeana_ts_report.pdf

16. Liu, B., Hu, B.: An evaluation of rdf storage systems for large data applications. In: First International Conference on Semantics, Knowledge and Grid, SKG 2005, p. 59 (November 2005)

17. Neumann, T., Weikum, G.: RDF-3X: a RISC-style engine for RDF. Proceedings of the VLDB Endowment (PVLDB) 1(1), 647–659 (2008)

18. Prud'hommeaux, E., Seaborne van Harmelen, A. (eds.): SPARQL Query Language for RDF. W3C Candidate Recommendation (April 2006), http://www.w3.org/TR/rdf-sparql-query/

19. Ramamurthy, R., DeWitt, D.J., Su, Q.: A case for fractured mirrors. In: CAiSE 2002 and VLDB 2002. VLDB Endowment, pp. 430–441 (2002)

20. Stonebraker, M., Abadi, D.J., Batkin, A., Chen, X., Cherniack, M., Ferreira, M., Lau, E., Lin, A., Madden, S.R., O'Neil, E., O'Neil, P., Rasin, A., Tran, N., Zdonik, S.: C-Store: A Column Oriented DBMS. In: International Conference on Very Large Data Bases, VLDB (2005)
21. Weiss, C., Karras, P., Bernstein, A.: Hexastore: sextuple indexing for semantic web data management. Proceeding of the VLDB Endowment (PVLDB) 1(1), 1008–1019 (2008)
22. Wilkinson, K., Sayers, C., Kuno, H.A., Reynolds, D.: Efficient rdf storage and retrieval in jena2. In: SWDB 2003, pp. 131–150 (2003)
23. Yan, Y., Wang, C., Zhou, A., Qian, W., Ma, L., Pan, Y.: Efficient indices using graph partitioning in rdf triple stores. In: Proceedings of the 2009 IEEE International Conference on Data Engineering, pp. 1263–1266. IEEE Computer Society, Washington, DC, USA (2009)
24. Zou, L., Mo, J., Chen, L., Oezsu, M.T., Zhao, D.: gstore: Answering sparql queries via subgraph matching. PVLDB 4(8) (2011)

Extending Functional Dependency to Detect Abnormal Data in RDF Graphs

Yang Yu and Jeff Heflin

Department of Computer Science and Engineering, Lehigh University
19 Memorial Drive West, Bethlehem, PA 18015
{yay208,heflin}@cse.lehigh.edu

Abstract. Data quality issues arise in the Semantic Web because data
is created by diverse people and/or automated tools. In particular, erroneous triples may occur due to factual errors in the original data source,
the acquisition tools employed, misuse of ontologies, or errors in ontology alignment. We propose that the degree to which a triple deviates
from similar triples can be an important heuristic for identifying errors.
Inspired by functional dependency, which has shown promise in database
data quality research, we introduce *value-clustered graph functional dependency* to detect abnormal data in RDF graphs. To better deal with
Semantic Web data, this extends the concept of functional dependency
on several aspects. First, there is the issue of scale, since we must consider the whole data schema instead of being restricted to one database
relation. Second, it deals with multi-valued properties without explicit
value correlations as specified as tuples in databases. Third, it uses clustering to consider classes of values. Focusing on these characteristics, we
propose a number of heuristics and algorithms to efficiently discover the
extended dependencies and use them to detect abnormal data. Experiments have shown that the system is efficient on multiple data sets and
also detects many quality problems in real world data.

Keywords: value-clustered graph functional dependency, abnormal data
in RDF graphs.

1 Introduction

Data quality (DQ) research has been intensively applied to traditional forms
of data, e.g. databases and web pages. The data are deemed of high quality if
they correctly represent the real-world construct to which they refer. In the last
decade, data dependencies, e.g. functional dependency (FD) [1] and conditional
functional dependency (CFD) [2, 3], have been used in promising DQ research
efforts on databases. Data quality is also critically important for Semantic Web
data. A large amount of heterogeneous data is converted into RDF/OWL format
by a variety of tools and then made available as Linked Data[1]. During the
creation or conversion of this data, numerous data quality problems can arise.

---

[1] http://linkeddata.org

L. Aroyo et al. (Eds.): ISWC 2011, Part I, LNCS 7031, pp. 794–809, 2011.
© Springer-Verlag Berlin Heidelberg 2011

Some works [4–6] began to focus on the quality of Semantic Web data, but such research is still in its very early stages. No previous work has utilized the fact that RDF data can be viewed as a graph database, therefore we can benefit from traditional database approaches, but we must make special considerations for RDF's unique features. Since the Semantic Web represents many points of view, there is no objective measure of correctness for all Semantic Web data. Therefore, we focus on the detection of abnormal triples, i.e., triples that violate certain data dependencies. This in turn is used as a heuristic of a potential data quality problem. We recognize that not all abnormal data is incorrect (in fact, in some scenarios the abnormal data may be the most interesting data) and thus leave it up to the application to determine how to use the heuristic.

A typical data dependency in databases is functional dependency [7]. Given a relation R, a set of attributes X in R is said to functionally determine another attribute Y, also in R, (written $X \to Y$), if and only if each X value is associated with precisely one Y value. An example FD $zipCode \to state$ means, for any tuple, the value of $zipCode$ determines the value of $state$.

RDF data also has various dependencies. But RDF data has a very different organization and FD cannot be directly applied because RDF data is not organized into relations with a fixed set of attributes. We propose *value-clustered graph functional dependency* (VGFD) based on the following thoughts. First, FD is formally defined over one entire relation. However RDF data can be seen as extremely decomposed tables where each table is a set of triples for a single property. Thus we must look for dependencies that cross these extremely decomposed tables and extend the concept of dependency from a single database relation to a whole data set. Second, the correlation between values is trivially determined in a database of relational tuples. But in RDF data, it is non-trivial to determine the correlation, especially for multi-valued properties. For example, in DBPedia, the properties *city* and *province* do not have cardinality restrictions, and thus instances can have multiple values for each property. This makes sense, considering that some organizations can have multiple places. Yet finding the correlation between the different values of *city* and *province* becomes non-trivial. Third, traditionally value equality is used to determine FD. However, this is not appropriate for real world, distributed data. Consider (1) for floating point numbers, rounding and measurement errors must be considered. (2) Sometimes dependencies are probabilistic in nature, and one-to-one value correspondences are inappropriate. For example, the days needed for processing an order before shipping for a certain product is usually limited to a small range but not an exact value. (3) Sometimes certain values can be grouped to form a more abstract value.

In sum, our work makes the following contributions.

- we automatically find optimal clusters of values
- we efficiently discover VGFDs over clustered values
- we use the clusters and VGFDs to detect outliers and abnormal data
- we conducts experiments on three data sets that validate the system

The rest of the paper is as follows. Section 2 discusses related work. Section 3 describes how to efficiently discover VGFDs while Section 4 discusses categorizing property values for their use. Sections 5 and 6 give the experimental results and the conclusion.

2   Related Work

Functional dependencies are by far the most common integrity constraints for databases in the real world. They are very important when designing or analyzing relational databases. Most approaches to find FD [8–10] are mainly based on the concept of an agree set [11]. Given a pair of tuples, the agree set is all the attributes for which these tuples have the same values. Since the search for FDs occurs over a given relation and each tuple has at most one value for each attribute, then each tuple can be placed into exactly one cluster where all tuples in the cluster have the same agree set with all other tuples. Agree sets are not very useful when applied to the extensions of RDF properties, which are equivalent to relations with only two attributes (i.e. the subject and object of the triple). Furthermore, many properties in RDF data are multi-valued and so the correlation between values of different properties becomes more complex. Finally, since most RDF properties are designed just for a subset of instances in the data set, an agree set-based approach will partition many instances based on null values is common.

RDF graphs are more like graph database models. The value functional dependency (VFD) [12] defined for the object-oriented data model can have multi-valued properties on the right-hand side, e.g. $title \rightarrow authors$. However the dependencies we envision can have multi-valued properties on both sides and our system can determine the correlation between each value in both sets. The path functional dependency (PFD) [13] defined for semantic data models considered multiple attributes on a path, however the PFD did not consider multi-valued attributes. FD$_{XML}$ is the FD's counterpart in XML [14] where its left-hand side is a path starting from the XML document root which essentially is another form of a record in a database. Hartmann et al. [15] generalized the definitions of several previous FDs in XML from a graph matching perspective.

As mentioned previously, the basic equality comparison of values used in FD is limited in many situations. Algebraic constraints [16, 17] in database relations are about the algebraic relation between two columns in a database and are often used for query optimization. The algebraic relation can be $+, -, \times, /$. However these works are limited to numerical attribute values and the mapping function can only be defined using several algebraic operators. The reason is that numerical columns are more often indexed and queried over as selective conditions in databases than strings. In contrast, we try to find a general mapping function between the values of different properties, both numbers and strings. Additionally, for the purpose of query optimization, they focus on pairs of columns with top ranked relational significance, the major parts in each of these pairs and the data related to dependencies that is often queried over, rather than all possible pairs of properties and all pairs of values existing in the data set.

Data dependencies have recently shown promise for data quality management in databases. Bohannon et al. [1] focuses on repairing inconsistencies based on standard FDs and inclusion dependencies, that is, to edit the instance via minimal value modification such that the updated instance satisfies the constraints. A CFD [2, 3] is more expressive than a FD because it can describe a dependency that only holds for a subset of the tuples in a relation, i.e., those that satisfy some condition. Fan et. al [2] gave a theoretical analysis and algorithms for computing implications and minimal cover of CFDs; Cong et al. [3], similar to Bohannon et al., focused on repairing inconsistent data. The CFD discovery problem has high complexity; it is known to be more complex than the implication problem, which is already coNP-complete [2]. In contrast to them, we are trying to both automatically find fuzzy constraints, i.e. those that hold for most of the data, and report on exceptional data for applications. Our work incorporates advantages from both FD and CFD, i.e. fast execution and the ability to tolerate exceptions.

With respect to data quality on the Semantic Web, Sabou et al. [4] evaluate semantic relations between concepts in ontologies by counting the similar axioms (both explicit and entailed) in online ontologies and their derivation length. For instance data, previous evaluations mainly focused on two types of errors: explicit inconsistency with the syntax of the ontologies and logical inconsistency that can be checked by DL reasoners. However, many Linked Data ontologies do not fully specify the semantics of the terms defined, and OWL cannot specify axioms that only hold part of the time. Our work focuses on detecting abnormal semantic data by automatically discovering probabilistic integrity constraints (IC). Tao et al. [6] proposed an IC semantics for ontology modelers and suggested that it is useful for data validation purposes. But the precursor problem of how to discover these ICs is not addressed. Furber et al. [5] also noticed that using FD could be helpful for data quality management on the Semantic Web, but do not give an automatic algorithm to find such FDs and, more importantly, direct application of FD to RDF data may not capture the unique characteristics of RDF data.

3   Discovering VGFDs

We begin with some definitions.

Definition 1. *An RDF graph is $G := \langle I, L, R, E \rangle$, where three sets I, L and R are instance, literal and relation identifiers and the set of directional edges is $E \subseteq I \times R \times (I \cup L)$. Let $\mathcal{G}$ be the set of all possible graphs and $G \in \mathcal{G}$. Let $R^- = \{r^- | r \in R\}$.*

Definition 2. *A Path c in graph G is a tuple $\langle I_0, r_1, I_1, ..., r_n, I_n \rangle$ where $I_i \in I, r_i \in R \cup R^-$, and $\forall i, 0 \leqslant i < n$, if $r_i \in R$ then $(I_i, r_{i+1}, I_{i+1}) \in E$ or if $r_{i+1} \in R^-$ then $(I_{i+1}, r_{i+1}, I_i) \in E$; $\forall j$, if $i \neq j$ then $I_i \neq I_j$.*

Paths are acyclic and directional, but can include inverted relations of the form r^-.

Definition 3. *A Composite Property (P_{comp}) $r°$ in graph G is $r_1 \circ r_2...r_n$, where $\exists I_0, ..., I_n$ and $\langle I_0, r_1, I_1, ..., r_n, I_n \rangle$ is a Path in G. Let $\mathcal{R}°$ be all possible $P_{comp}s$. Given $r° \in \mathcal{R}°$, $Inst(r°, G) = \{\langle I_0, r°, I_n \rangle | \langle I_0, r_1, I_1, r_2, I_2, ..., r_n, I_n \rangle$ is a Path in $G\}$. $Length(r°) = n$. $\forall r \in R$, $r \in \mathcal{R}°$ and $Length(r) = 1$.*

Definition 4. *A Conjunctive Property (P_{conj}) r^+ in graph G is a set $\{r_1, r_2, ..., r_n\}$ (written $r_1 + r_2 + ... + r_n$), where $\forall i, r_i \in \mathcal{R}°$ and $\exists I'$ s.t. $\forall 1 \leq i \leq n$, $\langle I', r_i, I_i \rangle \in Inst(r_i, G)$. Let $\mathcal{R}^+$ be all possible $P_{conj}s$. $Size(r^+) = \sum_{r_i \in r^+} Length(r_i)$.*

A Composite Property is a sequence of edges on a Path. The subject and object of a P_{comp} are the first and last objects on the Paths consisting of this sequence of edges. Every property is a special case of P_{comp}. A Conjunctive Property groups a set of $P_{comp}s$ that have a common subject I'. Note, each original $r \in R$ is also $r \in \mathcal{R}°$ and each $r° \in \mathcal{R}°$ is also $r° \in \mathcal{R}^+$.

Definition 5. *Given $i \in I$ and $r° \in \mathcal{R}°$, $V°(i, r°) = \{i' | \exists \langle i, r°, i' \rangle \in Inst(r°, G)\}$. Given $r^+ \in \mathcal{R}^+$, $V^+(i, r^+)$ is a tuple $\langle V°(i, r_1), ..., V°(i, r_1) \rangle$ where $\forall j, r_j \in \mathcal{R}^+$.*

Given a P_{comp}, value function $V°$ returns the objects connected with a subject through P_{comp}, and given a P_{conj}, the value function V^+ returns the set of objects connected with a subject through P_{conj}.

Definition 6. *Given $i, j \in I$ and $r° \in \mathcal{R}°$, Dependency Equality (DE) between i and j on $r°$ is: $V(i, r°) \doteq V(j, r°) \iff (\forall x \in V°(i, r°) \iff \exists y \in V°(j, r°), C(x) = C(y))$, where $C(x)$ is the dependency cluster of x (discussed in Section 4). With a slight abuse of notation for DE, given $r^+ \in \mathcal{R}^+$, $V^+(i, r^+) \doteq V^+(j, r^+) \iff \forall r_k \in r^+, V°(i, r_k) \doteq V°(j, r_k)$.*

Definition 7. *A value-clustered graph functional dependency (VGFD) s in graph G is $X \rightarrow Y$, where $X \in \mathcal{R}^+$, $Y \in \mathcal{R}°$ and $\forall i, j \in I$, if $V^+(i, X) \doteq V^+(j, X)$ then $V°(i, Y) \doteq V°(j, Y)$.*

These definitions state that for all instances, if the values of the left-hand side (LHS) P_{comp} of a given VGFD satisfy Dependency Equality (DE), then there is a DE on the right-hand side (RHS) P_{conj}. Note, due to the union rule of Armstrong's axioms used to infer all the functional dependencies, if $\alpha \rightarrow \beta$ and $\alpha \rightarrow \gamma$ hold, then $\alpha \rightarrow \beta\gamma$ holds. Therefore, it is enough to define the VGFD whose right-hand side (RHS) is each single element of a set of properties. In this work, DE includes basic equality for both object and datatype property values, transitivity of the sameAs relation for instances and clustering for datatype values.

Shown in Algorithm 1, this section introduces the VGFD search (line 8-15) and the next section introduces value clustering (line 2-5) which is used to detect dependencies. To efficiently discover a minimum set of VGFDs which is a cover of the whole set of VGFDs, our approach essentially is computed level-wise. Each level L_i consists of VGFDs with LHS of size i (Fig. 1 gives an example). The computation of VGFDs with smaller sets of LHS properties can be used

Algorithm 1. $Search\_VGFDs(G, \alpha, \beta, \gamma)$, $G = (I, L, R, E)$ is a graph; α is the confidence threshold for a VGFD; β is the sampling size; γ is the threshold for pre-clustering.

```
1:  S ← ∅, C ← ∅
2:  for each r ∈ R s.t. r is a datatype property do
3:      groups ← Preclustering(Range(r),γ)
4:      Cr ← Optimal_Kmeans(Range(r),groups)
5:      C ← C ∪ Cr
6:  i = 0
7:  Li ← ∅
8:  repeat
9:      i = i + 1
10:     Li ← Generate_Level_with_Static_Pruning(Li−1,E)
11:     for each s ∈ Li do
12:         if Runtime_Pruning(s,α,β,E,C) = FALSE then
13:             if (M ← Compute_VGFD(s,α,E,C)) ≠ ∅ then
14:                 S ← S ∪ (s,M)              //M is the set of value mappings of s.
15: until Li = ∅ or i >= DEPTH_LIMIT
16: return  S
```

when computing children VGFDs that have a superset of those LHS properties. A similar level-wise search was proposed for the Tane algorithm [9], but each node in Tane corresponds to a subset of our nodes whose LHS is based on single properties instead of P_{comp}s. In contrast, our nodes are finer grained which leads to more opportunities for pruning. Our algorithm starts with level 0. On each new level, it first generates possible VGFDs on this level based on the results of previous levels and it also eliminates many potential VGFDs from further consideration based on some easily computed heuristics (Section 3.1). Then, a runtime pruning (Section 3.3) and a detailed computation (Section 3.2) are conducted on each candidate VGFD. The whole process can terminate at a specified level, or after all levels, although the latter is usually unnecessary and unrealistic. The process returns each VGFD and its value mappings which is used for detecting violations.

3.1   Heuristics for Static Pruning

We first define the discriminability for a property as the number of distinct values divided by the sum of the property extension, and when it is compared between properties, it is over the instances they have in common. Then, static pruning heuristics used to eliminate potential VGFDs from further consideration are:

1. insufficient subject overlap between the LHS and the RHS,
2. the LHS or RHS has too high a discriminability,
3. the discriminability of the LHS is less than that of the RHS.

The information for rule 1 can be acquired from an ontology (e.g. using domain and range information) or a simple relational join on data. Here insufficient overlap means too few common subjects, e.g. 20. For rule 2, if the discriminability is close to one, e.g. 0.95 which means 95%, the property functions like a superkey in a database. Since such keys identify an individual, they are not useful for detecting abnormal data. For rule 3, if there is a mapping between two such properties, some values of the property with smaller discriminability must be mapped to different values on the RHS which would not be a functional mapping. In order to apply these heuristics, we make the additional observations:

1. The discriminability of a P_{comp} (P_{conj} resp.) is generally no greater than (no less than resp.) that of each property involved.
2. A P_{conj} (P_{comp} resp.) cannot be based on two properties that have few common subjects (objects and subjects resp.).
3. All children of a true VGFD on the level-wise search graph are also true VGFDs, but are not minimal.

For example, given a P_{comp} $A \circ B$, its values all come from the values of B and its extension is a subset of the Cartesian product between objects of A and subjects of B, then its discriminability, i.e. the distinct values divided by the usages, should be no greater than that of each component. A similar explanation applies for P_{conj} in observation 1. An extension of the observation 2 is that a P_{conj} cannot be followed by other properties in a property chain, e.g. $(A+B) \circ C$, since its values are tuples (e.g. the values of $A + B$) as opposed to the instances and literals that are the subjects of another property (e.g. subjects of C).

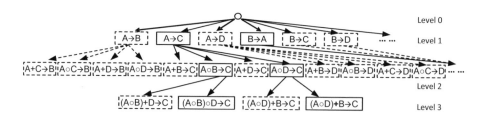

Fig. 1. An example of level-wise discovering process. We suppose that (1) property A and B have few common subjects, (2) the discriminability of B is less than that of C and (3) D has a high discriminability.

Fig.1 is an example showing how these heuristics are useful in the level-wise searching. Each edge is from a VGFD to a VGFD with an LHS that is a superset of the parent LHS and the two VGFDs have the same RHS. Note, our current algorithm does not support the use of composite properties on the RHS. The VGFDs pruned by the above heuristics are in dotted boxes and dotted lines pointing to the children pruned. For this example, we make assumptions typical of real RDF data. For instance, in DBPedia less than 2% of all possible pairs of properties share sufficient common instances. So following our heuristics, four VGFDs on level 1 are pruned: $A \rightarrow B$ is due to heuristic rule 1, $B \rightarrow C$ is due to rule 3 and the other two are due to rule 2. Then the children of $A \rightarrow B$ and

$A \rightarrow D$ are pruned due to the same reason as their parents. $A + B \rightarrow C$ on level 2 and $(A \circ D) + B \rightarrow C$ on level 3 are pruned due to the first assumption plus the observation 2. Finally, $A + D \rightarrow C$ on level 2 and $(A \circ B) + D \rightarrow C$ on level 3 are pruned due to the observation 1 and heuristic rule 2. From this example, we can see simple conditions can reduce the level-wise search space greatly based on these heuristics.

3.2 Handling Multi-valued Properties

The fundamental difference between VGFD and FD when computing VGFD is that we consider multi-valued properties. When finding FDs in databases, the multi-valued attributes either are not considered (if they are not in the same relation), or the correlation of their values is given by having separate tuples for each value. RDF frequently has multi-valued properties without any explicit correlation of values, e.g. in DBPedia, more than 60% properties are multi-valued. When computing a VGFD, we try to find a functional mapping

Table 1. The left table is the triple list. The right table is mapping count.

deptNo		deptName	
subject	object	subject	object
A	1	A	CS
A	2	A	EE
B	1	B	EE
C	2	C	CS
D	2	D	EE

Candidate Value Mapping	Count
$1 \rightarrow EE$	2
$2 \rightarrow EE$	2
$2 \rightarrow CS$	2
$1 \rightarrow CS$	1

from each LHS value to an RHS value such that this mapping maximizes the number of correlations. We consider any two values for a pair of multi-valued properties that share a subject to be a possible mapping. Then we greedily select the LHS value that has the most such mappings and remove all other possible mappings for this LHS value. If multiple RHS values are tied for the largest number of mappings, then we pick the one that appears in the fewest mappings so far. Consider Table 1 which analyzes the dependency $deptNo \rightarrow deptName$. The triples are given to the left and each possible value mapping and its maximal possible count are listed in descending order to the right. The maximal count of $1 \rightarrow EE$ is 2, because these two values co-occur for instances A and B once for each. We first greedily choose mapping $1 \rightarrow EE$, because it has the largest count among all mappings for $depNo = 1$. After this selection, the mapping $1 \rightarrow CS$ is removed since $deptNo = 1$ has been mapped. Then for $deptNo = 2$, to maximize the number of distinct values being matched on both sides, we choose $(2, CS)$ since CS has been mapped to by fewer LHS values than EE. Note the basic equality used here is a special case of cluster-based Dependency Equality. For example, if CS and EE are clustered together, then the mappings will be $1 \rightarrow EECS$ and $2 \rightarrow EECS$, where $EECS$ is the cluster.

Our confidence in a VGFD depends on how often the data agree with it, i.e., the total matches divided by the sum of the LHS's extension, e.g. the VGFD above has the confidence of $4/5 = 0.8$. In this work, we set the confidence threshold $\alpha = 0.9$ to ensure that patterns are significant, while allowing for some variation due to noise, input errors, and exceptions.

3.3   Run-time Pruning

In the worst case, the expensive full scan of value pairs must occur $|R^+| \cdot |R^\circ|$ times. So we propose to use *mutual information* (MI) computed over sampled value pairs for estimating the degree of dependency. In Algorithm 2, given a candidate VGFD s $X \rightarrow Y$, we start by randomly selecting a percentage β of the instances. In line 2, for each instance i, we randomly pick a pair of values from $V^+(i, X)$ and $V^\circ(i, Y)$. *Distribution*() also applies the clusters C_X and C_Y and returns these pairs in lieu of the actual values. In information theory, a MI I_{XY} of two random variables X and Y is formally defined as $I_{XY} = \sum_i \sum_j p_{i,j} \log (p_{i,j}/p_i p_j)$, where p_i, p_j are the marginal probability distribution functions of X and Y, and $p_{i,j}$ is the joint probability distribution function of X and Y respectively. Intuitively, MI measures how much knowing one of these variables reduces our uncertainty about the other. Furthermore, the *entropy coefficient* (EC), using MI, measures the percentage reduction in uncertainty in predicting the dependent variable based on knowledge of the independent variable. When it is zero, the independent variable is of no help in predicting the dependent variable; and when it is one, there is a full dependency. The EC is directional and $EC(X|Y)$ for predicting the variable X with respect to variable Y is defined as I_{XY}/E_Y, where E_Y is the entropy of variable Y, formally $\sum_j p_j \log 1/p_j$. Because I_{XY} also can be expressed as $E_X + E_Y - E_{XY}$ which has a easier form to compute.

Algorithm 2. *Runtime_Pruning*(s, α, β, E, C), s is a candidate VGFD $X \rightarrow Y$; α is the confidence threshold for a VGFD; β is the sampling size as a percentage; E is a set of triples. C is a set of cluster sets for each property.

1: $I \leftarrow Sampling\_Subjects(s, \beta, E)$ // Sampled subjects shared by the LHS and RHS.
2: $\{(X_i, Y_i)\} \leftarrow Distribution(s, I, E, C)$ //A list of value pairs where each pair consists of two single sampled values of LHS and RHS for the same subject.
3: $E_X = -\sum_{distinct\ x \in \{X_i\}} \frac{|\{X_i | X_i = x\}|}{|\{X_i\}|} \cdot log \frac{|\{X_i | X_i = x\}|}{|\{X_i\}|}$
4: $E_Y = -\sum_{distinct\ y \in \{Y_i\}} \frac{|\{Y_i | Y_i = y\}|}{|\{Y_i\}|} \cdot log \frac{|\{Y_i | Y_k = i\}|}{|\{Y_i\}|}$
5: $E_{XY} = -\sum_{distinct\ (x,y) \in \{(X_i, Y_i)\}} \frac{|\{(X_i, Y_i) | X_i = x \wedge Y_i = y\}|}{|\{(X_i, Y_i)\}|} \cdot log \frac{|\{(X_i, Y_i) | X_i = x \wedge Y_i = y\}|}{|\{(X_i, Y_i)\}|}$
6: **if** $(E_X + E_Y - E_{XY})/E_X < \alpha - 0.2$ **then**
7:    **return** TRUE
8: **return** FALSE

We note that Paradies et al. [18] also used entropy to estimate the dependency between two columns in databases. Since they want to determine attribute pairs that can be estimated with high certainty, i.e. focusing on precision of the positives, they need a complex statistical estimator. In contrast, our aim is a fast filter that is good enough to remove most negatives, i.e. independent pairs, thus a statistical estimator is not necessary. We can avoid missing positives by setting a low enough threshold. In our experiments, the difference between EC for a 20% sample and EC of full data is less than 0.15 on average and the estimated values typically have higher ECs. For example, it is very rare that a VGFD estimated lower than 0.7 has an actual value above 0.9. Therefore, a threshold of 0.2 less than α (line 6) is a reasonable lower bound for filtering out independent pairs.

4   Clustering Property Values

As introduced in Section 1, we must cluster property values in order to discover dependencies that allow for rounding errors, measurement errors, and distributions of values. For object property values, clustering groups all identifiers that stand for the same real world object by computing the transitive closure of sameAs. The rest of this section discusses clustering the values for each datatype property. This is used to determine Dependency Equality (Definition 6) between two objects.

4.1   Pre-clustering

The pre-clustering process is a light-weight computation that provides two benefits for finer clustering later: the minimum number of clusters and reserves expensive distance calculations for pairs of points within the same pre-cluster. Since the pre-clustering is used for VGFD discovery, there are three thoughts. First, the values to be clustered are from various properties and have very different features. So the clustering process needs to be generic in two aspects: (1) a pair-wise distance metric that is suitable for different types of values and multiple feature dimensions, and (2) suitable for the most common distribution in real world, i.e. the normal distribution. Second, we prefer a comparatively larger number of clusters where elements are really close (if not, they may not be clustered). The reason is that the clusters will be used as class types for detecting dependencies. Larger values of k generate finer-grained class types, which in turn allow us to generate more precise VGFDs, albeit at the risk of bluring boundaries between classes and making it harder to discover some dependencies. This point also makes our approach different from many other pre-clustering approaches, e.g. [19], because their results of pre-clustering can be overlapped and rigid clustering later could merge these groups into fewer clusters.

Based on the above thoughts, specifically, given a list of values, the process first selects a value that is closest to the center (we choose the mean for numeric values and discuss strings in the next paragraph), and then moves it from the list to be the centroid of a new group. Second, for each value on the list, if the distance to this centroid is within the threshold (we use the standard deviation), it will be moved from the list to a new group. Finally, the above process is repeated if the list is not empty. Thus the process generally finds the cluster around the original center first, and then the clusters further away from the center. This is much better than random selection, because if an outlier is selected, then most instances remain on the list for clustering after this round of computation.

To compute the center and distance of string values, we compute the weight of each token in a string according to its frequency in values for the property. Then we pick the string that has the largest sum of weights divided by the number of tokens in it as the center and the distance between two strings is the sum of weights of the different tokens in them. The intuition is that by taking these strings as a class, the most representative one is the one with the most common words. For example, the property *color* in DBPedia has values "light green", "lime green", etc. Then, the representative of these two strings is the common word "green". For "light green", the distance to "lime green" will be less than that to "light red", since "red" and "green" are more common and have larger weights.

4.2   Optimal k-Means Clustering

There are several popular clustering methods, e.g. k-Means, Greedy Agglomerative Clustering, etc. However most of them need a parameter for the number of resulting clusters. To automatically find the most natural/best clusters, we designed the following unsupervised method of finding optimal clusters.

The approach is inspired by the gap statistic [20] which is used to cluster numeric values with a gradually increasing number of clusters. The idea is that when we increase k to above the optimum, e.g. adding a cluster center in the middle of an already ideal cluster, the pooled within-cluster sum of squares around the cluster mean decreases more slowly than its expected rate. Thus the gap between the expectation and actual improvement over different k will be in a shape with an inflexion which indicates the best k. Our approach improves upon this idea in three aspects: we start at the number of pre-clusters instead of 1; in each round of k-Means, the initial centroids are selected according to pre-clusters; and the distance computation is only made among points within the same pre-cluster.

Our *Optimal_kMeans* algorithm is presented as Algorithm 3. At first, k is set to the number of pre-clusters. At each iteration, we increment k and select k random estimated centroids m_i, each of which starts a new cluster c_i. *Init()* selects the centroids from the pre-clusters in proportion to their sizes. In each inner loop (line 8-13), every value is labeled as a member of the cluster whose centroid has the shortest distance to this instance among all centroids that are within the same pre-cluster as that value (line 10). Then each centroid is recomputed based on the cheap distance metric until the centroid does not change. After each round of modified k-Means clustering, we compute the difference on $Gap(k)$ and stop the process if it is an inflexion point. Since the clustering is used to detect abnormal data in which string values are expected to be caused by accidental input or data conversion, in this clustering, we use edit distance as the distance metric for string values as opposed to the above pre-clustering.

Algorithm 3. $Optimal\_kMeans(L, groups)$, L is a set of literal values; $groups$ is a set of pre-clustered groups of L.

1: $k = |groups|$
2: $Gap(k) = Gap\_Statistic(groups)$
3: $tmpC \leftarrow groups$
4: **repeat**
5:     $k = k + 1$, $C \leftarrow tmpC$, $tmpC \leftarrow \emptyset$ //$tmpC$ is the set of k clusters
6:     **for** each $i \leq k$ **do**
7:         $Init(m_i), c_i \leftarrow c_i \cup m_i, tmpC \leftarrow tmpC \cup c_i$ //m_i is the center of each cluster
8:     **repeat**
9:         **for** each $x \in L$ **do**
10:            $c_i \leftarrow c_i \cup \arg\min_{m_i \in Group(x)} Distance(x, m_i)$
11:        **for** each $i \leq k$ **do**
12:            $m_i = Mean(c_i)$
13:     **until** $\forall i \leq k, m_i$ converges
14:     $Gap(k) = Gap\_Statistic(tmpC)$
15: **until** $Gap(k) < Gap(k-1)$
16: **return** C

---

5   Experimental Results

For our experiments, we selected the Semantic Web Research Corpus[2] (SWRC), DBPedia and RKB[3] data sets. All of them are widely used subsets of Linked Data that cover different domains. Experiments were conducted on a Sun workstation with 8 Xeon 2.8G cores and 6G memory. We observed that there are few dependencies with an LHS size larger than four and that such dependencies tend to have less plausible meanings. For this reason, we set the maximal size of a VGFD to four in our experiments. Based on clusters and VGFDs, abnormal data has two types: one is far away from other clusters and the other is a violation of VGFDs. Specifically, in this work, a triple is reported as an outlier if its value is the only element of some cluster whose distance to the nearest cluster centroid is above twice of the average distances between all clusters for this property. A triple is reported as abnormal due to violation of VGFDs only when its value conflicts with a value mapping determined by some VGFD and this value mapping is confirmed by other triple usages more than twice.

In our first experiment, we compared the overall performance of the system on three data sets. The sampling size β used in runtime pruning is 20%. In Table 2, we can see that the running time appears to be more heavily influenced by the number of properties than the data set size. Note that RKB has more triples but fewer properties than DBPedia, and thus has more triples per property. This leads to a longer clustering time, but thanks to static and runtime pruning, the total time to find VGFDs is less.

Table 3 gives some VGFDs from the three data sets and their short descriptions. In DBPedia, among 200 samples out of 2868 abnormal triples, 173 of them

---

[2] http://data.semanticweb.org/
[3] http://www.rkbexplorer.com/data/

Table 2. System overall performance on SWRC, DBPedia and RKB data sets

	SWRC	DBPedia	RKB
Number of Triples (M) / Properties	0.07 / 112	10 / 1114	38 / 54
Discovered VGFDs on Level 1	12	228	6
Discovered VGFDs on Level 2	37	304	3
Discovered VGFDs on Level 3	2	126	0
Discovered VGFDs on Level 4	0	53	0
Time for Clustering (s)	18	114	396
Time for Level 1 (s)	11	172	67
Time for Level 2 (s)	20	246	44
Time for Level 3 (s)	4	108	0
Time for Level 4 (s)	1	47	0
Total Time (s) / Discovered VGFDs	54 / 51	687 / 721	507 / 9
Reported Abnormal Triples	75	2868	227

Table 3. Some VGFDs from the three data sets. The first and second group of VGFDs are of size 1 and 2. The third group is a set of VGFDs with clustered values.

VGFD	Description
genus→family	Organisms in the same genus also have the same family.
writer→genre	A work's writer determines the work's genre.
teamOwner→chairman	The teams with the same owner also have the same chairman.
composer→mediaType	The works by the same composer have the same media type.
militaryRank→title	The people of the same military rank also have the same title.
location→nearestCity	The things on the same location have the same nearest city.
topic→primaryTopic	The papers with the same topic have the same primary topic.
manufacturer+oilSystem →compressionRatio	The manufacturer and oil system determine the engine's compression ratio.
publisher o country →language	The publisher's country determines the language of that published work.
article-of-journal+has-volume →has_date	The volume number of a journal where an article is published determines the published date of this article.
faculty→budget	The size of the faculty determines the budget range.
militaryRank→salary	The military rank determines the range of salary.
occupation→salary	The occupation determines the range of salary.
type→upperAge	A school's type determines the range of upper age.

(86.5%) are confirmed to be true errors in the original data. The correctness of 10 of the remaining triples was difficult to judge. SWRC and RKB have 51% and 62% precision respectively. We believe the lower precision for SWRC is because it has a higher initial data quality and its properties have a much smaller set of possible values than those of DBPedia.

We list a number of confirmed erroneous triples in Table 4. For example, the first triple is reported as an outlier after automatic clustering. The second triple violates the VGFD that a school's type determines the cluster of its upper age, because the triple's subject is a certain type of school while its value is not in the cluster of values for the same type of schools.

Next, to check the impact of our pruning algorithms, we performed an ablation study using DBPedia that removes these steps. The left part of Table 5 shows that using static and runtime pruning respectively saves over 62% and

Table 4. Some confirmed erroneous triples in the three data sets, where r, o, i, p, s are prefixes for http://www.dbpedia.org/resource/, http://www.dbpedia.org/ontology/, http://acm.rkbexplorer.com/id/, http://www.aktors.org/ontology/portal/ and http://data.semanticweb.org/

1	<r:Shanghai_Jiao_Tong_University, o:university/undergrad, 194323445>
2	<r:Harrow_College, o:School/upperAge, 2009.0>
3	<r:Melbourne_Grammar_School, o:School/ranking, 2006.0>
4	<r:Google_Maps, o:Work/language, r:Coverage_details_of_Google_Maps>
5	<r:Wiktionary, o:Work/language, r:History_and_development>
6	<r:Dembela, o:Place/coordinates, coord|N|W>
7	<r:Hutt_Valley_High_School, o:EducationalInstitution/principal, r:2008>
8	<r:Wake_Island, o:Island/country, r:United_States_Air_Force>
9	<r:Albuquerque_Plaza, o:Building/floorCount, 2221>
10	<r:varedo, o:City/province, r:Province_of_Milan>
11	<i:796511, p:has-date, to-10-01>
12	<i:journals/jair/DarwicheP97, p:has-date, 1998>
13	<s:person/bastian-quilitz, s:ns/swc/ontology#affiliation, research assistant>
14	<s:person/ulf-leser, s:ns/swc/ontology#affiliation, professor>

55% of time compared to using neither. Because they utilize different characteristics, using them together saves 85% over neither. When we do not prune, the few additional VGFDs discovered lead to fewer abnormal triples than those discovered with pruning (on average 2.2 per VGFD vs. 3.97 per VGFD). Thus the pruning techniques not only save time but do not affect the abnormality detection much.

Table 5. The left table is showing the impact of our pruning techniques. The right table is comparing our preclustering with an alternative called SortSeq on VGFDs using the clusters and abnormal data found based on these VGFDs.

	None	Static	Runtime	Both
Time (s)	4047	1529	1817	687
VGFDs	746	741	729	721
Abnormal	2923	2915	2887	2868

	Preclustering	SortSeq
Time (s)	114	83
VGFDs	42	23
Abnormal	625	391

Besides pruning, we also checked the impact of our pre-clustering. Because our approach is based on a generic pair-wise distance, we wanted to compare it with a simpler one based on the linear ordering of values where the distance is just the difference between numbers. After each iteration of clustering around the mean, this alternative, referred to as SortSeq, recursively clusters on two remaining value sets: one is above the mean and the other below the mean. To handle strings in this approach, we sort them alphabetically and assign each a sequence number. The right of Table 5 shows that VGFDs and abnormal data that are based on the baseline clustering are both less than that of our approach. Among the VGFDs not found by the SortSeq, most are for string values. SortSeq finds fewer VGFDs and less abnormal data, because it naively assumes that the more common leading characters two strings have, the more similar they are. Thus, our pre-clustering using cheap and generic computation captures the characteristics of different property values.

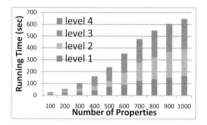 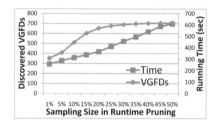

Fig. 2. The left is the effect of number of properties on the VGFD searching time. The right is the effect of sampling size in runtime pruning on the VGFD searching time.

Knowing that pre-clustering and pruning are useful for the system, we systematically checked the trend of system performance, especially time, by using these techniques. To be comparable on data set size, we picked subsets of properties from DBPedia. For each size, we randomly draw 10 different groups of this size and average the time over 10 runs. The left of Fig. 2 shows that the time for every level almost follows a linear trend. The right of Fig. 2 shows the effect of sampling size β used in runtime pruning on the system. We see that the running time is in linear proportion to the sampling size. As the VGFD curve shows, $\beta = 0.2$ is sufficient to find most dependencies for DBPedia.

6   Conclusion

We have presented a system to detect Semantic Web data that are abnormal and thus likely to be incorrect. Inspired by functional dependency in databases, we introduce *value-clustered graph functional dependency* which has three fundamental differences with functional dependency in order to better deal with Semantic Web data. First, the properties involved in a VGFD are across the whole data set schema instead of a single relation. Second, property value correlations, especially for multi-valued properties, are not explicitly given in RDF data. Third, using clusters for values greatly extends the detection of dependencies. Focusing on these unique characteristics, our system efficiently discovers VGFDs and effectively detects abnormal data, as shown in experiments on three Linked Data sets. In the future we plan to use subclass relationships to further generalize object property values. We also would like to take into account other features when clustering, for example the string patterns.

References

1. Bohannon, P., Fan, W., Flaster, M., Rastogi, R.: A cost-based model and effective heuristic for repairing constraints by value modification. In: SIGMOD 2005, pp. 143–154. ACM, New York (2005)
2. Fan, W., Geerts, F., Jia, X., Kementsietsidis, A.: Conditional functional dependencies for capturing data inconsistencies. ACM Trans. Database Syst. 33, 6:1–6:48 (2008)
3. Cong, G., Fan, W., Geerts, F., Jia, X., Ma, S.: Improving data quality: consistency and accuracy. In: VLDB 2007, pp. 315–326. VLDB Endowment (2007)
4. Sabou, M., Fernandez, M., Motta, E.: Evaluating semantic relations by exploring ontologies on the Semantic Web, pp. 269–280 (2010)
5. Fürber, C., Hepp, M.: Using SPARQL and SPIN for Data Quality Management on the Semantic Web. In: Abramowicz, W., Tolksdorf, R. (eds.) BIS 2010. LNBIP, vol. 47, pp. 35–46. Springer, Heidelberg (2010)
6. Tao, J., Sirin, E., Bao, J., McGuinness, D.L.: Integrity constraints in OWL. In: Fox, M., Poole, D. (eds.) AAAI. AAAI Press (2010)
7. Codd, E.F.: Relational completeness of data base sublanguages. In: Database Systems, pp. 65–98. Prentice-Hall (1972)
8. Mannila, H., Räihä, K.J.: Algorithms for inferring functional dependencies from relations. Data Knowl. Eng. 12(1), 83–99 (1994)
9. Huhtala, Y., Krkkinen, J., Porkka, P., Toivonen, H.: Tane: An efficient algorithm for discovering functional and approximate dependencies. The Computer Journal 42(2), 100–111 (1999)
10. Lopes, S., Petit, J.M., Lakhal, L.: Efficient Discovery of Functional Dependencies and Armstrong Relations. In: Zaniolo, C., Grust, T., Scholl, M.H., Lockemann, P.C. (eds.) EDBT 2000. LNCS, vol. 1777, pp. 350–364. Springer, Heidelberg (2000)
11. Beeri, C., Dowd, M., Fagin, R., Statman, R.: On the structure of Armstrong relations for functional dependencies. J. ACM 31, 30–46 (1984)
12. Levene, M., Poulovanssilis, A.: An object-oriented data model formalised through hypergraphs. Data Knowl. Eng. 6(3), 205–224 (1991)
13. Weddell, G.E.: Reasoning about functional dependencies generalized for semantic data models. ACM Trans. Database Syst., 32–64 (1992)
14. Li Lee, M., Ling, T.W., Low, W.L.: Designing functional dependencies for XML. In: Jensen, C.S., Jeffery, K., Pokorný, J., Šaltenis, S., Hwang, J., Böhm, K., Jarke, M. (eds.) EDBT 2002. LNCS, vol. 2287, pp. 124–141. Springer, Heidelberg (2002)
15. Hartmann, S., Link, S., Kirchberg, M.: A subgraph-based approach towards functional dependencies for XML. In: Computer Science and Engineering: II. SCI, vol. IX, pp. 200–211. IIIS (2003)
16. Brown, P.G., Hass, P.J.: Bhunt: automatic discovery of fuzzy algebraic constraints in relational data. In: VLDB 2003, pp. 668–679. VLDB Endowment (2003)
17. Haas, P.J., Hueske, F., Markl, V.: Detecting attribute dependencies from query feedback. In: VLDB 2007, pp. 830–841. VLDB Endowment (2007)
18. Paradies, M., Lemke, C., Plattner, H., Lehner, W., Sattler, K.U., Zeier, A., Krueger, J.: How to juggle columns: an entropy-based approach for table compression. In: IDEAS 2010, pp. 205–215. ACM, New York (2010)
19. McCallum, A., Nigam, K., Ungar, L.H.: Efficient clustering of high-dimensional data sets with application to reference matching. In: KDD 2000, pp. 169–178. ACM, New York (2000)
20. Tibshirani, R., Walther, G., Hastie, T.: Estimating the number of clusters in a data set via the gap statistic. Journal Of The Royal Statistical Society Series B 63(2), 411–423 (2001)

Author Index